# Differential Manifolds

## A Basic Approach for Experimental Physicists

# Differential Manifolds

## A Basic Approach for Experimental Physicists

**Paul Baillon**
CERN, Switzerland

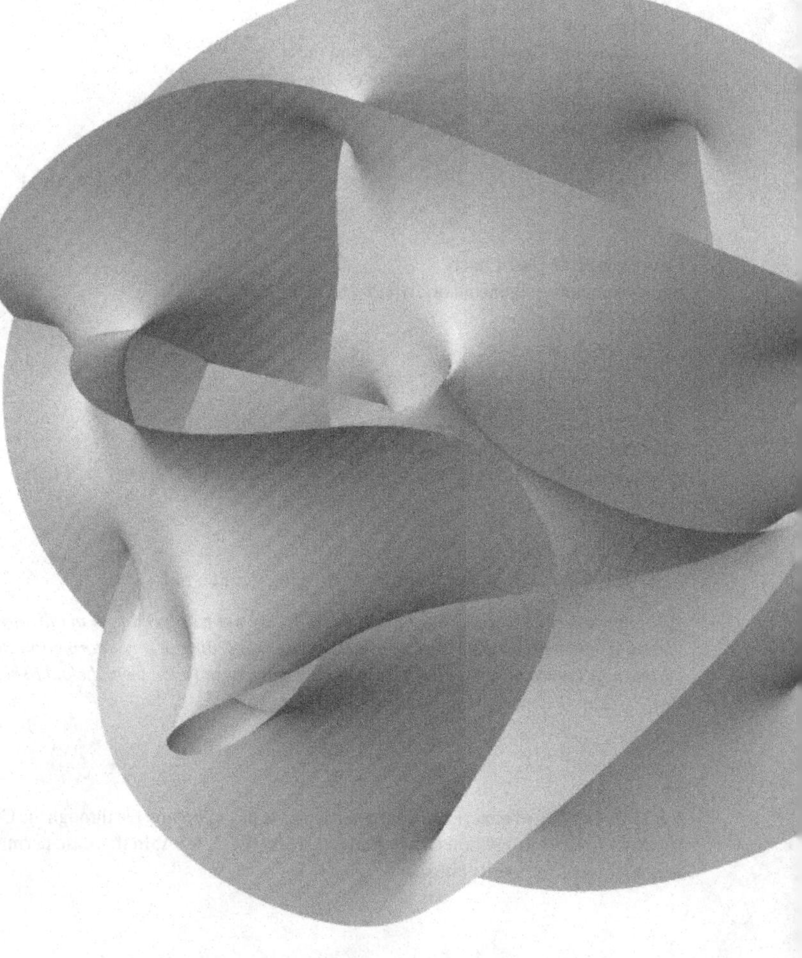

World Scientific

NEW JERSEY • LONDON • SINGAPORE • BEIJING • SHANGHAI • HONG KONG • TAIPEI • CHENNAI

*Published by*

World Scientific Publishing Co. Pte. Ltd.
5 Toh Tuck Link, Singapore 596224
*USA office:* 27 Warren Street, Suite 401-402, Hackensack, NJ 07601
*UK office:* 57 Shelton Street, Covent Garden, London WC2H 9HE

**Library of Congress Cataloging-in-Publication Data**
Baillon, Paul.
Differential manifolds : a basic approach for experimental physicists / by Paul Baillon (CERN, Switzerland).
pages cm
Includes bibliographical references and index.
ISBN 978-981-4449-56-4 (hardcover : alk. paper)
1. Differentiable manifolds. 2. Manifolds (Mathematics) I. Title.
QA614.3.B35 2013
516.3'6--dc23
2013010435



**British Library Cataloguing-in-Publication Data**
A catalogue record for this book is available from the British Library.



Printed in Singapore by B & Jo Enterprise Pte Ltd

# Contents

# Preface

Differential manifold is now a major frame for describing our physical world. There are several lecture notes on the subject dedicated to physicists. Those books, though they give a complete view of the subject, they often neglect to give the precise mathematical frame in which they do their development.
They just accept important theorems and think probably that physicists do not have enough background in mathematics for them. Very often for less important propositions they let to the reader the burden of finding a correct proof. For non-experts, it is a problem because those propositions are sometimes not that easy to prove. Nevertheless in spite of those critics, I am grateful to those lecture notes because they are the sources of my book on differential manifold [2], [3]. They also provide an important bibliography on the subject.
On the other hand, lecture notes for mathematicians on the same subject are unfortunately very obscure for non experts.
This book was written to fill the gap between those two kinds of lecture notes. It is aimed for students in physics and to physicists who want to enter in the framework of differential Manifold.
Differential manifolds deal with continuous maps which are indefinitely derivable. They deal also with integrals, differential equations and change of variables acting on various topological spaces. For that you need a very precise description of the various mathematical items used. That is why this book includes a description of standard general mathematics limited to what is needed for the differential manifolds. In order to have a precise language, you need a precise logical framework. We include it also in the book. It comes mainly from [1]. Strange enough when limited to what is really needed, those mathematics are not difficult and it is much easier to understand in a solid mathematical frame than in an artistic fuzzy.

I want to thank Robert Barate, Raymond Stora and Daniel Treille, for the time that they spent to read this book and for their advices and corrections. I also thank Philippe Bloch for his encouragements and comments. I am very grateful to CERN for its hospitality.

# Chapter 1

# Manifold

See section 1.19 (p. 128) for a quick view of the various notations.

## 1.1 Differentiable manifold

**Definition 1 Manifold.**

A topological set (or space) (see section 2.1.2 (p. 151)) **M** is said to be a manifold if[1]:

1. It is covered by a finite set of opens (definition 132 (p. 151)):
   Calling $\mathcal{O} \subset \mathcal{P}(\mathbf{M})$ (definition 350 (p. 453)), the set of opens of **M** (for the notations, see definitions 112 (p. 131) -equation (2.3), 390 (p. 499), 400 (p. 520)):

   $$(\exists \mathbf{K})\,(\mathbf{Card}(\mathbf{K}) \in \mathcal{Z}^+, \exists(\mathbf{U_i} : \mathbf{K} \to \mathcal{O})(\mathbf{M} = \bigcup_{\mathbf{i}\in\mathbf{K}} \mathbf{U_i}))$$

   very often, we shall set $\mathbf{K} = [\mathbf{1}, \mathbf{k}] = [\mathbf{1}, \mathbf{Card}(\mathbf{K})]$ (proposition 934 (p. 523)).

2. For any $\mathbf{i} \in \mathbf{K}$, there is an $\mathbf{n} \in \mathcal{Z}^+$ such that there is an isomorphism (definition 142 (p. 166)) $(\phi_\mathbf{i} : \mathbf{U_i} \leftrightarrow \mathbf{O} \subset \mathcal{R}^\mathbf{n})$ (definitions 143 (p. 170), 241 (p. 266)) between $\mathbf{U_i}$ and an open subset **O** of $\mathcal{R}^\mathbf{n}$ [2].

**Definition 2 Dimension of a covering set of a Manifold.**

Keeping the notation of definition 1 (p. 1), one calls dimension of a covering part $\mathbf{U_i}$ of a manifold **M** the dimension of the associated set $\mathcal{R}^\mathbf{n}$ (definitions 143 (p. 170), 241 (p. 266)) that is to say $\mathbf{n} \in \mathcal{Z}^+$ (definition 400 (p. 520)).

**Remark**

This definition makes sense because the isomorphisms $(\phi_\mathbf{i} : \mathbf{U_i} \leftrightarrow \mathbf{O} \subset \mathcal{R}^\mathbf{n})$ have a continuous inverse (definition 141 (p. 165)) which means that the image $\phi_\mathbf{i}(\mathbf{O_i}), \mathbf{i} \in [\mathbf{1}, \mathbf{k}]$ is open so any points of it is center of an open ball (definitions 260 (p. 285), 150 (p. 177)) fully in it (proposition 266 (p. 179)). According to definition 209 (p. 240), any open ball of $\mathcal{R}^\mathbf{n}$ can be considered as a subset of dimension **n** of $\mathcal{R}^\mathbf{n}$.

**Definition 3 Differentiable manifold.**

[1]For the definition of the symbols ($\subset$, $\cup$, $\cap$, $\exists$, $\forall$...) goes to appendix A (p. 433) ( section A.5.1 (p. 447) and definitions 333 (p. 441), 334 (p. 441). . . )

[2]Very often in this manual, we shall, for a point $p \in \mathbf{U_i} \subset \mathbf{M}$, quote as $\mathbf{x} \in \mathcal{R}^\mathbf{n}$ the value of $\phi_\mathbf{i} \in \mathcal{R}^\mathbf{n}$ for $p$: $\mathbf{x}(p) = \phi_\mathbf{i}(p)$ which are the coordinates of $p$ in the local chart or in the local reference system given by couple $\mathbf{U_i}, \phi_\mathbf{i}$ around $p$.

With the notations of definition 1 (p. 1), a manifold is said differentiable, if for any $\mathbf{i} \in [\mathbf{1}, \mathbf{k}]$ and $\mathbf{j} \in [\mathbf{1}, \mathbf{k}]$ such that:

$$U_i \cap U_j \neq \emptyset$$

considering the map $(f_{i,j} : \phi_i(U_i \cap U_j) \leftrightarrow \phi_j(U_i \cap U_j))$ (see definition 112 (p. 131) and section 2.1.1 (p. 131)) defined as:

$$f_{i,j} = \phi_j \circ \phi_i^{-1}$$

this map $f_{i,j}(x)$ belongs to $C^\infty\{\mathcal{R}^n \rightarrow \mathcal{R}^n\}$ (see definition 166 (p. 200)).

**Proposition 1** Given a differential manifold $\mathbf{M}$ if two covering elements $U_i, U_j$ are overlapping: $U_i \cap U_j \neq \emptyset$, they have the same dimension ( definition 2 (p. 1)).

**Proof**

From definitions 1 (p. 1) point 1 and 132 (p. 151) equation (2.18), $U_i \cap U_j$ is a non empty open. From definitions 1 (p. 1) point 2, 141 (p. 165), 3 (p. 1) and 142 (p. 166), from propositions 241 (p. 167), 177 (p. 142) and 193 (p. 149), the subsets $\phi_i(U_i \cap U_j)$ and $\phi_j(U_i \cap U_j)$ are open with an isomorphism $\phi_j \circ \phi_i^{-1} \in C^\infty$ between them. Then from proposition 642 (p. 373), we have proposition 1.

**Proposition 2** Using notations of definition 1 (p. 1), given an open subset $\mathbf{O}$ path connected (definition 285 (p. 327)) of a differential manifold $\mathbf{M}$, all covering elements $U_i$ of $\mathbf{M}$ which have a common part with $\mathbf{O}$ have the same dimension (definition 2 (p. 1)).

**Proof**

It is a consequence of proposition 1 (p. 2) and of proposition 596 (p. 330):
One defines $K' \subset K$ by $i \in K' \Leftrightarrow O_i = O \cap U_i \neq \emptyset$. Since $O \subset \bigcup_{i \in K} O_i = M$, we have:

$$(x \in O \Rightarrow (\exists j \in K)(x \in O_j)) \Rightarrow (j = \tau_i(i \in K, x \in O_i) \Rightarrow x \in O_j \cap O \neq \emptyset$$
$$\Rightarrow j \in K') \Rightarrow x \in \bigcup_{i \in K'} O_i) \Rightarrow O \subset \bigcup_{i \in K'} O_i$$

(see definitions 333 (p. 441), 349 (p. 451) and 362 (p. 462)). Using proposition 596 (p. 330), one may define a map (definition 112 (p. 131)) $(j(i) : [1, m] \rightarrow K')$ such that $O_{j(i)} \cap O_{j(i+1)} \neq \emptyset$. We call $i_m$ the biggest $i$ such that $(\forall \ell \in [1, i])(\text{dimension}(O_{j(\ell)}) = \text{dimension}(O_{j(1)})$. It exists from proposition 924 (p. 519). If $i_m \neq m$, since $O_{j(i_m)} \cap O_{j(i+1)} \neq \emptyset$, $[1, i_m + 1]$ would be such that $(\forall \ell \in [1, i+1])(\text{dimension}(O_{j(\ell)}) = \text{dimension}(O_{j(1)})$ and $i_m$ would not be the biggest element. Then $i_m = m$. All the sets $O_{j(i)}$ such that $i \in [1, m]$ have the same dimension $\mathbf{n}$. Since any $O_i$ with $i \in K'$ has in common a point in $\mathbf{O}$ so there is an $\ell \in [1, m]$ such that $O_i \cap O_{j(\ell)} \neq \emptyset$ which means $\text{dimension}(O_i) = n$.

**Proposition 3** If a differential manifold $\mathbf{M}$ (definition 3 (p. 1)) is a path connected set (definition 285 (p. 327)), all its covering subsets (see point 1 (p. 1) of definition 1 (p. 1)) have the same dimension.

**Proof**

It is a direct consequence of proposition 2 (p. 2).

**Definition 4 Dimension of a differential manifold.**

If all covering sets of a differential manifold (definition 3 (p. 1)) have the same dimension (definition 2 (p. 1)), that dimension is called the dimension of the differential manifold.

**Proposition 4** For all $n \in \mathcal{Z}^+$ (see definition 400 (p. 520)), $\mathcal{R}^n$ (definitions 143 (p. 170), 241 (p. 266)) is a differential manifold (definition 3 (p. 1)).

**Proof**
$\mathcal{R}^n$ being an open (definitions 132 (p. 151), 260 (p. 285)), the identity map (definition 113 (p. 132)) being in $C^\infty$ (definitions 276 (p. 304), 275 (p. 304), propositions 309 (p. 194), 308 (p. 194)), $\mathcal{R}^n$ is a differential manifold.

**Definition 5 Restriction of a differentiable manifold (definition 3 (p. 1)).**

Keeping the notations of definitions 1 (p. 1) and 3 (p. 1), O is an open subset of M. In the induced topology (definition 133 (p. 151)), the opens of O are also opens in the M topology (proposition 200 (p. 152)). The set of opens $V_i = O \cap U_i$ (equation (2.18) (p. 151)) is a covering set of O. From proposition 238 (p. 166), any restricted $\phi_i$ map forms an isomorphism. With the new couple $V_i, \phi_i$, O can also be considered as a differential manifold.

## 1.2 Smooth maps

**Definition 6 Smooth maps on differentiable manifolds.**

Given two differentiable manifolds (definition 3 (p. 1)) M and N of dimension m and n, with the continuous associated maps (definitions 112 (p. 131), 141 (p. 165)) $(\phi_i : U_i \to \phi_i(U_i) \subset \mathcal{R}^m)$ and $(\psi_j : V_j \to \psi_j(V_j) \subset \mathcal{R}^n)$ with $i \in [1, k_M]$, $j \in [1, k_N]$, $k_M \in \mathcal{Z}^+$, $k_M \in \mathcal{Z}^+$ (see definition 400 (p. 520)), $U_i$ and $V_j$ being the sets of opens covering M and N, a map $(f : M \to N)$ is a smooth map if:

$$(\forall i \in [1, k_M])(\forall j \in [1, k_N])((f(U_i) \cap V_j \neq \emptyset, O_M = \phi_i(f^{-1}(V_j) \cap U_i) \subset \mathcal{R}^m,$$
$$O_N = \psi_j(f(U_i) \cap V_j) \subset \mathcal{R}^n) \Rightarrow (\psi_j \circ f \circ \phi_i^{-1} : O_M \to O_N) \in C^\infty)$$

(see definitions 117 (p. 134), 278 (p. 306), 276 (p. 304)).

**Proposition 5** Keeping the above notations, given an open O of M, the smooth maps of M to N (definition 6 (p. 3)) are smooth maps of the restricted manifold O to N (see definition 118 (p. 135)).

**Proof**
It is a consequence of proposition 556 (p. 305).

**Definition 7 $\mathcal{F}$ is the set of smooth maps from M to $\mathcal{R}$.**

Given a differentiable manifold (definition 3 (p. 1)) M of dimension n with the covering set of opens: $(U_i : [1, k] \to M), M = \bigcup_{i \in [1,k]} U_i$ and with the continuous associated maps $(\phi_i : U_i \to \mathcal{R}^n)$, we call $\mathcal{F}$, the set of smooth maps from M to $\mathcal{R}$ (definition 6 (p. 3)). Here $\mathcal{R}$ is considered as a differentiable manifold (proposition 4 (p. 2)).

**Proposition 6**

$$(\forall f \in \mathcal{F})(\forall i \in [1, k])(f \circ \phi_i^{-1} \in C^\infty\{\phi_i(U_i) \subset \mathcal{R}^n \to \mathcal{R}\})$$

see definition 276 (p. 304).

**Proof**
It is a direct consequence of definition 6 (p. 3) and of proposition 563 (p. 311).

**Definition 8 $\mathcal{F}_O$.**

O being an open of a differential manifold (definition 3 (p. 1)) M, $\mathcal{F}_O$ is the set of smooth maps from O to $\mathcal{R}$ where O is the restricted manifold induced by M (definition 5 (p. 3)).

**Definition 9 Restricted smooth map $\mathcal{F}_{O}^{M}$.**

$O \subset M$ being an open, $\mathcal{F}_{O}^{M}$ is the set of maps from $M$ to $\mathcal{R}$ which are, when restricted to $O$, smooth maps from the manifold $O \subset M$ to $\mathcal{R}$ (see definition 5 (p. 3)). (In most of the case they are not in $\mathcal{F}$, since they are not necessarily smooth outside $O$. )

**Proposition 7** With the above notation of definition 9 (p. 4), we have $\mathcal{F} \subset \mathcal{F}_{O}^{M}$.

**Proof**
It is a direct consequence of proposition 5 (p. 3).

**Proposition 8** Any smooth map of the restricted manifold $O \subset M$ can be extended to be in $\mathcal{F}_{O}^{M}$.

**Proof**
Given a map $(f' : O \to \mathcal{R})$, it can be extended to $(f : M \to \mathcal{R})$ by any map $(f'' : \mathcal{C}_M(O) \to \mathcal{R})$ (definition 353 (p. 454)) for instance by $f'' = 0$ which gives:

$$p \in O \Rightarrow f(p) = f'(p), \; p \notin O \Rightarrow f(p) = 0$$

and we have $f \in \mathcal{F}_{O}^{M}$.

**Definition 10 $\mathcal{F}_{O}^{0}$.**

$O$ being an open of $M$, we shall call $\mathcal{F}_{O}^{0} \subset \mathcal{F}_{O}^{M}$ the set of maps of $\mathcal{F}_{O}^{M}$ which are zero outside $O$. Whenever we consider a smooth map of the restricted manifold attached to an open subset $O \subset M$, we can extend it to a map $M \to \mathcal{R}$ zero outside $O$. It belongs to $\mathcal{F}_{O}^{M}$, not necessarily to $\mathcal{F}$. This correspondence between the set $\mathcal{F}_{O}^{0}$ and the set of smooth maps of $O \to \mathcal{R}$ is a bijection (definition 124 (p. 139)).

**Proposition 9** When restricted to $O$ (definition 118 (p. 135)), the maps of $\mathcal{F}_{O}^{M}$, $\mathcal{F}_{O}^{0}$ (definitions 9 (p. 4), 10 (p. 4)) are both identical to the set $\mathcal{F}_{O}$ (definition 8 (p. 3)).

**Proof**
It is a direct consequence of the quoted definitions.

**Definition 11 $\mathcal{F}_k$, $\mathcal{F}_{k}^{M}$, $\mathcal{F}_{k}^{0}$ .**

When the open $O$ in the definitions 8 (p. 3), 9 (p. 4), 10 (p. 4) is one of the covering set $U_k$ of the manifold $M$ (see definition 1 (p. 1)), we shall replace the letter $O$ by $k$ in the quoting of the corresponding set of maps.

**Proposition 10** Given a differentiable manifold $M$ of dimension $n$, the covering set of opens: $(U_k : [1, K] \to M)$ and the continuous associated maps $(\phi_k : U_k \to \mathcal{R}^n)$, for any $k \in [1, K]$, and for any $i \in [1, n]$, we have $\phi_k^i \in \mathcal{F}_k$ (see definition 8 (p. 3)).

**Proof**
We call $\prod^{e}_{i \in [1,n]} x^i = (x^i : [1, n] \to \mathcal{R}) \in \mathcal{R}^n$, a point of $\mathcal{R}^n$ (definitions 241 (p. 266), 372 (p. 473)). With that and definition 3 (p. 1), we have:

$$p \in U_k \Rightarrow \prod^{e}_{i \in [1,n]} x^i = x = \phi_k(p) = \prod^{e}_{i \in [1,n]} \phi_k^i(p)$$

then we have:

$$(\forall i \in [1, n])(\phi_k^i \circ \phi_k^{-1}(x) = x^i)$$

which gives, using notation defined in definition 276 (p. 304):

$$(\forall j \in [1, n])(\forall i \in [1, n])(\partial_j x^i = \partial_j \phi^i_k \circ \phi_k^{-1}(x) = \delta^i_j,$$
$$\ell > 1 \Rightarrow (\forall \partial^\ell \in D^\ell)(\partial^\ell \phi^i_k \circ \phi_k^{-1}(x) = 0)).$$

Then $\phi^i_k \circ \phi_k^{-1} \in C^\infty$ and from definitions 6 (p. 3), 7 (p. 3), 11 (p. 4), we have:

$$(\forall i \in [1, n])(\phi^i_k \in \mathcal{F}_k).$$

**Proposition 11** $\mathcal{F}$ (and also for $\mathcal{F}_O, \mathcal{F}_O^M, \mathcal{F}_O^0$ replacing $\mathcal{F}$) is a commutative ring (definition 180 (p. 215)) and a vector space (definition 199 (p. 235)) with $\mathcal{R}$ as a field (definition 181 (p. 217)). The neutral element is the map where all elements of $\mathbf{M}$ go to zero.

**Proof**

From proposition 399 (p. 237) and from definitions 7 (p. 3), $\mathcal{F}$, $\mathcal{F}_O, \mathcal{F}_O^M, \mathcal{F}_O^0$ are vectorial spaces with $\mathcal{R}$ as a field. For the same reason, they form an Abelian group for the sum of maps. Since the product of two maps to $\mathcal{R}$ is a map to $\mathcal{R}$, since also:

$$(f_1 \in \mathcal{F}, f_2 \in \mathcal{F}, f_3 \in \mathcal{F}) \Rightarrow f_1 \times (f_2 + f_3) = f_1 \times f_2 + f_1 \times f_3$$

and the same replacing $\mathcal{F}$ by $\mathcal{F}_O, \mathcal{F}_O^M, \mathcal{F}_O^0$. $\mathcal{F}$, $\mathcal{F}_O, \mathcal{F}_O^M, \mathcal{F}_O^0$ are commutative rings (definition 180 (p. 215)).

**Proposition 12** Calling $\mathbf{E}$, the set of linear maps (definition 212 (p. 244)) of the vectorial space (definition 199 (p. 235), proposition 11 (p. 5)) $\mathcal{F}$ to the vectorial space $\mathcal{F}$ with $\mathcal{R}$ as field, $\mathbf{E}$ is a vectorial space (definition 199 (p. 235)) with $\mathcal{R}$ as a field (definition 181 (p. 217)). It is also a module (definition 182 (p. 218)) with $\mathcal{F}$ as commutative ring (definition 180 (p. 215)). The neutral element is, in both cases, the map where all elements of $\mathcal{F}$ go to zero.

**Proof**

From definition 212 (p. 244):

$$(u_1 \in E, u_2 \in E, f_1 \in \mathcal{F}, f_2 \in \mathcal{F}, \lambda_1 \in \mathcal{R}, \lambda_2 \in \mathcal{R}) \Rightarrow (f_1 \cdot u_1 + f_2 \cdot u_2 \in E,$$
$$u_1(\lambda_1 f_1 + \lambda_2 f_2) = \lambda_1 u_1(f_1) + \lambda_2 u_1(f_2)) \quad (1.1)$$

we also have

$$(\lambda \in \mathcal{R}, \mu \in \mathcal{R}, u_1 \in E, u_2 \in E) \Rightarrow \lambda u_1 + \mu u_2 \in E. \quad (1.2)$$

**Proposition 13** The same text as proposition 12 (p. 5), $\mathcal{F}$ being replaced by one of the sets: $\mathcal{F}_O$, $\mathcal{F}_O^M$, $\mathcal{F}_O^0$.

**Proof**

For the same reasons as for proposition 12 (p. 5).

**Definition 12 Constant smooth map.**

Given a differentiable manifold $\mathbf{M}$ of dimension $\mathbf{n}$ with for any $\mathbf{k} \in \mathbf{K}$ a couple $\mathbf{U_k}$, $\phi_k$, calling $\mathcal{F}$ its set of smooth maps from $\mathbf{M}$ to $\mathcal{R}$, a constant smooth map associated to an element $\lambda \in \mathcal{R}$ and quoted as $\lambda$ is the element of $\mathcal{F}$ such that:

$$(\forall p \in M)(\lambda(p) = \lambda).$$

**Proposition 14** $(\boldsymbol{\lambda}(p) : \mathrm{M} \to \mathcal{R})$ being a constant smooth map (definition 12 (p. 5)). We have:

$$(\forall \mathrm{k} \in \mathrm{K})(\forall \mathrm{x} \in \phi(\mathrm{U_k}))(\boldsymbol{\lambda}(\phi_{\mathrm{k}}^{-1}(\mathrm{x})) \in \mathrm{C}^{\infty})$$

see definition 276 (p. 304).

**Proof**
From proposition 234 (p. 165), $\boldsymbol{\lambda}(\phi_{\mathrm{k}}^{-1}(\mathrm{x}))$ is continuous; all the derivatives of $\boldsymbol{\lambda}(\phi_{\mathrm{k}}^{-1}(\mathrm{x}))$ are $\mathbf{0}$ then they exist.

**Definition 13** ${}^{\mathrm{i}}\mathbf{g}_{p'}(\mathrm{x},\mathrm{a})$, ${}^{\mathrm{i}}\mathbf{g}'_{p'}(p,\mathrm{a})$, $\mathbf{g}_{p'}(\mathrm{x},\mathrm{a})$, $\mathbf{g}'_{p'}(p,\mathrm{a})$.

We call $\mathbf{O}$ an open of $\mathrm{M}$. Taking a $p' \in \mathrm{O}$ and $\mathbf{i} \in \mathbf{K}$ with $p' \in \mathrm{U_i}$, we have $\mathrm{U_i} \cap \mathrm{O}$ open and not empty. We take $\mathbf{a} \in \mathcal{R}$ such that the closed cube (definition 152 (p. 179) and proposition 514 (p. 288)) $\overline{\mathrm{C}}(\phi_{\mathrm{i}}(p'),\mathrm{a})$ verifies $\overline{\mathrm{C}}(\phi_{\mathrm{i}}(p'),\mathrm{a}) \subset \mathrm{U_i} \cap \mathrm{O}$. $\mathbf{a}$ exists from propositions 269 (p. 179) and 507 (p. 286). From proposition 714 (p. 427), there is a map $(\mathbf{g}(\mathrm{y} - \phi_{\mathrm{i}}(p'),\mathrm{a}) : \mathcal{R}^{\mathrm{n}} \to \mathcal{R})$ of $\mathrm{C}^{\infty}$ (see proposition 712 (p. 426), definition 276 (p. 304)) which is $\mathbf{1}$ at $\phi_{\mathrm{i}}(p')$ and $\mathbf{0}$ in $\mathcal{C}(\mathrm{C}(\phi_{\mathrm{i}}(p'),\mathrm{a}))$ (see definition 353 (p. 454)). It allows to define $({}^{\mathrm{i}}\mathbf{g}_{p'}(\mathrm{x},\mathrm{a}) = \mathbf{g}(\mathrm{x} - \phi_{\mathrm{i}}(p'),\mathrm{a}) : \phi_{\mathrm{i}}(\mathrm{O}) \to \mathcal{R})$. We define also ${}^{\mathrm{i}}\mathbf{g}'_{p'}(p,\mathbf{a})$ by:

$$p \in \phi^{-1}(\mathrm{C}(\phi_{\mathrm{i}}(p'),\mathrm{a})) \Rightarrow {}^{\mathrm{i}}\mathbf{g}'_{p'}(p,\mathrm{a}) = \mathbf{g}(\phi_{\mathrm{i}}(p) - \phi_{\mathrm{i}}(p'),\mathrm{a}),$$
$$p \notin \phi^{-1}(\mathrm{C}(\phi_{\mathrm{i}}(p'),\mathrm{a})) \Rightarrow {}^{\mathrm{i}}\mathbf{g}'_{p'}(p,\mathrm{a}) = 0\,.$$

In most of the cases, the upper index $\mathbf{i}$ is clear from the surrounding, so to simplify the writing, we shall not always mention it. We shall use instead $\mathbf{g}_{p'}(\mathrm{x},\mathrm{a})$, $\mathbf{g}'_{p'}(p,\mathrm{a})$.

**Proposition 15** With the above notations of definition 13 (p. 6) we have:

$$(p' \in \mathrm{U_i}, \mathrm{C}(\phi(p'),\mathrm{a}) \subset \phi(\mathrm{U_i})) \Rightarrow \mathbf{g}'_{p'}(p,\mathrm{a}) \in \mathcal{F}_{\mathrm{i}}^{0} \cap \mathcal{F}\,.$$

**Proof**
It is a direct consequence of proposition 712 (p. 426) and of definition 10 (p. 4).

**Definition 14** ${}^{\mathrm{i}}\mathbf{g}_{p'}(\mathrm{x},\mathrm{a},\mathrm{b})$, ${}^{\mathrm{i}}\mathbf{g}'_{p'}(p,\mathrm{a},\mathrm{b})$, $\mathbf{g}_{p'}(\mathrm{x},\mathrm{a},\mathrm{b})$, $\mathbf{g}'_{p'}(p,\mathrm{a},\mathrm{b})$.

$\mathrm{O} \neq \emptyset$ is an open of $\mathrm{M}$, $p' \in \mathrm{O}$ and $\mathbf{i} \in \mathbf{K}$ with $p' \in \mathrm{U_i}$, we have $\mathrm{U_i} \cap \mathrm{O}$ open and not empty. We take $\mathbf{a} \in \mathcal{R}$, $\mathrm{b} \in \mathcal{R}$, $\mathrm{b} > \mathbf{a}$ such that the closed cubes (definition 152 (p. 179) and proposition 514 (p. 288)) $\overline{\mathrm{C}}(\phi_{\mathrm{i}}(p'),\mathrm{a})$ and $\overline{\mathrm{C}}(\phi_{\mathrm{i}}(p'),\mathrm{b})$ verify $\overline{\mathrm{C}}(\phi_{\mathrm{i}}(p'),\mathrm{a}) \subset \overline{\mathrm{C}}(\phi_{\mathrm{i}}(p'),\mathrm{b}) \subset \phi_{\mathrm{i}}(\mathrm{U_i} \cap \mathrm{O})$. $\mathbf{a}$ and $\mathbf{b}$ exist from propositions 269 (p. 179) and 507 (p. 286). We may consider the map $\mathcal{R}^{\mathrm{n}} \to \mathcal{R}$ of $\mathrm{C}^{\infty}$ (see definition 276 (p. 304)) with $\mathbf{x} \in \mathcal{R}^{\mathrm{n}}$ as variable:

$$(\mathbf{g}(\mathrm{x} - \phi_{\mathrm{i}}(p'),\mathrm{a},\mathrm{b}) : \mathcal{R}^{\mathrm{n}} \to \mathcal{R})$$

as defined in definition 715 (p. 428). All those maps are zero when: $\mathbf{min}_{\mathrm{j}\in[1,\mathrm{n}]} |\mathrm{x_j} - \phi_{\mathrm{i}}(p')_{\mathrm{j}}| \geq \mathrm{b}$ and equal to $\mathbf{1}$ when $\mathbf{max}_{\mathrm{j}\in[1,\mathrm{n}]} |\mathrm{x_j} - \phi_{\mathrm{i}}(p')_{\mathrm{j}}| \leq \mathbf{a}$. On the differentiable manifold, we can define[3]:

$${}^{\mathrm{i}}\mathbf{g}_{p'}(\mathrm{x},\mathrm{a},\mathrm{b}) = \mathbf{g}(\mathrm{x} - \phi_{\mathrm{i}}(p'),\mathrm{a},\mathrm{b}) \in \mathrm{C}_{\mathrm{x}}^{\infty}$$

and ${}^{\mathrm{i}}\mathbf{g}'_{p'}(p,\mathrm{a},\mathrm{b}) \in \mathcal{F}_{\mathrm{i}}^{0} \cap \mathcal{F}$:

$$p \in \mathrm{U_i} \cap \mathrm{O} \Rightarrow {}^{\mathrm{i}}\mathbf{g}'_{p'}(p,\mathrm{a},\mathrm{b}) = \mathbf{g}(\phi_{\mathrm{i}}(p) - \phi_{\mathrm{i}}(p'),\mathrm{a},\mathrm{b}),$$
$$p \notin \mathrm{U_i} \cap \mathrm{O} \Rightarrow {}^{\mathrm{i}}\mathbf{g}'_{p'}(p,\mathrm{a},\mathrm{b}) = 0\,. \quad (1.3)$$

[3]We put $\mathbf{x} \in \mathrm{C}_{\mathrm{x}}^{\infty}$ to indicate that the variable on which one derives is $\mathbf{x}$.

It means:

$$\begin{aligned} \phi_i(p) \in C(\phi_i(p'), a) \Rightarrow {}^i g'_{p'}(p, a, b) = 1, \\ \phi_i(p) \notin C(\phi_i(p'), b) \Rightarrow {}^i g'_{p'}(p, a, b) = 0 . \end{aligned} \tag{1.4}$$

As for definition 13 (p. 6), in most of the cases, the upper index **i** is clear from the surrounding, we shall then often forget it.
**Notice**: From the properties of $\mathcal{F}$ (proposition 11 (p. 5)) we have $g'^2_{p'} \in \mathcal{F}$. $g'^2_{p'}$ verifies also equation (1.4) (p. 7).

**Proposition 16** We have a differential manifold **M** and an open **O**. We consider an open covering couple $U_k, \phi_k$ (see definition 1 (p. 1) point 1). If $\phi_k(U_k \cap O) \neq \mathcal{R}^n$, at any point $p \in O \cap U_k$, there is $x_m \in \mathcal{R}^+$ such that (see definitions 353 (p. 454), 355 (p. 456)):

$$\begin{aligned} 0 < a < x_m \Rightarrow \overline{C}(\phi_k(p), a) \subset \phi_k(O \cap U_k), \\ 0 < x_m \leq a \Rightarrow \overline{C}(\phi_k(p), a) \cap \mathcal{C}(\phi_k(O \cap U_k)) \neq \emptyset . \end{aligned} \tag{1.5}$$

If $x_m$ does not exist that is to say $\phi_k(O \cap U_k) = \mathcal{R}^n$, equation (1.5) (p. 7) becomes:

$$0 < a \Rightarrow \overline{C}(\phi_k(p), a) \subset \phi_k(O \cap U_k) = \mathcal{R}^n . \tag{1.6}$$

**Proof**
It is a direct consequence of propositions 269 (p. 179) and 525 (p. 291).

**Proposition 17** We are given an open **O** of a manifold **M** with its set of smooth map $\mathcal{F}^0_O$ (definition 10 (p. 4)). By definition 1 (p. 1), for any $p \in O$, $(\exists K_p \subset K)(i \in K_p \Leftrightarrow p \in U_i)$. Taking an $i \in K_p$ $(i = \tau_j(j \in K_p))$, we define:

$$O' = O \cap U_i$$

$O'$ is open (definition 132 (p. 151), 2.18) and not empty. Since $\phi_i^{-1}$ is continuous, $\phi_i(O')$ is an open of $\mathcal{R}^n$ containing $\phi_i(p)$. We consider $g'_p(p', a, b) \in \mathcal{F} \cap \mathcal{F}^0_O)$ with **a** and **b** defined such that $\overline{C}(\phi_i(p), a) \subset \overline{C}(\phi_i(p), b) \subset \phi_i(O')$. **a** and **b** exist from propositions 269 (p. 179) and 507 (p. 286) (see definition 14 (p. 6)). For any $f \in \mathcal{F}^M_O$ (definition 9 (p. 4)) we have:

$${}^i g'_p(p', a, b) f(p') \in \mathcal{F} \cap \mathcal{F}^0_O .$$

**Proof**
From definition 14 (p. 6),

$$\begin{aligned} & p' \notin O' \Rightarrow {}^i g'_p(p', a, b) f(p') = 0 \\ & p' \in O' \Rightarrow (\forall k \in K_p)(x = \phi_k(p') \\ & \Rightarrow {}^i g'_p(p', a, b) = g(\phi_i(\phi_k^{-1}(x) - \phi_i(p), a, b) \in C^\infty)) \\ & p' \in M \Rightarrow ((p' \in U_k, x = \phi_k(p')) \Rightarrow (f(\phi_k^{-1}(x)) \in C^\infty)) . \end{aligned}$$

This means (see proposition 559 (p. 306)):

$$(\forall k \in K)(\forall x \in \phi_k(U_k))(g'_p(\phi_k^{-1}(x), a, b) f(\phi_k^{-1}(x)) \in C^\infty)$$

then $g'_p(p', a, b) f(p') \in \mathcal{F}$.

## 1.3 Vector fields on a differentiable manifold

### 1.3.1 Definition of a vector field

**Definition 15 Vector fields.**

We have a differentiable manifold $\mathbf{M}$ of dimension $\mathbf{n}$. We call $\mathcal{O}$ the set of opens of $\mathbf{M}$ with $(\mathbf{U_k} : \mathbf{K} \rightarrow \mathcal{O})$ the covering set of $\mathbf{M}$ together with the associated continuous maps $(\phi_k : \mathbf{U_k} \rightarrow \mathcal{R}^n)$. We call $\mathcal{F}$ the set of smooth maps from $\mathbf{M}$ to $\mathcal{R}$. One calls $\mathbf{E}$, the set of linear maps of $\mathcal{F}$ to $\mathcal{F}$ as defined in proposition 12 (p. 5). $\mathbf{E}$ is a module (definition 182 (p. 218)) with $\mathcal{F}$ as commutative ring (definition 180 (p. 215)), it is also a vectorial space (definition 199 (p. 235)) with $\mathcal{R}$ as a field (definition 181 (p. 217)). One calls a vector field $\mathbf{u}$ on $\mathbf{M}$, an element of $\mathbf{E}$ which has the following properties:

$$(\forall f \in \mathcal{F})(\forall g \in \mathcal{F})(u(fg) = u(f)g + fu(g)) \ . \tag{1.7}$$

Since $\mathbf{E}$ is a set of linear maps, we also have (definition 212 (p. 244), proposition 12 (p. 5)):

$$\begin{aligned} &(\forall f \in \mathcal{F})(\forall g \in \mathcal{F})(u(f+g) = u(f) + u(g)) \\ &(\forall f \in \mathcal{F})(\forall \lambda \in \mathcal{R})(u(\lambda f) = \lambda u(f)) \ . \end{aligned} \tag{1.8}$$

We use here the notation given in equation (2.3) (p. 131).
Equation (1.7) (p. 8) is called the Leibniz rule. It is similar to the rule for the derivative of the product of functions. We shall see in the following parts how to construct such vector fields.

**Definition 16 Sum of vector fields.**

Keeping the above notations of definition 15 (p. 8), by definition:

$$(u \in \mathcal{V}, v \in \mathcal{V}, f \in \mathcal{F}) \Rightarrow (u+v)(f) = u(f) + v(f) \in \mathcal{F} \ .$$

**Proposition 18** The sum of vector fields (definition 16 (p. 8)) is a vector field.

**Proof**
Applying definition 16 (p. 8), we verify equations (1.7) (p. 8), (1.8) (p. 8) of definition 15 (p. 8):

$$\begin{aligned} &(u \in \mathcal{V}, v \in \mathcal{V}, f \in \mathcal{F}, g \in \mathcal{F}, \lambda \in \mathcal{R}) \Rightarrow \{(u+v)(fg) = u(fg) + v(fg) \\ &= u(f)g + fu(g) + v(f)g + fv(g) = (u+v)(f) \times g + f \times (u+v)(g), \\ &(u+v)(f+g) = u(f+g) + v(f+g) = uf + ug + vf + vg \\ &= (u+v)(f) + (u+v)(g), \\ &(u+v)(\lambda f) = u(\lambda f) + v(\lambda f) = \lambda u(f) + \lambda v(f) = \lambda(u+v)(f)\} \ . \end{aligned}$$

**Definition 17 Product of a vector field by a smooth map over a differential manifold.**

Keeping the notations of definition 15 (p. 8), we set by definition:

$$(u \in \mathcal{V}, \lambda \in \mathcal{F}, f \in \mathcal{F}) \Rightarrow (\lambda u)(f) = \lambda u(f) \in \mathcal{F}$$

**Proposition 19** The product of a vector field (definition 17 (p. 8)) by a smooth map is a vector field.

**Proof**

Applying definition 17 (p. 8), we verify equations (1.7) (p. 8), (1.8) (p. 8) of definition 15 (p. 8):

$$\begin{aligned}&(\mathbf{u} \in \mathcal{V}, \mathbf{v} \in \mathcal{V}, \lambda \in \mathcal{F}, \mathbf{f} \in \mathcal{F}, \mathbf{g} \in \mathcal{F}, \mu \in \mathcal{R}) \Rightarrow \{(\lambda \mathbf{u})(\mathbf{fg}) = \lambda \mathbf{u}(\mathbf{fg}) = \\ &\lambda \mathbf{u}(\mathbf{f})\mathbf{g} + \lambda \mathbf{f}\mathbf{u}(\mathbf{g}) = \lambda \mathbf{u}(\mathbf{f}) \times \mathbf{g} + \mathbf{f}\lambda \mathbf{u}(\mathbf{g}) = (\lambda \mathbf{u})(\mathbf{f}) \times \mathbf{g} + \mathbf{f}(\lambda \mathbf{u})(\mathbf{g}), \\ &(\lambda \mathbf{u})(\mathbf{f}+\mathbf{g}) = \lambda \mathbf{u}(\mathbf{f}+\mathbf{g}) = \lambda \mathbf{u}(\mathbf{f}) + \lambda \mathbf{u}(\mathbf{g}) = (\lambda \mathbf{u})(\mathbf{f}) + (\lambda \mathbf{u})(\mathbf{g}), \\ &(\lambda \mathbf{u})(\mu \mathbf{f}) = \lambda \mathbf{u}(\mu \mathbf{f}) = \lambda \mu \mathbf{u}(\mathbf{f}) = \mu \lambda \mathbf{u}(\mathbf{f}) = \mu(\lambda \mathbf{u})(\mathbf{f})\} \, .\end{aligned}$$

**Proposition 20** With the above internal laws (definitions 16 (p. 8), 17 (p. 8)), the set of vector fields (definition 15 (p. 8)) $\mathcal{V}$, associated to a differential manifold **M** (definition 1 (p. 1)), is a module (definition 182 (p. 218)) over the commutative ring (definition 180 (p. 215)) $\mathcal{F}$. $\mathcal{F}$ is the set of smooth maps over **M** (definition 6 (p. 3)).

**Proof**

With definition 15 (p. 8) and definitions 16 (p. 8), 17 (p. 8), $\mathcal{V}$, the set of vector fields on **M**, satisfies the conditions of a module with $\mathcal{F}$ as commutative ring (see definition 180 (p. 215)).

## 1.3.2 Properties of vector fields

To help the understanding of the properties of vector fields, in what follows we suppose that we have defined:

- A differentiable manifold **M** of dimension **n**, with its covers $\mathbf{U_i}, \mathbf{i} \in \mathbf{K}$ and their associated bijections $(\phi_i : \mathbf{U_i} \leftrightarrow \mathcal{R}^n)$.
- Its set of smooth maps $\mathcal{F}$ of **M** to $\mathcal{R}$
- Its set of vector fields $\mathcal{V}$ associated to that differentiable manifold.

**Proposition 21** With those notations, and using the same term for an element of $\mathcal{R}$ and its associated map of $\mathcal{F}$ as defined in 12 (p. 5), we have:

$$(\forall \lambda \in \mathcal{R})((\lambda : \mathbf{M} \to \{\lambda\}) \in \mathcal{F}, (\forall \mathbf{u} \in \mathcal{V})(\mathbf{u}(\lambda) = 0) \, .$$

**Proof**

Using equation (1.1) of proposition 12 (p. 5), using proposition 399 (p. 237), definition 15 (p. 8), equation (1.7) and definition 12 (p. 5), we have:

$$\begin{aligned}&\lambda \in \mathcal{R} \Rightarrow \mathbf{u}(\lambda^2) = \lambda \mathbf{u}(\lambda) \\ &\lambda \in \mathcal{F} \Rightarrow \mathbf{u}(\lambda^2) = \lambda \mathbf{u}(\lambda) + \mathbf{u}(\lambda)\lambda = 2\lambda \mathbf{u}(\lambda) \\ &\Rightarrow 2\lambda \mathbf{u}(\lambda) = \lambda \mathbf{u}(\lambda) \Rightarrow \lambda \mathbf{u}(\lambda) = 0 \Rightarrow (\lambda \neq 0 \Rightarrow \mathbf{u}(\lambda) = 0) \qquad (1.9) \\ &\mathbf{u}(0+2) = \mathbf{u}(2) = \mathbf{u}(0) + \mathbf{u}(2) \Rightarrow \mathbf{u}(0) = 0 \\ &(\forall \lambda \in \mathcal{R})(\lambda \in \mathcal{F}, \mathbf{u}(\lambda) = 0) \, .\end{aligned}$$

**Proposition 22** Given a vector field $\mathbf{v} \in \mathcal{V}$ of a differential manifold **M**, we consider an open $\mathbf{O} \subset \mathbf{M}$, **f** and **h** being two maps both in $\mathcal{F}$, we have:

$$(\forall p \in \mathbf{O})(\mathbf{f}(p) = \mathbf{h}(p)) \Rightarrow (\forall p \in \mathbf{O})(\mathbf{v}(\mathbf{f})(p) = \mathbf{v}(\mathbf{h})(p)) \, . \qquad (1.10)$$

**Proof**
We call $(\mathrm{U_i} : \mathrm{K} \to \mathcal{P}(\mathrm{M}))$, the set of covering opens of $\mathrm{M}$ with its associated set of maps $(\phi_\mathrm{i} : \mathrm{U} - \mathrm{i} \to \mathcal{R}^\mathrm{n})$ as defined in definition 1 (p. 1). If $\mathrm{O} = \emptyset$, proposition 22 is trivially true. Considering $p \in \mathrm{O}$, following proposition 17 (p. 7), definition 1 (p. 1) gives: $(\exists \mathrm{K_p} \subset \mathrm{K})(\mathrm{i} \in \mathrm{K_p} \Leftrightarrow p \in \mathrm{U_i})$ (definition 333 (p. 441)). Taking an $\mathrm{i} \in \mathrm{K_p}$ $(\mathrm{i} = \tau_\mathrm{j}(\mathrm{j} \in \mathrm{K_p}))$, we define:

$$\mathrm{O}' = \mathrm{O} \cap \mathrm{U_i}$$

$\mathrm{O}'$ is open (definition 132 (p. 151), 2.18) and not empty. Since $\phi_\mathrm{i}^{-1}$ is continuous, using definition 141 (p. 165), $\phi_\mathrm{i}(\mathrm{O}')$ is an open of $\mathcal{R}^\mathrm{n}$ containing $\phi_\mathrm{i}(p)$. From propositions 269 (p. 179) and 507 (p. 286), there is a closed cube (definition 152 (p. 179) and proposition 514 (p. 288)) $\overline{\mathrm{C}}(\phi_\mathrm{i}(p), \mathrm{a}) \subset \phi_\mathrm{i}(\mathrm{O}')$ centered on $\phi_\mathrm{i}(p)$. From definition 13 (p. 6) and proposition 15 (p. 6), there is a map $(\mathrm{g}'_p(q), \mathrm{a}) : \mathrm{M} \to \mathcal{R}) \in \mathcal{F}_\mathrm{i}^0 \cap \mathcal{F}$ such that, for $q \in \mathrm{M}$, we have:

$$q \notin \mathrm{O} \cap \mathrm{U_i} \Rightarrow \mathrm{g}'_p(q, \mathrm{a}) = 0$$
$$\mathrm{g}'_p(p, \mathrm{a}) = 1 \; .$$

It gives:

$$(\forall q \in \mathrm{M})(\mathrm{g}'_p(q, \mathrm{a}) \times (\mathrm{f}(q) - \mathrm{h}(q)) = 0) \; .$$

Because for any point $q \in \mathrm{M}$ where $\mathrm{f}(q) \neq \mathrm{h}(q)$, $\mathrm{g}'_p(q, \mathrm{a}) = 0$. Then since $\mathrm{g}'_p(q, \mathrm{a}) \in \mathcal{F}$ we have from proposition 21 (p. 9):

$$\mathrm{v} \in \mathcal{V} \Rightarrow (\forall q \in \mathrm{M})(\mathrm{v}(\mathrm{g}'_p(x, \mathrm{a}) \times (\mathrm{f} - \mathrm{h}))(q) = 0)$$

which gives:

$$\begin{aligned}&(\forall q \in \mathrm{M})(\mathrm{v}(\mathrm{g}'_p(x, \mathrm{a}) \times (\mathrm{f} - \mathrm{h}))(q) = 0 \\ &= (\mathrm{f} - \mathrm{h})(q) \times \mathrm{v}(\mathrm{g}'_p(x, \mathrm{a}))(q) + \mathrm{g}'_p(q, \mathrm{a})(q) \times \mathrm{v}(\mathrm{f} - \mathrm{h})(q) = 0) \; .\end{aligned} \tag{1.11}$$

From what we suppose, we have $(\mathrm{f} - \mathrm{h})(p) = 0$ and $\mathrm{g}'_p(p, \mathrm{a}) = 1$. Then the above equation (1.11) (p. 10) gives when $q = p$:

$$\mathrm{v}(\mathrm{f} - \mathrm{h})(p) = \mathrm{v}(\mathrm{f})(p) - \mathrm{v}(\mathrm{h})(p) = 0 \; .$$

**Proposition 23** Keeping the notations of section 1.3.2 and of definition 14 (p. 6), we have:

$$\begin{aligned}&(\forall \mathrm{a} \in \mathcal{R}^+)(\forall \mathrm{b} > \mathrm{a} > 0)(\forall \mathrm{v} \in \mathcal{V})(\forall \mathrm{i} \in \mathrm{K})((p \in \mathrm{U_i}, \overline{\mathrm{C}(\phi_\mathrm{i}(\mathrm{p}), \mathrm{b})} \subset \phi_\mathrm{i}(\mathrm{U_i})) \\ &\Rightarrow (((\mathrm{h} : \mathrm{M} \to \mathcal{R}^\mathrm{n}), p' \notin \mathrm{U_i} \Rightarrow \mathrm{h}(p') = 0, p' \in \mathrm{U_i} \Rightarrow \mathrm{h}(p') = \mathrm{g}_p(\phi_\mathrm{i}(p'), \mathrm{a}, \mathrm{b})) \\ &\Rightarrow (\mathrm{h} \in \mathcal{F}_\mathrm{i}^0 \cap \mathcal{F}, p' \in \phi_\mathrm{i}^{-1}(\mathrm{C}(\phi_\mathrm{i}(p), \mathrm{a})) \Rightarrow \mathrm{v}(\mathrm{h})(p') = 0))) \; .\end{aligned}$$

**Proof**
It is a direct consequence of propositions 21 (p. 9) and 22 (p. 9) equation (1.10) (p. 9): Keeping the above notations, we have $\mathrm{C}(\phi_\mathrm{i}(p), \mathrm{a})$ open (definition 261 (p. 285)). Then from equation (1.10) (p. 9) we have:

$$q \in \phi_\mathrm{i}^{-1}(\mathrm{C}(\phi_\mathrm{i}(p), \mathrm{a})) \Rightarrow \mathrm{g}(\phi_\mathrm{i}(q) - \phi_\mathrm{i}(p), \mathrm{a}, \mathrm{b}) = 1$$
$$p' \in \phi_\mathrm{i}^{-1}(\mathrm{C}(\phi_\mathrm{i}(p), \mathrm{a})) \Rightarrow \mathrm{v}(\mathrm{h})(p') = \mathrm{v}(1) = 0 \; .$$

**Proposition 24** $\mathbf{O} \subset \mathbf{M}$ being an open, we have:

$$(\mathbf{f} \in \mathcal{F}, \mathbf{v} \in \mathcal{V}, p \in \mathbf{U_k} \cap \mathbf{O}, \mathbf{C}(\phi_\mathbf{k}(p), \mathbf{a}) \subset \mathbf{C}(\phi_\mathbf{k}(p), \mathbf{b}) \subset \phi_\mathbf{k}(\mathbf{U_k} \cap \mathbf{O})) \Rightarrow \mathbf{v}(\mathbf{g}'_p(q, \mathbf{a}, \mathbf{b})\mathbf{f}) \in \mathcal{F}^0_\mathbf{O} \cap \mathcal{F} \,.$$

**Proof**
It is a direct consequence of definition 14 (p. 6) and proposition 23 (p. 10).

**Proposition 25** Keeping the notations of section 1.3.2 and of definition 14 (p. 6), we have:

$$(\forall p' \in \mathbf{M})(\forall \mathbf{k} \in \mathbf{K})(p' \in \mathbf{U_k} \Rightarrow (\exists(\mathbf{a}, \mathbf{b}))(\mathbf{a} \in \mathcal{R}, \mathbf{b} \in \mathcal{R}, 0 < \mathbf{a} < \mathbf{b}, \mathbf{C}(\phi_\mathbf{k}(p), \mathbf{a}) \subset \mathbf{C}(\phi_\mathbf{k}(p), \mathbf{b}) \subset \phi_\mathbf{k}(\mathbf{O} \cap \mathbf{U_k}))) \,. \tag{1.12}$$

That defines, for a $\mathbf{k} \in \mathbf{K}$ such that $p' \in \mathbf{U_k}$, two maps (definition 112 (p. 131) point 4) $\mathbf{a}_{p'}$, $\mathbf{b}_{p'}$ which satisfies (1.12) for any $p'$ and any associated $\mathbf{k}$. Taking those associated values now called $\mathbf{a}$, $\mathbf{b}$ for any $p' \in \mathbf{M}$ and any associated $\mathbf{k}$, we have:

$$(\forall \mathbf{f} \in \mathcal{F})(\forall \mathbf{v} \in \mathcal{V})(\forall p \in \phi_\mathbf{k}^{-1}(\mathbf{C}(\phi_\mathbf{k}(p'), \mathbf{a})))(\mathbf{v}(\mathbf{f})(p) = \mathbf{v}(\mathbf{g}'_{p'}(q, \mathbf{a}, \mathbf{b})\mathbf{f})(p), (\mathbf{x} \in \phi_\mathbf{i}(\mathbf{U_i}) \Rightarrow \mathbf{v}(\mathbf{g}'_{p'}(q, \mathbf{a}, \mathbf{b})\mathbf{f})(\phi_\mathbf{i}^{-1}(\mathbf{x})) \in \mathbf{C}^\infty)) \,.$$

See definition 276 (p. 304).

**Proof**
Apart from the last line, it is a direct consequence of proposition 22 (p. 9) and proposition 23 (p. 10). The last line $\mathbf{v}(\mathbf{g}_{p'}(q, \mathbf{a}, \mathbf{b})\mathbf{f})(\phi^{-1}(\mathbf{x})) \in \mathbf{C}^\infty$ comes from $\mathbf{g}'_{p'}(q, \mathbf{a}, \mathbf{b})\mathbf{f} \in \mathcal{F}$ and from definition 15 (p. 8).

**Proposition 26** Given an open subset $\mathbf{O}$ of a differential manifold $\mathbf{M}$, given a vector field $\mathbf{v}$ of $\mathcal{V}$ and a map $\mathbf{f} \in \mathcal{F}_\mathbf{O}$ (see definition 8 (p. 3)). Using equation (1.12) (p. 11) of proposition 25 (p. 11), one defines for a $\mathbf{k} \in \mathbf{K}$ with $p \in \mathbf{U_k}$ two values $\mathbf{a}_p$, $\mathbf{b}_p$ in $\mathcal{R}^+$. Using them, we define a map $(\mathbf{v_O}(\mathbf{f})(p) : \mathbf{M} \to \mathcal{R})$ by (see definition 353 (p. 454)):

$$\begin{aligned} &p \in \mathcal{C}_\mathbf{M}(\mathbf{O}) \Rightarrow \mathbf{v_O}(\mathbf{f})(p) = 0, \\ &p \in \mathbf{O} \Rightarrow \mathbf{v_O}(\mathbf{f})(p) = \mathbf{v}(\mathbf{g}'_p(q, \mathbf{a}_p, \mathbf{b}_p)\mathbf{f})(p) \,. \end{aligned} \tag{1.13}$$

This map belongs to $\mathcal{F}_\mathbf{O}$ and does not vary if $\mathbf{k}$, $\mathbf{a}_p$, $\mathbf{b}_p$ are replaced by other values satisfying equation (1.12) (p. 11):

$$\begin{aligned} &(\forall \ell \in \mathbf{K})(\forall(\mathbf{a}, \mathbf{b}))((\mathbf{a} \in \mathcal{R}, \mathbf{b} \in \mathcal{R}, 0 < \mathbf{a} < \mathbf{b}, p \in \mathbf{O} \cap \mathbf{U}_\ell, \\ &\mathbf{C}(\phi_\ell(p), \mathbf{a}) \subset \mathbf{C}(\phi_\ell(p), \mathbf{b}) \subset \phi_\ell(\mathbf{O} \cap \mathbf{U}_\ell)) \\ &\Rightarrow \mathbf{v_O}(\mathbf{f})(p) = \mathbf{v}({}^\mathbf{k}\mathbf{g}'_p(q, \mathbf{a}_p, \mathbf{b}_p)\mathbf{f})(p) = \mathbf{v}({}^\ell\mathbf{g}'_p(q, \mathbf{a}, \mathbf{b})\mathbf{f})(p)) \,. \end{aligned} \tag{1.14}$$

**Proof**
From proposition 17 (p. 7), $\mathbf{g}'_p(q, \mathbf{a}_p, \mathbf{b}_p)\mathbf{f} \in \mathcal{F}$ then $\mathbf{v}(\mathbf{g}'_p(q, \mathbf{a}_p, \mathbf{b}_p)\mathbf{f}) \in \mathcal{F}$ is defined and its restriction to $\mathbf{O}$ is in $\mathcal{F}_\mathbf{O}$ (definition 118 (p. 135)).
We also have the relation (1.14):
We choose $\ell, \mathbf{a}, \mathbf{b}$ as defined by the first member of relation (1.14). Since $\phi_\mathbf{k}$ and $\phi_\ell$ are continuous, from definition 141 (p. 165), $\mathbf{O}' = \phi_\mathbf{k}^{-1}(\mathbf{C}(\phi_\mathbf{k}(p), \mathbf{a}_p)) \cap \phi_\ell^{-1}(\mathbf{C}(\phi_\ell(p), \mathbf{a}_p)) \subset \mathbf{O}$, is open and not empty since $p \in \mathbf{O}'$. Both ${}^\mathbf{k}\mathbf{g}'_p(q, \mathbf{a}_p, \mathbf{b}_p)$, ${}^\ell\mathbf{g}'_p(q, \mathbf{a}_p, \mathbf{b}_p)$ are equal to $\mathbf{1}$ when $q \in \mathbf{O}'$. We have then:

$$q \in \mathbf{O}' \Rightarrow {}^\mathbf{k}\mathbf{g}'_p(q, \mathbf{a}_p, \mathbf{b}_p)\mathbf{f}(q) = {}^\ell\mathbf{g}'_p(q, \mathbf{a}_p, \mathbf{b}_p)\mathbf{f}(q) \,.$$

From proposition 22 (p. 9), we shall have:

$$(p' \in \mathrm{O}' \Rightarrow \mathrm{v}({}^{\mathrm{k}}\mathrm{g}'_p(q,\mathrm{a},\mathrm{b})\mathrm{f})(p') = \mathrm{v}({}^{\mathrm{k}}\mathrm{g}'_p(q,\mathrm{a},\mathrm{b})\mathrm{f})(p') = \mathrm{v}_\mathrm{o}(\mathrm{f})(p') \; .$$

Since $p \in \mathrm{O}'$, the above is also valid for $p' = p$.
Let us show now that, in any $\mathrm{U_k}$ covering $\mathrm{O}$, we have:

$$(\mathrm{v}_\mathrm{o}(\mathrm{f})(\phi_\mathrm{k}^{-1}(\mathrm{x})) : \phi_\mathrm{k}(\mathrm{O} \cap \mathrm{U_k}) \subset \mathcal{R}^\mathrm{n} \to \mathcal{R}) \in \mathrm{C}^\infty \; .$$

From propositions 269 (p. 179) and 507 (p. 286), in the open cube (definition 260 (p. 285))

$$\mathrm{C}(\phi_\mathrm{k}(p), \mathrm{a}_p) \subset \mathrm{U_k} \cap \mathrm{O}$$

in which $\mathrm{g}'_p(q, \mathrm{a}_p, \mathrm{b}_p) = 1$, any point $\phi_\mathrm{k}(q) \in \mathrm{C}(\phi_\mathrm{k}(p), \mathrm{a}_p)$ is the center of two closed cubes (definition 152 (p. 179) and proposition 514 (p. 288)):

$$\overline{\mathrm{C}}(\phi_\mathrm{k}(q), \mathrm{a}_q) \subset \overline{\mathrm{C}}(\phi_\mathrm{k}(q), \mathrm{b}_q) \subset \overline{\mathrm{C}}(\phi_\mathrm{k}(p), \mathrm{a}_p) \subset \mathrm{U_k} \cap \mathrm{O} \; .$$

From proposition 22 (p. 9), we have:

$$\begin{aligned} &(\forall q \in \phi_\mathrm{k}^{-1}(\mathrm{C}(\phi_\mathrm{k}(p), \mathrm{a}_p)))(\mathrm{v}(\mathrm{g}'_q(q', \mathrm{a}_q, \mathrm{b}_q)\mathrm{f}(q'))(q) = \mathrm{v}_\mathrm{o}(\mathrm{f})(q) \\ &= \mathrm{v}(\mathrm{g}'_p(p', \mathrm{a}_p, \mathrm{b}_p)\mathrm{f}(p'))(q)) \; . \end{aligned} \tag{1.15}$$

Since (proposition 24 (p. 11)) $\mathrm{v}(\mathrm{g}'_p(p', \mathrm{a}_p, \mathrm{b}_p)\mathrm{f}(p'))(q) \in \mathcal{F}^0_\mathrm{k} \cap \mathcal{F}^0_\mathrm{O} \cap \mathcal{F}$, we have (see footnote 3 (p. 6)):

$$\mathrm{x} \in \mathrm{U_k} \Rightarrow \mathrm{v}(\mathrm{g}'_p(p', \mathrm{a}_p, \mathrm{b}_p)\mathrm{f}(p'))(\phi_\mathrm{k}^{-1}(\mathrm{x})) \in \mathrm{C}^\infty_\mathrm{x} \; . \tag{1.16}$$

Using proposition 556 (p. 305) with equations (1.15) (p. 12) and (1.16) (p. 12), we have:

$$\mathrm{x} = \phi_\mathrm{k}^{-1}(p) \Rightarrow \mathrm{v}_\mathrm{o}(\mathrm{f})(\phi_\mathrm{k}^{-1}(\mathrm{x})) \in \mathrm{C}^\infty_\mathrm{x} \; .$$

It means:

$$\mathrm{f} \in \mathcal{F}_\mathrm{O} \Rightarrow \mathrm{v}_\mathrm{o}(\mathrm{f}) \in \mathcal{F}_\mathrm{O} \; .$$

**Proposition 27** Using notations of proposition 26 (p. 11), the map $(\mathrm{v}_\mathrm{o} : \mathcal{F}_\mathrm{O} \to \mathcal{F}_\mathrm{O})$ verifies equations (1.1), (1.2) of proposition 12 (p. 5) and (1.7), (1.8) of the definition of a vector field (definition 15 (p. 8)) on $\mathrm{O}$.

**Proof**
From the definition of $\mathrm{g}'_p(p', \mathrm{a}_p, \mathrm{b}_p)$, for $\mathrm{f} \in \mathcal{F}_\mathrm{O}$ and $\mathrm{h} \in \mathcal{F}_\mathrm{O}$ we have:

$$\begin{aligned} &(\mathrm{fh} \in \mathcal{F}_\mathrm{O}, p \in \mathrm{U_k} \cap \mathrm{O}, p' \in \mathrm{C}(\phi(p), \mathrm{a}_p)) \Rightarrow \mathrm{f}(p')\mathrm{h}(p') = (\mathrm{g}'_p\mathrm{fh})(p') \\ &= (\mathrm{g}'_p\mathrm{f})(p') \times (\mathrm{g}'_p\mathrm{h})(p') \; . \end{aligned}$$

From proposition 22 (p. 9), we have:

$$\begin{aligned} &\mathrm{v}_\mathrm{o}(\mathrm{fh})(p) = \mathrm{v}(\mathrm{g}'_p\mathrm{fh})(p) = \mathrm{v}(\mathrm{g}'_p\mathrm{f}\mathrm{g}'_p\mathrm{h})(p) \\ &= (\mathrm{g}'_p\mathrm{h}\mathrm{v}(\mathrm{g}'_p\mathrm{f}))(p) + (\mathrm{g}'_p\mathrm{f}\mathrm{v}(\mathrm{g}'_p\mathrm{h}))(p) \\ &= (\mathrm{h}\mathrm{v}_\mathrm{o}(\mathrm{f}))(p) + (\mathrm{f}\mathrm{v}_\mathrm{o}(\mathrm{h}))(p) \Rightarrow \mathrm{v}_\mathrm{o}(\mathrm{fh}) = \mathrm{h}\mathrm{v}_\mathrm{o}(\mathrm{f}) + \mathrm{f}\mathrm{v}_\mathrm{o}(\mathrm{h}) \; . \end{aligned} \tag{1.17}$$

Then $\mathrm{v}_\mathrm{o} \in \mathcal{V}_\mathrm{O}$.

**Definition 18 Restriction of a vector field to an open subset of a differentiable manifold.**

Keeping the notation of proposition 26 (p. 11), one calls restriction of $\mathbf{v} \in \mathcal{V}$ to an open O of M, the vector field of $\mathbf{v}_o \in \mathcal{V}_O$ defined by:

$$(\forall \mathrm{f} \in \mathcal{F}_O)(\forall p \in O)(\mathbf{v}_o(\mathrm{f})(p) = \mathbf{v}(\mathrm{g}'_p(q, \mathrm{a}_p, \mathrm{b}_p)\mathrm{f}(q))(p) .$$

**Proposition 28** Keeping notation of proposition 5 (p. 3), given an open $O \in M$, a map $\mathrm{f} \in \mathcal{F}$, a vector field $\mathbf{v} \in \mathcal{V}$, its restriction (definition 18 (p. 13)) $\mathbf{v}_o \in \mathcal{V}_o$ of $\mathbf{v}$ to O, we have:

$$p \in O \Rightarrow \mathbf{v}(\mathrm{f})(p) = \mathbf{v}_o(\mathrm{f})(p) .$$

**Proof**
From proposition 5 (p. 3), $\mathrm{f} \in \mathcal{F} \Rightarrow \mathrm{f} \in \mathcal{F}_O$. From definition 18 (p. 13) and proposition 22 (p. 9) we have:

$$\mathbf{v}_o(\mathrm{f})(p) = \mathbf{v}(\mathrm{g}'_p\mathrm{f})(p) = \mathbf{v}(\mathrm{f})(p) .$$

**Proposition 29** Given an open O of a manifold M, considering O as a restricted manifold of M with its set of smooth map $\mathcal{F}_O$ and its set of vector field $\mathcal{V}_o$, for any $p \in O$ we consider $\mathrm{g}'_p(p', \mathrm{a}, \mathrm{b}) \in \mathcal{F} \cap \mathcal{F}^0_O)$ with a and b defined as in definition 14 (p. 6) (see proposition 17 (p. 7)). A vector field $\mathbf{v}_o \in \mathcal{V}_o$ multiplied by $\mathrm{g}'_p(p', \mathrm{a}, \mathrm{b})$ can be extended to a vector field on M written as $\mathrm{g}'_p(p', \mathrm{a}, \mathrm{b})\mathbf{v}_o$.

**Proof**
An $\mathrm{f} \in \mathcal{F}$ can be restricted to O (definition 118 (p. 135)). It allows to define $\mathbf{v}_o(\mathrm{f}) \in \mathcal{F}^0_O$ by equation (1.13) (p. 11) of proposition 25 (p. 11) and from proposition 17 (p. 7), we have:

$$\mathrm{g}'_p(p', \mathrm{a}, \mathrm{b})\mathbf{v}_o(\mathrm{f}) \in \mathcal{F} .$$

It satisfies definition 15 (p. 8):

- equation (1.7) (p. 8).

$$(\mathrm{f} \in \mathcal{F}, \mathrm{g} \in \mathcal{F}) \Rightarrow \mathrm{g}'_p(p', \mathrm{a}, \mathrm{b})\mathbf{v}_o(\mathrm{fg})(p') =$$
$$\mathrm{g}'_p(p', \mathrm{a}, \mathrm{b})\mathbf{v}_o(\mathrm{f})(p')\mathrm{g}(p') + \mathrm{f}(p')\mathrm{g}'_p(p', \mathrm{a}, \mathrm{b})\mathbf{v}_o(\mathrm{g})(p') .$$

- equation (1.8) (p. 8)

$$(\mathrm{f} \in \mathcal{F}, \mathrm{g} \in \mathcal{F}) \Rightarrow \mathrm{g}'_p(p', \mathrm{a}, \mathrm{b})\mathbf{v}_o(\mathrm{f}+\mathrm{g})(p') =$$
$$\mathrm{g}'_p(p', \mathrm{a}, \mathrm{b})\mathbf{v}_o(\mathrm{f})(p') + \mathrm{g}'_p(p', \mathrm{a}, \mathrm{b})\mathbf{v}_o(\mathrm{g})(p') .$$

**Proposition 30** Keeping notations of proposition 29 (p. 13) with $p \in U_k \cap O$ and $C(\phi_k(p), \mathrm{a}) \subset C(\phi_k(p), \mathrm{b}) \subset U_k \cap O$

$$p' \in \phi_k^{-1}(C(\phi_k(p), \mathrm{a})) \Rightarrow \mathrm{g}'_p(p', \mathrm{a}, \mathrm{b})\mathbf{v}_o(\mathrm{f})(p') = \mathbf{v}_o(\mathrm{f})(p') .$$

**Proof**
It is a direct consequence of proposition 23 (p. 10) and equation (1.7) of definition 15 (p. 8).

**Definition 19 $\mathbf{v_k}$, the restriction of a vector field $\mathbf{v} \in \mathcal{V}$ to $\mathbf{U_k}$.**

With the notation of proposition 25 (p. 11), given a manifold $\mathbf{M}$ and a vector field $\mathbf{v} \in \mathcal{V}$, one usually calls $\mathbf{v_k}$ the restriction of $\mathbf{v}$ to one of the covering sets $\mathbf{U_k}$ entering in the definition of $\mathbf{M}$.

**Proposition 31** With the notations given at the beginning of section 1.3.1 (p. 8) and in definition 15 (p. 8), we have:

$$(\forall \mathbf{v} \in \mathcal{V})(\forall \mathbf{i} \in \mathbf{K})(\exists \mathbf{h} \in (\mathcal{F}_\mathbf{i})^\mathbf{n})((\mathbf{h}^\mu : [1,\mathbf{n}] \to \mathcal{F}_\mathbf{i}), (\forall \mathbf{f} \in \mathcal{F})(\forall \mathbf{x} \in \phi_\mathbf{i}(\mathbf{U_i})) \\ (\mathbf{v}(\mathbf{f})(\phi_\mathbf{i}^{-1}(\mathbf{x})) = \frac{\partial \mathbf{f}(\phi_\mathbf{i}^{-1}(\mathbf{x}))}{\partial \mathbf{x}^\mu}\mathbf{h}^\mu(\mathbf{x}))) \; . \tag{1.18}$$

**Proof**

We take a $\mathbf{v} \in \mathcal{V}$ and an $\mathbf{i} \in \mathbf{K}$. If $\mathbf{U_i}$ is empty, 1.18 is trivially true. We suppose now $\mathbf{U_i} \neq \emptyset$ then $\exists p' \in \mathbf{U_i}$. We shall use now that $p'$ and set $\mathbf{x}' = \phi_\mathbf{i}(p')$. $\phi_\mathbf{i}^{-1}$ is continuous, $\mathbf{U_i}$ is open. From definition 141 (p. 165) $\phi_\mathbf{i}(\mathbf{U_i})$ is open. From definition 152 (p. 179), proposition 514 (p. 288), propositions 269 (p. 179) and 507 (p. 286), we have:

$$(\exists(\mathbf{a},\mathbf{b}))(0 < \mathbf{a} < \mathbf{b}, \overline{\mathbf{C}}(\phi_\mathbf{i}(p'),\mathbf{a}) \subset \overline{\mathbf{C}}(\phi_\mathbf{i}(p'),\mathbf{b}) \subset \phi_\mathbf{i}(\mathbf{U_i})) \; .$$

We use definition 14 (p. 6) to define:

$$\mathbf{g}'_{p'}(q) = \mathbf{g}'_{p'}(q,\mathbf{a},\mathbf{b}) = \mathbf{g}(\phi_\mathbf{i}(q) - \phi_\mathbf{i}(p'),\mathbf{a},\mathbf{b}) \; .$$

From definition 15 (p. 8), equation (1.7) and proposition 23 (p. 10), we have[4]:

$$(\forall \mathbf{v} \in \mathcal{V})(p \in \phi_\mathbf{i}^{-1}(\mathbf{C}(\mathbf{x}',\mathbf{a})) \Rightarrow \mathbf{v}\{\mathbf{g}'_{p'}\}(p) = \mathbf{v}\{\mathbf{g}'^{\,2}_{p'}\}(p) = \mathbf{v}\{1\}(p) = 0) \; . \tag{1.19}$$

Considering a smooth map $\mathbf{f} \in \mathcal{F}$, for any point $\mathbf{x} \in \mathbf{C}(\mathbf{x}',\mathbf{a})$, we can perform the Taylor expansion using proposition 658 (p. 401) (see definition 117 (p. 134)):

$$\mathbf{f} \circ \phi_\mathbf{i}^{-1}(\mathbf{x}) = \mathbf{f} \circ \phi_\mathbf{i}^{-1}(\mathbf{x}') + (\mathbf{x}^\mu(\mathbf{x}) - \mathbf{x}'^\mu)\frac{\partial \mathbf{f} \circ \phi_\mathbf{i}^{-1}(\mathbf{x}')}{\partial \mathbf{x}^\mu} \\ + 0.5(\mathbf{x}^\mu(\mathbf{x}) - \mathbf{x}'^\mu)(\mathbf{x}^\nu(\mathbf{x}) - \mathbf{x}'^\nu)(\frac{\partial^2 \mathbf{f} \circ \phi_\mathbf{i}^{-1}(\mathbf{x}')}{\partial \mathbf{x}^\mu \partial \mathbf{x}^\nu} + \eta_{\mu\nu}(\mathbf{x})) \; . \tag{1.20}$$

According to proposition 658 (p. 401), the functions $\eta_{\mu\nu}(\mathbf{x})$ belong to $\mathbf{C}^\infty$ (see definition 276 (p. 304)) defined in $\mathcal{R}^\mathbf{n}$. $\mathbf{x}'$ is here a fixed value, $\mathbf{x}$ is the variable.
$(\mathbf{x}^\mu(\mathbf{x}) : \mathcal{R}^\mathbf{n} \to \mathcal{R})$ gives the coordinate $\mu$ of $\mathbf{x}$: $\mathbf{x}^\mu(\mathbf{x}) = \mathbf{x}^\mu$ [5]. From proposition 554 (p. 305), it belongs to $\mathbf{C}^\infty$ in $\mathcal{R}^\mathbf{n}$. We have:

$$\mathbf{g}'_{p'}(p)\mathbf{x}^\mu(\phi_\mathbf{i}(p)) \in \mathcal{F}, \; \mathbf{g}'^{\,2}_{p'}(p)\mathbf{x}^\mu(\phi_\mathbf{i}(p)) \in \mathcal{F} \; .$$

We have with $p = \phi_\mathbf{i}^{-1}(\mathbf{x}) \in \mathbf{U_i}$:

$$\mathbf{f}(p)\mathbf{g}'^{\,2}_{p'}(p) = \mathbf{f} \circ \phi_\mathbf{i}^{-1}(\mathbf{x})\mathbf{g}^2(\mathbf{x} - \mathbf{x}',\mathbf{a},\mathbf{b}) = \\ \mathbf{g}^2(\mathbf{x} - \mathbf{x}',\mathbf{a},\mathbf{b})\mathbf{f} \circ \phi_\mathbf{i}^{-1}(\mathbf{x}') + \mathbf{g}^2(\mathbf{x} - \mathbf{x}',\mathbf{a},\mathbf{b})(\mathbf{x}^\mu(\mathbf{x}) - \mathbf{x}'^\mu)\frac{\partial \mathbf{f} \circ \phi_\mathbf{i}^{-1}(\mathbf{x}')}{\partial \mathbf{x}^\mu} + \\ 0.5\mathbf{g}(\mathbf{x} - \mathbf{x}',\mathbf{a},\mathbf{b})(\mathbf{x}^\mu(\mathbf{x}) - \mathbf{x}'^\mu)\mathbf{g}(\mathbf{x} - \mathbf{x}',\mathbf{a},\mathbf{b})(\mathbf{x}^\nu(\mathbf{x}) - \mathbf{x}'^\nu) \\ (\frac{\partial^2 \mathbf{f} \circ \phi_\mathbf{i}^{-1}(\mathbf{x}')}{\partial \mathbf{x}^\mu \partial \mathbf{x}^\nu} + \eta_{\mu\nu}(\mathbf{x})) \tag{1.21}$$

---

[4] To be clearer, we write for a while $\mathbf{v}(\mathbf{f})$ as $\mathbf{v}\{\mathbf{f}\}$ .

[5] $\mathbf{x}^\mu(\mathbf{x}) = \mathbf{e}^\mu(\mathbf{x})$, as defined in definition 213 (p. 244) when $\mathbf{e}_\mu = \prod^{\mathbf{e}}_{\nu \in [1,\mathbf{n}]} \delta^\nu_\mu = (\delta^\nu_\mu : [1,\mathbf{n}] \to \mathcal{R}) \in \mathcal{R}^\mathbf{n}$, is the reference basis of $\mathcal{R}^\mathbf{n}$ (see definition 242 (p. 267))

and when $p \notin U_i$: $f(p)g'^{\,2}_{p'}(p) = 0$ .
Applying $v \in \mathcal{V}$ to $f(p)g'^{\,2}_{p'}(p)$ and evaluating at $p'$, from equation (1.19) (p. 14) and (1.7) (p. 8) in definition 15 (p. 8), we have:

$$v\{f(p)g'^{\,2}_{p'}(p)\}(p') = v\{f\}(p') .$$

Now we apply $v$ to the second member of equation (1.21) (p. 14). We do it first on line 2 of (1.21). On that line, with $x = (\phi_i(p))$, remembering that $x'$ is a fixed term which means for instance (see proposition 21 (p. 9)): $v\{f(\phi^i(x'))\} = 0$, applying equations (1.7) (p. 8) and (1.19) (p. 14), we have:

$$v\{g^2(x-x',a,b)f\circ\phi_i^{-1}(x') + g^2(x-x',a,b)(x^\mu(x) - x'^\mu)\frac{\partial f\circ\phi_i^{-1}(x')}{\partial x^\mu}\}(p') =$$
$$= \frac{\partial f\circ\phi_i^{-1}(x')}{\partial x^\mu} v\{g^2(x-x',a,b)x^\mu(x)\} .$$

We use for that with $x'$ fixed (see equation (1.19) (p. 14)):

$$\begin{aligned} &v\{g^2(\phi_i(p)-x',a,b)f\circ\phi_i^{-1}(x')\}(p')\\ &= f\circ\phi_i^{-1}(x')v\{g^2(\phi_i(p))-x',a,b)\}(p') = 0\\ &v\{g^2(x-x',a,b)x'^\mu\}(p') = 0 .\end{aligned}$$

From equation (1.7) in definition 15 (p. 8), we have:

$$\begin{aligned} &v\{g^2(\phi_i(p)-x',a,b)x^\mu(\phi_i(p))\}(p') = x^\mu(x')g(0,a,b)v\{g(\phi_i(p)-x',a,b)\}(p')\\ &+ g(0,a,b)v\{g(\phi_i(p)-x',a,b)x^\mu(p)\}(p')\\ &= v\{g(\phi_i(p)-x',a,b)x^\mu(p)\}(p') = v\{g'_{p'}x^\mu(p)\} .\end{aligned}$$

The second line of (1.21) reduced to

$$v\{g'_{p'}x^\mu(p)\}\frac{\partial f\circ\phi_i^{-1}(x')}{\partial x^\mu} .$$

The last two lines of (1.21) can be considered as the product of three terms of $\mathcal{F}$:

$$\begin{aligned} &0.5g(x-x',a,b)(x^\mu(x) - x'^\mu)\\ &\times g(x-x',a,b)(x^\nu(x) - x'^\nu)\\ &\times (\frac{\partial^2 f\circ\phi_i^{-1}(x')}{\partial x^\mu \partial x^\nu} + \eta_{\mu\nu}(x)) .\end{aligned}$$

When $v \in \mathcal{V}$ is acting on that product of three terms of $\mathcal{F}$, from definition 15 (p. 8) and equation (1.7) (p. 8), it is the sum of the action on each term multiplied by the other two. When $p$ is set to $p'$, one of the two terms is zero. The last two lines give zero when $p = p'$. Setting:

$$h^\mu(x') = v\{g'_{p'}x^\mu(p)\}(\phi_i^{-1}(x')) = v\{g'_{p'}x^\mu\}(x') \tag{1.22}$$

we have proposition 1.18 (p. 14). From proposition 25 (p. 11), we have $h(x) \in C^\infty$ (see definition 276 (p. 304)).

**Definition 20 Coordinate maps of $\mathcal{F}_k$.**

Keeping the notations of proposition 31 (p. 14), for any $\mathbf{k} \in \mathbf{K}$, we consider $\mathbf{U_k}$ as a manifold restriction of $\mathbf{M}$ (definition 5 (p. 3)) with $\mathcal{F}_k$ its set of smooth maps (definition 8 (p. 3)) $((\mathbf{f_k} : \mathbf{U_k} \to \mathcal{R}) \in \mathcal{F}_k)$. We call coordinate maps $\mathbf{x}_k^\mu \in \mathcal{F}_k$, the map:

$$(\mathbf{x}_k^\mu = \phi_k^\mu(\mathbf{p}) : \mathbf{U_k} \to \mathcal{R})\ .$$

In the notation of definition 213 (p. 244), if $\mathbf{e}_\mu = \prod^{\mathbf{e}}_{\nu \in [1,\mathbf{n}]} \delta_\mu^\nu = (\delta_\mu^\nu : [1,\mathbf{n}] \to \mathcal{R}) \in \mathcal{R}^\mathbf{n}$ is the reference basis of $\mathcal{R}^\mathbf{n}$ (see definition 242 (p. 267)), we have:

$$\mathbf{x}_k^\mu = \mathbf{e}^\mu,\ \mathbf{x} \in \mathbf{U_k} \subset \mathcal{R}^\mathbf{n} \Rightarrow \mathbf{x}_k^\mu(\mathbf{x}) = \mathbf{e}^\mu(\mathbf{x}) \Rightarrow \mathbf{x} = \mathbf{x}_k^\mu \mathbf{e}_\mu$$

$(\mathbf{e}^\mu : [1,\mathbf{n}] \to (\mathcal{R}^\mathbf{n})^* \subset \mathcal{R}^\mathbf{n} \times \mathcal{R})$ is the dual basis of the reference basis of $\mathcal{R}^\mathbf{n}$ that is to say the reference basis of the vectorial space $(\mathcal{R}^\mathbf{n})^*$ generated by the linear maps on $\mathcal{R}^\mathbf{n}$.

**Definition 21 Basic vector field maps $(\partial_\mu : \mathcal{F}_k \to \mathcal{F}_k)$.**

Keeping the notations of proposition 31 (p. 14) and definition 20 (p. 15), in a covering set $\mathbf{U_k}$ of a differential manifold $\mathbf{M}$, one calls $(\partial_\mu : \mathcal{F}_k \to \mathcal{F}_k)$, the maps defined by:

$$(\forall \mathbf{f} \in \mathcal{F}_k)(\forall p \in \mathbf{U_k})\ (\partial_\mu\{\mathbf{f}\}(p) = \frac{\partial \mathbf{f} \circ \phi_k^{-1}}{\partial \mathbf{x}^\mu}(\phi_k(p)))$$

$\mathcal{F}_k$ is the set of smooth maps of $\mathbf{U_k}$ (see definition 8 (p. 3)).

**Proposition 32** Given a partial covering $\mathbf{U_k}$ of a differential manifold $\mathbf{M}$ of dimension $\mathbf{n}$, given its associated map $(\phi_k : \mathbf{U_k} \to \mathcal{R}^\mathbf{n})$, we have (see definition 18 (p. 13)):

$$(\forall \mu \in [1,\mathbf{n}])\ (\partial_\mu \in \mathcal{V}_k)\ .$$

**Proof**

$(\mathbf{f_k} \circ \phi_k^{-1} : \phi(\mathbf{U_k}) \subset \mathcal{R}^\mathbf{n} \to \mathcal{R})$ is a $\mathbf{C}^\infty$ map (see definition 276 (p. 304)) between an open of $\mathcal{R}^\mathbf{n}$ and $\mathcal{R}$ which has partial derivatives for any point of $\phi(\mathbf{U_k}) \subset \mathcal{R}^\mathbf{n}$. Any of such partial derivative $\partial_\mu(\mathbf{f_k} \circ \phi_k^{-1})$, $\mu \in [1,\mathbf{n}]$ is also a $\mathbf{C}^\infty$ map $(\partial_\mu(\mathbf{f_k} \circ \phi_k^{-1}) : \phi(\mathbf{U_k}) \subset \mathcal{R}^\mathbf{n} \to \mathcal{R})$ which belongs then also to $\mathcal{F}_k$ (see footnote 4 (p. 14)):

$$(\partial_\mu\{\mathbf{f_k}\} = \partial_\mu(\mathbf{f_k} \circ \phi_k^{-1})(\phi_k(\mathbf{p})) : \mathbf{U_k} \to \mathcal{R}) \in \mathcal{F}_k\ .$$

The map $\partial_\mu$ associated to $\phi_k$ is a map $(\partial_\mu : \mathcal{F}_k \to \mathcal{F}_k)$ which verifies (see proposition 325 (p. 201)):

$$\begin{aligned}(\mathbf{f} \in \mathcal{F}_k, \mathbf{h} \in \mathcal{F}_k) &\Rightarrow \partial_\mu\{\mathbf{fh}\}(\mathbf{p}) = \partial_\mu(\mathbf{f} \circ \phi_k^{-1} \times \mathbf{h} \circ \phi_k^{-1})(\phi_k(\mathbf{p}))\\ &= (\partial_\mu(\mathbf{f} \circ \phi_k^{-1}) \times \mathbf{h} \circ \phi_k^{-1})(\phi_k(\mathbf{p})) + (\mathbf{f} \circ \phi_k^{-1} \times \partial_\mu(\mathbf{h} \circ \phi_k^{-1}))(\phi_k(\mathbf{p}))\\ &= \partial_\mu\{\mathbf{f}\}\mathbf{h}(\mathbf{p}) + \mathbf{f}\partial_\mu\{\mathbf{h}\}(\mathbf{p})\ .\end{aligned}$$

Then $\partial_\mu \in \mathcal{V}_k$.

**Proposition 33** With the above notations, in any covering set $\mathbf{U_k}$ of $\mathbf{M}$, we have:

$$(\forall \mathbf{v} \in \mathcal{V}_k)(\exists(\mathbf{v}^\mu : [1,\mathbf{n}] \to \mathcal{F}_k))((\forall \mu \in [1,\mathbf{n}])(\mathbf{v}^\mu = \mathbf{v}\{\phi_k^\mu(q)\}), \mathbf{v} = \mathbf{v}^\mu \partial_\mu)\ .$$

**Proof**
It is a direct consequence of proposition 31 (p. 14) equation (1.18) and equation (1.22) (p. 15).

**Proposition 34** With the above notations, in any covering set $\mathbf{U_k}$ of $\mathbf{M}$, for any $\mathbf{v} \in \mathcal{V}$, we may consider its restriction $\mathbf{v_k} \in \mathcal{V}_\mathbf{k}$ to $\mathbf{U_k}$ and for it we have:

$$(\exists(\mathbf{v}^\mu_\mathbf{k} : [1,\mathbf{n}] \to \mathcal{F}_\mathbf{k}))((\forall\mu \in [1,\mathbf{n}])(\mathbf{v}^\mu_\mathbf{k} = \mathbf{v_k}\{\phi^\mu_\mathbf{k}(q)\}), \mathbf{v_k} = \mathbf{v}^\mu_\mathbf{k}\partial_\mu) \ .$$

Which means that for all $\mathbf{f} \in \mathcal{F}$, calling $\mathbf{f_k}$ the restriction of $\mathbf{f}$ to $\mathbf{U_k}$ (definition 118 (p. 135)), we have:

$$(\forall p \in \mathbf{U_k})\ (\mathbf{v}(\mathbf{f})(p) = \mathbf{v_k}(\mathbf{f_k})(p) = \mathbf{v}^\mu_\mathbf{k}\partial_\mu\mathbf{f}(p)) \ .$$

**Proof**
It is a consequence of definitions 18 (p. 13), 19 (p. 14), propositions 28 (p. 13), 33 (p. 16).

**Proposition 35** With the above notations, in any covering set $\mathbf{U_k}$ of $\mathbf{M}$, we have:

$$((\lambda^\mu : [1,\mathbf{n}] \to \mathcal{F}_\mathbf{k}), \lambda^\mu\partial_\mu = 0) \Rightarrow (\forall\nu \in [1,\mathbf{n}])(\lambda^\nu = 0) \ .$$

**Proof**
Since $(\forall p \in \mathbf{U_k})(\mathbf{x}^\nu(p) = \phi^\nu_\mathbf{k}(p) \in \mathcal{F}_\mathbf{k})$ we have:

$$\lambda^\mu\partial_\mu = 0 \Rightarrow (\forall\nu \in [1,\mathbf{n}])(0 = \lambda^\mu\partial_\mu\mathbf{x}^\nu = \lambda^\mu\delta^\nu_\mu = \lambda^\nu) \Rightarrow (\forall\nu \in [1,\mathbf{n}])(\lambda^\nu = 0) \ .$$

**Definition 22 Finite basis of a set of vector field.**

Since a set of vector fields is a module (definition 182 (p. 218)) over the commutative ring $\mathcal{F}$, we call a finite basis of a set of vector fields, a set of vector fields which satisfy the definition 184 (p. 219).

**Definition 23 $(\partial_\mu : [1,\mathbf{n}] \to \mathcal{V}_\mathbf{k})$, canonical basic vectors in the local frame.**

From proposition 33 (p. 16) and definition 22 (p. 17), $(\partial_\mu : [1,\mathbf{n}] \to \mathcal{V}_\mathbf{k})$ is a local basis of $\mathcal{V}_\mathbf{k}$. We call it canonical basis. It is related to the map $(\phi_\mathbf{k} : \mathbf{U_k} \to \mathcal{R}^\mathbf{n})$. Most of the time, one changes the basis by changing the map $\phi_\mathbf{k}$ and the term canonical is not really necessary, see convention 1 (p. 18).

**Proposition 36** A canonical basis (definition 23 (p. 17)) is an independent basis (definition 185 (p. 219)).

**Proof**
It is a direct consequence of propositions 33 (p. 16) and 35 (p. 17).

**Definition 24 Components of a vector field of a local manifold $\mathbf{U_k} \subset \mathbf{M}$.**

Given a basis $(\mathbf{h_i} : \mathbf{I} \to \mathcal{V}_\mathbf{k})$ of a local covering $\mathbf{U_k} \subset \mathbf{M}$, the map $\mathbf{v^i}$ associated to a vector $\mathbf{v} \in \mathcal{V}_\mathbf{k}$ in definition 22 (p. 17) is called components of the vector field in that basis. The $\mathbf{v}^\mu \in \mathcal{F}_\mathbf{k}$ of the above proposition 33 are called the components in the canonical basis. Those components will be practically the only one used, see convention 1 (p. 18).

**Definition 25 Local components of a vector field.**

Given a vector field $\mathbf{v} \in \mathcal{V}$, its local components relative to a cover $\mathbf{U_k}$ of $\mathbf{M}$ are the components (definition 24 (p. 17)) of its restriction $\mathbf{v_k}$ in $\mathbf{U_k}$ (definition 18 (p. 13)).

**Convention 1** $(\partial_\mu : [1,n] \to \mathcal{V}_k)$.
When it is not specified otherwise, the local basis used for a covering set $U_k \subset M$ is the canonical basis $(\partial_\mu : [1,n] \to \mathcal{V}_k)$.

**Proposition 37** In the above definition 25 (p. 17), we take $I = [1,n]$ and consider $(h_i : [1,n] \to \mathcal{V}_k)$ a basis of $\mathcal{V}_k$. From proposition 33 (p. 16), we may write: $h_i = h_i^j \partial_j$ with $(h_i^j : [1,n]^2 \to \mathcal{F}_k^0)$ and we have:

$$(\forall p \in U_k)\ (\det(h_i^j(p)) \neq 0)\ .$$

**Proof**
From definition 22 (p. 17) and proposition 35 (p. 17), we should have: $(h'^j_i : [1,n]^2 \to \mathcal{F}_k^0)$ such that

$$\partial_i = h'^j_i h_j = h'^j_i h_j^k \partial_k \Rightarrow h'^j_i h_j^k = \delta_i^k\ .$$

Then we have, from proposition 448 (p. 257):

$$(\forall p \in U_k)\ (h'^j_i(p) h_j^k(p) = \delta_i^k \Rightarrow \det(h'^j_i(p))\det(h_i^j(p)) = 1$$
$$\Rightarrow (\det(h'^j_i(p))) \neq 0, \det(h_i^j(p)) \neq 0)\ .$$

**Proposition 38** Taking again the premise of proposition 37 (p. 18), with $(h_i : [1,n] \to \mathcal{V}_k)$ considered as a set of vector fields of $\mathcal{V}_k$ with $h_i = h_i^j \partial_j$, $(h_i^j : [1,n]^2 \to \mathcal{F}_k^0)$ but verifying now:

$$(\forall p \in U_k)\ (\det(h_i^j(p)) \neq 0)$$

then $(h_i : [1,n] \to \mathcal{V}_k)$ is a finite independent basis of the local set of vector field definitions 184 (p. 219), 185 (p. 219).

**Proof**
From proposition 452 (p. 258) [6]

$$\det(h_i^j(p)) \neq 0 \Rightarrow (\exists (h'^i_j(p) : [1,n]^2 \to \mathcal{R}))(h'^\ell_j(p) h_\ell^i(p) = \delta_j^i)\ .$$

From definitions 221 (p. 250), 227 (p. 252), equation (2.93) (p. 252), the **det** map is a sum of products of a subset of its coefficients which are in $C^\infty$. From proposition 331 (p. 203), it is also a $C^\infty$ map:

$$(h_i^j : [1,n]^2 \to \mathcal{F}_k^0) \Rightarrow (\det(h_i^j(p)) \in \mathcal{F}_k^0)\ .$$

From proposition 455 (p. 259), any maps $h'^i_k$ are sums of products of subsets of set of maps $h_k^i \in C^\infty$ multiplied by $\frac{1}{\det(h_k^l)}$.
$\det(h_i^j)$ is different from zero and is in $C^\infty$. From proposition 332 (p. 205), we have $\frac{1}{\det(h_k^l)} \in C^\infty$. Then from proposition 331 (p. 203) we have:

$$(\forall p \in U_k)\ (\det(h_i^j(p)) \neq 0) \Rightarrow (\forall i \in [1,n])(\forall j \in [1,n])(h'^i_j \in \mathcal{F}_k^0)$$

then $(h_i : [1,n] \to \mathcal{V}_k)$ is a basis of the local set of vector field because we have:

$$(\forall v \in \mathcal{V}_k)(\exists (v^i : [1,n] \to \mathcal{F}_k^0))\ (v = v^i \partial_i = v^i \delta_i^j \partial_j = v^i h'^\ell_i h_\ell^j \partial_j = v^i h'^\ell_i h_\ell)\ .$$

Since $\partial_j : [1,n] \to \mathcal{V}_k$ are an independent basis (proposition 36 (p. 17)), using proposition 453 (p. 259), we have:

$$\lambda^i h_i = 0 \Rightarrow \lambda^i h_i^j \partial_j = 0 \Rightarrow (\forall j \in [1,n])\ (\lambda^i h_i^j = 0)$$
$$\Rightarrow (\forall j \in [1,n])(\forall p \in U_k)\ (\lambda^i(p) h_i^j(p) = 0)$$
$$\Rightarrow (\forall i \in [1,n])(\forall p \in U_k)\ (\lambda^i(p) = 0)\ .$$

[6] Throughout this book, we always have summation over repeated indices when they do not have special meaning defined earlier. See also convention 12 (p. 128).

## 1.4 Conventions

As usual, we shall adopt the following conventions otherwise the writing becomes too heavy:

**Convention 2 $\mathbf{f}$, $\partial_\mu \mathbf{f}$.**

When one deals with a map $\mathbf{f} \in \mathcal{F}$ and we are considering a covering set $\mathbf{U_k} \subset \mathbf{M}$, we call by the same name $\mathbf{f}$ the restriction of $\mathbf{f}$ to $\mathbf{U_k}$ so in $\mathcal{F}_\mathbf{k}$ (strictly speaking it is not the same element). That allows to write $\partial_\mu \mathbf{f}$ with $\partial_\mu \in \mathcal{V}_\mathbf{k}$ although it is a map $\mathcal{F}_\mathbf{k} \to \mathcal{F}_\mathbf{k}$.

**Convention 3 $\mathbf{f} \in \mathcal{F}_\mathbf{k}$, $\mathbf{v(f)}$.**

When one deals with a vector field of $\mathcal{V}$, we shall use often the same name for its restriction to a covering element $\mathbf{U_k}$ of $\mathbf{M}$ for instance we may write, with $\mathbf{f} \in \mathcal{F}_\mathbf{k}$, $\mathbf{v(f)}$ for $\mathbf{v_k(f)}$ where $\mathbf{v_k}$ is the restriction of $\mathbf{v}$ in $\mathcal{V}_\mathbf{k}$ (see definition 18 (p. 13)).

**Convention 4 $\mathbf{U_k} \to \mathbf{U}$, $\mathcal{F}_\mathbf{k} \to \mathcal{F}_\mathbf{U}$, $\mathcal{V}_\mathbf{k} \to \mathcal{V}_\mathbf{U}$.**

When one deals with a cover $\mathbf{U_k} \subset \mathbf{M}$, without caring which one in particular, it is clearer to suppress the index $\mathbf{k}$ and to consider instead a covering set $\mathbf{U} \subset \mathbf{M}$ and to replace $\mathbf{k}$ by $\mathbf{U}$:

$$\mathbf{U_k} \to \mathbf{U},\ \mathcal{F}_\mathbf{k} \to \mathcal{F}_\mathbf{U},\ \mathcal{V}_\mathbf{k} \to \mathcal{V}_\mathbf{U} .$$

## 1.5 Tangent spaces and tangent vectors

**Definition 26 $\mathbf{u} \sim_p \mathbf{v}$ in $\mathcal{V}$.**

We have a differential manifold $\mathbf{M}$, $\mathcal{F}$ its set of smooth maps, $\mathcal{V}$ its set of vector fields, a point $p \in \mathbf{M}$. We define $\sim_p$ in $\mathcal{V}$ by:

$$\mathbf{u} \sim_p \mathbf{v} \Leftrightarrow (p \in \mathbf{M}, \mathbf{u} \in \mathcal{V}, \mathbf{v} \in \mathcal{V}, (\forall \mathbf{f} \in \mathcal{F})(\mathbf{u(f)}(p) = \mathbf{v(f)}(p))) . \tag{1.23}$$

**Proposition 39** We have an open $\mathbf{O} \subset \mathbf{M}$ where $\mathbf{M}$ is a differential manifold, $\mathcal{V}$, $\mathcal{V}_\mathbf{O}$ their sets of vector field, $p \in \mathbf{O}$. Calling $\sim_p^\mathbf{O}$ the relation defined in definition 26 (p. 19) on the restricted manifold $\mathbf{O}$ from $\mathbf{M}$ and $\sim_p$ the relation defined on the manifold $\mathbf{M}$ also by definition 26, two vector fields $\mathbf{u} \in \mathcal{V}$ and $\mathbf{v} \in \mathcal{V}$ and $\mathbf{u_o} \in \mathcal{V}_\mathbf{O}$ and $\mathbf{v_o} \in \mathcal{V}_\mathbf{O}$ their restriction to the manifold $\mathbf{O}$ (definition 18 (p. 13)), we have:

$$\mathbf{u_o} \sim_p^\mathbf{O} \mathbf{v_o} \Leftrightarrow \mathbf{u} \sim_p \mathbf{v} .$$

**Proof**

It is a direct consequence of propositions 5 (p. 3), 22 (p. 9) and 30 (p. 13): any map $\mathbf{f}$ of $\mathcal{F}$ or $\mathcal{F}_\mathbf{O}$ can be transformed to a map $\mathbf{h}$ of $\mathcal{F} \cap \mathcal{F}_\mathbf{O}$: We call $\mathbf{U}$, an open member the set of opens covering $\mathbf{M}$ with its continuous bijection $(\phi : \mathbf{U} \to \phi(\mathbf{U}) \subset \mathcal{R}^\mathbf{n})$. From propositions 266 (p. 179), 507 (p. 286), definitions 141 (p. 165), 260 (p. 285), 261 (p. 285), 263 (p. 286), there is $\mathbf{a} \in \mathcal{R}^+$, $\mathbf{b} \in \mathcal{R}^+, 0 < \mathbf{a} < \mathbf{b}$ such that $\phi^{-1}(\mathbf{C}(\mathbf{a}, p)) \subset \phi^{-1}(\mathbf{C}(\mathbf{b}, p)) \subset \mathbf{O} \subset \mathbf{M}$ and there is a $\mathbf{C}^\infty$ map $(\mathbf{g}'_p(\phi^{-1}(\mathbf{x}), \mathbf{a}, \mathbf{b}) : \phi(\mathbf{M}) \to \mathcal{R}^\mathbf{n})$ (definition 14 (p. 6)) which is $\mathbf{0}$ for $\mathbf{x} \notin \mathbf{C}(\mathbf{b}, p)$ and $\mathbf{1}$ for $\mathbf{x} \in \mathbf{C}(\mathbf{a}, p)$. Using definition 18 (p. 13), for any $\mathbf{v} \in \mathcal{V}$, we have:

$$\begin{aligned}(\forall \mathbf{f} \in \mathcal{F}_\mathbf{O})\ &(\mathbf{v_o(f)}(p) = \mathbf{v}(\mathbf{g}'_p(q, \mathbf{a}, \mathbf{b})\mathbf{f}(q))(p) \\ &= \mathbf{v}(\mathbf{g}'_p(q, \mathbf{a}, \mathbf{b}))(p)\mathbf{f}(p) + \mathbf{g}'_p(p, \mathbf{a}, \mathbf{b})\mathbf{v}(\mathbf{f}(q))(p) = \mathbf{v}(\mathbf{f}(q))(p))\end{aligned}$$

because $\mathbf{v}(\mathbf{g}'_p(q,\mathrm{a},\mathrm{b}))(p) = \mathbf{0}$ (proposition 24 (p. 11)) and $\mathbf{g}'_p(p,\mathrm{a},\mathrm{b}) = \mathbf{1}$. Then

$$\mathbf{u} \sim_p \mathbf{v} \Rightarrow \mathbf{u}_\mathrm{O} \sim_p^\mathrm{O} \mathbf{v}_\mathrm{O} \,.$$

We have as well:

$$(\forall \mathrm{f} \in \mathcal{F})\ (\mathbf{v}(\mathrm{f}(q))(p) = \mathbf{v}(\mathbf{g}'_p(q,\mathrm{a},\mathrm{b})\mathrm{f}(q))(p) = \mathbf{v}_\mathrm{o}(\mathbf{g}'_p(q,\mathrm{a},\mathrm{b})\mathrm{f}(q))(p)) \,.$$

Then:

$$\mathbf{u}_\mathrm{O} \sim_p^\mathrm{O} \mathbf{v}_\mathrm{O} \Rightarrow \mathbf{u} \sim_p \mathbf{v} \,.$$

**Proposition 40** For all $p \in \mathrm{M}$, the relation (1.23) (p. 19) is an equivalence relation (definition 336 (p. 445)).

**Proof**

Relation $\mathbf{R}(\mathbf{u},\mathbf{v})$, defined by (1.23) (p. 19), verifies all terms of definition 336 (p. 445) setting $\mathbf{x} = \mathbf{u}, \mathbf{y} = \mathbf{v}$.

**Definition 27 Tangent space at $p$.**

We have a differentiable manifold M of dimension $\mathbf{n}$ with $\mathcal{V}$ its set of vector fields (definition 3 (p. 1)) and $\mathcal{F}$ the set of smooth maps $\mathrm{M} \to \mathcal{R}$ (definition 7 (p. 3)). A tangent space, ${}^{[p]}\mathrm{T}$ at $p \in \mathrm{M}$, is the set of equivalence classes (see definition 366 (p. 469)) generated in the set of vector fields $\mathcal{V}$ by the equivalence relation (1.23) of definition 26 (p. 19).

From definitions 366 (p. 469), 350 (p. 453), we have:

$$({}^{[p]}\mathrm{T} \subset \mathcal{P}(\mathcal{V}), \mathbf{u} \in {}^{[p]}\mathrm{T}) \Rightarrow \mathbf{u} \subset \mathcal{V} \tag{1.24}$$

$$(\mathbf{u} \in {}^{[p]}\mathrm{T}, \mathbf{v} \in {}^{[p]}\mathrm{T}, \mathbf{u} \cap \mathbf{v} \neq \emptyset) \Rightarrow \mathbf{u} = \mathbf{v} \tag{1.25}$$

$$\bigcup_{\mathbf{u} \in {}^{[p]}\mathrm{T}} \mathbf{u} = \mathcal{V} \tag{1.26}$$

$$(\mathbf{u} \in {}^{[p]}\mathrm{T}, \mathbf{x} \in \mathbf{u}, \mathbf{y} \in \mathbf{u}) \Rightarrow \mathbf{x} \sim_p \mathbf{y} \tag{1.27}$$

$$\mathbf{x} \sim_p \mathbf{y} \Rightarrow (\exists \mathbf{u} \in {}^{[p]}\mathrm{T})(\mathbf{x} \in \mathbf{u}, \mathbf{y} \in \mathbf{u}) \tag{1.28}$$

$$(\mathbf{x} \sim_p \mathbf{y}, \mathbf{x} \in \mathbf{u} \in {}^{[p]}\mathrm{T}) \Rightarrow \mathbf{y} \in \mathbf{u} \,. \tag{1.29}$$

**Proposition 41** We have a differentiable manifold M and $p \in \mathrm{M}$. We call $\mathcal{O}_p$, the set of opens of M containing $p$ and ${}^{[p]}\mathrm{T}$, the tangent space at $p$. We have:

$$(\forall \mathbf{u} \in {}^{[p]}\mathrm{T})(\forall \mathrm{O} \in \mathcal{O}_p)\ (\mathbf{u} \cap \mathcal{V}_\mathrm{O} \neq \emptyset) \,.$$

**Proof**

Using notation of proposition 39 (p. 19), at $p$ we define $\mathrm{a}, \mathrm{b}$ and the map $\mathbf{g}'_p(q,\mathrm{a},\mathrm{b})$ such that $\overline{\mathrm{C}}(\phi(p),\mathrm{a}) \subset \overline{\mathrm{C}}(\phi(p),\mathrm{b}) \subset \phi(\mathrm{O})$ (definition 14 (p. 6)). Taking $\mathbf{u} \in {}^{[p]}\mathrm{T}$ and $\mathbf{x} \in \mathbf{u}$, we have $\mathbf{x}\mathbf{g}'_p(q,\mathrm{a},\mathrm{b}) \in \mathcal{V}_\mathrm{O} \cap \mathcal{V}$ because

$$(\forall \mathrm{f} \in \mathcal{F} \cup \mathcal{F}_\mathrm{O})\ (\mathbf{g}'_p(q,\mathrm{a},\mathrm{b})\mathrm{f} \in \mathcal{F} \cap \mathcal{F}_\mathrm{O}, \phi(q) \notin \mathrm{C}(\phi(p),\mathrm{b}) \subset \phi(\mathrm{O})$$
$$\Rightarrow \mathbf{x}(\mathbf{g}'_p(q,\mathrm{a},\mathrm{b})\mathrm{f})(q) = 0) \,.$$

It means $\mathbf{x}(\mathbf{g}'_p(q,\mathrm{a},\mathrm{b})\mathrm{f}) \in \mathcal{F}_\mathrm{O}$ then $\mathbf{x}\mathbf{g}'_p(q,\mathrm{a},\mathrm{b}) \in \mathcal{V}_\mathrm{O}$.

Since $\mathbf{g}'_p(q,\mathrm{a},\mathrm{b}) \in \mathcal{F}$ we have: $\mathbf{x}\mathbf{g}'_p(q,\mathrm{a},\mathrm{b}) \in \mathcal{V}$. Moreover (proposition 22 (p. 9)):

$$\mathrm{f} \in \mathcal{F} \Rightarrow \mathbf{x}(\mathbf{g}'_p(q,\mathrm{a},\mathrm{b})\mathrm{f})(p) = \mathbf{x}(\mathrm{f})(p)$$

which means $\mathbf{x}\mathbf{g}'_p(q,\mathrm{a},\mathrm{b}) \in \mathbf{u}$.

**Definition 28 Operations on tangent space.**

With the notations of the above definition 27 (p. 20), considering the tangent space $^{[p]}\mathbf{T}$ at $p \in \mathrm{M}$, $\mathbf{u} \in {}^{[p]}\mathbf{T}$, $\mathbf{v} \in {}^{[p]}\mathbf{T}$, $\boldsymbol{\lambda} \in \mathcal{R}, \boldsymbol{\mu} \in \mathcal{R}$ using proposition 41 (p. 20) (see also definitions 16 (p. 8), 17 (p. 8), 12 (p. 5) and equation (1.1) of proposition 12 (p. 5)):

$$\mathbf{x} \in \lambda\mathbf{u} + \mu\mathbf{v} \Leftrightarrow (\exists \mathbf{x_u} \in \mathbf{u} \cap \mathcal{V})(\exists \mathbf{x_v} \in \mathbf{v} \cap \mathcal{V})(\mathbf{x} \sim_p \lambda\mathbf{x_u} + \mu\mathbf{x_v}) .$$

**Proposition 42** With the notations of the above definitions 27 (p. 20), 28 (p. 21)

$$\lambda\mathbf{u} + \mu\mathbf{v} \in {}^{[p]}\mathbf{T} .$$

**Proof**
We have from proposition 41 (p. 20) and definition 28 (p. 21):

$$\mathbf{x} \in \lambda\mathbf{u} + \mu\mathbf{v} \Rightarrow (\exists \mathbf{x_u} \in \mathbf{u} \cap \mathcal{V})(\exists \mathbf{x_v} \in \mathbf{v} \cap \mathcal{V})\ (\mathbf{x} \sim_p \lambda\mathbf{x_u} + \mu\mathbf{x_v}) .$$

We take those $\mathbf{x_u}, \mathbf{x_v}$ associated to any $\mathbf{x} \in \lambda\mathbf{u} + \mu\mathbf{v}$ (see definition 112 (p. 131) point 4) and the similar $\mathbf{y_u}, \mathbf{y_v}$ for any $\mathbf{y} \in \lambda\mathbf{u} + \mu\mathbf{v}$ we have:

$$\begin{aligned}&(\forall \mathbf{f} \in \mathcal{F})\ (\mathbf{x_u}(\mathbf{f})(p) = \mathbf{y_u}(\mathbf{f})(p), \mathbf{x_v}(\mathbf{f})(p) = \mathbf{y_v}(\mathbf{f})(p),\\ &\lambda\mathbf{x_u}(\mathbf{f})(p) + \mu\mathbf{x_v}(\mathbf{f})(p) = \lambda\mathbf{y_u}(\mathbf{f})(p) + \mu\mathbf{y_v}(\mathbf{f})(p)\\ &= \mathbf{x}(\mathbf{f})(p) = \mathbf{y}(\mathbf{f})(p)) .\end{aligned}$$

Then $\lambda\mathbf{u} + \mu\mathbf{v}$ as defined in definition 28 (p. 21) is an equivalence class of $\sim_\mathrm{p}$ then:

$$\lambda\mathbf{u} + \mu\mathbf{v} \in {}^{[p]}\mathbf{T} .$$

**Proposition 43** Keeping the notations of the definitions 27 (p. 20), 28 (p. 21), we have:

$$(\mathbf{x} \in \lambda\mathbf{u} + \mu\mathbf{v} \in {}^{[p]}\mathbf{T}, \mathbf{y} \sim_p \mathbf{x}) \Rightarrow (\exists \mathbf{y_u} \in \mathbf{u})(\exists \mathbf{y_v} \in \mathbf{v})(\mathbf{y} = \lambda\mathbf{y_u} + \mu\mathbf{y_v}) .$$

**Proof**
By definition 28 (p. 21), we have:

$$(\exists \mathbf{x_u} \in \mathbf{u})(\exists \mathbf{x_v} \in \mathbf{v})(\mathbf{y} \sim_p \mathbf{x} \sim_p \lambda\mathbf{x_u} + \mu\mathbf{x_v}) .$$

Since (definition 27 (p. 20)) $\mathbf{f} \in \mathcal{F} \Rightarrow (\mathbf{y}(\mathbf{f})(p) = \lambda\mathbf{x_u}(\mathbf{f})(p) + \mu\mathbf{x_v}(\mathbf{f})(p))$, we can choose:

- If $\lambda = \mu = 0$, $\mathbf{y_u} = \mathbf{y} + \mathbf{x_u}$, $\mathbf{y_v} = \mathbf{y} + \mathbf{x_v}$, we have for $\mathbf{f} \in \mathcal{F}$, $\mathbf{y}(\mathbf{f})(p) = \mathbf{0}$ then, $\mathbf{y_u}(\mathbf{f})(p) = \mathbf{x_u}(\mathbf{f})(p)$, $\mathbf{y_v}(\mathbf{f})(p) = \mathbf{x_v}(\mathbf{f})(p)$
- If $\lambda + \mu \neq 0$

$$\begin{aligned}&(\mathbf{y_u} = \frac{1}{\lambda+\mu}(\mathbf{y} - (\lambda\mathbf{x_u} + \mu\mathbf{x_v})) + \mathbf{x_u},\ \mathbf{y_v} = \frac{1}{\lambda+\mu}(\mathbf{y} - (\lambda\mathbf{x_u} + \mu\mathbf{x_v})) + \mathbf{x_v})\\ &\Rightarrow (\mathbf{y} = \lambda\mathbf{y_u} + \mu\mathbf{y_v}, (\mathbf{f} \in \mathcal{F} \Rightarrow (\mathbf{y_u}(\mathbf{f})(p) = \mathbf{x_u}(\mathbf{f})(p), \mathbf{y_v}(\mathbf{f})(p) = \mathbf{x_v}(\mathbf{f})(p))))\\ &\Rightarrow (\mathbf{y} = \lambda\mathbf{y_u} + \mu\mathbf{y_v}, \mathbf{y_u} \in \mathbf{u}, \mathbf{y_v} \in \mathbf{v})\end{aligned}$$

- If $\lambda - \mu \neq 0$

$$
\begin{aligned}
&(y_u = \frac{1}{\lambda - \mu}(y - (\lambda x_u + \mu x_v)) + x_u,\ y_v = \frac{-1}{\lambda - \mu}(y - (\lambda x_u + \mu x_v)) + x_v) \\
&\Rightarrow (y = \lambda y_u + \mu y_v, (f \in \mathcal{F} \Rightarrow (y_u(f)(p) = x_u(f)(p), y_v(f)(p) = x_v(f)(p)))) \\
&\Rightarrow (y = \lambda y_u + \mu y_v, y_u \in u, y_v \in v)\ .
\end{aligned}
$$

Then from proposition 727 (p. 441), we have proposition 43 (p. 21).

**Proposition 44** With the notations of the above definition 27 (p. 20), a tangent space at $p \in M$ is a vector space (definition 199 (p. 235)) over the field $\mathcal{R}$ (definition 181 (p. 217)).

**Proof**
From proposition 42 (p. 21), all elements of definition 199 (p. 235) are true.

**Proposition 45** With the notations of the above definition 27 (p. 20), we have:

$$(u \in \mathcal{V}, v \in \mathcal{V}, p \in U_k) \Rightarrow ((\forall \mu \in [1, n])(u_k^\mu(p) = v_k^\mu(p)) \Leftrightarrow u \sim_p v)\ . \tag{1.30}$$

Where $u_k = u_k(x^\mu)\partial_\mu = u_k^\mu \partial_\mu$, $v_k = v_k(x^\mu)\partial_\mu = u_k^\mu \partial_\mu$ are the restrictions of $u$ and $v$ to $U_k$ (definition 366 (p. 469)) and $x^\mu = \phi_k^\mu \in \mathcal{F}_k$ the coordinate maps in $U_k$ (definition 20 (p. 15)).

**Proof**
From proposition 31 (p. 14), we may write:

$$u \in \mathcal{V} \Rightarrow ((f \in \mathcal{F}, p \in U_k) \Rightarrow u(f)(p) = u_k^\mu(p)\partial_\mu f(p))\ . \tag{1.31}$$

Then:

$$(\forall \mu \in [1, n])\ (u_k^\mu(p) = v_k^\mu(p)) \Rightarrow u \sim_p v\ .$$

For the other relation, we remark that $u \sim_p v \Rightarrow (p \in M, u \in \mathcal{V}, v \in \mathcal{V})$ and using the same notations as in definition 18 (p. 13) and when $p \in U_k$ we have:

$$(\exists(a, b))\ (0 < a < b, C(\phi_k(p), a) \subset C(\phi_k(p), b) \subset \phi_k(U_k))$$

(definition 14 (p. 6), propositions 269 (p. 179) and 507 (p. 286)). We may define $x_p^\mu$ (see definition 20 (p. 15)) by:

$$x_p^\mu(p') = g_p'(p', a, b)\phi_k^\mu(p') \in \mathcal{F}\ .$$

From propositions 25 (p. 11), 26 (p. 11), 33 (p. 16), definition 18 (p. 13), we have:

$$(\forall u \in \mathcal{V})\ (u_k^\mu(p) = u(x_p^\mu)(p))\ . \tag{1.32}$$

Since $u \sim_p v \Rightarrow u(x_p^\mu)(p) = v(x_p^\mu)(p)$, equation (1.32) (p. 22) gives the $\Leftarrow$ part of relation (1.30) (p. 22).

**Definition 29 Tangent space basis vector ${}^{[p]}\partial_\mu$ of ${}^{[p]}T$.**

Keeping the notations of this section, given a point $p \in M$ and a $k \in K$ such that $p \in U_k$, one calls ${}^{[p]}\partial_\mu \in {}^{[p]}T$, the equivalence class of $\partial_\mu \in \mathcal{V}_k$ in $\mathcal{V}$ (definition 27 (p. 20)). It does depend on $k$ since $\partial_\mu$ depends on it. On the opposite, from proposition 39 (p. 19), ${}^{[p]}T$ is independent of $k$.

**Proposition 46** Keeping the notations of this section, we have a tangent space ${}^{[p]}\mathrm{T}$. ${}^{[p]}\mathrm{T}$, when considered as a vector space (definition 199 (p. 235)) built on $\mathcal{R}$ associated to $p \in \mathrm{M}$, has $({}^{[p]}\partial_\mu : [1, \mathrm{n}] \to \mathcal{V}_\mathrm{K})$ (definition 29 (p. 22)) as an independent basis of dimension $\mathrm{n}$ (definitions 205 (p. 239), 206 (p. 239)) when $p \in \mathrm{U_k} \subset \mathrm{M}$.

**Proof**

From proposition 44 (p. 22), ${}^{[p]}\mathrm{T}$ is a vector space on $\mathcal{R}$.
If $p \in \mathrm{U_k}$, from definition 29 (p. 22), we have:

$$(\mathrm{v} \in \mathcal{V}, p \in \mathrm{U_k}, \mathrm{f} \in \mathcal{F}) \Rightarrow (\mathrm{v}(\mathrm{f})(p) = \mathrm{v_k}(\phi_\mathrm{k}^\mu(p))\partial_\mu \mathrm{f} \circ \phi_\mathrm{k}^{-1}(\phi_\mathrm{k}(p))$$
$$= \mathrm{v}_\mathrm{k}^\mu(p)\partial_\mu(\mathrm{f})(p) \Rightarrow \mathrm{v} \sim_p \mathrm{v}_\mathrm{k}^\mu(p)\,{}^{[p]}\partial_\mu)$$

then the equivalence classes ${}^{[p]}\partial_\mu$ of $\partial_\mu$ are a basis of ${}^{[p]}\mathrm{T}$ (see definition 79 (p. 87)). It is an independent basis, because (see definition 20 (p. 15)):

$$((\lambda^\mu : [1, \mathrm{n}] \to \mathcal{R}), \lambda^\mu\,{}^{[p]}\partial_\mu = 0) \Rightarrow (\forall \nu \in [1, \mathrm{n}])\ (\lambda^\mu\,{}^{[p]}\partial_\mu \mathrm{x}^\nu(p) = \lambda^\mu \partial_\mu \mathrm{x}^\nu(p)$$
$$= \lambda^\mu \delta_\mu^\nu = \lambda^\nu = 0)\ .$$

## 1.6 Coordinate changes

**Definition 30 Change of coordinates in a differentiable manifold.**

Having a differentiable manifold $\mathrm{M}$ covered by a finite $\mathrm{K} \in \mathcal{Z}^+$ number of opens $\mathrm{U_k}$, $\mathrm{k} \in \mathrm{K}$ with their isomorphism $\phi_\mathrm{k}$ to $\mathcal{R}^\mathrm{n}$. Any isomorphism (definition 142 (p. 166)) $(\psi_\mathrm{k} : \mathcal{R}^\mathrm{n} \to \mathcal{R}^\mathrm{n})$ belonging to $\mathrm{C}^\infty$ (see definition 276 (p. 304)) which has also an inverse in $\mathrm{C}^\infty$ and defined at least on $\phi_\mathrm{k}(\mathrm{U_k})$ of $\mathcal{R}^\mathrm{n}$ can generate a change of coordinates. The isomorphisms $(\phi_\mathrm{k} : \mathrm{U_k} \to \mathcal{R}^\mathrm{n})$ are replaced by the isomorphisms (see propositions 195 (p. 149), 241 (p. 167)):

$$(\psi_\mathrm{k} \circ \phi_\mathrm{k} : \mathrm{U_k} \to \mathcal{R}^\mathrm{n})$$

with:

$$(\psi_\mathrm{k} \circ \phi_\mathrm{k})^{-1} = \phi_\mathrm{k}^{-1} \circ \psi_\mathrm{k}^{-1}\ .$$

Since from definition 3 (p. 1) $\phi_{\mathrm{k}'} \circ \phi_\mathrm{k}^{-1} \in \mathrm{C}^\infty$ (definition 276 (p. 304)), from propositions 562 (p. 309) and 563 (p. 311), we have:

$$(\forall \mathrm{k}' \in \mathrm{K})(x \in \mathrm{U_k} \cap \mathrm{U_{k'}} \neq \emptyset \Rightarrow \psi_{\mathrm{k}'} \circ \phi_{\mathrm{k}'} \circ (\psi_\mathrm{k} \circ \phi_\mathrm{k})^{-1}(x)$$
$$= \psi_{\mathrm{k}'} \circ \phi_{\mathrm{k}'} \circ \phi_\mathrm{k}^{-1} \circ \psi_\mathrm{k}^{-1}(x) \in \mathrm{C}^\infty)\ .$$

Then definition 30 (p. 23) satisfies definition 3 (p. 1).

**Proposition 47** Using the above notations, forgetting the subscript $\mathrm{k}$ in order to simplify, using definitions 25 (p. 17) and 23 (p. 17), any vector field $\mathrm{v} \in \mathcal{V}$ can be written in the two local frames as:

$$\mathrm{v} = \mathrm{v}^{\psi\phi \mathrm{j}}\partial_\mathrm{j}^{\psi\phi} = \mathrm{v}^{\phi \mathrm{i}}\partial_\mathrm{i}^\phi$$

with $(\partial^{\psi\phi} : [1, \mathrm{n}] \to \mathcal{V})$ and $(\partial^\phi : [1, \mathrm{n}] \to \mathcal{V})$ being the two bases of vector fields corresponding to the two different isomorphisms. Calling (see proposition 10 (p. 4)):

$$\mathrm{x}^\phi = \phi \in \mathcal{F}^\mathrm{n},\ \mathrm{x}^{\psi\phi} = \psi \circ \phi = \psi(\mathrm{x}^\phi) \in \mathcal{F}^\mathrm{n}$$

those two basic vector fields are related to each other by:

$$\partial_\mathrm{i}^\phi = \mathrm{H}_\mathrm{i}^\mathrm{j}\partial_\mathrm{j}^{\psi\phi} \tag{1.33}$$

where $\mathbf{H}_{\mathrm{i}}^{\mathrm{j}} \in \mathcal{F}$ is given by:

$$\mathbf{H}_{\mathrm{i}}^{\mathrm{j}} = \frac{\partial}{\partial \mathbf{x}^{\phi \mathrm{i}}} \mathbf{x}^{\psi\phi \mathrm{j}} = \frac{\partial}{\partial \mathbf{x}^{\phi \mathrm{i}}} (\psi^{\mathrm{j}}(\mathbf{x}^{\phi})) \ .$$

From definition 297 (p. 358)

$$(\mathbf{H}(p) = \mathbf{jac}(\psi)(\phi(p)) : \mathbf{U} \to \mathcal{R}^{2\mathrm{n}}) \ . \tag{1.34}$$

It also gives:

$$\mathbf{v}^{\psi\phi \mathrm{j}} = \mathbf{jac}(\psi)_{\mathrm{i}}^{\mathrm{j}} \mathbf{v}^{\phi \mathrm{i}} \ . \tag{1.35}$$

**Proof**
In what follows we have suppressed the index $\mathbf{k}$ as said above. Everything that we shall say, is only valid in $\mathbf{U} = \mathbf{U}_{\mathbf{k}}$ and $\mathcal{F} = \mathcal{F}_{\mathbf{k}}$.
We have the following equalities and definitions for any $\mathbf{f} \in \mathcal{F}$, with $p \in \mathbf{U}$ and

$$\begin{aligned} \mathbf{f}(p) &= \mathbf{f}(\phi^{-1}(\phi(p))) = \mathbf{f}^{\phi}(\phi(p)) = \mathbf{f}^{\phi}(\mathbf{x}^{\phi}) \\ &= \mathbf{f}(\phi^{-1}(\psi^{-1}(\psi(\phi(p))))) = \mathbf{f}^{\psi\phi}(\psi(\phi(p))) = \mathbf{f}^{\psi\phi}(\mathbf{x}^{\psi\phi}) \ . \end{aligned}$$

From proposition 10 (p. 4) combined with propositions 562 (p. 309) and 563 (p. 311), we have:

$$\frac{\partial}{\partial \mathbf{x}^{\psi\phi \mathrm{j}}} \mathbf{f}^{\psi\phi}(\mathbf{x}^{\psi\phi}) \frac{\partial}{\partial \mathbf{x}^{\phi \mathrm{i}}} \mathbf{x}^{\psi\phi \mathrm{j}} = \frac{\partial}{\partial \mathbf{x}^{\phi \mathrm{i}}} \mathbf{f}^{\psi\phi}(\mathbf{x}^{\psi\phi}) = \frac{\partial}{\partial \mathbf{x}^{\phi \mathrm{i}}} \mathbf{f}^{\phi}(\mathbf{x}^{\phi}) \ . \tag{1.36}$$

Setting:

$$\mathbf{H}_{\mathrm{i}}^{\mathrm{j}}(p) = \frac{\partial}{\partial \mathbf{x}^{\phi \mathrm{i}}} \mathbf{x}^{\psi\phi \mathrm{j}} = \frac{\partial}{\partial \mathbf{x}^{\phi \mathrm{i}}} (\psi^{\mathrm{j}}(\mathbf{x}^{\phi}))$$

we have:

$$\mathbf{v}^{\phi \mathrm{i}} \frac{\partial}{\partial \mathbf{x}^{\phi \mathrm{i}}} \mathbf{f}^{\phi}(\mathbf{x}^{\phi}) \ = \mathbf{v}^{\phi \mathrm{i}} \mathbf{H}_{\mathrm{i}}^{\mathrm{j}}(p) \frac{\partial}{\partial \mathbf{x}^{\psi\phi \mathrm{j}}} \mathbf{f}^{\psi\phi}(\mathbf{x}^{\psi\phi}) = \mathbf{v}^{\psi\phi \mathrm{j}} \frac{\partial}{\partial \mathbf{x}^{\psi\phi \mathrm{j}}} \mathbf{f}^{\psi\phi}(\mathbf{x}^{\psi\phi}) \ .$$

Then:

$$\mathbf{v}^{\psi\phi \mathrm{j}}(p) = \mathbf{H}_{\mathrm{i}}^{\mathrm{j}}(p) \mathbf{v}^{\phi \mathrm{i}}(p)$$

from

$$\mathbf{v} = \mathbf{v}^{\psi\phi \mathrm{j}} \partial_{\mathrm{j}}^{\psi\phi} = \mathbf{v}^{\phi \mathrm{i}} \partial_{\mathrm{i}}^{\phi}$$

we get

$$\partial_{\mathrm{i}}^{\phi} = \mathbf{H}_{\mathrm{i}}^{\mathrm{j}}(p) \partial_{\mathrm{j}}^{\psi\phi} \ .$$

In other words, using proposition 33 (p. 16), we may write:

$$\partial_{\mathrm{i}}^{\phi} = \partial_{\mathrm{i}}^{\phi}(\psi^{\mathrm{j}}(\mathbf{x}^{\phi})) \partial_{\mathrm{j}}^{\psi\phi} \ . \tag{1.37}$$

**Proposition 48** With the same notations, $\mathbf{det}(\mathbf{H}_{\mathrm{i}}^{\mathrm{j}}(p)) \neq 0$ .

**Proof**
It is a consequence of proposition 628 (p. 360).

## 1.7 Metric on a differentiable manifold

**Definition 31 Metric on a differentiable manifold.**

If $\mathcal{V}$ is the space of vector fields on a differentiable manifold M, a metric g is a map from $\mathcal{V} \times \mathcal{V}$ to $\mathcal{F}$ with the following properties:
With $\lambda \in \mathcal{F}, \mu \in \mathcal{F}, \mathrm{u} \in \mathcal{V}, \mathrm{v} \in \mathcal{V}, \mathrm{w} \in \mathcal{V}$ we have:

$$\mathrm{g}(\mathrm{u},\mathrm{v}) \Rightarrow (\mathrm{u} \in \mathcal{V}, \mathrm{v} \in \mathcal{V}) \tag{1.38}$$
$$\mathrm{g}(\mathrm{u},\mathrm{v}) = \mathrm{g}(\mathrm{v},\mathrm{u}) \tag{1.39}$$
$$\mathrm{g}(\lambda\mathrm{u} + \mu\mathrm{v}, \mathrm{w}) = \lambda\mathrm{g}(\mathrm{u},\mathrm{w}) + \mu\mathrm{g}(\mathrm{v},\mathrm{w}) \tag{1.40}$$
$$(\forall p \in \mathrm{M})((\forall \mathrm{w} \in \mathcal{V})\ (\mathrm{g}(\mathrm{v},\mathrm{w})(p) = 0) \Rightarrow (\forall \mathrm{f} \in \mathcal{F})\ (\mathrm{v}(\mathrm{f})(p) = 0))\ . \tag{1.41}$$

**Proposition 49** With the notations of definition 31 (p. 25), given metric g on a differentiable manifold M, we have:

$$(\forall p \in \mathrm{M})(\forall \mathrm{v} \in \mathcal{V})(\mathrm{g}(\mathrm{u},\mathrm{v})(p) = \mathrm{g}(\mathrm{u}'\mathrm{v})(p)) \Rightarrow \mathrm{u} = \mathrm{u}' \in \mathcal{V}\ . \tag{1.42}$$

**Proof**
From relation (1.38) (p. 25), $\mathrm{u} \in \mathcal{V},\ \mathrm{u}' \in \mathcal{V}$ We have from relation (1.40) (p. 25):

$$(\mathrm{g}(\mathrm{u},\mathrm{v})(p) = \mathrm{g}(\mathrm{u}'\mathrm{v})(p)) \Rightarrow \mathrm{g}(\mathrm{u},\mathrm{v})(p) - \mathrm{g}(\mathrm{u}'\mathrm{v})(p) = \mathrm{g}(\mathrm{u}-\mathrm{u}',\mathrm{v})(p) = 0\ .$$

From relation 1.41 (p. 25),

$$(\forall p \in \mathrm{M})(\forall \mathrm{v} \in \mathcal{V})\ (\mathrm{g}(\mathrm{u},\mathrm{v})(p) = \mathrm{g}(\mathrm{u}'\mathrm{v})(p)) \Rightarrow (\forall \mathrm{f} \in \mathcal{F})\ (\mathrm{u}(\mathrm{f}) = \mathrm{u}'(\mathrm{f})) \Rightarrow \mathrm{u} = \mathrm{u}'\ .$$

**Proposition 50** With the notations of definition 31 (p. 25), given metric g on a differentiable manifold M, we consider a covering element $\mathrm{U_k} \subset \mathrm{M}$ with its continuous map $(\phi_\mathrm{k} : \mathrm{U_k} \to \mathcal{R}^\mathrm{n})$, two open cubes $\mathrm{C}(\phi_\mathrm{k}(p),\mathrm{a}) \subset \phi_\mathrm{k}(\mathrm{U_k})$, $\mathrm{C}(\phi_\mathrm{k}(p),\mathrm{b}) \subset \phi_\mathrm{k}(\mathrm{U_k})$, $0 < \mathrm{a} < \mathrm{b}$ centered at $p \in \mathrm{U_k}$ and the associated map (definition 14 (p. 6)) $(\mathrm{g}'_p(q,\mathrm{a},\mathrm{b}) : \mathrm{M} \to \mathcal{R}) \in \mathcal{F} \subset \mathcal{F}_\mathrm{k}$ we have:

$$(\mathrm{u} \in \mathcal{V}, \mathrm{v} \in \mathcal{V}) \Rightarrow (\forall q \in \phi_\mathrm{k}^{-1}(\mathrm{C}(\phi_\mathrm{k}(p),\mathrm{a})))\ (\mathrm{g}(\mathrm{g}'_p(q',\mathrm{a},\mathrm{b})\mathrm{u},\mathrm{v})(q) = \mathrm{g}(\mathrm{u},\mathrm{v})(q))\ .$$

**Proof**
It is a direct consequence of equation (1.40) (p. 25).

**Definition 32 Restriction of a metric over an open.**

We keep the notations of definition 31 (p. 25) and of proposition 49 (p. 25). We consider an open O, two vector fields $\mathrm{u} \in \mathcal{V}_\mathrm{O}$ and $\mathrm{v} \in \mathcal{V}_\mathrm{O}$, $\mathcal{V}_\mathrm{O}$ being the set of vector fields of the manifold O restriction of the manifold M (definition 5 (p. 3)). There is a covering element $\mathrm{U_k}$ of M such that $\mathrm{U_k} \cap \mathrm{O} \neq \emptyset$ and a continuous map $(\phi_\mathrm{k} : \mathrm{U_k} \to \mathcal{R}^\mathrm{n})$. From definition 132 (p. 151), $\mathrm{U_k} \cap \mathrm{O}$ is also a non-empty open subset of M. From proposition 266 (p. 179), we may consider as above two open cubes $\mathrm{C}(\phi_\mathrm{k}(p),\mathrm{a_p}) \subset \mathrm{C}(\phi_\mathrm{k}(p),\mathrm{b_p}) \subset \phi_\mathrm{k}(\mathrm{U_k} \cap \mathrm{O})$, $0 < \mathrm{a_p} < \mathrm{b_p}$ centered at $p \in \mathrm{U_k} \cap \mathrm{O}$, and the associated map (definition 14 (p. 6)) $(\mathrm{g}'_p(q,\mathrm{a_p},\mathrm{b_p}) : \mathrm{M} \to \mathcal{R}) \in \mathcal{F} \subset \mathcal{F}_\mathrm{k}$. Using $\mathrm{g}'_p$ as in definition 14 (p. 6), one sets:

$$\mathrm{g_O}(\mathrm{u},\mathrm{v})(p) = \mathrm{g}(\mathrm{g}'_p(q,\mathrm{a_p},\mathrm{b_p})\mathrm{u}, \mathrm{g}'_p(q,\mathrm{a_p},\mathrm{b_p})\mathrm{v})(p)\ . \tag{1.43}$$

**Proposition 51** $\mathbf{g_O(u,v)}$ as defined in definition 32 (p. 25) equation (1.43) does not depend on $\mathbf{a}$ and $\mathbf{b}$, keeping the notation of 32 (p. 25).

**Proof**
In equation (1.43) (p. 25), $\mathbf{g}'_p(q,\mathbf{a},\mathbf{b})\mathbf{u}$, $\mathbf{g}'_p(q,\mathbf{a},\mathbf{b})\mathbf{v}$ are considered as a vector field of $\mathbf{M}$ (see proposition 29 (p. 13)) so the second member of equation (1.43) (p. 25) is defined. It does not depend on $\mathbf{a}$ and $\mathbf{b}$:
We take another couple $0 < \mathbf{a}' < \mathbf{b}'$ with the same requirement. We may suppose $\mathbf{b}' > \mathbf{b}$ otherwise we exchange $\mathbf{a},\mathbf{b}$ and $\mathbf{a}',\mathbf{b}'$. We define $\mathbf{x}^p_{\mathrm{m}} > 0$ as in proposition 16 (p. 7) at any point $p \in \mathbf{O} \cap \mathbf{U_k}$: $\mathbf{x}^p_{\mathrm{m}} \in \mathcal{R}^+$ is the biggest value of $\mathbf{x} \in \mathcal{R}^+$ such that for any values $0 < \mathbf{a_p} < \mathbf{b_p} < \mathbf{x}$, we have $\mathbf{C}(\phi_{\mathrm{k}}(p),\mathbf{a_p}) \subset \mathbf{C}(\phi_{\mathrm{k}}(p),\mathbf{b_p}) \subset \phi_{\mathrm{k}}(\mathbf{U_k} \cap \mathbf{O})$, $0 < \mathbf{a_p} < \mathbf{b_p}$. If $\mathbf{x}^p_{\mathrm{m}}$ does not exist, any positive values are possible. If $\mathbf{x}^p_{\mathrm{m}}$ exists, since, from proposition 524 (p. 291), $\mathbf{x}^p_{\mathrm{m}} > \mathbf{b}'$, we can choose $\mathbf{a}'',\mathbf{b}''$ such that $\mathbf{x}^p_{\mathrm{m}} > \mathbf{b}'' > \mathbf{a}'' > \mathbf{b}'$. If $\mathbf{x}^p_{\mathrm{m}}$ does not exist, we take $\mathbf{b}'' > \mathbf{a}'' > \mathbf{b}'$. Then, since $\mathbf{g}'_p(q,\mathbf{a}'',\mathbf{b}'') = 1$ when $\mathbf{g}'_p(q,\mathbf{a},\mathbf{b}) \neq 0$ or when $\mathbf{g}'_p(q,\mathbf{a}',\mathbf{b}') \neq 0$, we have:

$$\mathbf{g}'_p(q,\mathbf{a},\mathbf{b}) = \mathbf{g}'_p(q,\mathbf{a},\mathbf{b})\mathbf{g}'_p(q,\mathbf{a}'',\mathbf{b}''),\ \mathbf{g}'_p(q,\mathbf{a}',\mathbf{b}') = \mathbf{g}'_p(q,\mathbf{a}',\mathbf{b}')\mathbf{g}'_p(q,\mathbf{a}'',\mathbf{b}'')\ .$$

All those maps are in $\mathcal{F}$.
Applying proposition 49 (p. 25) together with $\mathbf{g}'_p(p,\mathbf{a},\mathbf{b}) = \mathbf{g}'_p(p,\mathbf{a}',\mathbf{b}') = 1$, we have:

$$\begin{aligned}
&\mathbf{g}(\mathbf{g}'_p(q,\mathbf{a},\mathbf{b})\mathbf{u},\mathbf{g}'_p(q,\mathbf{a},\mathbf{b})\mathbf{v})(p) = \mathbf{g}'_p(p,\mathbf{a},\mathbf{b})^2\mathbf{g}(\mathbf{g}'_p(q,\mathbf{a}'',\mathbf{b}'')\mathbf{u},\mathbf{g}'_p(q,\mathbf{a}'',\mathbf{b}'')\mathbf{v})(p)\\
&= \mathbf{g}(\mathbf{g}'_p(q,\mathbf{a}'',\mathbf{b}'')\mathbf{u},\mathbf{g}'_p(q,\mathbf{a}'',\mathbf{b}'')\mathbf{v})(p)\\
&= \mathbf{g}'_p(p,\mathbf{a}',\mathbf{b}')^2\mathbf{g}(\mathbf{g}'_p(q,\mathbf{a}'',\mathbf{b}'')\mathbf{u},\mathbf{g}'_p(q,\mathbf{a}'',\mathbf{b}'')\mathbf{v})(p)\\
&= \mathbf{g}(\mathbf{g}'_p(q,\mathbf{a}',\mathbf{b}')\mathbf{u},\mathbf{g}'_p(q,\mathbf{a}',\mathbf{b}')\mathbf{v})(p)\ .
\end{aligned}$$

Then:

$$\mathbf{g}(\mathbf{g}'_p(q,\mathbf{a},\mathbf{b})\mathbf{u},\mathbf{g}'_p(q,\mathbf{a},\mathbf{b})\mathbf{v})(p) = \mathbf{g}(\mathbf{g}'_p(q,\mathbf{a}',\mathbf{b}')\mathbf{u},\mathbf{g}'_p(q,\mathbf{a}',\mathbf{b}')\mathbf{v})(p)\ .$$

**Proposition 52** With the notations of definition 32 (p. 25), the map $(\mathbf{g_O} : \mathcal{V}_\mathbf{O} \times \mathcal{V}_\mathbf{O} \to \mathbf{O} \times \mathcal{R})$ given by equation (1.43) (p. 25) is a metric of the restricted manifold $\mathbf{O} \subset \mathbf{M}$.

**Proof**

- definition 31 (p. 25) $(\mathbf{u} \in \mathcal{V}_\mathbf{O}, \mathbf{v} \in \mathcal{V}_\mathbf{O}) \Rightarrow \mathbf{g_o(u,v)} \in \mathcal{F}_\mathbf{O}$. We use the same arguments as in proposition 26 (p. 11):
From proposition 269 (p. 179), definitions 152 (p. 179), 507 (p. 286) and proposition 514 (p. 288), we have $0 < \mathbf{a_p} < \mathbf{b_p}$ such that

$$\overline{\mathbf{C}}(\phi_{\mathrm{k}}(p),\mathbf{a_p}) \subset \overline{\mathbf{C}}(\phi_{\mathrm{k}}(p),\mathbf{b_p}) \subset \phi_{\mathrm{k}}(\mathbf{U_k} \cap \mathbf{O})$$

with: $\phi_{\mathrm{k}}(q) \in \mathbf{C}(\phi_{\mathrm{k}}(p),\mathbf{a_p}) \Rightarrow \mathbf{g}'_p(q,\mathbf{a_p},\mathbf{b_p}) = 1$.
From definition 243 (p. 267), propositions 473 (p. 268), 474 (p. 269), 475 (p. 270), a point $\mathbf{x} = \phi_{\mathrm{k}}(x) \in \mathbf{C}(\phi_{\mathrm{k}}(p),\mathbf{a_p}/4)$ is the center of two open cubes (definition 260 (p. 285))

$$\mathbf{C}(\phi_{\mathrm{k}}(x),\mathbf{a_p}/2) = \mathbf{C}(\mathbf{x},\mathbf{a_p}/2) \subset \mathbf{C}(\phi_{\mathrm{k}}(p),\mathbf{a_p}) \subset \phi_{\mathrm{k}}(\mathbf{U_k} \cap \mathbf{O})$$
$$\mathbf{C}(\phi_{\mathrm{k}}(x),\mathbf{b_p}/2) = \mathbf{C}(\mathbf{x},\mathbf{b_p}/2) \subset \mathbf{C}(\phi_{\mathrm{k}}(p),\mathbf{b_p}) \subset \phi_{\mathrm{k}}(\mathbf{U_k} \cap \mathbf{O})$$

with $\phi_k(p) \in C(\phi_k(x), a_p/2)$. From definition 32 (p. 25) equation (1.43) and proposition 51 (p. 26), we have:

$$\begin{aligned} x \in C(\phi_k(p), a/4) &\Rightarrow g_O(u, v)(\phi_k^{-1}(x)) \\ &= g(g_x(q, a_p/2, b_p/2)u, g_x(q, a_p/2, b_p/2)v)(\phi_k^{-1}(x)) \\ &= g(g'_p(q, a_p, b_p)g'_x(q, a/2, b/2)u, g'_p(q, a_p, b_p)g'_x(q, a/2, b/2)v)(\phi_k^{-1}(x)) \\ &= g'^2_x(x, a_p/2, b_p/2)g(g'_p(q, a_p, b_p)u, g'_p(q, a_p, b_p)v)(\phi_k^{-1}(x)) \\ &= g(g'_p(q, a_p, b_p)u, g'_p(q, a_p, b_p)v)(\phi_k^{-1}(x)) \in C^\infty . \end{aligned}$$

Since $p \in C(\phi_k(p), a_p/4)$, the above relation is true for $x = p$. That is valid for any $p \in O$. We also have:

$$g(g'_p(q, a_p, b_p)u, g'_p(q, a_p, b_p)v)(p) = g_O(u, v)(p) .$$

Altogether, it gives $g_O(u, v) \in \mathcal{F}_O$

- definition 31 (p. 25) equations (1.39), (1.40). It is clearly verified.
- definition 31 (p. 25) equation (1.41). If the first part of (1.41) is verified for any $w \in \mathcal{V}_O$, for any $w \in \mathcal{V}$, we have:

$$\begin{aligned} (\forall w \in \mathcal{V})\ (g(g'_p(p', a, b)u, w)(p) &= g'_p(p, a, b)g(g'_p(p', a, b)u, w)(p) \\ &= g(g'_p(p', a, b)u, g'_p(p', a, b)w)(p) = 0) \end{aligned}$$

because from proposition 28 (p. 13) and definition 18 (p. 13), $g'_p(p', a, b)w \in \mathcal{V}_O$. Since $g'_p(p', a, b)f \in \mathcal{F}$ applying equation (1.41) (p. 25) to $g'_p(p', a, b)f$ we have:

$$(\forall f \in \mathcal{F}_O)(u(f)(p) = g'_p(p, a, b)u(f)(p) = u(g'_p(p', a, b)f)(p) = 0) .$$

**Proposition 53** With the notations of definition 31 (p. 25), given a metric $g$ on a differentiable manifold $M$, two vector fields $u \in \mathcal{V}$, $v \in \mathcal{V}$, a point $p \in M$, we have (see definition 26 (p. 19)):

$$u \sim_p v \Rightarrow (\forall w \in \mathcal{V})g(u, w)(p) = g(v, w)(p) .$$

**Proof**
$U_k$ with its map $(\phi_k : U_k \to \mathcal{R}^n)$ is one open of the coverings of $M$ containing $p \in M$. We call $(x = \phi_k(p) : [1, n] \to \mathcal{R})$. We have from propositions 34 (p. 17), 45 (p. 22) and section 1.4 (p. 19):

$$g(u, w)(p) = g(u^i_k \partial_i, w)(p) = u^i_k(p)g(\partial_i, w)(p) = v^i_k(p)g(\partial_i, w) = g(v, w)(p) .$$

**Proposition 54** With the above notation, calling ${}^{[p]}T$ the tangent space at $p$ (definition 27 (p. 20)), we have the following relations which allow to define a metric on ${}^{[p]}T$:

$$\begin{aligned} &(t_1 \in {}^{[p]}T, t_2 \in {}^{[p]}T, u_1 \in t_1, v_1 \in t_1, u_2 \in t_2, v_2 \in t_2) \\ &\Rightarrow g(u_1, u_2)(p) = g(v_1, v_2)(p) = {}^{[p]}g(t_1, t_2) . \end{aligned}$$

**Proof**
In $\mathbf{U}$ with its map $(\phi : \mathbf{U} \to \mathcal{R}^n)$, one of the coverings of $\mathbf{M}$ with $(\mathbf{x} : [1,n] \to \mathcal{R}^n)$ and $x = \phi^{-1}(\mathbf{x}) \in \mathbf{U}$, from proposition 34 (p. 17) and section 1.4 (p. 19) we have:

$$u_1 = u_1^i(x)\partial_i, u_2 = u_2^i(x)\partial_i, v_1 = v_1^i(x)\partial_i, v_2 = v_2^i(x)\partial_i \; .$$

Where $(x^i : \mathcal{R}^n \to \mathcal{R})$, which is also a map $(x^i : [1,n] \to \mathcal{F}_u)$, is the coordinate $\mathbf{i}$ of the point $x$ in the local coordinate system of $\mathbf{U}$. From definition 27 (p. 20), we have:

$$(\forall i \in [1,n]) \; (u_1^i(p) = v_1^i(p), u_2^i(p) = v_2^i(p)) \; .$$

Applying equation (1.40) (p. 25) we have:

$$\begin{aligned} g(u_1,u_2)(p) &= u_1^i(p)u_2^j(p)g(\partial_i,\partial_j)(p) \\ &= v_1^i(p)v_2^j(p)g(\partial_i,\partial_j)(p) = g(v_1,v_2)(p) \; . \end{aligned}$$

**Proposition 55** We have a differentiable manifold $\mathbf{M}$ with a metric $\mathbf{g}$. That metric applied to a tangent space at the point $p$ defines a metric (definition 236 (p. 262)) on the vectorial space ${}^{[p]}\mathbf{T}$ (definition 199 (p. 235)).
**Proof**
From proposition 54 (p. 27), it satisfies definition 236 (p. 262).
**Definition 33** $g_{i,j}(p)$.

Having a differentiable manifold $\mathbf{M}$ with its vector field $\mathcal{V}$ and a metric $\mathbf{g}$, we have $\partial_i \in \mathcal{V}_k$ (definition 21 (p. 16)). Using definition 27 (p. 20), definition 31 (p. 25) relations (1.39)-(1.40) and definition 32 (p. 25), we have:

$$\begin{aligned} p \in U_k \Rightarrow g(u,v)(p) &= g_{U_k}(\partial_i,\partial_j)(p)u^i(p)v^j(p) = {}^{[p]}g({}^{[p]}\partial_i, {}^{[p]}\partial_j))u^i(p)v^j(p) \\ &= g_{i,j}(p)u^i(p)v^j(p) \; . \end{aligned} \tag{1.44}$$

With those equations, one may define:

$$g_{i,j}(p) = g_{U_k}(\partial_i,\partial_j)(p) = g(\partial_i,\partial_j)(p) = {}^{[p]}g({}^{[p]}\partial_i, {}^{[p]}\partial_j) \in \mathcal{R}$$

$g_{i,j}(p) = {}^{[p]}g({}^{[p]}\partial_i, {}^{[p]}\partial_j)$ can be considered as the metric in the tangent space ${}^{[p]}\mathbf{T}$ (see definition 237 (p. 263)).

**Proposition 56**

$$(\forall k)(\forall p \in U_k) \; (\det(g_{i,j})(p) \neq 0) \; . \tag{1.45}$$

**Proof**
If we have, a $p \in \mathbf{M}$ and a $\mathbf{k}$ with $p \in U_k$, $\det((g_{i,j})(p)) = 0$, at that $p$ we have (proposition 430 (p. 250)):

$$(\exists(\lambda^i : [1,n] \to \mathcal{R}))((\forall j \in [1,n])(\lambda^i g_{i,j}(p) = 0), (\exists k \in [1,n])(\lambda_k \neq 0)) \; .$$

Using this map $(\lambda^i : [1,n] \to \mathcal{R})$ (convention 26 (p. 441)) and the vector field $u = \lambda^i\partial_i$, we should have from definition 33 (p. 28):

$$(\forall v \in \mathcal{V}) \; (g(u,v)(p) = g_{i,j}(p)\lambda^i v^j(p) = 0) \; .$$

Then from equation (1.41) (p. 25), we should have:

$$(\forall j \in [1,n]) \; (u(x^j)(p) = 0 = \lambda^i\partial_i x^j(p) = \lambda^i\delta_i^j = \lambda^j) \; .$$

Which means $(\forall j \in [1,n])(\lambda^j = 0)$. It is in contradiction with what we supposed. From proposition 927 (p. 520), $\det(g_{i,j}(p)) \neq 0$.

**Proposition 57**

$$(\forall i \in [1,n])(\forall i \in [1,n]) \ (\mathbf{g_{i,j}}(\phi_k^{-1}(x)) \in C^\infty)$$

(see definition 276 (p. 304)).

**Proof**
It is a consequence of definitions 31 (p. 25), 32 (p. 25), proposition 52 (p. 26).

**Proposition 58**

$$\det(\mathbf{g_{i,j}}(\phi_k^{-1}(x))) \in C^\infty$$

(see definition 276 (p. 304)).

**Proof**
From proposition 57 (p. 29), each $\mathbf{g_{i,j}}(\phi_k^{-1}(x))$ is in $C^\infty$. Since, from proposition 442 (p. 254), the determinant is the sum of product of $\mathbf{g_{i,j}}(\phi_k^{-1}(x))$, it also belongs to $C^\infty$ (proposition 559 (p. 306)).

**Definition 34 $\mathbf{g^{i,j}}(p)$.**

With notation of definition 33 (p. 28), from proposition 56 (p. 28), we have:

$$(\forall p \in M) \ (\det(\mathbf{g_{i,j}}(p)) \neq 0) \ .$$

Then from proposition 444 (p. 255), $\mathbf{g_{i,j}}(p)$ has an inverse. The elements of $\mathbf{g}^{-1}(p)$ are named $\mathbf{g^{i,j}}(p)$ and we have:

$$(\forall p \in M) \ (\mathbf{g_{i,\ell}}(p)\mathbf{g^{\ell,j}}(p) = \delta_i^j) \ .$$

**Proposition 59**

$$(\forall i \in [1,n])(\forall i \in [1,n]) \ (\mathbf{g^{i,j}}(\phi_k^{-1}(x)) \in C^\infty)$$
$$\det(\mathbf{g^{i,j}}(\phi_k^{-1}(x))) \in C^\infty$$

(see definition 276 (p. 304)).

**Proof**
It is a direct consequence of propositions 57 (p. 29), 58 (p. 29), 444 (p. 255) and 559 (p. 306).

**Proposition 60** We have a covering open $\mathbf{U}$ with its map $(\phi : \mathbf{U} \to \mathcal{R}^n)$ of a differential manifold $\mathbf{M}$ of dimension $\mathbf{n}$ with a metric $\mathbf{g}$. We consider the determinant of the metric evaluated in the local canonical basis (definition 23 (p. 17)) of $\mathbf{U}$ at a point $p \in \mathbf{U}$ (see definitions 33 (p. 28), 221 (p. 250)). This determinant has the same sign at any point of a connected subset (definition 285 (p. 327)) of $\mathbf{U}$.

**Proof**
Given a connected subset $\mathbf{C} \subset \mathbf{U}$ and two points $p \in \mathbf{C}$, $q \in \mathbf{C}$ from definition 285 (p. 327) we have, $\mathbf{f}$ being a continuous map:

$$(\exists(\mathbf{f} : [0,1] \to \mathbf{C})) \ (\mathbf{f}(0) = p, \mathbf{f}(1) = q) \ .$$

Then from proposition 241 (p. 167), the map $\mathbf{h} = (\det(\mathbf{g}(\partial_i, \partial_j)) \circ \phi^{-1} \circ \phi \circ \mathbf{f} : \mathcal{R} \to \mathcal{R})$ is continuous. From proposition 299 (p. 190), $[\mathbf{h}(0), \mathbf{h}(1)] \subset \mathbf{h}([0,1])$. But:

$$\begin{aligned}&(\mathbf{h}(0) < 0 < \mathbf{h}(1) \vee \mathbf{h}(1) < 0 < \mathbf{h}(0)) \Rightarrow (0 \in [\mathbf{h}(0), \mathbf{h}(1)] \subset \mathbf{h}([0,1]))\\&\Rightarrow 0 \in \mathbf{h}([0,1]) \Rightarrow (\exists \mathbf{a} = \mathbf{h}^{-1}(0))(\mathbf{a} \in [0,1])\\&\Rightarrow (\mathbf{f}(\mathbf{a}) = \mathbf{f}(\mathbf{h}^{-1}(0)) \in \mathbf{U}, \det(\mathbf{g}(\partial_i, \partial_j)) \circ \phi^{-1} \circ \phi \circ \mathbf{f}(\mathbf{h}^{-1}(0)) =\\&\det(\mathbf{g}(\partial_i, \partial_j))(\mathbf{f}(\mathbf{a})) = \mathbf{h}(\mathbf{h}^{-1}(0)) = 0)\ .\end{aligned}$$

The last line is forbidden by proposition 56 (p. 28). Then, from section A.1.4 (p. 435) point 8, proposition 60 (p. 29) is true.

**Proposition 61** Given a differential manifold $\mathbf{M}$ of dimension $\mathbf{n}$ with one of its open covering sets $\mathbf{U_k}$, $\mathbf{k} \in \mathbf{K}$, we suppose that $\mathbf{M}$ has a metric $\mathbf{g}$. In $\mathbf{U_k}$, we consider the associated continuous map $(\phi_\mathbf{k} : \mathbf{M} \to \mathcal{R}^\mathbf{n})$ and an isomorphism (definition 142 (p. 166)) $(\psi : \phi(\mathbf{U_k}) \to \mathcal{R}^\mathbf{n}) \in \mathbf{C}^\infty$ (see definition 276 (p. 304) and section 1.6 (p. 23)) which generates another associated continuous map $(\psi \circ \phi_\mathbf{k} : \mathbf{M} \to \mathcal{R}^\mathbf{n})$. We call $(\partial_i^\phi : [1, \mathbf{n}] \to \mathcal{V}_\mathbf{k})$ the canonical basis associated to $\phi$ of the set of vector fields $\mathcal{V}_\mathbf{k}$ of the restricted manifold $\mathbf{U_k}$. We call $(\partial_i^{\psi\phi} : [1, \mathbf{n}] \to \mathcal{V}_\mathbf{k})$ the one associated to $\psi \circ \phi_\mathbf{k}$. We have in $\mathbf{U_k}$ the following relation:

$$\begin{aligned}\mathbf{g}_{ij}^\phi &= \mathbf{g}(\partial_i^\phi, \partial_j^\phi) = \mathbf{g}(\mathbf{jac}(\psi)_i^k \partial_k^{\psi\phi}, \mathbf{jac}(\psi)_j^\ell \partial_\ell^{\psi\phi}) = \mathbf{jac}(\psi)_i^k \mathbf{jac}(\psi)_j^\ell \mathbf{g}(\partial_k^{\psi\phi}, \partial_\ell^{\psi\phi})\\&= \mathbf{jac}(\psi)_i^k \mathbf{jac}(\psi)_j^\ell \mathbf{g}_{k\ell}^{\psi\phi}\ .\end{aligned}$$

The map $\mathbf{jac}$ is defined in definition 297 (p. 358).

**Proof**

It is a direct consequence of definition 31 (p. 25), equation (1.40) and equations (1.33), (1.34) of proposition 47 (p. 23).

**Proposition 62** Keeping the notation of proposition 61 (p. 30), the two determinants of the metric expressed in two local bases deduced from each other by an isomorphism as in proposition 61 (p. 30) have the same sign.

**Proof**

From propositions 61 (p. 30) and 448 (p. 257) we have:

$$\det(\mathbf{g}^\phi) = \det(\mathbf{g}_{ij}^\phi) = \det(\mathbf{jac}(\psi)_i^k)\det(\mathbf{jac}(\psi)_j^\ell)\det(\mathbf{g}_{k\ell}^{\psi\phi}) = \det(\mathbf{jac}(\psi)^2)\det(\mathbf{g}^{\psi\phi})\ .$$

From proposition 48 (p. 24), we have $\det(\mathbf{jac}(\psi)) \neq 0$ as it should.

**Proposition 63** In any tangent space, the metric is given by a real symmetric matrix. When this matrix is considered as a local self adjoint transformation (definition 324 (p. 429)) with the Euclidean norm (definition 237 (p. 263)), its eigenvalues (definition 323 (p. 428)) are all real and different from zero.

**Proof**

From equation (1.39) (p. 25), we have $\mathbf{g}_{i,j} = \mathbf{g}_{j,i} \in \mathcal{F}$. For any $p \in \mathbf{M}$, $\mathbf{g}_{i,j}(p) \in \mathcal{R}$ can be considered as a linear transformation on a vector space with $\mathbf{n}$ dimensions with a Euclidean norm (the inner product given by $\delta_j^i$ the Kronecker symbol (definition 215 (p. 245))). With that norm, it is a self adjoint transformation. Then, from proposition 720 (p. 429), the eigenvalues of $\mathbf{g}_{i,j}$ at each point $p$ are real. There are $\mathbf{n}$ roots of the eigenvalue polynomial (proposition 717 (p. 428)). The coefficients of that polynomial are also product and sum of $\phi_\mathbf{k}(p)$ and are in $\mathbf{C}^\infty$ (see definition 276 (p. 304) and proposition 718 (p. 428)). From proposition 720 (p. 429), those $\mathbf{n}$ roots are real. The product of the eigenvalues is $\det(\mathbf{g}_{i,j})$ (proposition 724 (p. 431)). From equation (1.45) (p. 28), none of the eigenvalues can be zero.

**Proposition 64** M is a differentiable manifold of dimension $\mathbf{n}$ with a metric $\mathbf{g}$. We consider at any point of M, the matrix given by the metric in the associated tangent space (definitions 27 (p. 20), 33 (p. 28)) and its eigenvalue polynomial. We consider the map (see proposition 704 (p. 421)) $(\mathbf{r(i) = r_i : [1,n] \to \mathcal{R}})$ associated to the eigenvalues (definition 323 (p. 428)) of that symmetric matrix. The two cardinals (see definitions 378 (p. 484), 174 (p. 141)) $\mathbf{n_+ = Card(r^{-1}(]0, \to [))}$, (number of $\mathbf{r_i}$ with a + sign) and $\mathbf{n_- = Card(r^{-1}(] \leftarrow, 0[))}$, (number of $\mathbf{r_i}$ with a − sign) are the same along all path connected subsets of M (definition 285 (p. 327)).

**Proof**

Each $\mathbf{g_{i,j}} \in \mathcal{R}$ of the matrix associated to the metric $\mathbf{g}$ is a map $(\mathbf{g_{i,j}}(p) : \mathbf{M} \to \mathcal{R})$ which belongs to $\mathcal{F}$. Then, all the coefficients of the eigenvalue polynomial are in $\mathcal{F}$. From proposition 63 (p. 30), the eigenvalues are all real (proposition 720 (p. 429)) and different from zero. The degree of that polynomial, being equal to the dimension $\mathbf{n}$ of the differentiable manifold M, is the same in the connected part. The sum of the multiplicity (definition 322 (p. 421)) of the eigenvalues (roots of the eigenvalue polynomial) is $\mathbf{n = n_+ + n_-}$ and remains constant in the connected part.

From proposition 707 (p. 423), the roots can be ordered such that they are continuous maps of $p \in \mathbf{M}$:

There are $\mathbf{n}$ continuous maps $(\mathbf{r_i : [1,n]} \to F)$. We call here $F \subset \mathcal{P}(\mathbf{M} \times \mathcal{R})$ the set of continuous map from M to $\mathcal{R}$:

$$(\forall \mathbf{i} \in [\mathbf{1,n}]) \; ((\mathbf{r_i}(p) : \mathbf{M} \to \mathcal{R}) \in F) \; .$$

Given two points $p_0 \in \mathbf{M}$ and $p_1 \in \mathbf{M}$ belonging to the same path connected subset of M, there exists (definition 285 (p. 327)) a continuous map $(\mathbf{h} : [\mathbf{0,1}] \to \mathbf{M})$ such as $\mathbf{h(0)} = p_0$ and $\mathbf{h(1)} = p_1$, if the number of positive eigenvalues at $p_0$ were not the same as at $p_1$, there would be an $\mathbf{i} \in [\mathbf{1,n}]$ such that $\mathbf{r_i}(p_0) < \mathbf{0} < \mathbf{r_i}(p_1)$ and $\mathbf{r_i(h(0))} < \mathbf{0} < \mathbf{r_i(h(1))}$. From proposition 241 (p. 167), $(\mathbf{r_i(h(t))} : [\mathbf{0,1}] \to \mathcal{R})$ is continuous. From proposition 299 (p. 190), there is a $\theta \in [\mathbf{0,1}]$ such that $\mathbf{r_i(h(\theta))} = \mathbf{0}$ which is forbidden at any $p \in \mathbf{M}$ because there is no root equal to zero.

**Proposition 65** M is a differentiable manifold of dimension $\mathbf{n}$ with a metric $\mathbf{g}$, with its covering set of opens $\mathbf{U_k}$ with their map $(\phi_\mathbf{k} : \mathbf{U_k} \to \mathcal{R}^\mathbf{n})$, $\mathbf{k} \in \mathbf{K}$. For any $\mathbf{k} \in \mathbf{K}$, there is a set of $\mathbf{n^2}$ maps $(\psi_\mathbf{i}^\mathbf{j} : \phi_\mathbf{k}(\mathbf{U_k}) \to \mathcal{R}^\mathbf{n})$ such that in any tangent space:

$$|\mathbf{g}(\psi_\mathbf{i}^\mathbf{m} \partial_\mathbf{m}, \psi_\mathbf{j}^{\mathbf{m}'} \partial_{\mathbf{m}'})(p)| = \delta_\mathbf{i}^\mathbf{j} \; .$$

The number of $\mathbf{i} \in [\mathbf{1,n}]$ for which $\mathbf{g}(\psi_\mathbf{i}^\mathbf{m} \partial_\mathbf{m}, \psi_\mathbf{i}^{\mathbf{m}'} \partial_{\mathbf{m}'})(p) = \mathbf{1}$ and the number for which it is $-\mathbf{1}$ are the same in any path connected subset of M (definition 285 (p. 327)). (We have to be careful, the $\psi$'s are not necessarily continuous maps like the ones generated by the eigen-vectors of the metric (proposition 707 (p. 423)). ) [7]

**Proof**

It is a direct application of propositions 64 (p. 31), 489 (p. 275). This set is not unique. One of them is given by the eigenvectors of $\mathbf{g_{i,j}}$ in each tangent space ${}^{[p]}\mathbf{T}$: From proposition 723 (p. 430), there is a set of orthogonal vectors in the Euclidean metric generated by the Euclidean norm (quoted as $< \mathbf{u}|\mathbf{v} >_\mathbf{e}$). Those vectors are also orthogonal in the semi Riemannian

[7] For instance taking as M, $\mathcal{R}^\mathbf{2}$, labeling a point of M by $(\mathbf{x,y}) \in \mathbf{M} = \mathcal{R}^\mathbf{2}$, one can imagine a manifold $\mathbf{n = 2}$ with the metric $\mathbf{g}$ given by $\mathbf{x \leq 0 \Rightarrow g_j^i(x,y) = \delta_i^j(1+x)}$, $\mathbf{x \geq 0 \Rightarrow g_j^i(x,y) = \delta_i^j + (1 - \delta_i^j)x}$. $\mathbf{g}$ is continuous for all $(\mathbf{x,y}) \in \mathbf{M}$. The eigenvectors of $\mathbf{g}$ are when $\mathbf{x \leq 0}$, $(\mathbf{1,0})$; $(\mathbf{0,1})$ and when $\mathbf{x > 0}$, $\frac{\mathbf{1}}{\sqrt{\mathbf{2}}}(\mathbf{1,1})$; $\frac{\mathbf{1}}{\sqrt{\mathbf{2}}}(\mathbf{1,-1})$. Their components are not continuous at $\mathbf{x = 0}$.

metric generated by $\mathbf{g}$ (quoted as $<\mathrm{u}|\mathrm{v}>_{\mathbf{g}}$) in each ${}^{[p]}\mathrm{T}$:
Considering two different eigenvectors of the above set: $\mathbf{v_i}$ $\mathbf{v_j}$ with their eigenvalues: $\mathbf{r_i}$, $\mathbf{r_j}$ we have:

$$<\mathrm{v_i}|\mathrm{v_j}>_{\mathbf{g}}=<\mathrm{v_i}|\mathbf{g}\mathrm{v_j}>_{\mathrm{e}}=\mathrm{r_j}<\mathrm{v_i}|\mathrm{v_j}>_{\mathrm{e}}=\mathrm{r_j}\delta^{\mathrm{i}}_{\mathrm{j}}\ .$$

Since none of the eigenvalue are zero (proposition 63 (p. 30)), each vector $\mathbf{v_i}$ can be renormalized as $\mathbf{v}'_{\mathrm{i}}=\frac{\mathbf{v_i}}{\sqrt{|\mathbf{r_i}|}}$. It gives now:

$$<\mathrm{v}'_{\mathrm{i}}|\mathrm{v}'_{\mathrm{j}}>_{\mathbf{g}}=<\mathrm{v}'_{\mathrm{i}}|\mathbf{g}\mathrm{v}'_{\mathrm{j}}>_{\mathrm{e}}=\frac{\mathrm{r_j}}{\sqrt{\mathrm{r}_{\mathrm{j}}^2}}\delta^{\mathrm{i}}_{\mathrm{j}}=\delta^{\mathrm{i}}_{\mathrm{j}}\ .$$

From the preceding proposition and from proposition 489 (p. 275), in every tangent space and in any orthogonal basis (according to $\mathbf{g}$), the number of $+$ and the number of $-$ sign in the length (definition 251 (p. 274)) of the basic vectors are the same.

**Definition 35 Signature of a metric.**

The numbers of positive and negative eigenvalues of the metric $\mathbf{g}$ are called its signature.

## 1.8 One-form field and differential

**Definition 36 One-form field.**

A **one-form** field or a **one-form** ( ${}_{1\triangleleft}\mathrm{f}:\mathcal{V}\rightarrow\mathcal{F}$) is a map of the set of vector fields $\mathcal{V}$ on $\mathcal{F}$ which is linear with $\mathcal{F}$ acting as a ring (definition 180 (p. 215)):
If ${}_{1\triangleleft}\mathrm{f}$ is **one-form**, $\mathbf{u}$ and $\mathbf{v}$ two vector fields, $\boldsymbol{\lambda}$ and $\boldsymbol{\mu}$ two elements of $\mathcal{F}$, one has:

$$\begin{aligned}&(\mathrm{u}\in\mathcal{V},\mathrm{v}\in\mathcal{V})\Rightarrow\\&(\ {}_{1\triangleleft}\mathrm{f}(\mathrm{u})\in\mathcal{F},\ {}_{1\triangleleft}\mathrm{f}(\mathrm{v})\in\mathcal{F},\ {}_{1\triangleleft}\mathrm{f}(\lambda\mathrm{u}+\mu\mathrm{v})=\lambda\times_{1\triangleleft}\mathrm{f}(\mathrm{u})+\mu\times_{1\triangleleft}\mathrm{f}(\mathrm{v})\in\mathcal{F})\ .\end{aligned}\tag{1.46}$$

**Definition 37 $\mathcal{V}^*$, dual of the set of vector fields $\mathcal{V}$ on a differentiable manifold.**

The dual $\mathcal{V}^*$ of the set of vector fields $\mathcal{V}$ is the set of **one-form** built on $\mathcal{V}$. It is also linear on $\mathcal{F}$.

**Definition 38 $\mathcal{V}^*_{\mathrm{O}}$, dual of the set of vector fields $\mathcal{V}_{\mathrm{O}}$ on the restricted manifold O open subset of a manifold M (definition 5 (p. 3)).**

One calls $\mathcal{V}^*_{\mathrm{O}}$ the dual defined by definition 37 (p. 32) of the restricted vector field $\mathcal{V}_{\mathrm{O}}$ as defined in definition 5 (p. 3).

**Proposition 66** $\mathcal{V}^*$ as $\mathcal{V}$ is a module over the commutative ring $\mathcal{F}$ (see definitions 180 (p. 215), 182 (p. 218)).

**Proof**
It comes directly from definition 36 (p. 32) which satisfies definition 180 (p. 215).

**Proposition 67** As in proposition 22 (p. 9), having a point $p\in\mathrm{M}$ and one of the covering sets $\mathrm{U_i}$ containing it, for any open $\mathrm{O}$ of $\mathrm{U_i}$ containing $p$: $p\in\mathrm{O}\subset\mathrm{U_i}\subset\mathrm{M}$, $\mathbf{u}$ and $\mathbf{v}$ being two vector fields of $\mathcal{V}$, we have:

$$\begin{aligned}&(\forall q\in\mathrm{O})(\forall\mathrm{f}\in\mathcal{F})\ (\mathrm{v}(\mathrm{f})(q)=\mathrm{u}(\mathrm{f})(q))\\&\Rightarrow(\forall p\in\mathrm{O})(\forall\mathrm{h}\in\mathcal{V}^*)(\mathrm{h}(\mathrm{u})(p)=\mathrm{h}(\mathrm{v})(p))\ .\end{aligned}$$

**Proof**
We proceed as in proposition 22 (p. 9). With $\mathbf{a} > 0, \overline{\mathrm{C}}(p,\mathrm{a}) \subset \phi_{\mathrm{i}}(\mathrm{O})$, we consider the map $(\mathrm{g}'_p(\phi(x),\mathrm{a}) = \mathrm{g}(\phi(x) - \phi(p),\mathrm{a}) : \mathrm{O} \to \mathcal{R})$ of proposition 714 (p. 427) giving $\mathrm{g}'_p(\phi_{\mathrm{i}}(p),\mathrm{a}) = 1$ and we introduce the same $\mathrm{g}^{\mathrm{M}} \in \mathcal{F}$, such that, for $x \in \mathrm{M}$, we have (see definition 353 (p. 454)):

$$x \in \mathrm{O} \Rightarrow \mathrm{g}^{\mathrm{M}}(x) = \mathrm{g}'_p(\phi_{\mathrm{i}}(x),\mathrm{a}),\ x \notin \mathrm{O} \Rightarrow \mathrm{g}^{\mathrm{M}}(x) = 0\ .$$

We have for any $\mathbf{h} \in \mathcal{V}^*$ (see definition 17 (p. 8)):

$$\begin{aligned}&(\forall x \in \mathrm{M})(\forall \mathrm{f} \in \mathcal{F})\ (\mathrm{g}^{\mathrm{M}}(\mathrm{u}-\mathrm{v})(\mathrm{f})(x) = \mathrm{g}^{\mathrm{M}}(x)(\mathrm{u}(\mathrm{f})(x) - \mathrm{v}(\mathrm{f})(x)) = 0)\\ &\Rightarrow \mathrm{g}^{\mathrm{M}}(\mathrm{u}-\mathrm{v}) = 0 \Rightarrow \mathrm{h}(\mathrm{g}^{\mathrm{M}}(\mathrm{u}-\mathrm{v})) = 0\\ &\Rightarrow \mathrm{h}(\mathrm{g}^{\mathrm{M}}\mathrm{u}) = \mathrm{h}(\mathrm{g}^{\mathrm{M}}\mathrm{v})\\ &\Rightarrow \mathrm{g}^{\mathrm{M}}(p)\mathrm{h}(\mathrm{u})(p) = \mathrm{g}^{\mathrm{M}}(p)\mathrm{h}(\mathrm{v})(p) \Rightarrow \mathrm{h}(\mathrm{u})(p) = \mathrm{h}(\mathrm{v})(p)\ .\end{aligned}$$

**Proposition 68** Given an open O of a differential manifold, any **one-form** (definition 36 (p. 32)) of $\mathcal{V}^*$, dual of the set of vector fields $\mathcal{V}$ of M, can be restricted as a **one-form** of $\mathcal{V}^*_{\mathrm{O}}$ (definition 38 (p. 32)) by

$$\begin{aligned}&(\forall\ {}_{1\lhd}\mathrm{f} \in \mathcal{V}^*)(\exists\ {}_{1\lhd}\mathrm{f}_{\mathrm{O}} \in \mathcal{V}^*_{\mathrm{O}})\\ &((p \in \mathrm{O}, \mathrm{v} \in \mathcal{V}_{\mathrm{O}}) \Rightarrow\ {}_{1\lhd}\mathrm{f}_{\mathrm{O}}(\mathrm{v})(p) =\ {}_{1\lhd}\mathrm{f}(\mathrm{g}'_p(q,\mathrm{a}_p,\mathrm{b}_p)\mathrm{v})(p))\end{aligned}$$

where as usual, if $\mathrm{U}_p, \phi_p$ are a local covering and local map around $p \in \mathrm{O} \subset \mathrm{M}$: $p \in \mathrm{U}_p$ (see definition 1 (p. 1) point 1).
$(\mathrm{a}_p : \mathrm{O} \to \mathcal{R}^+)$, $(\mathrm{b}_p : \mathrm{O} \to \mathcal{R}^+)$ are such that $0 < \mathrm{a}_p < \mathrm{b}_p$ and $\overline{\mathrm{C}}(\phi(p),\mathrm{a}_p) \subset \overline{\mathrm{C}}(\phi(p),\mathrm{b}_p) \subset \phi(\mathrm{U} \cap \mathrm{O})$ (see proposition 269 (p. 179)).
**Proof**
From proposition 29 (p. 13), $\mathrm{g}'_p(q,\mathrm{a}_p,\mathrm{b}_p)\mathrm{v}$ is a vector field in $\mathcal{V}$ that is to say it is a map $\mathcal{F} \to \mathcal{F}$ satisfying definition 15 (p. 8).
But we also have to prove that ${}_{1\lhd}\mathrm{f}_{\mathrm{O}}(\mathrm{v})(p)$ does not depend on the particular choice of $\mathrm{U}_p, \phi_p, \mathrm{a}_p, \mathrm{b}_p$. In any $\mathrm{U}_p$ if we have two sets of $\mathrm{a}_p, \mathrm{b}_p,\ \mathrm{a}'_p, \mathrm{b}'_p$, since $\mathrm{g}'_p(p,\mathrm{a}_p,\mathrm{b}_p) = \mathrm{g}'_p(p,\mathrm{a}'_p,\mathrm{b}'_p) = 1$, from definition 36 (p. 32) equation (1.46) we have:

$$\begin{aligned}&{}_{1\lhd}\mathrm{f}(\mathrm{g}'_p(q,\mathrm{a}_p,\mathrm{b}_p)\mathrm{v})(p) = \mathrm{g}'_p(p,\mathrm{a}'_p,\mathrm{b}'_p)\ {}_{1\lhd}\mathrm{f}(\mathrm{g}'_p(q,\mathrm{a}_p,\mathrm{b}_p)\mathrm{v})(p)\\ &=\ {}_{1\lhd}\mathrm{f}(\mathrm{g}'_p(q,\mathrm{a}'_p,\mathrm{b}'_p)\mathrm{g}'_p(q,\mathrm{a}_p,\mathrm{b}_p)\mathrm{v})(p) = \mathrm{g}'_p(p,\mathrm{a}_p,\mathrm{b}_p)\ {}_{1\lhd}\mathrm{f}(\mathrm{g}'_p(q,\mathrm{a}'_p,\mathrm{b}'_p)\mathrm{v})(p)\\ &=\ {}_{1\lhd}\mathrm{f}(\mathrm{g}'_p(q,\mathrm{a}'_p,\mathrm{b}'_p)\mathrm{v})(p)\ .\end{aligned}$$

If we have two covering elements of M $\mathrm{U}, \phi,\ \mathrm{U}_1, \phi_1$ at $p$, we consider the open (definition 132 (p. 151) equation (2.18)) $\mathrm{U} \cap \mathrm{U}_1 \cap \mathrm{O}$ and a couple $0 < \mathrm{a} < \mathrm{b}$ such that (see proposition 269 (p. 179)): $\overline{\mathrm{C}}(\phi(p),\mathrm{a}) \subset \overline{\mathrm{C}}(\phi(p),\mathrm{b}) \subset \phi(\mathrm{U} \cap \mathrm{U}_1 \cap \mathrm{O})$. $\phi_1(\phi^{-1}(\mathrm{C}(\phi(p),\mathrm{a}))))$ is open. In it, we also have $0 < \mathrm{a}_1 < \mathrm{b}_1$ such that (see proposition 269 (p. 179)): $\overline{\mathrm{C}}(\phi_1(p),\mathrm{a}_1) \subset \overline{\mathrm{C}}(\phi_1(p),\mathrm{b}_1) \subset \phi_1(\phi^{-1}(\overline{\mathrm{C}}(\phi(p),\mathrm{a})))) \subset \phi_1(\mathrm{U} \cap \mathrm{U}_1 \cap \mathrm{O})$. From the definition of $\mathbf{g}'$ (definition 14 (p. 6)), we have:

$$\mathrm{g}'_{1p}(q,\mathrm{a}_1,\mathrm{b}_1) = \mathrm{g}'_{1p}(q,\mathrm{a}_1,\mathrm{b}_1)\mathrm{g}'_p(q,\mathrm{a},\mathrm{b})\ .$$

When one of the subscripts of $\mathbf{g}'$ is $\mathbf{1}$ it indicates that the definitions of the balls are given using $\phi_1$ and $\phi$ otherwise. From that we have:

$$\begin{aligned}&{}_{1\lhd}\mathrm{f}(\mathrm{g}'_{1p}(q,\mathrm{a}_1,\mathrm{b}_1)\mathrm{v})(p) =\ {}_{1\lhd}\mathrm{f}(\mathrm{g}'_p(q,\mathrm{a},\mathrm{b})\mathrm{g}'_{1p}(q,\mathrm{a}_1,\mathrm{b}_1)\mathrm{v})(p)\\ &= \mathrm{g}'_{1p}(p,\mathrm{a}_1,\mathrm{b}_1)\ {}_{1\lhd}\mathrm{f}(\mathrm{g}'_p(q,\mathrm{a},\mathrm{b})\mathrm{v})(p) =\ {}_{1\lhd}\mathrm{f}(\mathrm{g}'_p(q,\mathrm{a},\mathrm{b})\mathrm{v})(p)\ .\end{aligned}$$

**Definition 39 Restricted one-form field to an open.**

Given an open $\mathbf{O}$ of a differential manifold $\mathbf{M}$ with its set of vector fields $\mathcal{V}$. We consider the set of vector fields $\mathcal{V}_\mathbf{O}$ on the restricted manifold $\mathbf{O}$ (definition 5 (p. 3)). Given ${}_{1\triangleleft}\mathbf{f}$ a **one-form** on $\mathcal{V}$, its restricted **one-form** field on O, $({}_{1\triangleleft}\mathbf{f}_\mathbf{O} : \mathcal{V}_\mathbf{O} \to \mathcal{F})$ is a **one-form** on $\mathcal{V}_\mathbf{O}$ defined as follows:
For all $\mathbf{x} \in \mathbf{O}$, we associate $\mathbf{U}$, an element of the covering of $\mathbf{M}$ with its map $\phi : \mathbf{U} \to \mathcal{R}^\mathrm{n}$ and with $\mathbf{x} \in \mathbf{U}$. We associate also $\mathbf{a} > 0,\ \mathbf{b} > 0$ such that $\mathbf{C}(\mathbf{x},\mathbf{a}) \subset \mathbf{C}(\mathbf{x},\mathbf{b}) \subset \phi(\mathbf{U}\cap\mathbf{O})$ (proposition 266 (p. 179)). It allows to write:

$$\mathbf{u} \in \mathcal{V}_\mathbf{O} \Rightarrow_{1\triangleleft} \mathbf{f}_\mathbf{O}(\mathbf{u})(\mathbf{x}) =_{1\triangleleft} \mathbf{f}(\mathbf{g}'_x(y,\mathbf{a},\mathbf{b})\mathbf{u})(x)$$

$1\triangleleft\mathbf{f}_\mathbf{O}(\mathbf{u})$ satisfies definition 36 (p. 32) and from proposition 68 (p. 33), it does not depend on the exact value of $\mathbf{a},\ \mathbf{b}$ and $\mathbf{U}$.

**Definition 40 Differential.**

In a differentiable manifold $\mathbf{M}$ (definition 3 (p. 1)), with its set of smooth maps $\mathcal{F}$ and its set of vector fields $\mathcal{V}$, one calls differential a map $(\mathbf{d} : \mathcal{F} \to \mathcal{V}^*)$ defined by:

$$(\forall \mathbf{f} \in \mathcal{F})(\forall \mathbf{u} \in \mathcal{V})\ (\mathbf{df}(\mathbf{u}) = \mathbf{u}(\mathbf{f}) \in \mathcal{F})\ . \tag{1.47}$$

**Definition 41 Restricted differential to an open of M.**

Given a map $\mathbf{f} \in \mathcal{F}$ and an open $\mathbf{O} \subset \mathbf{M}$, we call the restricted differential of $\mathbf{f}$ to the open O, the map $(\mathbf{d_o f} : \mathcal{F}_\mathrm{o} \to \mathcal{V}^*_\mathrm{o})$ defined by:

$$(\forall \mathbf{v} \in \mathcal{V}_\mathrm{o})\ (\mathbf{d_o f}(\mathbf{v}) = \mathbf{df_o}(\mathbf{v}) = \mathbf{v}(\mathbf{f_o}))\ ,$$

where $\mathbf{f_o}$ is the restriction to $\mathbf{O}$ of $\mathbf{f}$ (definition 118 (p. 135)) and $\mathcal{V}_\mathrm{o}$ the set of vector field of the restricted manifold $\mathbf{O}$ (definition 5 (p. 3)).

**Proposition 69** The restriction of a differential to an open $\mathbf{O}$ (definition 41 (p. 34)) subset of a differential manifold $\mathbf{M}$ is the same as the restriction of the associated **one-form** (definition 39 (p. 34)) to the open $\mathbf{O}$.

**Proof**
From definition 39 (p. 34), we have with $\mathbf{v} \in \mathcal{V}_\mathrm{o}\ x \in \mathbf{O}$:

$$\begin{aligned}(\mathbf{df})_\mathrm{o}(\mathbf{v})(x) &= \mathbf{df}(\mathbf{g}'_x(y,\mathbf{a},\mathbf{b})\mathbf{v})(x) = \mathbf{v}(\mathbf{g}'_x(y,\mathbf{a},\mathbf{b})\mathbf{f}(y))(x)\\ &= \mathbf{v}(\mathbf{g}'_x(y,\mathbf{a},\mathbf{b})\mathbf{f_o}(y))(x) = \mathbf{v}(\mathbf{f_o}(y))(x) = \mathbf{df_o}(\mathbf{v})(\mathbf{x})\ .\end{aligned}$$

**Proposition 70** In any covering set $\mathbf{U_i}$ of $\mathbf{M}$ having a local map $(\phi_\mathrm{i} : \mathbf{U_i} \to \mathcal{R}^\mathrm{n})$ (see definition 5 (p. 3)), the restricted differential to $\mathbf{U_i}$ of a map $\mathbf{f} \in \mathcal{F}$ can be written as[8]:

$$\mathbf{d_i f} = \mathbf{df_i} = (\partial_\mu \mathbf{f})\mathbf{dx}^\mu = \partial_\mu \mathbf{f} \mathbf{dx}^\mu\ , \tag{1.48}$$

where $\mathbf{f_i}$ is the restriction of $\mathbf{f}$ to $\mathbf{U_i}$ (see definitions 118 (p. 135), 20 (p. 15) and 23 (p. 17)). In equation (1.48) (p. 34), we apply the next convention 5 (p. 35).

[8]We always have summation over repeated indices when they do not have special meaning defined earlier.

**Proof**
Keeping the above notations, in the manifold $\mathbf{U_i}$ (see definition 5 (p. 3)), we have the functions $\mathbf{x}^\mu \in \mathcal{F}_\mathrm{i}$ (definitions 9 (p. 4), 20 (p. 15)) and the basic vector fields $\boldsymbol{\partial}_\nu \in \mathcal{V}_\mathrm{i}$ (definition 23 (p. 17)). Then we have (definitions 19 (p. 14), 112 (p. 131) propositions 28 (p. 13), 31 (p. 14), equation (1.18) (p. 14)):

$$(\forall \mathrm{v} \in \mathcal{V}_\mathrm{i})(\forall \mathrm{f} \in \mathcal{F})((\mathrm{f_i} = \mathrm{f} \cap \mathrm{U_i} \times \mathcal{R} \in \mathcal{F}_\mathrm{i}, p \in \mathrm{U_i}) \Rightarrow \mathrm{d_i f(v)}(p) = \mathrm{df_i(v)}(p)$$
$$= \mathrm{v(f_i)}(p) = \{\mathrm{v(x^\mu)}\partial_\mu(\mathrm{f} \circ \phi_\mathrm{i}^{-1}) \circ \phi_\mathrm{i}\}(p) = (\partial_\mu \mathrm{f}(p))\mathrm{v(x^\mu)}(p) = \{\partial_\mu \mathrm{f dx^\mu(v)}\}(p)) .$$

**Proposition 71** In any covering set $\mathbf{U_i}$ of $\mathbf{M}$ having a local map $(\phi_\mathrm{i} : \mathrm{U_i} \to \mathcal{R}^\mathrm{n})$ (see definition 5 (p. 3)), any one form defined on the restricted manifold $\mathbf{U_i}$ can be expressed as:

$${}_{1\triangleleft}\mathrm{f} \in \mathcal{V}_\mathrm{i}^* \Rightarrow {}_{1\triangleleft}\mathrm{f} = {}_{1\triangleleft}\mathrm{f}(\partial_\mu)\mathrm{dx}^\mu$$

**Proof**
From definition 36 (p. 32), equation (1.46) (p. 32) and definition 24 (p. 17) with the following definition of $\mathbf{x}^\mu$

$$q \in \mathrm{U_i} \Rightarrow \mathrm{x}^\mu = \phi_\mathrm{i}^\mu(q) \in \mathcal{F}_\mathrm{i}$$

we have:

$${}_{1\triangleleft}\mathrm{f} \in \mathcal{V}_\mathrm{i}^* \Rightarrow (\forall \mathrm{v} \in \mathcal{V}_\mathrm{i})(\ {}_{1\triangleleft}\mathrm{f(v)} = {}_{1\triangleleft}\mathrm{f}(\mathrm{v(x^\mu)}\partial_\mu) = \mathrm{v(x^\mu)}\ {}_{1\triangleleft}\mathrm{f}(\partial_\mu) = {}_{1\triangleleft}\mathrm{f}(\partial_\mu)\mathrm{dx^\mu(v)}$$

then we have proposition 71 (p. 35).

**Convention 5** $(\boldsymbol{\partial}_\mu \mathbf{f})\mathbf{dx}^\mu = \boldsymbol{\partial}_\mu \mathbf{f dx}^\mu$.

By convention to simplify the writing, we assume that $\boldsymbol{\partial}$ does not go over $\mathbf{d}$ when we are dealing with differentials.

**Proposition 72** Given a manifold $\mathbf{M}$ of dimension $\mathbf{n}$, in a local covering $\mathbf{U}$ with its map $(\phi : \mathrm{U} \to \mathcal{R}^\mathrm{n})$, its set of smooth maps $\mathcal{F}_\mathrm{u}$ and its set of vector fields $\mathcal{V}_\mathrm{U}$, we have the following relationship between the differential of the coordinate maps $\mathbf{x}^\mu$ and the basic vector field $\boldsymbol{\partial}_\nu$:

$$\mathrm{dx}^\mu(\partial_\nu) = \partial_\nu \mathrm{x}^\mu = \frac{\partial \mathrm{x}^\mu}{\partial \mathrm{x}^\nu} = \delta^\mu_\nu \tag{1.49}$$

where $\delta^\nu_\mu$ is the Kronecker symbol (definition 215 (p. 245)).

**Proof**
It is a direct consequence of equation (1.47) (p. 34) and definition 20 (p. 15) with proposition 31 (p. 14) equation (1.18) (p. 14), propositions 308 (p. 194) and 309 (p. 194).

**Proposition 73** Keeping the above notations, any ${}_{1\triangleleft}$**form h** combined with a vector field gives with no summation over $\mathbf{k}$ but over $\boldsymbol{\mu}$ and $\boldsymbol{\nu}$:

$$\begin{gathered}(\forall \mathrm{h} \in \mathcal{V}^*)(\forall \mathrm{k} \in \mathrm{K})(\exists(\mathrm{h_{k,\mu}} : [1, \mathrm{n}] \to \mathcal{F}_\mathrm{k}))(\forall \mathrm{v} \in \mathcal{V}) \\ (p \in \mathrm{U_k} \Rightarrow \mathrm{h(v)}(p) = \mathrm{h_{k,\mu}v_k^\mu}(p) = \mathrm{h_{k,\mu}dx^\mu}(\mathrm{v_k^\nu}\partial_\nu)(p))\end{gathered} \tag{1.50}$$

where $\mathbf{v}_\mathbf{k}^\mu$ is defined in definition 25 (p. 17) and $\mathbf{h}_{\mathbf{k},\mu} \in \mathcal{F}_{\mathrm{UK}}$.

**Proof**
From proposition 269 (p. 179), when $p \in \mathrm{U_k}$ there are $\mathrm{a} \in \mathcal{R}$, $\mathrm{b} \in \mathcal{R}$ with $0 < \mathrm{a} < \mathrm{b}$ such that $\overline{\mathrm{C}}(\phi_{\mathrm{k}}(p), \mathrm{a}) \subset \overline{\mathrm{C}}(\phi_{\mathrm{k}}(p), \mathrm{b}) \subset \phi_{\mathrm{k}}(\mathrm{U_k})$. It allows to define $\mathrm{g}'_p(\phi_{\mathrm{k}}(p'), \mathrm{a}, \mathrm{b}) \in \mathcal{F}$ with $p' \in \mathrm{M}$ as in definition 14 (p. 6). Then using definition 36 (p. 32) and applying proposition 67 (p. 32) several times:

$$p \in \mathrm{U_k} \Rightarrow \mathrm{h(v)}(p) = \mathrm{g}'_p(\phi_{\mathrm{k}}(p), \mathrm{a}, \mathrm{b})\mathrm{h(v)}(p) = \mathrm{h}(\mathrm{g}'_p\mathrm{v})(p)$$
$$= \mathrm{h}(\mathrm{g}'_p\mathrm{v}^\mu_{\mathrm{k}}\partial_\mu)(p) = \mathrm{v}^\mu_{\mathrm{k}}(p) \times \mathrm{h}(\mathrm{g}'_p\partial_\mu)(p) = \mathrm{h}_{\mathrm{k},\mu}(p)\mathrm{v}^\mu_{\mathrm{k}}(p)$$

Since from proposition 29 (p. 13) $\mathrm{g}'_p\partial_\mu \in \mathcal{V}$, we have:

$$\mathrm{h}(\mathrm{g}'_p\partial_\mu) \in \mathcal{F} .$$

From proposition 67 (p. 32) $\mathrm{h}(\mathrm{g}'_p\partial_\mu)$ does not depend on $0 < \mathrm{a} < \mathrm{b}$ as soon as:

$$\overline{\mathrm{C}}(\phi_{\mathrm{k}}(p), \mathrm{a}) \subset \overline{\mathrm{C}}(\phi_{\mathrm{k}}(p), \mathrm{b}) \subset \phi_{\mathrm{k}}(\mathrm{U_k}) .$$

As in definition 18 (p. 13), for any $q \in \mathrm{C}(\phi_{\mathrm{k}}(p), \mathrm{a})$, there are $\mathrm{a}_q, \mathrm{b}_q$ such that:

$$\overline{\mathrm{C}}(\phi_{\mathrm{k}}(q), \mathrm{a}_q) \subset \overline{\mathrm{C}}(\phi_{\mathrm{k}}(q), \mathrm{b}_q) \subset \overline{\mathrm{C}}(\phi_{\mathrm{k}}(p), \mathrm{a}) \subset \overline{\mathrm{C}}(\phi_{\mathrm{k}}(p), \mathrm{b}) \subset \phi_{\mathrm{k}}(\mathrm{U_k}) .$$

Setting $\mathrm{g}'_p = \mathrm{g}'_p(x, \mathrm{a}, \mathrm{b}), \mathrm{g}'_q = \mathrm{g}'_q(x, \mathrm{a}_q, \mathrm{b}_q)$, proposition 30 (p. 13) gives:

$$(\forall \mathrm{x} \in \mathrm{C}(\phi_{\mathrm{k}}(q), \mathrm{a}_q))(\forall \mathrm{f} \in \mathcal{F})(\mathrm{g}'_p\partial_\mu(\mathrm{f})(\phi^{-1}_{\mathrm{k}}(\mathrm{x}))$$
$$= \mathrm{g}'_q\partial_\mu(\mathrm{f})(\phi^{-1}_{\mathrm{k}}(\mathrm{x})) = \partial_\mu\mathrm{f}(\phi^{-1}_{\mathrm{k}}(\mathrm{x}))) .$$

From proposition 67 (p. 32)

$$\mathrm{h}(\mathrm{g}'_p\partial_\mu)(q) = \mathrm{h}(\mathrm{g}'_q\partial_\mu)(q)$$

since $\mathrm{h}(\mathrm{g}'_p\partial_\mu)(q) \in \mathrm{C}^\infty$ (see definition 276 (p. 304)), from proposition 556 (p. 305), we have $\mathrm{h}_{\mathrm{k},\mu} = \mathrm{h}(\mathrm{g}'_q\partial_\mu)(q) \in \mathrm{C}^\infty$ and in particular for $q = p$. The last equality of equation (1.50) (p. 35):

$$\mathrm{h}_{\mathrm{k},\mu}\mathrm{v}^\mu_{\mathrm{k}}(p) = \mathrm{h}_{\mathrm{k},\mu}\mathrm{dx}^\mu(\mathrm{v}^\nu_{\mathrm{k}}\partial_\nu)(p)$$

comes from $\mathrm{dx}^\mu(\partial_\nu) = \delta^\mu_\nu$ given by equation (1.49) (p. 35).

**Definition 42 Local coordinates of a $_{1\triangleleft}$form.**

Keeping the above notations, any $_{1\triangleleft}$**form** $\mathrm{h}$ can be written in a local cover $\mathrm{U_k} \subset \mathrm{M}$ as: $\mathrm{h} = \mathrm{h}_\mu\mathrm{dx}^\mu$ with $\mathrm{h}_\mu \in \mathcal{F}_{\mathrm{k}}$. It will give on a vector field $\mathrm{v} = \mathrm{v}^\nu\partial_\nu \in \mathcal{V}_{\mathrm{k}}, \mathrm{v}^\nu \in \mathcal{F}_{\mathrm{k}}$ (it can be the restriction in $\mathrm{U_k}$ of a vector field acting on M):

$$\mathrm{h(v)} = \mathrm{h}_\mu\mathrm{v}^\mu$$

$\mathrm{h}_\mu$ are called the local coordinates of $\mathrm{h}$ in $\mathcal{V}^*_{\mathrm{U_k}}$ and $\mathrm{dx}^\mu$ is considered as the basis of $\mathcal{V}^*_{\mathrm{U_k}}$ as $\partial_\nu$ is considered as the basis of $\mathcal{V}_{\mathrm{U_k}}$.

**Proposition 74** With the same notations, we suppose that one has, defined on $\phi(\mathrm{U_k}) \subset \mathcal{R}^{\mathrm{n}}$, a bijection $(\psi : \phi(\mathrm{U_k}) \leftrightarrow \psi\phi(\mathrm{U_k}))$. The local coordinates $\mathrm{h}^{\psi\phi}_\mu$ of the $_{1\triangleleft}$**form** $\mathrm{h}$ obtained with the map $(\psi\phi : U_k \to \mathcal{R}^n)$ are related to the $h^\phi_\mu$ obtained with the map $(\phi : U_k \to \mathcal{R}^n)$ by (see definition 297 (p. 358) and proposition 444 (p. 255))[9]:

$$h = h^{\psi\phi}_\mu dx^{\psi\phi\mu} = h^\phi_\nu dx^{\phi\nu} \Rightarrow h^{\psi\phi}_\mu \frac{\partial\psi^\mu(x^\phi)}{\partial x^{\phi\nu}} = jac(\psi)^\mu_\nu h^{\psi\phi}_\mu = h^\phi_\nu$$
$$dx^{\psi\phi\mu} = \frac{\partial\psi^\mu(x^\phi)}{\partial x^{\phi\nu}} dx^{\phi\nu} = \partial_\nu\psi^\mu(x^\phi) dx^{\phi\nu} \tag{1.51}$$
$$det(\frac{\partial\psi^\mu(x^\phi)}{\partial x^{\phi\nu}}) \neq 0, dx^{\phi\nu} = \{\partial_j\psi^i(x^\phi)\}^{-1\nu}_\mu dx^{\psi\phi\mu} = jac^{-1}(\psi)^\nu_\mu dx^{\psi\phi\mu}.$$

**Proof**
We have by definition $x^{\psi\phi} = \psi(x^\phi)$. Then its differential is (proposition 70 (p. 34)):

$$dx^{\psi\phi\mu} = \frac{\partial\psi^\mu(x^\phi)}{\partial x^{\phi\nu}} dx^{\phi\nu} = \partial_\nu\psi^\mu(x^\phi) dx^{\phi\nu}$$

The same formula can be obtained with proposition 47 (p. 23): $\mathbf{h}$ being a $_{1\triangleleft}$**form**, we have:

$$h^\phi_\mu = h(\partial^\phi_\mu) = \frac{\partial\psi^\nu(x^\phi)}{\partial x^{\phi\mu}} h(\partial^{\psi\phi}_\nu) = \frac{\partial\psi^\nu(x^\phi)}{\partial x^{\phi\mu}} h^{\psi\phi}_\nu$$

from proposition 48 (p. 24) $det(\frac{\partial\psi^\mu(x^\phi)}{\partial x^{\phi\nu}}) \neq 0$. Then $\{\partial\psi^\mu_\nu(x^\phi)\}$ has an inverse for any $p \in U = U_k$ (see proposition 446 (p. 256)). From proposition 455 (p. 259), $\{\partial\psi^i_j(x^\phi)\}^{-1\nu}_\mu \in C^\infty$ (see definition 276 (p. 304)).

**Proposition 75** Defined as in definition 36 (p. 32), the set of linear forms, built on $\mathcal{V}^*$ with $\mathcal{F}$ as a commutative ring (definition 180 (p. 215)), is $\mathcal{V}$.

**Proof**
We call $\mathcal{V}^{**}$ the set of linear forms on $\mathcal{V}^*$:

$$u \in \mathcal{V}^{**} \Rightarrow (u : \mathcal{V}^* \to \mathcal{F}) .$$

We can define a map $(h : \mathcal{V}^{**} \to \mathcal{V})$:
To any $u \in \mathcal{V}^{**}$, we associate a vector field $h(u) \in \mathcal{V}$ by:

$$(\forall f \in \mathcal{F}) (h(u)(f) = u(df)) \tag{1.52}$$

also to any $v \in \mathcal{V}$ we may associate an element of $\mathcal{V}^{**}$, $h^{inv}(v) \in \mathcal{V}^{**}$ by:

$$(\forall {}_{1\triangleleft}f \in \mathcal{V}^*) (h^{inv}(v)({}_{1\triangleleft}f) = {}_{1\triangleleft}f(v))$$

we have

$$(\forall f \in \mathcal{F})(\forall v \in \mathcal{V}) (h(h^{inv}(v))(f) = h^{inv}(v)(df) = df(v) = v(f))$$

then $h(\mathcal{V}^{**}) = \mathcal{V}$ (definition 114 (p. 132)) because any element of $\mathcal{V}$ has at least one inverse $h^{inv}(v)$. $\mathbf{h}$ is a surjective map (definition 123 (p. 139)).
Let us prove now that $\mathbf{h}$ is a bijection: We suppose that

$$u_1 \in \mathcal{V}^{**}, u_2 \in \mathcal{V}^{**}, h(u_1) = h(u_2)$$

[9] In order to be clearer when A is a matrix, we often refer to it by adding indices like $A^i_j$ which are meaningless: we may write $det(A^i_j)$ for $det(A)$ and $\{A^i_j\}^{-1\mu}_\nu$ for $A^{-1\mu}_\nu$.

We have to show that it implies $\mathbf{u_1} = \mathbf{u_2}$.
For any $p \in \mathbf{M}$, there is at least an $\mathbf{U_k} \subset \mathbf{M}$ with its $(\phi_k : \mathbf{U_k} \to \mathcal{R}^n)$ such that $p \in \mathbf{U_k}$. In those conditions (see proposition 269 (p. 179)), there are two closed cubes $\overline{\mathbf{C}}(\phi_k(p), \mathbf{a}) \subset \overline{\mathbf{C}}(\phi_k(p), \mathbf{b}) \subset \phi_k(\mathbf{U_k}) \subset \mathcal{R}^n$. It allows to define a map $\mathbf{g}'_p(q, \mathbf{a}, \mathbf{b}) \in \mathcal{F} \cap \mathcal{F}^0_k$ as in definition 14 (p. 6). We have $\mathbf{g}'_p(p, \mathbf{a}, \mathbf{b}) = \mathbf{1}$ (definition 14 (p. 6)). From the linearity of $\mathbf{u}_1$, $\mathbf{u_2}$ with $\mathcal{F}$ as a commutative ring, using equation (1.52) (p. 37) which defines $\mathbf{h}$, using propositions 71 (p. 35) and 73 (p. 35) applied to $\mathbf{U_k}$, we have for any ${}_{1\lhd}\mathbf{f} \in \mathcal{V}^*$:[10]

$$\begin{aligned}
&\mathbf{u}_1({}_{1\lhd}\mathbf{f})(p) = \mathbf{g}'_p(p, \mathbf{a}, \mathbf{b})\mathbf{u}_1({}_{1\lhd}\mathbf{f})(p) = \mathbf{u}_1(\mathbf{g}'_p(q, \mathbf{a}, \mathbf{b})_{1\lhd}\mathbf{f})(p)\\
&= \mathbf{g}'_p(p, \mathbf{a}, \mathbf{b})_{1\lhd}\mathbf{f}({}^{[k]}\partial_i)(p)\mathbf{u}_1({}^{[k]}\mathbf{dx}^i)(p) = \mathbf{g}'_p(p, \mathbf{a}, \mathbf{b})_{1\lhd}\mathbf{f}({}^{[k]}\partial_i)(p)\mathbf{h}(\mathbf{u}_1)({}^{[k]}\mathbf{x}^i)(p)\\
&= \mathbf{g}'_p(p, \mathbf{a}, \mathbf{b})_{1\lhd}\mathbf{f}({}^{[k]}\partial_i)(p)\mathbf{h}(\mathbf{u}_2)({}^{[k]}\mathbf{x}^i)(p) = \mathbf{g}'_p(p, \mathbf{a}, \mathbf{b})_{1\lhd}\mathbf{f}({}^{[k]}\partial_i)(p)\mathbf{u}_2({}^{[k]}\mathbf{dx}^i)(p)\\
&= \mathbf{u}_2(\mathbf{g}'_p(q, \mathbf{a}, \mathbf{b})_{1\lhd}\mathbf{f})(p) = \mathbf{g}'_p(p, \mathbf{a}, \mathbf{b})\mathbf{u}_2({}_{1\lhd}\mathbf{f})(p)\\
&= \mathbf{u}_2({}_{1\lhd}\mathbf{f})(p)
\end{aligned}$$

with:

$$ {}^{[k]}\mathbf{x}^i, {}^{[k]}\mathbf{dx}^i, {}^{[k]}\partial_i $$

standing for:

$$ \mathbf{g}'_p(q, \mathbf{a}, \mathbf{b})\mathbf{x}^i(q), \mathbf{g}'_p(q, \mathbf{a}, \mathbf{b})\mathbf{dx}^i, \mathbf{g}'_p(q, \mathbf{a}, \mathbf{b})\partial_i $$

with $q \in \phi_k^{-1}(\mathbf{C}(\phi_k(p), \mathbf{a})) \subset \mathbf{U_k} \subset \mathbf{M}$[11].
Then we have:

$$ (\forall p \in \mathbf{M})(\forall_{1\lhd}\mathbf{f} \in \mathcal{V}^*)\,(\mathbf{u}_1({}_{1\lhd}\mathbf{f})(p) = \mathbf{u}_2({}_{1\lhd}\mathbf{f})(p)) \,. $$

Then $\mathbf{u_1} = \mathbf{u_2}$. $\mathbf{h}$ is a bijection (definition 124 (p. 139)) as injective (definition 122 (p. 139)) and surjective.
As it is commonly done when there is a bijection, we call by the same name the elements which in fact are linked each other by a bijection: $\mathbf{u} \in \mathcal{V}^{**}$ and $\mathbf{h}(\mathbf{u}) \in \mathcal{V}$ are both called $\mathbf{u}$ and then we say abusively $\mathcal{V}^{**} = \mathcal{V}$.

**Proposition 76** A metric on a differential manifold defines a linear map between $\mathcal{V}$ and $\mathcal{V}^*$, $(\mathbf{v}^* : \mathcal{V} \to \mathcal{V}^*)$ with $\mathcal{F}$ as a commutative ring. It should satisfy the following relation setting $\mathbf{v}^*$:

$$ (\forall \mathbf{u} \in \mathcal{V})\,(\mathbf{v}^*(\mathbf{u}) = \mathbf{g}(\mathbf{v}, \mathbf{u})) \,. $$

**Proof**
Given a differentiable manifold $\mathbf{M}$ of dimension $\mathbf{n}$ with a metric $\mathbf{g}$ on its set of vector fields, one calls $\mathbf{v}^* \in \mathcal{V}^*$ dual of $\mathbf{v} \in \mathcal{V}$, the element of $\mathcal{V}^*$ defined by:

$$ (\forall \mathbf{u} \in \mathcal{V})\,(\mathbf{v}^*(\mathbf{u}) = \mathbf{g}(\mathbf{v}, \mathbf{u})) $$

$\mathbf{v}^*$ such defined is a map: $(\mathbf{v}^* : \mathcal{V} \to \mathcal{V}^*)$ which is linear with $\mathcal{F}$ as a commutative ring as $\mathbf{g}(\mathbf{v}, \mathbf{u})$ is (definition 31 (p. 25)).

**Proposition 77** A metric on a differential manifold defines also a linear map with $\mathcal{F}$ as a commutative ring between $\mathcal{V}^*$ and $\mathcal{V}$: $(\mathbf{u}^* : \mathcal{V}^* \to \mathcal{V})$ which satisfied the following relation setting $\mathbf{u}^*$

$$ \mathbf{u} \in \mathcal{V}^* \Rightarrow (\forall \mathbf{v} \in \mathcal{V})\,(\mathbf{u}(\mathbf{v}) = \mathbf{g}(\mathbf{u}^*, \mathbf{v})) \,. \tag{1.53} $$

[10] As usual $\mathbf{g}'_p(q, \mathbf{a}, \mathbf{b})$ means the map $\mathbf{M} \to [0, 1]$, and $\mathbf{g}'_p(p, \mathbf{a}, \mathbf{b})$ means the value of that map for $q = p$.
[11] All this gymnastic is due to the fact that the maps $\mathbf{x}^i(q)$ are valid only in a $\mathbf{U}$ covering $\mathbf{M}$ and cannot be easily extended to the whole manifold.

**Proof**
In an open cover $\mathbf{U_k}$, $\phi_k$ of the differential manifold $\mathbf{M}$ of dimension $\mathbf{n}$, $^{[k]}\mathbf{g}$ is defined at any point $p \in \mathbf{U_k}$ as a matrix $^{[k]}\mathbf{g}_{\mu,\nu}(p) \in \mathcal{F}_k$ with $\det(^{[k]}\mathbf{g}_{\mu,\nu}(p)) \neq 0$. $^{[k]}\mathbf{g}_{\mu,\nu}$ gets an inverse at $p$: $^{[k]}\mathbf{g}^{-1}_{\mu,\nu}(p) = {}^{[k]}\mathbf{g}^{\mu,\nu}(p) \in \mathcal{F}_k$ (see definition 34 (p. 29)) such that

$$^{[k]}\mathbf{g}_{\mu,\nu}(p)\,^{[k]}\mathbf{g}^{\nu,\tau}(p) = \delta^{\tau}_{\mu} \,.$$

One has the following in $\mathbf{U_k}$:

$$\mathbf{u}(\mathbf{v}) = {}^{[k]}\mathbf{u_j}\,^{[k]}\mathbf{v}^{\mathbf{j}} = \delta^{\ell}_{\mathbf{j}}\,^{[k]}\mathbf{u}_\ell\,^{[k]}\mathbf{v}^{\mathbf{j}} = {}^{[k]}\mathbf{g_{i,j}}\,^{[k]}\mathbf{u}_\ell\,^{[k]}\mathbf{g}^{\ell,\mathbf{i}}\,^{[k]}\mathbf{v}^{\mathbf{j}} \,.$$

Setting:

$$\mathbf{u} = {}^{[k]}\mathbf{u_i}\,^{[k]}\mathbf{dx}^{\mathbf{i}}, \;\; \mathbf{u}^* = {}^{[k]}\mathbf{u_i}\,^{[k]}\mathbf{g}^{\mathbf{i,j}}\,^{[k]}\partial_{\mathbf{j}} \tag{1.54}$$

one has in $\mathbf{U_k}$:

$$(\forall \mathbf{v} \in \mathcal{V})\,(\mathbf{u}(\mathbf{v}) = {}^{[k]}\mathbf{u_i}\,^{[k]}\mathbf{v}^{\mathbf{i}} = \delta^{\ell}_{\mathbf{j}}\,^{[k]}\mathbf{u}_\ell\,^{[k]}\mathbf{v}^{\mathbf{j}} = {}^{[k]}\mathbf{g_{i,j}}\,^{[k]}\mathbf{g}^{\ell,\mathbf{i}}\,^{[k]}\mathbf{u}_\ell\,^{[k]}\mathbf{v}^{\mathbf{j}} = \mathbf{g}(\mathbf{u}^*, \mathbf{v})$$

$\mathbf{u}^*$ verifies equation 1.53 (p. 38). From proposition 49 (p. 25), $\mathbf{u}^*$ is unique then $\mathbf{u}^*$ is a map (see definition 112 (p. 131)).
**We have to verify that $\mathbf{u}^*$, which is defined in each $\mathbf{U_k}$, can be considered as in $\mathcal{V}$.**
The map $(\mathbf{u}^* : \mathcal{V}^*_k \to \mathcal{V}_k)$ defined by equation (1.54) (p. 39) in $\mathbf{U_k}$ is linear in $\mathcal{V}^*_k$ with $\mathcal{F}_k$ as commutative ring. We have to prove that it does not depend on the map $(\phi_k : \mathbf{U_k} \to \mathcal{R}^n)$ chosen for defining the maps $^{[k]}\mathbf{x}$ and the differential $^{[k]}\mathbf{dx}$ so we have the same development for $\mathbf{U_m}$ and when $\mathbf{U_m} \cap \mathbf{U_k}$ is not empty, we get the same answer.
Considering in an open $\mathbf{O}$, surrounding $p \in \mathbf{O} \subset \mathbf{U_k}$, a bijection $(\psi : \phi_k(\mathbf{O}) \to \mathcal{R}^n)$, we have (see proposition 61 (p. 30)) [12]:

$$\begin{aligned}\mathbf{g}(\mathbf{v}', \mathbf{v}) = \mathbf{v}'^{\mathbf{i}}_{\phi}\mathbf{v}^{\mathbf{j}}_{\phi}\mathbf{g}(\partial^{\phi}_{\mathbf{i}}\partial^{\phi}_{\mathbf{j}}) = \mathbf{v}'^{\mathbf{i}}_{\phi}\mathbf{v}^{\mathbf{j}}_{\phi}\mathbf{g}^{\phi}_{\mathbf{i,j}} = \mathbf{v}'^{\mathbf{i}}_{\psi\phi}\mathbf{v}^{\mathbf{j}}_{\psi\phi}\mathbf{g}^{\psi\phi}_{\mathbf{i,j}} &= \mathbf{v}'^{\mathbf{i}}_{\psi\phi}\mathbf{v}^{\mathbf{j}}_{\psi\phi}\mathbf{g}(\partial^{\psi\phi}_{\mathbf{i}}\partial^{\psi\phi}_{\mathbf{j}}) \\ &= \mathbf{v}'^{\mathbf{i}}_{\psi\phi}\mathbf{v}^{\mathbf{j}}_{\psi\phi}\mathbf{jac}^{-1}(\psi)^{\lambda}_{\mathbf{i}}\mathbf{jac}^{-1}(\psi)^{\mu}_{\mathbf{j}}\mathbf{g}^{\phi}_{\lambda,\mu}\end{aligned}$$

which is consistent with proposition 47 (p. 23) equation (1.35) where we have:

$$\begin{aligned}\mathbf{v}^{\mathbf{i}}_{\psi\phi} &= \mathbf{jac}(\psi)^{\mathbf{i}}_{\mu}\mathbf{v}^{\mu}_{\phi} \\ \mathbf{jac}(\psi)^{\mathbf{i}}_{\mu}\partial^{\psi\phi}_{\mathbf{i}} &= \partial^{\phi}_{\mu} \\ \mathbf{jac}^{-1}(\psi)^{\mu}_{\mathbf{i}}\partial^{\phi}_{\mu} &= \partial^{\psi\phi}_{\mathbf{i}} \,.\end{aligned}$$

From that we have:

$$\mathbf{g}^{\mathbf{i,j}}_{\psi\phi} = \mathbf{jac}(\psi)^{\mathbf{i}}_{\lambda}\mathbf{jac}(\psi)^{\mathbf{j}}_{\mu}\mathbf{g}^{\lambda,\mu}_{\phi} \,.$$

From proposition 74 (p. 36) equation (1.51), with

$$\mathbf{h} = \mathbf{h}^{\phi}_{\nu}\mathbf{dx}^{\phi\nu} = \mathbf{h}^{\psi\phi}_{\nu}\mathbf{dx}^{\psi\phi\nu} = \mathbf{h}^{\psi\phi}_{\nu}\frac{\partial\psi^{\nu}(\mathbf{x}^{\phi})}{\partial\mathbf{x}^{\phi\mu}}\mathbf{dx}^{\phi\mu} \in \mathcal{V}^*$$

we also have:

$$\mathbf{jac}(\psi)^{\mu}_{\nu}\mathbf{h}^{\psi\phi}_{\mu} = \mathbf{h}^{\phi}_{\nu},\, \mathbf{h}^{\psi\phi}_{\mu} = \mathbf{jac}^{-1}(\psi)^{\nu}_{\mu}\mathbf{h}^{\phi}_{\nu} \,.$$

---

[12] In the following when a symbol has $\phi$ (resp. $\psi\phi$) in upper or lower position, it means that we use the map $\phi$ (resp. $\psi\phi$) to go from $\mathbf{U} \subset \mathbf{M}$ to $\mathcal{R}^n$ local system. By definition we have $\mathbf{x}^{\psi\phi} = \psi(\mathbf{x}^{\phi})$ and $\partial^{\phi}_{\mu} = \frac{\partial}{\partial \mathbf{x}^{\phi\mu}} = \frac{\partial \mathbf{x}^{\psi\phi\nu}}{\partial \mathbf{x}^{\phi\mu}}\frac{\partial}{\partial \mathbf{x}^{\psi\phi\nu}} = \mathbf{jac}(\psi)^{\nu}_{\mu}\partial^{\psi\phi}_{\nu}$.

All those relations give:

$$\mathbf{u}^{\psi\phi*} = \mathbf{u}_i^{\psi\phi}\mathbf{g}_{\psi\phi}^{i,j}\partial_j^{\psi\phi} = \mathrm{jac}^{-1}(\psi)_i^{\ell}\mathbf{u}_{\ell}^{\phi}\mathrm{jac}^{-1}(\psi)_j^{\ell'}\partial_{\ell'}^{\phi}\mathrm{jac}(\psi)_{\lambda}^{i}\mathrm{jac}(\psi)_{\mu}^{j}\mathbf{g}_{\phi}^{\lambda,\mu}$$
$$= \mathbf{u}_{\lambda}^{\phi}\partial_{\mu}^{\phi}\mathbf{g}_{\phi}^{\lambda,\mu} = \mathbf{u}^{\phi*} = \mathbf{u}^*$$

$\mathbf{u}^*$ transforms as a vector field in any transition region when the opens covering of $\mathbf{M}$ overlap. It allows to write:

$$(\forall f \in \mathcal{F})\,(\mathbf{u}^*(f) \in \mathcal{F}) \Rightarrow \mathbf{u}^* \in \mathcal{V}\ .$$

**Definition 43 Metric induced on $\mathcal{V}^*$ by the metric g on $\mathcal{V}$.**

Propositions 76 (p. 38) and 77 (p. 38) allow to set a metric or inner product $\mathbf{g}^*$ on $\mathcal{V}^*$. Having $\mathbf{u} \in \mathcal{V}$, $\mathbf{v} \in \mathcal{V}$:$\mathbf{u} = \mathbf{u}^i\partial_i$, $\mathbf{v} = \mathbf{v}^j\partial_j$ and their dual in $\mathcal{V}^*$: $\mathbf{u}^* = \mathbf{g}_{ik}\mathbf{u}^i\mathbf{dx}^k$ and $\mathbf{v}^* = \mathbf{g}_{j\ell}\mathbf{v}^j\mathbf{dx}^{\ell}$, we want to have:

$$\mathbf{g}^*(\mathbf{u}^*, \mathbf{v}^*) = \mathbf{g}(\mathbf{u}, \mathbf{v})\ .$$

Setting $\mathbf{g}^*(\mathbf{dx}^k, \mathbf{dx}^{\ell}) = \mathbf{g}^{*k,\ell}$, we should have:

$$\mathbf{g}^*(\mathbf{u}^*, \mathbf{v}^*) = \mathbf{g}_{ij}\mathbf{u}^i\mathbf{u}^j = \mathbf{g}^{*k\ell}\mathbf{g}_{ik}\mathbf{g}_{j\ell}\mathbf{u}^i\mathbf{u}^j\ .$$

Then $\mathbf{g}^{*k\ell}\mathbf{g}_{ik} = \delta_i^{\ell}$. From definition 34 (p. 29) we have:

$$\mathbf{g}^* = \mathbf{g}^{-1},\ \mathbf{g}^{*ij} = \mathbf{g}_{ij}^{-1} = \mathbf{g}^{ij}\ .$$

From now on, we call $\mathbf{g}^*$ as $\mathbf{g}$ since their set of definition is not the same:

$$(\mathbf{g} : \mathcal{V} \times \mathcal{V} \cup \mathcal{V}^* \times \mathcal{V}^* \to \mathcal{R})$$

with that definition

$$(\forall \mathbf{u}^* \in \mathcal{V}^*)(\forall \mathbf{v}^* \in \mathcal{V}^*)\,(\mathbf{g}(\mathbf{u}^*, \mathbf{v}^*) = \mathbf{g}(\mathbf{u}, \mathbf{v}))\ .$$

## 1.9 Tensorial field

We shall extend the multilinear form and the tensorial product defined in vector spaces to vector fields. The extension is rather direct but one has to keep in mind that the commutative ring $\mathcal{F}$ which plays the role of the field $\mathcal{RC}$ is not a field.

**Definition 44 Tensorial product of several sets which are either the set linear forms or the set of vector fields of a differential manifold.**

One can define tensor fields on a differential manifold $\mathbf{M}$. One considers the set of two elements $\{\mathcal{V}, \mathcal{V}^*\}$, $\mathbf{p} \in \mathcal{Z}^+$ and a map of $(\mathbf{i_k} : [\mathbf{1}, \mathbf{p}] \to \{\mathcal{V}, \mathcal{V}^*\})$ which sets $(\forall \mathbf{k} \in [\mathbf{1}, \mathbf{p}])(\mathbf{i_k} = \mathcal{V} \vee \mathbf{i_k} = \mathcal{V}^*)$. We consider the product of sets generated by $\mathbf{i_k}$ (definition 372 (p. 473)): $\prod_{k\in[1,p]} \mathbf{i_k}$[13].

[13] The following can be directly extended to the multimanifold case: Given a set $\mathbf{X}$ and a map $(\mathbf{M_k} : [\mathbf{1}, \mathbf{p}] \to \mathcal{P}(\mathbf{X}))$ each $\mathbf{M_k}$ being a differential manifold, we can as well define a map $(\mathbf{i_k} : [\mathbf{1}, \mathbf{p}] \to \bigcup_{\kappa\in[1,p]} \{\mathcal{V}_{\kappa}, \mathcal{V}_{\kappa}^*\})$ such that $(\forall \mathbf{k} \in [\mathbf{1}, \mathbf{p}])(\mathbf{i_k} \in \{\mathcal{V}_k, \mathcal{V}_k^*\})$ and everything nearby remains afterward identical.
It is also not really necessary to limit the tensorial product to a finite set. One could as well put, instead of $[\mathbf{1}, \mathbf{p}]$, any set of indices finite or not (see footnote 16 (p. 224)).

One calls tensor field associated to the map $\mathbf{i_k}$, a multilinear map between $\prod_{\mathbf{k}\in[1,\mathbf{p}]} \mathbf{i_k}$ and $\mathcal{F}$ (definition 191 (p. 223)). As before, the multilinear property is relative to the commutative ring $\mathcal{F}$. The set of tensors associated to $\mathbf{i_k}$ is written as:

$$\prod_{\mathbf{k}\in[1,\mathbf{p}]}^{\otimes} \mathbf{i}^*_\mathbf{k} = \bigotimes_{\mathbf{k}\in[1,\mathbf{p}]} \mathbf{i}^*_\mathbf{k} \,.$$

From proposition 75 (p. 37), we have $(\mathcal{V}^*)^* = \mathcal{V}$ then we also have $(\mathbf{i}^*_\mathbf{k} : [1,\mathbf{p}] \to \{\mathcal{V}, \mathcal{V}^*\})$ and $(\forall \mathbf{k} \in [1,\mathbf{p}])(\mathbf{i}^*_\mathbf{k} = \mathcal{V} \vee \mathbf{i}^*_\mathbf{k} = \mathcal{V}^*)$.

**Definition 45 Tensorial product of a mixed set of vector fields and of linear forms.**

Given a manifold $\mathbf{M}$, its set $\mathcal{V}$ of vector field with its dual $\mathcal{V}^*$, a map $(\mathbf{i_k} : [1,\mathbf{p}] \to \{\mathcal{V}, \mathcal{V}^*\})$ where $\mathbf{p} \in \mathcal{Z}^+$, given an element of $\mathbf{t} \in \prod_{\mathbf{k}\in[1,\mathbf{p}]} \mathbf{i}^*_\mathbf{k}$ (see definition 372 (p. 473)) where $\mathbf{i}^*_\mathbf{k}$ is the dual of $\mathbf{i_k}$, one calls $\otimes_{\mathbf{k}\in[1,\mathbf{p}]}\mathbf{t_k}$ the multilinear map:

$$(\bigotimes_{\mathbf{k}\in[1,\mathbf{p}]} \mathbf{t_k} : \prod_{\mathbf{k}\in[1,\mathbf{p}]} \mathbf{i_k} \to \mathcal{F})$$

defined by: [14]

$$(\forall \mathbf{t} \in \prod_{\mathbf{k}\in[1,\mathbf{p}]} \mathbf{i}^*_\mathbf{k})(\forall \mathbf{v} \in \prod_{\mathbf{k}\in[1,\mathbf{p}]} \mathbf{i_k})(\forall p \in \mathbf{M})((\bigotimes_{\mathbf{k}\in[1,\mathbf{p}]} \mathbf{t_k})(\mathbf{v})(p) = \prod_{\mathbf{k}\in[1,\mathbf{p}]}^{\times} \mathbf{t_k}(\mathbf{v_k})(p)) \,. \quad (1.55)$$

From definition 373 (p. 474), calling $\mathbf{I} = \prod_{\mathbf{k}\in[1,\mathbf{p}]} \mathbf{i_k}$

$$(\bigotimes_{\mathbf{k}\in[1,\mathbf{p}]} \mathbf{t_k} : \mathbf{I} \to \mathcal{F}) \in \mathcal{F}^\mathbf{I}$$

equation (1.55) defines a map which can be written as:

$$(\bigotimes_{\mathbf{k}\in[1,\mathbf{p}]} \mathbf{t_k}(\mathbf{v}) : \prod_{\mathbf{k}\in[1,\mathbf{p}]} \mathbf{i}^*_\mathbf{k} \to \mathcal{F}^\mathbf{I}) \,.$$

In order to simplify the writing, we shall often write $\bigotimes_{\mathbf{k}\in[1,\mathbf{p}]} \mathbf{t_k}$ as $\mathbf{t}$:

$$(\forall \mathbf{t} \in \prod_{\mathbf{k}\in[1,\mathbf{p}]} \mathbf{i}^*_\mathbf{k})(\forall \mathbf{v} \in \prod_{\mathbf{k}\in[1,\mathbf{p}]} \mathbf{i_k})(\mathbf{t}(\mathbf{v}) = \prod_{\mathbf{k}\in[1,\mathbf{p}]} \mathbf{t_k}(\mathbf{v_k}) : \prod_{\mathbf{k}\in[1,\mathbf{p}]} \mathbf{i_k} \to \mathcal{F}) \,.$$

**Proposition 78** Keeping notations of definition 45 (p. 41), we consider $\mathbf{t} = \bigotimes_{\mathbf{k}\in[1,\mathbf{p}]} \mathbf{t_k}$ as it is defined in equation (1.55) (p. 41), with $(\forall \mathbf{k} \in [1,\mathbf{p}])(\mathbf{t_k} \in \mathbf{i}^*_\mathbf{k})$.
Such a $\mathbf{t}$ is an element of $\bigotimes_{\mathbf{k}\in[1,\mathbf{p}]} \mathbf{i}^*_\mathbf{k}$. The terms $\mathbf{i_k}$, $\mathbf{i}^*_\mathbf{k}$, $\bigotimes_{\mathbf{k}\in[1,\mathbf{p}]} \mathbf{i}^*_\mathbf{k}$ are defined in definition 44 (p. 40).

[14] The term $\prod$ has different meaning in the literature: see definitions 372 (p. 473) ($\prod$), 126 (p. 144) ($\prod^{e}$), 128 (p. 145) ($\prod^{\times}$), 131 (p. 146) ($\prod^{\circ}$) and here ($\prod^{\otimes} = \bigotimes$). When it is obvious from the text, we shall forget the upper sign.

**Proof**
In equation (1.55) (p. 41), which defines $\bigotimes_{k\in[1,p]} t_k$, for any $k \in [1,p]$, $t_k(v_k)$ is linear in $v_k$ (definition 37 (p. 32), proposition 75 (p. 37)). From definition 180 (p. 215) point 2, $\prod^{\times}_{k\in[1,p]} t_k(v_k)$ satisfies the definition of a multilinear in $v_k$ (definition 191 (p. 223)).

**Definition 46 Tensorial product of several linear forms (versus vector fields).**

In the above definition we set $(\forall k \in [1,p])(i_k = \mathcal{V})$ (versus $(\forall k \in [1,p])(i_k = \mathcal{V}^*)$) and considering an element $f$ of $(\mathcal{V}^*)^p$ (versus $v \in (\mathcal{V})^p$): $f = (f_i : [1,p] \to \mathcal{V}^*)$ (versus $v = (v_i : [1,p] \to \mathcal{V})$), one calls:

$$\prod^{\otimes}_{i\in[1,p]} f_i = \bigotimes_{i\in[1,p]} f_i \ \ (\textbf{versus} \ \prod^{\otimes}_{i\in[1,p]} v_i = \bigotimes_{i\in[1,p]} v_i\ )$$

the linear map $((\mathcal{V})^p \to \mathcal{F})$ (versus $((\mathcal{V}^*)^p \to \mathcal{F})$) defined by (see definition 126 (p. 144)):

$$(\forall v \in (\mathcal{V})^p)\,((\prod^{\otimes}_{i\in[1,p]} f_i)(v) = \prod^{\times}_{i\in[1,p]} f_i(v_i))$$

$$(\textbf{versus}\ \ (\forall f \in (\mathcal{V}^*)^p)\,((\prod^{\otimes}_{i\in[1,p]} v_i)(f) = f(\prod^{e}_{i\in[1,p]} v_i) = \prod^{\times}_{i\in[1,p]} v_i(f_i)\ ))$$

**Definition 47 $_p$linear_form.**

The tensorial product of several linear forms is also called $_p$**linear_form**

**Proposition 79** We use the notation of the above definition 44 (p. 40). We consider the product of space $\prod_{k\in[1,p]} i_k$ and the tensor space $\bigotimes_{k\in[1,p]} i^*_k$. At any covering open $U \subset M$ with a continuous map $(\phi : U \to \mathcal{R}^n)$, we consider in $U$ the associated maps ${}^u i_k : [1,p] \to \{\mathcal{V}_U, \mathcal{V}^*_U)$ with the following convention $i_k = \mathcal{V} \Rightarrow {}^u i_k = \mathcal{V}_U$, $i_k = \mathcal{V}^* \Rightarrow {}^u i_k = \mathcal{V}^*_U$ see definition 18 (p. 13). For any element $v \in \prod_{k\in[1,p]} i_k$, we may consider its restriction to $U$ $v_u \in \prod_{k\in[1,p]} {}^u i_k$ written as:

$$v_u = \prod^{e}_{k\in[1,p]} \sum_{j\in[1,n]} v^j_{[k]} e(i_k, j)\ ,$$

where $e(i_k, j)$ is defined by:

$$
\begin{aligned}
{}^u i_k = \mathcal{V}_u &\Rightarrow e(i_k, j) = \partial_j\\
{}^u i_k = \mathcal{V}^*_u &\Rightarrow e(i_k, j) = dx^j
\end{aligned}
$$

$v^j_{[k]} e(i_k, j)$ is a vector field of ${}^u i_k$ if ${}^u i_k = \mathcal{V}_u$ or a linear form of ${}^u i_k$ if ${}^u i_k = \mathcal{V}^*_u$. We have from definitions 40 (p. 34) and 23 (p. 17):

$$dx^\mu(\partial_\nu) = \partial_\nu x^\mu = \delta^\mu_\nu \tag{1.56}$$

$\delta^\mu_\nu$ is the Kronecker symbol (definition 215 (p. 245)).
In $\mathbf{U}$, we define also for any $\mathbf{j} \in [\mathbf{1}, \mathbf{n}]^\mathbf{p}$ (see definitions 372 (p. 473), 126 (p. 144) and 128 (p. 145)) the map:

$$(\forall q \in \mathrm{U})(\forall \mathrm{j} \in [1, \mathrm{n}]^\mathrm{p})\, (\mathrm{v}(\mathrm{j})(q) = \prod_{\mathrm{k}\in[1,\mathrm{p}]}^{\times} \mathrm{v}^{\mathrm{j_k}}_{[\mathrm{k}]}(q) \in \mathcal{F}^0_\mathrm{u}) \,.$$

For each $t \in \bigotimes_{\mathrm{k}\in[1,\mathrm{p}]} \mathrm{i}^*_\mathrm{k}$ (definition 44 (p. 40)), there is a map $(\mathrm{t}(\mathrm{j}) : [1, \mathrm{n}]^\mathrm{p} \to \mathcal{F}^0_\mathrm{u})$ such as:

$$q \in \mathrm{U} \Rightarrow t(v)(q) = \sum_{\mathrm{j}\in[1,\mathrm{n}]^\mathrm{p}} \mathrm{t}(\mathrm{j})(q)\mathrm{v}(\mathrm{j})(q) \in \mathcal{F}^0_\mathrm{u} \,. \tag{1.57}$$

We may also write in $\mathbf{U}$:

$$^{\mathrm{u}}t = \sum_{\mathrm{j}\in[1,\mathrm{n}]^\mathrm{p}} \mathrm{t}(\mathrm{j}) \prod_{\mathrm{k}\in[1,\mathrm{p}]}^{\otimes} \mathrm{e}(\mathrm{i_k}, \mathrm{j_k})^* = \sum_{\mathrm{j}\in[1,\mathrm{n}]^\mathrm{p}} \mathrm{t}(\mathrm{j}) \bigotimes_{\mathrm{k}\in[1,\mathrm{p}]} \mathrm{e}(\mathrm{i_k}, \mathrm{j_k})^* \tag{1.58}$$

$$^{\mathrm{u}}v = \sum_{\mathrm{j}\in[1,\mathrm{n}]^\mathrm{p}} \mathrm{v}(\mathrm{j}) \prod_{\mathrm{k}\in[1,\mathrm{p}]}^{\otimes} \mathrm{e}(\mathrm{i_k}, \mathrm{j_k}) = \sum_{\mathrm{j}\in[1,\mathrm{n}]^\mathrm{p}} \mathrm{v}(\mathrm{j}) \bigotimes_{\mathrm{k}\in[1,\mathrm{p}]} \mathrm{e}(\mathrm{i_k}, \mathrm{j_k}) \tag{1.59}$$

such that:

$$q \in \mathrm{U} \Rightarrow t(v)(q) = {}^{\mathrm{u}}t({}^{\mathrm{u}}v)(q) = \sum_{\mathrm{j}\in[1,\mathrm{n}]^\mathrm{p}} \mathrm{t}(\mathrm{j})(q) \times \mathrm{v}(\mathrm{j})(q) \,. \tag{1.60}$$

**Proof**
We may write:

$$v = \prod_{\mathrm{k}\in[1,\mathrm{p}]}^{\mathrm{e}} v_{[\mathrm{k}]} \in \prod_{\mathrm{k}\in[1,\mathrm{p}]} \mathrm{i_k}$$

with $v_{[\mathrm{k}]} \in \mathrm{i_k}$.
From definitions 261 (p. 285) and 14 (p. 6), we have:

$$\overline{\mathrm{C}}(q, \mathrm{a}) \subset \overline{\mathrm{C}}(q, \mathrm{b}) \subset \mathrm{U} \subset \mathrm{M} \,.$$

As in proposition 73 (p. 35), using it with definition 47 (p. 42), from multilinearity, since $\mathrm{g'}_q(\phi(q), \mathrm{a}, \mathrm{b}) = 1$, we have:

$$\begin{aligned} t(v)(q) &= \mathrm{g'}^{\mathrm{p}}_q(\phi(q), \mathrm{a}, \mathrm{b}) t(v)(q) = t(\mathrm{g'}^{\mathrm{p}}_q(\phi(q'), \mathrm{a}, \mathrm{b}) v)(q) \\ &= t(\prod_{\mathrm{k}\in[1,\mathrm{p}]}^{\mathrm{e}} \mathrm{g'}_q(\phi(q'), \mathrm{a}, \mathrm{b}) v_{[\mathrm{k}]})(q) \,. \end{aligned}$$

We have since $(q' \notin \mathrm{C}(q, \mathrm{b}) \subset \mathrm{U}) \Rightarrow \mathrm{g'}_q(\phi(q'), \mathrm{a}, \mathrm{b}) = 0$:

$$\mathrm{g'}_q(\phi(q'), \mathrm{a}, \mathrm{b}) v_{[\mathrm{k}]} \in {}^{\mathrm{u}}\mathrm{i_k}$$

where ${}^{\mathrm{u}}\mathrm{i_k}$ or ${}^{\mathrm{u}}\mathrm{i}^*_\mathrm{k}$ are the set of vector field or its dual associated to the restricted manifold $\mathbf{U}$. $\mathrm{i_k}$ or $\mathrm{i}^*_\mathrm{k}$ are either the vector field of $\mathbf{M}$ or its dual. We set $\mathrm{e}(\mathrm{i}, \mathrm{j}) = \partial_\mathrm{j}$ when $\mathrm{i} = \mathcal{V}$ and $\mathrm{e}(\mathrm{i}, \mathrm{j}) = \mathrm{d}^\mathrm{j}$ when $\mathrm{i} = \mathcal{V}^*$. We may write from propositions 31 (p. 14) and 73 (p. 35):

$$\mathrm{g'}_q(\phi(q'), \mathrm{a}, \mathrm{b}) v_{[\mathrm{k}]} = \sum_{\mathrm{j}\in[1,\mathrm{n}]} {}_q\mathrm{v}^\mathrm{j}_{[\mathrm{k}]}(q')\mathrm{e}(\mathrm{i_k}, \mathrm{j}) = \sum_{\mathrm{j}\in[1,\mathrm{n}]} \mathrm{g'}_q(\phi(q'), \mathrm{a}, \mathrm{b}) \mathrm{v}^\mathrm{j}_{[\mathrm{k}]}(q')\mathrm{e}(\mathrm{i_k}, \mathrm{j})$$

by definition ${}_q\mathrm{v}^{\mathrm{j}}_{[\mathrm{k}]}(q) = \mathrm{v}^{\mathrm{j}}_{[\mathrm{k}]}(q)$ and ${}_q\mathrm{v}^{\mathrm{j}}_{[\mathrm{k}]}(q') \in \mathcal{F}$ (definitions 18 (p. 13), 19 (p. 14) and 25 (p. 17)).
When $\phi(q') \in \mathrm{C}(\phi(q),\mathrm{a})$, $\mathrm{g}'_q(\phi(q'),\mathrm{a},\mathrm{b}) = 1$, ${}_q\mathrm{v}^{\mathrm{j}}_{[\mathrm{k}]}(q') = \mathrm{v}^{\mathrm{j}}_{[\mathrm{k}]}(q')$. Then using the multilinearity, we get from proposition 289 (p. 186) (with $\mathbf{L} = [\mathbf{1},\mathbf{p}]$, $\mathbf{J}_\lambda = [\mathbf{1},\mathbf{n}]$):

$$t(v)(q) = t(\mathrm{g}'^{\mathrm{p}}_q(\phi(q'),\mathrm{a},\mathrm{b})v)(q) = t(\prod^{\mathrm{e}}_{\mathrm{k}\in[1,\mathrm{p}]}\sum_{\mathrm{j}\in[1,\mathrm{n}]} {}_q\mathrm{v}^{\mathrm{j}}_{[\mathrm{k}]}(q')\mathrm{e}(\mathrm{i}_\mathrm{k},\mathrm{j}))(q)$$

$$= t(\prod^{\mathrm{e}}_{\mathrm{k}\in[1,\mathrm{p}]}\sum_{\mathrm{j}\in[1,\mathrm{n}]} \mathrm{g}'_q(\phi(q'),\mathrm{a},\mathrm{b})\mathrm{v}^{\mathrm{j}}_{[\mathrm{k}]}(q')\mathrm{e}(\mathrm{i}_\mathrm{k},\mathrm{j}))(q)$$

$$= \sum_{\mathrm{j}\in[1,\mathrm{n}]^\mathrm{p}} t(\prod^{\mathrm{e}}_{\mathrm{k}\in[1,\mathrm{p}]} \mathrm{g}'_q(\phi(q'),\mathrm{a},\mathrm{b})\mathrm{v}^{\mathrm{j}_\mathrm{k}}_{[\mathrm{k}]}(q')\mathrm{e}(\mathrm{i}_\mathrm{k},\mathrm{j}_\mathrm{k}))(q)$$

$$= \sum_{\mathrm{j}\in[1,\mathrm{n}]^\mathrm{p}} (\prod^{\times}_{\mathrm{k}\in[1,\mathrm{p}]} \mathrm{v}^{\mathrm{j}_\mathrm{k}}_{[\mathrm{k}]}(q))t(\prod^{\mathrm{e}}_{\mathrm{k}\in[1,\mathrm{p}]} \mathrm{g}'_q(\phi(q'),\mathrm{a},\mathrm{b})\mathrm{e}(\mathrm{i}_\mathrm{k},\mathrm{j}_\mathrm{k}))(q)$$

and we may set:

$$\mathrm{t(j)(q)} = t(\prod^{\mathrm{e}}_{\mathrm{k}\in[1,\mathrm{p}]} \mathrm{g}'_q(\phi(q'),\mathrm{a},\mathrm{b})\mathrm{e}(\mathrm{i}_\mathrm{k},\mathrm{j}_\mathrm{k}))(\mathrm{q}) \in \mathcal{F}$$

which does not depend on $\mathbf{a}$ and $\mathbf{b}$. $\mathbf{t}(\mathbf{j})(q)$ belongs to $\mathcal{F}$ by the same argument as in proposition 73 (p. 35) and in definition 18 (p. 13). Using that definition of $t(\mathbf{j})$, we get equation (1.57) (p. 43).
Equations (1.58) (p. 43) and (1.59) (p. 43) are a direct application of equation (1.56) (p. 42) which specifies:

$$\mathrm{e}(\mathrm{i}_\mathrm{k},\mathrm{j})^*(\mathrm{e}(\mathrm{i}_\mathrm{k},\mathrm{j}')) = \delta^{\mathrm{j}}_{\mathrm{j}'} \;.$$

**Proposition 80** Taking the notation of the above definition 44 (p. 40), the set of multilinear forms on $\bigotimes_{\mathrm{k}\in[1,\mathrm{p}]} \mathrm{i}^*_\mathrm{k}$ is $\bigotimes_{\mathrm{k}\in[1,\mathrm{p}]} \mathrm{i}_\mathrm{k}$.

**Proof**
Since $\mathbf{i}_\mathbf{k} = \mathbf{i}^{**}_\mathbf{k}$ (proposition 75 (p. 37)), $\bigotimes_{\mathrm{k}\in[1,\mathrm{p}]} \mathrm{i}_\mathrm{k}$ is the set of multilinear forms on $\prod_{\mathrm{k}\in[1,\mathrm{p}]} \mathrm{i}^*_\mathrm{k}$ (definition 44 (p. 40)). In any covering open subset $\mathbf{U}$ of a differential manifold $\mathbf{M}$ with its map $(\phi : \mathrm{U} \to \mathcal{R}^\mathrm{n})$ (definitions 1 (p. 1) and 3 (p. 1)), at any point $q \in \mathbf{U}$ we can consider:

$$0 < \mathrm{a}_q < \mathrm{b}_q, \overline{\mathrm{C}}(\phi(q),\mathrm{a}_q) \subset \overline{\mathrm{C}}(\phi(q),\mathrm{b}_q) \subset \phi(\mathrm{U}) \subset \mathcal{R}^\mathrm{n} \;.$$

Those elements exist from proposition 269 (p. 179). For any $\mathbf{v} \in \bigotimes_{\mathrm{k}\in[1,\mathrm{p}]} \mathrm{i}_\mathrm{k}$, since:

$$(\forall \mathrm{j} \in [1,\mathrm{n}]^\mathrm{p})\,(\bigotimes_{\mathrm{k}\in[1,\mathrm{p}]} \mathrm{g}'_q(q',\mathrm{a}_q,\mathrm{b}_q)\mathrm{e}(\mathrm{i}^*_\mathrm{k},\mathrm{j}_\mathrm{k}) \in \prod_{\mathrm{k}\in[1,\mathrm{p}]} \mathrm{i}^*_\mathrm{k})$$

we may write using definition 14 (p. 6):

$$(\forall \mathrm{j} \in [1,\mathrm{n}]^\mathrm{p})\,(\mathrm{v}(\mathrm{j})(q) = \mathrm{v}(\bigotimes_{\mathrm{k}\in[1,\mathrm{p}]} \mathrm{g}'_q(q',\mathrm{a}_q,\mathrm{b}_q)\mathrm{e}(\mathrm{i}^*_\mathrm{k},\mathrm{j}_\mathrm{k}))(q))$$

and we may extend $\mathbf{v}$ to $\bigotimes_{k\in[1,p]} \mathrm{i}_k^*$ by setting for any $\mathbf{t} \in \bigotimes_{k\in[1,p]} \mathrm{i}_k^*$ (proposition 79 (p. 42), equation (1.58))

$$\begin{aligned}\mathbf{v}(\mathbf{t})(q) &= \mathbf{g}'^{\mathrm{p}}_q(q, \mathbf{a}_q, \mathbf{b}_q)\mathbf{v}(\mathbf{t})(q) = \mathbf{v}(\mathbf{g}'^{\mathrm{p}}_q(q', \mathbf{a}_q, \mathbf{b}_q)\mathbf{t})(q) \\ &= \mathbf{v}(\sum_{j\in[1,n]^p} \mathbf{t}(\mathbf{j}) \bigotimes_{k\in[1,p]} \mathbf{g}'_q(q', \mathbf{a}_q, \mathbf{b}_q)\mathbf{e}(\mathrm{i}_k^*, \mathrm{j}_k))(q) = \sum_{j\in[1,n]^p} \mathbf{t}(\mathbf{j})(q)\mathbf{v}(\mathbf{j})(q) \,. \end{aligned} \tag{1.61}$$

**Convention 6** Notation for tensorial product.

Following equation (1.58) (p. 43), a common notation for a tensorial product in a local covering set $\mathbf{U}$ of a manifold $\mathbf{M}$ is:

$$t = \mathrm{t}^{k_1,k_2,..,k_p}_{j_1,j_2,...,j_q} \partial_{k_1} \otimes \partial_{k_2} ... \otimes \partial_{k_p} \otimes \mathrm{d}^{j_1} \otimes \mathrm{d}^{j_2} ... \otimes \mathrm{d}^{j_q}$$

$\partial$ and $\mathbf{d}$ are not necessarily in the order indicated. In fact, they should be mixed using a map $\mathbf{i}$ as explained in definition 44 (p. 40). This notation is used commonly but it is sometimes heavy to manipulate.

**Definition 48 Metric on a tensorial product.**

We are given a metric $(\mathbf{g} : \mathcal{V}^2 \cup (\mathcal{V}^*)^2 \to \mathcal{F})$ (definitions 31 (p. 25), 43 (p. 40)) where $\mathcal{V}$ is the set of vector fields of a differential manifold M and $\mathcal{V}^*$ its dual (definition 37 (p. 32)). Given a tensor field settled by $(\mathrm{i}_k : [1,p] \to \{\mathcal{V}, \mathcal{V}^*\})$, one considers the map $\mathbf{g}_t$ deduced from $\mathbf{g}$:

$$(\mathbf{g}_t : \prod_{\ell\in[1,2]} \prod_{k\in[1,p]} \mathrm{i}_k = (\prod_{k\in[1,p]} \mathrm{i}_k)^2 \to \mathcal{F})$$

defined by (see definition 43 (p. 40)):

$$\begin{aligned}&(\forall \mathbf{u} = \prod_{k\in[1,p]} \mathbf{u}_k \in \prod_{k\in[1,p]} \mathrm{i}_k)(\forall \mathbf{v} = \prod_{k\in[1,p]} \mathbf{v}_k \in \prod_{k\in[1,p]} \mathrm{i}_k) \\ &(\mathbf{g}_t(\mathbf{u}, \mathbf{v}) = \prod_{k\in[1,p]} \mathbf{g}(\mathrm{i}_k)(\mathbf{u}_k, \mathbf{v}_k) = \prod_{k\in[1,p]} \mathbf{g}(\mathbf{u}_k, \mathbf{v}_k)) \,.\end{aligned}$$

From proposition 80 (p. 44), it can be extended to be an element of $\prod_{\ell\in[1,2]} \bigotimes_{k\in[1,p]} \mathrm{i}_k^*$:

$$(\mathbf{g}_t(\mathbf{u}, \mathbf{v}) : \prod_{\ell\in[1,2]} \bigotimes_{k\in[1,p]} \mathrm{i}_k \to \mathcal{F})$$

which is multilinear in the two arguments (see footnote 13 (p. 40)).

## 1.10 Wedge product of 1-linear forms (versus vector fields)

**Definition 49 Permutation of $(\mathcal{V}^*)^{\mathrm{p}}$ (versus $(\mathcal{V})^{\mathrm{p}}$).**

We have a differential manifold $\mathbf{M}$ and $\mathcal{V}$ the set of vector field. $(\otimes\mathcal{V}^*)^{\mathrm{p}}$ is the set of $_{\mathrm{p}}$**linear_form** (definition 44 (p. 40)). For any permutation $(\mathbf{P} : [1,p] \leftrightarrow [1,p])$ (see section

2.9.1), we have the corresponding map $(\mathbf{P}_{\otimes^*} : (\otimes\mathcal{V}^*)^\mathrm{p} \to (\otimes\mathcal{V}^*)^\mathrm{p})$ such that (see proposition 79 (p. 42)):

$$(\forall t \in (\otimes\mathcal{V}^*)^\mathrm{p})(\forall v = \prod^{\mathrm{e}}_{\mathrm{i}\in[1,\mathrm{p}]} v_\mathrm{i} \in (\mathcal{V})^\mathrm{p})(\mathbf{P}_{\otimes^*}(t)(\prod^{\mathrm{e}}_{\mathrm{i}\in[1,\mathrm{p}]} v_\mathrm{i}) = t(\prod^{\mathrm{e}}_{\mathrm{i}\in[1,\mathrm{p}]} v_{\mathrm{P}^{-1}(\mathrm{i})})) . \quad (1.62)$$

Using the definition given by equation (1.62) (p. 46) and definition 46 (p. 42) we can define a map $(\mathbf{P}_{\otimes} : (\otimes\mathcal{V})^\mathrm{p} \to (\otimes\mathcal{V})^\mathrm{p})$ setting:

$$(\forall u \in (\otimes\mathcal{V})^\mathrm{p})(\forall t \in (\otimes\mathcal{V}^*)^\mathrm{p})\,(\mathbf{P}_{\otimes}u(t) = u(\mathbf{P}^{-1}_{\otimes^*}(t))$$

with

$$u \in (\otimes\mathcal{V})^\mathrm{p} \Rightarrow (u(t) : (\otimes\mathcal{V}^*)^\mathrm{p} \to \mathcal{F}) .$$

Since all the maps implied in the definition are linear, the $\mathbf{P}_{\otimes}$, $\mathbf{P}_{\otimes^*}$ maps associated to $\mathbf{P}$ are linear. From now on, we shall call a permutation operator $\mathbf{P}$ by the same symbol independently of the set on which it acts.

**Proposition 81** Given the tensorial product of $\mathbf{p}$ linear forms:$\bigotimes_{\mathrm{k}\in[1,\mathrm{p}]} t_\mathrm{k}$ (versus vectors $\bigotimes_{\mathrm{k}\in[1,\mathrm{p}]} v_\mathrm{k}$ ) and a permutation $(\mathbf{P} : [1,\mathrm{p}] \to [1,\mathrm{p}])$ we have:

$$\mathrm{P}(\bigotimes_{\mathrm{k}\in[1,\mathrm{p}]} t_\mathrm{k}) = \bigotimes_{\mathrm{k}\in[1,\mathrm{p}]} t_{\mathrm{P}(\mathrm{k})}$$

versus

$$\mathrm{P}(\bigotimes_{\mathrm{k}\in[1,\mathrm{p}]} v_\mathrm{k}) = \bigotimes_{\mathrm{k}\in[1,\mathrm{p}]} v_{\mathrm{P}(\mathrm{k})} .$$

**Proof**
Proposition 997 (p. 556) applied to $t_\mathrm{k}(v_{\mathrm{P}^{-1}(\mathrm{k})})$ gives with the above definitions 49 (p. 45) and 46 (p. 42):

$$\mathrm{P}(\bigotimes_{\mathrm{k}\in[1,\mathrm{p}]} t_\mathrm{k})(\prod^{\mathrm{e}}_{\mathrm{i}\in[1,\mathrm{p}]} v_\mathrm{i}) = \prod^{\times}_{\mathrm{k}\in[1,\mathrm{p}]} t_\mathrm{k}(v_{\mathrm{P}^{-1}(\mathrm{k})}) = \prod^{\times}_{\mathrm{k}\in[1,\mathrm{p}]} t_{\mathrm{P}(\mathrm{k})}(v_{\mathrm{P}^{-1}(\mathrm{P}(\mathrm{k}))})$$
$$= \bigotimes_{\mathrm{k}\in[1,\mathrm{p}]} t_{\mathrm{P}(\mathrm{k})}(\prod^{\mathrm{e}}_{\mathrm{i}\in[1,\mathrm{p}]} v_\mathrm{i})$$

versus

$$(\forall t \in (\otimes\mathrm{V}^*)^\mathrm{p})\,(\mathrm{P}(\bigotimes_{\mathrm{k}\in[1,\mathrm{p}]} v_\mathrm{k})(t) = \mathrm{P}^{-1}(\mathrm{t})(\bigotimes_{\mathrm{k}\in[1,\mathrm{p}]} v_\mathrm{k}) = \mathrm{t}(\bigotimes_{\mathrm{k}\in[1,\mathrm{p}]} v_{\mathrm{P}(\mathrm{k})}))$$
$$= \bigotimes_{\mathrm{k}\in[1,\mathrm{p}]} v_{\mathrm{P}(\mathrm{k})}(\mathrm{t})) .$$

**Definition 50 Wedge projection on $(\otimes\mathcal{V}^*)^\mathrm{p}$ (versus $(\otimes\mathcal{V})^\mathrm{p}$ ).**

Calling $\mathbf{PG}_\mathrm{p}$ the permutation group of $[1,\mathrm{p}]$ (see definition 172 (p. 207)) and $(\mathbf{s}(\mathbf{P}) : \mathbf{PG}_\mathrm{p} \to \{-1,1\})$ the signature of the permutations (see definition 175 (p. 208)), one calls wedge projection or antisymmetric projection, the following linear map $(\mathbf{W} : (\otimes\mathcal{V}^*)^\mathrm{p} \to (\otimes\mathcal{V}^*)^\mathrm{p})$ (versus $(\mathbf{W} : (\otimes\mathcal{V})^\mathrm{p} \Rightarrow (\otimes\mathcal{V})^\mathrm{p})$) defined by:

$$(\forall t \in (\otimes\mathcal{V}^*)^\mathrm{p})\,(\mathbf{W}(t) = \sum_{\mathrm{P}\in\mathrm{PG}_\mathrm{p}} \mathrm{s}(\mathrm{P})\mathrm{P}(t))$$

versus:

$$(\forall v \in (\otimes\mathcal{V})^{\mathbf{p}})(\mathbf{W}(v) = \sum_{\mathbf{P}\in\mathbf{PG_p}} \mathbf{s}(\mathbf{P})\mathbf{P}(v)) .$$

Since the **P** are linear, **W** is linear and $\mathbf{W}((\otimes\mathcal{V})^{\mathbf{p}})$ is a linear vectorial subspace of $(\otimes\mathcal{V})^{\mathbf{p}}$ with $\mathcal{F}$ as a commutative ring.

**Definition 51 Wedge product of p 1-linear forms (versus vector fields).**

Wedge product of **p** 1-linear forms (versus vector fields) is equivalent to the wedge projection (definition 50 (p. 46)) of **p** 1-linear forms (versus vector fields):
Given $t = (t_{\mathbf{i}} : [1,\mathbf{p}] \to \mathcal{V}^*)$ (versus $v = (v_{\mathbf{i}} : [1,\mathbf{p}] \to \mathcal{V})$):

$$\bigwedge_{\mathbf{i}\in[1,\mathbf{p}]} t_{\mathbf{i}} = \mathbf{W}(\bigotimes_{\mathbf{i}\in[1,\mathbf{p}]} t_{\mathbf{i}}) = \sum_{\mathbf{P}\in\mathbf{PG_p}} \mathbf{s}(\mathbf{P}) \bigotimes_{\mathbf{i}\in[1,\mathbf{p}]} t_{\mathbf{P}(\mathbf{i})}$$

versus

$$\bigwedge_{\mathbf{i}\in[1,\mathbf{p}]} v_{\mathbf{i}} = \mathbf{W}(\bigotimes_{\mathbf{i}\in[1,\mathbf{p}]} v_{\mathbf{i}}) = \sum_{\mathbf{P}\in\mathbf{PG_p}} \mathbf{s}(\mathbf{P}) \bigotimes_{\mathbf{i}\in[1,\mathbf{p}]} v_{\mathbf{P}(\mathbf{i})}$$

where $\mathbf{s}(\mathbf{P})$ is the signature of **P**.

**Definition 52 p_form.**

An element of $\mathbf{W}((\mathcal{V}^*)^{\mathbf{p}})$ is called **p_form**.

**Proposition 82** If, in a wedge product of **p** 1-linear forms (versus vector field), there are two identical forms (versus vectors), the wedge product is zero.

**Proof**
We treat here only the case of two identical forms since the transposition to wedge product of vector fields is straightforward. Keeping the notation of definition 51 (p. 47), we suppose that there are two identical forms in a wedge product of **p** 1-linear forms. It means that if one considers the map $(t_{\mathbf{i}} : [1,\mathbf{p}] \to \mathcal{V}^*)$, in the set $[1,\mathbf{p}]$, there is $\mathbf{s_2} \subset [1,\mathbf{p}]$ with $\mathbf{Card}(\mathbf{s_2}) = 2$ such that $\mathbf{Card}(t_{\mathbf{s_2}}) = 1$. Then there are two sets of permutations of $[1,\mathbf{p}]$: The one with positive signature $\mathbf{PG_p^+}$ and the one with negative signature $\mathbf{PG_p^-}$. The transposition $\mathbf{T_{s_2}}$ of $\mathbf{s_2}$ defines a bijection between the two sets (see definition 175 (p. 208), proposition 183 (p. 143)). Then we have:

$$\mathbf{PG_p} = \mathbf{PG_p^+} \cup \mathbf{PG_p^-} = \mathbf{PG_p^+} \cup \mathbf{T_{s_2}}(\mathbf{PG_p^+}) = \bigcup_{\mathbf{P}\in\mathbf{PG_p^+}} \{\mathbf{P}, \mathbf{T_{s_2}P}\}$$

which gives since $\mathbf{S}(\mathbf{P}) = -\mathbf{S}(\mathbf{T_{s_2}P})$:

$$\sum_{\mathbf{P}\in\mathbf{PG_p}} \mathbf{S}(\mathbf{P}) \bigotimes_{\mathbf{i}\in[1,\mathbf{p}]} t_{\mathbf{P}(\mathbf{i})} = \sum_{\mathbf{P}\in\mathbf{PG_p^+}} \mathbf{S}(\mathbf{P}) \bigotimes_{\mathbf{i}\in[1,\mathbf{p}]} t_{\mathbf{P}(\mathbf{i})} + \mathbf{S}(\mathbf{T_{s_2}P}) \bigotimes_{\mathbf{i}\in[1,\mathbf{p}]} t_{\mathbf{T_{s_2}}(\mathbf{P}(\mathbf{i}))} = 0$$

since $(\forall \mathbf{i} \in [1,\mathbf{p}])(t_{\mathbf{i}} = t_{\mathbf{T_{s_2}}(\mathbf{i})})$ and $\mathbf{S}(\mathbf{P}) = -\mathbf{S}(\mathbf{T_{s_2}P})$.

**Proposition 83** In a manifold **M** of dimension **n** with $\mathcal{V}$ as set of vector field, when $\mathbf{p} > \mathbf{n}$, the wedge projection of $(\otimes\mathcal{V}^*)^{\mathbf{p}}$ (versus $(\otimes\mathcal{V})^{\mathbf{p}}$) is the zero element of each corresponding tensorial product.

**Proof**
When restricted to any of the covering set $\mathbf{U}$ of $\mathbf{M}$ each element of $(\otimes\mathcal{V}^*)^{\mathrm{p}}$ (versus $(\otimes\mathcal{V})^{\mathrm{p}}$) acts as an element $t$ of $(\otimes\mathcal{V}^*_{\mathrm{u}})^{\mathrm{p}}$ (versus $v$ of $(\otimes\mathcal{V}_{\mathrm{u}})^{\mathrm{p}}$). By definition (see definition 46 (p. 42) and proposition 79 (p. 42)) equations (1.58), (1.59) give:

$$\mathrm{t} \in (\otimes\mathcal{V}^*)^{\mathrm{p}} \Rightarrow (\exists(\mathrm{t_j} : [1,\mathrm{n}]^{\mathrm{p}} \to \mathcal{F}_{\mathrm{u}}))(t = \sum_{\mathrm{j}\in[1,\mathrm{n}]^{\mathrm{p}}} \mathrm{t_j} \bigotimes_{\ell\in[1,\mathrm{p}]} \mathrm{dx}^{\mathrm{j}(\ell)})$$

versus

$$\mathrm{v} \in (\otimes\mathcal{V})^{\mathrm{p}} \Rightarrow (\exists(\mathrm{v^j} : [1,\mathrm{n}]^{\mathrm{p}} \to \mathcal{F}_{\mathrm{u}}))(v = \sum_{\mathrm{j}\in[1,\mathrm{n}]^{\mathrm{p}}} \mathrm{v^j} \bigotimes_{\ell\in[1,\mathrm{p}]} \partial_{\mathrm{j}(\ell)})$$

then

$$\mathbf{W}(t) = \sum_{\mathrm{j}\in[1,\mathrm{n}]^{\mathrm{p}}} \mathrm{t_j}\mathbf{W}(\bigotimes_{\ell\in[1,\mathrm{p}]} \mathrm{dx}^{\mathrm{j}(\ell)})$$

versus

$$\mathbf{W}(v) = \sum_{\mathrm{j}\in[1,\mathrm{n}]^{\mathrm{p}}} \mathrm{v^j}\mathbf{W}(\bigotimes_{\ell\in[1,\mathrm{p}]} \partial_{\mathrm{j}(\ell)})$$

with $\mathbf{p} > \mathbf{n}$ a map $(\mathbf{j} : [\mathbf{1},\mathbf{p}] \to \mathbf{j}([\mathbf{1},\mathbf{p}]) \subset [\mathbf{1},\mathbf{n}])$ cannot be bijective otherwise $\mathbf{p} = \mathbf{Card}([\mathbf{1},\mathbf{p}]) = \mathbf{Card}(\mathbf{j}([\mathbf{1},\mathbf{p}])) \leq \mathbf{n}$ (definition 390 (p. 499)). It is forbidden because $\mathbf{p} > \mathbf{n}$. The map $\mathbf{j}$ is surjective (definition 123 (p. 139), proposition 170 (p. 139)), it cannot be injective otherwise it would be bijective (definition 124 (p. 139)). Then $(\exists \mathbf{l} \in [\mathbf{1},\mathbf{n}])(\mathbf{Card}(\mathbf{j}^{-1}(\mathbf{l})) \geq \mathbf{2})$ (definition 122 (p. 139)). From proposition 83 (p. 47) restricted in $\mathbf{U}$, $\mathbf{W}(t) = \mathbf{0}$ and for the same reason $\mathbf{W}(v) = \mathbf{0}$. Since those relations are true for any $\mathbf{U}$, they are true for the unrestricted elements [15].

**Definition 53** $\mathcal{I}_{\mathrm{p}}$.

We call $\mathcal{I}_{\mathrm{p}}$, the set of strictly increasing maps (definition 386 (p. 486)) $[\mathbf{1},\mathbf{p}] \to [\mathbf{1},\mathbf{n}]$.

**Proposition 84** There is a bijection between $\mathcal{I}_{\mathrm{p}}$ and the set of subsets of $[\mathbf{1},\mathbf{n}]$ with $\mathbf{p}$ elements. This bijection is given by the map $(\mathbf{f} : \mathcal{I}_{\mathrm{p}} \leftrightarrow \mathcal{P}([\mathbf{1},\mathbf{n}]))$:

$$(\forall \mathbf{h} \in \mathcal{I}_{\mathrm{p}})(\mathbf{f}(\mathbf{h}) = \mathbf{h}([\mathbf{1},\mathbf{p}]) \subset [\mathbf{1},\mathbf{n}]) \ .$$

**Proof**
It is a consequence of propositions 867 (p. 495) and 886 (p. 504).

**Proposition 85** $\mathbf{p} \in \mathcal{Z}^+, \mathbf{n} \in \mathcal{Z}^+, \mathbf{p} \leq \mathbf{n}$, there is a bijection (definition 124 (p. 139)) between the set of injective map (definition 122 (p. 139)) $[\mathbf{1},\mathbf{p}] \to [\mathbf{1},\mathbf{n}]$ and the product of the set $\mathcal{I}_{\mathrm{p}}$ and the set of permutation of $[\mathbf{1},\mathbf{p}]$. This bijection is given by

$$(\mathbf{f} : [\mathbf{1},\mathbf{p}] \leftrightarrow \mathbf{f}([\mathbf{1},\mathbf{p}]) \subset [\mathbf{1},\mathbf{n}]) \Rightarrow (\exists \mathbf{h} \in \mathcal{I}_{\mathrm{p}})(\exists \mathbf{P} \in \mathbf{PG_p})(\mathbf{f} = \mathbf{h} \circ \mathbf{P})$$

**Proof**
Given an injective map $(\mathbf{f} : [\mathbf{1},\mathbf{p}] \leftrightarrow \mathbf{f}([\mathbf{1},\mathbf{p}]) \subset [\mathbf{1},\mathbf{n}])$, from proposition 84 (p. 48), there is an injection $\mathbf{h} \in \mathcal{I}_{\mathrm{p}}$ such that $\mathbf{f}([\mathbf{1},\mathbf{p}]) = \mathbf{h}([\mathbf{1},\mathbf{p}])$. With that $\mathbf{h}$, $\mathbf{h}^{-1} \circ \mathbf{f}$ is an injection $[\mathbf{1},\mathbf{p}] \leftrightarrow [\mathbf{1},\mathbf{p}]$ as product of two injections. Since it is surjective, it is a bijection, an element $\mathbf{P}$ of $\mathbf{PG_p}$ (definition 172 (p. 207)). Then we have $\mathbf{h}^{-1} \circ \mathbf{f} = \mathbf{P}$. It means $\mathbf{f} = \mathbf{h} \circ \mathbf{P}$. Any product of a map of $\mathbf{h} \in \mathcal{I}_{\mathrm{p}}$ by a permutation of is an injection $[\mathbf{1},\mathbf{p}]$ to $[\mathbf{1},\mathbf{n}]$ as product

[15] In order to be correct, we should have used the auxiliary map coming from definition 14 (p. 6) and used it the same way as for the other proofs like in proposition 75 (p. 37). Since it is tedious although not difficult, we shall now on give only the sketch of the demonstration omitting the fact that we have $\phi_{\mathrm{U}}(p) \notin \mathcal{F}$ for the mapping to $\mathcal{R}^{\mathrm{n}}$ although in $\mathcal{F}_{\mathrm{u}}$.

of two injections (proposition 191 (p. 148)). Two different $\mathbf{h} \in \mathcal{I}_\mathbf{p}$ give two different sets in $[\mathbf{1}, \mathbf{n}]$ (proposition 84 (p. 48)). Moreover two different permutations $\mathbf{P}_1$, $\mathbf{P}_2$ of $\mathbf{PG_p}$ cannot lead to the same map $\mathbf{f}$ with the same $\mathbf{h} \in \mathcal{I}_\mathbf{p}$: $\mathbf{P_1(k) \neq P_2(k), k \in [1, p] \Rightarrow h(P_1(k)) \neq h(P_2(k))}$ because $\mathbf{h}$ is an injection. The correspondence $\mathbf{f}$ to $\mathbf{h} \circ \mathbf{P}$ is one to one.

**Proposition 86** $\mathcal{I}_\mathbf{n}$ has only one element which is the identity map.

**Proof**
It is a direct consequence of proposition 84 (p. 48).

**Proposition 87** With the notations of proposition 83 (p. 47) and definition 53 (p. 48), any term of $\mathbf{W}((\otimes\mathcal{V}_\mathbf{u}^*)^\mathbf{p})$ (versus $\mathbf{W}((\otimes\mathcal{V}_\mathbf{u})^\mathbf{p})$), $\mathbf{p} \leq \mathbf{n}$, can be written as:

$$\sum_{\mathbf{i}\in\mathcal{I}_\mathbf{p}} \mathbf{t_i} \sum_{\mathbf{P}\in\mathbf{PG_p}} \mathbf{s(P)} \bigotimes_{\ell\in[\mathbf{1},\mathbf{p}]} \mathbf{dx}^{\mathbf{i}(\mathbf{P}(\ell))} \in \mathbf{W}((\otimes\mathcal{V}_\mathbf{u}^*)^\mathbf{p})$$

versus

$$\sum_{\mathbf{i}\in\mathcal{I}_\mathbf{p}} \mathbf{v^i} \sum_{\mathbf{P}\in\mathbf{PG_p}} \mathbf{s(P)} \bigotimes_{\ell\in[\mathbf{1},\mathbf{p}]} \partial_{\mathbf{i}(\mathbf{P}(\ell))} \in \mathbf{W}((\otimes\mathcal{V}_\mathbf{u})^\mathbf{p}) \,.$$

**Proof**
It is a direct consequence of propositions 82 (p. 47) and 83 (p. 47).

**Proposition 88** We use the notations of proposition 87 (p. 49) for $\mathbf{p} = \mathbf{n}$. We have:

$$(\forall \mathbf{T} \in \mathbf{W}((\otimes\mathcal{V}_\mathbf{u}^*)^\mathbf{n}))(\exists \mathbf{t} \in \mathcal{F}_\mathbf{u})(\mathbf{T} = \mathbf{t} \sum_{\mathbf{P}\in\mathbf{PG_n}} \mathbf{s(P)} \bigotimes_{\ell\in[\mathbf{1},\mathbf{n}]} \mathbf{dx}^{\mathbf{P}(\ell)})$$

$$(\forall \mathbf{T} \in \mathbf{W}((\otimes\mathcal{V}_\mathbf{u})^\mathbf{n}))(\exists \mathbf{t} \in \mathcal{F}_\mathbf{u})(\mathbf{T} = \mathbf{t} \sum_{\mathbf{P}\in\mathbf{PG_n}} \mathbf{s(P)} \bigotimes_{\ell\in[\mathbf{1},\mathbf{n}]} \partial_{\mathbf{P}(\ell)}) \,.$$

**Proof**
It is a direct consequence of propositions 87 (p. 49) and 86 (p. 49).

**Definition 54 0_form.**

By definition an element of $\mathcal{F}$ is also called a **0_form**.

## 1.11 Exterior differential

**Definition 55** $\mathbf{\Omega^0}$**.**

By definition $\mathbf{\Omega^0} = \mathcal{F}$. It is the set of **0_form** (definition 54 (p. 49)).

**Definition 56** $\mathbf{\Omega^p}$**,** $\mathbf{p} \in \mathcal{Z}^+$**.**

$\mathbf{\Omega^p}$ is the vector space $\mathbf{W}((\otimes\mathcal{V}^*)^\mathbf{p}) \subset (\otimes\mathcal{V}^*)^\mathbf{p}$ (see definition 182 (p. 218)) with $\mathcal{F}$ as a commutative ring. $\mathbf{W}$ is the wedge projection as defined in 50 (p. 46). It is the set of **p_form** (definition 52 (p. 47)).

**Definition 57** $\mathbf{\Omega}$**.**

$$\mathbf{\Omega} = \bigcup_{\mathbf{p}\in\mathcal{Z}^{0+}} \mathbf{\Omega^p} \,.$$

**Definition 58** $\mathbf{a \otimes b, a \wedge b, \; a \in \Omega, b \in \Omega}$**.**

Following definitions 44 (p. 40), 45 (p. 41), considering $\mathbf{a} \in \Omega^{\mathrm{p}}, \mathbf{b} \in \Omega^{\mathrm{q}}$, we may define a map $(\mathbf{a} \otimes \mathbf{b} : (\otimes \mathcal{V})^{\mathrm{p+q}} \rightarrow \mathcal{F})$ by defining first a multilinear map:

$$(\mathbf{a} \otimes \mathbf{b} : (\mathcal{V})^{\mathrm{p+q}} = \prod_{\ell \in [1,\mathrm{p}]} \mathcal{V} \times \prod_{\ell \in [\mathrm{p}+1,\mathrm{p}+\mathrm{q}]} \mathcal{V} \rightarrow \mathcal{F}) .$$

as it follows:
For any element of $v \in (\mathcal{V})^{\mathrm{p+q}}$ given by $(v_{\mathrm{k}} : [1, \mathrm{p}+1] \rightarrow \mathcal{V})$ we have:

$$v = \prod_{\mathrm{k}=[1,\mathrm{p}+\mathrm{q}]}^{\mathrm{e}} v_{\mathrm{k}} = \bigotimes_{\mathrm{k}=[1,\mathrm{p}+\mathrm{q}]} v_{\mathrm{k}} \Rightarrow \mathbf{a} \otimes \mathbf{b}(v) = \mathbf{a}(\bigotimes_{\mathrm{j} \in [1,\mathrm{p}]} v_{\mathrm{j}})\mathbf{b}(\bigotimes_{\mathrm{j} \in [\mathrm{p}+1,\mathrm{p}+\mathrm{q}]} v_{\mathrm{j}}) \in \mathcal{F} .$$

From definition 44 (p. 40), $\mathbf{a} \otimes \mathbf{b} \in \bigotimes_{\mathrm{k} \in [1,\mathrm{p}+\mathrm{q}]} \mathcal{V}^*$. Using proposition 79 (p. 42), we can extend that definition to any element of $(\otimes \mathcal{V})^{\mathrm{p+q}}$.
Having set $\mathbf{a} \otimes \mathbf{b}$, we may set also (see definition 417 (p. 545)):

1. $\mathbf{a} \in \Omega^0, \mathbf{b} \in \Omega \Rightarrow \mathbf{a} \wedge \mathbf{b} = \mathbf{b} \wedge \mathbf{a} = \mathbf{ab}$

2. $(\mathrm{p} \in \mathcal{Z}^+, \mathrm{q} \in \mathcal{Z}^+, \mathbf{a} \in \Omega^{\mathrm{p}}, \mathbf{b} \in \Omega^{\mathrm{q}})$
$\Rightarrow \mathbf{a} \wedge \mathbf{b} = \mathbf{W}(\mathbf{a} \otimes \mathbf{b}) = (-1)^{\mathrm{pq}} \mathbf{W}(\mathbf{b} \otimes \mathbf{a}) = (-1)^{\mathrm{pq}} \mathbf{b} \wedge \mathbf{a} \in \Omega^{\mathrm{p+q}})$

**Proposition 89** In the differential manifold $\mathbf{M}$ of dimension $\mathbf{n}$, we consider a covering set $\mathbf{U_k}$ of $\mathbf{M}$ with its continuous injection $(\phi : \mathbf{U_k} \rightarrow \mathcal{R}^{\mathrm{n}})$ (definition 122 (p. 139)) to $\mathcal{R}^{\mathrm{n}}$. In an open $\mathbf{U}$ of $\mathbf{U_k}$, we consider two other continuous injections to $\mathcal{R}^{\mathrm{n}}$ $(\mathbf{x} : \mathbf{U} \rightarrow \mathcal{R}^{\mathrm{n}})$, $(\mathbf{y} : \mathbf{U} \rightarrow \mathcal{R}^{\mathrm{n}})$ such that $(\mathbf{x} \circ \phi^{-1} : \phi(\mathbf{U}) \rightarrow \mathbf{x}(\mathbf{U})) \in \mathbf{C}^{\infty}$, $(\phi \circ \mathbf{x}^{-1} : \mathbf{x}(\mathbf{U}) \rightarrow \phi(\mathbf{U})) \in \mathbf{C}^{\infty}$, $(\mathbf{y} \circ \phi^{-1} : \phi(\mathbf{U}) \rightarrow \mathbf{y}(\mathbf{U})) \in \mathbf{C}^{\infty}$ and $(\phi \circ \mathbf{y}^{-1} : \mathbf{y}(\mathbf{U}) \rightarrow \phi(\mathbf{U})) \in \mathbf{C}^{\infty}$. It means $\mathbf{x} \circ \mathbf{y}^{-1} = \mathbf{x} \circ \phi^{-1} \circ \phi \circ \mathbf{y}^{-1} \in \mathbf{C}^{\infty}$ and $\mathbf{y} \circ \mathbf{x}^{-1} \in \mathbf{C}^{\infty}$ and the determinants of all possible Jacobians different from $\mathbf{0}$ (proposition 628 (p. 360)). Given $p \in \mathbf{U}$ as in footnote 2 (p. 1), we set $\mathbf{x} = \mathbf{x}(p)$, $\mathbf{y} = \mathbf{y}(p)$ and $\mathbf{x}(\mathbf{y}) = \mathbf{x} \circ \mathbf{y}^{-1}$, $\mathbf{y}(\mathbf{x}) = \mathbf{y} \circ \mathbf{x}^{-1}$. $\mathbf{dx}^{\ell}$ stands for the differential of the map $(\mathbf{x}^{\ell} : \mathbf{U} \rightarrow \mathcal{R})$ when $(\mathbf{x}(p) : \mathbf{U} \rightarrow \mathcal{R}^{\mathrm{n}})$ is considered as the mapping to $\mathcal{R}^{\mathrm{n}}$ of the restriction to $\mathbf{U}$ (definition 41 (p. 34)) of the manifold $\mathbf{M}$. As in section 1.6 (p. 23) $\partial_{\mathrm{j}} \mathrm{y}^{\mathrm{i}} = \frac{\partial \mathrm{y}^{\mathrm{i}}(\mathrm{x})}{\partial \mathrm{x}^{\mathrm{j}}}$ (see definition 275 (p. 304)). Using proposition 87 (p. 49), we have with no implicit product or sum(See definition 50 (p. 46) and 53 (p. 48) for the notations like $\mathbf{PG_p}, \mathcal{I}_{\mathrm{p}}$):

$$\mathrm{i} \in \mathcal{I}_{\mathrm{p}} \Rightarrow \bigwedge_{\ell \in [1,\mathrm{p}]} \mathrm{dy}^{\mathrm{i}(\ell)} = \sum_{\mathrm{j} \in \mathcal{I}_{\mathrm{p}}} ( \sum_{\mathrm{P} \in \mathrm{PG_p}} \mathrm{s}(\mathrm{P}) \prod_{\ell \in [1,\mathrm{p}]} \partial_{\mathrm{j}(\mathrm{P}(\ell))} \mathrm{y}^{\mathrm{i}(\ell)}) \times \bigwedge_{\ell' \in [1,\mathrm{p}]} \mathrm{dx}^{\mathrm{j}(\ell')} .$$

**Proof**
We suppose $\mathbf{p} \in [1, \mathbf{n}]$, $q \in \mathbf{U}$, $\mathbf{i} \in \mathcal{I}_{\mathrm{p}}$, $v = \prod_{\ell \in [1,\mathrm{p}]} \mathrm{v}_{\ell} \in \mathcal{V}_{\mathrm{U}}^{\mathrm{p}}$. From proposition 352 (p. 209), we have $\mathbf{s}(\mathbf{P}) = \mathbf{s}(\mathbf{P}^{-1})$ because $\mathbf{s}(\mathbf{P})\mathbf{s}(\mathbf{P}^{-1}) = \mathbf{s}(\mathbf{P} \circ \mathbf{P}^{-1}) = \mathbf{1}$. Then applying

propositions 289 (p. 186), 291 (p. 187), 292 (p. 187), 293 (p. 188), we have:

$$\bigwedge_{\ell\in[1,p]} dy^{i(\ell)}(v_\ell)(q) = \sum_{P\in PG_p}\prod_{\ell\in[1,p]} s(P)dy^{i(P(\ell))}(v_\ell)(q)$$
$$= \sum_{P\in PG_p}\prod_{\ell\in[1,p]} s(P)\sum_{j\in[1,n]} \partial_j y^{i(P(\ell))}dx^j(v_\ell)(q)$$
$$= \sum_{P\in PG_p}\sum_{j\in[1,n]^p}\prod_{\ell\in[1,p]} s(P)\partial_{j(\ell)} y^{i(P(\ell))}dx^{j(\ell)}(v_\ell)(q)$$
$$= \sum_{P\in PG_p}\sum_{j\in[1,n]^p}\prod_{\ell\in[1,p]} s(P)\partial_{j(\ell)} y^{i(P(\ell))}\prod_{\ell'\in[1,p]} dx^{j(\ell')}(v_{\ell'})(q)$$
$$= \sum_{P\in PG_p}\sum_{j\in[1,n]^p}\prod_{\ell\in P([1,p])} s(P)\partial_{j(P^{-1}(\ell))} y^{i(P(P^{-1}(\ell)))}\prod_{\ell'\in[1,p]} dx^{j(\ell')}(v_{\ell'})(q)$$
$$= \sum_{j\in[1,n]^p}\sum_{P\in PG_p}\prod_{\ell\in[1,p]} s(P)\partial_{j(P^{-1}(\ell))} y^{i(\ell)}\prod_{\ell'\in[1,p]} dx^{j(\ell')}(v_{\ell'})(q) .$$

From proposition 359 (p. 213), given a bijection $(\mathbf{h} : \mathbf{B} \leftrightarrow \mathbf{A})$ with $\mathbf{Card(A)} = \mathbf{Card(B)} \in \mathcal{Z}^+$ and a map $(\mathbf{H} : \mathbf{A} \to \mathcal{R})$, we have:

$$\sum_{\ell\in A} H(\ell) = \sum_{\ell\in B} H(h(\ell));\ \prod_{\ell\in A} H(\ell) = \prod_{\ell\in B} H(h(\ell)) .$$

From propositions 198 (p. 150), 342 (p. 207), 343 (p. 207), 358 (p. 212) and definition 178 (p. 212), we have as bijection $\mathbf{PG_p} \leftrightarrow \mathbf{PG_p}$ with $\mathbf{P} \in \mathbf{PG_p}$ the variable and $\mathbf{Q} \in \mathbf{PG_p}$ fixed :

$$(Q\circ P : PG_p \leftrightarrow PG_p);\ (P\circ Q : PG_p \leftrightarrow PG_p);\ (P^{-1} : PG_p \leftrightarrow PG_p)$$

Since in the sum $\sum_{j\in[1,n]^p}$, all terms such that $\mathbf{j(\ell) = j(\ell'), \ell \neq \ell'}$ have a zero contribution (proposition 82 (p. 47)), the map $(\mathbf{j(\ell)} : [1,p] \to [1,n])$ can be restricted to an injection (definition 122 (p. 139)). We can then apply the following relation coming from proposition 85 (p. 48):

$$\{(H(j) : [1,n]^p \to \mathcal{R}), ((\exists i_1 \in [1,p])(\exists i_2 \in [1,p])$$
$$(i_1 \neq i_2, j(i_1) = j(i_2)) \Rightarrow H(j) = 0)\}$$
$$\Rightarrow \sum_{j\in[1,n]^p} H(j) = \sum_{j\in\mathcal{I}_p}\sum_{Q\in PG_p} H(j\circ Q) .$$

With all that we may write:

$$
\begin{aligned}
&\bigwedge_{\ell\in[1,p]} dy^{i(\ell)}(v_\ell)(q) \\
&= \sum_{j\in\mathcal{I}_p}\sum_{Q\in PG_p}\sum_{P\in PG_p}\prod_{\ell\in[1,p]} s(P^{-1})\partial_{j\circ Q(P^{-1}(\ell))}y^{i(\ell))}\prod_{\ell'\in[1,p]} dx^{j(Q(\ell'))}(v_{\ell'})(q) \\
&= \sum_{j\in\mathcal{I}_p}\sum_{Q\in PG_p}\sum_{P\in PG_p}\prod_{\ell\in[1,p]} s(P)\partial_{j\circ Q(P(\ell))}y^{i(\ell))}\prod_{\ell'\in[1,p]} dx^{j(Q(\ell'))}(v_{\ell'})(q) \\
&= \sum_{j\in\mathcal{I}_p}\sum_{Q\in PG_p}\sum_{P\in PG_p}\prod_{\ell\in[1,p]} s(Q^{-1}P)\partial_{j(P(\ell))}y^{i(\ell)}\prod_{\ell'\in[1,p]} dx^{j(Q(\ell'))}(v_{\ell'})(q) \\
&= \sum_{j\in\mathcal{I}_p}\sum_{Q\in PG_p} s(Q)\sum_{P\in PG_p} s(P)\prod_{\ell\in[1,p]} \partial_{j(P(\ell))}y^{i(\ell)}\prod_{\ell'\in[1,p]} dx^{j(Q(\ell'))}(v_{\ell'})(q) \\
&= \sum_{j\in\mathcal{I}_p}(\sum_{P\in PG_p} s(P)\prod_{\ell\in[1,p]} \partial_{j(P(\ell))}y^{i(\ell)}) \times (\sum_{Q\in PG_p} s(Q)\prod_{\ell'\in[1,p]} dx^{j(Q(\ell'))}(v_{\ell'})(q)) \\
&= \sum_{j\in\mathcal{I}_p}(\sum_{P\in PG_p} s(P)\prod_{\ell\in[1,p]} \partial_{j(P(\ell))}y^{i(\ell)}) \times \bigwedge_{\ell'\in[1,p]} dx^{j(\ell')}(v)(q)\ .
\end{aligned}
$$

**Proposition 90** We keep the notations and premises of proposition 89 (p. 50). For $\mathbf{a} \in \Omega_U$ which means $(\exists p \in \mathcal{Z}^{0+})(a \in \Omega_U^p)$, with no implicit product or sum we have:

$$
\begin{aligned}
&(\exists({}^x a_i : \mathcal{I}_p \to \mathcal{F}_k)) \\
&((q \in U, v \in \mathcal{V}_U^p) \Rightarrow a(v)(q) = \sum_{i\in\mathcal{I}_p} {}^x a_i \bigwedge_{j\in[1,p]} dx^{i(j)}(v)(q))
\end{aligned}
\tag{1.63}
$$

and the same thing exchanging $\mathbf{x}$ with $\mathbf{y}$.

$$
\begin{aligned}
&(\exists({}^y a_i : \mathcal{I}_p \to \mathcal{F}_k)) \\
&((q \in U, v \in \mathcal{V}_U^p) \Rightarrow a(v)(q) = \sum_{i\in\mathcal{I}_p} {}^y a_i \bigwedge_{\ell\in[1,p]} dy^{i(\ell)}(v)(q))\ .
\end{aligned}
\tag{1.64}
$$

We have the following relationship:

$$
(\forall i \in \mathcal{I}_p)\ ({}^x a_i = \sum_{j\in\mathcal{I}_p} {}^y a_j(\sum_{P\in PG_p} s(P)\prod_{\ell\in[1,p]} \partial_{i(P(\ell))}y^{j(\ell)}))\ .
\tag{1.65}
$$

**Proof**

Using proposition 89 (p. 50) and expanding equation (1.64) (p. 52), we have :

$$
\begin{aligned}
&(\exists({}^y a_i : \mathcal{I}_p \to \mathcal{F}_k)) \\
&((q \in U, v \in \mathcal{V}_U^p) \Rightarrow a(v)(q) = \sum_{i\in\mathcal{I}_p} {}^y a_i \bigwedge_{\ell\in[1,p]} dy^{i(\ell)}(v)(q) \\
&= \sum_{i\in\mathcal{I}_p} {}^y a_i \sum_{j\in\mathcal{I}_p}(\sum_{P\in PG_p} s(P)\prod_{\ell\in[1,p]} \partial_{j(P(\ell))}y^{i(\ell)}) \times \bigwedge_{\ell'\in[1,p]} dx^{j(\ell')}(v)(q) \\
&= \sum_{j\in\mathcal{I}_p}\sum_{i\in\mathcal{I}_p} {}^y a_i(\sum_{P\in PG_p} s(P)\prod_{\ell\in[1,p]} \partial_{j(P(\ell))}y^{i(\ell)}) \times \bigwedge_{\ell'\in[1,p]} dx^{j(\ell')}(v)(q) \\
&= \sum_{i\in\mathcal{I}_p}\sum_{j\in\mathcal{I}_p} {}^y a_j(\sum_{P\in PG_p} s(P)\prod_{\ell\in[1,p]} \partial_{i(P(\ell))}y^{j(\ell)}) \times \bigwedge_{\ell'\in[1,p]} dx^{i(\ell')}(v)(q))\ .
\end{aligned}
$$

Combining with equation (1.63) (p. 52), we get equation (1.65) (p. 52).

**Definition 59 The differential operator d as a map $(\mathbf{d} : \Omega \to \Omega)$.**

Setting $\mathbf{a} \in \Omega, \mathbf{b} \in \Omega$ we should have:

1. $(\mathbf{d} : \Omega^0 \to \Omega^1)$
   From definition 40 (p. 34)
$$\mathbf{a} \in \Omega^0 = \mathcal{F} \Rightarrow (\forall \mathbf{v} \in \mathcal{V})(\mathbf{da}(\mathbf{v}) = \mathbf{v}(\mathbf{a})) \Rightarrow \mathbf{da} \in \Omega^1$$
2. $(\mathbf{p} \in \mathcal{Z}^{0+}, \mathbf{q} \in \mathcal{Z}^{0+}, \mathbf{a} \in \Omega^{\mathbf{p}}, \mathbf{b} \in \Omega^{\mathbf{q}}) \Rightarrow \mathbf{d}(\mathbf{a} \wedge \mathbf{b}) = \mathbf{da} \wedge \mathbf{b} + (-1)^{\mathbf{p}} \mathbf{a} \wedge \mathbf{db} \in \Omega$
3. $(\mathbf{a} \in \Omega^{\mathbf{p}}, \mathbf{b} \in \Omega^{\mathbf{p}}) \Rightarrow \mathbf{d}(\mathbf{a} + \mathbf{b}) = \mathbf{da} + \mathbf{db}$
4. $(\forall \mathbf{a} \in \Omega)(\mathbf{dda} = \mathbf{0})$.

**Proposition 91** There is a map $(\mathbf{d} : \Omega \to \Omega)$ as in definition 59 (p. 53). This map is unique. This map does not depend on the covering of the manifold.

**Proof**
One considers a manifold $\mathbf{M}$ of dimension $\mathbf{n}$ and one of its covers defined by $\mathbf{k} \in \mathbf{K} = [\mathbf{1}, \mathbf{Card}(\mathbf{K})]$, $\mathbf{U_k}$, $(\phi_\mathbf{k} : \mathbf{U_k} \to \mathcal{R}^\mathbf{n})$ all elements $\mathbf{a} \in \Omega$ can be expressed in any $\mathbf{U_k}$ as:

$$(\exists \mathbf{p} \in [\mathbf{1}, \mathbf{n}])(\exists (\mathbf{a}^\mathbf{k}_\mathbf{i} : \mathcal{I}_\mathbf{p} \to \mathcal{F}_\mathbf{k}))(\mathbf{a}^\mathbf{k} = \sum_{\mathbf{i} \in \mathcal{I}_\mathbf{p}} \mathbf{a}^\mathbf{k}_\mathbf{i} \bigwedge_{\ell \in [\mathbf{1},\mathbf{p}]} \mathbf{dx}^{\mathbf{i}(\ell)}) \ . \tag{1.66}$$

In that cover $\mathbf{U_k}$, the map $\mathbf{d}$ is fully defined on $\mathbf{a^k}$ in equation (1.66) (p. 53):

- by definition 59 (p. 53) point 3 applied on the sum defining $\mathbf{a^k}$,
- by definition 59 (p. 53) point 2 applied on each term $\mathbf{a}^\mathbf{k}_\mathbf{i} \bigwedge_{\ell \in [\mathbf{1},\mathbf{p}]} \mathbf{dx}^{\mathbf{i}(\ell)}$ of that sum,
- by definition 59 (p. 53) points 1, 2 applied to any $\mathbf{a}^\mathbf{k}_\mathbf{i}$,
- by definition 59 (p. 53) point 4 applied to $\mathbf{dx}^\nu$ for all $\nu \in [\mathbf{1}, \mathbf{n}]$.

The $\mathbf{d}$ is unique. We have to prove that this map verifies definition 59 (p. 53) in any of its terms:

1. $\mathbf{f} \in \mathcal{F} \Rightarrow \mathbf{ddf} = \mathbf{0}$
$$\mathbf{ddf} = \partial_\mu \partial_\nu \mathbf{f} \mathbf{dx}^\mu \wedge \mathbf{dx}^\nu = \sum_{\mu \in [\mathbf{1},\mathbf{n}-\mathbf{1}]} \sum_{\nu \in [\mu+\mathbf{1},\mathbf{n}]} \partial_\mu \partial_\nu \mathbf{f} \mathbf{dx}^\mu \wedge \mathbf{dx}^\nu + \partial_\nu \partial_\mu \mathbf{f} \mathbf{dx}^\nu \wedge \mathbf{dx}^\mu = \mathbf{0}$$
   because $\partial_\mu \partial_\nu = \partial_\nu \partial_\mu$ (proposition 652 (p. 396)) and $\mathbf{dx}^\mu \wedge \mathbf{dx}^\nu = -\mathbf{dx}^\nu \wedge \mathbf{dx}^\mu$ (definition 51 (p. 47)).
2. $(\mathbf{a} \in \Omega^{\mathbf{p}}, \mathbf{b} \in \Omega^{\mathbf{p}}) \Rightarrow \mathbf{d}(\mathbf{a} + \mathbf{b}) = \mathbf{da} + \mathbf{db}$
   Using expression (1.66) (p. 53) for $\mathbf{a}$ and $\mathbf{b}$, together with definition 40 (p. 34) and definition 15 (p. 8), from definition 59 (p. 53) point 2 and from $\mathbf{ddx}^\nu = \mathbf{0}$, we have:
$$\mathbf{d}(\mathbf{a}+\mathbf{b})^\mathbf{k} = \mathbf{d}(\sum_{\mathbf{i} \in \mathcal{I}_\mathbf{p}} (\mathbf{a}+\mathbf{b})^\mathbf{k}_\mathbf{i}) \wedge \bigwedge_{\ell \in [\mathbf{1},\mathbf{p}]} \mathbf{dx}^{\mathbf{i}(\ell)} = \sum_{\mathbf{i} \in \mathcal{I}_\mathbf{p}} (\mathbf{da}^\mathbf{k}_\mathbf{i} + \mathbf{db}^\mathbf{k}_\mathbf{i}) \wedge \bigwedge_{\ell \in [\mathbf{1},\mathbf{p}]} \mathbf{dx}^{\mathbf{i}(\ell)}$$
$$= \sum_{\mathbf{i} \in \mathcal{I}_\mathbf{p}} \mathbf{da}^\mathbf{k}_\mathbf{i} \wedge \bigwedge_{\ell \in [\mathbf{1},\mathbf{p}]} \mathbf{dx}^{\mathbf{i}(\ell)} + \sum_{\mathbf{i} \in \mathcal{I}_\mathbf{p}} \mathbf{db}^\mathbf{k}_\mathbf{i} \wedge \bigwedge_{\ell \in [\mathbf{1},\mathbf{p}]} \mathbf{dx}^{\mathbf{i}(\ell)} = \mathbf{da}^\mathbf{k} + \mathbf{db}^\mathbf{k}$$

3. $\mathbf{a} \in \Omega^p, \mathbf{b} \in \Omega^q \Rightarrow \mathbf{d}(\mathbf{a} \wedge \mathbf{b}) = \mathbf{da} \wedge \mathbf{b} + (-1)^p \mathbf{a} \wedge \mathbf{db} \in \Omega$
Using again expression (1.66) (p. 53) for **a** and **b** we have:

$$\begin{aligned}
d(a \wedge b)^k &= d(\sum_{i \in \mathcal{I}_p} a_i^k \bigwedge_{\ell \in [1,p]} dx^{i(\ell)} \wedge \sum_{j \in \mathcal{I}_q} b_j^k \bigwedge_{l \in [1,q]} dx^{j(\ell)}) \\
&= \sum_{i \in \mathcal{I}_p} \sum_{j \in \mathcal{I}_q} d(a_i^k b_j^k) \wedge \bigwedge_{\ell \in [1,p]} dx^{i(\ell)} \wedge \bigwedge_{l \in [1,q]} dx^{j(\ell)} \\
&= \sum_{i \in \mathcal{I}_p} \sum_{j \in \mathcal{I}_q} (b_j^k da_i^k + a_i^k db_j^k) \wedge \bigwedge_{\ell \in [1,p]} dx^{i(\ell)} \wedge \bigwedge_{l \in [1,q]} dx^{j(\ell)} \\
&= \sum_{i \in \mathcal{I}_p} \sum_{j \in \mathcal{I}_q} (b_j^k da_i^k \wedge \bigwedge_{\ell \in [1,p]} dx^{i(\ell)} \wedge \bigwedge_{l \in [1,q]} dx^{j(\ell)} \\
&+ a_i^k db_j^k \wedge \bigwedge_{\ell \in [1,p]} dx^{i(\ell)} \wedge \bigwedge_{l \in [1,q]} dx^{j(\ell)}) \\
&= da^k \wedge b^k + \sum_{i \in \mathcal{I}_p} \sum_{j \in \mathcal{I}_q} a_i^k db_j^k \wedge \bigwedge_{\ell \in [1,p]} dx^{i(\ell)} \wedge \bigwedge_{l \in [1,q]} dx^{j(\ell)} \\
&= da^k \wedge b^k + (-1)^p \sum_{i \in \mathcal{I}_p} a_i^k \bigwedge_{\ell \in [1,p]} dx^{i(\ell)} \wedge (\sum_{j \in \mathcal{I}_q} db_j^k \wedge \bigwedge_{l \in [1,q]} dx^{j(\ell)}) \\
&= da^k \wedge b^k + (-1)^p a^k \wedge db^k \,.
\end{aligned}$$

**Proposition 92** Given a basic cover $\mathbf{U_k}$ of a differential manifold **M** of dimension **n** (definition 1 (p. 1)), an element $\omega \in \Omega^n$ we have:

$$(\exists f \in \mathcal{F}_k)(\forall v \in (\otimes \mathcal{V}_k)^n)(\forall p \in U_k)(\omega(v)(p) = f(p) \bigwedge_{i \in [1,n]} dx^i(v)(p) \,.$$

Proof
It is a direct consequence of propositions 86 (p. 49) and 87 (p. 49).

## 1.12 Volume and integral in differential manifold

### 1.12.1 Volume form

**Definition 60 Oriented volume form.**

In a differential manifold of dimension **n** with its cover: $(\mathbf{U_k} : \mathbf{K} = [\mathbf{1}, \mathbf{Card(K)}] \to \mathcal{P}(\mathbf{M}))$ (definition 1 (p. 1)) with $(\forall \mathbf{k} \in \mathbf{K})(\exists(\phi_\mathbf{k}(p) : \mathbf{U_k} \to \mathcal{R}^\mathbf{n}))$, **Vol** is called an oriented volume form if it is an **n_form** (definition 52 (p. 47)) on **M** [16] different from zero at any point of **M**:

$$\mathrm{Vol(M)} \subset \mathcal{R} - \{0\} = ]\leftarrow, 0[\cup]0, \rightarrow[$$

**Proposition 93** Keeping notations of definition 60 (p. 54), when an oriented volume form **Vol** is restricted (definition 41 (p. 34)) on any cover subset $\mathbf{U_k}$ of **M**, it gives the wedge product of the **n** basic vector fields of $\mathbf{U_k}$ (definition 21 (p. 16)) multiplied by a map of $\mathcal{F}_\mathbf{k}$ which is different from zero at any point of $\mathbf{U_k}$ (see definition 8 (p. 3)).

[16] One has to keep in mind that those volume forms as the vector fields on **M** are not guaranteed to exist. They are guaranteed to exist only when **M** is reduced to one cover: $\mathbf{K} = [\mathbf{1}, \mathbf{1}]$ otherwise one has to find them.

**Proof**
It is a consequence of proposition 88 (p. 49).

**Proposition 94** Keeping above notations, calling $\mathbf{C(x,a)}$ an open cube (definition 260 (p. 285)) centered on $\mathbf{x}$ in $\mathcal{R}^n$, $\mathbf{g}'_p(\mathbf{x},\mathbf{a},\mathbf{b})$ the map defined in definition 14 (p. 6) and $\mathbf{Vol_k}$ the restriction of **Vol** to $\mathbf{U_k}$ (definition 41 (p. 34)), we should have when **Vol** is an oriented volume form (definition 60 (p. 54)):

$$\begin{aligned}
&\mathbf{Vol} \in \Omega^n \Rightarrow (\forall k \in K)(\exists \mathrm{vol_k} \in \mathcal{F}_k)\\
&((p \in U_k, C(\phi_k(p), a) \subset C(\phi_k(p), b) \subset \phi_k(U_k) \subset \mathcal{R}^n)\\
&\Rightarrow \mathrm{Vol}(g'_p(p', a, b) \bigotimes_{i\in[1,n]} \partial_i)(p) = \mathrm{Vol_k}(\bigotimes_{i\in[1,n]} \partial_i)(p)\\
&= \mathrm{vol_k}(p) \bigwedge_{j\in[1,n]} dx^j(\bigotimes_{i\in[1,n]} \partial_i)(p) = \mathrm{vol_k}(p) \neq 0)
\end{aligned}$$

or more directly in any $\mathbf{U_k}$ the restriction $\mathbf{Vol_k}$ of **Vol** is such as

$$(\exists \mathrm{vol_k} \in \mathcal{F}_k)\,(\mathrm{Vol_k} = \mathrm{vol_k} \bigwedge_{\mu\in[1,n]} dx^\mu_k, (\forall p \in U_k)(\mathrm{vol_k}(p) \neq 0)) \,.$$

**Proof**
It a consequence of proposition 93 (p. 54) and definition 60 (p. 54).

**Proposition 95** Using the notation of the preceding definition 60 (p. 54), in the differential manifold **M** of dimension **n** with an oriented volume form **Vol**, we consider a local cover $\mathbf{U_k}$ and $\mathbf{Vol_k}$ the restriction of **Vol** to $\mathbf{U_k}$ (definition 41 (p. 34)). With that, we have:

$$\mathrm{Vol_k} = \mathrm{vol_k} \bigwedge_{i\in[1,n]} dx^i \tag{1.67}$$

with $\mathbf{vol_k} \in \mathcal{F}_k$ and if $\mathbf{U_k}$ is path connected (definition 285 (p. 327)) for all points in $\mathbf{U_k}$ $\mathbf{vol_k}$ has the same sign:

$$\mathrm{vol_k}(U_k) \subset ]0, \rightarrow [\vee \mathrm{vol_k}(U_k) \subset ] \leftarrow, 0[ \tag{1.68}$$

**Proof**
Equation (1.67) (p. 55) with $\mathbf{vol} \in \mathcal{F}_k$ is a direct consequence of propositions 88 (p. 49), 93 (p. 54) and definition 60 (p. 54).
If equation (1.68) (p. 55) were false, there would be two points $p_0, p_1$ of $\mathbf{U_k}$ such that:

$$\mathrm{vol}(p_0) < 0 < \mathrm{vol}(p_1)$$

since from (definition 285 (p. 327)) there is a continuous map $(\mathbf{h} : [0,1] \rightarrow \mathbf{U_k})$ with $\mathbf{h}(0) = p_0$ $\mathbf{h}(1) = p_1$ then we have a continuous map $(\mathbf{vol} \circ \mathbf{h} : [0,1] \rightarrow \mathcal{R})$ such that $\mathbf{vol}(\mathbf{h}(0)) < 0 < \mathbf{vol}(\mathbf{h}(1))$. From proposition 299 (p. 190), we would have

$$0 \in [\mathrm{vol}(h(0)), \mathrm{vol}(h(1))] \subset \mathrm{vol}(h([0,1])) \subset \mathrm{vol}(U_k) \,.$$

That would imply that $(\exists q \in \mathbf{U_k})(\mathbf{vol}(q) = 0)$ which is forbidden. We conclude by A.1.4 (p. 435) point 8.

**Proposition 96** We are given a differential manifold $\mathbf{M}$ of dimension $\mathbf{n}$ and a local cover $\mathbf{U_k}$ with $(\forall \mathbf{k} \in \mathbf{K})(\exists(\phi_\mathbf{k}(p) : \mathbf{U_k} \to \mathcal{R}^\mathbf{n}))$ and two oriented volume forms $\mathbf{Vol^1}$, $\mathbf{Vol^2}$. At any point $p$ of a connected part $\mathbf{C} \subset \mathbf{M}$ (definition 286 (p. 327)), for any set of $\mathbf{n}$ vector fields of $\mathcal{V}$ (or of $\mathbf{n}$ local vector fields of $\mathcal{V}_\mathbf{k}$ if $p \in \mathbf{U_k}$)), the two oriented volume forms yield two numbers of $\mathcal{R}$ which are either both zero or both with the same relative sign (The product of them has the same sign everywhere if not zero).

**Proof**

1. From proposition 95 (p. 55), equation (1.68) (p. 55), in any $\mathbf{U_k}$, we have

$$\mathbf{i} \in [1,2] \Rightarrow (\mathbf{Vol^i_k} = \mathbf{vol^i_k} \wedge_{\mathbf{j}\in[1,\mathbf{n}]} \mathbf{dx^j}, \mathbf{vol^i_k} \in \mathcal{F}_\mathbf{k})$$

$\mathbf{vol^1_k}$ and $\mathbf{vol^2_k}$ are never $\mathbf{0}$ and keep their sign over $\mathbf{U_k}$. Any set of $\mathbf{n}$ vector fields either in $\mathcal{V}$ or in $\mathcal{V}_\mathbf{k}$ can be restricted (definition 18 (p. 13)) in $\mathbf{U_k}$ to $(\mathbf{v}_\ell = \mathbf{v}^\mathbf{j}_\ell \partial_\mathbf{j} : [1,\mathbf{n}] \to \mathcal{V}_\mathbf{k})$ with $(\mathbf{v}^\mathbf{j}_\ell : [1,\mathbf{n}] \to \mathcal{F}_\mathbf{k})$ (see proposition 28 (p. 13) and definition 21 (p. 16)). At any $p \in \mathbf{U_k}$, we have:

$$\mathbf{Vol^i}(\bigotimes_{\ell\in[1,\mathbf{n}]} \mathbf{v}_\ell)(p) = \det(\mathbf{v}^\mathbf{j}_\ell)(p)\mathbf{vol^i_k}(p)$$

$$\prod_{\mathbf{i}\in[1,2]} \mathbf{Vol^i}(\bigotimes_{\ell\in[1,\mathbf{n}]} \mathbf{v}_\ell)(p) = \det{}^2(\mathbf{v}^\mathbf{j}_\ell)(p) \prod_{\mathbf{i}\in[1,2]} \mathbf{vol^i_k}(p)$$

if $\det(\mathbf{v}^\mathbf{j}_\ell) = 0$ both $\mathbf{Vol^i}(\bigotimes_{\ell\in[1,\mathbf{n}]} \mathbf{v}_\ell)(p)$ are zero and both of the same sign over $\mathbf{U_k}$ if $\det(\mathbf{v}^\mathbf{j}_\ell) \neq 0$.

2. If two different $\mathbf{U_k}$, $\mathbf{U_{k'}}$ $(\mathbf{k} \neq \mathbf{k'})$ have a common point $p$, they have in common an open $\mathbf{U_k} \cap \mathbf{U_{k'}} \neq \emptyset$ in that open there are the two reference vector fields $\partial^\mathbf{k}$, $\partial^\mathbf{k'}$. From proposition 47 (p. 23), equation (1.33) (p. 23) and proposition 48 (p. 24), we have:

$$\partial^\mathbf{k}{}_\ell = \frac{\partial(\phi_\mathbf{k'} \circ \phi_\mathbf{k}^{-1})^\mathbf{j}}{\partial \mathbf{x}^\ell_\mathbf{k}} \partial^\mathbf{k'}_\mathbf{j},$$

$$(\forall p \in \mathbf{U_k} \cap \mathbf{U_{k'}})(\det(\frac{\partial(\phi_\mathbf{k'} \circ \phi_\mathbf{k}^{-1})^\mathbf{j}}{\partial \mathbf{x}^\ell_\mathbf{k}})(p) \neq 0)$$

$$(\forall p \in \mathbf{U_k} \cap \mathbf{U_{k'}})(\prod_{\mathbf{i}\in[1,2]} \mathbf{vol^i_k}(p) = \det{}^2(\frac{\partial(\phi_\mathbf{k'} \circ \phi_\mathbf{k}^{-1})^\mathbf{j}}{\partial \mathbf{x}^\ell_\mathbf{k}}) \prod_{\mathbf{i}\in[1,2]} \mathbf{vol^i_{k'}}(p) .$$

Then using the preceding point 1 (p. 56), proposition 96 (p. 56) is true for $\mathbf{U_k} \cup \mathbf{U_{k'}}$.

3. From definition 60 (p. 54), both volume forms are not zero for the full set of reference vector fields associated to any $\mathbf{U_k}$. We can then split $\mathbf{K}$ into two sets $\mathbf{K}^+$ and $\mathbf{K}^-$ (see proposition 749 (p. 450)) defined as follows:

$$\mathbf{k} \in \mathbf{K}^- \subset \mathbf{K} \Leftrightarrow ((\mathbf{k} \in \mathbf{K}, p \in \mathbf{U_k}) \Rightarrow \prod_{\mathbf{i}\in[1,2]} \mathbf{vol^i_k}(p) < 0),$$

$$\mathbf{k} \in \mathbf{K}^+ \subset \mathbf{K} \Leftrightarrow ((\mathbf{k} \in \mathbf{K}, p \in \mathbf{U_k}) \Rightarrow \prod_{\mathbf{i}\in[1,2]} \mathbf{vol^i_k}(p) > 0) .$$

From proposition 598 (p. 331), for any point $\mathbf{a} \in \mathbf{C} \cap \mathbf{M}$, the connected set $\mathbf{C_a}$ of elements connected to $\mathbf{a}$ is open. From definition of $\mathbf{a}$ and $\mathbf{C_a}$, $\mathbf{a} \in \mathbf{C} \cap \mathbf{C_a}$ then

$\mathbf{C} \cap \mathbf{C_a} \neq \emptyset$. From proposition 809 (p. 469), we have $\mathbf{C_a} = \mathbf{C}$. Then $\mathbf{C}$ is open as $\mathbf{C_a}$. For any $\mathbf{k} \in \mathbf{K}$, $\mathbf{U_k} \cap \mathbf{C}$ is open (definition 132 (p. 151)). From definition 132 (p. 151), (2.17) and proposition 798 (p. 465), calling $\mathcal{O}^{\mathbf{M}}$ the set of opens of $\mathbf{M}$. We have:

$$\bigcup_{\mathbf{k}\in\mathbf{K}^-} \mathbf{U_k} \cap \mathbf{C} = \mathbf{C} \cap \bigcup_{\mathbf{k}\in\mathbf{K}^-} \mathbf{U_k} \in \mathcal{O}^{\mathbf{M}}, \ \bigcup_{\mathbf{k}\in\mathbf{K}^+} \mathbf{U_k} \cap \mathbf{C} = \mathbf{C} \cap \bigcup_{\mathbf{k}\in\mathbf{K}^+} \mathbf{U_k} \in \mathcal{O}^{\mathbf{M}} .$$

From propositions 798 (p. 465) and 797 (p. 465), we have:

$$(\mathbf{C} \cap \bigcup_{\mathbf{k}\in\mathbf{K}^+} \mathbf{U_k}) \cup (\mathbf{C} \cap \bigcup_{\mathbf{k}\in\mathbf{K}^-} \mathbf{U_k}) = \mathbf{C} \cap \mathbf{M} = \mathbf{C} .$$

Then from proposition 595 (p. 329), if $\mathbf{C} \cap \bigcup_{\mathbf{k}\in\mathbf{K}^+} \mathbf{U_k} \neq \emptyset, \mathbf{C} \cap \bigcup_{\mathbf{k}\in\mathbf{K}^-} \mathbf{U_k} \neq \emptyset$

$$(\mathbf{C} \cap \bigcup_{\mathbf{k}\in\mathbf{K}^+} \mathbf{U_k}) \cap (\mathbf{C} \cap \bigcup_{\mathbf{k}\in\mathbf{K}^-} \mathbf{U_k}) \neq \emptyset .$$

In that case, from point 2 (p. 56), definition 362 (p. 462), definition 363 (p. 463), the following statements would be true:

$$(\exists p \in (\mathbf{C} \cap \bigcup_{\mathbf{k}\in\mathbf{K}^+} \mathbf{U_k}) \cap (\mathbf{C} \cap \bigcup_{\mathbf{k}\in\mathbf{K}^-} \mathbf{U_k}))$$
$$(\exists p)(p \in \mathbf{C}, p \in \bigcup_{\mathbf{k}\in\mathbf{K}^+} \mathbf{U_k}, p \in \bigcup_{\mathbf{k}\in\mathbf{K}^-} \mathbf{U_k})$$
$$(\exists p)(p \in \mathbf{C}, (\exists \mathbf{k} \in \mathbf{K}^+)(\exists \mathbf{k}' \in \mathbf{K}^-)(p \in \mathbf{U_k} \cap \mathbf{U'_k}))$$
$$(\exists \mathbf{k} \in \mathbf{K}^+)(\exists \mathbf{k}' \in \mathbf{K}^-)(\forall p \in \mathbf{U_k})(\forall p' \in \mathbf{U_{k'}})$$
$$(\mathbf{vol}^1_{\mathbf{k}}(p) \times \mathbf{vol}^2_{\mathbf{k}}(p) = \mathbf{vol}^1_{\mathbf{k'}}(p') \times \mathbf{vol}^2_{\mathbf{k'}}(p'))$$

that is impossible then $\neg(\mathbf{C} \cap \bigcup_{\mathbf{k}\in\mathbf{K}^+} \mathbf{U_k} \neq \emptyset, \mathbf{C} \cap \bigcup_{\mathbf{k}\in\mathbf{K}^-} \mathbf{U_k} \neq \emptyset)$ is true (see A.1.4 (p. 435), 8 (p. 435)). It means $\mathbf{C} \cap \bigcup_{\mathbf{k}\in\mathbf{K}^+} \mathbf{U_k} = \emptyset \vee \mathbf{C} \cap \bigcup_{\mathbf{k}\in\mathbf{K}^-} \mathbf{U_k} = \emptyset$.

From points 1 (p. 56) to 3 (p. 56), proposition 96 (p. 56) is true.

**Proposition 97** If a manifold $\mathbf{M}$ of dimension $\mathbf{n}$ has an oriented volume form $\mathbf{Vol}$, all elements of $\mathbf{\Omega^n}$ (definition 56 (p. 49)) are the product of a map $\mathbf{f}$ belonging to $\mathcal{F}$ the set of smooth map of $\mathbf{M}$ (definition 7 (p. 3)) and $\mathbf{Vol}$.

**Proof**
We call $\mathbf{U_k}$, $(\phi_{\mathbf{k}} : \mathbf{U_k} \to \mathcal{R}^{\mathbf{n}})$ the cover system of $\mathbf{M}$. We consider $\omega \in \mathbf{\Omega^n}$. For any $\mathbf{v} \in \mathcal{V}^{\mathbf{n}}$, any $p \in \mathbf{M}$, for any $\mathbf{k}$ such that $p \in \mathbf{U_k}$, we have:

$$(\exists(\mathbf{a}, \mathbf{b}) \in \mathcal{R}^2)\,(0 < \mathbf{a} < \mathbf{b}, \mathbf{C}(\phi_{\mathbf{k}}(p), \mathbf{a}) \subset \mathbf{C}(\phi_{\mathbf{k}}(p), \mathbf{b}) \subset \phi_{\mathbf{k}}(\mathbf{U_k}) \subset \mathcal{R}^{\mathbf{n}})$$

$\mathbf{C}(\mathbf{x}, \mathbf{c})$ is an open cube of $\mathcal{R}^{\mathbf{n}}$ (definition 261 (p. 285)).
Using $(\forall p \in \mathbf{U_k})(\mathbf{vol_k}(p) \neq 0)$ together with proposition 92 (p. 54) and definition 14 (p. 6), we get:

$$\omega(\mathbf{v})(p) = \omega(\mathbf{g}'_p(p', \mathbf{a}, \mathbf{b})\mathbf{v})(p) = \omega_{\mathbf{k}}(p) \bigwedge_{\mathbf{i}\in[1,\mathbf{n}]} \mathbf{dx^i}(\mathbf{g}'_p(p', \mathbf{a}, \mathbf{b})\mathbf{v})(p)$$
$$= \frac{\omega_{\mathbf{k}}(p)}{\mathbf{vol_k}(p)} \mathbf{Vol}((\mathbf{g}'_p(p', \mathbf{a}, \mathbf{b})\mathbf{v}))(p)$$

$(\mathbf{Vol}(\mathbf{g}'_p(p',\mathrm{a},\mathrm{b})\mathrm{v})(\phi_k^{-1}(\mathrm{x})) : \phi_k(\mathrm{U}_k) \subset \mathcal{R}^n \to \mathcal{R})$ is not zero and belongs to $\mathbf{C}^\infty$ (see definition 276 (p. 304)) when $\|\phi_k(p'') - \phi_k(p)\| < \mathrm{a}$. We also get:

$$\omega(\mathrm{v})(p) = \omega(\mathrm{g}'_p(p',\mathrm{a},\mathrm{b})\mathrm{v})(p) = \omega_k(p) \bigwedge_{i\in[1,n]} \mathrm{dx}^i(\mathrm{g}'_p(p',\mathrm{a},\mathrm{b})\mathrm{v})(p)$$
$$= \frac{\omega_k(p)}{\mathrm{vol}_k(p)}\mathrm{Vol}(\mathrm{g}'_p(p',\mathrm{a},\mathrm{b})\mathrm{v})(p) = \frac{\omega_k(p)}{\mathrm{vol}_k(p)}\mathrm{Vol}(\mathrm{v})(p) \ .$$

Since $\omega(\mathrm{v})$ and $\mathbf{Vol}(\mathrm{v})$ are continuous map $\mathbf{M} \to \mathcal{R}$, when coverings overlap, we have:

$$p \in \mathrm{U}_k \cap \mathrm{U}_{k'} \Rightarrow \frac{\omega_k(p)}{\mathrm{vol}_k(p)} = \frac{\omega_{k'}(p)}{\mathrm{vol}_{k'}(p)} = \frac{\omega(\mathrm{v})(p)}{\mathrm{Vol}(\mathrm{v})(p)}$$

then $\frac{\omega_k(p)}{\mathrm{vol}_k(p)}$ does not depend on $\mathbf{k}$ in the overlapping regions like $\mathrm{U}_k \cap \mathrm{U}_{k'}$. $(\mathrm{h} = \frac{\omega_k(p)}{\mathrm{vol}_k(p)} : \mathrm{M} \to \mathcal{R})$ belongs to $\mathcal{F}$.

**Definition 61 Volume form attached to a metric.**

We consider with its associated map $(\phi_k : \mathrm{U}_k \to \mathcal{R}^n)$, an open $\mathrm{U}_k$ of the covering set of a differentiable manifold $\mathbf{M}$ of dimension $\mathbf{n}$ with a metric $\mathbf{g}$ (see section 1.19 (p. 128) and definition 31 (p. 25)). Acting on its associated set of smooth maps $\mathcal{F}_k$, we consider the associated vector field set $\mathcal{V}_k$. One considers a basic set of $\mathbf{n}$ vector fields of $\mathcal{V}_k$: $(\mathrm{e}_i : [1,\mathrm{n}] \to \mathcal{V}_k)$ which are linked to the canonical basis $\partial_i$ by $\mathrm{e}_i = \mathrm{e}_i^j\partial_j$ with $(\mathrm{e}_j^i : [1,\mathrm{n}]^2 \to \mathcal{F}_k)$. It can be $(\partial_i : [1,\mathrm{n}] \to \mathcal{V}_k)$ with $\mathrm{e}_j^i = \delta_j^i$. As indicated in proposition 65 (p. 31), in each tangent frame vectorial space ${}^{[p]}\mathrm{T}$ attached to any point $p \in \mathrm{U}_k$, we have basic sets of vector linked either to the $\mathrm{e}_i$ or to the canonical $\partial_i$: ${}^{[p]}\mathrm{e}_i$, ${}^{[p]}\partial_i$. In the vector space ${}^{[p]}\mathrm{T}$ with the basis ${}^{[p]}\mathrm{e}_i$, one has, generated by $\mathbf{g}$, a local metric given by $\mathrm{g}_{ij}(p) = \mathrm{g}(\mathrm{e}_i,\mathrm{e}_j)(p)$. This metric of ${}^{[p]}\mathrm{T}$ has an orthogonal set of vectors of norm $\pm 1$ (see proposition 64 (p. 31)) generated using the eigenvector of the matrix ${}^{[p]}\mathrm{T}$. This set can also be used as a local basis (proposition 463 (p. 264)): $(\mathrm{e}'_k : [1,\mathrm{n}] \to {}^{[p]}\mathrm{T})$. When one manages to have the local volume of a box supported by the unit vectors $\mathrm{e}'_i$ to be $\mathbf{1}$, the volume definition is said to be defined by the metric. Then the volume of a box supported by the basis ${}^{[p]}\mathrm{e}_i = \eta_i^j\mathrm{e}'_j$ is from proposition 626 (p. 357) $|\det(\eta_i^j)|$. With $\mathrm{r}_k = \mathrm{g}(\mathrm{e}'_k,\mathrm{e}'_k) = \pm 1$, we have:

$$\mathrm{g}(\mathrm{e}_i,\mathrm{e}_j)(p) = \mathrm{g}(\eta_i^k\mathrm{e}'_k, \eta_j^\ell\mathrm{e}'_\ell) = \eta_i^k\eta_j^\ell\mathrm{g}(\mathrm{e}'_k,\mathrm{e}'_\ell) = \mathrm{r}_k\eta_i^k\eta_j^\ell\delta_\ell^k$$

(see footnote 20 (p. 239)). From proposition 448 (p. 257), we also have:

$$\det(\mathrm{g}(\mathrm{e}_i,\mathrm{e}_j)(p)) = (\det(\eta_i^j))^2\det(\mathrm{r}_k\delta_\ell^k) = (\det(\eta_i^j))^2 \prod_{k\in[1,n]}^{\times} \mathrm{r}_k = \pm(\det(\eta_i^j))^2 \ .$$

Let us suppose now, we have two sets of basic vector fields described in $\mathrm{U}_k$ by:$(\mathrm{f}_i : [1,\mathrm{n}] \to \mathcal{V}_k)$ with $\mathrm{f}_i = \mathrm{f}_i^j\partial_j$, $\mathrm{f}_j^i \in \mathcal{F}_k$ and $\mathrm{e}_i = \mathrm{e}_i^j\partial_j$, $\mathrm{e}_j^i \in \mathcal{F}_k$. We consider also their duals: $\mathrm{f}^i = (\mathrm{f}^{-1})_j^i\mathrm{dx}^j$, $\mathrm{e}^i = (\mathrm{e}^{-1})_j^i\mathrm{dx}^j$ with $(\mathrm{f}^{-1})_k^j\mathrm{f}_i^k = \delta_i^j$, $(\mathrm{e}^{-1})_k^j\mathrm{e}_i^k = \delta_i^j$ (see propositions 47 (p. 23), 74 (p. 36) and its equation (1.33)). We consider the volume form $\mathbf{Vol} =$

$\sqrt{|\det(g(e_i, e_j))|} \bigwedge_{i\in[1,n]} e^i$. We have the following properties:

$$\begin{aligned}
&\det(g(e_i, e_j)) = (\det(e_i^j))^2 \det(g(\partial_i, \partial_j)) \\
&\det(g(f_i, f_j)) = (\det(f_i^j))^2 \det(g(\partial_i, \partial_j)) \\
&\mathrm{Vol} = \sqrt{|\det(g(e_i, e_j))|} \bigwedge_{i\in[1,n]} e^i = \sqrt{|\det(g(e_i, e_j))|} \bigwedge_{i\in[1,n]} (e^{-1})^i_j dx^j = \\
&\sqrt{\det(e^i_j)^2 |\det(g(\partial_i, \partial_j))| \det((e^{-1})^i_j)} \bigwedge_{i\in[1,n]} dx^i = \pm\sqrt{|\det(g(\partial_i, \partial_j))|} \bigwedge_{i\in[1,n]} dx^i \\
&= \pm\sqrt{|\det(g(f_i, f_j))|} \bigwedge_{i\in[1,n]} f^i .
\end{aligned} \tag{1.69}$$

If one uses the map $(v^* : \mathcal{V} \to \mathcal{V}^*)$ as defined in proposition 76 (p. 38), we may also write:

$$\begin{aligned}
\mathrm{Vol} &= \frac{\bigwedge_{i\in[1,n]} e_i^*}{\sqrt{|\det(g(e_i, e_j))|}} = \frac{\det(g_{\nu,j} e_i^\nu) \bigwedge_{i\in[1,n]} dx^i}{\sqrt{|\det(g(e_i, e_j))|}} = \frac{\det(g_{\nu,j}) \det(e_i^\nu) \bigwedge_{i\in[1,n]} dx^i}{\sqrt{|\det(g(\partial_i, \partial_j)) \det^2(e_i^\nu)|}} \\
&= \pm\frac{\det(g_{\nu,j}) \bigwedge_{i\in[1,n]} dx^i}{\sqrt{|\det(g(\partial_i, \partial_j))|}} = \pm\frac{\bigwedge_{i\in[1,n]} \partial_i^*}{\sqrt{|\det(g(\partial_i, \partial_j))|}} = \pm\sqrt{|\det(g(\partial_i, \partial_j))|} \bigwedge_{i\in[1,n]} dx^i
\end{aligned} \tag{1.70}$$

**Vol** is the form which defines volumes associated to the metric **g** by the integral as we shall define in section 1.12.2 (p. 61). Since $\det(g(\partial_i, \partial_j))$ is different from zero and always keeping the same sign for any connected subset of $U_k$ from propositions 330 (p. 203) and 321 (p. 198), we have: $\sqrt{|\det(g(\partial_i, \partial_j))|} \in \mathcal{F}_k$.

**Definition 62 Hodge * operator.**

Given a differential manifold **M** of dimension **n** with $\mathcal{V}$ as its set of vector fields, we suppose that $\mathcal{V}$ has a metric $(g : \mathcal{V}^2 \to \mathcal{F})$ (definition 31 (p. 25)), which induces another metric $(g^* : (\mathcal{V}^*)^2 \to \mathcal{F})$ (definition 43 (p. 40)). One can also put both as $(g : \mathcal{V}^2 \cup (\mathcal{V}^*)^2 \to \mathcal{F})$ as explained in definition 43 (p. 40). The metric allows to define a metric on any tensorial product defined on **M**, see definition 48 (p. 45). We quote that metric also by **g**.
One defines the Hodge operator as a bijection between $\Omega^p, p \in [1, n]$ and $\Omega^{n-p}$. If we take one element of $v \in \Omega^p$ the Hodge * operator maps this element into an element $*v \in \Omega^{n-p}$ asking the following formula (see equation (1.69) (p. 59) for the definition of **Vol**) to be true:

$$(\forall u \in \Omega^p)(u \wedge (*v) = g(u, v)\mathrm{Vol} = g(u, v)\mathrm{vol} \bigwedge_{\ell\in[1,n]} dx^\ell) \tag{1.71}$$

**Proposition 98** Taking the notation of the above definition 62 (p. 59), there is a unique solution to equation (1.71) (p. 59).

**Proof**
We consider a cover $U \subset M$ (we drop the index **k**) with its map $(\phi : U \to \mathcal{R}^n)$, the set $\mathcal{V}$ of vector field on the restricted manifold **U**.
For any $i \in \mathcal{I}_p$ (see definition 53 (p. 48)), we call $i^c \in \mathcal{I}_{n-p}$ the map such that $i([1, p]) \cup i^c([1, n-p]) = [1, n]$ and we call $s_{i,i^c}$ the signature of the permutation defined by the above

$i(j)$ when $j \in [1,p]$ and by $i^c(j-p)$ when $j \in [p+1,n]$. In U, we set $Vol = vol \bigwedge_{\ell\in[1,n]} dx^\ell$. We set:

$$v = \sum_{i\in\mathcal{I}_p} v^i \bigwedge_{\ell\in[1,p]} dx^{i(\ell)} = \sum_{i\in\mathcal{I}_p} v^i \sum_{P_v\in PG_p} s(P_v) \bigotimes_{\ell\in[1,p]} dx^{i(P_v(\ell))}$$
$$u = \sum_{j\in\mathcal{I}_p} u^j \bigwedge_{\ell\in[1,p]} dx^{j(\ell)} = \sum_{j\in\mathcal{I}_p} u^j \sum_{P_u\in PG_p} s(P_u) \bigotimes_{\ell\in[1,p]} dx^{j(P_u(\ell))} .$$

With that, we have:

$$g(u,v) = \sum_{i\in\mathcal{I}_p}\sum_{j\in\mathcal{I}_p} v^i u^j \sum_{P_v\in PG_p}\sum_{P_u\in PG_p} s(P_v)s(P_u)g(\bigotimes_{\ell\in[1,p]} dx^{i(P_v(\ell))}, \bigotimes_{\ell\in[1,p]} dx^{j(P_u(\ell))})$$

$$g(u,v) = \sum_{i\in\mathcal{I}_p}\sum_{j\in\mathcal{I}_p} v^i u^j \sum_{P_v\in PG_p}\sum_{P_u\in PG_p} s(P_v)s(P_u) \prod_{\ell\in[1,p]}^{\times} g(dx^{i(P_v(\ell))}, dx^{j(P_u(\ell))})$$
$$= \sum_{i\in\mathcal{I}_p}\sum_{j\in\mathcal{I}_p} v^i u^j \sum_{P_v\in PG_p}\sum_{P_u\in PG_p} s(P_v P_u^{-1}) \prod_{\ell\in[1,p]}^{\times} g(dx^{i(P_v(P_u^{-1}(\ell)))}, dx^{j(\ell)})$$
$$= \sum_{i\in\mathcal{I}_p}\sum_{j\in\mathcal{I}_p} v^i u^j \sum_{P_v\in PG_p}\sum_{P\in PG_p} s(P) \prod_{\ell\in[1,p]}^{\times} g(dx^{i(P(\ell))}, dx^{j(\ell)})$$
$$= \sum_{i\in\mathcal{I}_p}\sum_{j\in\mathcal{I}_p} v^i u^j p! \sum_{P\in PG_p} s(P) \prod_{\ell\in[1,p]}^{\times} g(dx^{i(P(\ell))}, dx^{j(\ell)})$$
$$u \wedge (*v) = \sum_{j\in\mathcal{I}_p} u^j (*v)^{j_c} s(j,j_c) \bigwedge_{\ell\in[1,n]} dx^{j(\ell)}$$
$$= g(u,v)Vol = vol \bigwedge_{\ell\in[1,n]} dx^\ell \sum_{i\in\mathcal{I}_p}\sum_{j\in\mathcal{I}_p} v^i u^j p! \sum_{P\in PG_p} s(P) \prod_{\ell\in[1,p]}^{\times} g(dx^{i(P(\ell))}, dx^{j(\ell)}) .$$

We used the fact that:

$$P \in PG_p \Rightarrow (\prod_{\ell\in[1,p]} a_\ell = \prod_{\ell\in[1,p]} a_{P(\ell)}, \sum_{\ell\in[1,p]} a_\ell = \sum_{\ell\in[1,p]} a_{P(\ell)})$$
$$P \in PG_p \Rightarrow PG_p = P(PG_p)$$
$$Card(PG_p) = p!$$
$$(P \in PG_p, P' \in PG_p) \Rightarrow (s(P)s(P') = s(P \circ P') = s(P \circ P'^{-1})) .$$

Because $\sum$ and $\prod$ are invariant under a permutation: see propositions 996 (p. 555), 997 (p. 556) and section 2.2.3 (p. 173) and because since it is a group, the set of permutation is invariant by translation. The last statement comes from the definition of the signature of a permutation 175 (p. 208). Then we have one solution to equation (1.71) (p. 59):

$$(\forall j \in \mathcal{I}_{n-p})((*v)^j = vol \times s(j^c,j) \sum_{i\in\mathcal{I}_p} v^i p! \sum_{P\in PG_p} s(P) \prod_{\ell\in[1,p]}^{\times} g(dx^{i(P(\ell))}, dx^{j^c(\ell)})) \quad (1.72)$$

$$\mathbf{u} \wedge (*\mathbf{v}) = \sum_{j \in \mathcal{I}_p} u^j \bigwedge_{\ell \in [1,p]} dx^{j(\ell)} \wedge \mathrm{vol} \sum_{j \in \mathcal{I}_p} s_{j,j^c} \bigwedge_{\ell \in [1,n-p]} dx^{j^c(\ell)} \sum_{i \in \mathcal{I}_p} v^i p! \sum_{P \in PG_p} s(P) \prod_{\ell \in [1,p]}^{\times} g(dx^{i(P(\ell))}, dx^{(j(\ell))})$$

$$\mathbf{u} \wedge (*\mathbf{v}) = \mathrm{Vol} \sum_{j \in \mathcal{I}_p} u^j \sum_{i \in \mathcal{I}_p} v^i p! \sum_{P \in PG_p} s(P) \prod_{\ell \in [1,p]}^{\times} g(dx^{i(P(\ell))}, dx^{j(\ell)}) = g(\mathbf{u}, \mathbf{v}) \mathrm{Vol}\,.$$

**This solution is unique.**

Let us suppose that equation (1.71) (p. 59) has two solution one $*\mathbf{v}$ given by equation (1.72) (p. 60), the other one being $*\mathbf{v}' \neq *\mathbf{v}$ the $_p$**linear_form** $\mathbf{w} = *\mathbf{v}' - *\mathbf{v} \neq 0$ is such that:

$$(\forall \mathbf{u} \in \Omega^p)\,(\mathbf{u} \wedge (*\mathbf{v} - *\mathbf{v}') = g(\mathbf{u}, \mathbf{v} - \mathbf{v})\mathrm{Vol} = g(\mathbf{u}, \mathbf{v} - \mathbf{v})\mathrm{vol} = 0)\,.$$

Since $\mathbf{w}$ is not a null $_p$**linear_form**, we have $p \in U, i \in \mathcal{I}_{n-p}, 0 < a < b$ with $C(\phi(p), b) \subset U$ such as (see definition 14 (p. 6)):

$$(\phi_k^{-1}(x) \in U_k \Rightarrow \mathbf{w}(g'_p(q, a, b) \bigotimes_{\ell \in [1,n-p]} \partial_{i(\ell)})(p) \neq 0,$$

$$(\forall \mathbf{u} \in \Omega^p)((\mathbf{u} \wedge \mathbf{w})(g''^2_p(q, a, b) \bigotimes_{\ell \in [1,n]} \partial_\ell)(p) = 0))\,.$$

Taking $\mathbf{u} = g'_p(q, a, b) \bigwedge_{\ell \in [1,p]} dx^{i^c(\ell)}$, it gives:

$$\begin{aligned}
&(\mathbf{u} \wedge \mathbf{w})(g'_p(q, a, b) \bigotimes_{\ell \in [1,n]} \partial_\ell)(p) \\
&= s_{i^c,i}(\mathbf{u} \wedge \mathbf{w})(g'^2_p(q, a, b) \bigotimes_{\ell \in [1,p]} \partial_{i_c(\ell)} \otimes \bigotimes_{\ell \in [1,n-p]} \partial_{i(\ell)})(p) \\
&= s_{i^c,i}\mathbf{u}(g'_p(q, a, b) \bigotimes_{\ell \in [1,p]} \partial_{i_c(\ell)})(p) \times \mathbf{w}(g'_p(q, a, b) \bigotimes_{\ell \in [1,n-p]} \partial_{i(\ell)})(p) \\
&= s_{i^c,i}\mathbf{w}(g'_p(q, a, b) \bigotimes_{\ell \in [1,n-p]} \partial_{i(\ell)})(p) \neq 0
\end{aligned}$$

see definitions 45 (p. 41), 52 (p. 47) and 58 (p. 49). Then from A.1.4 (p. 435), 8, we have $*\mathbf{v} = *\mathbf{v}'$.

## 1.12.2 Oriented integral

**Definition 63 Oriented Manifold.**

A differentiable manifold M is said to be oriented if there exists an oriented volume form (definition 60 (p. 54)) defined on M. We shall assume also that for any element of the covering set of M: U, the associated map $(\phi : U \to \mathcal{R}^n)$ is such that for any $x \in \mathcal{R}^n$ and for any $r \in \mathcal{R}$, $\phi(U) \cap B(x, r)$ has a volume. (This condition may be dropped if we consider the Lebesgue integral instead of the Riemann integral (see definition 306 (p. 382)). But, in that case, some continuous maps cannot be integrated. )

**Definition 64 Two local bases with the same orientation.**

In an oriented manifold (definition 1.12.2 (p. 61)), we consider two local bases $(\partial_\ell : [1,n] \to \mathcal{V}_u)$, $(\partial'_\ell : [1,n] \to \mathcal{V}_{u'})$ of two covering subsets U, U′ which are connected. Those local bases are said to have the same orientation relative to the volume form $\mathbf{Vol} \in (\wedge \mathcal{V}^*)^n$ if:

$$(\forall p \in U)(\forall p' \in U')\,(\mathbf{Vol}(\bigotimes_{\ell\in[1,n]} \partial_\ell)(p) \times \mathbf{Vol}(\bigotimes_{\ell\in[1,n]} \partial'_\ell)(p') > 0)\;.$$

**Proposition 99** All the local bases of a connected part of an oriented differential manifold can be oriented (see definition 64 (p. 61)) by an oriented volume form.

**Proof**

We have an oriented differential manifold **M** with an oriented volume form **Vol** with a set of covers $(U_k : K = [1, \mathrm{Card}(K)] \to \mathcal{P}(M))$ with its set of charts $(\phi_k = U_k \to \mathcal{R}^n)$ any transposition $T_{i,j}$ of 2 elements of $[1,n]$ defines also a map $(T_{i,j} : \mathcal{R}^n \to \mathcal{R}^n)$ (see definition 126 (p. 144)):

$$(\forall x \in \mathcal{R}^n)\,(x = \prod_{\ell\in[1,n]}^{e} x^\ell \Rightarrow T_{ij}(x) = \prod_{\ell\in[1,n]}^{e} x^{T_{ij}(\ell)})$$

with the following definition of $T_{ij}$:

$$(\forall \ell \in [1,n])\,(\ell \notin \{i,j\} \Rightarrow T_{ij}(\ell) = \ell), T_{ij}(i) = j, T_{ij}(j) = i$$

$T_{ij}$ generates a new chart $(T_{ij} \circ \phi_k : U_k \to \mathcal{R}^n)$. For each $k \in K$, we have a new basis $(\partial^T_\ell : [1,n] \to \mathcal{V}_k)$ corresponding to $(T_{ij} \circ \phi_k : U_k \to \mathcal{R}^n)$. This basis and the basis $(\partial_\ell : [1,n] \to \mathcal{V}_k)$, corresponding to the map $(\phi_k : U_k \to \mathcal{R}^n)$, are linked by:

$$(\forall \ell \in [1,n])\,(\partial^T_\ell = \partial_{T_{ij}(\ell)})\;.$$

Since $\mathbf{Vol} \in (\bigwedge \mathcal{V}^*)^n$

$$(\forall p \in U_k)\,(\mathbf{Vol}(\bigotimes_{\ell\in[1,n]} \partial^T_\ell)(p)) = -\mathbf{Vol}(\bigotimes_{\ell\in[1,n]} \partial_\ell)(p))\;.$$

Since $\mathbf{Vol}(\bigotimes_{\ell\in[1,n]} \partial_\ell))$ has the same sign for any $p$ in any connected part of $U_k$, one can choose to apply or not a $T_{i,j}$ in any $U_k$ in order to have a basis $(\partial'_\ell : [1,n] \to \mathcal{V}_k$ such that $\mathbf{Vol}(\bigotimes_{\ell\in[1,n]} \partial'_\ell))$ has the same sign in any connected part of **M**.

**Proposition 100** We have an oriented differentiable manifold **M** with an oriented volume form **Vol**. In one of the covers **U** of **M** for which we have a map $(\phi : U \to \mathcal{R}^n)$, we choose a subset **V** of **U** such that $\phi(V)$ is in a cube of $\mathcal{R}^n$ and has a volume in $\mathcal{R}^n$ (see definition 295 (p. 342)). From proposition 266 (p. 179) it is always possible since $\phi(U)$ is open. We call $\Phi_V$ the characteristic map of $V \subset U$ (definition 293 (p. 341)). Given any continuous map $(f : U \to \mathcal{R})$, the map:

$$(\Phi_V f \mathbf{Vol}(\bigotimes_{\mu\in[1,n]} \partial^\phi_\mu) \circ \phi^{-1} : \mathcal{R}^n \to \mathcal{R})$$

has a Riemann integral ${}^{\mathrm{Vol}}J^\phi_V$ (see definition 290 (p. 332), propositions 241 (p. 167) and 620 (p. 352)). Given another map $(\psi : \phi(U) \to \mathcal{R}^n)$ of $C^\infty$ (definitions 278 (p. 306), 276 (p. 304)) which defines a coordinate change on **U** (definition 30 (p. 23)). It allows to define another map:

$$(\Phi_V f \mathbf{Vol}(\bigotimes_{\mu\in[1,n]} \partial^{\psi\phi}_\mu) \circ \phi^{-1} \circ \psi^{-1} : \mathcal{R}^n \to \mathcal{R})\;.$$

It defines another Riemann integral ${}^{\mathrm{Vol}}J^{\psi\phi}_V$. Those two integrals are equal within a sign.

**Proof**
With the above notation, we have using proposition 47 (p. 23):

$$(\forall p \in \mathbf{U})\ ((\mathbf{x}^\phi = \phi(p), \mathbf{x}^{\psi\phi} = \psi(\phi(p))) \Rightarrow \mathbf{Vol}(\bigotimes_{\mu\in[1,n]} \frac{\partial}{\partial(\mathbf{x}^\phi)^\mu})(p)$$

$$= \mathbf{Vol}(\bigotimes_{\mu\in[1,n]} \frac{\partial(\mathbf{x}^{\psi\phi})^\nu}{\partial(\mathbf{x}^\phi)^\mu}\frac{\partial}{\partial(\mathbf{x}^{\psi\phi})^\nu})(p)$$

$$= \det(\frac{\partial(\mathbf{x}^{\psi\phi})^\nu}{\partial(\mathbf{x}^\phi)^\mu}(\phi(p)))\mathbf{Vol}(\bigotimes_{\mu\in[1,n]} \frac{\partial}{\partial(\mathbf{x}^{\psi\phi})^\nu})(p))$$

then the integral using $\phi$ is

$${}^{\mathbf{Vol}}\mathbf{J}^\phi_{\mathbf{V}} = \mathbf{J}(\mathbf{f}\Phi_{\mathbf{V}}(\phi^{-1}(\mathbf{x}^\phi))\mathbf{Vol}(\bigotimes_{\mu\in[1,n]} \frac{\partial}{\partial(\mathbf{x}^\phi)^\mu})(\phi^{-1}(\mathbf{x}^\phi)))$$

$$= \mathbf{J}(\mathbf{f}\Phi_{\mathbf{V}}(\phi^{-1}(\mathbf{x}^\phi))\det(\frac{\partial(\mathbf{x}^{\psi\phi})^\nu}{\partial(\mathbf{x}^\phi)^\mu})(\phi^{-1}(\mathbf{x}^\phi))\mathbf{Vol}(\bigotimes_{\mu\in[1,n]} \frac{\partial}{\partial(\mathbf{x}^{\psi\phi})^\nu})(\phi^{-1}(\mathbf{x}^\phi)))$$

using $\psi\phi$

$${}^{\mathbf{Vol}}\mathbf{J}^{\psi\phi}_{\mathbf{V}} = \mathbf{J}(\mathbf{f}\Phi_{\mathbf{V}}(\phi^{-1}\circ\psi^{-1}(\mathbf{x}^{\psi\phi}))\mathbf{Vol}(\bigotimes_{\mu\in[1,n]} \frac{\partial}{\partial(\mathbf{x}^{\psi\phi})^\mu})(\phi^{-1}\circ\psi^{-1}(\mathbf{x}^{\psi\phi}))) .$$

We have $\mathbf{x}^{\psi\phi} = \psi(\mathbf{x}^\phi)$. Using proposition 629 (p. 360) and the fact that in $\mathbf{U}$ neither $\mathbf{Vol}(\bigotimes_{\mu\in[1,n]}\partial_\mu)$ nor $\det(\frac{\partial(\mathbf{x}^{\psi\phi})^\nu}{\partial(\mathbf{x}^\phi)^\mu})$ can get to zero, we can evaluate ${}^{\mathbf{Vol}}\mathbf{J}^{\psi\phi}_{\mathbf{V}}$ in $\mathbf{x}^\phi$ coordinates with the $\mathbf{g}$ of equation (2.180) (p. 360) set to $\psi^{-1}$:

$${}^{\mathbf{Vol}}\mathbf{J}^{\psi\phi}_{\mathbf{V}} = \mathbf{J}(\mathbf{f}\Phi_{\mathbf{V}}(\phi^{-1}\circ\psi^{-1}(\psi(\mathbf{x}^\phi)))\mathbf{Vol}(\bigotimes_{\mu\in[1,n]} \frac{\partial}{\partial(\mathbf{x}^{\psi\phi})^\mu}))$$

$$(\phi^{-1}\circ\psi^{-1}(\psi(\mathbf{x}^\phi)))|\det(\mathrm{jac}(\psi))(\mathbf{x}^\phi)|)$$

$$= \mathbf{J}(\mathbf{f}\Phi_{\mathbf{V}}(\phi^{-1}(\mathbf{x}^\phi))\mathbf{Vol}(\bigotimes_{\mu\in[1,n]} \frac{\partial}{\partial(\mathbf{x}^\phi)^\mu})(\phi^{-1}(\mathbf{x}^\phi))|\det(\frac{\partial(\mathbf{x}^{\psi\phi})^\nu}{\partial(\mathbf{x}^\phi)^\mu}(\mathbf{x}^\phi))|)$$

$$= \mathbf{J}(\mathbf{f}\Phi_{\mathbf{V}}(\phi^{-1}(\mathbf{x}^\phi))\frac{\mathbf{Vol}(\bigotimes_{\mu\in[1,n]}\frac{\partial}{\partial(\mathbf{x}^\phi)^\mu})(\phi^{-1}(\mathbf{x}^\phi))}{\det(\frac{\partial(\mathbf{x}^{\psi\phi})^\nu}{\partial(\mathbf{x}^\phi)^\mu}(\mathbf{x}^\phi))}|\det(\frac{\partial(\mathbf{x}^{\psi\phi})^\nu}{\partial(\mathbf{x}^\phi)^\mu}(\mathbf{x}^\phi))|)$$

$$= \mathbf{J}(\mathbf{f}\Phi_{\mathbf{V}}(\phi^{-1}(\mathbf{x}^\phi))\mathbf{Vol}(\bigotimes_{\mu\in[1,n]} \frac{\partial}{\partial(\mathbf{x}^\phi)^\mu})(\phi^{-1}(\mathbf{x}^\phi))\frac{|\det(\frac{\partial(\mathbf{x}^{\psi\phi})^\nu}{\partial(\mathbf{x}^\phi)^\mu}(\mathbf{x}^\phi))|}{\det(\frac{\partial(\mathbf{x}^{\psi\phi})^\nu}{\partial(\mathbf{x}^\phi)^\mu}(\mathbf{x}^\phi))})$$

$$= {}^{\mathbf{Vol}}\mathbf{J}^\phi_{\mathbf{V}} \times \epsilon .$$

With $\epsilon = \pm 1$ because (see definition 154 (p. 181)) $\mathbf{U}$ being connected, the determinant $\det(\frac{\partial(\mathbf{x}^{\psi\phi})^\nu}{\partial(\mathbf{x}^\phi)^\mu}(\mathbf{x}^\phi))$ cannot change sign within $\mathbf{U}$. Both integrals, ${}^{\mathbf{Vol}}\mathbf{J}^\phi_{\mathbf{V}}$ and ${}^{\mathbf{Vol}}\mathbf{J}^{\psi\phi}_{\mathbf{V}}$, are equal within a sign.

**Proposition 101** With the same premises as proposition 100 (p. 62), we suppose that the bases defined in $\mathbf{U}$ by $\phi$ and $\psi\phi$ have the same orientation according to the volume form $\mathbf{Vol}$ (definition 64 (p. 61)) then the integrals of $\mathbf{f}$: ${}^{\mathbf{Vol}}\mathbf{J}^\phi_{\mathbf{V}}$ and ${}^{\mathbf{Vol}}\mathbf{J}^{\psi\phi}_{\mathbf{V}}$ are equal.

**Proof**
By definition:

$$(\forall p \in \mathrm{U})\,(\mathrm{Vol}(\bigotimes_{\mu\in[1,n]} \partial_\mu^{\psi\phi})(p) = \det(\partial_j\psi^i)(p)\mathrm{Vol}(\bigotimes_{\mu\in[1,n]} \partial_\mu^{\phi})(p))$$

and $\mathrm{Vol}(\bigotimes_{\mu\in[1,n]} \partial_\mu^{\psi\phi})(p)$ has the same sign as $\mathrm{Vol}(\bigotimes_{\mu\in[1,n]} \partial_\mu^{\phi})(p)$ (definition 64 (p. 61)). Then $\det(\partial_j\psi^i)(p) > 0$ and from proposition 100 (p. 62), ${}^{\mathrm{Vol}}\mathrm{J}_{\mathrm{V}}^{\phi} = {}^{\mathrm{Vol}}\mathrm{J}_{\mathrm{V}}^{\psi\phi}$.

**Proposition 102** Given an oriented differentiable manifold $\mathbf{M}$ with its covering sets $(\mathrm{U}_k : \mathrm{K} = [1, \mathrm{Card}(\mathrm{K})] \to \mathcal{P}(\mathrm{M}))$ with their covering chart $\phi_k$. We suppose that we have a connected subset (definition 285 (p. 327) of $\mathbf{M}$ where the $\phi_k$ are oriented by the volume form $\mathbf{Vol}$. We suppose that, $\forall \mathrm{K}_s \subset \mathrm{K}$, the set $\mathrm{V} \cap \bigcap_{k\in \mathrm{K}_s} \mathrm{U}_k \cap \mathcal{C}(\bigcup_{k\in\mathcal{C}(\mathrm{K}_s)} \mathrm{U}_k)$ (see definition 353 (p. 454)) has a volume. Then for any continuous map $(\mathrm{f} : \mathrm{M} \to \mathcal{R})$ an integral ${}^{\mathrm{Vol}}\mathrm{J}_{\mathrm{V}}(\mathrm{f})$ can be defined:

$$ {}^{\mathrm{Vol}}\mathrm{J}_{\mathrm{V}}(\mathrm{f}) = \sum_{\mathrm{K}_s\in\mathcal{P}(\mathrm{K})} \mathrm{J}_{\mathrm{V}\cap\bigcap_{k\in\mathrm{K}_s}\mathrm{U}_k\cap\mathcal{C}(\bigcup_{k\in\mathcal{C}(\mathrm{K}_s)}\mathrm{U}_k)}(\mathrm{f}\mathrm{Vol}(\bigotimes_{\mu\in[1,n]} \partial_\mu^{\phi_{k'(\mathrm{K}_s)}}))\,, \tag{1.73}$$

where $(\mathrm{k}'(\mathrm{K}_s) : \mathcal{P}(\mathrm{K}) \to \mathrm{K})$ is a map such that: $\mathrm{k}'(\mathrm{K}_s) \in \mathrm{K}_s$. ${}^{\mathrm{Vol}}\mathrm{J}_{\mathrm{V}}(\mathrm{f})$ remains the same when any change of coordinates (definition 30 (p. 23)) are applied to the charts $\phi_k$ and does not depend on the map $\mathrm{k}'(\mathrm{K}_s)$ used.

**Proof**

$$\mathrm{M} = \bigcup_{k\in\mathrm{K}} \mathrm{U}_k = \bigcup_{\mathrm{K}_s\in\mathcal{P}(\mathrm{K})} (\bigcap_{k\in\mathrm{K}_s} \mathrm{U}_k \cap \mathcal{C}(\bigcup_{k\in\mathcal{C}(\mathrm{K}_s)} \mathrm{U}_k)) \tag{1.74}$$

because:

$$(\forall p \in \mathrm{M})(\exists \mathrm{K}_s \subset \mathrm{K})(k \in \mathrm{K}_s \Leftrightarrow p \in \mathrm{U}_k)$$

according to definition 112 (p. 131) point 2, it defines a map $(\mathrm{K}_s(p) : \mathrm{M} \to \mathcal{P}(\mathrm{K}) \cap \{\emptyset\})$. Since (section A.1.4 -7 (p. 435))

$$p \in \mathrm{U}_k \Rightarrow k \in \mathrm{K}_s(p) \Leftrightarrow k \notin \mathrm{K}_s(p) \Rightarrow p \notin \mathrm{U}_k$$

and $p \in \mathcal{C}(\bigcup_{k\in\mathcal{C}(\mathrm{K}_s(p))} \mathrm{U}_k)$ and

$$p \in \bigcap_{k\in\mathrm{K}_s(p)} \mathrm{U}_k \cap \mathcal{C}(\bigcup_{k\in\mathcal{C}(\mathrm{K}_s(p))} \mathrm{U}_k)$$

which gives relation (1.74) (p. 64). But we also have:

$$\mathrm{K}_s \neq \mathrm{K}'_s \Rightarrow (\exists k)(k \in \mathrm{K}_s, k \notin \mathrm{K}'_s) \vee (\exists k)(k \in \mathrm{K}'_s, k \notin \mathrm{K}_s)$$

let us suppose $(\exists k)(k \in \mathrm{K}_s, k \notin \mathrm{K}'_s)$, then:

$$\bigcap_{k\in\mathrm{K}_s} \mathrm{U}_k \cap \mathcal{C}(\bigcup_{k\in\mathcal{C}(\mathrm{K}_s)} \mathrm{U}_k) \cap \bigcap_{k\in\mathrm{K}'_s} \mathrm{U}_k \cap \mathcal{C}(\bigcup_{k\in\mathcal{C}(\mathrm{K}'_s)} \mathrm{U}_k) = \emptyset$$

because we have:

$$k_1 \in \mathrm{K}_s \Rightarrow \bigcap_{k\in\mathrm{K}_s} \mathrm{U}_k \cap \mathcal{C}(\bigcup_{k\in\mathcal{C}(\mathrm{K}_s)} \mathrm{U}_k) \subset \mathrm{U}_{k_1}$$

and because

$$\mathcal{C}(\bigcup_{k\in\mathcal{C}(K_s)} U_k) = \bigcap_{k\in\mathcal{C}(K'_s)} \mathcal{C}(U_k)$$

we also have

$$k_1 \in \mathcal{C}(K'_s) \Rightarrow \bigcap_{k\in K'_s} U_k \cap \mathcal{C}(\bigcup_{k\in\mathcal{C}(K'_s)} U_k) \subset \mathcal{C}(U_{k_1})$$

then

$$\bigcap_{k\in K_s} U_k \cap \mathcal{C}(\bigcup_{k\in\mathcal{C}(K_s)} U_k) \cap \bigcap_{k\in K'_s} U_k \cap \mathcal{C}(\bigcup_{k\in\mathcal{C}(K'_s)} U_k) \subset U_{k_1} \cap \mathcal{C}(U_{k_1}) = \emptyset$$

One can set:

$${}^{Vol}J_V(f) = \sum_{K_s\in\mathcal{P}(K)} J_{V\cap\bigcap_{k\in K_s} U_k\cap\mathcal{C}(\bigcup_{k\in\mathcal{C}(K_s)} U_k)}(fVol(\bigotimes_{\mu\in[1,n]} \partial_\mu^{\phi_{k'(K_s)}}))$$

where $(k' : \mathcal{P}(K) \to K)$ is a map defined by $k'(K_s) = \tau_k(k \in K_s)$ (see definition 112 (p. 131), 2) which exists if $K_s \neq \emptyset$ when $K_s = \emptyset$, we have $\bigcap_{k\in K_s} U_k \cap \mathcal{C}(\bigcup_{k\in\mathcal{C}(K_s)} U_k) = \emptyset$ and $k'(K_s)$ does need to be defined.
From proposition 101 (p. 63), ${}^{Vol}J_V(f)$ remains the same when any change of coordinates (definition 30 (p. 23)) are applied to the charts $\phi_k$ and does not depend on the map $k'(K_s)$ used.

**Definition 65 Oriented integral of an element of $\mathcal{F}$.**

Given an oriented manifold **M** with its volume form **Vol**, with the notation of proposition 102 (p. 64), ${}^{Vol}J_V(f)$ is called the oriented integral of **f** on the volume **V** with its volume form **Vol** (definitions 60 (p. 54), 63 (p. 61)).

**Definition 66 Oriented integral of an element of $\Omega^n$ (see definition 56 (p. 49)).**

Given an oriented manifold **M** with a volume form **Vol** and using the notations of definition 65 (p. 65), any element of $\omega \in \Omega^n$ can be written (proposition 97 (p. 57)) as $\omega = fVol$ with $f \in \mathcal{F}$. The integral of $\omega$ on a volume **V** is the integral of **f** as given by the definition 65 (p. 65). Setting in **U**, $(\phi : U \to \mathcal{R}^n)$ one of the covering sets of **M**: $\omega = h(x) \bigwedge_{i\in[1,n]} dx^i$, $h \in \mathcal{F}_U$, one uses to write if $V \subset U$ although it is not fully correct:

$$\int_V h(x) \bigwedge_{i\in[1,n]} dx^i = \int_V \omega = {}^{Vol}J_{\phi(V)}(f) = J_{\phi(V)}(f(\phi_k^{-1}(x))Vol(\bigotimes \frac{\partial}{\partial x^\mu}))$$
$$= J_V(fVol) \ .$$

**Definition 67 Oriented surface integral.**

We suppose that we have a differentiable manifold **S** of dimension $n-1$ embedded in a differentiable manifold **M** of dimension **n**: $S \subset M$, a map $f \in \mathcal{F}(M)$: $(f : M \to \mathcal{R})$ and a volume form **Vol** defined on **S**. If it exists, the oriented Riemann integral of **f** with the volume form **Vol** on **S** is called the surface integral of **f** on **S**.

**Definition 68 Smooth surface.**

We have a differential manifold $\mathbf{M}$ of dimension $\mathbf{n}$, with its covering open subsets $(\mathbf{U_k} : \mathbf{K} \to \mathcal{P}(\mathbf{M}))$ with their injections $(\phi_\mathbf{k} : \mathbf{U_k} \to \mathcal{R}^\mathbf{n})$. In the differential manifold $\mathbf{M}$, we have a subset $\mathbf{S} \subset \mathbf{M}$ which is a differential manifold of dimension $\mathbf{n-1}$. We suppose that $\mathbf{S}$ has the induced topology of $\mathbf{M}$ (definition 133 (p. 151)). We label the covering opens of $\mathbf{S}$ by $(\mathbf{U^s_k} : [\mathbf{1}, \mathbf{K^s}] \to \mathcal{P}(\mathbf{S}))$ and their injections by $(\phi^\mathbf{s}_\mathbf{k} : \mathbf{U^s_k} \to \mathcal{R}^{\mathbf{n-1}})$. The set $\mathbf{S}$ is said to be a smooth surface if (see definition 276 (p. 304)):

$$(\forall \mathbf{k} \in \mathbf{K})(\forall \mathbf{k^s} \in [\mathbf{1}, \mathbf{K^s}])\,((\phi_\mathbf{k} \circ {\phi^\mathbf{s}_{\mathbf{k^s}}}^{-1} : \phi^\mathbf{s}_{\mathbf{k^s}}(\mathbf{U^s_{k^s}} \cap \mathbf{U_k}) \to \mathcal{R}^\mathbf{n}) \in \mathbf{C}^\infty)\;.$$

(From definition 133 (p. 151), $\mathbf{U^s_{k^s}} \cap \mathbf{U_k}$ is an open of $\mathbf{S}$. )

**Proposition 103** Given a smooth surface $\mathbf{S}$ (definition 68 (p. 65)) of an oriented differential manifold $\mathbf{M}$ of dimension $\mathbf{n}$ (definition 68 (p. 65)). $(\mathbf{U^s_k} : [\mathbf{1}, \mathbf{K^s}] \to \mathcal{P}(\mathbf{S}))$, $(\phi^\mathbf{s}_\mathbf{k} : \mathbf{U^s_k} \to \mathcal{R}^{\mathbf{n-1}})$ are the covering opens and injections of $\mathbf{S}$. $(\mathbf{U_k} : [\mathbf{1}, \mathbf{K}] \to \mathcal{P}(\mathbf{M}))$, $(\phi_\mathbf{k} : \mathbf{U_k} \to \mathcal{R}^\mathbf{n})$ are those for $\mathbf{M}$. We call $\mathcal{V}^\mathbf{s}$, $\mathcal{F}^\mathbf{s}$, $\mathcal{V}$, $\mathcal{F}$ the associated sets of vector fields and of smooth maps (definitions 1.3.1 (p. 8), 6 (p. 3)). We consider two covering sets $\mathbf{U^s_{k^s}}$ and $\mathbf{U_k}$ of respectively $\mathbf{S}$ and $\mathbf{M}$ such that $\mathbf{U^s_{k^s}} \cap \mathbf{U_k} \neq \emptyset$. The action of a vector field $\mathbf{v^s} \in \mathcal{V}^\mathbf{s}$ on the restriction to $\mathbf{S}$ of a smooth map $\mathbf{f} \in \mathcal{F}$ is given in $\mathbf{U^s_{k^s}} \cap \mathbf{U_k}$ by:

$$\begin{aligned}
&(\forall p \in \mathbf{U^s_{k^s}} \cap \mathbf{U_k})\;(\mathbf{v^s}(\mathbf{f})(p) = \mathbf{v^s}(\mathbf{f}) \circ \phi^{-1} \circ \phi \circ (\phi^\mathbf{s})^{-1} \circ \phi^\mathbf{s}(p)\\
&= \mathbf{v^s}(\mathbf{f})(\phi^{-1} \circ \mathbf{x}(\mathbf{x^s})) = \mathbf{v}^{\mathbf{s}\nu}\frac{\partial}{\partial {\mathbf{x_s}}^\nu}\mathbf{f} \circ \phi^{-1} \circ \mathbf{x}(\mathbf{x^s}) = \mathbf{v}^{\mathbf{s}\nu}\frac{\partial \mathbf{x}^\mu}{\partial {\mathbf{x_s}}^\nu}\frac{\partial}{\partial \mathbf{x}^\mu}\mathbf{f}(\phi(p)))
\end{aligned}$$

where $\mathbf{x} = \phi(p), \mathbf{x_s} = \phi^\mathbf{s}(p), \mathbf{x}(\mathbf{x_s}) = \phi \circ (\phi^\mathbf{s})^{-1}(\mathbf{x_s})$, $\mathbf{v}^{\mathbf{s}\nu}$ are the local component of $\mathbf{v^s}$ in the covering set $\mathbf{U^s_{k^s}}$ (definition 25 (p. 17)).

**Proof**

From definition 133 (p. 151), $\mathbf{U^s_{k^s}} \cap \mathbf{U_k}$ is an open of $\mathbf{S}$, then proposition 103 (p. 66) is a direct consequence of proposition 563 (p. 311).

**Proposition 104** With the same premise as in proposition 103 (p. 66), calling $\mathcal{V}^*_\mathbf{M}$ the set of differential (definition 40 (p. 34)) of $\mathbf{M}$ and $\mathcal{V}^*_\mathbf{S}$ for $\mathbf{S}$, there is a map $(\mathbf{V} : \mathcal{V}^*_\mathbf{M} \to \mathcal{V}^*_\mathbf{S})$ defined in any open $\mathbf{U^s_{k^s}} \cap \mathbf{U_k}$ of $\mathbf{S}$:

$$\begin{aligned}
&(\omega = \omega_\mathbf{i}\mathbf{dx^i} \in \mathcal{V}^*_\mathbf{M}, p \in \mathbf{U^s_{k^s}} \cap \mathbf{U_k}) \Rightarrow (\forall \mathbf{v} \in \mathcal{V}_\mathbf{S})(\mathbf{V}(\omega)(\mathbf{v})(p)\\
&= \omega_\mathbf{i}\frac{\partial \mathbf{x^i}}{\partial \mathbf{x^j_s}}\mathbf{dx^j_s}(\mathbf{v})(p))\;.
\end{aligned}$$

**Proof**

Since $\frac{\partial \mathbf{x^i}}{\partial \mathbf{x^j_s}}$ belongs to $\mathbf{C}^\infty$ and it is smoothed in any transition region this map $\mathbf{V}$ generates from a differential on $\mathbf{M}$, a differential on $\mathbf{S}$.

**Convention 7** Notation for a differential on a surface.

Keeping notations of proposition 104 (p. 66), we shall use, when it is clear enough, the same name for the differential defined on $\mathbf{M}$ and its image by $\mathbf{V}$ on the surface $\mathbf{S}$.

**Definition 69 Integral of an $\mathbf{n-1}$ differential form on a smooth surface.**

Given a differential manifold $\mathbf{M}$ (definition 60 (p. 54)) and a smooth oriented surface $\mathbf{S} \subset \mathbf{M}$ (definitions 68 (p. 65), 63 (p. 61)), given an $\mathbf{n-1}$ differential form $\omega$, using the map defined in proposition 104 (p. 66) and the local coordinates in any covering element $\mathbf{U^s_{k^s}} \cap \mathbf{U_k}$ of $\mathbf{S}$, $\omega$ can be considered as an $\mathbf{n-1}$ differential form on $\mathbf{S}$ and then we may use definition 66 (p. 65) to define $\int_\mathbf{S} \omega$:

In any covering set $\mathbf{U} = \mathbf{U}^{s}_{k^s} \cap \mathbf{U}_k$ of $\mathbf{S}$ with its coordinate system $(\mathbf{x}_s : \mathbf{U} \to \mathcal{R}^{n-1})$ and oriented by a volume form of $\mathbf{S}$ (definition 63 (p. 61)), it gives:

$$\int_U \omega = J_U(\omega(\bigotimes_{\mu\in[1,n-1]} \frac{\partial}{\partial x_s^{\mu}}))$$

With $\omega = \sum_{i\in[1,n]} \omega_i \bigwedge_{k\in[1,n]-i} dx^k$ we have using proposition 104 (p. 66) and convention 7 (p. 66):

$$\begin{aligned}\int_U \omega &= J_U(\sum_{i\in[1,n]} \omega_i \bigwedge_{k\in[1,n]-i} \frac{\partial x^k}{\partial x_s^{\mu}} dx_s^{\mu}(\bigotimes_{\mu\in[1,n-1]} \frac{\partial}{\partial x_s^{\mu}})) \\ &= J_U(\sum_{i\in[1,n]} \omega_i \det_{k\in[1,n]-i,\mu\in[1,n-1]} (\frac{\partial x^k}{\partial x_s^{\mu}}))\end{aligned} \tag{1.75}$$

**Definition 70 Flux of a linear $_1$form through a smooth surface (definition 68 (p. 65)) which is the boundary (definition 138 (p. 157)) of a convex differentiable manifold (definition 71 (p. 67)).**

Keeping the notations of definition 68 (p. 65), given an oriented differential manifold $\mathbf{M}$, with a volume form **Vol** (definition 60 (p. 54)), we suppose that we also have a metric $\mathbf{g}$ (definition 31 (p. 25)), and a smooth oriented surface $\mathbf{S} \subset \mathbf{M}$ (definitions 68 (p. 65), 63 (p. 61)). The metric and the volume form define a Hodge transformation of any linear $_1$**form t**: $*\mathbf{t}$ (see definition 62 (p. 59)). The flux of $\mathbf{t}$ through $\mathbf{S}$ is (definition 69 (p. 66)):

$$\int_S *t$$

## 1.12.3 Convex differentiable manifold

We shall consider now surfaces generated by the boundary of convex differentiable manifold.

**Definition 71 Convex differentiable manifold of dimension n.**

Given a differential manifold $\mathbf{M}$, a subset $\mathbf{C}$ we shall call here a convex differentiable manifold a subset of $\mathbf{M}$ which is equivalent to a convex set of $\mathcal{R}^n$ of dimension $\mathbf{n}$ (see definition 256 (p. 280)) include in a ball. That is to say, it belongs to a covering subset $\mathbf{U}$ of $\mathbf{M}$ with a map $\phi$ and its image by $\phi$ is a convex set of $\mathcal{R}^n$ (definition 255 (p. 277)) which is included in a ball of $\mathcal{R}^n$.

**Proposition 105** Considering a differential manifold $\mathbf{M}$ of dimension $\mathbf{n}$, a covering subset $\mathbf{U}$ of $\mathbf{M}$ with its map $\phi$ to $\mathcal{R}^n$, we suppose that embedded in $\mathbf{U}$, we have an open convex differential manifold (definition 71 (p. 67)) $\mathbf{C}$ of dimension $\mathbf{n}$ (see definition 256 (p. 280)). Its image, $\phi(\mathbf{C})$, is a convex (definition 255 (p. 277)) subset of an open ball $\mathbf{B}(\mathbf{a},\mathbf{r})$ $(\mathbf{a} \in \mathcal{R}^n, \mathbf{r} \in \mathcal{R}^+)$. We call $\mathcal{B}(\mathbf{C})$ the boundary of $\mathbf{C}$. From proposition 239 (p. 166), we have $\phi(\mathcal{B}(\mathbf{C})) = \mathcal{B}(\phi(\mathbf{C}))$. We assume that $\mathcal{B}(\mathbf{C})$ is also a differential manifold of dimension $\mathbf{m}$ with its covering open sets $(\mathbf{U}_i^{\mathcal{B}(C)} : [1,\mathbf{K}] \to \mathcal{P}(\mathcal{B}(\mathbf{C})))$ together with their associated map $(\mathbf{y}' = \phi_i^{\mathcal{B}(C)} : \mathbf{U}_i^{\mathcal{B}(C)} \to \mathcal{R}^m)$. We suppose that the maps $(\phi \circ (\phi_i^{\mathcal{B}(C)})^{-1} : \mathcal{R}^m \to \mathcal{R}^n)$ are in $\mathbf{C}^{\infty}$ (see definition 276 (p. 304)). We suppose also that the set of $\mathbf{m}$ vectors $(\mathbf{v}_j : [1,\mathbf{m}] \to \mathcal{R}^m)$ defined at any point $\mathbf{y}' \in \phi_i^{\mathcal{B}(C)}(\mathbf{U}^i)$ by $(\mathbf{v}_j = \partial_j \phi \circ (\phi_i^{\mathcal{B}(C)})^{-1}(\mathbf{y}'))$ is an independent set of vectors (definition 206 (p. 239)). For a point $p \in \mathbf{C}$, taking into account proposition 572 (p. 316), we have a continuous map:

$$(\lambda(p,\mathrm{x}) : \mathcal{R}^{\mathrm{n}} - \{\phi(p)\} \to \mathcal{R}^{+})$$

such as:

$$\phi(p) + \lambda(p,\mathrm{x})\frac{\mathrm{x} - \phi(p)}{\|\mathrm{x} - \phi(p)\|} \in \mathcal{B}(\phi(\mathrm{C})) = \phi(\mathcal{B}(\mathrm{C})) \; .$$

For this point $p$, using the map $\lambda(p,\mathrm{x})$ and the maps $(\phi_{\mathrm{i}}^{\mathcal{B}(\mathrm{C})} : \mathrm{U}_{\mathrm{i}}^{\mathcal{B}(\mathrm{C})} \to \mathcal{R}^{\mathrm{m}})$ together with the map $(\phi : \mathrm{M} \Leftrightarrow \mathcal{R}^{\mathrm{n}})$, we can generate a bijective map for a given $\mathbf{i} \in [\mathbf{1},\mathbf{K}]$:

$$(\mathrm{x}(\mathrm{y}',\mathrm{y}^{\mathrm{m}+1}) : \mathcal{R}^{\mathrm{m}} \times \mathcal{R}^{+} \leftrightarrow \mathcal{R}^{\mathrm{n}} - \{\phi(p)\})$$

with:

$$\mathrm{x}(\mathrm{y}',\mathrm{y}^{\mathrm{m}+1}) = \mathrm{y}^{\mathrm{m}+1} \times (\phi((\phi_{\mathrm{i}}^{\mathcal{B}(\mathrm{C})})^{-1}(\mathrm{y}')) - \phi(p)) + \phi(p) \; .$$

It allows to define the inverse map (for a given $\mathbf{i} \in [\mathbf{1},\mathbf{K}]$):

$$((\mathrm{y}',\mathrm{y}^{\mathrm{m}+1})(\mathrm{x}) : \mathcal{R}^{\mathrm{n}} - \{\phi(p)\} \to \mathcal{R}^{\mathrm{m}} \times \mathcal{R}^{+})$$

with

$$\mathrm{y}^{\mathrm{m}+1} = \|\mathrm{x} - \phi(p)\|/\lambda(p,\mathrm{x}),\; \mathrm{y}' = \phi_{\mathrm{i}}^{\mathcal{B}(\mathrm{C})} \circ \phi^{-1}(\frac{\mathrm{x} - \phi(p)}{\mathrm{y}^{\mathrm{m}+1}} + \phi(p)) \; .$$

We have the following properties:

1.
$$(\forall \mathrm{i} \in [\mathbf{1},\mathbf{K}])((\mathrm{x}_{\mathrm{i}}(\mathrm{y}',\mathrm{y}^{\mathrm{m}+1}) : \phi_{\mathrm{i}}^{\mathcal{B}(\mathrm{C})}(\mathrm{U}_{\mathrm{i}}) \times \mathcal{R}^{+} \leftrightarrow \mathcal{R}^{\mathrm{n}} - \{\phi(p)\}) \in \mathrm{C}^{\infty})$$

   This map and its inverse are in $\mathrm{C}^{\infty}$ (definitions 276 (p. 304), 166 (p. 200)).

2. $\mathbf{m} = \mathbf{n} - \mathbf{1}$

3. At any $\mathbf{x} \in \mathcal{R}^{\mathrm{n}} - \{\phi(p)\}$, considering an $\mathbf{i} \in [\mathbf{1},\mathbf{K}]$ such as

$$(\exists \mathrm{y}' \in \phi_{\mathrm{i}}^{\mathcal{B}(\mathrm{C})}(\mathrm{U}_{\mathrm{i}}))\,(\frac{\mathrm{x} - \phi(p)}{\|\mathrm{x} - \phi(p)\|} = \frac{\phi((\phi_{\mathrm{i}}^{\mathcal{B}(\mathrm{C})})^{-1}(\mathrm{y}')) - \phi(p)}{\|\phi((\phi_{\mathrm{i}}^{\mathcal{B}(\mathrm{C})})^{-1}(\mathrm{y}')) - \phi(p)\|})$$

   or setting $\mathbf{y} = (\mathrm{y}',\mathrm{y}^{\mathrm{n}}) \in \mathcal{R}^{\mathrm{n}-1} \times \mathcal{R}^{+}$ such that:

$$(\exists \mathrm{y} \in \mathcal{R}^{\mathrm{n}})\,(\mathrm{x} = \mathrm{x}_{\mathrm{i}}(\mathrm{y}))$$

   the determinants of the Jacobian of $\mathbf{x}_{\mathrm{i}}(\mathbf{y})$ and of $\mathbf{x}_{\mathrm{i}}^{-1}(\mathbf{x}) = \mathbf{y}(\mathbf{x})$ (definition 297 (p. 358)) are different from zero for $\mathbf{x}(\mathbf{y})$ and $\mathbf{y} = \mathbf{y}(\mathbf{x})$.

**Proof**

We keep the above notations. To simplify the writing we suppose that the null vector (origin) $\mathbf{0}$ of $\mathcal{R}^{\mathrm{n}}$ belongs to $\phi(\mathbf{C})$ and it is $\phi(p)$. Then we have $\mathbf{0} = \phi(p)$. Since $\phi(\mathbf{C})$ is open, we have from proposition 266 (p. 179):

$$(\exists \epsilon)(\mathrm{B}(0,\epsilon) \subset \phi(\mathrm{C})) \; .$$

From proposition 572 (p. 316), we have:

$$(\exists(\lambda(\mathrm{x}) : \mathcal{R}^{\mathrm{n}} - \{0\} \to \mathcal{R}^{+}))\,(\mathrm{x} \in \mathcal{R}^{\mathrm{n}} - \{0\} \Rightarrow \lambda(\mathrm{x})\frac{\mathrm{x}}{\|\mathrm{x}\|} \in \phi(\mathcal{B}(\mathrm{C}))) \; .$$

We have since $\mathbf{B(0,\epsilon)} \subset \phi(\mathbf{C})$, $\lambda(\mathbf{x}) = \|\lambda(\mathbf{x})\frac{\mathbf{x}}{\|\mathbf{x}\|}\| > \epsilon$.
From proposition 574 (p. 317), $\lambda(\mathbf{x})$ is a continuous map (see definition 141 (p. 165)): $\lambda(\mathbf{x}) \in \mathbf{C^0}$ (see definition 166 (p. 200)). From proposition 285 (p. 185), $\mathbf{1/\lambda(x)}$ is also a continuous map. From propositions 516 (p. 288), 286 (p. 185) and 285 (p. 185), $\lambda(\mathbf{x})\frac{\mathbf{x}}{\|\mathbf{x}\|}$ is continuous as $\phi^{-1}(\lambda(\mathbf{x})\frac{\mathbf{x}}{\|\mathbf{x}\|}) \in \mathbf{U_i}$ (proposition 241 (p. 167)). Then

$$(\phi_i^{\mathcal{B}(C)}(\phi^{-1}(\lambda(x)\frac{x}{\|x\|})) \times \lambda(x) : \mathcal{R}^n \to \mathcal{R}^m \times \mathcal{R}^+)$$

is continuous and from proposition 572 (p. 316), it is an injection (definition 122 (p. 139)).
Then $\mathbf{y} = (\mathbf{y}', \mathbf{y}^{m+1}) \in \phi_i^{\mathcal{B}(C)}(\mathbf{U_i}) \times \mathcal{R}^+$,

$$(x(y) = y^{m+1}\phi((\phi_i^{\mathcal{B}(C)})^{-1}(y')) : \phi_i^{\mathcal{B}(C)}(U_i) \times \mathcal{R}^+ \to \mathcal{R}^n - \{0\})$$

is a bijection (definition 124 (p. 139)) which is in $\mathbf{C^\infty}$ (see definition 276 (p. 304)) because:

$$\frac{\partial x}{\partial y^{m+1}} = \phi((\phi_i^{\mathcal{B}(C)})^{-1}(y')) \in C^\infty,$$

$$i \in [1,m] \Rightarrow \frac{\partial x}{\partial y'^i} = y^{m+1}\frac{\partial \phi((\phi_i^{\mathcal{B}(C)})^{-1}(y'))}{\partial y'^i} \in C^\infty$$

$(\mathbf{x(y)} : \mathcal{R}^m \times \mathcal{R}^+ \leftrightarrow \mathcal{R}^n - \{0\}) \in \mathbf{C^\infty}$ with the inverse map continuous. Then, from proposition 642 (p. 373), $\mathbf{m+1=n}$: point 2 is true.

**Let us prove that we have $\det(\mathrm{Jac}(\frac{\partial x^i}{\partial y^j})) \neq 0$ (point 3).**

We call $\mathbf{e_i}$, the basic vectors of $\mathcal{R}^n$ ($\mathbf{e_i}$ is the map $(\mathbf{e_i} : [1,n] \to \mathcal{R}) = \delta_i^j$ where $\delta_i^j$ is the Kronecker symbol (definition 215 (p. 245)) and $\mathbf{j}$ the variable of the map).

- For $\nu \in [1, n-1]$, the set $\frac{\partial x^i}{\partial y^\nu}e_i = y^n\partial_\nu\phi^i \circ (\phi_k^{\mathcal{B}(C)})^{-1}e_i$ is supposed to be an independent set.

- $\mathcal{B}(\mathbf{C})$ does not contain $\boldsymbol{p}$. $\boldsymbol{p}$ gives, by $\phi$, the origin of $\mathcal{R}^n$ because we suppose that at the beginning of the proof. Then $\|\phi^i \circ (\phi_k^{\mathcal{B}(C)})^{-1}(y'_k)\| \neq 0$. Let us prove that we cannot have: [17]

$$\frac{\partial x}{\partial y^{m+1}} = \partial_n x^i = \phi^i \circ (\phi_k^{\mathcal{B}(C)})^{-1}(y'_k) = a^\nu(y'_k)\partial_\nu^{y'}\phi^i \circ (\phi_k^{\mathcal{B}(C)})^{-1}(y'_k)$$

with $(\exists\nu \in [1,n-1])(a^\nu \neq 0)$ which means $\|a^\nu e_\nu\| = \|a\| > 0$. Otherwise, following proposition 560 (p. 307), given $\Delta \in \mathcal{R}$ and $a = (a^\nu : [1,n-1] \to \mathcal{R})$, we may write:

$$\begin{aligned}\phi^i \circ (\phi_k^{\mathcal{B}(C)})^{-1}(y'_k + a(y'_k)\Delta) =&\phi^i \circ (\phi_k^{\mathcal{B}(C)})^{-1}(y'_k)\\ &+ (\partial_\nu^y\phi^i \circ (\phi_k^{\mathcal{B}(C)})^{-1}(y'_k) + \eta_\nu^i(\Delta))a^\nu(y'_k)\Delta\end{aligned}$$

which leads to:

$$\begin{aligned}&(\forall\alpha > 0)(\exists\epsilon_0(\alpha) > 0)\,(\|a(y'_k)\Delta\| = \|a(y'_k)\| \times |\Delta| \leq \epsilon_0(\alpha)\\ &\Rightarrow \|\eta^i(\Delta)\| < \alpha)\\ &\Rightarrow (\forall\alpha > 0)(\exists\epsilon_1(\alpha) > 0)(|\Delta| < \epsilon_1(\alpha) = \epsilon_0(\frac{\alpha}{\|a\|})\\ &\Rightarrow \|\eta_\nu^i(\Delta)a^\nu\| \leq \|\eta^i(\Delta)\| \times \|a\| < \frac{\alpha}{\|a\|}\|a\| = \alpha)\,.\end{aligned} \tag{1.76}$$

[17]On a sphere, which is a typical convex surface (proposition 496 (p. 279)), the following equation would mean that a line going from a point **A** on the surface to a point **B** inside the surface could not be in the tangent plane to the surface at **A**.

But, from definition 141 (p. 165), since $\mathbf{U_k^{\mathcal{B}(C)}}$ is open and $(\phi_k^{\mathcal{B}(C)})^{-1}$ continuous, $\phi_k^{\mathcal{B}(C)}(\mathbf{U_k^{\mathcal{B}(C)}}) \subset \mathcal{R}^{n-1}$ is open and we have:

$$(\exists r_y > 0)\ (B(y'_k, r_y) \subset \phi_k^{\mathcal{B}(C)}(U_k^{\mathcal{B}(C)}))$$

which means:

$$|\Delta| < \frac{r_y}{\|a(y'_k)\|} \Rightarrow \phi \circ (\phi_k^{\mathcal{B}(C)})^{-1}(y'_k + a(y'_k)\Delta) \in \phi \circ (\phi_k^{\mathcal{B}(C)})^{-1}(B(y'_k, r_y))$$
$$\subset \phi \circ (\phi_k^{\mathcal{B}(C)})^{-1}(\phi_k^{\mathcal{B}(C)}(U_k^{\mathcal{B}(C)})) = \phi(U_k^{\mathcal{B}(C)}) \subset \phi(\mathcal{B}(C)) = \mathcal{B}(\phi(C)) \tag{1.77}$$

Setting $\mathbf{P(t)} : \mathcal{R} \to \mathcal{R}^n$ by:

$$P(t) = (\phi^i \circ (\phi_k^{\mathcal{B}(\phi(C))})^{-1}(y'_k)$$
$$+ (\partial_\nu \phi^i \circ (\phi_k^{\mathcal{B}(\phi(C))})^{-1}(y'_k) + \eta^i_\nu(\Delta))((1+\Delta)t - 1)a^\nu(y'_k))e_i$$

we have:

$$P(0) = -(\eta^i_\nu(\Delta)a^\nu(y'_k))e_i$$
$$P(\frac{1}{1+\Delta}) = (\phi^i \circ (\phi_k^{\mathcal{B}(C)})^{-1}(y'_k))e_i \in \mathcal{B}(\phi(C))$$
$$P(1) = \phi^i \circ (\phi_k^{\mathcal{B}(C)})^{-1}((y'_k + a(y'_k)\Delta)))e_i\ . = \phi \circ (\phi_k^{\mathcal{B}(C)})^{-1}(y'_k + a(y'_k)\Delta)$$

From equation (1.77) (p. 70)

$$|\Delta| < \frac{r_y}{\|a(y'_k)\|} \Rightarrow P(1) \in \mathcal{B}(\phi(C))\ .$$

Since $\phi(\mathbf{C})$ is open and since $\mathbf{0} \in \phi(\mathbf{C})$, we have:

$$(\exists \epsilon_3 > 0)\ (\|P\| < \epsilon_3 \Rightarrow P \in \phi(C))\ .$$

From equation (1.76) (p. 69), we have an $\epsilon_1(\epsilon_3) > 0$ such that:

$$|\Delta| \le \epsilon_1(\epsilon_3) \Rightarrow \|\eta^i_\nu(\Delta)a^\nu e_i\| < \epsilon_3\ .$$

We get for $0 < \Delta < \min(\epsilon_1(\epsilon_3), \frac{r_y}{\|a(y'_k)\|})$

$$P(0) = -\eta^i_\nu(\Delta)a^\nu e_i \in \phi(C) = \mathcal{I}(\phi(C))\ .$$

It is forbidden by proposition 570 (p. 316): together with $\mathbf{P(1)} \in \mathcal{B}(\phi(\mathbf{C}))$ this proposition implies $\mathbf{P}(\frac{1}{1+\Delta}) \in \phi(\mathbf{C}) = \mathcal{I}(\phi(\mathbf{C}))$ which is contradiction with $\mathbf{P}(\frac{1}{1+\Delta}) \in \mathcal{B}(\phi(\mathbf{C}))$.

Then combining both above points, from proposition 429 (p. 249), we have:

$$\det(\mathrm{Jac}(x)) = \det(\frac{\partial x^i}{\partial y_j}) \neq 0\ .$$

Since $\det(\mathbf{Jac(x)}) \neq 0$, from proposition 627 (p. 358), $\mathbf{x^{-1}}$ has a continuous Jacobian. Since

$$\mathrm{Jac}(x^{-1})(x(y))\mathrm{Jac}(x)(y) = 1$$

its determinant is not zero.
Since all derivatives of $\mathbf{x^{-1}}$ are given by a product of derivatives of $\mathbf{x}$ at some order divided by $\det(\mathbf{jac(x)})$ at some power, $\mathbf{x^{-1}} \in \mathbf{C^\infty}$ (see definition 276 (p. 304)). Point 1 is true.

**Definition 72 Convex smooth surface of dimension $\mathbf{n-1}$.**

A Convex smooth surface is the boundary of a convex differentiable manifold $\mathbf{C}$ of dimension $\mathbf{n}$ (definition 71 (p. 67)).
It has to be also a smooth surface (definition 68 (p. 65)).
We call $\mathbf{M}$, the differential manifold $\mathbf{M}$ of dimension $\mathbf{n}$, in which $\mathbf{C}$ is a subset of a covering element $\mathbf{U}$. The associated map to $\mathbf{U}$ is $(\phi : \mathbf{U} \to \phi(\mathbf{U}) \subset \mathbf{M})$. Calling $\mathbf{S} \subset \mathbf{M}$ its convex smooth surface, we call the covering open sets of $\mathbf{S} = \mathcal{B}(\mathbf{C})$:

$$(\mathbf{U}_i^{\mathcal{B}(\mathbf{C})} : [1, \mathbf{K}] \to \mathcal{P}(\mathcal{B}(\mathbf{C})))$$

and their associated map $(\mathbf{y}' = \phi_i^{\mathcal{B}(\mathbf{C})} : \mathbf{U}_i^{\mathcal{B}(\mathbf{C})} \to \mathcal{R}^{n-1})$. We suppose also that the set of $\mathbf{n-1}$ vectors $(\mathbf{v_j} : [1, \mathbf{n}-1] \to \mathcal{R}^n)$ defined at any point $\mathbf{y}' \in \phi_i^{\mathcal{B}(\mathbf{C})}(\mathbf{U}^i)$ by $\mathbf{v_j} = \partial_j\phi \circ (\phi_i^{\mathcal{B}(\mathbf{C})})^{-1}(\mathbf{y}')$ is an independent set of vectors (definition 206 (p. 239)).

**Definition 73 Interior of a convex smooth surface.**

The convex open set (see definition 137 (p. 156)) which defines a convex smooth surface (definition 72 (p. 71)) is called the interior of the smooth surface.

**Proposition 106** A convex smooth surface, definition 72 (p. 71), is fully path connected (definition 285 (p. 327)).

**Proof**
We uses the notations of proposition 105 (p. 67). Considering two points $q_1$ $q_2$ of $\mathcal{B}(\mathbf{C})$ with $\mathbf{q_1} = \phi(q_1)$ , $\mathbf{q_2} = \phi(q_2)$. $\mathbf{C} \neq \emptyset$ and $\phi(\mathbf{C})$ being an open, there is an open ball in $\phi(\mathbf{C})$ and a point $p$ inside that open ball not on the line $\mathbf{q_1}, \mathbf{q_2}$[18]. We set $p = \phi^{-1}(\mathbf{p}) \in \mathbf{C}$. Since $\phi(\mathbf{C})$ is open there is an open ball $\mathbf{B}(\phi(p), 2\mathbf{r})$ subset of $\phi(\mathbf{C})$. From proposition 105 (p. 67), one can define a continuous bijective map $\psi$ between $\mathbf{B}(\phi(p), 2\mathbf{r}) - \mathbf{p}$ and $\mathcal{B}(\mathbf{C}) \times ]0, 1]$. Considering two points $q_1' = \phi^{-1}(\mathbf{p} + \mathbf{r}\frac{\mathbf{q_1}-\mathbf{p}}{\|\mathbf{q_1}-\mathbf{p}\|})$, $q_2' = \phi^{-1}(\mathbf{p} + \mathbf{r}\frac{\mathbf{q_1}-\mathbf{p}}{\|\mathbf{q_2}-\mathbf{p}\|})$ those two points are in $\mathbf{B}(\phi(p), 2\mathbf{r}) - \mathbf{p}$. The map

$$(\mathbf{f}(\mathbf{m}) = \mathbf{m}\mathbf{q}_1' + (\mathbf{m}-1)\mathbf{q}_2' : [0, 1] \to \mathbf{B}(\phi(p), 2\mathbf{r}) - 1)$$

---

[18] $\mathbf{B}(\mathbf{u}, \rho)$ being that open ball, if $\mathbf{u}$ is not aligned with $\mathbf{q_1}$, $\mathbf{q_2}$ that is to say:

$$(\forall \mu)(\mathbf{u} \neq \mu \mathbf{q_1} + (1-\mu)\mathbf{q_2})$$

we set $\mathbf{p} = \mathbf{u}$. In the opposite, that is to say if:

$$(\exists \mu)(\mathbf{u} = \mu \mathbf{q_1} + (1-\mu)\mathbf{q_2})$$

since $\mathbf{q_1} \neq \mathbf{q_2} \Rightarrow (\exists \mathbf{j} \in [1, \mathbf{n}])(\mathbf{q}_2^j - \mathbf{q}_1^j \neq 0)$ with that using $\mathbf{k} \in [1, \mathbf{n}] - \mathbf{j}$, instead of $\mathbf{u}$ we choose

$$\mathbf{p} = \mathbf{u} + 0.5\frac{\rho}{\sqrt{(\mathbf{q}_2^j - \mathbf{q}_1^j)^2 + (\mathbf{q}_2^k - \mathbf{q}_1^k)^2}}((\mathbf{q}_2^j - \mathbf{q}_1^j)\mathbf{e_k} - (\mathbf{q}_2^k - \mathbf{q}_1^k))\mathbf{e_j} \in \mathbf{B}(\mathbf{u}, \rho) \subset \phi(\mathbf{C})$$

with that we clearly have:

$$(\forall \nu \in \mathcal{R})(\mathbf{p} \neq \nu \mathbf{q_1} + (1-\nu)\mathbf{q_2})$$

because if not it would mean with $\alpha = 0.5\frac{\rho}{\sqrt{(\mathbf{q}_2^j - \mathbf{q}_1^j)^2 + (\mathbf{q}_2^k - \mathbf{q}_1^k)^2}} > 0$

$$(\mathbf{q_2} - \mathbf{q_1})^k(\mu - \nu) = (\mathbf{q_2} - \mathbf{q_1})^j\alpha$$
$$(\mathbf{q_2} - \mathbf{q_1})^j(\mu - \nu) = -(\mathbf{q_2} - \mathbf{q_1})^k\alpha$$

It would mean, if $(\mathbf{q_2} - \mathbf{q_1})^k = 0$, $\alpha = 0$ and, if not, $\frac{\alpha}{\mu-\nu} = -\frac{\mu-\nu}{\alpha}$ and $\alpha^2 = -(\mu-\nu)^2$. Both are impossible since $\alpha \neq 0$.

is a continuous map (see propositions 591 (p. 328), 537 (p. 297)). $\psi \circ \mathbf{f}$ is also continuous calling $\mathbf{proj}_\lambda$ the projection map $\mathcal{B}(\mathrm{C})\times]0,1] \to \mathcal{B}(\mathrm{C})$ as defined in definition 121 (p. 136) is continuous (proposition 237 (p. 166)) then $(\mathbf{g} = \phi^{-1} \circ \mathbf{proj}_\lambda \circ \psi \circ \mathbf{f} : [0,1] \to \mathcal{B}(\mathrm{C}))$ is continuous (proposition 241 (p. 167)) with $\mathbf{g}(0) = q_1$, $\mathbf{g}(0) = q_2$, $\mathcal{B}(\mathrm{C})$ is path connected.

**Proposition 107** Keeping the notations of proposition 105 (p. 67), given a manifold $\mathbf{M}$ with a convex open subset $\mathrm{C} \neq \emptyset$ inside a covering element $\mathrm{U}$ of $\mathbf{M}$ with its mapping $\phi$ to $\mathcal{R}^n$, we suppose that $\phi(\mathrm{U})$ is within a ball of $\mathcal{R}^n$. Given a point $p \in \mathrm{C}$, we set $\mathbf{x} = \phi(x) - \phi(p)$ when $x \in \mathrm{U} - \{p\}$. We consider the positive map $(\lambda(\mathrm{x}) : \phi(\mathrm{C}) - \{\phi(p)\} \to \mathcal{R}^+)$ as defined in proposition 574 (p. 317). It means that $\lambda(\mathrm{x})$ is such that:

$$\phi(p) + \lambda(\mathrm{x})\frac{\mathrm{x}}{||\mathrm{x}||} \in \phi(\mathcal{B}(\mathrm{C}))$$

We have the following:

- $\lambda(\mathrm{x})$ is a $\mathrm{C}^\infty$ map (see definition 276 (p. 304)).
- We also have: $\frac{\lambda(\mathrm{x})}{||\mathrm{x}||} > 1$ if $x \in \mathrm{C}$, $0 < \frac{\lambda(\mathrm{x})}{||\mathrm{x}||} < 1$ if $x \notin \mathrm{C}$, $\frac{\lambda(\mathrm{x})}{||\mathrm{x}||} = 1$ if $\mathrm{x} \in \mathcal{B}(\mathrm{C})$.

**Proof**
$\lambda(\mathrm{x}) \in \mathrm{C}^\infty$ is a direct consequence of proposition 105 (p. 67). Because if we set

$$\mathrm{m} = \frac{\lambda(\mathrm{x})}{||\mathrm{x}||},\ \mathrm{q} = \phi(p) + \lambda(\mathrm{x})\frac{\mathrm{x}}{||\mathrm{x}||} = \phi(p)(1-\mathrm{m}) + \mathrm{m}\phi(x)$$

we have:

$$\mathrm{q} \in \phi(\mathcal{B}(\mathrm{C})) = \mathcal{B}(\phi(\mathrm{C})), \phi(p) \in \phi(\mathrm{C})\ .$$

Then from proposition 569 (p. 315), proposition 572 (p. 316) and definition 255 (p. 277) we have:

$$\begin{aligned}
&\phi(x) = \phi(p) + \frac{1}{\mathrm{m}}(\mathrm{q} - \phi(p)) = \frac{\mathrm{m}-1}{\mathrm{m}}\phi(p) + \frac{1}{\mathrm{m}}\mathrm{q}\\
&0 < \mathrm{m} < 1 \Leftrightarrow \phi(x) \notin \phi(\overline{\mathrm{C}})\\
&\mathrm{m} = 1 \Leftrightarrow \phi(x) \in \phi(\mathcal{B}(\mathrm{C}))\\
&\mathrm{m} \geq 1 \Leftrightarrow \phi(x) \in \phi(\overline{\mathrm{C}})\ .
\end{aligned}$$

**Proposition 108** Keeping the notations and the conditions of propositions 105 (p. 67) and 107 (p. 72), a point $q \in \mathcal{B}(\mathrm{C})$ with $\mathrm{q} = \phi(q)$, there is a basic vector $\mathrm{e_i}$ of $\mathcal{R}^n$ such as:

$$(\exists\mu)(\mathrm{q} + \mu\mathrm{e_i} \in \phi(\mathrm{C}))\ .$$

**Proof**
We set $\mathrm{p} = \phi(p) = 0$ and a new variable $\mathrm{y^n} = \frac{||\mathrm{x}||}{\lambda(\mathrm{x})}$ (see proposition 574 (p. 317)). It allows to define for any point of $\mathcal{R}^n - \{\mathrm{p}\} = \mathcal{R}^n - 0$, a new coordinate system (see proposition 105 (p. 67)): $\mathrm{y}(\mathrm{x})$ such as:

$$\begin{aligned}
&\mathrm{i} \in [1,\mathrm{n}] \Rightarrow \mathrm{x^i}(\mathrm{y}) = \mathrm{y^n} \times \phi^{\mathrm{i}}((\phi^{\mathcal{B}})^{-1}(\prod^{\mathrm{e}}_{\mathrm{j}\in[1,\mathrm{n}-1]} \mathrm{y^j}))\\
&\mathrm{y^n}(\mathrm{x}) = \frac{||\mathrm{x}||}{\lambda(\mathrm{x})},\\
&(\mathrm{i} \in [1,\mathrm{n}-1] \Rightarrow \mathrm{y^i} = \frac{\lambda(\mathrm{x})||\mathrm{x}||}{||\mathrm{x}||}(\phi^{\mathcal{B}})^{\mathrm{i}}(\phi^{-1}(\mathrm{x})) = \frac{(\phi^{\mathcal{B}})^{\mathrm{i}}(\phi^{-1}(\mathrm{x}))}{\mathrm{y^n}(\mathrm{x})})\ .
\end{aligned} \tag{1.78}$$

When $\mathbf{x} = \mathbf{x}^i\mathbf{e}_i = \phi(q) = \mathbf{q}$, we have: $\mathbf{y}^i = (\phi^{\mathcal{B}})^i(q)$ and $\mathbf{y}^n = \frac{\|\mathbf{q}\|}{\lambda(\mathbf{q})} = 1$ since $\mathbf{q} \in \phi(\mathcal{B}(\mathbf{C}))$.
We set for $\mathbf{x} = \mathbf{q}$, $\mathbf{i} \in [\mathbf{1}, \mathbf{n}-\mathbf{1}] \Rightarrow \mathbf{v}_i = \frac{\partial(\mathbf{x}^j)}{\partial \mathbf{y}^i}\mathbf{e}_j$ and $\mathbf{v}_n = \mathbf{q} - \mathbf{p} = \mathbf{q}$, $\mathbf{p} = \phi(p) = 0$. Applying item 3 of proposition 105 (p. 67), we have: $\mathbf{det}(\mathbf{jac}(\mathbf{x}(\mathbf{y}))) = \mathbf{det}(\mathbf{v}) \neq \mathbf{0}$. Then $\mathbf{v}_n = \mathbf{q} - \mathbf{p} = \mathbf{q}$ is independent of the $\mathbf{v}_i, \mathbf{i} \in [\mathbf{1}, \mathbf{n}-\mathbf{1}]$.
There is an $\mathbf{e}_i$ of the standard basis of $\mathcal{R}^n$ not dependent of the $\mathbf{v}_i, \mathbf{i} \in [\mathbf{1}, \mathbf{n}-\mathbf{1}]$ otherwise the $\mathbf{e}$ would not be independent (proposition 412 (p. 242), A.1.4-8 (p. 435)). Let us suppose that it is $\mathbf{e}_1$.
From proposition 105 (p. 67) $\mathbf{det}(\mathbf{jac}(\mathbf{x}(\mathbf{y}))) \neq \mathbf{0}$, we have (see proposition 444 (p. 255), definition 417 (p. 545)):

$$\frac{\partial \mathbf{y}^n}{\partial \mathbf{x}^1} = \mathbf{jac}(\mathbf{y}(\mathbf{x}))^n_1 = (\mathbf{jac}(\mathbf{x}(\mathbf{y}))^{-1})^1_n = (-1)^{n+1}\frac{\mathbf{det}(\mathbf{Minor}(\mathbf{x}(\mathbf{y}))^1_n)}{\mathbf{det}(\mathbf{jac}(\mathbf{x}(\mathbf{y})))} .$$

From definition 228 (p. 254):

$$\{\mathbf{Minor}(\mathbf{x}(\mathbf{y}))^1_n\}^i_j = \frac{\partial \mathbf{x}^{i+1}}{\partial \mathbf{y}^j}, \mathbf{i} \in [\mathbf{1}, \mathbf{n}-\mathbf{1}], \mathbf{j} \in [\mathbf{1}, \mathbf{n}-\mathbf{1}]$$

If $\mathbf{u} = \mathbf{u}^i\mathbf{e}_i$ together with the $\mathbf{v}_i$, $\mathbf{i} \in [\mathbf{1}, \mathbf{n}-\mathbf{1}]$ is making an independent set of vector, we have (proposition 442 (p. 254)):

$$\begin{aligned}\mathbf{det}(\mathbf{u} \wedge \bigwedge_{j\in[1,n-1]} \mathbf{v}_j) &= \mathbf{u}^i(-1)^{i+1}\mathbf{det}(\mathbf{Minor}(\mathbf{u} \wedge \bigwedge_{j\in[1,n-1]} \mathbf{v}_j)^i_1) \\ &= \mathbf{u}^i(-1)^{i+1} \det_{\ell\in[1,n]-i, m\in[1,n-1]} (\mathbf{v}^\ell_m) \neq \mathbf{0} .\end{aligned}$$

Since we suppose that $\mathbf{e}_1 = \mathbf{e}^i_1 = \delta^i_1\mathbf{e}_i$ (definition 242 (p. 267)), $\mathbf{e}_1$ is making with the $\mathbf{v}_i$, $\mathbf{i} \in [\mathbf{1}, \mathbf{n}-\mathbf{1}]$ an independent set, we have (proposition 430 (p. 250)):

$$\frac{\partial \mathbf{y}^n}{\partial \mathbf{x}^1}(\mathbf{q}) = (-1)^{n+1} \det_{\ell\in[2,n], m\in[1,n-1]} (\mathbf{v}^\ell_m) = (-1)^{n+1}\mathbf{det}(\mathbf{e}_1 \wedge \bigwedge_{j\in[1,n-1]} \mathbf{v}_j) \neq \mathbf{0} .$$

Then $\frac{\partial \mathbf{y}^n}{\partial \mathbf{x}^1}(\mathbf{q}) \neq \mathbf{0}$.
From definitions 275 (p. 304) and 161 (p. 192), equation (2.51) (p. 192), we have:

$$(\forall \alpha)(\exists \epsilon(\alpha) > 0)(|\mu| < \epsilon(\alpha) \Rightarrow |\frac{\mathbf{y}^n(\mathbf{q} + \mu\mathbf{e}_1) - \mathbf{y}^n(\mathbf{q}) - \mu\frac{\partial \mathbf{y}^n}{\partial \mathbf{x}^1}(\mathbf{q})}{\mu}| < \alpha) .$$

Choosing $\mu$ with the opposite sign of $\frac{\partial \mathbf{y}^n}{\partial \mathbf{x}^1}$ ($\mu\frac{\partial \mathbf{y}^n}{\partial \mathbf{x}^1} < \mathbf{0}$) and such as:

$$|\mu| < \min(\epsilon(0.5|\frac{\partial \mathbf{y}^n}{\partial \mathbf{x}^1}|), \frac{1}{1.5|\frac{\partial \mathbf{y}^n}{\partial \mathbf{x}^1}|}) .$$

Since $\mathbf{y}^n(\mathbf{q}) = \mathbf{1}$, we have:

$$\begin{aligned}&|\mathbf{y}^n(\mathbf{q} + \mu\mathbf{e}_1) - 1 - \mu\frac{\partial \mathbf{y}^n}{\partial \mathbf{x}^1}| < 0.5|\mu||\frac{\partial \mathbf{y}^n}{\partial \mathbf{x}^1}| = -0.5\mu\frac{\partial \mathbf{y}^n}{\partial \mathbf{x}^1} \\ &0.5\mu\frac{\partial \mathbf{y}^n}{\partial \mathbf{x}^1} < |\mathbf{y}^n(\mathbf{q} + \mu\mathbf{e}_1) - 1 - \mu\frac{\partial \mathbf{y}^n}{\partial \mathbf{x}^1}| < -0.5\mu\frac{\partial \mathbf{y}^n}{\partial \mathbf{x}^1} \\ &0 < 1 + 1.5\mu\frac{\partial \mathbf{y}^n}{\partial \mathbf{x}^1} < \mathbf{y}^n(\mathbf{q} + \mu\mathbf{e}_1) < 1 + 0.5\mu\frac{\partial \mathbf{y}^n}{\partial \mathbf{x}^1} < 1 .\end{aligned}$$

From the definition of $\mathbf{y}$ given by equation (1.78) (p. 72), we have:

$$\frac{q+\mu e_1}{y^n(q+\mu e_1)} = \lambda(q+\mu e_1)\frac{q+\mu e_1}{\|q+\mu e_1\|} \in \phi(\mathcal{B}(C)) \ .$$

It gives for $x = q + \mu e_1$

$$\frac{\lambda(x)}{\|x\|} = \frac{1}{y^n(q+\mu e_1)} > 1 \ .$$

Then from proposition 107 (p. 72),

$$q + \mu e_1 \in \phi(C) \ .$$

Proposition 1.78 (p. 72) is true.

### 1.12.4 Stokes Kelvin-Thomson theorem on a convex smooth surface

**Proposition 109** Given a differential manifold of dimension $\mathbf{n}$, in a covering open set $U \subset M$ having a continuous injection $\phi$ to $\mathcal{R}^n$, we have an open convex subset $C \subset M$ (definition 71 (p. 67)) not empty and of dimension $\mathbf{n}$ (definition 256 (p. 280)) with a smooth surface as boundary (definition 68 (p. 65)). Moreover $\phi(C)$ is in a closed ball of $\mathcal{R}^n$. Considering the projection map $(\mathrm{proj}_i : \mathcal{R}^n \to \mathcal{R}^{n-1})$ (see definitions 121 (p. 136), 244 (p. 270)) along the coordinate $i \in [1, n]$, we have (see definitions 166 (p. 200), 276 (p. 304)):

$$\begin{aligned}&(\exists(h_1^i : \mathrm{proj}_i(\phi(C)) \to \mathcal{R}) \in C^\infty)(\exists(h_2^i : \mathrm{proj}_i(\phi(C)) \to \mathcal{R}) \in C^\infty)\\&((x \in \mathcal{R}^n, \mathrm{proj}_i(x) \in \mathrm{proj}_i(\phi(C))) \Rightarrow (h_1^i(\mathrm{proj}_i(x)) \le h_2^i(\mathrm{proj}_i(x))),\\&(x^i = h_1^i(\mathrm{proj}_i(x)) \vee x^i = h_2^i(\mathrm{proj}_i(x))) \Leftrightarrow x \in \phi(\mathcal{B}(C))) \ .\end{aligned}$$

**Proof**

We keep the above notation, calling $(e_i : [1, n] \to \mathcal{R}^n)$, the reference basis of $\mathcal{R}^n$ such that $e_i = \prod^e_{j\in[1,n]} \delta^j_i = (\delta^j_i : [1, n] \to \mathcal{R}) \in \mathcal{R}^n$ (see equation (2.106) (p. 267) in definition 242 (p. 267)), we have from proposition 572 (p. 316): [19]

$$\begin{aligned}&(\exists(u_1(x) : \mathrm{proj}_i(\phi(C)) \to \phi(\mathcal{B}(C))))(\exists(u_2(x) : \mathrm{proj}_i(\phi(C)) \to \phi(\mathcal{B}(C))))\\&(x_p \in \phi(C) \Rightarrow (u_1(\mathrm{proj}_i(x_p)) = x_p + (h_1^i - x_p^i)e_i,\\&u_2(\mathrm{proj}_i(x_p)) = x_p + (x_p^i - h_2^i)(-e_i)) \Rightarrow (h_1^i > x_p^i, h_2^i < x_p^i)) \ .\end{aligned}$$

From definition 112 (p. 131), 4, it defines two maps $h_1^i, h_2^i$. From proposition 572 (p. 316), those two maps do not depend on i component: $\mathrm{proj}_i(x) = \mathrm{proj}_i(x') \Rightarrow (h_1^i(x) = h_1^i(x'), h_2^i(x) = h_2^i(x'))$. Then we may write:

$$(h_1^i : \mathrm{proj}_i(\phi(C)) \to \mathcal{R}), (h_2^i : \mathrm{proj}_i(\phi(C)) \to \mathcal{R}) \ .$$

From proposition 105 (p. 67), those two maps are in $C^\infty$ (see 276 (p. 304)). It splits $\mathcal{B}(C)$ in two subsets $\mathcal{B}(C)^+$ and $\mathcal{B}(C)^-$ which do not overlap since $h_1^i - x_p^i$ and $x_p^i - h_2^i$ are strictly positive.

**Proposition 110** With the same notations as in proposition 109 (p. 74), for any $i \in [1, n]$, the two subsets of $\mathcal{B}(C)$: ${}^C\mathcal{B}_\ell^i$, $\ell \in [1, 2]$ such that:

$$x \in {}^C\mathcal{B}_\ell^i \Leftrightarrow (\mathrm{proj}_i(\phi(x)) \in \mathrm{proj}_i(\phi(C)), \phi(x)^i = h_\ell^i(\mathrm{proj}_i(\phi(x))))$$

are open (in the induced topology by the M topology on $\mathcal{B}(C)$).

[19] As usual when an index has some significance, there is no summation over it even when repeated.

**Proof**
Calling:

$$\mathbf{hm}(\phi(x)) = \frac{\mathbf{h}^i_1(\mathbf{proj_i}(\phi(x))) + \mathbf{h}^i_2(\mathbf{proj_i}(\phi(x)))}{2}$$

$\mathbf{hm}(\phi(x))$ is continuous (propositions 237 (p. 166), 234 (p. 165) and 534 (p. 296)). We define the set $\mathbf{S}^i_\ell$ by:

$$\mathbf{x} \in \mathbf{S}^i_\ell \Leftrightarrow (\mathbf{x} \in \phi(\mathbf{C}) \subset \mathcal{R}^n, \mathbf{proj_i}(\mathbf{x}) \in \mathbf{proj_i}(\phi(\mathbf{C})), (-1)^\ell \mathbf{x}^i > (-1)^\ell \mathbf{hm}(\mathbf{x})) .$$

Since $\mathbf{hm(x)}$ is continuous, $\mathbf{f(x)} = \mathbf{x}^i - \mathbf{hm(x)}$ is also continuous. From definition 141 (p. 165), $\mathbf{f}^{-1}(]\mathbf{0}, \rightarrow [)$, $\mathbf{f}^{-1}(] \leftarrow, \mathbf{0}[)$ are open because $]\mathbf{0}, \rightarrow [$, $] \leftarrow, \mathbf{0}[$ are open (proposition 275 (p. 181)). We have

$$\mathbf{S}^i_1 = \mathbf{f}^{-1}(]\mathbf{0}, \rightarrow [) \cap \phi(\mathbf{C}),\ \mathbf{S}^i_2 = \mathbf{f}^{-1}(] \leftarrow, \mathbf{0}[) \cap \phi(\mathbf{C})$$

$\phi(\mathbf{C})$ being open, from definition 132 (p. 151), $\mathbf{S}^i_1$, $\mathbf{S}^i_2$ are open. Since $\phi$ and $\phi^{-1}$ are continuous (definitions 1 (p. 1), 2 and 141 (p. 165)), from definition 133 (p. 151), the subsets $\phi({}^\mathbf{C}\mathcal{B}^i_\ell) = \mathbf{S}^i_\ell \cap \mathcal{B}(\phi(\mathbf{C})) = \mathbf{S}^i_\ell \cap \phi(\mathcal{B}(\mathbf{C}))$ are open in the $\phi(\mathcal{B}(\mathbf{C}))$ topology as the subsets $\phi^{-1} \circ \phi({}^\mathbf{C}\mathcal{B}^i_\ell) = {}^\mathbf{C}\mathcal{B}^i_\ell$ in the $\mathcal{B}(\mathbf{C})$ topology.

**Proposition 111** Keeping notation of proposition 110 (p. 74), we have

$$(\forall \ell \in [1,2])(\mathbf{proj_i}(\phi({}^\mathbf{C}\mathcal{B}^i_\ell)) = \mathbf{proj_i}(\mathbf{C}))$$

and the two maps $\ell \in [1,2]$, $(\mathbf{proj_i} : \phi({}^\mathbf{C}\mathcal{B}^i_\ell) \leftrightarrow \mathbf{proj_i}(\phi(\mathbf{C})))$ are isomorphisms (definition 142 (p. 166)).

**Proof**
Calling $\mathbf{e_i} = \prod^e_{j\in[1,n]} \delta^j_i = (\delta^j_i : [1,n] \rightarrow \mathcal{R}) \in \mathcal{R}^n$, the reference basis of $\mathcal{R}^n$ (see equation (2.106) (p. 267) in definition 242 (p. 267)), from proposition 109 (p. 74), we have:

$$(\ell \in [1,2], \mathbf{i} \in [1,n]) \Rightarrow (\mathbf{h}^i_1 : \mathbf{proj_i}(\phi(\mathbf{C})) \rightarrow \mathcal{R}) .$$

Then we have:

$$(\ell \in [1,2], \mathbf{x} = \mathbf{x}^j\mathbf{e}_j \in \mathcal{R}^n) \Rightarrow$$
$$((\sum_{j\in[1,n]-i} \mathbf{x}^j\mathbf{e}_j \in \mathbf{proj_i}(\phi(\mathbf{C})), \mathbf{x}^i = \mathbf{h}^i_\ell(\sum_{j\in[1,n]-i} \mathbf{x}^j\mathbf{e}_j)) \Leftrightarrow \mathbf{x} \in \phi({}^\mathbf{C}\mathcal{B}^i_\ell))$$

then $(\forall \ell \in [1,2])(\mathbf{proj_i}(\phi({}^\mathbf{C}\mathcal{B}^i_\ell)) = \mathbf{proj_i}(\phi(\mathbf{C})))$. The map $(\mathbf{proj_i} : \phi({}^\mathbf{C}\mathcal{B}^i_\ell) \rightarrow \mathbf{proj_i}(\phi(\mathbf{C})))$ is surjective (definition 123 (p. 139)). From proposition 572 (p. 316), we have:

$$(\ell \in [1,2], \mathbf{x} \in \phi({}^\mathbf{C}\mathcal{B}^i_\ell)) \Rightarrow \mathbf{Card}(\phi({}^\mathbf{C}\mathcal{B}^i_\ell) \cap \mathbf{proj}_i^{-1}(\mathbf{proj_i}(\mathbf{x}))) = \mathbf{1}$$

then the maps $(\mathbf{proj_i} : \phi({}^\mathbf{C}\mathcal{B}^i_\ell) \leftrightarrow \mathbf{proj_i}(\phi(\mathbf{C})))$, for $\ell \in [1,2]$, are injective (definition 122 (p. 139)) and from definition 124 (p. 139) are bijective.
The $\phi({}^\mathbf{C}\mathcal{B}^i_\ell)$ are open in the $\mathcal{B}(\phi(\mathbf{C}))$ induced topology (proposition 110 (p. 74)), any open of $\mathbf{proj_i}(\phi(\mathbf{C}))$ has an inverse in $\mathcal{R}^n$ which is open (proposition 237 (p. 166), definition 141 (p. 165)). Then its intersection with $\mathcal{B}(\phi(\mathbf{C}))$ is also open in the $\mathcal{B}(\phi(\mathbf{C}))$ induced topology (definition 132 (p. 151), 2.18) then its intersection with $\phi({}^\mathbf{C}\mathcal{B}^i_\ell)$ is open in the $\mathcal{B}(\phi(\mathbf{C}))$ induced topology. We conclude with definition 141 (p. 165): the maps $(\mathbf{proj_i} : \phi({}^\mathbf{C}\mathcal{B}^i_\ell) \rightarrow \mathbf{proj_i}(\phi(\mathbf{C})))$ are continuous in $\mathcal{B}(\phi(\mathbf{C}))$ induced topology.
$\mathbf{h}^i_\ell$ are continuous maps (proposition 109 (p. 74)), then, from propositions 233 (p. 165) and 242 (p. 168), the maps $(\mathbf{proj}_i^{-1} : \mathbf{proj_i}(\phi(\mathbf{C})) \rightarrow \phi({}^\mathbf{C}\mathcal{B}^i_\ell))$ are continuous. They are isomorphisms (definition 142 (p. 166)).

**Proposition 112** We consider the differential manifold associated to the boundary $\mathcal{B}(\mathbf{C})$ of a convex differentiable manifold $\mathbf{C}$ (definition 71 (p. 67)) subset of a covering open $\mathbf{U}$ of a differential manifold, $\mathbf{M}$, $\phi$ being the associated bijection $(\phi : \mathbf{U} \to \mathcal{R}^{\mathbf{n}})$. We suppose that $\phi(\mathbf{C})$ is in a closed ball of $\mathcal{R}^{\mathbf{n}}$. We suppose also that $\mathcal{B}(\mathbf{C})$ is a smooth surface of dimension $n-1$ and has a covering set (see definition 72 (p. 71)) $(\mathbf{U}_{\mathrm{i}}^{\mathcal{B}(\mathbf{C})} : [1, \mathbf{K}] \to \mathcal{P}(\mathcal{B}(\mathbf{C})))$ and their associated map $(\phi_{\mathrm{i}}^{\mathcal{B}(\mathbf{C})} : \mathbf{U}_{\mathrm{i}}^{\mathcal{B}(\mathbf{C})} \to \mathcal{R}^{\mathbf{n}-1})$ (proposition 105 (p. 67)). The set of opens ${}^{\mathbf{C}}\mathcal{B}_{\ell}^{\mathbf{i}}$ is another covering set of $\mathcal{B}(\mathbf{C})$:

$$\mathcal{B}(\mathbf{C}) = \bigcup_{\mathrm{i}\in[1,\mathrm{n}],\ell\in[1,2]} {}^{\mathbf{C}}\mathcal{B}_{\ell}^{\mathbf{i}} \,.$$

The bijections $(\mathbf{proj}_{\mathrm{i}} \circ \phi : \phi({}^{\mathbf{C}}\mathcal{B}_{\ell}^{\mathbf{i}}) \leftrightarrow \mathcal{R}^{\mathbf{n}-1})$ are the associated maps (see definitions 1 (p. 1), 3 (p. 1), 124 (p. 139)).

**Proof**

From proposition 110 (p. 74), the ${}^{\mathbf{C}}\mathcal{B}_{\ell}^{\mathbf{i}}$, are open. From proposition 108 (p. 72),

$$(\forall q \in \mathcal{B}(\mathbf{C}))(\exists \mathrm{i} \in [1, \mathrm{n}])(\exists \mu)(\phi(q) + \mu \mathrm{e}_{\mathrm{i}} \in \phi(\mathbf{C})) \,.$$

From definition 121 (p. 136), taking that $\mathbf{i}$ and $\mu$ for any $\mathbf{q}$ such as:

$$\mathrm{q} = \phi(q) = \prod_{\mathrm{j}\in[1,\mathrm{n}]}^{\mathrm{e}} \mathrm{q}^{\mathrm{j}} = \mathrm{q}^{\mathrm{j}}\mathrm{e}_{\mathrm{j}}, q \in \mathcal{B}(\mathbf{C})$$

we have:

$$\mathbf{proj}_{\mathrm{i}}(\prod_{\mathrm{j}\in[1,\mathrm{n}]}^{\mathrm{e}} \mathrm{q}^{\mathrm{j}}) = \prod_{\mathrm{j}\in[1,\mathrm{n}]-\{\mathrm{i}\}}^{\mathrm{e}} \mathrm{q}^{\mathrm{j}} = \mathbf{proj}_{\mathrm{i}}(\mathrm{q} + \mu \mathrm{e}_{\mathrm{i}}) \in \mathbf{proj}_{\mathrm{i}}(\phi(\mathbf{C})) \,. \tag{1.79}$$

Since for any point in $\mathbf{x} \in \mathbf{proj}_{\mathrm{i}}(\phi(\mathbf{C}))$ there are two and only two points of $\mathcal{B}(\phi(\mathbf{C}))$, one in ${}^{\mathbf{C}}\mathcal{B}_{1}^{\mathbf{i}}$ one in ${}^{\mathbf{C}}\mathcal{B}_{2}^{\mathbf{i}}$ which project along $\mathbf{i}$ in $\mathbf{x}$ (proposition 572 (p. 316)). Then combining that with equation (1.79) (p. 76), we have:

$$(\forall q \in \mathcal{B}(\mathbf{C}))(\exists \mathrm{i} \in [1, \mathrm{n}])(\phi(q) \in {}^{\mathbf{C}}\mathcal{B}_{1}^{\mathbf{i}} \cup {}^{\mathbf{C}}\mathcal{B}_{2}^{\mathbf{i}})$$

then (see definition 349 (p. 451), 362 (p. 462))

$$\mathcal{B}(\mathbf{C}) \subset \bigcup_{\mathrm{i}\in[1,\mathrm{n}],\ell\in[1,2]} {}^{\mathbf{C}}\mathcal{B}_{\ell}^{\mathbf{i}} \,.$$

Since $(\forall \mathrm{i} \in [1, \mathrm{n}])(\forall \ell \in [1, 2])({}^{\mathbf{C}}\mathcal{B}_{\ell}^{\mathbf{i}} \subset \mathcal{B}(\mathbf{C}))$ we have:

$$\bigcup_{\mathrm{i}\in[1,\mathrm{n}],\ell\in[1,2]} {}^{\mathbf{C}}\mathcal{B}_{\ell}^{\mathbf{i}} \subset \mathcal{B}(\mathbf{C})$$

then from proposition 754 (p. 451),

$$\bigcup_{\mathrm{i}\in[1,\mathrm{n}],\ell\in[1,2]} {}^{\mathbf{C}}\mathcal{B}_{\ell}^{\mathbf{i}} = \mathcal{B}(\mathbf{C}) \tag{1.80}$$

From definition 121 (p. 136), propositions 241 (p. 167), 235 (p. 165) and 237 (p. 166), the map $(\mathbf{proj}_{\mathrm{i}} \circ \phi : \phi({}^{\mathbf{C}}\mathcal{B}_{\ell}^{\mathbf{i}}) \to \mathcal{R}^{\mathbf{n}-1})$ is continuous. From proposition 111 (p. 75) it has an inverse which is also continuous. We have, by what we suppose for $\mathcal{B}(\mathbf{C})$, $\bigcup_{\mathrm{k}\in[1,\mathrm{K}]} \mathbf{U}_{\mathrm{k}}^{\mathcal{B}(\mathbf{C})} = \mathcal{B}(\mathbf{C})$.

If we have $q \in {}^{C}\mathcal{B}^{i'}_{\ell'} \cap {}^{C}\mathcal{B}^{i}_{\ell}$, the three sets ${}^{C}\mathcal{B}^{i}_{\ell}$, ${}^{C}\mathcal{B}^{i'}_{\ell'}$, $U^{\mathcal{B}(C)}_{k}$ which are open have a non-empty intersection. This intersection is open and different of zero (definition 132 (p. 151)). Then $\phi({}^{C}\mathcal{B}^{i}_{\ell} \cap {}^{C}\mathcal{B}^{i'}_{\ell'} \cap U^{\mathcal{B}(C)}_{k})$ contains an open ball $B(\phi(q), r_q) \cap \phi(\mathcal{B}(C))$ (proposition 266 (p. 179)). In that open, from proposition 111 (p. 75) and definition 68 (p. 65), calling $\mathbf{proj}^{\ell}_{i}$ the projection along $e_i$ limited to $\phi({}^{C}\mathcal{B}^{i}_{\ell})$, we have:

$$(\mathbf{proj}^{\ell}_{i} \circ \phi \circ (\phi^{\mathcal{B}(C)}_{k})^{-1} : \mathcal{R}^{n-1} \to \mathcal{R}^{n-1}) \in C^{\infty}$$
$$(\mathbf{proj}^{\ell'}_{i'} \circ \phi \circ (\phi^{\mathcal{B}(C)}_{k})^{-1} : \mathcal{R}^{n-1} \to \mathcal{R}^{n-1}) \in C^{\infty}$$

their inverses are also in $C^{\infty}$. Then we have:

$$\mathbf{proj}^{\ell}_{i} \circ \phi \circ (\phi^{\mathcal{B}(C)}_{k})^{-1} \circ \phi^{\mathcal{B}(C)}_{k} \circ \phi^{-1} \circ (\mathbf{proj}^{\ell'}_{i'})^{-1} = \mathbf{proj}^{\ell}_{i} \circ \phi \circ (\mathbf{proj}^{\ell'}_{i'} \circ \phi)^{-1} \in C^{\infty}$$

All requirements of definitions 1 (p. 1), 3 (p. 1), 124 (p. 139) are satisfied. Proposition 112 (p. 76) is true.

**Proposition 113** Keeping the notations of proposition 112 (p. 76), ($e_i = \delta^{j}_{i} : [1, n] \to \mathcal{R}^{n}$) being the reference basis of $\mathcal{R}^{n}$ (definition 242 (p. 267)), we have the following relations:

$$(\forall i_a \in [1, n])(\forall i_b \in [1, n] - \{i_a\})(\forall \ell_a \in [1, 2])(\forall \ell_b \in [1, 2])$$
$$((x \in {}^{C}\mathcal{B}^{i_a}_{\ell_a} \cap {}^{C}\mathcal{B}^{i_b}_{\ell_b}, x = \phi(x) = x^{i}e_i) \Rightarrow ((x^{i_a}(\prod^{e}_{j \in [1,n]-i_a} x^{j})$$
$$= h^{i_a}_{\ell_a}(\prod^{e}_{j \in [1,n]-i_a} x^{j}) : \mathcal{R}^{n-1} \to \mathcal{R}) \in C^{\infty}, \frac{\partial x^{i_a}}{\partial x^{i_b}} = (-1)^{\ell_a + \ell_b + 1}|\frac{\partial x^{i_a}}{\partial x^{i_b}}|)) \tag{1.81}$$

$$(x \in {}^{C}\mathcal{B}^{i_a}_{\ell_a} \cap \mathcal{C}_{\mathcal{B}(C)}({}^{C}\mathcal{B}^{i_b}_{1} \cup {}^{C}\mathcal{B}^{i_b}_{2}), x = \phi(x) = x^{i}e_i)$$
$$\Rightarrow ((x^{i_a}(\prod^{e}_{j \in [1,n]-i_a} x^{j}) : \mathcal{R}^{n-1} \to \mathcal{R}) \in C^{\infty}, \frac{\partial x^{i_a}}{\partial x^{i_b}} = 0) \tag{1.82}$$

(see definition 353 (p. 454))

**Proof**

**Let us demonstrate equation (1.81) (p. 77)**

From proposition 109 (p. 74), we have:

$$(p \in {}^{C}\mathcal{B}^{i}_{\ell}, \phi(p) = x^{j}e_j) \Rightarrow x^{i} = h^{i}_{\ell}(\mathbf{proj}_i(\phi(p))) \in C^{\infty}$$

Setting

$$x^{i}_{m} = 0.5(h^{i}_{1}(\mathbf{proj}_i(\phi(p))) + h^{i}_{2}(\mathbf{proj}_i(\phi(p))))$$

from proposition 109 (p. 74), we have[20]:

$$\mathbf{proj}_i(\phi(p)) + x^{i}_{m}e_i \in \phi(C)$$

and from proposition 572 (p. 316)

$$(-1)^{\ell}x^{i}_{m} < (-1)^{\ell}x^{i}.$$

[20]No summation here on repeated indices

Using proposition 572 (p. 316), at $p \in {}^{C}\mathcal{B}^{i_b}_{\ell_b} \cap {}^{C}\mathcal{B}^{i_a}_{\ell_a}$, it gives:

$$(\exists\alpha > 0)\,(\mathrm{proj}_{i_b}(\phi(p)) + x^{i_b}_{m}e_{i_b} + 2\alpha(-1)^{\ell_a}e_{i_a} \in \phi({}^{C}\mathcal{B}^{i_a}_{\ell_a}) \subset \phi(\mathcal{B}(C)))\ .$$

Then, taking that $\alpha > 0$, we also have:

$$\phi(p) + (x^{i_b}_{m} - x^{i_b})e_{i_b} + \alpha(-1)^{\ell_a}e_{i_a} \in \phi(C)\ .$$

Then, again from proposition 572 (p. 316):

$$\mu \in ]0,1] \Rightarrow (q = \phi(p) + \mu(x^{i_b}_{m} - x^{i_b})e_{i_b} + \mu\alpha(-1)^{\ell_a}e_{i_a} \in \phi(C))\ .$$

From proposition 572 (p. 316), we have:

$$(\mu \in ]0,1] \Rightarrow (\exists\lambda(\mu) > 0)(\phi(p) + \mu(x^{i_b}_{m} - x^{i_b})e_{i_b} + (\mu\alpha + \lambda(\mu))(-1)^{\ell_a}e_{i_a}$$
$$\in \phi({}^{C}\mathcal{B}^{i_b}_{\ell_b} \cap {}^{C}\mathcal{B}^{i_a}_{\ell_a}))\ .$$

Then taking this $\lambda(\mu)$ and calling $(h^{i_a}_{\ell_a}({}^{\Delta}x^{i_b}) : ]0, x^{i_b}_{m} - x^{i_b}[ \rightarrow \mathcal{R})$ the value of $h^{i_a}_{\ell_a}$ when the $i_b$ coordinate of $\mathrm{proj}_{i_a}$ is moved to ${}^{\Delta}x^{i_b} = x^{i_b} + \Delta x^{i_b}$ , we have:

$$(\Delta x^{i_b} = \mu(x^{i_b}_{m} - x^{i_b}), \mu \in ]0,1[)$$
$$\Rightarrow \Delta x^{i_a} = h^{i_a}_{\ell_a}(x^{i_b} + \Delta x^{i_b}) - h^{i_a}_{\ell_a}(x^{i_b}) = (\mu\alpha + \lambda(\mu))(-1)^{\ell_a}$$

It means

$$\frac{\Delta x^{i_a}(\Delta x^{i_b})}{\Delta x^{i_b}} = \frac{(\mu\alpha + \lambda(\mu))(-1)^{\ell_b}}{\mu(x^{i_b}_{m} - x^{i_b})}$$

From proposition 572 (p. 316), we have $(x^{i_b}_{m} - x^{i_b})(-1)^{\ell_b} < 0$. From what we just suppose, $\mu$, $\alpha$, $\lambda(\mu)$ are all positive. We have:

$$\mu \in ]0,1] \Rightarrow \frac{\Delta x^{i_a}(\Delta x^{i_b})}{\Delta x^{i_b}}(-1)^{\ell_a+\ell_b+1} \geq 0\ .$$

By exchanging $\mathbf{a}$ and $\mathbf{b}$ we go to the other side of the derivative where the sign of $\Delta x^{i_a}$ becomes $-\ell_a$ and $\Delta x^{i_b}$, $\ell_b$ keeping the ratio at the same sign. From definitions 161 (p. 192), 275 (p. 304) and propositions 275 (p. 181), 512 (p. 288), we have:

$$\begin{aligned}\frac{\partial x^{i_a}}{\partial x^{i_b}} &= (-1)^{\ell_a+\ell_b+1}\left|\frac{\partial x^{i_a}}{\partial x^{i_b}}\right| \\ \frac{\partial x^{i_b}}{\partial x^{i_a}} &= (-1)^{\ell_a+\ell_b+1}\left|\frac{\partial x^{i_b}}{\partial x^{i_a}}\right|\ .\end{aligned} \tag{1.83}$$

**Let us demonstrate equation (1.82) (p. 77)**

Given $i_a \in [1,n]$, $i_b \in [1,n] - \{i_a\}$, from propositions 109 (p. 74) and 110 (p. 74), we have:

$$\begin{aligned}&\mathrm{proj}_{i_b} \circ \phi(C) = \mathrm{proj}_{i_b} \circ \phi({}^{C}\mathcal{B}^{i_b}_{1}) = \mathrm{proj}_{i_b} \circ \phi({}^{C}\mathcal{B}^{i_b}_{2}) \\ &\mathrm{proj}_{i_b} \circ \phi(\mathcal{B}(C)) = \mathrm{proj}_{i_b}\mathcal{B}(\phi(C)) = \overline{\mathrm{proj}_{i_b} \circ \phi(C)} \\ &\mathcal{B}(\mathrm{proj}_{i_b} \circ \phi(C)) = \overline{\mathrm{proj}_{i_b} \circ \phi(C)} \cap \mathcal{C}(\mathrm{proj}_{i_b} \circ \phi(C)) \\ &\mathcal{B}(\mathrm{proj}_{i_b} \circ \phi(C)) = \overline{\mathrm{proj}_{i_b} \circ \phi({}^{C}\mathcal{B}^{i_b}_{1})} \cap \mathcal{C}(\mathrm{proj}_{i_b} \circ \phi({}^{C}\mathcal{B}^{i_b}_{1})) \\ &= \overline{\mathrm{proj}_{i_b} \circ \phi({}^{C}\mathcal{B}^{i_b}_{2})} \cap \mathcal{C}(\mathrm{proj}_{i_b} \circ \phi({}^{C}\mathcal{B}^{i_b}_{2})) \\ &\mathcal{B}(\mathrm{proj}_{i_b} \circ \phi(C)) = \mathrm{proj}_{i_b}(\mathcal{B}(\phi({}^{C}\mathcal{B}^{i_b}_{1}))) = \mathrm{proj}_{i_b}(\mathcal{B}(\phi({}^{C}\mathcal{B}^{i_b}_{2})))\ .\end{aligned} \tag{1.84}$$

The last line comes from proposition 184 (p. 143) and the fact that for $\ell \in [1,2]$ the maps $(\mathrm{proj}_{i_b} \circ \phi : {}^{C}\mathcal{B}_{\ell}^{i_b} \to \overline{\phi(C)})$ are bijections. From propositions 164 (p. 138), 165 (p. 138) and 184 (p. 143), we have:

$$\begin{aligned}
&\mathrm{proj}_{i_b}^{-1}\mathcal{B}(\mathrm{proj}_{i_b} \circ \phi(C)) = \mathrm{proj}_{i_b}^{-1}(\overline{\mathrm{proj}_{i_b} \circ \phi(C)}) \cap \mathrm{proj}_{i_b}^{-1}(\mathcal{C}(\mathrm{proj}_{i_b} \circ \phi(C)))\\
&= \mathrm{proj}_{i_b}^{-1}(\overline{\mathrm{proj}_{i_b} \circ \phi(C)}) \cap \mathcal{C}_{\phi(\mathcal{B}(C))}(\mathrm{proj}_{i_b}^{-1} \circ \mathrm{proj}_{i_b} \circ \phi(C) \cap \phi(\mathcal{B}(C)))\\
&= \mathrm{proj}_{i_b}^{-1}(\overline{\mathrm{proj}_{i_b} \circ \phi(C)}) \cap \mathcal{C}_{\phi(\mathcal{B}(C))}(\phi({}^{C}\mathcal{B}_1^{i_b}) \cup \phi({}^{C}\mathcal{B}_2^{i_b}))
\end{aligned}$$

$$x \in \phi(\mathcal{C}_{\mathcal{B}(C)}({}^{C}\mathcal{B}_1^{i_b} \cup {}^{C}\mathcal{B}_2^{i_b})) \Rightarrow \mathrm{proj}_{i_b}(x) \in \mathcal{B}(\mathrm{proj}_{i_b}\phi(C)) \ .$$

From what we suppose (see premise of proposition 112 (p. 76)) $\phi(C)$ is an open convex in a closed ball of $\mathcal{R}^n$, $\overline{\phi(C)}$ is also in that closed ball (proposition 211 (p. 155)) and is convex (proposition 569 (p. 315)). From proposition 494 (p. 278), $\mathrm{proj}_{i_b}^{-1} \circ \mathrm{proj}_{i_b}(x) \cap \overline{\phi(C)}$ is convex. From proposition 571 (p. 316) we have:

$$\begin{aligned}
&x \in \phi({}^{C}\mathcal{B}_{\ell_a}^{i_a} \cap \mathcal{C}_{\mathcal{B}(C)}({}^{C}\mathcal{B}_1^{i_b} \cup {}^{C}\mathcal{B}_2^{i_b})) \Rightarrow\\
&(\exists \alpha \in \mathcal{R})(\exists \beta \in \mathcal{R})(\alpha \le \beta,\ \mathrm{proj}_{i_b}^{-1} \circ \mathrm{proj}_{i_b}(x) \cap \overline{\phi(C)}\\
&\quad = \mathrm{set}_{x'}\{(\exists \lambda \in [\alpha,\beta])(x' = \mathrm{proj}_{i_b}(x) + \lambda e_{i_b})\})
\end{aligned}$$

- $x = \alpha = \beta$, we have:

$$x \in \phi({}^{C}\mathcal{B}_{\ell_a}^{i_a} \cap \mathcal{C}_{\mathcal{B}(C)}({}^{C}\mathcal{B}_1^{i_b} \cup {}^{C}\mathcal{B}_2^{i_b})),\ \mathrm{proj}_{i_b}^{-1} \circ \mathrm{proj}_{i_b}(x) \cap \overline{\phi(C)} = \{x\}$$

  Since we suppose that the derivatives are continuous and in $C^\infty$ for any value of $x^{i_b}$, from the last line of equation (1.84) (p. 78), $x \in \mathcal{B}({}^{C}\mathcal{B}_1^{i_b}) \cap \mathcal{B}({}^{C}\mathcal{B}_2^{i_b})$ and from preceding paragraph:

$$\frac{\partial x^{i_a}}{\partial x^{i_b}} = |\frac{\partial x^{i_a}}{\partial x^{i_b}}|(-1)^{\ell_a+1+1} = |\frac{\partial x^{i_a}}{\partial x^{i_b}}|(-1)^{\ell_a+2+1} = 0$$

- $x \in [\alpha,\beta]$, $\alpha < \beta$, we have:

$$x \in \phi({}^{C}\mathcal{B}_{\ell_a}^{i_a} \cap \mathcal{C}_{\mathcal{B}(C)}({}^{C}\mathcal{B}_1^{i_b} \cup {}^{C}\mathcal{B}_2^{i_b})),\ \mathrm{proj}_{i_b}^{-1} \circ \mathrm{proj}_{i_b}(x) \cap \overline{\phi(C)} = [\alpha,\beta]$$

  In that the component $x^{i_a}$ of $x \in \mathcal{B}(C)$ does not depend on $x^{i_b}$ when $x^{i_b} \in ]\alpha,\beta[$ then its derivative $\frac{\partial x^{i_a}}{\partial x^{i_b}} = 0$. It is also true at the limits. Then $x^{i_b} \in [\alpha,\beta] \Rightarrow \frac{\partial x^{i_a}}{\partial x^{i_b}} = 0$.

**Proposition 114** Keeping the notations of proposition 112 (p. 76), if an oriented volume form (definition 60 (p. 54)) **Vol** has $(-1)^n$ as sign in ${}^{C}\mathcal{B}_2^n$ along the basis $\bigwedge_{j\in[1,n-1]} dx^j$ that is to say:[21]

$$x \in {}^{C}\mathcal{B}_2^n \Rightarrow \mathrm{Vol}(x) = \mathrm{vol}_2^n(x) \bigwedge_{j\in[1,n-1]} dx^j = (-1)^n|\mathrm{vol}_2^n(x)| \bigwedge_{j\in[1,n-1]} dx^j \qquad (1.85)$$

Then, in the manifold $\mathcal{B}(C)$ oriented by **Vol** (definition 63 (p. 61)), it is written as:

$$\begin{aligned}
&(i \in [1,n], \ell \in [1,2], x \in {}^{C}\mathcal{B}_\ell^i) \Rightarrow \mathrm{Vol}(x) = \mathrm{vol}_\ell^i(x) \bigwedge_{j\in[1,n]-i} dx^j\\
&= |\mathrm{vol}_\ell^i(x)|(-1)^{\ell+i} \bigwedge_{j\in[1,n]-i} dx^j \ .
\end{aligned} \qquad (1.86)$$

[21]This condition is not restrictive because we can always reshuffle the indices to make it true.

**Proof**

We keep the notations of proposition 112 (p. 76). Since we have (see definition 112 (p. 131), 121 (p. 136)):

$$\mathbf{C} \neq \emptyset,\ \phi(\mathbf{C}) \neq \emptyset,\ (\forall \mathbf{i} \in [1,\mathbf{n}])(\mathbf{proj_i}(\phi(\mathbf{C})) \neq \emptyset),$$
$$(\forall \mathbf{i} \in [1,\mathbf{n}])(\forall \ell \in [1,2])(\mathbf{proj_i}{}^{\mathbf{C}}\mathcal{B}^{\mathbf{i}}_{\ell} = \mathbf{proj_i}(\phi(\mathbf{C})))\ .$$

Then we have:

$$(\forall \mathbf{i} \in [1,\mathbf{n}])(\forall \ell \in [1,2])(\mathbf{proj_i}({}^{\mathbf{C}}\mathcal{B}^{\mathbf{i}}_{\ell}) \neq \emptyset, {}^{\mathbf{C}}\mathcal{B}^{\mathbf{i}}_{\ell} \neq \emptyset)\ .$$

Let us suppose that ${}^{\mathbf{C}}\mathcal{B}^{\mathbf{i_a}}_{\ell_a} \cap {}^{\mathbf{C}}\mathcal{B}^{\mathbf{i_b}}_{\ell_b} \neq \emptyset$. Since ${}^{\mathbf{C}}\mathcal{B}^{\mathbf{i_a}}_{\ell_a}$ and ${}^{\mathbf{C}}\mathcal{B}^{\mathbf{i_b}}_{\ell_b}$ are open, their intersections are open and $\mathbf{proj_{i_a}}({}^{\mathbf{C}}\mathcal{B}^{\mathbf{i_a}}_{\ell_a} \cap {}^{\mathbf{C}}\mathcal{B}^{\mathbf{i_b}}_{\ell_b})$ is open as $\mathbf{proj_{i_b}}({}^{\mathbf{C}}\mathcal{B}^{\mathbf{i_a}}_{\ell_a} \cap {}^{\mathbf{C}}\mathcal{B}^{\mathbf{i_b}}_{\ell_b})$ and both have in it an open ball centered on the projection of the same point $\phi(p)$ with the same radius $\mathbf{r}$. At that point $p \in {}^{\mathbf{C}}\mathcal{B}^{\mathbf{i_a}}_{\ell_a} \cap {}^{\mathbf{C}}\mathcal{B}^{\mathbf{i_b}}_{\ell_b}$, $\mathcal{B}$ has two coordinate systems $\prod^{\mathbf{e}}_{\mathbf{j}\in[1,\mathbf{n}]-\mathbf{i_a}} \mathbf{x^j}$ for ${}^{\mathbf{C}}\mathcal{B}^{\mathbf{i_a}}_{\ell_a}$ and $\prod^{\mathbf{e}}_{\mathbf{j}\in[1,\mathbf{n}]-\mathbf{i_b}} \mathbf{x^j}$ for ${}^{\mathbf{C}}\mathcal{B}^{\mathbf{i_b}}_{\ell_b}$. Let us write $\mathbf{Vol}$ in any ${}^{\mathbf{C}}\mathcal{B}^{\mathbf{i}}_{\ell}$ as

$$\mathbf{Vol} = \mathbf{vol}^{\mathbf{i}}_{\ell} \bigwedge_{\mathbf{j}\in[1,\mathbf{n}]-\mathbf{i}} \mathbf{dx^j}$$

with that we have in any overlapping region as $p \in {}^{\mathbf{C}}\mathcal{B}^{\mathbf{i_a}}_{\ell_a} \cap {}^{\mathbf{C}}\mathcal{B}^{\mathbf{i_b}}_{\ell_b}$ with $\mathbf{i_a} < \mathbf{i_b}$:

$$\begin{aligned}
\mathbf{Vol} &= \mathbf{vol}^{\mathbf{i_a}}_{\ell_a} \bigwedge_{\mathbf{j}\in[1,\mathbf{n}]-\mathbf{i_a}} \mathbf{dx^j} = \mathbf{vol}^{\mathbf{i_a}}_{\ell_a}(-1)^{\mathbf{i_b}+\mathbf{n}}(\bigwedge_{\mathbf{j}\in[1,\mathbf{n}]-\mathbf{i_a}-\mathbf{i_b}} \mathbf{dx^j}) \wedge \mathbf{dx^{i_b}} \\
&= \mathbf{vo}^{\mathbf{i_a}}_{\ell_a}(-1)^{\mathbf{i_b}+\mathbf{n}}(\bigwedge_{\mathbf{j}\in[1,\mathbf{n}]-\mathbf{i_a}-\mathbf{i_b}} \mathbf{dx^j}) \wedge \sum_{\mathbf{k}\in[1,\mathbf{n}]-\mathbf{i_b}} \frac{\partial \mathbf{x^{i_b}}}{\partial \mathbf{x^k}} \mathbf{dx^k} \\
&= \mathbf{vol}^{\mathbf{i_a}}_{\ell_a}(-1)^{\mathbf{i_b}+\mathbf{n}}(\bigwedge_{\mathbf{j}\in[1,\mathbf{n}]-\mathbf{i_a}-\mathbf{i_b}} \mathbf{dx^j}) \wedge \frac{\partial \mathbf{x^{i_b}}}{\partial \mathbf{x^{i_a}}} \mathbf{dx^{i_a}} \\
&= \mathbf{vol}^{\mathbf{i_a}}_{\ell_a} \frac{\partial \mathbf{x^{i_b}}}{\partial \mathbf{x^{i_a}}}(-1)^{\mathbf{i_b}+\mathbf{n}+\mathbf{i_a}+\mathbf{n}-1}(\bigwedge_{\mathbf{j}\in[1,\mathbf{n}]-\mathbf{i_b}} \mathbf{dx^j}) = \mathbf{vol}^{\mathbf{i_a}}_{\ell_a} \frac{\partial \mathbf{x^{i_b}}}{\partial \mathbf{x^{i_a}}}(-1)^{\mathbf{i_b}+\mathbf{i_a}-1}(\bigwedge_{\mathbf{j}\in[1,\mathbf{n}]-\mathbf{i_b}} \mathbf{dx^j}) \\
&= \mathbf{vol}^{\mathbf{i_b}}_{\ell_b} \bigwedge_{\mathbf{j}\in[1,\mathbf{n}]-\mathbf{i_b}} \mathbf{dx^j}\ .
\end{aligned} \tag{1.87}$$

Using equation (1.81) (p. 77) from proposition 113 (p. 77) and equation (1.87) (p. 80), we get:

$$\begin{aligned}
\mathbf{Vol} &= \mathbf{vol}^{\mathbf{i_a}}_{\ell_a} \bigwedge_{\mathbf{j}\in[1,\mathbf{n}]-\mathbf{i_a}} \mathbf{dx^j} = \mathbf{vol}^{\mathbf{i_a}}_{\ell_a} \frac{\partial \mathbf{x^{i_b}}}{\partial \mathbf{x^{i_a}}}(-1)^{\mathbf{i_b}+\mathbf{n}+\mathbf{i_a}+\mathbf{n}-1}(\bigwedge_{\mathbf{j}\in[1,\mathbf{n}]-\mathbf{i_b}} \mathbf{dx^j}) \\
&= \mathbf{vol}^{\mathbf{i_a}}_{\ell_a} \left|\frac{\partial \mathbf{x^{i_b}}}{\partial \mathbf{x^{i_a}}}\right|(-1)^{\mathbf{i_b}+\mathbf{i_a}+\ell_a+\ell_b}(\bigwedge_{\mathbf{j}\in[1,\mathbf{n}]-\mathbf{i_b}} \mathbf{dx^j}) \\
&= \mathbf{vol}^{\mathbf{i_b}}_{\ell_b} \bigwedge_{\mathbf{j}\in[1,\mathbf{n}]-\mathbf{i_b}} \mathbf{dx^j}\ .
\end{aligned}$$

Then we get:

$$\mathbf{vol}^{\mathbf{i_a}}_{\ell_a} \mathbf{vol}^{\mathbf{i_b}}_{\ell_b}(-1)^{\mathbf{i_b}+\mathbf{i_a}+\ell_a+\ell_b} > 0\ . \tag{1.88}$$

From proposition 112 (p. 76)

$$\mathcal{B}(\mathrm{C}) = \bigcup_{i\in[1,n],\ell\in[1,2]} {}^{\mathrm{C}}\mathcal{B}^{i}_{\ell} .$$

Since $\mathcal{B}(\mathrm{C})$ is fully path connected (proposition 106 (p. 71), definition 72 (p. 71)), from proposition 596 (p. 330), we have:

$$(\exists l_p \in \mathcal{Z}^+)(\exists(bp : [1,l_p] \to [1,n]))(\exists(bs : [1,l_p] \to [1,2]))(bp(1) = n, bs(1) = 2,$$
$$(k \in [1,l_p - 1] \Rightarrow {}^{\mathrm{C}}\mathcal{B}^{bp(k)}_{bs(k)} \cap {}^{\mathrm{C}}\mathcal{B}^{bp(k+1)}_{bs(k+1)} \neq \emptyset), \bigcup_{k\in[1,l_p]} {}^{\mathrm{C}}\mathcal{B}^{bp(k)}_{bs(k)} = \mathcal{B}(\mathrm{C})) .$$

Since $\bigcup_{k\in[1,l_p]} {}^{\mathrm{C}}\mathcal{B}^{bp(k)}_{bs(k)} = \mathcal{B}(\mathrm{C})$

$$(i \in [1,n], \ell \in [1,2]) \Rightarrow ({}^{\mathrm{C}}\mathcal{B}^{i}_{\ell} \neq \emptyset,$$
$$(\exists q \in {}^{\mathrm{C}}\mathcal{B}^{i}_{\ell})(\exists k \in [1,l_p])(q \in {}^{\mathrm{C}}\mathcal{B}^{bp(k)}_{bs(k)}, {}^{\mathrm{C}}\mathcal{B}^{i}_{\ell} \cap {}^{\mathrm{C}}\mathcal{B}^{bp(k)}_{bs(k)} \neq \emptyset)) .$$

We take that $k$, we call $l'_p = k + 1$ and we generate two new maps:

$$(u : [1,l'_p] \to [1,n]), (d : [1,l'_p] \to [1,2]),$$
$$j \in [1,l'_p[ \Rightarrow (u(j) = bp(j), d(j) = bs(j)),$$
$$u(l'_p) = i, d(l'_p) = \ell .$$

Supposing equation (1.86) (p. 79) is true for ${}^{\mathrm{C}}\mathcal{B}^{u(k')}_{d(k')}$ when $k' \in [1,j] \subset [1,l'_p[$, let us show that equation (1.86) (p. 79) is then also true for $k' \in [1,j+1]$:
Since ${}^{\mathrm{C}}\mathcal{B}^{u(j)}_{d(j)} \cap {}^{\mathrm{C}}\mathcal{B}^{u(j+1)}_{d(j+1)} \neq \emptyset$, we can apply equation (1.88) (p. 80) with (1.86) (p. 79) for $j$. It gives:

$$\mathrm{vol}^{u(j)}_{d(j)} \mathrm{vol}^{u(j)+1}_{d(j+1)} (-1)^{u(j)+u(j+1)+d(j)+d(j+1)}$$
$$= (-1)^{u(j)+d(j)} \mathrm{vol}^{u(j)+1}_{d(j+1)} (-1)^{u(j)+u(j+1)+d(j)+d(j+1)}$$
$$= \mathrm{vol}^{u(j+1)}_{d(j+1)} (-1)^{u(j+1)+d(j+1)} > 0,$$
$$\mathrm{vol}^{u(j+1)}_{d(j+1)} = |\mathrm{vol}^{u(j+1)}_{d(j+1)}| (-1)^{u(j+1)+d(j+1)} .$$

We impose by equation (1.85) (p. 79) that 1.86 (p. 79) is true for $j = 1$, 1.86 (p. 79) is then true for any $j \in [1,l'_p]$ (proposition 927 (p. 520)) and in particular for $j = l'_p$. It gives:

$$\mathrm{vol}^{i}_{\ell}(-1)^{i+\ell} > 0 \text{ or } \mathrm{vol}^{i}_{\ell} = |\mathrm{vol}^{i}_{\ell}|(-1)^{i+\ell} .$$

**Proposition 115** Given a differential manifold $\mathrm{M}$ of dimension $n$ and a convex open $\mathrm{C} \subset \mathrm{M}$ of dimension $n$ included in a ball (definition 71 (p. 67)) with an oriented (definition 63 (p. 61)) differentiable manifold smooth surface $\mathcal{B}(\mathrm{C})$ of dimension $n - 1$ (definition 68 (p. 65)), we have:

$$\int_{\mathrm{C}} d\omega = \int_{\mathcal{B}(\mathrm{C})} \omega .$$

See definition 69 (p. 66).

**Proof**

We use the notation explained in definitions 71 (p. 67) and 63 (p. 61). We also set $\mathbf{e_i} = \prod^{e}_{j\in[1,n]} \delta^j_i = (\delta^j_i : [1,n] \to \mathcal{R}) \in \mathcal{R}^n$ as the reference basis of $\mathcal{R}^n$ (see equation (2.106) (p. 267) in definition 242 (p. 267)). We suppose for simplicity $\phi^{-1}(0) \in \mathbf{C}$. We take as a volume form:

$$\mathrm{Vol} = \bigwedge_{j\in[1,n]} dx^j$$

and we set $p = \phi^{-1}(0) \in \mathbf{C}$. Since $\omega \in \Omega^{n-1}$, we have (see definition 56 (p. 49)):

$$\begin{aligned}
\omega &= \sum_{i\in[1,n]} f_i(x) \bigwedge_{j\in[1,n]-i} dx^j = \sum_{i\in[1,n]} \omega_i, \quad \omega_i = f_i(x) \bigwedge_{j\in[1,n]-i} dx^j \\
d\omega &= \sum_{i\in[1,n]} d\omega_i = \sum_{i\in[1,n]} \sum_{\ell\in[1,n]} \partial_\ell f_i(x) dx^\ell \wedge \bigwedge_{j\in[1,n]-i} dx^j \\
&= \sum_{i\in[1,n]} \sum_{\ell\in[1,n]} \partial_\ell f_i(x) \delta^\ell_i dx^i \wedge \bigwedge_{j\in[1,n]-i} dx^j \\
&= \sum_{i\in[1,n]} \partial_i f_i(x) dx^i \wedge \bigwedge_{j\in[1,n]-i} dx^j = \sum_{i\in[1,n]} (-1)^{i-1} \partial_i f_i \bigwedge_{j\in[1,n]} dx^j
\end{aligned} \tag{1.89}$$

with $\omega_i = f_i(x) \bigwedge_{j\in[1,n]-i} dx^j$, $d\omega_i = \partial_i f_i(x) dx^i \wedge \bigwedge_{j\in[1,n]-i} dx^j$. (definitions 121 (p. 136), 244 (p. 270), 292 (p. 337), proposition 616 (p. 349)). Using equation (1.89) (p. 82), we have (see propositions 624 (p. 357), 618 (p. 351)):

$$\begin{aligned}
&\int_{\overline{C}} d\omega = \int_C d\omega = \sum_{i\in[1,n]} \int_{\overline{C}} d\omega_i = \sum_{i\in[1,n]} J_{\overline{\phi(C)}} d\omega_i (\bigotimes_{j\in[1,n]} \frac{\partial}{\partial x^j}) \\
&= \sum_{i\in[1,n]} (-1)^{i-1} J_{\overline{\phi(C)}} (\partial_i f_i(x) \bigwedge_{j\in[1,n]} dx^j (\bigotimes_{j\in[1,n]} \frac{\partial}{\partial x^j})) = \sum_{i\in[1,n]} (-1)^{i-1} J_{\overline{\phi(C)}} (\partial_i f_i(x)) .
\end{aligned}$$

Since $\phi$ is an isomorphism (definition 142 (p. 166)), from proposition 239 (p. 166), we have $\overline{\phi(C)} = \phi(\overline{C})$. Calling $\Psi_{\mathbf{A}}(x)$ the characteristic map of the set $\mathbf{A}$ (definition 293 (p. 341)), we have from propositions 109 (p. 74), 568 (p. 315), 569 (p. 315):

$$\Psi_{\overline{\phi(C)}}(x) = \Psi_{\mathrm{proj}_i(\overline{\phi(C)})}(\mathrm{proj}_i x) \times \Psi_{[h^i_1(\mathrm{proj}_i x), h^i_2(\mathrm{proj}_i x)]}(x^i) .$$

Then

$$\begin{aligned}
&J_{\overline{\phi(C)}}(\partial_i f_i(x)) = J(\Psi_{\overline{\phi(C)}}(x) \partial_i f_i(x)) \\
&= J_{\mathcal{R}^n}(\Psi_{\mathrm{proj}_i(\overline{\phi(C)})}(\mathrm{proj}_i x) \times \Psi_{[h^i_1(\mathrm{proj}_i x), h^i_2(\mathrm{proj}_i x)]}(x^i) \times \partial_i f_i(x)) .
\end{aligned}$$

Applying propositions 620 (p. 352), 618 (p. 351), 606 (p. 338), we have:

$$\begin{aligned}
&J_{\mathcal{R}^n}(\Psi_{\mathrm{proj}_i(\overline{\phi(C)})}(\mathrm{proj}_i x) \times \Psi_{[h^i_1(\mathrm{proj}_i x), h^i_2(\mathrm{proj}_i x)]}(x^i) \times \partial_i f_i(x)) = \\
&J_{\mathcal{R}^{n-1}}(\Psi_{\mathrm{proj}_i(\overline{\phi(C)})}(\mathrm{proj}_i x) J_{\mathcal{R}}(\Psi_{[h^i_1(\mathrm{proj}_i x), h^i_2(\mathrm{proj}_i x)]}(x^i) \times \partial_i f_i(x))) = \\
&J_{\mathcal{R}^{n-1}}(\Psi_{\mathrm{proj}_i(\overline{\phi(C)})}(\mathrm{proj}_i x)(f_i(x^i_2) - f_i(x^i_1))
\end{aligned}$$

$$\int_{\overline{C}} d\omega = \sum_{i\in[1,n]} (-1)^{i-1} J_{\mathrm{proj}_i(\overline{\phi(C)})} ( \bigwedge_{j\in[1,n]-i} dx^j ( \bigotimes_{j\in[1,n]-i} \frac{\partial}{\partial x^j})(f_i(x^i_2) - f_i(x^i_1))) \tag{1.90}$$

with:

$$\mathbf{x}^{\mathbf{i}}_{\ell} = (\prod^{\mathbf{e}}_{\mathbf{k}\in[1,\mathbf{i}-1]} \mathbf{x}^{\mathbf{k}}, \mathbf{h}^{\mathbf{i}}_{\ell}(\prod^{\mathbf{e}}_{\mathbf{k}\in[1,\mathbf{n}]-\mathbf{i}} \mathbf{x}^{\mathbf{k}}), \prod^{\mathbf{e}}_{\mathbf{k}\in[\mathbf{i}+1,\mathbf{n}]} \mathbf{x}^{\mathbf{k}}) = \sum_{\mathbf{k}\in[1,\mathbf{n}]-\mathbf{i}} \mathbf{x}^{\mathbf{k}}\mathbf{e}_{\mathbf{k}} + \mathbf{h}^{\mathbf{i}}_{\ell}(\sum_{\mathbf{k}\in[1,\mathbf{n}]-\mathbf{i}} \mathbf{x}^{\mathbf{k}}\mathbf{e}_{\mathbf{k}})\mathbf{e}_{\mathbf{i}} \,.$$

(See definition 378 (p. 484) and proposition 841 (p. 484). )
$\mathbf{C}$ being open, $\boldsymbol{\phi}$ being an isomorphism (definition 142 (p. 166)), $\boldsymbol{\phi}(\mathbf{C})$ is open (definition 141 (p. 165)), $\overline{\boldsymbol{\phi}(\mathbf{C})}$ is convex (definition 71 (p. 67)). From proposition 575 (p. 319) $\mathbf{proj_i}(\boldsymbol{\phi}(\mathbf{C})) = \mathbf{proj_i}(\overline{\boldsymbol{\phi}(\mathbf{C})})$. Then from propositions 624 (p. 357) and 618 (p. 351):

$$\begin{aligned}&\int_{\overline{\mathbf{C}}} \mathbf{d}\omega = \int_{\mathbf{C}} \mathbf{d}\omega = \\ &\sum_{\mathbf{i}\in[1,\mathbf{n}]} (-1)^{\mathbf{i}-1}\mathbf{J}_{\mathbf{proj_i}(\boldsymbol{\phi}(\mathbf{C}))}(\bigwedge_{\mathbf{j}\in[1,\mathbf{n}]-\mathbf{i}} \mathbf{dx^j}(\bigotimes_{\mathbf{j}\in[1,\mathbf{n}]-\mathbf{i}} \frac{\partial}{\partial \mathbf{x^j}})(\mathbf{f_i}(\mathbf{x}^{\mathbf{i}}_{2}) - \mathbf{f_i}(\mathbf{x}^{\mathbf{i}}_{1}))) \,.\end{aligned} \tag{1.91}$$

From proposition 112 (p. 76), the set of opens ${}^{\mathbf{C}}\boldsymbol{\mathcal{B}}^{\mathbf{i}}_{\ell}$ is a covering set of $\boldsymbol{\mathcal{B}}(\mathbf{C})$ with $(\mathbf{proj_i} \circ \boldsymbol{\phi} : \boldsymbol{\phi}({}^{\mathbf{C}}\boldsymbol{\mathcal{B}}^{\mathbf{i}}_{\ell}) \leftrightarrow \boldsymbol{\mathcal{R}}^{\mathbf{n}-1})$ as associated bijections. We also have:

$$\begin{aligned}&(\forall \ell \in [1,2])(\forall \mathbf{i} \in [1,\mathbf{n}])\,(\mathbf{proj_i}(\boldsymbol{\phi}(\mathbf{C})) = \mathbf{proj_i}(\boldsymbol{\phi}({}^{\mathbf{C}}\boldsymbol{\mathcal{B}}^{\mathbf{i}}_{\ell})),\\ &{}^{\mathbf{C}}\boldsymbol{\mathcal{B}}^{\mathbf{i}}_{\ell} = \boldsymbol{\mathcal{I}}({}^{\mathbf{C}}\boldsymbol{\mathcal{B}}^{\mathbf{i}}_{\ell}),\\ &\boldsymbol{\mathcal{B}}({}^{\mathbf{C}}\boldsymbol{\mathcal{B}}^{\mathbf{i}}_{\ell}) \cap \boldsymbol{\mathcal{I}}({}^{\mathbf{C}}\boldsymbol{\mathcal{B}}^{\mathbf{i}}_{\ell}) = \emptyset,\\ &\boldsymbol{\mathcal{B}}({}^{\mathbf{C}}\boldsymbol{\mathcal{B}}^{\mathbf{i}}_{\ell}) \cup \boldsymbol{\mathcal{I}}({}^{\mathbf{C}}\boldsymbol{\mathcal{B}}^{\mathbf{i}}_{\ell}) = \overline{{}^{\mathbf{C}}\boldsymbol{\mathcal{B}}^{\mathbf{i}}_{\ell}})\end{aligned}$$

where $\overline{{}^{\mathbf{C}}\boldsymbol{\mathcal{B}}^{\mathbf{i}}_{\ell}}$ is the closure (definition 136 (p. 155)) of ${}^{\mathbf{C}}\boldsymbol{\mathcal{B}}^{\mathbf{i}}_{\ell}$. From proposition 219 (p. 158), $\overline{{}^{\mathbf{C}}\boldsymbol{\mathcal{B}}^{\mathbf{i}}_{\ell}}$ is closed.
We choose for the surface $\boldsymbol{\mathcal{B}}(\mathbf{C})$, the covering sets

$$\mathbf{i} \in [1,\mathbf{n}], \ell \in [1,2], {}^{\mathbf{C}}\boldsymbol{\mathcal{B}}^{\mathbf{i}}_{\ell}$$

the maps associated to each $\mathbf{i} \in [1,\mathbf{n}], \ell \in [1,2]$ are $\mathbf{proj_i} \circ \boldsymbol{\phi}$ with $\prod^{\mathbf{e}}_{\mathbf{j}\in[1,\mathbf{n}]-\mathbf{i}} \mathbf{x^j}$ as variables defining the position of an element of ${}^{\mathbf{C}}\boldsymbol{\mathcal{B}}^{\mathbf{i}}_{\ell}$. The orientation of the bases of each covering element is given by an oriented volume form of $\boldsymbol{\mathcal{B}}(\mathbf{C})$ multiplied by $\pm\mathbf{1}$ in order to have the sign + for $\bigotimes_{\mathbf{j}\in[2,\mathbf{n}]} \mathbf{dx^j}$ in ${}^{\mathbf{C}}\boldsymbol{\mathcal{B}}^{1}_{2}$. Proposition 114 (p. 79) gives $\mathbf{Vol(x)} = -|\mathbf{Vol(x)}|$ when $\mathbf{l} = \mathbf{2},\ \mathbf{i} = \mathbf{1}$ and $\mathbf{Vol(x)} = (-1)^{\mathbf{n}}|\mathbf{Vol(x)}|$ when $\mathbf{l} = \mathbf{2},\ \mathbf{i} = \mathbf{n}$ so we have reverse the sign of the volume. Doing that we get in each covering element an integration volume equal to $(-1)^{\mathbf{i}+\ell-1}$ for $\bigotimes_{\mathbf{j}\in[1,\mathbf{n}]-\mathbf{i}} \mathbf{dx^j}$. As explained in section 1.12.2 (p. 61) and in definitions 63 (p. 61), 67 (p. 65), 65 (p. 65), $\mathbf{g}$ being a map $(\mathbf{g} : {}^{\mathbf{C}}\boldsymbol{\mathcal{B}}^{\mathbf{i}}_{\ell} \to \boldsymbol{\mathcal{R}})$, when we do an integration of $\mathbf{G} = \mathbf{g} \bigwedge_{\mathbf{j}\in[1,\mathbf{n}]-\mathbf{i}} \mathbf{dx^j}$ on the surface ${}^{\mathbf{C}}\boldsymbol{\mathcal{B}}^{\mathbf{i}}_{\ell}$, we get:

$$\int_{{}^{\mathbf{C}}\boldsymbol{\mathcal{B}}^{\mathbf{i}}_{\ell}} \mathbf{G} = \mathbf{J}_{{}^{\mathbf{C}}\boldsymbol{\mathcal{B}}^{\mathbf{i}}_{\ell}}(\mathbf{sign}(\mathbf{Vol}(\bigotimes_{\mathbf{j}\in[1,\mathbf{n}]-\mathbf{i}} \partial_{\mathbf{j}}))\mathbf{G}(\prod^{\mathbf{e}}_{\mathbf{j}\in[1,\mathbf{n}]-\mathbf{i}} \mathbf{x^j})(\bigotimes_{\mathbf{j}\in[1,\mathbf{n}]-\mathbf{i}} \partial_{\mathbf{j}})) \,.$$

It gives from formula (1.91) (p. 83):

$$\int_{{}^{\mathbf{C}}\boldsymbol{\mathcal{B}}^{\mathbf{i}}_{\ell}} \mathbf{G} = (-1)^{\mathbf{i}+\ell-1}\mathbf{J}_{{}^{\mathbf{C}}\boldsymbol{\mathcal{B}}^{\mathbf{i}}_{\ell}}(\mathbf{g}(\prod^{\mathbf{e}}_{\mathbf{j}\in[1,\mathbf{n}]-\mathbf{i}} \mathbf{x^j}) \bigwedge_{\mathbf{j}\in[1,\mathbf{n}]-\mathbf{i}} \mathbf{dx^j}(\bigotimes_{\mathbf{j}\in[1,\mathbf{n}]-\mathbf{i}} \partial_{\mathbf{j}})) \,.$$

From that we get:

$$\sum_{\ell\in[1,2]} \int_{{}^{C}\mathcal{B}^{i}_{\ell}} f_i \bigwedge_{j\in[1,n]-i} dx^j$$
$$= \sum_{\ell\in[1,2]} (-1)^{i+\ell-1} J_{\mathrm{proj}_i(\overline{\phi(C)})}\Big(\bigwedge_{k\in[1,n]-i} dx^k\Big(\bigotimes_{j\in[1,n]-i}\frac{\partial}{\partial x^j}\Big) f_i(x^i_\ell)\Big)$$
$$= (-1)^{i-1} J_{\mathrm{proj}_i(\overline{\phi(C)})}\Big(\bigwedge_{k\in[1,n]-i} dx^k\Big(\bigotimes_{j\in[1,n]-i}\frac{\partial}{\partial x^j}\Big)(f_i(x^i_2) - f_i(x^i_1))\Big)\ .$$

Applying that to formula (1.91) (p. 83), we have:

$$\int_C d\omega = \sum_{i\in[1,n]}\sum_{\ell\in[1,2]} \int_{{}^{C}\mathcal{B}^{i}_{\ell}} f_i \bigwedge_{j\in[1,n]-i} dx^j = \sum_{i\in[1,n]}\sum_{\ell\in[1,2]} \int_{{}^{C}\mathcal{B}^{i}_{\ell}} \omega_i\ .$$

Using proposition 110 (p. 74), we have for any $\mathbf{i} \in [\mathbf{1}, \mathbf{n}]$:

$${}^{C}\mathcal{B}^{i}_{1} \cap {}^{C}\mathcal{B}^{i}_{2} = \emptyset$$
$$(\forall \ell \in [1,2])(\mathrm{proj}_j\phi({}^{C}\mathcal{B}^{i}_{\ell}) = \mathrm{proj}_j\phi(C))$$
$$(\forall \ell \in [1,2])(\overline{\mathrm{proj}_j\phi({}^{C}\mathcal{B}^{i}_{\ell})} = \mathrm{proj}_j\overline{\phi({}^{C}\mathcal{B}^{i}_{\ell})} = \overline{\mathrm{proj}_j(\phi(C))} = \mathrm{proj}_j\mathcal{B}(\phi(C)))$$

$$\sum_{\ell\in[1,2]} \int_{{}^{C}\mathcal{B}^{i}_{\ell}} \omega_i = \int_{\bigcup_{\ell\in[1,2]} {}^{C}\mathcal{B}^{i}_{\ell}} \omega_i = \int_{\mathcal{B}(C)} \omega_i - \int_{\mathcal{C}_{\mathcal{B}(C)}(\bigcup_{\ell\in[1,2]} {}^{C}\mathcal{B}^{i}_{\ell})} \omega_i\ .$$

**Let us prove that:**

$$(\forall i \in [1,n])(x \in \mathcal{B}(C) - \bigcup_{\ell\in[1,2]} \overline{{}^{C}\mathcal{B}^{i}_{\ell}})$$
$$\Rightarrow (j \neq i \Rightarrow \omega_i(\bigotimes_{\nu\in[1,n]-j}\frac{\partial}{\partial x^\nu})(x) = 0)$$

We have:

$$\omega_i(\bigotimes_{\nu\in[1,n]-j}\frac{\partial}{\partial x^\nu})(x) = f_i(x) \bigwedge_{k\in[1,n]-i} dx^k(\bigotimes_{\nu\in[1,n]-j}\frac{\partial}{\partial x^\nu})(x)$$
$$= f_i \det_{k\in[1,n]-i,\mu\in[1,n]-j}(\frac{\partial x^k}{\partial x^\mu})$$

Since at any point in $x^i e_i \in \phi({}^{C}\mathcal{B}^{j}_{\ell})$ corresponding to $\prod^{e}_{k\in[1,n]-j} x^k$ with $x^j = h^j_\ell(\prod^{e}_{k\in[1,n]-j} x^k)$, we have:

$$k \in [1,n] - i - j \Rightarrow \frac{\partial x^k}{\partial x^i} = 0\ . \tag{1.92}$$

From proposition 108 (p. 72), we have:

$$(\exists j \in [1,n])(\exists \ell \in [1,2])(x \in {}^{C}\mathcal{B}^{j}_{\ell})\ .$$

Then:

$$x \in (\mathcal{B}(C) - \bigcup_{\ell\in[1,2]} \overline{{}^{C}\mathcal{B}^{i}_{\ell}}) \Rightarrow x \in {}^{C}\mathcal{B}^{j}_{\ell} \cap \mathcal{C}(\bigcup_{\ell\in[1,2]} {}^{C}\mathcal{B}^{i}_{\ell})\ .$$

Using that with proposition 113 (p. 77), equation (1.82), we have:

$$\frac{\partial x^j}{\partial x^i} = \frac{\partial h^j_\ell(\prod^e_{k\in[1,n]-j} x^k)}{\partial x^i} = 0 \tag{1.93}$$

then from equation (1.92) (p. 84), equation (1.93) (p. 85) and proposition 443 (p. 255), we have:

$$\det_{k\in[1,n]-i,\mu\in[1,n]-j}(\frac{\partial x^k}{\partial x^\mu}) = 0 \ .$$

Proposition 115 (p. 81) is true when the covering of $\mathcal{B}(C)$ is the one chosen: $({}^C\mathcal{B}^i_\ell : [1, n] \times [1, 2] \to \mathcal{P}(\mathcal{B}(C)))$ for the opens covering and $(proj_i \circ \phi : {}^C\mathcal{B}^i_\ell \to \mathcal{R}^{n-1})$ for the associated mapping to $\mathcal{R}^{n-1}$. From proposition 101 (p. 63), the integral remains the same when using another open covering $(U_k : [1, Card(K)] \to \mathcal{PB}(C))$ with another set of associated maps $({}^{\mathcal{B}}\phi_k : U_k \to \mathcal{R}^{n-1})$ such that $(\phi \circ {}^{\mathcal{B}}\phi_k^{-1} : {}^{\mathcal{B}}\phi_k^{-1}(U_k) \to \mathcal{R}^{n-1}) \in C^\infty$. Proposition 115 (p. 81) is true in the general case.

**Proposition 116** Keeping notations of proposition 115 (p. 81), $C$ being in a ball $\bar{B}(0, r_1)$ and $\psi$ a map $(\psi : B(0, r_1) \to \mathcal{R}^n) \in C^\infty$. Then with the new map $(\psi\phi : C \to \mathcal{R}^n)$ defining the integral over $C$ we have:

$$\int_C d\omega = \int_{\mathcal{B}(C)} \omega \ .$$

It justifies the fact that we do not specify the mapping used in the above formula and in proposition 115 (p. 81).

**Proof**

$\bar{B}(0, r_1)$ is compact (definition 136 (p. 155), 140 (p. 160), propositions 219 (p. 158), 521 (p. 290)), $\psi$ is a continuous map, $\psi(\bar{B}(0, r_1))$ is compact (proposition 240 (p. 167)) then is also in a ball (proposition 523 (p. 291)). Then proposition 116 (p. 85) is a consequence of propositions 115 (p. 81) and 101 (p. 63).

## 1.13 Lie bracket

**Definition 74 Lie bracket of two vector fields.**

One calls Lie bracket of two vector fields $u$ and $v$ on a differentiable manifold $M$, the map $[u, v] : \mathcal{F} \to \mathcal{F}$ which for any mapping $f \in \mathcal{F}$ $(f : M \to \mathcal{R})$ gives the following element of $\mathcal{F}$:

$$[u, v](f) = u(v(f)) - v(u(f)) \ . \tag{1.94}$$

**Proposition 117** The Lie bracket of two vector fields is a vector field.

**Proof**

Given two vector fields $u$ and $v$ on a differentiable manifold $M$, their Lie bracket, two elements $f \in \mathcal{F}$, $g \in \mathcal{F}$ (definitions 6 (p. 3), 15 (p. 8)), we have:

$$\begin{aligned}[u, v](fg) &= u(v(fg)) - v(u(fg)) = u(v(f)g + fv(g)) - v(u(f)g + fu(g)) \\ &= u(v(f))g + v(f)u(g) + u(f)v(g) + fu(v(g)) \\ &- v(u(f))g - u(f)v(g) - v(f)u(g) - fv(u(g)) \\ &= u(v(f))g - v(u(f))g + fu(v(g)) - fv(u(g)) = [u, v](f)g + f[u, v](g) \ .\end{aligned}$$

**Proposition 118** Given two vector fields $\mathbf{u}$ and $\mathbf{v}$ on a differentiable manifold $\mathbf{M}$, their Lie bracket, two elements $\mathbf{f} \in \mathcal{F}$, $\lambda \in \mathcal{F}$ (definitions 6 (p. 3), 15 (p. 8)), we have:

$$[\mathbf{u}, \mathbf{v}] = -[\mathbf{v}, \mathbf{u}] \tag{1.95}$$

$$[\mathbf{u}, \lambda\mathbf{v}](\mathbf{f}) = \lambda[\mathbf{u}, \mathbf{v}](\mathbf{f}) + \mathbf{u}(\lambda)\mathbf{v}(\mathbf{f}) \tag{1.96}$$

$$[\lambda\mathbf{u}, \mathbf{v}](\mathbf{f}) = \lambda[\mathbf{u}, \mathbf{v}](\mathbf{f}) - \mathbf{v}(\lambda)\mathbf{u}(\mathbf{f}) \ . \tag{1.97}$$

**Proof**
From definition 74 (p. 85) we have:

$$[\mathbf{u}, \mathbf{v}]\mathbf{f} = \mathbf{u}(\mathbf{v}(\mathbf{f})) - \mathbf{v}(\mathbf{u}(\mathbf{f})) = -(\mathbf{v}(\mathbf{u}(\mathbf{f})) - \mathbf{u}(\mathbf{v}(\mathbf{f}))) = -[\mathbf{v}, \mathbf{u}]$$

$$[\mathbf{u}, \lambda\mathbf{v}](\mathbf{f}) = \mathbf{u}(\lambda\mathbf{v}(\mathbf{f})) - \lambda\mathbf{v}(\mathbf{u}(\mathbf{f})) = \lambda[\mathbf{u}, \mathbf{v}](\mathbf{f}) + \mathbf{u}(\lambda)\mathbf{v}(\mathbf{f})$$

and

$$[\lambda\mathbf{u}, \mathbf{v}] = -[\mathbf{v}, \lambda\mathbf{u}] = -\lambda[\mathbf{v}, \mathbf{u}] + \mathbf{u}(\lambda)\mathbf{v} = \lambda[\mathbf{u}, \mathbf{v}] - \mathbf{u}(\lambda)\mathbf{v} \ .$$

**Proposition 119** Given an open cover $\mathbf{U}$ of a manifold $\mathbf{M}$ of dimension $\mathbf{n}$, considering the restricted manifold $\mathbf{U}$ with its set of restricted vector fields $\mathcal{V}$, given the standard basis of $\mathcal{V}$: $(\partial_i : [1, \mathbf{n}] \to \mathcal{V})$ (definition 21 (p. 16)), we have:

$$(\forall(i, j) \in [1, \mathbf{n}]^2)([\partial_i, \partial_j] = 0) \tag{1.98}$$

$$(\mathbf{u} = \mathbf{u}^i\partial_i \in \mathcal{V}, \mathbf{v} = \mathbf{v}^i\partial_i \in \mathcal{V}) \Rightarrow [\mathbf{u}, \mathbf{v}] = \mathbf{u}^i\partial_i(\mathbf{v}^j)\partial_j - \mathbf{v}^j\partial_j(\mathbf{u}^i)\partial_i = \mathbf{u}(\mathbf{v}^j)\partial_j - \mathbf{v}(\mathbf{u}^i)\partial_i \ . \tag{1.99}$$

**Proof**
The first equation comes directly from proposition 652 (p. 396):

$$[\partial_i, \partial_j]\mathbf{f} = \partial_i\partial_j\mathbf{f} - \partial_j\partial_i\mathbf{f} = 0 \ .$$

Then from that and definition 74 (p. 85), we have:

$$[\mathbf{u}, \mathbf{v}] = [\mathbf{u}^i\partial_i, \mathbf{v}^j\partial_j] = \mathbf{v}^j[\mathbf{u}^i\partial_i, \partial_j] + \mathbf{u}^i\partial_i(\mathbf{v}^j)\partial_j = \mathbf{u}^i\mathbf{v}^j[\partial_i, \partial_j] - \mathbf{v}^j\partial_j(\mathbf{u}^i)\partial_i + \mathbf{u}^i\partial_i(\mathbf{v}^j)$$
$$= \mathbf{u}^i\partial_i(\mathbf{v}^j)\partial_j - \mathbf{v}^j\partial_j(\mathbf{u}^i)\partial_i = \mathbf{u}(\mathbf{v}^j)\partial_j - \mathbf{v}(\mathbf{u}^i)\partial_i \ .$$

## 1.14 Bundles and differentiable manifold

### 1.14.1 Bundles and sections of vector bundle

**Definition 75 Bundle.**

One calls bundle, a set of two differentiable manifolds $\mathbf{E}$ and $\mathbf{M}$ with a surjective (definition 123 (p. 139)) continuous map $(\pi : \mathbf{E} \to \mathbf{M})$ (definition 141 (p. 165)). We suppose also that the covering sets of opens of $\mathbf{E}$ and $\mathbf{M}$ are matching each other by $\pi$.

**Definition 76 Fibers of a bundle.**

For any element $\mathbf{x} \in \mathbf{M}$, one calls fiber $\mathbf{F}_\mathbf{x} = \pi^{-1}(\mathbf{x})$ the inverse by $\pi$ of $\mathbf{x}$ in $\mathbf{E}$.

**Definition 77 Vector bundle.**

A bundle is called vector bundle when each fiber $\mathbf{F_x}$ of that bundle is isomorphic to a common set $\mathbf{F}$. We suppose also that $\mathbf{F}$ is a vector space of dimension $\mathbf{n'}$ on a field ($\mathcal{R}$ or $\mathcal{C}$) isomorphic to $\mathcal{R}^{\mathbf{n'}}$ or $\mathcal{C}^{\mathbf{n'}}$. In a vector bundle, we have:

$$\mathbf{E} = \mathbf{F} \times \mathbf{M} \,.$$

We suppose that the topology of $\mathbf{F}$ is generated by a metric of the type:

$$(\mathbf{x} \in \mathbf{F}, \mathbf{y} \in \mathbf{F}, \mathbf{F} = \mathcal{R}^{\mathbf{n'}} \vee \mathbf{F} = \mathcal{C}^{\mathbf{n'}}) \Rightarrow \mathbf{d}(\mathbf{y}, \mathbf{x}) = \|\mathbf{x} - \mathbf{y}\|_{\mathbf{p}}$$
$$= (\sum_{\mathbf{i} \in [1, \mathbf{n'}]} (\|\mathbf{x_i} - \mathbf{y_i}\|)^{\mathbf{p}})^{1/\mathbf{p}} \,.$$

We shall suppose that $\mathbf{p} = 2$ and

$$\mathbf{F} = \mathcal{R}^{\mathbf{n'}} \Rightarrow \|\mathbf{x_i} - \mathbf{y_i}\| = |\mathbf{x_i} - \mathbf{y_i}|$$
$$\mathbf{F} = \mathcal{C}^{\mathbf{n'}} \Rightarrow \|\mathbf{x_i} - \mathbf{y_i}\| = \sqrt{(\mathbf{Real}(\mathbf{x_i} - \mathbf{y_i}))^2 + (\mathbf{Im}(\mathbf{x_i} - \mathbf{y_i}))^2} \,.$$

We suppose also that the topology of $\mathbf{E}$ is the one generated by product $\mathbf{F} \times \mathbf{M}$ with definition 134 (p. 153). The mapping $(\pi : \mathbf{E} \to \mathbf{M})$ is the projection along $\mathbf{F}$ of $\mathbf{E}$ on $\mathbf{M}$ according to definition 121 (p. 136). From proposition 237 (p. 166) this map is continuous. With our simplifications, to define $\mathbf{F}$ as a manifold, we use $\mathbf{F}$ as the unique covering element with the identity map $(\mathbf{y}(\mathbf{x}) = \mathbf{x} : \mathbf{F} \to \mathcal{R}^{\mathbf{n'}})$ as the map for connecting $\mathbf{F}$ and $\mathcal{R}^{\mathbf{n'}}$ when $\mathbf{F} = \mathcal{R}^{\mathbf{n'}}$ and the map

$$\mathbf{y} = \prod_{\mathbf{i} \in [1, \mathbf{n'}]}^{\mathbf{e}} (\mathbf{Real}(\mathbf{x^i}), \mathbf{Im}(\mathbf{x^i}))$$

to connect $\mathbf{F}$ and $\mathcal{R}^{2\mathbf{n'}}$ when $\mathbf{F} = \mathcal{C}^{\mathbf{n'}}$.
To define $\mathbf{E} = \mathbf{F} \times \mathbf{M}$ as a manifold, $\mathbf{U_k}, \phi_{\mathbf{k}}, \mathbf{k} \in [1, \mathbf{K}]$ being the covering element of the manifold $\mathbf{M}$ of dimension $\mathbf{n}$, $\mathbf{F} \times \mathbf{U_k}$ are taken as the covering elements of $\mathbf{E} = \mathbf{F} \times \mathbf{M}$. The map $\mathbf{y}(\mathbf{x})$ together with the maps $\phi_{\mathbf{k}}$ associated to the covering $\mathbf{U_k}$ of the manifold $\mathbf{M}$ generates the maps needed for $\mathbf{E} = \mathbf{F} \times \mathbf{M}$ to be a differential manifold. For $\mathbf{k} \in [1, \mathbf{K}]$ those maps are:

$$(\mathbf{y}(\mathbf{x}) \times \phi_{\mathbf{k}}(p) : \mathbf{E} = \mathbf{F} \times \mathbf{M} \to \mathcal{R}^{\mathbf{n''} + \mathbf{n}})$$

$\mathbf{n''} = \mathbf{n'}$ when $\mathbf{F} = \mathcal{R}^{\mathbf{n'}}$ and $2\mathbf{n'}$ when $\mathbf{F} = \mathcal{C}^{\mathbf{n'}}$. In all what follows we shall deal only with vector bundle (definition 77 (p. 87)) even if not specified. Most of the time it shall be on real number see convention 8 (p. 87).

**Definition 78 Sections of a vector bundle (definition 77 (p. 87)), section module.**

Given a vector bundle $\mathbf{E} = \mathbf{F} \times \mathbf{M}$ (definition 77 (p. 87)) with the above notation and definitions, one calls section $\mathbf{s}$, a smooth map $(\mathbf{s} : \mathbf{M} \to \mathbf{F})$. The set of sections is a module with $\mathcal{F}$ as ring (definition 182 (p. 218)). It is called in the following $\mathbf{V_s}$.

**Definition 79 Basis of the section module.**

Having a basis $(\mathbf{f_i} : [1, \mathbf{n''}] \to \mathbf{F})$ of the vectorial space $\mathbf{F}$, for any $\mathbf{i} \in [1, \mathbf{n''}]$, we define a section $(\mathbf{e_i}(p) = \mathbf{f_i} : \mathbf{M} \to \mathbf{F})$ of $\mathbf{E}$. With those $\mathbf{n''}$ sections, we may express any section $\mathbf{s}$ of $\mathbf{E}$ by $\mathbf{s} = \mathbf{s^i}(\mathbf{x})\mathbf{e_i}$ where $\mathbf{s^i}(\mathbf{x}) \in \mathcal{F}$, $\mathcal{F}$ being the set of smooth map on $\mathbf{M}$ (definition 7 (p. 3)). Those $\mathbf{n''}$ sections $\mathbf{e_i}$ are making a basis of the section module.

**Convention 8** Vectorial space on $\mathcal{C}$.

A vectorial space $\mathbf{F}$ built on $\mathcal{C}$ with dimension $\mathbf{n}'$ can also be considered as a vectorial space on $\mathcal{R}$ with dimension $\mathbf{n}'' = 2\mathbf{n}'$. In order to be simpler, we shall consider that $\mathbf{F}$ is a vectorial space on $\mathcal{R}$ with dimension $\mathbf{n}'$. The transposition to a vectorial space built on $\mathcal{C}$ is most of the time straightforward.

**Definition 80 Dual of the section module.**

Given a bundle $\mathbf{E} = \mathbf{F} \times \mathbf{M}$, calling $\mathbf{V_s}$ the section module over $\mathcal{F}$ (definition 78 (p. 87)), the dual $\mathbf{V^*_s}$ of $\mathbf{V_s}$ is the set of linear map $(\mathbf{t} : \mathbf{V_s} \to \mathcal{F})$ such as:

$$\mathbf{t} \in \mathbf{V^*_s} \Leftrightarrow ((\mathbf{t} : \mathbf{V_s} \to \mathcal{F}), (\forall \lambda_1 \in \mathcal{F})(\forall \lambda_2 \in \mathcal{F})(\forall \mathbf{s}_1 \in \mathbf{V_s})(\forall \mathbf{s}_2 \in \mathbf{V_s})$$
$$(\mathbf{t}(\lambda_1 \mathbf{s}_1 + \lambda_2 \mathbf{s}_2) = \lambda_1 \mathbf{t}(\mathbf{s}_1) + \lambda_2 \mathbf{t}(\mathbf{s}_2))) \ .$$

See definition 187 (p. 221).

**Definition 81 Dual of a basis of the section module.**

We are given a bundle $\mathbf{E} = \mathbf{F} \times \mathbf{M}$. Calling $\mathbf{V_s}$ the section module over $\mathcal{F}$, we call $\mathbf{V^*_s}$ its dual. We suppose that we have a basis (definition 79 (p. 87)) $(\mathbf{e_i} : [\mathbf{1}, \mathbf{m}] \to \mathbf{V_s})$ of $\mathbf{V_s}$ such that any element $\mathbf{s}$ of $\mathbf{V_s}$ can be written as $\mathbf{s} = \mathbf{s^i e_i}$, $\mathbf{s^i} \in \mathcal{F}$. One calls dual basis of $(\mathbf{e_i} : [\mathbf{1}, \mathbf{m}] \to \mathbf{V_s})$, the map $(\mathbf{e^i} : [\mathbf{1}, \mathbf{m}] \to \mathbf{V^*_s})$ such as $\mathbf{e^j}(\mathbf{e_i}) = \delta^j_i$ ($\delta^j_i$ is the Kronecker symbol: definition 215 (p. 245)). It gives for any section $\mathbf{s} = \mathbf{s^i e_i}$: $\mathbf{e^j}(\mathbf{s}) = \mathbf{s^j}$ establishing a map belonging to $\mathbf{V^*_s}$. For any element $\mathbf{t} \in \mathbf{V^*_s}$, we have $\mathbf{t} = \mathbf{t}(\mathbf{e_i})\mathbf{e^i} = \mathbf{t_i e^i}$. It leads to (see footnote 20 (p. 239)):

$$\mathbf{t}(\mathbf{s}) = \mathbf{t}(\mathbf{s^i e_i}) = \mathbf{s^i t}(\mathbf{e_i}) = \mathbf{t_i s^i} = \mathbf{t_i e^i}(\mathbf{s}) \ .$$

**Definition 82 Standard basis of the section module and of its dual.**

Given a bundle $\mathbf{E} = \mathbf{F} \times \mathbf{M}$, the dimension of $\mathbf{M}$ being $\mathbf{n}$ and that of $\mathbf{F}$ being $\mathbf{n}'$, according to convention 8 (p. 87), we suppose that $\mathbf{F}$ is a vectorial space with $\mathcal{R}$ as a commutative ring. Any point $p_\mathbf{E}$ of $\mathbf{E}$ can be written as $p_\mathbf{F} \times p_\mathbf{M}, p_\mathbf{F} \in \mathbf{F}, p_\mathbf{M} \in \mathbf{M}$. Since $\mathbf{F} = \mathcal{R}^{\mathbf{n}'}$, if we consider $\mathbf{k} \in [\mathbf{1}, \mathbf{K}]$ such that $p_\mathbf{M} \in \mathbf{U_k}$, the couple $\mathbf{U_k}, (\phi_\mathbf{k} : \mathbf{U_k} \to \mathcal{R}^\mathbf{n})$ being a covering element of $\mathbf{M}$, we have $\mathbf{F} \times \mathbf{U_k} = \mathcal{R}^{\mathbf{n}'} \times \mathbf{U_k}$ as a covering element of $\mathbf{E}$ with for local map:

$$(\phi^\mathbf{E}_\mathbf{k}(p_\mathbf{E}) = p_\mathbf{F} \times \phi_\mathbf{k}(p_\mathbf{M}) : \mathbf{F} \times \mathbf{U_k} \to \mathcal{R}^{\mathbf{n}'+\mathbf{n}}) \ .$$

Taking $(\mathbf{f_i} = \delta^j_i : [\mathbf{1}, \mathbf{n}'] \to \mathcal{R}^{\mathbf{n}'})$ as the standard basis of $\mathbf{F}$ (see equation (2.106) (p. 267) in definition 242 (p. 267)), from the definitions 78 (p. 87), and 6 (p. 3), we may call $\mathbf{e_i}$ the section of $\mathbf{E}$ defined by $\mathbf{e_i} = (\mathbf{f_i} : \mathbf{M} \to \mathbf{F} = \mathcal{R}^{\mathbf{n}'}), \mathbf{i} \in [\mathbf{1}, \mathbf{n}']$. From the same definitions, any section $(\mathbf{s}(p) : \mathbf{M} \to \mathbf{F} = \mathcal{R}^{\mathbf{n}'})$ is such that, writing $\mathbf{s}(p) = \mathbf{s^i}(p)\mathbf{f_i}$, we have:

$$(\forall \mathbf{i} \in [\mathbf{1}, \mathbf{n}']) \, (\mathbf{x} \in \phi_\mathbf{k}(\mathbf{U_k}) \Rightarrow \mathbf{s^i}(\phi^{-1}_\mathbf{k}(\mathbf{x})) \in \mathbf{C}^\infty)$$

(see definition 276 (p. 304)). Then any section of $\mathbf{E}$ can be written as: $\mathbf{s}(p) = (\mathbf{s^i}(p)\mathbf{e_i}(p) : \mathbf{M} \to \mathbf{F})$, the $\mathbf{s^i}$ being smooth maps. The sections $\mathbf{e_i}$ defined that way are the standard basis of the set $\mathbf{V_s}$ of the sections of $\mathbf{E}$. The dual of that basis (definition 81 (p. 88)) is the standard basis of $\mathbf{V^*_s}$.

**Proposition 120** We have a vector bundle (definition 77 (p. 87)) $\mathbf{F} \times \mathbf{M}$ where $\mathbf{F}$ is the fiber and $\mathbf{M}$ the main manifold (see definition 77 (p. 87)). The dual of the section module of $\mathbf{F} \times \mathbf{M}$ is the section module of the vector bundle (definition 77 (p. 87)) $\mathbf{F}^* \times \mathbf{M}$ where $\mathbf{F}^*$ is the dual of $\mathbf{F}$.

**Proof**
We have a vector bundle $\mathbf{E} = \mathbf{F} \times \mathbf{M}$ (see definition 77 (p. 87) for the notations). $\mathbf{n}'$ is the dimension of the vectorial space $\mathbf{F}$ (fiber) built on $\mathcal{R}^{\mathbf{n}'}$, $\mathbf{n}$ the dimension of the differential manifold $\mathbf{M}$. $(\mathbf{f_i} = \delta_i^j : [1, \mathbf{n}'] \to \mathcal{R}^{\mathbf{n}'} = \mathbf{F})$ the standard basis of $\mathbf{F}$ (see equation (2.106) (p. 267) in definition 242 (p. 267)), $(\mathcal{F} : \mathbf{M} \to \mathcal{R})$ the set of smooth maps on $\mathbf{M}$ to $\mathcal{R}$ (definition 6 (p. 3)), $(\mathbf{e_i} : \mathbf{M} \to \{\mathbf{f_i}\})$ the standard basis of the section module $\mathbf{V_s}$ (definition 78 (p. 87)). $\mathbf{V_s^*}$ being the dual of $\mathbf{V_s}$, we call $(\mathbf{e^i} : [1, \mathbf{n}'] \to \mathbf{V_s^*})$ the dual basis of the standard basis of $\mathbf{V_s}$ which is also a basis of $\mathbf{V_s^*}$ (see definitions 81 (p. 88), 82 (p. 88)). From those definitions, $(\mathbf{e^i}(\mathbf{e_j}) : \mathbf{M} \to \{\delta_j^i\}) \in \mathcal{F}$. Then calling $(\mathbf{f^i} : [1, \mathbf{n}'] \to \mathbf{F}^* = \mathcal{R}^{\mathbf{n}'})$ the standard basis of $\mathbf{F}^*$ with $\mathbf{f^i}(\mathbf{f_j}) = \delta_j^i$, we may write:

$$(\mathbf{e^i} : \mathbf{M} \to \{\mathbf{f^i}\} \subset \mathbf{F}^*)$$

then $\mathbf{e^i}$ is a section of $\mathbf{F}^* \times \mathbf{M}$. Since any term $\mathbf{t}$ of $\mathbf{V_s^*}$ can be written as $\mathbf{t} = \mathbf{t_i}\mathbf{e^i}$, $\mathbf{t_i} \in \mathcal{F}$ (definition 7 (p. 3)), $\mathbf{t}$ is a smooth map $(\mathbf{t} : \mathbf{M} \to \mathbf{F}^*)$ (definition 6 (p. 3)) $\mathbf{F}^*$ being considered as a manifold $\approx \mathcal{R}^{\mathbf{n}'}$. $\mathbf{t}$ is a section of $\mathbf{F}^* \times \mathbf{M}$ considered as a vector bundle (definition 77 (p. 87)).

**Definition 83 Linear map or linear transformation on the sections of a vector bundle (definition 77 (p. 87)).**

The linear maps (linear transformations) on the section module are defined in definition 186 (p. 220).

**Definition 84 End_E.**

Given a vector bundle (definition 77 (p. 87)) $\mathbf{E} = \mathbf{F} \times \mathbf{M}$ built on the differential manifold $\mathbf{M}$ with a fiber $\mathbf{F} = \mathcal{R}^{\mathbf{n}'}$, one calls **End_E** the vector bundle built also on $\mathbf{M}$ but with the set $\mathcal{L}_\mathbf{F}$ of linear transformation on $\mathbf{F}$ (definition 186 (p. 220)) as the fiber:

$$\mathbf{End\_E} = \mathcal{L}_\mathbf{F} \times \mathbf{M} \,.$$

**Proposition 121** The set of linear transformations (definition 186 (p. 220)) on the section module of a vector bundle (definition 77 (p. 87)) $\mathbf{F} \times \mathbf{M}$ associated to a differential manifold $\mathbf{M}$ can also be identified as the set $\mathbf{V_s^{End\_E}}$ of sections (definition 78 (p. 87)) of a vector bundle **End_E** associated to the same manifold with, as fiber, the set of linear transformations on the fiber $\mathbf{F}$. If we call $\mathbf{V_s^E}$ the section module associated to $\mathbf{E}$ (definition 78 (p. 87)), from proposition 121 (p. 89), we have:

$$\mathbf{V_s^{End\_E}} = \mathbf{V_s^E} \otimes (\mathbf{V_s^E})^* \,.$$

**Proof**
We have a vector bundle (definition 77 (p. 87)) $\mathbf{E} = \mathbf{F} \times \mathbf{M}$. $\mathbf{F}$ is a vectorial space of dimension $\mathbf{n}'$ with a standard basis $\mathbf{f_i}$. We call $\mathbf{e_i}$ the standard basis associated to the $\mathbf{f_i}$ for the section module $\mathbf{V_s}$. For a section $\mathbf{s} \in \mathbf{V_s}$ , we set $\mathbf{s} = \mathbf{s^i}\mathbf{e_i}$. We call $\mathcal{F}$, the set of smooth maps $\mathbf{M} \to \mathcal{R}$ (definition 7 (p. 3)). For any linear transformation $\mathbf{A}$ on $\mathbf{V_s}$, we set $\mathbf{A_j^i} \in \mathcal{F}$ such that $\mathbf{A}(\mathbf{e_i}) = \mathbf{A_i^j}\mathbf{e_j}$. One may write $\mathbf{A} = \mathbf{A_i^j}\mathbf{e_j}\mathbf{e^i} \in \mathbf{V_s} \otimes \mathbf{V_s^*}$ (see definition 80 (p. 88)). From definition 186 (p. 220), we have:

$$\mathbf{s} = \mathbf{s^k}\mathbf{e_k} \Rightarrow \mathbf{A}(\mathbf{s}) = \mathbf{s^k}\mathbf{A}(\mathbf{e_k}) = \mathbf{s^k}\mathbf{A_i^j}\mathbf{e_j}\mathbf{e^i}(\mathbf{e_k}) = \mathbf{s^k}\delta_k^i\mathbf{A_i^j}\mathbf{e_j} = \mathbf{s^k}\mathbf{A_k^j}\mathbf{e_j} \,.$$

Using $(\mathbf{e_i} = \delta_i^k \mathbf{f_k} : \mathbf{M} \rightarrow \mathbf{F})$, $(\mathbf{e^i} = \delta_k^i \mathbf{f^k} : \mathbf{M} \rightarrow \mathbf{F^*})$, we have:

$$(\mathbf{e_j e^i} = \delta_i^k \mathbf{f_k} \delta_{k'}^j \mathbf{f^{k'}} : \mathbf{M} \rightarrow \mathbf{F} \otimes \mathbf{F^*}) .$$

So **A** is defined by a smooth map of **M** (definitions 6 (p. 3), 7 (p. 3), 278 (p. 306)) on the linear transformation set of **F**, $\mathcal{L}_\mathbf{F} = \mathbf{F} \otimes \mathbf{F^*}$. $\mathcal{L}_\mathbf{F}$ is a vectorial space of dimension $\mathbf{n'} \times \mathbf{n'}$. The basic vectors of $\mathcal{L}_\mathbf{F}$ are the $\mathbf{n'^2}$ linear maps $(\mathbf{f_i f^j} : \mathbf{F} \rightarrow \mathbf{F})$ which gives $\mathbf{f_i}$ for $\mathbf{f_j}$ and 0 for the other. From that, a linear transformation **A** on the module of sections is a map $(\mathbf{A} : \mathbf{M} \rightarrow \mathcal{L}_\mathbf{F})$.

**Definition 85 Multilinear map or $_\mathbf{p}$linear_form on a bundle.**

We have a vector bundle $\mathbf{E} = \mathbf{F} \times \mathbf{M}$ (see definition 77 (p. 87)), **F** being a fiber vectorial set of dimension $\mathbf{n'}$ and **M** a differential manifold of dimension **n**. We call as usual $\mathbf{V_s}$ the set of sections of **E** (definition 78 (p. 87)), $\mathcal{F}$ the set of smooth maps $\mathbf{M} \rightarrow \mathcal{R}$ (definition 7 (p. 3)). As defined in definition 191 (p. 223), given an integer $\mathbf{p} \in \mathcal{Z}^+$, we call a map $({}_{\mathbf{p}\triangleleft}\mathbf{f} : \mathbf{V_s^p} \rightarrow \mathcal{F})$ a **p** multilinear map or $_\mathbf{p}$**linear_forms** on the bundle if it is a **p** multilinear form on $\prod_{\mathbf{i} \in [1,\mathbf{p}]} \mathbf{V_s}$ as defined in definition 217 (p. 246) with definitions 119 (p. 135) and 372 (p. 473). The set of $_\mathbf{p}$**linear_forms** may be written as:

$$\bigotimes_{\mathbf{i} \in [1,\mathbf{p}]} \mathbf{V_s^*} = (\bigotimes \mathbf{V_s^*})^\mathbf{p} .$$

**Proposition 122** The set of $_\mathbf{p}$**linear_forms** of a bundle is a module with $\mathcal{F}$ (definition 7 (p. 3)) as a commutative ring.

**Proof**

$$\begin{aligned}&(\lambda_1 \in \mathcal{F}, \lambda_2 \in \mathcal{F}, \mathbf{u_1} \in (\bigotimes \mathbf{V_s^*})^\mathbf{p}, \mathbf{u_2} \in (\bigotimes \mathbf{V_s^*})^\mathbf{p}) \Rightarrow \\ &(\mathbf{s} = \prod_{\mathbf{i} \in [1,\mathbf{p}]} \mathbf{s^i} \in \mathbf{V_s^p} \Rightarrow \lambda_1 \mathbf{u_1}(\mathbf{s}) + \lambda_2 \mathbf{u_2}(\mathbf{s}) \in \mathcal{F})\end{aligned}$$

$\mathbf{u_1}(\mathbf{s})$ and $\mathbf{u_2}(\mathbf{s})$ verifying definition 191 (p. 223), from the properties of $\mathcal{F}$, $\lambda_1 \mathbf{u_1} + \lambda_2 \mathbf{u_2}$ verifies also 191 (p. 223). Then $\lambda_1 \mathbf{u_1} + \lambda_2 \mathbf{u_2} \in (\bigotimes \mathbf{V_s^*})^\mathbf{p}$.

### 1.14.2 Covariant derivative

We keep in section 1.14.2, the notations of the previous section 1.14.1 (p. 86).

**Definition 86 Covariant derivative on a vector bundle (definition 77 (p. 87)).**

We suppose that we have a vector bundle (definition 77 (p. 87)) $\mathbf{E} = \mathbf{F} \times \mathbf{M}$ with its set of sections $\mathbf{V_s}$ (definition 78 (p. 87)) and its commutative ring (definitions 179 (p. 214) and 180 (p. 215)) $\mathcal{F}$: the set of smooth maps $\mathbf{M} \rightarrow \mathcal{R}$ (definition 7 (p. 3)). We call as usual $\mathcal{V}$ the set of vector fields on the manifold **M** (definition 15 (p. 8)). We shall call covariant derivative $\mathbf{D_v}$ along a vector field $\mathbf{v} \in \mathcal{V}$, a map $(\mathbf{D_v} : \mathbf{V_s} \rightarrow \mathbf{V_s})$ with the following Leibniz rules:

$$\begin{aligned}&(\mathbf{v} \in \mathcal{V}, \mathbf{f} \in \mathcal{F}, \mathbf{s} \in \mathbf{V_s}, \mathbf{t} \in \mathbf{V_s}) \Rightarrow \\ &(\mathbf{D_v}(\mathbf{s}+\mathbf{t}) = \mathbf{D_v}(\mathbf{s}) + \mathbf{D_v}(\mathbf{t}), \\ &\mathbf{D_v}(\mathbf{f} \cdot \mathbf{s}) = \mathbf{v}(\mathbf{f}) \cdot \mathbf{s} + \mathbf{f} \cdot \mathbf{D_v}(\mathbf{s}))\end{aligned} \tag{1.100}$$

$\mathbf{D_v}$ can be seen as the action of **v** on $\mathbf{V_s}$ and we may also set $\mathbf{v}(\mathbf{f}) = \mathbf{D_v}(\mathbf{f})$ for consistency.

**Definition 87 Connection on a bundle.**

Given a vector bundle as defined in definition 77 (p. 87): $\mathbf{E} = \mathbf{F} \times \mathbf{M}$, calling $\boldsymbol{\mathcal{M}}$ the set of maps $\mathbf{V_s} \to \mathbf{V_s}$, $\mathbf{V_s}$ being the set of sections of $\mathbf{E}$, $\boldsymbol{\mathcal{M}}$ is a module (proposition 370 (p. 219)). Calling $\boldsymbol{\mathcal{V}}$ the set of vector fields of $\mathbf{M}$, one calls connection $\mathbf{D}$ on the bundle $\mathbf{E}$, a linear map (definition 186 (p. 220)) of $\boldsymbol{\mathcal{V}}$ to $\boldsymbol{\mathcal{M}}$: $(\mathbf{D} : \boldsymbol{\mathcal{V}} \to \boldsymbol{\mathcal{M}})$ such that each image of $\mathbf{v} \in \boldsymbol{\mathcal{V}}$: $\mathbf{D(v)} = (\mathbf{D_v} : \boldsymbol{\mathcal{V}} \times \mathbf{V_s} \to \mathbf{V_s})$ is a covariant derivative along $\mathbf{v}$ (definition 86 (p. 90)) as defined in definition 86 (p. 90). We should then have:

$$\begin{aligned}
&(\mathbf{v} \in \boldsymbol{\mathcal{V}}, \mathbf{w} \in \boldsymbol{\mathcal{V}}, \mathbf{f} \in \boldsymbol{\mathcal{F}}, \mathbf{s} \in \mathbf{V_s}, \mathbf{t} \in \mathbf{V_s}) \Rightarrow \\
&(\mathbf{D_{v+w}(s)} = \mathbf{D_v(s)} + \mathbf{D_w(s)}, \\
&\mathbf{D_{fv}(s)} = \mathbf{f} \cdot \mathbf{D_v(s)}, ) \\
&\mathbf{D_v(fs)} = \mathbf{v(f)s} + \mathbf{f} \cdot \mathbf{D_v(s)}), \\
&\mathbf{D_v(t+s)} = \mathbf{D_v(t)} + \mathbf{D_v(s)})\ .
\end{aligned} \tag{1.101}$$

**Convention 9** $\mathbf{D}_\mu = \mathbf{D}_{\partial_\mu}$.

With notation of definition 87 (p. 91), in a cover $\mathbf{U}$ of the manifold $\mathbf{M}$, taking the local canonical basis (definition 23 (p. 17)) of $\boldsymbol{\mathcal{V}}_\mathbf{U}$: $(\partial_\mu : [\mathbf{1}, \mathbf{n}] \to \boldsymbol{\mathcal{V}}_\mathbf{U})$ such that we have (see section 1.4 (p. 19)):

$$(\forall \mathbf{f} \in \boldsymbol{\mathcal{F}}_\mathbf{U})\ (\partial_\mu(\mathbf{f}) = \partial_\mu \mathbf{f} \in \boldsymbol{\mathcal{F}}_\mathbf{U})$$

we set, to simplified the writing, $\mathbf{D}_\mu = \mathbf{D}_{\partial_\mu}$ in the restricted bundle $\mathbf{F} \times \mathbf{U}$.

### 1.14.3 Vector potential

**Proposition 123** $\mathbf{D}$ and $\mathbf{D'}$ are two connections on a vector bundle (definition 77 (p. 87)) $\mathbf{E} = \mathbf{F} \times \mathbf{M}$. Having $\mathbf{D_v} \in \mathbf{D}$ and $\mathbf{D'_v} \in \mathbf{D'}$, $\mathbf{D_v} - \mathbf{D'_v}$ is a linear transformation (definition 186 (p. 220)) on the section module $\mathbf{V_s}$ (definition 78 (p. 87)).

**Proof**
If $\mathbf{f}$ and $\mathbf{g} \in \boldsymbol{\mathcal{F}}$ are two smooth maps from $\mathbf{M}$ to $\boldsymbol{\mathcal{R}}$, $\mathbf{s}$ and $\mathbf{t}$ two sections, one has:

$$\begin{aligned}
(\mathbf{D_v} - \mathbf{D'_v})(\mathbf{fs} + \mathbf{gt}) = &\ \mathbf{v(f)s} + \mathbf{v(g)t} + \mathbf{fD_v(s)} + \mathbf{gD_v(t)} \\
&- \mathbf{v(f)s} - \mathbf{v(g)t} - \mathbf{fD'_v(s)} - \mathbf{gD'_v(t)} \\
= &\ \mathbf{f(D_v - D'_v)s} + \mathbf{g(D_v - D'_v)t}\ .
\end{aligned}$$

**Definition 88** $\mathbf{D^0}$.

Given a vector bundle (definition 77 (p. 87)) $\mathbf{E} = \mathbf{F} \times \mathbf{M}$ with the above notations, given a basis $\mathbf{e_i}$ for the sections, one uses to call $\mathbf{D^0}$, the connection such that for any section $\mathbf{s} \in \mathbf{V_s}, \mathbf{s} = \mathbf{s^i e_i}, (\mathbf{s^i} : [\mathbf{1}, \mathbf{n'}] \to \boldsymbol{\mathcal{F}})$ ($\mathbf{n'}$ is the dimension of the fiber $\mathbf{F}$):

$$\mathbf{D^0_v(s)} = \mathbf{D^0_v(s^i e_i)} = \mathbf{v(s^i)e_i}\ . \tag{1.102}$$

Since $(\forall \mathbf{v} \in \boldsymbol{\mathcal{V}})(\mathbf{v(1)} = \mathbf{0})$ (proposition 21 (p. 9)), it means:

$$\mathbf{D^0_v(e_i)} = \mathbf{0}\ . \tag{1.103}$$

With equation (1.102) (p. 91), $\mathbf{D^0_v(s)}$ satisfies the axioms of the definition of a covariant derivative 86 (p. 90) and $\mathbf{D^0}$ the axioms of the definition of a connection 87 (p. 91).

**Definition 89 Vector potential.**

One can write any connection as

$$\mathbf{D_v} = \mathbf{D^0_v} + \mathbf{A_v} \tag{1.104}$$

where $\mathbf{A_v}$ is a linear transformation on the section module $\mathbf{V_s}$ (definitions 186 (p. 220), 78 (p. 87), proposition 123 (p. 91)). $\mathbf{A_v}$ depends on the bases of the sections. It is called, in gauge theories like electromagnetism, vector potential.

**Proposition 124** Keeping the above notations, given a vector bundle (definition 77 (p. 87)) $\mathbf{E = F \times M}$, calling $(\mathbf{e_i} : [1, \mathbf{n'}] \to \mathbf{V_s})$, $(\mathbf{f_i} : [1, \mathbf{n'}] \to \mathbf{F})$ the standard basis of the section module (definition 78 (p. 87)) and of the fiber $\mathbf{F}$, we consider another basis $(\mathbf{u_i} = \mathbf{u_i^j e_j} : [1, \mathbf{n'}] \to \mathbf{V_s})$ of the sections such that $\mathbf{0 \notin det(u_i^j(M))}$[22] and a linear transformation $\mathbf{T}$ (definition 186 (p. 220)) on the sections expressed in that basis by:

$$\begin{aligned} &p \in \mathrm{M} \Rightarrow \mathrm{T}(\mathrm{u_j})(p) = \mathrm{T^i_j}(p)\mathrm{u_i}(p) \\ &0 \notin \det(\mathrm{T^i_j}(\mathrm{M})) \end{aligned}$$

The sections $\mathbf{u^t_j} = \mathbf{Tu_j} = \mathbf{T^i_j u^k_i e_k}$: $\mathbf{u^t_j}(p) = \mathbf{u^t{}^i_j}(p)\mathbf{f_i} = \mathbf{T^i_j}(p)\mathbf{u_i}(p) = \mathbf{T^i_j}(p)\mathbf{u^k_i}(p)\mathbf{f_k}$ form a new basis of $\mathbf{V_s}$ with $\mathbf{u_i} = (\mathbf{T^{-1}})^j_i(p)\mathbf{u^t_i}(p)$, where, for each value of $p \in \mathbf{M}$, $\mathbf{T^{-1}}(p)$ is the inverse matrix of $\mathbf{T}(p)$. $\mathbf{T^{-1}}(p)$ exists for each value a $p \in \mathbf{M}$ since $\mathbf{0 \notin det(u^j_i)(M)}$ (see definitions 225 (p. 250), 227 (p. 252) and proposition 444 (p. 255)). Taking a connection $\mathbf{D}$ which can be described in the basis $\mathbf{u_j}$ by

$$(\mathrm{s} = \mathrm{s^i u_i}, \mathrm{v} \in \mathcal{V}) \Rightarrow \mathrm{D_v(s)} = \mathrm{v(s^i)u_i} + \mathrm{s^i D_v(u_i)^j u_j}$$

and in the basis $\mathbf{u^t_j} = \mathbf{T^i_j u_i}$ by:

$$(\mathrm{s} = \mathrm{s^i u_i} = \mathrm{s^t{}^i u^t{}_i}, \mathrm{v} \in \mathcal{V}) \Rightarrow \mathrm{D_v(s)} = \mathrm{v(s^{tj})u^t_j} + \mathrm{s^{tj} D_v(u^t{}_j)^i u^t_i}$$

we have:

$$\mathrm{D_v(u_i)^\ell} = \mathrm{D_v{}^\ell_i} = \mathrm{v(T^{-1}{}^j_i)T^\ell_j} + \mathrm{T^{-1}{}^j_i D_v(u^t_j)^k T^\ell_k} = \mathrm{v(T^{-1}{}^j_i)T^\ell_j} + \mathrm{T^{-1}{}^j_i D^t_{vj}{}^k T^\ell_k} \, .$$

**Proof**
With $\mathbf{u_i} = \mathbf{T^j_i u^t_j}$, $\mathbf{u^t_j} = (\mathbf{T^{-1}})^i_j \mathbf{u_i}$, we have:

$$\begin{aligned} &\mathrm{D_v(u_i)} = \mathrm{D_v(u_i)^\ell u_\ell} = \mathrm{D_v(T^j_i u^t_j)} = \mathrm{v(T^j_i)(T^{-1})^\ell_j u_\ell} + \mathrm{T^j_i D_v(u^t_j)^k u^t_k} \\ &= \mathrm{v(T^j_i)(T^{-1})^\ell_j u_\ell} + \mathrm{T^j_i D_v(u^t_j)^k (T^{-1})^\ell_k u_\ell} \, . \end{aligned}$$

Then:

$$\mathrm{D^\ell_{v_i}} = \mathrm{v(T^{-1}{}^j_i)T^\ell_j} + \mathrm{T^{-1}{}^j_i D^t_{vj}{}^k T^\ell_k} \, .$$

**Definition 90 Connection $\mathbf{D^{End\_E}}$ induced by a connection D on a vector bundle E.**

[22] See definition 114 (p. 132) which means $\mathbf{a \notin f(A) \Leftrightarrow (\forall x \in A)(f(x) \neq a)}$.

We consider (definition 77 (p. 87)) a vector bundle $\mathbf{E} = \mathbf{F} \times \mathbf{M}$ where $\mathbf{F}$ is the vectorial space representing the fiber. We consider the vector bundle (definitions 84 (p. 89), 77 (p. 87)) $\mathbf{End\_E} = \mathcal{L}_\mathbf{F} \times \mathbf{M}$ where $\mathcal{L}_\mathbf{F}$ is the set of linear transformations on $\mathbf{F}$ (definition 186 (p. 220)). From proposition 121 (p. 89), $\mathbf{End\_E}$ is also the set of linear transformations on the module $\mathbf{V_s}$, the set of sections of $\mathbf{E}$ (definitions 83 (p. 89), 186 (p. 220)). We call $\mathbf{V_s^{End\_E}}$, the set of sections of $\mathbf{End\_E}$ (definition 78 (p. 87)). We call $\mathbf{C(E)}$ the set of connections of $\mathbf{E}$ (definition 87 (p. 91)) and $\mathbf{C(End\_E)}$ the set of connections of $\mathbf{End\_E}$.
We define a map $(\mathbf{D^{End\_E}} : \mathbf{C(E)} \to \mathbf{C(End\_E)})$:
For any $\mathbf{D} \in \mathbf{C(E)}$, we call:

$$(\mathbf{D_v^{End\_E}(H)} : \mathcal{V} \times \mathbf{V_s^{End\_E}} \to \mathbf{V_s^{End\_E}}) \in \mathbf{C(End\_E)}$$

the connection defined as follows:

$$\begin{gathered}(\forall \mathbf{H} \in \mathbf{V_s^{End\_E}})(\forall \mathbf{v} \in \mathcal{V})(\forall \mathbf{s} \in \mathbf{V_s}) \\ (\{\mathbf{D_v^{End\_E}(H)}\}(\mathbf{s}) = \mathbf{D_v(H(s))} - \mathbf{H(D_v(s))}) \,.\end{gathered} \tag{1.105}$$

**Proposition 125** Keeping the above notations of definition 90 (p. 92), the formula (1.105) defined a connection on the vector bundle $\mathbf{End\_E}$.

**Proof**
With formula (1.105), since $\mathbf{H}$ is a linear transformation on section with $\mathcal{F}$ as commutative ring (definition 180 (p. 215)), $\mathbf{D_v^{End\_E}}$ is linear in $\mathbf{v} \in \mathcal{V}$ because $\mathbf{D_v}$ is linear in $\mathbf{v}$ and $\mathbf{H}$ is linear in $\mathbf{s} \in \mathbf{V_s}$. $\mathbf{D_v^{End\_E}}$ is a covariant derivative of a linear transformation $\mathbf{H}$ on $\mathbf{V_s}$. It verifies definition 86 (p. 90) for $\mathbf{H}$ which is also a section of $\mathbf{End\_E}$, $\mathbf{H} \in \mathbf{V_s^{End\_E}}$:

$$\begin{aligned}&(\mathbf{v} \in \mathcal{V}, \mathbf{f} \in \mathcal{F}, \mathbf{H} \in \mathbf{V_s^{End\_E}}, \mathbf{G} \in \mathbf{V^{End\_E_s}}, \mathbf{s} \in \mathbf{V_s}) \Rightarrow \\ &(\mathbf{D_v^{End\_E}(H+G)(s)} = \mathbf{D_v(H(s)+G(s))} - \mathbf{(H+G)(D_v(s))} \\ &= \mathbf{D_v(H(s))} + \mathbf{D_v(G(s))} - \mathbf{H(D_v(s))} - \mathbf{G(D_v(s))} = \mathbf{D_v^{End\_E}(H)} + \mathbf{D_v^{End\_E}(G)}, \\ &\mathbf{D_v^{End\_E}(f \cdot H)(s)} = \mathbf{D_v(f \cdot H(s))} - \mathbf{f \cdot H(D_v(s))} \\ &= \mathbf{v(f) \cdot H(s)} + \mathbf{f \cdot D_v(H(s))} - \mathbf{f \cdot H(D_v(s))} = \mathbf{v(f) \cdot H(s)} + \mathbf{f \cdot D_v^{End\_E}(H)(s)}, \\ &\mathbf{D_v^{End\_E}(H)(fs)} = \mathbf{D_v(H(fs))} - \mathbf{H(D_v(fs))} = \mathbf{D_v(fH(s))} - \mathbf{H(D_v(fs))} \\ &= \mathbf{v(f)H(s)} + \mathbf{fD_v(H(s))} - \mathbf{H(v(f)s)} - \mathbf{H(fD_v(s))} \\ &= \mathbf{fD_v(H(s))} - \mathbf{fH(D_v(s))} = \mathbf{fD_v^{End\_E}(H)(s))} \,.\end{aligned} \tag{1.106}$$

With that definition, we have extended the Leibniz law equation (1.7) in definition 15 (p. 8).

$$\mathbf{D_v(Hs)} = \mathbf{D_v^{End\_E}(H)\ s} + \mathbf{HD_v(s)} = \mathbf{D_v(H)\ s} + \mathbf{HD_v(s)} \,. \tag{1.107}$$

**Definition 91 Derivative and differential of linear transformations on a bundle.**

We have a differentiable manifold $\mathbf{M}$ of dimension $\mathbf{n}$ with its set of smooth maps $\mathcal{F}$ and its set of vector fields $\mathcal{V}$. We also have a vectorial space $\mathbf{F}$ of dimension $\mathbf{m}$. We can consider a linear transformation $\mathbf{H}$ on a vector bundle (definition 77 (p. 87)) $\mathbf{F} \times \mathbf{M}$ as an element of $\mathcal{F}^{\mathbf{m}\times\mathbf{m}}$: $(\mathbf{H} = \mathbf{H_j^i} : [\mathbf{1}, \mathbf{m}] \times [\mathbf{1}, \mathbf{m}] \to \mathcal{F})$ and consider for $\mathbf{v} \in \mathcal{V}$ the corresponding map $(\mathbf{v} : \mathcal{F}^{\mathbf{m}\times\mathbf{m}} \to \mathcal{F}^{\mathbf{m}\times\mathbf{m}})$ (see definition 373 (p. 474)):

$$(\mathbf{v(H)} = \mathbf{v(H)_j^i} = \mathbf{v(H_j^i)} : [\mathbf{1}, \mathbf{m}] \times [\mathbf{1}, \mathbf{m}] \to \mathcal{F}) \in \mathcal{F}^{\mathbf{m}\times\mathbf{m}}$$

and the corresponding differential $\mathbf{dH} : \mathcal{V} \to \mathcal{F}^{m\times m}$

$$(\mathbf{dH} = \mathbf{dH(v)}^i_j = \mathbf{dH}^i_j(\mathbf{v}) = \mathbf{v}(\mathbf{H}^i_j) : \mathcal{V} \to \mathcal{F}^{m\times m}) .$$

In a local reference frame we have

$$(\partial_\mu \mathbf{H})^i_j = \partial_\mu \mathbf{H}^i_j, \ \mathbf{dH} = \partial_\mu \mathbf{H} \mathbf{dx}^\mu .$$

This allows to generate also external differential as in section 1.11 replacing $\mathcal{F}$ by $\mathcal{F}^{m\times m}$. We may then consider also the sets $\Omega_{\mathcal{F}^{m\times m}}, \Omega^0_{\mathcal{F}^{m\times m}} = \mathcal{F}^{m\times m}, \mathcal{F}^{m\times mP}$. See also section 2.10.1.

**Proposition 126** The covariant derivatives of linear transformations on a bundle associated to a connection on $\mathbf{V_s}$ defines a connection on the bundle $\mathbf{E}^{\mathbf{End\_E}} = \mathcal{L}_\mathbf{F} \times \mathbf{M}$ (see definition 90 (p. 92)). For any section written as $\mathbf{s} = \mathbf{s}^i\mathbf{f}_i$, with:

$$\{\mathbf{D}^{\mathbf{End\_E}}_\mathbf{v}(\mathbf{H})\}(\mathbf{f}_i) = \{\mathbf{D}^{\mathbf{End\_E}}_\mathbf{v}(\mathbf{H})\}^k_i \mathbf{f}_k, \quad \mathbf{D}_\mathbf{v}(\mathbf{f}_i) = \mathbf{D}_{\mathbf{v}i}{}^k \mathbf{f}_k$$

we have:

$$\{\mathbf{D}^{\mathbf{End\_E}}_\mathbf{v}(\mathbf{H})\}^k_i = \mathbf{v}(\mathbf{H}^k_i) + \mathbf{H}^j_i \mathbf{D}_{\mathbf{v}j}{}^k - \mathbf{D}_{\mathbf{v}i}{}^j \mathbf{H}^k_j .$$

**Proof**
With the notation of definition 90 (p. 92), $\{\mathbf{D}^{\mathbf{End\_E}}_\mathbf{v}(\mathbf{H})(\mathbf{H})\}(\mathbf{s}) = \mathbf{D_v}(\mathbf{Hs}) - \mathbf{HD_v}(\mathbf{s})$ is a connection on the space of $\mathbf{V}^{\mathbf{End\_E}}_\mathbf{s}$. Let us call $(\mathbf{f}_i : [1, \mathbf{m}] \to \mathbf{V_s})$ a basis of the section module. Any section can be written as $\mathbf{s} = \mathbf{s}^i\mathbf{f}_i$ with $\mathbf{i} \in [1, \mathbf{m}] \Rightarrow \mathbf{s}^i \in \mathcal{F}$. We may write using the linearity of $\mathbf{H}$ and of $\mathbf{D}^{\mathbf{End\_E}}_\mathbf{v}(\mathbf{H})$ (equation (1.106) (p. 93)):

$$\begin{aligned}
&\{\mathbf{D}^{\mathbf{End_E}}_\mathbf{v}(\mathbf{H})\}(\mathbf{s}) = \{\mathbf{D}^{\mathbf{End_E}}_\mathbf{v}(\mathbf{H})\}(\mathbf{s}^i\mathbf{f}_i) = \mathbf{s}^i\{\mathbf{D}^{\mathbf{End_E}}_\mathbf{v}(\mathbf{H})\}(\mathbf{f}_i) = \mathbf{s}^i\{\mathbf{D}^{\mathbf{End_E}}_\mathbf{v}(\mathbf{H})\}^k_i\mathbf{f}_k \\
&= \mathbf{D_v}(\mathbf{H}(\mathbf{s}^i\mathbf{f}_i)) - \mathbf{HD_v}(\mathbf{s}^i\mathbf{f}_i) = \mathbf{D_v}(\mathbf{s}^i\mathbf{H}(\mathbf{f}_i)) - \mathbf{HD_v}(\mathbf{s}^i\mathbf{f}_i) = \mathbf{D_v}(\mathbf{s}^i\mathbf{H}^k_i\mathbf{f}_k) - \mathbf{HD_v}(\mathbf{s}^i\mathbf{f}_i) \\
&= \mathbf{v}(\mathbf{s}^i\mathbf{H}^k_i)\mathbf{f}_k + \mathbf{s}^i\mathbf{H}^k_i\mathbf{D_v}(\mathbf{f}_k) - \mathbf{v}(\mathbf{s}^i)\mathbf{H}\mathbf{f}_i - \mathbf{s}^i\mathbf{HD_v}(\mathbf{f}_i) \\
&= \mathbf{v}(\mathbf{s}^i)\mathbf{H}^k_i\mathbf{f}_k + \mathbf{s}^i\mathbf{v}(\mathbf{H}^k_i)\mathbf{f}_k + \mathbf{s}^i\mathbf{H}^k_i\mathbf{D_v}(\mathbf{f}_k) - \mathbf{v}(\mathbf{s}^i)\mathbf{H}^k_i\mathbf{f}_k - \mathbf{s}^i\mathbf{HD_v}(\mathbf{f}_i) \\
&= \mathbf{s}^i\mathbf{v}(\mathbf{H}^k_i)\mathbf{f}_k + \mathbf{s}^i\mathbf{H}^k_i\mathbf{D_v}(\mathbf{f}_k) - \mathbf{s}^i\mathbf{HD}_{\mathbf{v}i}{}^j\mathbf{f}_j \\
&= \mathbf{s}^i\mathbf{v}(\mathbf{H}^k_i)\mathbf{f}_k + \mathbf{s}^i\mathbf{H}^k_i\mathbf{D}_{\mathbf{v}k}{}^j\mathbf{f}_j - \mathbf{s}^i\mathbf{H}^k_j\mathbf{D}_{\mathbf{v}i}{}^j\mathbf{f}_k \\
&= \mathbf{s}^i(\mathbf{v}(\mathbf{H}^k_i) + \mathbf{H}^j_i\mathbf{D}_{\mathbf{v}j}{}^k - \mathbf{D}_{\mathbf{v}i}{}^j\mathbf{H}^k_j)\mathbf{f}_k .
\end{aligned}$$

**Definition 92 Covariant derivative on the dual of a bundle.**

We have a vector bundle (definition 77 (p. 87)) $\mathbf{E} = \mathbf{F} \times \mathbf{M}$ and $\mathbf{E}^* = \mathbf{F}^* \times \mathbf{M}$ its dual which is also a bundle (definition 80 (p. 88), proposition 120 (p. 88)). We define a bijection (definition 124 (p. 139)) between the set of connection on the vector bundle $\mathbf{E} = \mathbf{F} \times \mathbf{M}$ and the set of connection on $\mathbf{E}^* = \mathbf{F}^* \times \mathbf{M}$. We call $\mathbf{s}$ a section of $\mathbf{E} = \mathbf{F} \times \mathbf{M}$ (definition 78 (p. 87)), $\mathbf{t}$ a section of $\mathbf{E}^* = \mathbf{F}^* \times \mathbf{M}$. We consider $\mathbf{D}$ a connection on the section module and $\mathbf{v} \in \mathcal{V}$ a vector field (definition 1.3.1 (p. 8)). By definition the corresponding element $\mathbf{D}^*$ of $\mathbf{D}$ is given by:

$$(\mathbf{t} \in \mathbf{V}^*_\mathbf{s}, \mathbf{s} \in \mathbf{V_s}) \Rightarrow \{\mathbf{D}^*_\mathbf{v}(\mathbf{t})\}(\mathbf{s}) = \mathbf{v}(\mathbf{t}(\mathbf{s})) - \mathbf{t}(\mathbf{D_v}(\mathbf{s})) \in \mathcal{F} \tag{1.108}$$

$\mathbf{D}^*$ is a map $(\mathbf{V}^*_\mathbf{s} \to \mathbf{V}^*_\mathbf{s})$.

**Proposition 127** In the definition 92 (p. 94), equation (1.108) (p. 94) defines a bijection (definition 124 (p. 139)).

**Proof**
Using proposition 420 (p. 245), $\mathbf{F^{**} = F}$, $\mathbf{E^{**} = E}$. Then equation (1.108) (p. 94) transposes to:

$$(\mathbf{t} \in \mathbf{V}^*_{\mathbf{s}}, \mathbf{s} \in \mathbf{V_s}) \Rightarrow \mathbf{s}(\mathbf{D}^*_{\mathbf{v}}(\mathbf{t})) = \mathbf{v}(\mathbf{s}(\mathbf{t})) - \{\mathbf{D_v}(\mathbf{s})\}(\mathbf{t}) \in \mathcal{F}$$

it gives

$$(\mathbf{t} \in \mathbf{V}^*_{\mathbf{s}}, \mathbf{s} \in \mathbf{V_s}) \Rightarrow \{\mathbf{D}^{**}_{\mathbf{v}}(\mathbf{s})\}(\mathbf{t}) = \{\mathbf{D_v}(\mathbf{s})\}(\mathbf{t}) = \mathbf{v}(\mathbf{s}(\mathbf{t})) - \mathbf{s}(\mathbf{D}^*_{\mathbf{v}}(\mathbf{t}))$$

Then the map $(\mathbf{D} \to \mathbf{D}^*)$ has an inverse. From proposition 171 (p. 140) it is a bijection because the map inverse is defined for all $\mathbf{D}^*$ as the direct for all $\mathbf{D}$.

**Proposition 128** Using notation of definition 92 (p. 94), the map $(\mathbf{D}^* : \mathbf{V}^*_{\mathbf{s}} \to \mathbf{V}^*_{\mathbf{s}})$ as defined by definition 92 (p. 94) and equation (1.108) (p. 94) is a connection on $\mathbf{E}^*$.

**Proof**
$\mathbf{V}^*_{\mathbf{s}}$ dual of $\mathbf{V_s}$ is the set of sections of $\mathbf{E^* = F^* \times M}$ (proposition 120 (p. 88)). $(\mathbf{D} : \mathbf{V_s} \to \mathbf{V_s})$ being a connection on $\mathbf{E}$, $(\mathbf{D}^* : \mathbf{V}^*_{\mathbf{s}} \to \mathbf{V}^*_{\mathbf{s}})$ is defined by definition 92 (p. 94) and equation (1.108) (p. 94).
$\mathbf{v(t(s))}$ and $\mathbf{tD_v(s)}$ are well defined so $\mathbf{D}^*_{\mathbf{v}}(\mathbf{t})\mathbf{s}$ is also defined and then $\mathbf{D}^*_{\mathbf{v}}(\mathbf{t})$. It is a connection (definition 87 (p. 91)):

- $\{\mathbf{D}^*_{\mathbf{v}}(\mathbf{t})\}$ is a linear map on $\mathbf{V_s}$:

$$\begin{aligned}
&\{\mathbf{D}^*_{\mathbf{v}}(\mathbf{t})\}(\mathbf{fs}) = \mathbf{vt(fs) - tD_v(fs) = v(ft(s)) - t(v(f)s) - t(fD_v(s))}\\
&= \mathbf{fv(t(s)) + v(f)t(s) - v(f)t(s) - ft(D_v(s))}\\
&= \mathbf{f(v(t(s)) - t(D_v(s)) = f\{D^*_v(t)\}(s)}
\end{aligned} \tag{1.109}$$

$$\begin{aligned}
&\{\mathbf{D}^*_{\mathbf{v}}(\mathbf{t})\}(\mathbf{s+u}) = \mathbf{vt(s+u) - tD_v(s+u)}\\
&= \mathbf{vt(s) - tD_v(s) + vt(u) - tD_v(u) = \{D^*_v(t)\}(s) + \{D^*_v(t)\}(u)}
\end{aligned}$$

- We also have:

$$\begin{aligned}
&\{\mathbf{D}^*_{\mathbf{v}}(\mathbf{ft})\}(\mathbf{s}) = \mathbf{v(ft(s)) - ftD_v(s) = v(f)t(s) + fv(t(s)) - ftD_v(s)}\\
&= \mathbf{\{v(f)t + fD^*_v(t)\}(s)}\\
&\mathbf{D^*_v(ft) = v(f)t + fD^*_v(t)}
\end{aligned} \tag{1.110}$$

- It is a linear map in $\mathbf{v}$ because $\mathbf{v(t(s))}$ and $\mathbf{D_v(s)}$ are linear in $\mathbf{v}$ (definition 87 (p. 91)).

**Definition 93 Covariant derivative on multilinear forms on a bundle.**

Generalizing the same argument, $\mathbf{v}$ acting on a multilinear form ${}_{\mathbf{p}\triangleleft}\mathbf{f}$ acting on the sections:

$$\mathbf{v}({}_{\mathbf{p}\triangleleft}\mathbf{f}(\prod_{\mathbf{i}\in[1,\mathbf{p}]} \mathbf{s_i})) = (\mathbf{D}^*_{\mathbf{v}}\, {}_{\mathbf{p}\triangleleft}\mathbf{f})(\prod_{\mathbf{i}\in[1,\mathbf{p}]} \mathbf{s_i}) + \sum_{\mathbf{l}\in[1,\mathbf{p}]} {}_{\mathbf{p}\triangleleft}\mathbf{f}(\prod_{\mathbf{i}\in[1,\mathbf{l}-1]} \mathbf{s_i D_v s_l} \prod_{\mathbf{i}\in[\mathbf{l}+1,\mathbf{p}]} \mathbf{s_i}) \tag{1.111}$$

equation (1.111) defines $\mathbf{D}^*_{\mathbf{v}}\, {}_{\mathbf{p}\triangleleft}\mathbf{f}$.

**Definition 94 Covariant derivative on tensorial product of bundles.**

Applying the same technique, we may define the transformed of the tensorial product setting (see definition 222 (p. 250), and section 44 (p. 40)):

$$\mathrm{v}(\ _{p\triangleleft}f(\prod_{i\in[1,p]} s_i)) = \mathrm{v}(\bigotimes_{i\in[1,p]} s_i(\ _{p\triangleleft}f)) = \bigotimes_{i\in[1,p]} s_i(D^*_v\ _{p\triangleleft}f) + D_v \bigotimes_{i\in[1,p]} s_i(\ _{p\triangleleft}f) \tag{1.112}$$

combined to equation (1.111) (p. 95) it gives:

$$D_v \bigotimes_{i\in[1,p]} s_i = \sum_{l\in[1,p]} \bigotimes_{i\in[1,l-1]} s_i \otimes D_v s_l \bigotimes_{i\in[l+1,p]} s_i \ . \tag{1.113}$$

**Definition 95 Covariant derivative of a metric.**

A metric on the section space is a bi-linear mapping of $\mathbf{V_s} \otimes \mathbf{V_s}$ on $\mathcal{F}$. It is a symmetric element of $\mathbf{V^*_s} \otimes \mathbf{V^*_s}$ and from equation (1.112) (p. 96) and (1.113) (p. 96):

$$g = \sum g_{i,j} s^{*i} \otimes t^{*j}\ ,\ s^{*i}\&t^{*j} \in V^*_s$$
$$v(g(s,t)) = (D_v g)(s,t) + g(D_v s, t) + g(s, D_v t)$$
$$(D_v g)(s,t) = v(g(s,t)) - g(D_v s, t) - g(s, D_v t)\ .$$

### 1.14.4 Differential on a bundle

**Definition 96 Differential of a section associated to a connection.**

We have a manifold $\mathbf{M}$ of dimension $\mathbf{n}$, a vector bundle (definition 77 (p. 87)) $\mathbf{F} \times \mathbf{M}$ built on $\mathbf{M}$ with the vectorial space $\mathbf{F}$ of dimension $\mathbf{m}$ built on $\mathcal{R}$ as the fiber. $(e_j : [1,m] \to F)$ is a basis of $\mathbf{F}$ and $(f_\ell = e^i_\ell(x) e_i : M \to F)$ a basis of the sections with $e^i_\ell(x) \in \mathcal{F}$. We have a connection $\mathbf{D}$. Considering the module $\mathbf{V_s}$ of section of $\mathbf{F} \times \mathbf{M}$, one calls differential $\mathbf{d_D s}$ of a section $\mathbf{s} \in \mathbf{V_s}$ associated to the connection $\mathbf{D}$, the following linear map:

$$(d_D s(v) = D_v s : \mathcal{V} \to V_s)$$

$\mathcal{V}$ is the set of vector fields of $\mathbf{M}$.

### 1.14.5 Exterior differential on a bundle

**Proposition 129** With notations of definition 96 (p. 96),

$$(f \in \mathcal{F}, s \in V_s) \Rightarrow d_D fs = s df + f d_D s\ .$$

See definition 40 (p. 34).

**Proof**
From definitions 96 (p. 96), 40 (p. 34), we have:

$$d_D fs(v) = D_v(fs) = v(f)s + fD_v(s) = s df(v) + f d_D s(v)\ .$$

**Proposition 130** $\mathbf{d_D s} \in \mathcal{V}^*_{V_s} = V_s \otimes \mathcal{V}^* = V_s \otimes \Omega^1$ (see section 1.11 (p. 49) and convention 20 (p. 221)).

**Proof**
Any $\mathbf{d_D s}$ is a linear map of $\mathcal{V}$ to $\mathbf{V_s}$, so an element of $\mathcal{V}^*_{\mathbf{V_s}}$ as defined in definition 188 (p. 221). It is also an element of $\mathbf{V_s} \otimes \mathcal{V}^*$ (proposition 393 (p. 232)). So an element of $\mathbf{V_s} \otimes \mathbf{\Omega^1}$.

**Definition 97 Wedge product of a term of $\mathbf{V_s} \otimes \Omega^p$ and a term of $\mathbf{V_s} \otimes \Omega^{p'}$.**

Given $\mathbf{t} = \mathbf{t^\ell_i f_\ell a^i} \in \mathbf{V_s} \otimes \Omega^p$ and $\mathbf{t'} = \mathbf{t'^\ell_j f_\ell a'^j} \in \mathbf{V_s} \otimes \Omega^{p'}$, we set:

$$\mathbf{t} \wedge \mathbf{t'} = \mathbf{t^\ell_i t'^\ell_j f_\ell a^i} \wedge \mathbf{a'^j} \ .$$

**Definition 98 Definition of $(\mathbf{d_D t} : \mathbf{V_s} \otimes \Omega^p \to \mathbf{V_s} \otimes \Omega^{p+1})$, a map of $\mathbf{V_s} \otimes \Omega^p$ to $\mathbf{V_s} \otimes \Omega^{p+1}$ (See section 1.11 (p. 49)).**

With notations of definition 96 (p. 96), given a term of $\mathbf{t} = \mathbf{t^\ell_i f_\ell} \otimes \mathbf{a^i} = \mathbf{t^\ell_i f_\ell a^i} \in \mathbf{V_s} \otimes \Omega^p$ with $(\mathbf{t^\ell} : [\mathbf{1}, \mathbf{m}] \to \mathcal{F})$, one states:

$$\mathbf{d_D t} = (\mathbf{f_\ell dt^\ell_i} + \mathbf{t^\ell_i d_D f_\ell}) \wedge \mathbf{a^i} + \mathbf{t^\ell_i f_\ell da^i} \in \mathbf{V_s} \otimes \Omega^{p+1} \tag{1.114}$$

$\mathbf{da^i}$ is defined in definitions 59 (p. 53) and 56 (p. 49).

**Proposition 131** Using notation of definition 98 (p. 97), we call $\mathcal{I}_\mathbf{p}$ (definition 53 (p. 48)), the set of increasing bijections $\boldsymbol{\sigma}$ (definitions 124 (p. 139), 384 (p. 486)) of $[\mathbf{1}, \mathbf{p}]$ on a subset of $[\mathbf{1}, \mathbf{n}]$ with cardinal $\mathbf{p}$: $(\boldsymbol{\sigma} : [\mathbf{1}, \mathbf{p}] \leftrightarrow \boldsymbol{\sigma}([\mathbf{1}, \mathbf{p}]) \subset [\mathbf{1}, \mathbf{n}])$. In a covering element $\mathbf{U}$ of $\mathbf{M}$ an element $\mathbf{t}$ of $\mathbf{V_s} \otimes \Omega^p$ can be written as $\mathbf{t} = \sum_{\sigma \in \mathcal{I}_p} \mathbf{t^\ell_\sigma f_\ell} \bigwedge_{i \in [1,p]} \mathbf{dx^{\sigma(i)}}$. With that, the formula (1.114) gives in local coordinates:

$$\begin{aligned}
\mathbf{d_D t} &= \sum_{\sigma \in \mathcal{I}_p} \mathbf{dt^\ell_\sigma f_\ell} \wedge \Big( \bigwedge_{i \in [1,p]} \mathbf{dx^{\sigma(i)}} \Big) \\
&= \sum_{\sigma \in \mathcal{I}_p} \sum_{j \in \mathcal{C}(\sigma([1,p]))} (\mathbf{t^\ell_\sigma (A_j)^k_\ell f_k} + \mathbf{f_\ell \partial_j t^\ell_\sigma}) \mathbf{dx^j} \wedge \Big( \bigwedge_{i \in [1,p]} \mathbf{dx^{\sigma(i)}} \Big) \ .
\end{aligned}$$

**Proof**
It is an application of propositions 129 (p. 96), 130 (p. 96) and definition 98 (p. 97).
**Remark** The property $\mathbf{a} \in \Omega \Rightarrow \mathbf{dda} = \mathbf{d_D d_D a} = \mathbf{0}$ only applies to elements of $\Omega$, see definition 59 (p. 53) and the next proposition 132.

**Proposition 132** $\mathbf{t} \in \mathbf{V_s} \otimes \Omega^p \Rightarrow \mathbf{d_D d_D t} = \mathbf{t^\ell_i} (\mathbf{d_D d_D f_\ell}) \wedge \mathbf{a^i}$ .

**Proof**
Given $\mathbf{t} = \mathbf{t^\ell_i f_\ell a^i} \in \mathbf{V_s} \otimes \Omega^p$, from definition 98 (p. 97)

$$\begin{aligned}
\mathbf{d_D d_D t} &= \mathbf{d_D}((\mathbf{f_\ell dt^\ell_i} + \mathbf{t^\ell_i d_D f_\ell}) \wedge \mathbf{a^i} + \mathbf{t^\ell_i f_\ell da^i}) = \mathbf{d_D f_\ell} \wedge \mathbf{dt^\ell_i} + \mathbf{dt^\ell_i} \wedge \mathbf{d_D f_\ell} \\
&+ \mathbf{t^\ell_i d_D d_D f_\ell} \wedge \mathbf{a^i} - (\mathbf{f_\ell dt^\ell_i} + \mathbf{t^\ell_i d_D f_\ell}) \wedge \mathbf{da^i} + (\mathbf{f_\ell dt^\ell_i} + \mathbf{t^\ell_i d_D f_\ell}) \wedge \mathbf{da^i} \\
&= \mathbf{t^\ell_i} (\mathbf{d_D d_D f_\ell}) \wedge \mathbf{a^i} \ .
\end{aligned}$$

In local coordinates, setting $\mathbf{d_D f_\ell} = \mathbf{A_{\mu\ell}{}^i f_i dx^\mu}$ we have:

$$\begin{aligned}
\mathbf{d_D d_D f_\ell} &= \mathbf{dA_{\mu\ell}{}^i f_i dx^\mu} + \mathbf{A_{\mu\ell}{}^k A_{\nu k}{}^i f_i dx^\nu} \wedge \mathbf{dx^\mu} \\
&= (\partial_\nu \mathbf{A_{\mu\ell}{}^i} + \mathbf{A_{\mu\ell}{}^k A_{\nu k}{}^i}) \mathbf{f_i dx^\nu} \wedge \mathbf{dx^\mu} = \frac{\mathbf{1}}{\mathbf{2}} (\partial_\nu \mathbf{A_{\mu\ell}{}^i} - \partial_\mu \mathbf{A_{\nu\ell}{}^i} + [\mathbf{A_{\nu\ell}{}^k}, \mathbf{A_{\mu k}{}^i}]) \mathbf{f_i dx^\nu} \wedge \mathbf{dx^\mu} \ .
\end{aligned}$$

## 1.15 Parallel transport

Let us consider a differentiable manifold $\mathbf{M}$ of dimension $\mathbf{n}$ and a map of $[0, \mathbf{t}_{\max}]$ into the set of smooth bijective map from $\mathbf{M}$ to $\mathbf{M}$ (definition 6 (p. 3)). Having for a $\mathbf{t}$ the map $(\mathbf{g}(\mathbf{t}, p) : \mathbf{M} \leftrightarrow \mathbf{M})$, for simplicity, we may replace it by $\mathbf{f}(\mathbf{t}, p) = \mathbf{g}(\mathbf{t}, p) \circ \mathbf{g}^{-1}(0, p)$ so we have $(\forall p \in \mathbf{M})(\mathbf{f}(0, p) = p)$. $\mathbf{g}^{-1}(\mathbf{t}, p)$ is the inverse of the map $(\mathbf{g}(\mathbf{t}, p) : \mathbf{M} \leftrightarrow \mathbf{M})$ with $\mathbf{g}(\mathbf{t}, \mathbf{g}^{-1}(\mathbf{t}, p)) = p$.
We write $\mathbf{O}_q$ one of the covering opens containing $q \in \mathbf{M}$ with its associated covering map $(\phi_q \mathbf{O}_q \to \mathcal{R}^{\mathbf{n}})$, see definition 1 (p. 1). We shall also suppose that for any $\mathbf{t} \in [0, \mathbf{t}_{\max}]$, the map $(\mathbf{f}(\mathbf{t}, p) : \mathbf{M} \leftrightarrow \mathbf{M})$ is such that we have $(\phi_{\phi_p(p)}(\mathbf{f}(\mathbf{t}, p)) : [0, \mathbf{t}_{\max}] \to \mathcal{R}^{\mathbf{n}}) \in \mathbf{C}^{\infty}$. From propositions 561 (p. 308) point 2, 562 (p. 309) and 563 (p. 311), it implies that:

$$(\phi_{\phi_p(p)}(\mathbf{f}(\mathbf{t}, \phi_p^{-1}(\mathbf{p}))) : [0, \mathbf{t}_{\max}] \times \phi_p(\mathbf{O}_p) \to \mathcal{R}^{\mathbf{n}}) \in \mathbf{C}^1$$

We shall suppose that it is in fact $\mathbf{C}^{\infty}$ instead of $\mathbf{C}^1$.
At each point $p \in \mathbf{M}$, it defines a curve $(\gamma_p(\mathbf{t}) = \mathbf{f}(\mathbf{t}, p) : [0, \mathbf{t}_{\max}] \to \mathbf{M})$ with $\gamma_p(0) = p$. For each value of $\mathbf{t}$ and for each point $q \in \mathbf{M}$, we have a curve $\gamma$ and a point $p = \mathbf{f}^{-1}(\mathbf{t}, q) \in \mathbf{M}$[23]. It gives $q = \gamma_p(\mathbf{t})$. We suppose that at any $q$ we have the associated tangent space (definition 27 (p. 20)) ${}^{[q]}\mathbf{T}$. When taking the classical derivative of $\gamma_p(\mathbf{t})$ at $\mathbf{t}$, we define for each $\mathbf{t}$ a classical vector at $q$:

$$\dot{\gamma}(\mathbf{t}, q) = \frac{\partial \gamma_{\mu}}{\partial \mathbf{t}} \partial_{\mu} \in {}^{[q]}\mathbf{T}$$

which is tangent to the curve $\gamma_p(\mathbf{t})$ at $q = \gamma_p(\mathbf{t})$. The $\dot{\gamma}(\mathbf{t}, q)$ defines a vector field $\dot{\gamma}(\mathbf{t})$ on $\mathbf{M}$. Having a bundle associated to $\mathbf{M}$ with a vectorial space fiber $\mathbf{F} = \mathcal{R}^{\mathbf{m}}$ and a connection $\mathbf{D}$. We can now consider a continuous map of $[0, 1]$ to section space $\mathbf{V}_s$: $(\mathbf{s}(\mathbf{t}) : [0, 1] \to \mathbf{V}_s)$. We say that those sections are parallel transport for the vector fields $\dot{\gamma}(\mathbf{t})$ and the connection $\mathbf{D}$ if:

$$\mathbf{D}_{\dot{\gamma}(\mathbf{t})}(\mathbf{s}(\mathbf{t})) = 0 \ . \tag{1.115}$$

Remarks: In practical applications, one may consider only the region we are interested in. For instance one can consider only the curve $\gamma_p(\mathbf{t})$ associated to a point $p$ and to the value on the fiber at $\mathbf{t}$.

**Definition 99 Parallel transport.**

We have a vector bundle $\mathbf{E} = \mathbf{F} \times \mathbf{M}$ (definition 77 (p. 87)) defined on a differentiable manifold (definition 3 (p. 1)) $\mathbf{M}$ with a fiber $\mathbf{F}$. $\mathbf{V}_s$ is the set of sections (definition 78 (p. 87)), $\mathcal{V}$ the set of vector fields (definition 15 (p. 8)). We have two continuous maps $(\mathbf{v}(\mathbf{t}) : [0, 1] \to \mathcal{V})$, $(\mathbf{s}(\mathbf{t}) : [0, 1] \to \mathbf{V}_s)$ and a connection $\mathbf{D}$ (definition 87 (p. 91)). Those elements define a parallel transport if we have:

$$(\forall \mathbf{t} \in [0, 1])(\mathbf{D}_{\mathbf{v}(\mathbf{t})}(\mathbf{s}(\mathbf{t})) = 0) \ .$$

## 1.16 Curvature

**Definition 100 Curvature.**

Given a vector bundle (definition 77 (p. 87)) $\mathbf{F} \times \mathbf{M}$ built on a differential manifold (definition 3 (p. 1)) $\mathbf{M}$, given two vector fields (definition 15 (p. 8)) $\mathbf{u}$ and $\mathbf{v}$, given a connection (definition

[23] Here $\mathbf{f}(\mathbf{t}, \mathbf{f}^{-1}(\mathbf{t}, q)) = q$

87 (p. 91)) $\mathbf{D}$ on the bundle $\mathbf{F} \times \mathbf{M}$, one calls curvature, the following transformation on the section module (definition 182 (p. 218)) $\mathbf{V_s}$:

$$\boldsymbol{R}(\mathbf{u},\mathbf{v})\mathbf{s} = \mathbf{D_u}(\mathbf{D_v}(\mathbf{s})) - \mathbf{D_v}(\mathbf{D_u}(\mathbf{s})) - \mathbf{D}_{[\mathbf{u},\mathbf{v}]}(\mathbf{s}) \tag{1.116}$$

$\mathbf{D_u}$ is the covariant derivative associated to $\mathbf{u}$ in the connection $\mathbf{D}$, $[\mathbf{u},\mathbf{v}]$ is the Lie bracket (definitions 1.94 (p. 85)).

**Proposition 133** $\boldsymbol{R}$ is antisymmetric in the two vector fields dependence.

**Proof**

Using the notations of definition 100 (p. 98), since the connection $\mathbf{D_u}$ is linear in $\mathbf{u}$ (definition 87 (p. 91)) and since $[\mathbf{u},\mathbf{v}] = -[\mathbf{v},\mathbf{u}]$ (definition 1.94 (p. 85)) we have $\mathbf{D}_{[\mathbf{u},\mathbf{v}]} = \mathbf{D}_{-[\mathbf{v},\mathbf{u}]} = -\mathbf{D}_{[\mathbf{v},\mathbf{u}]}$. Then

$$\begin{aligned} &\boldsymbol{R}(\mathbf{u},\mathbf{v})(\mathbf{s}) = \mathbf{D_u}(\mathbf{D_v}(\mathbf{s})) - \mathbf{D_v}(\mathbf{D_u}(\mathbf{s})) - \mathbf{D}_{[\mathbf{u},\mathbf{v}]}(\mathbf{s}) = \\ &- \mathbf{D_v}(\mathbf{D_u}(\mathbf{s})) + \mathbf{D_u}(\mathbf{D_v}(\mathbf{s})) + \mathbf{D}_{[\mathbf{v},\mathbf{u}]}(\mathbf{s}) = -\boldsymbol{R}(\mathbf{v},\mathbf{u})(\mathbf{s}) \ . \end{aligned}$$

**Proposition 134** Given two vector fields $\mathbf{u}$ and $\mathbf{v}$ and a section $\mathbf{s}$ in a section module (definition 182 (p. 218)) $\mathbf{V_s}$ the curvature $\boldsymbol{R}(\mathbf{u},\mathbf{v})\mathbf{s}$ is linear on $\mathbf{u},\mathbf{v},\mathbf{s}$.

**Proof**

For $\boldsymbol{\lambda} \in \boldsymbol{\mathcal{F}}, \mathbf{u} \in \boldsymbol{\mathcal{V}}, \mathbf{v} \in \boldsymbol{\mathcal{V}}, \mathbf{w} \in \boldsymbol{\mathcal{V}}, \mathbf{s} \in \mathbf{V_s}, \mathbf{t} \in \mathbf{V_s}$

$$\begin{aligned} \boldsymbol{R}(\mathbf{u},\mathbf{v})\lambda\mathbf{s} &= \mathbf{D_u}(\mathbf{D_v}(\lambda\mathbf{s})) - \mathbf{D_v}(\mathbf{D_u}(\lambda\mathbf{s})) - \mathbf{D}_{[\mathbf{u},\mathbf{v}]}(\lambda\mathbf{s}) \\ &= \mathbf{D_u}(\mathbf{v}(\lambda)\mathbf{s}) + \mathbf{D_u}(\lambda\mathbf{D_v}(\mathbf{s})) - \mathbf{D_v}(\mathbf{u}(\lambda)\mathbf{s}) - \mathbf{D_v}(\lambda\mathbf{D_u}(\mathbf{s})) \\ &- [\mathbf{u},\mathbf{v}](\lambda)\mathbf{s} - \lambda\mathbf{D}_{[\mathbf{u},\mathbf{v}]}(\mathbf{s}) \\ &= \mathbf{u}(\mathbf{v}(\lambda))\mathbf{s} + \mathbf{v}(\lambda)\mathbf{D_u}(\mathbf{s}) + \mathbf{u}(\lambda)\mathbf{D_v}(\mathbf{s}) + \lambda\mathbf{D_u}(\mathbf{D_v}(\mathbf{s})) \\ &- \mathbf{v}(\mathbf{u}(\lambda))\mathbf{s} - \mathbf{u}(\lambda)\mathbf{D_v}(\mathbf{s}) - \mathbf{v}(\lambda)\mathbf{D_u}(\mathbf{s}) - \lambda\mathbf{D_v}(\mathbf{D_u}(\mathbf{s})) \\ &- (\mathbf{u}(\mathbf{v}(\lambda)) - \mathbf{v}(\mathbf{u}(\lambda)))\mathbf{s} - \lambda\mathbf{D}_{[\mathbf{u},\mathbf{v}]}(\mathbf{s}) \\ &= \lambda(\mathbf{D_u}(\mathbf{Dv}(\mathbf{s})) - \mathbf{Dv}(\mathbf{Du}(\mathbf{s})) - \mathbf{D}_{[\mathbf{u},\mathbf{v}]}(\mathbf{s})) \\ \boldsymbol{R}(\mathbf{u},\lambda\mathbf{v})\mathbf{s} &= \mathbf{D_u}(\lambda\mathbf{D_v}(\mathbf{s})) - \lambda\mathbf{D_v}(\mathbf{D_u}(\mathbf{s})) - \mathbf{D}_{[\mathbf{u},\lambda\mathbf{v}]}(\mathbf{s}) \\ &= \mathbf{u}(\lambda)\mathbf{D_v}(\mathbf{s}) + \lambda(\mathbf{D_u}(\mathbf{Dv}(\mathbf{s})) - \mathbf{D_v}(\mathbf{Du}(\mathbf{s}))) - \mathbf{D}_{(\mathbf{u}(\lambda)\mathbf{v}+\lambda[\mathbf{u},\mathbf{v}])}(\mathbf{s}) \\ &= \mathbf{u}(\lambda)\mathbf{D_v}(\mathbf{s}) + \lambda(\mathbf{D_u}(\mathbf{Dv}(\mathbf{s})) - \mathbf{D_v}(\mathbf{Du}(\mathbf{s}))) - \mathbf{u}(\lambda)\mathbf{D_v}(\mathbf{s}) - \lambda\mathbf{D}_{[\mathbf{u},\mathbf{v}]}(\mathbf{s}) \\ &= \lambda\boldsymbol{R}(\mathbf{u},\mathbf{v})\mathbf{s} \end{aligned} \tag{1.117}$$

see proposition 118 (p. 86). From definition 87 (p. 91), we also have:

$$\begin{aligned} \boldsymbol{R}(\mathbf{u}+\mathbf{w},\mathbf{v})\mathbf{s} &= \boldsymbol{R}(\mathbf{u},\mathbf{v})\mathbf{s} + \boldsymbol{R}(\mathbf{w},\mathbf{v})\mathbf{s} \\ \boldsymbol{R}(\mathbf{u},\mathbf{v})(\mathbf{s}+\mathbf{t}) &= \boldsymbol{R}(\mathbf{u},\mathbf{v})\mathbf{s} + \boldsymbol{R}(\mathbf{u},\mathbf{v})\mathbf{t} \ . \end{aligned} \tag{1.118}$$

**Proposition 135** In the coordinate reference frame of the section module (see definitions 88 (p. 91), 182 (p. 218) and 89 (p. 92)), we have, using conventions 1 (p. 18), 9 (p. 91):

$$\boldsymbol{R}(\partial_\mu,\partial_\nu)\mathbf{s} = [\mathbf{D}_\mu,\mathbf{D}_\nu]\mathbf{s} = (\partial_\mu\mathbf{A}_\nu - \partial_\nu\mathbf{A}_\mu + [\mathbf{A}_\mu,\mathbf{A}_\nu])\mathbf{s} \ . \tag{1.119}$$

We also have:

$$\boldsymbol{R} = \boldsymbol{R}_{\mu,\nu}\mathbf{dx}^\mu \wedge \mathbf{dx}^\nu = (\partial_\mu\mathbf{A}_\nu + \mathbf{A}_\mu\mathbf{A}_\nu)\mathbf{dx}^\mu \wedge \mathbf{dx}^\nu \ . \tag{1.120}$$

**Proof**
We call $(\mathbf{e_i} : [\mathbf{1}, \mathbf{m}] \rightarrow \mathbf{V_s})$, a basis of the section module (definition 182 (p. 218)), $\mathbf{m}$ being the dimension of the fiber vectorial space. We write (conventions 1 (p. 18), 9 (p. 91)) $\frac{\partial}{\partial x^\mu} = \partial_\mu, \mathrm{D}_{\partial_\mu} = \mathrm{D}_\mu$. We have from definition 89 (p. 92):

$$\mathbf{s} = \mathbf{s^i e_i} \in \mathbf{V_s} \Rightarrow \mathbf{D_\mu s} = \partial_\mu \mathbf{s^i e_i} + \mathbf{A_{\mu j}}^{\mathbf{i}} \mathbf{s^j e_i}$$
$$\Rightarrow \mathbf{D_\nu D_\mu s} = (\partial_\nu(\partial_\mu \mathbf{s^i} + \mathbf{A_{\mu j}}^{\mathbf{i}}\mathbf{s^j}) + \mathbf{A_{\nu k}}^{\mathbf{i}}\mathbf{A_{\mu j}}^{\mathbf{k}}\mathbf{s^j})\mathbf{e_i}$$
$$\Rightarrow [\mathbf{D_\nu}, \mathbf{D_\mu}]\mathbf{s} = ([\partial_\nu, \partial_\mu]\mathbf{s^i} + (\partial_\nu \mathbf{A_{\mu j}}^{\mathbf{i}} - \partial_\mu \mathbf{A_{\nu j}}^{\mathbf{i}})\mathbf{s^j} + (\mathbf{A_{\nu k}}^{\mathbf{i}}\mathbf{A_{\mu j}}^{\mathbf{k}} - \mathbf{A_{\mu k}}^{\mathbf{i}}\mathbf{A_{\nu j}}^{\mathbf{k}})\mathbf{s^j})\mathbf{e_i}$$

using

$$[\partial_\mu, \partial_\nu] = 0$$

we get equation (1.119) (p. 99). Since from propositions 133 (p. 99) and 134 (p. 99), $\boldsymbol{R}(\partial_\mu, \partial_\nu)^{\mathbf{i}}_{\mathbf{j}}$ is an antisymmetric multilinear map in $\mathbf{u}, \mathbf{v}$ from $\boldsymbol{\mathcal{V}} \times \boldsymbol{\mathcal{V}} \rightarrow \boldsymbol{\mathcal{F}}$ (definition 40 (p. 34), 50 (p. 46)) given $\mathbf{i}, \mathbf{j}$, we may write:

$$\boldsymbol{R}(\partial_\mu, \partial_\nu)^{\mathbf{i}}_{\mathbf{j}} = \boldsymbol{R}(\partial_{\mu'}, \partial_{\nu'})^{\mathbf{i}}_{\mathbf{j}} \mathbf{dx}^{\mu'} \otimes \mathbf{dx}^{\nu'}(\partial_\mu \otimes \partial_\nu) = \boldsymbol{R}_{\mu',\nu'}{}^{\mathbf{i}}_{\mathbf{j}} \mathbf{dx}^{\mu'} \otimes \mathbf{dx}^{\nu'}(\partial_\mu \otimes \partial_\nu) \ .$$

From that we have:

$$\begin{aligned}\boldsymbol{R}\mathbf{s} &= \boldsymbol{R}_{\mu,\nu}(\mathbf{s})\mathbf{dx}^\mu \otimes \mathbf{dx}^\nu = (\partial_\mu \mathbf{A}_\nu - \partial_\nu \mathbf{A}_\mu + [\mathbf{A}_\mu, \mathbf{A}_\nu])\mathbf{s}\mathbf{dx}^\mu \otimes \mathbf{dx}^\nu \\ &= (\partial_\mu \mathbf{A}_\nu + \mathbf{A}_\mu, \mathbf{A}_\nu)\mathbf{s}\mathbf{dx}^\mu \otimes \mathbf{dx}^\nu - (\partial_\mu \mathbf{A}_\nu + \mathbf{A}_\mu, \mathbf{A}_\nu)\mathbf{s}\mathbf{dx}^\nu \otimes \mathbf{dx}^\mu \\ &= (\partial_\mu \mathbf{A}_\nu + \mathbf{A}_\mu \mathbf{A}_\nu)\mathbf{s}\mathbf{dx}^\mu \wedge \mathbf{dx}^\nu\end{aligned}$$

we may also write:

$$\boldsymbol{R}(\mathbf{s^i e_i}) = \mathbf{R}_{\mu,\nu}{}^{\mathbf{j}}_{\mathbf{i}} \mathbf{s^i e_j} \mathbf{dx}^\mu \wedge \mathbf{dx}^\nu = (\partial_\mu \mathbf{A}_{\nu}{}^{\mathbf{j}}_{\mathbf{i}} + \mathbf{A}_{\mu}{}^{\mathbf{j}}_{\mathbf{k}} \mathbf{A}_{\nu}{}^{\mathbf{k}}_{\mathbf{i}})\mathbf{s^i e_j}\mathbf{dx}^\mu \wedge \mathbf{dx}^\nu$$

or:

$$\boldsymbol{R} = \boldsymbol{R}_{\mu,\nu}\mathbf{dx}^\mu \wedge \mathbf{dx}^\nu = (\partial_\mu \mathbf{A}_\nu + \mathbf{A}_\mu \mathbf{A}_\nu)\mathbf{dx}^\mu \wedge \mathbf{dx}^\nu \ .$$

**Proposition 136** $\boldsymbol{R} = \mathbf{dA} + \mathbf{A} \wedge \mathbf{A}$.

**Proof**
As in definition 91 (p. 93), we may consider $(\mathbf{A} : \boldsymbol{\mathcal{V}} \rightarrow \boldsymbol{\mathcal{F}}^{\mathbf{m}\times\mathbf{m}} \in \boldsymbol{\Omega}^1_{\boldsymbol{\mathcal{F}}^{\mathbf{m}\times\mathbf{m}}})$. In the local frame we have $\mathbf{A} = \mathbf{A}_\mu \mathbf{dx}^\mu$ then $\mathbf{dA} = \partial_\nu \mathbf{A}_\mu \mathbf{dx}^\nu \wedge \mathbf{dx}^\mu$. So in any local frame: $\boldsymbol{R} = \mathbf{dA} + \mathbf{A} \wedge \mathbf{A}$ so it is true everywhere.

**Proposition 137** Using notations of propositions 135 (p. 99) and 136 (p. 100), given $\mathbf{t} = \mathbf{t}^\ell_{\mathbf{i}} \mathbf{f}_\ell \mathbf{a}^{\mathbf{i}} \in \mathbf{V_s} \otimes \boldsymbol{\Omega}^{\mathbf{p}}$ we have:

$$\mathbf{d_D d_D t} = \mathbf{t}^\ell_{\mathbf{i}} \boldsymbol{R}^{\mathbf{k}}_{\mu,\nu,\ell} \mathbf{f_k} \mathbf{dx}^\mu \mathbf{dx}^\nu \wedge \mathbf{a}^{\mathbf{i}} \ .$$

**Proof**
From propositions 132 (p. 97) and 136 (p. 100), we have:

$$\mathbf{d_D d_D t} = \mathbf{t}^\ell_{\mathbf{i}}(\mathbf{d_D d_D f}_\ell) \wedge \mathbf{a}^{\mathbf{i}} = \mathbf{t}^\ell_{\mathbf{i}} \boldsymbol{R}^{\mathbf{k}}_{\mu,\nu,\ell} \mathbf{f_k} \mathbf{dx}^\mu \mathbf{dx}^\nu \wedge \mathbf{a}^{\mathbf{i}} \ .$$

**Proposition 138**

$$\mathbf{d_D}\boldsymbol{R} = 0 \ . \tag{1.121}$$

See definition 96 (p. 96) and sections 1.14.5 (p. 96), 2.10.1 (p. 232).

**Proof**
From equation (1.105) (p. 93) in definition 90 (p. 92) and definition 98 (p. 97), we have:

$$\mathbf{d_D}\boldsymbol{R} = (\mathbf{D}_\lambda \boldsymbol{R}_{\mu,\nu} - \boldsymbol{R}_{\mu,\nu}\mathbf{D}_\lambda)\mathbf{dx}^\lambda \wedge \mathbf{dx}^\mu \wedge \mathbf{dx}^\nu$$
$$\mathbf{d_D}\boldsymbol{R} = [\mathbf{D}_\lambda, [\mathbf{D}_\mu, \mathbf{D}_\mathrm{v}]]\mathbf{dx}^\lambda \wedge \mathbf{dx}^\mu \wedge \mathbf{dx}^\nu$$
$$\mathbf{d_D}\boldsymbol{R} = 1/3([\mathbf{D}_\lambda, [\mathbf{D}_\mu, \mathbf{D}_\mathrm{v}]] + [\mathbf{D}_\mu, [\mathbf{D}_\mathrm{v}, \mathbf{D}_\lambda]] + [\mathbf{D}_\nu, [\mathbf{D}_\lambda, \mathbf{D}_\mu]])\mathbf{dx}^\lambda \wedge \mathbf{dx}^\mu \wedge \mathbf{dx}^\nu \,.$$

From Bianchi (Jacobi) identity (proposition 363 (p. 215)), we have:

$$[\mathbf{D}_\lambda, [\mathbf{D}_\mu, \mathbf{D}_\mathrm{v}]] + [\mathbf{D}_\mu, [\mathbf{D}_\mathrm{v}, \mathbf{D}_\lambda]] + [\mathbf{D}_\nu, [\mathbf{D}_\lambda, \mathbf{D}_\mu]] = 0$$

$$\mathbf{d_D}\boldsymbol{R} = 0 \,.$$

This expression means for instance (see definition 90 (p. 92)):

$$\mathbf{s} \in \mathbf{V_s} \Rightarrow \mathbf{d_D}(\boldsymbol{R}\mathbf{s}) = \mathbf{d_D}(\boldsymbol{R})\mathbf{s} + \boldsymbol{R}\mathbf{d_D}\mathbf{s} = \boldsymbol{R}\mathbf{d_D}\mathbf{s} \,.$$

## 1.17 Lagrangian of the electro-weak interactions

We shall show here, as an example, how one uses the covariant derivatives of a bundle to build the Lagrangian of the electro-weak interactions (left-handed).
Let us suppose that we have a principal vector bundle $\mathbf{E} = \mathbf{F} \times \mathbf{M}$ (definition 77 (p. 87)) built on a manifold $\mathbf{M}$ of dimension $\mathbf{n} = \mathbf{4}$ and a fiber $\mathbf{F}$ which is a normal vectorial space of dimension $\mathbf{m} = \mathbf{16}$ (definition 199 (p. 235)), we call as usual $\mathcal{F}$ (definition 7 (p. 3)) the commutative ring (definition 180 (p. 215)) generated by the smooth map $\mathbf{M} \to \mathcal{R}$ (definition 6 (p. 3)), $\mathcal{V}$ the module (definitions 182 (p. 218), 15 (p. 8)) made of vector fields built on $\mathbf{M}$, $\mathbf{V_s}$ the module of the section (definition 78 (p. 87)) of $\mathbf{E}$. $\mathbf{D}$ is a connection (definition 87 (p. 91)) and $\boldsymbol{R}$ the associated curvature (definition 100 (p. 98)).
We could have taken as fiber $\mathbf{F^c}$, the product of two 4 dimension vectorial spaces on complex numbers (definition 245 (p. 271)) $\mathcal{C}^4$. Having a basic vector set $\mathbf{F^c}$, $(\mathbf{f_i^c} : [1,2] \times [1,4] \to \mathcal{C}^8 = \mathbf{F^c})$, a section $\mathbf{s^c}$ would have to be written $(\mathbf{s^c(x)} = \mathbf{s^i(x)f_i^c} : \mathbf{M} \to \mathbf{F^c})$ and a standard basis, as usual, given by $(\mathbf{f_i} = \delta_\mathrm{i}^\mathrm{j} : [1,2] \times [1,4] \to \mathbf{F})$ where $\delta_\mathrm{i}^\mathrm{j}$ is the map on j: $([1,2] \times [1,4] \to \mathcal{C})$ which is zero when $\mathrm{i} \neq \mathrm{j}$ and 1 when $\mathrm{i} = \mathrm{j}$ (see definition 242 (p. 267)). That is possible according to definitions 75 (p. 86) and 77 (p. 87).
But since the manifold theory is built mainly on $\mathcal{R}$, in order to be clearer, we shall use as a fiber $(\mathcal{R}^4)^4 = \mathbf{F}$ which can carry the same properties as $(\mathcal{C}^4)^2$. The vectors of $\mathbf{F}$ are the maps $([1,2] \times [1,2] \times [1,4] \to \mathcal{R})$. The standard basis is now $(\mathbf{f_i} = \delta_\mathrm{i}^\mathrm{j} : [1,2] \times [1,2] \times [1,4] \to \mathbf{F})$. $\delta_\mathrm{i}^\mathrm{j}$ is the map on j, $([1,2] \times [1,2] \times [1,4] \to \mathcal{R})$, which is zero when $\mathrm{i} \neq \mathrm{j}$ and 1 when $\mathrm{i} = \mathrm{j}$. In that standard basis a section $\mathbf{s}$ is written as $(\mathbf{s(x)} = \mathbf{s^i(x)f_i} : \mathbf{M} \to \mathbf{F})$. Splitting i in $\mathrm{i}_1 \times \mathrm{i}_2 \times \mathrm{i}_3$ with the following meaning:

1. $(\mathbf{s}^{1,1,\mathrm{i}_3}(\mathbf{x}) : [1,4] \times \mathbf{M} \to \mathbf{F})$ represents the real part of the Dirac spinor left-handed electronic field,

2. $(\mathbf{s}^{2,1,\mathrm{i}_3}(\mathbf{x}) : [1,4] \times \mathbf{M} \to \mathbf{F})$ its imaginary part,

3. $(\mathbf{s}^{1,2,\mathrm{i}_3}(\mathbf{x}) : [1,4] \times \mathbf{M} \to \mathbf{F})$ the real part of the Dirac spinor left-handed neutrino field,

4. $(\mathbf{s}^{2,2,\mathrm{i}_3}(\mathbf{x}) : [1,4] \times \mathbf{M} \to \mathbf{F})$ the corresponding imaginary part.

As explained in definition 245 (p. 271), we shall associate to any complex number of $\mathcal{C}$, a linear transformation acting only on the first index of $\mathbf{s}$. We call $\mathbb{1}$ the identity transformation $\mathbb{1}^i_j = \delta^i_j$ and $\imath$, the rotation matrix

$$(\imath^i_j : \mathcal{R}^2 \to \mathcal{R}^2)$$

defined by:

$$(i \in [1,2], j \in [1,2]) \Rightarrow (i = j \to \imath^i_j = 0, \imath^1_2 = -\imath^2_1 = 1) \,. \tag{1.122}$$

Considering a complex number $\mathbf{w}$, we write it as $\mathbf{w} = \mathbf{u}\mathbb{1} + \mathbf{v}\imath$ where $\mathbf{u}$ is the real part of $\mathbf{w}$ and $\mathbf{v}$ its imaginary part. The operation $(\mathbf{ws} : \mathbf{F} \to \mathbf{F})$ becomes:

$$\mathbf{w}s^{i1,i2,i3}f_{i1,i2,i3} = s^{i1,i2,i3}(u\mathbb{1} + v\imath)f_{i1,i2,i3} = s^{i1,i2,i3}(u\mathbb{1}^i_{i_1} + v\imath^i_{i_1})f_{i,i2,i3}$$

$(\mathbf{s} : \mathbf{M} \to \mathbf{F})$ being a section, we call $(\mathbf{s}^\dagger : \mathbf{M} \to \mathbf{F}^*)$ a map such that:

$$(\forall x \in M)(i \in [1,2] \times [1,2] \times [1,4], s(x) = s^i(x)f_i, s^\dagger(x) = s^\dagger_i(x)f^i \Rightarrow s^\dagger_i(x) = s^i(x))$$

$s^\dagger_i f^i$ is the adjoint of $\mathbf{s}$ under the metric $g(f_i, f_j) = \delta^j_i$ ( definition 234 (p. 261), 235 (p. 262) and 236 (p. 262)).
$f_i, f^i$ are the standard basis of $\mathbf{F}$ and $\mathbf{F}^*$ (definition 242 (p. 267)): $f^i(f_j) = \delta^i_j$, $\delta$ being the Kronecker symbol (definition 215 (p. 245)).
We call $\mathbf{T}^\dagger$ linear transformation on $\mathbf{F}^*$ such that $T^\dagger(s^\dagger) = (T(s))^\dagger$ which means $T^{\dagger j}_{\ i} = T^i_j$ in order to have $T^{\dagger i}_{\ j} s^\dagger_i e^j = T^{\dagger i}_{\ j} s^i e^j = T^j_i s^i e^j = T^j_i s^\dagger_i e^j$.
We may write the index as $\mathbf{i} = (i_1, i_2, i_3) \in [1,2] \times [1,2] \times [1,4]$.
$\mathbf{M}$ is considered as a Minkowski space that is to say with a metric $(\mathcal{V} \times \mathcal{V} \to \mathcal{F})$ which as a signature: (1-1-1-1) (definition 35 (p. 32)) in any local system. To satisfy the common use, in a local system, we shall take the $\mathbf{4}$ index of the basic vector field of $\mathbf{M}$ in $[0,3]$ and we shall write $(\partial_\mu : [0,3] \to \mathcal{V})$ the basic vector fields of $\mathcal{V}$ in the local system. The metric is given by $g^{\mu\nu} = g(\partial_\mu, \partial_\nu) \in \mathcal{F}$: $(g^{\mu\nu} : M \times [0,3] \times [0,3] \to \mathcal{R})$. Let us consider in the local system, a free Lagrangian for two left-handed leptons of zero masses given by:

$$\mathcal{L}^L = s^\dagger \imath \gamma^0 \gamma^\mu D_\mu s \tag{1.123}$$

or if one wants to keep right and left fermions, one can write also $\mathcal{L}^L = s^\dagger \imath \gamma^0 \gamma^\mu (1 - \gamma^5) D_\mu s$, $(1 - \gamma^5)$ selecting left handed fermions. But since we are dealing with only left handed fermions we can forget about the $\gamma^5$. $\gamma_\ell$ are the $\mathbf{4} \times \mathbf{4}$ Dirac matrices acting separately on the $\mathbf{2}$ different leptonic sections:

$$s = s^{j_1 j_2 j_3} f_{j_1 j_2 j_3} \Rightarrow \gamma^\mu s = (\gamma^\mu)^{i_3}_{j_3} s^{j_1 j_2 j_3} f_{j_1 j_2, i_3} \,.$$

Since we do not use complex number but their equivalent transformation matrix, we write $\gamma^\mu = \gamma^{(r)\mu} + \imath \gamma^{(i)\mu}$ where $\imath$ is defined in equation (1.122) (p. 102). $D_\mu = \partial_\mu + A_\mu$ is a covariant derivative (definition 86 (p. 90)). $\gamma^0$ and $\gamma^5$ are real. We want a total Lagrangian $\mathcal{L}^L$ invariant under a unitary $\mathbf{T}$ in the leptonic space (different) at any point. A transformation $\mathbf{T}(x)$ such as:

$$(\forall x \in M)(T^{\dagger i}_{\ k}(x) T^k_j(x) = \delta^i_j)$$

is called unitary transformation on the section. We want $\mathbf{T}$ to act only on the leptonic part of the fiber space: the $\mathcal{R}^2 \otimes \mathcal{R}^2$ part excluding the spinor part $\mathcal{R}^4$. We have then:

$$s = s^{i,\ell,k} f_{i,\ell,k} \Rightarrow Ts = T^{i_1, i_2}_{j_1 j_2} s^{j_1 j_2 j_3} f_{i_1, i_2 j_3} \,.$$

Since it is a unitary transformation:

$$\mathbf{T}^{i_1,i_2}_{j_1,j_2}\mathbf{T}^{\ell_1,\ell_2}_{j_1,j_2} = \delta^{i_1}_{\ell_1}\delta^{i_2}_{\ell_2}$$

We want also that $\mathbf{T}^{i_1,i_2}_{j_1,j_2}$ commutes with $\imath$ which means that it is equivalent to a complex number (definition 245 (p. 271)) for the first up and down indices. Then:

$$(\forall j_1 \in [1,2)(\forall j_2 \in [1,2])(\mathbf{T}^{1,i_2}_{1,j_2} = \mathbf{T}^{2,i_2}_{2,j_2}, \mathbf{T}^{1,i_2}_{2,j_2} = -\mathbf{T}^{2,i_2}_{1,j_2})$$

with that we have $\imath\mathbf{T} = \mathbf{T}\imath$ (see section 2.13.1 (p. 272)). We also have:

$$\mathbf{T}\imath\gamma^\mu\mathbf{s} = \imath\mathbf{T}\gamma^\mu\mathbf{s} = \imath\gamma^\mu\mathbf{T}\mathbf{s}$$

because:

$$\begin{aligned}
\mathbf{s} = \mathbf{s}^{j_1 j_2 j_3}\mathbf{f}_{j_1,j_2,j_3} &\Rightarrow \mathbf{T}\gamma^\mu\mathbf{s} = \mathbf{T}(\gamma^{(r)\mu} + \imath\gamma^{(i)\mu})\mathbf{s} = \mathbf{T}\gamma^{(r)\mu}\mathbf{s} + \mathbf{T}\imath\gamma^{(i)\mu}\mathbf{s} \\
&= \mathbf{T}\gamma^{(r)\mu}\mathbf{s} + \imath\mathbf{T}\gamma^{(i)\mu}\mathbf{s} = (\mathbf{T}^{i_1,i_2}_{j_1,j_2}(\gamma^{(r)\mu})^{i_3}_{j_3} + (\imath\mathbf{T})^{i_1,i_2}_{j_1,j_2}(\gamma^{(i)\mu})^{i_3}_{j_3})\mathbf{s}^{j_1 j_2 j_3}\mathbf{f}_{i_1,i_2,i_3} \\
&= ((\gamma^{(r)\mu})^{i_3}_{j_3}\mathbf{T}^{i_1,i_2}_{j_1,j_2} + (\gamma^{(i)\mu})^{i_3}_{j_3}(\imath\mathbf{T})^{i_1,i_2}_{j_1,j_2})\mathbf{s}^{j_1 j_2 j_3}\mathbf{f}_{i_1,i_2,i_3} = (\gamma^{(r)\mu} + \imath\gamma^{(i)\mu})\mathbf{T}\mathbf{s} = \gamma^\mu\mathbf{T}\mathbf{s}\ .
\end{aligned}$$

Let us consider now, $\mathcal{L}^L(\mathbf{s})$ as defined by equation (1.123) (p. 102). From proposition 325 (p. 201) $\partial_\mu\mathbf{T}\mathbf{s} = \mathbf{T}\partial_\mu\mathbf{s} + (\partial_\mu\mathbf{T})\mathbf{s}$ then we have:

$$\begin{aligned}
\mathcal{L}^L(\mathbf{T}\mathbf{s}) &= \mathbf{s}^\dagger\mathbf{T}^\dagger\imath\gamma^0\gamma^\mu\mathbf{D}_\mu\mathbf{T}\mathbf{s} = \mathbf{s}^\dagger\mathbf{T}^\dagger\imath\gamma^0\gamma^\mu\mathbf{T}\partial_\mu\mathbf{s} + \mathbf{s}^\dagger(\mathbf{T}^\dagger\mathbf{A}_\mu\mathbf{T} + \partial_\mu\mathbf{T})\mathbf{s} \\
&= \mathbf{s}^\dagger\imath\gamma^0\gamma^\mu\partial_\mu\mathbf{s} + \mathbf{s}^\dagger(\mathbf{T}^\dagger\mathbf{A}_\mu\mathbf{T} + \partial_\mu\mathbf{T})\mathbf{s}\ .
\end{aligned}$$

Using proposition 136 (p. 100), we have:

$$R(\mathbf{A}) = \mathbf{d}(\mathbf{A}_\mu\mathbf{d}\mathbf{x}^\mu) + \mathbf{A}_\mu\mathbf{d}\mathbf{x}^\mu \wedge \mathbf{A}_\nu\mathbf{d}\mathbf{x}^\nu\ . \tag{1.124}$$

But (see section 1.11 (p. 49)):

$$\mathbf{d}(\mathbf{T}^\dagger\mathbf{d}\mathbf{T}) = \mathbf{d}\mathbf{T}^\dagger \wedge \mathbf{d}\mathbf{T} + \mathbf{T}^\dagger\mathbf{d}\mathbf{d}\mathbf{T} = \mathbf{d}\mathbf{T}^\dagger \wedge \mathbf{d}\mathbf{T} \tag{1.125}$$

since $\mathbf{T}^\dagger\mathbf{T} = \mathbf{T}\mathbf{T}^\dagger = \mathbf{1}$ we also have:

$$\mathbf{d}(\mathbf{T}^\dagger\mathbf{T}) = \mathbf{d}(\mathbf{T}\mathbf{T}^\dagger) = \mathbf{d}\mathbf{T}^\dagger\mathbf{T} + \mathbf{T}^\dagger\mathbf{d}\mathbf{T} = \mathbf{d}\mathbf{T}\mathbf{T}^\dagger + \mathbf{T}\mathbf{d}\mathbf{T}^\dagger = \mathbf{0}\ . \tag{1.126}$$

Moreover, if $\mathbf{A}$ is a differential of order 1 ($\mathbf{A} \in \mathbf{\Omega}^1$) and ($\boldsymbol{\lambda} : \mathbf{M} \to \mathcal{R} \in \mathcal{F}$), $\mathbf{d}\boldsymbol{\lambda}\mathbf{A} = \mathbf{d}\boldsymbol{\lambda} \wedge \mathbf{A} + \boldsymbol{\lambda}\mathbf{d}\mathbf{A} = -\mathbf{A} \wedge \mathbf{d}\boldsymbol{\lambda} + \boldsymbol{\lambda}\mathbf{d}\mathbf{A}$. If we have matrices, we get the following:

$$\mathbf{d}(\mathbf{T}\mathbf{A}) = \mathbf{d}\mathbf{T} \wedge \mathbf{A} + \mathbf{T}\mathbf{d}\mathbf{A},\quad \mathbf{d}\mathbf{A}\mathbf{T} = \mathbf{d}\mathbf{A}\mathbf{T} - \mathbf{A} \wedge \mathbf{d}\mathbf{T} \tag{1.127}$$

since the matrices do not generally commute. Moreover, $\mathbf{A},\ \mathbf{B}$ being two differential of $\mathbf{\Omega}^1$, T a matrix acting on the sections, let us prove that we have:

$$\mathbf{A} \wedge \mathbf{T}\mathbf{B} = \mathbf{A}\mathbf{T} \wedge \mathbf{B}\ . \tag{1.128}$$

Indeed, going to the local system, we get:

$$\mathbf{A} \wedge \mathbf{T}\mathbf{B} = \mathbf{A}_{\mu}{}^{i}_{j}\mathbf{d}\mathbf{x}^\mu \wedge \mathbf{T}^{j}_{k}\mathbf{B}_{\nu}{}^{k}_{\ell}\mathbf{d}\mathbf{x}^\nu = \mathbf{A}_{\mu}{}^{i}_{j}\mathbf{T}^{j}_{k}\mathbf{B}_{\nu}{}^{k}_{\ell}\mathbf{d}\mathbf{x}^\mu \wedge \mathbf{d}\mathbf{x}^\nu = \mathbf{A}\mathbf{T} \wedge \mathbf{B}\ .$$

Using equation (1.127) (p. 103), we have:

$$\begin{aligned}
&\boldsymbol{R}(\mathbf{T}^\dagger\mathbf{A}\mathbf{T}+\mathbf{T}^\dagger\mathbf{d}\mathbf{T}) = \boldsymbol{R}(\mathbf{T}^\dagger\mathbf{A}_\mu\mathbf{T}\mathbf{d}\mathbf{x}^\mu+\mathbf{T}^\dagger\partial_\mu\mathbf{T}\mathbf{d}\mathbf{x}^\mu)\\
&= \mathbf{d}(\mathbf{T}^\dagger\mathbf{A}\mathbf{T}+\mathbf{T}^\dagger\mathbf{d}\mathbf{T}) + (\mathbf{T}^\dagger\mathbf{A}_\mu\mathbf{T}+\mathbf{T}^\dagger\partial_\mu\mathbf{T})\mathbf{d}\mathbf{x}^\mu \wedge (\mathbf{T}^\dagger\mathbf{A}_\nu\mathbf{T}+\mathbf{T}^\dagger\partial_\nu\mathbf{T})\mathbf{d}\mathbf{x}^\nu\\
&= \mathbf{d}\mathbf{T}^\dagger\wedge\mathbf{A}\mathbf{T}+\mathbf{T}^\dagger\mathbf{d}\mathbf{A}\mathbf{T}-\mathbf{T}^\dagger\mathbf{A}\wedge\mathbf{d}\mathbf{T}+\mathbf{d}(\mathbf{T}^\dagger\mathbf{d}\mathbf{T})+\mathbf{T}^\dagger\mathbf{A}\mathbf{T}\wedge\mathbf{T}^\dagger\mathbf{A}\mathbf{T}\\
&+\mathbf{T}^\dagger\mathbf{A}\mathbf{T}\wedge\mathbf{T}^\dagger\mathbf{d}\mathbf{T}+\mathbf{T}^\dagger\mathbf{d}\mathbf{T}\wedge\mathbf{T}^\dagger\mathbf{A}\mathbf{T}+\mathbf{T}^\dagger\mathbf{d}\mathbf{T}\wedge\mathbf{T}^\dagger\mathbf{d}\mathbf{T}\ .
\end{aligned}$$

From equations (1.126) (p. 103) and (1.128) (p. 103), we have:

$$\mathbf{T}^\dagger\mathbf{d}\mathbf{T}\wedge\mathbf{T}^\dagger\mathbf{A}\mathbf{T} = -\mathbf{d}\mathbf{T}^\dagger\mathbf{T}\wedge\mathbf{T}^\dagger\mathbf{A}\mathbf{T} = -\mathbf{d}\mathbf{T}^\dagger\wedge\mathbf{A}\mathbf{T}\ .$$

From equation (1.128) (p. 103) we have:

$$\mathbf{T}^\dagger\mathbf{A}\mathbf{T}\wedge\mathbf{T}^\dagger\mathbf{d}\mathbf{T} = \mathbf{T}^\dagger\mathbf{A}\wedge\mathbf{d}\mathbf{T}\ .$$

Then with equations (1.125) (p. 103) and (1.128) (p. 103), we get:

$$\begin{aligned}
&\boldsymbol{R}(\mathbf{T}^\dagger\mathbf{A}\mathbf{T}+\mathbf{T}^\dagger\mathbf{d}\mathbf{T}) = \mathbf{T}^\dagger\mathbf{d}\mathbf{A}\mathbf{T}+\mathbf{d}(\mathbf{T}^\dagger\mathbf{d}\mathbf{T})+\mathbf{T}^\dagger\mathbf{A}\wedge\mathbf{A}\mathbf{T}+\mathbf{T}^\dagger\mathbf{d}\mathbf{T}\wedge\mathbf{T}^\dagger\mathbf{d}\mathbf{T}\\
&= \mathbf{T}^\dagger\mathbf{d}\mathbf{A}\mathbf{T}+\mathbf{T}^\dagger\mathbf{A}\wedge\mathbf{A}\mathbf{T}+\mathbf{d}\mathbf{T}^\dagger\wedge\mathbf{d}\mathbf{T}-\mathbf{d}\mathbf{T}^\dagger\mathbf{T}\wedge\mathbf{T}^\dagger\mathbf{d}\mathbf{T}\\
&= \mathbf{T}^\dagger\boldsymbol{R}(\mathbf{A})\mathbf{T}\ .
\end{aligned}$$

Then since $(\mathbf{T}^\dagger\mathbf{T})^{\mathbf{j}_1,\mathbf{j}_2}_{\mathbf{j}'_1,\mathbf{j}'_2} = \delta^{\mathbf{j}_1}_{\mathbf{j}'_1}\delta^{\mathbf{j}_2}_{\mathbf{j}'_2}$, $\mathbf{T}^{\dagger\,\ell_1,\ell_2}_{\ \ \mathbf{i}_1,\mathbf{i}_2}\mathbf{T}^{\mathbf{i}''_1,\mathbf{i}''_2}_{\ell_1,\ell_2} = \delta^{\mathbf{i}''_1}_{\mathbf{i}_1}\delta^{\mathbf{i}''_2}_{\mathbf{i}_2}$, we have:

$$\boldsymbol{R}_{\mu\nu}{}^{\mathbf{i}_1,\mathbf{i}_2,\mathbf{i}_3}_{\mathbf{j}_1,\mathbf{j}_2,\mathbf{j}_3}(\mathbf{A})\boldsymbol{R}_{\mu'\nu'}{}^{\mathbf{j}_1,\mathbf{j}_2,\mathbf{j}_3}_{\mathbf{i}_1,\mathbf{i}_2,\mathbf{i}_3}(\mathbf{A}) = \mathbf{T}^{\dagger\,\ell_1,\ell_2}_{\ \ \mathbf{i}_1,\mathbf{i}_2}\boldsymbol{R}_{\mu\nu}{}^{\mathbf{i}'_1,\mathbf{i}'_2,\mathbf{i}_3}_{\mathbf{j}_1,\mathbf{j}_2,\mathbf{j}_3}(\mathbf{A})(\mathbf{T}^\dagger\mathbf{T})^{\mathbf{j}_1,\mathbf{j}_2}_{\mathbf{j}'_1,\mathbf{j}'_2}\boldsymbol{R}_{\mu'\nu'}{}^{\mathbf{j}'_1,\mathbf{j}'_2,\mathbf{j}_3}_{\mathbf{i}''_1,\mathbf{i}''_2,\mathbf{i}_3}(\mathbf{A})\mathbf{T}^{\mathbf{i}''_1,\mathbf{i}''_2}_{\ell_1,\ell_2}\ .$$

Then

$$\mathbf{g}^{\mu\mu'}\mathbf{g}^{\nu\nu'}\mathbf{T}^\dagger\boldsymbol{R}_{\mu\nu}(\mathbf{A})\mathbf{T}\mathbf{T}^\dagger\boldsymbol{R}_{\mu'\nu'}(\mathbf{A})\mathbf{T} = \mathbf{g}^{\mu\mu'}\mathbf{g}^{\nu\nu'}\boldsymbol{R}_{\mu\nu}(\mathbf{A})\boldsymbol{R}_{\mu'\nu'}(\mathbf{A})\ .$$

Now if we take as a general Lagrangian with $\Bbbk \in \mathcal{R}$ as an arbitrary constant:

$$\boldsymbol{\mathcal{L}}(\mathbf{A},\mathbf{s}) = \mathbf{g}^{\mu\mu'}\mathbf{g}^{\nu\nu'}\boldsymbol{R}_{\mu\nu}{}^{\mathbf{i}_1,\mathbf{i}_2,\mathbf{i}_3}_{\mathbf{j}_1,\mathbf{j}_2,\mathbf{j}_3}(\mathbf{A})\boldsymbol{R}_{\mu'\nu'}{}^{\mathbf{j}_1,\mathbf{j}_2,\mathbf{j}_3}_{\mathbf{i}_1,\mathbf{i}_2,\mathbf{i}_3}(\mathbf{A}) + \Bbbk\times\mathbf{s}^\dagger_{\mathbf{j}_1,\mathbf{j}_2,\mathbf{j}_3}(\imath\gamma^0\gamma^\mu\mathbf{D}_\mu(\mathbf{A}))^{\mathbf{j}_1,\mathbf{j}_2,\mathbf{j}_3}_{\mathbf{i}_1,\mathbf{i}_2,\mathbf{i}_3}\mathbf{s}^{\mathbf{i}_1,\mathbf{i}_2,\mathbf{i}_3}\ . \tag{1.129}$$

Replacing **s** by **Ts** we have:

$$\begin{aligned}
&\boldsymbol{\mathcal{L}}(\mathbf{A},\mathbf{T}\mathbf{s})\\
&= \mathbf{g}^{\mu\mu'}\mathbf{g}^{\nu\nu'}\boldsymbol{R}_{\mu\nu}(\mathbf{A})\boldsymbol{R}_{\mu'\nu'}(\mathbf{A}) + \mathbf{s}^\dagger\mathbf{T}^\dagger\imath\gamma^\mu(\partial_\mu+\mathbf{A})\mathbf{T}\mathbf{s}\\
&= \mathbf{g}^{\mu\mu'}\mathbf{g}^{\nu\nu'}\boldsymbol{R}_{\mu\nu}(\mathbf{A})\boldsymbol{R}_{\mu'\nu'}(\mathbf{A}) + \mathbf{s}^\dagger\imath\gamma^\mu(\mathbf{T}^\dagger\mathbf{T}\partial_\mu+\mathbf{T}^\dagger\partial_\mu\mathbf{T}+\mathbf{T}^\dagger\mathbf{A}\mathbf{T})\mathbf{s}\\
&= \mathbf{g}^{\mu\mu'}\mathbf{g}^{\nu\nu'}\boldsymbol{R}_{\mu\nu}(\mathbf{T}^\dagger\partial_\mu\mathbf{T}+\mathbf{T}^\dagger\mathbf{A}\mathbf{T})\boldsymbol{R}_{\mu'\nu'}(\mathbf{T}^\dagger\partial_\mu\mathbf{T}+\mathbf{T}^\dagger\mathbf{A}\mathbf{T})\\
&+ \Bbbk\times\mathbf{s}^\dagger\imath\gamma^0\gamma^\mu(\partial_\mu+\mathbf{T}^\dagger\partial_\mu\mathbf{T}+\mathbf{T}^\dagger\mathbf{A}\mathbf{T})\mathbf{s}\\
&\boldsymbol{\mathcal{L}}(\mathbf{T}^\dagger\partial_\mu\mathbf{T}+\mathbf{T}^\dagger\mathbf{A}\mathbf{T},\mathbf{s}) = \boldsymbol{\mathcal{L}}(\mathbf{A},\mathbf{T}\mathbf{s})\ .
\end{aligned} \tag{1.130}$$

Then if **A′**, **s** satisfy the action principle with a Lagrangian $\boldsymbol{\mathcal{L}}(\mathbf{A}',\mathbf{s}')$ it will also satisfy for **A** and $\mathbf{s} = \mathbf{T}\mathbf{s}'$ such that

$$\mathbf{T}^\dagger\partial_\mu\mathbf{T}+\mathbf{T}^\dagger\mathbf{A}\mathbf{T} = \mathbf{A}' \tag{1.131}$$

(The integral of $\boldsymbol{\mathcal{L}}(\mathbf{T}^\dagger\partial_\mu\mathbf{T}+\mathbf{T}^\dagger\mathbf{A}\mathbf{T},\mathbf{s}')$ shall be extremal[24] for **A** and **s** when $\mathbf{T}^\dagger\partial_\mu\mathbf{T} + \mathbf{T}^\dagger\mathbf{A}\mathbf{T} = \mathbf{A}'$ and $\mathbf{s} = \mathbf{T}\mathbf{s}'$ with the transformed limits if it is extremal for $\boldsymbol{\mathcal{L}}(\mathbf{A}',\mathbf{s}')$). Then **A** and $\mathbf{s} = \mathbf{T}\mathbf{s}'$ verify the action principle, if $\boldsymbol{\mathcal{L}}(\mathbf{A}',\mathbf{s}')$ verifies it with $\mathbf{A}' = \mathbf{T}^\dagger\partial_\mu\mathbf{T}+\mathbf{T}^\dagger\mathbf{A}\mathbf{T}$.

[24]It means that $\psi$ is minimal or maximal at the physical value as in proposition 312 (p. 195).

### 1.17.1 Applications

On the opposite of general relativity which was developed only in the manifold frame, particle physics uses manifold frame only as a starting point. The rest of it is developed in the frame of quantum mechanics which is far beyond the scope of that book. We shall not extend the use of the Lagrangian (1.129) (p. 104). We shall only show that it leads to classical electrodynamics when one deals with the potential vector.

**Euler-Lagrange equations**

Having a Lagrangian, $\mathcal{L}(\psi, \frac{\partial\psi}{\partial x^\mu}) = \mathcal{L}(\psi, \partial_\mu\psi)$, the general principle of classical mechanics says that given an open convex $C \in \mathcal{R}^n$, setting the value of $\psi$ at the boundary (definition 138 (p. 157)) of that convex C, the physical evolution, inside that open convex, is such that the variation of the oriented integral (definitions 65 (p. 65), 67 (p. 65)) of $\mathcal{L}$ over C is extremal (see footnote 24 (p. 104)) when $\psi$ is varying from its physical values. That principle can be written as (proposition 312 (p. 195)):

$$(\forall\Delta\psi \in C^\infty)(\{(x \in \mathcal{C}(C) \Rightarrow \Delta\psi = 0), (f : \mathcal{R} \to \mathcal{R}),$$
$$f(v) = \int_C (\mathcal{L}(\psi + v\Delta\psi, \partial_\mu(\psi + v\Delta\psi))) \bigwedge_{i\in[0,3]} dx^i\} \Rightarrow \frac{df(0)}{dv} = f'(0) = 0) .$$

From propositions 622 (p. 356), 321 (p. 198) and 309 (p. 194), we have:

$$f'(v) = \int_C (\frac{\partial(\mathcal{L})}{\partial\psi}(\psi + v\Delta\psi, \partial_\mu(\psi + v\Delta\psi))\Delta\psi +$$
$$\frac{\partial(\mathcal{L})}{\partial(\partial_\mu\psi)}(\psi + v\Delta\psi, \partial_\mu(\psi + v\Delta\psi))\Delta\partial_\mu\psi) \bigwedge_{i\in[0,3]} dx^i$$
$$f'(0) = \int_C (\frac{\partial(\mathcal{L})}{\partial\psi}(\psi, \partial_\mu\psi)\Delta\psi + \frac{\partial(\mathcal{L})}{\partial(\partial_\mu\psi)}(\psi, \partial_\mu\psi)\Delta\partial_\mu\psi) \bigwedge_{i\in[0,3]} dx^i$$
$$= \int_C (\frac{\partial(\mathcal{L})}{\partial\psi}(\psi, \partial_\mu\psi)\Delta\psi + \partial_\mu(\frac{\partial(\mathcal{L})}{\partial(\partial_\mu\psi)}\Delta\psi) - \partial_\mu\frac{\partial(\mathcal{L})}{\partial(\partial_\mu\psi)}\Delta\psi) \bigwedge_{i\in[0,3]} dx^i .$$

We suppose that we have a set $\bigcup_{k\in[1,K]}\{U_k\}$ of K disjoint covers of $\mathcal{B}(C)$ (see equation (1.73) in proposition 102 (p. 64) for the definition of such a cover) with a $\prod^e_{j\in[1,3]} u^j_k$ local reference frame coordinates which are linked to the $\mathcal{R}^4$ reference frame of C by:

$$(\prod^e_{i\in[0,3]} x^i_k(\prod^e_{j\in[1,3]} u^j_k) : \mathcal{R}^3 \to \mathcal{R}^4) \in C^\infty .$$

We use proposition 115 (p. 81) ( see definition 67 (p. 65)) and we get:

$$\int_C (\partial_\mu(\frac{\partial(\mathcal{L})}{\partial(\partial_\mu\psi)}\Delta\psi) \bigwedge_{i\in[0,3]} dx^i = \sum_{\mu\in[0,3]} \int_C (\partial_\mu(\frac{\partial(\mathcal{L})}{\partial(\partial_\mu\psi)}\Delta\psi)(-1)^\mu dx^\mu \wedge \bigwedge_{i\in[0,3]-\mu} dx^i$$
$$= \sum_{\nu\in[0,3]} \sum_{\mu\in[0,3]} \int_C d(\frac{\partial(\mathcal{L})}{\partial(\partial_\mu\psi)}\Delta\psi(-1)^\mu \bigwedge_{i\in[0,3]-\nu} dx^i)$$
$$= \sum_{k\in[1,K]} \sum_{\mu\in[0,3]} \int_{U_k} \frac{\partial(\mathcal{L})}{\partial(\partial_\mu\psi)}\Delta\psi(-1)^\mu \det_{i\in[0,3]-\mu, j\in[1,3]} (\frac{\partial(x^i_k)}{\partial u^j_k}) \bigwedge_{\ell\in[1,3]} du^\ell_k = 0$$

because $\Delta\psi = 0$ on any $U_k \subset \mathcal{B}(C)$. In the above development, we use the relation (see definition 215 (p. 245) for $\delta^\mu_\nu$ ) :

$$dx^\mu \wedge \bigwedge_{i\in[0,3]-\nu} dx^i = \delta^\mu_\nu dx^\mu \wedge \bigwedge_{i\in[0,3]-\mu} dx^i \,.$$

From that, we get the relation:

$$(\forall\Delta\psi \in C^\infty)((x \in \mathcal{C}(C) \Rightarrow \Delta\psi = 0) \Rightarrow \int_C (\frac{\partial(\mathcal{L})}{\partial\psi} - \partial_\mu \frac{\partial(\mathcal{L})}{\partial(\partial_\mu\psi)})\Delta\psi \bigwedge_{i\in[0,3]} dx^i = 0) \,.$$

Applying proposition 148 (p. 117), we get:

$$\frac{\partial(\mathcal{L})}{\partial\psi} - \partial_\mu \frac{\partial(\mathcal{L})}{\partial(\partial_\mu\psi)} = 0 \tag{1.132}$$

**Maxwell equations**

We take, as a gauge transformation, the unitary transformation with positive determinant on $\mathcal{R}^2$: the rotations in $\mathcal{R}^2$. $\mathcal{R}^2$ corresponds to the complex numbers set (definition 245 (p. 271)) associated to the real part and the imaginary of the bundle with which we are dealing (see section 2.13 (p. 271)). They depend on only one parameter and commute: If the angle of rotation is $(\alpha : \mathcal{R}^4 \to \mathcal{R}) \in C^\infty$, that gauge transformation is given by:

$$T^1_1 = T^2_2 = \cos\alpha, T^1_2 = -T^2_1 = \sin\alpha, T = \mathbb{1}\cos\alpha + \imath\sin\alpha = e^{\imath\alpha} \,.$$

According to propositions 679 (p. 413), 321 (p. 198), equations (1.131) (p. 104) and (1.122) (p. 102), we have:

$$\begin{aligned}
&\partial_\mu T^1_1 = \partial_\mu T^2_2 = -\sin\alpha\partial_\mu\alpha, \partial_\mu T^1_2 = -\partial_\mu T^2_1 = \cos\alpha\partial_\mu\alpha \\
&A^t_\mu = T^\dagger\partial_\mu T + T^\dagger A_\mu T \\
&\partial_\mu T = T\imath\partial_\mu\alpha \\
&T^\dagger = T^{-1} \,, \imath T = T\imath \\
&T^\dagger\partial_\mu T = T^\dagger T\imath\partial_\mu\alpha = \imath\partial_\mu\alpha \,.
\end{aligned}$$

Then if $A_\mu$ is restricted to a map $(A_\mu : \mathcal{R}^4 \to \mathcal{R})$ multiplied by the $\imath e$ matrix (see equation (1.122) (p. 102)) $e \in \mathcal{R}$ being the coupling of the electrical field to the photon field : $A_\mu = \imath e A_\mu$, it stays like that in any gauge transformation:

$$\begin{aligned}
&A^t_\mu = T^\dagger\partial_\mu T + T^\dagger A T_\mu = \imath\partial_\mu\alpha + T^\dagger\imath e A_\mu T_\mu = \imath\partial_\mu\alpha + \imath e A_\mu \\
&A^t_\mu = \frac{1}{e}\partial_\mu\alpha + A_\mu;\ A^t = \imath e A^t
\end{aligned}$$

With that we have $[A_\mu, A_\nu] = 0$. Furthermore dealing only with photons, left handed and right handed electrons and positrons have to be treated on equal footing: we can suppress the $1 - \gamma^5$ in the definition of $\mathcal{L}$. Using equation (1.129) (p. 104) and supposing $\mu \neq \nu \Rightarrow$

$g^{\mu\nu} = 0$, we have (equation (1.119) (p. 99))[25]:

$$\begin{aligned}
&F_{\mu\nu} = \partial_\mu A_\nu - \partial_\nu A_\mu \\
&R = dA + A \wedge A \Rightarrow R_{\mu\nu}(A) = \partial_\mu A_\nu - \partial_\nu A_\mu + [A_\mu, A_\nu] = \imath e(\partial_\mu A_\nu - \partial_\nu A_\mu) \\
&= \imath e F_{\mu\nu} \\
&dA = \imath e \frac{F_{\mu\nu}}{2} dx^\mu \wedge dx^\nu \\
&\frac{\partial R_{\mu\nu}(\imath e A)}{\partial(\partial_\tau A_\lambda)} = \imath e(\delta^\mu_\tau \delta^\nu_\lambda - \delta^\nu_\tau \delta^\mu_\lambda) \\
&\mathbb{k} \times s^\dagger(\imath\gamma^\mu D_\mu(\imath e A))s = \mathbb{k} \times s^\dagger(\imath\gamma^0\gamma^\mu(\mathbb{1}\partial_\mu + \imath e A_\mu)s \\
&\mathbb{k} \times \frac{\partial s^\dagger(\imath\gamma^\mu D_\mu(\imath e A))s}{\partial A_\lambda} = -\mathbb{k}e \times \delta^\mu_\lambda s^\dagger\gamma^0\gamma^\mu s \\
&\frac{\partial(\mathcal{L})}{\partial A_\nu} = -\mathbb{k}e \times s^\dagger\gamma^0\gamma^\nu s \\
&\partial_\mu \frac{\partial(\mathcal{L})}{\partial(\partial_\mu A_\nu)} = \mathrm{Trace}\{2\imath e \partial_\mu g^{\mu\mu} g^{\nu\nu} R_{\mu'\nu'}(\delta^\mu_{\mu'}\delta^\nu_{\nu'} - \delta^\mu_{\nu'}\delta^\nu_{\mu'})\} \\
&= 2\mathrm{Trace}\{\imath e \partial_\mu g^{\mu\mu} g^{\nu\nu}(R_{\mu\nu} - R_{\nu\mu})\} = 2e^2 \partial_\mu g^{\mu\mu} g^{\nu\nu}(F_{\mu\nu} - F_{\nu\mu})\mathrm{Trace}\mathbb{1} \\
&= -8e^2 g^{\mu\mu} g^{\nu\nu} \partial_\mu F_{\mu\nu} .
\end{aligned}$$

Putting the above expressions in equation (1.132) (p. 106) we get:

$$-8e^2 g^{\mu\mu} g^{\nu\nu} \partial_\mu F_{\mu\nu} + \mathbb{k}e \times s^\dagger\gamma^0\gamma^\nu s = 0 .$$

From that, setting $\mathbb{k} = 8e^2$, we get the usual formula:

$$\sum_{\mu\in[0,3]} g^{\mu\mu} g^{\nu\nu} \partial_\mu F_{\mu\nu} = e s^\dagger\gamma^0\gamma^\nu s . \tag{1.133}$$

Calling $x^0 = ct$, the time coordinate[26] and $x^1, x^2, x^3$, the spatial coordinates ($\delta^j_i$ is the Kronecker symbol (definition 215 (p. 245))), we set:

$$\begin{aligned}
&g^{00} = 1, \mu \times \nu \in ([0,3] \times [0,3] - \{0 \times 0\}) \Rightarrow g^{\mu\nu} = -\delta^\mu_\nu \\
&B_1 = F_{32},\ B_2 = F_{13},\ B_3 = F_{21} \\
&E_1 = F_{01},\ E_2 = F_{02},\ E_3 = F_{03} .
\end{aligned}$$

Equation (1.133) gives four of the Maxwell relations.
For $\nu = 0$, (1.133) gives:

$$\begin{aligned}
&g^{11}g^{00}\partial_1 F_{10} + g^{22}g^{00}\partial_2 F_{20} + g^{33}g^{00}\partial_3 F_{30} = \partial_1 E_1 + \partial_2 E_2 + \partial_3 E_3 = e s^\dagger\gamma^0\gamma^0 s \\
&\partial_1 E_1 + \partial_2 E_2 + \partial_3 E_3 = e s^\dagger\gamma^0\gamma^0 s .
\end{aligned}$$

For $\nu = 1$:

$$\begin{aligned}
&g^{00}g^{11}\partial_0 F_{01} + g^{22}g^{11}\partial_2 F_{21} + g^{33}g^{11}\partial_3 F_{31} = -\partial_0 E_1 + \partial_2 B_3 - \partial_3 B_2 = e s^\dagger\gamma^0\gamma^1 s \\
&\partial_2 B_3 - \partial_3 B_2 - \partial_0 E_1 = e s^\dagger\gamma^0\gamma^1 s .
\end{aligned}$$

[25] $T \in \mathcal{R}^{2n} \Rightarrow \mathrm{Trace}(T) = \sum_{i=[1,n]} T^i_i$

[26] $c$ is the speed of light

And similarly for the other $\nu$. Then we have:

$$\begin{aligned}
\partial_1 E_1 + \partial_2 E_2 + \partial_3 E_3 &= e s^\dagger \gamma^0 \gamma^0 s \\
\partial_2 B_3 - \partial_3 B_2 - \partial_0 E_1 &= e s^\dagger \gamma^0 \gamma^1 s \\
\partial_3 B_1 - \partial_1 B_3 - \partial_0 E_2 &= e s^\dagger \gamma^0 \gamma^2 s \\
\partial_1 B_2 - \partial_2 B_1 - \partial_0 E_3 &= e s^\dagger \gamma^0 \gamma^3 s \; .
\end{aligned}$$

The other Maxwell relations are given by $\imath dF = ddA = \imath \partial_\tau F_{\mu\nu} dx^\tau \wedge dx^\mu \wedge dx^\nu = 0$ (see definition 59 (p. 53) point 4, proposition 91 (p. 53) point 1). It gives:

$$\begin{aligned}
&(\forall \lambda \in [0,3])((\{\tau,\mu,\nu\} = [0,3] - \lambda, \tau < \mu < \nu) \\
&\Rightarrow \sum_{P \in PG_{\{\tau,\mu,\nu\}}} s(P) \partial_{P(\tau)} F_{P(\mu)P(\nu)} = 0) \; .
\end{aligned}$$

For $\lambda = 0$, we get since $F_{\mu\nu} = -F_{\nu\mu}$:

$$\begin{aligned}
&\partial_1 F_{23} + \partial_2 F_{31} + \partial_3 F_{12} - \partial_1 F_{32} - \partial_2 F_{13} - \partial_3 F_{21} = 2(\partial_1 F_{23} + \partial_2 F_{31} + \partial_3 F_{12}) \\
&= -2(\partial_1 B_1 + \partial_2 B_2 + \partial_3 B_3) = 0 \\
&\sum_{i \in [1,3]} \partial_i B_i = 0
\end{aligned}$$

for $\lambda = 1$:

$$2(\partial_0 F_{23} + \partial_2 F_{30} + \partial_3 F_{02}) = 2(-\partial_0 B_1 - \partial_2 E_3 + \partial_3 E_2) = 0 \; .$$

Then:

$$\begin{aligned}
&\sum_{i \in [1,3]} \partial_i B_i = 0 \\
&\partial_2 E_3 - \partial_3 E_2 + \partial_0 B_1 = 0 \\
&\partial_3 E_1 - \partial_1 E_3 + \partial_0 B_2 = 0 \\
&\partial_1 E_2 - \partial_2 E_1 + \partial_0 B_3 = 0 \; .
\end{aligned}$$

## 1.18 General relativity

**Definition 101 Metric in general relativity.**

In general relativity, we consider a differentiable manifold $M$ of 4 dimensions fully path connected (definition 285 (p. 327)). This manifold is given with a metric which has a signature (+ - - -) (definition 35 (p. 32) and proposition 64 (p. 31)). It is called a Minkowski space. $\mathcal{V}$ is the set of vector fields on $M$.

**Convention 10** $g_{00} = 1$, $g_{11} = g_{22} = g_{33} = -1$.

Given a local cover $U \subset M$ with its map $(\phi : U \to \mathcal{R}^4)$, we call $\mathcal{V}_U$ the set of vector fields of the manifold $U$ considered as the restriction of the manifold $M$ (definition 5 (p. 3)). We consider the matrix $g_{ij} = g(\partial_i, \partial_j)(p)$ at $p \in M$. If $g_{ij} = \pm\delta^i_j$, from proposition 486 (p. 274) there are three values of $i$ for which $g_{ij} = -\delta^i_j$ and one for which $g_{ij} = \delta^i_j$. Then by convention we label the coordinates such that:

$$g_{00} = 1, \; g_{11} = g_{22} = g_{33} = -1$$

In what will follow most of the time the dimension of the manifold is $\mathbf{n}$, the indices are in $[\mathbf{1}, \mathbf{n}]$ and the metric is not specified. It is only when we shall write down the Einstein Lagrangian that the dimension will be set to 4 and the index in $[\mathbf{0}, \mathbf{3}]$ and the signature of the metric set to $(+ - --)$.

**Definition 102 Tangent space bundle, vector bundle (definition 77 (p. 87)) in general relativity in dimension n.**

We suppose that we have a manifold $\mathbf{M}$ of dimension $\mathbf{n}$ with a finite set of vector fields $(\mathbf{n})$ forming a basis of $\mathcal{V}$. Each point $\boldsymbol{p}$ of $\mathbf{M}$ has a tangent space ${}^{[p]}\mathbf{T}$ which is an $\mathbf{n}$ dimensional vector space, $\mathbf{n} = \mathbf{4}$ in our space. The set of those tangent spaces forms a vector bundle (definition 77 (p. 87)) $\mathbf{E} = \mathbf{F} \times \mathbf{M}$ with $\mathbf{F} = \mathcal{R}^{\mathbf{n}}$. At each $\boldsymbol{p} \in \mathbf{M}$ each ${}^{[p]}\mathbf{T}$ is a fiber attached to $\boldsymbol{p}$ (see definitions 75 (p. 86) and 76 (p. 86)).

**Definition 103 Section attached to a vector field.**

From the definition of tangent space (definition 27 (p. 20)), each vector field defines at each point $\boldsymbol{p}$ of $\mathbf{M}$, a vector in the tangent space ${}^{[p]}\mathbf{T}$. Then $(\forall \mathbf{v} \in \mathcal{V})(\mathbf{v} : \mathbf{M} \to \mathbf{F} = \mathcal{R}^{\mathbf{n}})$, $\mathbf{v}$ is a section (definition 78 (p. 87)) of $\mathbf{E} = \mathbf{F} \times \mathbf{M}$.

**Definition 104 Levi-Civita connection $\nabla$.**

We call $\mathcal{V}$ the set of vector fields on $\mathbf{M}$. A Levi-Civita connection is a connection (definition 87 (p. 91)) defined on the sections (definitions 78 (p. 87), 103 (p. 109)) of the tangent space bundle (definitions 77 (p. 87), 102 (p. 109)) that is to say on the vector fields. This connection $\nabla$ is such that $\nabla_{\mathbf{u}}(\mathbf{g})(\mathbf{v}, \mathbf{w})$, defined in definition 95 (p. 96), is zero:

$$\begin{aligned} &(\forall \mathbf{u} \in \mathcal{V})(\forall \mathbf{v} \in \mathcal{V})(\forall \mathbf{w} \in \mathcal{V}) \\ &(\nabla_{\mathbf{u}}(\mathbf{g})(\mathbf{v}, \mathbf{w}) = \mathbf{u}(\mathbf{g}(\mathbf{v}, \mathbf{w})) - \mathbf{g}(\nabla_{\mathbf{u}}(\mathbf{v}), \mathbf{w}) - \mathbf{g}(\mathbf{v}, \nabla_{\mathbf{u}}(\mathbf{w})) = 0) \; . \end{aligned} \tag{1.134}$$

Moreover, one also asks:

$$(\forall \mathbf{u} \in \mathcal{V})(\forall \mathbf{v} \in \mathcal{V})([\mathbf{u}, \mathbf{v}] = \nabla_{\mathbf{u}}(\mathbf{v}) - \nabla_{\mathbf{v}}(\mathbf{u})) \; , \tag{1.135}$$

where $[\mathbf{u}, \mathbf{v}]$ is the Lie bracket (see definition 74 (p. 85)).

**Proposition 139** Given differential manifold $\mathbf{M}$ (definition 3 (p. 1)), its set of vector field $\mathcal{V}$ (definition 15 (p. 8)) with its associated tangent space bundle (definitions 77 (p. 87), 102 (p. 109)), we suppose that $\mathbf{M}$ has a metric $\mathbf{g} : \mathcal{V}^2 \to \mathcal{F}$ (definitions 31 (p. 25), 101 (p. 108)), if a connection (definition 87 (p. 91)) on $\mathbf{M}$ verifies the Levi-Civita conditions (definition 104 (p. 109)) on a basic set (definition 21 (p. 16)) of the restricted vector fields (definition 19 (p. 14)) of a covering open $\mathbf{U}$ (definition 1 (p. 1)), the connection is a Levi-Civita condition on the restricted vector fields of $\mathbf{U}$.

**Proof**
Calling $\mathcal{F}_{\mathbf{U}}$ the restricted set of smooth maps on $\mathbf{U}$ (definitions 7 (p. 3), 8 (p. 3), given six vector fields of $\mathcal{V}_{\mathbf{U}}$: $(\mathbf{u_i} : [\mathbf{1}, \mathbf{3}] \to \mathcal{V}_{\mathbf{U}})$, $(\mathbf{v_i} : [\mathbf{1}, \mathbf{3}] \to \mathcal{V}_{\mathbf{U}})$ one smooth map of $\mathcal{F}_{\mathbf{U}}$:

($\lambda \in \mathcal{F}_U$), applying definitions, supposing definitions 104 (p. 109) for the $\mathbf{u}$ and $\mathbf{v}$, we have:

$$\begin{aligned}
&\nabla_{u_1+v_1}(g)(u_2,u_3) = u_1(g(u_2,u_3)) + v_1(g(u_2,u_3)) - g(\nabla_{u_1}(u_2),u_3)\\
&-g(\nabla_{v_1}(u_2),u_3) - g(u_2,\nabla_{u_1}(u_3)) - g(\nabla_{v_1}(u_2),u_3) - g(u_2,\nabla_{v_1}(u_3)) = 0\\
&\nabla_{\lambda u_1}g(u_2,u_3) = \lambda u_1(g(u_2,u_3)) - g(\nabla_{\lambda u_1}(u_2),u_3) - g(u_2,\nabla_{\lambda u_1}(u_3))\\
&= \lambda u_1(g(u_2,u_3)) - \lambda g(\nabla_{u_1}(u_2),u_3) - \lambda g(u_2,\nabla_{u_1}(u_3)) = 0\\
&\nabla_{u_1}g(u_2+v_2,u_3) = u_1 g(u_2,u_3) + u_1 g(v_2,u_3) - g(\nabla_{u_1}(u_2),u_3)\\
&-g(\nabla_{u_1}(v_2),u_3) - g(u_2,\nabla_{u_1}(u_3)) - g(v_2,\nabla_{u_1}(u_3)) = 0\\
&\nabla_{u_1}(g)(u_2,u_3+v_3) = u_1 g(u_2,u_3) + u_1 g(u_2,v_3) - g(\nabla_{u_1}(u_2),u_3)\\
&-g(\nabla_{u_1}(u_2),v_3) - g(u_2,\nabla_{u_1}(u_3)) - g(u_2,\nabla_{u_1}(v_3)) = 0\\
&\nabla_{u_1}(g)(\lambda u_2,u_3) = u_1(\lambda g(u_2,u_3)) - g(\lambda\nabla_{u_1}(u_2) + u_1(\lambda)u_2,u_3) - \lambda g(u_2,\nabla_{u_1}(u_3))\\
&= u_1(\lambda)g(u_2,u_3) + \lambda u_1(g(u_2,u_3))\\
&-\lambda g(\nabla_{u_1}(u_2),u_3) - u_1(\lambda)g(u_2,u_3) - \lambda g(u_2,\nabla_{u_1}(u_3)) = 0\,.
\end{aligned}$$

Moreover

$$\begin{aligned}
&[u_1+u_2,u_3] = (u_1+u_2)u_3 - u_3(u_1+u_2) = [u_1,u_3] + [u_2,u_1]\\
&= \nabla_{u_1}u_3 + \nabla_{u_2}u_3 - \nabla_{u_3}u_1 - \nabla_{u_3}u_2 = \nabla_{u_1+u_2}u_3 - \nabla_{u_3}u_1 + u_2\\
&[u_3,u_1+u_2] = u_3(u_1+u_2) - (u_1+u_2)u_3 = [u_3,u_1] + [u_3,u_2]\\
&= \nabla_{u_3}u_1 + \nabla_{u_3}u_2 - \nabla_{u_1}u_3 - \nabla_{u_2}u_3 = \nabla_{u_3}(u_1+u_2) - \nabla_{u_1+u_2}u_3\\
&[\lambda u_1,u_2] = \lambda u_1 u_2 - u_2(\lambda u_1) = \lambda u_1 u_2 - \lambda u_2 u_1 - u_2(\lambda)u_2 u_1\\
&= \lambda\nabla_{u_1}u_2 - \lambda\nabla_{u_2}u_1 - u_2(\lambda)u_1 = \nabla_{\lambda u_1}u_2 - \nabla_{u_2}(\lambda u_1)\\
&[u_1,\lambda u_2] = u_1(\lambda)u_2 + \lambda u_1 u_2 - \lambda u_2 u_1 = \nabla_{u_1}(\lambda u_2) - \nabla_{\lambda u_2}u_1
\end{aligned}$$

so if definition 104 (p. 109) is true for a set of vector fields, it is also true for any linear combination of them with element of $\mathcal{F}_U$. [27]

**Definition 105 Christoffel symbols.**

Given a connection $\mathbf{D}$ (definition 87 (p. 91)) defined on the sections (the vector fields) of the tangent space bundle, using definitions 88 (p. 91), 89 (p. 92), proposition 123 (p. 91), one can write

$$D_v(u) = D^0_v(u) + A_v u$$

where $\mathbf{D^0}$ is a connection which is by convention attached to the basis chosen for the sections here $(\partial_i : [1,n] \to \mathcal{V} = V_s)$. We have by definition:

$$D^0_v(u^j\partial_j) = v(u^j)\partial_j = v^i\partial_i(u^j)\partial_j \tag{1.136}$$

$\mathbf{A_v}$ is a linear transformation on the sections which is also linear in $\mathbf{v}$ (see proposition 123 (p. 91)). Then, in the basis of $\mathcal{V}$ which is also the basis of the sections, we can write:

$$A_v = v^i A^k_{i,j},\ A^k_{i,j}\partial_k = D_{\partial_i}(\partial_j) - D^0_{\partial_i}(\partial_j)\,.$$

From equation (1.136) (p. 110) and proposition 21 (p. 9), in the basis $(\partial_i : [1,n] \to \mathcal{V} = V_s)$, we have $(\forall i \in [1,n])(\forall j \in [1,n])(D^0_{\partial_i}(\partial_j) = 0)$. From that we also have:

$$\begin{aligned}
&D_{v^i\partial_i}(u^j\partial_j) = D^0_{v^i\partial_i}(u^j\partial_j) + v^i A^k_{i,j}u^j\partial_k = v^i\partial_i(u^j)\partial_j + v^i u^j A^k_{i,j}\partial_k\\
&= (v^i\partial_i(u^j) + v^i u^k A^j_{i,k})\partial_j\,.
\end{aligned}$$

[27] We have to be careful here: At any point of $\mathbf{M}$, on the tangent space there is a linear transformation which can transform $\mathbf{g}$ into a unitary matrix. But as it can be seen on practical examples although the coefficients of that matrix can be in $\mathcal{F}_U$, in most of the cases there is no coordinate change as defined in 1.6 (p. 23) such that the Jacobian of that transformation is that matrix.

One calls Christoffel symbols the above potential $\mathbf{A}$ written in the local basis $(\partial_i : [1,n] \to \mathcal{V} = V_s)$ when $D = \nabla$ is a Levi-Civita connection. Those symbols are written with $\Gamma$.

$$\begin{aligned} \nabla_{\partial_i}(\partial_j) &= A^k_{i,j}\partial_k = \Gamma^k_{i,j}\partial_k \\ \Gamma^k_{i,j} &= A^k_{i,j} = dx^k(\nabla_{\partial_i}(\partial_j)) = \nabla_{\partial_i}(\partial_j)(x^k) \end{aligned} \tag{1.137}$$

and we have with that definition:

$$\nabla_{v^i\partial_i}(u^j\partial_j) = v^i\partial_i(u^j)\partial_j + v^iu^j\Gamma^k_{i,j}\partial_k \,.$$

**Convention 11** Conventions in general relativity.

In order to simplify the writing we shall adopt a usual convention in general relativity:
In any set of elements of $\mathcal{F}$ labeled by several indices taken in $[1,n]$ where $n$ is the dimension of the differentiable manifold (4 in the real world general relativity) any upper index can be lowered using the metric

$$T_{ijk} = g_{im}T^m_{jk}$$

and setting

$$g^{ij} = g^{-1}_{ij} \,.$$

We also have

$$T^i_{jk} = g^{im}T_{mjk}$$

$g^{ij}$ is then used to raise an index.

**Proposition 140** In the Levi-Civita connection $\Gamma^k_{ij} = \Gamma^k_{ji}$

**Proof**
From equations (1.135) (p. 109) and (1.98) (p. 86) we have:

$$\nabla_{\partial_i}(\partial_j) - \nabla_{\partial_j}(\partial_i) = [\partial_i, \partial_j] = 0$$

but from definition 105 (p. 110)

$$\nabla_{\partial_i}(\partial_j) - \nabla_{\partial_j}(\partial_i) = (\Gamma^k_{ij} - \Gamma^k_{ji})\partial_k$$

then from equations (2.139) (p. 305) and (1.49) (p. 35) plus the definitions 40 (p. 34) and 41 (p. 34), we have:

$$\begin{aligned} &(\forall \ell \in [1,n])(0 = dx^\ell((\Gamma^k_{ij} - \Gamma^k_{ji})\partial_k) = (\Gamma^k_{ij} - \Gamma^k_{ji})\partial_k(x^\ell) = (\Gamma^k_{ij} - \Gamma^k_{ji})\delta^\ell_k = \Gamma^\ell_{ij} - \Gamma^\ell_{ji}) \\ &(\forall \ell \in [1,n])(\Gamma^\ell_{ij} = \Gamma^\ell_{ji}) \end{aligned}$$

$\delta^i_j$ is the Kronecker symbol (definition 215 (p. 245)).

**Proposition 141** In the Levi-Civita connection, the Christoffel symbols are fully defined.

**Proof**
Equation (1.134) (p. 109) gives:

$$\partial_i(g(\partial_j, \partial_k)) = g(\nabla_{\partial_i}(\partial_j), \partial_k) + g(\partial_j, \nabla_{\partial_i}(\partial_k)) \tag{1.138}$$

but $g(\partial_j, \partial_k) = g_{jk}$ and

$$g(\nabla_{\partial_i}(\partial_j), \partial_k) = g(\Gamma^\ell_{ij}\partial_\ell, \partial_k) = \Gamma^\ell_{ij}g_{\ell,k} \tag{1.139}$$

then

$$\partial_i(g(\partial_j,\partial_k)) = \Gamma^\ell_{ij} g_{\ell k} + \Gamma^\ell_{ik} g_{j\ell} \,. \tag{1.140}$$

Since $g_{ij} = g_{ji}$ and $\Gamma^\ell_{ij} = \Gamma^\ell_{ji}$, setting:

$$T_{ijk} = \partial_i(g(\partial_j,\partial_k)) + \partial_j(g(\partial_i,\partial_k)) - \partial_k(g(\partial_i,\partial_j)) \tag{1.141}$$

we have:

$$T_{ijk} = \Gamma^\ell_{ij} g_{\ell k} + \Gamma^\ell_{ik} g_{j\ell} + \Gamma^\ell_{ji} g_{\ell k} + \Gamma^\ell_{jk} g_{i\ell} - \Gamma^\ell_{kj} g_{\ell i} - \Gamma^\ell_{ik} g_{j\ell} = 2\Gamma^\ell_{ij} g_{\ell k} \,. \tag{1.142}$$

The set of equations (1.142) (p. 112) gives for each $i, j$ and at each $p \in M$ the linear system:

$$(\forall p \in M)(\Gamma^\ell_{ij} g_{\ell k}(p) = T_{ijk}(p)/2) \,.$$

Since $(\forall p \in M)(\det(g_{\ell k}) \neq 0)$, from proposition 452 (p. 258), it has for each value of $p$ a unique solution. From proposition 455 (p. 259), since by definition $g_{\ell k}(p) \in \mathcal{F}$, this solution is a linear combination of elements which are maps of $M \to \mathcal{R}$ belonging to $\mathcal{F}$. The Christoffel symbols $\Gamma$ solution of equation (1.141) (p. 112) define a connection.
This connection verifies equation (1.134) (p. 109):
By definition (equation (1.141) (p. 112))

$$(T_{ijk} + T_{ikj})/2 = \partial_i(g(\partial_j,\partial_k))$$

$\Gamma$ being solution of equation (1.142) (p. 112), we also have:

$$(T_{ijk} + T_{ikj})/2 = \Gamma^\ell_{ij} g_{\ell k} + \Gamma^\ell_{ik} g_{j\ell} = \partial_i(g(\partial_j,\partial_k)) = g(\nabla_{\partial_i}(\partial_j),\partial_k) + g(\partial_j,\nabla_{\partial_i}(\partial_k)) \,.$$

We have, from equation (1.141) (p. 112), $T_{ijk} = T_{jik}$. Then the solution of equation (1.142) (p. 112) verifies $\Gamma^k_{ij} = \Gamma^k_{ji}$. Using that, it verifies also equation (1.135) (p. 109) (see definitions 74 (p. 85), 24 (p. 17) and propositions 31 (p. 14), 119 (p. 86)):

$$\nabla_u(v) - \nabla_v(u) = \nabla_u(v^i\partial_i) - \nabla_v(u^j\partial_j) = u(v^i)\partial_i - v(u^j)\partial_j + v^i\nabla_u(\partial_i) - u^j\nabla_v(\partial_j)$$
$$= [u,v] + v^i u^j \Gamma^k_{ij}\partial_k - u^j v^i \Gamma^k_{ji}\partial_k = [u,v] \,.$$

**Definition 106 Riemann curvature.**

One calls Riemann curvature the curvature map $(R : \mathcal{V}^3 \to \mathcal{V})$ (see definition 100 (p. 98)) associated with Levi-Civita connection in the Minkowski differentiable manifold $M$:

$$(u,v,w) \in \mathcal{V}^3 \Rightarrow R(u,v,w) = \nabla_u(\nabla_v(w)) - \nabla_v(\nabla_u(w)) - \nabla_{[u,v]}(w) \,.$$

**Proposition 142** With the above notation $R(u,v,w) = -R(v,u,w)$

**Proposition 143** With the above notation $R(u,v,w)$ is linear in all the arguments.

**Proof**
It is a rephrasing of proposition 134 (p. 99).

**Definition 107 Riemann curvature tensor $R^i_{\ell jk}$.**

From proposition 143 (p. 112), in a cover $\mathbf{U}$ of a manifold $\mathbf{M}$ (see definition 1 (p. 1)), we may define:

$$\begin{aligned}\mathbf{R}(\partial_j, \partial_k, \partial_\ell) &= \mathbf{R}^i_{jk\ell}\partial_i \\ &= \nabla_{\partial_j}(\nabla_{\partial_k}(\partial_\ell)) - \nabla_{\partial_k}(\nabla_{\partial_j}(\partial_\ell)) - \nabla_{[\partial_j,\partial_k]}(\partial_\ell)\end{aligned}$$

$\nabla_{[\partial_j,\partial_k]}(\partial_j) = 0$ since $[\partial_j, \partial_k] = 0$ (equation (1.98) (p. 86)) which means

$$\nabla_{[\partial_j,\partial_k]} = \nabla_0 = 0$$

because $\nabla_\mathbf{v}$ is linear in $\mathbf{v}$ (definition 87 (p. 91), equation (1.101) (p. 91)). Then we have:

$$\mathbf{R}^i_{jk\ell} = dx^i(\mathbf{R}(\partial_j, \partial_k, \partial_\ell)) = \mathbf{R}(\partial_j, \partial_k, \partial_\ell)(x^i) = dx^i(\nabla_{\partial_j}(\nabla_{\partial_k}(\partial_\ell)) - \nabla_{\partial_k}(\nabla_{\partial_j}(\partial_\ell))) .$$

See definitions 40 (p. 34) and 41 (p. 34) and proposition 72 (p. 35).

**Proposition 144**

$$\mathbf{R}^i_{jk\ell} = -\mathbf{R}^i_{kj\ell} .$$

**Proof**
It is a direct consequence of definition 107 (p. 112).

**Definition 108 Ricci tensor.**

One calls Ricci tensor $\mathbf{Ric}_{ij} = \mathbf{R}^k_{ikj}$.

**Definition 109 Scalar curvature.**

One calls scalar curvature $\mathbf{R} \in \mathcal{F}$ the following term (see proposition 144 (p. 113)):

$$\mathbf{R} = \mathbf{R}^j_{ijk}\mathbf{g}^{ik} = -\mathbf{R}^j_{jik}\mathbf{g}^{ik} = \mathbf{Ric}_{ij}\mathbf{g}^{ij} .$$

**Proposition 145** $\mathbf{R}^i_{jk\ell} = \partial_j\Gamma^i_{k\ell} - \partial_k\Gamma^i_{j\ell} + \Gamma^i_{jm}\Gamma^m_{k\ell} - \Gamma^i_{km}\Gamma^m_{j\ell}$.

**Proof**

$$\nabla_{\partial_j}(\nabla_{\partial_k}(\partial_\ell)) = \nabla_{\partial_j}(\Gamma^m_{k\ell}\partial_m) = \Gamma^i_{jm}\Gamma^m_{k\ell}\partial_i + (\partial_j\Gamma^m_{k\ell})\partial_m = \Gamma^i_{jm}\Gamma^m_{k\ell}\partial_i + (\partial_j\Gamma^i_{k\ell})\partial_i$$

then

$$\begin{aligned}\mathbf{R}(\partial_j, \partial_k, \partial_\ell) &= (\partial_j\Gamma^i_{k\ell} - \partial_k\Gamma^i_{j\ell} + \Gamma^i_{jm}\Gamma^m_{k\ell} - \Gamma^i_{km}\Gamma^m_{j\ell})\partial_i \\ \mathbf{R}^i_{jk\ell} &= \partial_j\Gamma^i_{k\ell} - \partial_k\Gamma^i_{j\ell} + \Gamma^i_{jm}\Gamma^m_{k\ell} - \Gamma^i_{km}\Gamma^m_{j\ell} .\end{aligned}$$

**Proposition 146** $\mathbf{g}_{\ell m}\mathbf{R}^m_{ijk} = \mathbf{R}_{\ell ijk} = -\mathbf{R}_{kij\ell}$ (see convention 11 (p. 111)).

**Proof**
We have from relation (1.134) (p. 109):

$$\begin{aligned}\partial_i\partial_j\mathbf{g}(\partial_k, \partial_\ell) &= \partial_i\mathbf{g}(\nabla_{\partial_j}\partial_k, \partial_\ell) + \partial_i\mathbf{g}(\partial_k, \nabla_{\partial_j}\partial_\ell) \\ &= \mathbf{g}(\nabla_{\partial_i}\nabla_{\partial_j}\partial_k, \partial_\ell) + \mathbf{g}(\nabla_{\partial_j}\partial_k, \nabla_{\partial_i}\partial_\ell) + \mathbf{g}(\nabla_{\partial_i}\partial_k, \nabla_{\partial_j}\partial_\ell) + \mathbf{g}(\partial_k, \nabla_{\partial_i}\nabla_{\partial_j}\partial_\ell)\end{aligned}$$

then

$$\begin{aligned}[\partial_i, \partial_j]\mathbf{g}(\partial_k, \partial_\ell) = 0 &= \mathbf{g}([\nabla_{\partial_i}\nabla_{\partial_j}]\partial_k, \partial_\ell) + \mathbf{g}(\partial_k, [\nabla_{\partial_i}\nabla_{\partial_j}]\partial_\ell) \\ &= \mathbf{g}(\mathbf{R}^m_{ijk}\partial_m, \partial_\ell) + \mathbf{g}(\partial_k, \mathbf{R}^m_{ij\ell}\partial_m) \\ &= \mathbf{g}_{\ell m}\mathbf{R}^m_{ijk} + \mathbf{g}_{km}\mathbf{R}^m_{ij\ell} = \mathbf{R}_{\ell ijk} + \mathbf{R}_{kij\ell} .\end{aligned}$$

**Proposition 147** $\mathbf{R}^{\ell}_{ijk} + \mathbf{R}^{\ell}_{jki} + \mathbf{R}^{\ell}_{kij} = 0$ also called Bianchi first identity.

**Proof**

It comes from proposition 145 (p. 113) coupled to proposition 140 (p. 111):

$$\begin{aligned}
\mathbf{R}^{\ell}_{ijk} &= \partial_i \Gamma^{\ell}_{jk} - \partial_j \Gamma^{\ell}_{ik} + \Gamma^{\ell}_{im}\Gamma^{m}_{jk} - \Gamma^{\ell}_{jm}\Gamma^{m}_{ik} \\
\mathbf{R}^{\ell}_{jki} &= \partial_j \Gamma^{\ell}_{ki} - \partial_k \Gamma^{\ell}_{ji} + \Gamma^{\ell}_{jm}\Gamma^{m}_{ki} - \Gamma^{\ell}_{km}\Gamma^{m}_{ji} \\
\mathbf{R}^{\ell}_{kij} &= \partial_k \Gamma^{\ell}_{ij} - \partial_i \Gamma^{\ell}_{kj} + \Gamma^{\ell}_{km}\Gamma^{m}_{ij} - \Gamma^{\ell}_{im}\Gamma^{m}_{kj} \, .
\end{aligned}$$

Applying proposition 140 (p. 111), we get $\mathbf{R}^{\ell}_{ijk} + \mathbf{R}^{\ell}_{jki} + \mathbf{R}^{\ell}_{kij} = 0$.

**Example**

Taking $\mathbf{M}$ as $\mathcal{R}^3$ with the Euclidean metric (see definition 243 (p. 267) and section 2.15 (p. 285)) without the third axis: $\mathbf{x} \in \mathbf{M} \Leftrightarrow (\mathbf{x} = (\mathbf{x}^1, \mathbf{x}^2, \mathbf{x}^3), \mathbf{x} \in \mathcal{R}^3, \mathbf{x}^1 \neq 0, \mathbf{x}^2 \neq 0)$ $\mathbf{M}$ is an open subspace of $\mathcal{R}^3$ (definition 132 (p. 151) equation (2.17) and definition 150 (p. 177)):

$$\mathbf{M} = \bigcup_{\mathbf{x} \in \mathbf{M}} \mathbf{B}(\mathbf{x}, \sqrt{(\mathbf{x}^1)^2 + (\mathbf{x}^2)^2}/2)$$

$\mathbf{M}$ defined a manifold with a metric (definition 31 (p. 25)) which gives the Euclidean metric of $\mathcal{R}^3$ in any tangent space (section 1.5 (p. 19)):

$$\mathbf{g}^{\mathbf{x}}_{ij} = \mathbf{g}^{\mathbf{x}}(\frac{\partial}{\partial \mathbf{x}^i}, \frac{\partial}{\partial \mathbf{x}^j}) = \delta^i_j \, .$$

Going to spherical coordinates when $\mathbf{x} \in \mathbf{M}$ (see section 2.29.2 (p. 413)) with (see definition 378 (p. 484)):

$$\begin{aligned}
&\mathbf{u}^3 = \mathbf{r} = \sqrt{(\mathbf{x}^1)^2 + (\mathbf{x}^2)^2 + (\mathbf{x}^3)^2} \in ]0, \rightarrow [ \\
&\mathbf{u}^1 = \theta = \cos^{-1}(\mathbf{x}^3/\mathbf{r}), \mathbf{u}^1 \in ]0, \pi[, \\
&\mathbf{x}^2 < 0 \Rightarrow (\mathbf{u}^2 = \cos^{-1}(\frac{\mathbf{x}^1}{\mathbf{r}\sin(\theta)}), \mathbf{u}^2 \in ] - \pi, 0[), \\
&\mathbf{x}^2 \geq 0 \Rightarrow (\mathbf{u}^2 = \cos^{-1}(\frac{\mathbf{x}^1}{\mathbf{r}\sin(\theta)}), \mathbf{u}^2 \in [0, \pi]) \\
&\mathbf{x}^1 = \mathbf{u}^3\sin(\mathbf{u}^1)\cos(\mathbf{u}^2) \\
&\mathbf{x}^2 = \mathbf{u}^3\sin(\mathbf{u}^1)\sin(\mathbf{u}^2) \\
&\mathbf{x}^3 = \mathbf{u}^3\cos(\mathbf{u}^1) \, .
\end{aligned}$$

It defines a map $(\mathbf{u}(\mathbf{x}) : \mathbf{M} \rightarrow \mathcal{R}^3)$. $\mathbf{F} = \mathbf{u}(\mathbf{M})$ has the induced topology by $\mathcal{R}^3$ (see definition 133 (p. 151)). One can avoid the problem of having $[0, \pi]$ in the definition of $\mathbf{F}$ by excluding also the first axis but it is not really necessary. That defines a bijection (definition 124 (p. 139)) $\mathbf{M} \leftrightarrow \mathbf{F} = ]0 \rightarrow [\times]0, \pi[\times] - \pi, \pi] \subset \mathcal{R}^3$. Any point satisfying $\mathbf{u}^3 = \mathbf{r}$ is on a sphere of radius $\mathbf{r}$ since $(\mathbf{x}^1)^2 + (\mathbf{x}^2)^2 + (\mathbf{x}^3)^2 = (\mathbf{u}^3)^2 = \mathbf{r}^2$. In the $\mathbf{u}$ system (in $\mathbf{F}$) (see definition 30 (p. 23)), the metric is given by (proposition 61 (p. 30))(see also definition 215 (p. 245) for Kronecker symbol):

$$\mathbf{g}_{ij} = \mathbf{g}^{\mathbf{u}}(\frac{\partial}{\partial \mathbf{u}^i}, \frac{\partial}{\partial \mathbf{u}^j}) = \frac{\partial \mathbf{x}^k}{\partial \mathbf{u}^i}\frac{\partial \mathbf{x}^\ell}{\partial \mathbf{u}^j}\mathbf{g}^{\mathbf{x}}_{k\ell} = \frac{\partial \mathbf{x}^k}{\partial \mathbf{u}^i}\frac{\partial \mathbf{x}^\ell}{\partial \mathbf{u}^j}\delta_{k\ell} = \frac{\partial \mathbf{x}^k}{\partial \mathbf{u}^i}\frac{\partial \mathbf{x}^k}{\partial \mathbf{u}^j} \, .$$

Then since:

$$\begin{array}{llll}
\frac{\partial x^1}{\partial u^i} \underset{i\in[1,3]}{=} & u^3\cos(u^1)\cos(u^2) & , -u^3\sin(u^1)\sin(u^2) & , \sin(u^1)\cos(u^2) \\
\frac{\partial x^2}{\partial u^i} \underset{i\in[1,3]}{=} & u^3\cos(u^1)\sin(u^2) & , u^3\sin(u^1)\cos(u^2) & , \sin(u^1)\sin(u^2) \\
\frac{\partial x^3}{\partial u^i} \underset{i\in[1,3]}{=} & -u^3\sin(u^1) & , 0 & , \cos(u^1)\ .
\end{array}$$

We have from that with $r = u^3$:

$$g_{ij} = \frac{\partial x^k}{\partial u^i}\frac{\partial x^\ell}{\partial u^j}\delta_{k\ell} = \frac{\partial x^k}{\partial u^i}\frac{\partial x^k}{\partial u^j} = \begin{pmatrix} r^2 & , 0 & , 0 \\ 0 & , r^2\sin^2(u^1) & , 0 \\ 0 & , 0 & , 1 \end{pmatrix}.$$

Fixing the radius $r$, we define a smooth surface of $\mathcal{R}^3$: a sphere $S_r$ of radius $r$ minus the points (poles) on the third axis. It is also a differential manifold (see definition 68 (p. 65)). The mapping to $\mathcal{R}^2$ is given by the two $u$: $u^1(x), u^2(x)$. We have on $S_r$ the ordinary metric given by:

$$g_{11} = r^2,\ g_{22} = r^2\sin^2(u^1),\ g_{21} = g_{12} = 0\ .$$

With the Levi-Civita connection (definition 104 (p. 109)) associated to that metric, using the formulae of propositions 141 (p. 111) and 145 (p. 113)), we get the following expressions for the $T$ of proposition 141 (p. 111) together with the Christoffel symbols (definition 105 (p. 110)):

$$\begin{aligned}
&T_{111} = \partial_1 g_{11} + \partial_1 g_{11} - \partial_1 g_{11} = 0\\
&T_{122} = \partial_1 g_{22} + \partial_2 g_{12} - \partial_2 g_{12} = 2r^2\sin(u^1)\cos(u^1)\\
&T_{112} = \partial_1 g_{12} + \partial_1 g_{12} - \partial_2 g_{22} = 0\\
&T_{121} = \partial_1 g_{12} + \partial_2 g_{22} - \partial_1 g_{12} = 0\\
&T_{211} = \partial_2 g_{11} + \partial_1 g_{21} - \partial_1 g_{21} = 0\\
&T_{222} = \partial_2 g_{22} = 0\\
&T_{212} = \partial_2 g_{12} + \partial_1 g_{22} - \partial_2 g_{12} = 2r^2\sin(u^1)\cos(u^1)\\
&T_{221} = \partial_2 g_{12} + \partial_2 g_{21} - \partial_1 g_{22} = -2r^2\sin(u^1)\cos(u^1)\\
&\Gamma^i_{11} g_{i1} = \Gamma^1_{11} r^2 = T_{111}/2 = 0 = \Gamma^1_{11}\\
&\Gamma^i_{11} g_{i2} = \Gamma^2_{11} r^2\sin^2(u^1) = T_{112}/2 = 0 = \Gamma^2_{11}\\
&\Gamma^i_{12} g_{i1} = \Gamma^1_{12} r^2 = T_{121}/2 = 0 = \Gamma^1_{12}\\
&\Gamma^1_{21} = 0\\
&\Gamma^i_{12} g_{i2} = \Gamma^2_{12} r^2\sin^2(u^1) = T_{122}/2 = r^2\sin(u^1)\cos(u^1),\ \Gamma^2_{12} = \cos(u^1)/\sin(u^1)\\
&\Gamma^2_{21} = \cos(u^1)/\sin(u^1)\\
&\Gamma^i_{22} g_{i1} = \Gamma^1_{22} r^2 = T_{221}/2 = -r^2\sin(u^1)\cos(u^1),\ \Gamma^1_{22} = -\sin(u^1)\cos(u^1)\\
&\Gamma^i_{22} g_{i2} = \Gamma^2_{22} r^2\sin^2(u^1) = T_{222}/2 = 0 = \Gamma^2_{22} = 0\ .
\end{aligned}$$

Using all that, we get the Christoffel symbols (definition 105 (p. 110)), the curvature (definitions 100 (p. 98), 106 (p. 112)) and the scalar curvature $\mathbf{R}$ (definition 109 (p. 113)):

$$\Gamma^i_{11} \underset{i\in[1,2]}{=} 0,\ 0$$
$$\Gamma^i_{12} = \Gamma^i_{21} \underset{i\in[1,2]}{=} 0,\ \cos(u^1)/\sin(u^1)$$
$$\Gamma^i_{22} \underset{i\in[1,2]}{=} -\sin(u^1)\cos(u^1),\ 0$$
$$R^i_{jk\ell} = \partial_j\Gamma^i_{k\ell} - \partial_k\Gamma^i_{j\ell} + \Gamma^i_{jm}\Gamma^m_{k\ell} - \Gamma^i_{km}\Gamma^m_{j\ell}$$
$$g^{11} = 1/r^2,\ g^{22} = 1/(r^2\sin^2(u^1)),\ g^{21} = g^{12} = 0$$
$$R = R^j_{ijk}g^{ik} = g^{11}(R^1_{111} + R^2_{121}) + g^{22}(R^1_{212} + R^2_{222})$$
$$R^1_{111} = R^2_{222} = 0$$
$$R^2_{121} = \partial_1\Gamma^2_{21} - \partial_2\Gamma^2_{11} + \Gamma^2_{1m}\Gamma^m_{21} - \Gamma^2_{2m}\Gamma^m_{11} = \partial(\cos(u^1)/\sin(u^1)) + \cos^2(u^1)/\sin^2(u^1)$$
$$= -1$$
$$R^1_{212} = \partial_2\Gamma^1_{12} - \partial_1\Gamma^1_{22} + \Gamma^1_{2m}\Gamma^m_{12} - \Gamma^1_{1m}\Gamma^m_{22} = \partial(\sin(u^1)\cos(u^1)) - \cos^2(u^1)$$
$$= -\sin^2(u^1)$$
$$R = -2/r^2\ .$$

The minus sign obtained here looks troublesome. To get a plus sign, some authors use as scalar curvature, minus the one used here. It gives:

$$R' = -R^i_{ji\ell}g^{j\ell} = R^i_{ij\ell}g^{j\ell} = (\partial_i\Gamma^i_{j\ell} - \partial_j\Gamma^i_{i\ell} + \Gamma^i_{im}\Gamma^m_{j\ell} - \Gamma^i_{jm}\Gamma^m_{i\ell})g^{j\ell}\ .$$

### Some useful formulae

We suppose that $\mathbf{det(g)} < 0$ because in the real world the signature of $\mathbf{g}$ is $(+ - --)$ (definition 35 (p. 32)).

1. $g_{ik}\partial_\ell g^{kj} + \partial_\ell g_{ik}g^{kj} = 0$.
   From convention 11 (p. 111), $\delta^i_j$ being the Kronecker symbol (definition 215 (p. 245)): $g_{ik}g^{kj} = \delta^i_j$. Then
   $$\partial_\ell(g_{ik}g^{kj}) = g_{ik}\partial_\ell g^{kj} + \partial_\ell g_{ik}g^{kj} = 0\ .$$
2. $\partial_\ell g^{ij} = -\partial_\ell g_{mk}g^{mi}g^{kj} = -g^{im}\partial_\ell g_{mk}g^{kj}$.
   From previous item 1 (p. 116)
   $$\partial_\ell g^{ij} = \delta^i_k\partial_\ell g^{kj} = g^{im}g_{mk}\partial_\ell g^{kj} = -\partial_\ell g_{mk}g^{im}g^{kj} = -g^{im}\partial_\ell g_{mk}g^{kj}\ .$$
3. $g_{ij} = g_{ji}$ (definition 31 (p. 25)).
4. $\partial_\mu\det(g_{\ell m}) = \det(g_{\ell m})g^{ij}\partial_\mu g_{ij}$.
   Using definition 228 (p. 254) and proposition 442 (p. 254), we have [28]:
   $$\frac{\partial\det(g_{\mu\nu})}{\partial g_{ij}} = g^{ji}\det(g_{\mu\nu})\ .$$
   Then from proposition 560 (p. 307)
   $$\partial_\mu\det(g_{\mu\nu}) = \frac{\partial\det(g_{\mu\nu})}{\partial g_{ij}}\partial_\mu g_{ij} = g^{ji}\det(g_{\mu\nu})\partial_\mu g_{ij}$$
   $\partial_\mu\det(g_{\ell m}) = \det(g_{\ell m})g^{ij}\partial_\mu g_{ij}$.

[28] $g_{ik}\frac{\partial\det(g_{\mu\nu})}{\partial g_{jk}} = \delta^i_j\det(g_{\mu\nu})$

5. $\partial_\ell\sqrt{|\det(g_{\ell m})|} = \partial_\ell\sqrt{-\det(g_{\ell m})} = \frac{1}{2}\sqrt{-\det(g_{\ell m})}g^{ji}\partial_\ell g_{ji}$.
As we suppose that at the beginning of the paragraph, we have $\det(g) < 0$. Then:

$$-\det(g_{\ell m}) = |\det(g_{\ell m})| \,.$$

From the above item 4 (p. 116) we have

$$\partial_\mu\sqrt{-\det(g_{\ell m})} = \partial_\mu(-\det(g_{\ell m}))^{1/2} = \frac{-\partial_\mu \det(g_{\ell m})}{2\sqrt{-\det(g_{\ell m})}} = \frac{-\det(g_{\ell m})g^{ij}\partial_\mu g_{ij}}{2\sqrt{-\det(g_{\ell m})}}$$
$$= \frac{1}{2}\sqrt{-\det(g_{\ell m})}g^{ij}\partial_\mu g_{ij} \,.$$

6. From equation (1.139) (p. 111):

$$\begin{aligned}\Gamma^\ell_{ij} &= \frac{1}{2}g^{\ell k}(\partial_i(g(\partial_j,\partial_k)) + \partial_j(g(\partial_i,\partial_k)) - \partial_k(g(\partial_i,\partial_j)))\\ &= \frac{1}{2}g^{\ell k}(\partial_i g_{jk} + \partial_j g_{ik} - \partial_k g_{ij})\end{aligned} \,. \quad (1.143)$$

7. We have:

$$\partial_k g^{ij} = -\Gamma^i_{k\ell}g^{\ell j} - \Gamma^j_{k\ell}g^{i\ell} \,. \quad (1.144)$$

We have from point 2 and equation (1.140) (p. 112) of proposition 141 (p. 111):

$$\begin{aligned}\partial_k g^{ij} &= -g^{i\ell}g^{mj}\partial_k g_{\ell m} = -g^{i\ell}g^{mj}(\Gamma^\mu_{k\ell}g_{\mu m} + \Gamma^\mu_{km}g_{\ell\mu})\\ &= -g^{i\ell}\Gamma^j_{k\ell} - g^{mj}\Gamma^i_{km} = -\Gamma^i_{k\ell}g^{\ell j} - \Gamma^j_{k\ell}g^{i\ell} \,.\end{aligned}$$

8. Using equation (1.143) (p. 117) and item 4 (p. 116), we have:

$$\Gamma^i_{i\ell} = \frac{1}{2}g^{im}(\partial_i g_{\ell m} + \partial_\ell g_{im} - \partial_m g_{i\ell}) = \frac{1}{2}g^{im}\partial_\ell g_{im} = \frac{1}{2}\frac{\partial_\ell \det(g)}{\det(g)} \,. \quad (1.145)$$

**Definition 110 Einstein Hilbert Lagrangian.**

In general relativity the field variables is $\mathbf{g}$. The associated Lagrangian is called the Einstein Hilbert Lagrangian given by (see definition 61 (p. 58) and equation (1.70) (p. 59)):

$$\mathcal{L}(g) = \int_C R\sqrt{-\det(g)} \bigwedge_{i\in[0,3]} dx^i = \int_C L(g) \bigwedge_{i\in[0,3]} dx^i \quad (1.146)$$

$\mathbf{C}$ is a convex subset of $\mathcal{R}^n$ in open cube $\mathbf{C}(0,a)$ of $\mathcal{R}^n$ (definition 261 (p. 285)). It defines the characteristic map needed for calculating the integral. This Lagrangian is independent of the system used to calculate it. Applying the standard variational treatment to $\mathcal{L}$ one gets the evolution of the world. To do so, we shall use the following proposition (fundamental lemma of the variational principle):

**Proposition 148** Given a convex open set $\mathbf{B}$ of $\mathcal{R}^n$, $f \in C^\infty(\mathcal{R}^n \to \mathcal{R})$ we have:

$$\left((\forall g \in C^\infty(\mathcal{R}^n \to \mathcal{R}))(g(\mathcal{B}(B)) = \{0\} \Rightarrow \int_{\bar{B}} fg \bigwedge_{i\in[1,n]} dx^i = 0)\right) \Rightarrow f(\bar{B}) = \{0\}$$

$\mathcal{B}(\mathbf{B})$ is the boundary of $\mathbf{B}$, see definition 138 (p. 157).

**Proof**
Let us suppose $(\exists p \in \bar{B})(f(p) = u \neq 0)$ with $u > 0$. Taking that $p$, $f$ being continuous (definition 141 (p. 165)), it is continuous at $p$ (definition 159 (p. 183), proposition 281 (p. 184)). From proposition 157 (p. 182), equation (2.41), there is an open cube $C(p, s_1)$ such that $f^{-1}(C(p, s_1)) \subset ]u/2, 3u/2[$. From definition 132 (p. 151) equation (2.18), $C(p, u_1) \cap B$ is an open of $B$. Since $\bar{B}$ is the closure of $B$, from definition 136 (p. 155), $C(p, u_1) \cap B$ contains a point $q$ and, from proposition 266 (p. 179), an open cube $C(q, s_2) \subset C(p, s_1) \cap B$ and we have $f^{-1}(C(q, s_2)) \subset ]u/2, 3u/2[$. Then using proposition 715 (p. 428) and the map $h(x) = g(x - p, s_1/2, s_1) \in C^\infty(\mathcal{R}^n \to \mathcal{R})$ which verifies ($g$ is defined in definition 14 (p. 6) and $\mathcal{C}$ the complement of a set, is defined in definition 353 (p. 454)):

$$h(\mathcal{R}^n) \subset \mathcal{R}^{+0}, h(C(q, s_1/2)) = \{1\}, h(\mathcal{C}(C(q, s_1))) = \{0\} .$$

We have then:

$$0 = \int_{\bar{B}} fh \bigwedge_{i \in [1,n]} dx^i > u/2 \int_{\bar{B}} h \bigwedge_{i \in [1,n]} dx^i .$$

Since $0 < \frac{s_1^n}{2^n} \leq \int_{\bar{B}} h \bigwedge_{i \in [1,n]} dx^i \leq s_1^n$, $u$ cannot be positive. If $u < 0$, applying what precedes to $-f$, it is also impossible. Thus $u = 0$.

### 1.18.1 Einstein equations in vacuum

One shall derive the Einstein equations from the variational principle when the Lagrangian given by equation (1.146) (p. 117) is alone (vacuum case): We have an $n$ dimensional manifold $M$ ($n = 4$ in real world and the index of the variables goes from 0 to 3 in order to separate the time component from the space component) and in a cover $O$ with its map $(\phi : O \to \mathcal{R}^n)$, an open convex subset $C \subset \phi(O)$ which can be a cube. We suppose that $(\exists a > 0)(C \subset C(0, a) \subset \mathcal{R}^n)$. $C(0, a)$ is an open cube of $\mathcal{R}^n$ (definition 261 (p. 285)). $\mathcal{B}(C)$ (definition 138 (p. 157)) is called the convex surface. A metric $g$ can be described by a map $(g : [1, n] \times [1, n] \to \mathcal{F})$ such that $(\mu \in [1, n], \nu \in [1, n]) \Rightarrow (g_{\mu\nu}(\phi^{-1}(x)) : \mathcal{R}^n \to \mathcal{R}) \in C^\infty$. To be physical $g$ should be such as[29]:

$$\begin{aligned}&(\forall \alpha > 0)(\forall (h : [1, n] \times [1, n] \to C^\infty(\mathcal{R}^n \to \mathcal{R})))((x \notin C \Rightarrow h(x) = 0) \Rightarrow \\ &(\exists \epsilon_{\mathcal{L}}(\alpha, h) > 0)(0 < \tau < \epsilon_{\mathcal{L}}(\alpha, h) \Rightarrow |\mathcal{L}(g + \tau h) - \mathcal{L}(g)| < \alpha\tau)) .\end{aligned} \quad (1.147)$$

In the following, we write $\Delta g = \tau h$ where $h$ is made of $n \times n$ $C^\infty$ maps $\mathcal{R}^n \to \mathcal{R}$. In the classical way of doing the calculation to find such a $g$, one writes:

$$\begin{aligned}\Delta L = {} & \Delta g^{ik} R^j_{ijk} \sqrt{-\det(g)} + g^{ik} \Delta R^j_{ijk} \sqrt{-\det(g)} + g^{ik} R^j_{ijk} \Delta \sqrt{-\det(g)} \\ & + g^{ik} \Delta R^j_{ijk} \Delta \sqrt{-\det(g)} + R^j_{ijk} \Delta g^{ik} \Delta \sqrt{-\det(g)} + \sqrt{-\det(g)} \Delta g^{ik} \Delta R^j_{ijk} \\ & + \Delta g^{ik} \Delta R^j_{ijk} \Delta \sqrt{-\det(g)} .\end{aligned}$$

Since all the maps which we consider are defined and continuous in $\bar{C}$ which is closed (proposition 219 (p. 158)) and compact (proposition 521 (p. 290)), from proposition 297 (p. 189), their absolute values have a maximum $A$ for all of them. All expressions, which we consider, are $C^\infty$ maps of the type $\mathcal{R}^q \to \mathcal{R}$. In those maps, each of the variables are replaced by either $g$ or their partial derivatives at any finite order.

$$(\forall \lambda \in I_\lambda)(\exists q_\lambda \in \mathcal{Z}^+)(\exists (H_\lambda \in C^\infty(\mathcal{R}^{q_\lambda} \to \mathcal{R})))(\exists (\mu_\ell \nu_\ell : [1, q_\lambda] \to [1, n]^2))$$

$$(\exists (\partial^{[\ell]} : [1, q_\lambda] \to \mathcal{D}))(H_\lambda(g) = H_\lambda( \prod^{e}_{\ell \in [1, q_\lambda]} \partial^{[\ell]} g_{\mu_\ell \nu_\ell}(x)) : \mathcal{R}^n \to \mathcal{R})$$

[29]In order to simplify, from now on we refer $g$ for $(g \circ \phi^{-1} : \mathcal{R}^n \to \mathcal{R})$ and the same for $h$.

where $\mathcal{D}$ is the set of the multiorder partial derivatives, (see definition 312 (p. 398)). Since we have $\mathbf{A}$ such that (we set $\partial^{[0]}\mathbf{f} = \mathbf{f}$):

$$(\forall \ell \in \mathcal{Z}^{0+})(\forall \lambda \in \mathrm{I}_\lambda)(\forall x \in \bar{\mathrm{C}})(\forall \ell\mu\nu \in [1, q_\lambda] \times [1, n]^2)(|\partial^{[\ell]} g_{\mu\nu}(x)| < A,$$
$$|\partial^{[\ell]} h_{\mu\nu}(x)| < A) .$$

Applying proposition 560 (p. 307) point 2, we have:

$$(\forall \alpha)(\exists \epsilon_\lambda(\alpha))(|\tau| < \epsilon_\lambda(\alpha), H_\lambda(g + \tau h) - H_\lambda(g) = \Delta H_\lambda$$
$$= \sum_{\ell \in [1, q_\lambda]} (\partial_\ell H_\lambda + \eta_\ell)\tau \partial^{[\ell]} h \Rightarrow \|\eta\| < \alpha) . \tag{1.148}$$

For that we take the map $\epsilon(\alpha)$ of equation (2.143) (p. 307) of proposition 560 (p. 307) point 2 and we set $\epsilon_\lambda(\alpha) = \epsilon(\alpha/A)$.
For $\lambda_1\lambda_2 \in \mathrm{I}_\lambda^2$, we have:

$$\Delta H_{\lambda_1} \Delta H_{\lambda_2} = \tau^2 (\sum_{\ell \in [1, q_{\lambda_1}]} (\partial_\ell H_{\lambda_1} + \eta_\ell)\partial^{[\ell]} h) \times (\sum_{\ell \in [1, q_{\lambda_2}]} (\partial_\ell H_{\lambda_2} + \eta_\ell)\partial^{[\ell]} h) .$$

Then for $|\tau| < \epsilon^{12} = \min(\epsilon_{\lambda_1}(A), \epsilon_{\lambda_2}(A))$ and $u(x) \in C^\infty(\mathcal{R}^n \to \mathcal{R})$ such that $x \in \bar{C} \Rightarrow |u(x)| < A$ (we can always increase $\mathbf{A}$ so it is true see propositions 297 (p. 189) and 521 (p. 290)).

$$|\int_C u(x)\Delta H_{\lambda_1} \Delta H_{\lambda_2} \bigwedge_{i \in [1,n]} dx^i| < \int_C |u(x)| \times |\Delta H_{\lambda_1}| \times |\Delta H_{\lambda_2}| \bigwedge_{i \in [1,n]} dx^i$$
$$< \tau^2 q_{\lambda_1} q_{\lambda_2} A(2A)^2(2a)^n$$

then

$$(\forall \alpha > 0)(\exists \epsilon'^{12}(\alpha) = \min(\frac{\alpha}{q_{\lambda_1} q_{\lambda_2} A(2A)^2(2a)^n}, \epsilon^{12}))(|\tau| < \epsilon'^{12}$$
$$\Rightarrow |\int_C u(x)\Delta H_{\lambda_1} \Delta H_{\lambda_2} \bigwedge_{i \in [1,n]} dx^i| < \alpha\tau) . \tag{1.149}$$

With the same reasoning:

$$(\forall \alpha > 0)(\exists \epsilon'^{123}(\alpha))(|\tau| < \epsilon'^{123}(\alpha)) \Rightarrow |\int_C u(x)\Delta H_{\lambda_1} \Delta H_{\lambda_2} \Delta H_{\lambda_3} \bigwedge_{i \in [1,n]} dx^i| < \alpha\tau \tag{1.150}$$

**Some more relations**
In the following, by convention, for all $\eta^{\text{index}}$ we have the relation:

$$(\forall \alpha > 0)(\exists \epsilon^{\text{index}}(\alpha) > 0)(|\tau| < \epsilon^{\text{index}}(\alpha) \Rightarrow |\int_C \eta^{\text{index}} \bigwedge_{i \in [1,n]} dx^i| < \alpha\tau) . \tag{1.151}$$

Using formula 2 (p. 116) and equation (1.148) (p. 119) we have:

$$R^j_{ijk}\sqrt{-\det(g)}\Delta g^{ik} = R^j_{ijk}\sqrt{-\det(g)}\tau h_{m\ell}\partial g^{ik}/\partial_{g_{m\ell}} + \eta^{\text{Inv}}$$
$$= -g^{im} R^j_{ijk}\sqrt{-\det(g)}\tau h_{m\ell} g^{\ell k} + \eta^{\text{Inv}} . \tag{1.152}$$

Using equation (1.148) (p. 119), definition 228 (p. 254), propositions 442 (p. 254), 444 (p. 255) we have:

$$\Delta \det(g) = (-1)^{i+j}\det(\mathrm{Minor}_{ij}(g))\Delta g_{ij} = \det(g)g^{ij}\Delta g_{ij} = \tau \det(g)g^{ij}h_{ij} \,. \tag{1.153}$$

From proposition 329 (p. 203) we have:

$$(\forall\alpha)(\exists\epsilon^{sq}(\alpha))((\Delta\sqrt{cy} = \frac{c\Delta y}{2\sqrt{cy}} + \zeta\Delta y, |\Delta y| < \epsilon^{sq}(\alpha)) \Rightarrow |\zeta| < \alpha) \,. \tag{1.154}$$

The above two relations 1.153 (p. 120), 1.154 (p. 120) give:

$$g^{ik}R^j_{ijk}\Delta\sqrt{-\det(g)} = \tau g^{ik}R^j_{ijk}\frac{\sqrt{-\det(g)}g^{\ell m}}{2}h_{\ell m} + \eta^{sqd} \tag{1.155}$$

$$\begin{aligned}
&\Delta R^j_{ijk} = \partial_i\Delta\Gamma^j_{jk} - \partial_j\Delta\Gamma^j_{ik}\\
&+ \Gamma^m_{jk}\Delta\Gamma^j_{im} - \Gamma^m_{ik}\Delta\Gamma^j_{jm} + \Gamma^j_{im}\Delta\Gamma^m_{jk} - \Gamma^j_{jm}\Delta\Gamma^m_{ik} + \Delta\Gamma^j_{im}\Delta\Gamma^m_{jk} - \Delta\Gamma^j_{jm}\Delta\Gamma^m_{ik}\\
&\sqrt{-\det(g)}g^{ik}\Delta R^j_{ijk} = \sqrt{-\det(g)}g^{ik}(\partial_i\Delta\Gamma^j_{jk} - \partial_j\Delta\Gamma^j_{ik}\\
&+ \Gamma^m_{jk}\Delta\Gamma^j_{im} - \Gamma^m_{ik}\Delta\Gamma^j_{jm} + \Gamma^j_{im}\Delta\Gamma^m_{jk} - \Gamma^j_{jm}\Delta\Gamma^m_{ik}) + \eta^R
\end{aligned} \tag{1.156}$$

$$\begin{aligned}
&\Delta L = \Delta g^{ik}\, R^j_{ijk}\sqrt{-\det(g)} + g^{ik}\Delta R^j_{ijk}\sqrt{-\det(g)} + g^{ik}R^j_{ijk}\Delta\sqrt{-\det(g)}\\
&+ g^{ik}\Delta R^j_{ijk}\Delta\sqrt{-\det(g)} + \Delta g^{ik}R^j_{ijk}\Delta\sqrt{-\det(g)} + \Delta g^{ik}\Delta R^j_{ijk}\sqrt{-\det(g)}\\
&+ \Delta g^{ik}\Delta R^j_{ijk}\Delta\sqrt{-\det(g)}\\
&\Delta L = \Delta g^{ik}\, R^j_{ijk}\sqrt{-\det(g)} + g^{ik}\Delta R^j_{ijk}\sqrt{-\det(g)} + g^{ik}R^j_{ijk}\Delta\sqrt{-\det(g)} + \eta^{\mathcal{L}} \,.
\end{aligned} \tag{1.157}$$

All the above calculations are to show that we can neglect second and third order terms which are summed in the $\eta$'s. We apply for that formulae like equation (1.151) (p. 119) which we prove to be true.

**Computation of the second term:** $\mathbf{I} = \int_C g^{ik}\Delta R^j_{ijk}\sqrt{-\det(g)} \bigwedge_{\ell\in[1,n]} dx^\ell$

We shall use the Stokes theorem 115 (p. 81) which states:

$$\int_C \partial_i f \bigwedge_{k\in[1,n]} dx^k = (-1)^i \int_C d(f \bigwedge_{k\in[1,n]-i} dx^k) = (-1)^i \int_{\mathcal{B}(C)} f \bigwedge_{k\in[1,n]-i} dx^k \,.$$

The integral is then 0 if $f = 0$ on the boundary $\mathcal{B}(C)$.
In the following expressions, we shall relate two maps by

$$=_{int}$$

if the difference of the two maps has a null integral on $C$ using the above relationship. From equations (1.145) (p. 117) and (1.144) (p. 117), with the integration by part[30], we have:

$$\begin{aligned}
&\sqrt{-\det(g)}g^{ik}\Delta R^j_{ijk} = \sqrt{-\det(g)}g^{ik}(\partial_i\Delta\Gamma^j_{jk} - \partial_j\Delta\Gamma^j_{ik}\\
&+ \Gamma^m_{jk}\Delta\Gamma^j_{im} - \Gamma^m_{ik}\Delta\Gamma^j_{jm} + \Gamma^j_{im}\Delta\Gamma^m_{jk} - \Gamma^j_{jm}\Delta\Gamma^m_{ik}) + \eta^R \,.
\end{aligned}$$

[30] It uses $a\partial b = \partial(ab) - (\partial a)b$.

Setting

$$\mathbf{T_1 = \sqrt{-\det(g)}g^{ik}(\partial_i\Delta\Gamma^j_{jk} - \partial_j\Delta\Gamma^j_{ik})}$$
$$\mathbf{T_2 = \sqrt{-\det(g)}\{g^{ik}\Gamma^m_{jk}\Delta\Gamma^j_{im} - g^{ik}\Gamma^m_{ik}\Delta\Gamma^j_{jm} + g^{ik}\Gamma^j_{im}\Delta\Gamma^m_{jk} - g^{ik}\Gamma^j_{jm}\Delta\Gamma^m_{ik}\}}\ .$$

We have

$$\mathbf{\sqrt{-\det(g)}g^{ik}\Delta R^j_{ijk} = T_1 + T_2 + \eta^R}\ .$$

But we also have:

$$\begin{aligned}
&\mathbf{T_1 = \partial_i(\sqrt{-\det(g)}g^{ik}\Delta\Gamma^j_{jk}) - \partial_j(\sqrt{-\det(g)}g^{ik}\Delta\Gamma^j_{ik})}\\
&\mathbf{- \partial_i(\sqrt{-\det(g)}g^{ik})\Delta\Gamma^j_{jk} + \partial_j(\sqrt{-\det(g)}g^{ik})\Delta\Gamma^j_{ik}}\\
&\mathbf{\partial_j(\sqrt{-\det(g)}g^{ik}) = \partial_j(\sqrt{-\det(g)})g^{ik} + \sqrt{-\det(g)}\partial_j g^{ik}}\\
&\mathbf{\partial_j(\sqrt{-\det(g)}) = \frac{-1}{2\sqrt{-\det(g)}}2\det(g)\Gamma^\ell_{\ell j} = \sqrt{-\det(g)}\Gamma^\ell_{\ell j}}\\
&\mathbf{\partial_j g^{ik} = -g^{i\mu}\partial_j g_{\mu\nu}g^{\nu k} = -g^{i\mu}(\Gamma^\ell_{j\mu}g_{\ell\nu} + \Gamma^\ell_{j\nu}g_{\mu\ell})g^{\nu k} = -g^{i\mu}\Gamma^k_{j\mu} - \Gamma^i_{j\nu}g^{\nu k}}\\
&\mathbf{T_1 =_{int} \sqrt{-\det(g)}\{(-g^{ik}\Gamma^\ell_{\ell i} + g^{i\mu}\Gamma^k_{i\mu} + g^{\nu k}\Gamma^i_{i\nu})\Delta\Gamma^j_{jk}}\\
&\mathbf{+ (g^{ik}\Gamma^\ell_{\ell j} - g^{i\mu}\Gamma^k_{j\mu} - g^{\nu k}\Gamma^i_{j\nu})\Delta\Gamma^j_{ik}\}}\\
&\mathbf{= \sqrt{-\det(g)}\{(-g^{ik}\Gamma^\ell_{\ell i} + g^{i\mu}\Gamma^k_{i\mu} + g^{ik}\Gamma^\ell_{\ell i})\Delta\Gamma^j_{jk} + (g^{ik}\Gamma^\ell_{\ell j} - g^{i\mu}\Gamma^k_{j\mu} - g^{\nu k}\Gamma^i_{j\nu})\Delta\Gamma^j_{ik}\}}\\
&\mathbf{= \sqrt{-\det(g)}\{g^{i\mu}\Gamma^k_{i\mu}\Delta\Gamma^j_{jk} + (g^{ik}\Gamma^\ell_{\ell j} - g^{i\mu}\Gamma^k_{j\mu} - g^{\nu k}\Gamma^i_{j\nu})\Delta\Gamma^j_{ik}\}}
\end{aligned}$$

and

$$\begin{aligned}
&\mathbf{T_2 = \sqrt{-\det(g)}\{g^{i\mu}\Gamma^k_{j\mu}\Delta\Gamma^j_{ik} - g^{i\mu}\Gamma^k_{i\mu}\Delta\Gamma^j_{jk} + g^{\nu k}\Gamma^i_{\nu j}\Delta\Gamma^j_{ik} - g^{ik}\Gamma^\ell_{\ell j}\Delta\Gamma^j_{ik}\}}\\
&\mathbf{= \sqrt{-\det(g)}\{-g^{i\mu}\Gamma^k_{i\mu}\Delta\Gamma^j_{jk} + (-g^{ik}\Gamma^\ell_{\ell j} + g^{i\mu}\Gamma^k_{j\mu} + g^{\nu k}\Gamma^i_{\nu j})\Delta\Gamma^j_{ik}\} =_{int} -T_1}\ .
\end{aligned}$$

Then $\mathbf{T_1 + T_2 =_{int} 0}$ which gives:

$$\mathbf{\sqrt{-\det(g)}g^{ik}\Delta R^j_{ijk} = T_1 + T_2 + \eta^R =_{int} \eta^R}\ .$$

**Einstein equation**

We now have:

$$\begin{aligned}
\mathbf{\Delta\mathcal{L}} &= \mathbf{\int_C (\Delta g^{ik}\, R^j_{ijk}\sqrt{-\det(g)} + g^{ik}R^j_{ijk}\Delta\sqrt{-\det(g)}) \bigwedge_{\mu\in[1,n]} dx^\mu}\\
&\mathbf{+ \int_C (\eta^{\mathcal{L}} + \eta^R) \bigwedge_{\mu\in[1,n]} dx^\mu}\ .
\end{aligned}$$

From equations (1.152) (p. 119), (1.155) (p. 120) we have:

$$\begin{aligned}
\mathbf{\Delta\mathcal{L}} &= \mathbf{\tau\int_C (-g^{im}h_{m\ell}g^{\ell k}R^j_{ijk}\sqrt{-\det(g)} + g^{ik}R^j_{ijk}\frac{\sqrt{-\det(g)}g^{\ell m}}{2}h_{\ell m}) \bigwedge_{\mu\in[1,n]} dx^\mu}\\
&\mathbf{+ \int_C (\eta^{\mathcal{L}} + \eta^R + \eta^{sqd} + \eta^{Inv}) \bigwedge_{\mu\in[1,n]} dx^\mu}\ .
\end{aligned}$$

Calling:

$$\mathcal{I} = \int_C h_{m\ell}\sqrt{-\det(g)}(-g^{im}g^{\ell k}R^j_{ijk} + g^{ik}R^j_{ijk}\frac{g^{m\ell}}{2}) \bigwedge_{\mu\in[1,n]} dx^\mu$$

We have from equations (1.147) (p. 118), (1.155) (p. 120), (1.156) (p. 120) , (1.157) (p. 120) with convention of equation (1.151) (p. 119):

$$(\forall\alpha)(\exists\epsilon(\alpha) = \min(\epsilon_{\mathcal{L}}(\alpha/5, h), \epsilon^{\mathcal{L}}(\alpha/5), \epsilon^{R}(\alpha/5), \epsilon^{sqd}(\alpha/5)), \epsilon^{Inv}(\alpha/5)) > 0)$$
$$(|\tau| < \epsilon(\alpha) \Rightarrow |\tau||\mathcal{I}| < |\Delta\mathcal{L}| + \int_C (|\eta^{\mathcal{L}}| + |\eta^{R}| + |\eta^{sqd}| + |\eta^{Inv}|) \bigwedge_{i\in[1,n]} dx^i < \alpha|\tau|$$
$$\Rightarrow |\mathcal{I}| < \alpha) .$$

Then we have:

$$(\forall h)(((h : [1,n]\times[1,n] \to C^\infty(\mathcal{R}^n \to \mathcal{R})), x \notin C \Rightarrow h(x) = 0) \Rightarrow |\mathcal{I}| = 0)$$

otherwise if $|\mathcal{I}| \neq 0$ one can set $\alpha = |\mathcal{I}|$ and we have an $\epsilon > 0$ such that $|\tau| < \epsilon$ gives $|\mathcal{I}| < \alpha$ which is absurd. By A.1.4 (p. 435), 8 (p. 435) we have $|\mathcal{I}| = 0$.
Applying now proposition 148 (p. 117), we have:

$$(\forall x \in \bar{C})(\forall m\times\ell \in [1,n]^2)(\{\sqrt{-\det(g)}(-g^{im}g^{\ell k}R^j_{ijk} + g^{ik}R^j_{ijk}\frac{g^{m\ell}}{2})\}(x) = 0) .$$

Since $\det(g) \neq 0$, it gives:

$$g^{mi}g^{\ell k}R^j_{ijk} = \frac{1}{2}g^{m\ell}g^{ik}R^j_{ijk} = g^{m\ell}\frac{R}{2}$$

$$g_{\mu m}g_{\nu\ell}g^{mi}g^{\ell k}R^j_{ijk} = R^\lambda_{\mu\lambda\nu} = g_{\mu m}g_{\nu\ell}g^{m\ell}\frac{R}{2} = g_{\mu\nu}\frac{R}{2} .$$

We then get:

$$R^\lambda_{\mu\lambda\nu} = \mathrm{Ric}_{\mu\nu} = g_{\mu\nu}\frac{R}{2} \tag{1.158}$$

$\mathrm{Ric}_{\mu\nu}$ is the Ricci tensor (definition 108 (p. 113)). By definitions 109 (p. 113), 34 (p. 29), 215 (p. 245), we have:

$$R = g^{\mu\nu}R^\lambda_{\mu\lambda\nu},\ g^{\mu\lambda}g_{\lambda\nu} = g^{\mu\lambda}g_{\nu\lambda} = \delta^\mu_\nu,\ \delta^\mu_\mu = 4 .$$

The above equation (1.158) (p. 122) gives $R = \delta^\mu_\mu R/2 = 4R/2 = 2R$ which means $R = 0$ so in vacuum $R^\lambda_{\mu\lambda\nu} = 0$. But when not in vacuum one describes the interaction of the matter with the metric by adding to Einstein Hilbert Lagrangian, a Lagrangian which also depends on the metric and on the matter we want to deal with. In that frame, the term $G_{\mu\nu} = R^\lambda_{\mu\lambda\nu} - g_{\mu\nu}\frac{R}{2}$ is not zero everywhere.

**Definition 111 Einstein tensor.**

The tensor $G_{\mu\nu} = R^\lambda_{\mu\lambda\nu} - g_{\mu\nu}\frac{R}{2}$ is called the Einstein tensor.

**Example 1: The Schwarzschild metric.**

Schwarzschild proposed as solution of the Einstein equation, the following metric:

$$
\begin{aligned}
g_{00} &= g_{tt} = (1 - \frac{r_s}{r})c^2 \\
g_{11} &= g_{\theta,\theta} = -r^2 \\
g_{22} &= g_{\phi\phi} = -r^2 \sin^2\theta \\
g_{33} &= g_{rr} = -(1 - \frac{r_s}{r})^{-1}
\end{aligned}
\tag{1.159}
$$

with $x^0 = t,\ x^1 = \theta,'\ x^2 = \phi,\ x^3 = r$. All the other terms are zero.
$t$ being the time component and $\theta, \phi, r$ the traditional polar coordinates. We shall use now the following relations (see proposition 141 (p. 111), equations (1.141) (p. 112), (1.142) (p. 112) and proposition 145 (p. 113)):

$$
\begin{aligned}
R^i_{jk\ell} &= \partial_j \Gamma^i_{k\ell} - \partial_k \Gamma^i_{j\ell} + \Gamma^i_{jm}\Gamma^m_{k\ell} - \Gamma^i_{km}\Gamma^m_{j\ell} \\
T_{ijk} &= \partial_i g_{jk} + \partial_j g_{ik} - \partial_k g_{ij} \\
\Gamma^k_{ij} &= \frac{1}{2} T_{ijm} g^{mk} \,.
\end{aligned}
$$

With that at any point where $r = x^3 \neq r_s$ and $r = x^3 \neq 0$ we have:

- The Christoffel symbols (definition 105 (p. 110)):

$$
\begin{array}{lll}
\Gamma^3_{00} = \frac{1}{2}\frac{r_s c^2 (r-r_s)}{r^3} & \partial_3\Gamma^3_{00} = \frac{r_s c^2}{2}(\frac{1}{r^3} - 3\frac{r-r_s}{r^4}) & \\
\Gamma^0_{30} = \frac{1}{2}\frac{r_s}{r(r-r_s)} & \partial_3\Gamma^0_{30} = \frac{-1}{2}\frac{r_s(2r-r_s)}{r^2(r-r_s)^2} & \\
\Gamma^0_{03} = \frac{1}{2}\frac{r_s}{r(r-r_s)} & \partial_3\Gamma^0_{03} = \frac{-1}{2}\frac{r_s(2r-r_s)}{r^2(r-r_s)^2} & \\
\Gamma^3_{11} = -(r - r_s) & \partial_3\Gamma^3_{11} = -1 & \\
\Gamma^1_{31} = \frac{1}{r} & \partial_3\Gamma^1_{31} = \frac{-1}{r^2} & \\
\Gamma^1_{13} = \frac{1}{r} & \partial_3\Gamma^1_{13} = \frac{-1}{r^2} & \\
\Gamma^3_{22} = -\sin^2\theta(r - r_s) & \partial_1\Gamma^3_{22} = -2\sin\theta\cos\theta(r - r_s) & \partial_3\Gamma^3_{22} = -\sin^2\theta \\
\Gamma^2_{32} = \frac{1}{r} & \partial_3\Gamma^2_{32} = \frac{-1}{r^2} & \\
\Gamma^2_{23} = \frac{1}{r} & \partial_3\Gamma^2_{23} = \frac{-1}{r^2} & \\
\Gamma^1_{22} = -\cos\theta\sin\theta & \partial_1\Gamma^1_{22} = (\sin^2\theta - \cos^2\theta) & \\
\Gamma^2_{12} = \frac{\cos\theta}{\sin\theta} & \partial_1\Gamma^2_{12} = \frac{-1}{\sin^2\theta} & \\
\Gamma^2_{21} = \frac{\cos\theta}{\sin\theta} & \partial_1\Gamma^2_{21} = \frac{-1}{\sin^2\theta} & \\
\Gamma^3_{33} = -\frac{1}{2}\frac{r_s}{r(r-r_s)} & \partial_3\Gamma^3_{33} = \frac{1}{2}\frac{r_s(2r-r_s)}{r^2(r-r_s)^2} \,. &
\end{array}
$$

  All the other terms are zero.

- Curvature tensor, calculus of the 16 terms $R^\lambda_{\mu\lambda\nu}$ which have to be zero.
  We give here all the components of the curvature tensor which are different from zero and we sum them according to the above formula and we verify that they are null. This was done using a computer program which assembles the $\Gamma$ expressions. It rejects all the expressions which do not have any contribution from the Christoffel symbols.

1. $\mathbf{R}^{\mathrm{i}}_{0\mathrm{i}0}$

$$R^0_{000} = \frac{1}{2}\frac{r_s}{r(r-r_s)} \times \frac{1}{2}\frac{r_s c^2(r-r_s)}{r^3} - \frac{1}{2}\frac{r_s}{r(r-r_s)} \times \frac{1}{2}\frac{r_s c^2(r-r_s)}{r^3}$$
$$R^1_{010} = -\frac{1}{r} \times \frac{1}{2}\frac{r_s c^2(r-r_s)}{r^3}$$
$$R^2_{020} = -\frac{1}{r} \times \frac{1}{2}\frac{r_s c^2(r-r_s)}{r^3}$$
$$R^3_{030} = -\frac{r_s c^2}{2}\left(\frac{1}{r^3} - 3\frac{r-r_s}{r^4}\right) + \frac{1}{2}\frac{r_s c^2(r-r_s)}{r^3} \times \frac{1}{2}\frac{r_s}{r(r-r_s)}$$
$$+\frac{1}{2}\frac{r_s}{r(r-r_s)} \times \frac{1}{2}\frac{r_s c^2(r-r_s)}{r^3}$$

$$\begin{aligned}
&R^i_{0i0} = \\
&\frac{1}{2}\frac{r_s}{r(r-r_s)} \times \frac{1}{2}\frac{r_s c^2(r-r_s)}{r^3} - \frac{1}{2}\frac{r_s}{r(r-r_s)} \times \frac{1}{2}\frac{r_s c^2(r-r_s)}{r^3} \\
&-\frac{1}{r} \times \frac{1}{2}\frac{r_s c^2(r-r_s)}{r^3} \\
&-\frac{1}{r} \times \frac{1}{2}\frac{r_s c^2(r-r_s)}{r^3} \\
&-\frac{r_s c^2}{2}\left(\frac{1}{r^3} - 3\frac{r-r_s}{r^4}\right) + \frac{1}{2}\frac{r_s c^2(r-r_s)}{r^3} \times \frac{1}{2}\frac{r_s}{r(r-r_s)} \\
&+\frac{1}{2}\frac{r_s}{r(r-r_s)} \times \frac{1}{2}\frac{r_s c^2(r-r_s)}{r^3} = 0
\end{aligned}$$

2. $\mathbf{R}^{\mathrm{i}}_{1\mathrm{i}1}$

$$R^0_{101} = \frac{1}{2}\frac{r_s}{r(r-r_s)} \times (r-r_s)$$
$$R^1_{111} = -\frac{1}{r} \times (r-r_s) + \frac{1}{r} \times (r-r_s)$$
$$R^2_{121} = \frac{-1}{\sin^2\theta} + \frac{\cos\theta}{\sin\theta} \times \frac{\cos\theta}{\sin\theta} + \frac{1}{r} \times (r-r_s)$$
$$R^3_{131} = 1 - (r-r_s) \times \frac{1}{r} - \frac{1}{2}\frac{r_s}{r(r-r_s)} \times (r-r_s)$$

$$\begin{aligned}
&R^i_{1i1} = \\
&\frac{1}{2}\frac{r_s}{r(r-r_s)} \times (r-r_s) \\
&-\frac{1}{r} \times (r-r_s) + \frac{1}{r} \times (r-r_s) \\
&+\frac{-1}{\sin^2\theta} + \frac{\cos\theta}{\sin\theta} \times \frac{\cos\theta}{\sin\theta} + \frac{1}{r} \times (r-r_s) \\
&+1 - (r-r_s) \times \frac{1}{r} - \frac{1}{2}\frac{r_s}{r(r-r_s)} \times (r-r_s) = 0
\end{aligned}$$

3. $\mathbf{R^{i}_{1i3}}$

$$\mathbf{R^{2}_{123} = \frac{\cos\theta}{\sin\theta} \times \frac{1}{r} - \frac{\cos\theta}{\sin\theta} \times \frac{1}{r}}$$

$$\mathbf{R^{i}_{1i3} = \frac{\cos\theta}{\sin\theta} \times \frac{1}{r} - \frac{\cos\theta}{\sin\theta} \times \frac{1}{r} = 0}$$

4. $\mathbf{R^{i}_{2i2}}$

$$\mathbf{R^{0}_{202} = \frac{1}{2}\frac{r_s}{r(r - r_s)} \times \sin^2\theta(r - r_s)}$$
$$\mathbf{R^{1}_{212} = -(\sin^2\theta - \cos^2\theta) - \cos\theta\sin\theta \times \frac{\cos\theta}{\sin\theta} + \frac{1}{r} \times \sin^2\theta(r - r_s)}$$
$$\mathbf{R^{2}_{222} = -\frac{\cos\theta}{\sin\theta} \times \cos\theta\sin\theta - \frac{1}{r} \times \sin^2\theta(r - r_s) + \frac{\cos\theta}{\sin\theta} \times \cos\theta\sin\theta}$$
$$\mathbf{+\frac{1}{r} \times \sin^2\theta(r - r_s)}$$
$$\mathbf{R^{3}_{232} = \sin^2\theta - \sin^2\theta(r - r_s) \times \frac{1}{r} - \frac{1}{2}\frac{r_s}{r(r - r_s)} \times \sin^2\theta(r - r_s)}$$

$$\mathbf{R^{i}_{2i2} =}$$
$$\mathbf{\frac{1}{2}\frac{r_s}{r(r - r_s)} \times \sin^2\theta(r - r_s)}$$
$$\mathbf{-(\sin^2\theta - \cos^2\theta) - \cos\theta\sin\theta \times \frac{\cos\theta}{\sin\theta} + \frac{1}{r} \times \sin^2\theta(r - r_s)}$$
$$\mathbf{-\frac{\cos\theta}{\sin\theta} \times \cos\theta\sin\theta - \frac{1}{r} \times \sin^2\theta(r - r_s) + \frac{\cos\theta}{\sin\theta} \times \cos\theta\sin\theta}$$
$$\mathbf{+\frac{1}{r} \times \sin^2\theta(r - r_s)}$$
$$\mathbf{+\sin^2\theta - \sin^2\theta(r - r_s) \times \frac{1}{r} - \frac{1}{2}\frac{r_s}{r(r - r_s)} \times \sin^2\theta(r - r_s) = 0}$$

5. $\mathbf{R^{i}_{3i1}}$

$$\mathbf{R^{2}_{321} = \frac{1}{r} \times \frac{\cos\theta}{\sin\theta} - \frac{\cos\theta}{\sin\theta} \times \frac{1}{r}}$$
$$\mathbf{R^{i}_{3i1} = \frac{1}{r} \times \frac{\cos\theta}{\sin\theta} - \frac{\cos\theta}{\sin\theta} \times \frac{1}{r} = 0}$$

6. $\mathbf{R}^{i}_{3i3}$

$$R^0_{303} = \frac{-1}{2}\frac{r_s(2r - r_s)}{r^2(r - r_s)^2} + \frac{1}{2}\frac{r_s}{r(r - r_s)} \times \frac{1}{2}\frac{r_s}{r(r - r_s)}$$
$$+ \frac{1}{2}\frac{r_s}{r(r - r_s)} \times \frac{1}{2}\frac{r_s}{r(r - r_s)}$$
$$R^1_{313} = \frac{-1}{r^2} + \frac{1}{r} \times \frac{1}{r} + \frac{1}{r} \times \frac{1}{2}\frac{r_s}{r(r - r_s)}$$
$$R^2_{323} = \frac{-1}{r^2} + \frac{1}{r} \times \frac{1}{r} + \frac{1}{r} \times \frac{1}{2}\frac{r_s}{r(r - r_s)}$$
$$R^3_{333} = \frac{1}{2}\frac{r_s(2r - r_s)}{r^2(r - r_s)^2} - \frac{1}{2}\frac{r_s(2r - r_s)}{r^2(r - r_s)^2} + \frac{1}{2}\frac{r_s}{r(r - r_s)} \times \frac{1}{2}\frac{r_s}{r(r - r_s)}$$
$$- \frac{1}{2}\frac{r_s}{r(r - r_s)} \times \frac{1}{2}\frac{r_s}{r(r - r_s)}$$

$$R^i_{3i3} =$$
$$\frac{-1}{2}\frac{r_s(2r - r_s)}{r^2(r - r_s)^2} + \frac{1}{2}\frac{r_s}{r(r - r_s)} \times \frac{1}{2}\frac{r_s}{r(r - r_s)}$$
$$+ \frac{1}{2}\frac{r_s}{r(r - r_s)} \times \frac{1}{2}\frac{r_s}{r(r - r_s)}$$
$$+ \frac{-1}{r^2} + \frac{1}{r} \times \frac{1}{r} + \frac{1}{r} \times \frac{1}{2}\frac{r_s}{r(r - r_s)}$$
$$+ \frac{-1}{r^2} + \frac{1}{r} \times \frac{1}{r} + \frac{1}{r} \times \frac{1}{2}\frac{r_s}{r(r - r_s)}$$
$$+ \frac{1}{2}\frac{r_s(2r - r_s)}{r^2(r - r_s)^2} - \frac{1}{2}\frac{r_s(2r - r_s)}{r^2(r - r_s)^2} + \frac{1}{2}\frac{r_s}{r(r - r_s)} \times \frac{1}{2}\frac{r_s}{r(r - r_s)}$$
$$- \frac{1}{2}\frac{r_s}{r(r - r_s)} \times \frac{1}{2}\frac{r_s}{r(r - r_s)} = 0$$

Then the metric defined by equation (1.159) (p. 123) verifies the Einstein equation.
We take a convex such that the closure of that convex does not have a point at a spacial distance $\mathbf{r_s}$ from the origin and does not contain the origin. From proposition 148 (p. 117), if we suppose than the metric on the boundary of that convex is given by equation (1.159) (p. 123), then inside the convex, the metric is also given by equation (1.159) (p. 123).
The metric defined by equation (1.159) (p. 123) is usually taken for a stable star.

**Example 2 Einstein Tensor**

Let us take a simple metric:

$$g_{00} = c^2,\ g_{11} = -a(t),\ g_{22} = -a(t),\ g_{33} = -a(t)$$

where $\mathbf{a(t)}$ is a (positive) map $(\mathbf{a} : [\mathbf{1}, \rightarrow [\rightarrow \mathcal{R}^+)$ (see definition 378 (p. 484)). We have for the Christoffel symbols which are not zero:

$$\Gamma^0_{11} = \Gamma^0_{22} = \Gamma^0_{33} = \frac{a'(t)}{2c^2}; \quad \partial_0\Gamma^0_{11} = \partial_0\Gamma^0_{22} = \partial_0\Gamma^0_{33} = \frac{a''(t)}{2c^2}$$
$$\Gamma^1_{01} = \Gamma^1_{10} = \Gamma^2_{02} = \Gamma^2_{02} = \Gamma^3_{03} = \Gamma^3_{30} = \frac{a'(t)}{2a(t)}$$
$$\partial_0\Gamma^1_{01} = \partial_0\Gamma^1_{10} = \partial_0\Gamma^2_{02} = \partial_0\Gamma^2_{20} = \partial_0\Gamma^3_{03} = \partial_0\Gamma^3_{30} = \frac{a''(t)}{2a(t)} - \frac{a'(t)^2}{2a(t)^2}\,.$$

For the Ricci curvature tensor (definition 108 (p. 113)) we have:

$$\mathrm{Ric}_{00} = R^i_{0i0} = 3R^1_{010} = 3\frac{a''(t)}{2a(t)} - 3\frac{a'(t)^2}{4a(t)^2}\,.$$

For $\mathbf{j = 1, 2, 3}$ with summation on $\mathbf{i}$ we have:

$$j \in [1,3] \Rightarrow \mathrm{Ric}_{jj} = R^i_{jij} = -\frac{a''(t)}{2c^2} - \frac{a'(t)^2}{4c^2a(t)}$$

and zero for the other terms. The scalar curvature $\mathbf{R}$ (definition 109 (p. 113)) is then:

$$R = g^{\mu\nu}\mathrm{Ric}_{\mu\nu} = 3\frac{a''(t)}{2c^2a(t)} - 3\frac{a'(t)^2}{4c^2a(t)^2} + 3\frac{a''(t)}{2c^2a(t)} + 3\frac{a'(t)^2}{4c^2a(t)^2}$$
$$= 6\frac{a''(t)}{2c^2a(t)}\,.$$

In that case the Einstein tensor (definition 111 (p. 122)) $\mathbf{G_{\mu\nu} = R^\lambda_{\mu\lambda\nu} - g_{\mu\nu}\frac{R}{2}}$ is given by:

$$G_{00} = 3\frac{a''(t)}{2a(t)} - 3\frac{a'(t)^2}{4a(t)^2} - 3\frac{a''(t)}{2a(t)} = -3\frac{a'(t)^2}{4a(t)^2}$$
$$j \in [1,3] \Rightarrow G_{jj} = -\frac{a''(t)}{2c^2} - \frac{a'(t)^2}{4c^2a(t)} + 3\frac{a''(t)}{2c^2} = \frac{a''(t)}{c^2} - \frac{a'(t)^2}{4c^2a(t)}\,. \tag{1.160}$$

All the other terms are zero. We have as it should:

$$g^{\mu\nu}G_{\mu\nu} = g^{\mu\nu}\mathrm{Ric}_{\mu\nu} - g^{\mu\nu}g_{\mu\nu}R/2 = R - 2R = -R\,.$$

See definitions 109 (p. 113), 108 (p. 113), 34 (p. 29), 215 (p. 245). If we set $\mathbf{c^2 = 1}$ and $\mathbf{b^2(t) = a(t)}$ as it is usually done, then equation (1.160) (p. 127) becomes:

$$G_{00} = -3\frac{(2b(t)b'(t))^2}{4b(t)^4} = -3\frac{b'(t)^2}{b(t)^2}$$
$$j \in [1,3] \Rightarrow G_{jj} = (2b(t)b'(t))' - \frac{(2b(t)b'(t))^2}{4b(t)^2} = 2b'(t)^2 + 2b(t)b''(t) - b'(t)^2$$
$$= b'(t)^2 + 2b(t)b''(t)$$
$$R = 6\frac{b'(t)^2}{b(t)^2} + 6\frac{b''(t)}{b(t)}\,.$$

## 1.19 Notations

1. **Convention 12** Einstein convention.

   When two letters are repeated in a formula, there is an implicit summation over them (Einstein convention footnote 20 (p. 239)) and, in principle, they should be once in lower position (covariant) and once in upper position (contravariant). However, if a letter has a meaning; for instance if there is an equation and the letter appears once in the first member then there is no more implicit summation in the second member. See also convention 27 (p. 445).

2. The expressions $(\mathbf{f(x) : A \to B})$ or $(\mathbf{x^i : A \to B})$ design maps of $\mathbf{A}$ on $\mathbf{B}$ as defined in section 2.1.1 definition 112 (p. 131) [31].

3. In the following, $\mathbf{K}$ is a finite set of indices. To be clearer, we take it as a set of integers: $\mathbf{K} \subset \mathcal{Z}^+$ with $\mathbf{Card(K)} \in \mathcal{Z}^+$ and most of the time we suppose: $\mathbf{K} = [\mathbf{1}, \mathbf{Card(K)}]$ (definitions 378 (p. 484), 390 (p. 499), 400 (p. 520)).
   $\mathbf{n} \in \mathcal{Z}^+$ is the dimension of the manifold except when it is specified otherwise.
   Except also when specified otherwise, Greek letters like $\boldsymbol{\lambda}, \boldsymbol{\mu}, \boldsymbol{\nu}$ or Latin letters like $\mathbf{i}, \mathbf{j}$, adjacent in upper or lower position to another letter, are elements of $[\mathbf{1}, \mathbf{n}]$. Together with the adjacent letters, they design either an element of $\mathcal{R}$ (2.2 (p. 170)) or an element of another set like $\mathcal{V}$ or $\mathcal{F}$ (see below):

   $$(\mathbf{a^i : [1, n] \to \mathcal{R}}) \ .$$

4. $\mathbf{M} = \bigcup_{\mathbf{k} \in \mathbf{K}} \mathbf{U_k}$ with $(\boldsymbol{\phi}_\mathbf{k}(\mathbf{x}) : \mathbf{U_k} \to \mathcal{R}^\mathbf{n})$ is a differentiable manifold of dimension $\mathbf{n}$ (definition 3 (p. 1)) (or alternatively $\mathbf{N}\ \mathbf{V}\ \boldsymbol{\psi}$ in place of $\mathbf{M}\ \mathbf{U}\ \boldsymbol{\phi}$).

5. $\mathcal{F}$ is the set of smooth maps $(\mathbf{f : M \to \mathcal{R}})$ (definition 7 (p. 3)). It has to be understood as the commutative ring associated to the vector fields set $\mathcal{V}$.

6. Open Cube $\mathbf{C(x, a)} \subset \mathcal{R}^\mathbf{n}$, $(\mathbf{x^i : [1, n] \to \mathcal{R}}) \in \mathcal{R}^\mathbf{n}$, $(\mathbf{y^i : [1, n] \to \mathcal{R}}) \in \mathcal{R}^\mathbf{n}$, $\mathbf{a} \in \mathcal{R}^+$, $\mathbf{C(0, a)} = ]-\mathbf{a}, \mathbf{a}[^\mathbf{n}$, definition 261 (p. 285):

   $$\mathbf{y} \in \mathbf{C(x, a)} \Leftrightarrow \mathbf{y^i} \in ]\mathbf{x^i} - \mathbf{a}, \mathbf{x^i} + \mathbf{a}[ \ .$$

7. Closed Cube $\bar{\mathbf{C}}(\mathbf{x}, \mathbf{a})$. It is the closure of $\mathbf{C(x, a)}$ (definitions 136 (p. 155), 262 (p. 286)).

8. Special maps in $\mathbf{C}^\infty$ $\mathcal{R} \to \mathcal{R}$ with $\mathbf{a} \in \mathcal{R}, \mathbf{x} \in \mathcal{R}$

   (a) Slope map $\mathcal{U}(\mathbf{x})$ (proposition 710 (p. 426)):
   $(\mathcal{U}(\mathbf{x}) : \mathcal{R} \Rightarrow [\mathbf{0}, \mathbf{1}]) \in \mathbf{C}^\infty$

   $$\mathbf{x} > \mathbf{0} \Rightarrow \mathcal{U}(\mathbf{x}) = \mathbf{e}^{\frac{-1}{\mathbf{x}^2}}, \ \mathbf{x} \leq \mathbf{0} \Rightarrow \mathcal{U}(\mathbf{x}) = \mathbf{0} \ .$$

   (b) Bell map $\mathbf{h(x, a)}$ (proposition 712 (p. 426)):

   $$(\mathbf{h} : \mathcal{R} \times \mathcal{R}^+ \to \mathcal{R}), \ \ \mathbf{x} \in \mathcal{R}, \ \mathbf{a} \in \mathcal{R}^+, \ \mathbf{h(x, a)} = \frac{\mathcal{U}(\mathbf{a}^2 - \mathbf{x}^2)}{\mathcal{U}(\mathbf{a}^2)}$$

   $$(\mathbf{x} \notin [-\mathbf{a}, \mathbf{a}] \Rightarrow \mathbf{h(x, a)} = \mathbf{0}),$$
   $$(\mathbf{x} \in [-\mathbf{a}, \mathbf{a}] \Rightarrow \mathbf{0} \leq \mathbf{h(x, a)} \leq \mathbf{1}, \mathbf{h(0, a)} = \mathbf{1}) \ .$$

[31] 112 (p. 131) means 112 on page 131.

(c) Jump map $\mathbf{Y(x,a)}$ (proposition 713 (p. 427)):

$$\mathbf{x \leq a \Rightarrow Y(x,a) = 0;\ x \in [-a,a] \Rightarrow Y(x,a) = \frac{\int_{-a}^{x} h(x,a)}{\int_{-a}^{a} h(x,a)}}$$
$$\mathbf{x \geq a \Rightarrow Y(x,a) = 1\ .}$$

9. Point continuous map $(\mathbf{g(x,a)} : \mathcal{R}^n \to \mathcal{R})$, $\mathbf{a} \in \mathcal{R}^+$, proposition 714 (p. 427):

$$\mathbf{x \in C(0,a) \Rightarrow g(x,a) = \prod_{i \in [1,n]} \frac{\mathcal{U}(a^2 - (x^i)^2)}{\mathcal{U}(a^2)},\quad x \notin C(0,a) \Rightarrow g(x,a) = 0.}$$

10. Cube continuous map, $(\mathbf{g(x,a,b)} : \mathcal{R}^n \to \mathcal{R})$, $\mathbf{a} \in \mathcal{R}^+$, $\mathbf{b} \in \mathcal{R}^+$, $\mathbf{0 < a < b}$, proposition 715 (p. 428):

$$\mathbf{x \in C(0,a) \Rightarrow g(x,a,b) = 1,}$$
$$\mathbf{(x \notin C(0,a),\ x \in C(0,b)) \Rightarrow (g(x,a,b) \in [0,1],}$$
$$\mathbf{g(x,a,b) = \prod_{i \in [1,n]} Y(-|y_i - x_i| + (a+b)/2, (b-a)/2)),}$$
$$\mathbf{x \notin C(0,b) \Rightarrow g(x,a,b) = 0\ .}$$

11. In order to simplify the writing, when we have a map $(\phi : \mathbf{U} \to \mathcal{R}^n)$ which is clear from the surrounding, we set:

$$\mathbf{g}_p(\mathbf{x},\mathbf{a}) = \mathbf{g}(\mathbf{x} - \phi(p), \mathbf{a});\quad \mathbf{g}_p(\mathbf{x},\mathbf{a},\mathbf{b}) = \mathbf{g}(\mathbf{x} - \phi(p), \mathbf{a}, \mathbf{b})$$
$$\mathbf{g}'_p(q,\mathbf{a}) = \mathbf{g}(\phi(q) - \phi(p), \mathbf{a});\quad \mathbf{g}'_p(q,\mathbf{a},\mathbf{b}) = \mathbf{g}(\phi(q) - \phi(p), \mathbf{a}, \mathbf{b})\ .$$

12. $\mathcal{F}_k$ is the set of maps $(\mathbf{f} : \mathbf{U_k} \to \mathcal{R})$ which are smooth on $\mathbf{U_k}$ which is considered as a manifold, (definitions 5 (p. 3), 8 (p. 3), 11 (p. 4)).

13. $\mathcal{F}_k^M$ is the set of maps $(\mathbf{f} : \mathbf{M} \to \mathcal{R})$ which are smooth when acting on $\mathbf{U_k}$: $\mathcal{F} \subset \mathcal{F}_k^M$ (definitions 9 (p. 4), 11 (p. 4)).

14. $\mathcal{F}_k^0 \subset \mathcal{F}_k^M$ are maps $(\mathbf{f} : \mathbf{M} \to \mathcal{R})$ which are zero for any points outside $\mathbf{U_k}$ and smooth when in $\mathbf{U_k}$ (definitions 10 (p. 4), 11 (p. 4)).

15. $\mathcal{F}_O$, $\mathcal{F}_O^M$, $\mathcal{F}_O^0$. Here the open $\mathbf{O} \subset \mathbf{M}$ replace $\mathbf{U_k}$ in the three preceding items: 12 (p. 129), 13 (p. 129), 14 (p. 129).

16. $\mathcal{V}$ is the set of vector fields on $\mathbf{M}$ (definition 15 (p. 8)).
Given a vector field $\mathbf{v} \in \mathcal{V}$: $(\mathbf{v(f)} : \mathcal{F} \to \mathcal{F})$ [32] its action is given by:

$$\mathbf{(v \in \mathcal{V}, f \in \mathcal{F}, g \in \mathcal{F}) \Rightarrow (v(f) \in \mathcal{F}, v(fg) = gv(f) + fv(g) \in \mathcal{F})\ .}$$

17. $\mathcal{V}_k$ is the set of vector fields restricted to $\mathbf{U_k}$ (definition 15 (p. 8)).

18. $\mathcal{V}^*$ is the dual of the set of vector fields $\mathcal{V}$ (definition 37 (p. 32)).
An element of $\mathcal{V}^*$ is called **one-form** field $_{1\triangleleft}\mathbf{f}$ (definition 36 (p. 32)) : $(\ _{1\triangleleft}\mathbf{f} : \mathcal{V} \to \mathcal{F})$
The action of a **one-form** field is given by:

$$_{1\triangleleft}\mathbf{f} \in \mathcal{V}^*, \mathbf{v} \in \mathcal{V} \Rightarrow {}_{1\triangleleft}\mathbf{f}(\mathbf{v}) \in \mathcal{F}\ .$$

[32] See definition 112 (p. 131) equation (2.3) for the notation.

19. $\mathbf{d}$ is the differential operator (definition 40 (p. 34)). It is a map $(\mathbf{d} : \mathcal{F} \rightarrow \mathcal{V}^*)$. Its action is given by:
$$\mathrm{f} \in \mathcal{F}, \mathrm{v} \in \mathcal{V} \Rightarrow \mathrm{df}(\mathrm{v}) = \mathrm{v}(\mathrm{f}) \ .$$

20. $\delta^\nu_\mu$ is the Kronecker symbol (definition 215 (p. 245)): $\mu = \nu \Rightarrow \delta^\nu_\mu = 1, \mu \neq \nu \Rightarrow \delta^\nu_\mu = 0$.

21. For any $\mathrm{k} \in \mathrm{K}$, $\mathrm{x}^\mu$: $(\mathrm{x}^\mu : [1, \mathrm{n}] \rightarrow \mathcal{F}^{\mathrm{M}}_{\mathrm{k}})$ is defined in $\mathrm{U_k}$ by (definition 126 (p. 144)):
$$(p \in \mathrm{U_k}, \mathrm{x} = \phi_{\mathrm{k}}(p) \in \mathcal{R}^{\mathrm{n}}, \mu \in [1, \mathrm{n}]) \Rightarrow \mathrm{x}^\mu = \phi^\mu_{\mathrm{k}}(p) \ .$$

22. For any $\mathrm{k} \in \mathrm{K}$, $\partial_\mu \in \mathcal{V}_{\mathrm{k}}$ is defined in each $\mathrm{U_k}$ by:
$$(p \in \mathrm{U_k}, \mathrm{f} \in \mathcal{F}, \mathrm{x} = \phi_{\mathrm{k}}(p), \mathrm{f_k}(\mathrm{x}) = \mathrm{f}(\phi^{-1}_{\mathrm{k}}(\mathrm{x}))) \Rightarrow \partial_\mu \mathrm{f}(p) = \frac{\partial \mathrm{f_k}(\mathrm{x})}{\partial \mathrm{x}^\mu} \ .$$

23. In each $\mathrm{U_k}$, the coordinates (definition 23 (p. 17)) $(\mathrm{v}^\mu : [1, \mathrm{n}] \rightarrow \mathcal{F}^{\mathrm{M}}_{\mathrm{k}})$ of the restriction (definition 18 (p. 13)) of a vector field $\mathrm{v} \in \mathcal{V}$ to $\mathrm{U_k}$ is defined by proposition 1.18 (p. 14) [33]:
$$(\mathrm{f} \in \mathcal{F}, \mathrm{v} \in \mathcal{V}, p \in \mathrm{U_k}, \mathrm{x} = \phi_{\mathrm{k}}(p), \mathrm{v}^\mu = \mathrm{v_k}(\mathrm{x}^\mu))$$
$$\Rightarrow \mathrm{v}(\mathrm{f})(\phi^{-1}_{\mathrm{k}}(\mathrm{x})) = \mathrm{v}^\mu \frac{\partial \mathrm{f}(\phi^{-1}_{\mathrm{k}}(\mathrm{x}))}{\partial \mathrm{x}_\mu} \ .$$

24. $\mathrm{dx}^\mu$ defined in any $\mathrm{U_k}$ as $(\mathrm{dx}^\mu : [1, \mathrm{n}] \rightarrow \mathcal{V}^*_{\mathrm{k}})$ by:
$$(p \in \mathrm{U_k}, \mathrm{x} = \phi_{\mathrm{k}}(p), \mathrm{v} \in \mathcal{V}_{\mathrm{k}}) \Rightarrow (\mathrm{dx}^\mu(\mathrm{v}) = \mathrm{v}(\mathrm{x}^\mu) = \mathrm{v}^\mu = \mathrm{v}^\nu \partial_\nu \mathrm{x}^\mu)$$
then we have by definition in any $\mathrm{U_k}$:
$$\partial_\mu \mathrm{dx}^\nu = \mathrm{dx}^\nu(\partial_\mu) = \partial_\mu(\mathrm{x}^\nu) = \delta^\nu_\mu$$
$$p \in \mathrm{U_k}, \mathrm{x} = \phi_{\mathrm{k}}(p) \Rightarrow \mathrm{df}(\mathrm{v})(\phi^{-1}_{\mathrm{k}}(\mathrm{x})) = \mathrm{v}^\mu \frac{\partial \mathrm{f_k}(\phi^{-1}_{\mathrm{k}}(\mathrm{x}))}{\partial \mathrm{x}_\mu}$$
$$\Rightarrow \mathrm{df_k} = \frac{\partial \mathrm{f_k}(\phi^{-1}_{\mathrm{k}}(\mathrm{x}))}{\partial \mathrm{x}_\mu} \mathrm{dx}^\mu \ .$$

[33] As a general rule, when it is clear enough we shall forget extra reference index: here we write $\mathrm{v^i}$ the coordinate relative to $\mathrm{U_k}$ of $\mathrm{v} \in \mathcal{V}$. If it is ambiguous we shall write ${}^{[\mathrm{k}]}\mathrm{v}^{\mathrm{i}}$ instead of $\mathrm{v^i}$.

# Chapter 2

# Some basic mathematics needed for manifolds

## 2.1 General concepts

### 2.1.1 Maps and functions

**Definition 112 Map.**

1. See definition 357 (p. 460). Let us consider two sets (see footnote 3 (p. 133)) $\mathbf{A}$ and $\mathbf{B}$ and a subset $\mathbf{f}$ of the product $\mathbf{A}$ by $\mathbf{B}$ (see section A.5.1 (p. 447) and definitions 342 (p. 447), 344 (p. 447), 356 (p. 459)): $\mathbf{f} \subset \mathbf{A} \times \mathbf{B}$. $\mathbf{f}$ defines a map (function, application) of $\mathbf{A}$ on $\mathbf{B}$, $(\mathbf{f} : \mathbf{A} \to \mathbf{B})$ if (see section A.3 (p. 441)):
$$(\forall \mathbf{x} \in \mathbf{A})(\exists \mathbf{y} \in \mathbf{B})((\mathbf{x}, \mathbf{y}) \in \mathbf{f}) \tag{2.1}$$
and if [1]:
$$((\mathbf{x}, \mathbf{y}) \in \mathbf{f}, (\mathbf{x}, \mathbf{y}') \in \mathbf{f}) \Rightarrow \mathbf{y} = \mathbf{y}' \tag{2.2}$$

2. Any member of $\mathbf{f}$ can be written as $(\mathbf{x}, \mathbf{f}(\mathbf{x})) \in \mathbf{f}$ with $\mathbf{f}(\mathbf{x}) = \tau_{\mathbf{y}}((\mathbf{x}, \mathbf{y}) \in \mathbf{f})$ (see definitions 332 (p. 440), 333 (p. 441), 358 (p. 461)) since $\mathbf{f}(\mathbf{x})$ is unique for any $\mathbf{x}$. One uses to also write [2]:
$$(\mathbf{f}(\mathbf{x}) : \mathbf{A} \to \mathbf{B}) \text{ or } (\mathbf{f} : \mathbf{A} \to \mathbf{B}) \tag{2.3}$$
with $\mathbf{x} \in \mathbf{A}$ and $\mathbf{f}(\mathbf{x}) \in \mathbf{B}$.

3. There are other popular notations for a map $\mathbf{x}$ between a set $\mathbf{I}$ and a set $\mathbf{X}$. One writes $\mathbf{x}$ as $(\mathbf{x}^{\mathbf{i}} : \mathbf{I} \to \mathbf{X})$ or $(\mathbf{x}_{\mathbf{i}} : \mathbf{I} \to \mathbf{X})$.
An element of the set $\mathbf{x}$ is then $(\mathbf{i}, \mathbf{x}^{\mathbf{i}})$ or $(\mathbf{i}, \mathbf{x}_{\mathbf{i}})$. When $\mathbf{X} = \mathcal{R}$ or $\mathbf{X} = \mathcal{C}$ (definitions 143 (p. 170), 245 (p. 271)) and when $\mathbf{I}$ is finite ($\mathbf{Card}(\mathbf{I}) \in \mathcal{Z}^+$), making use of the Einstein notation (convention 12 (p. 128)), one often writes $\mathbf{x} = \mathbf{x}^{\mathbf{i}}\mathbf{e}_{\mathbf{i}}$ where the $\mathbf{e}_{\mathbf{k}} = (\delta^{\mathbf{i}}_{\mathbf{k}} : \mathbf{I} \to \mathcal{R} \subset \mathcal{C})$ are called the basic maps (vectors) (definition 242 (p. 267)). $(\delta^{\mathbf{i}}_{\mathbf{j}} : \mathbf{I} \times \mathbf{I} \to \{0, 1\} \subset \mathcal{R} \subset \mathcal{C})$ is the Kronecker symbol: $\mathbf{i} \neq \mathbf{j} \Rightarrow \delta^{\mathbf{i}}_{\mathbf{j}} = 0$, $\mathbf{i} = \mathbf{j} \Rightarrow \delta^{\mathbf{i}}_{\mathbf{j}} = 1$ (definitions 215 (p. 245), 345 (p. 447)).

---

[1] According to convention 27 (p. 445), we shall use for simplicity a generalization of the Einstein notation for indices: Whenever a letter appears in a true relation without any relation with a known term, it means that the relation is true for any term put in the place of the letter: $\mathbf{R}(\mathbf{x})$ true means $(\forall \mathbf{x})\mathbf{R}(\mathbf{x})$ true.

[2] The notation is not always very precise mainly when $\mathbf{A}$ is a product of several sets as for instance $\mathbf{A} = [1, \mathbf{K}] \times \mathcal{R}$, $\mathbf{K} \in \mathcal{Z}^+$. Very often the two components of $\mathbf{A}$ will not be treated on the same footing. We shall sometimes write a set of finite maps from $\mathcal{R}$ to $\mathcal{R}$ as $(\mathbf{f}_{\mathbf{i}}(\mathbf{x}) : \mathcal{R} \to \mathcal{R})$ or as $(\mathbf{f}_{\mathbf{i}} : \mathcal{R} \to \mathcal{R})$ instead of $(\mathbf{f}_{\mathbf{i}}(\mathbf{x}) : [1, \mathbf{K}] \times \mathcal{R} \to \mathcal{R})$.

4. Map given by a relation.
   $\mathbf{A}$ and $\mathbf{B}$ being set, $\mathbf{R(x,y)}$ a relation, $\mathbf{(\forall x \in A)(\exists y \in B)(R(x,y))}$ defines a map $\mathbf{(f : A \to B)}$ with $\mathbf{f(x) = \tau_y R(x,y)}$ see definition 332 (p. 440), section A.4.1 (p. 445). It is a map because from the axioms defined by equations (A.2) (p. 441) and (A.3) (p. 441), applying relation (A.2) (p. 441) and the logical rules A.1.3 (p. 435), 6 (p. 435), we have:

$$\mathbf{x_1 = x_2 \Rightarrow (R(x_1,z) \Leftrightarrow R(x_2,z)) \Rightarrow \tau_y R(x_1,y) = \tau_z(R(x_2,z))} \ .$$

**Convention 13** $\mathbf{(f : A \to B)}$.

By convention a map $\mathbf{f \subset A \times B}$ is written $\mathbf{(f : A \to B)}$. See equation (2.3) (p. 131).

**Definition 113 Identity map.**

Given a set $\mathbf{E}$, the identity map $\mathbf{(I : E \to E)}$ is defined by: $\mathbf{(\forall x \in E)(f(x) = x)}$

**Proposition 149** $\mathbf{A}$**,** $\mathbf{B}$ are two sets, $\mathbf{(h_1 : A \to B)}$, $\mathbf{(h_2 : A \to B)}$ two maps, we have:

$$\mathbf{h_1 \neq h_2 \Leftrightarrow (\exists x)((x, h_1(x)) \neq (x, h_2(x)))} \ .$$

**Proof**
From proposition 757 (p. 452) and definition 112 (p. 131), we have:

$$\mathbf{h_1 \neq h_2 \Rightarrow (\exists u)(u = (x,y), ((u \in h_1, y = h_1(x), u \notin h_2)} \\ \mathbf{\vee (u \in h_2, y = h_2(x), u \notin h_1)))} \ .$$

We use the first option $\mathbf{(u \in h_1, y = h_1(x), u \notin h_2)}$ and that $\mathbf{u = (x, h_1(x)) = (x,y)}$ (convention 42 (p. 560)). For the second option, we exchange 1 and 2 in what follows and we get the same conclusion. We have from equation (2.1) (p. 131):

$$\mathbf{(\exists y' \in B)((x,y') \in h_2, y' = h_2(x))} \ .$$

Since $\mathbf{(x, h_1(x)) \notin h_2}$, using that $\mathbf{y' = h_2(x)}$, we have $\mathbf{h_1(x) \neq h_2(x)}$. Then (see equation (A.15) (p. 448), proposition 727 (p. 441)), we also have:

$$\mathbf{h_1 \neq h_2 \Rightarrow (\exists x)((x, h_1(x)) \neq (x, h_2(x)))} \ .$$

We suppose $\mathbf{(\exists x)((x, h_1(x)) \neq (x, h_2(x)))}$. Taking that $\mathbf{x}$ (convention 42 (p. 560)), we cannot have $\mathbf{(x, h_1(x)) \in h_2}$ because from equation (2.2) (p. 131), it would mean $\mathbf{h_1(x) = h_2(x)}$. Then from proposition 756 (p. 451), section A.1.4 point 7 (p. 435) and section A.1.3 point 6 (p. 435), we have:

$$\mathbf{(\exists x)((x, h_1(x)) \neq (x, h_2(x))) \Rightarrow h_1 \neq h_2} \ .$$

Proposition 149 (p. 132) is true.

**Proposition 150** The Identity map (definition 113 (p. 132)) is a map

**Proof**
It comes directly from definition 112 (p. 131).

**Definition 114 Image of a subset given by a map.**

For any map $(\mathbf{f}:\mathbf{A}\longrightarrow\mathbf{B})$ and for any subset $\mathbf{C}$ of $\mathbf{A}$, we can define a subset $\mathbf{f}(\mathbf{C})$ of $\mathbf{B}$, called the image in $\mathbf{B}$ of the subset $\mathbf{C}$:

$$\mathbf{f}(\mathbf{C})=\mathbf{set_y}\{(\mathbf{y}\in\mathbf{B}),(\exists\mathbf{x}\in\mathbf{C})((\mathbf{x},\mathbf{y})\in\mathbf{f})\}\,. \tag{2.4}$$

This set exists because of the proposition 749 (p. 450).

**Proposition 151** Having a map $(\mathbf{f}:\mathbf{A}\longrightarrow\mathbf{B})$, we have $\mathbf{f}(\mathbf{A})\subset\mathbf{B}$.

**Proof**
It is a direct consequence of definition 114 (p. 132) equation (2.4) (p. 133), definition 112 (p. 131) equations (2.1) (p. 131) and (349) (p. 451).

**Proposition 152** Having a map $(\mathbf{f}:\mathbf{E}\longrightarrow\mathbf{F})$ and $\mathbf{A}\subset\mathbf{B}\subset\mathbf{E}$, we have:

$$\mathbf{f}(\mathbf{A})\subset\mathbf{f}(\mathbf{B})\subset\mathbf{f}(\mathbf{E})\subset\mathbf{F}\,.$$

**Proof**
It is also a direct consequence of definition 114 (p. 132) equation (2.4) (p. 133), definition 112 (p. 131) equation (2.1) (p. 131) and (349) (p. 451).

**Proposition 153** Given three sets $\mathbf{A},\ \mathbf{B},\ \mathbf{C}$ [3], two maps $(\mathbf{f}:\mathbf{A}\rightarrow\mathbf{C})$, $(\mathbf{g}:\mathbf{B}\rightarrow\mathbf{C})$. If $\mathbf{A}\cap\mathbf{B}=\emptyset$, $\mathbf{f}\cup\mathbf{g}$ is also a map. $(\mathbf{f}\cup\mathbf{g}:\mathbf{A}\cup\mathbf{B}\rightarrow\mathbf{C})$.

**Proof**
Using A.1.4 (p. 435) - 25 (p. 436), 18 (p. 436) and A.1.3 (p. 435) -3 (p. 435), we have:

$$\begin{aligned}&((\mathbf{x},\mathbf{y})\in\mathbf{f}\cup\mathbf{g},(\mathbf{x},\mathbf{y}')\in\mathbf{f}\cup\mathbf{g})\Rightarrow\\&(\{(\mathbf{x},\mathbf{y})\in\mathbf{f}\vee((\mathbf{x},\mathbf{y})\in\mathbf{g}\},\{(\mathbf{x},\mathbf{y}')\in\mathbf{f}\vee(\mathbf{x},\mathbf{y}')\in\mathbf{g}\})\\&\Leftrightarrow(\{(\mathbf{x},\mathbf{y})\in\mathbf{f},(\mathbf{x},\mathbf{y}')\in\mathbf{f}\}\vee\{(\mathbf{x},\mathbf{y})\in\mathbf{g},(\mathbf{x},\mathbf{y}')\in\mathbf{g}\}\\&\vee\{(\mathbf{x},\mathbf{y})\in\mathbf{f},(\mathbf{x},\mathbf{y}')\in\mathbf{g}\}\vee\{(\mathbf{x},\mathbf{y})\in\mathbf{g},(\mathbf{x},\mathbf{y}')\in\mathbf{f}\})\,.\end{aligned}$$

Along the last logical expression, if one of the two first items is true, since $\mathbf{f}$ and $\mathbf{g}$ are maps, we should have $\mathbf{y}=\mathbf{y}'$. The last two cannot be true since it would imply $\mathbf{x}\in\mathbf{A}\cap\mathbf{B}$ which cannot be true since $\mathbf{A}\cap\mathbf{B}=\emptyset$.

**Definition 115 Inverse image of a subset given by a map.**

Given a map $(\mathbf{f}:\mathbf{A}\longrightarrow\mathbf{B})$, one defines the inverse $\mathbf{f}^{-1}(\mathbf{D})\subset\mathbf{A}$ of a subset $\mathbf{D}$ of $\mathbf{B}$ as:

$$\mathbf{f}^{-1}(\mathbf{D})=\mathbf{set_x}\{(\mathbf{x}\in\mathbf{A}),(\exists\mathbf{y})(\mathbf{y}\in\mathbf{D},(\mathbf{x},\mathbf{y})\in\mathbf{f})\}\,. \tag{2.5}$$

This set exists from proposition 750 (p. 450) in A.5.1.
For convenience, one sometimes suppresses equation (2.1) (p. 131) in the map definition $(\mathbf{f}:\mathbf{A}\longrightarrow\mathbf{B})$ and we define by $\mathbf{f}^{-1}(\mathbf{B})\subset\mathbf{A}$, the subset of $\mathbf{A}$ for which $\mathbf{f}$ is defined: $(\mathbf{f}:\mathbf{f}^{-1}(\mathbf{B})\longrightarrow\mathbf{B})$.

**Proposition 154** $(\mathbf{f}:\mathbf{A}\longrightarrow\mathbf{B})$ being a map[4] $\mathbf{f}^{-1}(\mathbf{f}(\mathbf{A}))=\mathbf{A}$.

---

[3] "Given three sets" is in fact a useless expression, since any term is set: there is only one term $\mathbf{A}$ for which $\neg(\exists\mathbf{x})(\mathbf{x}\in\mathbf{A})$ is the empty set $\emptyset$ which is a set. We shall nevertheless use that sentence to make the reader aware that we are using the set theory. See proposition 774 (p. 457).

[4] One has to be cautious when using that proposition because we have to keep in mind that $\mathbf{f}$ should only act on $\mathbf{A}$ to be true: we do not have $\mathbf{C}\subset\mathbf{A},\mathbf{f}^{-1}\mathbf{f}(\mathbf{C})=\mathbf{C}$.

**Proof**

$$\mathbf{x \in A \Rightarrow (\exists y = f(x))(f(x) = y) \Rightarrow x \in f^{-1}(f(A))}$$

then $\mathbf{A \subset f^{-1}(f(A))}$ but since $\mathbf{f \subset A \times B}$

$$\mathbf{x \in f^{-1}(f(A)) \Rightarrow (\exists y \in B)((x, y) \in f \subset A \times B) \Rightarrow x \in A}$$

then $\mathbf{f^{-1}(f(A)) \subset A}$. We conclude by proposition 754 (p. 451) (extensionality principle).

**Definition 116 Inverse image of an element given by a map.**

Given a map $\mathbf{(f : A \rightarrow B)}$, for any element $\mathbf{b \in f(A)}$, one defines its inverse set: $\mathbf{f^{-1}(b) = f^{-1}(\{b\}) \subset A}$ (see definition 345 (p. 447)).

**Proposition 155** $\mathbf{f^{-1}(b)}$ defines a partition of $\mathbf{A}$ (definition 370 (p. 471)).

**Proof**
It is a direct consequence of proposition 816 (p. 471): From that proposition, we have:

$$\mathbf{A = \bigcup_{b \in f(A)} f^{-1}(b), \quad b \neq b' \Rightarrow f^{-1}(b) \cap f^{-1}(b') = \emptyset} .$$

**Definition 117 Product of maps.**

Given three sets $\mathbf{A}$, $\mathbf{B}$, $\mathbf{C}$ and two maps: $\mathbf{(f : A \rightarrow B)}$, $\mathbf{(g : B \rightarrow C)}$, one defines the map $\mathbf{(g \circ f : A \rightarrow C)}$ as (see definition 344 (p. 447)):

$$\mathbf{g \circ f = set_{(x,z)}\{(\exists y)((x, y) \in f), (y, z) \in g\}} .$$

**Proposition 156** Using notations of definition 117 (p. 134), $\mathbf{g \circ f}$ is a map.

**Proof**
For all $\mathbf{x}$ in $\mathbf{A}$, there is a $\mathbf{f(x)}$ in $\mathbf{B}$, since there is a $\mathbf{z = g(y)}$ for any $\mathbf{y}$ in $\mathbf{B}$, there is a $\mathbf{z = g(f(x))}$ in $\mathbf{C}$ for any $\mathbf{x}$ in $\mathbf{A}$. $\mathbf{g \circ f}$ verifies equation (2.1) (p. 131).
Moreover, we have:

$$\mathbf{((x, z) \in g \circ f, (x, z') \in g \circ f) \Rightarrow (\exists y)((x, y) \in f) \Rightarrow (x, \tau_y((x, y) \in f)) \in f}$$

but, from definition 112 (p. 131), we have:

$$\mathbf{((x, y) \in f, (x, y') \in f) \Rightarrow y = y' = \tau_y((x, y) \in f)} .$$

Then, from definition 117 (p. 134), if we have $\mathbf{(x, z) \in f \circ g}$, $\mathbf{(x, z') \in f \circ g}$, we have:

$$\mathbf{(x, \tau_y((x, y) \in f)) \in f, \ (\tau_y((x, y) \in f), z) \in g, (\tau_y((x, y) \in f), z') \in g} .$$

From definition 112 (p. 131) $\mathbf{z = z'}$. $\mathbf{g \circ f}$ verifies equation (2.2) (p. 131) and then definition 112 (p. 131).

**Proposition 157** $\mathbf{A}$, $\mathbf{B}$ two sets, $\mathbf{f}$ a map $\mathbf{(f : A \rightarrow B)}$.

$$\mathbf{U \subset V \subset A \Rightarrow f(U) \subset f(V)} .$$

**Proof**

$$y \in f(U) \Rightarrow (\exists x \in U)(y = f(x))$$
$$U \subset V \Rightarrow (x \in U \Rightarrow x \in V) \Rightarrow (y \in f(U) \Rightarrow (\exists x \in V)(y = f(x)) \Rightarrow y \in f(V))$$
$$\Rightarrow f(U) \subset f(V)$$

See definition 349 (p. 451).

**Definition 118 Restriction of a map to a subset.**

For any map $(f : A \to B)$ and for any subset $C$ of $A$, we can define also the map $(f_C : C \to B)$ from $f$. We can consider $f$ as $f \subset A \times B$ and $f_C$ is given by:

$$f_C = f \cap (C \times B) \subset A \times B$$

$f_C$ verifies equation (2.1) (p. 131) and equation (2.2) (p. 131).

**Definition 119 Product of sets.**

The product of sets is defined in definition 372 (p. 473): Given two sets (see footnote 3 (p. 133)) $I$, $X$ and a map $(X_i : I \to \mathcal{P}(X))$ (definition 350 (p. 453)), we have:

$$\prod_{i \in I} X_i = \text{set}_u\{(u : I \to \bigcup_{i \in I} X_i), (\forall i \in I)(u(i) \in X_i)\} .$$

See definition 362 (p. 462). We use to write such a $u$: $\prod^{e}_{i \in I} u(i)$.

**Definition 120 Product of maps induced by product of sets.**

Given three sets $I$, $X$, $Y$, two maps $(X_i : I \to \mathcal{P}(X))$, $(Y_i : I \to \mathcal{P}(Y))$, a set of maps $i \in I \Rightarrow (f_i : X_i \to Y_i)$, we call:

$$(\prod^{e}_{i \in I} f_i = \prod^{e}_{i \in I} (f_i(x_i)) : \prod_{i \in I} X_i \to \prod_{i \in I} Y_i)$$

**Proposition 158** Using the same notations as in definition 120 (p. 135), we have (see definition 115 (p. 133)):

$$(\prod^{e}_{i \in I} f_i)^{-1} = \prod^{e}_{i \in I} f_i^{-1}$$

**Proof**

From definitions 115 (p. 133) and 372 (p. 473), given $y \in \prod_{i \in I} Y_i$, we have:

$$x \in (\prod^{e}_{i \in I} f_i)^{-1}(y) \Leftrightarrow (x \in \prod_{i \in I} X_i, y = \prod^{e}_{i \in I} f_i(x_i)) \Leftrightarrow (\forall i \in I)(x_i \in f_i^{-1}(y_i))$$
$$\Leftrightarrow x \in \prod^{e}_{i \in I} f_i^{-1}(y_i) .$$

We conclude by proposition 754 (p. 451) (extensionality principle).

**Convention 14** $\prod_{i \in [1,n]}$.

See also convention 18 (p. 146). Keeping notations of definition 119 (p. 135), if $\mathbf{I} = [\mathbf{1}, \mathbf{n}], \mathbf{n} \in \mathcal{Z}^+$ (see proposition 926 (p. 519)) and $(\forall \mathbf{i} \in [1, \mathbf{n}])(\mathbf{X_i} = \mathbf{X})$, one also uses to write:

$$\prod_{i \in [1,n]} X_i = \prod_{i \in [1,n]} X = X^n .$$

From that we have:

$$u \in X^n \Rightarrow u = (u : [1, n] \to X) = \prod_{i \in [1,n]}^{e} u(i) .$$

**Definition 121 Projection on product of sets.**

Taking definition 372 (p. 473) of product of sets, calling $\mathbf{I}$ a set of indices, $\mathbf{X}$ a set and having a map (see definition 350 (p. 453))

$$(X^i : I \to \mathcal{P}(X)) \tag{2.6}$$

it allows to define (see definition 372 (p. 473)) $\mathbf{Y^I} = \prod_{i \in I} \mathbf{X^i}$. Let us have $\mathbf{K} \subset \mathbf{I}$, and its complement $\mathbf{K_C} = \mathcal{C}_\mathbf{I}(\mathbf{K})$:
We can consider the restriction (definition 118 (p. 135)) to $\mathbf{K}$ and to $\mathbf{K_C}$ of the map given by equation (2.6).
It allows to define:

$$Y^K = \prod_{i \in K} X^i$$
$$Y^{K_C} = \prod_{i \in K_C} X^i .$$

One calls $(\mathbf{Pr_K} : \mathbf{Y^I} \to \mathbf{Y^K})$ the projection of $\mathbf{Y^I}$ on $\mathbf{Y^K}$ along the set $\mathbf{Y^{K_C}}$, the restriction to $\mathbf{K}$ of the set of maps defined as $(\mathbf{x^i} : \mathbf{I} \to \mathbf{X})$ with $(\forall \mathbf{i} \in \mathbf{I})(\mathbf{x^i} \in \mathbf{X^i})$ which are the elements of $\mathbf{Y^I}$.
**For example:**
Taking $\mathbf{I} = [1, 3]$, $\mathbf{X} = \{\mathcal{R}\}$ with $(\forall \mathbf{i} \in \mathbf{I})(\mathbf{X^i} = \mathcal{R})$ and $\mathbf{K} = [1, 2]$,
we have for $\mathbf{x} = (\mathbf{x^1}, \mathbf{x^2}, \mathbf{x^3}) \in \mathcal{R}^3$, $\mathbf{Pr_K(x)} = (\mathbf{x^1}, \mathbf{x^2})$.

**Proposition 159** $(\mathbf{f} : \mathbf{A} \to \mathbf{B})$ being a map,

$$(C \subset A, D \subset A) \Rightarrow f(C) \cup f(D) = f(C \cup D) .$$

**Proof**
Directly from the definition of the image of a set.
In a more general way:

**Proposition 160** Given $(\mathbf{f} : \mathbf{A} \to \mathbf{B})$, $(\mathbf{O_i} : \mathbf{I} \to \mathcal{P}(\mathbf{A}))$ we have

$$f(\bigcup_{i \in I} O_i) = \bigcup_{i \in I} f(O_i) .$$

See definitions 350 (p. 453) and 362 (p. 462).

**Proof**
From propositions 737 (p. 444) and 735 (p. 442):

$$\begin{aligned} y \in \bigcup_{i\in I} f(O_i) \Leftrightarrow (\exists j \in I)(y \in f(O_j)) &\Leftrightarrow (\exists j)(j \in I, (\exists x)(x \in O_j, y = f(x))) \\ &\Leftrightarrow (\exists j)(\exists x)(j \in I, x \in O_j, y = f(x)) \\ &\Leftrightarrow (\exists x)(\exists j)(j \in I, x \in O_j, y = f(x)) \\ &\Leftrightarrow (\exists x)((\exists j)(j \in I, x \in O_j), y = f(x)) \\ &\Leftrightarrow (\exists x)(x \in \bigcup_{i\in I} O_i, y = f(x)) \\ &\Leftrightarrow y \in f(\bigcup_{i\in I} O_i) \ . \end{aligned}$$

Then proposition 160 (p. 136) is true from definition 349 (p. 451) and proposition 754 (p. 451).

**Proposition 161** Given $(f : A \to B)$, we have:

$$C \subset A \Rightarrow C \subset f^{-1}(f(C)) \ .$$

**Proof**

$$(x \in C \Rightarrow f(x) \in f(C) \Rightarrow x \in f^{-1}(f(C))) \Rightarrow C \subset f^{-1}(f(C)) \ .$$

**Proposition 162** Given $(f : A \to B)$, we have:

$$C \subset D \subset B \Rightarrow f^{-1}(C) \subset f^{-1}(D) \ .$$

**Proof**

$$\begin{aligned} &((x \in f^{-1}(C), C \subset D) \Rightarrow (f(x) \in C, f(x) \in D) \Rightarrow x \in f^{-1}(D)) \\ &\Rightarrow f^{-1}(C) \subset f^{-1}(D) \ . \end{aligned}$$

**Proposition 163** Given $(f : A \to B)$, $(O_i : I \to \mathcal{P}(B))$, we have:

$$f^{-1}(\bigcup_{i\in I} O_i) = \bigcup_{i\in I} f^{-1}(O_i) \ .$$

See definitions 350 (p. 453) and 362 (p. 462).

**Proof**
From propositions 737 (p. 444) and 735 (p. 442):

$$\begin{aligned} x \in \bigcup_{i\in I} f^{-1}(O_i) &\Leftrightarrow (\exists j)(j \in I, x \in f^{-1}(O_j)) \\ &\Leftrightarrow (\exists j)(j \in I, (\exists y)(y \in O_j, y = f(x))) \\ &\Leftrightarrow (\exists j)(\exists y)(j \in I, y \in O_j, y = f(x)) \\ &\Leftrightarrow (\exists y)(\exists j)(j \in I, y \in O_j, y = f(x)) \\ &\Leftrightarrow (\exists y)((\exists j)(j \in I, y \in O_j), y = f(x)) \\ &\Leftrightarrow (\exists y)(y \in \bigcup_{i\in I} O_i, y = f(x)) \\ &\Leftrightarrow x \in f^{-1}(\bigcup_{i\in I} O_i) \ . \end{aligned}$$

Then proposition 163 (p. 137) is true from definition 349 (p. 451) and proposition 754 (p. 451) in A.5.1.

**Proposition 164** $((\mathbf{f}:\mathbf{A}\to\mathbf{B}),\mathbf{D}\subset\mathbf{B})\Rightarrow \mathbf{f}^{-1}(\mathcal{C}(\mathbf{D}))=\mathcal{C}(\mathbf{f}^{-1}(\mathbf{D}))$[5].

**Proof**
$\mathcal{C}(\mathbf{D})$ is such that $\mathcal{C}(\mathbf{D})\cup\mathbf{D}=\mathbf{B}$ and $\mathcal{C}(\mathbf{D})\cap\mathbf{D}=\emptyset$:
From proposition 163 (p. 137) and relation (2.1) (p. 131):
$\mathbf{f}^{-1}(\mathbf{D})\cup\mathbf{f}^{-1}(\mathcal{C}(\mathbf{D}))=\mathbf{f}^{-1}(\mathbf{B})=\mathbf{A}$
and from relation (2.2) (p. 131):

$$\begin{aligned}&\mathbf{f}^{-1}(\mathbf{D})\cap\mathbf{f}^{-1}(\mathcal{C}(\mathbf{D}))\neq\emptyset\Rightarrow(\exists\mathbf{x})(\mathbf{f}^{-1}(\mathbf{D})\cap\mathbf{f}^{-1}(\mathcal{C}(\mathbf{D})))\\&\Rightarrow(\exists\mathbf{x})(\mathbf{f}(\mathbf{x})\in\mathbf{D},\mathbf{f}(\mathbf{x})\in\mathcal{C}(\mathbf{D}))\Rightarrow\mathbf{D}\cap\mathcal{C}(\mathbf{D})\neq\emptyset\,.\end{aligned}$$

It is absurd. We conclude with proposition 8 (p. 435).

**Proposition 165** Given $(\mathbf{f}:\mathbf{A}\to\mathbf{B})$, $(\mathbf{O_i}:\mathbf{I}\to\mathcal{P}(\mathbf{B}))$, we have:

$$\mathbf{f}^{-1}(\bigcap_{\mathbf{i}\in\mathbf{I}}\mathbf{O_i})=\bigcap_{\mathbf{i}\in\mathbf{I}}\mathbf{f}^{-1}(\mathbf{O_i})\,.$$

**Proof**
Using propositions 163 (p. 137), 164 (p. 138), 767 (p. 455), 801 (p. 466) and 802 (p. 466), we have:

$$\begin{aligned}&\mathbf{f}^{-1}(\bigcup_{\mathbf{i}\in\mathbf{I}}\mathcal{C}(\mathbf{O_i}))=\bigcup_{\mathbf{i}\in\mathbf{I}}\mathbf{f}^{-1}(\mathcal{C}(\mathbf{O_i}))\\&\mathcal{C}(\mathbf{f}^{-1}(\bigcup_{\mathbf{i}\in\mathbf{I}}\mathcal{C}(\mathbf{O_i})))=\mathcal{C}(\bigcup_{\mathbf{i}\in\mathbf{I}}\mathbf{f}^{-1}(\mathcal{C}(\mathbf{O_i})))=\bigcap_{\mathbf{i}\in\mathbf{I}}\mathcal{C}(\mathbf{f}^{-1}(\mathcal{C}(\mathbf{O_i})))=\bigcap_{\mathbf{i}\in\mathbf{I}}\mathbf{f}^{-1}(\mathbf{O_i})\\&\mathcal{C}(\mathbf{f}^{-1}(\bigcup_{\mathbf{i}\in\mathbf{I}}\mathcal{C}(\mathbf{O_i})))=\mathbf{f}^{-1}(\mathcal{C}(\bigcup_{\mathbf{i}\in\mathbf{I}}\mathcal{C}(\mathbf{O_i})))=\mathbf{f}^{-1}(\bigcap_{\mathbf{i}\in\mathbf{I}}\mathcal{C}(\mathcal{C}(\mathbf{O_i})))\\&=\mathbf{f}^{-1}(\bigcap_{\mathbf{i}\in\mathbf{I}}\mathbf{O_i})=\bigcap_{\mathbf{i}\in\mathbf{I}}\mathbf{f}^{-1}(\mathbf{O_i})\,.\end{aligned}$$

**Proposition 166** Given $(\mathbf{f}:\mathbf{A}\to\mathbf{B})$, we have:

$$\mathbf{C}\subset\mathbf{D}\subset\mathbf{B}\Rightarrow\mathbf{f}^{-1}(\mathbf{C})\subset\mathbf{f}^{-1}(\mathbf{D})\,.$$

**Proof**
From proposition 763 (p. 453), we have $\mathbf{C}\subset\mathbf{D}\Leftrightarrow\mathbf{C}\cap\mathbf{D}=\mathbf{C}$, from proposition 165 (p. 138) we have:

$$\mathbf{f}^{-1}(\mathbf{C}\cap\mathbf{D})=\mathbf{f}^{-1}(\mathbf{C})\cap\mathbf{f}^{-1}(\mathbf{D})=\mathbf{f}^{-1}(\mathbf{C})\,.$$

From proposition 763 (p. 453),

$$\mathbf{f}^{-1}(\mathbf{C})\cap\mathbf{f}^{-1}(\mathbf{D})=\mathbf{f}^{-1}(\mathbf{C})\Leftrightarrow\mathbf{f}^{-1}(\mathbf{C})\subset\mathbf{f}^{-1}(\mathbf{D})\,.$$

Then proposition 166 (p. 138) is true.

**Proposition 167** Given $(\mathbf{f}:\mathbf{A}\to\mathbf{B})$, from definition 349 (p. 451), we have

$$\mathbf{C}\subset\mathbf{f}(\mathbf{A})\Rightarrow\mathbf{f}(\mathbf{f}^{-1}(\mathbf{C}))=\mathbf{C}\,.$$

[5] see definition 353 (p. 454).

**Proof**

$$\begin{aligned}&(C \subset f(A), y \in C) \Rightarrow (\exists x \in f^{-1}(C))(y = f(x), x \in f^{-1}(y))\\&\Rightarrow y \in f(f^{-1}(y)) \Rightarrow y \in f(f^{-1}(C))\\&C \subset f(f^{-1}(C))\\&y \in f(f^{-1}(C)) \Rightarrow (\exists x \in f^{-1}(C))(y = f(x)) \Rightarrow y \in C\\&f(f^{-1}(C)) \subset C\,.\end{aligned}$$

We conclude by proposition 754 (p. 451).

**Proposition 168** Given $(f : A \to B)$, we have

$$C \subset B \Rightarrow f(f^{-1}(C)) \subset C\,.$$

**Proof**

$$f^{-1}(C) = f^{-1}(C \cap f(A))$$

from proposition 167 (p. 138)

$$C \cap f(A) \subset f(A) \Rightarrow f(f^{-1}(C \cap f(A))) = C \cap f(A)$$

then combining the two:

$$f(f^{-1}(C)) = C \cap f(A) \subset C\,.$$

**Definition 122 Injective map, one-to-one map.**

See definition 359 (p. 461). If a map $f$ is such that:

$$f(x) = f(x') \Rightarrow x = x' \tag{2.7}$$

the map is said to be injective or one-to-one.

**Proposition 169** If a map $(f : A \to B)$ is injective, it has an inverse when it is restricted to the map $(f : A \to f(A))$.

**Proof**
It is a direct consequence of definition 122 together with definitions 174 (p. 141) and 114 (p. 132).

**Definition 123 Surjective map, onto map.**

(See definition 360 (p. 461)) If a map $(f : A \to B)$ is such that

$$(\forall y \in B)(\exists x \in A)((x, y) \in f) \Leftrightarrow B = f(A) \tag{2.8}$$

the map is said to be surjective or onto.

**Proposition 170** Any map $(f : A \to B)$ between two sets $A$ and $B$ is surjective when it is considered as a map $(f : A \to f(A) \subset B)$.

**Proof**
It is a direct consequence of the two definitions 123 (p. 139) and 114 (p. 132).

**Definition 124 Bijective map or bijection.**

See definition 361 (p. 461). If a map $(\mathbf{f}:\mathbf{A}\rightarrow\mathbf{B})$ is both injective and surjective, it is said to be bijective.

**Convention 15** $(\mathbf{f}:\mathbf{A}\leftrightarrow\mathbf{B})$.

By convention a bijective map $(\mathbf{f}:\mathbf{A}\rightarrow\mathbf{B})$ is written as $(\mathbf{f}:\mathbf{A}\leftrightarrow\mathbf{B})$.

**Definition 125 Inverse map of a map.**

A map $(\mathbf{f}:\mathbf{A}\rightarrow\mathbf{B})$ has a map $(\mathbf{g}:\mathbf{B}\rightarrow\mathbf{A})$ as an inverse, if $(\mathbf{g}\circ\mathbf{f}:\mathbf{A}\rightarrow\mathbf{A})$ is such that (see definition 117 (p. 134)):

$$\mathbf{x}\in\mathbf{A}\Rightarrow\mathbf{g}\circ\mathbf{f}(\mathbf{x})=\mathbf{x}\,.$$

The map $\mathbf{g}$ is quoted as $\mathbf{f}^{-1}$.

**Convention 16** Quotation of the inverse of a map.

The inverse of a map $(\mathbf{f}:\mathbf{A}\rightarrow\mathbf{B})$ when it exists is written as $(\mathbf{f}^{-1}:\mathbf{B}\rightarrow\mathbf{A})$.

**Proposition 171** If a surjective map (definition 123 (p. 139)) $(\mathbf{f}:\mathbf{A}\rightarrow\mathbf{B})$ has an inverse $\mathbf{g}$, $\mathbf{f}$ is the inverse of $\mathbf{g}$.

**Proof**
Since $\mathbf{f}$ is surjective

$$(\forall\mathbf{b}\in\mathbf{B})(\exists\mathbf{a_b}\in\mathbf{A})(\mathbf{b}=\mathbf{f}(\mathbf{a_b}),\mathbf{g}(\mathbf{b})=\mathbf{g}\circ\mathbf{f}(\mathbf{a_b})=\mathbf{a_b},\mathbf{f}\circ\mathbf{g}(\mathbf{b})=\mathbf{f}(\mathbf{a_b})=\mathbf{b})\,.$$

**Proposition 172** If a surjective map (definition 123 (p. 139)) $(\mathbf{f}:\mathbf{A}\rightarrow\mathbf{B})$ has an inverse, this inverse is unique.

**Proof**
Let us suppose that a surjective map $\mathbf{f}:\mathbf{A}\rightarrow\mathbf{B}$ has two inverses $\mathbf{g}$ and $\mathbf{g}'$. Let us suppose $(\mathbf{b},\mathbf{a})\in\mathbf{g}$. From proposition 171 (p. 140), we have:

$$\mathbf{g}(\mathbf{b})=\mathbf{a},\mathbf{f}\circ\mathbf{g}(\mathbf{b})=\mathbf{f}(\mathbf{a})=\mathbf{b}\,.$$

Since

$$\mathbf{g}'\circ\mathbf{f}(\mathbf{A})=\mathbf{A}=\mathbf{g}'(\mathbf{f}(\mathbf{A})),\mathbf{f}(\mathbf{A})\subset\mathbf{B}$$

we have $\mathbf{g}'(\mathbf{B})=\mathbf{A}$. Then we have[6]: $(\exists\mathbf{b}'_{\mathbf{a}})((\mathbf{b}'_{\mathbf{a}},\mathbf{a})\in\mathbf{g}')$ and for the same reason as before we have $\mathbf{b}'_{\mathbf{a}}=\mathbf{f}(\mathbf{a})=\mathbf{b}$, then we have $(\mathbf{b},\mathbf{a})\in\mathbf{g}\Rightarrow(\mathbf{b},\mathbf{a})\in\mathbf{g}'$ and $\mathbf{g}\subset\mathbf{g}'$. With the same reasoning, we get $\mathbf{g}'\subset\mathbf{g}$. From proposition 754 (p. 451), $\mathbf{g}=\mathbf{g}'$.

**Proposition 173** The identity map (definition 113 (p. 132)) has an inverse map which is the identity map.

**Proof**
$\mathbf{A}$ being a set with $\mathbf{f}$ as its identity map, from definition 113 (p. 132) $(\forall\mathbf{y}\in\mathbf{A})(\mathbf{x}\in\mathbf{f}(\mathbf{y})\Rightarrow\mathbf{y}=\mathbf{x})$ then:

$$\mathbf{x}\in\mathbf{A}\Rightarrow\mathbf{f}(\mathbf{f}(\mathbf{x}))=\mathbf{f}\circ\mathbf{f}(\mathbf{x})=\mathbf{x}$$

$$\mathbf{x}\in\mathbf{A}\Rightarrow(\mathbf{x},\mathbf{x})\in\mathbf{f}\Rightarrow(\exists\mathbf{y})((\mathbf{x},\mathbf{y})\in\mathbf{f},(\mathbf{y},\mathbf{x})\in\mathbf{f})$$

we satisfy definitions 125 (p. 140) and 117 (p. 134).

[6]As usual, we apply here for $\exists\mathbf{b}'_{\mathbf{a}}$ the convention 26 (p. 441).

**Proposition 174** We have a map $(\mathbf{f} : \mathbf{A} \to \mathbf{B})$.

1. if $\mathbf{f}$ is such that (see definitions 390 (p. 499)):

$$(\forall \mathbf{b} \in \mathbf{f}(\mathbf{A}))(\mathbf{f}(\mathbf{a}) = \mathbf{f}(\mathbf{a}') = \mathbf{b} \Rightarrow \mathbf{a} = \mathbf{a}')$$

   $\mathbf{f}$ has an inverse $\mathbf{f}^{-1}$.

2. If the map $\mathbf{f}$ has an inverse we have:

$$(\forall \mathbf{b} \in \mathbf{f}(\mathbf{A}))(\mathbf{f}(\mathbf{a}) = \mathbf{f}(\mathbf{a}') = \mathbf{b} \Rightarrow \mathbf{a} = \mathbf{a}') .$$

In both cases we have:

$$(\mathbf{f}^{-1} : \mathbf{f}(\mathbf{A}) \to \mathbf{A}) = \mathbf{set_a}\{(\exists \mathbf{x} \in \mathbf{A})(\exists \mathbf{y} \in \mathbf{B})((\mathbf{y}, \mathbf{x}) = \mathbf{a}, (\mathbf{x}, \mathbf{y}) \in \mathbf{f})\} .$$

**Proof**

1. We call

$$\mathbf{g} = \mathbf{set_a}\{(\exists \mathbf{x} \in \mathbf{A})(\exists \mathbf{y} \in \mathbf{B})((\mathbf{y}, \mathbf{x}) = \mathbf{a}, (\mathbf{x}, \mathbf{y}) \in \mathbf{f})\} .$$

   Since from definition 356 (p. 459) and proposition 780 (p. 458), we have:

$$(\exists \mathbf{x} \in \mathbf{A})(\exists \mathbf{y} \in \mathbf{B})((\mathbf{y}, \mathbf{x}) = \mathbf{a}, (\mathbf{x}, \mathbf{y}) \in \mathbf{f}) \Rightarrow \mathbf{a} \in \mathbf{B} \times \mathbf{A} ,$$

   $\mathbf{g}$ exists from proposition 750 (p. 450).
   We suppose

$$(\forall \mathbf{b} \in \mathbf{f}(\mathbf{A}))(\mathbf{f}(\mathbf{a}) = \mathbf{f}(\mathbf{a}') = \mathbf{b} \Rightarrow \mathbf{a} = \mathbf{a}') .$$

   We have:

$$(\forall \mathbf{b} \in \mathbf{f}(\mathbf{A}))(\exists \mathbf{a_b} \in \mathbf{A})(\mathbf{f}(\mathbf{a_b}) = \mathbf{b}, (\mathbf{b}, \mathbf{f}(\mathbf{a_b})) \in \mathbf{g})$$
$$(\forall \mathbf{b} \in \mathbf{B})(((\mathbf{b}, \mathbf{x}) \in \mathbf{g}, (\mathbf{b}, \mathbf{x}') \in \mathbf{g}) \Rightarrow ((\mathbf{x}, \mathbf{b}) \in \mathbf{f}, (\mathbf{x}', \mathbf{b}) \in \mathbf{f})$$
$$\Rightarrow (\mathbf{x} \in \mathbf{f}^{-1}(\{\mathbf{b}\}), \mathbf{x}' \in \mathbf{f}^{-1}(\{\mathbf{b}\})) \Rightarrow \mathbf{x} = \mathbf{x}') .$$

   Then we verify in definitions 112 (p. 131), equations (2.1) (p. 131) and (2.2) (p. 131) that $\mathbf{g}$ is a map. Since we have:

$$(\forall \mathbf{a} \in \mathbf{A})(\exists \mathbf{b} = \mathbf{f}(\mathbf{a}) \in \mathbf{B})((\mathbf{a}, \mathbf{b}) \in \mathbf{f}, (\mathbf{b}, \mathbf{a}) \in \mathbf{g})$$

   from definition 117 (p. 134), $\mathbf{g} \circ \mathbf{f}(\mathbf{a}) = \mathbf{a}$, $\mathbf{g}$ is the inverse $\mathbf{f}^{-1}$ of $\mathbf{f}$ (definition 125 (p. 140)).

2. $\mathbf{f}(\mathbf{a}) = \mathbf{f}(\mathbf{a}') \Rightarrow \mathbf{f}^{-1}(\mathbf{f}(\mathbf{a})) = \mathbf{a} = \mathbf{f}^{-1}(\mathbf{f}(\mathbf{a}')) = \mathbf{a}' \Rightarrow \mathbf{a} = \mathbf{a}'$ .

Since in point 2 we verify also the conditions of point 1 we have:

$$(\mathbf{f}^{-1} : \mathbf{f}(\mathbf{A}) \to \mathbf{A}) = \mathbf{set_a}\{(\exists \mathbf{x} \in \mathbf{A})(\exists \mathbf{y} \in \mathbf{B})((\mathbf{y}, \mathbf{x}) = \mathbf{a}, (\mathbf{x}, \mathbf{y}) \in \mathbf{f})\} .$$

**Proposition 175** $(\mathbf{f} : \mathbf{A} \to \mathbf{B})$ being a map with an inverse, we have:

$$(\mathbf{x}, \mathbf{y}) \in (\mathbf{f} : \mathbf{A} \to \mathbf{B}) \Leftrightarrow (\mathbf{y}, \mathbf{x}) \in (\mathbf{f}^{-1} : \mathbf{f}(\mathbf{A}) \to \mathbf{A}) . \tag{2.9}$$

**Proof**
It is a direct consequence of proposition 174 (p. 141).

**Proposition 176** If a map (definition 123 (p. 139)) $(\mathbf{f}:\mathbf{A}\longrightarrow\mathbf{B})$ has an inverse $\mathbf{g}$, it is such that (see definition 390 (p. 499)):

$$(\forall \mathbf{b}\in\mathbf{B})((\mathbf{x}\in\mathbf{f}^{-1}(\{\mathbf{b}\}),\mathbf{x}'\in\mathbf{f}^{-1}(\{\mathbf{b}\}))\Rightarrow\mathbf{x}=\mathbf{x}') \tag{2.10}$$

or using cardinals (section A.5.10 (p. 499)), we may write:

$$(\forall \mathbf{b}\in\mathbf{B})(\mathbf{Card}(\mathbf{f}^{-1}(\{\mathbf{b}\}))\leq 1)\ .$$

If $\mathbf{f}$ is surjective (definition 123 (p. 139)), with equation (2.10), we have:

$$(\forall \mathbf{b}\in\mathbf{B})(\exists\mathbf{x})(\mathbf{x}\in\mathbf{f}^{-1}(\{\mathbf{b}\})) \tag{2.11}$$

or using cardinals combining equations (2.10) and (2.11), we may write:

$$(\forall \mathbf{b}\in\mathbf{B})(\mathbf{Card}(\mathbf{f}^{-1}(\{\mathbf{b}\}))=1)\ .$$

**Proof**
We have $\mathbf{b}\in\mathbf{f}(\mathbf{A})\vee\mathbf{b}\notin\mathbf{f}(\mathbf{A})$ (point 4 (p. 435) in A.1.4).

1. $\mathbf{b}\in\mathbf{f}(\mathbf{A})\subset\mathbf{B}$, in that case we have:

$$(\exists\mathbf{a}\in\mathbf{A})(\mathbf{b}=\mathbf{f}(\mathbf{a})),\quad(\exists\mathbf{a}\in\mathbf{A})(\mathbf{a}\in\mathbf{f}^{-1}(\{\mathbf{b}\}))\ .$$

   From propositions 878 (p. 501), 1, $\mathbf{card}(\mathbf{f}^{-1}(\{\mathbf{b}\}))\neq\mathbf{0}$. If we have:

$$\mathbf{b}=\mathbf{f}(\mathbf{a}'),\mathbf{a}'\in\mathbf{f}^{-1}(\{\mathbf{b}\})$$

   we have $\mathbf{a}'=\mathbf{g}(\mathbf{b})=\mathbf{g}\circ\mathbf{f}(\mathbf{a})=\mathbf{a}$. From proposition 878 (p. 501), 2, we have

$$\mathbf{card}(\mathbf{f}^{-1}(\{\mathbf{b}\}))=\mathbf{1}\ .$$

2. $\mathbf{b}\notin\mathbf{f}(\mathbf{A})\subset\mathbf{B}$ in that case we have:

$$\neg(\exists\mathbf{a})(\mathbf{b}=\mathbf{f}(\mathbf{a})),\quad\neg(\exists\mathbf{a})(\mathbf{a}\in\mathbf{f}^{-1}(\{\mathbf{b}\}))\ .$$

   From propositions 773 (p. 457), 878 (p. 501), 1, $\mathbf{Card}(\mathbf{f}^{-1}(\{\mathbf{b}\}))=\mathbf{0}$.

In the case where $\mathbf{f}$ is surjective (definition 123 (p. 139)), only point 1 applies.

**Proposition 177** A bijection has an inverse map which is a bijection.

**Proof**
It is a direct consequence of definition 124 (p. 139) and proposition 174 (p. 141).

**Proposition 178** A surjective map (definition 123 (p. 139)) which has an inverse is a bijection.

**Proof**
It is a direct consequence of definitions 124 (p. 139), 122 (p. 139), 123 (p. 139), 172 (p. 140).

**Proposition 179** The identity map (definition 113 (p. 132)) is a bijection.

It is a direct consequence of propositions 173 (p. 140) and 178 (p. 142).

**Proposition 180** Given an injective map $(\mathbf{f}:\mathbf{A}\to\mathbf{B})$ (definition 122 (p. 139)) between two sets $\mathbf{A}$ and $\mathbf{B}$, $\mathbf{f}$ is a bijective map between $\mathbf{A}$ and $\mathbf{f(A)}$: $(\mathbf{f}:\mathbf{A}\leftrightarrow\mathbf{f(A)})$.

**Proof**
It is a direct consequence of propositions 178 (p. 142) and 169 (p. 139).

**Proposition 181** Given a bijective map $(\mathbf{f}:\mathbf{A}\to\mathbf{C}\subset\mathbf{B})$ (definition 122 (p. 139)) between two sets $\mathbf{A}$ and $\mathbf{C}\subset\mathbf{B}$, $\mathbf{f}$ is an injective map between $\mathbf{A}$ and $\mathbf{B}$: $(\mathbf{f}:\mathbf{A}\to\mathbf{B})$.

**Proof**
It is a direct consequence of definitions 124 (p. 139) and 122 (p. 139).

**Proposition 182** Given a map $\mathbf{f}$ between two sets $\mathbf{A}$ and $\mathbf{B}$: $(\mathbf{f}:\mathbf{A}\to\mathbf{B})$, one can define from $\mathbf{f}$ an injective map $\mathbf{g}$ (definition 122 (p. 139)) between $\mathbf{f(A)}$ and $\mathbf{A}$, $(\mathbf{g}:\mathbf{f(A)}\to\mathbf{A})$.

**Proof**
Since from definition 114 (p. 132), one has

$$(\forall \mathbf{y}\in\mathbf{f(A)})(\exists \mathbf{x}\in\mathbf{A})(\mathbf{f(x)}=\mathbf{y})\,.$$

Using the remark in definition 112 (p. 131), 4, we define $(\mathbf{g}:\mathbf{f(A)}\to\mathbf{A})$ such that $\mathbf{g(f(A))}=\mathbf{A}$: $\mathbf{g(y)}=\tau_{\mathbf{x}}(\mathbf{f(x)}=\mathbf{y})$. Then:

$$(\forall \mathbf{y}\in\mathbf{f(A)})(\mathbf{f}\circ\mathbf{g(y)}=\mathbf{y})$$

so $\mathbf{f}=\mathbf{g}^{-1}$. $\mathbf{g}$ is a bijection between $\mathbf{f(A)}$ and $\mathbf{A}$.

**Proposition 183** Given two sets $\mathbf{A}$, $\mathbf{B}$, a bijection $(\mathbf{f}:\mathbf{A}\leftrightarrow\mathbf{B})$, $\mathbf{C}\subset\mathbf{A}$, the restricted map (definition 118 (p. 135)) $\mathbf{f_C}$ of $\mathbf{f}$ to $\mathbf{C}$ is a bijection on $\mathbf{f(C)}$.

**Proof**
The restricted map verifies all the points of definition 124.

**Proposition 184** Given two sets $\mathbf{A}$, $\mathbf{B}$, a bijection $(\mathbf{f}:\mathbf{A}\leftrightarrow\mathbf{B})$,

$$\mathbf{C}\subset\mathbf{A}\Rightarrow\mathbf{f}(\mathcal{C}(\mathbf{C}))=\mathcal{C}\mathbf{f(C)}\,.$$

**Proof**
From proposition 164 (p. 138), applied to $\mathbf{f}^{-1}$, which is also a map since $\mathbf{f}$ is bijective, we have:

$$(\mathbf{f}^{-1})^{-1}(\mathcal{C}(\mathbf{C}))=\mathcal{C}((\mathbf{f}^{-1})^{-1}(\mathbf{C}))\,.$$

It gives proposition 184, since for a bijection $\mathbf{f}$, we have $(\mathbf{f}^{-1})^{-1}=\mathbf{f}$.

**Proposition 185** $\mathbf{f}$ being a bijection $(\mathbf{f}:\mathbf{A}\leftrightarrow\mathbf{B})$, $\mathbf{f}(\emptyset)=\emptyset$ (definition 355 (p. 456)).

**Proof**
We have $\emptyset=\mathcal{C}(\mathbf{A})$ (definition 355 (p. 456)). Then, since $\mathbf{B}=\mathbf{f(A)}$ (definition 124 (p. 139)), from proposition 184 (p. 143), $\emptyset=\mathcal{C}(\mathbf{B})=\mathbf{f}(\mathcal{C}(\mathbf{A}))=\mathbf{f}(\emptyset)$.

**Proposition 186** Given two sets $\mathbf{A}$, $\mathbf{B}$, a bijection $(\mathbf{f}:\mathbf{A}\leftrightarrow\mathbf{B})$, a set of index $\mathbf{I}$ and a map $(\mathbf{O_i}:\mathbf{I}\to\mathcal{P}(\mathbf{A}))$ we have:

$$\mathbf{f}(\bigcap_{\mathbf{i}\in\mathbf{I}}\mathbf{O_i})=\bigcap_{\mathbf{i}\in\mathbf{I}}\mathbf{f}(\mathbf{O_i})\,.$$

**Proof**
f being a bijection, the map $(\mathbf{f}^{-1} : \mathbf{B} \to \mathbf{A})$ exists. Applying proposition 165 (p. 138) to $\mathbf{f}^{-1}$ we get:

$$(f^{-1})^{-1}(\bigcap_{i \in I} O_i) = \bigcap_{i \in I} (f^{-1})^{-1}(O_i)$$

since $(\mathbf{f}^{-1})^{-1} = \mathbf{f}$ we have proposition 186.

**Definition 126** $(\mathbf{x}^1, \mathbf{x}^2, ..., \mathbf{x}^n)$, $(\mathbf{x}_1, \mathbf{x}_2, ..., \mathbf{x}_n)$, $\prod_{i \in I}^{e} \mathbf{x}^i$.

Given a set $\mathbf{X}$, $\mathbf{n} \in \mathcal{Z}^+$ an integer, a map $(\mathbf{x} : [1, \mathbf{n}] \to \mathbf{X})$ one uses to call this map $(\mathbf{x}^1, \mathbf{x}^2, ..., \mathbf{x}^n)$ or $(\mathbf{x}_1, \mathbf{x}_2, ..., \mathbf{x}_n)$.
It may also be called $\prod_{i \in [1,n]} x^i$ or $\prod_{i \in [1,n]} x_i$. When it is not obvious, in order to avoid confusion with other type of product like product of numbers (definition 128 (p. 145)) or product of maps (definition 117 (p. 134)), to avoid confusion we may use $\prod_{i \in [1,n]}^{e} x^i$ or $\prod_{i \in [1,n]}^{e} x^i$.
The notation holds for any set $\mathbf{I}$, called the set of indices, associated to any other set $\mathbf{X}$: Any map $(\mathbf{x}^i : \mathbf{I} \to \mathbf{X})$ defines a term $\prod_{i \in I}^{e} x^i$:

$$\prod_{i \in I}^{e} x^i = (x^i : I \to X) \ .$$

See the definition 372 (p. 473).

**Proposition 187** With $\mathbf{I}_1 \cap \mathbf{I}_2 = \emptyset$, $\mathbf{X}$ being a set, we consider the elements $\prod_{i \in I_1}^{e} x^i \in \prod_{i \in I_1}^{e} X$, $\prod_{i \in I_2}^{e} x^i \in \prod_{i \in I_2}^{e} X$ and $\prod_{i \in I_1 \cup I_2}^{e} x^i \in \prod_{i \in I_1 \cup I_2}^{e} X$. There is a bijection between $\prod_{i \in I_1}^{e} x^i \overset{e}{\times} \prod_{i \in I_2}^{e} x^i$ and $\prod_{i \in I_1 \cup I_2}^{e} x^i$.

**Proof**
We have $\mathbf{X}, \mathbf{I}_1, \mathbf{I}_2$ considered as set with $\mathbf{I}_1 \cap \mathbf{I}_2 = \emptyset$ and $(\mathbf{x}^i : \mathbf{I}_1 \cup \mathbf{I}_2 \to \mathbf{X})$ maps which are elements of $\prod_{i \in I_1 \cup I_2} X$. Calling $\mathbf{I} = \mathbf{I}_1 \cup \mathbf{I}_2$, from proposition 836 (p. 480), we have a bijection between $\prod_{i \in I_1} X \times \prod_{i \in I_2} X$ and $\prod_{i \in I} X$. Given a map $\prod_{i \in I}^{e} x^i = (x^i : I \to X)$ for which we have two restricted maps $\prod_{i \in I_1}^{e} x^i = (x^i : I_1 \to X)$ and $\prod_{i \in I_2}^{e} x^i = (x^i : I_1 \to X)$. To those maps correspond by the bijection of proposition 836 (p. 480): $\prod_{i \in I}^{e} x^i = (x^i : I \to X)$.

**Convention 17** $\prod_{i \in I_1 \cup I_2}^{e} x^i$.

If $\mathbf{I}_1 \cap \mathbf{I}_2 = \emptyset$, by convention although it is not really correct, we set:

$$\prod_{i \in I_1}^{e} x^i \overset{e}{\times} \prod_{i \in I_2}^{e} x^i = \prod_{i \in I_1 \cup I_2}^{e} x^i \ .$$

**Definition 127** $\sum_{i\in[1,n]} x^i, n \in \mathcal{Z}^+$.

In the same conditions and with the same notations as in the definition just above (definition 126), if the set $X$ has an internal law $(x+y : X\times X \to X)$, the map $(\sum_{i\in[1,n]} x^i : X^n \to X)$ is defined by the relation

$$\sum_{i\in[1,n]} x^i = \sum_{i\in[1,n-1]} x^i + x^n . \tag{2.12}$$

It is defined for $n = 1$, if it is defined for $n = p$, it is also defined by equation (2.12) for $n = p + 1$ then by the recurrence principle it is defined for any $n \in \mathcal{Z}^+$ (proposition 927 (p. 520)). If we also have $x + y = y + x$, it does not depend on the order as we shall see in proposition 347 (p. 208).

**Definition 128** $\prod^{\times}_{i\in[1,n]} x^i$.

When the internal law is quoted as $(x \times y : X \times X \to X)$, one replaces the symbol $\sum$ by $\prod$. There is a potential ambiguity with the product of sets or of elements, but in most of the cases, it is obvious which situation one deals with. When confusion is possible, we shall call

$$\prod^{e}_{i\in[1,n]} x^i = (x^1, x^2, ..., x^n) \text{ and } \prod^{\times}_{i\in[1,n]} x^i = x^1 \times x^2 \times ... \times x^n .$$

**Definition 129 Restriction of a map defined on a product of sets.**

Given two sets $A$ and $B$, a set of index $I$, a map $(C_i : I \to \mathcal{P}(A))$, a map $(f(x) : \prod_{i\in I} C_i \to B)$ (see definition 372 (p. 473)), a $k \in I$ and a term $y \in \prod_{i\in I-\{k\}} C_i$[7], one calls the restricted map of $f$ to $y$, the map $(f_y(x) : C_k \to B)$ such that:

$$(\forall x \in C_k)((z = (z_i : I \to B) \in \prod_{i\in I} C_i, z_k = x,$$
$$(\forall j \in I - \{k\})(z_j = y_j) \Rightarrow f_y(x) = f(z)) .$$

**Definition 130 Transformed of a product of sets by a map.**

Given two sets $I$ and $X$, a map $(X_i : I \to \mathcal{P}(X))$, using definition 119 (p. 135), it defines: $\prod^{e}_{i\in I} X_i$. To any map $(P : I \to I)$ (if $P$ is a bijection (definition 124 (p. 139)), we may call it permutation of $I$, see definition 172 (p. 207)), we set a corresponding map $(P_\Pi : \prod_{i\in I} X_i \to \prod_{i\in I} X_i)$ defined by:

$$(P : I \to I) \Rightarrow (\exists (P_\Pi : \prod_{i\in I} X_i \to \prod_{i\in I} X_i))((u \in \prod_{i\in I} X_i \Rightarrow P_\Pi(u) = u \circ P \in \prod_{i\in I} X_i),$$
$$(\prod^{e}_{i\in I} x^i \in \prod_{i\in I} X_i \Rightarrow P_\Pi(\prod^{e}_{i\in I} x^i) = \prod_{i\in I} x^{P(i)} \in \prod_{i\in I} X_i)) . \tag{2.13}$$

[7] $I - \{k\} = \mathcal{C}_I(\{k\})$

**Proposition 188** Given a set $\mathbf{A}$, we call $\mathcal{F} \subset \mathbf{A} \times \mathbf{A}$, the set of maps $\mathbf{A} \rightarrow \mathbf{A}$ (it is a set from proposition 749 (p. 450)). Given a map $(\mathbf{f_i} : \mathcal{Z}^+ \rightarrow \mathcal{F})$ (see definitions 397 (p. 516), 400 (p. 520)), setting $(\mathbf{h_0} : \mathbf{A} \rightarrow \mathbf{A})$ such that $\mathbf{x} \in \mathbf{A} \Rightarrow \mathbf{h_0(x)} = \mathbf{x}$ (identity map), we have the following statement:

$$(\forall \mathbf{n} \in \mathcal{Z}^+)(\exists \mathbf{h_k} = ( \overset{\circ}{\prod_{\mathbf{j} \in [1,\mathbf{k}]}} \mathbf{f_j} : [1,\mathbf{n}] \rightarrow \mathcal{F}))(\forall \mathbf{k} \in [1,\mathbf{n}])(\mathbf{h_k} = \mathbf{f_k} \circ \mathbf{h_{k-1}}) \ . \qquad (2.14)$$

**Proof**

The part of relation (2.14) controlled by $(\forall \mathbf{n} \in \mathcal{Z}^+)$ is true for $\mathbf{n} = 1$. If it is true for $\mathbf{n} = \mathbf{n}'$, it is also true for $\mathbf{n} = \mathbf{n}' + 1$. From propositions 927 (p. 520), 2.14 is then true. $(\mathbf{h_k} : \mathcal{Z} \rightarrow \mathcal{F})$ is unique when $\mathbf{k} = 1$. If it is unique up to $\mathbf{n}$, it is unique for $\mathbf{n} + 1$ then it is unique for any $\mathbf{n}$ (proposition 927 (p. 520)).

**Definition 131 Product of several maps $\overset{\circ}{\prod}$.**

relation (2.14) of proposition 188 (p. 146) allows to define:

$$\overset{\circ}{\prod_{\mathbf{j} \in [1,\mathbf{n}]}} \mathbf{f_j}$$

for any positive integer $\mathbf{n} > 0$. We can expand the definition of the product of several maps to the case where we have a well-ordered set of indices (definition 388 (p. 487)) $\mathbf{I}$ such that $\mathbf{Card(I)} \in \mathcal{Z}^+$. From proposition 947 (p. 532), there is a unique isomorphism (definition 387 (p. 487)) $\mathbf{i(j)} : [1, \mathbf{Card(I)}] \rightarrow \mathbf{I}$. We use it to expand the definition of the product of several maps:

$$\overset{\circ}{\prod_{\mathbf{i} \in \mathbf{I}}} \mathbf{f_i} = \overset{\circ}{\prod_{\mathbf{j} \in [1,\mathbf{Card(I)}]}} \mathbf{f_{i(j)}} \ .$$

**Convention 18** Sum and product over a finite set of indices.

See also convention 14 (p. 135). Having a set $\mathbf{E}$ with a map $(\mathbf{h(x,y)} : \mathbf{E} \times \mathbf{E} \rightarrow \mathbf{E})$, a finite set $\mathbf{I}$ with a total order (definition 377 (p. 483)) and a map $(\mathbf{x_k} : \mathbf{I} \rightarrow \mathbf{E})$ with its associated unique isomorphism $(\mathbf{i(j)} : [1, \mathbf{Card(I)}] \rightarrow \mathbf{I})$ (proposition 947 (p. 532)), as above (equation (2.14)) we have[8]:

$$(\exists(\mathbf{H(k)} : [1, \mathbf{Card(I)}] \rightarrow \mathbf{E}))(\forall \mathbf{k} \in [2, \mathbf{Card(I)}])$$
$$(\mathbf{H(k)} = \mathbf{h(H(k-1), x_{i(k)})}, \ \mathbf{H(1)} = \mathbf{x_{i(1)}}) \ .$$

Quoting $\mathbf{h(x,y)}$ as $\mathbf{x} \times \mathbf{y}$ or $\mathbf{x} + \mathbf{y}$, by convention we define respectively:

$$\prod_{\mathbf{k} \in \mathbf{I}} \mathbf{x_k} = \mathbf{H(Card(I))}, \ \sum_{\mathbf{k} \in \mathbf{I}} \mathbf{x_k} = \mathbf{H(Card(I))} \ .$$

[8]Unfortunately, there is no universal convention for the order in which the operations are quoted. Sometimes, one prefers to write successive operations from left to right as one writes or reads in European languages. It is the case when one does product of spaces or of elements. Sometimes nevertheless, one prefers the opposite as for the product of map where one follows the habit for writing numbers. Most of the times the order to be chosen is obvious from the surrounding.

In other words being less precise and more intuitive, any times we have an operation implying an ordered set like $\overset{\circ}{\prod\limits_{i\in I}} f_i$ where the order in which the operations are done matters, the increasing order is taken. For instance:

$$\overset{\circ}{\prod_{i\in[1,3]}} f_i = f_3(f_2(f_1(x))) = f_3 \circ f_2 \circ f_1 \ .$$

For expressions like:

$$\overset{e}{\prod_{i\in[1,n]}} x_i = (x_1, x_2, ..., x_n) \in \prod_{i\in[1,n]} \mathcal{R} = \mathcal{R}^n$$

$\overset{e}{\prod\limits_{i\in[1,n]}}$ represents a map of $[1,n] \to \mathcal{R}$. There is no ambiguity and the set of indices does not need to be ordered. The same remark applies to expressions like: $\bigotimes\limits_{i\in[1,n]} v_i$, $\bigwedge\limits_{i\in[1,n]} df_i$ which are defined from a product of elements (see definition 44 (p. 40)). But on the opposite, we need a convention for terms like $x \times y \times z \in \mathcal{R}^3$: By convention, this term represents a map $[1,3] \to \mathcal{R}$ such as $1 \to x$, $2 \to y$, $3 \to z$ we follow from left to right the order of $1, 2, 3$.

**Proposition 189** The product of map is associative: $f, g, h$ being three maps we have:

$$(f \circ g) \circ h = f \circ (g \circ h) = f \circ g \circ h \ .$$

**Proof**
From definition 117 (p. 134), we have:

$$\begin{aligned}(f \circ g) \circ h &= set_{(x,y)}\{(\exists z_2)((x, z_2) \in h, (z_2, y) \in (f \circ g))\} \\ &= set_{(x,y)}\{(\exists z_1)(\exists z_2)((x, z_2) \in h, (z_2, z_1) \in g, (z_1, y) \in f)\} \\ &= set_{(x,y)}\{(\exists z_1)((x, z_1) \in g \circ h, (z_1, y) \in f)\} \\ &= f \circ (g \circ h) \ .\end{aligned}$$

**Proposition 190 Associativity of the product of a finite set of maps.**
With the notations of definition 131 (p. 146), given a map $(f_i : \mathcal{Z}^+ \to \mathcal{F})$, a map $(k_p : [1,n] \to \mathcal{Z}^+)$ (see definition 112 (p. 131), 3, it defines a set of $n$ positive integers labeled from $1$ to $n$), we set (see definition 393 (p. 506)):

$$I = \bigcup_{p\in[1,n]} [1, k_p] \times \{p\} \ .$$

It is the union of $n$ intervals (definition 378 (p. 484)) of positive integers labeled from $1$ to $n$. Given two elements of $I$: $(k, p) \in I, (k', p') \in I$, we set the good order on $I$ by:

$$(k, p) \leq (k', p') \Leftrightarrow p < p' \vee (p = p', k \leq k') \ .$$

It is a good order because the order on cardinals is a good order (proposition 886 (p. 504)): Given a subset $A$ of $I$, one takes its projections (definition 121 (p. 136)) on the second coordinates. It makes a subset of $\mathcal{Z}^+$ which has a smallest element. Taking all elements of $A$ having that second coordinate, the set of their first coordinates is a subset of $\mathcal{Z}^+$ and has

then a smallest element. The corresponding element is the smallest element of $\mathbf{A}$.
Calling $\mathbf{m} = \sum_{\mathbf{p}\in[\mathbf{1},\mathbf{n}]} \mathbf{k_p}$, we have $\mathbf{m} = \mathbf{Card(I)}$ (definition 393 (p. 506), proposition 942 (p. 527)). We have an isomorphism (definition 387 (p. 487)) between $[\mathbf{1},\mathbf{m}]$ and $\mathbf{I}$ (definition 387 (p. 487), propositions 869 (p. 496), 923 (p. 518)):

$$(\mathbf{i(j)} : [\mathbf{1},\mathbf{m}] \to \mathbf{I}) \ .$$

Using that isomorphism, we have (definition 131 (p. 146), convention 18 (p. 146)):

$$\overset{\circ}{\prod_{\mathbf{i}=(\mathbf{k},\mathbf{p})\in\mathbf{I}}} \mathbf{f_i} = \overset{\circ}{\prod_{\mathbf{j}\in[\mathbf{1},\mathbf{m}]}} \mathbf{f_{i(j)}} = \overset{\circ}{\prod_{\mathbf{p}\in[\mathbf{1},\mathbf{n}]}} \overset{\circ}{\prod_{\mathbf{k}\in[\mathbf{1},\mathbf{k_p}]}} \mathbf{f_{(k,p)}} \ . \tag{2.15}$$

**Proof**

Equation (2.15) is true for $\mathbf{n} = \mathbf{1}$ and any $\mathbf{k_1}$ (definition 131 (p. 146)). Let us suppose that it is true for $\mathbf{n} = \mathbf{n'}$ then from definition 131 (p. 146) it is true for $\mathbf{n} = \mathbf{n'} + \mathbf{1}$ and $\mathbf{k_{n'+1}} = \mathbf{1}$. Let us suppose that it is true for $\mathbf{n} = \mathbf{n'} + \mathbf{1}$ and $\mathbf{k_{n'+1}} = \mathbf{k'_{n'+1}}$.
Setting:

$$\mathbf{I'} = [\mathbf{1}, \mathbf{k'_{n'+1}}] \times \{\mathbf{n'} + \mathbf{1}\} \cup \bigcup_{\mathbf{p}\in[\mathbf{1},\mathbf{n'}]} [\mathbf{1}, \mathbf{k_p}] \times \{\mathbf{p}\}$$

$$\mathbf{I} = [\mathbf{1}, \mathbf{k'_{n'+1}} + \mathbf{1}] \times \{\mathbf{n'} + \mathbf{1}\} \cup \bigcup_{\mathbf{p}\in[\mathbf{1},\mathbf{n'}]} [\mathbf{1}, \mathbf{k_p}] \times \{\mathbf{p}\} \ .$$

Calling $\mathbf{m'} = \mathbf{Card(I')}, \mathbf{n} = \mathbf{n'} + \mathbf{1}, \mathbf{m} = \mathbf{Card(I)}, \mathbf{k_{n'+1}} = \mathbf{k'_{n'+1}} + \mathbf{1}$. Using what we assume, plus proposition 189 (p. 147) and definition 131 (p. 146) (see convention 18 (p. 146)), we have (see convention 18 (p. 146)):

$$\overset{\circ}{\prod_{\mathbf{j}\in[\mathbf{1},\mathbf{m}]}} \mathbf{f_{i(j)}} = \mathbf{f_{(k'_{n'+1}+1,n'+1)}} \circ \overset{\circ}{\prod_{\mathbf{i}\in\mathbf{I'}}} \mathbf{f_i}$$

$$= \mathbf{f_{(k'_{n'+1}+1,n'+1)}} \circ \overset{\circ}{\prod_{\mathbf{k}\in[\mathbf{1},\mathbf{k'_{n'+1}}]}} \mathbf{f_{(k,n'+1)}} \circ \overset{\circ}{\prod_{\mathbf{p}\in[\mathbf{1},\mathbf{n'}]}} \overset{\circ}{\prod_{\mathbf{k}\in[\mathbf{1},\mathbf{k_p}]}} \mathbf{f_{(k,p)}} \ .$$

From proposition 189 (p. 147):

$$\mathbf{f_{(k'_{n'+1}+1,n'+1)}} \circ \overset{\circ}{\prod_{\mathbf{k}\in[\mathbf{1},\mathbf{k'_{n'+1}}]}} \mathbf{f_{(k,n'+1)}} \circ \overset{\circ}{\prod_{\mathbf{p}\in[\mathbf{1},\mathbf{n'}]}} \overset{\circ}{\prod_{\mathbf{k}\in[\mathbf{1},\mathbf{k_p}]}} \mathbf{f_{(k,p)}}$$

$$= \overset{\circ}{\prod_{\mathbf{k}\in[\mathbf{1},\mathbf{k'_{n'+1}}+\mathbf{1}]}} \mathbf{f_{(k,n'+1)}} \circ \overset{\circ}{\prod_{\mathbf{p}\in[\mathbf{1},\mathbf{n'}]}} \overset{\circ}{\prod_{\mathbf{k}\in[\mathbf{1},\mathbf{k_p}]}} \mathbf{f_{(k,p)}} = \overset{\circ}{\prod_{\mathbf{p}\in[\mathbf{1},\mathbf{n}]}} \overset{\circ}{\prod_{\mathbf{k}\in[\mathbf{1},\mathbf{k_p}]}} \mathbf{f_{(k,p)}}$$

equation (2.15) is now true for $\mathbf{n} = \mathbf{n'} + \mathbf{1}$ and $\mathbf{k_{n'+1}} = \mathbf{k'_{n'+1}} + \mathbf{1}$. Using proposition 927 (p. 520) proposition 190 is true for $\mathbf{n} = \mathbf{n'} + \mathbf{1}$ and any $\mathbf{k_{n'+1}}$. Using proposition 927 (p. 520) again, proposition 190 is true for any $\mathbf{n}$ and any associated set of $\mathbf{k_p}$.

**Proposition 191** The product of two injective maps is an injective map (definition 122 (p. 139)), (see proposition 193 (p. 149)).

**Proof**

We have three sets $\mathbf{A}$, $\mathbf{B}$, $\mathbf{C}$ and two injective maps: $(\mathbf{f} : \mathbf{A} \to \mathbf{B})$, $(\mathbf{g} : \mathbf{B} \to \mathbf{C})$. $\mathbf{f}$ and $\mathbf{g}$ being injective, we have:

$$(\mathbf{g(f(x))} = \mathbf{g(f(x'))}) \Rightarrow (\mathbf{f(x)} = \mathbf{f(x')}) \Rightarrow \mathbf{x} = \mathbf{x'}$$

proposition 191 (p. 148) is true.

**Proposition 192** The product of two surjective maps is a surjective map (definition 123 (p. 139)), (see proposition 193 (p. 149)).

**Proof**

We have three sets $\mathbf{A}$, $\mathbf{B}$, $\mathbf{C}$ and two surjective maps: $(\mathbf{f} : \mathbf{A} \rightarrow \mathbf{B})$, $(\mathbf{g} : \mathbf{B} \rightarrow \mathbf{C})$. $\mathbf{f}$ and $\mathbf{g}$ being surjective, we have: $\mathbf{B} = \mathbf{f}(\mathbf{A})$ and $\mathbf{C} = \mathbf{g}(\mathbf{B})$. Then $\mathbf{g} \circ \mathbf{f}(\mathbf{A}) = \mathbf{g}(\mathbf{B}) = \mathbf{C}$. Proposition 192 (p. 149) is true.

**Proposition 193** The product of a finite number of bijective (resp. injective; surjective) maps is a bijective (resp. injective; surjective) map.

**Proof**

We call $\mathcal{F}$, the set of bijective maps (resp. injective; surjective) $\mathbf{A} \rightarrow \mathbf{B}$. This set exists from proposition 750 (p. 450) since $\mathbf{f} \in \mathcal{F} \Rightarrow \mathbf{f} \in \mathcal{P}(\mathbf{A} \times \mathbf{B})$. Having a map $(\mathbf{f_i} : [\mathbf{1}, \mathbf{m}] \rightarrow \mathcal{F})$, if

$$1 \leq n < m \Rightarrow \overset{\circ}{\prod_{i \in [1,n]}} f_i \in \mathcal{F}$$

from definitions 124 (p. 139), 131 (p. 146) and propositions 191 (p. 148), 192 (p. 149), we have:

$$\overset{\circ}{\prod_{i \in [1,n]}} f_i \circ f_{n+1} = \overset{\circ}{\prod_{i \in [1,n+1]}} f_i \in \mathcal{F}$$

$\mathbf{n} = \mathbf{1} \Rightarrow \overset{\circ}{\prod_{i \in [1,n]}} f_i \in \mathcal{F}$, then from proposition 927 (p. 520), $\overset{\circ}{\prod_{i \in [1,n]}} f_i \in \mathcal{F}$ is true for any $\mathbf{n} \in [\mathbf{1}, \mathbf{m}]$.

**Proposition 194** If the product of two maps is a bijective map (definition 124 (p. 139)) with the first of them surjective (definition 123 (p. 139)), those two maps are bijective.

**Proof**

Having three sets $\mathbf{A}$, $\mathbf{B}$, $\mathbf{C}$ two maps $(\mathbf{f} : \mathbf{A} \rightarrow \mathbf{B})$ $(\mathbf{g} : \mathbf{B} \rightarrow \mathbf{C})$ such that $\mathbf{f}(\mathbf{A}) = \mathbf{B}$, since $\mathbf{g} \circ \mathbf{f}(\mathbf{A}) = \mathbf{C}$, then $\mathbf{g}(\mathbf{f}(\mathbf{A})) = \mathbf{g}(\mathbf{B}) = \mathbf{C}$, $\mathbf{g}$ is surjective.
Since $\mathbf{f}(\mathbf{A}) = \mathbf{B}$, we have:

$$(x_1 \in B, x_2 \in B) \Rightarrow (\exists y_1 \in A)(\exists y_2 \in A)(f(y_1) = x_1, f(y_2) = x_2) \,.$$

Taking those values corresponding to $\mathbf{x_1}, \mathbf{x_2}$, since $\mathbf{gf}$ is a bijection we have $\mathbf{g}(\mathbf{x_1}) = \mathbf{g}(\mathbf{x_2}) \Rightarrow \mathbf{gf}(\mathbf{y_1}) = \mathbf{gf}(\mathbf{y_2}) \Rightarrow \mathbf{y_1} = \mathbf{y_2} \Rightarrow \mathbf{x_1} = \mathbf{x_2}$ (see definition 112 (p. 131), equation (2.2)). $\mathbf{g}$ is injective. From definition 124 (p. 139), since $\mathbf{g}$ is also surjective, $\mathbf{g}$ is bijective as $\mathbf{g}^{-1}$ (propositions 177 (p. 142), 178 (p. 142)). Since $\mathbf{g}^{-1} \circ (\mathbf{g} \circ \mathbf{f}) = (\mathbf{g}^{-1} \circ \mathbf{g}) \circ \mathbf{f} = \mathbf{f}$ (proposition 189 (p. 147)), $\mathbf{f}$ is the product of two bijections and from proposition 193 (p. 149) $\mathbf{f}$ is a bijection.

**Proposition 195** $\mathbf{f}$ and $\mathbf{g}$ bijective gives $(\mathbf{g} \circ \mathbf{f})^{-1} = \mathbf{f}^{-1} \circ \mathbf{g}^{-1}$ .

**Proof**

From proposition 193 (p. 149) $\mathbf{g} \circ \mathbf{f}$ is bijective map as $(\mathbf{g} \circ \mathbf{f})^{-1}$, $\mathbf{f}^{-1}$ and $\mathbf{g}^{-1}$ (proposition 177 (p. 142)). From the definitions of a map (definition 112 (p. 131)), a bijective map (definition 124 (p. 139)) and the product of maps (definition 117 (p. 134)), we have:

$$\begin{aligned}(x, z) \in g \circ f &\Rightarrow (\exists y)((x, y) \in f, (y, z) \in g) \\ &\Rightarrow (\exists y)((y, x) \in f^{-1}, (z, y) \in g^{-1}) \\ &\Rightarrow (z, x) \in f^{-1} \circ g^{-1} \,.\end{aligned}$$

**Proposition 196** $\mathbf{h_1}$, $\mathbf{h_2}$ are two injective map $\mathbf{A} \longrightarrow \mathbf{B}$, $\mathbf{f}$ an injective map $\mathbf{B} \longrightarrow \mathbf{C}$. We have:

$$\mathbf{h_1 \neq h_2 \Rightarrow f \circ h_1 \neq f \circ h_2} \, .$$

**Proof**

From proposition 149 (p. 132),

$$\mathbf{h_1 \neq h_2 \Leftrightarrow (\exists x)((x, h_1(x)) \neq (x, h_2(x)))} \, .$$

Then calling $\mathbf{x} = \tau_{\mathbf{x}}(\mathbf{h_1(x) \neq h_2(x)})$, we have:

$$\mathbf{f \circ h_1(x) = f(h_1(x)) \neq f(h_2(x)) = f \circ h_2(x)}$$

then from propositions 149 (p. 132), 196 (p. 150) is true.

**Proposition 197** $\mathbf{h_1}$, $\mathbf{h_2}$ are two injective maps $\mathbf{B} \longrightarrow \mathbf{C}$, $\mathbf{f}$ a bijective map $\mathbf{A} \longrightarrow \mathbf{B}$. We have:

$$\mathbf{h_1 \neq h_2 \Rightarrow h_1 \circ f \neq h_2 \circ f} \, .$$

**Proof**

From proposition 149 (p. 132),

$$\mathbf{h_1 \neq h_2 \Leftrightarrow (\exists x)((x, h_1(x)) \neq (x, h_2(x)))} \, .$$

Then calling $\mathbf{x} = \tau_{\mathbf{x}}(\mathbf{h_1(x) \neq h_2(x)})$, from definition 124 (p. 139) there is a map $(\mathbf{f^{-1} : B \longrightarrow A})$ such that $(\mathbf{f \circ f^{-1} : B \longrightarrow B})$ is the identity map (definition 113 (p. 132)). From definition 113 (p. 132), we have:

$$\mathbf{h_1 \circ f(f^{-1}(x)) = h_1(x) \neq h_2(x) = h_1 \circ f(f^{-1}(x))} \, .$$

Then from proposition 727 (p. 441), we have: $(\exists \mathbf{y = f^{-1}(x)})(\mathbf{h_1 \circ f(y) \neq h_2 \circ f(y)})$. From propositions 149 (p. 132), 197 (p. 150) is true.

**Proposition 198** Given a set $\mathbf{E}$ and the set of bijections $\boldsymbol{\mathcal{E}}$ of $\mathbf{E}$ on $\mathbf{E}$, a map $\mathbf{a} \in \boldsymbol{\mathcal{E}}$, the following maps $\boldsymbol{\mathcal{E}} \longrightarrow \boldsymbol{\mathcal{E}}$ are bijective:

1. $$\mathbf{h_1(s) = (s^{-1} : \boldsymbol{\mathcal{E}} \leftrightarrow \boldsymbol{\mathcal{E}})}$$

2. $$\mathbf{h_2^a(s) = (a \circ s : \boldsymbol{\mathcal{E}} \leftrightarrow \boldsymbol{\mathcal{E}})}$$

3. $$\mathbf{h_3^a(s) = (s \circ a : \boldsymbol{\mathcal{E}} \leftrightarrow \boldsymbol{\mathcal{E}})} \, .$$

**Proof**

1. From definition 124 (p. 139) and proposition 177 (p. 142), $\mathbf{s} \in \boldsymbol{\mathcal{E}} \Rightarrow \mathbf{s^{-1}} \in \boldsymbol{\mathcal{E}}$. From proposition 171 (p. 140) $\mathbf{t} \in \boldsymbol{\mathcal{E}}$, $\mathbf{s = t^{-1} \Rightarrow s^{-1} = t}$ $\mathbf{h_1(s)}$ is surjective. Then from proposition 178 (p. 142) it is bijective.

2. If $\mathbf{a} \in \boldsymbol{\mathcal{E}}$, $\mathbf{s} \in \boldsymbol{\mathcal{E}}$, from proposition 193 (p. 149), we have: $\mathbf{a \circ s} \in \boldsymbol{\mathcal{E}}$, $\mathbf{s \circ a} \in \boldsymbol{\mathcal{E}}$. $\mathbf{h_2^a}$ and $\mathbf{h_3^a}$ defined maps $\boldsymbol{\mathcal{E}} \longrightarrow \boldsymbol{\mathcal{E}}$. From proposition 171 (p. 140), we have:

   $$(\forall \mathbf{s} \in \boldsymbol{\mathcal{E}})(\exists \mathbf{s_2 = a^{-1} os})(\exists \mathbf{s_3 = soa^{-1}})(\mathbf{aos_2 = aoa^{-1}os = s, s_3oa = soa^{-1}oa = s})$$

   $\mathbf{h_2^a}$, $\mathbf{h_3^a}$ are surjective. $\mathbf{h_2^{a^{-1}}}$, $\mathbf{h_3^{a^{-1}}}$ are the inverse of $\mathbf{h_2^a}$, $\mathbf{h_3^a}$:

   $$(\forall \mathbf{s} \in \boldsymbol{\mathcal{E}})(\mathbf{h_2^a \circ h_2^{a^{-1}}(s) = a \circ a^{-1} \circ s = s, h_3^a \circ h_3^{a^{-1}}(s) = s \circ a \circ a^{-1} \circ s = s})$$

   then from proposition 178 (p. 142), the maps $\mathbf{h_2^a}$, $\mathbf{h_3^a}$ are bijections $\boldsymbol{\mathcal{E}} \leftrightarrow \boldsymbol{\mathcal{E}}$.

### 2.1.2 Topology

**Definition 132 Open subset.**

The set **E** is called topological set if there is a subset $\mathcal{O}$ of the set of subsets of **E** ($\mathcal{O} \subset \mathcal{P}(\mathbf{E})$). cf. definition 350 (p. 453) such that (see definitions 355 (p. 456), 351 (p. 453), 352 (p. 453)):

$$\mathbf{E} \in \mathcal{O}, \emptyset \in \mathcal{O} \tag{2.16}$$

and

$$(\forall \mathcal{O}_s \subset \mathcal{O})(\bigcup_{O \in \mathcal{O}_s} O \in \mathcal{O}) \tag{2.17}$$

and[9]

$$(\forall \mathcal{O}_s \subset \mathcal{O})(\mathbf{Card}(\mathcal{O}_s) \in \mathcal{Z}^+ \Rightarrow \bigcap_{O \in \mathcal{O}_s} O \in \mathcal{O}) . \tag{2.18}$$

The elements of $\mathcal{O}$ are called open subsets of **E**.

**Definition 133 Induced topology on subset.**

Given a topological space **E** and a subset **F** of **E**, the induced topology from **E** to **F** is obtained by taking as open subsets of **F** the intersection of the open subsets of **E** by **F**.

**Proposition 199** An induced topology, defined a topological subset, the opens of which verify definition 132 (p. 151).

**Proof**

Given a topological space **E** and a subset **F** of **E**, calling $\mathcal{O}$, the set of opens of **E**, $\mathcal{O}_F$, the set of opens of **F** as defined in 133 (p. 151), we have from that definition:

$$O_F \in \mathcal{O}_F \Leftrightarrow (\exists O \in \mathcal{O})(O_F = O \cap F) . \tag{2.19}$$

Let us prove that $\mathcal{O}_F$ verifies definition 132 (p. 151).
To ease the demonstration, using (2.19) (p. 151) and point 4 of definition 112 (p. 131), we define the injective map (definition 122 (p. 139)) $(O_E(O) : \mathcal{O}_F \to \mathcal{O})$ such that $O \in \mathcal{O}_F \Rightarrow O = O_E(O) \cap F$, $O_E(O) \in \mathcal{O}$. It is injective because $O_E(O) = O_E(O') \Rightarrow O_E(O) \cap F = O_E(O') \cap F = O = O'$.

- Equation (2.16) (p. 151) is verified.
  From propositions 763 (p. 453) and 805 (p. 467), we have:
  $$(\emptyset = \emptyset \cap F, \emptyset \in \mathcal{O}) \Rightarrow \emptyset \in \mathcal{O}_F$$
  $$F = F \cap E, E \in \mathcal{O}) \Rightarrow F \in \mathcal{O}_F .$$
- Equation (2.17) (p. 151) is verified.
  From propositions 798 (p. 465), 797 (p. 465), we have:
  $$S \subset \mathcal{O}_F \Rightarrow \bigcup_{s \in S} s = \bigcup_{s \in S} O_E(s) \cap F = F \cap \bigcup_{u \in O_E(S)} u .$$
  Since from propositions 151 (p. 133) and 157 (p. 134), we have $O_E(S) \subset \mathcal{O}$. From equation (2.17) (p. 151), we also have $\bigcup_{u \in O_E(S)} u \in \mathcal{O}$. From definition 133 (p. 151), $F \cap \bigcup_{u \in O_E(S)} u \in \mathcal{O}_F$ then $\bigcup_{s \in S} s \in \mathcal{O}_F$.

[9] "$\mathbf{Card}(\mathcal{O}_s) \in \mathcal{Z}$" is equivalent to "the number of elements in $\mathcal{O}_s$ is finite" (see definition 390 (p. 499)).

- Equation (2.18) (p. 151).
  From propositions 800 (p. 466), 797 (p. 465), we have:

$$\mathbf{S} \subset \mathcal{O}_\mathbf{F}, \mathbf{Card(s)} \in \mathcal{Z}^+ \Rightarrow \bigcap_{\mathbf{s}\in\mathbf{S}} \mathbf{s} = \bigcap_{\mathbf{s}\in\mathbf{S}} \mathbf{O_E(s)} \cap \mathbf{F} = \mathbf{F} \cap \bigcap_{\mathbf{s}\in\mathbf{S}} \mathbf{O_E(s)} = \mathbf{F} \cap \bigcap_{\mathbf{u}\in\mathbf{O_E(S)}} \mathbf{u}\,.$$

From equation (2.18) (p. 151), we have: $\bigcap_{\mathbf{u}\in\mathbf{O_E(S)}} \mathbf{u} \in \mathcal{O}$. Then, from definition 133 (p. 151), we have: $\bigcap_{\mathbf{s}\in\mathbf{S}} \mathbf{s} \in \mathcal{O}_\mathbf{F}$.

**Proposition 200** Given a topological space **E** and an open subset **O** of **E**, any open of the induced topology by **E** on **O** is also an open of **E**.

**Proof**

It is a consequence of equation (2.18) (p. 151) combined to definition 133 (p. 151).

**Proposition 201** Given a topological set **E** with $\mathcal{O} \subset \mathcal{P}(\mathbf{E})$ the set of opens, a set **I** and a map $(\mathbf{O_i} : \mathbf{I} \to \mathcal{O})$, we have:

$$\bigcup_{\mathbf{i}\in\mathbf{I}} \mathbf{O_i} \in \mathcal{O}$$

$$\mathbf{Card(I)} \in \mathcal{Z}^+ \Rightarrow \bigcap_{\mathbf{i}\in\mathbf{I}} \mathbf{O_i} \in \mathcal{O}\,.$$

**Proof**

Setting (see proposition 760 (p. 453)) $\mathcal{O}_\mathbf{s} = \mathbf{O_I} \subset \mathcal{P}(\mathbf{E})$, applying, in equations (2.17) (p. 151) and (2.18) (p. 151), propositions 797 (p. 465) and 804 (p. 467), setting in those propositions $\mathbf{A} = \mathcal{O}_\mathbf{s}$, we get proposition 201.

**Proposition 202** Given a set **E**, a subset $\mathcal{O}_\mathbf{r} \subset \mathcal{P}(\mathbf{E})$, calling $\mathcal{O}$ the set constructed from $\mathcal{O}_\mathbf{r}$ as:

$$\mathcal{O} = \mathbf{set}_\omega\{(\exists \mathbf{P} \subset \mathcal{O}_\mathbf{r})(\omega = \bigcup_{\mathbf{p}\in\mathbf{P}} \mathbf{p})\}\,. \tag{2.20}$$

The set $\mathcal{O} \subset \mathcal{P}(\mathbf{E})$ verifies equation (2.17) (p. 151).

**Proof**

**E** a set, $\mathcal{O}_\mathbf{r} \subset \mathcal{P}(\mathbf{E})$ a subset of $\mathcal{P}(\mathbf{E})$ with $\mathcal{O}$ defined by equation (2.20). Equation (2.20) says (see definition 349 (p. 451)):

$$(\forall \mathcal{O}_\mathbf{s} \subset \mathcal{O})(\mathbf{x} \in \mathcal{O}_\mathbf{s} \Rightarrow \mathbf{x} \in \mathcal{O} \Rightarrow (\exists \mathbf{P(x)} \subset \mathcal{O}_\mathbf{r})(\mathbf{x} = \bigcup_{\mathbf{p}\in\mathbf{P(x)}} \mathbf{p}))\,. \tag{2.21}$$

From that, using the definition 112 (p. 131), item 4 with the same notations, for any $\mathcal{O}_\mathbf{s} \subset \mathcal{O}$, equation (2.21) defines a map $(\mathbf{P(x)} : \mathcal{O}_\mathbf{s} \to \mathcal{P}(\mathcal{O}_\mathbf{r}))$ such that: $\mathbf{x} = \bigcup_{\mathbf{p}\in\mathbf{P(x)}} \mathbf{p}$. From proposition 796 (p. 464), equation (A.25) (p. 464), we have:

$$\bigcup_{\mathbf{x}\in\mathcal{O}_\mathbf{s}} \mathbf{x} = \bigcup_{\mathbf{x}\in\mathcal{O}_\mathbf{s}} \bigcup_{\mathbf{p}\in\mathbf{P(x)}} \mathbf{p} = \bigcup_{\mathbf{p}\in\bigcup_{\mathbf{x}\in\mathcal{O}_\mathbf{s}} \mathbf{P(x)}} \mathbf{p}\,.$$

From proposition 797 (p. 465), we have:

$$\bigcup_{\mathbf{x}\in\mathcal{O}_\mathbf{s}} \mathbf{P(x)} = \bigcup_{\mathbf{p}\in\mathbf{P}(\mathcal{O}_\mathbf{s})} \mathbf{p}\,.$$

Since, from proposition 151 (p. 133), we have: $\mathbf{P}(\mathcal{O}_\mathbf{s}) \subset \mathcal{P}(\mathcal{O}_\mathbf{r})$, from proposition 799 (p. 465), we have: $\mathbf{S} = \bigcup_{\mathbf{p}\in\mathbf{P}(\mathcal{O}_\mathbf{s})} \mathbf{p} \in \mathcal{P}(\mathcal{O}_\mathbf{r})$ and we may write (see proposition 727 (p. 441)):

$$(\exists \mathbf{S} \in \mathcal{P}(\mathcal{O}_\mathbf{r}))(\bigcup_{\mathbf{x}\in\mathcal{O}_\mathbf{s}} \mathbf{x} = \bigcup_{\mathbf{p}\in\mathbf{S}} \mathbf{p})$$

then $\mathcal{O}$ verifies equation (2.17) (p. 151).

**Proposition 203** $\mathbf{E}$ a set, $\mathcal{O} \subset \mathcal{P}(\mathbf{E})$ a subset of $\mathcal{P}(\mathbf{E})$ if we have:

$$\mathrm{A} \in \mathcal{O}, \mathrm{B} \in \mathcal{O} \Rightarrow \mathrm{A} \cap \mathrm{B} \in \mathcal{O} \tag{2.22}$$

then $\mathcal{O}$ verifies equation (2.18) (p. 151).

**Proof**
$\mathbf{E}, \mathcal{O}$ defined as in proposition 203, if equation (2.18) (p. 151) is verified for any $\mathcal{O}_\mathbf{s} \subset \mathcal{O}$ with $\mathbf{Card}(\mathcal{O}_\mathbf{s}) = \mathbf{n}$, from equation (2.22), it is true for any $\mathcal{O}_\mathbf{s} \subset \mathcal{O}$ with $\mathbf{Card}(\mathcal{O}_\mathbf{s}) = \mathbf{n}+1$. Since it is true for $\mathbf{n} = 1$ and $\mathbf{n} = 2$, from the recurrence principle (proposition 927 (p. 520)) it is true for any integer.

**Proposition 204** Given a set $\mathbf{E}$, a subset $\mathcal{O}_\mathbf{r} \subset \mathcal{P}(\mathbf{E})$, the set $\mathcal{O}$ constructed from $\mathcal{O}_\mathbf{r}$ by equation (2.20) (p. 152), if $\mathcal{O}_\mathbf{r}$ is such that:

$$(\forall \mathrm{A} \in \mathcal{O}_\mathrm{r})(\forall \mathrm{B} \in \mathcal{O}_\mathrm{r})(\mathrm{A} \cap \mathrm{B} \in \mathcal{O}) \,,$$

then set $\mathcal{O}$ verifies the equation (2.18) (p. 151).

**Proof**
Keeping the term defined as in the proposition, $\mathbf{O_1}$ and $\mathbf{O_2}$ being two elements of $\mathcal{O}$ we have from relation (2.20) (p. 152):

$$\mathrm{i} \in \{1,2\} \Rightarrow (\exists \mathrm{p_i} \subset \mathcal{O}_\mathrm{r})(\mathrm{O_i} = \bigcup_{\mathrm{u} \in \mathrm{p_i}} \mathrm{u}) \,.$$

Using those $\mathbf{p}$, one has from proposition 830 (p. 476):

$$\mathrm{O_1} \cap \mathrm{O_2} = \bigcap_{\mathrm{i} \in \{1,2\}} \bigcup_{\mathrm{u} \in \mathrm{p_i}} \mathrm{u} = \bigcup_{p \in \mathrm{p_1} \times \mathrm{p_2}} \bigcap_{\mathrm{i} \in \{1,2\}} p(\mathrm{i}) \,.$$

Since $\mathbf{i} \in \{1,2\} \Rightarrow p(\mathbf{i}) \in \mathcal{O}_\mathbf{r}$ we have $\bigcap_{\mathrm{i} \in \{1,2\}} p(\mathrm{i}) \in \mathcal{O}$. Then from equation (2.17) (p. 151)

$$\mathrm{O_1} \bigcap \mathrm{O_2} \in \mathcal{O} \,.$$

Then one concludes using proposition 203 (p. 153).

**Definition 134 Induced topology on product of spaces (definition 119 (p. 135)).**

$\mathbf{A}$ being a set, $(\mathbf{A_i} : [1,\mathbf{n}] \subset \mathcal{Z}^+ \to \mathcal{P}(\mathbf{A}))$ a set of topological sets with $\mathcal{O}_\mathbf{i}$ as their set of opens. [10] We define the induced set of opens $\mathcal{O}^\pi$ of $\prod_{\mathrm{i} \in [1,\mathrm{n}]} \mathbf{A_i}$ (see definition 372 (p. 473) and proposition 826 (p. 475)) as:

$$\mathcal{O}^\pi = \mathbf{set}_{\mathrm{O}^\pi}\{(\exists \mathrm{P} \in \mathcal{P}(\prod_{\mathrm{i} \in [1,\mathrm{n}]} \mathcal{O}_\mathrm{i}))(\mathrm{O}^\pi = \bigcup_{\mathrm{x} \in \mathrm{P}} \mathrm{x})\} \,. \tag{2.23}$$

The set $\mathcal{O}^\pi$ is a set of opens because it satisfies the conditions of definition 132 (p. 151) (see A.1.2 (p. 434)):

[10] We could also generate a topology on the product of sets generated by $(\mathbf{A_i} : \mathbf{I} \to \mathcal{P}(\mathbf{A}))$ where $\mathbf{I}$ is whatever set. In what follows, it is enough to replace $[1,\mathbf{n}]$ by $\mathbf{I}$ and every statement remains true, but it is not the topology used when $\mathbf{I}$ is not finite: if $\mathbf{I} = \mathcal{R}$, $(\forall \mathbf{i} \in \mathcal{R})(\mathbf{A_i} = \mathcal{R})$ the product considered is the set of maps $(\mathcal{R} \to \mathcal{R})$ and in that case this induced topology is never used.

- relation (2.16) (p. 151)

$$(\forall i \in [1,n])(A_i \in \mathcal{O}_i)$$
$$(\forall i \in [1,n])(A_i \in \mathcal{O}_i) \Rightarrow \{ \prod_{i\in[1,n]} A_i\} \in \mathcal{P}( \prod_{i\in[1,n]} \mathcal{O}_i) \Rightarrow \prod_{i\in[1,n]} A_i \in \mathcal{O}^\pi$$
$$\prod_{i\in[1,n]} A_i \in \mathcal{O}^\pi .$$

From proposition 827 (p. 475) $\emptyset = \prod_{i\in[1,n]} \emptyset$. Then:

$$(\forall i \in [1,n])(\emptyset \in \mathcal{O}_i)$$
$$(\forall i \in [1,n])(\emptyset \in \mathcal{O}_i) \Rightarrow \{\emptyset\} = \{ \prod_{i\in[1,n]} \emptyset\} \in \mathcal{P}( \prod_{i\in[1,n]} \mathcal{O}_i) \Rightarrow \emptyset \in \mathcal{O}^\pi$$
$$\emptyset \in \mathcal{O}^\pi$$

- relation (2.17) (p. 151)
  It is an application of proposition 202 (p. 152)

- relation (2.18) (p. 151)
  Calling $\mathbf{u}$ and $\mathbf{v}$ two elements of $\prod_{i\in[1,n]} \mathcal{O}_i$. From definitions 126 (p. 144) and 372 (p. 473), one can write $\mathbf{u}$ as

$$(u_1, u_2, ..., u_n),\ (\mathbf{u} : [1,n] \to \prod_{i\in[1,n]} \mathcal{O}_i), (\forall i \in [1,n])(u_i \in \mathcal{O}_i)$$

and also, from our convention at the beginning of this demonstration, $\mathbf{u} = \prod_{i\in[1,n]} u_i$ with $u_i \in \mathcal{O}_i$. We can write the same for $\mathbf{v}$. From proposition 829 (p. 476):

$$\mathbf{u} \cap \mathbf{v} = \prod_{i\in[1,n]} u_i \cap v_i . \tag{2.24}$$

From the definition of $\mathcal{O}_i$. $u_i \cap v_i \in \mathcal{O}_i$. Then we have $\mathbf{u} \cap \mathbf{v} \in \prod_{i\in[1,n]} \mathcal{O}_i$. We conclude using proposition 204 (p. 153)

**Proposition 205** Given a product **P** of a finite number of topological spaces (definition 119 (p. 135)), the projection (definition 121 (p. 136)) of an open subset of **P** in the induced topology of **P** is also open in the induced topology on the projection of **P**.

**Proof**

It s a straightforward consequence of definitions 134 (p. 153) and 121 (p. 136).

**Definition 135 Closed subset.**

**E** is a topological space and **C** a subset of **E**. **C** is said **closed** if its completing set $\mathcal{C}(\mathbf{C})$ is open (see definition 353 (p. 454)). The total set and the empty set $\emptyset$ are both closed and open.

**Proposition 206** Calling $\mathcal{O}^c$ the set of closed subsets of a topological set E. Given a set $\mathcal{O}^c_s \subset \mathcal{O}^c$ we have:

$$\bigcap_{C\in\mathcal{O}^c_s} C \in \mathcal{O}^c \tag{2.25}$$

$$\mathrm{Card}(\mathcal{O}^c_s) \in \mathcal{Z}^+ \Rightarrow \bigcup_{C\in\mathcal{O}^c_s} C \in \mathcal{O}^c . \tag{2.26}$$

**Proof**
Propositions 801 (p. 466) and 802 (p. 466) applied to proposition 201 and equations (2.17) (p. 151), (2.18) (p. 151) give proposition 206 (p. 154).

**Proposition 207** Given a topological space **E** and a closed subset **C** of **E**, any closed subset of the induced topology by **E** on **C** is also a closed subset of **E**.

**Proof**
**B** being a closed subset of the induced topology on **C**, from definition 135 (p. 154), there is an open **O** of **E** which verifies:

$$\mathcal{C}_C(B) = O \cap C$$

then from proposition 767 (p. 455) and from proposition 768 (p. 455):

$$B = \mathcal{C}_C(\mathcal{C}_C(B)) = \mathcal{C}_C(O \cap C) = \mathcal{C}_E(O) \cap C \,.$$

Since, from definition 135 (p. 154), $\mathcal{C}_E(O)$ is closed. **C** being closed, from proposition 206 (p. 154) $\mathcal{C}_E(O) \cap C$ is closed in the **E** topology. Then **B** is closed in the **E** topology.

**Proposition 208** Given a topological space **E** and a closed subset **C** of **E**, any subset of **C** closed in the **E** topology is also a closed subset in the induced topology by **E** on **C**.

**Proof**
**B** being a subset of **C** closed subset the **E** topology, we have

$$\mathcal{C}_C(B) = \mathcal{C}(B)_E \cap C \,.$$

Since $\mathcal{C}(B)_E$ is open in the **E** topology (definition 135 (p. 154)), $\mathcal{C}(B)_E \cap C$ is open in the induced **C** topology (definition 133 (p. 151)). Then from definition 135 (p. 154) **B** is closed in the **C** induced topology.

**Definition 136 Closure of a subset.**

**E** is a topological space with $\mathcal{O}$ as the set of open sets and **A** a subset of **E**. One calls **closure** of **A**, the set $\bar{A}$ of any element for which any open including it includes an element of **A**.

$$\bar{A} = \mathrm{set}_x\{(x \in E), (\forall O \in \mathcal{O})(x \in O \Rightarrow O \cap A \neq \emptyset)\} \,.$$

**Proposition 209** With the same notations as in definition 136 (p. 155) and with $A \subset E$, $B \subset E$ we have:

$$A \subset B \Rightarrow \bar{A} \subset \bar{B} \,.$$

**Proof**
From definition 349 (p. 451)

$$O \cap A \neq \emptyset \Rightarrow (\exists x)(x \in O, x \in A \subset B) \Rightarrow (\exists x)(x \in O, x \in B) \Rightarrow O \cap B \neq \emptyset$$

then $x \in \bar{A} \Rightarrow x \in \bar{B}$ and proposition 209 (p. 155) is true.

**Proposition 210** The closure of a set includes the set.

**Proof**
It is a direct consequence of the definition of the closure of a set.

**Proposition 211** Any closed set includes the closure of any of its subsets.

**Proof**
Given a topological set $\mathbf{E}$ and a closed subset $\mathbf{C} \subset \mathbf{E}$. Any element of $\mathcal{C}(\mathbf{C})$ is in an open: $\mathcal{C}(\mathbf{C})$ (see definitions 135 (p. 154) and 353 (p. 454)). This open ($\mathcal{C}(\mathbf{C})$) does not contain any point of $\mathbf{C}$ and then any point of any subset of $\mathbf{C}$. From definition 136 (p. 155), it cannot be in the closure of any subset of $\mathbf{C}$.

**Proposition 212** The closure of union of a finite set of subsets is the union of their closure.
**Proof**
$\mathbf{E}$ is a topological set with $\mathcal{O}$ its set of opens. Given a finite set $\mathbf{I}$ ($\mathbf{Card(I)} \in \mathcal{Z}^+$) a map ($\mathbf{A_i} : \mathbf{I} \to \mathcal{P}(\mathbf{E})$), from propositions 754 (p. 451), 801 (p. 466), 802 (p. 466), 767 (p. 455), definition 132 (p. 151) equation (2.18), we have:

$$\mathrm{x} \in \mathcal{C}(\overline{\bigcup_{\mathrm{i}\in\mathrm{I}} \mathrm{A_i}}) \Leftrightarrow (\exists \mathrm{O} \in \mathcal{O})(\mathrm{x} \in \mathrm{O}, \mathrm{O} \subset \mathcal{C}(\bigcup_{\mathrm{i}\in\mathrm{I}} \mathrm{A_i}) = \bigcap_{\mathrm{i}\in\mathrm{I}} \mathcal{C}(\mathrm{A_i}))$$
$$\Rightarrow (\forall \mathrm{i} \in \mathrm{I})(\exists \mathrm{O} \in \mathcal{O})(\mathrm{x} \in \mathrm{O}, \mathrm{O} \subset \mathcal{C}(\mathrm{A_i})) \Rightarrow (\forall \mathrm{i} \in \mathrm{I})(\mathrm{x} \in \mathcal{C}(\overline{\mathrm{A_i}}))$$
$$\mathcal{C}(\overline{\bigcup_{\mathrm{i}\in\mathrm{I}} \mathrm{A_i}}) \subset \bigcap_{\mathrm{i}\in\mathrm{I}} \mathcal{C}(\overline{\mathrm{A_i}}) = \mathcal{C}(\bigcup_{\mathrm{i}\in\mathrm{I}} \overline{\mathrm{A_i}})$$
$$\mathrm{x} \in \bigcap_{\mathrm{i}\in\mathrm{I}} \mathcal{C}(\overline{\mathrm{A_i}}) \Leftrightarrow (\forall \mathrm{i} \in \mathrm{I})(\exists \mathrm{O_i} \in \mathcal{O})(\mathrm{x} \in \mathrm{O_i}, \mathrm{O_i} \subset \mathcal{C}(\mathrm{A_i}))$$
$$\Rightarrow (\exists \mathrm{O} = \bigcap_{\mathrm{i}\in\mathrm{I}} \mathrm{O_i} \in \mathcal{O})(\mathrm{x} \in \mathrm{O}, (\forall \mathrm{i} \in \mathrm{I})(\mathrm{O} \subset \mathcal{C}(\mathrm{A_i})))$$
$$\Rightarrow (\exists \mathrm{O} \in \mathcal{O})(\mathrm{x} \in \mathrm{O}, \mathrm{O} \subset \bigcap_{\mathrm{i}\in\mathrm{I}} \mathcal{C}(\mathrm{A_i}) = \mathcal{C}(\bigcup_{\mathrm{i}\in\mathrm{I}} \mathrm{A_i})) \Rightarrow \mathrm{x} \in \mathcal{C}(\overline{\bigcup_{\mathrm{i}\in\mathrm{I}} \mathrm{A_i}})$$
$$\bigcap_{\mathrm{i}\in\mathrm{I}} \mathcal{C}(\overline{\mathrm{A_i}}) \subset \mathcal{C}(\overline{\bigcup_{\mathrm{i}\in\mathrm{I}} \mathrm{A_i}}) \Leftrightarrow \mathcal{C}(\bigcup_{\mathrm{i}\in\mathrm{I}} \overline{\mathrm{A_i}}) \subset \mathcal{C}(\overline{\bigcup_{\mathrm{i}\in\mathrm{I}} \mathrm{A_i}})$$
$$\mathcal{C}(\bigcup_{\mathrm{i}\in\mathrm{I}} \overline{\mathrm{A_i}}) = \mathcal{C}(\overline{\bigcup_{\mathrm{i}\in\mathrm{I}} \mathrm{A_i}})$$
$$\bigcup_{\mathrm{i}\in\mathrm{I}} \overline{\mathrm{A_i}} = \overline{\bigcup_{\mathrm{i}\in\mathrm{I}} \mathrm{A_i}} \,.$$

**Definition 137 Interior of a subset.**

The interior part $\mathcal{I}(\mathbf{A})$ of a subset $\mathbf{A}$ of a topological space $\mathbf{E}$ is the subset of $\mathbf{A}$ which contains the elements for which there is an open subset of $\mathbf{A}$ including that element. $\mathcal{O}$ being the set of opens of $\mathbf{E}$, we should have:

$$\mathrm{x} \in \mathcal{I}(\mathrm{A}) \Leftrightarrow (\exists \mathrm{O} \in \mathcal{O})(\mathrm{x} \in \mathrm{O}, \mathrm{O} \subset \mathrm{A}) \,.$$

**Proposition 213** Having a subset $\mathbf{A}$ of a topological space $\mathbf{E}$, we have the identity (see definition 353 (p. 454)):

$$\mathcal{I}(\mathrm{A}) = \mathcal{C}(\overline{\mathcal{C}(\mathrm{A})}) \,.$$

See definitions 137 (p. 156), 136 (p. 155).
**Proof**
Setting $\mathbf{B} = \mathcal{C}(\overline{\mathcal{C}(\mathbf{A})})$, $\mathcal{O}$ the set of opens of $\mathbf{E}$ we have:

$$\mathrm{b} \in \mathrm{B} \Leftrightarrow (\exists \mathrm{O} \in \mathcal{O})(\mathrm{b} \in \mathrm{O}, \mathrm{O} \cap \mathcal{C}(\mathrm{A}) = \emptyset) \Leftrightarrow (\exists \mathrm{O} \in \mathcal{O})(\mathrm{b} \in \mathrm{O}, \mathrm{O} \subset \mathrm{A}) \,.$$

**Proposition 214** In a topological set, the interior of a subset is open.

**Proof**

Calling $\mathcal{I}(\mathbf{A})$ the interior part of a subset **A** of a topological space **E**, calling $\mathcal{O} \subset \mathcal{P}(\mathbf{E})$ the set of opens of **E**, we have from definition 137:

$$(\mathbf{p} \in \mathcal{I}(\mathbf{A})) \Leftrightarrow (\exists \mathbf{B} \in \mathcal{O})(\mathbf{B} \subset \mathbf{A}, \mathbf{p} \in \mathbf{B}) .$$

From definition 137, we also have:

$$(\forall \mathbf{B} \in \mathcal{O})(\mathbf{B} \subset \mathbf{A} \Rightarrow (\forall \mathbf{x} \in \mathbf{B})(\mathbf{x} \in \mathcal{I}(\mathbf{A})) \Rightarrow \mathbf{B} \subset \mathcal{I}(\mathbf{A})) .$$

It allows to define a map, according to definition 112 (p. 131), 4:

$$(\mathbf{B}(\mathbf{p}) : \mathcal{I}(\mathbf{A}) \rightarrow \mathcal{P}(\mathcal{I}(\mathbf{A}))) .$$

From definition 362 (p. 462) we have:

$$\bigcup_{\mathbf{p} \in \mathcal{I}(\mathbf{A})} \mathbf{B}(\mathbf{p}) \subset \mathcal{I}(\mathbf{A}) .$$

By definition we have:

$$(\forall \mathbf{p} \in \mathcal{I}(\mathbf{A}))(\mathbf{p} \in \mathbf{B}(\mathbf{p})) \Rightarrow \mathcal{I}(\mathbf{A}) \subset \bigcup_{\mathbf{p} \in \mathcal{I}(\mathbf{A})} \mathbf{B}(\mathbf{p}) .$$

Then applying proposition 754 (p. 451)

$$\mathcal{I}(\mathbf{A}) = \bigcup_{\mathbf{p} \in \mathcal{I}(\mathbf{A})} \mathbf{B}(\mathbf{p}) .$$

Then from definition 132 equation (2.17) and proposition 201 (p. 152) $\mathcal{I}(\mathbf{A})$ is open.

**Proposition 215** In a topological set, the interior of an open is the open itself.

**Proof**

**A** being an open, any point of **A** is in an open included in **A**: It is **A** itself.

**Proposition 216** **A** being a subset of a topological set, we have:

$$\mathcal{I}(\mathcal{I}(\mathbf{A})) = \mathcal{I}(\mathbf{A}) .$$

**Proof**

Since from proposition 214 $\mathcal{I}(\mathbf{A})$ is open, from proposition 215 we have proposition 216.

**Definition 138 Boundary of a subset.**

**A** being a subset of a topological set, the boundary $\mathcal{B}(\mathbf{A})$ of **A** is the subset of $\bar{\mathbf{A}}$, the closure of **A** (definition 136 (p. 155)), which contains all elements in $\bar{\mathbf{A}}$ and outside the inner part $\mathcal{I}(\mathbf{A})$ (definition 137 (p. 156)) of **A**. It is the complementary of the interior part in $\bar{\mathbf{A}}$ (see definition 353 (p. 454) and proposition 213 (p. 156))

$$\mathcal{B}(\mathbf{A}) = \mathcal{C}_{\bar{\mathbf{A}}}(\mathcal{I}(\mathbf{A})) = \bar{\mathbf{A}} - \mathcal{I}(\mathbf{A}) = \bar{\mathbf{A}} \cap \mathcal{C}(\mathcal{I}(\mathbf{A})) = \bar{\mathbf{A}} \cap \overline{\mathcal{C}(\mathbf{A})} .$$

**Proposition 217** $\mathcal{B}(\mathbf{A})$ being the boundary of a subset $\mathbf{A}$ of a topological set, any open containing a point of $\mathcal{B}(\mathbf{A})$ contains a point of $\mathbf{A}$ and a point of $\mathcal{C}(\mathbf{A})$ (definition 353 (p. 454)).

**Proof**
From definition 138, any point $\mathbf{p}$ in $\mathcal{B}(\mathbf{A})$ is in $\bar{\mathbf{A}}$ closure of $\mathbf{A}$ then from definition 136 (p. 155) any open containing $\mathbf{p}$ contains a point of $\mathbf{A}$. From definition 138, $\mathbf{p}$ cannot be in $\mathcal{I}(\mathbf{A})$ interior of $\mathbf{A}$. From definition 137, any open containing $\mathbf{p}$ is not subset of $\mathbf{A}$ then it has a point in $\mathcal{C}(\mathbf{A})$.

**Proposition 218** $\mathbf{A}$ being a subset of a topological set $\mathbf{E}$, any point of $\mathbf{E}$ for which any open containing it contains a point of $\mathbf{A}$ and a point of $\mathcal{C}(\mathbf{A})$ is in the boundary of $\mathbf{A}$.

**Proof**
If $\mathbf{p} \in \mathbf{E}$ is such that any open containing $\mathbf{p}$ contains a point of $\mathbf{A}$, we have, from definition 136 (p. 155), $\mathbf{p} \in \bar{\mathbf{A}}$. But if there is no open containing $\mathbf{p}$ which is a subset of $\mathbf{A}$ $\mathbf{p} \notin \mathcal{I}(\mathbf{A})$ (definition 137 (p. 156)). Then from definition 138 (p. 157) $\mathbf{p} \in \mathcal{B}(\mathbf{A})$.

**Proposition 219** The closure of a set is closed.

**Proof**
$\mathbf{E}$ is a topological space with $\mathcal{O}$ as the set of open sets and $\mathbf{A}$ a subset of $\mathbf{E}$. From the relation (2.16) (p. 151) and from the definition 136 (p. 155) any element $\mathbf{c}$ of $\mathcal{C}(\bar{\mathbf{A}})$ is contained in an open $\mathbf{O_c}$ with no element of $\mathbf{A}$. $\mathbf{O_c}$ does not contain any element of $\bar{\mathbf{A}}$ because any open containing an element of $\bar{\mathbf{A}}$ contains an element of $\mathbf{A}$.
In a more logical way, see A.3 (p. 441) and, in A.1 page 433, A.1.4, 5, A.1.4, 20, A.1.4, 23, definitions 329 (p. 434) and 330 (p. 434):

$$\begin{aligned}
\mathrm{c} \in \mathcal{C}(\bar{\mathbf{A}}) &\Leftrightarrow (\mathrm{c} \in \mathbf{E}, \neg(\mathrm{c} \in \bar{\mathbf{A}})) \\
&\Leftrightarrow (\mathrm{c} \in \mathbf{E}, \neg(\forall \mathrm{O} \in \mathcal{O})(\mathrm{c} \in \mathrm{O} \Rightarrow \neg(\mathrm{O} \cap \mathbf{A} = \emptyset))) \\
&\Leftrightarrow (\mathrm{c} \in \mathbf{E}, \neg\neg(\exists \mathrm{O} \in \mathcal{O})\neg(\neg(\mathrm{c} \in \mathrm{O}) \vee \neg(\mathrm{O} \cap \mathbf{A} = \emptyset))) \\
&\Leftrightarrow (\exists \mathrm{O_c} \in \mathcal{O})(\mathrm{c} \in \mathrm{O_c}, \mathrm{O_c} \cap \mathbf{A} = \emptyset) \ .
\end{aligned}$$

So:

$$(\forall \mathrm{c} \in \mathcal{C}(\bar{\mathbf{A}}))(\exists \mathrm{O_c})(\mathrm{c} \in \mathrm{O_c}, \mathrm{O_c} \subset \mathcal{C}(\bar{\mathbf{A}}))$$

and from proposition 754 (p. 451) and using convention 26 (p. 441):

$$\mathcal{C}(\bar{\mathbf{A}}) = \bigcup_{\mathrm{c} \in \mathcal{C}(\bar{\mathbf{A}})} \mathrm{O_c}$$

$\mathcal{C}(\bar{\mathbf{A}})$ is the union of opens and then, from equation (2.17) (p. 151), it is open. From definition 132 (p. 151), $\mathcal{C}(\bar{\mathbf{A}})$ open means that $\bar{\mathbf{A}}$ is closed.

**Proposition 220** A closed set is its own closure.

**Proof**
$\mathbf{A}$ is a closed set and $\bar{\mathbf{A}}$ its closure. $\mathcal{C}(\mathbf{A})$ is open and does not contain any point of $\mathbf{A}$. Any element of $\mathcal{C}(\mathbf{A})$ is then not in $\bar{\mathbf{A}}$ and $\mathcal{C}(\mathbf{A}) \subset \mathcal{C}(\bar{\mathbf{A}}) \Rightarrow \bar{\mathbf{A}} \subset \mathbf{A}$ (proposition 769 (p. 455)). From proposition 210 (p. 155) $\mathbf{A} \subset \bar{\mathbf{A}}$ then $\mathbf{A} = \bar{\mathbf{A}}$ from proposition 754 (p. 451) (see also proposition 211 (p. 155)).

**Proposition 221** In a topological set, the boundary of the union of a finite number of disjoint subsets (definition 369 (p. 470)) is the union of their boundary.

**Proof**
$\mathbf{E}$ is a topological set with $\mathcal{O}$ its set of opens. $\mathbf{A} \in \mathcal{O}$ and $\mathbf{B} \in \mathcal{O}$ with $\mathbf{A} \cap \mathbf{B} = \emptyset$, we have:

$$\mathbf{A} \cap \bar{\mathbf{B}} = \emptyset,\ \mathbf{B} \cap \bar{\mathbf{A}} = \emptyset,\ \mathcal{C}(\mathbf{A}) \cap \bar{\mathbf{B}} = \bar{\mathbf{B}},\ \mathcal{C}(\mathbf{B}) \cap \bar{\mathbf{A}} = \bar{\mathbf{A}}$$

because from definition 136 (p. 155)

$$(\mathbf{x} \in \mathbf{A} \in \mathcal{O}, \mathbf{A} \cap \mathbf{B} = \emptyset) \Rightarrow \mathbf{x} \notin \bar{\mathbf{B}}$$

the same for $\mathbf{B}$ then from propositions 798 (p. 465), 800 (p. 466), we have:

$$\begin{aligned}&\mathcal{B}(\mathbf{A}) \cup \mathcal{B}(\mathbf{B}) = (\bar{\mathbf{A}} \cap \mathcal{C}(\mathbf{A})) \cup (\bar{\mathbf{B}} \cap \mathcal{C}(\mathbf{B})) = (\bar{\mathbf{A}} \cap \mathcal{C}(\mathbf{A}) \cap \mathcal{C}(\mathbf{B})) \cup (\bar{\mathbf{B}} \cap \mathcal{C}(\mathbf{B}))\\ &= \mathcal{C}(\mathbf{B}) \cap ((\bar{\mathbf{A}} \cap \mathcal{C}(\mathbf{A})) \cup \bar{\mathbf{B}})\\ &= \mathcal{C}(\mathbf{B}) \cap ((\bar{\mathbf{A}} \cap \mathcal{C}(\mathbf{A})) \cup \bar{\mathbf{B}} \cap \mathcal{C}(\mathbf{A})) = (\mathcal{C}(\mathbf{B}) \cap \mathcal{C}(\mathbf{A})) \cap (\bar{\mathbf{A}} \cup \bar{\mathbf{B}})\\ &= \mathcal{C}(\mathbf{A} \cup \mathbf{B}) \cap (\bar{\mathbf{A}} \cup \bar{\mathbf{B}}) = \mathcal{B}(\mathbf{A} \cup \mathbf{B})\ .\end{aligned}$$

Now given a finite subset $\mathbf{F}$ of $\mathcal{P}(\mathbf{E})$, if proposition 221 is true for $\mathbf{Card}(\mathbf{F}) = \mathbf{n} \in \mathcal{Z}^+$ it is true for $\mathbf{Card}(\mathbf{F}) = \mathbf{n}+1$: Taking $\mathbf{F}$ such that $\mathbf{Card}(\mathbf{F}) = \mathbf{n}+1$, we have since $\mathbf{F} \neq \emptyset$, $\exists \mathbf{B} \in \mathbf{F}$ and taking that $\mathbf{B} = \tau_{\mathbf{x}}(\mathbf{x} \in \mathbf{F})$ (see section A.3 (p. 441) convention 26 (p. 441)), we have from proposition 796 (p. 464) since $\mathbf{F} = (\mathbf{F} - \{\mathbf{B}\}) \cup \{\mathbf{B}\}$:

$$\bigcup_{\mathbf{x} \in \mathbf{F}} \mathbf{x} = \bigcup_{\mathbf{x} \in \mathbf{F} - \{\mathbf{B}\}} \mathbf{x} \cup \bigcup_{\mathbf{x} \in \{\mathbf{B}\}} \mathbf{x} = \bigcup_{\mathbf{x} \in \mathbf{F} - \{\mathbf{B}\}} \mathbf{x} \cup \mathbf{B}\ .$$

From what we suppose and what we just demonstrated, we have (proposition 796 (p. 464)):

$$\mathcal{B}(\bigcup_{\mathbf{x} \in \mathbf{F}} \mathbf{x}) = \bigcup_{\mathbf{x} \in \mathbf{F} - \{\mathbf{B}\}} \mathcal{B}(\mathbf{x}) \cup \mathcal{B}(\mathbf{B}) = \bigcup_{\mathbf{x} \in \mathbf{F} - \{\mathbf{B}\}} \mathcal{B}(\mathbf{x}) \cup \bigcup_{\mathbf{x} \in \{\mathbf{B}\}} \mathcal{B}(\mathbf{x}) = \bigcup_{\mathbf{x} \in \mathbf{F}} \mathcal{B}(\mathbf{x})\ .$$

Then proposition 221 is true for $\mathbf{Card}(\mathbf{F}) = \mathbf{n}+1$. We conclude using proposition 927 (p. 520) (recurrence principle).

**Definition 139 Separated topology (Hausdorff space).**

$\mathcal{O}$ being the set of opens of a topological set $\mathbf{E}$, the set $\mathbf{E}$ is said to be **separated**, if for any couple of elements $\mathbf{a}$ and $\mathbf{b}$ of $\mathbf{E}$, there are two disjoint opens containing them:

$$\begin{aligned}&(\forall \mathbf{u} \in \mathbf{E})(\forall \mathbf{v} \in \mathbf{E})((\mathbf{u} \neq \mathbf{v})\\ &\Rightarrow (\exists \mathbf{O}_\mathbf{u} \in \mathcal{O})(\exists \mathbf{O}_\mathbf{v} \in \mathcal{O})(\mathbf{u} \in \mathbf{O}_\mathbf{u}, \mathbf{v} \in \mathbf{O}_\mathbf{v}, \mathbf{O}_\mathbf{u} \cap \mathbf{O}_\mathbf{v} = \emptyset))\ .\end{aligned} \tag{2.27}$$

**Proposition 222** Any subset of a separated topological set $\mathbf{E}$ with cardinal $\mathbf{1}$ is closed.

**Proof**
Given the set $\mathcal{O} \subset \mathcal{P}(\mathbf{E})$ of opens of $\mathbf{E}$ and a subset $\{\mathbf{x}\} \in \mathbf{E}$, from definition 139 (p. 159) we have a map:

$$(\mathbf{O}_\mathbf{y} : \mathbf{E} - \{\mathbf{x}\} = \mathcal{C}(\{\mathbf{x}\}) \to \mathcal{O})$$

such as $\mathbf{y} \in \mathbf{O}_\mathbf{y}, \mathbf{x} \notin \mathbf{O}_\mathbf{y}$). It gives (see definition 362 (p. 462), proposition 754 (p. 451)):

$$\begin{aligned}&(\forall \mathbf{y} \in \mathcal{C}(\{\mathbf{x}\}))(\mathbf{y} \in \mathbf{O}_\mathbf{y}, \mathbf{x} \notin \mathbf{O}_\mathbf{y}) \Rightarrow (\bigcup_{\mathbf{y} \in \mathcal{C}(\{\mathbf{x}\})} \mathbf{O}_\mathbf{y} \subset \mathcal{C}(\{\mathbf{x}\}), \mathbf{x} \notin \bigcup_{\mathbf{y} \in \mathcal{C}(\{\mathbf{x}\})} \mathbf{O}_\mathbf{y})\\ &\Rightarrow \bigcup_{\mathbf{y} \in \mathcal{C}(\{\mathbf{x}\})} \mathbf{O}_\mathbf{y} = \mathcal{C}(\{\mathbf{x}\})\end{aligned}$$

then, from definition 132 (p. 151) - equation (2.17), $\mathcal{C}(\{\mathbf{x}\}) \in \mathcal{O}$.

**Definition 140 Compact set.**

**E** is a topological set with $\mathcal{O}$ as the set of the open subsets of **E**. **C** subset of **E** is said to be compact if:

$$(\forall \mathcal{O}_s \subset \mathcal{O})(\mathrm{C} \subset \bigcup_{\mathrm{O} \in \mathcal{O}_s} \mathrm{O} \Rightarrow \exists \mathrm{B}_s(\mathrm{B}_s \subset \mathcal{O}_s, \mathrm{Card}(\mathrm{B}_s) \in \mathcal{Z}^+, \mathrm{C} \subset \bigcup_{\mathrm{O} \in \mathrm{B}_s} \mathrm{O})) . \quad (2.28)$$

In other words, **C** is compact if from any cover of **C** by a set of opens, one can extract a finite number of opens covering **C**.

**Proposition 223** In a topological set, any closed subset of a compact set is compact.

**Proof**
**C** being a compact subset (definition 140 (p. 160)) of a topological set **E** (definition 132 (p. 151)) with $\mathcal{O}$ as the set of the open subsets of **E**, $\mathbf{A} \subset \mathbf{C}$ being closed (definition 135 (p. 154)), let us suppose that we have $\mathbf{S} \subset \mathcal{O}$ such that: $\mathbf{A} \subset \bigcup_{s \in \mathbf{S}} s$. From definitions 353 (p. 454), 362 (p. 462), we have: $\mathbf{E} = \mathbf{A} \cup \mathcal{C}(\mathbf{A})$ and $\mathbf{E} = \bigcup_{s \in \mathbf{S}} s \cup \mathcal{C}(\mathbf{A})$. Then $\mathbf{C} \subset \bigcup_{s \in \mathbf{S}} s \cup \mathcal{C}(\mathbf{A}) = \mathbf{E}$. Since **A** is closed from definition 135 (p. 154), $\mathcal{C}(\mathbf{A})$ is open. **C** being compact, from definition 140 (p. 160), we have $\mathbf{S}_1 \subset \mathbf{S} \cup \{\mathcal{C}(\mathbf{A})\} \subset \mathcal{O}$ with $\mathbf{Card}(\mathbf{S}_1) \in \mathcal{Z}^{0+}$ and $\mathbf{C} \subset \bigcup_{s \in \mathbf{S}_1} s$ Calling $\mathbf{S}_2 = \mathcal{C}_{\mathbf{S}_1}(\{\mathcal{C}(\mathbf{A})\}) = \mathbf{S}_1 \cap \mathcal{C}(\{\mathcal{C}(\mathbf{A})\}) \subset \mathbf{S}$, from propositions 761 (p. 453), 762 (p. 453), 763 (p. 453), 764 (p. 454), we have:

$$\mathbf{C} \subset \bigcup_{s \in \mathbf{S}_2} s \cup \mathcal{C}(\mathbf{A})$$

then

$$\mathbf{A} = \mathbf{C} \cap \mathbf{A} \subset (\bigcup_{s \in \mathbf{S}_2} s \cup \mathcal{C}(\mathbf{A})) \cap \mathbf{A} = \bigcup_{s \in \mathbf{S}_2} s \cap \mathbf{A} \subset \bigcup_{s \in \mathbf{S}_2} s .$$

Since $\mathbf{S}_2 \subset \mathbf{S}, \mathbf{S}_2 \subset \mathbf{S}_1$ we have $\mathbf{Card}(\mathbf{S}_2) \leq \mathbf{Card}(\mathbf{S}_1)$. Then from proposition 918 (p. 516), $\mathbf{Card}(\mathbf{S}_2) \in \mathcal{Z}^{0+}$ and proposition 223 (p. 160) is true.

**Proposition 224** The empty set is compact.

**Proof**
Since the empty set is subset of any set (proposition 775 (p. 457)), it is trivially compact.

**Proposition 225** In a topological set, the union of a finite (integer) number of compact subsets is compact.

**Proof**
It is a direct application of proposition 941 (p. 526) which says that the union of a finite number of finite sets is finite.

**Proposition 226** In a separated topological set (definition 139 (p. 159)), any compact subset is closed.

**Proof**
**E** is a topological separated set, $\mathcal{O}$ the set of the opens of **E**, **C** a compact subset of **E**. Let us first prove the following statement:

**For any element x of E not in C, there is an open $\mathrm{O}^1_x$ including x with no element of C:**

Since $\mathbf{E}$ is separated, there is a map $\mathbf{O}_{\mathbf{y}}^{\mathbf{k},\mathbf{x}}$ ($\mathbf{k} \in \{1,2\}$, $\mathbf{x} \in \mathcal{C}(\mathbf{C})$, $\mathbf{y} \in \mathbf{C}$):

$$(\mathbf{O}_{\mathbf{y}}^{\mathbf{k},\mathbf{x}} : \{1,2\} \times \mathbf{C} \times \mathcal{C}(\mathbf{C}) \to \mathcal{O})$$

because we have (definition 139 (p. 159)):

$$(\forall \mathbf{y} \in \mathbf{C})(\forall \mathbf{x} \in \mathcal{C}(\mathbf{C}))(\exists \mathbf{O}_{\mathbf{y}}^{1,\mathbf{x}})(\exists \mathbf{O}_{\mathbf{y}}^{2,\mathbf{x}})(\mathbf{x} \in \mathbf{O}_{\mathbf{y}}^{1,\mathbf{x}} \in \mathcal{O}, \mathbf{y} \in \mathbf{O}_{\mathbf{y}}^{2,\mathbf{x}} \in \mathcal{O}, \mathbf{O}_{\mathbf{y}}^{1,\mathbf{x}} \cap \mathbf{O}_{\mathbf{y}}^{2,\mathbf{x}} = \emptyset) .$$

Then

$$(\forall \mathbf{x} \in \mathcal{C}(\mathbf{C}))(\mathbf{C} \subset \bigcup_{\mathbf{y} \in \mathbf{C}} \mathbf{O}_{\mathbf{y}}^{2,\mathbf{x}})$$

since

$$(\forall \mathbf{x} \in \mathcal{C}(\mathbf{C}))(\forall \mathbf{y} \in \mathbf{C})(\mathbf{x} \in \mathbf{O}_{\mathbf{y}}^{1,\mathbf{x}}, \mathbf{y} \in \mathbf{O}_{\mathbf{y}}^{2,\mathbf{x}}) .$$

See section A.5.1 for the definition and the properties of the various symbols used here. Then since $\mathbf{C}$ is compact and since all $\mathbf{O}_{\mathbf{y}}^{2,\mathbf{x}}$ are open:

$$(\forall \mathbf{x} \in \mathcal{C}(\mathbf{C}))(\exists \mathbf{B}_{\mathbf{x}} \subset \mathbf{C})(\mathbf{Card}(\mathbf{B}_{\mathbf{x}}) \in \mathcal{Z}^+)(\mathbf{C} \subset \bigcup_{\mathbf{y} \in \mathbf{B}_{\mathbf{x}}} \mathbf{O}_{\mathbf{y}}^{2,\mathbf{x}})$$

and we have:

$$(\forall \mathbf{x} \in \mathcal{C}(\mathbf{C}))(\forall \mathbf{y} \in \mathbf{B}_{\mathbf{x}})(\bigcap_{\mathbf{z} \in \mathbf{B}_{\mathbf{x}}} \mathbf{O}_{\mathbf{z}}^{1,\mathbf{x}} \subset \mathbf{O}_{\mathbf{y}}^{1,\mathbf{x}}, \mathbf{O}_{\mathbf{y}}^{1,\mathbf{x}} \bigcap \mathbf{O}_{\mathbf{y}}^{2,\mathbf{x}} = \emptyset)$$

$$\Rightarrow (\forall \mathbf{x} \in \mathcal{C}(\mathbf{C}))((\bigcap_{\mathbf{y} \in \mathbf{B}_{\mathbf{x}}} \mathbf{O}_{\mathbf{y}}^{1,\mathbf{x}}) \bigcap \bigcup_{\mathbf{y} \in \mathbf{B}_{\mathbf{x}}} \mathbf{O}_{\mathbf{y}}^{2,\mathbf{x}} = \emptyset)$$

$$\Rightarrow (\forall \mathbf{x} \in \mathcal{C}(\mathbf{C}))(\bigcap_{\mathbf{y} \in \mathbf{B}_{\mathbf{x}}} \mathbf{O}_{\mathbf{y}}^{1,\mathbf{x}}) \bigcap \mathbf{C} = \emptyset) .$$

Since $\mathbf{B}_{\mathbf{x}}$ is finite, $\bigcap_{\mathbf{y} \in \mathbf{B}_{\mathbf{x}}} \mathbf{O}_{\mathbf{y}}^{1,\mathbf{x}}$ is open from the definition of open set: (equation (2.18) (p. 151)). Then calling:

$$\mathbf{O}^{1,\mathbf{x}} = \bigcap_{\mathbf{y} \in \mathbf{B}_{\mathbf{x}}} \mathbf{O}_{\mathbf{y}}^{1,\mathbf{x}} \in \mathcal{O}$$

$$(\forall \mathbf{x} \in \mathcal{C}(\mathbf{C}))(\exists \mathbf{O}^{1,\mathbf{x}} \in \mathcal{O})(\mathbf{x} \in \mathbf{O}^{1,\mathbf{x}}, \mathbf{O}^{1,\mathbf{x}} \bigcap \mathbf{C} = \emptyset) .$$

**Let us prove now proposition 226 (p. 160).**
Calling $\mathbf{O}^1 = \bigcup_{\mathbf{x} \in \mathcal{C}(\mathbf{C})} \mathbf{O}^{1,\mathbf{x}}$, we have:

$$\mathcal{C}(\mathbf{C}) \subset \bigcup_{\mathbf{x} \in \mathcal{C}(\mathbf{C})} \mathbf{O}^{1,\mathbf{x}} \Rightarrow \mathcal{C}(\mathbf{C}) \subset \mathbf{O}^1$$

and

$$\mathbf{O}^1 \bigcap \mathbf{C} = \emptyset \Rightarrow \mathbf{O}^1 \subset \mathcal{C}(\mathbf{C})$$

from the axiom given by proposition 754 (p. 451) in A.5.1:

$$\mathbf{O}^1 = \mathcal{C}(\mathbf{C}) .$$

Since $\mathbf{O}^{1,\mathbf{x}}$ are open, $\mathbf{O}^1$ is open from (equation (2.17) (p. 151)) so $\mathcal{C}(\mathbf{C})$ is also open. Then $\mathbf{C}$ is closed from definition 135 (p. 154).

**Proposition 227** **A** being a topological set, **B** a subset of **A**, **C** a subset of **B**, if **C** is compact in the **A** topology, **C** is also compact in the induced topology (definition 133 (p. 151)) of **B** and vice versa: If **C** is compact in the induced topology of **B**, **C** is also compact in the **A** topology.

**Proof**
We call $\mathcal{O}_A$ the set of opens of **A**, $\mathcal{O}_B$ the set of opens of **B**. By definition (see also 112 (p. 131), 4) we have:

$$O_b \in \mathcal{O}_B \Leftrightarrow (\exists O_a(O_b) \in \mathcal{O}_A)(O_b = O_a \cap B) .$$

If **C** is compact, from definition 140 (p. 160) and since $C = C \cap B$

$$(E \subset \mathcal{O}_A, C \subset \bigcup_{O \in E} O) \Rightarrow (\exists n \in \mathcal{Z}^+)(\exists O_i : [1,n] \to E)(C \subset \bigcup_{i \in [1,n]} O_i) .$$

Then, from propositions 794 (p. 464), 798 (p. 465), 762 (p. 453), we have:

$$E_B \subset \mathcal{O}_B, C \subset \bigcup_{O_B \in E_B} O_B = \bigcup_{O_B \in E_B} O_a(O_B) \cap B =$$

$$B \cap \bigcup_{O_B \in E_B} O_a(O_B) \subset \bigcup_{O_B \in E_B} O_a(O_B) \Rightarrow (\exists (O_i : [1,n] \to E_B))(C \subset \bigcup_{i \in [1,n]} O_a(O_i))$$

$$\Rightarrow C = C \cap B \subset \bigcup_{i \in [1,n]} O_a(O_i) \cap B = \bigcup_{i \in [1,n]} O_i .$$

Then, if **C** is compact in the **A** topology, it is also compact in the **B** topology. As well we have:

$$(E_A \subset \mathcal{O}_A, C \subset \bigcup_{O_A \in E_A} O_A) \Rightarrow C = C \cap B \subset \bigcup_{O_A \in E_A} O_A \cap B$$

$$\Rightarrow (\exists (O_i : [1,n] \to E_A))(C \subset \bigcup_{i \in [1,n]} O_i \cap B = B \cap \bigcup_{i \in [1,n]} O_i \subset \bigcup_{i \in [1,n]} O_i) .$$

Then, if **C** is compact in the **B** induced topology, it is also compact in the **A** topology.

**Proposition 228** Given a non-empty set of compact subsets of a topological separated set (definition 139 (p. 159)), their intersection is compact.

**Proof**
Given a topological separated set **A** and a subset $S \neq \emptyset$ of $\mathcal{P}(A)$ such that any element of **S** is compact, let us consider the subset

$$B = \bigcap_{C \in S} C \subset A .$$

From propositions 206 (p. 154) and 226 (p. 160), **B** is closed. Since $S \neq \emptyset$, $(\exists D \in S)$ with $D = \tau_x x \in S$, we have from definition 363 (p. 463) $B \subset D$. Then from proposition 223 (p. 160), we have proposition 228.

**Proposition 229** The product of two compact sets is compact.

**Proof**

$\mathbf{A}$ and $\mathbf{B}$ are two topological sets, $\mathcal{O}_a$ and $\mathcal{O}_b$ their sets of opens and $\mathcal{O}^\pi$ the set of opens of $\mathbf{A} \times \mathbf{B}$:

$$\mathcal{O}^\pi = \mathrm{set}_{O^\pi}\{(\exists P \in \mathcal{P}(\mathcal{O}_a \times \mathcal{O}_b))(O^\pi = \bigcup_{x \in P} x)\}$$

see definition 134 (p. 153). Given $\mathbf{A}^c \subset \mathbf{A}$ and $\mathbf{B}^c \subset \mathbf{B}$ two compact subsets of $\mathbf{A}$ and $\mathbf{B}$. Let us consider a subset $\mathbf{S} \subset \mathcal{O}^\pi$ such that $\mathbf{A}^c \times \mathbf{B}^c \subset \bigcup_{x \in S} x$.
By definition of $\mathbf{x}$

$$x \in \mathcal{O}^\pi \Leftrightarrow (\exists P(x) \subset \mathcal{O}_a \times \mathcal{O}_b)(x = \bigcup_{y \in P(x)} y \subset A \times B) \tag{2.29}$$

(here $\mathbf{y}$ is an element of $\mathcal{O}_a \times \mathcal{O}_b$ ).
From proposition 796 (p. 464), one has, calling $\mathbf{S}' = \bigcup_{x \in S} P(x) \subset (\mathcal{O}_a \times \mathcal{O}_b)$:

$$\bigcup_{x \in S} x = \bigcup_{x \in S}(\bigcup_{y \in P(x)} y) = \bigcup_{y \in \bigcup_{x \in S} P(x)} y = \bigcup_{y \in S'} y\,. \tag{2.30}$$

One has $\mathbf{A}^c \times \mathbf{B}^c \subset \bigcup_{y \in S'} y$ and

$$(\forall u \in A^c)(\{u\} \times B^c \subset \bigcup_{y \in S'} y). \tag{2.31}$$

Since $\mathbf{y} = \mathbf{y}_a \times \mathbf{y}_b$ with $\mathbf{y}_a \in \mathcal{O}_a$, $\mathbf{y}_b \in \mathcal{O}_b$, we can restrict equation (2.31) (p. 163) to the union when $\mathbf{u} \in \mathbf{y}_a$. Calling:

$$S'^a_u = \mathrm{set}_y\{y = (y_a, y_b) \in \mathcal{O}_a \times \mathcal{O}_b, y \in S', u \in y_a\}\,.$$

We have (see propositions 798 (p. 465), 829 (p. 476), 763 (p. 453), 829 (p. 476), 826 (p. 475), 827 (p. 475) and 806 (p. 468)):

$$(\{u\} \times B^c \subset A^c \times B^c \subset \bigcup_{y \in S'} y,\ \ \{u\} \times B^c \subset \{u\} \times B)$$
$$\Rightarrow \{u\} \times B^c \subset (\{u\} \times B) \cap \bigcup_{y \in S'} y\,.$$

We also have:

$$(\{u\} \times B) \cap \bigcup_{y \in S'} y = \bigcup_{y \in S'} ((\{u\} \times B) \cap y)$$
$$= \bigcup_{(y_a \times y_b) \in S'} ((\{u\} \times B) \cap (y_a \times y_b)) = \bigcup_{(y_a \times y_b) \in S'} ((y_a \cap \{u\}) \times (y_b \cap B))$$
$$= \bigcup_{(y_a \times y_b) \in S'} ((y_a \cap \{u\}) \times y_b) \subset \bigcup_{(y_a \times y_b) \in S'^a_u} (y_a \times y_b)\,.$$

It gives:

$$(\forall u \in A^c)(\{u\} \times B^c \subset \bigcup_{y \in S'^a_u} y)\,.$$

All the associated $\mathbf{y}_b$ cover $\mathbf{B}^c$:

$$B^c = \bigcup_{i=(\alpha,\beta) \in S'^a_u} \beta\,.$$

From $\mathbf{S'^a_u}$, one can extract a finite set ${}^f\mathbf{S'^a_u}$ which also covers $\mathbf{B^c}$.
The union ${}^{\cup}\mathbf{y^u_a} = \bigcup_{\mathbf{i}=(\alpha,\beta)\in {}^f\mathbf{S'^a_u}} \alpha$ of the finite sets of corresponding $\mathbf{y_a}$ is open (definition 132 (p. 151) equation (2.17)) and from definition 351 (p. 453), it contains $\mathbf{u}$ because $(\forall(\alpha,\beta) \in {}^f\mathbf{S'^a_u})(\mathbf{u} \in \alpha)$. We have then:

$$(\forall \mathbf{u} \in \mathbf{A^c})(\mathbf{u} \in {}^{\cup}\mathbf{y^u_a} \in \mathcal{O}_\mathbf{a}, \mathbf{B^c} \subset \bigcup_{\mathbf{i}=(\alpha,\beta)\in {}^f\mathbf{S'^a_u}} \beta, \mathbf{Card}({}^f\mathbf{S'^a_u}) \in \mathcal{Z}^+,$$
$$((\alpha,\beta) \in {}^f\mathbf{S'^a_u} \Rightarrow (\alpha \in \mathcal{O}_\mathbf{a}, \beta \in \mathcal{O}_\mathbf{b})))$$

we have (see definitions 362 (p. 462), 349 (p. 451)):

$$\mathbf{A^c} \subset \bigcup_{\mathbf{u}\in\mathbf{A^c}} {}^{\cup}\mathbf{y^u_a}\,.$$

Since $\mathbf{A^c}$ is compact (definitions 140 (p. 160)), we can extract from it a finite subset ${}^f\mathbf{A_c}$ of $\mathbf{A^c}$ such that

$$\mathbf{A^c} \subset \bigcup_{\mathbf{u}\in{}^f\mathbf{A_c}} {}^{\cup}\mathbf{y^u_a}, \mathbf{Card}({}^f\mathbf{A_c}) \in \mathcal{Z}^+, (\forall \mathbf{u} \in {}^f\mathbf{A_c})({}^{\cup}\mathbf{y^u_a} = \bigcup_{(\alpha,\beta)\in {}^f\mathbf{S'^a_u}} \alpha \in \mathcal{O}_\mathbf{a},$$
$$\mathbf{B^c} \subset \bigcup_{(\alpha,\beta)\in {}^f\mathbf{S'^a_u}} \beta)$$

then from proposition 796 (p. 464):

$$\mathbf{A^c} \times \mathbf{B^c} \subset \bigcup_{\mathbf{u}\in{}^f\mathbf{A_c}} \bigcup_{(\alpha,\beta)\in {}^f\mathbf{SCard'^a_u}} \alpha \times \beta = \bigcup_{(\alpha,\beta)\in\bigcup_{\mathbf{u}\in{}^f\mathbf{A_c}} {}^f\mathbf{S'^a_u}} \alpha \times \beta\,.$$

We have from propositions 894 (p. 508), 941 (p. 526) and definition 393 (p. 506):

$$\mathbf{Card}(\bigcup_{\mathbf{u}\in{}^f\mathbf{A_c}} {}^f\mathbf{S'^a_u}) \leq \sum_{\mathbf{u}\in{}^f\mathbf{A_c}} \mathbf{Card}({}^f\mathbf{S'^a_u}) \in \mathcal{Z}^+\,.$$

From equation (2.30) (p. 163) and (2.29) (p. 163):

$$((\alpha,\beta) \in {}^f\mathbf{S'^a_u} \subset \mathbf{S'} = \bigcup_{\mathbf{x}\in\mathbf{S}} \mathbf{P}(\mathbf{x})) \Rightarrow (\exists \mathbf{x}(\alpha,\beta) \in \mathbf{S} \subset \mathcal{O}^\pi)((\alpha,\beta) \in \mathbf{P}(\mathbf{x}(\alpha,\beta)))\,.$$

Then because $\mathbf{x} = \bigcup_{(\lambda,\mu)\in\mathbf{P}(\mathbf{x})} \lambda \times \mu$, we have:

$$\alpha \times \beta \subset \mathbf{x}(\alpha,\beta)\,.$$

Using that $\mathbf{x}(\alpha,\beta)$ we have:

$$\mathbf{A^c} \times \mathbf{B^c} \subset \bigcup_{(\alpha,\beta)\in\bigcup_{\mathbf{u}\in{}^f\mathbf{A_c}} {}^f\mathbf{S'^a_u}} \mathbf{x}(\alpha,\beta)\,.$$

Then $\mathbf{A^c} \times \mathbf{B^c}$ is compact because it is a subset of a finite union of sets of $\mathbf{S}$.

**Proposition 230** The product of a finite number (integer) of compact sets is compact.

**Proof**

From proposition 229 (p. 162) if the product $\mathbf{P}$ of $\mathbf{n} \in \mathcal{Z}^+$ compact subsets is compact, the product of $\mathbf{P}$ by another compact set is compact. Then, if 230 (p. 164) is true up to $\mathbf{n}$, it is true for $\mathbf{n}+\mathbf{1}$ so, by the recurrence principle, it is true for any $\mathbf{n}$ (proposition 927 (p. 520)).

### 2.1.3 Continuous map

**Definition 141 Continuous map.**

A map $\mathbf{f}$ between two topological sets $\mathbf{A}$ and $\mathbf{B}$:$(\mathbf{f}:\mathbf{A}\to\mathbf{B})$ is said to be **continuous** if the inverse of an open set is open.

**Proposition 231** If a map $\mathbf{f}$ between two topological sets $\mathbf{A}$ and $\mathbf{B}$:$(\mathbf{f}:\mathbf{A}\to\mathbf{B})$ is such that the inverse of a closed set is closed, this map $\mathbf{f}$ is continuous. The reverse is also true: If a map $\mathbf{f}$ between two topological sets $\mathbf{A}$ and $\mathbf{B}$:$(\mathbf{f}:\mathbf{A}\to\mathbf{B})$ is continuous then the inverse of a closed set is closed.

**Proof**

Since $(\forall\mathbf{D}\subset\mathbf{B})(\mathbf{f}^{-1}(\mathcal{C}(\mathbf{D}))=\mathcal{C}\mathbf{f}^{-1}(\mathbf{D}))$, the condition: the inverse of a closed set is closed is equivalent to the inverse of an open set is open.

**Proposition 232** $(\mathbf{f}:\mathbf{A}\to\mathbf{B})$ is a continuous map between two topological sets $\mathbf{A}$, $\mathbf{B}$, we have

$$\mathbf{C}\subset\mathbf{A}\Rightarrow\mathbf{f}(\overline{\mathbf{C}})\subset\overline{\mathbf{f}(\mathbf{C})}$$

$\overline{\mathbf{C}}$, $\overline{\mathbf{f}(\mathbf{C})}$ being the closure of $\mathbf{C}$, $\mathbf{f}(\mathbf{C})$ (definition 136 (p. 155)).

**Proof**

$\mathcal{O}_\mathbf{A}$, $\mathcal{O}_\mathbf{B}$ being the set of opens of $\mathbf{A}$, $\mathbf{B}$, (see proposition 165 (p. 138)) we have:

$$\mathbf{p}\in\mathbf{f}(\overline{\mathbf{C}})\Rightarrow(\mathbf{p}\in\mathbf{O}\in\mathcal{O}_\mathbf{B}\Rightarrow\mathbf{f}^{-1}(\mathbf{p})\subset\mathbf{f}^{-1}(\mathbf{O})\in\mathcal{O}_\mathbf{A},\mathbf{f}^{-1}(\mathbf{p})\cap\overline{\mathbf{C}}\neq\emptyset$$
$$\Rightarrow\mathbf{C}\cap\mathbf{f}^{-1}(\mathbf{O})\neq\emptyset\Rightarrow\mathbf{f}(\mathbf{C})\cap\mathbf{O}\neq\emptyset)\Rightarrow\mathbf{p}\in\overline{\mathbf{f}(\mathbf{C})}\Rightarrow\mathbf{f}(\overline{\mathbf{C}})\subset\overline{\mathbf{f}(\mathbf{C})}\,.$$

**Proposition 233** The identity map (definition 113 (p. 132)) between a topological set $\mathbf{E}$ and itself is continuous.

**Proof**

From definition 113 (p. 132) the identity map $(\mathbf{f}(\mathbf{x})=\mathbf{x}:\mathbf{E}\to\mathbf{E})$ is such that, for any subset $\mathbf{O}$, we have $\mathbf{O}=\mathbf{f}(\mathbf{O})=\mathbf{f}^{-1}(\mathbf{O})$. Then if $\mathbf{O}$ is open $\mathbf{f}^{-1}(\mathbf{O})=\mathbf{O}$ is open.

**Proposition 234** Given two topological sets $\mathbf{A}$ and $\mathbf{B}$, given $\boldsymbol{\lambda}\in\mathbf{B}$, the map $(\boldsymbol{\lambda}(\mathbf{x}):\mathbf{A}\to\mathbf{B})$ such as: $(\forall\mathbf{x}\in\mathbf{A})(\boldsymbol{\lambda}(\mathbf{x})=\boldsymbol{\lambda})$, is continuous.

**Proof**

For any open $\mathbf{O}\subset\mathbf{B}$, we have $\boldsymbol{\lambda}\in\mathbf{O}\vee\boldsymbol{\lambda}\notin\mathbf{O}$ and we have:

$$\boldsymbol{\lambda}\in\mathbf{O}\Rightarrow\boldsymbol{\lambda}^{-1}(\mathbf{O})=\mathbf{A},\ \boldsymbol{\lambda}\notin\mathbf{O}\Rightarrow\boldsymbol{\lambda}^{-1}(\mathbf{O})=\emptyset\,.$$

Since $\mathbf{A}$ and $\emptyset$ are both open (definition 132 (p. 151) relation (2.16)), $\boldsymbol{\lambda}^{-1}(\mathbf{O})$ is open (A.1.4, 16). $\boldsymbol{\lambda}(\mathbf{x})$ is continuous.

**Proposition 235** Given two topological sets $\mathbf{A}$ and $\mathbf{B}$, a subset $\mathbf{C}\subset\mathbf{A}$ and a continuous map $(\mathbf{f}:\mathbf{A}\to\mathbf{B})$, the restriction of $\mathbf{f}$ to $\mathbf{C}$ is also continuous if $\mathbf{C}$ has the induced topology from the topology of $\mathbf{A}$.

**Proof**

It is a direct consequence of proposition 165 (p. 138) and definition 133 (p. 151).

**Proposition 236** Given two topological sets $\mathbf{A}$ and $\mathbf{B}$, a subset $\mathbf{C}\subset\mathbf{B}$ and a continuous map $(\mathbf{f}:\mathbf{A}\to\mathbf{B})$, such that $\mathbf{f}(\mathbf{A})\subset\mathbf{C}\subset\mathbf{B}$, the map $(\mathbf{f}:\mathbf{A}\to\mathbf{C})$ is also continuous if $\mathbf{C}$ has the induced topology from the topology of $\mathbf{B}$. The opposite is also true:
If $(\mathbf{f}:\mathbf{A}\to\mathbf{C})$ is continuous, $(\mathbf{f}:\mathbf{A}\to\mathbf{B})$ is continuous.

**Proof**
Since $\mathbf{f(A) \subset C}$, we have $\mathbf{f^{-1}(C) = A}$ because $(\forall \mathbf{x} \in \mathbf{A})(\exists \mathbf{y} = \mathbf{f(x)} \in \mathbf{C})(\mathbf{y} = \mathbf{f(x)})$. $\mathbf{O}$ being an open of $\mathbf{C}$, from definition 133 (p. 151), $(\exists \mathbf{O'})$ open of $\mathbf{B}$ such that $\mathbf{O} = \mathbf{O'} \cap \mathbf{C}$. From proposition 165 (p. 138), we have:

$$\mathbf{f^{-1}(O) = f^{-1}(O') \cap f^{-1}(C) = f^{-1}(O') \cap A = f^{-1}(O')} .$$

Since $\mathbf{f^{-1}(O')}$ is open, $\mathbf{f^{-1}(O)}$ is open. $(\mathbf{f : A \to C})$ is continuous. If $(\mathbf{f : A \to C})$ is continuous $\mathbf{f^{-1}(O)}$ is open then $\mathbf{f^{-1}(O')}$ is open and $(\mathbf{f : A \to B})$ is continuous.

**Proposition 237** The projection maps defined (definition 121 (p. 136)) on product of topological sets are continuous.

**Proof**
Keeping the notations used in definition 121 (p. 136), an open $\mathbf{O^K}$ of $\mathbf{Y^K} = \prod_{\mathbf{k \in K}} \mathbf{X^k}$ is given by an element of $\mathbf{H} \in \mathcal{P}(\prod \mathcal{O}^\mathbf{k})$ where $\mathcal{O}^\mathbf{k}$ is the set of opens of $\mathbf{X^k}$:

$$\mathbf{O^K} = \bigcup_{\mathbf{h \in H}} \prod_{\ell \in \mathbf{K}} \mathbf{h}(\ell)$$

defining for any $\mathbf{h}$, the map $(\mathbf{U^i : I \to \mathcal{P}(X_i)})$ such that:

$$\mathbf{i \in K \Rightarrow U^i = h(\ell), i \in K_C = \mathcal{C}_I(K) \Rightarrow U^i = X^i} .$$

We have from definition 121 (p. 136),

$$\mathbf{Pr_K^{-1}}(\prod_{\mathbf{l \in K}} \mathbf{h(l)}) = \prod_{\ell \in \mathbf{I}} \mathbf{U}^\ell$$

which is open from definitions 134 (p. 153) and 132 (p. 151) equation (2.16). Then from definition 132 (p. 151) equation (2.17), $\bigcup_{\mathbf{h \in H}} \mathbf{Pr_K^{-1}}(\prod_{\ell \in \mathbf{K}} \mathbf{h}(\ell))$ is open. From proposition 160 (p. 136)

$$\bigcup_{\mathbf{h \in H}} \mathbf{Pr_K^{-1}}(\prod_{\ell \in \mathbf{K}} \mathbf{h}(\ell)) = \mathbf{Pr_K^{-1}}(\bigcup_{\mathbf{h \in H}} (\prod_{\ell \in \mathbf{K}} \mathbf{h}(\ell)) = \mathbf{Pr_K^{-1}(O^k)}$$

$\mathbf{Pr_K^{-1}(O^k)}$ is open and $\mathbf{Pr_K^{-1}}$ is continuous.

**Definition 142 Isomorphism.**

If a continuous map is bijective (definition 124 (p. 139)) and if its inverse is also continuous, the map is said to be isomorph or the map defines an **isomorphism.**

**Proposition 238** Given two sets $\mathbf{A}$, $\mathbf{B}$, an isomorphism $(\mathbf{f : A \leftrightarrow B})$, $\mathbf{C \subset A}$, the restricted map (definition 118 (p. 135)) $\mathbf{f_C}$ of $\mathbf{f}$ to $\mathbf{C}$ is also an isomorphism on $\mathbf{f(C)}$ using the induced topology in $\mathbf{C}$ and $\mathbf{f_C}$.

**Proof**
It is a direct consequence of propositions 183 (p. 143) and 235 (p. 165).

**Proposition 239** Given two topological sets $\mathbf{A}$ and $\mathbf{B}$, a subset $\mathbf{C \subset A}$ and an isomorphism $(\mathbf{f : A \leftrightarrow B})$, we have (see definition 136 (p. 155)):

$$\mathbf{f(\overline{C}) = \overline{f(C)}} .$$

**Proof**
We call $\mathcal{O}_A$, $\mathcal{O}_B$ the sets of open of A and B. From definitions 115 (p. 133), 116 (p. 134), 174 (p. 141) and from propositions 154 (p. 133), 157 (p. 134), we have:

$$p \in f(\overline{C}) \Rightarrow ((O \in \mathcal{O}_B, p \in O) \Rightarrow (f^{-1}(p) \in f^{-1}(O) \cap \overline{C}, f^{-1}(O) \in \mathcal{O}_A))$$
$$\Rightarrow (\exists x \in f^{-1}(O) \cap C)$$

but

$$x \in f^{-1}(O) \cap C \Rightarrow f(x) \in O \cap f(C) \Rightarrow \exists y \in O \cap f(C)$$

then $p \in \overline{f(C)}$. Since f is an isomorphism, we have from the above applied to $f^{-1}$:

$$p \in \overline{f(C)} \Rightarrow f^{-1}(p) \in f^{-1}(\overline{f(C)}) \Rightarrow f^{-1}(p) \in \overline{C} \Rightarrow p \in f(\overline{C}) \; .$$

**Proposition 240** A, B are two topological sets, C a compact subset of A, $(f : A \rightarrow B)$ a continuous map. Then f(C) is compact.

**Proof**
$\mathcal{O}$ is the set of opens of B. If there is a cover of f(C) by opens of B, we have:

$$(\exists \mathcal{O}^c \subset \mathcal{O})(f(C) \subset \bigcup_{O \in \mathcal{O}^c} O) \; .$$

We also have: (propositions 161 (p. 137), 162 (p. 137), 163 (p. 137))

$$C \subset \bigcup_{O \in \mathcal{O}^c} f^{-1}(O) \; .$$

Since f is continuous, the $f^{-1}(O)$ are open. C being compact, we have:

$$(\exists \mathcal{O}_z \subset \mathcal{O}^c)(Card(\mathcal{O}_z) \in \mathcal{Z}^+)(C \subset \bigcup_{O \in \mathcal{O}_z} f^{-1}(O)) \; .$$

We have $f(f^{-1}(O)) \subset O$ proposition 168 (p. 139). Then

$$(\exists \mathcal{O}_z \subset \mathcal{O}^c)(Card(\mathcal{O}_z) \in \mathcal{Z}^+, f(C) \subset \bigcup_{O \in \mathcal{O}_z} f(f^{-1}(O)) \subset \bigcup_{O \in \mathcal{O}_z} O)$$

f(C) is compact.

**Proposition 241** If A, B, C are three topological sets, $(F_{A,B} : B \rightarrow A)$ a continuous map from B to A, $(F_{B,C} : C \rightarrow B)$ a continuous map from C to B, the composite map (see definition 117 (p. 134)) from C to B and to A: $F_{A,C} = F_{A,B}(F_{B,C}) = F_{A,B} \circ F_{B,C}$ is a continuous map from C to A.

**Proof**
If $O_A$ is an open of A, from the definition of continuous maps, $F^{-1}_{A,B}(O_A)$ is open and then $F^{-1}_{B,C}(F^{-1}_{A,B}(O_A))$ is also open. Then $F^{-1}_{A,C}(O_A)$ is open.

**Proposition 242** Given $\mathbf{2n}$ topological sets:
$\mathbf{A_1, ..., A_n}$: $(\mathbf{A_i : [1, n] \to \mathcal{P}(X)})$,
$\mathbf{B_1, ..., B_n}$: $(\mathbf{B_i : [1, n] \to \mathcal{P}(Y)})$,
$\mathbf{X, Y}$ being two topological sets and $\mathbf{n}$ maps $(\forall \mathbf{i} \in [\mathbf{1, n}])(\exists(\mathbf{g_i(x) : A_i \to B_i}))$ all continuous then the map

$$(\mathbf{g} = \prod_{\mathbf{i}\in[\mathbf{1,n}]} \mathbf{g_i(x^i)} : \prod_{\mathbf{i}\in[\mathbf{1,n}]} \mathbf{A_i} \to \prod_{\mathbf{i}\in[\mathbf{1,n}]} \mathbf{B_i})$$

is continuous.

**Proof**
Any open $\mathbf{O^{\pi b}}$ of $\prod_{\mathbf{i}\in[\mathbf{1,n}]} \mathbf{B_i}$ is such that $\mathbf{O^{\pi b}} = \bigcup_{\mathbf{j}\in\mathbf{J}} \prod_{\mathbf{i}\in[\mathbf{1,n}]} \mathbf{O^b_{i,j}}$ where the $\mathbf{O^b_{i,j}}$ are open of $\mathbf{B_i}$ (see definition 134 (p. 153)). Then from proposition 163 (p. 137),

$$(\prod_{\mathbf{i}\in[\mathbf{1,n}]} \mathbf{g_i})^{-1}(\mathbf{O^{\pi b}}) = \bigcup_{\mathbf{j}\in\mathbf{J}}(\prod_{\mathbf{i}\in[\mathbf{1,n}]} \mathbf{g_i})^{-1}(\prod_{\mathbf{i}\in[\mathbf{1,n}]} \mathbf{O^b_{i,j}}) = \bigcup_{\mathbf{j}\in\mathbf{J}}(\prod_{\mathbf{i}\in[\mathbf{1,n}]} \mathbf{g_i^{-1}}(\mathbf{O^b_{i,j}}))\ .$$

Since all $\mathbf{g_i}$ are continuous, all $\mathbf{g_i^{-1}(O^b_{i,j})}$ are open and, from proposition 158 (p. 135), $(\prod_{\mathbf{i}\in[\mathbf{1,n}]} \mathbf{g_i})^{-1}(\mathbf{O^{\pi b}})$ is open as union of opens. The map $\prod_{\mathbf{i}\in[\mathbf{1,n}]} \mathbf{g_i}$ is continuous.

**Proposition 243** We keep the premise of proposition 242 (p. 168) and we suppose we have another set $\mathbf{C}$ and a continuous map $(\mathbf{f} : \prod_{\mathbf{i}\in[\mathbf{1,n}]} \mathbf{B_i} \to \mathbf{C})$. As above we call $\mathbf{g}$ the map:

$$(\mathbf{g} = \prod_{\mathbf{i}\in[\mathbf{1,n}]} \mathbf{g_i(x^i)} : \prod_{\mathbf{i}\in[\mathbf{1,n}]} \mathbf{A_i} \to \prod_{\mathbf{i}\in[\mathbf{1,n}]} \mathbf{B_i})\ .$$

Then the map $(\mathbf{f \circ g} : \prod_{\mathbf{i}\in[\mathbf{1,n}]} \mathbf{A_i} \to \mathbf{C})$ is continuous.

**Proof**
It is a direct consequence of propositions 241 (p. 167) and 242 (p. 168).

**Proposition 244** $\mathbf{A}$ being a set, $(\mathbf{A_i} : [\mathbf{1, n}] \to \mathcal{P}(\mathbf{A}))$ are $\mathbf{n}$ subsets of $\mathbf{A}$ which are topological sets, $\mathbf{B}$ is a topological set, $(\mathbf{F_{A_i,B} : B \to A_i})$, $\mathbf{n}$ continuous maps from $\mathbf{B}$ to $\mathbf{A_i}$. The map $(\mathbf{F : B} \to \prod_{\mathbf{i}\in[\mathbf{1,n}]} \mathbf{A_i})$ from $\mathbf{B}$ to $\prod_{\mathbf{i}\in[\mathbf{1,n}]} \mathbf{A_i}$ defined by:

$$\mathbf{F} = \mathbf{set}_{(\mathbf{x,y})}\{(\mathbf{x} \in \mathbf{B}), (\mathbf{y} = \prod_{\mathbf{i}\in[\mathbf{1,n}]} \mathbf{F_{A_i,B}(x)} \in \prod_{\mathbf{i}\in[\mathbf{1,n}]} \mathbf{A_i})\}$$

is continuous.

**Proof**
Any open $\mathbf{O^\pi}$ of $\prod_{\mathbf{i}\in[\mathbf{1,n}]} \mathbf{A_i}$ is a union of opens such that $\mathbf{O} = \prod_{\mathbf{i}\in[\mathbf{1,n}]} \mathbf{O_i}$ where $\mathbf{O_i}$ are opens of $\mathbf{A_i}$:
$\mathcal{O}_\mathbf{i} \subset \mathcal{P}(\mathbf{A_i})$ being the set of opens of $\mathbf{A_i}$, $\mathcal{O}^\pi = \mathcal{P}(\prod_{\mathbf{i}\in[\mathbf{1,n}]} \mathcal{O}_\mathbf{i})$ defined the set of opens of $\prod_{\mathbf{i}\in[\mathbf{1,n}]} \mathbf{A_i}$ (see definition 134 (p. 153)). For any open $\mathbf{O^\pi} \in \mathcal{O}^\pi$, there is $\mathbf{U^\pi} \subset \prod_{\mathbf{i}\in[\mathbf{1,n}]} \mathcal{O}_\mathbf{i}$ such that $\mathbf{O^\pi} = \bigcup_{\mathbf{O}\in\mathbf{U^\pi}} \mathbf{O}$ where $\mathbf{O} = \prod_{\mathbf{i}\in[\mathbf{1,n}]} \mathbf{O_i} \in \prod_{\mathbf{i}\in[\mathbf{1,n}]} \mathcal{O}_\mathbf{i}$ which means $(\forall \mathbf{i} \in [\mathbf{1, n}])(\mathbf{O_i} \in \mathcal{O}_\mathbf{i})$. An element $\mathbf{b}$ of $\mathbf{B}$ having an image in an open $\mathbf{O}$ has an image by $\mathbf{F_{A_i,B}}$ in every corresponding $\mathbf{O_i}$. Then:

$$\mathbf{F(b)} \in \mathbf{O} \Leftrightarrow \mathbf{b} \in \bigcap_{\mathbf{i}\in[\mathbf{1,n}]} \mathbf{F^{-1}_{A_i,B}(O_i)}\ .$$

Then we have:

$$\mathbf{F^{-1}(O)} = \bigcap_{\mathbf{i}\in[\mathbf{1,n}]} \mathbf{F^{-1}_{A_i,B}(O_i)}\ .$$

All $F^{-1}_{A_i,B}(O_i)$ are open so $\bigcap_{i\in[1,n]} F^{-1}_{A_i,B}(O_i)$ is also open since $n$ is finite. Using proposition 163 (p. 137) we also have:

$$F^{-1}(O^\pi) = \bigcup_{O\in U^\pi} F^{-1}(O)$$

we conclude proposition 244 using relation (2.17) (p. 151).

**Proposition 245** Conversely $A$ being a set, $(A_i : [1,n] \to \mathcal{P}(A))$ are $n$ subsets of $A$ which are topological sets, $B$ is a topological set, $(F_{A_i,B} : B \to A_i)$ $n$ maps from $B$ to $A_i$. The map $(F : B \to \prod_{i\in[1,n]} A_i)$ from $B$ to $\prod_{i\in[1,n]} A_i$ defined by:

$$F = set_{(x,y)}\{(x \in B), (y = \prod_{i\in[1,n]} F_{A_i,B}(x) \in \prod_{i\in[1,n]} A_i)\}$$

being continuous, each individual map $(F_{A_i,B}, B \to A_i)$, $i \in [1,n]$ is continuous.

**Proof**

Given an open $O_i$ of $A_i$, any point of $B$ having an image by $F_{A_i,B}$ in $O_i$ also has an image by $F_{A_j,B}$, $j \in [1,n] - i$ in $A_j$ so we have:

$$F^{-1}_{A_i,B}(O_i) \subset (\prod_{j\in[1,n]} F_{A_j,B})^{-1}(\prod_{j\in[1,n]} O_j) = F^{-1}(\prod_{j\in[1,n]} O_j) .$$

With $(\forall j \in [1,n] - i)(O_j = A_j)$. But any point of $B$, which has an image by $F = \prod_{j\in[1,n]} F_{A_j,B}$ in $\prod_{j\in[1,n]} O_j$, is in $F^{-1}_{A_i,B}(O_i)$. Then:

$$F^{-1}(\prod_{j\in[1,n]} O_j) \subset F^{-1}_{A_i,B}(O_i)$$

and by proposition 754 (p. 451):

$$F^{-1}_{A_i,B}(O_i) = F^{-1}(\prod_{j\in[1,n]} O_j) .$$

Since $O_j = A_j$, $j \neq i$ are open (definition 132 (p. 151), equation (2.16)) $\prod_{j\in[1,n]} O_j$ is open. Since $F$ is continuous, $F^{-1}(\prod_{j\in[1,n]} O_j)$ is open then $F^{-1}_{A_i,B}(O_i)$ is open $F_{A_i,B}$ is continuous (definition 141 (p. 165)).

**Proposition 246** $A$ being a set, $(A_i : [1,n] \to \mathcal{P}(A))$ are $n$ subsets of $A$ which are topological sets, $B$ is a topological set, $(f : \prod_{i\in[1,n]} A_i \to B)$ a continuous map from $\prod_{i\in[1,n]} A_i$ to $B$. $K$ a subset of $[1,n]$ and $(a \in \prod_{i\in K} A_i)$ a term which defines a map

$$(x'(x,a) : \prod_{i\in[1,n]-K} A_i \to \prod_{i\in[1,n]} A_i)$$

such that $i \in K \Rightarrow x'_i = a_i, i \in [1,n] - K \Rightarrow x'_i = x_i$. The restricted map

$$g(x) = (f(x'(x,a)) : \prod_{i\in[1,n]-K} A_i \to B)$$

is continuous.

**Proof**
Let us call $\mathcal{O}_i$ and $\mathcal{O}_B$ the sets of open sets of $\mathbf{A_i}$ and $\mathbf{B}$. $\mathbf{O_B}$ being an open of $\mathbf{B}$ and $\mathbf{f}$ being continuous, $\mathbf{f^{-1}(O_B)} \subset \prod_{i\in[1,n]} \mathbf{A_i}$ is open. From the definition 134 (p. 153) of open sets of $\prod_{i\in[1,n]} \mathbf{A_i}$, we have:

$$(\exists U^\pi \in \mathcal{P}(\prod \mathcal{O}_i))(f^{-1}(O_b) = \bigcup_{O\in U^\pi} O = \bigcup_{O\in U^\pi} \prod_{i\in[1,n]} O_i^o)$$

with $\mathbf{O} = \prod_{i\in[1,n]} \mathbf{O_i^o}$.
For any $\mathbf{x}$ in $\mathbf{g^{-1}(O_B)}$, we have

$$x'(x,a) \in f^{-1}(O_B) = \bigcup_{O\in U^\pi} \prod_{i\in[1,n]} O_i^o \Leftrightarrow x \in g^{-1}(O_B)\ .$$

Let us consider $\mathbf{V^\pi} \subset \mathbf{U^\pi}$ such that:

$$v \in V^\pi \Leftrightarrow (v \in U^\pi, (\forall i \in K)(a_i \in v_i))\ .$$

We have

$$g^{-1}(O_B) = \bigcup_{v\in V^\pi} \prod_{j\in[1,n]-K} v_i\ .$$

Since $(\forall \mathbf{i} \in [1,n])(\mathbf{v_i} \in \mathcal{O}_i)$, from definition 134 (p. 153), $\mathbf{g^{-1}(O_B)}$ is open.

## 2.2 Real numbers, set $\mathcal{R}$

**Definition 143 Real numbers, $\mathcal{R}$.**

Integers $\mathcal{Z}$ and rational (fractional) $\mathcal{Q}$ numbers are supposed to be defined (definition 397 (p. 516) and 418 (p. 551)) and in appendix A (p. 433). Real numbers are the segments of $\mathcal{Q}$ which included their **upper_limit** if they have one in $\mathcal{Q}$, which are not empty and not equal to $\mathcal{Q}$ (see definitions 389 (p. 488) and 381 (p. 485)). From that definition of $\mathcal{R}$ we have: [11]

$$\begin{aligned}\mathcal{R} = \text{set}_A\{\ & A \subset \mathcal{Q}, A \neq \mathcal{Q}, A \neq \emptyset, \\ & (a \in \mathcal{Q}, (\forall b \in \mathcal{C}_Q(A)), a \leq b) \Rightarrow a \in A\}\ .\end{aligned} \tag{2.32}$$

See definitions 338 (p. 446) and 374 (p. 482). From definition 350 (p. 453) and proposition 749 (p. 450), this set exists.
One uses to write an element of $\mathbf{A} \in \mathcal{R}$ as $\mathbf{A}|\mathcal{C}(\mathbf{A})$.

**Proposition 247** There is an injective map (definition 122 (p. 139) between $\mathcal{Q}$ and $\mathcal{R}$ defined as (see definition 378 (p. 484)):

$$(] \leftarrow, q] : \mathcal{Q} \to \mathcal{R})\ .$$

**Proof**
$] \leftarrow, \mathbf{q}]$ is a segment which satisfy definition 143 (p. 170). If there is another rational $\mathbf{q'}$ such that $] \leftarrow \mathbf{q'}] =] \leftarrow, \mathbf{q}]$ we would have:

$$q' \in ] \leftarrow, q],\ q \in ] \leftarrow, q']$$

which means from definition 378 (p. 484), $\mathbf{q'} \leq \mathbf{q}, \mathbf{q} \leq \mathbf{q'}$ which gives from definition 338 (p. 446) $\mathbf{q'} = \mathbf{q}$ and most of the time in order to be simpler one writes $] \leftarrow, \mathbf{q}]$ as $\mathbf{q}$. The map such defined is, as we shall see, an isomorphism for the order and the four operations. Since integers are also rational numbers, they can also be considered as real. Then we can say:

[11]We apply as all along this book convention 27 (p. 445) in appendix A.3 (p. 441).

**Proposition 248** Rational and Integers numbers are also real numbers.

**Definition 144** $\pm\infty$, $\mathcal{R}_\infty$.

In Lebesgue integration, $\mathcal{R}$ is not big enough, to definition 143 (p. 170), we need to add two points at $\infty$. They correspond to the two excluded segments of $\mathcal{Q}$:

$$\emptyset = -\infty,\ \mathcal{Q} = \infty\ .$$

We shall call that extended $\mathcal{R}$, $\mathcal{R}_\infty$ and similarly. $\mathcal{R}^+_\infty = \mathcal{R}^+ \cup \{\infty\}$, $\mathcal{R}^{0+}_\infty = \mathcal{R}^+ \cup \{\infty\} \cup \{0\}$. We have to do those distinctions because $\mathcal{R}_\infty$ is not a group for the $\times$ and $+$ internal laws as $\mathcal{R}$ is (see section 2.9 (p. 205)).

### 2.2.1 Example of a real number

$$\mathbf{A} = \mathrm{set}_\mathbf{x}\{(\mathbf{x} \in \mathcal{Q}), (\mathbf{x} < 0 \text{ or } \mathbf{x}^2 \le 2)\}; \tag{2.33}$$

$$\mathbf{B} = \mathcal{C}(\mathbf{A}) = \mathrm{set}_\mathbf{x}\{(\mathbf{x} \in \mathcal{Q}), (\mathbf{x} > 0), (\mathbf{x}^2 > 2)\} \tag{2.34}$$

$$\sqrt{2} = \mathbf{A}|\mathbf{B} \tag{2.35}$$

as it is well known, the real number $\sqrt{2}$ cannot be rational.

### 2.2.2 Order on $\mathcal{R}$

**Definition 145 Order on $\mathcal{R}_\infty$.**

We set $\mathbf{A} \in \mathcal{R}_\infty$, $\mathbf{B} \in \mathcal{R}_\infty$ as defined in (2.32) (p. 170) and in 144 (p. 171) and we define:

$$\mathbf{A} \subset \mathbf{B} \Leftrightarrow (\mathbf{A}|\mathcal{C}(\mathbf{A}) \le \mathbf{B}|\mathcal{C}(\mathbf{B}))\ . \tag{2.36}$$

From proposition 838 (p. 483) equation (2.36) defines an order on $\mathcal{R}_\infty$. $\mathbf{A}$ and $\mathbf{B}$ being segments of $\mathcal{Q}$.

**Proposition 249** $(\forall \mathbf{r} \in \mathcal{R}) - \infty < \mathbf{r} < \infty$ **.**

**Proof**
From definitions 143 (p. 170) and 144 (p. 171), we have

$$\mathbf{r} \in \mathcal{R} \Rightarrow (\mathbf{r} \ne \emptyset,\ \mathbf{r} \ne \mathcal{Q}, \emptyset \subset \mathbf{r} \subset \mathcal{Q}) \Rightarrow -\infty < \mathbf{r} < \infty\ .$$

**Proposition 250** The order defined by definition 145 is a total order (see definition 377 (p. 483)).

**Proof**
From proposition 858 (p. 489) that order is total as the order on $\mathcal{Q}$ is.

**Proposition 251** Any rational $\mathbf{q}$ in the segment definition of a real $\mathbf{r}$ is smaller than or equal to $\mathbf{r}$ when $\mathbf{q}$ is considered as real.

**Proof**
When $\mathbf{q}$ is considered as real, its segment of definition is $]\leftarrow, \mathbf{q}] \subset \mathcal{Q}$ (definition 378 (p. 484), proposition 248 (p. 171)). From 858 (p. 489) we have either $]\leftarrow, \mathbf{q}] \subset \mathbf{r}$ or $\mathbf{r} \subset ]\leftarrow, \mathbf{q}]$. Since $\mathbf{q} \in \mathbf{r}$ only $]\leftarrow, \mathbf{q}] \subset \mathbf{r}$ is possible and from 145 (p. 171)

$$\mathbf{q} \leq \mathbf{r} .$$

**Proposition 252** When considered as real, any rational $\mathbf{q}$ smaller than a real $\mathbf{r}$, it belongs to the segment definition of $\mathbf{r}$.
**Proof**
From definition 143 (p. 170), 378 (p. 484), propositions 247 (p. 170) and 248 (p. 171), the real number associated to $\mathbf{q}$ is $]\leftarrow, \mathbf{q}]$. From definition 145 (p. 171), if $\mathbf{q} \leq \mathbf{r}$ we have:

$$]\leftarrow, \mathbf{q}] \subset \mathbf{r} \subset \mathcal{Q} .$$

From definition 349 (p. 451), since (definition 378 (p. 484)) $\mathbf{q} \in ]\leftarrow, \mathbf{q}]$ we have $\mathbf{q} \in \mathbf{r}$.

**Proposition 253** Given two different extended reals $\mathbf{r_1}$, $\mathbf{r_2}$ of $\mathcal{R}_\infty$ (definition 144 (p. 171)) $\mathbf{r_1} < \mathbf{r_2} \Rightarrow (\exists \mathbf{q} \in \mathcal{Q})(\mathbf{r_1} < \mathbf{q} < \mathbf{r_2}) \Rightarrow (\exists \mathbf{r} \in \mathcal{R})(\mathbf{r_1} < \mathbf{r} < \mathbf{r_2})$.
**Proof**
From definition 145 (p. 171) and the fact that the order on reals is total (proposition 250 (p. 171)), we have $\mathbf{r_1} < \mathbf{r_2}$. It means that there is $\mathbf{q} \in \mathcal{Q}$ such that $\mathbf{q} \in \mathbf{r_2}, \mathbf{q} \notin \mathbf{r_1}$ then from proposition 251 equivalencing $\mathbf{q}$ and $]\leftarrow, \mathbf{q}]$ we have $\mathbf{r_1} < \mathbf{q} < \mathbf{r_2}$. Then setting $\mathbf{r} = ]\leftarrow, \mathbf{q}]$ we have $\mathbf{r_1} < \mathbf{r} < \mathbf{r_2}$.

**Proposition 254** We have $\mathbf{r_1}$ and $\mathbf{r_2}$ being real:

$$\mathbf{r_1} \leq \mathbf{r_2} \Leftrightarrow ((\mathbf{q} \in \mathcal{Q}, \mathbf{q} \leq \mathbf{r_1}) \Rightarrow \mathbf{q} \leq \mathbf{r_2}) .$$

**Proof**
From definition 145 (p. 171) $\mathbf{r_1} \leq \mathbf{r_2} \Leftrightarrow \mathbf{r_1} \subset \mathbf{r_2}$, we conclude using proposition 251 (p. 171) and proposition 252 (p. 172).

**Proposition 255** Any set of reals which has, as a cardinal, an integer larger than 0, (a finite set) has a biggest and a smallest element which are also its **upper_limit** and its **lower _limit** (definition 381 (p. 485)).
**Proof**
It is a direct consequence of propositions 250 (p. 171), 945 (p. 531) and 845 (p. 485).

**Proposition 256** On any finite set of real numbers $\mathbf{A}$ of cardinal $\mathbf{n} \in \mathcal{Z}^+, \mathbf{n} = \mathbf{Card(A)}$, there is a strictly increasing bijection (definition 386 (p. 486)) (isomorphism definition 387 (p. 487)):

$$(\mathbf{x^i} : [\mathbf{1}, \mathbf{n}] \leftrightarrow \mathbf{A}), (\mathbf{y^j} : [\mathbf{0}, \mathbf{n} - \mathbf{1}] \leftrightarrow \mathbf{A}) .$$

**Proof**
From proposition 255, the order on any finite set of real is a good order (definition 388 (p. 487)). Then from proposition 869 (p. 496) and definition 390 (p. 499), we have:

$$(\exists \mathbf{x})(\mathbf{x} = (\mathbf{x^i} : [\mathbf{1}, \mathbf{n}] \leftrightarrow \mathbf{A})) .$$

For the same reasons, we have:

$$(\exists \mathbf{y})(\mathbf{y} = (\mathbf{y^j} : [\mathbf{0}, \mathbf{n} - \mathbf{1}] \leftrightarrow \mathbf{A})) .$$

**Proposition 257** Given $\mathbf{p} \in \mathcal{Z}$, $\mathbf{q} \in \mathcal{Z}$, $\mathbf{p} \leq \mathbf{q}$ and $(\mathbf{f} : [\mathbf{p}, \mathbf{q}] \to \mathcal{R})$ a strictly increasing map (definition 386 (p. 486)), we have:

$$\bigcup_{i \in [p,q[} [f(i), f(i+1)[= [f(p), f(q)[ \; .$$

**Proof**

Let us suppose that it is true for $\mathbf{q} \in [\mathbf{p}, \mathbf{q}']$. Let us prove that in that case, it is also true for $\mathbf{q} = \mathbf{q}' + 1$ we have:

$$\bigcup_{i \in [p,q[} [f(i), f(i+1)[= [f(p), f(q)[= \bigcup_{i \in [p,q'[} [f(i), f(i+1)[\cup[f(q'), f(q)[$$
$$= [f(p), f(q')[\cup[f(q'), f(q)[$$

but $\mathbf{x} \in [f(p), f(q')[ \Rightarrow \mathbf{x} \in [f(p), f(q)[$ since $f(q') < f(q), x < f(q') \Rightarrow x < f(q)$ as well as $\mathbf{x} \in [f(q'), f(q)[ \Rightarrow \mathbf{x} \in [f(p), f(q)[$ since $f(p) \leq f(q'), f(q') \leq x \Rightarrow f(p) \leq x$ then from section A.1.3 (p. 435) point 16

$$x \in [f(p), f(q')[\cup[f(q'), f(q)[\Rightarrow x \in [f(p), f(q')[\vee x \in [f(q'), f(q)[\Rightarrow x \in [f(p), f(q)[$$

similarly

$$x \in [f(p), f(q)[\Rightarrow f(p) \leq x < f(q') \vee f(q') \leq x < f(q)$$
$$\Rightarrow x \in [f(p), f(q')[\cup[f(q'), f(q)[ \; .$$

Then we conclude by proposition 754 (p. 451) (extensionality principle) and 927 (p. 520) (recurrence principle).

### 2.2.3 Operations on $\mathcal{R}$

As for rational numbers (see section A.8.2 (p. 553)), we shall not elaborate too much because the exact definition of elementary operations on $\mathcal{R}$ are not difficult to implement but it is tedious to make it rigorous. One can define addition and multiplication of real numbers as maps of $\mathcal{R} \times \mathcal{R} \to \mathcal{R}$:

$$(u = A|\mathcal{C}(A)), (v = B|\mathcal{C}(B)) \Rightarrow u + v = +(A, B)|\mathcal{C}(+(A, B)) \; .$$

With

$$+(A, B) = \text{set}_x\{(\exists a \in A)(\exists b \in B)(x = a + b)\} \; .$$

For the multiplication and division, one has to care about 0 and the sign of the numbers. The easiest thing is to define the cut of the product with element of the same sign. For instance with $\mathbf{u} = (\mathbf{A}|\mathcal{C}(\mathbf{A})$ and $\mathbf{v} = (\mathbf{B}|\mathcal{C}(\mathbf{B}))$ in $\mathcal{R}^+$, one defines the cut of $\mathbf{uv}$ taking:

$$C = \mathcal{R}^- \cup \{0\} \cup \times(A \cap \mathcal{R}^+, B \cap \mathcal{R}^+)$$

and making $\mathbf{uv} = (\mathbf{C}|\mathcal{C}(\mathbf{C}))$.
Similarly, taking:

$$D = \mathcal{R}^- \cup \{0\} \cup /(A \cap \mathcal{R}^+, \mathcal{C}(B))$$

and making $\mathbf{u}/\mathbf{v} = (\mathbf{D}|\mathcal{C}(\mathbf{D}))$, it is easy to prove that the properties defined in section A.8.2 (p. 553) can be extended to $\mathcal{R}$.

### 2.2.4 Upper_limit and lower_limit of a subset of $\mathcal{R}$

**Proposition 258** Any $\mathbf{C} \subset \mathcal{R}$, not empty and with a majoring (resp. minoring) element $\mathbf{r_m}$ (see definition 380 (p. 484)), has an **upper_limit** (resp. **lower_limit**) (see definition 381 (p. 485)). They are unique numbers from proposition 844 (p. 485).

**Proof**
Let us call $\mathbf{M} \subset \mathcal{Q}$, the set of the majoring (resp. minoring) elements of $\mathbf{C}$ in $\mathcal{Q}$.

- $\mathbf{M} \neq \emptyset$ because $\mathbf{r_m}$ is majoring (resp. minoring) $\mathbf{C}$ and from proposition 253 (p. 172) $(\exists \mathbf{q} \in ]\mathbf{r_m}, \rightarrow [\cap \mathcal{Q})$ (resp. $(\exists \mathbf{q} \in ] \leftarrow, \mathbf{r_m}] \cap \mathcal{Q})$ where $\mathbf{q}$ is identified to the real $] \leftarrow, \mathbf{q}]$
- $\mathbf{M} \neq \mathcal{Q}$ because from proposition 253 (p. 172)
$$(\exists \mathbf{x} \in \mathbf{C}) \Rightarrow (\exists \mathbf{y} \in \mathcal{Q})(\mathbf{y} < (\textbf{resp.} >)\mathbf{x}, \mathbf{y} \notin \mathbf{M})$$
(propositions 251 (p. 171), 1007 (p. 559), definition 143 (p. 170)).

We call $\mathbf{L} \subset \mathcal{Q}$, the set of the minoring elements in $\mathcal{Q}$ of $\mathbf{M}$ **(resp.** $\mathbf{L} = \mathbf{M}$**)**. $\mathbf{L}$ is a segment of $\mathcal{Q}$.
All elements of $\mathbf{M}$ is a majoring element of $\mathbf{L}$. If $\mathbf{L}$ has an **upper_limit** $\boldsymbol{\ell}$, from the definition 381 (p. 485), $\boldsymbol{\ell}$ is a minoring element of $\mathbf{M}$, it belongs to $\mathbf{L}$ from the definition of $\mathbf{L}$.
**(Resp.** Calling $\mathbf{L'}$ set of elements of $\mathcal{Q}$ such that $\mathbf{p} \in \mathbf{L'} \Leftrightarrow (\exists \mathbf{x} \in \mathbf{C})(\mathbf{x} \leq \mathbf{p})$, we have (see proposition 253 (p. 172) and definitions 329 (p. 434), 330 (p. 434), 333 (p. 441), 334 (p. 441)):

$$\begin{aligned}
&\mathbf{q} \in \mathbf{L} \Rightarrow (\forall \mathbf{p} \in \mathbf{L'})(\exists \mathbf{x} \in \mathbf{C})(\mathbf{q} \leq \mathbf{x} \leq \mathbf{p}) \Rightarrow (\forall \mathbf{p} \in \mathbf{L'})(\mathbf{q} \leq \mathbf{p}) \\
&\{(\exists \mathbf{x})(\mathbf{x} \in \mathbf{C}, \mathbf{x} < \mathbf{q}) \Rightarrow (\exists \mathbf{x} \in \mathbf{C})(\exists \mathbf{p'} \in \mathcal{Q})(\mathbf{x} < \mathbf{p'} < \mathbf{q}) \Rightarrow (\exists \mathbf{p'} \in \mathbf{L'})(\mathbf{q} > \mathbf{p'})\} \\
&(\exists \mathbf{x})(\mathbf{x} \in \mathbf{C}, \mathbf{x} < \mathbf{q}) \Rightarrow (\exists \mathbf{p'} \in \mathbf{L'})(\mathbf{q} > \mathbf{p'}) \\
&\neg\neg(\exists \mathbf{p'})(\mathbf{p'} \in \mathbf{L'}, \neg(\mathbf{q} \leq \mathbf{p'})) \vee \neg(\exists \mathbf{x})(\mathbf{x} \in \mathbf{C}, \mathbf{x} < \mathbf{q}) \\
&\forall \mathbf{p'}(\neg \mathbf{p'} \in \mathbf{L'} \vee \mathbf{q} \leq \mathbf{p'}) \Rightarrow \neg(\exists \mathbf{x})\neg(\neg \mathbf{x} \in \mathbf{C} \vee \neg \mathbf{x} < \mathbf{q}) \\
&\forall \mathbf{p'}(\mathbf{p'} \in \mathbf{L'} \Rightarrow \mathbf{q} \leq \mathbf{p'}) \Rightarrow (\forall \mathbf{x})(\mathbf{x} \in \mathbf{C} \Rightarrow \neg \mathbf{x} < \mathbf{q} = \mathbf{q} \leq \mathbf{x}) \Rightarrow \mathbf{q} \in \mathbf{L}
\end{aligned}$$

then
$$\mathbf{q} \in \mathbf{L} \Leftrightarrow (\forall \mathbf{p} \in \mathbf{L'})(\mathbf{q} \leq \mathbf{p})$$
From the definition 381 (p. 485) and proposition 1006 (p. 559), if $\mathbf{L}$ has an **upper_limit**, it belongs to $\mathbf{L}$. )
From what we said on $\mathbf{M}$, $\mathbf{L}$ is not empty and not $\mathcal{Q}$. Then $\mathbf{L} \in \mathcal{R}$ from definition 143 (p. 170).
Following proposition 253 (p. 172), equivalencing $\mathbf{q}$ and $] \leftarrow, \mathbf{q}]$ when $\mathbf{q} \in \mathcal{Q}$ (proposition 247 (p. 170)), we have:
By definition of $\mathbf{L}$ and $\mathbf{M}$, since $\mathbf{M} \neq \emptyset$

$$\begin{aligned}
&\mathbf{q} \in \mathbf{L} \Leftrightarrow (\forall \mathbf{p} \in \mathbf{M})(\mathbf{q} \leq \mathbf{p}) \\
&\mathbf{p} \in \mathbf{M} \Leftrightarrow (\forall \mathbf{x} \in \mathbf{C})(\mathbf{x} \leq \mathbf{p}) \\
&(\mathbf{p} \in \mathcal{Q}, (\exists \mathbf{x} \in \mathbf{C})(\mathbf{p} \leq \mathbf{x})) \Rightarrow \mathbf{p} \in \mathbf{L}\,.
\end{aligned}$$

From proposition 253 (p. 172), we have with $\mathbf{r} \in \mathcal{R}$:

$$(\exists \mathbf{x} \in \mathbf{C})(\mathbf{x} > \mathbf{r}) \Rightarrow (\exists \mathbf{x} \in \mathbf{C})(\exists \mathbf{q} \in \mathcal{Q})(\mathbf{x} > \mathbf{q} > \mathbf{r}, \mathbf{q} \in \mathbf{L})\,. \tag{2.37}$$

From definitions 143 (p. 170) and 389 (p. 488), equation (2.37) (p. 174) gives $\mathbf{r} \subset \mathbf{L}, \mathbf{r} \neq \mathbf{L}$. It implies (definition 145 (p. 171)) $\mathbf{r} < \mathbf{L}$. We also have (see A.3 (p. 441), definition 334 (p. 441)):

$$(\mathbf{r} \in \mathcal{R}, (\forall \mathbf{x} \in \mathbf{C})(\mathbf{r} \geq \mathbf{x})) \Rightarrow \mathbf{r} \geq \mathbf{L}\ .$$

Moreover $\mathbf{x} \in \mathbf{C}, \mathbf{x} > \mathbf{L}$ is impossible because we would have $(\exists \mathbf{q} \in \mathcal{Q})(\mathbf{x} > \mathbf{q} > \mathbf{L})$ which means $(\forall \mathbf{p} \in \mathbf{M})(\mathbf{p} > \mathbf{x} > \mathbf{q})$ and, from the definition of $\mathbf{L}$, $\mathbf{q} \in \mathbf{L}$. It means $\mathbf{q} \leq \mathbf{L}$. Then we have:

$$((\forall \mathbf{x} \in \mathbf{C})(\mathbf{x} \leq \mathbf{L}), \mathbf{r} \geq \mathbf{L}) \Rightarrow (\forall \mathbf{x} \in \mathbf{C})(\mathbf{x} \leq \mathbf{r})\ .$$

Then L is the smallest of all majoring elements of **C**. It is its **upper_limit** (definition 381 (p. 485)). **(Resp.** From proposition 254 (p. 172), since $(\forall \mathbf{q} \in \mathbf{L})(\forall \mathbf{x} \in \mathbf{C})(\mathbf{q} \leq \mathbf{x})$, we have:

$$(\forall \mathbf{x} \in \mathbf{C})(\mathbf{L} \leq \mathbf{x})\ .$$

From proposition 253 (p. 172) and definition 145 (p. 171), we have:

$$\mathbf{y} < \mathbf{L} \Rightarrow (\exists \mathbf{q} \in \mathcal{Q})(\mathbf{y} < \mathbf{q} < \mathbf{L}) \Rightarrow (\forall \mathbf{x} \in \mathbf{C})(\mathbf{y} < \mathbf{x}),$$
$$\mathbf{y} = \mathbf{L} \Rightarrow (\forall \mathbf{x} \in \mathbf{C})(\mathbf{y} \leq \mathbf{x})\ .$$

Then $\mathbf{y} \leq \mathbf{L} \Rightarrow (\forall \mathbf{x} \in \mathbf{C})(\mathbf{y} \leq \mathbf{x})$. From propositions 253 (p. 172) and 738 (p. 444), we have:

$$(\forall \mathbf{x} \in \mathbf{C})(\mathbf{y} \leq \mathbf{x}) \Rightarrow (\forall \mathbf{q} \in \mathbf{y})(\forall \mathbf{x} \in \mathbf{C})(\mathbf{q} \leq \mathbf{x}) \Rightarrow (\forall \mathbf{q} \in \mathbf{y})(\mathbf{q} \in \mathbf{L}) \Rightarrow \mathbf{y} \leq \mathbf{L}$$

**L** is the **lower_limit** of **C** ).

**Definition 146** $\min_{\mathbf{i} \in \mathbf{I}} \mathbf{a}(\mathbf{i}),\ \max_{\mathbf{i} \in \mathbf{I}} \mathbf{a}(\mathbf{i})$.

Having a set **I** and a map $\mathbf{a}(\mathbf{i}) : \mathbf{I} \to \mathcal{R}$ with $\mathbf{C} = \mathbf{a}(\mathbf{I})$, for simplification in case of $\mathcal{R}$, one calls $\min_{\mathbf{i} \in \mathbf{I}} \mathbf{a}(\mathbf{i})$ and $\max_{\mathbf{i} \in \mathbf{I}} \mathbf{a}(\mathbf{i})$, the **lower_limit** of **C** and the **upper_limit** of **C** in contradiction with definition 382 (p. 485).

**Proposition 259** For any $\mathbf{C} \subset \mathcal{R}$ with an **upper_limit** (resp. **lower_limit**) $\boldsymbol{\ell}$, we have:

$$\mathbf{u} \geq \boldsymbol{\ell} \Rightarrow (\forall \mathbf{v} \in \mathbf{C})(\mathbf{u} \geq \mathbf{v}) \tag{2.38}$$

resp.

$$\mathbf{u} \leq \boldsymbol{\ell} \Rightarrow (\forall \mathbf{v} \in \mathbf{C})(\mathbf{u} \leq \mathbf{v})\ .$$

**Proof**
It comes from the definition of $\boldsymbol{\ell}$ (definition 381 (p. 485)).

**Proposition 260** For any $\mathbf{C} \subset \mathcal{R}$ with an **upper_limit** (resp. **lower_limit**) $\boldsymbol{\ell}$, we have:

$$\mathbf{u} < \boldsymbol{\ell} \Rightarrow (\exists \mathbf{y} \in \mathbf{C})(\mathbf{u} < \mathbf{y} \leq \boldsymbol{\ell}) \tag{2.39}$$

resp.

$$\mathbf{u} > \boldsymbol{\ell} \Rightarrow (\exists \mathbf{y} \in \mathbf{C})(\mathbf{u} > \mathbf{y} \geq \boldsymbol{\ell})\ .$$

**Proof**
Otherwise **u** will be a majoring (resp. minoring) element of **C** which is in contradiction with the definition of $\boldsymbol{\ell}$.

## 2.3 Euclidean Metric

**Definition 147 Euclidean Metric, Euclidean Metric space.**

A set $\mathbf{A}$ has a Euclidean metric defined on it or is a Euclidean metric space, if there is a map:

$$(\mathbf{d}(\mathbf{x},\mathbf{y}) : \mathbf{A} \times \mathbf{A} \to \mathcal{R}^{+o})$$

with the following properties:

1. $\mathbf{x} \neq \mathbf{y} \Rightarrow \mathbf{d}(\mathbf{x},\mathbf{y}) > 0$
2. $\mathbf{d}(\mathbf{x},\mathbf{x}) = 0$
3. $\mathbf{d}(\mathbf{x},\mathbf{y}) = \mathbf{d}(\mathbf{y},\mathbf{x})$
4. $\mathbf{d}(\mathbf{x},\mathbf{y}) + \mathbf{d}(\mathbf{y},\mathbf{z}) \geq \mathbf{d}(\mathbf{x},\mathbf{z})$

Unfortunately, the term metric does not always in the literature refer to those properties. For instance the Minkowski metric does not satisfy those while the Riemannian and the Euclidean metric do. In that book, except if it is specified clearly otherwise, when we speak about metric, we refer to the above Euclidean metric.

**Definition 148 Minimum distance of a point to a subset of a metric set.**

Given a metric set $\mathbf{A}$, a point $\mathbf{x} \in \mathbf{A}$, a subset $\mathbf{B} \subset \mathbf{A}$, one calls $(\mathbf{md_B}(\mathbf{x}) : \mathbf{A} \to \mathcal{R}^{+0})$, the following map:

$$\mathbf{x} \in \mathbf{A} \Rightarrow \mathbf{md_B}(\mathbf{x}) = \min_{\mathbf{y} \in \mathbf{B}} ||\mathbf{x} - \mathbf{y}||$$

**Proposition 261** The map of definition 148 (p. 176) exists.

**Proof**
We keep notation of definition 148 (p. 176), for any $\mathbf{x} \in \mathbf{A}$, we can define $\mathbf{H_x} \subset \mathcal{R}^{+0}$ by:

$$\mathbf{d} \in \mathbf{H_x} \Leftrightarrow (\exists \mathbf{y} \in \mathbf{B})(\mathbf{d} = ||\mathbf{x} - \mathbf{y}||)$$

because we have $(\exists \mathbf{y} \in \mathbf{B})(\mathbf{d} = ||\mathbf{x} - \mathbf{y}||) \Rightarrow \mathbf{d} \in \mathcal{R}$ (proposition 750 (p. 450)). Since from definition 147 (p. 176) $\mathbf{d} \in \mathbf{H_x} \Rightarrow \mathbf{d} \geq 0$, $\mathbf{H_x}$ has a **lower_limit** (proposition 258 (p. 174), definition 381 (p. 485)) which defines $\mathbf{md_B}(\mathbf{x})$.

**Definition 149 Induced Euclidean Metric.**

$\mathbf{A}$ being a Euclidean metric space $(\mathbf{d}(\mathbf{x},\mathbf{y}) : \mathbf{A} \times \mathbf{A} \to \mathcal{R}^{+o})$, $\mathbf{d}$ when it is restricted to $\mathbf{B} \subset \mathbf{A}$, defines also a restricted map (definition 118 (p. 135)), $(\mathbf{d}(\mathbf{x},\mathbf{y}) : \mathbf{B} \times \mathbf{B} \to \mathcal{R}^{+o})$ which verifies all the above properties when $\mathbf{x}$ and $\mathbf{y}$ are in $\mathbf{B}$. It is called the induced metric of $\mathbf{A}$ in $\mathbf{B}$.

**Proposition 262** $\mathbf{A}$ being a Euclidean metric space,

$$(\mathbf{x} \in \mathbf{A}, \mathbf{y} \in \mathbf{A}, \mathbf{z} \in \mathbf{A}) \Rightarrow \mathbf{d}(\mathbf{x},\mathbf{y}) - \mathbf{d}(\mathbf{y},\mathbf{z}) \leq \mathbf{d}(\mathbf{x},\mathbf{z}) .$$

**Proof**
From the properties of definition 147, we have:

$$\mathbf{d}(\mathbf{x},\mathbf{z}) + \mathbf{d}(\mathbf{y},\mathbf{z}) = \mathbf{d}(\mathbf{x},\mathbf{z}) + \mathbf{d}(\mathbf{z},\mathbf{y}) \geq \mathbf{d}(\mathbf{x},\mathbf{y}) \Rightarrow \mathbf{d}(\mathbf{x},\mathbf{y}) - \mathbf{d}(\mathbf{y},\mathbf{z}) \leq \mathbf{d}(\mathbf{x},\mathbf{z}) .$$

**Definition 150 Open ball on a Euclidean metric space.**

$\mathbf{A}$ being a Euclidean metric space, one calls an open ball of center $\mathbf{x} \in \mathbf{A}$ and radius $\rho \in \mathcal{R}$, $\rho \geq 0$, the subset $\mathbf{B}(\mathbf{x}, \rho) = \mathbf{set}_{\mathbf{y}}\{\mathbf{y} \in \mathbf{A}, \mathbf{d}(\mathbf{x}, \mathbf{y}) < \rho\}$.
We have $\mathbf{B}(\mathbf{x}, \rho) \subset \mathbf{A}$. From definition 147 (p. 176), 2, when $\rho > 0$, $\mathbf{x} \in \mathbf{B}(\mathbf{x}, \rho)$. From definition 147 (p. 176), 1, 2 $(\forall \mathbf{x} \in \mathbf{A})(\mathbf{B}(\mathbf{x}, 0) = \emptyset)$.

**Definition 151 Open subsets on a Euclidean metric set.**

$\mathbf{A}$ being a Euclidean metric space, any subset $\mathbf{S} \subset \mathbf{A} \times \mathcal{R}^+$, defined an open set $\mathbf{O}$ of $\mathbf{A}$:

$$\mathbf{O} = \bigcup_{(\mathbf{x},\rho) \in \mathbf{S}} \mathbf{B}(\mathbf{x}, \rho) .$$

**Proposition 263** A Euclidean metric space with the above open subsets is a topological space.

**Proof**
The set $\mathcal{O}$ of $\mathbf{A}$ defined as

$$\mathcal{O} = \mathbf{set}_{\mathbf{O}}\{(\exists \mathbf{S} \subset \mathbf{A} \times \mathcal{R}^+), \mathbf{O} = \bigcup_{(\mathbf{x},\rho) \in \mathbf{S}} \mathbf{B}(\mathbf{x}, \rho)\}$$

is a set of opens because:

- The empty space is open, it corresponds to the empty space of the ball $\mathbf{B}(\mathbf{x}, 0)$ associated to any $\mathbf{x} \in \mathbf{A}$.
  $\mathbf{A}$ is $\mathcal{O}$ because since $\rho > 0 \Rightarrow \mathbf{x} \in \mathbf{B}(\mathbf{x}, \rho)$ and $(\forall \mathbf{x})(\mathbf{B}(\mathbf{x}, \rho) \subset \mathbf{A})$:

$$\mathbf{A} = \bigcup_{\mathbf{x} \in \mathbf{A}} \mathbf{B}(\mathbf{x}, 1)$$

  so equation (2.16) (p. 151) is verified.
- It verifies equation (2.17) (p. 151) according to proposition 202 (p. 152).
- The intersection of two open balls $\mathbf{B}(\mathbf{x}^1, \rho_1), \mathbf{B}(\mathbf{x}^2, \rho_2)$ is an open ball because:

$$\begin{aligned}&(\forall \mathbf{x} \in \mathbf{B}(\mathbf{x}^1, \rho_1) \cap \mathbf{B}(\mathbf{x}^2, \rho_2))(0 \leq \mathbf{d}(\mathbf{x}, \mathbf{x}^1) < \rho_1, 0 \leq \mathbf{d}(\mathbf{x}, \mathbf{x}^2) < \rho_2)\\ &\Rightarrow ((\exists \mathbf{r})(0 < \mathbf{r} \leq \min\{\rho_1 - \mathbf{d}(\mathbf{x}, \mathbf{x}^1), \rho_2 - \mathbf{d}(\mathbf{x}, \mathbf{x}^2)\},\\ &\mathbf{B}(\mathbf{x}, \mathbf{r}) \subset \mathbf{B}(\mathbf{x}^1, \rho_1), \mathbf{B}(\mathbf{x}, \mathbf{r}) \subset \mathbf{B}(\mathbf{x}^2, \rho_2))\end{aligned}$$

  which defines a map (see 4 (p. 132)) $(\mathbf{r}(\mathbf{x}) : \mathbf{B}(\mathbf{x}^1, \rho_1) \cap \mathbf{B}(\mathbf{x}^2, \rho_2)) \to \mathcal{R}^+)$. We have then used that map:

$$\mathbf{B}(\mathbf{x}^1, \rho_1) \cap \mathbf{B}(\mathbf{x}^2, \rho_2) = \bigcup_{\mathbf{x} \in \mathbf{B}(\mathbf{x}^1, \rho_1) \cap \mathbf{B}(\mathbf{x}^2, \rho_2))} \mathbf{B}(\mathbf{x}, \mathbf{r}(\mathbf{x})) .$$

  Then using definition 151, $\mathbf{B}(\mathbf{x}^1, \rho_1) \cap \mathbf{B}(\mathbf{x}^2, \rho_2))$ is open. One concludes using proposition 204 (p. 153).

**Proposition 264** $\mathbf{A}$ being a Euclidean metric topological space, $\mathbf{C}$ a subset of $\mathbf{A}$, the induced topology by $\mathbf{A}$ on $\mathbf{C}$ is the same as the topology of $\mathbf{C}$ generated by the induced metric by $\mathbf{A}$ on $\mathbf{C}$.

**Proof**
Calling $\mathcal{O}$ the set of opens of **A**, we have from definition 151:

$$O \in \mathcal{O} \Leftrightarrow (\exists S \subset A \times \mathcal{R}^+)(O = \bigcup_{(x,\rho)\in S} B(x,\rho)) .$$

Calling $\mathcal{O}_C$ the set of opens of **C** induced by the topology of **A**, we have from definition 133 (p. 151):

$$\begin{aligned} &O_C \in \mathcal{O}_C \Leftrightarrow (\exists O \in \mathcal{O})(O_C = O \cap B) \\ &\Leftrightarrow (\exists S \subset A \times \mathcal{R}^+)(O_C = \bigcup_{(x,\rho)\in S\times\mathcal{R}+} B(x,\rho) \cap C) . \end{aligned}$$

From definition 149 $x \in C \Rightarrow B(x,\rho) \cap C = B_C(x,\rho)$ where $B_C(x,\rho)$ is an open ball of **C** with the induced metric. We have from definition 147:

$$\begin{aligned} &y \in B(x,\rho) \Rightarrow d(y,x) < \rho \Rightarrow \\ &(u \in B(y,\rho - d(x,y)) \Rightarrow d(u,x) < d(u,y) + d(y,x) \le \rho \Rightarrow u \in B(x,\rho)) . \end{aligned}$$

Then

$$B(x,\rho) \cap C \subset \bigcup_{y\in B(x,\rho)\cap C} B(y,\rho - d(x,y)) \cap C,$$

$$\bigcup_{y\in B(x,\rho)\cap C} B(y,\rho - d(x,y)) \cap C \subset B(x,\rho) \cap C$$

we have

$$B(x,\rho) \cap C = \bigcup_{y\in B(x,\rho)\cap C} B(y,\rho - d(x,y)) \cap C .$$

Then $O_C$ is an open of the topology of the open balls of **C** as union of open balls (proposition 263 (p. 177), definition 132 (p. 151) equation (2.17)) If $O_C = \bigcup_{(x,\rho)\in S_C} B_C(x,\rho)$, we have

$$O_C = \bigcup_{(x,\rho)\in S_C} B(x,\rho) \cap C = C \cap \bigcup_{(x,\rho)\in S_C} B(x,\rho)$$

with $S_c \subset A \times \mathcal{R}^+$ then $O_C$ is an open in the induced topology.

**Proposition 265** Given a Euclidean metric topological space **A**, any finite subset of **A** different from **A** is closed.

**Proof**
Given **D** with $Card(D) \in \mathcal{Z}^{+0}$, if $Card(D) = 0$, $D = \emptyset$, $\mathcal{C}(D) = A$. $\mathcal{C}(D)$ is open by definition 132 (p. 151), 2.16.
$D \neq \emptyset, D \neq A$ means $\mathcal{C}(D) \neq \emptyset$ and we may consider for any $x \in \mathcal{C}(D)$ the set $S \subset \mathcal{R}^+$ of strictly positive reals defined by:

$$s \in S \Leftrightarrow (\exists y \in D)(s = d(x,y)) .$$

Since **S** is finite being equivalent to a subset of **D** which is finite, **S** has a lower limit $r(x) \in S$ proposition 255 (p. 172). Then $B(x, 0.5r(x)) \cap D = \emptyset$ which means $B(x, 0.5r(x)) \subset \mathcal{C}(D)$ and

$$\bigcup_{x\in\mathcal{C}(D)} B(x, 0.5r(x)) = \mathcal{C}(D) .$$

From definition 132 (p. 151), $\mathcal{C}(D)$ is open. $\mathcal{C}(D)$ being open from definition 135 (p. 154), **D** is closed.

**Proposition 266** Any element of an open $\mathbf{O}$ of a Euclidean metric topological space is the center of an open ball with a nonzero radius subset of $\mathbf{O}$.
**Proof**
$\mathbf{A}$ being a Euclidean metric topological space with the distance quoted as $\mathbf{d(x,y)}$ for $\mathbf{x} \in \mathbf{A}, \mathbf{y} \in \mathbf{A}$, from the definition 151 (p. 177), it follows that any element $\mathbf{u}$ of an open $\mathbf{O}$ of $\mathbf{A}$ is in an open ball $\mathbf{B}$ fully in $\mathbf{O}$ ($\mathbf{B} \subset \mathbf{O}$). Let us call $\mathbf{c}$ the center of the ball and $\boldsymbol{\rho}$ its radius. Since $\mathbf{u} \in \mathbf{B}(\mathbf{c}, \boldsymbol{\rho})$, we have $\boldsymbol{\rho} > \mathbf{d(u,c)} \geq 0$. The open ball $\mathbf{B}(\mathbf{u}, \boldsymbol{\rho} - \mathbf{d(u,c)})$ centered on $\mathbf{u}$ is fully in $\mathbf{B}$ so fully in $\mathbf{O}$ and has a nonzero radius (see definition 147 (p. 176)).

**Proposition 267** A Euclidean metric topological space is a separated set (definition 139 (p. 159)).
**Proof**
$\mathbf{A}$ being a Euclidean metric topological space, with the distance quoted as $\mathbf{d(x,y)}$ for $\mathbf{x} \in \mathbf{A}, \mathbf{y} \in \mathbf{A}$, from 4 of definition 147 (p. 176), there is no common point $\mathbf{z}$ to the balls $\mathbf{B(x, d(x,y)/2)}$ and $\mathbf{B(y, d(x,y)/2)}$ because otherwise we would have:

$$\mathbf{d(x,z) + d(z,y) < d(x,y)}$$

It would violate 4 of definition 147 (p. 176).

**Definition 152 Closed ball on a Euclidean metric space.**

$\mathbf{A}$ being a Euclidean metric space, one calls a closed ball of center $\mathbf{x} \in \mathbf{A}$ and radius $\boldsymbol{\rho} \in \mathcal{R}$, $\boldsymbol{\rho} \geq 0$, the subset $\bar{\mathbf{B}}(\mathbf{x}, \boldsymbol{\rho}) = \mathbf{set_y\{y \in A, d(x,y) \leq} \boldsymbol{\rho}\}$.
We have: $\bar{\mathbf{B}}(\mathbf{x}, \boldsymbol{\rho}) \subset \mathbf{A}$. From definition 147 (p. 176), 2, $\mathbf{x} \in \bar{\mathbf{B}}(\mathbf{x}, \boldsymbol{\rho})$. From definition 147 (p. 176), 1, 2 $(\forall \mathbf{x} \in \mathbf{A})(\bar{\mathbf{B}}(\mathbf{x}, 0) = \{\mathbf{x}\})$.

**Proposition 268** $\boldsymbol{\rho} > 0$, $\bar{\mathbf{B}}(\mathbf{x}, \boldsymbol{\rho})$ (definition 152 (p. 179)) is closed (definition 135 (p. 154)). For any $\mathbf{x} \in \mathbf{A}$, $\bar{\mathbf{B}}(\mathbf{x}, 0) = \{\mathbf{x}\}$ is closed. See proposition 514 (p. 288) for the case of $\mathcal{R}^{\mathbf{n}}$.
**Proof**
We have $\bar{\mathbf{B}}(\mathbf{x}, \boldsymbol{\rho})$ with $\boldsymbol{\rho} > 0$. From definition 152 (p. 179) we have (see definition 353 (p. 454) for $\mathcal{C}$):

$$\begin{aligned}&\mathbf{y} \in \mathcal{C}(\bar{\mathbf{B}}(\mathbf{x}, \boldsymbol{\rho})) \Rightarrow \mathbf{d(x,y)} > \boldsymbol{\rho} \Rightarrow \boldsymbol{\rho}_{\mathbf{y}} = \mathbf{d(x,y)} - \boldsymbol{\rho} > 0 \Rightarrow (\mathbf{d(z,y)} < \boldsymbol{\rho}_{\mathbf{y}} \\ &\Rightarrow \mathbf{d(z,x) > d(y,x) - d(z,y) > d(y,x)} - \boldsymbol{\rho}_{\mathbf{y}} = \boldsymbol{\rho} \Rightarrow \mathbf{z} \notin \bar{\mathbf{B}}(\mathbf{x}, \boldsymbol{\rho})) \\ &\Rightarrow \mathbf{B}(\mathbf{y}, \boldsymbol{\rho}_{\mathbf{y}}) \subset \mathcal{C}(\bar{\mathbf{B}}(\mathbf{x}, \boldsymbol{\rho}))\ .\end{aligned}$$

We have then:

$$\mathcal{C}(\bar{\mathbf{B}}(\mathbf{x}, \boldsymbol{\rho})) = \bigcup_{\mathbf{y} \in \mathcal{C}(\bar{\mathbf{B}}(\mathbf{x}, \boldsymbol{\rho}))} \mathbf{B}(\mathbf{y}, \boldsymbol{\rho}_{\mathbf{y}})$$

$\mathcal{C}(\bar{\mathbf{B}}(\mathbf{x}, \boldsymbol{\rho}))$ is open then $\bar{\mathbf{B}}(\mathbf{x}, \boldsymbol{\rho})$ is closed (definition 135 (p. 154)). If $\boldsymbol{\rho} = 0$, $\bar{\mathbf{B}}(\mathbf{x}, \boldsymbol{\rho})$ is closed from definition 147 (p. 176) point 1, propositions 222 (p. 159) and 265 (p. 178).

**Proposition 269** Any element of an open $\mathbf{O}$ of a Euclidean metric topological space is the center of a closed ball with a nonzero radius (definition 152 (p. 179)).
**Proof**
From proposition 266 (p. 179), any point $\mathbf{c} \in \mathbf{O}$ is center of an open ball $\mathbf{B}(\mathbf{c}, \boldsymbol{\rho}) \subset \mathbf{O}$. $\boldsymbol{\rho} \in \mathcal{R}^+$. Since $\boldsymbol{\rho} > 0$ we have from proposition 253 (p. 172):

$$(\exists \boldsymbol{\rho}' \in ]0, \boldsymbol{\rho}[)(0 < \boldsymbol{\rho}' < \boldsymbol{\rho})$$

Using that $\boldsymbol{\rho}'$, we have (definitions 349 (p. 451), 152 (p. 179), 150 (p. 177)):

$$\begin{aligned}&(\mathbf{x} \in \bar{\mathbf{B}}(\mathbf{c}, \boldsymbol{\rho}') \Leftrightarrow 0 < \mathbf{d(c,x)} \leq \boldsymbol{\rho}' < \boldsymbol{\rho} \Rightarrow \mathbf{x} \in \mathbf{B}(\mathbf{c}, \boldsymbol{\rho}) \subset \mathbf{O}) \\ &\Rightarrow \bar{\mathbf{B}}(\mathbf{c}, \boldsymbol{\rho}') \subset \mathbf{B}(\mathbf{c}, \boldsymbol{\rho}) \subset \mathbf{O}\ .\end{aligned}$$

**Proposition 270** $\mathbf{A}$ and $\mathbf{B}$ being two subsets of a Euclidean metric topological space $\mathbf{E}$, we have

$$\bar{\mathbf{A}} \cup \bar{\mathbf{B}} = \overline{\mathbf{A} \cup \mathbf{B}}$$

where $\bar{\mathbf{X}}$ means closure of $\mathbf{X}$ (definition 136 (p. 155)).

**Proof**
$\mathcal{O}$ is the set of opens of $\mathbf{E}$. From definition 136 (p. 155),

$$\begin{aligned}(\mathbf{p} \in \bar{\mathbf{A}} \vee \mathbf{p} \in \bar{\mathbf{B}}) &\Rightarrow (\forall \mathbf{O} \in \mathcal{O})(\exists \mathbf{q} \in \mathbf{O} \cap \mathbf{A} \vee \exists \mathbf{q} \in \mathbf{O} \cap \mathbf{B}) \\ &\Rightarrow (\forall \mathbf{O} \in \mathcal{O})(\exists \mathbf{q} \in \mathbf{O} \cap \mathbf{A} \cup \mathbf{B}) \Rightarrow \mathbf{p} \in \overline{\mathbf{A} \cup \mathbf{B}}\end{aligned}$$

then $\bar{\mathbf{A}} \cup \bar{\mathbf{B}} \subset \overline{\mathbf{A} \cup \mathbf{B}}$. From definitions 136 (p. 155), 150 (p. 177), 151 (p. 177) and proposition 263 (p. 177),

$$\begin{aligned}\mathbf{p} \in \overline{\mathbf{A} \cup \mathbf{B}} &\Rightarrow (\forall \mathbf{r} > 0)(\exists \mathbf{q} \in \mathbf{B}(\mathbf{p}, \mathbf{r}) \cap (\mathbf{A} \cup \mathbf{B})) \\ &\Rightarrow (\forall \mathbf{r} > 0)(\exists \mathbf{q} \in \mathbf{B}(\mathbf{p}, \mathbf{r}) \cap \mathbf{A}) \vee (\exists \mathbf{q} \in \mathbf{B}(\mathbf{p}, \mathbf{r}) \cap \mathbf{B})\ .\end{aligned}$$

We can define two sets of $\mathcal{R}^+$ $\mathbf{R_a}$ (resp. $\mathbf{R_b}$) for which $\mathbf{r} \in \mathbf{R_a} \Rightarrow (\exists \mathbf{q} \in \mathbf{B}(\mathbf{p}, \mathbf{r}) \cap \mathbf{A})$ (resp. $\mathbf{r} \in \mathbf{R_b} \Rightarrow (\exists \mathbf{q} \in \mathbf{B}(\mathbf{p}, \mathbf{r}) \cap \mathbf{B})$). Since 0 is a minoring element of both sets, $\mathbf{R_a}$ (resp. $\mathbf{R_b}$) has a lower limit $\mathbf{r_a} \geq \mathbf{0}$ (resp. $\mathbf{r_b} \geq \mathbf{0}$). One of them has to be zero, otherwise if we have $\mathbf{0} < \mathbf{r} < \mathbf{min}(\mathbf{r_a}, \mathbf{r_b})$, $\mathbf{B}(\mathbf{p}, \mathbf{r})$ would have no point either in $\mathbf{A}$ or in $\mathbf{B}$ which is against $\mathbf{p} \in \overline{\mathbf{A} \cup \mathbf{B}}$. If let us say $\mathbf{r_a}$ is zero then for any $\mathbf{r}$ we have $\mathbf{r}' < \mathbf{r}, \mathbf{r}' \in \mathbf{R_a}$. Then

$$\exists \mathbf{q} \in \mathbf{B}(\mathbf{p}, \mathbf{r}') \cap \mathbf{A} \subset \mathbf{B}(\mathbf{p}, \mathbf{r}) \cap \mathbf{A} \subset \mathbf{B}(\mathbf{p}, \mathbf{r}) \cap (\mathbf{A} \cup \mathbf{B})$$

$\mathbf{p} \in \overline{\mathbf{A}} \cup \overline{\mathbf{B}}$. We conclude by the extensionality principle 754 (p. 451).

**Proposition 271** Any compact subset of a metric space $\mathbf{E}$ is a closed set included in a ball.

**Proof**
Setting $\mathbf{C}$ as a compact subset of $\mathbf{E}$, from 267 (p. 179) $\mathbf{E}$ is separated, from 226 (p. 160) $\mathbf{C}$ is closed. Given $\mathbf{r} \in \mathcal{R}^+$ and the set of opens:

$$\mathbf{S} = \mathbf{set_O}\{(\exists \mathbf{x} \in \mathbf{C})(\mathbf{O} = \mathbf{B}(\mathbf{x}, \mathbf{r}))\}$$

we have

$$\mathbf{C} \subset \bigcup_{\mathbf{x} \in \mathbf{C}} \mathbf{B}(\mathbf{x}, \mathbf{r}) = \bigcup_{\mathbf{O} \in \mathbf{S}} \mathbf{O}\ .$$

Since $\mathbf{C}$ is compact there is a finite subset $\mathbf{I}$ of $\mathbf{S}$ such that

$$\mathbf{C} \subset \bigcup_{\mathbf{O} \in \mathbf{I}} \mathbf{O}\ .$$

Using the triangle inequality, we have:

$$\mathbf{B}(\mathbf{x}, \mathbf{r}) \subset \mathbf{B}(\mathbf{0}, \|\mathbf{x}\| + \mathbf{r})\ .$$

Since the set $\mathbf{A} = \mathbf{set_z}(\exists \mathbf{x} \in \mathbf{I})(\mathbf{z} = \|\mathbf{x}\| + \mathbf{r} \in \mathcal{R})$ is finite, it gets an **upper_limit**: $\mathbf{R} = \mathbf{upper\_limit}(\mathbf{A})$ (proposition 255 (p. 172)) and we have:

$$\mathbf{C} \subset \mathbf{B}(\mathbf{0}, \mathbf{R})\ .$$

## 2.4 Metric and topology on $\mathcal{R}$

**Definition 153 Sign of a real number.**

The sign of a real number $\mathbf{u} \in \mathcal{R}$ is a map $(\mathbf{sign(u)} : \mathcal{R} \to \{-1,+1\})$ such that $\mathbf{sign}(\mathcal{R}^-) = -1, \mathbf{sign}(\mathcal{R}^{0+}) = +1$.

**Definition 154 Absolute value of a real number.**

It is the same definition as definition 407 (p. 532), it is a map $(|\mathbf{x}| : \mathcal{R} \to \mathcal{R}^+)$ defined by

$$\mathbf{x} \in \mathcal{R}^{0+} \Rightarrow |\mathbf{x}| = \mathbf{x},$$
$$\mathbf{x} \in \mathcal{R}^{-} \Rightarrow |\mathbf{x}| = -\mathbf{x}$$

or by $|\mathbf{x}| = \mathbf{sign(x)x}$.

**Proposition 272** The absolute value defines a metric and a topology on $\mathcal{R}$.

**Proof**
Given $\mathbf{x} \in \mathcal{R}$, $\mathbf{y} \in \mathcal{R}$, $\mathbf{d(x,y)} = |\mathbf{x} - \mathbf{y}|$ defines a metric on $\mathcal{R}$: The definition 147 (p. 176) points 1 to 3 are clearly verified. From proposition 965 (p. 541) extended to $\mathcal{Q}$ as for proposition 995 and extended to $\mathcal{R}$ as in section 2.2.3 (p. 173), we have with $\mathbf{z} \in \mathcal{R}$:

$$|\mathbf{x} - \mathbf{z}| = |\mathbf{x} - \mathbf{y} + \mathbf{y} - \mathbf{z}| \leq |\mathbf{x} - \mathbf{y}| + |\mathbf{y} - \mathbf{z}| \tag{2.40}$$

it is the definition 147 (p. 176) point 4.

**Proposition 273** The absolute value verifies also the axioms of a norm see definition 203 (p. 237):

$$(\mathbf{x},\mathbf{y}) \in \mathcal{R}^2 \Rightarrow |\mathbf{x}| - |\mathbf{y}| \leq |\mathbf{x} + \mathbf{y}| \leq |\mathbf{x}| + |\mathbf{y}| \,.$$

**Proof**
Setting $\mathbf{y} = \mathbf{0}$ in equation (2.40) (p. 181) and in proposition 262 (p. 176) and changing $\mathbf{z}$ to $-\mathbf{z}$.

**Definition 155 Open interval (see definition 378 (p. 484)).**

An open ball of $\mathcal{R}$, $\mathbf{B(x,r)}$, is the interval $]\mathbf{x} - \mathbf{r}, \mathbf{x} + \mathbf{r}[$ which is also called open interval (see definition 378 (p. 484)).

**Definition 156 Closed interval.**

The closure of a ball of $\mathcal{R}$, $\bar{\mathbf{B}}(\mathbf{x},\mathbf{r})$, is the interval $\bar{\mathbf{B}}(\mathbf{x},\mathbf{r}) = [\mathbf{x} - \mathbf{r}, \mathbf{x} + \mathbf{r}]$ (see proposition 514 (p. 288)). It is also called closed interval (see definition 378 (p. 484)).

**Proposition 274** $\bar{\mathbf{B}}(\mathbf{x},\mathbf{r}) = [\mathbf{x} - \mathbf{r}, \mathbf{x} + \mathbf{r}]$ is closed and is the closure of $]\mathbf{x} - \mathbf{r}, \mathbf{x} + \mathbf{r}[$.

**Proof**
From proposition 268 (p. 179), $[\mathbf{x} - \mathbf{r}, \mathbf{x} + \mathbf{r}]$ is closed. Given $\mathbf{y} \in [\mathbf{x} - \mathbf{r}, \mathbf{x} + \mathbf{r}]$ and an open $\mathbf{O_y}$ containing $\mathbf{y}$, from proposition 266 (p. 179), we have $\mathbf{r_y} > \mathbf{0}$ such as $]\mathbf{y} - \mathbf{r_y}, \mathbf{y} + \mathbf{r_y}[ \subset \mathbf{O_y}$. It means, calling $\mathbf{r'_y} = \mathbf{min(r_y, d(x,y))/2}$ and $\mathbf{z} = (\mathbf{x} - \mathbf{y})\mathbf{r'_y}/|\mathbf{x} - \mathbf{y}| + \mathbf{y}$, that $|\mathbf{z} - \mathbf{y}| < \mathbf{r_y}, |\mathbf{z} - \mathbf{x}| < \mathbf{d(x,y)} \leq \mathbf{r}$. Then $\mathbf{O_y}$ contains a point of $]\mathbf{x} - \mathbf{r}, \mathbf{x} + \mathbf{r}[$. Since $[\mathbf{x} - \mathbf{r}, \mathbf{x} + \mathbf{r}]$ is closed, any point not in it is contained in an open with no point of $]\mathbf{x} - \mathbf{r}, \mathbf{x} + \mathbf{r}[$. $[\mathbf{x} - \mathbf{r}, \mathbf{x} + \mathbf{r}]$ is the closure of $]\mathbf{x} - \mathbf{r}, \mathbf{x} + \mathbf{r}[$.

**Proposition 275** $\mathbf{a} \in \mathcal{R}$, $\mathbf{b} \in \mathcal{R}$, any interval (definition 378 (p. 484)) of type: $]\mathbf{a}, \mathbf{b}[$ or $]\mathbf{a}, \rightarrow [$, $] \leftarrow, \mathbf{a}[$ are open.
Any interval of type $[\mathbf{a}, \mathbf{b}]$ or $[\mathbf{a}, \rightarrow [$, $] \leftarrow, \mathbf{a}]$ are closed.

**Proof**
From definitions 150 (p. 177), 154 (p. 181) and proposition 272 (p. 181), we have:

$$\mathbf{B}(\frac{\mathbf{a}+\mathbf{b}}{2}, \frac{|\mathbf{a}-\mathbf{b}|}{2}) = ]\mathbf{a},\mathbf{b}[\ .$$

Then $]\mathbf{a},\mathbf{b}[$ is open and we also have:

$$]\mathbf{a},\rightarrow[ = \bigcup_{\mathbf{x}\in]\mathbf{a},\rightarrow[} \mathbf{B}(\mathbf{x},\mathbf{x}-\mathbf{a})\ .$$

Then from definition 132 (p. 151), $]\mathbf{a},\rightarrow[$ is open. For the same reason $]\leftarrow,\mathbf{a}[$ is open. $[\mathbf{a},\mathbf{b}]$ is closed from proposition 274 (p. 181). $[\mathbf{a},\rightarrow[ = \mathcal{C}(]\leftarrow,\mathbf{a}[),\ ]\leftarrow,\mathbf{a}] = \mathcal{C}(]\mathbf{a},\rightarrow[)$ are closed from definition 135 (p. 154) and first part of proposition 275 (p. 181).

**Proposition 276** Given $\mathbf{A} \subset \mathcal{R}$, if it exists, its **upper_limit** (resp. **lower_limit**) belongs to its closure (definition 136 (p. 155)) $\bar{\mathbf{A}}$.

**Proof**
Calling $\boldsymbol{\ell} = \mathbf{upper\_limit}(\mathbf{A})$ (resp. $\boldsymbol{\ell} = \mathbf{lower\_limit}(\mathbf{A})$) (definition 381 (p. 485), see section 2.2.4), from proposition 266 (p. 179), for any open $\mathbf{O}$ containing $\boldsymbol{\ell}$ there is $\epsilon > 0$ such that $]\boldsymbol{\ell}-\epsilon,\boldsymbol{\ell}+\epsilon[\subset \mathbf{O}$. From proposition 260 (p. 175), there is an element of $\mathbf{A}$ in $]\boldsymbol{\ell}-\epsilon,\boldsymbol{\ell}+\epsilon[$ then in $\mathbf{O}$. From definition 136 (p. 155), $\boldsymbol{\ell} \in \bar{\mathbf{A}}$.

**Proposition 277** Given a Euclidean metric space $\mathbf{E}$ the map $(\mathbf{d}(\mathbf{x},\mathbf{y}) : \mathbf{E} \rightarrow \mathcal{R})$ is continuous in $\mathbf{x}$ and $\mathbf{y}$.

**Proof**
Any open of $\mathcal{R}$ is the union of opens of the type $]\mathbf{a},\mathbf{b}[$ (definition 151 (p. 177)). From definition 132 (p. 151) equation (2.17) and proposition 362 (p. 462), it is enough to prove that the set $\mathbf{O}$ defined by

$$\mathbf{y} \in \mathbf{O} \Leftrightarrow \mathbf{d}(\mathbf{x},\mathbf{y}) \in ]\mathbf{a},\mathbf{b}[$$

is open.

- If $\mathbf{a} \leq 0$ and $\mathbf{b} \geq 0$, $\mathbf{O_y} = \mathbf{B}(\mathbf{x},\mathbf{b})$. $\mathbf{B}(\mathbf{x},\mathbf{b})$ is open from definitions 150 (p. 177) and 151 (p. 177).
- If $\mathbf{a} \leq 0$ and $\mathbf{b} \leq 0$ $\mathbf{O} = \emptyset$ so it is open from definition 132 (p. 151) equation (2.16).
- If $\mathbf{a}$ and $\mathbf{b}$ are both positive $\mathbf{O} = \mathcal{C}(\bar{\mathbf{B}}(\mathbf{x},\mathbf{a})) \cap \mathbf{B}(\mathbf{x},\mathbf{b})$. $\bar{\mathbf{B}}(\mathbf{x},\mathbf{a})$ is closed (proposition 268 (p. 179)). $\mathcal{C}(\bar{\mathbf{B}}(\mathbf{x},\mathbf{a}))$ is open (definition 135 (p. 154)), then $\mathbf{O}$ is open from definition 132 (p. 151), (2.18).

Proposition 277 (p. 182) is true from definition 141 (p. 165).

## 2.5 Behavior at a point

**Definition 157 Limit of a map at a point.**

Given two topological spaces $\mathbf{A}$ and $\mathbf{B}$ and a point $\mathbf{x} \in \mathbf{A}$, a map $(\mathbf{f} : \mathbf{A} - \{\mathbf{x}\} \rightarrow \mathbf{B})$[12] is said to have a **limit** $\mathbf{y}$ for $\mathbf{x}$, if for any open $\mathbf{O_y} \subset \mathbf{B}$ containing $\mathbf{y}$ there is an open $\mathbf{O_x} \subset \mathbf{A}$ containing $\mathbf{x}$ such that $\mathbf{f}(\mathbf{O_x} - \{\mathbf{x}\}) \subset \mathbf{O_y}$.

[12] If $\mathbf{f}$ is defined for all $\mathbf{A}$, one restricts $\mathbf{f}$ to $\mathbf{A} - \{\mathbf{x}\} = \mathcal{C}_{\mathbf{A}}(\{\mathbf{x}\})$ (see definitions 118 (p. 135), 353 (p. 454)).

**Proposition 278** Keeping the above notations, if $\mathbf{A}$ and $\mathbf{B}$ are Euclidean metric topological spaces (definition 147 (p. 176)) with a distance quoted as $\mathbf{d_A(u,v)}$ or $\mathbf{d_B(u,v)}$, definition 157 (p. 182) is equivalent to:

$$(\forall \alpha \in \mathcal{R}^+)(\exists \epsilon \in \mathcal{R}^+)(\forall \mathbf{u} \in \mathbf{A})(0 < \mathbf{d_A(u,x)} < \epsilon \Rightarrow \mathbf{d_B(f(u),y)} < \alpha) \, . \tag{2.41}$$

**Proof**

- In any open $\mathbf{O_y}$ (proposition 266 (p. 179)), there is an open ball (definition 150 (p. 177)) centered on $\mathbf{y}$ the radius of which can be called $\alpha$. If equation (2.41) is true, the open of definition 157 can be the open ball of radius $\epsilon$ centered on $\mathbf{x}$.
- If definition 157 is true to any open ball $\mathbf{B(y,\alpha)} \subset \mathbf{B}$ of radius $\alpha$ centered on $\mathbf{y}$, we have an open $\mathbf{O} \subset \mathbf{A}$ containing $\mathbf{x}$ such that $\mathbf{f(O) - \{x\} \subset B(y,\alpha)}$. $\mathbf{O}$ contains an open ball $\mathbf{B(x,\epsilon)} \subset \mathbf{A}$ of radius $\epsilon$ centered on $\mathbf{x}$ (proposition 266 (p. 179)) and we have $\mathbf{f(B(x,\epsilon) - \{x\}) \subset B(y,\alpha)}$. $\epsilon$ satisfies expression (2.41).

**Proposition 279** When a map $(\mathbf{f(x) : A \to B})$, between two topological spaces $\mathbf{A}$ and $\mathbf{B}$, has a limit for $\mathbf{u}$, this limit is unique.

**Proof**
If there were more than one limit, we should have at least two limits $\mathbf{y_1}$ and $\mathbf{y_2}$. Taking $\alpha = \mathbf{d_B(y_1,y_2)/3}$ we have two corresponding $\epsilon_1$ and $\epsilon_2$ satisfying relation (2.41). Since $\mathbf{d(u,u) = 0 \leq \min(\epsilon_1,\epsilon_2)}$, (2.41) with point 4 of definition 147 (p. 176), it gives

$$\mathbf{d(f(u),y_1) \leq d_B(y_1,y_2)/3, d(f(u),y_2) \leq d_B(y_1,y_2)/3}$$
$$\Rightarrow \mathbf{d_B(y_1,y_2) \leq d(f(u),y_1) + d(f(u),y_2) \leq \frac{2}{3} d_B(y_1,y_2)}$$

which is impossible to satisfy except if $\mathbf{d_B(y_1,y_2) = 0}$. That gives from 2 and 3 of definition 147 (p. 176) $\mathbf{y_1 = y_2}$.

**Definition 158 $\mathbf{\lim_{x\to u} f(x)}$.**

With notations of proposition 279 (p. 183), the unique limit of $\mathbf{f}$ for $\mathbf{u} \in \mathbf{A}$ is written as:

$$\lim_{\mathbf{x}\to\mathbf{u}} \mathbf{f(x)} \, . \tag{2.42}$$

**Proposition 280** Given two topological spaces $\mathbf{A}$ & $\mathbf{B}$, a map $(\mathbf{f : A \to B})$, an open $\mathbf{O} \subset \mathbf{A}$, if $\mathbf{x} \in \mathbf{O}$, the limits of $\mathbf{f}$ at $\mathbf{x}$ are equal to the limits at $\mathbf{x}$ of the map $\mathbf{f}$ restricted to $\mathbf{O}$ (definition 118 (p. 135)).

**Proof**
The induced topology of $\mathbf{A}$ on $\mathbf{O}$ (definition 133 (p. 151)) is the same as $\mathbf{A}$ (proposition 200 (p. 152)). Then if $\mathbf{y}$ is limit of $\mathbf{f}$ restricted to $\mathbf{O}$, any open $\mathbf{O_x} \subset \mathbf{O}$ such that $\mathbf{f(O_x - x) \subset O_y}$ verify also $\mathbf{O_x} \subset \mathbf{A}$ and open on $\mathbf{A}$. If $\mathbf{y}$ is limit for $\mathbf{f}$ to any $\mathbf{O_y}$ there is $\mathbf{O_x \cap O}$ which is open of $\mathbf{O}$ (definition 133 (p. 151)) and such that $\mathbf{f(O_x \cap O) \in O_y}$ (proposition 157 (p. 134)).

**Definition 159 Continuous map at a point.**

If the limit of a map at a point $\mathbf{x}$ is the image by the map of the point $\mathbf{x}$ the map is said to be continuous at $\mathbf{x}$.

**Proposition 281** If a map is continuous, it is continuous at any point.

**Proof**

it is enough to take $\mathbf{O_x = f^{-1}(O_{f(x)})}$.

**Proposition 282** If a map $\mathbf{(f : A \longrightarrow B)}$ is continuous at any point of a subset $\mathbf{C \subset A}$, it is continuous on $\mathbf{C}$.

**Proof**

For any open $\mathbf{O \subset B}$, we have:

$$f^{-1}(O) \cap C = \bigcup_{x \in f^{-1}(O) \cap C} O_x \cap C$$

where $\mathbf{O_x}$ is an open around $\mathbf{x}$ such that $\mathbf{f(O_x) \subset O}$ which implies $\mathbf{O_x \subset f^{-1}(O)}$. It exists since $\mathbf{f(x) \in O}$ and $\mathbf{f}$ continuous at $\mathbf{x}$ when $\mathbf{x \in f^{-1}(O) \cap C}$. Then $\mathbf{f^{-1}(O) \cap C}$ is open as a union of open subsets in the induced topology on $\mathbf{C}$.

**Definition 160 Uniformly continuous map defined between a subset C of A and a set B, the topology being induced by a Euclidean metric.**

The map $\mathbf{f : C \longrightarrow B}$ is uniformly continuous if we have:

$$(\forall \alpha \in \mathcal{R}^+)(\exists \epsilon \in \mathcal{R}^+)(\forall u \in C)(\forall v \in C)(d_A(u,v) < \epsilon \Rightarrow d_B(f(u), f(v)) < \alpha) \, .$$

**Proposition 283** A continuous map defined between two Euclidean metric topological spaces $\mathbf{A}$ and $\mathbf{B}$ on a compact subset $\mathbf{C}$ of $\mathbf{A}$ is uniformly continuous on the compact subset $\mathbf{C}$.

**Proof**

$\mathbf{A}$ and $\mathbf{B}$ being Euclidean metric topological sets, given a map $\mathbf{(f : A \longrightarrow B)}$ continuous on the compact $\mathbf{C \subset B}$, from proposition 236 $\mathbf{(f : A \longrightarrow C)}$ is continuous. From proposition 281 and from equation (2.41) in definition 157 (p. 182), we have

$$(\forall \alpha \in \mathcal{R}^+)(\forall x \in C)(\exists \epsilon_x \in \mathcal{R}^+)((u \in C, d_A(u,x) < 2\epsilon_x) \Rightarrow d_B(f(u), f(x)) < \alpha/2) \, .$$

It defines a map $\mathbf{(\epsilon_x : C \longrightarrow \mathcal{R}^+)}$ (see point 4 in definition 112 (p. 131)). Since $\mathbf{C}$ is compact (see definition 140 (p. 160)), we have :

$$C \subset \bigcup_{x \in C} B(x, \epsilon_x) \Rightarrow (\exists I)(Card(I) \in \mathcal{Z}^+, I \subset C, C \subset \bigcup_{x \in I} B(x, \epsilon_x))$$

calling $\mathbf{\epsilon = \min_{x \in I} \epsilon_x}$ which is bigger than zero according to proposition 255 (p. 172). Then we have:

$$\begin{aligned}
&(\forall u \in C)(\forall v \in C)(d_A(u,v) < \epsilon \Rightarrow \\
&(\exists x \in I)(d_A(u,x) \leq \epsilon_x \quad \Rightarrow d_A(v,x) < d_A(u,v) + d_A(u,x) < 2\epsilon_x \\
&\qquad\qquad\qquad\qquad\quad \Rightarrow d_B(f(u), f(x)) < \alpha/2, d_B(f(v), f(x)) < \alpha/2 \\
&\qquad\qquad\qquad\qquad\quad \Rightarrow d_B(f(u), f(v)) < d_B(f(u), f(x)) + d_B(f(v), f(x)) < \alpha) \\
&\Rightarrow d_B(f(u), f(v)) < \alpha)
\end{aligned}$$

then

$$(\forall \alpha \in \mathcal{R}^+)(\exists \epsilon \in \mathcal{R}^+)(\forall u \in C)(\forall v \in C)(d(u,v) < \epsilon \Rightarrow d_B(f(u), f(v)) < \alpha) \, .$$

**Proposition 284** Given 3 topological sets $\mathbf{A}$, $\mathbf{B}$, $\mathbf{C}$, a map $\mathbf{F_{B,A}}$ from $\mathbf{A}$ to $\mathbf{B}$ continuous at a point $\mathbf{x} \in \mathbf{A}$, a map $\mathbf{F_{C,B}}$ from $\mathbf{B}$ to $\mathbf{C}$ continuous at $\mathbf{F_{B,A}(x)} \in \mathbf{B}$. The map $\mathbf{F_{C,B}(F_{B,A})}$ is continuous at $\mathbf{x}$ .

**Proof**

Applying twice the definition of a continuous function at a point for any open around $\mathbf{O_z}$, $\mathbf{z} = \mathbf{F_{C,B}(y)}$ with $\mathbf{y} = \mathbf{F_{B,A}(x)}$:

$$\begin{aligned}&(\forall \mathbf{O_z} \subset \mathbf{C})(\exists \mathbf{O_y})((\mathbf{F_{C,B}(O_y)} \subset \mathbf{O_z}),\\&(\exists \mathbf{O_x})(\mathbf{F_{B,A}(O_x)} \subset \mathbf{O_y}))\\&\Rightarrow (\forall \mathbf{O_z} \subset \mathbf{C})(\exists \mathbf{O_x})(\mathbf{F_{C,B}(F_{B,A}(O_x))} \subset \mathbf{O_z})\,.\end{aligned}$$

## 2.6 Some properties of continuous maps from $\mathcal{R}$ to $\mathcal{R}$

**Proposition 285** The map $(\mathbf{1/x} : \mathcal{R} - \{\mathbf{0}\} \to \mathcal{R})$ is continuous for $\mathbf{x} \neq \mathbf{0}$.

**Proof**

$$|\frac{1}{\mathbf{x}+\Delta\mathbf{x}} - \frac{1}{\mathbf{x}}| = \frac{|\Delta\mathbf{x}|}{|\mathbf{x}^2 + \mathbf{x}\Delta\mathbf{x}|}$$

$$|\Delta\mathbf{x}| < |\mathbf{x}|/2 \Rightarrow |\mathbf{x}^2 + \mathbf{x}\Delta\mathbf{x}| > \mathbf{x}^2 - |\mathbf{x}||\Delta\mathbf{x}| > \mathbf{x}^2/2$$

then

$$(\forall \alpha > 0)(\exists \epsilon = \min(0.5\alpha\mathbf{x}^2, |\mathbf{x}|/2) > 0)(|\Delta\mathbf{x}| < \epsilon \Rightarrow |\frac{1}{\mathbf{x}+\Delta\mathbf{x}} - \frac{1}{\mathbf{x}}| < \frac{\alpha/2}{\mathbf{x}^2/2} = \alpha)\,.$$

**Proposition 286** The maps $\mathcal{R}^2$ to $\mathcal{R}$ defined by the sum and product are continuous.

**Proof**

$$\begin{aligned}&(\forall \alpha \in \mathcal{R}^+)(\forall \mathbf{x} \in \mathcal{R}^2)(\forall \mathbf{y} \in \mathcal{R}^2)(\exists \epsilon = \alpha/\sqrt{2})(\|\mathbf{x}-\mathbf{y}\| < \epsilon\\&\Rightarrow |(\mathbf{x_1}+\mathbf{x_2}) - (\mathbf{y_1}+\mathbf{y_2})| < \alpha)\,.\end{aligned}$$

Then the map $\mathbf{x}+\mathbf{y}$ is continuous from proposition 532 (p. 296). We also have:

$$\begin{aligned}&(\forall \alpha \in \mathcal{R}^+)(\forall \mathbf{x} \in \mathcal{R}^2)(\forall \mathbf{y} \in \mathcal{R}^2)(\exists \epsilon = \min(\alpha/\sqrt{2}, \tfrac{\alpha^2}{4|\mathbf{x_1}|}, \tfrac{\alpha^2}{4|\mathbf{x_2}|}))\\&(\|\mathbf{x}-\mathbf{y}\| < \epsilon \Rightarrow |\mathbf{x_1x_2} - \mathbf{y_1y_2}| < \alpha)\,.\end{aligned} \tag{2.43}$$

Calling $\Delta_1 = \mathbf{x_1} - \mathbf{y_1}$ and $\Delta_2 = \mathbf{x_2} - \mathbf{y_2}$, we have:

$$\|\mathbf{x}-\mathbf{y}\|^2 = \Delta_1^2 + \Delta_2^2 \Rightarrow (\|\mathbf{x}-\mathbf{y}\| < \epsilon \Rightarrow |\Delta_1| < \epsilon, |\Delta_2| < \epsilon)$$

$$|\mathbf{x_1x_2} - \mathbf{y_1y_2}|^2 = |\mathbf{x_1x_2} - (\mathbf{x_1}+\Delta_1)(\mathbf{x_2}+\Delta_2)| = |\mathbf{x_1}\Delta_2 + \mathbf{x_2}\Delta_1 + \Delta_1\Delta_2|$$

then

$$|\mathbf{x_1x_2} - \mathbf{y_1y_2}|^2 < |\mathbf{x_1}|\epsilon + |\mathbf{x_2}|\epsilon + \epsilon^2$$

and using equation (2.43) (p. 185), we have $|\mathbf{x_1x_2} - \mathbf{y_1y_2}|^2 < \alpha^2$ and $|\mathbf{x_1x_2} - \mathbf{y_1y_2}| < \alpha$.

**Proposition 287** If $\mathbf{f(x)}$ and $\mathbf{g(x)}$ are two continuous functions from $\mathcal{R}$ to $\mathcal{R}$ the sum $\mathbf{f(x)}+\mathbf{g(x)}$ and product $\mathbf{f(x)g(x)}$ are continuous functions from $\mathcal{R}$ to $\mathcal{R}$.

**Proof**

$\mathbf{f(x)}$ and $\mathbf{g(x)}$ are defining a map $((\mathbf{f},\mathbf{g}) : \mathcal{R} \to \mathcal{R}^2)$ since $\mathbf{x} \in \mathcal{R}^\mathbf{n}$ and $(\mathbf{f(x)}, \mathbf{g(x)}) \in \mathcal{R}^2$. $(\mathbf{f},\mathbf{g})$ is continuous according to proposition 244 (p. 168). $\mathbf{f(x)}+\mathbf{g(x)}$ and $\mathbf{f(x)g(x)}$ are composite maps of $(\mathbf{f},\mathbf{g})$ and respectively of the maps sum and product $(\mathcal{R}^2 \to \mathcal{R})$ which are continuous, proposition 286 (p. 185). From proposition 241 (p. 167), $\mathbf{f(x)}+\mathbf{g(x)}$ and $\mathbf{f(x)g(x)}$ are continuous.

**Proposition 288** The product and sum of finite sets of continuous functions are continuous.

**Proof**
Calling $\mathcal{F}$ the set of continuous function $\mathcal{R} \to \mathcal{R}$ [13], having two finite sets $\mathbf{I_p}$ and $\mathbf{J_s}$ and two maps $(\mathbf{f_i} : \mathbf{I_p} \to \mathcal{F})$, $(\mathbf{g_j} : \mathbf{J_s} \to \mathcal{F})$, proposition 288 (p. 186) means:

$$\sum_{j \in J_s} g_j \times \prod_{i \in I_p} f_j \in \mathcal{F}, \ \prod_{i \in I_p} f_j \times \sum_{j \in J_s} g_j \in \mathcal{F} . \tag{2.44}$$

From proposition 287 (p. 185), equation (2.44) is true for $\mathbf{Card(I_p) = Card(J_s) = 1}$. From the same proposition, if it is true for $\mathbf{n_i = Card(I_p), n_j = Card(J_s)}$, it is also true when $\mathbf{Card(I_p) = n_i + 1}$, $\mathbf{Card(J_s) = n_j}$ or when $\mathbf{Card(I_p) = n_i}$, $\mathbf{Card(J_s) = n_j + 1}$. Then from proposition 927 (p. 520) it is true for any $\mathbf{Card(I_p)} \in \mathcal{Z}^+, \mathbf{Card(J_s)} \in \mathcal{Z}^+$ as soon as they are integers (definition 397 (p. 516)).

### 2.6.1 Distributivity of product with the sum

**Proposition 289** $\mathbf{J}$ and $\mathbf{L}$ are two sets with $\mathbf{Card(J)} \in \mathcal{Z}^+$ and $\mathbf{Card(L)} \in \mathcal{Z}^+$. We suppose that we have a partition (definition 370 (p. 471)) of $\mathbf{J}$: $(\mathbf{J_\lambda} : \mathbf{L} \to \mathcal{P}(\mathbf{J}))$ with $(\forall \lambda \in \mathbf{L})(\mathbf{Card(J_\lambda)} \in \mathcal{Z}^+)$ and a set of maps over $\mathcal{R}$: $(\mathbf{a_{\lambda,\iota}} : \mathbf{J_\lambda} \to \mathcal{R})$. We set $\mathbf{I} = \prod_{\lambda \in L}^{\times} \mathbf{J_\lambda}$. We have:

$$\prod_{\lambda \in L}^{\times} \sum_{\iota \in J_\lambda} a_{\lambda,\iota} = \sum_{f \in I} \prod_{\lambda \in L}^{\times} a_{\lambda, f(\lambda)} . \tag{2.45}$$

**Proof**
Calling $\mathcal{I} = \prod_{\lambda \in L}^{\times} \mathbf{Card(J_\lambda)} = \mathbf{Card(I)}$, from proposition 941 (p. 526), we have $\mathcal{I} \in \mathcal{Z}^+$. There is a bijection from $(\lambda, \iota)_i : [1, \mathcal{I}] \to \mathbf{I}$ (see definition 390 (p. 499)) and we can also label any $\mathbf{a}$ by $(\mathbf{a'_i} = \mathbf{a}_{(\lambda,\iota)_i} : [1, \mathcal{I}] \to \mathcal{R})$. We use that to generate two maps $\mathcal{R}^{\mathcal{I}} \to \mathcal{R}$:

$$(F_{ps} = (\prod_{\lambda \in L}^{\times} \sum_{\iota \in J_\lambda} a_{\lambda,\iota} : \mathcal{R}^{\mathcal{I}} \to \mathcal{R})), \ (F_{sp} = (\sum_{f \in I} \prod_{\lambda \in L}^{\times} a_{\lambda,f(\lambda)} : \mathcal{R}^{\mathcal{I}} \to \mathcal{R})) .$$

From proposition 1000 (p. 557), $\mathbf{F_{ps}(x) = F_{sp}(x)}$ when $\mathbf{x} \in \mathcal{Q}^{\mathcal{I}} \subset \mathcal{R}^{\mathcal{I}}$. $\mathbf{F_{ps}(x)}$ and $\mathbf{F_{sp}(x)}$ are continuous maps when restricted to $\mathbf{x_i}$ (see definition 129 (p. 145) and proposition 288 (p. 186)). Using that, we have[14]:

$$\begin{aligned} &(\forall i \in [1, \mathcal{I}])(\forall \alpha \in \mathcal{R}^+)(\forall x \in \mathcal{R}^{\mathcal{I}})(\forall x' \in \mathcal{R}^{\mathcal{I}})(\forall t \in \{ps, sp\})(\exists \epsilon_t^i(\alpha)) \\ &(((\forall j \in [1, \mathcal{I}] - i)(x'_j = x_j), |x'_i - x_i| < \epsilon_t^i(\alpha)) \Rightarrow |F_t(x') - F_t(x)| < \alpha) . \end{aligned} \tag{2.46}$$

From proposition 253 (p. 172),

$$(\forall u \in \mathcal{R})(\forall \epsilon \in \mathcal{R}^+)(\exists r \in \mathcal{Q})(|u - r| < \epsilon)$$

so in formula (2.46), taking $\mathbf{r_i}$ such that $|\mathbf{r_i} - \mathbf{x_i}| < \min(\epsilon_{ps}^i(\alpha/\mathcal{I}), \epsilon_{sp}^i(\alpha/\mathcal{I}))$ and $\mathbf{x'_i = r_i}$ we have:

$$(\forall t \in \{ps, sp\})(|F_t(x') - F_t(x)| < \alpha/\mathcal{I}) .$$

---

[13] It is a set because from definition 112 (p. 131) any map $\mathcal{R} \to \mathcal{R}$ belongs to $\mathcal{R}^2$ which is a set as explained in definition 356 (p. 459) and from proposition 749 (p. 450), it means that $\mathcal{F}$ is a set.

[14] $\forall \mathbf{t} \in \{\mathbf{ps,sp}\}$ means that we replace the letter $\mathbf{t}$ either by $\mathbf{ps}$ or by $\mathbf{sp}$.

We can then change $\mathbf{x_1}$ to $\mathbf{r_1}$ rational with $|\mathbf{F_{sp}} - \mathbf{F_{ps}}| < \alpha/\mathcal{I}$. If we have $\mathbf{q} \in \mathcal{Z}^+$ rational numbers $(\mathbf{r_k} : [1, \mathbf{q}] \to \mathcal{Q})$ such as:

$$(\mathbf{i} \in [1, \mathbf{q}] \Rightarrow \mathbf{x}'_\mathbf{i} = \mathbf{r_i}, \mathbf{i} \in [\mathbf{q}+1, \mathcal{I}] \Rightarrow \mathbf{x}'_\mathbf{i} = \mathbf{x_i}) \Rightarrow (\forall \mathbf{t} \in \{\mathbf{ps}, \mathbf{sp}\}) \\ (|\mathbf{F_t}(\mathbf{x}') - \mathbf{F_t}(\mathbf{x})| < \frac{\mathbf{q}\alpha}{\mathcal{I}}) . \tag{2.47}$$

From what we just say (2.47) is also true for $\mathbf{p}+\mathbf{1}$. From proposition 927 (p. 520), it is true for $\mathbf{p} = \mathcal{I}$. Then changing that way $(\mathbf{x_i} : [1, \mathcal{I}] \to \mathcal{R})$ to $(\mathbf{x}'_\mathbf{i} : [1, \mathcal{I}] \to \mathcal{Q})$, we have

$$(\forall \mathbf{t} \in \{\mathbf{ps}, \mathbf{sp}\})(|\mathbf{F_t}(\mathbf{x}') - \mathbf{F_t}(\mathbf{x})| < \alpha) .$$

From proposition 1000 (p. 557), $\mathbf{F_{sp}}(\mathbf{x}') = \mathbf{F_{ps}}(\mathbf{x}')$. Then we have (proposition 272 (p. 181))

$$(\forall \alpha > 0)(|\mathbf{F_{sp}}(\mathbf{x}) = \mathbf{F_{ps}}(\mathbf{x})| < 2\alpha) .$$

$0 \leq |\mathbf{F_{sp}}(\mathbf{x}) - \mathbf{F_{ps}}(\mathbf{x})|$ (definition 154 (p. 181)) is a minoring element (definition 380 (p. 484)) of $\mathcal{R}^+$. From definition 154 (p. 181) $\mathbf{0}$ is the **lower_limit** (definition 380 (p. 484)) of $\mathcal{R}^+$. $0 \geq |\mathbf{F_{sp}}(\mathbf{x}) - \mathbf{F_{ps}}(\mathbf{x})|$ then $\mathbf{F_{sp}}(\mathbf{x}) = \mathbf{F_{ps}}(\mathbf{x})$.

**Proposition 290** Given $\mathbf{u} \in \mathcal{R}$, an integer $\mathbf{n}$ and a map $(\mathbf{v_i} : [1, \mathbf{n}] \to \mathcal{R})$, we have:

$$\sum_{\mathbf{i} \in [1,\mathbf{n}]} \mathbf{a}\mathbf{v_i} = \mathbf{a} \sum_{\mathbf{i} \in [1,\mathbf{n}]} \mathbf{v_i} .$$

**Proof**
In proposition 289 (p. 186), we set $\mathbf{J} = [1, \mathbf{n}]$ $\mathbf{L} = \{1, 2\}$ $\mathbf{J_1} = \{1\}$, $\mathbf{J_2} = \mathbf{J}$ $\mathbf{a_{1,1}} = \mathbf{u}$, $\mathbf{k} \in \mathbf{J} \Rightarrow \mathbf{a_{2,k}} = \mathbf{v_k}$. Then proposition 289 (p. 186) gives:

$$\mathbf{u}\sum_{\mathbf{i} \in \mathbf{J}} \mathbf{v_i} = \prod_{\lambda \in [1,2]} \sum_{\iota \in \mathbf{J}_\lambda} \mathbf{a}_{\lambda,\iota} = \sum_{\mathbf{f} \in [1,1] \times [1,\mathbf{n}]} \prod_{\lambda \in [1,2]} \mathbf{a}_{\lambda,\mathbf{f}(\lambda)} = \sum_{\mathbf{f} \in [1,1] \times [1,\mathbf{n}]} \mathbf{a_{1,1}}\mathbf{a_{2,f(2)}} = \sum_{\mathbf{k} \in \mathbf{J}} \mathbf{u}\mathbf{v_k} .$$

## 2.6.2 Commutativity of sum and product

**Proposition 291** Given two finite sets of indices $\mathbf{I}$, $\mathbf{J}$ with $\mathbf{Card}(\mathbf{I}) = \mathbf{Card}(\mathbf{J})$ and a map $(\mathbf{a_i} : \mathbf{I} \to \mathcal{R})$ calling $(\mathbf{i}(\mathbf{j}) : \mathbf{J} \leftrightarrow \mathbf{I})$ and $(\mathbf{j}(\mathbf{i}) : \mathbf{I} \leftrightarrow \mathbf{J})$ two bijective maps with $\mathbf{i}(\mathbf{j}(\mathbf{k})) = \mathbf{k}$ which exist since $\mathbf{I} \sim \mathbf{J}$, we have:

$$\sum_{\mathbf{i} \in \mathbf{I}} \mathbf{a_i} = \sum_{\mathbf{j} \in \mathbf{J}} \mathbf{a_{i(j)}} .$$

**Proof**
See proposition 996 (p. 555) and apply the same argument as for proposition 289 (p. 186).

**Proposition 292** Given two finite sets of index $\mathbf{I} \sim \mathbf{J}$ with two associated bijections $(\mathbf{i}(\mathbf{j}) : \mathbf{J} \leftrightarrow \mathbf{I})$, $(\mathbf{j}(\mathbf{i}) : \mathbf{I} \leftrightarrow \mathbf{J})$, $\mathbf{i}(\mathbf{j}(\mathbf{k})) = \mathbf{k}$ and a map $(\mathbf{a_i} : \mathbf{I} \to \mathcal{R})$, we have:

$$\prod_{\mathbf{i} \in \mathbf{I}} \mathbf{a_i} = \prod_{\mathbf{j} \in \mathbf{J}} \mathbf{a_{i(j)}} .$$

**Proof**
See proposition 997 (p. 556) and apply the same argument as for proposition 289 (p. 186).

### 2.6.3 Associativity of sum and product

**Proposition 293** Given two finite sets $\mathbf{I}$ and $\mathbf{J}$, a map $(\mathbf{a_i} : \mathbf{I} \to \mathcal{R})$, a partition of $\mathbf{I}$ (definition 370 (p. 471)) $(\mathbf{I_j} : \mathbf{J} \to \mathcal{P}(\mathbf{I}))$, then:

$$\sum_{j \in J} \sum_{i \in I_j} a_i = \sum_{i \in \bigcup_{j \in J} I_j} a_i = \sum_{i \in I} a_i \,. \tag{2.48}$$

**Proof**
See proposition 998 (p. 556) and apply the same argument as for proposition 289 (p. 186).

**Proposition 294** Given two finite sets $\mathbf{L_1}$ and $\mathbf{L_2}$, a map $(\mathbf{a_\ell} : \mathbf{L_1} \times \mathbf{L_2} \to \mathcal{R})$ we have:

$$\sum_{\ell = \ell_1 \times \ell_2 \in L_1 \times L_2} a_\ell = \sum_{\ell_2 \in L_2} \sum_{\ell_1 \in L_1} a_{\ell_1 \times \ell_2} \,. \tag{2.49}$$

**Proof**
The map $(\mathbf{i} \times \mathbf{j} : \mathbf{L_2} \to \{\mathbf{i}\} \times \mathbf{L_2})$ is a bijection then from proposition 291 (p. 187),

$$(\forall \ell_1 \in L_1)(\sum_{\ell_2 \in L_2} a_{\ell_1 \times \ell_2} = \sum_{\ell \in \{\ell_1\} \times L_2} a_\ell)$$

then:

$$\sum_{\ell_1 \in L_1} \sum_{\ell_2 \in L_2} a_{\ell_1 \times \ell_2} = \sum_{\ell_1 \in L_1} \sum_{\ell \in \{\ell_1\} \times L_2} a_\ell \,.$$

From proposition 293 (p. 188):

$$\sum_{\ell_1 \in L_1} \sum_{\ell \in \{\ell_1\} \times L_2} a_\ell = \sum_{\ell \in \bigcup_{\ell_1 \in L_1} (\{\ell_1\} \times L_2)} a_\ell \,.$$

From proposition 830 (p. 476) with $\mathbf{L} = \{1, 2\}$, $\mathbf{J_1} = \mathbf{L_1}$, $\mathbf{J_2} = \{1\}$, $\mathbf{X_{1,\iota}} = \{\iota\}$, $\mathbf{X_{2,1}} = \mathbf{L_2}$ we have:

$$\bigcup_{\ell_1 \in L_1} (\{\ell_1\} \times L_2) = (\bigcup_{\ell_1 \in L_1} \{\ell_1\}) \times L_2 \,.$$

From proposition 795 (p. 464), we have $\mathbf{L_1} = \bigcup_{\ell_1 \in L_1} \{\ell_1\}$ then:

$$\bigcup_{\ell_1 \in L_1} (\{\ell_1\} \times L_2) = L_1 \times L_2 \,.$$

It gives equation (2.49) (p. 188).

## 2.7 Continuous maps from topological sets to $\mathcal{R}$

**Proposition 295** The maps $(\mathbf{min(x, y)} : \mathcal{R}^2 \to \mathcal{R})$, $(\mathbf{max(x, y)} : \mathcal{R}^2 \to \mathcal{R})$ are continuous.

**Proof**
Let us prove it for $\mathbf{max(x, y)}$. The proof is the same for $\mathbf{min(x, y)}$. Let us suppose $\mathbf{x} \geq \mathbf{y}, \mathbf{x} - \mathbf{y} > 0$ and $|\Delta \mathbf{x}| < \epsilon, |\Delta \mathbf{y}| < \epsilon$, we have two cases:

- $\Delta y \leq x + \Delta x - y$. Then $y + \Delta y \leq x + \Delta x, \max(x + \Delta x, y + \Delta y) = x + \Delta x$ which gives:

$$|\max(x + \Delta x, y + \Delta y) - \max(x, y)| = |x + \Delta x - x| = \Delta x < \epsilon$$

- $\Delta y > x + \Delta x - y$. Then $\max(x + \Delta x, y + \Delta y) = y + \Delta y, 0 < x - y < \Delta y - \Delta x$ which gives:

$$|\max(x + \Delta x, y + \Delta y) - \max(x, y)| = |y + \Delta y - x| = |x - y - \Delta y|,$$
$$-\Delta y < x - y - \Delta y < -\Delta x$$

since $-\epsilon < -\Delta y < \epsilon, -\epsilon < -\Delta x < \epsilon$ because $|\Delta x| < \epsilon$ and $|\Delta y| < \epsilon$, we have:

$$|\max(x + \Delta x, y + \Delta y) - \max(x, y)| < \epsilon$$

and the same for $y \geq x$. In any cases, we have:

$$(\forall \alpha)(\exists \epsilon = \alpha)((|\Delta x| < \epsilon, |\Delta y| < \epsilon) \Rightarrow \max(x + \Delta x, y + \Delta y) - \max(x, y) < \alpha)$$

then $\max(x, y)$ is uniformly continuous in the box metric (see definitions 157 (p. 182), 159 (p. 183), 243 (p. 267) proposition 282 (p. 184) and proposition 506 (p. 286)) so it is continuous in the product of space topology.

**Proposition 296** Given two topological spaces $\mathbf{A}$ and $\mathbf{B}$ and two continuous maps $(f : A \to \mathcal{R}), (g : B \to \mathcal{R})$ the maps $\max(f, g) : A \times B \to \mathcal{R}), \min(f, g) : A \times B \to \mathcal{R})$ are continuous.

**Proof**

It is a direct consequence of proposition 295 (p. 188) together with propositions 242 (p. 168) and 241 (p. 167).

**Proposition 297** Given a compact set (definition 140 (p. 160)) $\mathbf{C}$ subset of a topological separated space (definition 139 (p. 159)) and a continuous map or function $(f : C \to \mathcal{R})$, $f(C)$ has an **upper_limit** and a **lower_limit** (definition 381 (p. 485)) which belongs to $f(C)$.

**Proof**

Taking the above notation, from proposition 240 (p. 167), $f(C)$ is compact and from proposition 226 (p. 160), $f(C)$ is closed.

Any point of $f(C)$ is in an open interval (definition 378 (p. 484)):

$$(\forall r \in \mathcal{R})(f(C) \subset \bigcup_{x \in f(C)} ]x - r, x + r[)$$

then from definition 140 (p. 160), given an $r$

$$(\exists n \in \mathcal{Z}^+)(\exists x)(x = (x_i : [1, n] \to f(C)), f(C) \subset \bigcup_{i \in [1,n]} ]x_i - r, x_i + r[) .$$

From proposition 256 (p. 172) the map $x_i$ can be a strictly increasing map (definition 386 (p. 486)) and

$$(\forall i \in [1, n])(x_1 \leq x_i \leq x_n, x_i + r \leq x_n + r, x_1 - r$$
$$\leq x_i - r, ]x_i - r, x_i + r[ \subset ]x_1 - r, x_n + r[) .$$

Then $f(C) \subset ]x_1 - r, x_n + r[$ and from proposition 844 (p. 485), it has an **upper_limit** and a **lower_limit** $f_M, f_m$. Any open including $f_M$ (resp. $f_m$) contains an interval (definitions 378 (p. 484), 156 (p. 181)) centered on $f_M$ (resp. $f_m$) with a point of $f(C)$ (see proposition 260 (p. 175)). Then from definition 136 (p. 155) and proposition 220 (p. 158), $f_M \in f(C)$ (resp. $f_m \in f(C)$).

**Proposition 298** A closed interval of $\mathcal{R}$ is compact (definitions 378 (p. 484), 156 (p. 181)).

**Proof**
Let us call $[a, b]$ such a closed interval of $\mathcal{R}$. Let us call $\mathcal{O}$ the set of open sets of $\mathcal{R}$ and B a subset of $\mathcal{O}$ such that:

$$[a, b] \subset \bigcup_{O \in B} O .$$

Let us consider the set C defined as follows:

$$C = \text{set}_u\{u \in [a, b], (\exists I \subset B)(\text{Card}(I) \in \mathcal{Z}^+, [a, u] \subset \bigcup_{O \in I} O)\} \tag{2.50}$$

C is not empty because $a \in C$ ($[a, a]$ is covered by $\tau_x(x \in \mathcal{O}, a \in x)$ which exists and belongs to $\mathcal{O}$ see A.3 (p. 441) and definition 332 (p. 440)). C gets b as a majoring element. It has from proposition 258 (p. 174) an **upper_limit(C)** c which means that there is an open $O^c$ of $\mathcal{O}$ covering c. From proposition 266 (p. 179), any open containing c contains an open interval which contains a point of C (propositions 260 (p. 175)) so of $[a, b]$. Then c belongs to the closure (definition 136 (p. 155)) of $[a, b]$ which is $[a, b]$ (propositions 274 (p. 181), 220 (p. 158)). Then we have $c \in [a, b]$. From definition of B (see definition 362 (p. 462)), there is an open set $O_c \in B$ which contains c and an open ball $B(c, r) = ]c - r, c + r[$ (proposition 266 (p. 179)) in it. This ball contains a point $c' \in C, c - r < c' \leq c$ (see proposition 260 (p. 175)). From the definition of C, we have a finite covering $I_{c'} \subset \mathcal{O}$ of $[a, c']$. Adding $O_c$ to that covering we get a finite covering of $[a, c + r] \cap [a, b]$ then c cannot be the **upper_limit(C)** except if $c = b$. Proposition 298 (p. 190) is true.

**Proposition 299** a and b are two real numbers. If f is a continuous map from $[a, b]$ to $\mathcal{R}$, if $\text{Min}(f([a, b])$ and $\text{Max}(f([a, b])$ exist in $\mathcal{R}$ and we have:

$$f([a, b]) = [\text{Min}(f([a, b])), \text{Max}(f([a, b]))] .$$

**Proof**
f can be extended to a continuous map from $\mathcal{R}$ to $\mathcal{R}$ by setting $f(]\leftarrow, a]) = \{f(a)\}$, $f([b, \rightarrow[) = \{f(b)\}$. $C = [a, b]$ is a closed interval of $\mathcal{R}$, C is compact (proposition 298 (p. 190)). If f is a continuous map from $\mathcal{R}$ to $\mathcal{R}$, there exist a **upper_limit** and a **lower_limit** (definition 381 (p. 485)) of f(C) which belong to f(C) (proposition 297 (p. 189)).
Let us prove now that any point between those two limits belongs to f(C). Calling u and v two elements of $[a, b]$ such as:

$$u = f^{-1}(\text{lower\_lim}(f(C)), \quad v = f^{-1}(\text{upper\_lim}(f(C))$$

u and v exist according to what we have just written. From definition 160 (p. 184) and proposition 283 (p. 184), we have:

$$(\forall \alpha \in \mathcal{R}^+)(\exists \epsilon(\alpha) > 0)(|x - x'| < \epsilon(\alpha) \Rightarrow |f(x) - f(x')| < \alpha) .$$

Using that map $(\epsilon(\alpha) : \mathcal{R}^+ \rightarrow \mathcal{R}^+)$ (see definition 112 (p. 131) point 4), we have:

$$(\forall n > \frac{|v - u|}{\epsilon(\alpha)})(\exists (x_i^n : [0, n] \rightarrow C))((i \in [1, n] \Rightarrow x_i^n = u + (v - u) \times i/n),$$
$$(\forall p \in [0, n - 1])(x_{p+1}^n - x_p^n = (u - v)/n, |x_{p+1}^n - x_p^n| < \epsilon(\alpha))) .$$

For any $\mathbf{n} > \frac{|\mathbf{v}-\mathbf{u}|}{\epsilon(\alpha)}$, $\mathbf{n} \in \mathcal{Z}^+$ and any $\mathbf{y} \in [\mathbf{Min}(\mathbf{f}([\mathbf{a},\mathbf{b}])), \mathbf{Max}(\mathbf{f}([\mathbf{a},\mathbf{b}]))]$, using those $\mathbf{x}^\mathbf{n}$, from proposition 924 (p. 519), we have successively:

$$\mathbf{f}(\mathbf{x}^\mathbf{n}_0) \leq \mathbf{y}, \mathbf{f}(\mathbf{x}^\mathbf{n}_\mathbf{n}) \geq \mathbf{y}$$

$$(\exists \mathbf{p_y})(\mathbf{p_y} = \max_{\mathbf{p}\in[0,\mathbf{n}],\mathbf{f}(\mathbf{x}^\mathbf{n}_\mathbf{p})<\mathbf{y}} (\mathbf{p}), 0 \leq \mathbf{p_y} \leq \mathbf{n}-1)$$

$$\mathbf{f}(\mathbf{x}^\mathbf{n}_{\mathbf{p_y}}) \leq \mathbf{y} \leq \mathbf{f}(\mathbf{x}^\mathbf{n}_{\mathbf{p_y}+1}), |\mathbf{f}(\mathbf{x}^\mathbf{n}_{\mathbf{p_y}+1}) - \mathbf{f}(\mathbf{x}^\mathbf{n}_{\mathbf{p_y}})| < \alpha \Rightarrow |\mathbf{y} - \mathbf{f}(\mathbf{x}^\mathbf{n}_{\mathbf{p_y}})| < \alpha)$$

so:

$$\mathbf{Min}(\mathbf{f}(\mathbf{C})) \leq \mathbf{y} \leq \mathbf{Max}(\mathbf{f}(\mathbf{C})) \Rightarrow (\forall \alpha)(\exists \mathbf{x} \in \mathbf{C})(|\mathbf{y} - \mathbf{f}(\mathbf{x})| < \alpha) .$$

Since, from propositions 226 (p. 160) and 240 (p. 167), $\mathbf{f}(\mathbf{C})$ is closed, from definition 136 (p. 155), propositions 266 (p. 179) and 220 (p. 158), $\mathbf{y}$ belongs to $\mathbf{f}(\mathbf{C})$.

**Proposition 300** $[\mathbf{a},\mathbf{b}] \subset \mathcal{R}$, a strictly increasing continuous map or strictly decreasing map $\mathbf{f}$ is a continuous bijection (definition 124 (p. 139)) with a continuous inverse:

$$(\mathbf{f} : [\mathbf{a},\mathbf{b}] \leftrightarrow [\mathbf{f}(\mathbf{a}), \mathbf{f}(\mathbf{b})]) .$$

**Proof**

From proposition 299 (p. 190), $\mathbf{f}$ is a surjective map on $[\mathbf{f}(\mathbf{a}), \mathbf{f}(\mathbf{b})]$ (definition 123 (p. 139)). From proposition 846 (p. 486), it is an injective map it is then a bijection (definition 124 (p. 139)). Let us suppose that it is a strictly increasing map, given an $\mathbf{y_0} \in [\mathbf{f}(\mathbf{a}), \mathbf{f}(\mathbf{b})]$ and a $\alpha > 0$, we call $\mathbf{x_0} = \mathbf{f}^{-1}(\mathbf{y})$.
If $\mathbf{x_0} + \alpha \geq \mathbf{b}, \mathbf{x_0} - \alpha \leq \mathbf{a}$, we choose $\epsilon = \mathbf{f}(\mathbf{b}) - \mathbf{f}(\mathbf{a})$.
If $\mathbf{x_0} + \alpha \geq \mathbf{b}, \mathbf{x_0} - \alpha > \mathbf{a}$ we choose $\epsilon = \mathbf{y_0} - \mathbf{f}(\mathbf{x_0} - \alpha)$.
If $\mathbf{x_0} + \alpha < \mathbf{b}, \mathbf{x_0} - \alpha \leq \mathbf{a}$ we choose $\epsilon = \mathbf{f}(\mathbf{x_0} + \alpha) - \mathbf{y_0}$.
If $\mathbf{x_0} + \alpha < \mathbf{b}, \mathbf{x_0} - \alpha > \mathbf{a}$ we choose $\epsilon = \mathbf{Min}(\mathbf{f}(\mathbf{x_0} + \alpha) - \mathbf{y_0}, \mathbf{y_0} - \mathbf{f}(\mathbf{x_0} - \alpha))$. Since $\mathbf{f}$ is a strictly increasing map, in all the cases, we have $\epsilon > 0$ and

$$\mathbf{y} \in [\mathbf{f}(\mathbf{a}), \mathbf{f}(\mathbf{b})] \cap [\mathbf{y_0} - \epsilon, \mathbf{y_0} + \epsilon] \Rightarrow |\mathbf{f}^{-1}(\mathbf{y}) - \mathbf{x_0}| < \alpha .$$

Then $\mathbf{f}^{-1}$ is continuous for all $\mathbf{y} \in [\mathbf{f}(\mathbf{a}), \mathbf{f}(\mathbf{b})]$.

**Proposition 301** $[\mathbf{a},\mathbf{b}] \subset \mathcal{R}$, a continuous injective map $(\mathbf{f} : [\mathbf{a},\mathbf{b}] \rightarrow \mathcal{R})$ is either a strictly increasing or a strictly decreasing map (definition 386 (p. 486)).

**Proof**

Let us suppose $\mathbf{f}(\mathbf{a}) < \mathbf{f}(\mathbf{b})$ (resp. $\mathbf{f}(\mathbf{a}) > \mathbf{f}(\mathbf{b})$). $\mathbf{f}(\mathbf{a}) \neq \mathbf{f}(\mathbf{b})$ since $\mathbf{f}$ is injective (definition 122 (p. 139)). We cannot have a $\mathbf{c} \in [\mathbf{a},\mathbf{b}[$ with $\mathbf{f}(\mathbf{c}) = \mathbf{f}(\mathbf{b})$ because $\mathbf{f}$ is injective. We cannot have $\mathbf{f}(\mathbf{c}) > \mathbf{f}(\mathbf{b})$ (resp. $\mathbf{f}(\mathbf{c}) < \mathbf{f}(\mathbf{b})$) because if it was true, calling $\mathbf{M} = \max \mathbf{f}([\mathbf{a},\mathbf{b}])$ (resp. $\mathbf{M} = \min \mathbf{f}([\mathbf{a},\mathbf{b}])$) we should have $\mathbf{M} \geq \mathbf{f}(\mathbf{c}) > \mathbf{f}(\mathbf{b})$ (resp. $\mathbf{M} \leq \mathbf{f}(\mathbf{c}) < \mathbf{f}(\mathbf{b})$). From proposition 299, it means $\mathbf{f}^{-1}(\mathbf{M}) \in [\mathbf{a},\mathbf{b}]$ and $\mathbf{M} \neq \mathbf{b}$. We should also have $\mathbf{f}([\mathbf{a}, \mathbf{f}^{-1}(\mathbf{M})[) \cap \mathbf{f}([\mathbf{f}^{-1}(\mathbf{M})], \mathbf{b}) = \emptyset$ because $\mathbf{f}$ is injective but $\mathbf{f}(\mathbf{b}) \in \mathbf{f}([\mathbf{a}, \mathbf{f}^{-1}(\mathbf{M})])$ because from proposition 299 $[\mathbf{f}(\mathbf{a}), \mathbf{M}] \subset \mathbf{f}([\mathbf{a}, \mathbf{f}^{-1}(\mathbf{M})])$ and $\mathbf{f}(\mathbf{b}) > \mathbf{f}(\mathbf{a}), \mathbf{f}(\mathbf{b}) < \mathbf{M}$ (resp. $\mathbf{f}(\mathbf{b}) < \mathbf{f}(\mathbf{a}), \mathbf{f}(\mathbf{b}) > \mathbf{M}$) since $\mathbf{f}(\mathbf{b})$ is in $\mathbf{f}([\mathbf{f}^{-1}(\mathbf{M}), \mathbf{b}])$ $\mathbf{f}([\mathbf{a}, \mathbf{f}^{-1}(\mathbf{M})[) \cap \mathbf{f}([\mathbf{f}^{-1}(\mathbf{M})], \mathbf{b}) = \emptyset$ cannot be true and we conclude by point A.1.4 point 8 (p. 435).

**Proposition 302** Having a continuous increasing map, $\mathbf{a} \in \mathcal{R}, \mathbf{b} \in \mathcal{R}, \mathbf{a} < \mathbf{b}$, see definition 384 (p. 486) (resp. decreasing map, see definition 385 (p. 486)), defined on a semi open intervals $]\mathbf{a},\mathbf{b}]$ (resp. 2. $[\mathbf{a},\mathbf{b}[$) $(\mathbf{f} : ]\mathbf{a},\mathbf{b}] \rightarrow \mathcal{R}$ (resp. 2. $(\mathbf{f} : [\mathbf{a},\mathbf{b}[ \rightarrow \mathcal{R})$) setting $\mathbf{f}(\mathbf{a}) = \underset{\mathbf{x}\in]\mathbf{a},\mathbf{b}]}{\mathbf{lower\_limit}}(\mathbf{f}(\mathbf{x})$ (resp. $\mathbf{f}(\mathbf{a}) = \underset{\mathbf{x}\in]\mathbf{a},\mathbf{b}]}{\mathbf{upper\_limit}}(\mathbf{f}(\mathbf{x}))$ (resp. 2 $\mathbf{f}(\mathbf{b}) = \underset{\mathbf{x}\in[\mathbf{a},\mathbf{b}[}{\mathbf{upper\_limit}}(\mathbf{f}(\mathbf{x})$ (resp. $\mathbf{f}(\mathbf{b}) = \underset{\mathbf{x}\in[\mathbf{a},\mathbf{b}[}{\mathbf{lower\_limit}}(\mathbf{f}(\mathbf{x}))$), the obtained map $(\mathbf{f} : [\mathbf{a},\mathbf{b}] \rightarrow \mathcal{R})$ is continuous.

**Proof**
Let us demonstrate it for an increasing map (definition 384 (p. 486)) on $]\mathbf{a}, \mathbf{b}]$. Setting $\mathbf{f}(\mathbf{a}) = \underset{\mathbf{x}\in]\mathbf{a},\mathbf{b}]}{\mathbf{lower\_limit}}(\mathbf{f}(\mathbf{x})$, from proposition:

$$(\forall \alpha > 0)(\exists \epsilon(\alpha) > 0)(\mathbf{f}(\mathbf{a} + \epsilon(\alpha)) - \mathbf{f}(\mathbf{a}) > \alpha)$$

since $\mathbf{f}(\mathbf{x})$ is an increasing map (definition 384 (p. 486):

$$(\forall \alpha > 0)(\mathbf{x} - \mathbf{a} < \epsilon(\alpha) \Rightarrow \mathbf{f}(\mathbf{a}) \leq \mathbf{f}(\mathbf{x}) \leq \mathbf{f}(\mathbf{a} + \epsilon(\alpha)) \leq \mathbf{f}(\mathbf{a}) + \alpha$$

then

$$-\alpha < \mathbf{f}(\mathbf{x}) - \mathbf{f}(\mathbf{a}) < \alpha$$

then

$$0 < \mathbf{x} - \mathbf{a} < \epsilon(\alpha) \Rightarrow |\mathbf{f}(\mathbf{x}) - \mathbf{f}(\mathbf{a})| < \alpha$$

$\mathbf{f}(\mathbf{x})$ is continuous in $\mathbf{a}$ (definition 159 (p. 183)) since it is continuous in any point of $[\mathbf{a}, \mathbf{b}]$, it is continuous on $[\mathbf{a}, \mathbf{b}]$ (propositions 281 (p. 184), 282 (p. 184)).

## 2.8 Derivable function

**Definition 161 Derivative of a function at a point.**

Given a function $(\mathbf{f}(\mathbf{x}) : \mathbf{O} \to \mathcal{R})$, with $\mathbf{O} \subset \mathcal{R}$, if at a value of $\mathbf{x} \in \mathbf{O}$, there is an open interval $]\mathbf{x} - \mathbf{a}, \mathbf{x} + \mathbf{a}[\subset \mathbf{O}$, $\mathbf{a} \in \mathcal{R}^+$ and if the function $(\mathbf{h}(\Delta\mathbf{x}) :] - \mathbf{a}, 0[\cup]0, \mathbf{a}[\to \mathcal{R})$

$$\mathbf{h}(\Delta\mathbf{x}) = \frac{\mathbf{f}(\mathbf{x} + \Delta\mathbf{x}) - \mathbf{f}(\mathbf{x})}{\Delta\mathbf{x}}$$

has a limit at $\Delta\mathbf{x} = 0$, $\mathbf{f}$ is said to have a **derivative** at $\mathbf{x}$. The value of the limit (definition 157 (p. 182)) is the derivative of $\mathbf{f}$ at $\mathbf{x}$. One uses to write this derivative as $\mathbf{f}'(\mathbf{x})$. From definition 157 (p. 182), we have for $\mathbf{f}'(\mathbf{x})$:

$$(\forall \alpha > 0)(\exists \epsilon \in ]0, \mathbf{a}[)(|\Delta\mathbf{x}| < \epsilon \Rightarrow |\frac{\mathbf{f}(\mathbf{x} + \Delta\mathbf{x}) - \mathbf{f}(\mathbf{x})}{\Delta\mathbf{x}} - \mathbf{f}'(\mathbf{x})| < \alpha) \,. \tag{2.51}$$

**Proposition 303** $\mathbf{O} \in \mathcal{R}$, a function $(\mathbf{f}(\mathbf{x}) : \mathbf{O} \to \mathcal{R})$ which is derivable at $\mathbf{x} \in \mathbf{O}$ is continuous at $\mathbf{x}$.

**Proof**
Setting as above $\mathbf{h}(\Delta\mathbf{x})$ and $\mathbf{f}'(\mathbf{x})$ from definitions 157 (p. 182) and 161 (p. 192), we have:

$$(\forall \alpha \in \mathcal{R}^+)(\exists \epsilon \in \mathcal{R}^+)(|\Delta\mathbf{x}| < \epsilon \Rightarrow |\mathbf{h}(\Delta\mathbf{x}) - \mathbf{f}'(\mathbf{x})| < \alpha) \tag{2.52}$$

which defines a map (definition 112 (p. 131), 4) $(\epsilon(\alpha) : \mathcal{R}^+ \to \mathcal{R}^+)$. We have:

$$|\mathbf{h}(\Delta\mathbf{x}) - \mathbf{f}'(\mathbf{x})| < \alpha \Rightarrow |\mathbf{f}(\mathbf{x} + \Delta\mathbf{x}) - \mathbf{f}(\mathbf{x}) - \mathbf{f}'(\mathbf{x})\Delta\mathbf{x}| < \alpha|\Delta\mathbf{x}| \,.$$

Using proposition 272 (p. 181), we get:

$$\begin{aligned} |\Delta\mathbf{x}| < \epsilon(\alpha) &\Rightarrow ||\mathbf{f}(\mathbf{x} + \Delta\mathbf{x}) - \mathbf{f}(\mathbf{x})| - |\mathbf{f}'(\mathbf{x})\Delta\mathbf{x}|| \\ &\leq |\mathbf{f}(\mathbf{x} + \Delta\mathbf{x}) - \mathbf{f}(\mathbf{x}) - \mathbf{f}'(\mathbf{x})\Delta\mathbf{x}| < \alpha|\Delta\mathbf{x}| \end{aligned}$$

then:

$$|\Delta\mathbf{x}| < \epsilon(\alpha) \Rightarrow |\mathbf{f}(\mathbf{x} + \Delta\mathbf{x}) - \mathbf{f}(\mathbf{x})| < (\alpha + |\mathbf{f}'(\mathbf{x})|)|\Delta\mathbf{x}|$$

so we have:

$$(\forall \alpha)(\exists \epsilon' = \min\{\epsilon(\alpha), \alpha/(\alpha + |\mathbf{f}'(\mathbf{x})|)\})(|\Delta\mathbf{x}| < \epsilon' \Rightarrow |\mathbf{f}(\mathbf{x} + \Delta\mathbf{x}) - \mathbf{f}(\mathbf{x})| < \alpha) \,.$$

**Definition 162 Derivable function.**

If the derivative of a function exists for all values in a subset $\mathbf{A}$ of $\mathcal{R}$, the function is called derivable on $\mathbf{A}$.

**Definition 163** $\frac{d^p}{dx^p}$.

Having a derivable map $(\mathbf{f}:\mathcal{R}\to\mathcal{R})$ on a subset $\mathbf{A}\subset\mathcal{R}$, one writes $\mathbf{p}=\mathbf{1}\Rightarrow \mathbf{f}'(\mathbf{x})=\frac{d^p f}{dx^p}$. Let us suppose now $\frac{d^p}{dx^p}$ is defined for $\mathbf{f}$ and $\mathbf{p}=\mathbf{p}'$ on the subset $\mathbf{A}$ and that it is a map $(\frac{d^{p'}f}{dx^{p'}}:\mathcal{R}\to\mathcal{R})$ derivable on $\mathbf{A}$, one writes:

$$\frac{df}{dx}\frac{d^{p'}f}{dx^{p'}}=\frac{d^{p'+1}f}{dx^{p'+1}}\,.$$

Since $\frac{d^p}{dx^p}$ is defined for $\mathbf{p}=\mathbf{1}$ and since if it is defined for $\mathbf{p}=\mathbf{p}'$, it is also defined for $\mathbf{p}=\mathbf{p}'+\mathbf{1}$, from proposition 927 (p. 520), it is defined for any $\mathbf{p}\in\mathcal{Z}^{0+}$.

**Definition 164 Notations for the derivative of a derivable function.**

$(\mathbf{f}:\mathcal{R}\to\mathcal{R})$ being a derivable map on a subset $\mathbf{A}\subset\mathcal{R}$, the derivative of $\mathbf{f}$ for $\mathbf{x}\in\mathbf{A}$ is quoted as $\mathbf{f}'(\mathbf{x})$, $\frac{\partial f(x)}{\partial x}$ or $\partial \mathbf{f}(\mathbf{x})$ following definition 275 (p. 304) or as $\frac{df}{dx}$ following definition 40 (p. 34) and equation (1.48) (p. 34). We have:

$$f'(x)=\frac{\partial f(x)}{\partial x}=\partial f(x)=\frac{df}{dx}\,.$$

If the derivative of $\mathbf{f}$ $(\frac{\partial^1 f(x)}{\partial x^1}=\partial^1 f(x))$ also has a derivative $(\frac{\partial^2 f(x)}{\partial x^2}=\partial^2 f(x))$ which is also a derivable function and so on up to $\mathbf{n}\in\mathcal{Z}$, one writes for $\mathbf{p}\in[\mathbf{1},\mathbf{n}]$:

$$(\prod_{i\in[1,p]}\partial)f(x)=f^{[p]}(x)=\frac{\partial^p f(x)}{\partial x^p}=\partial^p f(x)=\frac{d^p f}{dx^p}\,.$$

**Proposition 304** A function $(\mathbf{f}:\mathcal{R}\to\mathcal{R})$ derivable on a segment $[\mathbf{a},\mathbf{b}]$ is continuous for all value of $[\mathbf{a},\mathbf{b}]$.

**Proof**
It is a direct application of propositions 303 (p. 192) and 281 (p. 184).

**Proposition 305** $\mathbf{O}$ being an open of $\mathcal{R}$, a function $(\mathbf{f}:\mathcal{R}\to\mathcal{R})$ derivable on any point of $\mathbf{O}$ is continuous.

**Proof**
It comes also from propositions 303 (p. 192) and 281 (p. 184).

**Proposition 306** $\mathbf{A}\subset\mathcal{R}$, $\mathbf{B}\subset\mathcal{R}$, $\mathbf{O}\subset\mathbf{A}\cap\mathbf{B}$ being an open of $\mathcal{R}$, two functions $(\mathbf{f}:\mathbf{A}\to\mathcal{R})$, $(\mathbf{g}:\mathbf{B}\to\mathcal{R})$, $\mathbf{f}$ derivable on any point of $\mathbf{O}$ with $(\forall\mathbf{x}\in\mathbf{O})(\mathbf{f}(\mathbf{x})=\mathbf{g}(\mathbf{x}))$, then for all points of $\mathbf{O}$ $\mathbf{g}$ is derivable and its derivative is equal to the derivative of $\mathbf{f}$.

**Proof**
It is a direct consequence of proposition 280 (p. 183) and definition 161 (p. 192).

**Proposition 307** Given a function $(\mathbf{f}:\mathcal{R}\to\mathcal{R})$ derivable on a segment $[\mathbf{a},\mathbf{b}]$, if that derivative is bigger (resp. smaller) than zero, the function is a strictly increasing (resp. decreasing) function (see definition 386 (p. 486)).

**Proof**
Let us consider the set $\mathbf{C}$ of $\mathbf{x} \in [\mathbf{a}, \mathbf{b}]$ for which the map $\mathbf{f}$ is a strictly increasing (resp. decreasing) map of $[\mathbf{a}, \mathbf{x}]$ on $\mathbf{f}([\mathbf{a}, \mathbf{x}])$ (definition 386 (p. 486)). The set $\mathbf{C}$ is not empty because $\mathbf{a} \in \mathbf{C}$. It has a majoring (resp. minoring) value because $\mathbf{b} \geq \mathbf{x}$. It has then an **upper_limit** $\mathbf{c} \leq \mathbf{b}$. If $\mathbf{c} \neq \mathbf{b}$ since $\mathbf{f}$ has a derivative at $\mathbf{c}$ bigger (resp. smaller) than zero, for $\alpha = |\mathbf{f}'(\mathbf{c})|/2 > 0$, $\exists \epsilon > 0$ such that

$$\begin{aligned}
|\Delta x| \leq \epsilon \Rightarrow |\frac{f(c+\Delta x)-f(c)}{\Delta x} - f'(c)| &< |f'(c)|/2 \\
\Rightarrow -|f'(c)|/2 < \frac{f(c+\Delta x)-f(c)}{\Delta x} &- f'(c) < |f'(c)|/2 \\
\Rightarrow f'(c)/2 < \frac{f(c+\Delta x)-f(c)}{\Delta x} &.
\end{aligned}$$

Then

$$\begin{aligned}
&(\forall \Delta x \in ]0, \epsilon[)(f(c+\Delta x) - f(c) > f'(c)\Delta x/2 > 0) \\
&\quad (\text{resp. } (f(c+\Delta x) - f(c) < f'(c)\Delta x/2 < 0))
\end{aligned}$$

and $\mathbf{c} + \epsilon \in \mathbf{C}$ which is impossible except if $\mathbf{c} = \mathbf{b}$.

**Proposition 308** The derivative of a constant function is zero.

**Proof**
In notation of definition 161 (p. 192), $\Delta x \neq 0 \Rightarrow (h(\Delta x) = \frac{a-a}{\Delta x} = 0)$.

**Proposition 309** The derivative of $\mathbf{f}(\mathbf{x}) = \mathbf{a}\mathbf{x}$ is $\mathbf{a}$ .

**Proof**
In notation of definition 161 (p. 192), $\mathbf{h}(\Delta \mathbf{x}) = \mathbf{a}$ for $\mathbf{f}(\mathbf{x}) = \mathbf{a}\mathbf{x}$.

**Proposition 310** The derivative of $1/\mathbf{x}$ is $-1/\mathbf{x}^2$ when $\mathbf{x} \neq 0$.

**Proof**
Since $1/\mathbf{x}$ is continuous when $\mathbf{x} \neq 0$ (proposition 285 (p. 185)), we have:

$$(\forall x \neq 0)(\exists \epsilon_1(\alpha, x))(|\Delta x| < \epsilon_1(\alpha) \Rightarrow |\frac{1}{x+\Delta x} - \frac{1}{x}| < \alpha)$$

with

$$\frac{1}{x+\Delta x} - \frac{1}{x} = -\frac{\Delta x}{x(x+\Delta x)} = -\frac{\Delta x}{x}(\frac{1}{x+\Delta x} - \frac{1}{x}) - \frac{\Delta x}{x^2}$$

we have:

$$\frac{\frac{1}{x+\Delta x} - \frac{1}{x}}{\Delta x} + \frac{1}{x^2} = -\frac{1}{x}(\frac{1}{x+\Delta x} - \frac{1}{x}) .$$

Setting $\epsilon = \epsilon_1(|\mathbf{x}|\alpha)$, we get:

$$|\Delta x| < \epsilon \Rightarrow |\frac{\frac{1}{x+\Delta x} - \frac{1}{x}}{\Delta x} + \frac{1}{x^2}| < \alpha .$$

**Proposition 311** $(\mathbf{f}(\mathbf{x}) + \mathbf{g}(\mathbf{x}))' = \mathbf{f}'(\mathbf{x}) + \mathbf{g}'(\mathbf{x})$ .

**Proof**

Let us suppose that $\mathbf{f(x)}$ and $\mathbf{g(x)}$ are two derivable maps $\mathcal{R}$ to $\mathcal{R}$ at any value of $\mathbf{x}$ with as derivatives $\mathbf{f'(x)}$ and $\mathbf{g'(x)}$. Then we can write:

$$\begin{aligned}&(\forall x),(\forall\alpha),(\exists\epsilon_f),(\exists\epsilon_g),\\&\quad(|\Delta x_f|<\epsilon_f\Rightarrow|\tfrac{f(x+\Delta x_f)-f(x)}{\Delta x_f}-f'(x)|<\alpha),\\&\quad(|\Delta x_g|<\epsilon_g\Rightarrow|\tfrac{g(x+\Delta x_g)-g(x)}{\Delta x_g}-g'(x)|<\alpha)\ .\end{aligned}\tag{2.53}$$

From proposition 273 (p. 181), one gets taking $\alpha = \alpha'/2$ in (2.53 (p. 195)) and $\epsilon = \mathrm{Min}(\epsilon_f, \epsilon_g)$:

$$\begin{aligned}&(\forall x),(\forall\alpha'),(\exists\epsilon),(|\Delta x|<\epsilon\\&\Rightarrow|\tfrac{f(x+\Delta x)+g(x+\Delta x)-f(x)-g(x)}{\Delta x}-f'(x)-g'(x)|<2\alpha=\alpha')\ .\end{aligned}$$

**Proposition 312** If $\mathbf{f(x)}$ is a continuous function on $\mathcal{R}$ and has a maximum (resp. minimum) on the interval $]\mathbf{a},\mathbf{b}[$, its derivative, if it exists, is zero at the maximum (resp. minimum).

**Proof**

We call $\mathbf{x_m}$ that maximum (resp. minimum) of $\mathbf{f(x)}$ when $\mathbf{x}\in]\mathbf{a},\mathbf{b}[$. Since the derivative of $\mathbf{f}$ exists at $\mathbf{x_m}$, $h(\Delta x) = \frac{f(x_m+\Delta x)-f(x_m)}{\Delta x}$ is continuous on $\Delta x$ for $\Delta x = 0$ (proposition 303 (p. 192)). The relation (2.51) of definition 161 (p. 192) gives:

$$(\forall\alpha>0)(\exists\epsilon>0)(|\eta|<\epsilon\Rightarrow|h(0)-h(\eta)|<\alpha)\ .\tag{2.54}$$

If $\mathbf{h(0)} \neq \mathbf{0}$, we can take in equation (2.54), $\alpha = |h(0)|/2$. Let us suppose $h(0) > 0$ with the corresponding $\epsilon$. We have:

$$\begin{aligned}&|\eta|<\epsilon'\Rightarrow|h(0)-h(\eta)|<h(0)/2\\&\quad\Rightarrow 3\alpha=3h(0)/2>h(\eta)\le h(0)=2\alpha\vee h(0)\le h(\eta)>h(0)/2\\&\quad\Rightarrow 3\alpha>h(\eta)>\alpha\end{aligned}$$

so $h(\eta)\eta$ would change sign with $\eta$ which is impossible because if we have a maximum for $\mathbf{x_m}$

$$0\le f(x_m+\eta)-f(x_m)=h(\eta)\eta\tag{2.55}$$

if we have a minimum for $\mathbf{x_m}$

$$0\ge f(x_m+\eta)-f(x_m)=h(\eta)\eta\ .\tag{2.56}$$

Choosing $\mathbf{u}$ as $0 < u < \mathrm{Min}(\epsilon, |b-x_m|, |a-x_m|)$, $\mathbf{u}$ exists and both $\mathbf{u}$ and $-\mathbf{u}$ are in $]\mathbf{a},\mathbf{b}[$. When going from $\eta = -u$ to $\eta = u$ the first members of equations (2.55) and (2.56) do not change sign since $\mathbf{f(x_m)}$ is maximum (resp. minimum) while the second member change sign. The same reasoning holds when $h(0) < 0$ which gives instead $3\alpha > -h(\eta) > \alpha$. Then $h(0) \neq 0$ leads to an absurd situation. From A.1.4, 8 $h(0) = 0$ and proposition 312 is true.

**Proposition 313** If $\mathbf{f(x)}$ is a continuous function on $[\mathbf{a},\mathbf{b}]$ and is such that $f(a) = f(b)$ and derivable on $]\mathbf{a},\mathbf{b}[$, there is a point of $]\mathbf{a},\mathbf{b}[$ for which $f'(x) = 0$.

**Proof**

Setting

$$M=\mathrm{Lim\_sup}_{x\in[a,b]}f(x)=\mathrm{upper\_limit}(f([a,b])$$

and

$$m=\mathrm{Lim\_inf}_{x\in[a,b]}f(x)=\mathrm{lower\_limit}(f([a,b])$$

(see definitions 381 (p. 485), 114 (p. 132))

- if $\mathbf{M} = \mathbf{m}$, one has:
  $\forall \mathbf{x} \in [\mathbf{a}, \mathbf{b}] \mathbf{f}(\mathbf{x}) = \mathbf{f}(\mathbf{a}) = \mathbf{m} = \mathbf{M}$ and $(\forall \mathbf{x} \in ]\mathbf{a}, \mathbf{b}[) \mathbf{f}'(\mathbf{x}) = 0$ and proposition 313 is true

- If $\mathbf{M} \neq \mathbf{m}$ one of them is not equal to $\mathbf{f}(\mathbf{a})$.
  Let us take $\mathbf{M} > \mathbf{f}(\mathbf{a})$ (resp. $\mathbf{m} < \mathbf{f}(\mathbf{a})$). From proposition 299 (p. 190) there is a value for $\mathbf{x}$ in $[\mathbf{a}, \mathbf{b}]$ for which $\mathbf{f}(\mathbf{x}) = \mathbf{M}$ (resp. $\mathbf{f}(\mathbf{x}) = \mathbf{m}$). It cannot be $\mathbf{a}$ or $\mathbf{b}$ because $\mathbf{f}(\mathbf{a}) = \mathbf{f}(\mathbf{b}), \mathbf{M} \neq \mathbf{f}(\mathbf{a})$ (resp. $\mathbf{f}(\mathbf{a}) = \mathbf{f}(\mathbf{b}), \mathbf{m} \neq \mathbf{f}(\mathbf{a})$). From proposition 312 (p. 195) the derivative is zero at that point and proposition 313 is true.

**Proposition 314** If $\mathbf{f}(\mathbf{x})$ is a continuous function on $\mathcal{R}$ and derivable on $]\mathbf{a}, \mathbf{b}[$, there is a point of $]\mathbf{a}, \mathbf{b}[$ for which $\mathbf{f}'(\mathbf{x}) = (\mathbf{f}(\mathbf{b}) - \mathbf{f}(\mathbf{a}))/(\mathbf{b} - \mathbf{a})$.

**Proof**
One applies proposition 313 (p. 195) to

$$\mathbf{g}(\mathbf{x}) = \mathbf{f}(\mathbf{x}) - \frac{\mathbf{f}(\mathbf{b})(\mathbf{x} - \mathbf{a}) - \mathbf{f}(\mathbf{a})(\mathbf{x} - \mathbf{b})}{\mathbf{b} - \mathbf{a}} \tag{2.57}$$

$\mathbf{g}(\mathbf{x})$ is a continuous function from proposition 287 (p. 185), it is zero for $\mathbf{x} = \mathbf{a}$ and $\mathbf{x} = \mathbf{b}$. From propositions 308 (p. 194), 309 (p. 194), 313 (p. 195) and

$$(\exists \mathbf{x}_\mathbf{m} \in ]\mathbf{a}, \mathbf{b}[)(\mathbf{g}'(\mathbf{x}_\mathbf{m}) = 0) \, .$$

From equation (2.57) we have:

$$\mathbf{g}'(\mathbf{x}_\mathbf{m}) = \mathbf{f}'(\mathbf{x}_\mathbf{m}) - \frac{\mathbf{f}(\mathbf{b}) - \mathbf{f}(\mathbf{a})}{\mathbf{b} - \mathbf{a}} \, .$$

Then

$$\mathbf{g}'(\mathbf{x}_\mathbf{m}) = 0 \Rightarrow \mathbf{f}'(\mathbf{x}_\mathbf{m}) = \frac{\mathbf{f}(\mathbf{b}) - \mathbf{f}(\mathbf{a})}{\mathbf{b} - \mathbf{a}} \, .$$

**Proposition 315** If a function which is derivable on a segment $]\mathbf{a}, \mathbf{b}[$ and continuous on $[\mathbf{a}, \mathbf{b}]$ has a derivative bigger than (resp. bigger than or equal to) zero on $]\mathbf{a}, \mathbf{b}[$ then it is a strictly increasing (resp. increasing) function on $[\mathbf{a}, \mathbf{b}]$ (definitions 384 (p. 486), 386 (p. 486)).

**Proof**
For $\mathbf{c} \in [\mathbf{a}, \mathbf{b}]$ and $\mathbf{d} \in [\mathbf{a}, \mathbf{b}]$ $\mathbf{c} < \mathbf{b}$ we have from proposition 314

$$(\exists \mathbf{x} \in ]\mathbf{c}, \mathbf{d}[)(\mathbf{f}'(\mathbf{x})(\mathbf{d} - \mathbf{c}) = \mathbf{f}(\mathbf{d}) - \mathbf{f}(\mathbf{c}))$$

then

$$\mathbf{f}'(\mathbf{x}) > 0 \Rightarrow (\mathbf{d} > \mathbf{c} \Leftrightarrow \mathbf{f}(\mathbf{d}) > \mathbf{f}(\mathbf{c}))$$

resp.

$$\mathbf{f}'(\mathbf{x}) \geq 0 \Rightarrow (\mathbf{d} > \mathbf{c} \Leftrightarrow \mathbf{f}(\mathbf{d}) \geq \mathbf{f}(\mathbf{c})) \, .$$

**Proposition 316** Given $\mathbf{a} \in \mathcal{R}^+$: $\mathbf{a} > 0$ and two numbers $\mathbf{b} \in \mathcal{R}$, $\mathbf{c} \in \mathcal{R}, \mathbf{b} < \mathbf{c}$, we have:

$$\mathbf{ab} < \mathbf{ac} \, .$$

**Proof**
It is a direct consequence of propositions 309 (p. 194) and 315 (p. 196).

**Proposition 317**

$$a < b \Rightarrow -a > -b \,.$$

**Proof**
It is a direct consequence of propositions 309 (p. 194) and 315 (p. 196).

**Proposition 318** We have:

$$a > 0 \Rightarrow ((x \in ]0,1[ \Rightarrow ax < a), (x > 1 \Rightarrow ax > a))$$

$$a \in \mathcal{R} \Rightarrow (x = 0 \Rightarrow ax = 0, x = 1 \Rightarrow ax = a) \,.$$

**Proof**
From proposition 309 (p. 194), the derivative of $f(x) = ax$ is $a$. From proposition 315 (p. 196):

$$(x < 1, a > 0) \Rightarrow ax < a \times 1 = a, \ (x > 1, a > 0) \Rightarrow ax > a \times 1 = a \,.$$

From the definition of **1** (definitions 170 (p. 206), 180 (p. 215)) and from proposition 364 (p. 215), we have:

$$a \in \mathcal{R} \Rightarrow (x = 0 \Rightarrow ax = 0, x = 1 \Rightarrow ax = a) \,.$$

**Definition 165** $r^n$, $r \in \mathcal{R}$, $n \in \mathcal{Z}$.

We extend first the definition 422 (p. 554) to $\mathcal{R}$ as explained in section 2.2.3 (p. 173). Then as in 422 (p. 554), setting a map $(v_i : [1,n] \to \{r\} \subset \mathcal{R})$, one calls:

$$\prod_{i \in [1,n]} v_i = r^n \,.$$

From that definition, we have:

$$r^p \times r^q = r^{p+q}$$
$$p > q \Rightarrow r^p \times (1/r)^q = r^{p-q}$$
$$r^0 = 1$$

We may write:

$$r^{-n} = (1/r)^n \,.$$

**Proposition 319**

$$(a \in ]0,1[, n \geq 0) \Rightarrow 1 > a^n > a^{n+1} > 0$$
$$(a > 1, n \geq 0) \Rightarrow 1 < a^n < a^{n+1}$$
$$a = 0 \Rightarrow a^n = 0, a = 1 \Rightarrow a^n = 1 \,.$$

**Proof**
From proposition 318 (p. 197) and definition 165 (p. 197), we have:

$$(a \in ]0,1[, n \geq 0) \Rightarrow a^n > a^{n+1}$$
$$(a > 1, n \geq 0) \Rightarrow a^n < a^{n+1} .$$

For $n = 1$ we have:

$$a \in ]0,1[ \Rightarrow a^n \in ]0,1[, a > 1 \Rightarrow a^n > 1, a = 0 \Rightarrow a^n = 0, a = 1 \Rightarrow a^n = 1 .$$

Let us suppose that proposition 319 is true for $n$, from proposition 318 (p. 197) and definition 165 (p. 197) and from the definition of $1$ (definition 167 (p. 205), 170 (p. 206), 180 (p. 215)) and from proposition 364 (p. 215), it is true also for $n + 1$. Then from proposition 927 (p. 520), it is true for any $n$.

**Proposition 320**

$$(\forall i \in \mathcal{Z}^+)(\forall j \in \mathcal{Z}^+)((a \in ]0,1[, i > j) \Rightarrow 0 < a^i < a^j < 1) . \tag{2.58}$$

**Proof**
From proposition 319 (p. 197), for $a \in ]0,1[$ and $j \in \mathcal{Z}^+$, we have:

$$0 < a^{j+1} < a^j < 1$$

let us suppose for $k \in \mathcal{Z}^+$

$$0 < a^{j+k} < a^j < 1 .$$

From proposition 319 (p. 197) again, we have:

$$0 < a^{j+k+1} < a^{j+k} < a^j < 1 .$$

We conclude using proposition 927 (p. 520).

**Proposition 321** $(f : \mathcal{R} \to \mathcal{R})$, $(g : \mathcal{R} \to \mathcal{R})$, being continuous maps at $x \in \mathcal{R}$, $f$ and $g$ having derivative at $x$: $f'$ and $g'$, then the derivative $h'$ of $h(x) = f \circ g(x)$ at $x$ exists and is given by:

$$h' = f'(g(x))g'(x) .$$

**Proof**
With the above notations, we may write:

$$\Delta g = g(x + \Delta x) - g(x)$$
$$\Delta h = f(g(x) + \Delta g) - f(g(x))$$

and from definition 161 (p. 192)

$$(\forall \alpha > 0)(\exists(\epsilon_g(\alpha) > 0, \epsilon_f(\alpha) > 0))(|\Delta x| < \epsilon_g(\alpha) \Rightarrow |\frac{\Delta g}{\Delta x} - g'(x)| < \alpha,$$
$$|\Delta g| < \epsilon_f(\alpha) \Rightarrow |\frac{\Delta h}{\Delta g} - f'(g(x))| < \alpha) .$$

Then since:

$$|\frac{\Delta h}{\Delta x} - f'(g(x))g'(x)| = |(\frac{\Delta h}{\Delta g} - f'(g(x)))\frac{\Delta g}{\Delta x} - f'(g(x))(g'(x) - \frac{\Delta g}{\Delta x})|$$
$$\leq |\frac{\Delta h}{\Delta g} - f'(g(x))| \times |\frac{\Delta g}{\Delta x}| + |f'(g(x))| \times |g'(x) - \frac{\Delta g}{\Delta x}|$$

calling $\alpha' = \frac{\alpha}{2(|f'(g(x))|+1)}$ and $\epsilon = \min(\epsilon_g(\alpha'), \epsilon_f(\frac{\alpha}{2(\alpha'+|g'(x)|)}))$, we have:

$$|\Delta x| < \epsilon \Rightarrow |\Delta x| < \epsilon_g(\alpha') \Rightarrow |g'(x) - \frac{\Delta g}{\Delta x}| < \alpha'$$
$$\Rightarrow |f'(g(x))| \times |g'(x) - \frac{\Delta g}{\Delta x}| < |f'(g(x))| \frac{\alpha}{2(|f'(g(x))| + 1)} < \alpha/2$$

and (see propositions 272 (p. 181), 273 (p. 181))

$$|\Delta x| < \epsilon \Rightarrow |\Delta x| < \epsilon_g(\alpha') \Rightarrow |\frac{\Delta g}{\Delta x}| - |g'(x)| \leq |\frac{\Delta g}{\Delta x} - g'(x)| < \alpha'$$
$$\Rightarrow |\frac{\Delta g}{\Delta x}| \leq \alpha' + |g'(x)|$$
$$\Rightarrow |\frac{\Delta h}{\Delta x} - f'(g(x))| \times |\frac{\Delta g}{\Delta x}| < |\frac{\Delta h}{\Delta x} - f'(g(x))|(\alpha' + |g'(x)|)$$

and also:

$$|\Delta x| < \epsilon \Rightarrow |\Delta x| < \epsilon_f(\frac{\alpha}{2(\alpha' + |g'(x)| + 1)})$$
$$\Rightarrow |\frac{\Delta h}{\Delta x} - f'(g(x))| < \frac{\alpha}{2(\alpha' + |g'(x)|)}$$
$$\Rightarrow |\frac{\Delta h}{\Delta x} - f'(g(x))| \times |\frac{\Delta g}{\Delta x}| < \alpha \frac{\alpha' + |g'(x)|}{2(\alpha' + |g'(x)|)} \leq \alpha/2 .$$

Combining everything we have

$$(\forall \alpha)(\exists \epsilon = \min(\epsilon_g(\alpha'), \epsilon_f(\frac{\alpha}{2(\alpha' + |g'(x)|)})) > 0)(|\Delta x| < \epsilon$$
$$\Rightarrow |\frac{f(g(x + \Delta x)) - f(g(x))}{\Delta x} - f'(g(x))g'(x)| < \alpha) .$$

**Proposition 322** ($f : \mathcal{R} \to \mathcal{R}$), ($g : \mathcal{R} \to \mathcal{R}$), ($h = f \circ g : \mathcal{R} \to \mathcal{R}$) being continuous maps at $x \in \mathcal{R}$, $f$ and $h$ having derivative at $x$: $f'$ and $h'$ with $f'(g(x)) \neq 0$, then $g(x)$ has a derivative $g'(x)$ at $x$ which is given by:

$$g'(x) = \frac{h'(x)}{f'(g(x))} .$$

**Proof**
With the above notations, we may write:

$$\Delta g = g(x + \Delta x) - g(x)$$
$$\frac{f(g(x) + \Delta g) - f(g(x))}{\Delta g} \frac{\Delta g}{\Delta x} = (f'(g(x)) + \eta_f(\Delta g)) \frac{\Delta g}{\Delta x}$$
$$\frac{h(x + \Delta x) - h(x)}{\Delta x} = h'(x) + \eta_h(\Delta x)$$

and

$$\frac{\Delta g}{\Delta x} = \frac{h'(x) + \eta_h(\Delta x)}{f'(g(x)) + \eta_f(\Delta g)} .$$

From what we suppose, we have (see propositions 241 (p. 167), 285 (p. 185), definition 161 (p. 192)), given $\mathbf{x}$, with $\mathbf{f'(g(x)) \neq 0}$:

$$(\forall\alpha > 0)(\exists\epsilon_g(\alpha))(|\Delta x| < \epsilon_g(\alpha) \Rightarrow |\Delta g| < \alpha)$$
$$(\forall\alpha > 0)(\exists\epsilon_f(\alpha))(|\Delta g| < \epsilon_f(\alpha)) \Rightarrow (|\frac{1}{f'(g(x)) + \eta_f(\Delta g)} - \frac{1}{f'(g(x))}| < \alpha)$$
$$(\forall\alpha > 0)(\exists\epsilon_\Delta(\alpha))(|\Delta g| < \epsilon_\Delta(\alpha) \Rightarrow |\eta_f(\Delta g)| < \alpha)$$
$$(\forall\alpha > 0)(\exists\epsilon_h(\alpha))(|\Delta x| < \epsilon_h(\alpha) \Rightarrow |\eta_h(\Delta x)| < \alpha)\ .$$

Then setting

$$\epsilon = \min\left(\epsilon_g(\min(\epsilon_f(\frac{0.5\alpha}{|h'(x)| + 1}), \epsilon_\Delta(0.5|f'(g(x))|))), \epsilon_h(0.25\alpha f'(g(x)))\right)$$

$|\Delta x| < \epsilon$ leads to:

$$|\frac{\Delta g}{\Delta x} - \frac{h'(x)}{f'(g(x))}| < |h'(x)||\frac{1}{f'(g(x)) + \eta_f(\Delta g)} - \frac{1}{f'(g(x))}| + \frac{|\eta_h(\Delta x)|}{||f'(g(x))| - |\eta_f(\Delta g)||}$$
$$< |h'(x)|\frac{0.5\alpha}{|h'(x)| + 1} + \frac{0.25\alpha f'(g(x))}{0.5f'(g(x))} = 0.5\alpha + \frac{0.5\alpha}{1 + 1/|h'(x)|} \leq \alpha$$

and then to

$$(\forall\alpha > 0)(\exists\epsilon > 0)(|\Delta x| < \epsilon \Rightarrow |\frac{\Delta g}{\Delta x} - \frac{h'(x)}{f'(g(x))}| < \alpha)\ .$$

Then the derivative of $\mathbf{g}$ at $\mathbf{x}$ is $\mathbf{g}' = \frac{h'(x)}{f'(g(x))}$.

**Proposition 323** Having a map $(\mathbf{f} : \mathcal{R} \subset\!\!\rightarrow \mathcal{R})$ with a nonzero derivative at $\mathbf{x}$ if its inverse exists and is continuous at $\mathbf{x}$, it has a derivative at $\mathbf{f(x)}$ which is $(f^{-1})'(f(x)) = \frac{1}{f'(x)}$.

**Proof**
In the above proposition 322 (p. 199), setting $\mathbf{g} = \mathbf{f}^{-1}$, one gets: $f(f^{-1}(x)) = x$. Then $(f^{-1})'(y) = \frac{1}{f'(f^{-1}y)}$. Setting $\mathbf{y} = \mathbf{f(x)}$, it gives proposition 323.

**Definition 166** $C^0$, $(C^i : \mathcal{Z}^+ \rightarrow \mathcal{P}(\mathcal{R}^2))$, $C^\infty$.

Derivative defines a derivative function which can also be continuous and derivable. If a function is derivable to order $\mathbf{i} \in \mathcal{Z}^+$, it is said to belong to $C^i$, at any order, it is said to belong to $C^\infty$. Then by convention, the set of continuous functions, $\mathbf{i}$ times derivable functions, infinitely derivable functions are called respectively $C^0$, $(C^i : \mathcal{Z}^+ \rightarrow \mathcal{P}(\mathcal{R}^2))$, $C^\infty$. Those are sets from proposition 749 (p. 450). Example:

$$f(x) = \frac{1}{\sqrt{\sigma}}e^{\frac{-(x-a)^2}{2\sigma}}\ .$$

We call functions of $C^\infty$ smooth functions.

**Proposition 324** Given two maps $(\mathbf{f} : \mathcal{R} \rightarrow \mathcal{R})$, $(\mathbf{g} : \mathcal{R} \rightarrow \mathcal{R})$ both in $C^\infty$ (definitions 166 (p. 200), 276 (p. 304)), the composite map is in $C^\infty$.

**Proof**
Let us suppose that for $\mathbf{n} \in \mathcal{Z}^+$ we have: $\exists P_n \in \mathcal{Z}^+$, $\exists Q_n \in \mathcal{Z}^+$ for which we have

a map $(C_i : [1,P_n] \to \mathcal{R})$, a map $(j^i_k : [1,P_n] \times [1,Q_n] \to \mathcal{Z}^{0+})$ and a map $(j'^i_k : [1,P_n] \times [2,Q_n] \to \mathcal{Z}^{0+})$ such as (see definition 164 (p. 193)):

$$\frac{d^n f\circ g(x)}{dx^n} = \sum_{i\in[1,P_n]} C_i f^{[j^i_1]} \prod_{k\in[2,Q_n]} g^{[j^i_k],j'^i_k} \tag{2.59}$$

where

$$f^{[j]}(x) = \frac{d^j f(x)}{dx^j},\ g^{[j],j'}(x) = (\frac{d^j g(x)}{dx^j})^{j'} .$$

We have used proposition 562 (p. 309):

$$\frac{d\frac{d^n f\circ g(x)}{dx^n}}{dx} = \frac{d^{n+1} f\circ g(x)}{dx^{n+1}} = \sum_{i\in[1,P_n]} C_i (f^{[j^i_1+1]} \prod_{k\in[2,Q_n]} g^{[j^i_k],j'^i_k} +$$
$$f^{[j^i_1]} \sum_{k\in[2,Q_n]} j'^i_k g^{[j^i_k+1]} \prod_{k'\in[2,Q_n]} g^{[j^i_{k'}],j'^i_{k'}-\delta^{k'}_k})$$

$j'^i_{k'} - \delta^{k'}_k \geq 0$ because if $j'^i_{k'} = 0$, the corresponding multiplicative term $j'^i_{k'}$ is zero and the whole term disappears. The sum is of the same type as before with $Q_{n+1} = Q_n + 1$ and with a set of terms having a cardinal $P_{n+1}$ below $P_n \times Q_n$ so $P_{n+1} \in \mathcal{Z}^{0+}$. From the definition 390 (p. 499), there are three news maps for $n+1$ giving the same equation (2.59) but for $n+1$. Since equation (2.59) (p. 201) is true for $n = 1$, it is true for any $n$ and from proposition 559 (p. 306) we have proposition 324.

**Proposition 325** u and v being two maps $(\mathcal{R} \to \mathcal{R})$ with derivative, the derivative of $u \times v$ is:

$$(u\times v)' = u'\times v + u\times v' .$$

**Proof**
From definition 161 (p. 192):

$$(\forall\alpha > 0)(\exists\epsilon_u(\alpha) > 0, \epsilon_v(\alpha) > 0)$$
$$(|\Delta x| < \epsilon_u(\alpha) \Rightarrow |\frac{u(x+\Delta x) - u(x)}{\Delta x} - u'(x)| < \alpha,$$
$$|\Delta x| < \epsilon_v(\alpha) \Rightarrow |\frac{v(x+\Delta x) - v(x)}{\Delta x} - v'(x)| < \alpha)$$

and

$$|\frac{u(x+\Delta x)v(x+\Delta x) - u(x)v(x)}{\Delta x} - u'(x)\times v(x) - u(x)\times v'(x)|$$
$$= |\frac{u(x+\Delta x)v(x+\Delta x) - u(x)v(x+\Delta x)}{\Delta x} - u'(x)\times v(x+\Delta x)$$
$$+ u'(x)\times v(x+\Delta x) - u'(x)\times v(x) + \frac{u(x)v(x+\Delta x) - u(x)v(x)}{\Delta x}$$
$$- u(x)\times v(x)'| .$$

We have, setting $\epsilon_1(\alpha) = \frac{\alpha}{3(|u'(x)|+1)(|v'(x)|+\alpha)}$:

$$|\Delta x| < \min(\epsilon_v(\alpha), \epsilon_1(\alpha))$$
$$\Rightarrow |v(x+\Delta x) - v(x)| \leq (|v'(x)| + \alpha)|\Delta x| \leq \frac{\alpha}{3(|u'(x)|+1)} \Rightarrow$$
$$(|(v(x+\Delta x) - v(x))u'(x)| < \alpha/3,$$
$$|v(x+\Delta x)| < |v(x)| + \frac{\alpha}{3(|u'(x)|+1)})$$

$$|\Delta x| < \min(\epsilon_u(\frac{\alpha}{3(|v(x)| + \frac{\alpha}{3(|u'(x)|+1)})}), \epsilon_1(\alpha)) \Rightarrow$$
$$|\frac{u(x+\Delta x)v(x+\Delta x) - u(x)v(x+\Delta x)}{\Delta x} - u'(x) \times v(x+\Delta x)| < \alpha/3$$

and finally

$$|\Delta x| < \epsilon_v(\frac{\alpha}{3(|u(x)|+1)}) \Rightarrow |\frac{u(x)v(x+\Delta x) - u(x)v(x)}{\Delta x} - u(x) \times v(x)'|$$
$$< |u(x)|\frac{\alpha}{3(|u(x)|+1)} < \alpha/3 .$$

Combining everything:

$$(\forall \alpha > 0)(\exists \epsilon =$$
$$\min(\epsilon_v(\alpha), \epsilon_1(\alpha), \epsilon_u(\frac{\alpha}{3(|v(x)| + \frac{\alpha}{3(|u'(x)|+1)})}), \epsilon_v(\frac{\alpha}{3(|u(x)|+1)})) > 0)$$
$$(|\Delta x| < \epsilon \Rightarrow |\frac{u(x+\Delta x)v(x+\Delta x) - u(x)v(x)}{\Delta x} - u(x)v'(x) - u'(x)v(x)| < \alpha) .$$

**Proposition 326** $w$ being a map $(w : \mathcal{R} \to \mathcal{R})$ and $n \geq 0$ an integer, the derivative of $w^n$ (see definition 165 (p. 197)) is:

$$(w^n)' = n \times w^{n-1}w' .$$

**Proof**
Applying proposition 325 with $u = w^n$ and $v = w$, if proposition 326 is true for $n \geq 0$ it is true for $n + 1$. Since it is true for $n = 0$, it is true for any $n$. See proposition 927 (p. 520).

**Proposition 327** $w$ being a map $(w : \mathcal{R} \to \mathcal{R})$ and $n < 0$ an integer, the derivative of $w^n$ when $w \neq 0$ is:

$$(w^n)' = n \times w^{n-1}w' .$$

**Proof**
It is a direct application of propositions 326 (p. 202), 563 (p. 311) and 285 (p. 185). We have $w^n = \frac{1}{w^{-n}}$. Then applying propositions 563 (p. 311) and 285 (p. 185) when $w \neq 0$ $w^{-n} \neq 0$, we have:

$$(w^n)' = -w^{2n}(w^{-n})' = -w^{2n} \times (-n)w^{-n-1}w' = n \times w^{n-1}w' .$$

**Proposition 328** The map (see definition 378 (p. 484)) $(x^n : [0, \to [\ \to\ [0, \to [)$ is a strictly increasing and continuous map. It has a continuous inverse called $x^{1/n}$ which is also a strictly increasing and continuous map.

**Proof**
It is a consequence of propositions 326 (p. 202), 307 (p. 193) and 300 (p. 191). Because $(\forall \epsilon > 0)$ and $(\forall A > \epsilon)$, $x^n$ has a positive derivative when restricted on $[\epsilon, A]$, $x^n$ is a

strictly increasing map on $[\epsilon, A]$. We also have (see definitions 396 (p. 510), 415 (p. 543), 422 (p. 554), section 2.2.3 and proposition 827 (p. 475)):

$$x = 0 \Rightarrow x^n = 0, \epsilon > 0 \Rightarrow \epsilon^n > 0$$

then (see definition 378 (p. 484))

$$(\forall x \in [0, \rightarrow [)(\forall y \in [0, \rightarrow [)(x < y \Rightarrow x^n < y^n) .$$

From proposition 300 (p. 191), it has a continuous inverse $x^{1/n}$ on $[0, \rightarrow [$ which is also a strictly increasing map (definition 386 (p. 486)).

**Proposition 329** When $x > 0.$, the derivative of $x^{1/n}$ is $\frac{x^{1/n-1}}{n}$.

**Proof**

From proposition 323, setting $f(x) = x^n$, $f'(x) = nx^{n-1}$ then

$$(f^{-1})'(y) = dy^{1/n}/dy = 1/f'(f^{-1}(y)) = \frac{1}{n(y^{1/n})^{n-1}} = \frac{y^{1/n}}{ny} = \frac{y^{1/n-1}}{n} .$$

Note that $x^{1/n}$ is not derivable at $x = 0$.

**Proposition 330** When $x > 0$ the derivative of $x^{p/q}$ is $\frac{px^{p/q-1}}{q}$.

**Proof**

It is a consequence of propositions 329 and 326.

**Proposition 331** Given a finite set of maps $(f_\ell : [1, n] \rightarrow C^i)$, $(i, n) \in (\mathcal{Z}^{+0})^2$, we have the following statement:

$$k \in [1, i] \Rightarrow \frac{d^k}{dx^k} \sum_{\ell \in [1,n]} f^\ell(x) = \sum_{\ell \in [1,n]} \frac{d^k}{dx^k} f^\ell(x) \in C^{i-k} . \tag{2.60}$$

Setting

$$j \in [1, n]^k \Leftrightarrow j = \prod_{\ell \in [1,k]}^{e} j_\ell \in n^k .$$

It means $(\forall \ell \in [1, k])(j_\ell \in [1, n])$, we have:

$$k \in [1, i] \Rightarrow \frac{d^k}{dx^k} \prod_{\ell \in [1,n]} f^\ell(x) = \sum_{j \in [1,n]^k} \{ \prod_{t \in [1,k]} \frac{d_{j_t}}{dx} ( \prod_{\ell \in [1,n]} f^\ell(x))\} \in C^{i-k} . \tag{2.61}$$

In equation (2.61) (p. 203) $\frac{d_t}{dx}$ acts only on $f^t$ and by convention here we have (see proposition 292 (p. 187)):

$$(p_\ell : [1, n] \in \mathcal{Z}^{0+}) \in (\mathcal{Z}^{0+})^n \Rightarrow \frac{d_t}{dx} \prod_{\ell \in [1,n]} (\frac{d^{p_\ell}}{dx^{p_\ell}} f^\ell) = \frac{d^{p_t+1}}{dx^{p_t+1}} f^t \prod_{\ell \in [1,n]-t} (\frac{d^{p_\ell}}{dx^{p_\ell}} f^\ell) .$$

**Proof**

If equation (2.60) (p. 203) is true for $n \in [1, n'], i \in [1, i']$, it is also true for $n \in [1, n' + 1]$: $\sum_{\ell \in [1,n'+1]} f^\ell(x) = \sum_{\ell [1,n']} f^\ell(x) + f^{n'+1}$. From propositions 311 (p. 194), 293 (p. 188), when $k \in [1, i]$, we have:

$$\frac{d^k}{dx^k} \sum_{\ell \in [1,n'+1]} f^\ell(x) = \frac{d^k}{dx^k} \sum_{\ell \in [1,n']} f^\ell(x) + \frac{d^k}{dx^k} f^{n'+1} = \sum_{\ell \in [1,n'+1]} \frac{d^k}{dx^k} f^\ell(x) \in C^{i-k} .$$

From proposition 927 (p. 520), since it is true for $\mathbf{n} = \mathbf{1}$, it is true for any $\mathbf{n}$. If equation (2.60) (p. 203) is true for $\mathbf{n} \in [\mathbf{1}, \mathbf{n'}], \mathbf{i} \in [\mathbf{1}, \mathbf{i'}]$ it is also true for $\mathbf{i} \in [\mathbf{1}, \mathbf{i'} + \mathbf{1}]$:
For $\mathbf{k} = \mathbf{i'} + \mathbf{1}$, from proposition 311 (p. 194), definition 163 (p. 193) and what we suppose, we have:

$$\frac{d^{i'+1}}{dx^{i'+1}} \sum_{\ell \in [1,n]} f^{\ell}(x) = \frac{d^{i'}}{dx^{i'}} \frac{d}{dx} \sum_{\ell \in [1,n]} f^{\ell}(x) = \frac{d^{i'}}{dx^{i'}} \sum_{\ell \in [1,n]} \frac{d}{dx} f^{\ell}(x) = \sum_{\ell \in [1,n]} \frac{d^{i'}}{dx^{i'}} \frac{d}{dx} f^{\ell}(x)$$
$$= \sum_{\ell \in [1,n]} \frac{d^{i'+1}}{dx^{i'+1}} f^{\ell}(x) .$$

Then from proposition 927 (p. 520), since it is true for $\mathbf{i} = \mathbf{0}$, it is true for any $\mathbf{i}$. From proposition 928 (p. 520), equation (2.60) (p. 203) is true.
From proposition 325 (p. 201), equation (2.61) (p. 203) is true for $\mathbf{k} = \mathbf{1},\ \mathbf{n} = \mathbf{2}$.
If equation (2.61) (p. 203) is true for $\mathbf{k} = \mathbf{1}$ and $\mathbf{n} = \mathbf{n'}$, it is also true for $\mathbf{k} = \mathbf{1}$ and $\mathbf{n} = \mathbf{n'} + \mathbf{1}$:

$$\frac{d}{dx} \prod_{\ell \in [1,n'+1]} f^{\ell}(x) = \frac{d}{dx}\{f^{n'+1}(x) \prod_{\ell \in [1,n']} f^{\ell}(x)\}$$
$$= \{\frac{d}{dx}(f^{n'+1}(x))\} \prod_{\ell \in [1,n']} f^{\ell}(x) + \sum_{j \in [1,n']} \frac{d_j}{dx}( \prod_{\ell \in [1,n'+1]-j} f^{\ell}(x))$$
$$= \sum_{j \in [1,n'+1]} \{\frac{d_j}{dx}( \prod_{\ell \in [1,n'+1]-j} f^{\ell}(x))\} \in C^{i-1} .$$

Then from proposition 927 (p. 520), equation (2.61) (p. 203) is true for $\mathbf{k} = \mathbf{1}$ and $\mathbf{n} \in [\mathbf{1}, \rightarrow[$. If equation (2.61) (p. 203) is true for $\mathbf{k} \in [\mathbf{1}, \mathbf{k'}]$ and $\mathbf{n}$, it is also true for $\mathbf{k} = \mathbf{k'} + \mathbf{1}$ and $\mathbf{n}$ because we have:

$$\frac{d^{k'+1}}{dx^{k'+1}} \prod_{\ell \in [1,n]} f^{\ell}(x) = \frac{d^{k'}}{dx^{k'}} \frac{d}{dx} \prod_{\ell \in [1,n]} f^{\ell}(x) = \frac{d^{k'}}{dx^{k'}} \sum_{j \in [1,n]} \frac{d}{dx} f^{j} \prod_{\ell \in [1,n]-j} f^{j}$$
$$= \sum_{j' \in [1,n]} \frac{d^{k'}}{dx^{k'}} \frac{d}{dx} f^{j'} \prod_{\ell \in [1,n]-j'} f^{j}$$
$$= \sum_{j' \in [1,n]} \sum_{j \in n^{k'}} \{( \prod_{t \in [1,k']} \frac{d_{j_t}}{dx})((\frac{d}{dx} f^{j'}) \prod_{\ell \in [1,n]-j'} f^{\ell}(x))$$
$$= \sum_{j' \in [1,n]} \sum_{j \in n^{k'}} \{( \prod_{t \in [1,k']} \frac{d_{j_t}}{dx})(\frac{d_{j'}}{dx} \prod_{\ell \in [1,n]} f^{\ell}(x))\} = \sum_{j \in n^{k'+1}} \{( \prod_{t \in [1,k'+1]} \frac{d_{j_t}}{dx})( \prod_{\ell \in [1,n]} f^{\ell}(x))\} .$$

From proposition 367 (p. 216), we have:

$$\sum_{j' \in [1,n]} \sum_{j \in n^{k'}} u(j' \times j) = \sum_{j \in [1,n]^{k'} \times [1,n]} u(j) .$$

From proposition 836 (p. 480), with $\mathbf{I} = \{\mathbf{k'} + \mathbf{1}\},\ \mathbf{J} = [\mathbf{1}, \mathbf{k'}],\ \mathbf{X} = \{[\mathbf{1}, \mathbf{n}]\}$ with:

$$(\forall j \in J) X_j = [1, n] \in X,\ X_{k'+1} = [1, n] \in X$$

we have:

$$\prod_{i \in I} X_i \prod_{j \in J} X_j = [1, n] \times [1, n]^{k'} = \prod_{i \in I \cup J} [1, n] = [1, n]^{k'+1} .$$

From proposition 927 (p. 520), equation (2.61) (p. 203) is true.

**Proposition 332** Given an open $O \subset \mathcal{R}$, $i \in \mathcal{Z}^+$ and a map $(f : O \to \mathcal{R}) \in C^i$, with $(\forall x \in O)(f(x) \neq 0)$, we have: $\frac{1}{f(x)} \in C^i$.

**Proof**

From proposition 327 (p. 202) we have:

$$\frac{d}{dx}\frac{1}{f(x)} = -\frac{\frac{d}{dx}f(x)}{f(x)^2} .$$

Let us suppose that for $n \in [1, n'] \subset [1, i[$, we have $\frac{d^{n'}}{dx^{n'}}\frac{1}{f(x)} \in C^{i-n'}$ and let us suppose that it is the finite sum of product of a map of $C^{i-n'}$, $g$, by $f^{-p}$ with $p \in \mathcal{Z}^+$.

Let us prove that, in that case, we have the same properties for $n = n' + 1$:

From propositions 325 (p. 201) and 327 (p. 202), since $g$ and $f$ are in $C^{i-n'}$, we have:

$$\frac{d}{dx}gf^{-p} = f^{-p}\frac{d}{dx}g - pf^{-p-1}g\frac{d}{dx}f \in C^{i-n'-1}$$

then we have those properties for $n \in [1, n' + 1]$. Those properties are verified for $n = 1$. From proposition 928 (p. 520), proposition 332 (p. 205) is true.

## 2.9 Group

**Definition 167 Group.**

A set $G$ is called a group if there is a map $(a \cdot b : G \times G \to G)$ such that:

1. $a \cdot (b \cdot c) = (a \cdot b) \cdot c$ (associativity)
2. $(\exists e \in G)(\forall a \in G)(a \cdot e = a)$ e is called unit or neutral element.
3. $(\forall a \in G)(\exists a^{-1} \in G)(a \cdot a^{-1} = e)$

**Proposition 333** $e \cdot a = a$

**Proof**

From definition 167 (p. 205), 2:

$$e \cdot e = e \Rightarrow e \cdot a \cdot a^{-1} = e \cdot e = e = a \cdot a^{-1} \tag{2.62}$$

but:

| | |
|---|---|
| $e \cdot a = e \cdot (a \cdot e)$ | from definition 167, 2 |
| $e \cdot a = e \cdot (a \cdot (a^{-1} \cdot (a^{-1})^{-1}))$ | from definition 167, 3 |
| $e \cdot a = (e \cdot a \cdot a^{-1}) \cdot (a^{-1})^{-1}$ | from definition 167, 1 |
| $e \cdot a = a \cdot a^{-1} \cdot (a^{-1})^{-1} = a \cdot e$ | from equation (2.62) and from definition 167, 1 |
| $e \cdot a = a.$ | |

**Proposition 334** $(a^{-1})^{-1} = a$, $a^{-1} \cdot a = e$.

**Proof**

$$\begin{aligned} a \cdot a^{-1} \cdot (a^{-1})^{-1} &= a \cdot (a^{-1} \cdot (a^{-1})^{-1}) = a \cdot e = a \\ &= (a \cdot a^{-1}) \cdot (a^{-1})^{-1} = e \cdot (a^{-1})^{-1} = (a^{-1})^{-1} = a \end{aligned}$$

$$a^{-1}(a^{-1})^{-1} = e = a^{-1}a .$$

**Proposition 335** $\mathbf{Card(set_e\{(\forall a \in G)(a \cdot e = a)\}) = 1}$.

**Proof**
The set of unit element is not empty from definition 167 (p. 205), 2. If there were more than 1 element, we may label $\mathbf{e}$ and $\mathbf{f}$ as two different unit elements. But from definition 167 (p. 205), 2 $\mathbf{f \cdot e = f}$ and from proposition 333 (p. 205) $\mathbf{f \cdot e = e}$ then $\mathbf{e = f}$ and the unit element is unique.

**Proposition 336** $\mathbf{(a \in G, b \in G, c \in G, ab = ac = e) \Rightarrow b = c = a^{-1}}$.

**Proof**
Using definition 167 (p. 205) and propositions 333 (p. 205), 334 (p. 205), 335 (p. 206), we have:

$$\mathbf{abc = ec = c}$$
$$\mathbf{abc = bac = be = b}$$
$$\mathbf{b = c}\ .$$

**Proposition 337** Given a set $\mathbf{E}$, the set of bijections of $\mathbf{E}$ on $\mathbf{E}$ is a group over the product of maps (see 117 (p. 134), 124 (p. 139)).

**Proof**
It is a direct application of the definition bijective maps and of propositions 189 (p. 147), 193 (p. 149) and 178 (p. 142).

**Definition 168 Subgroup.**

Given a group $\mathbf{G}$, a subset $\mathbf{G'}$ of $\mathbf{G}$ is a subgroup if it is stable for the internal law of $\mathbf{G}$. That is to say if:

$$\mathbf{G' \subset G, (\forall u \in G')(u^{-1} \in G', (\forall v \in G')(u \cdot v \in G'))}\ .$$

**Definition 169 Map conserving the internal law.**

Given two groups $\mathbf{G}$ and $\mathbf{G'}$ and a map $\mathbf{(f : G \to G')}$, the map $\mathbf{f}$ is said to keep the internal law of the groups if:

$$\mathbf{(\forall u \in G)(\forall v \in G)(f(u) \cdot f(v) = f(u \cdot v))}\ .$$

**Proposition 338** Given a group $\mathbf{G}$, its set of maps $(\boldsymbol{\mathcal{F}}\mathbf{: G \to G})$, from definition 167 (p. 205), any element of $\mathbf{G}$ is mapped by the internal law to an element of $\boldsymbol{\mathcal{F}}$. This map preserves the internal law.

**Proof**
Calling $\mathbf{F}$ that map: $(\mathbf{F_u : G \to} \boldsymbol{\mathcal{F}})$, we have $\mathbf{(u \in G, v \in G) \Rightarrow F_u(v) = u \cdot v}$ and from 1 of definition 167 (p. 205), we have:

$$\mathbf{w \in G \Rightarrow F_{u \cdot v}(w) = (u \cdot v) \cdot w = u \cdot v \cdot w = u \cdot F_v(w) = F_u(F_v(w)) = F_u \circ F_v(w)}\ .$$

**Definition 170 Commutative group or Abelian group.**

A group is commutative (Abelian) if $\mathbf{a \cdot b = b \cdot a}$. One used to note Abelian group by the sign $+$ instead of $\cdot$ and $\mathbf{a^{-1}}$ by $\mathbf{-a}$. In such notation $\mathbf{e}$ is often called $\mathbf{0}$.

**Proposition 339** Given a group $\mathbf{G}$, an integer $\mathbf{n} \in \mathcal{Z}^+$, a map $(\mathbf{g_i} : [\mathbf{1}, \mathbf{n}] \to \mathbf{G})$, we have

$$(\exists(\sigma_{\mathbf{k}}^{\mathbf{g}} : [\mathbf{1}, \mathbf{n}] \to \mathbf{G}))(\forall \mathbf{k} \in [\mathbf{2}, \mathbf{n}])(\sigma_{\mathbf{k}}^{\mathbf{g}} = \sigma_{\mathbf{k}-1}^{\mathbf{g}} \cdot \mathbf{g_k})$$

the map $(\sigma_{\mathbf{k}}^{\mathbf{g}} : [\mathbf{1}, \mathbf{n}] \to \mathbf{G})$ is unique and verifies:

$$((\mathbf{g} : [\mathbf{1}, \mathbf{n}] \to \mathbf{G}), (\mathbf{g}' : [\mathbf{1}, \mathbf{n}'] \to \mathbf{G}), (\mathbf{n}' \geq \mathbf{n}), (\forall \mathbf{k} \in [\mathbf{1}, \mathbf{n}])(\mathbf{g_k} = \mathbf{g}'_{\mathbf{k}}))$$
$$\Rightarrow (\forall \mathbf{k} \in [\mathbf{1}, \mathbf{n})(\sigma_{\mathbf{k}}^{\mathbf{g}} = \sigma_{\mathbf{k}'}^{\mathbf{g}'}) .$$

**Proof**
Let us suppose that it is true for $\mathbf{k} = \mathbf{k}' \in [\mathbf{1}, \mathbf{n} - \mathbf{1}]$ then it is true for $\mathbf{k} = \mathbf{k}' + \mathbf{1}$ because $\sigma_{\mathbf{k}+1}^{\mathbf{g}} = \sigma_{\mathbf{k}}^{\mathbf{g}} \cdot \mathbf{g_{k+1}}$. $\sigma_{\mathbf{k}+1}^{\mathbf{g}}$ exists and is unique. We conclude by proposition 927 (p. 520) since it is true for $\mathbf{k} = \mathbf{k}' = \mathbf{2}$.

**Definition 171** $\prod_{\ell=1}^{\ell=\mathbf{k}} \mathbf{g}_\ell, \sum_{\ell=1}^{\ell=\mathbf{k}} \mathbf{g}_\ell$.

The map $(\sigma_{\mathbf{k}}^{\mathbf{g}} : [\mathbf{1}, \mathbf{n}] \to \mathbf{G})$ of proposition 339 (p. 207) is written $\sigma_{\mathbf{k}}^{\mathbf{g}} = \prod_{\ell=1}^{\ell=\mathbf{k}} \mathbf{g}_\ell$ if the group is not Abelian and $\sigma_{\mathbf{k}}^{\mathbf{g}} = \sum_{\ell=1}^{\ell=\mathbf{k}} \mathbf{g}_\ell$ if it is Abelian (definition 170 (p. 206)).

## 2.9.1 Finite groups

**Definition 172 Permutation of $[\mathbf{1}, \mathbf{n}]$.**

One calls the permutation of $[\mathbf{1}, \mathbf{n}]$, $\mathbf{n} \in \mathcal{Z}^+$, a bijective map (definition 124 (p. 139)) $(\mathbf{P(i)} : [\mathbf{1}, \mathbf{n}] \to [\mathbf{1}, \mathbf{n}])$.

**Proposition 340** The identity map (definition 113 (p. 132)) of $[\mathbf{1}, \mathbf{n}] \to [\mathbf{1}, \mathbf{n}]$, $\mathbf{n} \in \mathcal{Z}^+$ is a permutation.

**Proof**
It is a direct consequence of proposition 179 (p. 142).

**Proposition 341** The set $\mathbf{PG_n}$ of permutations of $[\mathbf{1}, \mathbf{n}]$ with the map product (definition 117 (p. 134)) as an internal law, is a group (definition 167 (p. 205)).

**Proof**
$\mathbf{PG_n}$ is a group because the product of two bijective maps is a bijective map. The unit element is the identity map which is a bijective map. Since a permutation is a bijective map, it has an inverse which is the inverse map.

**Proposition 342** $\mathbf{PG_n}$ being the set of permutations of $[\mathbf{1}, \mathbf{n}]$, the map $(\mathbf{p(s)} = \mathbf{s}^{-1} : \mathbf{PG_n} \to \mathbf{PG_n})$ transforming any permutation on its inverse is a bijection on $\mathbf{PG_n}$.

**Proof**
It is a direct application of proposition 198 (p. 150), point 1.

**Proposition 343** Using the product of permutations, any element of the permutation group transforms the permutation group to itself as a bijection.

**Proof**
It is a direct application of proposition 198 (p. 150), points 2 and 3.

**Proposition 344** The number of permutations of set $\mathbf{A}$ with $\mathbf{n} = \mathbf{Card(A)} \in \mathcal{Z}^+$ is $\mathbf{n}!$.

**Proof**
It comes from proposition 944 (p. 529).

**Definition 173 Transposition.**

Given an integer subset $[1, n]$, a transposition $(T_{i,j}, i \in [1, n], j \in [1, n])$ is a map $(P : [1, n] \to [1, n])$ such that:

$$(l \neq i, l \neq j) \Rightarrow T_{i,j}(l) = l, T_{i,j}(i) = j, T_{i,j}(j) = i\ .$$

**Proposition 345** The product of two identical transpositions is the identity map.

**Proof**
It is a direct consequence of definition 173 (p. 208).

**Proposition 346** A transposition (definition 173 (p. 208)) is a permutation.

**Proof**
From proposition 345 (p. 208), a transposition is its own inverse. From proposition 124 (p. 139), it is a bijection then from definition 172 (p. 207) a permutation.

**Definition 174 Transposition of two consecutive terms.**

We use the same notation as definition 173. With $j \in [1, n-1]$, $T_{j,j+1}$ is a transposition of two consecutive terms

**Proposition 347** Any permutation is the product of transpositions.

**Proof**
Let us suppose that it is true for the permutations on $[1, n]$, let us take a permutation $P$ of $[1, n+1]$. The permutation $T_{p(n+1),n+1}P$ is a permutation of $[1, n+1]$ which leaves $n+1$ at its place so it is a permutation of $[1, n]$ then it is a product of transpositions: $(\exists J)(\prod_{j \in J} T_j = T_{p(n+1),n+1}P)$ by applying $T_{p(n+1),n+1}$ again we have:

$$(\exists J)(T_{p(n+1),n+1} \prod_{j \in J} T_j = T_{p(n+1),n+1} T_{p(n+1),n+1} P = P)\ .$$

Since it is true for $n = 2$, applying proposition 927 (p. 520), it is true for any $n$.

**Proposition 348** Any transposition is the product of transpositions of two consecutive terms (definition 174 (p. 208)).

**Proof**
We use the same notation as proposition 347, proposition 348 is true for $j \leq n-1, T_{j,j+1}$. We suppose that 348 is true for $j + n < n, T_{j,j+n}$. We have: for $j + n + 1 \leq n$:

$$T_{j,j+n+1} = T_{j+n,j+n+1} T_{j,j+n} T_{j+n,j+n+1}\ .$$

Applying proposition 927 (p. 520), proposition 348 is true for any $n$.

**Proposition 349** Any permutation is the product of transpositions of two consecutive terms.

**Proof**
It is the consequence of propositions 347 and 348.

**Definition 175 Signature of a permutation.**

$\mathbf{G}$ being a bijective map, $(\mathbf{G(i) : [1,n] \rightarrow [1,n]})$. Let us consider the set:

$$\mathbf{E(G) = set_{(k,l)}\{k \in [1,n], l \in [1,n], k < l, G(k) > G(l)\}} \ .$$

One calls the signature $\mathbf{s}$ of the map $\mathbf{G}$ (see definition 417 (p. 545)):

$$\mathbf{s(G) = (-1)^{Card(E(G))}} \ .$$

**Proposition 350** Keeping above notations, we call $\mathbb{1}$ the permutation such that

$$(\forall \mathbf{k} \in [\mathbf{1,n}])(\mathbb{1}(\mathbf{k}) = \mathbf{k}) \ .$$

We have

$$\mathbf{s}(\mathbb{1}) = \mathbf{1} \ .$$

**Proof**
We have $\mathbf{Card(E(G)) = 0}$ so $\mathbf{s}(\mathbb{1}) = \mathbf{1}$.

**Proposition 351** Given a permutation $(\mathbf{P}(\ell) : [\mathbf{1,n}] \rightarrow [\mathbf{1,n}])$ and a transposition $\mathbf{T_{i,j}}$ calling $\mathbf{s(Q)}$ the signature of a permutation $\mathbf{Q}$, we have:

$$\mathbf{s(P \circ T_{i,j}) = -s(P)} \ .$$

**Proof**
Let us suppose that $\mathbf{j = i + 1}$ with $\mathbf{j \leq n}$ and that we have a permutation $\mathbf{P}$ such that $\mathbf{P(i) < P(i+1)}$ (resp. $\mathbf{P(i) > P(i+1)}$). The element $\mathbf{(i,j)}$ is not (resp. is) on the set $\mathbf{E(P)}$ which enters in the evaluation of $\mathbf{s(P)}$. Applying $\mathbf{T_{i,j}}$ first then $\mathbf{P}$, the elements of $\mathbf{E(P)}$ remain elements of $\mathbf{E(PT)}$ but $\mathbf{(i,j)}$ is now an element of $\mathbf{E(PT)}$ (resp. the elements of $\mathbf{E(P)}$ except $\mathbf{(i,j)}$ make the elements of $\mathbf{E(PT)}$) because $\mathbf{T(i) = j}$ and $\mathbf{T(j) = i}$ which implies $\mathbf{P(T(i)) > P(T(i+1))}$ (resp. $\mathbf{P(T(i)) < P(T(i+1))}$). Then $\mathbf{Card(E(P)) = Card(E(PT)) - 1}$ (resp. $\mathbf{Card(E(P)) = Card(E(PT)) + 1}$). Proposition 351 is then true when $\mathbf{j = i \pm 1}$.
Let us suppose now we have demonstrated the proposition for $\mathbf{T_{i,i+q}}$ with $\mathbf{0 < q \leq p}$ and let us prove that it is then true for $\mathbf{q = p + 1}$. We have:

$$\mathbf{T_{i,i+p+1} = T_{i,i+p} T_{i+p,i+p+1} T_{i,i+p}}$$

Because the first transposition $\mathbf{T_{i,i+p}}$ puts the element $\mathbf{a}$ in $\mathbf{i}$ into $\mathbf{i+p}$ and $\mathbf{b}$ in $\mathbf{i+p}$ into $\mathbf{i}$, $\mathbf{T_{i+p,i+p+1}}$ puts the element in $\mathbf{i+p}$ (now $\mathbf{a}$) into its final position $\mathbf{i+p+1}$ and the element $\mathbf{c}$ from $\mathbf{i+p+1}$ into $\mathbf{i+p}$. $\mathbf{T_{i,i+p}}$ puts back $\mathbf{b}$ into $\mathbf{i+p}$ and $\mathbf{c}$ into $\mathbf{i}$ which is $\mathbf{T_{i,i+p+1}}$: $\mathbf{c}$ of $\mathbf{i+p+1}$ into $\mathbf{i}$ and $\mathbf{a}$ of $\mathbf{i}$ into $\mathbf{i+p+1}$, the rest keeping the same place. There are two transpositions $\mathbf{T_{i,i+q}}$ with $\mathbf{0 < q \leq p}$ for which proposition 351 (p. 209) is true and one $\mathbf{T_{i+p,i+p+1}}$ for which we just demonstrate that it is also true. The number of transpositions are odd (definition 403 (p. 522)) so proposition 351 is true from the recurrence principle (proposition 927 (p. 520)).

**Proposition 352** The signature of the product of a finite number of permutations is the product of their signature.
**Proof**
It is a direct consequence of propositions 347 (p. 208), 350 (p. 209) and 351 (p. 209).

**Proposition 353** The signature of the product of a finite number $\mathbf{n}$ of transpositions is $\mathbf{1}$ if $\mathbf{n}$ is even and $\mathbf{-1}$ if odd (see definition 403 (p. 522)).

**Proof**
It is a direct consequence of propositions 352 (p. 209), 350 (p. 209), 351 (p. 209), 975 (p. 545) and definition 417 (p. 545).

**Proposition 354** $\mathbf{P}$ being a permutation of $[\mathbf{1},\mathbf{n}]$, the signature of $\mathbf{P}$ is equal to the signature of $\mathbf{P}^{-1}$ (see definition 172 (p. 207)).

**Proof**
We have, calling $\mathbb{1}$ the identity map, $\mathbf{P}\circ\mathbf{P}^{-1}=\mathbb{1}$. Then from propositions 350 (p. 209) and 352 (p. 209), $\mathbf{s}(\mathbf{P}\circ\mathbf{P}^{-1})=\mathbf{s}(\mathbf{P})\mathbf{s}(\mathbf{P}^{-1})=1$. Since $\mathbf{s}(\mathbf{P})=\pm\mathbf{1}=\mathbf{1}/\mathbf{s}(\mathbf{P})$, we have proposition 354 (p. 210).

**Definition 176 Permutation with identical elements.**

Given a set $\mathbf{E}$ with an integer $\mathbf{n}$ as a cardinal and a set of index $\mathbf{I}$ ($\mathbf{Card}(\mathbf{I})\leq\mathbf{n}$) defining a partition on $\mathbf{E}$ (see definition 370 (p. 471)):$(\mathbf{F}_\mathbf{i}:\mathbf{I}\rightarrow\mathcal{P}(\mathbf{E}))$. One says that two permutations $\mathbf{u},\mathbf{v}$ of $\mathbf{E}$ are equivalent (see definition 336 (p. 445)) in the partition $\mathbf{F}_\mathbf{i}$ of $\mathbf{E}$, if: $\mathbf{u}\circ\mathbf{v}^{-1}=\prod_{\mathbf{i}\in\mathbf{I}}\circ\mathbf{P}_\mathbf{i}$ where $\mathbf{P}_\mathbf{i}$ are permutations of $\mathbf{E}$ changing only $\mathbf{F}_\mathbf{i}$ that is to say permutations of $\mathbf{E}$ keeping $\mathcal{C}_\mathbf{E}(\mathbf{F}_\mathbf{i})$ unchanged. The representatives of the sets of equivalent permutations (see definition 337 (p. 446)) form a group $\mathbf{H}$ with the following product law (see definition 332 (p. 440)):

$$\mathbf{x}\in\mathbf{H},\mathbf{y}\in\mathbf{H}\Rightarrow\mathbf{xy}=\tau_\mathbf{z}(\mathbf{z}\sim\mathbf{x}\circ\mathbf{y})$$

$\mathbf{H}$ is called the permutation group with identical elements associated to the partition $\mathbf{F}_\mathbf{i}$.

**Proposition 355** Given a finite set $\mathbf{E}$ and a finite set of indices $\mathbf{I}$ defining a partition $\mathbf{F}_\mathbf{i}$ of $\mathbf{E}$, the cardinal of the associated permutation group $\mathbf{H}$ with identical elements associated to the partition $\mathbf{F}_\mathbf{i}$ is:

$$\mathbf{Card}(\mathbf{H})=\frac{\mathbf{Card}(\mathbf{E})!}{\prod_{\mathbf{i}\in\mathbf{I}}\mathbf{Card}(\mathbf{F}_\mathbf{i})!}.$$

**Proof**
The number of $\mathbf{u}$ equivalent to any $\mathbf{x}\in\mathbf{H}$ is from proposition 344 (p. 207) $\prod_{\mathbf{i}\in\mathbf{I}}\mathbf{Card}(\mathbf{F}_\mathbf{i})!$. Calling $\mathbf{G}$ the set of permutations of $\mathbf{E}$ since the map: $(\mathbf{H}_\mathbf{u}=\mathbf{set}_\mathbf{x}\{\mathbf{x}\in\mathbf{G},\mathbf{x}\sim\mathbf{u}\}:\mathbf{H}\rightarrow\mathcal{P}(\mathbf{G}))$ is a partition of $\mathbf{G}$:

$$\mathbf{G}=\bigcup_{\mathbf{u}\in\mathbf{H}}\mathbf{set}_\mathbf{x}\{\mathbf{x}\in\mathbf{G},\mathbf{x}\sim\mathbf{u}\}$$

$$\mathbf{u}\in\mathbf{H},\mathbf{v}\in\mathbf{H},\mathbf{u}\neq\mathbf{v}\Rightarrow\mathbf{set}_\mathbf{x}\{\mathbf{x}\in\mathbf{G},\mathbf{x}\sim\mathbf{u}\}\cap\mathbf{set}_\mathbf{x}\{\mathbf{x}\in\mathbf{G},\mathbf{x}\sim\mathbf{v}\}=\emptyset$$

we have:

$$\mathbf{Card}(\mathbf{H})\times\prod_{\mathbf{i}\in\mathbf{I}}\mathbf{Card}(\mathbf{F}_\mathbf{i})!=\mathbf{Card}(\mathbf{G})=\mathbf{n}!$$

then

$$\mathbf{Card}(\mathbf{H})=\frac{\mathbf{n}!}{\prod_{\mathbf{i}\in\mathbf{I}}\mathbf{Card}(\mathbf{F}_\mathbf{i})!}$$

**Proposition 356** In the same condition as proposition 355, $\mathbf{I}$ and $\mathbf{E}$ being two finite sets, we consider the set of maps $(\mathbf{g}:\mathbf{E}\rightarrow\mathbf{I})$, $(\mathbf{g}\in\mathbf{I}^{\mathbf{Card}(\mathbf{E})})$ (see definitions 370 (p. 471), 373 (p. 474), propositions 815 (p. 471), 816 (p. 471)) defining partitions of $\mathbf{E}$ with $\mathbf{i}\in\mathbf{g}(\mathbf{E})\subset\mathbf{I}\Rightarrow\mathbf{F}_\mathbf{i}=\mathbf{g}^{-1}(\{\mathbf{i}\})\neq\emptyset$. We define an equivalence relation between two elements of $\mathbf{I}^{\mathbf{Card}(\mathbf{E})}$ by

$$\begin{aligned}&\mathbf{g}_1\sim\mathbf{g}_2\\&\Leftrightarrow(\mathbf{g}_1\in\mathbf{I}^{\mathbf{Card}(\mathbf{E})},\mathbf{g}_2\in\mathbf{I}^{\mathbf{Card}(\mathbf{E})},(\forall\mathbf{i}\in\mathbf{I})(\mathbf{Card}(\mathbf{g}_1^{-1}(\{\mathbf{i}\}))=\mathbf{Card}(\mathbf{g}_2^{-1}(\{\mathbf{i}\})))).\end{aligned}$$

For any partition defined by a map

$$\mathbf{g \in I^{Card(E)} = (g : E \to I)}$$

of **E** to **I**, the cardinal of the equivalence class $\mathcal{E}_F$ to **g** is also **Card(H)**, **H** being defined in definition 176 (p. 210):

$$\mathbf{Card(\mathcal{E}_F) = \frac{Card(E)!}{\prod_{i\in I} Card(F_i)!} = \frac{Card(E)!}{\prod_{i\in I} Card(g^{-1}(\{i\}))!}}\ .$$

We have used the definition 404 (p. 528) where $\mathbf{0! = Card(\emptyset)! = 1}$ which allows to extend the definition of **Card(H)** to the whole set **I** and all maps $\mathbf{E \to I}$.

**Proof**
Having a partition of **E** given by $\mathbf{(g : E \to I)}$ and $\mathbf{(F_i : I \to \mathcal{P}(E))}$ with

$$\mathbf{F_i = g^{-1}(\{i\}) \subset E}$$

considering among the permutation group, the set of equivalence class **H** as defined by definition 176 (p. 210) and the partition $\mathbf{F_i}$ which for simplicity can be empty, we have:

$$\begin{aligned}&\mathbf{(\forall a \in H)(\forall b \in H)(a \neq b \Rightarrow ((x_a \in a, y_a \in a, x_b \in b, y_b \in b) \Rightarrow}\\&\mathbf{((\forall i \in I)(x_a(F_i) = y_a(F_i), x_b(F_i) = y_b(F_i),}\\&\mathbf{Card(x_a(F_i)) = Card(x_b(F_i)) = Card(F_i)), (\exists i \in I)(x_a(F_i) \neq x_b(F_i)))))}\end{aligned}$$

so any element of **H** defines a new partition of **E** with the same number of elements as the $\mathbf{F_i}$. It is then a map $\mathbf{(u : H \to \mathcal{E}_F)}$. **u** is an injection because at two different elements of **H** correspond two different elements of $\mathcal{E}_F$. Any element $\mathbf{s \in \mathcal{E}_F}$ is the image of an element of $\mathbf{t \in H}$ because $\mathbf{s \in \mathcal{E}_F}$ gives rise from definition 390 (p. 499) to the following bijections:

$$\mathbf{(\forall i \in I)(Card(F_i) = Card(F_i^s)) \Rightarrow (\forall i \in I)(\exists p_i)(p_i : F_i \leftrightarrow F_i^s)}\ .$$

The $\mathbf{p_i}$ generate a permutation **p** of **E**: $\mathbf{p \in P(E)}$:

$$\mathbf{x \in E \Rightarrow x \in F_{g(x)} \Rightarrow p(x) = p_{g(x)}(x)}\ .$$

Since

$$\mathbf{i \neq j \leftrightarrow F_i \cap F_j = \emptyset \leftrightarrow p_i(F_i) \cap p_j(F_j) = \emptyset}\ .$$

Then $\mathbf{p(x)}$ is a bijection and belongs to $\mathbf{P(E)}$ and then defines a class of equivalence $\mathbf{t \in H}$ associated to $\mathbf{F_i}$ and we have $\mathbf{s = u(t)}$. Then **u** is a bijection. It gives (definition 390 (p. 499)):

$$\mathbf{Card(H) = Card(\mathcal{E}_F) = \frac{n!}{\prod_{i\in I} Card(F_i)!}}\ .$$

### 2.9.2 Operations on groups over a finite set of index

**Definition 177 Product $\prod$, Sum $\sum$ of a finite ordered subset of a group.**

Given a group **G**, an integer $\mathbf{n \in \mathcal{Z}^+}$, a map $\mathbf{(g_i : [1, n] \to G)}$ one calls $\mathbf{\prod_{\ell=1}^{\ell=k} g_\ell}$ or $\mathbf{\sum_{\ell=1}^{\ell=k} g_\ell}$ the associated $\mathbf{(\sigma_k : [1, n] \to G)}$ map quoted in proposition 339 (p. 207).

**Proposition 357** Using notations of definition 177 (p. 211), $\mathbf{p}$ and $\mathbf{n}$ being integers with $\mathbf{p} < \mathbf{n}$, given an increasing map (definition 384 (p. 486)) $(\mathbf{k_i} : [1, \mathbf{p}] \to [1, \mathbf{n}])$, we have set $\mathbf{k}(0) = 1$:

$$\prod_{i=1}^{i=n} g_i = \prod_{j=1}^{j=p} \prod_{i=k(j-1)}^{i=k(j)} g_i$$

see proposition 190 (p. 147).

**Proof**

Given $\mathbf{u} \in \mathbf{G}$, $\mathbf{u}$ can be considered as a map $(\mathbf{u} : \mathbf{G} \to \mathbf{G})$ with $(\forall \mathbf{v} \in \mathbf{G})(\mathbf{u}(\mathbf{v}) = \mathbf{u} \cdot \mathbf{v})$. From proposition 338 (p. 206), it keeps the internal law as the composition law of maps then we can apply proposition 190 (p. 147) to $\prod_{i=1}^{i=n} g_i$ and applying it to the unit element we get proposition 357 (p. 212).

**Proposition 358** Given an Abelian group $\mathbf{G}$, an integer $\mathbf{n} \in \mathcal{Z}^+$ a map $\mathbf{g_i} : [1, \mathbf{n}] \to \mathbf{G}$, $(\mathbf{P}(\ell) : [1, \mathbf{k}] \to [1, \mathbf{k}])$ a permutation of $[1, \mathbf{k}]$, $\mathbf{k} \in [1, \mathbf{n}]$ we have:

$$\sum_{\ell=1}^{\ell=k} g_\ell = \sum_{\ell=1}^{\ell=k} g_{P(\ell)}$$

where $\sum_{\ell=1}^{\ell=k}$ is defined in definition 177 (p. 211) from proposition 339 (p. 207).

**Proof**

For $\mathbf{k} = 2$ it is true from definition 170 (p. 206). Let us suppose that proposition 358 (p. 212) is true up to $\mathbf{k} - 1$. From definition 167 (p. 205), 1, we have:

$$\sum_{\ell=1}^{\ell=k} g_\ell = \sum_{\ell \in [1,k-2]} g_\ell + g_{k-1} + g_k = \sum_{\ell \in [1,k-2]} g_\ell + (g_{k-1} + g_k)$$

$$= \sum_{\ell \in [1,k-2]} g_\ell + (g_k + g_{k-1}) = \sum_{\ell=1}^{\ell=k} g_{T_{k-1,k}(\ell)} \ .$$

From proposition 349 (p. 208), any permutation of $[1, \mathbf{k}]$ is a permutation of $[1, \mathbf{k} - 1]$ or the product of a permutation of $[1, \mathbf{k} - 1]$ by a permutation of the $\mathbf{k} - 1$, $\mathbf{k}$ terms. Proposition 358 (p. 212) is true for $\mathbf{k}$. We conclude by proposition 927 (p. 520).

**Definition 178** $\sum_{\ell \in I} g_\ell$.

Given an Abelian group $\mathbf{G}$, a finite set $\mathbf{I}$: $\mathbf{Card}(\mathbf{I}) = \mathbf{n} \in \mathcal{Z}^+$, a map $(\mathbf{g_i} : \mathbf{I} \to \mathbf{G})$ from definition 390 (p. 499) and proposition 934 (p. 523), there is a bijection $(\mathbf{P}(\mathbf{i}) : \mathbf{I} \leftrightarrow [1, \mathbf{n}])$. One may write from proposition 339 (p. 207):

$$\sum_{\ell \in I} g_\ell = \sum_{\ell=1}^{\ell=n} g_{P^{-1}(\ell)} \ .$$

From proposition 358 (p. 212), it is not dependent of $\mathbf{P}$ since if we have another map $(\mathbf{Q}(\mathbf{i}) : \mathbf{I} \leftrightarrow [1, \mathbf{n}])$, $\mathbf{P}(\mathbf{Q}^{-1})$ is a permutation of $[1, \mathbf{n}]$ (see definitions 172 (p. 207), 124 (p. 139), proposition 193 (p. 149)). Then:

$$\sum_{\ell \in I} g_\ell = \sum_{\ell=1}^{\ell=n} g_{P^{-1}(\ell)} = \sum_{\ell=1}^{\ell=n} g_{P^{-1}(P(Q^{-1}(\ell)))} = \sum_{\ell=1}^{\ell=n} g_{Q^{-1}(\ell)} \ .$$

The order does not matter, so the notation $\sum_{\ell \in I}$ is fully adequate.

**Proposition 359** Given an Abelian group $\mathbf{G}$, given two finite sets $\mathbf{I}$ and $\mathbf{J}$ of the same cardinal (definition 390 (p. 499)) then with a bijection (definition 124 (p. 139)) $(\mathbf{h}: \mathbf{J} \leftrightarrow \mathbf{I})$, a map $(\mathbf{g_i} = \mathbf{g(i)} : \mathbf{I} \to \mathbf{G})$: we have:

$$\sum_{\ell \in I} g_\ell = \sum_{\ell \in J} g_{h(\ell)} .$$

**Proof**
From definition 390 (p. 499), there is a bijection $(\mathbf{P(i)} : [1, n] \leftrightarrow \mathbf{J})$ and a bijection $(\mathbf{h} \circ \mathbf{P(i)} : [1, n] \leftrightarrow \mathbf{I})$. From proposition 358 (p. 212) and definition 178 (p. 212), we have:

$$\sum_{\ell \in I} g_\ell = \sum_{\ell \in [1,n]} g_{h \circ P(\ell)} .$$

But from proposition 180 (p. 143), we also have:

$$\sum_{\ell \in J} g_{h(\ell)} = \sum_{\ell \in [1,n]} g_{h \circ P(\ell)}$$

then

$$\sum_{\ell \in I} g_\ell = \sum_{\ell \in J} g_{h(\ell)} .$$

**Proposition 360** Given an Abelian group $\mathbf{G}$, given two finite sets $\mathbf{I}$ and $\mathbf{J}$, two injective maps (definition 122 (p. 139)): $(\mathbf{g^I_i} = \mathbf{g^I(i)} : \mathbf{I} \to \mathbf{G})$, $(\mathbf{g^J_j} = \mathbf{g^J(j)} : \mathbf{J} \to \mathbf{G})$ such as $\mathbf{g^I(I)} = \mathbf{g^J(J)}$ then (see definition 178 (p. 212)) we have:

$$\sum_{i \in I} g^I_i = \sum_{j \in J} g^J_j .$$

**Proof**
Since we may write: $\mathbf{Card(J)} = \mathbf{Card(I)} = \mathbf{n} \in \mathcal{Z}^+$, $\exists (\mathbf{f} : [1, n] \leftrightarrow \mathbf{g^I(I)} = \mathbf{g^J(J)})$ (definition 390 (p. 499), proposition 934 (p. 523)). From proposition 180 (p. 143) $(\mathbf{g^I} : \mathbf{I} \leftrightarrow \mathbf{g^I(I)})$ and $(\mathbf{g^J} : \mathbf{J} \leftrightarrow \mathbf{g^J(J)})$ are bijections (definition 124 (p. 139)). From proposition 193 (p. 149), $(\mathbf{f^{-1}} \circ \mathbf{g^I} : \mathbf{I} \leftrightarrow [1, n])$ and $(\mathbf{f^{-1}} \circ \mathbf{g^J} : \mathbf{J} \leftrightarrow [1, n])$ are bijections. From definition 178 (p. 212) and proposition 358 (p. 212):

$$\sum_{i \in I} g^I(i) = \sum_{k=1}^{k=n} g^I \circ (f^{-1} \circ g^I)^{-1}(k) = \sum_{k=1}^{k=n} g^I \circ (g^I)^{-1} \circ f(k) = \sum_{k=1}^{k=n} f(k)$$

$$= \sum_{k=1}^{k=n} g^J \circ (f^{-1} \circ g^J)^{-1}(k) = \sum_{j \in J} g^J(j) .$$

**Proposition 361** Given an Abelian group $\mathbf{G}$, $\mathbf{p} \in \mathcal{Z}^+$, $\mathbf{q} \in \mathcal{Z}^+$, a map $(\mathbf{g_{i,j}} : [1, p] \times [1, q] \to \mathbf{G})$ we have:

$$\sum_{i \in [1,p]} \sum_{j \in [1,q]} g_{i,j} = \sum_{j \in [1,q]} \sum_{i \in [1,p]} g_{i,j} .$$

**Proof**
From proposition 358 (p. 212) and definition 178 (p. 212) and applying propositions 357 (p. 212) and 782 (p. 459) we have:

$$\sum_{i\in[1,p]}\sum_{j\in[1,q]} g_{i,j} = \sum_{i\times j\in\bigcup_{i\in[1,p]}\{i\}\times[1,q]} g_{i,j} = \sum_{i\times j\in[1,p]\times[1,q]} g_{i,j}$$
$$= \sum_{i\times j\in\bigcup_{j\in[1,q]}[1,p]\times\{j\}} g_{i,j} = \sum_{j\in[1,q]}\sum_{i\in[1,p]} g_{i,j} .$$

**Proposition 362** Given an Abelian group $\mathbf{G}$, two disjoint finite sets $\mathbf{I}$, $\mathbf{J}$ ($\mathbf{I}\cap\mathbf{J}=\emptyset$), two maps $(\mathbf{g}^{\mathbf{I}}_{\mathbf{i}}:\mathbf{I}\to\mathbf{G})$, $(\mathbf{g}^{\mathbf{J}}_{\mathbf{j}}:\mathbf{J}\to\mathbf{G})$, those maps generate a map $(\mathbf{g}_{\mathbf{k}}:\mathbf{I}\cup\mathbf{J}\to\mathbf{G})$ by $\mathbf{k}\in\mathbf{I}\Rightarrow\mathbf{g}_{\mathbf{k}}=\mathbf{g}^{\mathbf{I}}_{\mathbf{k}}$, $\mathbf{k}\in\mathbf{J}\Rightarrow\mathbf{g}_{\mathbf{k}}=\mathbf{g}^{\mathbf{J}}_{\mathbf{k}}$ and we have:

$$\sum_{i\in I} g^I_i + \sum_{j\in J} g^J_j = \sum_{k\in I\cup J} g_k .$$

**Proof**
We call $\mathbf{p}=\mathbf{Card}(\mathbf{I})$, $\mathbf{q}=\mathbf{Card}(\mathbf{J})$. We have since $\mathbf{i}\cap\mathbf{j}=\emptyset$, $\mathbf{p}+\mathbf{q}=\mathbf{Card}(\mathbf{I}\cup\mathbf{J})$ (definition 393 (p. 506), proposition 893 (p. 507)). Then we have three bijections (see proposition 934 (p. 523)):

$$(f^I:[1,p]\leftrightarrow I),\ (f^J:[1,q]\leftrightarrow J),\ (f^{I\cup J}:[1,p+q]\leftrightarrow I\cup J) .$$

We have since $\mathbf{Card}[1,q]=\mathbf{Card}[p+1,p+q]$:

$$\sum_{i\in I} g^I_i + \sum_{j\in J} g^J_j = \sum_{k\in[1,p]} g_{f^I(k)} + \sum_{k\in[1,q]} g_{f^J(k)} = \sum_{k\in[1,p]} g_k + \sum_{k\in[p+1,p+q]} g_{f^J(k-p)} .$$

From definition 178 (p. 212), we have:

$$\sum_{k\in[1,p]} g_k + \sum_{k\in[p+1,p+q]} g_{f^J(k-p)} = \sum_{k\in[1,p+q]} g_{f^{I\cup J}(k)} = \sum_{k\in I\cup J} g_k .$$

### 2.9.3 Rings, fields

**Definition 179 Ring F.**

One shall call ring, a set $\mathbf{F}$ with two internal laws $(\mathbf{F}\times\mathbf{F}\to\mathbf{F})$:

1. One law, which is quoted as $+$, defines a commutative (Abelian) group (definition 170 (p. 206)).

2. A second law indicated as a product with the sign $\cdot$ or $\times$ or the two elements of $\mathbf{F}$ side by side. This law which does not define necessarily a group (any element does not always have an inverse see definition 167 (p. 205)), has to be associative (definition 167 (p. 205) point 1) but need not be commutative. We should have:

$$(a\in F, b\in F, c\in F)$$
$$\Rightarrow ab = a\cdot b = a\times b, a(bc) = (ab)c = abc . \tag{2.63}$$

3. The multiplicative map has to be distributive relative to the addition group, that is to say:
$$(\forall(\mathbf{a},\mathbf{b},\mathbf{c}) \in \mathbf{F}^3)(\mathbf{a}\cdot(\mathbf{b}+\mathbf{c}) = \mathbf{a}\cdot\mathbf{b}+\mathbf{a}\cdot\mathbf{c}) \ . \tag{2.64}$$

4. We suppose also that it has a unit element:
$$(\exists \mathbf{e} \in \mathbf{F})(\forall \mathbf{a} \in \mathbf{F})(\mathbf{ae} = \mathbf{ea} = \mathbf{a}) \ . \tag{2.65}$$

The unit element $\mathbf{e}$ of the second law is unique:
Let us suppose that $\mathbf{e}, \mathbf{f}$ are two unit elements of the second law. We have from equation (2.65):
$$\mathbf{ef} = \mathbf{f} = \mathbf{fe} = \mathbf{ef} = \mathbf{e} \ . \tag{2.66}$$
This unit element is called $\mathbf{1}$. The quaternions form a ring.

**Proposition 363** Having a ring $\mathbf{F}$ (definition 179 (p. 214)) and defining the map $([\mathbf{a},\mathbf{b}] = \mathbf{ab}-\mathbf{ba} : \mathbf{F}\times\mathbf{F} \to \mathbf{F})$, we have the following Bianchi (Jacobi) identity:
$$(\mathbf{a}\in\mathbf{F},\mathbf{b}\in\mathbf{F},\mathbf{c}\in\mathbf{F}) \Rightarrow [\mathbf{a},[\mathbf{b},\mathbf{c}]]+[\mathbf{b},[\mathbf{c},\mathbf{a}]]+[\mathbf{c},[\mathbf{a},\mathbf{b}]] = \mathbf{0} \ . \tag{2.67}$$

**Proof**
From definition 179 (p. 214), we have:
$$(\mathbf{a}\in\mathbf{F},\mathbf{b}\in\mathbf{F},\mathbf{c}\in\mathbf{F}) \Rightarrow [\mathbf{a},[\mathbf{b},\mathbf{c}]]+[\mathbf{b},[\mathbf{c},\mathbf{a}]]+[\mathbf{c},[\mathbf{a},\mathbf{b}]] =$$
$$\mathbf{abc}+\mathbf{bca}+\mathbf{cab}-\mathbf{acb}-\mathbf{bac}-\mathbf{cba}-\mathbf{bca}-\mathbf{cab}-\mathbf{abc}+\mathbf{cba}+\mathbf{acb}+\mathbf{bac} = \mathbf{0} \ .$$

**Definition 180 Commutative ring F.**

One shall call a commutative ring, a ring $\mathbf{F}$ (definition 179 (p. 214)) with the two internal laws $(\mathbf{F}\times\mathbf{F}\to\mathbf{F})$ commutative: We should have:
$$\begin{aligned}&(\mathbf{a}\in\mathbf{F},\mathbf{b}\in\mathbf{F},\mathbf{c}\in\mathbf{F})\\ &\Rightarrow \mathbf{ab} = \mathbf{a}\cdot\mathbf{b} = \mathbf{a}\times\mathbf{b} = \mathbf{ba}, \mathbf{a}(\mathbf{bc}) = (\mathbf{ab})\mathbf{c} \ .\end{aligned} \tag{2.68}$$

We define also $\sum$ and $\prod$ as in definition 177 (p. 211). They have the properties 357 (p. 212). $\mathcal{R}$ (definition 143 (p. 170)) and $\mathcal{C}$ (definition 245 (p. 271)) can be considered as commutative rings. Apart from those trivial cases, the set of continuous maps $\mathbf{f}$ from a topological set $\mathbf{A}$ to $\mathcal{R}$: $(\mathbf{f}:\mathbf{A}\to\mathcal{R})$ is an example of a commutative ring $\mathbf{F}$.

**Proposition 364** Having a ring $\mathbf{F}$ (definition 180 (p. 215)), $\mathbf{0}$ being the neutral element of the $+$ Abelian group (point 1 of 180 (p. 215)), we have:
$$(\forall \mathbf{a}\in\mathbf{F})(\mathbf{a0} = \mathbf{0})$$
where $(\mathbf{ab} : \mathbf{F}\times\mathbf{F}\to\mathbf{F})$ represents the result of the multiplicative law (point 2).

**Proof**
Applying the rules of (definition 180 (p. 215)) we have:
$$\mathbf{a}\in\mathbf{F} \Rightarrow \mathbf{a0} = \mathbf{a}(\mathbf{0}+\mathbf{0}) = \mathbf{a0}+\mathbf{a0} \Rightarrow \mathbf{a0}+\mathbf{a0} = \mathbf{a0} \Rightarrow \mathbf{a0}+\mathbf{a0}-\mathbf{a0} = \mathbf{0}$$
$$\Rightarrow \mathbf{a0}+\mathbf{0} = \mathbf{0} \Rightarrow \mathbf{a0} = \mathbf{0} \ .$$

**Proposition 365** With the same conditions and notations as proposition 364 (p. 215) but with $\mathbf{F}$ as a commutative ring, we have

$$(\forall a \in F)(a0 = 0a = 0) .$$

**Proof**
It is a direct consequence of definition 180 (p. 215) and proposition 364 (p. 215). For simplicity since we shall deal only with commutative rings, we restrict ourselves now to commutative rings.

**Proposition 366** $\mathbf{F}$ is a commutative ring (definition 180 (p. 215)). Given $a \in F$, $n \in \mathcal{Z}^+$, a map $(c_i : [1, n] \to F)$, we have (definition 177 (p. 211)):

$$a \sum_{i \in [1,n]} c_i = \sum_{i \in [1,n]} ac_i .$$

**Proof**
Proposition 366 is true for $n \in [1, 2]$: point 3 of definition 180 (p. 215). Let us suppose that it is true for $n$ then it is true for $n + 1$. From definitions 177 (p. 211) and 180 (p. 215) point 3:

$$\begin{aligned} &a \sum_{i \in [1,n+1]} c_i \\ &= a(\sum_{i \in [1,n]} c_i + c_{n+1}) = a \sum_{i \in [1,n]} c_i + ac_{n+1} = \sum_{i \in [1,n]} ac_i + ac_{n+1} = \sum_{i \in [1,n+1]} ac_i . \end{aligned}$$

We conclude using proposition 927 (p. 520).

**Proposition 367** $\mathbf{F}$ is a commutative ring (definition 180 (p. 215)). Given a map $(n_i : [1, 2] \to \mathcal{Z}^+)$, two maps $(a_{1,i_1} : [1, n_1] \to F)$, $(a_{2,i_2} : [1, n_2] \to F)$, we have (definition 177 (p. 211)):

$$\sum_{i_1 \in [1,n_1]} \sum_{i_2 \in [1,n_2]} a_{i_1,i_2} = \sum_{f \in [1,n_1] \times [1,n_2]} a_{f(1),f(2)} = \sum_{f \in [1,n_1] \times [1,n_2]} a_f \quad (2.69)$$

$$\sum_{i_1 \in [1,n_1]} a_{1,i_1} \sum_{i_2 \in [1,n_2]} a_{2,i_2} = \sum_{f \in [1,n_1] \times [1,n_2]} a_{1,f(1)} a_{2,f(2)} . \quad (2.70)$$

**Proof**
Let us suppose proposition 367 (p. 216) is true whatever $n_2$ for an $n_1$, we shall demonstrate it for the same $n_2$ at $n_1 + 1$. Using points 1, 2 and 3 of definition 179 (p. 214), definition 177 (p. 211), propositions 366 (p. 216), 362 (p. 214) and 782 (p. 459), we have:

$$\begin{aligned} &\sum_{i_1 \in [1,n_1+1]} \sum_{i_2 \in [1,n_2]} a_{i_1,i_2} = (\sum_{i_1 \in [1,n_1]} \sum_{i_2 \in [1,n_2]} a_{i1,i_2}) + \sum_{i_2 \in [1,n_2]} a_{n_1+1,i_2} \\ &= \sum_{f \in [1,n_1] \times [1,n_2]} a_{f(1),f(2)} + \sum_{f \in \{n_1+1\} \times [1,n_2]} a_{f(1),f(2)} \\ &= \sum_{f \in [1,n_1] \times [1,n_2]} a_{f(1),f(2)} + \sum_{f \in [n_1+1,n_1+1] \times [1,n_2]} a_{f(1),f(2)} \\ &= \sum_{f \in [1,n_1] \times [1,n_2] \cup [n_1+1,n_1+1] \times [1,n_2]} a_{f(1),f(2)} = \sum_{f \in [1,n_1+1] \times [1,n_2]} a_{f(1),f(2)} \\ &= \sum_{f \in [1,n_1+1] \times [1,n_2]} a_f . \end{aligned}$$

Equation (2.69) (p. 216) is trivially true for $\mathbf{n_1} = \mathbf{1}$ (definition 171 (p. 207)) then from proposition 928 (p. 520) it is true for any $(\mathbf{n_1}, \mathbf{n_2}) = \mathbf{n_1} \times \mathbf{n_2} \in \mathcal{Z}^2$. Equation (2.70) (p. 216) is obtained by setting $\mathbf{a_{i_1,i_2}} = \mathbf{a_{1,i_1} a_{2,i_2}}$ and using proposition 366 (p. 216).

**Proposition 368** $\mathbf{F}$ is a commutative ring (definition 180 (p. 215)). $\mathbf{n} \in \mathcal{Z}^+$, a map $(\mathbf{j_\mu} : [\mathbf{1}, \mathbf{n}] \to \mathcal{Z}^+)$, a set of maps $(\mathbf{a_{\lambda,i}} : [\mathbf{1}, \mathbf{j_\lambda}] \to \mathbf{F})$, we have (definition 177 (p. 211)):

$$\prod_{\lambda \in [1,n]} \sum_{i \in [1,j_\lambda]} a_{\lambda,i} = \sum_{f \in \prod_{\mu \in [1,n]} [1,j_\mu]} \prod_{\lambda \in [1,n]} a_{\lambda,f(\lambda)} \,.$$

**Proof**
From proposition 367 (p. 216) equation (2.70) (p. 216), proposition 368 is true for $\mathbf{n} = \mathbf{2}$. Let us suppose that it is true for $\mathbf{n}$. From propositions 367 (p. 216), 833 (p. 479), definition 177 (p. 211) and proposition 362 (p. 214), setting $\mathbf{g} = \prod_{\nu \in [1,n+1]} \mathbf{g}(\nu) \in \prod_{\mu \in [1,n+1]} [\mathbf{1}, \mathbf{j_\mu}]$ and $\mathbf{f} \in \prod_{\mu \in [1,n]} [\mathbf{1}, \mathbf{j_\mu}]$, we have (see convention 30 (p. 463) for the index of summation):

$$\prod_{\lambda \in [1,n+1]} \sum_{i \in [1,j_\lambda]} a_{\lambda,i} = \Big(\prod_{\lambda \in [1,n]} \sum_{i \in [1,j_\lambda]} a_{\lambda,i}\Big) \sum_{i \in [1,j_{n+1}]} a_{n+1,i}$$
$$= \Big(\sum_{f \in \prod_{\mu \in [1,n]} [1,j_\mu]} \prod_{\lambda \in [1,n]} a_{\lambda,f(\lambda)}\Big) \sum_{i \in [1,j_{n+1}]} a_{n+1,i}$$
$$= \sum_{i \in [1,j_{n+1}]} \Big(\sum_{f \in \prod_{\mu \in [1,n]} [1,j_\mu]} \prod_{\lambda \in [1,n]} a_{\lambda,f(\lambda)}\Big) a_{n+1,i}$$
$$= \sum_{i \in [1,j_{n+1}]} \sum_{f \in \prod_{\mu \in [1,n]} [1,j_\mu]} \prod_{\lambda \in [1,n]} a_{\lambda,f(\lambda)} a_{n+1,i}$$
$$= \sum_{k \in [1,j_{n+1}]} \sum_{g \in \prod_{\mu \in [1,n]} [1,j_\mu] \times [k,k]} \prod_{\lambda \in [1,n]} a_{\lambda,g(\lambda)} a_{n+1,g(n+1)}$$
$$= \sum_{g \in \prod_{\mu \in [1,n+1]} [1,j_\mu]} \prod_{\lambda \in [1,n+1]} a_{\lambda,g(\lambda)} = \sum_{g \in \prod_{\mu \in [1,n+1]} [1,j_\mu]} \prod_{\lambda \in [1,n+1]} a_{\lambda,g(\lambda)} \,.$$

We conclude as above using proposition 928 (p. 520).

**Definition 181 Field.**

One calls field: a commutative ring (definition 180 (p. 215)) $\mathbf{A}$ such that any element of $\mathbf{A}$, except the unit element of the addition group, has an inverse over the multiplicative law:

$$(\forall a \in \mathcal{C}_A(\{0\}) = A - \{0\})(\exists a^{-1} \in A)(a \cdot a^{-1} = 1) \tag{2.71}$$

$\mathbf{0}$ is the neutral element of the addition. $\mathbf{1}$ is the neutral element of the multiplication. Examples: $\mathcal{R}$, $\mathcal{C}$ are fields.

**Proposition 369** Having $\mathbf{m} \in \mathcal{Z}^+$, $\mathbf{n} \in \mathcal{Z}^+$ and a map $(\mathbf{a_i} : [\mathbf{1}, \mathbf{m}] \to \mathcal{RC})$ where $\mathcal{RC}$ is a field, we have the following identity:

$$\begin{aligned}(\sum_{i\in[1,m]} a_i)^n &= \sum_{g\in[1,m]^n} \prod_{i\in[1,n]} a_{g(i)} = \sum_{g\in[1,m]^n} \prod_{i\in[1,m]} a_i^{Card(g^{-1}(\{i\}))} \\ &= \sum_{h\in[0,n]^m, \sum_{k\in[1,m]} h(k)=n} n! \prod_{k\in[1,m]} \frac{a_k^{h(k)}}{h(k)!} .\end{aligned} \tag{2.72}$$

$\mathbf{h} \in [\mathbf{0}, \mathbf{n}]^{\mathbf{m}}, \sum_{\mathbf{k}\in[\mathbf{1},\mathbf{m}]} \mathbf{h}(\mathbf{k}) = \mathbf{n}$ under the sign $\sum$ means that the sum is on the set of maps $\mathbf{h} \in [\mathbf{0}, \mathbf{n}]^{\mathbf{m}}$ which verifies $\sum_{\mathbf{k}\in[\mathbf{1},\mathbf{m}]} \mathbf{h}(\mathbf{k}) = \mathbf{n}$ see convention 30 (p. 463).

**Proof**

From definition of a field: definition 181 (p. 217) we have:

1. Since $\mathcal{RC}$ is also a commutative ring, from proposition 368 (p. 217), we have:

$$(\sum_{i\in[1,m]} a_i)^n = \sum_{g\in[1,m]^n} \prod_{i\in[1,n]} a_{g(i)} .$$

2. Since it is commutative, we can regroup all the terms with the same index:

$$\prod_{i\in[1,n]} a_{g(i)} = \prod_{k\in[1,m]} a_k^{Card(g^{-1}(\{k\}))} .$$

3. One put together all the terms for which the map $(\mathbf{Card}(\mathbf{g}^{-1}) = \mathbf{h} : [\mathbf{1}, \mathbf{m}] \to [\mathbf{1}, \mathbf{n}])$ is the same, they are equal. Their number is given by proposition 356 (p. 210). One gets by summing them:

$$\sum_{g\in[1,m]^n} \prod_{i\in[1,m]} a_i^{Card(g^{-1}(\{i\}))} = \sum_{h\in[0,n]^m, \sum_{k\in[1,m]} h(k)=n} n! \prod_{k\in[1,m]} \frac{a_k^{h(k)}}{h(k)!} .$$

## 2.10 Module over a commutative ring

**Definition 182 Module over a commutative ring F.**

We call a set $\mathbf{V}$, a module over the commutative ring $\mathbf{F}$ (definition 180 (p. 215)), if there are an Abelian group (definition 170 (p. 206)) map $(\mathbf{a} + \mathbf{b} : \mathbf{V} \times \mathbf{V} \to \mathbf{V})$ and a map $(\lambda \mathbf{a} : \mathbf{F} \times \mathbf{V} \to \mathbf{V})$ with the following relations:
With $\mathbf{a} \in \mathbf{V}, \mathbf{b} \in \mathbf{V}, \lambda \in \mathbf{F}, \mu \in \mathbf{F}$, we have:

1. $\lambda(\mu \mathbf{a}) = (\lambda\mu)\mathbf{a}$.

2. $\lambda(\mathbf{a} + \mathbf{b}) = \lambda \mathbf{a} + \lambda \mathbf{b}$.

3. $\lambda \mathbf{a} + \mu \mathbf{a} = (\lambda + \mu)\mathbf{a}$.

In manifold theory the commutative ring is, most of the time, the set of smooth maps $\mathcal{F}$ from the manifold $\mathbf{M}$ (definition 7 (p. 3)) to $\mathcal{R}$. Given such a map $(\mathbf{f} : \mathbf{M} \to \mathcal{R})$, we have a set of points of $\mathbf{M}$ for which the map is zero: $\mathbf{f}^{-1}(\mathbf{0}) \neq \emptyset$. It prevents to define a map $\frac{1}{\mathbf{f}}$. Under these conditions, some properties of section 2.11 do not apply.

**Proposition 370** We have a module $\mathbf{V}$ (definition 182 (p. 218)) over the commutative ring $\mathbf{F}$ (definition 180 (p. 215)). We have a set $\mathbf{A}$, we consider the set $\mathcal{M}$ of maps (definition 112 (p. 131)) $\mathbf{A} \to \mathbf{V}$.
$\mathcal{M}$ is a module over $\mathbf{F}$ with an internal law:

$$(\mathbf{f} \in \mathcal{M}, \mathbf{g} \in \mathcal{M}, \lambda \in \mathbf{F}, \mathbf{x} \in \mathbf{A}) \Rightarrow ((\mathbf{f}+\mathbf{g})(\mathbf{x}) = \mathbf{f}(\mathbf{x}) + \mathbf{g}(\mathbf{x}) \in \mathcal{M},$$
$$(\lambda \mathbf{f})(\mathbf{x}) = \lambda \mathbf{f}(\mathbf{x}) \in \mathcal{M}) .$$

**Proof**
It is a direct consequence of the definition.

**Definition 183 Submodule of a module.**

Given a module $\mathbf{V}$ definition 182 (p. 218), a subset $\mathbf{U}$ of $\mathbf{V}$ is a submodule of $\mathbf{V}$ if $\mathbf{U}$ is stable for the internal laws: The Abelian group of definition 182 (p. 218) and the relations 1, 2, 3 of definition 182 (p. 218) acting on elements of $\mathbf{U}$ give an element of $\mathbf{U}$.

**Proposition 371** A commutative ring is a module over itself.

**Proof**
If $\mathbf{F}$ is the commutative ring, it verifies all elements of definition 182 (p. 218) with $\mathbf{V} = \mathbf{F}$.

**Proposition 372** Given a module (definition 182 (p. 218)) $\mathbf{V}$ over a commutative ring $\mathbf{F}$ (definitions 182 (p. 218), 180 (p. 215)), we call $\mathbf{0}$ the null element of the Abelian group acting on $\mathbf{F}$, $\mathbf{0_V}$ the zero vector of the Abelian group acting on $\mathbf{V}$. We have:

$$(\forall \mathbf{a} \in \mathbf{V})(\mathbf{0a} = \mathbf{0_V}) .$$

**Proof**

$$\mathbf{a} \in \mathbf{V} \Rightarrow \mathbf{0a} + \mathbf{0a} = (\mathbf{0}+\mathbf{0})\mathbf{a} = \mathbf{0a}$$
$$\mathbf{a} \in \mathbf{V} \Rightarrow \mathbf{0a} + \mathbf{0a} = \mathbf{0a} \Rightarrow \mathbf{0a} + \mathbf{0a} - \mathbf{0a} = \mathbf{0_V} = \mathbf{0a} - \mathbf{0a} + \mathbf{0a} = \mathbf{0a} \Rightarrow \mathbf{0a} = \mathbf{0_V} .$$

**Definition 184 Finite basis in a module over a commutative ring F.**

The definition looks similar to definition 205 (p. 239), replacing $\mathcal{RC}$ by $\mathbf{F}$ but we have to be careful because the dimension of a module (definition 182 (p. 218)) over a commutative ring is not defined. Keeping the notations of definition 182 (p. 218), having $\mathbf{V}$ a module over a commutative ring $\mathbf{F}$, we say that a finite set $(\mathbf{e_i} : [\mathbf{1}, \mathbf{n}] \to \mathbf{V})$, $\mathbf{n} \in \mathcal{Z}^+$ is a basis of $\mathbf{V}$ (see footnote 20 (p. 239)) if:

$$\mathbf{v} \in \mathbf{V} \Leftrightarrow (\exists(\mathbf{v^i} : [\mathbf{1}, \mathbf{n}] \to \mathbf{F}))(\mathbf{v} = \mathbf{v^i e_i}) .$$

In manifold theory, such bases are not supposed to exist really. They are supposed to exist only when the manifold $\mathbf{M}$ and the set $\mathbf{F} = \mathcal{F}$ (definition 7 (p. 3)) are restricted (definition 9 (p. 4)) to one of the covering opens of the manifold $\mathbf{M}$.

**Definition 185 Independent finite basis in a module over a commutative ring.**

Having a module (definition 182 (p. 218)) $\mathbf{V}$ over a commutative ring $\mathbf{F}$ (definitions 182 (p. 218), 180 (p. 215)), a basis (definition 184 (p. 219)) of $\mathbf{V}$, $(\mathbf{e_i} : [\mathbf{1}, \mathbf{n}] \to \mathbf{V})$, $\mathbf{n} \in \mathcal{Z}^+$, is said to be independent if:

$$((\lambda_\mathbf{i} : [\mathbf{1}, \mathbf{n}] \to \mathbf{F}), \sum_{\mathbf{i} \in [\mathbf{1},\mathbf{n}]} \lambda_\mathbf{i} \mathbf{e_i} = \mathbf{0}) \Rightarrow (\forall \mathbf{i} \in [\mathbf{1}, \mathbf{n}])(\lambda_\mathbf{i} = \mathbf{0}) .$$

**Proposition 373** Having a module $\mathbf{V}$ over a commutative ring $\mathbf{F}$ (definitions 182 (p. 218), 180 (p. 215)) and an independent finite basis $(\mathbf{e_i} : [\mathbf{1}, \mathbf{n}] \to \mathbf{V})$, $\mathbf{n} \in \mathcal{Z}^+$, any vector of $\mathbf{V}$ is expressed in a unique way in terms of the basic vectors:

$$\sum_{i\in[1,n]} v_1^i e_i = \sum_{i\in[1,n]} v_2^i e_i \Rightarrow (\forall i \in [1,n])(v_1^i = v_2^i) . \tag{2.73}$$

**Proof**
We have from definition 185 (p. 219):

$$\sum_{i\in[1,n]} v_1^i e_i = \sum_{i\in[1,n]} v_2^i e_i \Rightarrow \sum_{i\in[1,n]} (v_1^i - v_2^i) e_i = 0 \Rightarrow (\forall i \in [1,n])(v_1^i - v_2^i = 0) .$$

**Definition 186 Linear maps, linear forms, linear transformations.**

Having two modules (definition 182 (p. 218)) $\mathbf{U}$ and $\mathbf{V}$ over a commutative ring $\mathbf{F}$ (definitions 182 (p. 218), 180 (p. 215)), a map $(\mathbf{f} : \mathbf{U} \to \mathbf{V})$ is linear if:

$$(p \in \mathcal{Z}^+, (u_i : [1,p] \to U), (\lambda^i : [1,p] \to F)) \Rightarrow f(\sum_{i\in[1,p]} \lambda^i u_i) = \sum_{i\in[1,p]} \lambda^i f(u_i) . \tag{2.74}$$

**Proposition 374** Using notation of definition 186 (p. 220), if for a map $(\mathbf{f} : \mathbf{U} \to \mathbf{V})$ equation (2.74) (p. 220) holds for $\mathbf{p} \in [\mathbf{1}, \mathbf{2}]$, it holds for any $\mathbf{p} \in \mathcal{Z}^+$.

**Proof**
It is true for $\mathbf{p} = \mathbf{1}$, let us suppose that it is true up to $\mathbf{p} \geq \mathbf{1}$. Setting

$$u_1' = \sum_{i\in[1,p]} \lambda^i u_i,\ u_2' = u_{p+1},\ \mu^1 = 1, \mu^2 = \lambda^{p+1}$$

we have:

$$\begin{aligned} f(\sum_{i\in[1,p+1]} \lambda^i u_i) &= f(\mu^1 u_1' + \mu^2 u_2') = \mu^1 f(u_1') + \mu^2 f(u_2') \\ &= \sum_{i\in[1,p]} \lambda^i f(u_i) + \lambda^{p+1} f(u_{p+1}) = \sum_{i\in[1,p+1]} \lambda^i f(u_i) . \end{aligned}$$

We conclude by proposition 927 (p. 520).

**Proposition 375** Having a module (definition 182 (p. 218)) $\mathbf{U}$ over a commutative ring $\mathbf{F}$ (definitions 182 (p. 218), 180 (p. 215)), an element $\boldsymbol{\lambda}$ of $\mathbf{F}$ defines a linear map $(\boldsymbol{\lambda}\mathbf{u} : \mathbf{U} \to \mathbf{U})$.

**Proof**
From definitions 180 (p. 215) and 182 (p. 218), we have:

$$((\mu_i : [1,2] \to F), (u_i : [1,2] \to U)) \Rightarrow \lambda \sum_{i\in[1,2]} \mu_i u_i = \sum_{i\in[1,2]} \mu_i \lambda u_i .$$

We conclude by proposition 374 (p. 220).

**Proposition 376** Having two modules (definition 182 (p. 218)) $\mathbf{U}$ and $\mathbf{V}$ over a commutative ring $\mathbf{F}$ (definitions 182 (p. 218), 180 (p. 215)), the set $\mathcal{L}$ of linear maps $\mathbf{U} \to \mathbf{V}$ forms a module with laws defined as in proposition 370 (p. 219).

**Proof**
$\mathcal{L}$ is a subset of $\mathcal{M}$ set of map $U \to V$ as defined in proposition 370 (p. 219). $\mathcal{M}$ is a module. We have to show that $\mathcal{L}$ is a submodule of $\mathcal{M}$ (definition 183 (p. 219)). From proposition 361 (p. 213), we have $p \in \mathcal{Z}^+, q \in \mathcal{Z}^+$:

$$((\lambda_i : [1,p] \to F), (u_i : [1,p] \to U), (\mu_j : [1,q] \to F), (f_j : [1,q] \to \mathcal{L}))$$
$$\Rightarrow (\sum_{j\in[1,q]} \mu_j f_j)(\sum_{i\in[1,p]} \lambda_i u_i) = \sum_{j\in[1,q]} \mu_j f_j(\sum_{i\in[1,p]} \lambda_i u_i) = \sum_{j\in[1,q]} \sum_{i\in[1,p]} \mu_j \lambda_i f_j(u_i)$$
$$= \sum_{i\in[1,p]} \lambda_i \sum_{j\in[1,q]} \mu_j f_j(u_i) = \sum_{i\in[1,p]} \lambda_i (\sum_{j\in[1,q]} \mu_j f_j)(u_i) .$$

**Proposition 377** The product map of two linear maps (definition 117 (p. 134)) is a linear map: Having three modules (definition 182 (p. 218)) **U**, **V** and **W** over a commutative ring **F** (definitions 182 (p. 218), 180 (p. 215)), two linear maps (definition 186 (p. 220)) $(f : U \to V)$, $(g : V \to W)$, the map $(g(f) = g \circ f : U \to W)$ is also a linear map.

**Proof**
Keeping the above notations of definition 186 (p. 220), we have:

$$(p \in \mathcal{Z}^+, (u_i : [1,p] \to U), (\lambda^i : [1,p] \to F)) \Rightarrow f(\sum_{i\in[1,p]} \lambda^i u_i) = \sum_{i\in[1,p]} \lambda^i f(u_i)$$
$$\Rightarrow g(f(\sum_{i\in[1,p]} \lambda^i u_i) = g(\sum_{i\in[1,p]} \lambda^i f(u_i)) = \sum_{i\in[1,p]} \lambda^i g(f(u_i)) .$$

**Definition 187 Dual of a module over a commutative ring.**

Having a module **V** (definition 182 (p. 218)) over a commutative ring **F**, the set of linear maps of **V** to the commutative ring **F** is called the dual of **V** (see also definition 213 (p. 244)).

**Definition 188 Dual of a module over another module both built on a commutative ring.**

One can extend the above definition to any set of linear maps:
Having two modules **U** and **V** (definition 182 (p. 218)) over a commutative ring **F**, the set of linear maps of **V** to **U** can be called dual of **V** relative to **U**.

**Proposition 378** The dual of a module over a commutative ring (definition 187 (p. 221)) or over another module (definition 188 (p. 221)) is also a module over the same commutative ring.

**Proof**
It is a consequence of proposition 376 (p. 220).

**Convention 19** $V^*$ dual of a module **V**.

Having a module **V** (definition 182 (p. 218)) over a commutative ring **F**, the dual (definition 187 (p. 221)) of the module **V** is written $V^*$.

**Convention 20** $V^*_U$ dual of a module over another module.

Having two modules **U** and **V** (definition 182 (p. 218)) over a commutative ring **F**, the dual of **V** relative to **U** (definition 188 (p. 221)) is written in this section as $V^*_U$: there is no commonly adopted notation for that. When we consider a submodule of $V^*_U$ (definition 183 (p. 219)), we denote it by $V^{*s}_U$, we shall add other indices if we want to distinguish them.

**Proposition 379** The dual $\mathbf{V}^*$ of a module (definition 182 (p. 218)) $\mathbf{V}$ over a commutative ring $\mathbf{F}$ with an independent finite basis $\mathbf{n} \in \mathcal{Z}^+, (\mathbf{e_i} : [\mathbf{1}, \mathbf{n}] \to \mathbf{V})$ (definition 185 (p. 219)) also has independent finite bases. One of them is $(\mathbf{e^i} : [\mathbf{1}, \mathbf{n}] \to \mathbf{V}^*)$ such as $(\forall \mathbf{i} \in [\mathbf{1}, \mathbf{n}])(\forall \mathbf{j} \in [\mathbf{1}, \mathbf{n}])(\mathbf{e^i}(\mathbf{e_j}) = \delta^\mathbf{i}_\mathbf{j})$ where $\delta^\mathbf{i}_\mathbf{j}$ is the Kronecker symbol (definition 215 (p. 245)).

**Proof**

Since $(\forall \mathbf{v} \in \mathbf{V})(\exists \mathbf{v^i} : [\mathbf{1}, \mathbf{n}] \to \mathbf{F})(\mathbf{v} = \sum_{\mathbf{i} \in [\mathbf{1},\mathbf{n}]} \mathbf{v^i e_i})$, the map $(\mathbf{v^i} : [\mathbf{1}, \mathbf{n}] \to \mathbf{F})$ (see point 4 (p. 132) of definition 112) defines an element of $\mathbf{F^n}$ (definition 373 (p. 474)). From propositions 370 (p. 219) and 371 (p. 219) $\mathbf{F^n}$ defines a module. The map

$$(\mathbf{e}(\mathbf{v}) = \prod_{\mathbf{j} \in [\mathbf{1},\mathbf{n}]}^{\mathbf{e}} \mathbf{v^j} : \mathbf{V} \to \mathbf{F^n})$$

is linear because:

$$\begin{aligned}&(\mathbf{u} = \mathbf{u^i e_i} \in \mathbf{V}, \mathbf{v} = \mathbf{v^i e_i} \in \mathbf{V}, \lambda \in \mathbf{F}, \mu \in \mathbf{F}) \Rightarrow \lambda\mathbf{u} + \mu\mathbf{v} = (\lambda\mathbf{u^i} + \mu\mathbf{v^i})\mathbf{e_i}\\ &(\lambda\mathbf{u^i} + \mu\mathbf{v^i} : [\mathbf{1}, \mathbf{n}] \to \mathbf{F}) = \lambda(\mathbf{u^i} : [\mathbf{1}, \mathbf{n}] \to \mathbf{F}) + \mu(\mathbf{v^i} : [\mathbf{1}, \mathbf{n}] \to \mathbf{F})\ .\end{aligned}$$

From proposition 373 (p. 220), this map is a bijection (definition 124 (p. 139)). Then we have:

$$(\forall \mathbf{j} \in [\mathbf{1}, \mathbf{n}])(\mathbf{e^j}(\sum_{\mathbf{i} \in [\mathbf{1},\mathbf{n}]} \mathbf{v^i e_i}) = \mathbf{v^j})\ .$$

From proposition 373 (p. 220), for any $\mathbf{j} \in [\mathbf{1}, \mathbf{n}]$, the corresponding map $(\mathbf{e^j}(\mathbf{v}) = \mathbf{v^j} : \mathbf{v} \to \mathbf{F})$ is a linear map:

$$\begin{aligned}&(\mathbf{v_1} = \mathbf{v^i_1 e_i}, \mathbf{v_2} = \mathbf{v^i_2 e_i}, \lambda_1 \in \mathbf{F}, \lambda_2 \in \mathbf{F}) \Rightarrow \mathbf{e^j}(\lambda_1\mathbf{v_1} + \lambda_2\mathbf{v_2})\\ &= \mathbf{e^j}((\lambda_1\mathbf{v^i_1} + \lambda_2\mathbf{v^i_2})\mathbf{e_i}) = \lambda_1\mathbf{v^j_1} + \lambda_2\mathbf{v^j_2} = \lambda_1\mathbf{e^j}(\mathbf{v_1}) + \lambda_2\mathbf{e^j}(\mathbf{v_2})\ .\end{aligned} \tag{2.75}$$

We were using, in equation (2.75), the Einstein notation (see footnote 20 (p. 239)). We also have $(\forall \mathbf{i} \in [\mathbf{1}, \mathbf{n}])(\mathbf{e_i} = \delta^\mathbf{j}_\mathbf{i}\mathbf{e_j})$, then from the above definition of $\mathbf{e^j}$ in proposition 379 and in the next definition 189 (p. 222),

$$(\forall \mathbf{i} \in [\mathbf{1}, \mathbf{n}])(\forall \mathbf{j} \in [\mathbf{1}, \mathbf{n}])(\mathbf{e^j}(\mathbf{e_i}) = \delta^\mathbf{j}_\mathbf{i}) \tag{2.76}$$

Any element $\mathbf{u}$ of $\mathbf{V}^*$ gives on $\mathbf{v}$:

$$\begin{aligned}\mathbf{u}(\mathbf{v}) &= \sum_{\mathbf{i} \in [\mathbf{1},\mathbf{n}]} \mathbf{v^i u}(\mathbf{e_i}) = \sum_{\mathbf{i} \in [\mathbf{1},\mathbf{n}]} \sum_{\mathbf{j} \in [\mathbf{1},\mathbf{n}]} \mathbf{v^i u}(\mathbf{e_j})\delta^\mathbf{j}_\mathbf{i} = \mathbf{v^i u}(\mathbf{e_j})\delta^\mathbf{j}_\mathbf{i}\\ &= \sum_{\mathbf{i} \in [\mathbf{1},\mathbf{n}]} \sum_{\mathbf{j} \in [\mathbf{1},\mathbf{n}]} \mathbf{v^i u}(\mathbf{e_j})\mathbf{e^j}(\mathbf{e_i}) = \mathbf{v^i u}(\mathbf{e_j})\mathbf{e^j}(\mathbf{e_i}) = \sum_{\mathbf{j} \in [\mathbf{1},\mathbf{n}]} \mathbf{u}(\mathbf{e_j})\mathbf{e^j}(\mathbf{v}) = \mathbf{u}(\mathbf{e_j})\mathbf{e^j}(\mathbf{v})\end{aligned}$$

which gives $\mathbf{u} = \mathbf{u}(\mathbf{e_j})\mathbf{e^j}$ and setting $\mathbf{u_j} = \mathbf{u}(\mathbf{e_j})$, $\mathbf{u} = \mathbf{u_j e^j}$.
The $(\mathbf{e^j} : [\mathbf{1}, \mathbf{n}] \to \mathbf{V}^*)$ are a basis of $\mathbf{V}^*$ from definition 184 (p. 219).
It is an independent basis (definition 185 (p. 219)) because:

$$\mathbf{u_i e^i} = 0 \Rightarrow (\forall \mathbf{j} \in [\mathbf{1}, \mathbf{n}])(\mathbf{u_i e^i}(\mathbf{e_j}) = \mathbf{u_i}\delta^\mathbf{i}_\mathbf{j} = \mathbf{u_j} = 0)\ .$$

Then we have proposition 379 (p. 222).

**Definition 189 Induced dual basis.**

We have a module (definition 182 (p. 218)) $\mathbf{V}$ over a commutative ring $\mathbf{F}$ with an independent finite basis $\mathbf{n} \in \mathcal{Z}^+, (\mathbf{e_i} : [1,\mathbf{n}] \to \mathbf{V})$ (definition 185 (p. 219)). The independent basis of $\mathbf{V}^*$ dual of $\mathbf{V}$: $(\mathbf{e^i} : [1,\mathbf{n}] \to \mathbf{V})$ such that: $(\forall \mathbf{i} \in [1,\mathbf{n}])(\forall \mathbf{j} \in [1,\mathbf{n}])(\mathbf{e^i}(\mathbf{e_j}) = \delta^{\mathbf{i}}_{\mathbf{j}})$ is called the induced dual basis of $\mathbf{V}^*$ (see proposition 379 (p. 222)).

**Proposition 380** The dual $\mathbf{V}^*_{\mathbf{U}}$ of a module (definition 182 (p. 218)) $\mathbf{V}$ over another module $\mathbf{U}$ (definition 188 (p. 221)) with the same commutative ring $\mathbf{F}$ both having an independent finite basis $\mathbf{n} \in \mathcal{Z}^+, (\mathbf{e_i} : [1,\mathbf{n}] \to \mathbf{V}),\ \mathbf{m} \in \mathcal{Z}^+, (\mathbf{f_i} : [1,\mathbf{m}] \to \mathbf{U}),$ (definition 185 (p. 219)) also has independent finite bases. One of them is $(\mathbf{e^i_k} : [1,\mathbf{n}] \times [1,\mathbf{m}] \to \mathbf{V}^*_{\mathbf{U}})$ such that $(\forall \mathbf{i} \in [1,\mathbf{n}])(\forall \mathbf{j} \in [1,\mathbf{n}])(\forall \mathbf{k} \in [1,\mathbf{m}])(\mathbf{e^i_k}(\mathbf{e_j}) = \delta^{\mathbf{i}}_{\mathbf{j}}\mathbf{f_k} = \mathbf{e^i}(\mathbf{e_j})\mathbf{f_k})$ where $\delta^{\mathbf{i}}_{\mathbf{j}}$ is the Kronecker symbol (definition 215 (p. 245)). We may write that basis as:

$$\mathbf{e^i_k} = \mathbf{e^i f_k}\ .$$

Where $(\mathbf{e^i} : [1,\mathbf{n}] \to \mathbf{V})$ is the induced dual basis of $\mathbf{V}^*$ as described in definition 189 (p. 222).

**Proof**
The proof is similar to the proof of proposition 379 (p. 222). Using again equation (2.75) (p. 222), $(\forall \mathbf{v} \in \mathbf{V})(\exists \mathbf{v^i} : [1,\mathbf{n}] \to \mathbf{V})(\mathbf{v} = \sum_{\mathbf{i}\in[1,\mathbf{n}]} \mathbf{v^i e_i})$, it defines an $\mathbf{n} \times \mathbf{m}$ linear map

$$(\mathbf{e^i_k}(\mathbf{v}) = \mathbf{v^i f_k} : \mathbf{V} \to \mathbf{U})$$

Using equation (2.76) (p. 222), we have:

$$(\mathbf{u} \in \mathbf{V}^*_{\mathbf{U}}, \mathbf{v} = \mathbf{v^i e_i} \in \mathbf{V}) \Rightarrow \mathbf{u}(\mathbf{v}) = \mathbf{v^i u}(\mathbf{e_i}) = \mathbf{v^i u^k_i f_k} = \mathbf{v^i} \delta^{\mathbf{j}}_{\mathbf{i}} \mathbf{u^k_j f_k} = \mathbf{v^i u^k_j e^j_k}(\mathbf{e_i})\ .$$

The $\mathbf{e^i_k}$ are a basis of $\mathbf{V}^*_{\mathbf{U}}$ (definition 184 (p. 219)). Since $\mathbf{f_k} : [1,\mathbf{m}] \to \mathbf{U}$ are an independent basis (definition 185 (p. 219)) of $\mathbf{U}$, the $\mathbf{e^i_k}$ form also an independent basis of $\mathbf{V}^*_{\mathbf{U}}$:

$$(\forall \mathbf{v} \in \mathbf{V})(\mathbf{u^k_i e^i_k}(\mathbf{v}) = 0) \Rightarrow (\forall \mathbf{j} \in [1,\mathbf{n}])(\mathbf{u^k_i e^i_k}(\mathbf{e_j}) = \mathbf{u^k_i} \delta^{\mathbf{i}}_{\mathbf{j}} \mathbf{f_k} = 0$$
$$\Rightarrow (\forall \mathbf{k} \in [1,\mathbf{m}])(\mathbf{u^k_j} = 0))\ .$$

**Definition 190 Induced basis of the dual of a module over another module.**

Keeping notation of proposition 380 (p. 223), one can also generalize definition 189 (p. 222) taking for induced basis of $\mathbf{V}^*_{\mathbf{U}}$, the $\mathbf{e^i_k} = \mathbf{e^i f_k}$ of proposition 380 (p. 223).

**Proposition 381** Using above notations (definition 190 (p. 223), proposition 380 (p. 223)), any element of $\mathbf{V}^*_{\mathbf{U}}$ can be written as $\mathbf{t^i_j e^j f_i}$:

$$(\forall \mathbf{t} \in \mathbf{V}^*_{\mathbf{U}})(\exists (\mathbf{t^i_j} : [1,\mathbf{m}] \times [1,\mathbf{n}] \to \mathbf{F}))(\mathbf{t} = \mathbf{t^i_j e^j f_i})\ .$$

**Proof**
It is a direct consequence of proposition 380 (p. 223).

**Definition 191 Multilinear maps or Multilinear form.**

We have a set $\mathbf{A}$ of modules over the same commutative ring $\mathbf{F}$ (definitions 182 (p. 218), 180 (p. 215)), a module $\mathbf{U}$ over $\mathbf{F}$ also. We also have with $\mathbf{K} \in \mathcal{Z}^+$, a map $(\mathbf{V_k} : [1,\mathbf{K}] \to \mathbf{A})$. One calls multilinear map on $\prod_{\mathbf{i}\in[1,\mathbf{K}]} \mathbf{V_k}$ (definitions 372 (p. 473) and 119 (p. 135)) a map

$(\mathbf{f}: \prod_{\mathbf{i}\in[\mathbf{1},\mathbf{K}]} \mathbf{V_k} \to \mathbf{U})$ verifying the following (see definition 126 (p. 144), 187 (p. 144)):
When $\mathbf{K} = \mathbf{1}$, it has to be an ordinary linear map (definition 186 (p. 220)). When $\mathbf{K} > \mathbf{1}$[15], we should have:

1. For any $\mathbf{x} \in \mathbf{V_K}$, considering the map $(\mathbf{g_1} : \prod_{\mathbf{k}\in[\mathbf{1},\mathbf{K}-\mathbf{1}]} \mathbf{V_k} \to \mathbf{U})$ defined by

$$(\forall \mathbf{u} \in \prod_{\mathbf{k}\in[\mathbf{1},\mathbf{K}-\mathbf{1}]} \mathbf{V_k})(\mathbf{g_1}(\mathbf{u}) = \mathbf{f}(\mathbf{u} \overset{\mathbf{e}}{\times} \mathbf{x}))$$

   $\mathbf{g_1}$ is multilinear along $\prod_{\mathbf{k}\in[\mathbf{1},\mathbf{K}-\mathbf{1}]} \mathbf{V_k}$.

2. For any $\mathbf{u} \in \prod_{\mathbf{k}\in[\mathbf{1},\mathbf{K}-\mathbf{1}]} \mathbf{V_k}$, the map $(\mathbf{g_2} : \mathbf{V_K} \to \mathbf{U})$ defined by

$$(\forall \mathbf{x} \in \mathbf{V_K})(\mathbf{g_2}(\mathbf{x}) = \mathbf{f}(\mathbf{u} \overset{\mathbf{e}}{\times} \mathbf{x}))$$

   is such that $\mathbf{g_2}$ is linear.

**Proposition 382** A multilinear map is linear in any of its components and vice versa[16].

**Proof**
Keeping the notations of definition 191 (p. 223), proposition 382 is true for $\mathbf{K} = \mathbf{1}$. If it is true up to $\mathbf{K} = \mathbf{K}'$ from point 191 (p. 223) -2, it is true when $\mathbf{K} = \mathbf{K}' + \mathbf{1}$ for the last coordinate. From point 191 (p. 223) -1, it is multilinear for the $\mathbf{K}'$ first component then linear in any of them.
If linear in any of its components, the reverse is true for $\mathbf{K} = \mathbf{1}$. If true up to $\mathbf{K}'$, 191 (p. 223) -1 is verified for $\mathbf{K} = \mathbf{K} + \mathbf{1}$, proposition 191 (p. 223) -2 is verified since it is linear in the last component.
In both cases we conclude by proposition 927 (p. 520).

**Definition 192** $\bigotimes_{\mathbf{k}\in[\mathbf{1},\mathbf{K}]} \mathbf{V^*_{U,k}}$, $\bigotimes_{\mathbf{k}\in[\mathbf{1},\mathbf{K}]} \mathbf{V^{*s}_{U,k}}$, $\bigotimes_{\mathbf{k}\in[\mathbf{1},\mathbf{K}]} \mathbf{V^*_k}$.

Using definition 191 (p. 223), one calls $\bigotimes_{\mathbf{k}\in[\mathbf{1},\mathbf{K}]} \mathbf{V^*_{U,k}}$, the set of multilinear maps $\prod_{\mathbf{k}\in[\mathbf{1},\mathbf{K}]} \mathbf{V_k} \to \mathbf{U}$. If $\mathbf{U} = \mathbf{F}$, one calls the set of multilinear maps $\prod_{\mathbf{k}\in[\mathbf{1},\mathbf{K}]} \mathbf{V_k} \to \mathbf{F}$:

$$\bigotimes_{\mathbf{k}\in[\mathbf{1},\mathbf{K}]} \mathbf{V^*_k}$$

One calls $\bigotimes_{\mathbf{k}\in[\mathbf{1},\mathbf{K}]} \mathbf{V^{*s}_{U,k}}$ a subset of $\bigotimes_{\mathbf{k}\in[\mathbf{1},\mathbf{K}]} \mathbf{V^*_{U,k}}$ such that when restricted to one of its components $\mathbf{k} \in [\mathbf{1}, \mathbf{K}]$, it is a linear map element of $\mathbf{V^{*s}_{U,k}}$, see convention 20 (p. 221).

---

[15] In what follows we shall always apply without referencing them propositions 836 (p. 480) and 837 (p. 481). They stipulate that $(\prod_{\mathbf{i}\in\mathbf{I}} \mathbf{X_i}) \times (\prod_{\mathbf{j}\in\mathbf{J}} \mathbf{X_j}) \sim \prod_{\mathbf{i}\in\mathbf{I}\cup\mathbf{J}} \mathbf{X_i}$. We shall use them as $(\prod_{\mathbf{i}\in\mathbf{I}} \mathbf{X_i}) \times (\prod_{\mathbf{j}\in\mathbf{J}} \mathbf{X_j}) = \prod_{\mathbf{i}\in\mathbf{I}\cup\mathbf{J}} \mathbf{X_i}$.

[16] One could take proposition 382 as the definition of multilinear maps and definition 191 (p. 223) as a proposition and use for the set $[\mathbf{1}, \mathbf{K}]$, any set of indices finite or not.

**Definition 193 Operations on multilinear maps.**

Keeping the notations of definition 191 (p. 223), having two multilinear maps $\mathbf{f}$ and $\mathbf{g}$ on $\prod_{i\in[1,K]} \mathbf{V_k}$, having $\lambda \in \mathbf{F}, \mu \in \mathbf{F}$, by definition:

$$\mathbf{v} \in \prod_{i\in[1,K]} \mathbf{V_k} \Rightarrow (\lambda\mathbf{f} + \mu\mathbf{g})(\mathbf{v}) = \lambda\mathbf{f}(\mathbf{v}) + \mu\mathbf{g}(\mathbf{v}) \in \mathbf{U}$$

**Proposition 383** Keeping the notations of definitions 191 (p. 223) and 193 (p. 225), if $\mathbf{f}$ and $\mathbf{g}$ belong to $\bigotimes_{k\in[1,K]} \mathbf{V^{*s}_{U,k}}$, $\lambda\mathbf{f} + \mu\mathbf{g}$ belongs also to $\bigotimes_{k\in[1,K]} \mathbf{V^{*s}_{U,k}}$. With that internal law, $\bigotimes_{k\in[1,K]} \mathbf{V^{*s}_{U,k}}$ is a module with $\mathbf{F}$ as a commutative ring.

**Proof**

We keep the above notations. From proposition 376 (p. 220), proposition 383 is true for $\mathbf{K = 1}$. If it is true for $\mathbf{K = K'}$, point 1 of definition 191 (p. 223) is true for $\mathbf{K = K' + 1}$. Point 2 of definition 191 (p. 223) is also true:
$\mathbf{u}$ being an element of $\bigotimes_{k\in[1,K']} \mathbf{V_k}$, from what we suppose $\mathbf{x} \in \mathbf{V_{K'+1}}$, $\mathbf{f(u \overset{e}{\times} x)}$ and $\mathbf{g(u \overset{e}{\times} x)}$ are linear in $\mathbf{x}$ and in $\mathbf{V^{*s}_{U,K'+1}}$ then $\lambda\mathbf{f(u \overset{e}{\times} x)} + \mu\mathbf{g(u \overset{e}{\times} x)}$ is also linear and in $\mathbf{V^{*s}_{U,K'+1}}$ (proposition 376 (p. 220), definition 183 (p. 219)). Then, if proposition 383 is true for $\mathbf{K = K'}$, it is true also for $\mathbf{K = K' + 1}$. From proposition 927 (p. 520), proposition 383 is true for any $\mathbf{K} \in \mathcal{Z}^+$.

**Proposition 384** We have a set $\mathbf{A}$ of modules over a commutative ring $\mathbf{F}$ (definitions 182 (p. 218), 180 (p. 215)), a map $(\mathbf{V_k} : [\mathbf{1, K}] \to \mathbf{A})$, $\mathbf{K} \in \mathcal{Z}^+$, another module $\mathbf{U}$ also over $\mathbf{F}$. $\mathbf{U}$ is supposed to be of finite dimension $\mathbf{m}$ and with an independent basis $(\mathbf{f_i} : [\mathbf{1, m}] \to \mathbf{U})$. We consider for each $\mathbf{V_k}$, a submodule of $\mathbf{V^{*s}_{U,k}}$ as defined in 188 (p. 221). Any element of $\prod_{k\in[1,K]} \mathbf{V^{*s}_{U,k}}$ can be transformed by a map $\mathbf{h}$ to a multilinear map belonging to $\bigotimes_{k\in[1,K]} \mathbf{V^{*s}_{U,k}}$.
This map $\mathbf{h}$ is defined by [17]:

$$(\mathbf{i} \in [\mathbf{1, m}], \mathbf{u} = \overset{e}{\prod_{k\in[1,K]}} \mathbf{u_k} = \overset{e}{\prod_{k\in[1,K]}} \mathbf{u^j_k f_j} \in \prod_{k\in[1,K]} \mathbf{V^{*s}_{U,k}}, \mathbf{v} = \overset{e}{\prod_{k\in[1,K]}} \mathbf{v_k} \in \prod_{k\in[1,K]} \mathbf{V_k})$$

$$\Rightarrow \mathbf{h^i(u)(v)} = \overset{\times}{\prod_{k\in[1,K]}} \mathbf{u^i_k(v_k)}$$

$\mathbf{h(u)}$ is an element of $\bigotimes_{i\in[1,K]} \mathbf{V^{*s}_{U,k}}$:

$$(\mathbf{h} : \prod_{k\in[1,K]} \mathbf{V^{*s}_{U,k}} \to \bigotimes_{i\in[1,K]} \mathbf{V^{*s}_{U,k}}) .$$

**Proof**

When $\mathbf{K = 1}$, it is clearly true. Let us suppose that it is true up to $\mathbf{K = K'}$, then point 1 of definition 191 (p. 223), is true for $\mathbf{K = K' + 1}$. By definition $\mathbf{u_{K'+1}}$ is linear, from definition 180 (p. 215) and point 3 of definition 179 (p. 214), point 2 of definition 191 (p. 223) is satisfied for $\mathbf{K = K' + 1}$. Then by proposition 927 (p. 520), proposition 384 (p. 225) is true.

[17] See definition 128 (p. 145) for $\overset{\times}{\prod}$.

**Definition 194** $\bigotimes_{k\in[1,K]} u_k$.

Keeping notations of proposition 384 (p. 225), one usually denotes $\bigotimes_{k\in[1,K]} u_k$ the map:

$$\prod_{k\in[1,K]} V^{*s}_{U,k} \to \bigotimes_{k\in[1,K]} V^{*s}_{U,k}$$

given by:[18]

$$(j \in [1,m], u \in \prod_{k\in[1,K]} V^{*s}_{U,k}, v \in \prod_{k\in[1,K]} V_k) \Rightarrow \bigotimes_{k\in[1,K]} u^j_k(v) = \prod^{\times}_{k\in[1,K]} u^j_k(v_k)$$

which is from proposition 384 (p. 225) an element of $\bigotimes_{k\in[1,K]} V^{*s}_{U,k}$.

**Proposition 385** Keeping the notations of proposition 384 (p. 225), the map

$$(\bigotimes_{k\in[1,K]} u_k : \prod_{k\in[1,K]} V^{*s}_{U,k} \to \bigotimes_{k\in[1,K]} V^{*s}_{U,k})$$

is multilinear.

**Proof**

It is trivially true for $K = 1$. If true for $K = K'$, the point 191 (p. 223), 1 of definition 191 (p. 223) is satisfied for $K = K' + 1$. We also have 191 (p. 223), 2 for $K = K' + 1$ because, if $u_{K+1} = \sum_{i\in[1,p]} \lambda^i x_i$, we have with summation over $j \in [1,m]$ (see footnote 15 (p. 224)):

$$\begin{aligned}
v = \prod_{k\in[1,K+1]} v_k \in \prod_{k\in[1,K+1]} V_k &\Rightarrow (\bigotimes_{k\in[1,K+1]} u_k)(v) \\
&= (\bigotimes_{k\in[1,K]} u_k \otimes \sum_{i\in[1,p]} \lambda^i x_i)(v) \\
&= \prod^{\times}_{k\in[1,K]} u^j_k(v_k)(\sum_{i\in[1,p]} \lambda^i x^j_i(v_{K+1}))f_j = \sum_{i\in[1,p]} \lambda^i (\prod^{\times}_{k\in[1,K]} u^j_k(v_k) \times x^j_i(v_{K+1}))f_j \\
&= \sum_{i\in[1,p]} \lambda^i (\bigotimes_{k\in[1,K]} u_k \otimes x_i)(v) .
\end{aligned}$$

Then 385 (p. 226) is true for $K = K' + 1$ if true for $K = K'$. We conclude by proposition 927 (p. 520).

**Proposition 386** We have a set $A$ of modules over the same commutative ring $F$ (definitions 182 (p. 218), 180 (p. 215)), $K \in \mathcal{Z}^+$, a map $(V_k : [1,K] \to A)$, a module $U$ over $F$ also, a multilinear map $(f : \prod_{i\in[1,K]} V_k \to U)$. We have the following relation (see proposition 368 (p. 217), definitions 372 (p. 473), 119 (p. 135) and convention 14 (p. 135)):

$$\begin{aligned}
&(p \in (\mathcal{Z}^+)^K, (\forall k \in [1,K])(\lambda_k \in F^{p(k)}, (v^k_i : [1,p(k)] \to V_k))) \\
&\Rightarrow f(\prod^{e}_{k\in[1,K]} \sum_{i\in[1,p(k)]} \lambda^i_k v^k_i) = \sum_{\sigma\in\prod_{k\in[1,K]}[1,p(k)]} f(\prod^{e}_{k\in[1,K]} v^k_{\sigma(k)}) \prod^{\times}_{k\in[1,K]} \lambda^{\sigma(k)}_k .
\end{aligned} \quad (2.77)$$

[18] We have to realize that the above definition depends on the basis chosen for the module $U$.

**Proof**
Equation (2.77) (p. 226) is trivially true for $\mathbf{K}=\mathbf{1}$. Let us suppose that equation (2.77) (p. 226) is true up to $\mathbf{K}=\mathbf{K}'$. Using notation of proposition 386, we have from definition 191 (p. 223) and propositions 361 (p. 213), 368 (p. 217):

$$\begin{aligned}
&f(\prod^{e}_{k\in[1,K+1]}\sum_{i\in[1,p(k)]}\lambda^{i}_{k}v^{k}_{i}) = \sum_{i\in[1,p(K+1)]}\lambda^{i}_{K+1}f(\prod^{e}_{k\in[1,K]}\sum_{j\in[1,p(k)]}\lambda^{j}_{k}v^{k}_{j}\times v^{K+1}_{i})\\
&=\sum_{i\in[1,p(K+1)]}\lambda^{i}_{K+1}\sum_{\sigma\in\prod_{k\in[1,K]}[1,p(k)]}f(\prod^{e}_{k\in[1,K]}v^{k}_{\sigma(k)}\times v^{K+1}_{i})\prod^{\times}_{k\in[1,K]}\lambda^{\sigma(k)}_{k}\\
&=\sum_{\sigma\in\prod_{k\in[1,K]}[1,p(k)]}\sum_{i\in[1,p(K+1)]}f(\prod^{e}_{k\in[1,K]}v^{k}_{\sigma(k)}\times v^{K+1}_{i})\prod^{\times}_{k\in[1,K]}\lambda^{\sigma(k)}_{k}\lambda^{i}_{K+1}\\
&=\sum_{\sigma\in\prod_{k\in[1,K]}[1,p(k)]}\sum_{\sigma(K+1)\in[1,p(K+1)]}\\
&f(\prod^{e}_{k\in[1,K]}v^{k}_{\sigma(k)}\times v^{K+1}_{\sigma(K+1)})\prod^{\times}_{k\in[1,K]}\lambda^{\sigma(k)}_{k}\lambda^{\sigma(K+1)}_{K+1}\\
&=\sum_{\sigma\in\prod_{k\in[1,K+1]}[1,p(k)]}f(\prod^{e}_{k\in[1,K+1]}v^{k}_{\sigma(k)})\prod^{\times}_{k\in[1,K+1]}\lambda^{\sigma(k)}_{k}
\end{aligned}$$

then it is also true for $\mathbf{K}=\mathbf{K}'+\mathbf{1}$. We conclude by proposition 927 (p. 520).

**Proposition 387** Keeping notation of proposition 386 (p. 226), if relation (2.77) (p. 226) is true then $(\mathbf{f}:\prod_{i\in[1,K]}\mathbf{V_k}\to\mathbf{U})$ is a multilinear map.

**Proof**
For $\mathbf{K}=\mathbf{1}$, it is the definition of a linear map (definitions 186 (p. 220)), 387 (p. 227) is true in that case. If 387 (p. 227) is true up to $\mathbf{K}=\mathbf{K}'$, then taking a map $(\mathbf{f}:\prod_{k\in[1,K+1]}\mathbf{V_k}\to\mathbf{U})$ satisfying 387 (p. 227), we have for any $\mathbf{x}\in\mathbf{V_{K+1}}$ a map $(\mathbf{g_1(u)}=\mathbf{f(u\times x)}:\prod_{k\in[1,K]}\mathbf{V_k}\to\mathbf{U})$. We have from relation (2.77) (p. 226):

$$\begin{aligned}
&(p\in(\mathcal{Z}^{+})^{K+1},p(K+1)=1,\\
&(\forall k\in[1,K+1])(\lambda_k\in F^{p(k)},(v^{k}_{i}:[1,p(k)]\to V_k)),\lambda^{1}_{K+1}=1,\\
&(\forall k\in[1,K])((u^{k}_{i}=v^{k}_{i}:[1,p(k)]\to V_k)),v^{K+1}_{1}=x)\\
&\Rightarrow f(\prod_{k\in[1,K]}\sum_{i\in[1,p(k)]}\lambda^{i}_{k}u^{k}_{i}\times x)=f(\prod_{k\in[1,K+1]}\sum_{i\in[1,p(k)]}\lambda^{i}_{k}v^{k}_{i})=\\
&\sum_{\sigma\in\prod_{k\in[1,K+1]}[1,p(k)]}f(\prod_{k\in[1,K+1]}v^{k}_{\sigma(k)})\prod^{\times}_{k\in[1,K+1]}\lambda^{\sigma(k)}_{k}=\\
&\sum_{\sigma\in\prod_{k\in[1,K]}[1,p(k)]}f(\prod_{k\in[1,K]}u^{k}_{\sigma(k)}\times x)\prod^{\times}_{k\in[1,K]}\lambda^{\sigma(k)}_{k}.
\end{aligned}\tag{2.78}$$

From proposition 387 (p. 227), which is supposed to be true up to $\mathbf{K}$, point 1 of definition 191 (p. 223) is verified.

Setting $\mathbf{u} = \prod_{i \in K}^{e} \mathbf{u}_i = \prod_{i \in K}^{e} \mathbf{v}_i^1 \in \prod_{k \in [1,K]} \mathbf{V}_k$ $\mathbf{x} \in \mathbf{V}_{K+1}$, any term of $\prod_{k \in [1,K]} \mathbf{V}_k$ can be written as

$$\mathbf{u} \times \mathbf{x} = (\prod_{i \in [1,K]}^{e} \mathbf{v}_i^1) \times \sum_{i \in [1,p(k+1)]} \lambda_{K+1}^i \mathbf{v}_i^{K+1} \in \prod_{k \in [1,K]} \mathbf{V}_k .$$

We can write formula 2.77 (p. 226) with

$$(\forall k \in [1,K])(p(k) = 1, \lambda_k^1 = 1),\ \mathbf{x} = \sum_{i \in [1,p(k+1)]} \lambda_{K+1}^i \mathbf{v}_i^{K+1} .$$

We then get for any $\mathbf{u} \in \prod_{k \in [1,K]} \mathbf{V}_k$ with $\mathbf{g}_2(\mathbf{x}) = \mathbf{f}(\mathbf{u} \times \mathbf{x})$:

$$\begin{aligned} \mathbf{g}_2(\mathbf{x}) &= \mathbf{f}(\mathbf{u} \times \mathbf{x}) = \mathbf{f}(\prod_{k \in [1,K+1]} \sum_{i \in [1,p(k)]} \lambda_k^i \mathbf{v}_i^k) \\ &= \sum_{\sigma \in \prod_{k \in [1,K+1]} [1,p(k)]} \mathbf{f}(\prod_{k \in [1,K+1]} \mathbf{v}_{\sigma(k)}^k) \prod_{k \in [1,K+1]}^{\times} \lambda_k^{\sigma(k)} \\ &= \sum_{\sigma \in [1,p(K+1)]} \mathbf{f}(\mathbf{u} \times \mathbf{v}_\sigma^{K+1}) \lambda_{K+1}^\sigma . \end{aligned}$$

Then $\mathbf{g}_2(\mathbf{x})$ is linear and point 2 of definition 191 (p. 223) is verified. $\mathbf{f}$ is the multilinear map on $\prod_{k \in [1,K+1]} \mathbf{V}_k$ if all maps $(\prod_{k \in [1,K]}^{\times} \mathbf{V}_k \to \mathbf{U})$ verifying relation (2.77) (p. 226) are multilinear maps. We conclude by proposition 927 (p. 520).

**Proposition 388** We have a module $\mathbf{U}$ with a finite basis $(\mathbf{f} : [1,m] \to \mathbf{U})$ (definition 192 (p. 224)). We consider the set $\bigotimes_{i \in [1,K]} \mathbf{V}_k^*$ (definition 187 (p. 221)) and the set $\bigotimes_{i \in [1,K]} \mathbf{V}_{U,k}^*$ (definition 188 (p. 221)) of the multilinear maps on $\mathbf{K}$ modules over the commutative ring $\mathbf{F}$ respectively to the commutative ring $\mathbf{F}$ and to the module $\mathbf{U}$ (definition 191 (p. 223), 192 (p. 224)). The modules are defined by a map $(\mathbf{V}_k : [1,K] \to \mathbf{A})$, $\mathbf{A}$ being a set of modules. Each of the $\mathbf{K}$ modules has a finite independent basis (definition 185 (p. 219))

$$(\forall k \in [1,K])(\exists (\mathbf{e}_{k;i} : [1,n_k] \to \mathbf{V}_k))$$

and a dual (definition 187 (p. 221)) $\mathbf{V}_k^*$ with its induced independent basis (proposition 379 (p. 222) and definition 189 (p. 222)):

$$(\forall k \in [1,K])(\exists (\mathbf{e}_k^j : [1,n_k] \to \mathbf{V}_k^*))(\mathbf{e}_k^j(\mathbf{e}_{k,i}) = \delta_i^j)$$

($\delta_i^j$ is the Kronecker symbol (definition 215 (p. 245))).
From proposition 380 (p. 223) and definition 190 (p. 223), $\mathbf{V}_{U,k}^*$ also has independent basis

$$(\forall \ell \in [1,m])(\forall k \in [1,K])(\exists (\mathbf{e}_k^{j\ell} : [1,n_k] \to \mathbf{V}_{U,k}^*))(\mathbf{e}_k^{j\ell}(\mathbf{e}_{k,i}) = \mathbf{f}^\ell \delta_i^j) .$$

Then using proposition 386 (p. 226), $\bigotimes_{i \in [1,K]} \mathbf{V}_{U,k}^*$ has finite independent bases. One of such bases is given by a map:

$$({}^{\otimes}\mathbf{e}_\ell^\sigma : [1,m] \times \prod_{k \in [1,K]} [1,n_k] \to \bigotimes_{i \in [1,K]} \mathbf{V}_{U,k}^*) \tag{2.79}$$

such that ($\delta^i_j$ is the Kronecker symbol (definition 215 (p. 245))):

$$(\forall \ell \in [1,m])(\forall \sigma \in \prod_{k\in[1,K]} [1,n_k])((\forall \tau \in \prod_{k\in[1,K]} [1,n_k])(^{\otimes}e^{\sigma}_{\ell}(\prod^{e}_{k\in[1,K]} e_{k;\tau(k)})$$
$$= f_\ell \prod^{\times}_{k\in[1,K]} \delta^{\sigma(k)}_{\tau(k)} = f_\ell \prod^{\times}_{k\in[1,K]} e^{\sigma(k)}_k(e_{k;\tau(k)})), \tag{2.80}$$
$$^{\otimes}e^{\sigma}_{\ell} = f_\ell \bigotimes_{k\in[1,K]} e^{\sigma(k)}_k) .$$

See definition 194 (p. 226).
For any $u \in \bigotimes_{k\in[1,K]} V^*_{U,k}$, calling $u^\ell_\sigma f_\ell = u(\prod_{k\in[1,K]} e_{k;\sigma(k)})$ we have (see (2.79) (p. 228)):

$$u = \sum_{\sigma \in \prod_{k\in[1,K]}[1,n_k]} u^\ell_\sigma {}^{\otimes}e^{\sigma}_{\ell} . \tag{2.81}$$

**Proof**
Keeping the notations of definition 191 (p. 223), we suppose that we have a set $\mathbf{A}$ of modules over the same commutative ring $\mathbf{F}$ (definitions 182 (p. 218), 180 (p. 215)), $K \in \mathcal{Z}^+$, a map $(V_k : [1,K] \to A)$. Moreover, for any $k \in [1,K]$, $V_k$ has a basis of $n_k \in \mathcal{Z}^+$ independent vectors $(e_{k;i} : [1,n_k] \to V_k)$ which defines an induced one (definition 189 (p. 222)) on $V^*_{U,k}$: $(e^i_k : [1,n_k] \to V^*_{U,k})$. We consider:

$$v \in \prod_{k\in[1,K]} V_k \Rightarrow v = \prod_{k\in[1,K]} \sum_{i\in[1,n_k]} v^{k;i} e_{k;i} .$$

Given a multilinear map $u \in \bigotimes_{k\in[1,K]} V^*_{U,k}$, we have from equation (2.77) (p. 226):

$$u(v) = \sum_{\sigma \in \prod_{k\in[1,K]}[1,n_k]} u(\prod_{k\in[1,K]} e_{k;\sigma(k)}) \prod^{\times}_{k\in[1,K]} v^{k;\sigma(k)} . \tag{2.82}$$

Given a term $\tau \in \prod_{k\in[1,K]} [1,n_k]$ (definition 372 (p. 473)), we define a multilinear map on $\prod_{k\in[1,K]} V_k$:

$$^{\otimes}e^{\tau}_{\ell} = f_\ell \bigotimes_{k\in[1,K]} e^{\tau(k)}_k \in \bigotimes_{k\in[1,K]} V^*_{U,k} .$$

From definition 194 (p. 226), the map is such that (see proposition 379 (p. 222)):

$$(\forall \sigma \in \prod_{k\in[1,K]} [1,n_k])(^{\otimes}e^{\tau}_{\ell}(\prod^{e}_{k\in[1,K]} e_{k;\sigma(k)}) = f_\ell \prod^{\times}_{k\in[1,K]} \delta^{\tau(k)}_{\sigma(k)})$$

where $\delta$ is the Kronecker symbol (definition 215 (p. 245)). From proposition 386 (p. 226), it gives for $\mathbf{v}$:

$$\begin{aligned}
{}^{\otimes}\mathbf{e}^{\tau}_{\ell}(\mathbf{v}) &= \sum_{\sigma\in\prod_{k\in[1,K]}[1,n_k]} {}^{\otimes}\mathbf{e}^{\tau}_{\ell}\Big(\prod_{k\in[1,K]} \mathbf{e}_{k;\sigma(k)}\Big) \prod^{\times}_{k\in[1,K]} \mathbf{v}^{k;\sigma(k)} \\
&= \mathbf{f}_{\ell} \sum_{\sigma\in\prod_{k\in[1,K]}[1,n_k]} \prod^{\times}_{k_1\in[1,K]} \delta^{\tau(k_1)}_{\sigma(k_1)} \prod^{\times}_{k_2\in[1,K]} \mathbf{v}^{k_2;\sigma(k_2)} \\
&= \mathbf{f}_{\ell} \sum_{\sigma\in\prod_{k\in[1,K]}[1,n_k]} \prod^{\times}_{k\in[1,K]} \delta^{\tau(k)}_{\sigma(k)} \mathbf{v}^{k;\sigma(k)} = \mathbf{f}_{\ell} \prod^{\times}_{k\in[1,K]} \mathbf{v}^{k;\tau(k)} = \mathbf{f}_{\ell} \prod^{\times}_{k\in[1,K]} \mathbf{e}^{\tau(k)}_{k}(\mathbf{v}) \,.
\end{aligned}$$

From proposition 384 (p. 225), it is a multilinear map. From proposition 379 (p. 222), we can rewrite equation (2.82) (p. 229) as:

$$\begin{aligned}
\mathbf{u}(\mathbf{v}) &= \sum_{\sigma\in\prod_{k\in[1,K]}[1,n_k]} \mathbf{u}\Big(\prod_{k\in[1,K]} \mathbf{e}_{k;\sigma(k)}\Big) \prod^{\times}_{k\in[1,K]} \mathbf{e}^{\sigma(k)}_{k}(\mathbf{v}) \\
&= \sum_{\sigma\in\prod_{k\in[1,K]}[1,n_k]} \mathbf{u}\Big(\prod_{k\in[1,K]} \mathbf{e}_{k;\sigma(k)}\Big) {}^{\otimes}\mathbf{e}^{\sigma}_{\ell}(\mathbf{v}) \,.
\end{aligned}$$

Calling $\mathbf{u}^{\ell}_{\sigma}\mathbf{f}_{\ell} = \mathbf{u}\Big(\prod_{k\in[1,K]} \mathbf{e}_{k;\sigma(k)}\Big)$ we have:

$$\mathbf{u}(\mathbf{v}) = \sum_{\sigma\in\prod_{k\in[1,K]}[1,n_k]} \mathbf{u}^{\ell}_{\sigma}\, {}^{\otimes}\mathbf{e}^{\sigma}_{\ell}(\mathbf{v}) \,.$$

So we may write:

$$(\forall \mathbf{u} \in \bigotimes_{k\in[1,K]} \mathbf{V}^{*}_{U,k})(\exists(\mathbf{u}^{\ell}_{\sigma} : [1,m] \times \prod_{k\in[1,K]} [1,n_k] \to \mathbf{F}))(\mathbf{u} = \sum_{\sigma\in\prod_{k\in[1,K]}[1,n_k]} \mathbf{u}^{\ell}_{\sigma}\, {}^{\otimes}\mathbf{e}^{\sigma}_{\ell}) \,.$$

The ${}^{\otimes}\mathbf{e}^{\sigma}_{\ell}$ are a basis of $\prod_{k\in[1,K]} \mathbf{V}^{*}_{k}$.

**Definition 195 Tensorial product.**

Using the notations of definition 192 (p. 224), one calls $\bigotimes_{k\in[1,K]} \mathbf{V}_{k}$, the set of linear map on $\bigotimes_{k\in[1,K]} \mathbf{V}^{*}_{k}$ or the dual of $\bigotimes_{i\in[1,K]} \mathbf{V}^{*}_{k}$.

**Proposition 389** The tensorial product is a module under the same commutative ring.

**Proof**

It is a direct consequence of proposition 378 (p. 221).

**Proposition 390** Any element of $\bigotimes_{k\in[1,K]} \mathbf{V}_{k}$ can be considered as a multilinear map on $\prod_{i\in[1,K]} \mathbf{V}^{*}_{k}$.

**Proof**
It is a direct consequence of proposition 385 (p. 226), proposition 383 (p. 225) and definition 195 (p. 230).

**Proposition 391** Keeping the notations of definition 191 (p. 223), for any element of $\prod_{k\in[1,K]} V_k$ there is a multilinear map between $\prod_{k\in[1,K]} V_k$ and $\bigotimes_{k\in[1,K]} V_k$ quoted for: $u = \prod^{e}_{k\in[1,K]} u_k$ as $\bigotimes_{k\in[1,K]} u_k$. It is defined by:

$$(\forall u = \prod^{e}_{k\in[1,K]} u_k \in \prod_{k\in[1,K]} V_k)(\forall v \in \bigotimes_{k\in[1,K]} V^*_k)((\bigotimes_{k\in[1,K]} u_k)(v) = v(u)) .$$

**Proof**
For $K = 1$ for simplification we write:

$$v_1 \in V^*_1 \bigotimes u_1(v_1) = u_1(v_1) = v_1(u_1) .$$

We may write for any element of $[1, K]$:

$$v_k \in V^*_k \bigotimes u_k(v_k) = u_k(v_k) = v_k(u_k)$$

and we have:

$$(\bigotimes_{k\in[1,K]} u_k)(v) = v(u) = \prod^{\times}_{k\in[1,K]} v_k(u_k) = \prod^{\times}_{k\in[1,K]} u_k(v_k) .$$

Written that way proposition 390 (p. 230) is the same as proposition 385 (p. 226).

**Proposition 392** Having a set $A$ of modules with $F$ as commutative ring (definitions 182 (p. 218), 180 (p. 215)), $K \in \mathcal{Z}^+$, a map $(V_k : [1, K] \to A)$. Each of the $K$ modules, has a finite independent basis (definition 185 (p. 219)) $(\forall k \in [1, K])(\exists(e_{k;i} : [1, n_k] \to V_k))$. Then $\bigotimes_{k\in[1,K]} V_k$ has a finite independent basis. Setting $I = \prod_{k\in[1,K]} [1, n_k]$, we have one of them $(e_\sigma : I \to \bigotimes_{k\in[1,K]} V_k)$ which is given by:

$$(\forall \sigma \in I)(e_\sigma = \bigotimes_{k\in[1,K]} e_{k;\sigma(k)}) .$$

**Proof**
Using definition 189 (p. 222) and proposition 388 (p. 228), we have an induced basis of $\bigotimes_{k\in[1,K]} V^*_k$: For any $V^*_k$ we have a basis:

$$(\forall k \in [1, K])(\exists(e^{k;i} : [1, n_k] \to V^*_k))$$

and for $\bigotimes_{k\in[1,K]} V^*_k$ a basis:

$$(\forall \tau \in I)(e^\tau = \bigotimes_{k\in[1,K]} e^{k;\tau(k)})$$

such that:

$$\mathbf{v} = \prod_{\mathbf{k}\in[1,\mathbf{K}]} \mathbf{v}^{\mathbf{k}} \in \prod_{\mathbf{k}\in[1,\mathbf{K}]} \mathbf{V}^*_{\mathbf{k}}$$

gives:

$$(\forall \tau \in \mathbf{I})(\mathbf{e}^{\tau}(\mathbf{v}) = \prod^{\times}_{\mathbf{k}\in[1,\mathbf{K}]} \mathbf{e}^{\mathbf{k};\tau(\mathbf{k})}(\mathbf{v}^{\mathbf{k}})) \; .$$

From proposition 391 (p. 231), we have:

$$\mathbf{e}_{\sigma} \in \bigotimes_{\mathbf{k}\in[1,\mathbf{K}]} \mathbf{V}_{\mathbf{k}} \; .$$

From propositions 391 (p. 231) and 388 (p. 228): Given $\mathbf{u} \in \bigotimes_{\mathbf{k}\in[1,\mathbf{K}]} \mathbf{V}_{\mathbf{k}}$, $\mathbf{v} \in \bigotimes_{\mathbf{k}\in[1,\mathbf{K}]} \mathbf{V}^*_{\mathbf{k}}$, we have (definition 184 (p. 219), proposition 388 (p. 228)):

$$(\exists(\mathbf{v}_{\sigma} : \mathbf{I} \to \mathbf{F}))(\mathbf{v} = \sum_{\sigma\in\mathbf{I}} \mathbf{v}_{\sigma}\mathbf{e}^{\sigma}) \; .$$

Then from definitions 187 (p. 221) and 195 (p. 230), we have:

$$\begin{aligned}
\mathbf{u}(\mathbf{v}) &= \sum_{\sigma\in\mathbf{I}} \mathbf{v}_{\sigma}\mathbf{u}(\mathbf{e}^{\sigma}) \\
&= \sum_{\sigma\in\mathbf{I}}\sum_{\tau\in\mathbf{I}} \mathbf{v}_{\tau}\mathbf{u}(\mathbf{e}^{\sigma}) \prod^{\times}_{\mathbf{k}\in[1,\mathbf{K}]} \delta^{\tau(\mathbf{k})}_{\sigma(\mathbf{k})} \\
&= \sum_{\sigma\in\mathbf{I}}\sum_{\tau\in\mathbf{I}} \mathbf{v}_{\tau}\mathbf{u}(\mathbf{e}^{\sigma})\mathbf{e}_{\sigma}(\mathbf{e}^{\tau}) \\
&= \sum_{\sigma\in\mathbf{I}} \mathbf{u}(\mathbf{e}^{\sigma})\mathbf{e}_{\sigma}(\sum_{\tau\in\mathbf{I}} \mathbf{v}_{\tau}\mathbf{e}^{\tau}) \\
&= \sum_{\sigma\in\mathbf{I}} \mathbf{u}(\mathbf{e}^{\sigma})\mathbf{e}_{\sigma}(\mathbf{v}) \; .
\end{aligned}$$

We then have:

$$\mathbf{u} = \sum_{\sigma\in\mathbf{I}} \mathbf{u}(\mathbf{e}^{\sigma})\mathbf{e}_{\sigma} \; .$$

**Proposition 393** The dual of module **V** on a commutative ring **F** over another module **U** on the same commutative ring is isomorph to $\mathbf{V}^* \otimes \mathbf{U}$.

**Proof**

From propositions 391 (p. 231), 392 (p. 231) and 388 (p. 228), an element of $\mathbf{V}^*_{\mathbf{U}}$ can be expressed as an element of $\mathbf{V}^* \otimes \mathbf{U}$.

### 2.10.1 Wedge projection

**Definition 196 Permutation of $(\mathbf{V}^*_{\mathbf{U}})^{\mathbf{p}}$ .**

Given two modules **U** and **V** over the same commutative ring **F**, we consider (see definition 192 (p. 224)):

$$\bigotimes_{\mathbf{k}\in[1,\mathbf{K}]} \mathbf{V}^*_{\mathbf{U}} = (\otimes\mathbf{V}^*_{\mathbf{U}})^{\mathbf{K}} \; .$$

For any permutation ($\leftrightarrow$ is definition in convention 15 (p. 140)) $(\mathbf{P} : [\mathbf{1},\mathbf{p}] \leftrightarrow [\mathbf{1},\mathbf{p}])$ (see section 2.9.1 (p. 207)), we have the corresponding map $(\mathbf{P}_{\otimes^*} : (\otimes\mathbf{V}^*_{\mathbf{U}})^{\mathbf{p}} \to (\otimes\mathbf{V}^*_{\mathbf{U}})^{\mathbf{p}})$ or, when it is clear enough, $(\mathbf{P} : (\otimes\mathbf{V}^*_{\mathbf{U}})^{\mathbf{p}} \to (\otimes\mathbf{V}^*_{\mathbf{U}})^{\mathbf{p}})$ defined by:

$$(\forall(\mathbf{P} : [\mathbf{1},\mathbf{p}] \leftrightarrow [\mathbf{1},\mathbf{p}]))(\forall t \in (\otimes\mathbf{V}^*_{\mathbf{U}})^{\mathbf{p}})(\forall v = \prod^{\mathbf{e}}_{\mathbf{i}\in[\mathbf{1},\mathbf{p}]} v_{\mathbf{i}} \in (\mathbf{V})^{\mathbf{p}})$$

$$(\mathbf{P}(t)(\prod^{\mathbf{e}}_{\mathbf{i}\in[\mathbf{1},\mathbf{p}]} v_{\mathbf{i}}) = \mathbf{P}_{\otimes^*}(t)(\prod^{\mathbf{e}}_{\mathbf{i}\in[\mathbf{1},\mathbf{p}]} v_{\mathbf{i}}) = t(\prod^{\mathbf{e}}_{\mathbf{i}\in[\mathbf{1},\mathbf{p}]} v_{\mathbf{P}^{-1}(\mathbf{i})})) .$$

**Definition 197 Wedge projection on $(\otimes\mathbf{V}^*_{\mathbf{U}})^{\mathbf{p}}$.**

Calling $\mathbf{PG_p}$ the permutation group of $[\mathbf{1},\mathbf{p}]$ (see definition 172 (p. 207)) and $(\mathbf{s(P)} : \mathbf{PG_p} \to \{-\mathbf{1},\mathbf{1}\})$ the signature of the permutations (see definition 175 (p. 208)), one calls wedge projection or the antisymmetric projection, the following linear map $(\mathbf{W} : (\otimes\mathbf{V}^*_{\mathbf{U}})^{\mathbf{p}} \to (\otimes\mathbf{V}^*_{\mathbf{U}})^{\mathbf{p}})$ defined by:

$$(\forall t \in (\otimes\mathbf{V}^*_{\mathbf{U}})^{\mathbf{p}})(\mathbf{W}(t) = \sum_{\mathbf{P}\in\mathbf{PG_p}} \mathbf{s(P)P}(t)) .$$

**Proposition 394** With notations of definition 197 (p. 233) and proposition 388 (p. 228) supposing that $\mathbf{V}$ has a finite basis $(\mathbf{e_i} : [\mathbf{1},\mathbf{n}] \to \mathbf{V})$, we have (with Einstein convention 12 (p. 128)):

$$t \in (\otimes\mathbf{V}^*_{\mathbf{U}})^{\mathbf{p}} \Rightarrow (\exists(\mathbf{t}^{\ell}_{\sigma} : [\mathbf{1},\mathbf{m}] \times [\mathbf{1},\mathbf{n}]^{\mathbf{p}} \to \mathcal{R}))$$
$$(\mathbf{t} = \sum_{\ell\in[\mathbf{1},\mathbf{m}]} \sum_{\sigma\in[\mathbf{1},\mathbf{n}]^{\mathbf{p}}} \mathbf{t}^{\ell}_{\sigma}\otimes\mathbf{e}^{\sigma}_{\ell} = \sum_{\ell\in[\mathbf{1},\mathbf{m}]} \sum_{\sigma\in[\mathbf{1},\mathbf{n}]^{\mathbf{p}}} \mathbf{t}^{\ell}_{\sigma}\mathbf{f}_{\ell} \bigotimes_{\mathbf{k}\in[\mathbf{1},\mathbf{p}]} \mathbf{e}^{\sigma(\mathbf{k})},$$
$$\mathbf{W(t)} = \sum_{\ell\in[\mathbf{1},\mathbf{m}]} \mathbf{f}_{\ell} \sum_{\mathbf{h}\in\mathcal{I}_{\mathbf{p}}} \mathbf{W}(\mathbf{t}^{\ell}_{\mathbf{h}})\mathbf{W}(\bigotimes_{\mathbf{k}\in[\mathbf{1},\mathbf{p}]} \mathbf{e}^{\mathbf{h(k)}})) .$$

Where $\mathcal{I}_{\mathbf{p}}$ is defined in 53 (p. 48), and setting (definition 117 (p. 134)):

$$\mathbf{h} \in \mathcal{I}_{\mathbf{p}} \Rightarrow \mathbf{W}(\mathbf{t_h}) = \sum_{\mathbf{P}\in\mathbf{PG_p}} \mathbf{S(P)t_{h\circ P}} .$$

See equation (2.13) (p. 145) for the notation.

**Proof**

Keeping the above notation, from definitions 196 (p. 232), 197 (p. 233), proposition 388 (p. 228) equation (2.81) (p. 229), we have:[19]

$$(\forall\sigma' \in [\mathbf{1},\mathbf{n}]^{\mathbf{p}})(\mathbf{W(t)}(\mathbf{e}_{\sigma'}) = \mathbf{W(t)}(\bigotimes_{\mathbf{k}\in[\mathbf{1},\mathbf{p}]} \mathbf{e}_{\sigma'(\mathbf{k})}) = \mathbf{t}(\sum_{\mathbf{P}\in\mathbf{PG_p}} \mathbf{S(P)e}_{\sigma'\circ\mathbf{P}^{-1}})$$
$$= \sum_{\ell\in[\mathbf{1},\mathbf{m}]} \sum_{\sigma\in[\mathbf{1},\mathbf{n}]^{\mathbf{p}}} \mathbf{t}^{\ell}_{\sigma}\mathbf{f}_{\ell}\mathbf{e}^{\sigma}(\sum_{\mathbf{P}\in\mathbf{PG_p}} \mathbf{S(P)e}_{\sigma'\circ\mathbf{P}^{-1}})$$
$$= \sum_{\ell\in[\mathbf{1},\mathbf{m}]} \sum_{\mathbf{P}\in\mathbf{PG_p}} \mathbf{S(P)} \sum_{\sigma\in[\mathbf{1},\mathbf{n}]^{\mathbf{p}}} \mathbf{t}^{\ell}_{\sigma}\mathbf{f}_{\ell}\mathbf{e}^{\sigma}(\mathbf{e}_{\sigma'\circ\mathbf{P}^{-1}})$$
$$= \sum_{\ell\in[\mathbf{1},\mathbf{m}]} \sum_{\mathbf{P}\in\mathbf{PG_p}} \mathbf{S(P)t}^{\ell}_{\sigma'\circ\mathbf{P}^{-1}}\mathbf{f}_{\ell} = \sum_{\ell\in[\mathbf{1},\mathbf{m}]} \sum_{\mathbf{P}\in\mathbf{PG_p}} \mathbf{S(P)} \sum_{\sigma\in[\mathbf{1},\mathbf{n}]^{\mathbf{p}}} \mathbf{t}^{\ell}_{\sigma}\mathbf{f}_{\ell}\delta^{\sigma\circ\mathbf{P}}_{\sigma'}$$
$$= \sum_{\ell\in[\mathbf{1},\mathbf{m}]} \sum_{\mathbf{P}\in\mathbf{PG_p}} \mathbf{S(P)} \sum_{\sigma\in[\mathbf{1},\mathbf{n}]^{\mathbf{p}}} \mathbf{t}^{\ell}_{\sigma}\mathbf{f}_{\ell}\mathbf{e}^{\sigma\circ\mathbf{P}}(\mathbf{e}_{\sigma'})) .$$

[19] In order to simplify the writing, we call $\mathbf{e}_{\sigma'} = \bigotimes_{\mathbf{k}\in[\mathbf{1},\mathbf{p}]} \mathbf{e}_{\sigma'(\mathbf{k})}$ and the same for $\mathbf{e}^{\sigma}$.

Then

$$\mathbf{W(t)} = \sum_{\ell\in[1,m]} \sum_{P\in PG_p} \mathbf{S(P)} \sum_{\sigma\in[1,n]^p} \mathbf{t}^{\ell}_{\sigma}\mathbf{f}_{\ell}\mathbf{e}^{\sigma\circ P} \ . \tag{2.83}$$

Since

$$\mathbf{S(P^{-1}\circ P) = S(P^{-1})S(P) = 1,\ S(P^{-1}) = S(P) = \pm 1} \tag{2.84}$$

and

$$(\forall \mathbf{P} \in \mathbf{PG_p})(\mathbf{P^{-1}} \in \mathbf{PG_p}, \mathbf{P}\circ\mathbf{PG_p} = \mathbf{P^{-1}}\circ\mathbf{PG_p} = \mathbf{PG_p}) \tag{2.85}$$

we also have:

$$\sum_{P\in PG_p} \mathbf{S(P)t}^{\ell}_{\sigma'\circ P^{-1}} = \sum_{P\in PG_p} \mathbf{S(P)t}^{\ell}_{\sigma'\circ P} = \mathbf{W(t}^{\ell}_{\sigma'}) \ .$$

Let us call $\mathbf{I_p}$, the set of injections (definition 122 (p. 139)) of $[\mathbf{1},\mathbf{p}]$ to $[\mathbf{1},\mathbf{n}]$. From proposition 869 (p. 496) and definition 397 (p. 516) we have successively:

$$(\forall \mathbf{x} \in \mathbf{I_p})(\mathbf{Card}(\mathbf{set_h}\{\mathbf{h} \in \mathcal{I}_p, (\mathbf{x}([1,p]) = \mathbf{h}([1,p]))\}) = 1)$$
$$(\forall \mathbf{x} \in \mathbf{I_p})(\exists \mathbf{h_x} \in \mathcal{I}_p)(\mathbf{x}([1,p]) = \mathbf{h_x}([1,p]), \mathbf{x}^{-1}\circ\mathbf{h_x} \in \mathbf{PG_p})$$
$$(\forall \mathbf{x} \in \mathbf{I_p})((\exists \mathbf{P_x} \in \mathbf{PG_p})(\mathbf{x}\circ\mathbf{P_x} \in \mathcal{I}_p), \mathbf{Card}(\mathbf{set_P}\{\mathbf{P} \in \mathbf{PG_p}, \mathbf{x}\circ\mathbf{P} \in \mathcal{I}_p\}) = 1)$$

$(\mathbf{h_x} : \mathbf{I_p} \to \mathcal{I}_p)$ is a map (definition 112 (p. 131)) and $\mathbf{h_x^{-1}}$ defines a partition (definition 370 (p. 471), proposition 816 (p. 471)) of $\mathbf{I_p}$ and equivalence classes (definition 336 (p. 445)) in $\mathbf{I_p}$ with the equivalence relation $\mathbf{x} \approx \mathbf{y} \Leftrightarrow \mathbf{h_x} = \mathbf{h_y}$ (proposition 814 (p. 471)). The map $(\mathbf{P_x} : \mathbf{I_p} \leftrightarrow \mathbf{PG_p})$ is a bijection (definition 124 (p. 139)). We also have (see definition 353 (p. 454)):

$$\mathbf{h} \in \mathcal{C}_{[1,n]^p}(\mathbf{I_p}) \Rightarrow (\exists \mathbf{i} \in [1,p])(\exists \mathbf{j} \in [1,p])(\mathbf{i} \neq \mathbf{j}, \mathbf{h(i)} = \mathbf{h(j)})$$

otherwise $\mathbf{h}$ would be an injection and being member of $\mathbf{I_p}$.
Then if $\mathbf{h} \in \mathcal{C}(\mathbf{I_p})_{[1,n]^p}$, we have a transposition $\mathbf{T_{ij}} \in \mathbf{PG_p}$ (definition 173 (p. 208)) such that $(\forall \mathbf{k} \in [1,p])(\mathbf{h(k)} = \mathbf{h(T_{ij}(k))})$. Since we have:
$\mathbf{T_{ij}}\circ\mathbf{PG_p} = \mathbf{PG_p}$ (equation (2.85) (p. 234)) and $\mathbf{S(T_{ij}\circ P)} = -\mathbf{S(P)}$ (proposition 351 (p. 209)), we have:

$$\mathbf{h} \in \mathcal{C}(\mathbf{I})_{[1,n]^p} \Rightarrow \mathbf{W(e^h)} = \bigwedge_{i\in[1,p]} \mathbf{e}^{\mathbf{h(i)}} = \sum_{P\in T_{ij}\circ PG_p} \mathbf{S(P)e}^{\mathbf{h(P(i))}}$$
$$= \sum_{P\in PG_p} \mathbf{S(T_{ij}P)e}^{\mathbf{h(P(i))}} = 0$$

then using equations (2.83) (p. 234), 2.84 (p. 234) and 2.85 (p. 234), we have:

$$\begin{aligned}
W(t) &= \sum_{\ell\in[1,m]} \sum_{P\in PG_p} S(P) \sum_{\sigma\in[1,n]^p} t^\ell_\sigma f_\ell e^{\sigma\circ P} \\
&= \sum_{\ell\in[1,m]} \sum_{\sigma\in[1,n]^p} \sum_{P\in PG_p} S(P) t^\ell_\sigma f_\ell e^{\sigma\circ P} \\
&= \sum_{\ell\in[1,m]} \sum_{\sigma\in I_p} \sum_{P\in PG_p} S(P) t^\ell_\sigma f_\ell e^{\sigma\circ P} \\
&= \sum_{\ell\in[1,m]} \sum_{h\in\mathcal{I}_p} \sum_{P'\in PG_p} \sum_{P\in PG_p} S(P) t^\ell_{h\circ P'} f_\ell e^{h\circ P'\circ P} \\
&= \sum_{\ell\in[1,m]} \sum_{h\in\mathcal{I}_p} \sum_{P'\in PG_p} S(P') t^\ell_{h\circ P'} \sum_{P'\circ P\in PG_p} S(P'\circ P) f_\ell e^{h\circ P'\circ P} \\
&= \sum_{\ell\in[1,m]} \sum_{h\in\mathcal{I}_p} \sum_{P'\in PG_p} S(P') t^\ell_{h\circ P'} \sum_{Q\in PG_p} S(Q) f_\ell e^{h\circ Q} \\
&= \sum_{\ell\in[1,m]} f_\ell \sum_{h\in\mathcal{I}_p} W(t^\ell_h) W(e^h)\ .
\end{aligned}$$

**Definition 198** $(\bigwedge_{i\in[1,p]} t_i : \prod^{e}_{i\in[1,p]} V^*_U = (V^*_U)^p \to \bigotimes_{i\in[1,p]} V^*_U)$.

One sets:

$$(\forall t = (t_i : [1,p] \to V^*_U) \in \prod^{e}_{i\in[1,p]} V^*_U)(\bigwedge_{i\in[1,p]} t_i = W(\bigotimes_{i\in[1,p]} t_i))\ .$$

**Proposition 395** Setting $(t_i : [1,p] \to V^*_U)$ as $(\forall i \in [1,p])(t_i = t^\ell_{i,j} f_\ell e^j \in V^*_U)$, with $(\forall i \in [1,p])(\forall \ell \in [1,m])(\forall j \in [1,n])(t^\ell_{i,j} \in F)$, we have:

$$\bigwedge_{i\in[1,p]} t_i = f_\ell \sum_{h\in\mathcal{I}_p} W(\prod^{\times}_{i\in[1,p]} t^\ell_{i,h(i)}) \bigwedge_{i\in[1,p]} e^{h(i)}\ . \tag{2.86}$$

**Proof**
From proposition 388 (p. 228) equation (2.81) (p. 229), we have:

$$\bigotimes_{i\in[1,p]} t_i = \sum_{\sigma\in[1,n]^p} \prod^{\times}_{i\in[1,p]} t^\ell_{i,\sigma(i)} f_\ell \bigotimes_{i\in[1,p]} e^{\sigma(i)}\ .$$

We conclude using proposition 394 (p. 233).

## 2.11 Vector spaces

### 2.11.1 Definition and properties of vector space

**Definition 199 Vector space.**

A module (definition 182 (p. 218)) $\mathbf{V}$ is a vector space if it is a module over a commutative ring (definition 180 (p. 215) which is also a field (definition 181 (p. 217)). So following our definition 182 (p. 218) we should have: an Abelian group (definition 170 (p. 206)) map $(\mathbf{a}+\mathbf{b} : \mathbf{V}\times\mathbf{V} \to \mathbf{V})$, a field (definition 181 (p. 217)) $\mathcal{RC}$ and a map of $(\boldsymbol{\lambda}\mathbf{a} : \mathcal{RC}\times\mathbf{V} \to \mathbf{V})$ such that: with $\mathbf{a} \in \mathbf{V}, \mathbf{b} \in \mathbf{V}, \boldsymbol{\lambda} \in \mathcal{RC}, \boldsymbol{\mu} \in \mathcal{RC}$ we have the following relations:

1. $\boldsymbol{\lambda}(\boldsymbol{\mu}\mathbf{a}) = (\boldsymbol{\lambda}\boldsymbol{\mu})\mathbf{a}$
2. $\boldsymbol{\lambda}(\mathbf{a}+\mathbf{b}) = \boldsymbol{\lambda}\mathbf{a}+\boldsymbol{\lambda}\mathbf{b}$
3. $\boldsymbol{\lambda}\mathbf{a}+\boldsymbol{\mu}\mathbf{a} = (\boldsymbol{\lambda}+\boldsymbol{\mu})\mathbf{a}$
4. $\mathbf{1}\boldsymbol{a} = \boldsymbol{a}$.

The null element of the Abelian group is the zero vector quoted as $\mathbf{0}$. It could be confused with the zero of $\mathcal{RC}$ but since (see proposition 398 (p. 236))

$$(\mathbf{a} \in \mathbf{V}, \boldsymbol{\lambda} \in \mathcal{RC}) \Rightarrow \mathbf{0a} = (\boldsymbol{\lambda}-\boldsymbol{\lambda})\mathbf{a} = \boldsymbol{\lambda}\mathbf{a}-\boldsymbol{\lambda}\mathbf{a} = \mathbf{0}$$

there is no confusion most of the time and we have:

**Proposition 396** $\mathbf{V}$ is a vector space with the field $\mathcal{RC}$, $\mathbf{0}$ the null element of $\mathcal{RC}$, $\mathbf{0_V}$ the zero vector. We have:

$$(\forall \mathbf{a} \in \mathbf{V})(\mathbf{0a} = \mathbf{0_V}) .$$

**Proof**

$$\begin{aligned}&\mathbf{a} \in \mathbf{V} \Rightarrow \mathbf{a}+\mathbf{0a} = \mathbf{1a}+\mathbf{0a} = (\mathbf{1}+\mathbf{0})\mathbf{a} = \mathbf{1a} = \mathbf{a}\\ &\Rightarrow \mathbf{a}+\mathbf{0a} = \mathbf{a} \Rightarrow \mathbf{a}-\mathbf{a}+\mathbf{0a} = \mathbf{0_V} \Rightarrow \mathbf{0_V}+\mathbf{0a} = \mathbf{0_V} \Rightarrow \mathbf{0a} = \mathbf{0_V} .\end{aligned}$$

**Proposition 397** With notation of proposition 396 (p. 236), $\boldsymbol{\mu} \in \mathcal{RC} \Rightarrow \boldsymbol{\mu}\mathbf{0_V} = \mathbf{0_V}$.
**Proof**
We have from definition 199 (p. 235):

$$(\mathbf{a} \in \mathbf{V}, \boldsymbol{\mu} \in \mathcal{RC}) \Rightarrow \mathbf{a}+\boldsymbol{\mu}\mathbf{0_v} = \boldsymbol{\mu}(\mathbf{a}/\boldsymbol{\mu}+\mathbf{0_v}) = \boldsymbol{\mu}\mathbf{a}/\boldsymbol{\mu} = \mathbf{a} .$$

Then from proposition 335 (p. 206), $\boldsymbol{\mu}\mathbf{0_v} = \mathbf{0_v}$.

**Proposition 398** With notation of proposition 396 (p. 236),

$$\begin{aligned}&(\boldsymbol{\mu} \neq \mathbf{0}, \boldsymbol{\mu}\mathbf{a} = \mathbf{0}) \Rightarrow \mathbf{a} = \mathbf{0}\\ &(\mathbf{a} \neq \mathbf{0}, \boldsymbol{\mu}\mathbf{a} = \mathbf{0}) \Rightarrow \boldsymbol{\mu} = \mathbf{0} .\end{aligned}$$

**Proof**
From definition 181 (p. 217) $\boldsymbol{\mu} \neq \mathbf{0} \Rightarrow (\exists \mathbf{1}/\boldsymbol{\mu})(\boldsymbol{\mu}/\boldsymbol{\mu} = \mathbf{1})$. Then we have from propositions 396 (p. 236), 397 (p. 236) and definition 199 (p. 235) and point 199 (p. 235), 1, 4:

$$(\boldsymbol{\mu} \neq \mathbf{0}, \boldsymbol{\mu}\mathbf{a} = \mathbf{0}) \Rightarrow \boldsymbol{\mu}/\boldsymbol{\mu}\mathbf{a} = \mathbf{a} = (\mathbf{1}/\boldsymbol{\mu})\boldsymbol{\mu}\mathbf{a} = \mathbf{1}/\boldsymbol{\mu}\mathbf{0} = \mathbf{0} \Rightarrow \mathbf{a} = \mathbf{0} .$$

From above and definition 330 (p. 434), point 5 (p. 435) of A.1.4 (p. 435), we have:

$$\begin{aligned}&((\boldsymbol{\mu} \neq \mathbf{0}, \boldsymbol{\mu}\mathbf{a} = \mathbf{0}) \Rightarrow \mathbf{a} = \mathbf{0})\\ &\Leftrightarrow (\neg\neg\mathbf{a} = \mathbf{0} \vee \neg(\boldsymbol{\mu} \neq \mathbf{0}, \boldsymbol{\mu}\mathbf{a} = \mathbf{0})) = (\mathbf{a} \neq \mathbf{0} \Rightarrow \neg(\boldsymbol{\mu} \neq \mathbf{0}, \boldsymbol{\mu}\mathbf{a} = \mathbf{0}))\end{aligned}$$

From section A.1.2 (p. 434), $\mathbf{a} \neq \mathbf{0} \Rightarrow \neg(\boldsymbol{\mu} \neq \mathbf{0}, \boldsymbol{\mu}\mathbf{a} = \mathbf{0})$ is true. From section A.1.4 (p. 435) point 7 (p. 435), 17 (p. 436), 23 (p. 436), 13 (p. 435), we have:

$$\begin{aligned}&(\boldsymbol{\mu}\mathbf{a} = \mathbf{0}, \mathbf{a} \neq \mathbf{0}) \Rightarrow (\boldsymbol{\mu}\mathbf{a} = \mathbf{0}, \neg(\boldsymbol{\mu} \neq \mathbf{0}, \boldsymbol{\mu}\mathbf{a} = \mathbf{0})) = (\boldsymbol{\mu}\mathbf{a} = \mathbf{0}, (\boldsymbol{\mu} = \mathbf{0} \vee \boldsymbol{\mu}\mathbf{a} \neq \mathbf{0}))\\ &= (\boldsymbol{\mu}\mathbf{a} = \mathbf{0}, (\boldsymbol{\mu}\mathbf{a} = \mathbf{0} \Rightarrow \boldsymbol{\mu} = \mathbf{0})) \Rightarrow \boldsymbol{\mu} = \mathbf{0} .\end{aligned}$$

**Definition 200** $\sum_{\ell\in I} v_\ell$.

Given a vectorial space **V**, a finite set **I**: $\mathbf{Card}(\mathbf{I}) = \mathbf{n} \in \mathcal{Z}^+$, a map $(\mathbf{v_i} : \mathbf{I} \to \mathbf{V})$, using definition 178 (p. 212), one can define $\sum_{\ell\in I} v_\ell$.

**Definition 201 Sub-vector space.**

Given a vector space **E**, a subset $\mathbf{F} \subset \mathbf{E}$ is said to be a sub-vectorial space of **E**, if it is a vectorial space according to the same internal combination law which make **E** a vectorial space:

$$((\lambda, \mu) \in \mathcal{RC}^2, a \in F, b \in F) \Rightarrow \lambda a + \mu b \in F\ .$$

**Definition 202 Direct sum of vectorial spaces $\oplus$.**

Given a set **E**, a set of index **I** and a map $(\mathbf{V_i} : \mathbf{I} \to \mathcal{P}(\mathbf{E}))$ (definition 350 (p. 453)) such that for all $\mathbf{i} \in \mathbf{I}$, $\mathbf{V_i}$ is a vectorial space on the same field $\mathcal{RC}$. One calls $\bigoplus_{i\in I} V_i \subset E$ the following product of set (see definition 372 (p. 473)):

$$\bigoplus_{i\in I} V_i = \prod_{i\in I} V_i$$

an element of $v = \prod^{e}_{i\in I} v_i \in \bigoplus_{i\in I} V_i$ is written as $v = \sum_{i\in I} v_i \in \bigoplus_{i\in I} V_i$.
The vectorial internal composition laws are defined by:

$$(\lambda \in \mathcal{RC}, v = \sum_{i\in I} v_i \in \bigoplus_{i\in I} V_i) \Rightarrow \lambda v = \sum_{i\in I} \lambda v_i$$

$$(u = \sum_{i\in I} u_i \in \bigoplus_{i\in I} V_i, v = \sum_{i\in I} v_i \in \bigoplus_{i\in I} V_i) \Rightarrow u + v = \sum_{i\in I} u_i + v_i$$

those laws clearly verify definition 199 (p. 235).

**Proposition 399** Given a set **E** and a vector space **V**, the set of maps $\mathcal{F}$ of **E** to **V** is a vector space.

**Proof**

Since any map of **E** to **V** is an element of $\mathbf{E} \times \mathbf{V}$, those maps form a set (proposition 749 (p. 450)). Let us call it $\mathcal{F}$. We have an Abelian group $\mathcal{F} \times \mathcal{F} \to \mathcal{F}$ which is given by:

$$(f \in \mathcal{F}, g \in \mathcal{F}, a \in E) \Rightarrow ((f + g)(a) = f(a) + g(a) \in V)\ .$$

We also have:

$$(\lambda \in \mathcal{RC}, f \in \mathcal{F}, a \in E) \Rightarrow ((\lambda f)(a) = \lambda f(a) \in V)\ .$$

From the properties of **V**, we have $\mathbf{f+g} \in \mathcal{F}$ and $\lambda \mathbf{f} \in \mathcal{F}$. The neutral element $(\mathbf{0} : \mathbf{E} \to \mathbf{V})$ of the Abelian group is given by $(\forall a \in E)(0(a) = 0 \in V)$. **0** is a map $\mathbf{E} \to \mathbf{V}$ then it is in $\mathcal{F}$. All the axioms of a vector space are clearly verified.

**Definition 203 Norm on a vector space.**

The norm on a vector space **V** is a map $(\|x\| : V \to \mathcal{R}^{+0})$ and a map of $(\|\lambda\| : \mathcal{RC} \to \mathcal{R}^{+0})$ with the following properties:

1. $||a|| \geq 0$

2. $||a|| = 0 \Leftrightarrow a = e = 0_V$

3. $||\lambda a|| = ||\lambda|| ||a||$

4. $||a + b|| \leq ||a|| + ||b||$

**Proposition 400** Given a normed vectorial space $\mathbf{V}$, a finite set $\mathbf{I}$: $\mathbf{Card(I)} = n \in \mathcal{Z}^+$, a map $(v_i : I \to V)$ using definition 200 (p. 237) we have:

$$||\sum_{i \in I} v_i|| \leq \sum_{i \in I} ||v_i||$$

**Proof**
$n > 0 \Rightarrow (\exists q \in I), \mathbf{Card}(I - \{q\}) = n - 1$. Then if proposition 400 is true for $n - 1$, it is true for $n$ because using properties of definition 200 (p. 237) we have:

$$||\sum_{i \in I} v_i|| = ||\sum_{i \in I - \{q\}} v_i + v_q|| \leq ||\sum_{i \in I - \{q\}} v_i|| + ||v_q|| \leq \sum_{i \in I - \{q\}} ||v_i|| + ||v_q||$$
$$= \sum_{i \in I} ||v_i|| .$$

**Proposition 401** A vector space with a norm is a Euclidean metric space (definition 147 (p. 176)).

**Proof**
Given a norm on a vector space $\mathbf{V}$, we have the map $||a - b|| : V \times V \to \mathcal{R}$ which defines a Euclidean metric $d(a, b) = ||a - b||$. This map gives:

1. $a \neq b \Rightarrow d(a, b) > 0$

2. $d(a, b) = 0 \Leftrightarrow a - b = e = 0_V \Leftrightarrow a - b + b = e + b \Leftrightarrow a = b$

3. $d(a, b) = d(b, a)$

4. $d(a, b) = ||a - b|| = ||a - z - (b - z)|| \leq ||a - z|| + ||b - z|| = d(a, z) + d(z, b)$.

(To avoid confusion between various $\mathbf{0}$, we have called here $\mathbf{0}$ the neutral element of the addition group of $\mathcal{RC}$ and $e = 0_V$ the neutral element of $\mathbf{V}$. $\mathbf{1}$ staying the neutral element for the multiplication group of $\mathcal{RC}$. When there is no confusion possible we shall call also $\mathbf{0}$ the neutral element or the zero vector of $\mathbf{V}$. )

**Proposition 402** A vector space $\mathbf{V}$, with a norm on the field $\mathcal{RC}$ and with vectors different from $0 = e$, defines a Euclidean metric on $\mathcal{RC}$.

**Proof**
By applying proposition 401 (p. 238) with $a = \lambda c$, $b = \mu c$ $c \in V$, $c \neq 0$, $\lambda \in \mathcal{RC}$, $\mu \in \mathcal{RC}$ and dividing the relations 1 to 4 by $||c||$, we get also the definition of a Euclidean metric: 147 (p. 176).

**Proposition 403** Given a normed vector space $\mathbf{V}$ on a field $\mathcal{RC}$, $\mathbf{V}$ and $\mathcal{RC}$ are both separated topological spaces with the open ball topology (definition 151 (p. 177)).

**Proof**
It is a direct consequence of propositions 263 (p. 177), 401 (p. 238), 402 (p. 238).

**Proposition 404** $\mathbf{a}, \mathbf{b}$ being two vectors of a normed vectorial space, we have:

$$\|\mathbf{a}+\mathbf{b}\| \geq \|\mathbf{a}\| - \|\mathbf{b}\| .$$

**Proof**

$$(\|\mathbf{a}\| = \|\mathbf{a}+\mathbf{b}-\mathbf{b}\| \leq \|\mathbf{a}+\mathbf{b}\| + \|\mathbf{b}\|) \Rightarrow \|\mathbf{a}\| - \|\mathbf{b}\| \leq \|\mathbf{a}+\mathbf{b}\| .$$

**Definition 204 Linear combination of vectors.**

Given a vectorial space $\mathbf{V}$, a finite subspace $\mathbf{E}$ of $\mathbf{V}$ with its bijective map $(\mathbf{e_i} : [\mathbf{1}, \mathbf{n}] \rightarrow \mathbf{E})$ (see definition A.5.10 (p. 499)) where $\mathbf{n} = \mathbf{Card(E)} \in (\mathbf{Z})^+$, one says that a vector $\mathbf{x} \in \mathbf{V}$ is a linear combination of vectors of $\mathbf{E}$ if there is a map $(\mathbf{x^i} : [\mathbf{1}, \mathbf{n}] \rightarrow \mathcal{RC})$ such that[20]:

$$\mathbf{x} = \mathbf{x^i e_i} = \sum_{\mathbf{i} \in [\mathbf{1},\mathbf{n}]} \mathbf{x^i e_i} .$$

**Definition 205 Bases of a vectorial space.**

A finite subset $\mathbf{E}$ of a vectorial $\mathbf{V}$ is called a basis of $\mathbf{V}$, if any vector $\mathbf{x}$ of $\mathbf{V}$ is a linear combination of vectors of $\mathbf{E}$ (see definition 204).

**Definition 206 Independent vectors set.**

A subset $\mathbf{U}$ of a vector space $\mathbf{V}$ is said to be a set of independent vectors if:

1. their cardinal $\mathbf{n}$ is integer,
2. none is the zero vector,
3. calling $\mathbf{u_i}$, a bijection $(\mathbf{u_i} : [\mathbf{1}, \mathbf{n}] \rightarrow \mathbf{U})$ (see definition 390 (p. 499)), if a map $(\boldsymbol{\lambda}^\mathbf{i} : [\mathbf{1}, \mathbf{n}] \rightarrow \mathcal{RC})$ is such that $\boldsymbol{\lambda}^\mathbf{i}\mathbf{u_i} = \mathbf{0}$ we have $(\forall \mathbf{i} \in [\mathbf{1}, \mathbf{n}])(\boldsymbol{\lambda}^\mathbf{i} = \mathbf{0})$.

**Proposition 405** Any single vector of a vectorial space, different from the null vector makes an independent set of vector.

**Proof**
It is a direct consequence of proposition 396 (p. 236).

**Proposition 406** If a finite set $\mathbf{U}$ of $\mathbf{n}$ vectors of a vectorial space $\mathbf{V}$ is not a set of independent vectors, one of them $\mathbf{u_i} \in \mathbf{U}$ is a linear combination of the vectors of $\mathcal{C}_\mathbf{U}(\{\mathbf{u_i}\})$.

**Proof**
From definition 390 (p. 499), there is a bijection (definition 124 (p. 139) $(\mathbf{u_i} : [\mathbf{1}, \mathbf{n}] \leftrightarrow \mathbf{U})$. From definition 206 (p. 239):

$$(\exists(\boldsymbol{\lambda}_\mathbf{j} : [\mathbf{1}, \mathbf{n}] \rightarrow \mathcal{RC})(\exists \mathbf{i} \in [\mathbf{1}, \mathbf{n}])(\boldsymbol{\lambda}_\mathbf{i} = \boldsymbol{\lambda} \neq \mathbf{0}, \sum_{\mathbf{j} \in [\mathbf{1},\mathbf{n}]} \boldsymbol{\lambda}^\mathbf{j}\mathbf{u_j} = \mathbf{0}) .$$

Then $\mathbf{u_i} = \sum_{\mathbf{j} \in [\mathbf{1},\mathbf{n}] - \mathbf{i}} -\frac{\boldsymbol{\lambda}^\mathbf{j}}{\boldsymbol{\lambda}}\mathbf{u_j}$ since $\mathbf{u_i} \neq \mathbf{0}$ there is, from proposition 396 (p. 236), a $\mathbf{k} \in [\mathbf{1}, \mathbf{n}] - \mathbf{i}$ with $\boldsymbol{\lambda}^\mathbf{k} \neq \mathbf{0}$. Then we have proposition 406 true.

[20] We shall use very often the Einstein notation to simplify formulae: When a not defined index is clearly in a set and is appearing several times in a formula (and most of the times up and down), it means that there is an implicit summation of that index over its set.

**Proposition 407** Given an independent subset **U** of a vector space **V**, to any vector of **U** we can add a linear combination of all the others. This new vector combined with the unchanged vectors of **U** is a new independent subset of vectors of **V**.

**Proof**
Taking **u**, a vector of **U** and a map

$$(\mu^{\mathrm{v}} : \mathrm{U} - \{\mathrm{u}\} \rightarrow \mathcal{RC})$$

we set

$$\lambda \in \mathcal{RC}, (\lambda^{\mathrm{w}} : \mathrm{U} - \{\mathrm{u}\} \rightarrow \mathcal{RC}) .$$

If we have:

$$\lambda(\mathrm{u} + \sum_{\mathrm{v} \in \mathrm{U}-\{\mathrm{u}\}} \mu^{\mathrm{v}}\mathrm{v}) + \sum_{\mathrm{v} \in \mathrm{U}-\{\mathrm{u}\}} \lambda^{\mathrm{v}}\mathrm{v} = 0 .$$

That expression may be written as:

$$\lambda \mathrm{u} + \sum_{\mathrm{v} \in \mathrm{U}-\{\mathrm{u}\}} \lambda'^{\mathrm{v}}\mathrm{v} = 0, \mathrm{v} \neq \mathrm{u} \Rightarrow \lambda'^{\mathrm{v}} = \lambda^{\mathrm{v}} + \lambda\mu^{\mathrm{v}} .$$

From definition 206 (p. 239), it gives

$$\lambda = 0, (\forall \mathrm{v} \in \mathrm{U} - \{\mathrm{u}\})(\lambda'^{\mathrm{v}} = 0) .$$

Then:

$$(\forall \mathrm{v} \in \mathrm{U} - \{\mathrm{u}\})(\lambda'^{\mathrm{v}} = \lambda^{\mathrm{v}} + \lambda\mu^{\mathrm{v}} = \lambda^{\mathrm{v}} = 0) .$$

Then proposition 407 is true.

**Definition 207 Vector space with finite dimension.**

A vector space **V** is said to be of finite dimension, if there is an integer **N** (definition 397 (p. 516)) such that there is no subset of independent vectors in **V** with a cardinal (definition 390 (p. 499)) bigger than **N**.

**Proposition 408** Having a finite dimension vectorial space **V**, there is a subset **b** of **V** with a cardinal (definition 390 (p. 499)) $\mathbf{n} = \mathbf{Card(b)} \in \mathcal{Z}^+$ (definitions 397 (p. 516), 401 (p. 520)) which is bigger than or equal to the cardinal of any independent subset of **V**.

**Proof**
Any basis of **V** belongs to $\mathcal{P}(\mathbf{V})$ which is a set (definition 350 (p. 453)). Then they form a set (proposition 749 (p. 450)) **B**. The cardinals of any of element of **B** form a set (proposition 884 (p. 503)) **Bc**. This set of cardinals has a majoring integer from definition 207 (p. 240). From proposition 918 (p. 516), it is a set of integers. From proposition 924 (p. 519), it has a biggest element **n**. From our definitions and proposition 885 (p. 504), $\mathbf{x} = \mathbf{card(s)}$ is a map $\mathbf{B} \rightarrow \mathbf{B_c}$. From definition 357 (p. 460), there is an element **b** of **B** which has cardinal **n**.

**Definition 208 Dimension of a vectorial space with finite dimension.**

One calls dimension of a vectorial space with finite dimension, the maximal value of the cardinal of a set of independent vectors (proposition 408 (p. 240)).

**Definition 209 Dimension of a subset of a vectorial space with finite dimension.**

Given a subspace **S** of a vectorial space with finite dimension **V**, one calls the dimension of **S** the maximum value of the cardinal of an independent set of vectors belonging to **S**.
Remark: **S** does not need to be a vectorial space in that definition.

**Proposition 409** Given a finite dimension vectorial space $\mathbf{V}$ of dimension $\mathbf{n} \in \mathcal{Z}^+$ and a set $\mathbf{E}$ of $\mathbf{n}$ independent vectors $(\mathbf{e_i} : [1, \mathbf{n}] \to \mathbf{V})$, for any vector $\mathbf{x} \in \mathbf{V}$ there is a set of $\mathbf{n}$ numbers from $\mathcal{RC}$: $(\mathbf{x^i} : [1, \mathbf{n}] \to \mathcal{RC})$ such that $\mathbf{x} = \sum_{\mathbf{i}\in[1,\mathbf{n}]} \mathbf{x^i e_i} = \mathbf{x^i e_i}$. This set of numbers is unique.

**Proof**
If the space $\mathbf{V}$ is of dimension $\mathbf{n}$, there is a set $\mathbf{E}$ of $\mathbf{n}$ vectors $\mathbf{e_i}$ $(\mathbf{e_i} : [1, \mathbf{n}] \to \mathbf{V})$ independent, any vector $\mathbf{x} \in \mathbf{V}$ added to the set $\mathbf{E}$ makes it a dependent set of $\mathbf{n+1}$ vectors so there is a set of $\mathbf{n+1}$ numbers $(\lambda^\mathbf{i} : [1, \mathbf{n+1}] \to \mathcal{RC})$ not all zero such that

$$\lambda^{\mathbf{n+1}}\mathbf{x} + \sum_{\mathbf{i}\in[1,\mathbf{n}]} \lambda^\mathbf{i}\mathbf{e_i} = 0$$

$\lambda_{\mathbf{n+1}} \neq 0$ otherwise the $\mathbf{e_i}$ would be dependent. Then by setting $\mathbf{x^i} = -\lambda^\mathbf{i}/\lambda^{\mathbf{n+1}}$ we have:

$$\mathbf{x} = \sum_{\mathbf{i}\in[1,\mathbf{n}]} \mathbf{x^i e_i} = \mathbf{x^i e_i}\ .$$

It is unique, otherwise if there were another set of numbers such that $\mathbf{x} = \mathbf{x^{i\prime} e_i}$, we would have $(\mathbf{x^{i\prime}} - \mathbf{x^i})\mathbf{e_i} = 0$ which implies since $\mathbf{E}$ is an independent set of vectors:

$$(\forall \mathbf{i} \in [1, \mathbf{n}])(\mathbf{x^{i\prime}} - \mathbf{x^i} = 0, \mathbf{x^{i\prime}} = \mathbf{x^i})\ .$$

**Proposition 410** In a vectorial space of finite dimension $\mathbf{n}$, any set of independent vectors with cardinal $\mathbf{n}$ is a basis of the vectorial space.

**Proof**
It is a direct consequence of proposition 409 and definition 205 (p. 239).

**Definition 210 Independent bases of a finite vector space.**

One calls independent basis of a finite vector space, an independent set of vectors the cardinal of which is maximal (definition 383 (p. 485)) as defined in proposition 408 (p. 240).

**Proposition 411** If in a vectorial space $\mathbf{V}$, a finite set $\mathbf{E}$ of $\mathbf{n}$ independent vectors is a linear combination of another finite set $\mathbf{F}$ of $\mathbf{n}$ vectors, then any vector of $\mathbf{F}$ is a linear combination of the vectors of $\mathbf{E}$ and $\mathbf{F}$ is an independent set of $\mathbf{n}$ vectors.

**Proof**
Let us call $\mathbf{e_1, ..., e_n}$ $(\mathbf{e_i} : [1, \mathbf{n}] \leftrightarrow \mathbf{E})$ the vectors of $\mathbf{E}$ and $\mathbf{f_1, ..., f_n}$ $(\mathbf{f_i} : [1, \mathbf{n}] \leftrightarrow \mathbf{F})$ the vectors of $\mathbf{F}$ and let us suppose that we have a map $(\lambda_\mathbf{i}^\mathbf{j} : [1, \mathbf{n}] \times [1, \mathbf{n}] \to \mathcal{RC})$ such that: $\mathbf{e_i} = \lambda_\mathbf{i}^\mathbf{j}\mathbf{f_j}$. Then, if proposition 411 is true, we should have a map $(\mu_\mathbf{i}^\mathbf{j} : [1, \mathbf{n}] \times [1, \mathbf{n}] \to \mathcal{RC})$ such that $\mathbf{f_i} = \mu_\mathbf{i}^\mathbf{j}\mathbf{e_j}$. It is true for $\mathbf{n} = 1$ with $\mu_1^1 = 1/\lambda_1^1$ since $\lambda_1^1$ cannot be zero otherwise $\mathbf{e_1} = 0$ which is forbidden. Let us prove now that, if it is true for $\mathbf{n}$, it is true for $\mathbf{n+1}$ and we conclude the first part of 411 by proposition 927 (p. 520).
We have $\mathbf{e_{n+1}} = \lambda_{\mathbf{n+1}}^\mathbf{j}\mathbf{f_j}$, $(\exists \mathbf{k} \in [1, \mathbf{n+1}])(\lambda_{\mathbf{n+1}}^\mathbf{k} \neq 0)$ otherwise $\mathbf{e_{n+1}} = 0$. Then

$$(\exists \mathbf{k})(\mathbf{f_k} = \frac{\mathbf{e_{n+1}}}{\lambda_{\mathbf{n+1}}^\mathbf{k}} - \sum_{\mathbf{i}\in[1,\mathbf{n+1}],\mathbf{i}\neq\mathbf{k}} \frac{\lambda_{\mathbf{n+1}}^\mathbf{i}}{\lambda_{\mathbf{n+1}}^\mathbf{k}}\mathbf{f_i})\ .$$

To simplify, let us suppose that $\mathbf{k} = \mathbf{n}+\mathbf{1}$ [21] and we replace $\mathbf{f_{n+1}}$ by its new value in any equation $\mathbf{e_i} = \lambda_i^j \mathbf{f_j}$ and we sum the vector with the same $\mathbf{f_i}$. We get a map $(\lambda'^j_i : [1, n]^2 \to \mathcal{RC})$ such as:

$$e_i = \frac{\lambda_i^{n+1}}{\lambda_{n+1}^{n+1}} e_{n+1} + \lambda'^j_i f_j$$

with $\mathbf{i} \in [\mathbf{1}, \mathbf{n}], \mathbf{j} \in [\mathbf{1}, \mathbf{n}]$.
Since the vectors of $\mathbf{E}$ are independent, the vectors $e_i - \frac{\lambda_i^{n+1}}{\lambda_{n+1}^{n+1}} e_{n+1}$ are also independent:

$$\sum_{i \in [1,n]} \lambda'_i (e_i - \frac{\lambda_i^{n+1}}{\lambda_{n+1}^{n+1}} e_{n+1}) = 0 \Rightarrow ((\forall i \in [1,n])(\lambda'_i = 0), \sum_{i \in [1,n]} \lambda'_i \frac{\lambda_i^{n+1}}{\lambda_{n+1}^{n+1}} = 0)$$

Then from what we have assumed, we have $(\mu'^j_i : [1, n]^2 \to \mathcal{RC})$ such that:

$$f_i = \mu'^j_i (e_j - \frac{\lambda_i^{n+1}}{\lambda_{n+1}^{n+1}} e_{n+1})$$

and

$$f_{n+1} = \frac{e_{n+1}}{\lambda_{n+1}^{n+1}} - \frac{\lambda_{n+1}^i}{\lambda_{n+1}^{n+1}} \mu'^j_i (e_j - \lambda_j^{n+1} e_{n+1})$$

and the first part of proposition 411 is true for $\mathbf{n}+\mathbf{1}$ if true for $\mathbf{n}$. It is then true for any $\mathbf{n}$.

For the second part of 411, let us suppose $\mathbf{Card(F)} = \mathbf{Card(E)} = \mathbf{n}$. The vectors of $\mathbf{F}$ have to be independent otherwise one of them, let us say $\mathbf{f_n}$, would be a linear combination of the others. Using that linear combination, the vectors of $\mathbf{E}$ would then be a linear combination of the $\mathbf{n}-\mathbf{1}$ first vectors of $\mathbf{F}$ and applying first part of proposition 411, those $\mathbf{n}-\mathbf{1}$ vectors of $\mathbf{F}$ would be a linear combination of the $\mathbf{n}-\mathbf{1}$ first vectors of $\mathbf{E}$ and the last one $\mathbf{e_n}$ through those linear combinations would be a linear combination of the $\mathbf{n}-\mathbf{1}$ first. It would mean that the vectors of $\mathbf{E}$ are dependent which is forbidden.

**Proposition 412** If in a vectorial space $\mathbf{V}$, a finite set $\mathbf{E}$ of $\mathbf{n}$ independent vectors is a linear combination of another finite set $\mathbf{F}$ of $\mathbf{p}$ vectors, then we have $\mathbf{p} \geq \mathbf{n}$. If $\mathbf{p} = \mathbf{n}$ the set $\mathbf{F}$ is a set of independent vector.

**Proof**
If $\mathbf{p} < \mathbf{n}$, since $\mathbf{Card(E)} = \mathbf{n}$, from definition 390 (p. 499), there is a bijection $(\mathbf{e_i} : [\mathbf{1}, \mathbf{n}] \leftrightarrow \mathbf{E})$ which induced another bijection $(\mathbf{e_i} : [\mathbf{1}, \mathbf{p}] \leftrightarrow \mathbf{e_{[1,p]}})$ (proposition 183 (p. 143)). The sets $[\mathbf{1}, \mathbf{p}]$ and $\mathbf{e_{[1,p]}}$ have $\mathbf{p}$ as cardinal (definition 390 (p. 499), proposition 934 (p. 523)). The set $\mathbf{e_{[1,p]}}$ is an independent set. From proposition 411 (p. 241), the vectors of $\mathbf{F}$ are linear combination of the vectors $\mathbf{e_{[1,p]}}$, $\mathbf{e_{p+1}}$ is a linear combination of vectors of $\mathbf{F}$. Then $\mathbf{e_{p+1}}$ is a linear combination of the vectors $\mathbf{e_{[1,p]}}$. The set $\mathbf{e_{[1,p+1]}}$ is not independent. From A.1.4 point 8 (p. 435), we have $\mathbf{p} \geq \mathbf{n}$.
If $\mathbf{p} = \mathbf{n}$, $\mathbf{F}$ is an independent set of vectors otherwise from proposition 406 (p. 239), one of them will be a linear combination of the $\mathbf{n}-\mathbf{1}$ remaining of $\mathbf{F}$ and $\mathbf{E}$ would be a linear combination of $\mathbf{n}-\mathbf{1}$ vector of $\mathbf{F}$ which is forbidden by the first part of this proposition.

**Definition 211 Line and plane in a vectorial space.**

We are given a vectorial space $\mathbf{V}$ built with the field $\mathcal{RC}$.

[21] We can always manage to do so when choosing the map $(\mathbf{f_i} : [\mathbf{1}, \mathbf{n}+\mathbf{1}] \to \mathbf{F})$ by replacing it by $\mathbf{f} \circ \mathbf{T_{k,n+1}}$ which is fully eligible for labeling the vectors of $\mathbf{F}$.

1. Taking two vectors $\mathbf{p}, \mathbf{u}$ of $\mathbf{V}$, one calls line $\mathcal{L}(\mathbf{p}, \mathbf{u})$ of direction $\mathbf{u}$ going through $\mathbf{p}$, the following subset of $\mathbf{V}$:

$$\mathbf{x} \in \mathcal{L}(\mathbf{p}, \mathbf{u}) \Leftrightarrow (\mathbf{p} \in \mathbf{V}, \mathbf{u} \in \mathbf{V}, (\exists \lambda \in \mathcal{RC})(\mathbf{x} = \mathbf{p} + \lambda \mathbf{u}))$$

A line can also be defined by two points $\mathbf{p} \in \mathbf{V}, \mathbf{q} \in \mathbf{V}$ by setting $\mathbf{u} = \mathbf{q} - \mathbf{p}$.

2. We have three vectors $\mathbf{p}, \mathbf{u}, \mathbf{v}$ of $\mathbf{V}$, the last two being independent vectors (see definition 206 (p. 239)). One calls plane $\Pi(\mathbf{p}, \mathbf{u}, \mathbf{v})$ along the directions $\mathbf{u}, \mathbf{v}$ going through $\mathbf{p}$, the following subset of $\mathbf{V}$:

$$\begin{aligned}&\mathbf{x} \in \Pi(\mathbf{p}, \mathbf{u}, \mathbf{v}) \Leftrightarrow (\mathbf{p} \in \mathbf{V},\ \mathbf{u} \in \mathbf{V}, \mathbf{v} \in \mathbf{V}, ((\lambda \in \mathcal{RC},\ \mu \in \mathcal{RC},\\ &\lambda \mathbf{u} + \mu \mathbf{v} = 0) \Rightarrow \lambda = \mu = 0), (\exists \lambda \in \mathcal{RC})(\exists \mu \in \mathcal{RC})(\mathbf{x} = \mathbf{p} + \lambda \mathbf{u} + \mu \mathbf{v}))\ .\end{aligned}$$

A plane can also be defined by three points $\mathbf{p} \in \mathbf{V},\ \mathbf{q} \in \mathbf{V},\ \mathbf{r} \in \mathbf{V}$ by setting $\mathbf{u} = \mathbf{q} - \mathbf{p}, \mathbf{v} = \mathbf{r} - \mathbf{p}$.

**Proposition 413** Having a line $\mathcal{L}(\mathbf{p}, \mathbf{u})$ (versus a plane $\Pi(\mathbf{p}, \mathbf{u}, \mathbf{v})$) (definition 211 (p. 242)), changing the point in the first argument of the definition by another one belonging to the line (versus plane) and replacing the next vector argument (versus two vector arguments) by a product of an element of $\mathcal{RC}$ by that vector (versus by a linear independent combination of those two vectors) one gets the same line (versus plane).

**Proof**

1. If $\mathbf{q} \in \mathcal{L}(\mathbf{p}, \mathbf{u})$, $\mu \neq 0$, we have:

$$\mathbf{q} = \mathbf{p} + \lambda_{\mathbf{q}} \mathbf{u} = \mathbf{p} + (\lambda_{\mathbf{q}}/\mu)\mu \mathbf{u} = \mathbf{p} + \mu_{\mathbf{q}} \mu \mathbf{u}\ .$$

Then:

$$\begin{aligned}&\mathbf{x} \in \mathcal{L}(\mathbf{p}, \mathbf{u}) \Leftrightarrow (\exists \lambda_{\mathbf{x}} \in \mathcal{RC})(\mathbf{x} = \mathbf{p} + \lambda_{\mathbf{x}} \mathbf{u})\\ &\Leftrightarrow (\exists \mu_{\mathbf{x}} = \lambda_{\mathbf{x}}/\mu - \mu_{\mathbf{q}} \in \mathcal{RC})(\mathbf{x} = \mathbf{q} + \mathbf{p} - \mathbf{q} + \mu \lambda_{\mathbf{x}}/\mu \mathbf{u} = \mathbf{q} + \mu_{\mathbf{x}}(\mu \mathbf{u}))\\ &\Leftrightarrow \mathbf{x} \in \mathcal{L}(\mathbf{q}, \mu \mathbf{u})\end{aligned}$$

2. If $\mathbf{q} \in \Pi(\mathbf{p}, \mathbf{u}, \mathbf{v})$ and if $\mathbf{u}'$ and $\mathbf{v}'$ are independent vector linear combination of $\mathbf{u}$ and $\mathbf{v}$ from proposition 411 (p. 241), we have $\mathbf{u} = \lambda_{\mathbf{u}} \mathbf{u}' + \mu_{\mathbf{u}} \mathbf{v}',\ \mathbf{v} = \lambda_{\mathbf{v}} \mathbf{u}' + \mu_{\mathbf{v}} \mathbf{v}'$. Then from definition 211 (p. 242), $\mathbf{q} = \mathbf{p} + \alpha_{\mathbf{q}} \mathbf{u} + \beta_{\mathbf{q}} \mathbf{v}$ and

$$\begin{aligned}&\mathbf{x} \in \Pi(\mathbf{p}, \mathbf{u}, \mathbf{v}) \Leftrightarrow (\exists \alpha_{\mathbf{x}} \in \mathcal{RC})(\exists \beta_{\mathbf{x}} \in \mathcal{RC})\\ &(\mathbf{x} = \mathbf{p} + \alpha_{\mathbf{x}} \mathbf{u} + \beta_{\mathbf{x}} \mathbf{v} = \mathbf{q} + (\alpha_{\mathbf{x}} - \alpha_{\mathbf{q}})\mathbf{u} + (\beta_{\mathbf{x}} - \beta_{\mathbf{q}})\mathbf{v}\\ &= \mathbf{q} + (\alpha_{\mathbf{x}} - \alpha_{\mathbf{q}})(\lambda_{\mathbf{u}} \mathbf{u}' + \mu_{\mathbf{u}} \mathbf{v}') + (\beta_{\mathbf{x}} - \beta_{\mathbf{q}})(\lambda_{\mathbf{v}} \mathbf{u}' + \mu_{\mathbf{v}} \mathbf{v}')\\ &= \mathbf{q} + ((\alpha_{\mathbf{x}} - \alpha_{\mathbf{q}})\lambda_{\mathbf{u}} + (\beta_{\mathbf{x}} - \beta_{\mathbf{q}})\lambda_{\mathbf{v}})\mathbf{u}' + ((\alpha_{\mathbf{x}} - \alpha_{\mathbf{q}})\mu_{\mathbf{u}} + (\beta_{\mathbf{x}} - \beta_{\mathbf{q}})\mu_{\mathbf{v}})\mathbf{v}')\\ &\Leftrightarrow \mathbf{x} \in \Pi(\mathbf{q}, \mathbf{u}', \mathbf{v}')\ .\end{aligned}$$

**Proposition 414** If a vectorial space $\mathbf{V}$ has a finite basis $\mathbf{E}$ composed of independent vectors, any set $\mathbf{F}$ of independent vectors has $\mathbf{Card(F)} \leq \mathbf{n} = \mathbf{Card(E)}$ and the dimension of the vectorial space $\mathbf{V}$ is $\mathbf{n} = \mathbf{Card(E)}$.

**Proof**
If a set $\mathbf{F} \subset \mathbf{V}$ of independent vectors has a cardinal bigger than $\mathbf{n}$, one can extract from $\mathbf{F}$ a subset $\mathbf{F}'$ of $\mathbf{n} = \mathbf{Card(E)}$ independent vectors. They are linear combinations of the vectors of $\mathbf{E}$ as any vectors of $\mathbf{V}$. Then from proposition 411, the vectors of $\mathbf{E}$ are also linear combinations of the vectors of $\mathbf{F}'$. Then all vectors of $\mathbf{V}$ are linear combinations of vectors of $\mathbf{F}'$ through their linear combination of vectors of $\mathbf{E}$. Then any vector of $\mathcal{C}_{\mathbf{F}}(\mathbf{F}')$ is a linear combination of vectors of $\mathbf{F}'$. If $\mathcal{C}_{\mathbf{F}}(\mathbf{F}') \neq \emptyset$, $\mathbf{F}$ would not be made of independent vectors. Then $\mathcal{C}_{\mathbf{F}}(\mathbf{F}') = \emptyset$ and the dimension of $\mathbf{V}$ is $\mathbf{Card(E)}$.

**Proposition 415** If a vectorial space $\mathbf{V}$ has a finite basis $\mathbf{E}$ composed of independent vectors, the dimension of the vectorial space $\mathbf{V}$ is $\mathbf{n} = \mathbf{Card(E)}$. Any set $\mathbf{F}$ of independent vectors of $\mathbf{V}$ can be completed up to $\mathbf{n}$ vectors of $\mathbf{E}$ such that to be a basis of $\mathbf{V}$ with $\mathbf{n}$ vectors.

**Proof**
Keeping the above notations, if one could not add to a finite set $\mathbf{F} \subset \mathbf{V}$ of independent vectors a new independent vector in $\mathbf{E}$, it would mean that any vector of $\mathbf{E}$ could be expressed as a linear combination of vector of $\mathbf{F}$ and $\mathbf{F}$ would be a basis of $\mathbf{V}$ and, from the above proposition 414, it should have as cardinal the dimension of the vectorial space. So if the cardinal of $\mathbf{F}$ is smaller than the dimension of $\mathbf{V}$, it can be completed by new vectors in order to make an independent basis of $\mathbf{V}$.

**Proposition 416** Given a vectorial space $\mathbf{V}$ of finite dimension $\mathbf{n} \in \mathcal{Z}^+$ and a set $\mathbf{E}$ of $\mathbf{n}$ independent vectors $(\mathbf{e_i} : [\mathbf{1}, \mathbf{n}] \rightarrow \mathbf{V})$, for any no zero vector $\mathbf{x} \in \mathbf{V}$, there is a subset $\mathbf{E}' \subset \mathbf{E}$ of cardinal $\mathbf{n} - \mathbf{1}$ such as $\{\mathbf{x}\} \cup \mathbf{E}'$ is a basis of $\mathbf{V}$.

**Proof**
Taking $\mathbf{F} = \{\mathbf{x}\}$, $\mathbf{F}$ is an independent set of vectors because $(\lambda \in \mathcal{RC}, \lambda\mathbf{x} = 0) \Rightarrow \lambda = 0$. Then proposition 415 (p. 244) gives proposition 416 (p. 244).

## 2.11.2 Linear forms

**Definition 212 Linear form or linear map on a vectorial space.**

Given a vectorial space $\mathbf{U}$, $\mathbf{U}$ is also a module on $\mathcal{RC}$. $\mathcal{RC}$ is also a commutative ring from definitions 182 (p. 218), 181 (p. 217)). $\mathcal{RC}$ can also be considered as a module $\mathbf{V}$ on itself (proposition 371 (p. 219)). With that we may use definition 186 (p. 220) and call a map $\mathbf{f}$ $(\mathbf{f} : \mathbf{U} \rightarrow \mathbf{V} = \mathcal{RC})$ a linear form or sometimes linear map if it fills the condition of 186 (p. 220). That is to say:

$$(\forall(\lambda, \mu) \in \mathcal{RC}^2)(\forall(\mathbf{u}, \mathbf{v}) \in \mathbf{U} \times \mathbf{U})(\mathbf{f}(\lambda\mathbf{u} + \mu\mathbf{v}) = \lambda\mathbf{f}(\mathbf{u}) + \mu\mathbf{f}(\mathbf{v}))\ .$$

If the vectorial space $\mathbf{U}$ is of finite dimension $\mathbf{n}$ with a basis $(\mathbf{e_i} : [\mathbf{1}, \mathbf{n}] \rightarrow \mathbf{U})$ and if we call $\mathbf{y_i} = \mathbf{f}(\mathbf{e_i})$, we have for a vector $\mathbf{v} = \mathbf{x^i e_i}$, $\mathbf{f}(\mathbf{v}) = \mathbf{y_i x^i}$.

**Definition 213 Dual space.**

It is the same definition as 187 (p. 221). Given a vectorial space $\mathbf{V}$, one calls dual space $\mathbf{V}^*$, the set of all the linear forms on $\mathbf{V}$ (definition 212 (p. 244)).

**Proposition 417** $\mathbf{V}^*$ is a vectorial space on $\mathcal{RC}$.

**Proof**
See also 378 (p. 221). Given:

$$\mathbf{v^1} \in \mathbf{V}^*,\ \mathbf{v^2} \in \mathbf{V}^*,\ \mathbf{u_1} \in \mathbf{V},\ \mathbf{u_2} \in \mathbf{V},\ \lambda_1 \in \mathcal{RC},\ \lambda_2 \in \mathcal{RC},\ \mu_1 \in \mathcal{RC},\ \mu_2 \in \mathcal{RC}$$

we may set by definition:

$$(\forall u \in V)((\{\lambda_1 v^1 + \lambda_2 v^2\}(u) = \lambda_1 v^1(u) + \lambda_2 v^2(u) : V \to \mathcal{RC}))$$

which is an internal law $V^* \times V^* \to V^*$ which verifies definition 199 (p. 235).

**Definition 214 Bases $e^j : [1, n] \to V^*$.**

$V$ is a vectorial space of finite dimension $n$ with an independent basis $(e_i : [1, n] \to V)$. $V^*$ being the dual of $V$. By definition, one calls $(e^j : [1, n] \to V^*)$, the basis of $V^*$ associated to the basis $(e_i : [1, n] \to V)$, the linear forms $e^j$ of $V^*$ such that for any vector $u = u^i e_i$ of $V$ we associate the number $u^i$. With that definition we have: $e^j(e_i) = \delta^i_j$. $\delta^i_j$ is the Kronecker symbol (see definition 215 (p. 245) and proposition 419 (p. 245)).

**Proposition 418** $V$ is a vectorial space of finite dimension $n$ with an independent basis $(e_i : [1, n] \to V)$. $V^*$ is its dual with its associated basis $(e^j : [1, n] \to V^*)$. Any $e^j$ defined by 214 (p. 245) is a linear map $V \to \mathcal{RC}$.

**Proof**

From proposition 409 (p. 241), for all $j \in [1, n]$, $e^j(u^i e_i) = uj$ is a map (see definition 112 (p. 131)) defined for all $u \in V$. Given:

$$u_1 = u^i_1 e_i \in V,\ u_2 = u^i_2 e_i \in V,\ \lambda_1 \in \mathcal{RC},\ \lambda_2 \in \mathcal{RC}$$

we have:

$$\begin{aligned}(\forall j \in [1, n])(e^j(\lambda_1 u_1 + \lambda_2 u_2) &= e^j(\lambda_1 u^i_1 e_i + \lambda_2 u^i_2 e_i) = e^j(\lambda_1 u^i_1 + \lambda_2 u^i_2) e_i \\ &= \lambda_1 u^j_1 + \lambda_2 u^j_2 = e^j(\lambda_1 u^i_1 e_i) + e^j(\lambda_2 u^i_2 e_i))\ .\end{aligned}$$

**Proposition 419** If the vectorial space $V$ is of finite dimension $n$ with an independent basis $(e_i : [1, n] \to V)$, the set of linear forms of $V^*$ given by the linear forms $e^i$ is an independent basis of $V^*$.

**Proof**

If for an element $f \in V^*$ we call $y_j = f(e_j)$, we have

$$(\forall u = u^i e_i \in V)(f(u) = u^i f(e_i) = u^i y_i = y_i e^i(u))\ .$$

The $e^i$ is an independent basis because $f = y_j e^j = 0 \Rightarrow (\forall k \in [1, n])(f(e_k) = y_k = 0)$.

**Proposition 420** Given $V$ a vectorial space of finite dimension $n$ with $(e_i : [1, n] \to V)$ as an independent basis, calling its dual, $V^*$ with $(e^i : [1, n] \to V)$ as an independent basis (definition 213 (p. 244)), the dual of $V^*$ is $V$ with $(e_i : [1, n] \to V)$ as an independent basis and we have (see definition 215 (p. 245)):

$$\begin{aligned}&(v = v^i e_i \in V = V^{**}, t = t_j e^j \in V^*) \\ &\Rightarrow t(v) = v(t) = t_j v^i e^j(e_i) = t_j v^i e_i(e^j) = t_j v^i \delta^i_j = t_i v^i\ .\end{aligned} \tag{2.87}$$

**Proof**

It is a direct consequence of definitions 213 (p. 244), 214 (p. 245) and propositions 417 (p. 244), 418 (p. 245), 419 (p. 245).

**Definition 215 Kronecker symbol $\delta^i_j = \delta_{ij} = \delta^{ij}$.**

The Kronecker symbol is a map $(\delta^i_j = \delta_{ij} = \delta^{ij} : \mathcal{Z} \times \mathcal{Z} \to (0, 1))$ such that:

$$i = j \Rightarrow \delta^i_j = \delta_{ij} = \delta^{ij} = 1, i \neq j \Rightarrow \delta^i_j = \delta_{ij} = \delta^{ij} = 0\ .$$

**Definition 216 Heaviside map $\mathcal{Y}(x)$.**

One calls Heaviside map, the map $(\mathcal{Y} : \mathcal{R} \to \{0, 1\})$ such that $x < 0 \Rightarrow \mathcal{Y}(x) = 0$ and $0 \leq x \Rightarrow \mathcal{Y}(x) = 1$ which is also a map $(\mathcal{Y} : \mathcal{Z} \to \{0, 1\})$ with the same definition.

### 2.11.3 Multilinear forms

**Definition 217 Multilinear forms on a vectorial space.**

Following definition 191 (p. 223), having a vectorial space $\mathbf{V}$, an integer $\mathbf{p} \in \mathcal{Z}^+$, one calls a $_{\mathbf{p}}$**linear_form** $_{\mathbf{p}\triangleleft}\mathbf{f}$, a $_{\mathbf{p}}$**linear_form** of definition 191 (p. 223). $\mathcal{RC}$ is considered as both a commutative ring and a module (proposition 371 (p. 219)) while $\mathbf{V}$ as a module definition 191 (p. 223):
a $_{\mathbf{p}}$**linear_form** $_{\mathbf{p}\triangleleft}\mathbf{f}$ is a map $(_{\mathbf{p}\triangleleft}\mathbf{f} : \mathbf{V}^{\mathbf{p}} \to \mathcal{RC})$ such that when restricted to $\prod_{1,p-1}\{x_i\} \times \mathbf{V}$, $(x_i : [1, p-1] \to \mathbf{V})$, it is a linear map $(\mathbf{V} \to \mathcal{RC})$ and when restricted to $\prod_{1,p-1} \mathbf{V} \times \{x_p\}$, $x_p \in \mathbf{V}$, it is a $_{\mathbf{p}-1}$**linear_form**.

**Proposition 421** A multilinear form is a map $(_{\mathbf{p}\triangleleft}\mathbf{f} : \mathbf{V}^{\mathbf{p}} \to \mathcal{RC})$ which is for any $\mathbf{i} \in [1, \mathbf{p}]$ linear in $\mathbf{x}^{\mathbf{i}}$ which means:
For any $\mathbf{i} \in [1, \mathbf{p}]$ and any element of $\mathbf{y} \in \prod_{j\in[1,p]-\{i\}} \mathbf{V}$, the restricted map (definition 129 (p. 145)) $(_{\mathbf{p}\triangleleft}\mathbf{f}_{\mathbf{y}} : \mathbf{V} \to \mathcal{RC})$ is linear in $\mathbf{x}^{\mathbf{i}}$.

**Proof**
It is the proposition 382 (p. 224).

**Definition 218 $(\otimes\mathbf{V}^*)^{\mathbf{p}}$.**

As in definition 192 (p. 224), the set of $_{\mathbf{p}}$**linear_form** is called $(\otimes\mathbf{V}^*)^{\mathbf{p}}$.

**Definition 219 $(\bigotimes_{k\in[1,p]} u_k : \prod_{k\in[1,p]}(\mathbf{V}^*) \to \bigotimes_{k\in[1,p]} \mathbf{V}^* = (\otimes\mathbf{V}^*)^{\mathbf{p}})$.**

It is the same as definition 194 (p. 226):

$$((u_k : [1,p] \to \mathbf{V}^*), (v_k : [1,p] \to \mathbf{V})) \Rightarrow (\bigotimes_{k\in[1,p]} u_k)(\prod_{k\in[1,p]} v_k) = \prod_{k\in[1,p]}^{\times} u_k(v_k) .$$

**Proposition 422** Given a vectorial space $\mathbf{V}$ of finite dimension $\mathbf{n}$ with an independent basis B: $(e_i : [1, n] \to \mathbf{V})$ (definition 210 (p. 241)), an integer $\mathbf{p} \in \mathcal{Z}^+$, the set $(\otimes\mathbf{V}^*)^{\mathbf{p}}$ of $_{\mathbf{p}}$**linear_form** is a vectorial space of dimension $\mathbf{n}^{\mathbf{p}}$. The set $[1, \mathbf{n}]^{\mathbf{p}}$, an element of which can be written as the $(i_k : [1,p] \to [1,n])$ (definition 372 (p. 473)) together with definition 219 (p. 246), allows to define a set of $\mathbf{n}^{\mathbf{p}}$ linear $_{\mathbf{p}}$**linear_form**: $(\bigotimes_{k\in[1,p]} e^{i_k} : [1,n]^p \to (\otimes\mathbf{V}^*)^{\mathbf{p}})$, where the $(e^i : [1,n] \to \mathbf{V}^*)$ is the independent basis of $\mathbf{V}^*$ associated to the basis $(e_i : [1,n] \to \mathbf{V})$ of $\mathbf{V}$ (definition 214 (p. 245)).
The set of those $_{\mathbf{p}}$**linear_form** forms an independent basis of $(\otimes\mathbf{V}^*)^{\mathbf{p}}$ and we have the following relations:

$$\begin{aligned}&(\forall i \in [1,n]^p)(\forall j \in [1,n]^p)(i = j \Rightarrow \bigotimes_{\ell\in[1,p]} e^{i_\ell}(\prod_{\ell'\in[1,p]}^{e} e_{i_{\ell'}}) = 1, i \neq j \\ &\Rightarrow \bigotimes_{\ell\in[1,p]} e^{i_\ell}(\prod_{\ell'\in[1,p]}^{e} e_{i_{\ell'}}) = 0)\end{aligned} \tag{2.88}$$

and we have:

$$_{p\triangleleft}f = \sum_{i\in[1,n]^p} {}_{p\triangleleft}f(\prod_{j\in[1,p]} e_{i(j)}) \bigotimes_{j\in[1,p]} e^{i(j)} . \tag{2.89}$$

**Proof**
The relations 2.88 are consequences of definition 219 (p. 246).
It is true for $\mathbf{p} = \mathbf{1}$ from definition 212 (p. 244). Let us suppose is true up to $\mathbf{p} = \mathbf{p}'$, let us show that then it is true for $\mathbf{p} = \mathbf{p}' + \mathbf{1}$. From definition 219 (p. 246) and proposition 367 (p. 216), we have:

$$(\mathrm{v_{p'+1}} = \mathrm{v^i_{p'+1}e_i} \in \mathrm{V},\ \mathrm{v} = (\mathrm{v_l} : [1,\mathrm{p}'] \to \mathrm{V})) \Rightarrow$$

$$\begin{aligned}({}_{\mathrm{p'+1}\triangleleft}\mathrm{f}(\mathrm{v} \times \mathrm{v_{p'+1}}) &= {}_{\mathrm{p'+1}\triangleleft}\mathrm{f}(\mathrm{v} \times \sum_{\mathrm{i}\in[1,\mathrm{n}]} \mathrm{v^i_{p'+1}e_i}) = \sum_{\mathrm{i}\in[1,\mathrm{n}]} \mathrm{v^i_{p'+1}}\{{}_{\mathrm{p'+1}\triangleleft}\mathrm{f}\}(\mathrm{v} \times \mathrm{e_i})\\
&= \sum_{\mathrm{i}\in[1,\mathrm{n}]} \mathrm{e^i}(\mathrm{v_{p'+1}}) \sum_{\ell\in[1,\mathrm{n}]^{\mathrm{p}'}} \prod^{\times}_{\mathrm{j}\in[1,\mathrm{p}']} \mathrm{v_j^{\ell(j)}}\{{}_{\mathrm{p'+1}\triangleleft}\mathrm{f}(\prod_{\mathrm{j}\in[1,\mathrm{p}']} \mathrm{e_{\ell(j)}} \times \mathrm{e_i})\}\\
&= \sum_{\mathrm{i}\in[1,\mathrm{n}]} \sum_{\ell\in[1,\mathrm{n}]^{\mathrm{p}'}} \prod^{\times}_{\mathrm{j}\in[1,\mathrm{p}']} \mathrm{e^{\ell(j)}}(\mathrm{v_j})\mathrm{e^i}(\mathrm{v_{p'+1}})\{{}_{\mathrm{p'+1}\triangleleft}\mathrm{f}(\prod_{\mathrm{j}\in[1,\mathrm{p}']} \mathrm{e_{\ell(j)}} \times \mathrm{e_i})\}\\
&= \sum_{\ell(\mathrm{p'+1})\in[1,\mathrm{n}]} \sum_{\ell\in[1,\mathrm{n}]^{\mathrm{p}'}} \prod^{\times}_{\mathrm{j}\in[1,\mathrm{p'+1}]} \mathrm{e^{\ell(j)}}(\mathrm{v_j})\{{}_{\mathrm{p'+1}\triangleleft}\mathrm{f}(\prod_{\mathrm{j}\in[1,\mathrm{p'+1}]} \mathrm{e_{\ell(j)}})\}\\
&= \sum_{\ell\in[1,\mathrm{n}]^{\mathrm{p'+1}}} \prod^{\times}_{\mathrm{j}\in[1,\mathrm{p'+1}]} \mathrm{e^{\ell(j)}}(\mathrm{v_j})\{{}_{\mathrm{p'+1}\triangleleft}\mathrm{f}(\prod_{\mathrm{j}\in[1,\mathrm{p'+1}]} \mathrm{e_{\ell(j)}})\})\,.\end{aligned}$$

We conclude, using the recurrence principle, for any $\mathbf{p}$ (proposition 927 (p. 520)).

### 2.11.4 Wedge projection, antisymmetric multilinear forms

**Definition 220 Wedge projection on $(\otimes \mathbf{V}^*)^{\mathbf{p}}$.**

It is the definition 197 (p. 233) with $\mathbf{U} = \mathcal{RC}$: Calling $\mathbf{PG_p}$ the permutation group of $[\mathbf{1},\mathbf{p}]$ (see definition 172 (p. 207)) and $(\mathbf{s}(\mathbf{P}) : \mathbf{PG_p} \to \{-\mathbf{1},\mathbf{1}\})$ the signature of the permutations (see definition 175 (p. 208)) and using the associated map $\mathbf{P}(\mathbf{t}) : (\mathbf{V}^*)^{\mathbf{p}} \leftrightarrow (\mathbf{V}^*)^{\mathbf{p}}$ to any perturbation $\mathbf{P}$ which is defined in definition 196 (p. 232), one calls wedge projection or the antisymmetric projection, the following linear map $(\mathbf{W} : (\otimes \mathbf{V}^*)^{\mathbf{p}} \to (\otimes \mathbf{V}^*)^{\mathbf{p}})$ defined by:

$$(\forall t \in (\otimes \mathrm{V}^*)^{\mathrm{p}})(\mathrm{W}(t) = \sum_{\mathrm{P}\in\mathrm{PG_p}} \mathrm{s}(\mathrm{P})\mathrm{P}(t))\,.$$

**Proposition 423** Using notations of definition 220 (p. 247) we also have:

$$(\forall \mathrm{x} \in \prod_{\mathrm{k}\in[1,\mathrm{p}]} \mathrm{V})(\forall t \in \bigotimes_{\mathrm{k}\in[1,\mathrm{p}]} \mathrm{V}^*)(\mathrm{W}(t)(\mathrm{x}) = t(\sum_{\mathrm{P}\in\mathrm{PG_p}} \mathrm{s}(\mathrm{P}) \prod^{\mathrm{e}}_{\mathrm{k}\in[1,\mathrm{p}]} \mathrm{x_{P(k)}}))\,.$$

**Proof**
It is a direct consequence of definition 196 (p. 232) used for definition 220 (p. 247)

**Proposition 424** Given a vectorial space $\mathbf{V}$, a map $(\mathbf{x_k} : [\mathbf{1},\mathbf{p}] \to \mathbf{V})$, a transposition $(\mathbf{T_{l,m}} : [\mathbf{1},\mathbf{p}] \to [\mathbf{1},\mathbf{p}])$ on $[\mathbf{1},\mathbf{p}]$ (definition 173 (p. 208)) and a $\mathbf{p}$ antisymmetric $_{\mathbf{p}}$**linear_form** $({}_{\mathbf{p}\triangleleft}\mathbf{f^a} : \mathbf{V^p} \to \mathcal{RC})$, one should have:

$$ {}_{\mathrm{p}\triangleleft}\mathrm{f^a}(\prod_{\mathrm{k}\in[1,\mathrm{n}]} \mathrm{x_k}) = (-1) \times {}_{\mathrm{p}\triangleleft}\mathrm{f^a}(\prod_{\mathrm{k}\in[1,\mathrm{n}]} \mathrm{x_{T_{l,m}(k)}})\,. \tag{2.90}$$

**Proof**
Using propositions 351 (p. 209) and 350 (p. 209), it is a peculiar case of proposition 423 (p. 247).

**Proposition 425** Given a vectorial space $\mathbf{V}$, a map $(\mathbf{x_k} : [1, p] \to \mathbf{V})$ and a $\mathbf{p}$ antisymmetric $_\mathbf{p}$**linear_form** $({}_{p\triangleleft}\mathbf{f}^a : \mathbf{V}^p \to \mathcal{RC})$, for any permutation $\mathbf{P}$ of $[1, p]$ one has:

$$_{p\triangleleft}f^a(\prod_{k\in[1,n]} x_k) = s(P)\ {}_{p\triangleleft}f^a(\prod_{k\in[1,n]} x_{P(k)})$$

where $\mathbf{s(P)}$ is the signature of the permutation (definition 175 (p. 208)).

**Proof**
It is an immediate consequence, proposition 424 (p. 247): $\mathbf{s(P)}$ the signature of the permutation (definition 175 (p. 208)) is given by the parity (odd or even) of the number of transpositions to which $\mathbf{P}$ is equivalent (proposition 351 (p. 209)).

**Proposition 426** Keeping notations and conditions of proposition 425 (p. 248), we have:

$$_{p\triangleleft}f^a(x_1, x_2, ..., x_p) \neq 0 \Rightarrow (\forall\lambda \in \mathcal{RC})(\forall i \in [1, p])(\forall j \in [1, p])(i \neq j \Rightarrow x_i \neq \lambda x_j) . \tag{2.91}$$

**Proof**
Otherwise, by exchanging the $\mathbf{i}$ and $\mathbf{j}$ for which $\mathbf{x_i} = \lambda\mathbf{x_j}$, ${}_{p\triangleleft}\mathbf{f}^a$ changes its sign. That it is impossible if ${}_{p\triangleleft}\mathbf{f}^a$ is not zero.

**Proposition 427** Let us suppose now that $\mathbf{V}$ is of finite dimension $\mathbf{n}$ with $(\mathbf{e_i} : [1, n] \to \mathbf{V})$ as an independent basis. Taking the set $\mathbf{B}$ of all possible subsets of $\mathbf{p}$ different vectors $\mathbf{e_i}$ (clearly we have $\mathbf{Card(B)} = \frac{n!}{p!}$ and $\mathbf{p} \leq \mathbf{n}$), calling $\mathbf{PG_p}$ the permutation group of $[1, p]$ (see definition 172 (p. 207)), we have:

$$(\forall b \in B)(\exists\ {}^b e_k)({}^b e_k : [1, p] \to b)$$

${}_{p\triangleleft}\mathbf{f}^a$ can be written as with $(\lambda_b : \mathbf{B} \to \mathcal{RC})$:

$$_{p\triangleleft}f^a = \sum_{b\in B} \lambda_b \sum_{P\in PG_p} s(P) \prod_{k\in[1,p]} {}^b e^{P(k)} . \tag{2.92}$$

**Proof**
We have:

$$_{p\triangleleft}f^a(\prod_{k\in[1,p]} {}^b e_k) = s(P)\ {}_{p\triangleleft}f^a(\prod_{k\in[1,p]} {}^b e_{P(i)})$$

each $\sum_P s(P) \prod^b_{k\in[1,p]} e^{P(k)}$ is a basis of the vectorial space of $\mathbf{p}$ antisymmetric multilinear forms.

**Let us show now that ${}_{p\triangleleft}f^a$, being like in equation (2.92), satisfies the equation (2.90)**

We can write for ${}_{p\triangleleft}\mathbf{f}^a$ which is also a $_\mathbf{p}$**linear_form**:

$$_{p\triangleleft}f^a = f_{i_1, i_2 ..., i_p} e^{i_1} e^{i_2} ... e^{i_p} .$$

Applied to a set of $\mathbf{p}$ vectors obtained by a map $(\mathbf{x_k} : [1, p] \to \mathbf{V})$ with $\mathbf{x_k} = \mathbf{x^i_k e_i}$, it gives:

$$_{p\triangleleft}f^a(x_1 ..., x_p) = f_{i_1, i_2 ..., i_p} x_1^{i_1} x_2^{i_2} ... x_p^{i_p} .$$

If we permute the set of vector $\mathbf{x}$ by a permutation $\mathbf{Q}$ of $[1,\mathbf{p}]$ we have:

$$ {}_{p\triangleleft}f^a(Q(x_1...x_p)) = {}_{p\triangleleft}f^a(x_{Q(1)}...x_{Q(p)}) = f_{i_1,i_2,...,i_p} x^{i_1}_{Q(1)} x^{i_2}_{Q(2)} ... x^{i_p}_{Q(p)} \,. $$

From the definition of $e^{i_1} \otimes e^{i_2} ..... \otimes e^{i_p}$ in definition 217 we can write:

$$ \begin{aligned} {}_{p\triangleleft}f^a \circ Q(x_1, x_2...x_p) &= f_{i_1,i_2,...,i_p} e^{i_{Q(1)}} e^{i_{Q(2)}} ... e^{i_{Q(p)}} (x_1, x_2...x_p) \\ &= f_{i_1,i_2,...,i_p} x^{i_1}_{Q(1)} x^{i_2}_{Q(2)} ... x^{i_p}_{Q(p)} \end{aligned} $$

then plugging it in equation (2.92):

$$ {}_{p\triangleleft}f^a \circ Q = \sum_{b \in B} \lambda_b \sum_P s(P) \prod_{k \in [1,p]} {}^b e^{Q \circ P(k)} \,. $$

Setting $\mathbf{P}' = \mathbf{Q} \circ \mathbf{P}$

$$ {}_{p\triangleleft}f^a \circ Q = \sum_{b \in B} \lambda_b \sum_P s(Q^{-1} \circ P') \prod_{k \in [1,p]} {}^b e^{P'(k)} \,. $$

Using proposition 343 (p. 207) which says that $\mathbf{Q} \circ \mathbf{P}$ is a bijection on the set of permutation $\mathbf{PG_p}$ and that the order in a sum does not count (see definition 393 (p. 506)), we also have:

$$ {}_{p\triangleleft}f^a \circ Q = \sum_{b \in B} \lambda_b \sum_P s(Q^{-1} \circ P) \prod_{k \in [1,p]} {}^b e^{P(k)} \,. $$

Since $s(Q^{-1} \circ P) = s(Q^{-1})s(P)$ and $s(Q^{-1} \circ Q) = s(Q^{-1})s(Q) = 1 = S(Q \circ Q) = s(Q)^2 \Rightarrow s(Q^{-1}) = s(Q)$ (proposition 352 (p. 209)) we have:

$$ {}_{p\triangleleft}f^a \circ Q = s(Q) \sum_{b \in B} \lambda_b \sum_P s(P) \prod_{k \in [1,p]} {}^b e^{P(k)} = s(Q) {}_{p\triangleleft}f^a $$

${}_{p\triangleleft}f^a$ verifies the definition of an antisymmetric multilinear form.

**Proposition 428** If a set of $\mathbf{p}$ vectors of a finite dimension $\mathbf{n}$ vectorial space $\mathbf{V}$ is an independent set, there is a nonzero antisymmetric ${}_p$**linear_form** when acting on those $\mathbf{p}$ vectors.

**Proof**

We have by definition 207 (p. 240) $\mathbf{p} \leq \mathbf{n}$. Calling $(x_i : [1,p] \to V)$ those $\mathbf{p}$ independent vectors, if $\mathbf{p} < \mathbf{n}$, we may add $\mathbf{n} - \mathbf{p}$ vectors of $V{:}x_{p+1}, ..., x_n$ so to form an independent basis $(x_i : [1,n] \to V)$ of $\mathbf{V}$ (proposition 415 (p. 244)) and

$$ {}_{p\triangleleft}f^a = \sum_P s(P) \prod_{k \in [1,p]} x^{P(k)} $$

is an antisymmetric ${}_p$**linear_form** on $\mathbf{V}$ which gives $\mathbf{1}$ for the set $x_1, ..., x_p$ and 0 for any subset $\mathbf{p}$ basis vectors containing one or more basic vector with an index bigger than $\mathbf{p}$. (see definition 213 (p. 244)): The linear form $x^k$ is such that $x^k(x_l) = \delta^k_l$.

**Proposition 429** If $\mathbf{p}$ vectors of a finite dimension $\mathbf{n}$ vectorial space $\mathbf{V}$ are dependent, all $\mathbf{p}$ antisymmetric multilinear forms acting on them are zero.

**Proof**

Since the vectors are dependent, one of them can be expressed as a linear combination of the others. Then any antisymmetric ${}_p$**linear_form**, ${}_{p\triangleleft}f^a$ acting on those $\mathbf{p}$ vectors, could be expanded into terms in which there are always two vectors $\mathbf{x}$ and $\mathbf{y}$ for which there is a $\lambda$ such that $\mathbf{x} = \lambda\mathbf{y}$. All those terms are null from equation (2.91) (p. 248). Their sum is also null.

**Definition 221 Determinant.**

If $\mathbf{n} = \mathbf{p}$, the set $\mathbf{B}$ in equation (2.92) (p. 248) has only one element. Then given a vectorial space $\mathbf{V}$ of dimension $\mathbf{n}$ and an independent basis $(\mathbf{e_i} : [\mathbf{1}, \mathbf{n}] \to \mathbf{V})$ in formula (2.92) (p. 248) there is only one arbitrary $\boldsymbol{\lambda}$. When that $\boldsymbol{\lambda} = \mathbf{1}$, the antisymmetric **n_linear** form is called determinant of a set of $\mathbf{n}$ vectors of $\mathbf{V}$ in the basis $(\mathbf{e_i} : [\mathbf{1}, \mathbf{n}] \to \mathbf{V})$.

**Proposition 430** When $\mathbf{n}$ vectors of a finite dimension $\mathbf{n}$ vectorial space are independent (resp. dependent), their determinant is not zero (resp. is zero).

**Proof**
It is a direct consequence of propositions 428 and 429.

### 2.11.5 Tensorial product and linear transformation

**Definition 222 Tensorial product.**

Given a vectorial space $\mathbf{V}$, one calls $(\otimes \mathbf{V})^{\mathbf{p}}$ (**p_tensorial** product of $\mathbf{V}$), the dual of the vectorial space formed by the **p_multilinear** maps on $\mathbf{V}$.

**Definition 223 Basis of a p_tensorial product of a finite dimension vector space.**

If $\mathbf{V}$ is of finite dimension $\mathbf{n}$, from what we said in definition 212, to the basis of the **p_multilinear** map corresponds a basis of $(\otimes \mathbf{V})^{\mathbf{p}}$. One writes

$$\mathbf{e_{i_1} \otimes e_{i_2} ... \otimes e_{i_p}}$$

the basic vector corresponding to $\mathbf{e^{i_1} \otimes e^{i_2} ... \otimes e^{i_p}}$. We have the relations of definition (see definition 213 (p. 244)):

$$\mathbf{e_{i_1} \otimes e_{i_2} ... \otimes e_{i_p}(e^{j_1} \otimes e^{j_2} ... \otimes e^{j_p})} = \delta^{\mathbf{j_1}}_{\mathbf{i_1}} ... \delta^{\mathbf{j_p}}_{\mathbf{i_p}} \ .$$

**Definition 224 Linear transformation.**

Given two vectorial spaces $\mathbf{U}$ and $\mathbf{V}$, one calls a linear transformation, a map of $\mathbf{U}$ over $\mathbf{V}$:$(\mathbf{T} : \mathbf{U} \to \mathbf{V})$ which keeps the internal law of the vectorial space:

$$(\forall(\boldsymbol{\lambda}, \boldsymbol{\mu}) \in \mathcal{RC}^2)(\forall(\mathbf{a}, \mathbf{b}) \in \mathbf{U})(\mathbf{T}(\boldsymbol{\lambda}\mathbf{a} + \boldsymbol{\mu}\mathbf{b}) = \boldsymbol{\lambda}\mathbf{T}(\mathbf{a}) + \boldsymbol{\mu}\mathbf{T}(\mathbf{b}))\ .$$

**Proposition 431** We have two vectorial spaces $\mathbf{U}$ and $\mathbf{V}$ of finite dimension $\mathbf{p}$, $\mathbf{q}$ with the corresponding independent bases: the set $\mathbf{E}$ $(\mathbf{e_i} : [\mathbf{1}, \mathbf{p}] \to \mathbf{U})$ and the set $\mathbf{F}$ $(\mathbf{f_i} : [\mathbf{1}, \mathbf{q}] \to \mathbf{V})$. Writing the transformation of the basis of $\mathbf{U}$ in terms of basis of $\mathbf{U}$:

$$\mathbf{T(e_i) = T^j_i f_j}$$

the transformed of any vector of $\mathbf{U}$ is defined in a unique way by:

$$\mathbf{u = u^i e_i \in U \Rightarrow T(u) = T(u^i e_i) = T^j_i u^i j_j}\ .$$

**Proof**
It comes directly from definition 224 (p. 250) and proposition 409 (p. 241).

**Definition 225 Matrix of a linear transformation.**

Given two vectorial spaces of finite dimension $\mathbf{p}$ and $\mathbf{q}$ and a linear transformation from the first to the second, one calls matrix of the transformation, the map $(\mathbf{T}^{\mathbf{i}}_{\mathbf{j}} : [\mathbf{1},\mathbf{p}]\times[\mathbf{1},\mathbf{q}] \rightarrow \mathcal{RC})$ which defines that transformation as described in definition 224. One calls column $\mathbf{j}$ of the matrix $\mathbf{T}$, the coordinates of the transformed vectors of the basic vector $\mathbf{j} \in [\mathbf{1},\mathbf{p}]$: $\mathbf{T}^{\mathbf{i}}_{\mathbf{j}}$. One calls row $\mathbf{i}$, the set of coordinates of the transformed first set of basic vectors along the second set of basic vectors: $\mathbf{i} \in [\mathbf{1},\mathbf{q}]$: $\mathbf{T}^{\mathbf{i}}_{\mathbf{j}}$.

**Proposition 432** Given three vectorial spaces $\mathbf{U},\mathbf{V},\mathbf{W}$ of finite dimensions respectively $\mathbf{p},\mathbf{q},\mathbf{r}$, given also three independent bases of respectively $\mathbf{U},\mathbf{V},\mathbf{W}$ (definition 210 (p. 241)):

$$(\mathbf{e_i} : [\mathbf{1},\mathbf{p}] \rightarrow \mathbf{U}),\ (\mathbf{f_i} : [\mathbf{1},\mathbf{q}] \rightarrow \mathbf{V}),\ (\mathbf{g_i} : [\mathbf{1},\mathbf{r}] \rightarrow \mathbf{W})$$

given also two linear transformations (definition 224 (p. 250)) $(\mathbf{S} : \mathbf{U} \rightarrow \mathbf{V})$, $(\mathbf{T} : \mathbf{V} \rightarrow \mathbf{W})$, the product of the two linear transformations (definition 117 (p. 134)) is also a linear transformation, the matrix of which is given by:

$$(\mathbf{T} \circ \mathbf{S})^{\mathbf{i}}_{\mathbf{j}} = (\mathbf{TS})^{\mathbf{i}}_{\mathbf{j}} = \mathbf{T}^{\mathbf{i}}_{\mathbf{k}}\mathbf{S}^{\mathbf{k}}_{\mathbf{j}}\ .$$

**Proof**
From 224 (p. 250) and proposition 431 (p. 250)

$$\mathbf{u} = \mathbf{u}^{\mathbf{j}}\mathbf{e_j} \Rightarrow \mathbf{S}(\mathbf{u}) = \mathbf{u}^{\mathbf{i}}\mathbf{S}(\mathbf{e_j}) = \mathbf{u}^{\mathbf{j}}\mathbf{S}^{\mathbf{k}}_{\mathbf{j}}\mathbf{f_k}$$
$$\Rightarrow \mathbf{T} \circ \mathbf{S}(\mathbf{u}) = \mathbf{u}^{\mathbf{j}}\mathbf{S}^{\mathbf{k}}_{\mathbf{j}}\mathbf{T}(\mathbf{f_k}) = \mathbf{u}^{\mathbf{j}}\mathbf{S}^{\mathbf{k}}_{\mathbf{j}}\mathbf{T}^{\mathbf{i}}_{\mathbf{k}}\mathbf{g_i} = \mathbf{T}^{\mathbf{i}}_{\mathbf{k}}\mathbf{S}^{\mathbf{k}}_{\mathbf{j}}\mathbf{g_i}\ .$$

**Proposition 433** With notations of proposition 431 (p. 250), if $\mathbf{p} = \mathbf{q} = \mathbf{n}$ and if the determinant of the set of vectors $\mathbf{T}(\mathbf{E})$: $\mathbf{det}(\mathbf{T}^{\mathbf{j}}_{\mathbf{i}}\mathbf{f_j})$ is not zero, the set $\mathbf{T}(\mathbf{E})$ is formed of n independent vectors and we have:

$$\mathbf{V} = \mathbf{T}(\mathbf{U})\ .$$

**Proof**
The set $\mathbf{T}(\mathbf{E})$ is a basis of $\mathbf{T}(\mathbf{U})$, the vectors of which are $\mathbf{T}^{\mathbf{j}}_{\mathbf{i}}\mathbf{f_j}$. If $\mathbf{p} = \mathbf{q} = \mathbf{n}$ and if the determinant of the set of vectors $\mathbf{T}(\mathbf{E})$: $\mathbf{det}(\mathbf{T}^{\mathbf{j}}_{\mathbf{i}}\mathbf{f_j})$ is not zero, the set $\mathbf{T}(\mathbf{E})$ is formed of n independent vectors (proposition 430 (p. 250)) so it is a basis of $\mathbf{V}$ and, from definition 205 (p. 239), we have:

$$\mathbf{V} = \mathbf{T}(\mathbf{U})\ .$$

**Proposition 434** We keep the notations of propositions 432 (p. 251) and 431 (p. 250) with $\mathbf{p} = \mathbf{q} = \mathbf{n}$. If $\mathbf{V} = \mathbf{T}(\mathbf{U})$, any vector of $\mathbf{V}$ has a unique inverse in $\mathbf{U}$.

**Proof**
$\mathbf{V} = \mathbf{T}(\mathbf{U})$ implies that any $\mathbf{v} \in \mathbf{V}$ has an inverse (definition 114 (p. 132)). If it was not unique, we would have for a $\mathbf{v} \in \mathbf{V}$, a $\mathbf{u_1} \in \mathbf{U}, \mathbf{u_1} = \mathbf{u}^{\mathbf{i}}_{\mathbf{1}}\mathbf{e_i}$ and a $\mathbf{u_2} \in \mathbf{U}, \mathbf{u_2} = \mathbf{u}^{\mathbf{i}}_{\mathbf{2}}\mathbf{e_i} \neq \mathbf{u_1}$ such as:

$$\mathbf{v} = \mathbf{T}(\mathbf{u_1}) = \mathbf{T}(\mathbf{u_2}) \Rightarrow \mathbf{T}(\mathbf{u_1} - \mathbf{u_2}) = (\mathbf{u}^{\mathbf{i}}_{\mathbf{1}} - \mathbf{u}^{\mathbf{i}}_{\mathbf{2}})\mathbf{T}^{\mathbf{j}}_{\mathbf{i}}\mathbf{f_j} = \mathbf{0}$$

and the vectors $\mathbf{T}^{\mathbf{j}}_{\mathbf{i}}\mathbf{f_j}$ would not be independent since not all the $\mathbf{u}^{\mathbf{i}}_{\mathbf{1}} - \mathbf{u}^{\mathbf{i}}_{\mathbf{2}}$ are zero.

**Proposition 435** Given two vector spaces, $\mathbf{U}$, $\mathbf{V}$ and a linear transformation $(\mathbf{T} : \mathbf{U} \rightarrow \mathbf{V})$ (definition 224 (p. 250)) the dimension of $\mathbf{T}(\mathbf{U})$ is smaller than or equal to the dimension of $\mathbf{U}$ and of $\mathbf{V}$.

**Proof**
Calling $\mathbf{p}$ and $\mathbf{q}$ the dimension of $\mathbf{U},\ \mathbf{V}$. Since $\mathbf{T(U)} \subset \mathbf{V}$, from definition 208 (p. 240), the dimension of $\mathbf{T(U)}$ is smaller than $\mathbf{q}$. If we have $\mathbf{p'} > \mathbf{p}$ vectors of $\mathbf{T(U)}$ which can be written as (see definition 112 (p. 131)) $(\mathbf{v_i} = \mathbf{T(u_i)} : [1, \mathbf{p'}] \to \mathbf{T(U)})$. Since the dimension of $\mathbf{U}$ is $\mathbf{p} < \mathbf{p'}$ from definition 208 (p. 240) we have:

$$(\exists(\lambda_i : [1, p'] \to \mathcal{R}))(\lambda_{[1,p']} \neq \{0\}, \sum_{i\in[1,p']} \lambda_i u_i = 0) .$$

It implies since $\mathbf{T}$ is linear: $\sum_{i\in[1,p']} \lambda_i \mathbf{T}(u_i) = \mathbf{T}(0) = 0$. Then from definition 208 (p. 240), $\mathbf{p'}$ cannot be the dimension of $\mathbf{T(U)}$.

**Proposition 436** We keep notations of propositions 432 (p. 251) and 431 (p. 250) with $\mathbf{p} = \mathbf{q} = \mathbf{n}$. If $\mathbf{V} = \mathbf{T(U)}$, $\mathbf{T(E)}$ is a basis of $\mathbf{V}$ and their determinant is not zero.

**Proof**
It is a direct consequence of proposition 411 (p. 241).

**Definition 226 Transposition of a matrix.**

$\mathbf{a} = (a^i_j : [1, n]^2 \to \mathcal{RC})$ being a matrix, setting $\mathbf{a^T} = ((a^T)^j_i = a^i_j : [1, n]^2 \to \mathcal{RC})$, $\mathbf{a^T}$ is called the transposed matrix of $\mathbf{a}$.

**Definition 227 Determinant of a matrix.**

In a vectorial space $\mathbf{V}$ of finite dimension $\mathbf{n}$ with its independent basis $(e_i : [1, n] \to V)$ which exists from definition 208 (p. 240), we take a nonzero antisymmetric **n_linear** form ${}_{n\triangleleft}f^a$. This form also exists from proposition 428 (p. 249). Having a set $\mathbf{v}$ of $\mathbf{n}$ vectors $\mathbf{v} = (v_i : [1, n] \to V)$, the ratio:

$$\det(v) = \frac{{}_{n\triangleleft}f^a(\prod_{i\in[1,n]} v_i)}{{}_{n\triangleleft}f^a(\prod_{i\in[1,n]} e_i)} \tag{2.93}$$

From definition 221 (p. 250), $\mathbf{det(v)}$ is the determinant of $\mathbf{v}$ relative to the basis $\mathbf{e_i}$.
From the definition of an antisymmetric multilinear form (proposition 424 (p. 247)) one has:

$$\det(v) = \sum_{P\in PG_n} s(P) \prod_{k\in[1,n]} v_k^{P(k)}$$

where $\mathbf{PG_n}$ is the permutation group of $[1, n]$ (see definition 172 (p. 207)) and $(s(P) : PG_n \to \{-1, 1\})$ is the signature of any element of $\mathbf{PG_n}$. This definition gives the same result as in definition 221 (p. 250) if one sets the determinant of the basic vectors to $\mathbf{1}$. Since this is a map of $(\mathcal{RC})^{n^2} \to \mathcal{RC}$, it is called determinant of the matrix of the $\mathbf{n}$ vectors coordinates. Those coordinates can be whatever element of $(\mathcal{RC})^{n^2}$. It can also be written as:

$$(\forall v \in (\mathcal{RC})^{n^2})(v = (v^i_j : [1, n]^2 \to \mathcal{RC}), \det(v) = \sum_{P\in PG_n} s(P) \prod_{k\in[1,n]} v_k^{P(k)}) .$$

There is an ambiguity between the definitions of the term determinant: We have the definition 227 (p. 252) and the definition 221 (p. 250). They may differ by the value given for the determinant of the basis. By convention if the determinant is a map of $\mathbf{V^n} \to \mathcal{RC}$ the definition 221 (p. 250) is used, if it is a map $(\mathcal{RC})^{n^2} \to \mathcal{RC}$ the definition 227 (p. 252) is used. In all cases used here, the distinction is always very clear.

**Proposition 437** Given a matrix $\mathbf{v} \in \mathcal{RC})^{\mathbf{n}^2}, \mathbf{v} = (\mathbf{v}^{\mathbf{i}}_{\mathbf{j}} : [\mathbf{1}, \mathbf{n}]^2 \to \mathcal{RC})$, we have:

$$\det(\mathbf{v}) = \sum_{\mathbf{P} \in \mathbf{PG_n}} \mathbf{s(P)} \prod_{\mathbf{k} \in [1,n]} \mathbf{v}^{\mathbf{P(k)}}_{\mathbf{k}} = \sum_{\mathbf{P} \in \mathbf{PG_n}} \mathbf{s(P)} \prod_{\mathbf{k} \in [1,n]} \mathbf{v}^{\mathbf{k}}_{\mathbf{P(k)}} .$$

**Proof**

Since $\mathcal{RC}$ is a field the product of number of $\mathcal{RC}$ does not depend on the order we have

$$\mathbf{v}_1^{\mathbf{P(1)}} \mathbf{v}_2^{\mathbf{P(2)}} \ldots \mathbf{v}_\mathbf{n}^{\mathbf{P(n)}} = \mathbf{v}_{\mathbf{P^{-1}(1)}}^{\mathbf{P(P^{-1}(1))}} \mathbf{v}_{\mathbf{P^{-1}(2)}}^{\mathbf{P(P^{-1}(2))}} \ldots \mathbf{v}_{\mathbf{P^{-1}(n)}}^{\mathbf{P(P^{-1}(n))}} = \mathbf{v}^1_{\mathbf{P^{-1}(1)}} \mathbf{v}^2_{\mathbf{P^{-1}(2)}} \ldots \mathbf{v}^\mathbf{n}_{\mathbf{P^{-1}(n)}} .$$

Since $(\mathbf{P^{-1}} : \mathbf{PG_n} \to \mathbf{PG_n})$ is a bijection we have $\mathbf{PG_n^{-1}} = \mathbf{PG_n}$. It means that the summation over $\mathbf{PG_n^{-1}}$ is the same as summing over $\mathbf{PG_n}$. Then, since from proposition 354 (p. 210) $\mathbf{s(P)} = \mathbf{s(P^{-1})}$, we have proposition 437.

**Proposition 438** $\mathbf{a}$ being a finite $\mathbf{n} \times \mathbf{n}$ matrix, $\mathbf{a^T}$ the transposed matrix (definition 228 (p. 254)), we have:

$$\det(\mathbf{a}) = \det(\mathbf{a^T}) .$$

**Proof**

$\mathbf{a} = (\mathbf{a^i_j} : [\mathbf{1}, \mathbf{n}]^2 \to \mathcal{RC})$ being a matrix, setting $\mathbf{a^T} = ((\mathbf{a^T})^\mathbf{j}_\mathbf{i} = \mathbf{a^i_j} : [\mathbf{1}, \mathbf{n}]^2 \to \mathcal{RC})$ as the transposed matrix, from proposition 437 (p. 253) we have

$$\det(\mathbf{a}) = \sum_{\mathbf{P} \in \mathbf{PG_n}} \mathbf{s(P)} \prod_{\mathbf{k} \in [1,n]} \mathbf{a}^{\mathbf{P(k)}}_{\mathbf{k}} = \sum_{\mathbf{P} \in \mathbf{PG_n}} \mathbf{s(P)} \prod_{\mathbf{k} \in [1,n]} \mathbf{a}^{\mathbf{k}}_{\mathbf{P(k)}}$$
$$= \sum_{\mathbf{P} \in \mathbf{PG_n}} \mathbf{s(P)} \prod_{\mathbf{k} \in [1,n]} (\mathbf{a^T})^{\mathbf{P(k)}}_{\mathbf{k}} = \det(\mathbf{a^T}) .$$

**Proposition 439** We consider the matrices of $\mathcal{RC}^{2\mathbf{n}}$ acting on vectorial spaces of finite dimension $\mathbf{n}$ built on a field $\mathcal{RC}$ with their reference bases written as $(\mathbf{e_i} : [\mathbf{1}, \mathbf{n}] \to \mathcal{RC}^\mathbf{n})$. Given such a matrix $\mathbf{u} = \mathbf{u^i_j}$, we call column vector $\mathbf{u_i} = \mathbf{u^j_i e_j}$ and given a map $\lambda^\mathbf{i} : [\mathbf{1}, \mathbf{n}] \to \mathcal{RC}$ we have:

$$(\mathbf{k} \in [\mathbf{1}, \mathbf{n}], (\forall \mathbf{i} \in ([\mathbf{1}, \mathbf{n}] - \mathbf{k}))(\mathbf{v_i} = \mathbf{u_i}), \mathbf{v_k} = \lambda^\mathbf{j} \mathbf{u_j}) \Rightarrow \det(\mathbf{v}) = \lambda^\mathbf{k} \det(\mathbf{u}) .$$

**Proof**

From the definition 227 (p. 252) and because

$$(\mathbf{j} \neq \mathbf{k} \Rightarrow \mathbf{j} \in [\mathbf{1}, \mathbf{k} - \mathbf{1}] \cup [\mathbf{k} + \mathbf{1}, \mathbf{n}]) \Rightarrow \det(\prod_{\mathbf{i} \in [1,k-1]} \mathbf{u_i} \times \mathbf{u_j} \times \prod_{\mathbf{i} \in [k+1,n]} \mathbf{u_i}) = 0 .$$

We have (no implicit summation on $\mathbf{k}$):

$$\det(\mathbf{v}) = \sum_{\mathbf{j} \in [1,n]} \lambda^\mathbf{j} \det(\prod_{\mathbf{i} \in [1,k-1]} \mathbf{u_i} \times \mathbf{u_j} \times \prod_{\mathbf{i} \in [k+1,n]} \mathbf{u_i})$$
$$= \lambda^\mathbf{k} \det(\prod_{\mathbf{i} \in [1,k-1]} \mathbf{u_i} \times \mathbf{u_k} \times \prod_{\mathbf{i} \in [k+1,n]} \mathbf{u_i}) = \lambda^\mathbf{k} \det(\mathbf{u}) .$$

**Proposition 440** As for proposition 439, we consider the matrices of $\mathcal{RC}^{2\mathbf{n}}$ acting on vectorial spaces of finite dimension $\mathbf{n}$ built on a field $\mathcal{RC}$ with their reference bases written as $(\mathbf{e^i} : [\mathbf{1}, \mathbf{n}] \to \mathcal{RC}^\mathbf{n})$. Given such a matrix $\mathbf{u} = \mathbf{u^i_j}$, we call row vector $\mathbf{u^i} = \mathbf{u^i_j e^j}$ and given a map $(\lambda_\mathbf{i} : [\mathbf{1}, \mathbf{n}] \to \mathcal{RC})$, we have:

$$(\mathbf{k} \in [\mathbf{1}, \mathbf{n}], (\forall \mathbf{i} \in ([\mathbf{1}, \mathbf{n}] - \mathbf{k}))(\mathbf{v^i} = \mathbf{u^i}), \mathbf{v^k} = \lambda_\mathbf{j} \mathbf{u^j}) \Rightarrow \det(\mathbf{v}) = \lambda_\mathbf{k} \det(\mathbf{u}) .$$

**Proof**

It is a direct consequence of propositions 439 and 437 (p. 253).

**Definition 228 Minor of a term of a matrix.**

Given a matrix $\mathbf{a} \in \mathcal{RC}^{\mathbf{n}^2}$, $\mathbf{i} \in [\mathbf{1}, \mathbf{n}]$, $\mathbf{j} \in [\mathbf{1}, \mathbf{n}]$ one calls the Minor of $\mathbf{a}$ relative to $\mathbf{i}$ and $\mathbf{j}$

$$\mathbf{A}^{\mathbf{i}}_{\mathbf{j}} \in \mathcal{RC}^{(\mathbf{n}-1)^2}, \mathbf{A}^{\mathbf{i}}_{\mathbf{j}} = \mathbf{Minor}^{\mathbf{i}}_{\mathbf{j}}(\mathbf{a}) \Leftrightarrow (\mathbf{A}^{\mathbf{i}}_{\mathbf{j}})^{\mathbf{k}}_{\mathbf{m}} = \mathbf{a}^{\mathbf{k}+\mathcal{Y}(\mathbf{i}-\mathbf{k})}_{\mathbf{m}+\mathcal{Y}(\mathbf{j}-\mathbf{m})} , \tag{2.94}$$

where $\mathcal{Y}(\mathbf{m})$ is the Heaviside map (definition 216 (p. 245)) $\mathbf{A}^{\mathbf{i}}_{\mathbf{j}}$ is the matrix $\mathbf{a}$ where the row or line $\mathbf{i}$ and the column $\mathbf{j}$ have been suppressed.

**Proposition 441** $\mathbf{a} = (\mathbf{a}^{\mathbf{i}}_{\mathbf{j}} : [\mathbf{1}, \mathbf{n}]^2 \to \mathcal{RC})$ being a matrix, setting $\mathbf{a}^{\mathbf{T}} = ((\mathbf{a}^{\mathbf{T}})^{\mathbf{j}}_{\mathbf{i}} = \mathbf{a}^{\mathbf{i}}_{\mathbf{j}} : [\mathbf{1}, \mathbf{n}]^2 \to \mathcal{RC})$ as the transposed matrix (definition 228 (p. 254)), calling $\mathbf{A}^{\mathbf{i}}_{\mathbf{j}}$ the minor of the element $\mathbf{a}^{\mathbf{i}}_{\mathbf{j}}$ and $(\mathbf{A}^{\mathbf{T}})^{\mathbf{i}}_{\mathbf{j}}$ the minor of the element $(\mathbf{a}^{\mathbf{T}})^{\mathbf{i}}_{\mathbf{j}}$, we have:

$$\mathbf{i} \times \mathbf{j} \in [\mathbf{1}, \mathbf{n}]^2 \Rightarrow (\mathbf{A}^{\mathbf{T}})^{\mathbf{i}}_{\mathbf{j}} = (\mathbf{A}^{\mathbf{j}}_{\mathbf{i}})^{\mathbf{T}} .$$

**Proof**

From definition 228 (p. 254) equation (2.94), we have:

$$((\mathbf{A}^{\mathbf{T}})^{\mathbf{i}}_{\mathbf{j}})^{\mathbf{k}}_{\mathbf{m}} = (\mathbf{a}^{\mathbf{T}})^{\mathbf{k}+\mathcal{Y}(\mathbf{i}-\mathbf{k})}_{\mathbf{m}+\mathcal{Y}(\mathbf{j}-\mathbf{m})} = \mathbf{a}^{\mathbf{m}+\mathcal{Y}(\mathbf{j}-\mathbf{m})}_{\mathbf{k}+\mathcal{Y}(\mathbf{i}-\mathbf{k})} = (\mathbf{A}^{\mathbf{j}}_{\mathbf{i}})^{\mathbf{m}}_{\mathbf{k}} .$$

**Proposition 442** $\mathbf{a}^{\mathbf{i}}_{\mathbf{j}} : [\mathbf{1}, \mathbf{n}]^2 \to \mathcal{RC}$ being a matrix, $\mathbf{A}^{\mathbf{i}}_{\mathbf{j}}$ the minor of the element $\mathbf{a}^{\mathbf{i}}_{\mathbf{j}}$ one has the following identity:

$$\det(\mathbf{a}) = \sum_{\mathbf{i} \in [\mathbf{1},\mathbf{n}]} (-1)^{\mathbf{i}+\mathbf{j}} \mathbf{a}^{\mathbf{i}}_{\mathbf{j}} \det(\mathbf{A}^{\mathbf{i}}_{\mathbf{j}})$$

and consequently, from proposition 439, we also have:

$$\det(\mathbf{a}) = \sum_{\mathbf{j} \in [\mathbf{1},\mathbf{n}]} (-1)^{\mathbf{i}+\mathbf{j}} \mathbf{a}^{\mathbf{i}}_{\mathbf{j}} \det(\mathbf{A}^{\mathbf{i}}_{\mathbf{j}}) .$$

**Proof**

From equation (2.94) we have:

$$\begin{aligned} \det(\mathbf{A}^{\mathbf{j}}_{\mathbf{i}}) &= \sum_{\mathbf{P} \in \mathbf{PG}_{\mathbf{n}-1}} \mathbf{s}(\mathbf{P}) \prod_{\mathbf{m} \in [\mathbf{1},\mathbf{n}-1]} (\mathbf{A}^{\mathbf{j}}_{\mathbf{i}})^{\mathbf{P}(\mathbf{m})}_{\mathbf{m}} \\ &= \sum_{\mathbf{P} \in \mathbf{PG}_{\mathbf{n}-1}} \mathbf{s}(\mathbf{P}) \prod_{\mathbf{m} \in [\mathbf{1},\mathbf{n}-1]} \mathbf{a}^{\mathbf{P}(\mathbf{m})+\mathcal{Y}(\mathbf{j}-\mathbf{P}(\mathbf{m}))}_{\mathbf{m}+\mathcal{Y}(\mathbf{i}-\mathbf{m})} . \end{aligned}$$

$\mathbf{PG}_{\mathbf{n}-1}$ is the permutation group of $\mathbf{n}-1$ elements (see definition 172 (p. 207)), $\mathbf{s}(\mathbf{P})$ is the signature of the permutation $\mathbf{P}$. Then we have

$$\mathbf{a}^{\mathbf{j}}_{\mathbf{i}} \det(\mathbf{A}^{\mathbf{j}}_{\mathbf{i}}) = \sum_{\mathbf{P} \in \mathbf{PG}_{\mathbf{n}-1}} \mathbf{s}(\mathbf{P}) \mathbf{a}^{\mathbf{j}}_{\mathbf{i}} \prod_{\mathbf{m} \in [\mathbf{1},\mathbf{n}-1]} \mathbf{a}^{\mathbf{P}(\mathbf{m})+\mathcal{Y}(\mathbf{j}-\mathbf{P}(\mathbf{m}))}_{\mathbf{m}+\mathcal{Y}(\mathbf{i}-\mathbf{m})} .$$

The permutation $\mathbf{P}' \in \mathbf{PG}_{\mathbf{n}}$ defined as

$$\mathbf{P}'(\mathbf{i}) = \mathbf{j}, \mathbf{k} \neq \mathbf{i} \Rightarrow \mathbf{P}'(\mathbf{k}) = \mathbf{P}(\mathbf{k} - \mathcal{Y}(\mathbf{i}-\mathbf{k})) + \mathcal{Y}(\mathbf{j} - \mathbf{P}(\mathbf{k} - \mathcal{Y}(\mathbf{i}-\mathbf{k}))) .$$

It is an injection of $\mathbf{PG}_{\mathbf{n}-1} \to \mathbf{PG}_{\mathbf{n}}$. Since $\mathbf{k} - \mathcal{Y}(\mathbf{i}-\mathbf{k})$ and $\mathbf{k} - \mathcal{Y}(\mathbf{j}-\mathbf{k})$ are injections of $[\mathbf{1}, \mathbf{n}-1] \to [\mathbf{1}, \mathbf{n}]$ they commute with a transposition. We then have $\mathbf{S}(\mathbf{P}') = \epsilon \mathbf{S}(\mathbf{P})$

where $\epsilon \in \{-1, 1\}$ does not depend on $\mathbf{P}$. We have to compare it only when $\mathbf{P}$ is the identity permutation. In that case $\mathbf{S(P')}$ is the number of transpositions to bring $\mathbf{j}$ to $\mathbf{i}$ keeping all the other at the relative place and in particular $\mathbf{i}$ so $\epsilon = (-1)^{i+j}$ and we have with no implicit summation:

$$a^j_i \det(A^j_i) = (-1)^{i+j} \sum_{P' \in PG_n} \delta^{P'(i)}_j \prod_{m \in [1,n]} a^{P'(m)}{}_m$$

and using

$$\sum_{i \in [1,n]} \delta^{P'(i)}_j = \sum_{j \in [1,n]} \delta^{P'(i)}_j = 1$$

$$\sum_{i \in [1,n]} (-1)^{i+j} a^j_i \det(A^j_i) = \sum_{P' \in PG_n} \prod_{m \in [1,n]} a^{P'(m)}{}_m = \det(a)$$

$$\sum_{j \in [1,n]} (-1)^{i+j} a^j_i \det(A^j_i) = \sum_{P' \in PG_n} \prod_{m \in [1,n]} a^{P'(m)}{}_m = \det(a) \ .$$

**Proposition 443** Given a matrix $\mathbf{a} \in \mathcal{RC}^{n^2}$, calling $(a^i_j : [1,n]^2 \to \mathcal{RC})$ the terms of that matrix, we have:

$$(\exists j \in [1,n])(\forall i \in [1,n])(a^i_j = 0) \Rightarrow \det(a) = 0,$$
$$(\exists i \in [1,n])(\forall j \in [1,n])(a^i_j = 0) \Rightarrow \det(a) = 0 \ .$$

**Proof**
It is a direct consequence of proposition 442 (p. 254).

**Proposition 444** Given a matrix $\mathbf{a} \in \mathcal{RC}^{n^2}$ with $\det(a) \neq 0$, calling $(A^j_i)^k_m = a^{k+\mathcal{Y}(j-k)}_{m+\mathcal{Y}(i-m)}$ the minor of the term $a^j_i$ (see definition 228) where $\mathcal{Y}(m)$ is the Heaviside map (definition 216 (p. 245)). We have:

$$(-1)^{i+k} a^k_j \det(A^k_i) = \delta^j_i \det(a) \ .$$

Using notations of proposition 441 (p. 254), we have as well:

$$(-1)^{i+k} (a^T)^k_j \det(A^i_k) = \delta^j_i \det(a) = \delta^j_i \det(a^T) \ .$$

If $\det(a) \neq 0$, the matrix $\mathbf{M}$ defined by $M^i_k = (-1)^{i+k} \det(A^k_i)/\det(a)$ is the inverse of $\mathbf{a}$ while the matrix $\mathbf{M^T}$ is the inverse of $\mathbf{a^T}$.

**Proof**
From proposition 442 we have:

$$(\forall k \in [1,n])(\sum_{j \in [1,n]} (-1)^{i+j} a^j_k \det(A^j_k) = \det(\prod_{i \in [1,n]} a_i))$$

$A^k_j$ does not depend on the vector $a_k$ then one has as well:

$$(\forall u \in \mathcal{RC}^n)(a'_k = u, (\forall i \in [1,n] - k)(a'_i = a_i))$$
$$\Rightarrow \sum_{j \in [1,n]} (-1)^{i+j} u^j \det(A^j_k) = \det(\prod_{k \in [1,n]} a'_k) \ .$$

If $u = a_m, m \neq k$ then there are two equal vectors in the argument of the determinant which has to be zero since it changes sign without changing value when exchanging those two equal vectors. We then have the first part of proposition 444. Using that we have:

$$(-1)^{i+k} (a^T)^k_j \det((A^T)^i_k) = \delta^j_i \det(a^T) = \delta^j_i \det(a) \ .$$

Using propositions 441 (p. 254) and 438 (p. 253), we have:

$$(-1)^{i+k}(a^T)^k_j \det((A^T)^i_k) = (-1)^{i+k}(a^T)^k_j \det((A^k_i)^T)$$
$$= (-1)^{i+k}(a^T)^k_j \det((A^k_i)) = \delta^j_i \det(a) .$$

**Proposition 445** Given a matrix $\mathbf{a} \in \mathcal{RC}^{n^2}$, with $\det(a) \neq 0$, we have $aa^{-1} = a^{-1}a = \mathbb{1}$ where $\mathbb{1}$ is the matrix of the identity transformation.

**Proof**

**a** generate a linear transformation on a vectorial space of dimension **n** the inverse of which is a linear transformation of matrix $a^{-1}$. From proposition 171 (p. 140), $(a^{-1})^{-1} = a$ then $aa^{-1} = a^{-1}a = \mathbb{1}$.

**Proposition 446** Given a topological set **E** and a normed vectorial space **V** (definition 203 (p. 237)) built on a field $\mathcal{RC}$ with a finite dimension **n** (definition 209 (p. 240)), we call $(e_i : [1, n] \to V)$ an independent basis of **V** (definitions 205 (p. 239), 206 (p. 239)). From definition 203 (p. 237), $\mathcal{RC}$ is a metric set which makes $\mathcal{RC}$ a topological separated set. We suppose that sum, subtraction, product and division (when the denominator is not **0** see definition A.8.2 (p. 553)) in $\mathcal{RC}$ are continuous maps of $(\mathcal{RC}^2 \to \mathcal{RC})$. Calling $\mathcal{F}$ the set of continuous maps $(E \to \mathcal{RC})$, given $n^2$ continuous maps $(a^i_j : [1, n]^2 \to \mathcal{F})$ defining a matrix **A**, if $\det(A) \neq 0$ in one point of **E** there is an open **O** in **E** containing that point where $A^{-1}$ exists and where $(\forall i \in [1, n])(\forall j \in [1, n])$ $((A^{-1})^i_j : O \to \mathcal{RC})$ are continuous maps.

**Proof**

$(\det(A) : E \to \mathcal{RC})$ is a continuous map as sum of products of continuous maps $(E \to \mathcal{RC})$ (see proposition 241 (p. 167)). $\mathcal{C}(0)$ is open since $\mathcal{C}(0) = \bigcup_{r \in \mathcal{C}(0)} B(r, \|r\|)$ (see definitions 132 (p. 151), 150 (p. 177), proposition 263 (p. 177)). Then from definition 141 (p. 165), $O = \det^{-1}(\mathcal{C}(0)) \subset E$ is open. From proposition 444 (p. 255), $(\forall i \in [1, n])(\forall j \in [1, n])$ $((A^{-1})^i_j : O \to \mathcal{RC})$ are continuous maps.

**Definition 229 Determinant of a linear transformation of a vectorial space of finite dimension on itself.**

We are given a finite dimension **n** vectorial space**V** with an independent basis **E** $(e_i : [1, n] \to V)$. As in definition 227 (p. 252), we take a nonzero antisymmetric **n_linear** form ${}_{n\triangleleft}f^a$. This form exists from proposition 428 (p. 249). One calls determinant of a linear transformation $(T : V \to V)$ relative to the basis $e_i$ which defines **T** by $(T : E \to E)$:

$$\det(T_{e_i}) = {}_{n\triangleleft}f^a(T(e_1), T(e_2), ..., T(e_n)) / {}_{n\triangleleft}f^a(e_1, e_2, ..., e_n)$$
$$= \sum_{P \in PG_n} s(P) T_1^{P(1)} T_2^{P(2)} ... T_n^{P(n)} \tag{2.95}$$

with $T^j_i = T(e_i)^j$. $PG_n$ is the set of permutation of $[1, n]$ (see definition 172 (p. 207)) and $s(P)$ the signature of the permutation **P** (definition 175 (p. 208)).

**Proposition 447** The determinant of a linear transformation does not depend on the basis used to defined it.

**Proof**

Since from equation (2.92) (p. 248), all fully antisymmetric **n_linear** forms are proportional to each other (see also definition 221 (p. 250)), using notations of definition 229 (p. 256), for any set of $\mathbf{n}$ vectors of $\mathbf{V}$ :$(\mathbf{v_i} : [1,\mathbf{n}] \to \mathbf{V})$, the following expression defines a fully antisymmetric map $(\mathbf{V^n} \to \mathcal{R})$:

$$_{\mathbf{n}\triangleleft}\mathbf{f^a}(\mathbf{T}(\mathbf{v_1}), \mathbf{T}(\mathbf{v_2}), ..., \mathbf{T}(\mathbf{v_n})) = {_{\mathbf{n}\triangleleft}\mathbf{f^a}}(\prod_{\mathbf{i}\in[1,\mathbf{n}]} (\mathbf{T}(\mathbf{v_i}))) .$$

Then we have $\boldsymbol{\lambda} \in \mathcal{R}$ such that:

$$_{\mathbf{n}\triangleleft}\mathbf{f^a}(\prod_{\mathbf{i}\in[1,\mathbf{n}]} \mathbf{T}(\mathbf{v_i})) = \boldsymbol{\lambda}\ {_{\mathbf{n}\triangleleft}\mathbf{f^a}}(\prod_{\mathbf{i}\in[1,\mathbf{n}]} \mathbf{v_i}) .$$

Then having two independent bases $(\mathbf{e_i} : [1,\mathbf{n}] \to \mathbf{V})$, $(\mathbf{f_i} : [1,\mathbf{n}] \to \mathbf{V})$ for which $_{\mathbf{n}\triangleleft}\mathbf{f^a}(\prod_{\mathbf{i}\in[1,\mathbf{n}]} \mathbf{e_i}) \neq 0$, $_{\mathbf{n}\triangleleft}\mathbf{f^a}(\prod_{\mathbf{i}\in[1,\mathbf{n}]} \mathbf{f_i}) \neq 0$, we have:

$$\mathbf{det}(\mathbf{T_{e_i}}) = {_{\mathbf{n}\triangleleft}\mathbf{f^a}}(\prod_{\mathbf{i}\in[1,\mathbf{n}]} \mathbf{T}(\mathbf{e_i}))/{_{\mathbf{n}\triangleleft}\mathbf{f^a}}(\prod_{\mathbf{i}\in[1,\mathbf{n}]} \mathbf{e_i}) = \boldsymbol{\lambda} = {_{\mathbf{n}\triangleleft}\mathbf{f^a}}(\prod_{\mathbf{i}\in[1,\mathbf{n}]} \mathbf{T}(\mathbf{f_i}))/{_{\mathbf{n}\triangleleft}\mathbf{f^a}}(\prod_{\mathbf{i}\in[1,\mathbf{n}]} \mathbf{f_i})$$

$\mathbf{det}(\mathbf{T_{e_i}})$ does not depend on the basis where the vectors are expressed.

**Proposition 448** Given two linear transformations on the same vectorial space of finite dimension, the determinant of the product of the two linear transformations in a basis, is the product of the determinants of the two transformations expressed in that basis.

**Proof**

Given two linear transformations $(\mathbf{S} : \mathbf{V} \to \mathbf{V})$ and $(\mathbf{T} : \mathbf{V} \to \mathbf{V})$ acting on the vectorial space $\mathbf{V}$ of finite dimension $\mathbf{n}$ with an independent basis $\mathbf{E}$: $(\mathbf{e_i} : [1,\mathbf{n}] \to \mathbf{E})$, if $\mathbf{det}(\mathbf{S}) = 0$, it means with notation of proposition 447 (p. 256), $_{\mathbf{n}\triangleleft}\mathbf{f^a}(\prod_{\mathbf{i}\in[1,\mathbf{n}]} \mathbf{S}(\mathbf{e_i})) = 0$ the set $\mathbf{S}(\mathbf{e_i})$ is not an independent set then $\mathbf{dim}(\mathbf{T}(\mathbf{V})) < \mathbf{n}$ then from proposition 435 (p. 251) $\mathbf{dim}(\mathbf{ST}(\mathbf{V})) < \mathbf{n}$ and $\mathbf{det}(\mathbf{ST}) = 0$. For the same reason if $\mathbf{det}(\mathbf{T}) \neq 0$ but $\mathbf{det}(\mathbf{S}) = 0$, $\mathbf{dim}(\mathbf{ST}(\mathbf{V})) < \mathbf{n}$ and $\mathbf{det}(\mathbf{ST}) = 0$. Let us suppose $\mathbf{det}(\mathbf{S}) \neq 0$ and $\mathbf{det}(\mathbf{T}) \neq 0$. Given on $\mathbf{V}$ an antisymmetric **n_linear** form $\mathbf{A}$, we have:

$$\begin{aligned}\mathbf{det}((\mathbf{ST})_\mathbf{E}) &= \frac{\mathbf{A}(\mathbf{ST}(\mathbf{e_1}), ..., \mathbf{ST}(\mathbf{e_n}))}{\mathbf{A}(\mathbf{e_1}, ..., \mathbf{e_n})} \\ &= \frac{\mathbf{A}(\mathbf{ST}(\mathbf{e_1}), ..., \mathbf{ST}(\mathbf{e_n}))}{\mathbf{A}(\mathbf{S}(\mathbf{e_1}), ..., \mathbf{S}(\mathbf{e_n}))} \frac{\mathbf{A}(\mathbf{S}(\mathbf{e_1}), ..., \mathbf{S}(\mathbf{e_n}))}{\mathbf{A}(\mathbf{e_1}, ..., \mathbf{e_n})}\end{aligned}$$

$\mathbf{A}(\mathbf{S}(\mathbf{v_1}), ..., \mathbf{S}(\mathbf{v_n}))$ with $(\mathbf{v_i} : [1,\mathbf{n}] \to \mathbf{V})$ is an antisymmetric **n_linear** form on $\mathbf{V}$. From proposition 447 (p. 256):

$$\frac{\mathbf{A}(\mathbf{ST}(\mathbf{e_1}), ..., \mathbf{ST}(\mathbf{e_n}))}{\mathbf{A}(\mathbf{S}(\mathbf{e_1}), ..., \mathbf{S}(\mathbf{e_n}))} = \frac{\mathbf{A}(\mathbf{T}(\mathbf{e_1}), ..., \mathbf{T}(\mathbf{e_n}))}{\mathbf{A}((\mathbf{e_1}), ..., (\mathbf{e_n}))} = \mathbf{det}(\mathbf{T_E})$$

then

$$\mathbf{det}((\mathbf{ST})_\mathbf{E}) = \mathbf{det}(\mathbf{T_E})\mathbf{det}(\mathbf{S_E})$$

**Proposition 449** Given a linear transformation $\mathbf{T}$ on a vectorial space $\mathbf{V}$ to a vectorial space $\mathbf{U}$: $(\mathbf{T} : \mathbf{V} \to \mathbf{U})$, $\mathbf{T}(\mathbf{V})$ is a vectorial space.

**Proof**

$$\begin{aligned}&(\forall(\mathbf{u}, \mathbf{v}) \in \mathbf{T}(\mathbf{V})^2)(\exists(\mathbf{a}, \mathbf{b}) \in \mathbf{V}^2)(\mathbf{u} = \mathbf{T}(\mathbf{a}), \mathbf{v} = \mathbf{T}(\mathbf{b})) \\ &\Rightarrow (\forall(\boldsymbol{\lambda}, \boldsymbol{\mu}) \in \mathcal{RC}^2)(\boldsymbol{\lambda}\mathbf{u} + \boldsymbol{\mu}\mathbf{v} = \boldsymbol{\lambda}\mathbf{T}(\mathbf{a}) + \boldsymbol{\mu}\mathbf{T}(\mathbf{b}) = \mathbf{T}(\boldsymbol{\lambda}\mathbf{a} + \boldsymbol{\mu}\mathbf{b}) \in \mathbf{T}(\mathbf{v})) .\end{aligned}$$

**Proposition 450** Given a linear transformation on a vectorial space of finite dimension. If the inverse exists that is to say if it is a bijection, this inverse is a linear transformation.

**Proof**
We call $\mathbf{T}$ a linear transformation on a vectorial space $\mathbf{V}$, $\mathbf{T(V)}$ is a vectorial space. If $\mathbf{T}$ has an inverse $\mathbf{T^{-1}}$, $\mathbf{(a, b) \in V^2}$, $(\lambda, \mu) \in \mathcal{RC}^2$

$$\lambda T^{-1}T(a) + \mu T^{-1}T(b) = \lambda a + \mu b = T^{-1}T(\lambda a + \mu b) \in T^{-1}T(V)$$

$$\lambda T^{-1}T(a) + \mu T^{-1}T(b) = T^{-1}T(\lambda a + \mu b) = T^{-1}(\lambda T(a) + \mu T(b)) \ .$$

**Proposition 451** Given a linear transformation on a vectorial space of finite dimension. The determinant of the transformation is nonzero when the transformation is a bijection (definition 124 (p. 139)) and zero when it is not.

**Proof**
We call $\mathbf{T}$ a linear transformation on a vectorial space $\mathbf{V}$ of finite dimension $\mathbf{n}$ and $(\mathbf{e_i} : [\mathbf{1}, \mathbf{n}] \to \mathbf{V})$ a basis of $\mathbf{V}$. If $\mathbf{T}$ is a bijection, $\mathbf{T^{-1}}$ exists and is a linear transformation (proposition 450), if the determinant of $\mathbf{T}$ was zero, the $\mathbf{n}$ vectors $\mathbf{T(e_i)}$ would be dependent and since:

$$(\forall i \in [1, n])(\lambda^i \in \mathcal{RC}), \lambda^i T(e_i) = 0 \Rightarrow \lambda^i T^{-1}T(e_i) = \lambda^i e_i = 0$$

the $\mathbf{e_i}$ would also be dependent then $\mathbf{det(T) \neq 0}$.
If $\mathbf{det(T) = 0}$, the $\mathbf{T(e_i)}$ are dependent and we have:

$$(\exists(\lambda^i : [1, n] \to \mathcal{RC}))(\exists j \in [1, n])(\lambda^j \neq 0, \lambda^i T(e_i) = T(\lambda^i e_i) = 0) \ .$$

The $\mathbf{e_i}$ being independent, we have: $\lambda^i \mathbf{e_i} \neq \mathbf{0}$.
We may have $(\forall i)(\mathbf{T(e_i) = 0})$ in that case $\mathbf{V = T^{-1}(0)}$ which means $\mathbf{Card(T^{-1}(0)) > 1}$ since $\mathbf{V}$ has more than one element:$\mathbf{e_1, e_1 + e_1}$ $\mathbf{T}$ is not a bijection (definition 124 (p. 139)).
If $(\exists j \in [1, n])(T(e_j) \neq 0)$, we have for that $\mathbf{j}$, $\mathbf{T(e_j} + \lambda^i \mathbf{e_i}) = \mathbf{T(e_j) + 0 = T(e_j) \neq 0}$. Then

$$e_j \in T^{-1}T(e_j), e_j + \lambda^i e_i \in T^{-1}T(e_j), e_j + \lambda^i e_i \neq e_j \ .$$

It means $\mathbf{Card(T^{-1}T(e_j)) \geq 2}$ then $\mathbf{T}$ is not a bijection.

**Definition 230 Linear equations.**

Given a matrix $(\mathbf{T^j_i} : [\mathbf{1}, \mathbf{p}] \times [\mathbf{1}, \mathbf{q}] \to \mathcal{RC})$, considering the maps $(\mathbf{x^i} : [\mathbf{1}, \mathbf{p}] \to \mathcal{RC})$ and having a map $(\mathbf{y^i} : [\mathbf{1}, \mathbf{q}] \to \mathcal{R})$, one calls a system of $\mathbf{p}$ linear equations the following expression:

$$T^j_i x^i = y^j \ .$$

The set of maps $\mathbf{x}$, for which the expression is true, is called the set of solutions of the linear equations. $\mathbf{p}$ is called the number of equations, $\mathbf{q}$ the number of variables.

**Proposition 452** With the above notation, given a set of $\mathbf{n}$ linear equations with $\mathbf{n}$ variables, if the determinant of the matrix $\mathbf{T}$ is not zero, the set of solutions has cardinal 1.

**Proof**
It is a direct application of proposition 451.

**Proposition 453** With the above notation, given the set of $\mathbf{n}$ linear equations with $\mathbf{n}$ variables

$$\mathbf{T^j_i x^i = 0} \ ,$$

if the determinant of the matrix $\mathbf{T}$ is not zero, the unique solution is:

$$(\forall \mathbf{i} \in [\mathbf{1}, \mathbf{n}])(\mathbf{x^i = 0}) \ .$$

**Proof**
We have:

$$(\forall \mathbf{i} \in [\mathbf{1}, \mathbf{n}])(\mathbf{x^i = 0}) \Rightarrow \mathbf{T^j_i x^i = 0}$$

then $(\forall \mathbf{i} \in [\mathbf{1}, \mathbf{n}])(\mathbf{x^i = 0})$ is a solution of $\mathbf{T^j_i x^i = 0}$ From proposition 452 (p. 258) it is unique. Then we have proposition 453 (p. 259).

**Proposition 454** . With the above notation, given the set of $\mathbf{n}$ linear equations with $\mathbf{n}$ variables

$$\mathbf{T^j_i x^i = 0} \ ,$$

if it has a solution with one $\mathbf{i} \in [\mathbf{1}, \mathbf{n}]$ such that $\mathbf{x^i \neq 0}$, then the determinant of the matrix $\mathbf{T}$ is zero.

**Proof**
It is a direct application of proposition 452 since there is already a solution with all $\mathbf{x^i = 0}$.

**Proposition 455** With the above notation, the unique solution of

$$\mathbf{T^j_i x^i = y^j}$$

when $\mathbf{det(T^j_i) \neq 0}$ is

$$\mathbf{x^i = {T^{-1}}^i_j y^j}$$

where

$$\mathbf{{T^{-1}}^j_i = (-1)^{i+k} det(Minor^k_i(T))/det(T)}$$

(see definition 228 (p. 254) and proposition 444 (p. 255)).

**Proof**
From proposition 444 (p. 255), we have $\mathbf{{T^{-1}}^j_k T^k_i = \delta^j_i}$. Then

$$\mathbf{(det(T) \neq 0) \Rightarrow (T^j_i x^i = y^j, {T^{-1}}^k_j T^j_i x^i = {T^{-1}}^k_j y^j, x^i = {T^{-1}}^k_j y^j)} \ .$$

### 2.11.6 Riemannian inner product (or simply inner product)

We will define here the Riemannian inner product which we shall simply call inner product. (We shall see later on (see 2.13.2 (p. 273)) a semi Riemannian inner product which is used in relativity and in manifold but in that case we specify it. ) We want to do it before introducing complex numbers (2.13 (p. 271)). In what follows, we suppose this new definition of $\mathcal{RC}$:

**Definition 231** $\mathcal{RC}$.

From now on, $\mathcal{RC}$ is restricted to a field (definition 181) which has $\mathcal{R}$ as a subset stable over the two internal laws of the field and which has two maps defined on it: the norm and the complex conjugate.

**Definition 232 Norm and complex conjugate on** $\mathcal{RC}$.

The norm written as $||\mathbf{a}||$ is a map

$$(||\mathrm{a}|| : \mathcal{RC} \to \mathcal{R}^{+0})$$

This map should verify the definition 203 (p. 237) (see definitions 154 (p. 181)):

$$(\mathrm{a} \in \mathcal{RC}, \mathrm{b} \in \mathcal{RC}) \Rightarrow$$

1. $\mathbf{a} \in \mathcal{R} \Rightarrow ||\mathrm{a}|| = |\mathrm{a}|$
2. $||\mathrm{a}|| = 0 \Rightarrow \mathrm{a} = 0$
3. $||\mathbf{ab}|| = ||\mathrm{a}||||\mathrm{b}||$.

Using that map one defines the complex conjugate map: $(\lambda^* : \mathcal{RC} \to \mathcal{RC})$

$$\lambda^* = \frac{||\lambda||^2}{\lambda}$$

then

$$\lambda^*\lambda = \lambda\lambda^* = ||\lambda||^2, ||\lambda|| = \sqrt{\lambda^*\lambda}\ .$$

That map should have the following properties:

$$\lambda \in \mathcal{RC}, \mu \in \mathcal{RC} \Rightarrow (\lambda + \mu)^* = \lambda^* + \mu^*\ . \tag{2.96}$$

**Definition 233 Norm and complex conjugate on $\mathcal{R}$.**

We reduce definitions 232 (p. 259) to the case where $\mathcal{RC} = \mathcal{R}$ by making $\mathbf{a} \in \mathcal{R} \Rightarrow (||\mathrm{a}|| = |\mathrm{a}|, \mathrm{a}^* = \mathrm{a})$. It is clear from definitions 232 (p. 259): Those are the maps which are also defined for $\mathcal{R}$ as a subspace of $\mathcal{RC}$.

**Proposition 456** With the above notations, we have the following properties for complex conjugate:

1. $||1/\lambda|| = 1/||\lambda||$
2. $(\lambda^*)^* = \lambda$
3. $(\lambda\mu)^* = \lambda^*\mu^*$
4. $0^* = 0$.

**Proof**

1 is given by definitions 232 (p. 259), 1, -3:

$$||1/\lambda||||\lambda|| = ||\lambda/\lambda|| = ||1|| = 1$$

then $||1/\lambda|| = 1/||\lambda||$.

2 is given by:

$$(\lambda^*)^* = \frac{(||\lambda||^2(1/||\lambda||))^2}{||\lambda||^2/\lambda} = \frac{||\lambda||^2}{||\lambda||^2/\lambda} = \lambda$$

3 is given by:

$$(\lambda\mu)^* = \frac{||\lambda\mu||^2}{\lambda\mu} = \frac{||\lambda||^2}{\lambda}\frac{||\mu||^2}{\mu} = \lambda^*\mu^*$$

4 is given by:

$$(\forall\lambda \in \mathcal{RC})(0 = \lambda + (-1)\lambda, 0^* = \lambda^* + (-1)^*\lambda^* = \lambda^* + (-1)\lambda^* = \lambda^* - \lambda^* = 0)\ .$$

**Proposition 457** The norm defined by $(\|\mathbf{a}\| : \mathcal{RC} \to \mathcal{R}^{+0})$ of definition 232 is also a norm on $\mathcal{RC}$ when $\mathcal{RC}$ is considered as a vectorial space (definition 199 (p. 235)).

**Proof**

From definition 232 (p. 259), it verifies points 1, 2 and 3 of definition 203 (p. 237). From properties defined in definition 232 (p. 259), we have also point 4:

$$\lambda \in \mathcal{RC}, \mu \in \mathcal{RC} \Rightarrow \|\lambda + \mu\| \leq \|\lambda\| + \|\mu\|$$

$$\begin{aligned}
&(\forall \lambda \in \mathcal{RC})( \\
&(\lambda + \lambda^*)(\lambda + \lambda^*)^* = (\lambda + \lambda^*)^2 \geq 0, \\
&(\lambda - \lambda^*)(\lambda - \lambda^*)^* = -(\lambda - \lambda^*)^2 \geq 0, \\
&(\lambda + \lambda^*)^2 - (\lambda - \lambda^*)^2 = 4\lambda\lambda^* \\
&\Rightarrow (\lambda + \lambda^*)^2 \leq 4\lambda\lambda^*)
\end{aligned} \tag{2.97}$$

we also have:

$$(\lambda - \mu)(\lambda^* - \mu^*) = \lambda\lambda^* + \mu\mu^* - \lambda\mu^* - \lambda^*\mu \in \mathcal{R}, \lambda\lambda^* \in \mathcal{R}, \mu\mu^* \in \mathcal{R}$$
$$\Rightarrow -\lambda\mu^* - \lambda^*\mu \in \mathcal{R} \Rightarrow -\lambda\mu^* - \lambda^*\mu \leq |\lambda\mu^* + \lambda^*\mu| = \sqrt{(\lambda\mu^* + \lambda^*\mu)^2}\,.$$

Since from equation (2.97):

$$(\lambda\mu^* + \lambda^*\mu)^2 \leq 4(\lambda\lambda^*\mu\mu^*) = 4\|\lambda\|^2\|\mu\|^2$$

then:

$$\begin{aligned}
0 &\leq (\lambda - \mu)(\lambda^* - \mu^*) \leq \lambda\lambda^* + \mu\mu^* + 2\|\lambda\|\|\mu\| \\
&= \|\lambda\|^2 + \|\mu\|^2 + 2\|\lambda\|\|\mu\| = (\|\lambda\| + \|\mu\|)^2
\end{aligned}$$

then:

$$\|\lambda - \mu\| \leq \|\lambda\| + \|\mu\|\,.$$

Since $\|-\mu\| = \|-1 \times \mu\| = |-1|\|\mu\| = \|\mu\|$ we also have:

$$\|\lambda + \mu\| = \|\lambda - (-\mu)\| \leq \|\lambda\| + \|\mu\|$$

and proposition 457 is true.

**Proposition 458** The map $(\mathbf{d}(\lambda, \mu) = \|\lambda - \mu\| : \mathcal{RC} \to \mathcal{R}^{+0})$ is a Euclidean metric (definition 147 (p. 176)).

**Proof**

With the properties of definitions 232 and proposition 457, $\mathbf{d}(\lambda, \mu) = \|\lambda - \mu\|$ verifies all requirement of the Euclidean metric (definitions 147 (p. 176)).

**Definition 234 Inner product (Riemannian inner product), adjoint of a vector.**

Given a vectorial space $\mathbf{V}$ on the field $\mathcal{RC}$ defined as in definitions 231 (p. 259) with the definitions 232 (p. 259) of a norm and of complex conjugate, one calls an inner product, a "semi-linear" map of $\mathbf{V}$ over its dual $\mathbf{V}^*$: [22]

$$(<\mathbf{v}| : \mathbf{V} \to \mathbf{V}^*)\,.$$

[22]We adopt here the physicist notation for the inner product called the bracket notation. It is supposed to make the equations clearer. But we shall not use all the time that notation which is rather heavy and we shall often only use the up and down positions of the indices to go from the vectorial space to its dual.

The element $\mathbf{v}^\dagger$ of $\mathbf{V}^*$ corresponding to $\mathbf{v}$ is also called the adjoint of $\mathbf{v}$:

$$\mathbf{v}^\dagger =< \mathbf{v}|,\ \mathbf{v}^\dagger(\mathbf{u}) =< \mathbf{v}|\mathbf{u} > \ .$$

Since $< \mathbf{v}| \in \mathbf{V}^*$, it is a linear map of $\mathbf{V}$ on $\mathcal{RC}$. Its action on a vector $|\mathbf{u} >\in \mathbf{V}$ is quoted as $< \mathbf{v}|\mathbf{u} >\in \mathcal{RC}$. In order to define an inner product, this map should have the following "semi-linear" properties:

$$(|\mathbf{u} >\in \mathbf{V}) \Rightarrow (< \mathbf{u}|\mathbf{u} >= 0 \Leftrightarrow |\mathbf{u} >= 0) \tag{2.98}$$

$$(|\mathbf{u} >\in \mathbf{V}) \Rightarrow (< \mathbf{u}|\mathbf{u} >\in \mathcal{R}^{+0}) \tag{2.99}$$

$$\begin{gathered}(|\mathbf{u} >\in \mathbf{V},\ |\mathbf{v} >\in \mathbf{V}) \Rightarrow (< \mathbf{u}|\mathbf{v} >=< \mathbf{v}|\mathbf{u} >^*) \\ |\mathbf{a} >= \lambda|\mathbf{u} > +\mu|\mathbf{v} >\Rightarrow< \mathbf{a}| = \lambda^* < \mathbf{u}| + \mu^* < \mathbf{v}|\ .\end{gathered} \tag{2.100}$$

**Definition 235 Matrix associated to an inner product.**

If the vectorial space has a finite dimension $\mathbf{n}$ with a basis $\mathbf{B_e}$ defined by $(|\mathbf{e_i} >: [1,\mathbf{n}] \to \mathbf{V}) \in \mathbf{B_e}$, we can consider the operator $(\mathbf{g} : \mathbf{V} \to \mathbf{V}^*)$ such that $\mathbf{g}(\mathbf{v}) =< \mathbf{v}|$ and the associated matrix $(\mathbf{g_{ij}} : [1,\mathbf{n}] \times [1,\mathbf{n}] \to \mathcal{RC})$ defined by:

$$\begin{gathered}\mathbf{g_{ij}} =< \mathbf{e_i}|\mathbf{e_j} >= \mathbf{g}(\mathbf{e_i})(\mathbf{e_j}) \\ < \mathbf{e_i}| = \mathbf{g}(\mathbf{e_i}) = \mathbf{g_{ij}}\mathbf{e^j}\end{gathered} \tag{2.101}$$

With $(\mathbf{e^i} : [1,\mathbf{n}] \to \mathbf{V}^*)$ the basis of the dual space $\mathbf{V}^*$ $(\mathbf{e^i}(\mathbf{e_j}) = \delta^i_j)$, a vector $< \mathbf{v}| = \mathbf{g}(\mathbf{v}) \in \mathbf{V}^*$ corresponding to $|\mathbf{v} >\in \mathbf{V}$ can be written as:

$$< \mathbf{v}| = \mathbf{g}(|\mathbf{v} >) = \mathbf{v}^{*\mathbf{i}} < \mathbf{e_i}| = \mathbf{v}^{*\mathbf{i}}\mathbf{g}(|\mathbf{e_i} >) = \mathbf{v}^{*\mathbf{i}}\mathbf{g_{ij}}\mathbf{e^j} \Leftrightarrow \mathbf{g_{ij}} =< \mathbf{e_i}|\mathbf{e_j} > \ .$$

That defines the map $(\mathbf{g} : \mathbf{V} \to \mathbf{V}^*) = (< \mathbf{v}| : \mathbf{V} \to \mathbf{V}^*)$ The above conditions imply:

$$(\forall \mathbf{i} \in [1,\mathbf{n}])(\forall \mathbf{j} \in [1,\mathbf{n}])(\mathbf{g_{ij}} = \mathbf{g^*_{ji}}) \tag{2.102}$$

and

$$\begin{gathered}(|\mathbf{u} >= \mathbf{u^i e_i} = \mathbf{u^i}|\mathbf{e_i} >, |\mathbf{v} >= \mathbf{v^i e_i} = \mathbf{v^i}|\mathbf{e_i} >), < \mathbf{e_i}| = \mathbf{g_{ij}}\mathbf{e^j} \\ \Rightarrow (< \mathbf{u}|\mathbf{v} >= \mathbf{u^{i*}}\mathbf{v^j} < \mathbf{e_i}|\mathbf{e_j} >= \mathbf{u^{i*}}\mathbf{g_{ij}}\mathbf{v^j})\ .\end{gathered}$$

To make the thing fully symmetric:

$$< \mathbf{e_i}| = \mathbf{g_{ij}}\mathbf{e^j} = \mathbf{g_{ij}} < \mathbf{e^j}| \Rightarrow \mathbf{g^*_{ij}}|\mathbf{e^j} >= |\mathbf{e_i} > \ . \tag{2.103}$$

**Definition 236 Inner product when $\mathcal{RC} = \mathcal{R}$.**

Definition 236 is definition 234 (p. 261) setting $\mathcal{RC} = \mathcal{R}$. It means according to definition 233 (p. 260), $\mathbf{a} \in \mathcal{R} \Rightarrow (||\mathbf{a}|| = |\mathbf{a}|, \mathbf{a}^* = \mathbf{a})$ and $\mathbf{g_{ij}} = \mathbf{g^*_{ij}} \in \mathcal{R}$.

**Proposition 459** Given a vectorial space $\mathbf{V}$ on the field $\mathcal{RC}$ of finite dimension $\mathbf{n}$ with an independent basis $(\mathbf{e_i} : [1,\mathbf{n}] \to \mathbf{V})$ with a Riemannian inner product $\mathbf{g}$. The determinant of the associated matrix (definition 235 (p. 262)) $\mathbf{g_{ij}} =< \mathbf{e_i}|\mathbf{e_j} >$ is different from zero.

**Proof**

We have $\mathbf{det(g_{ij}) \neq 0}$ because since the $\mathbf{e_j}$ are independent we have from definition 206 (p. 239), equation (2.98) and propositions 456 (p. 260), 4:

$$\begin{aligned}&\lambda^i g_{ij} e_j = 0 \Rightarrow (\forall j \in [1,n])((\lambda^i g_{ij} = 0) \Rightarrow < \lambda^{*i} e_i | \lambda^{*j} e_i >= \lambda^i g_{ij} \lambda^{*j} = 0)\\ &\Rightarrow \lambda^{*j} e_i = 0 \Rightarrow (\forall i \in [1,n])(\lambda^{*i} = \lambda^i = 0)\ .\end{aligned}$$

Then from definition 206 (p. 239), the $\mathbf{n}$ vectors $\mathbf{g_{ij} e_j}$ are independent. From proposition 430 (p. 250) $\mathbf{det(g_{ij}) \neq 0}$.
From propositions 450 (p. 258) and 451 (p. 258), there is a matrix $\mathbf{g^{ij}}$ such that $\mathbf{g^{ik} g_{kj}} = \mathbf{g^{ik} g_{kj}} = \delta^i_j$ (proposition 445 (p. 256)) then by convention:

$$\begin{aligned}e^i &=< e^i | = g^{ik} < e_k |, |e^i >= g^{*ik} |e_k >\\ e^i e_j &= g^{ik} < e_k | e_j >= g^{ik} g_{kj} = \delta^i_j\ .\end{aligned} \tag{2.104}$$

**Proposition 460** The inner product defines a norm through $\|\mathbf{u}\| = \sqrt{<\mathbf{u}|\mathbf{u}>}$ (see definition 203 (p. 237)).

**Proof**

Points 1 to 3 of definition 203 (p. 237) are verified.
Point 4 is verified with the use of the Cauchy-Schwartz inequality:

$$\begin{aligned}&|a > \in V, |b > \in V \Rightarrow\\ &(< a| < a|b >^* / \|a\|^2 - < b|)(< a|b > / \|a\|^2 |a > - |b >) \geq 0\\ &\Rightarrow < a|b >^* < a|b > / \|a\|^2 + < b|b > -2 < a|b >^* < a|b > / \|a\|^2 \geq 0\\ &\Rightarrow \|a\|^2 \|b\|^2 \geq < a|b >< a|b >^* \Rightarrow \|a\| \|b\| \geq | < a|b > |\ .\end{aligned}$$

Applying now equation (2.97) (p. 261), we get the triangular inequality:

$$\begin{aligned}&< a - b | a - b >=< a|a > + < b|b > - < a|b > - < b|a >\\ &\leq \|a\|^2 + \|b\|^2 + \sqrt{(< a|b > + < a|b >^*)^2}\\ &\leq \|a\|^2 + \|b\|^2 + 2\sqrt{< a|b >< a|b >^*}\\ &\leq \|a\|^2 + \|b\|^2 + 2| < a|b > |\\ &\leq \|a\|^2 + \|b\|^2 + 2\|a\| \|b\|\\ &\Rightarrow \|a - b\|^2 \leq (\|a\| + \|b\|)^2 \Rightarrow \|a - b\| < \|a\| + \|b\|\ .\end{aligned}$$

**Definition 237 Euclidean vector space, Euclidean norm.**

When a vectorial space $\mathbf{V}$ of finite dimension $\mathbf{n}$ with a basis $\mathbf{e_i : [1,n] \rightarrow V}$ has an inner product given $\mathbf{g^{ij}} =< \mathbf{e_i | e_j} >= \delta^i_j$, it is often called Euclidean and its norm Euclidean (see definition 215 (p. 245) for $\delta^i_j$). It introduces confusion with Euclidean metric. Euclidean space has a Euclidean metric but any norm (definition 203 (p. 237)) on a vector space leads also to a Euclidean metric (definition 147 (p. 176) proposition 401 (p. 238)).

**Definition 238 Orthogonal vectors.**

A set of vectors $\mathbf{O}$ of a vectorial space $\mathbf{V}$ with an inner product, is said to form a set of orthogonal vectors if the inner product of any two different vectors of the set is zero.

$$a \in O, b \in O, a \neq b \Rightarrow < a|b >= 0\ .$$

**Proposition 461** A set of orthogonal vectors $\mathbf{O}$ is an independent set.

**Proof**

If we have $\sum_{v\in O} \lambda_v v = 0$, then:

$$(\forall u \in O)(\sum_{v \in O} < \lambda_v v|u >= \lambda_u < u|u >= 0 \Rightarrow \lambda_u = 0) \ .$$

**Proposition 462** Given a set of orthogonal vector $\mathbf{O}$ from a finite dimension vectorial space $\mathbf{V}$ with an inner product, if $\mathbf{O}$ is not a basis of $\mathbf{V}$ there is a vector $\mathbf{v} \in \mathbf{V}$ which is orthogonal to all vectors of $\mathbf{O}$.

**Proof**

Since $\mathbf{O}$ is a set of independent vector (proposition 461) which is not a basis of $\mathbf{V}$ there is a vector $\mathbf{u}$ of $\mathbf{V}$ such that $O \cup \{u\}$ is an independent set of vectors. The vector:

$$v = u - \sum_{w \in O} \frac{< w|u >}{< w|w >} w$$

is orthogonal to all vectors of $\mathbf{O}$ and it is not zero because the set $O \cup \{u\}$ is an independent set. From proposition 461, $O \cup \{v\}$ is an independent set.

**Proposition 463** A finite dimension vectorial space with an inner product has an independent basis which is an orthogonal set of vectors.

**Proof**

Same demonstration as in proposition 415 (p. 244) using proposition 462.

**Proposition 464** $\mathbf{O}$ being an orthogonal basis of a normed vectorial space $\mathbf{V}$ of finite dimension $\mathbf{n}$, we have the following identity:

$$(\forall v \in V)(v = \sum_{o \in O} \frac{< o|v >}{< o|o >} |o >) \ .$$

**Proof**

$\mathbf{O}$ being an orthogonal basis with $\mathbf{Card(O) = n}$, for any $\mathbf{v} \in \mathbf{V}$, there is $(\lambda_o : O \to V)$ such that:

$$|v >= \sum_{o \in O} \lambda_o |o >$$

for any vector $\mathbf{o'} \in \mathbf{O}$ we have since $< o|o' >= \delta_o^{o'}$ (definition 215 (p. 245))

$$< o'|v >= \lambda'_o < o'|o' > \ .$$

This equality plugs into the preceding one gives proposition 464.

**Definition 239 Adjoint transformation over an inner product.**

Given a finite dimension vectorial space $\mathbf{V}$ on $\mathcal{RC}$ with an inner product $(g(u,v) : V^2 \to \mathcal{RC})$ and a linear transformation $(T : V \to V)$, one calls the adjoint $(T^\dagger : V \to V)$ of $\mathbf{T}$ a linear transformation such as:

$$(\forall u)(\forall v)(g(u, T(v)) = g(T^\dagger(u), v)) \ .$$

Setting $(\mathbf{e_i} : [\mathbf{1}, \mathbf{n}] \to \mathbf{V})$, an independent basis of $\mathbf{V}$, $\mathbf{u} = \mathbf{u^i e_i}$, $\mathbf{v} = \mathbf{v^i e_i}$, $\mathbf{g(u, v)} = \mathbf{g_{ij} u^{*i} v^j}$. One may consider $\mathbf{g_{ij}}$ as a linear transformation $\mathbf{g'}$ on $\mathbf{V}$ with $\mathbf{g'(e_i)} = \mathbf{g(e_j, e_i) e_j}$ and if $\mathbf{u} = \mathbf{u^i e_i}$, we have $\mathbf{g'(u)} = \mathbf{g(e_i, u) e_i} = \mathbf{g_{ij} u^j e_i}$. If $\mathbf{g_{ij}}$ has a nonzero determinant, $\mathbf{g'}$ has an inverse (proposition 459 (p. 262)) and there is a $\mathbf{g^{ij}}$ (equation (2.104)) such that $\mathbf{g^{ik} g_{kj}} = \delta^{\mathbf{i}}_{\mathbf{j}}$. The adjoint always exists since we should have:

$$\mathbf{u^{*i} g_{ik} T^k_j v^j} = (\mathbf{T^{\dagger k}_{\ \ i} u^i})^* \mathbf{g_{kj} v^j}$$

or:

$$\mathbf{g_{ik} T^k_j} = (\mathbf{T^{\dagger k}_{\ \ i}})^* \mathbf{g_{kj}}$$

we have:

$$(\mathbf{g_{ik} T^k_j g^{jl}})^* = \mathbf{T^{\dagger l}_{\ \ i}}$$

If $\mathbf{g^{il}} = \delta^{\mathbf{il}}$, $\mathbf{g_{kj}} = \delta_{\mathbf{kj}}$ (see definition 215 (p. 245)), we have

$$\mathbf{T^{*i}_j} = \mathbf{T^{\dagger j}_{\ \ i}} \,.$$

**Proposition 465** $\mathbf{S}$ and $\mathbf{T}$ being two linear transformations of a finite dimension vectorial space with an inner product. The adjoint of $\mathbf{ST}$ is $\mathbf{T^\dagger S^\dagger}$.

**Proof**
Calling $(\mathbf{g(u, v)} : \mathbf{V^2} \to \mathcal{R})$ the inner product, we have by definition:

$$\mathbf{g(u, S(T(v)))} = \mathbf{g(S^\dagger(u), T(v))} = \mathbf{g(T^\dagger S^\dagger(u), v)} \,.$$

**Definition 240 Unitary transformation.**

Given a vectorial space with finite dimension with an inner product defining a norm. A transformation is said to be unitary on that space with that inner product, if the inner product of any two vectors of that space is equal to the inner product of the two transformed vectors over that transformation.
If $\mathbf{n}$ is the dimension of a vectorial space $\mathbf{V}$ with an inner product $\mathbf{g}$ defining a norm, $(\mathbf{e_i} : [\mathbf{1}, \mathbf{n}] \to \mathbf{V})$ an independent basis and $(\mathbf{g_{ij}} : [\mathbf{1}, \mathbf{n}]^2 \to \mathcal{R})$, a unitary transformation $(\mathbf{T} : \mathbf{V} \to \mathbf{V})$ must verify:

$$(\forall \mathbf{i} \in [\mathbf{1}, \mathbf{n}])(\forall \mathbf{j} \in [\mathbf{1}, \mathbf{n}])(\mathbf{g(e_i, e_j)} = \mathbf{g(T(e_i), T(e_j))}) \,.$$

The matrix of the inner product in the basis $\mathbf{e_i}$ being $\mathbf{g_{ij}} = \mathbf{g(e_i, e_j)}$ and the unitary transformation $(\mathbf{T^i_j} : [\mathbf{1}, \mathbf{n}]^2 \to \mathcal{R}) \Rightarrow \mathbf{T(e_i)} = \mathbf{T^j_i e_j}$, we should have:

$$\mathbf{g(T^j_i e_j, T^l_k e_l)} = \mathbf{T^{*j}_i T^l_k g_{jl}} = \mathbf{g_{ik}} \,.$$

If $\mathbf{g_{il}} = \delta_{\mathbf{il}}$ (Euclidean norm) and if one calls $\mathbf{{}^t T^i_j} = \mathbf{T^j_i}$ one should have:

$$\mathbf{{}^t T^{*i}_j \delta^j_l T^l_k} = \mathbf{{}^t T^{*i}_j T^j_k} = \mathbf{T^{*j}_i T^j_k} = \mathbf{T^{\dagger i}_j T^j_k} = \delta^{\mathbf{i}}_{\mathbf{k}} \,.$$

**Proposition 466** The product of two unitary transformations is a unitary transformation.

**Proof**
It is a direct consequence of the definition and of the linearity properties.

**Proposition 467** Given a vectorial space **V**, the set of linear transformations is a vectorial space.

**Proof**
**S** and **T** being two linear transformations, $\boldsymbol{\lambda}$ and $\boldsymbol{\mu}$ two numbers from $\mathcal{RC}$. $\boldsymbol{\lambda}\mathbf{S} + \boldsymbol{\mu}\mathbf{T}$ is the linear transformation such that:

$$(\forall \mathbf{u} \in \mathbf{V})((\boldsymbol{\lambda}\mathbf{S} + \boldsymbol{\mu}\mathbf{T})\mathbf{u} = \boldsymbol{\lambda}\mathbf{S}\mathbf{u} + \boldsymbol{\mu}\mathbf{T}\mathbf{u}) .$$

The unit element (or the zero element) is the linear transformation $\mathbf{T_0}$ defined by:

$$(\forall \mathbf{u} \in \mathbf{V})(\mathbf{T_0}\mathbf{u} = \mathbf{0}) .$$

Since it is clear enough in most of the cases, one uses to write $\mathbf{T_0}$ as $\mathbf{0}$, $-\mathbf{T}$ is the inverse of $\mathbf{T}$ since $\mathbf{T} - \mathbf{T} = 0$.

**Proposition 468** We suppose that we have a set **A** of linear transformations with an inverse on a finite dimension vectorial space **V**. From propositions 448 (p. 257), 450 (p. 258) and 451 (p. 258), the set **A** forms a group over the product of transformations. We can define $\boldsymbol{\lambda}\mathbf{A}$ $\forall \boldsymbol{\lambda} \in \mathcal{RC}, \boldsymbol{\lambda} \neq \mathbf{0}$ by:

$$\boldsymbol{\lambda}\mathbf{A} = \mathbf{set}_{\mathbf{T}}\{(\exists \mathbf{S} \in \mathbf{A})(\mathbf{T} = \boldsymbol{\lambda}\mathbf{S})\} .$$

Then the set $\bigcup_{\boldsymbol{\lambda} \in \mathcal{RC}, \boldsymbol{\lambda} \neq 0} \boldsymbol{\lambda}\mathbf{A}$ is also a group.

**Proof**
Calling $\mathbf{A_U} = \bigcup_{\boldsymbol{\lambda} \in \mathcal{RC}, \boldsymbol{\lambda} \neq 0} \boldsymbol{\lambda}\mathbf{A}$,

$$\boldsymbol{\lambda}\mathbf{U} \in \mathbf{A_U}, \boldsymbol{\mu}\mathbf{V} \in \mathbf{A_U} \Rightarrow \boldsymbol{\lambda}\mathbf{U}\boldsymbol{\mu}\mathbf{V} = \boldsymbol{\lambda}\boldsymbol{\mu}\mathbf{U}\mathbf{V} \in \mathbf{A_U}$$

because $\boldsymbol{\lambda} \neq \mathbf{0}, \boldsymbol{\mu} \neq \mathbf{0} \Rightarrow \boldsymbol{\lambda}\boldsymbol{\mu} \neq \mathbf{0}$ and because $\boldsymbol{\lambda}\boldsymbol{\mu} \neq \mathbf{0} \Rightarrow (\boldsymbol{\lambda}\boldsymbol{\mu}\mathbf{U}\mathbf{V})^{-1} = \frac{1}{\boldsymbol{\lambda}\boldsymbol{\mu}}\mathbf{V}^{-1}\mathbf{U}^{-1}$.

## 2.12 $\mathcal{R}^{\mathbf{n}}$

**Definition 241** $\mathcal{R}^{\mathbf{n}}$**.**

Let us define $\mathcal{R}^{\mathbf{n}}$ as the set of maps $(\mathcal{R}^{\mathbf{n}} = \mathbf{set}_{\mathbf{x}}\{(\mathbf{x} : [\mathbf{1}, \mathbf{n}] \subset \mathcal{Z} \rightarrow \mathcal{R})\}$ which is the product of $\mathcal{R}$ **n** times see definition 372 (p. 473). An element of $\mathbf{x} \in \mathcal{R}^{\mathbf{n}}$ is written (see definition 126 (p. 144)) as:

$$\mathbf{x} = \prod_{\mathbf{i} \in [\mathbf{1},\mathbf{n}]}^{\mathbf{e}} \mathbf{x}^{\mathbf{i}} = (\mathbf{x}^{\mathbf{i}} : [\mathbf{1}, \mathbf{n}] \rightarrow \mathcal{R})$$

It is also often written as $\mathbf{x} = (\mathbf{x}^{\mathbf{1}}, \mathbf{x}^{\mathbf{2}}, ..., \mathbf{x}^{\mathbf{n}})$.

**Proposition 469** $\mathcal{R}^{\mathbf{n}}$ can be considered as a vectorial space (definition 199 (p. 235)) with (definition 202 (p. 237)):

$$\mathcal{R}^{\mathbf{n}} = \prod_{\mathbf{i} \in [\mathbf{1},\mathbf{n}]} \mathcal{R} = (\oplus\mathcal{R})^{\mathbf{n}} = \bigoplus_{\mathbf{i} \in [\mathbf{1},\mathbf{n}]} \mathcal{R} .$$

**Proof**
$\mathcal{R}^n$ is a vector space (definition 199 (p. 235)) with $\mathcal{R}$ as a field because we may write (see also definition 202 (p. 237)):

$$\lambda \in \mathcal{R}, \mu \in \mathcal{R}$$

$$\begin{aligned} x &= (x^1, x^2, ..., x^n) \\ y &= (y^1, y^2, ..., y^n) \\ \lambda x + \mu y &= (\lambda x^1 + \mu y^1, \lambda x^2 + \mu y^2, ..., \lambda x^n + \mu y^n) . \end{aligned} \tag{2.105}$$

**Definition 242 Reference or standard basis of $\mathcal{R}^n$.**

We call reference basis or standard basis, the $\mathbf{n}$ maps $(e_i : [1, n] \to \mathcal{R}^n)$ which are defined as:

$$(\forall i \in [1, n])(e_i = \prod_{j \in [1,n]}^{e} \delta_i^j \in \mathcal{R}^n) , \tag{2.106}$$

where $\delta_i^j$ is the Kronecker symbol (definition 215 (p. 245)). This basis could also be written as a map of $[1, n]$ to $\mathcal{R}^n$:

$$\begin{aligned} &(e_i : [1, n] \to \mathcal{R}^n) \\ &e_i = (\delta_i^j : [1, n] \to \mathcal{R}^n) \end{aligned}$$

$\mathbf{i}$ being the variable (see definition 372 (p. 473)) in the first line and $\mathbf{j}$ in the second line. We can write any term of $\mathcal{R}^n$ as:

$$x = x^i e_i = x^i \delta_i^j = (x^j : [1, n] \to \mathcal{R}) = \prod_{j \in [1,n]}^{e} x^j \in \mathcal{R}^n .$$

**Proposition 470** The reference basis of $\mathcal{R}^n$ is a basis of $\mathcal{R}^n$.

**Proof**
From definition 242 (p. 267) and proposition 469 (p. 266), any $x = (x^i : [1, n] \to \mathcal{R}) \in \mathcal{R}^n$ is such that

$$x = \sum_{j \in [1,n]} x^j (\delta_j^i : [1, n] \to \mathcal{R}) .$$

**Proposition 471** The reference basis of $\mathcal{R}^n$ is an independent set of vectors (definition 206 (p. 239)).

**Proof**
From definition 241 (p. 266) and proposition 469 (p. 266), we have:

$$x = (x^i : [1, n] \to \mathcal{R}) = \sum_{j \in [1,n]} x^j \delta_j^i \Rightarrow (x = 0 \Leftrightarrow (\forall i \in [1, n])(x^i = 0)) .$$

**Definition 243 p_norm in $\mathcal{R}^n$.**

$\mathcal{R}^n$ can be considered as a vector space with the **p_norm** ($\mathbf{p} \in \mathcal{R}, \mathbf{p} \geq 1$):

$$\|x\|_p = (|x^1|^p + |x^2|^p + ... + |x^n|^p)^{1/p} = (\sum_{i\in[1,n]} |x^i|^p)^{1/p} \tag{2.107}$$

for $\mathbf{p} = \mathbf{2}$, one has the Euclidean norm:

$$\|x\|_2 = (\sum_{i\in[1,n]} (x^i)^2)^{1/2} \tag{2.108}$$

for $\mathbf{p} = \infty$ it is the norm which leads to the cube topology: [23]

$$\|x\|_\infty = \lim_{p\to\infty} \|x\|_p = \lim_{p\to\infty} (\sum_{i\in[1,n]} |x^i|^p)^{1/p} = \max_{i\in[1,n]} |x_i| \,. \tag{2.109}$$

**Proposition 472** Given $\mathbf{x} \in \mathcal{R}^n$: $\mathbf{x} = (\mathbf{x}^i : [1,n] \to \mathcal{R})$ (definition 241 (p. 266)) and an element $\mathbf{A} \in \mathcal{R}^{n^2}$: $\mathbf{A} = (\mathbf{A}^i_j : [1,n]^2 \to \mathcal{R})$ we have the following inequality:

$$\| \sum_{j\in[1,n]} A^i_j x^j \|_p = \|Ax\|_p \leq \|A\|_p \|x\|_p \,.$$

**Proof**
Using propositions 273 (p. 181), 328 (p. 202) and 368 (p. 217), we have:

$$\|A\|_p = (\sum_{i\in[1,n]} \sum_{j\in[1,n]} |A^i_j|^p)^{1/p}$$

$$\| \sum_{j\in[1,n]} A^i_j x^j \|_p = (\sum_{i\in[1,n]} | \sum_{j\in[1,n]} A^i_j x^j|^p)^{1/p} \leq (\sum_{i\in[1,n]} (\sum_{j\in[1,n]} |A^i_j|^p) \sum_{j\in[1,n]} |x^j|^p)^{1/p}$$
$$\leq (\sum_{i\in[1,n]} \sum_{j\in[1,n]} |A^i_j|^p)^{1/p} \times (\sum_{j\in[1,n]} |x^j|^p)^{1/p} \,.$$

**Proposition 473** $\mathbf{x} \in \mathcal{R}^n, \mathbf{y} \in \mathcal{R}^n, \mathbf{p} \in \mathcal{Z}^+, \mathbf{p} \geq 1 \Rightarrow \|\mathbf{x}\|_p - \|\mathbf{y}\|_p \leq \|\mathbf{x}+\mathbf{y}\|_p \leq \|\mathbf{x}\|_p + \|\mathbf{y}\|_p$.
**Proof**

1. We want to prove first:

$$\|x+y\|_p \leq \|x\|_p + \|y\|_p \tag{2.110}$$

$p = \infty$

$$\max_{i\in[1,n]} |x^i + y^i| \leq \max_{i\in[1,n]} (|x^i| + |y^i|) \leq \max_{i\in[1,n]} |x^i| + \max_{j\in[1,n]} |y_j|$$

this leads to

$$\|x+y\|_\infty \leq \|x\|_\infty + \|y\|_\infty \tag{2.111}$$

let us now demonstrate the relation (2.110), for $\mathbf{p} \geq 1$

[23] Let us suppose $(\forall i \in [1,n])(x^i > 0)$ and $\max_{i\in[1,n]} x_i = x_1$. Since $x^y, x > 0, y > 0$ is a strictly increasing map in $\mathbf{x}$, we have $x_1 \leq (\sum_{i\in[1,n]} |x^i|^p)^{1/p} \leq x^1 n^{1/p}$. We also have $1 \leq n^{1/p}$. $n^{1/p}$ is a strictly decreasing map of $\mathbf{p}$. If its **lower_limit** was $1+\epsilon$, $\epsilon > 0$, we would have (proposition 369 (p. 217)) $(\forall p \in \mathcal{Z}^+)(n \geq (1+\epsilon)^p \geq 1 + p\epsilon)$ which is impossible (proposition 259 (p. 175)). Then $\epsilon = 0$.

- n=1

$$\|x+y\|_p = (|x+y|)^p)^{1/p} = |x+y| \leq |x| + |y|$$
$$\Rightarrow \|x+y\|_p \leq \|x\|_p + \|y\|_p$$

- Let us suppose (2.110) hold up to $\mathbf{n}$ let us prove it for $\mathbf{n+1}$.
  Let us prove it for $\mathbf{y^{n+1} = 0}$. Let us set:

$$\begin{aligned} f(|x^{n+1}|^p) = & (\textstyle\sum_{i\in[1,n]} |x^i + y^i|^p + |x^{n+1}|^p)^{p/p} \\ & -((\textstyle\sum_{i\in[1,n]} |x^i|^p + |x^{n+1}|^p)^{1/p} + (\textstyle\sum_{i\in[1,n]} |y^i|^p)^{1/p})^p \end{aligned}$$

from what we suppose $\mathbf{f(0) < 0}$. The derivative of $\mathbf{f}$ is for $\alpha = |x_{n+1}|^p > 0$

$$\begin{aligned} f'(\alpha) = & 1 - ((\textstyle\sum_{i\in[1,n]} |x_i|^p + \alpha)^{1/p} \\ & +(\textstyle\sum_{i\in[1,n]} |y^i|^p)^{1/p})^{p-1} \times p \times (\textstyle\sum_{i\in[1,n]} |x_i|^p + \alpha)^{1/p-1}/p \end{aligned}$$

since $\mathbf{p > 1}$ and since all the expressed terms are positive, one can drop the terms beside $\alpha$ after the minus sign and one gets a bigger term:

$$f'(\alpha) \leq 1 - (\alpha^{1/p})^{p-1}\alpha^{1/p-1} = 0\,.$$

From that, we have $\alpha \geq 0 \Rightarrow f'(\alpha) \leq 0$.
Applying proposition 307 (p. 193), we get:

$$\alpha \geq 0 \Rightarrow -f(\alpha) \geq -f(0) > 0 \Rightarrow f(\alpha) < 0$$

our relation is true for $\mathbf{y^{n+1} = 0}$ whatever $\mathbf{x^{n+1}}$. We pursue with a similar argument when $\mathbf{y^{n+1} \neq 0}$ using:

$$\begin{aligned} & f(y^{n+1}) = \textstyle\sum_{i\in[1,n]} |x^i + y^i|^p + |x^{n+1} + y^{n+1}|^p \\ & -((\textstyle\sum_{i\in[1,n]} |x^i|^p + |x^{n+1}|^p)^{1/p} + (\textstyle\sum_{i\in[1,n]} |y^i|^p + |y^{n+1}|^p)^{1/p})^p\,. \end{aligned}$$

We may suppose $\mathbf{x^{n+1} > 0}$ otherwise one may change the sign of both $\mathbf{x^{n+1}}$ and $\mathbf{y^{n+1}}$, $\mathbf{f(y^{n+1})}$ clearly will not change when doing that. It is also enough to demonstrate that $\mathbf{f(y^{n+1}) < 0}$ for $\mathbf{y^{n+1} > 0}$ since we have $f(-|y^{n+1}|) < f(|y^{n+1}|)$ because $|x^{n+1} - |y^{n+1}||^p < |x^{n+1} + |y^{n+1}||^p$. Setting $\alpha = y^{n+1} > 0$, we may write when $\mathbf{p > 1}$:

$$\begin{aligned} & f'(\alpha) = p(x^{n+1} + \alpha)^{p-1} \\ & -p/p((\textstyle\sum_{i\in[1,n]} |x^i|^p + |x^{n+1}|^p)^{1/p} + (\textstyle\sum_{i\in[1,n]} |y^i|^p + \alpha^p)^{1/p})^{p-1} \\ & (\textstyle\sum_{i\in[1,n]} |y^i|^p + \alpha^p)^{1/p-1} p\alpha^{p-1} \\ & \leq p(x^{n+1} + \alpha)^{p-1} - (x^{n+1} + \alpha)^{p-1}(\alpha^p)^{1/p-1} p\alpha^{p-1} = 0\,. \end{aligned}$$

Since $\mathbf{f'(\alpha) < 0}$ for $\alpha \geq 0$ our relation is true also for any $\mathbf{x^{n+1}}$ and $\mathbf{y^{n+1}}$. We conclude, using the recurrence principle, for any $\mathbf{n}$ (proposition 927 (p. 520)).

2. $|\,\|x\|_p - \|y\|_p\,| \leq \|x+y\|_p$.
   It is a consequence of proposition 404 (p. 239).

**Proposition 474**

$$x \in \mathcal{R}^n, y \in \mathcal{R}^n, p \in \mathcal{Z}^+ \Rightarrow |\,\|x\|_p - \|y\|_p\,| \leq \|x-y\|_p \leq \|x\|_p + \|y\|_p\,.$$

**Proof**
Setting $\mathbf{y}' = -\mathbf{y}$, applying proposition 473 (p. 268) to $\mathbf{y}'$ and $\mathbf{x}$, replacing $\mathbf{y}'$ by $-\mathbf{y}$ in the result one gets proposition 474 (p. 269) by using $||-\mathbf{y}|| = ||\mathbf{y}||$.

**Proposition 475** The **p_norm** (definition 243 (p. 267)) is a norm (see definition 203 (p. 237)).

**Proof**
It is a direct consequence of propositions 472 (p. 268) and 473 (p. 268).

**Proposition 476** $\mathbf{d_p(x,y) = ||x-y||_p, p \in \mathcal{R}, p \geq 1}$ defines a Euclidean metric on $\mathcal{R}^n$

**Proof**
Applying proposition 474 (p. 269), we verify all the axioms of a Euclidean metric. See definition 147 (p. 176).

**Definition 244 Projection of $\mathcal{R}^n$ on $\mathcal{R}^{n-1}$.**

As in definition 121 (p. 136), one calls projection along the axis $\mathbf{i}$ of $\mathcal{R}^n$ on $\mathcal{R}^{n-1}$ the map $\mathbf{Pr_i} : \mathcal{R}^n \to \mathcal{R}^{n-1}$ which is acting on any maps $(\mathbf{x^k} : [1,n] \to \mathcal{R}) \in \mathcal{R}^n$ defining $\mathcal{R}^n$ giving the restricted map $(\mathbf{x^k} : [1,n] - \{i\} \to \mathcal{R}) \in \mathcal{R}^{n-1}$ (see definitions 126 (p. 144), 129 (p. 145))

**Proposition 477** A projection (definition 244 (p. 270)) along an axis $\mathbf{i}$ of $\mathcal{R}^n$ on $\mathcal{R}^{n-1}$ conserves the vectorial space internal laws:

$$(\lambda \in \mathcal{R}, \mu \in \mathcal{R}, a \in \mathcal{R}^n, b \in \mathcal{R}^n) \Rightarrow \mathrm{proj_i}(\lambda a + \mu b) = \lambda \mathrm{proj_i}(a) + \mu \mathrm{proj_i}(b) \ .$$

**Proof**
It is straightforward from the definition 244 (p. 270).

**Proposition 478** We consider $\mathcal{R}^n$ with a $\mathbf{p}$ norm. We call $\mathbf{e_i} = \prod^{e}_{j \in [1,n]} \delta^j_i = (\delta^j_i : [1,n] \to \mathcal{R}) \in \mathcal{R}^n$ the reference basis of $\mathcal{R}^n$. We are given a partition (definition 370 (p. 471)) of $[1,n]$ defined by a map $(\mathbf{g} : [1,n] \to [1,\ell]),\ \ell \leq n$ (see proposition 815 (p. 471)). We call $\mathbf{e_i}, \mathbf{i} \in \mathbf{g^{-1}(k)}$ the reference basis of $\mathcal{R}^{\mathrm{Card}(g^{-1}(k))}$ (see equation (2.106) (p. 267) in definition 242 (p. 267)). Given a vector $\mathbf{v} = \mathbf{v^i e_i} \in \mathcal{R}^n$, we write $(\mathbf{v_{g^{-1}(k)}} = \sum_{i \in g^{-1}(k)} \mathbf{v^i e_i}$ the projection (definition 121 (p. 136)) of $\mathbf{v}$ on $\mathcal{R}^{\mathrm{Card}(g^{-1}(k))}$ along $\mathcal{C}_{[1,n]}(\mathbf{g^{-1}(k)})$ (see definition 117 (p. 134)):

$$\mathbf{v}_{g^{-1}(k)} = \mathrm{Pr}_{\mathcal{C}(g^{-1}(k))} \circ \mathbf{v} = \sum_{i \in g^{-1}(k)} \mathbf{v^i e_i} \ .$$

We have the following inequalities:

$$||\mathbf{v}_{g^{-1}(k)}|| = (\sum_{i \in g^{-1}(k)} |\mathbf{v}^i|^p)^{1/p} \leq ||\mathbf{v}|| = (\sum_{i \in [1,n]} |\mathbf{v}^i|^p)^{1/p}$$

$$||\mathbf{v}|| \leq \sum_{k \in [1,\ell]} ||\mathbf{v}_{g^{-1}(k)}|| \ .$$

**Proof**
The first inequality comes from propositions 307 (p. 193) and 330 (p. 203). The second from proposition 473 (p. 268).

## 2.13 Complex numbers

Let us consider a vectorial space $\mathbf{V}$ of dimension 2 built on the field $\mathcal{R}$ with an independent basis $(\mathbf{e_i} : [1,2] \rightarrow \mathbf{V})$. We take the inner product and the norm associated by setting $< \mathbf{e_j}| = \mathbf{e^j}$ (see definitions 234 (p. 261) and 213 (p. 244)) with a norm given in that basis by $< \mathbf{e_i}|\mathbf{e_j} >= \mathbf{e^i e_j} = \delta^j_i$. Let us consider in that vectorial space, the set of linear transformations which are the product of a real number by a unitary transformation of positive determinant which forms a group from proposition 468 (p. 266).
In order to be unitary, a linear transformation should be represented by a matrix $\mathbf{U}$ such that $\mathbf{U^k_i U^k_j} = \delta^i_j$ see definition 240 (p. 265). It gives:

$$\mathbf{U^1_1 U^1_1 + U^2_1 U^2_1 = 1} \tag{2.112}$$
$$\mathbf{U^1_2 U^1_2 + U^2_2 U^2_2 = 1} \tag{2.113}$$
$$\mathbf{U^1_1 U^1_2 + U^2_1 U^2_2 = 0}\ . \tag{2.114}$$

We have only 3 equations because $\mathbf{U^k_i U^k_j}$ is symmetric by the exchange of $\mathbf{i}$ and $\mathbf{j}$. Using those we get:

$$\mathbf{2.114 \Rightarrow U^2_2 = -U^1_1 U^1_2 / U^2_1}$$
$$\mathbf{2.113 \Rightarrow (U^1_2)^2 (1 + (U^1_1/U^2_1)^2) = 1}$$
$$\mathbf{2.112 \Rightarrow (U^1_2)^2 = (U^2_1)^2 \Rightarrow U^1_2 = \pm U^2_1}\ .$$

Then we have two choices using (2.112): $\mathbf{U^1_2 = -U^2_1, U^2_2 = U^1_1}$ or $\mathbf{U^1_2 = U^2_1, U^2_2 = -U^1_1}$. Since $\mathbf{det(U) = U^1_1 U^2_2 - U^1_2 U^2_1}$, the first choice gives $\mathbf{det(U) = 1}$ and the second $\mathbf{det(U) = -1}$. We have to use the first choice. The group of linear transformations $\mathbf{U}$ for which $\mathbf{U^1_2 = -U^2_1, U^2_2 = U^1_1}$ is called the set of complex numbers $\mathcal{C}$. Multiplying those matrices by any real number $\lambda \in \mathcal{R}$, we keep the sign of the determinant because it is multiplied by $\lambda^2$.
Then the set of unitary matrix with positive determinant multiplied by any real is the set of linear transformations represented by $\mathbf{2 \times 2}$ matrices $\mathbf{U}$ such that $\mathbf{U^1_2 = -U^2_1, U^2_2 = U^1_1}$. This condition is kept when we add those transformations and multiplied by a real number. The group is a vectorial space over the Abelian group of the addition of matrices. Since from their definitions, the product of linear transformations is distributive for the addition: $\mathbf{A, B, C}$ being linear transformations, we have:

$$\mathbf{A(B + C) = AB + AC}\ .$$

Calling $\mathbf{U, V}$ two complex numbers and setting $\mathbf{u_r = U^1_1 = U^2_2, u_i = U^1_2 = -U^2_1}$, $\mathbf{v_r = V^1_1 = V^2_2, v_i = V^1_2 = -V^2_1}$, we have:

$$\mathbf{(UV)^1_1 = (UV)^2_2 = U^1_1 V^1_1 + U^1_2 V^2_1 = u_r v_r - u_i v_i}$$
$$\mathbf{(UV)^1_2 = -(UV)^2_1 = U^1_1 V^1_2 + U^1_2 V^2_2 = u_r v_i + u_i v_r}\ .$$

Then $\mathbf{UV = VU}$. The multiplication is also commutative. $\mathcal{C}$ can be used for defining a vectorial space.
One uses to call $\mathbf{1}$, the identity transformation and $\mathbf{i}$ the unitary transformation which transforms $\mathbf{e_1}$ in $\mathbf{e_2}$ and $\mathbf{e_2}$ in $\mathbf{-e_1}$. Any complex number $\mathbf{z}$ can then be written as $\mathbf{z = 1x + iy = x + iy}$ we have clearly $\mathbf{i^2 = -1}$. It defines a bijection between $\mathcal{R}^2$ and $\mathcal{C}$.
**Definition 245 Complex numbers $\mathcal{C}$.**
Given a vectorial space $\mathbf{V}$ of dimension $\mathbf{2}$ on real numbers with an independent basis $(\mathbf{e_i} : [1,2] \rightarrow \mathbf{V})$. One calls complex numbers $\mathcal{C}$, the subset of linear transformations which can be written as $\mathbf{1x + yi}$ where $\mathbf{x, y} \in \mathcal{R}^2$, $\mathbf{1}$ is the identity linear transformation and $\mathbf{i}$ the transformation such that $\mathbf{i(e_1) = e_2, i(e_2) = -e_1}$.

### 2.13.1 Properties of complex numbers

**Definition 246 Multiplication and addition on complex numbers.**

As explained in section 2.13 (p. 271) and from definition 245 (p. 271), complex numbers are also elements of $\mathcal{C} = \mathcal{R}^2$ considered as a vectorial space. On $\mathcal{C}$ the above sum can also be considered as the vectorial sum which defines an addition Abelian group. One calls $x = (x, 0)$ when $x \in \mathcal{R}$ and we also have $i = (0, 1)$. Another group with $(0, 0)$ excluded is associated to the product of the above associated linear transformations. It is also Abelian and is called the multiplication group:

$$(z_1, z_2) \in \mathcal{C}^2, z_1 = (x_1, y_1) = x_1 + iy_1, z_2 = (x_2, y_2) = x_2 + iy_2$$
$$z_1 \times z_2 = (x_1x_2 - y_1y_2, x_1y_2 + x_2y_1) = x_1x_2 - y_1y_2 + i(x_1y_2 + x_2y_1) .$$

**Proposition 479** The set of complex numbers, with the Multiplication and addition as defined in 246 (p. 272), is a field (definition 181 (p. 217)).

**Proof**
From definition 246 (p. 272) and beginning of section 2.13 (p. 271), it is a commutative ring (definitions 180 (p. 215) and 179 (p. 214)). Any element $z = x + iy \neq 0 \in \mathcal{C}$ has an inverse for the multiplication law: $z^{-1} = \frac{x-iy}{x^2+y^2} \in \mathcal{C}$:

$$\frac{(x + iy)(x - iy)}{x^2 + y^2} = \frac{x^2 + y^2}{x^2 + y^2} = 1 .$$

Then $\mathcal{C}$ satisfies definition 181 (p. 217).

**Definition 247 Norm of a complex number.**

The determinant of the associated linear transformation $\det(x + iy) = x^2 + y^2$ is also the modulus squared of the associated vector:

$$||x + iy|| = \sqrt{\det(x + iy)} = \sqrt{x^2 + y^2} .$$

It is the 2-norm on the associated $\mathcal{R}^2 = \mathcal{C}$ (definition 243 (p. 267)).

**Proposition 480**

$$z_1 \in \mathcal{C}, z_2 \in \mathcal{C} \rightarrow ||z_1z_2|| = ||z_1||\,||z_2|| .$$

**Proof**
Since the determinant of the product of two transformations is the product of the determinants of each transformation, we have:

$$z_1 \in \mathcal{C}, z_2 \in \mathcal{C} \rightarrow ||z_1z_2|| = ||z_1||\,||z_2|| .$$

**Definition 248 Complex conjugate of a complex number.**

$z = x + iy$ being a complex number, one calls its complex conjugate $\bar{z}$ also called $z^*$ $\bar{z} = z^* = x - iy$:

$$\overline{x + iy} = (x + iy)^* = x - iy = (x)^* + (iy)^* .$$

**Proposition 481** $\mathbf{a} = \mathbf{a_1} + \mathbf{i}\mathbf{a_2}, \mathbf{b} = \mathbf{b_1} + \mathbf{i}\mathbf{b_2}$ being two complex numbers, $\overline{\mathbf{ab}} = \bar{\mathbf{a}}\bar{\mathbf{b}}$ and $\overline{\mathbf{a}+\mathbf{b}} = \bar{\mathbf{a}} + \bar{\mathbf{b}}$.

**Proof**

$\mathbf{ab} = \mathbf{a_1 b_1} - \mathbf{a_2 b_2} + \mathbf{i}(\mathbf{a_1 b_2} + \mathbf{a_2 b_1})$,

$$\bar{\mathbf{a}}\bar{\mathbf{b}} = (\mathbf{a_1} - \mathbf{i}\mathbf{a_2})(\mathbf{b_1} - \mathbf{i}\mathbf{b_2}) = \mathbf{a_1 b_1} - \mathbf{a_2 b_2} - \mathbf{i}(\mathbf{a_1 b_2} + \mathbf{a_2 b_1}) = \overline{\mathbf{ab}}$$

$$\overline{\mathbf{a}+\mathbf{b}} = \overline{\mathbf{a_1} + \mathbf{i}\mathbf{a_2} + \mathbf{b_1} + \mathbf{i}\mathbf{b_2}} = \mathbf{a_1} + \mathbf{b_1} - \mathbf{i}(\mathbf{a_2} + \mathbf{b_2}) = \bar{\mathbf{a}} + \bar{\mathbf{b}}\ .$$

**Proposition 482** $\mathbf{z} \in \mathcal{C} \Rightarrow \mathbf{z} = \bar{\bar{\mathbf{z}}}$.

**Proof**

Setting $\mathbf{z} = \mathbf{x} + \mathbf{iy}$, we have: $\bar{\mathbf{z}} = \mathbf{x} - \mathbf{iy},\ \bar{\bar{\mathbf{z}}} = \mathbf{x} + \mathbf{iy} = \mathbf{z}$.

**Proposition 483** $\mathbf{z} \in \mathcal{C} \Rightarrow \|\mathbf{z}\| = \sqrt{\mathbf{z}\bar{\mathbf{z}}} = \|\bar{\mathbf{z}}\|$, see also definition 247 (p. 272).

**Proof**

Setting $\mathbf{z} = \mathbf{x} + \mathbf{iy}, \bar{\mathbf{z}} = \mathbf{x} - \mathbf{iy} \Rightarrow \mathbf{z}\bar{\mathbf{z}} = \mathbf{x}^2 + \mathbf{y}^2 = \|\mathbf{z}\|^2 = \|\bar{\mathbf{z}}\|^2$ see also definition 247 (p. 272).

**Proposition 484** The map $(\bar{\mathbf{z}} : \mathcal{C} \to \mathcal{C})$ is continuous in the topology generated by the norm $\sqrt{\mathbf{z}\bar{\mathbf{z}}}$.

**Proof**

From proposition 483, we have:

$$(\forall \alpha \in \mathcal{R}^+)(\exists \epsilon = \alpha)(\forall z \in \mathcal{C})(\forall \Delta z \in \mathcal{C})(\|z + \Delta z - z\| < \epsilon \Rightarrow \|\overline{z + \Delta} - \bar{z}\| < \alpha)$$

see definition 157 (p. 182) and proposition 282 (p. 184).

**Proposition 485** $\overline{\mathbf{a}^{\mathbf{n}}} = \bar{\mathbf{a}}^{\mathbf{n}}$.

**Proof**

If it is true for $\mathbf{n}$, it is true for $\mathbf{n+1}$ using proposition 481 (p. 273) then it is true for any $\mathbf{n}$ using proposition 927 (p. 520).

### 2.13.2 Semi Riemannian metric

Since we have defined complex numbers, we shall use from now on for field $\mathcal{RC}$, only the complex numbers $\mathcal{C}$.

**Definition 249 Semi Riemannian inner product (or Semi Riemannian metric).**

A finite dimension vector space $\mathbf{V}$ built on a field $\mathcal{C}$ as is said to have a semi Riemannian metric, if one has defined a map $\mathbf{g}$ between $\mathbf{V}$ and its dual $\mathbf{V}^*$ quoted as:

$$(<\mathbf{u}| = \mathbf{g}(|\mathbf{u}>) : \mathbf{V} \to \mathbf{V}^*)$$

with the following properties (see definition 234 (p. 261)):

$$<\mathbf{u}|\mathbf{v}> = <\mathbf{v}|\mathbf{u}>^* \tag{2.115}$$

$$(\forall \lambda \in \mathcal{C})(\forall \mu \in \mathcal{C})(\forall |\mathbf{u}> \in \mathbf{V})(\forall |\mathbf{v}> \in \mathbf{V})(<\lambda\mathbf{u} + \mu\mathbf{v}| = \lambda^* <\mathbf{u}| + \mu^* <\mathbf{v}|)\ . \tag{2.116}$$

Then we have $< \mathbf{u}|\mathbf{u} >=< \mathbf{u}|\mathbf{u} >^* \in \mathcal{R}$ so it can also be negative and zero (in Riemannian inner product (definition 234 (p. 261)) $< \mathbf{u}|\mathbf{u} >=< \mathbf{u}|\mathbf{u} >^* > 0$). Given a basis of $\mathbf{V}$, $(\mathbf{e_i} : [1, \mathbf{n}] \to \mathbf{V})$ the semi Riemannian metric is given by:

$$\mathbf{g_{ij}} = \mathbf{g^*_{ji}} =< \mathbf{e_i}|\mathbf{e_j} > \ .$$

As in definition 234 (p. 261) one may set by convention:

$$\begin{aligned} &\mathbf{g_{ij}} =< \mathbf{e_i}|\mathbf{e_j} >= \mathbf{g}(\mathbf{e_i})(\mathbf{e_j}) \\ &< \mathbf{e_i}| = \mathbf{g}(\mathbf{e_i}) = \mathbf{g_{ij}}\mathbf{e^j} \\ &\mathbf{g^*_{ij}}|\mathbf{e^j} >= |\mathbf{e_i} > \end{aligned}$$

if $\mathbf{det}(\mathbf{g_{ij}}) \neq 0$ we can define $\mathbf{g^{ij}} = \mathbf{g_{ij}^{-1}}$ such that $\mathbf{g^{ik}}\mathbf{g_{kj}} = \delta^i_j$. Then by convention:

$$\mathbf{e^i} =< \mathbf{e^i}| = \mathbf{g^{*ik}}|\mathbf{e_k} >= \mathbf{g^{*ik}}\mathbf{e_k} \ .$$

**Definition 250 Set of orthogonal vectors.**

With the above definition, a subset $\mathbf{O}$ of a vectorial space $\mathbf{V}$ with a semi Riemannian inner product, is called orthogonal set of vectors if

$$(\mathbf{u} \in \mathbf{O}, \mathbf{v} \in \mathbf{O}, \mathbf{u} \neq \mathbf{v}) \Rightarrow < \mathbf{u}|\mathbf{v} >= 0 \ .$$

**Definition 251 Length of a vector.**

One shall call here length of a vector $\mathbf{v} \in \mathbf{V}$: $\frac{<\mathbf{v}|\mathbf{v}>}{\sqrt{|<\mathbf{v}|\mathbf{v}>|}}$ which can be any real number.

**Proposition 486** Given a vectorial space of dimension $\mathbf{n}$ built on $\mathcal{C}$ with a semi Riemannian inner product of nonzero determinant, any set of orthogonal vectors with a length different from zero is an independent set of vectors.

**Proof**

Let us call $(\mathbf{v_i} : [1, \mathbf{p}] \to \mathbf{V})$ with $1 \leq \mathbf{p} \leq \mathbf{n}$ such a set we have:

$$\begin{aligned} &(\lambda^i : [1, \mathbf{p}] \to \mathcal{C}), \lambda^i \mathbf{v_i} = 0 \Rightarrow \\ &(\forall \mathbf{j} \in [1, \mathbf{p}])(0 = \lambda^{*i} < \mathbf{v_i}|\mathbf{v_j} >= \lambda^{*j} < \mathbf{v_j}|\mathbf{v_j} > \Rightarrow \lambda^{*j} = \lambda^j = 0) \ . \end{aligned}$$

**Proposition 487** Given a vectorial space of dimension $\mathbf{n}$ built on $\mathcal{C}$ with a semi Riemannian inner product of nonzero determinant, there is no independent orthogonal basis with a vector of length zero.

**Proof**

We call $(\mathbf{f} : [1, \mathbf{n}] \to \mathbf{V})$ an independent orthogonal basis, $\mathbf{f_i} = \mathbf{A^j_i}\mathbf{e_j}$. We have:

$$< \mathbf{f_i}|\mathbf{f_j} >=< \mathbf{f_i}|\mathbf{f_i} > \delta^i_j = \mathbf{A^{*k}_i}\mathbf{g_{kl}}\mathbf{A^l_j} \ .$$

From propositions 448 (p. 257) and 430 (p. 250), since $\mathbf{det}(\mathbf{A}) \neq 0$ (proposition 430 (p. 250)) we also have:

$$\mathbf{det}(\mathbf{A^{*k}_i}\mathbf{g_{kl}}\mathbf{A^l_j}) = \mathbf{det}(\mathbf{A})^*\mathbf{det}(\mathbf{g})\mathbf{det}(\mathbf{A}) \neq 0$$

then

$$\mathbf{det}(< \mathbf{f_i}|\mathbf{f_j} >) = \mathbf{det}(< \mathbf{f_i}|\mathbf{f_i} > \delta^i_j) = \prod_{\mathbf{i} \in [1,\mathbf{n}]} < \mathbf{f_i}|\mathbf{f_i} >) \neq 0 \ .$$

Then we have proposition 487.

**Proposition 488** Given a vectorial space $\mathbf{V}$ of dimension $\mathbf{n}$ built on $\mathcal{C}$ with a semi Riemannian inner product of determinant nonzero we have the following true statements:

1. For any subset $\mathbf{O_1}$ of nonzero length orthogonal vectors of cardinal smaller than $\mathbf{n}$, there is a nonzero length vector orthogonal to all vectors of $\mathbf{O_1}$.

2. For any subset $\mathbf{O_1}$ of nonzero length orthogonal vectors of cardinal $\mathbf{p}$ smaller than $\mathbf{n}$ there is a subset $\mathbf{O_2}$ of $\mathbf{n-p}$ nonzero length orthogonal vectors all of them orthogonal to all vectors of $\mathbf{O_1}$ such that $\mathbf{O_1 \cup O_2}$ is an independent basis of $\mathbf{V}$.

**Proof**

Let us suppose we already have a set $\mathbf{O_1}$ of $\mathbf{p \in [0, n-1]}$ vectors with a nonzero length making an orthogonal set. Since $\mathbf{p < n}$, from proposition 415 (p. 244) there is a vector $\mathbf{v}$ making with $\mathbf{O_1}$ a set of independent vectors and one can generate:

$$\mathbf{v}' = \mathbf{v} - \sum_{\mathbf{w} \in \mathbf{O_1}} \frac{<\mathbf{w}|\mathbf{v}>}{<\mathbf{w}|\mathbf{w}>}\mathbf{w} \tag{2.117}$$

$\mathbf{v}'$ is orthogonal to all vectors of $\mathbf{O_1}$. If there is one $\mathbf{v}'$ such that $<\mathbf{v}'|\mathbf{v}'> \neq 0$ we can increase $\mathbf{p}$ by $\mathbf{1}$.

If not, all vectors built that way are of length zero. From proposition 415 (p. 244), we may find a set $\mathbf{O}'_2$ of $\mathbf{n-p}$ vectors making with $\mathbf{O_1}$ an independent basis of $\mathbf{V}$. Using, for each of them, the formula (2.117) adding a linear combination of vectors of $\mathbf{O_1}$, we generate another set $\mathbf{O_2}$ of $\mathbf{n-p}$ vectors of length $\mathbf{0}$ orthogonal to all vectors of $\mathbf{O_1}$ making with $\mathbf{O_1}$ an independent basis (see proposition 407 (p. 240)). It means that we can build a full set of $\mathbf{n-p}$ independent vectors $\mathbf{O_2}$ such that they are of length zero and orthogonal to all vectors of $\mathbf{O_1}$ and making with $\mathbf{O_1}$ a basis of $\mathbf{V}$. If two of them (in $\mathbf{O_2}$) were not orthogonal, we would have $\mathbf{u \in O_2}$ and $\mathbf{v \in O_2}$ with $<\mathbf{u}|\mathbf{v}> \neq 0$ then the vector $\mathbf{t} = \mathbf{u}+ <\mathbf{u}|\mathbf{v}>\mathbf{v}$ would have a nonzero length:

$$<\mathbf{t}|\mathbf{t}> = (<\mathbf{u}| + <\mathbf{v}|\mathbf{u}><\mathbf{v}|)(|\mathbf{u}> + <\mathbf{u}|\mathbf{v}>|\mathbf{v}>) = 2<\mathbf{v}|\mathbf{u}><\mathbf{v}|\mathbf{u}> \neq 0$$

$\mathbf{t}$ is orthogonal to all vectors of $\mathbf{O_1}$ and independent of them and then could be added to $\mathbf{O_1}$. The only remaining case is

$$(\mathbf{u} \in \mathbf{O_2}, \mathbf{v} \in \mathbf{O_2}) \Rightarrow <\mathbf{u}|\mathbf{u}> = <\mathbf{v}|\mathbf{v}> = <\mathbf{u}|\mathbf{v}> = 0\ .$$

Then $\mathbf{O_1 \cup O_2}$ is an orthogonal basis of V with zero length vectors. It is impossible for a semi Riemannian metric with a nonzero determinant (proposition 487). Then we can always add a nonzero length vector to a set of $\mathbf{p \in [0, n-1]}$ orthogonal vectors of $\mathbf{V}$ with non zero length such that to make a $\mathbf{p+1}$ set of orthogonal vectors of $\mathbf{V}$ with nonzero length. We then have proposition 487 from proposition 927 (p. 520).

**Proposition 489** Given a vectorial space of dimension $\mathbf{n}$ built on $\mathcal{C}$ with a semi Riemannian inner product of nonzero determinant, given two orthogonal bases, the number of vectors with a positive length is the same in each basis.

**Proof**

Let us call $(\mathbf{e_i} : [\mathbf{1}, \mathbf{n}] \to \mathbf{V})$ and $(\mathbf{f_i} : [\mathbf{1}, \mathbf{n}] \to \mathbf{V})$ the two orthogonal bases with

$$<\mathbf{e_i}|\mathbf{e_j}> = <\mathbf{e_j}|\mathbf{e_i}>^* = \mathbf{s^e_i}\delta^i_j,\ <\mathbf{f_i}|\mathbf{f_j}> = <\mathbf{f_j}|\mathbf{f_i}>^* = \mathbf{s^f_i}\delta^i_j\ .$$

If the number of vectors with a positive length in a basis is zero, it is the same for the other basis: We have if $(\forall \mathbf{i} \in [\mathbf{1}, \mathbf{n}])(\mathbf{s^e_i} < 0)$:

$$(\forall \mathbf{i} \in [\mathbf{1}, \mathbf{n}])(\mathbf{f_i} = \mathbf{x^j_i e_j} \Rightarrow <\mathbf{f_i}|\mathbf{f_i}> = \mathbf{s^f_i}\delta^i_j = \mathbf{x^{*j}_i x^k_i} <\mathbf{e_j}|\mathbf{e_k}> = \mathbf{x^{*j}_i x^k_i s^e_k}\delta^j_k = \mathbf{x^{*j}_i x^j_i s^e_j})\ .$$

Since, from definition 232 (p. 259), $x^{*j}_i x^j_i = ||x^j_i||^2 \geq 0$ we have:

$$(\forall i \in [1,n])(s^e_i < 0) \Rightarrow (\forall i \in [1,n])(s^f_i < 0)$$

if $p^e > 0$ (resp. $p^f > 0$) is the number of positive length of the $e$ (resp. $f$) basis, we can relabel the index such that to have:
$i \in [1,p^e] \Leftrightarrow < e_i|e_i >> 0$ and $i \in [1,p^f] \Leftrightarrow < f_i|f_i >> 0$ .
We then have:

$$v = \sum_{i\in[1,p^f]} \lambda_i f_i \Rightarrow < v|v >= \lambda_i \lambda^*_i < f_i|f_i > \tag{2.118}$$

but in that case, we also have if $(\forall i \in [1,n])(f_i = x^j_i e_j)$:

$$v = \sum_{i\in[1,p^f]} \lambda_i \sum_{j\in[1,p^e]} x^j_i e_j + \lambda_i \sum_{j\in[p^e+1,n]} x^j_i e_j \; . \tag{2.119}$$

If $p^e < p^f$, the $p^f$ vectors $\sum_{j\in[1,p^e]} x^j_i e_j, i \in [1,p^f]$ are not independent because they belong to a subspace of dimension $p^e < p^f$ (proposition 414 (p. 243)) then:

$$(\exists(\lambda^i : [1,p^f] \to \mathcal{C}))(\exists k \in [1,p^f])(\lambda^k \neq 0, \sum_{i\in[1,p^f]} \lambda^i \sum_{j\in[1,p^e]} x^j_i e_j = 0) \; .$$

But taking that map $(\lambda_i : [1,p^f] \to \mathcal{C})$, we also have $v = \sum_{i\in[1,p^f]} \lambda^i f_i \neq 0$ since the $f_i$ are independent and since the $\lambda$ are not all zero. From equation (2.118) (p. 276):

$$((\forall i \in [1,p_f])(< f_i|f_i >> 0), (\exists k \in [1,p_f])(\lambda^k \neq 0))$$
$$\Rightarrow < v|v >=< v|v >^* > \lambda^k \lambda^{*k} < f_k|f_k >> 0$$

but equation (2.119) (p. 276) also gives

$$< v|v >=< v|v >^* = \sum_{j\in[p^e+1,n]} (\sum_{i\in[1,p^f]} \lambda_i x^j_i)^* (\sum_{i\in[1,p^f]} \lambda_i x^j_i) < e_j|e_j > \leq 0 \; .$$

We have a contradiction. From A.1.4, 8, $p^f \leq p^e$. Since we can exchange $e$ and $f$ the only possibility remaining is $p^f = p^e$.

**Definition 252 Signature of a vectorial space with a semi Riemannian inner product (metric).**

In a vectorial space with a semi Riemannian inner product having a nonzero determinant of the associated metric, the couple $p, n$ of the numbers of vectors with a positive length and with a negative length in an independent orthogonal basis is called the signature of the inner product. It does not depend on the particular independent orthogonal basis (proposition 489 (p. 275)).

**Definition 253 Adjoint transformation over a semi Riemannian inner product.**

Given a finite dimension vectorial space $V$ on $\mathcal{C}$ with a semi Riemannian inner product $(< u|v >= g(u,v) : V^2 \to \mathcal{RC})$ and a linear transformation $(T : V \to V)$, as in definition 239 (p. 264), one calls the adjoint of $T$, a linear transformation $(T^\dagger : V \to V)$ such that:

$$(\forall u)(\forall v)(< u|Tv >= g(u,T(v)) =< T^\dagger u|v >= g(T^\dagger(u),v)) \; .$$

One also uses to write:

$$< T^\dagger u|v >=< uT^\dagger|v >=< v|T^\dagger u >^*=< u|Tv > .$$

Setting ($e_i : [1,n] \to V$) an independent basis of **V** and $g_{ij} =< e_i|e_j >$ we should have as in definition 239 (p. 264) with $Te_j = T^k_j e_k$:

$$< e_i|Te_j >= T^k_j < e_i|e_k >=< T^\dagger e_i|e_j >= T^{\dagger *k}_i < e_k|e_j >$$

then:

$$T^k_j g_{ik} = T^{\dagger *k}_i g_{kj} .$$

If $\det(g) \neq 0$, from propositions 451 (p. 258) and 450 (p. 258) $g^{-1}$ exists and it is a linear transformation. Then we may find $g^{ij}$ such that $g^{ik}g_{kj} = \delta^i_j$ giving:

$$T^{\dagger j}_i = (g^{lj}T^k_l g_{ik})^* .$$

**Definition 254 Unitary transformation within a semi Riemannian inner product.**

With the same notation as in definition 253, a unitary transformation should be such that:

$$(\forall i \in [1,n])(\forall j \in [1,n])(g(e_i,e_j) = g(T(e_i),T(e_j))) .$$

## 2.14 Convex subset

**Definition 255 Convex subset.**

A subset **C** of a vectorial space **E** built on real numbers such that:

$$(\forall x \in C)(\forall y \in C)(\forall t \in [0,1] \subset \mathcal{R})(xt + (1-t)y \in C)$$

is a convex subset.

**Proposition 490** Given a vectorial space **E** built on real numbers, **E** and any subvectorial space of **E** is convex.

**Proof**

It is a direct consequence of definitions 255 (p. 277), 199 (p. 235) and 201 (p. 237).

**Proposition 491** Given a vectorial space **E** built on real numbers, given $p \in \mathcal{Z}^+$, a map ($v_i : [1,p] \to E$) and a map ($A_i : [0,p] \to \mathcal{R}$), the set defined by:

$$x \in C \Leftrightarrow (x \in E, (\exists (x^i : [1,p] \to \mathcal{R}))(x = x^i v_i), \sum_{i\in[1,p]} A_i x^i = A_0)$$

is convex.

**Proof**

Given two elements of **C** $x = x^i v_i$, $y = y^i v_i$ we have (see propositions 289 (p. 186), 290 (p. 187) and 291 (p. 187)):

$$(\forall t \in [0,1])(\sum_{i\in[1,p]} A_i(tx^i + (1-t)y^i) = t \sum_{i\in[1,p]} A_i x^i + (1-t) \sum_{i\in[1,p]} A_i y^i$$
$$= tA_0 + (1-t)A_0 = A_0) .$$

**Proposition 492** Given a convex subset $\mathbf{C}$ of a vectorial space $\mathbf{E}$ built on real numbers. Given an integer $\mathbf{p}$, a map $(\mathbf{v_i} : [1,\mathbf{p}] \to \mathbf{C})$, we have:

$$((\mathbf{m^i} : [1,\mathbf{p}] \to \mathcal{R}^{0+}), \sum_{\mathbf{i}\in[1,\mathbf{p}]} \mathbf{m^i} = 1) \Rightarrow \sum_{\mathbf{i}\in[1,\mathbf{p}]} \mathbf{m^i v_i} \in \mathbf{C} .$$

**Proof**

Let us suppose that it is true for $\mathbf{p} = \mathbf{p}'$, let us prove then that it is true for $\mathbf{p} = \mathbf{p}'' = \mathbf{p}'+1$ We have $(\exists \mathbf{j_1} \in [1,\mathbf{p}''])(\mathbf{m^{j_1}} \neq 0)$ otherwise $\sum_{\mathbf{i}\in[1,\mathbf{p}'']} \mathbf{m^i} = 0$. We suppose $\mathbf{m^{j_1}} \neq 0$, if $\neg(\exists \mathbf{j_2} \in [1,\mathbf{p}''] - \{\mathbf{j_1}\})(\mathbf{m^{j_2}} \neq 0)$, $\mathbf{m^{j_1}} = 1$ proposition 492 is true.
We suppose $\mathbf{m^{j_1}} \neq 0$ and $\mathbf{m^{j_2}} \neq 0, \mathbf{j_2} \in [1,\mathbf{p}''] - \{\mathbf{j_1}\}$. Setting $\mathbf{M^1} = \sum_{\mathbf{i}\in[1,\mathbf{p}'']-\{\mathbf{j_1}\}} \mathbf{m^i}$, we have $\mathbf{M^1} > 0$ and since proposition 492 is true for $\mathbf{p} = \mathbf{p}'' - 1$ we have:

$$\mathbf{X} = \sum_{\mathbf{i}\in[1,\mathbf{p}'']-\{\mathbf{j_1}\}} \frac{\mathbf{m^i}}{\mathbf{M^1}}\mathbf{v_i} = \frac{\sum_{\mathbf{i}\in[1,\mathbf{p}'']-\{\mathbf{j_1}\}} \mathbf{m^i}}{\mathbf{M^1}}\mathbf{v_i} \in \mathbf{C} .$$

Since we have $\mathbf{M^1} + \mathbf{m^{j_1}} = \sum_{\mathbf{i}\in[1,\mathbf{p}'']} \mathbf{m^i} = 1$, from definition 255 (p. 277) we have:

$$\mathbf{M^1X} + \mathbf{m^{j_1}v_{j_1}} = \sum_{\mathbf{i}\in[1,\mathbf{p}'']} \mathbf{m^i v_i} \in \mathbf{C} .$$

Proposition 492 is true for $\mathbf{p} = 1$, when it is true for $\mathbf{p} = \mathbf{p}'$ it also true for $\mathbf{p} = \mathbf{p}' + 1$. We conclude using proposition 927 (p. 520).

**Proposition 493** The projection along an axis of $\mathcal{R}^{\mathbf{n}}$ (definition 244 (p. 270)) of a convex subset of $\mathcal{R}^{\mathbf{n}}$ is a convex subset of $\mathcal{R}^{\mathbf{n}-1}$.

**Proof**

It is a direct consequence of proposition 477 (p. 270).

**Proposition 494** Given a vectorial space $\mathbf{E}$ built on real numbers, a set of index $\mathbf{I}$, a family of convex subsets of $\mathbf{E}$ $(\mathbf{C_i} : \mathbf{I} \to \mathcal{P}(\mathbf{E}))$ (definition 350 (p. 453)), $\bigcap_{\mathbf{i}\in\mathbf{I}} \mathbf{C_i}$ is convex (see definition 363 (p. 463)).

**Proof**

From definition 363 (p. 463) and proposition 736 (p. 443) equation (A.6), we have:

$$\begin{aligned}
&(\mathbf{p} \in \bigcap_{\mathbf{i}\in\mathbf{I}} \mathbf{C_i}, \mathbf{q} \in \bigcap_{\mathbf{i}\in\mathbf{I}} \mathbf{C_i}) \Rightarrow ((\forall \mathbf{i} \in \mathbf{I})(\mathbf{p} \in \mathbf{C_i}), (\forall \mathbf{j} \in \mathbf{I})(\mathbf{q} \in \mathbf{C_j}))\\
&\Rightarrow (\forall \mathbf{i} \in \mathbf{I})(\mathbf{p} \in \mathbf{C_i}, \mathbf{q} \in \mathbf{C_i}) \Rightarrow (\forall \mathbf{i} \in \mathbf{I})(\forall \mathbf{t} \in [0,1])(\mathbf{tp} + (1-\mathbf{t})\mathbf{q} \in \mathbf{C_i})\\
&\Rightarrow (\forall \mathbf{t} \in [0,1])(\forall \mathbf{i} \in \mathbf{I})(\mathbf{tp} + (1-\mathbf{t})\mathbf{q} \in \mathbf{C_i})\\
&\Rightarrow (\forall \mathbf{t} \in [0,1])(\mathbf{tp} + (1-\mathbf{t})\mathbf{q} \in \bigcap_{\mathbf{i}\in\mathbf{I}} \mathbf{C_i}) .
\end{aligned}$$

See propositions 734 (p. 442), A.6, 738 (p. 444), 730 (p. 442).

**Proposition 495** Given a convex subset $\mathbf{C}$ of the vectorial space $\mathcal{R}^{\mathbf{n}}$, given a map $(\mathbf{y}(\mathbf{x}) : \mathcal{R}^{\mathbf{n}} \to \mathcal{R}^{\mathbf{n}})$ defined by $\mathbf{y^i} = \mathbf{a^i} + \mathbf{B^i_j x^j}$, the transformed set $\mathbf{y}(\mathbf{C})$ is convex.

**Proof**

Keeping the above notation,

$$\begin{aligned}
&(\forall \mathbf{y_1} = \mathbf{y}(\mathbf{x_1}) \in \mathbf{y}(\mathbf{C}))(\forall \mathbf{y_2} = \mathbf{y}(\mathbf{x_2}) \in \mathbf{y}(\mathbf{C}))(\forall \mathbf{t} \in [0,1])\\
&(\mathbf{ty_1} + (1-\mathbf{t})\mathbf{y_2} = \mathbf{a^i}(\mathbf{t} + 1 - \mathbf{t}) + \mathbf{tB^i_j x^j_1} + (1-\mathbf{t})\mathbf{B^i_j x^j_2} = \mathbf{a^i} + \mathbf{B^i_j}(\mathbf{tx^j_1} + (1-\mathbf{t})\mathbf{x^j_2}))
\end{aligned}$$

since $\mathbf{tx^j_1} + (1-\mathbf{t})\mathbf{x^j_2} \in \mathbf{C}$ because $\mathbf{C}$ is convex. We have $\mathbf{ty_1} + (1-\mathbf{t})\mathbf{y_2} \in \mathbf{y}(\mathbf{C})$ as the transformed of an element of $\mathbf{C}$ then $\mathbf{y}(\mathbf{C})$ is convex.

**Proposition 496** An open ball (definition 150 (p. 177)) in a normed (definition 203 (p. 237)) vectorial space $\mathbf{E}$ built on $\mathcal{R}$ is a convex.

**Proof**
Given an open ball $\mathbf{B(x,\rho)}$, $\mathbf{x \in E}$, $\rho \in \mathcal{R}$,

$$\begin{aligned}&(\mathbf{x_1 \in B(x,\rho), x_2 \in B(x,\rho), \lambda \in [0,1]})\\ &\Rightarrow (\|\mathbf{x_1 - x}\| < \rho, \|\mathbf{x_2 - x}\| < \rho, \lambda \in [0,1])\\ &\Rightarrow \|\lambda \mathbf{x_1} + (1-\lambda)\mathbf{x_2} - \mathbf{x}\| = \|\lambda(\mathbf{x_1 - x}) + (1-\lambda)(\mathbf{x_2 - x})\|\\ &\leq \|(\lambda(\mathbf{x_1 - x})\| + \|(1-\lambda)(\mathbf{x_2 - x})\|\\ &\leq \lambda\|(\mathbf{x_1 - x})\| + (1-\lambda)\|(\mathbf{x_2 - x})\| < (\lambda + (1-\lambda))\rho = \rho\end{aligned}$$

then $\mathbf{B(x,\rho)}$ is convex.

**Proposition 497** Any interval (definition 378 (p. 484)) is a convex set of $\mathcal{R}$ and vice versa.

**Proof**
From proposition 496 (p. 279), any open or closed intervals (definition 378 (p. 484)) are convex. Any semi-open interval $\mathbf{[a,b[}$ is convex because from proposition 496 (p. 279), :

$$\begin{aligned}&(\mathbf{x_1 \in [a,b[, x_2 \in [x_1,b[, t \in [0,1]}) \Rightarrow \mathbf{a \leq x_1 = tx_1 + (1-t)x_2 - (1-t)(x_2 - x_1)}\\ &\leq \mathbf{tx_1 + (1-t)x_2 \leq tx_1 + (1-t)x_2 + (x_2 - x_1)t = x_2 < b}\\ &\Rightarrow \mathbf{tx_1 + (1-t)x_2 \in [a,b[}\ .\end{aligned}$$

The same reasoning apply for:

$$\mathbf{]a,b], [a,\rightarrow[, ]\leftarrow,b], ]a,\rightarrow[, ]\leftarrow,b[}\ .$$

If $\mathbf{C \subset \mathcal{R}}$ is convex and has majoring and a minoring element, it has a **lower_limit**: $\mathbf{a}$ and an **upper_limit**: $\mathbf{b}$ (see definition 381 (p. 485), proposition 258 (p. 174)). We suppose first $\mathbf{a}$ and $\mathbf{b}$ are in $\mathbf{C}$. In that case, we have:

$$\begin{aligned}&\mathbf{t \in [0,1] \Rightarrow a = ta + (1-t)b - (1-t)(b-a) \leq ta + (1-t)b}\\ &\leq \mathbf{ta + (1-t)b + t(b-a) = b \Rightarrow ta + (1-t)b \in [a,b]}\ .\end{aligned} \tag{2.120}$$

Then $\mathbf{C \subset [a,b]}$. We also have:

$$\begin{aligned}&\mathbf{a \leq ta + (1-t)b \leq b \Rightarrow b - t(b-a) \leq b \Rightarrow t(b-a) \geq 0 \Rightarrow t \geq 0}\\ &\mathbf{a \leq ta + (1-t)b \leq b \Rightarrow a \leq a + (1-t)(b-a) \Rightarrow 0 \leq (1-t)(b-a) \Rightarrow t \leq 1}\end{aligned} \tag{2.121}$$

then $\mathbf{[a,b] \subset C}$. We have both $\mathbf{C \subset [a,b]}$ and $\mathbf{[a,b] \subset C}$ then $\mathbf{[a,b] = C}$ (proposition 754 (p. 451)). If $\mathbf{a \notin C}$ and $\mathbf{b \notin C}$ for any $\mathbf{x_1 < x_2, x_1 \in ]a,b[, x_2 \in ]a,b[}$, we have (proposition 259 (p. 175)):

$$(\exists \mathbf{x'_1 \in ]a,x_1[\cap C}), (\exists \mathbf{x'_2 \in ]x_2,b[\cap C})\ .$$

Using those $\mathbf{x'_1}$, $\mathbf{x'_2}$, we have $\mathbf{x_1 \in C}$, $\mathbf{x_2 \in C}$ because they are in $\mathbf{[x'_1,x'_2]}$ (see equation (2.120) (p. 279)) Then $\mathbf{]a,b[\subset C}$ since $\mathbf{\{a,b\} \cap C = \emptyset}$, we have $\mathbf{]a,b[= C}$ because any term in $\mathbf{x \in C}$ are such $\mathbf{a < x < b}$ so in $\mathbf{]a,b[}$. It means $\mathbf{C \subset ]a,b[}$ .
With the same reasoning, if we have only one minoring (resp. majoring) element and then only a **lower_limit** (resp. **upper_limit**) $\mathbf{a}$, $\mathbf{a \notin C \Rightarrow ]a \rightarrow [= C}$ (resp. $\mathbf{a \notin C \Rightarrow ]\leftarrow, a[= C}$. If there is no minoring and no majoring element, we get $\mathbf{C = \mathcal{R}}$.

**Proposition 498** Given a convex subset $\mathbf{C}$ of a vectorial space $\mathbf{E}$ built on real numbers, given two points $\mathbf{p} \in \mathbf{C}$ and $\mathbf{q} \in \mathbf{C}$, we call $\mathbf{C_p} \subset \mathbf{E}$ and $\mathbf{C_q} \subset \mathbf{E}$ the sets:

$$\mathbf{v_p} \in \mathbf{C_p} \Leftrightarrow (\exists \mathbf{v} \in \mathbf{C})(\mathbf{v_p} = \mathbf{v} - \mathbf{p})$$
$$\mathbf{v_q} \in \mathbf{C_q} \Leftrightarrow (\exists \mathbf{v} \in \mathbf{C})(\mathbf{v_q} = \mathbf{v} - \mathbf{q})$$

If the dimension (see definition 209 (p. 240)) of $\mathbf{C_p}$ is finite, it is equal to the dimension of $\mathbf{C_q}$.

**Proof**
Since $\mathbf{p} \in \mathbf{C}$ and $\mathbf{q} \in \mathbf{C}$, we have $\mathbf{q} - \mathbf{p} \in \mathbf{C_p}$ and $\mathbf{p} - \mathbf{q} \in \mathbf{C_q}$. We call $\mathbf{n_p}$, $\mathbf{n_q}$ the finite dimensions of respectively $\mathbf{C_p}$ and $\mathbf{C_q}$. Following the same argument as in proposition 416 (p. 244), since $\mathbf{q} - \mathbf{p} \in \mathbf{C_p}$, there is in $\mathbf{C_p}$, a set of $\mathbf{n_p}$ independent vectors $(\mathbf{v_{p,i}} : [\mathbf{1}, \mathbf{n_p}] \to \mathbf{V})$ of containing the vector $\mathbf{q} - \mathbf{p}$. Let us call them $\mathbf{v_{p,1}} = \mathbf{v_1} - \mathbf{p} = \mathbf{q} - \mathbf{p}$ and $(\mathbf{v_{p,i}} = \mathbf{v_i} - \mathbf{p} : [\mathbf{1}, \mathbf{n}] \to \mathbf{C_p})$ for the following. Using them, we get $\mathbf{n_p}$ vectors of $\mathbf{C_q}$ with $\mathbf{i} \in [\mathbf{2}, \mathbf{n_p}] \Rightarrow \mathbf{v_{q,i}} = \mathbf{v_i} - \mathbf{q} = \mathbf{v_{p,i}} - (\mathbf{q} - \mathbf{p})$, $\mathbf{v_{q,1}} = -\mathbf{v_{p,1}} = \mathbf{p} - \mathbf{q}$ they are also independent vectors because:

$$\lambda_i \mathbf{v_{q,i}} = \sum_{i \in [2, n_p]} \lambda_i (\mathbf{v_{p,i}} - \mathbf{v_{p,1}}) - \mathbf{v_{p,1}} \lambda_1 = \sum_{i \in [2, n_p]} \lambda_i \mathbf{v_{p,i}} - \mathbf{v_{p,1}} \sum_{i \in [1, n_p]} \lambda_i \, .$$

Then, since the $\mathbf{v_{p,i}}$ are independent, we have:

$$\lambda_i \mathbf{v_{q,i}} = 0 \Rightarrow ((\forall i \in [2, n])(\lambda_i = 0), \sum_{i \in [1, n_p]} \lambda_i = 0) \Rightarrow (\forall i \in [1, n])(\lambda_i = 0)$$

then there is a subset of $\mathbf{C_q}$ made of $\mathbf{n_p}$ independent vectors. Then from definition 209 (p. 240), $\mathbf{n_q} \geq \mathbf{n_p}$. Inverting $\mathbf{p}$ and $\mathbf{q}$ in the preceding argument we also have $\mathbf{n_p} \geq \mathbf{n_q}$ then by combining both we have $\mathbf{n_p} = \mathbf{n_q}$.

**Definition 256 Dimension of a convex subset.**

Given a convex subset $\mathbf{C}$ of a vectorial space $\mathbf{E}$ built on real numbers, given $\mathbf{p} \in \mathbf{C}$, calling $\mathbf{C_p} \subset \mathbf{E}$, the set such that

$$\mathbf{v_p} \in \mathbf{C_p} \Leftrightarrow (\exists \mathbf{v} \in \mathbf{C})(\mathbf{v_p} = \mathbf{v} - \mathbf{p}) \, .$$

If the dimension (definition 209 (p. 240)) of $\mathbf{C_p}$ is finite, it is called the dimension of $\mathbf{C}$. From proposition 498 (p. 280), it does not depend on $\mathbf{p} \in \mathbf{C}$.

**Proposition 499** Being in the same conditions as in definition 256 (p. 280), keeping the same notations, calling $\mathbf{n}$ the dimension of $\mathbf{C}$ as defined by 256, we consider a point $\mathbf{p_0} \in \mathbf{C}$ with the associated set $\mathbf{C_{p_0}}$ in which there are $\mathbf{n}$ independent vectors $\mathbf{w_i}$. We consider the set of $\mathbf{n}$ points of $\mathbf{C}$ defined by $\mathbf{i} \in [\mathbf{1}, \mathbf{n}] \Rightarrow \mathbf{p_i} = \mathbf{p_0} + \mathbf{w_i} \in \mathbf{C}$. We have the following relation:

$$\mathbf{x} \in \mathbf{C} \Rightarrow (\exists (\mathbf{x^i} : [\mathbf{1}, \mathbf{n}] \to \mathcal{R}))(\mathbf{x} - \mathbf{p_0} = \mathbf{x^i}(\mathbf{p_i} - \mathbf{p_0})) \, . \tag{2.122}$$

**Proof**
As in proposition 409 (p. 241), $\mathbf{x} - \mathbf{p_0} \in \mathbf{C_{p_0}}$ is a linear combination of the $\mathbf{n}$ $\mathbf{w_i}$ because, from definition 209 (p. 240), $\mathbf{n}$ is the maximum number of independent vectors of $\mathbf{C_{p_0}}$.

**Proposition 500** Given $\mathbf{k} \in \mathcal{Z}^+$ vectors of $\mathcal{R}^n$:$(\mathbf{v_i} : [1, \mathbf{k}] \to \mathcal{R}^n)$, the subset $\mathbf{A} \subset \mathcal{R}^n$ defined by:

$$\mathbf{A} = \mathbf{Set_x}\{\mathbf{x} \in \mathcal{R}^n, (\exists \mathbf{m})(\mathbf{m} = (\mathbf{m^i} : [1, \mathbf{k}] \to \mathcal{R}^{+0}), \sum_{\mathbf{i} \in [1,\mathbf{k}]} \mathbf{m^i} = 1, \mathbf{x} = \mathbf{m^i v_i})\}$$

is a convex set.

**Proof**
If we have two vectors $\mathbf{x}$, $\mathbf{x'}$ of $\mathbf{A}$: $\mathbf{x} = \mathbf{m^i v_i}$ and $\mathbf{x'} = \mathbf{m'^i v_i}$, from definition of $\mathbf{A}$ we may write:

$$\sum_{\mathbf{i} \in [1,\mathbf{k}]} \mathbf{m^i} = \sum_{\mathbf{i} \in [1,\mathbf{k}]} \mathbf{m'^i} = 1, \ (\forall \mathbf{i} \in [1, \mathbf{k}])(\mathbf{m^i} \geq 0, \mathbf{m'^i} \geq 0) .$$

From that we get:

$$\mathbf{t} \in [0, 1] \Rightarrow ((\forall \mathbf{i})(\mathbf{m^i t} + (1 - \mathbf{t})\mathbf{m'^i} > 0), \sum_{\mathbf{i} \in [1,\mathbf{k}]} \mathbf{m^i t} + (1 - \mathbf{t})\mathbf{m'^i} = \mathbf{t} + 1 - \mathbf{t} = 1)$$

$$\Rightarrow \mathbf{xt} + (1 - \mathbf{t})\mathbf{x'} = (\mathbf{m^i t} + (1 - \mathbf{t})\mathbf{m'^i})\mathbf{v_i} \in \mathbf{A} .$$

**Definition 257 The inner part of a finite set of vectors of a vectorial space.**

$\mathbf{V}$ with $\mathbf{Card(V)} \in \mathcal{Z}^+$, being a finite set of vectors in a vectorial space $\mathbf{E}$ built on $\mathcal{R}$, one calls the inner part $\overline{\mathbf{V}}$ of $\mathbf{V}$, the following set:

$$\mathbf{x} \in \overline{\mathbf{V}} \Leftrightarrow (\exists \mathbf{m^v})((\mathbf{m^v} : \mathbf{V} \to \mathcal{R}^{+0}), \sum_{\mathbf{v} \in \mathbf{V}} \mathbf{m^v} = 1, \mathbf{x} = \sum_{\mathbf{v} \in \mathbf{V}} \mathbf{m^v v}) .$$

**Proposition 501** The inner part of a finite set of vectors in a vectorial space is a convex set.

**Proof**
It is a direct consequence of proposition 500 (p. 281).

**Definition 258** $\sum_{\mathbf{i} \in \emptyset} \mathbf{i} = 0, \ \sum_{\mathbf{i} \in \emptyset} \mathbf{v(i)} = 0$ .

We are given a vectorial space $\mathbf{E}$ built on $\mathcal{R}$, a set $\mathbf{I}$ and a map $(\mathbf{v(i)} : \mathbf{I} \to \mathbf{E})$. One sets $\sum_{\mathbf{i} \in \emptyset} \mathbf{v(i)} = 0$. $\emptyset$ is the empty set (definition 355 (p. 456)), $\mathbf{0}$ is the null vector (definition 199 (p. 235)). If $\mathbf{I}$ is a subspace of $\mathbf{E}$ and $\mathbf{v(i)}$ the identity map we also have $\sum_{\mathbf{i} \in \emptyset} \mathbf{i} = 0$. Using proposition 362 (p. 214) and identifying $\mathbf{v} \in \mathbf{E}$ with $\sum_{\mathbf{i} \in \{\mathbf{v}\}} \mathbf{i}$ we have:

$$\sum_{\mathbf{i} \in \{\mathbf{v}\}} \mathbf{i} + \sum_{\mathbf{i} \in \emptyset} \mathbf{i} = \mathbf{v} + 0 = \sum_{\mathbf{i} \in \{\mathbf{v}\} \cup \emptyset} \mathbf{i} = \sum_{\mathbf{i} \in \{\mathbf{v}\}} \mathbf{i} = \mathbf{v} .$$

It is compatible with the definition of the vector $\mathbf{0}$ see definitions 199 (p. 235), 170 (p. 206).

**Definition 259 The inner part of a box defined by a basis.**

Given a set of independent vectors $\mathbf{V}$ of a vectorial space $\mathbf{E}$ built on $\mathcal{R}$: $\mathbf{V} = (\mathbf{v_i} : [1, \mathbf{n}] \to \mathbf{E})$. We call the inner part $\overline{\mathbf{V}}_\mathbf{b}$ of the box defined by $\mathbf{V}$, the inner part (definition 257 (p. 281)) of the set of vectors $\mathbf{P_V}$ defined by (see definition 350 (p. 453)):

$$\mathbf{x} \in \mathbf{P_V} \Leftrightarrow (\exists \mathbf{a} \in \mathcal{P}(\mathbf{V}))(\mathbf{x} = \sum_{\mathbf{v} \in \mathbf{a}} \mathbf{v}) .$$

If the vector $\mathbf{v_i}$ are along the standard basic vectors:

$$\mathbf{v_i = y^i \delta_i^j e_j, (y^i : [1, n] \to \mathcal{R})}$$

the inner part of $\mathbf{P_V}$ is the box $\overline{\mathbf{Bx}}(\mathbf{0}, \sum_{\mathbf{v \in V}} \mathbf{v})$ of definition 279 (p. 320) (see the following proposition 502 (p. 282)).

**Proposition 502** Given a vectorial space $\mathbf{E}$ built on $\mathcal{R}$ and a set of $\mathbf{n}$ independent vectors $\mathbf{V = (v_i : [1, n] \to E)}$ with $\overline{\boldsymbol{V}}_\mathbf{b}$ the inner part of the box generated by $\mathbf{V}$, we have the following relations:

$$\begin{aligned}
&\mathbf{x} \in \overline{\boldsymbol{V}}_\mathbf{b}\\
&\Leftrightarrow (\exists(\mathbf{m_i} : \mathcal{P}([1, n]) \to \mathcal{R}^{0+})), \sum_{i \in \mathcal{P}([1,n])} \mathbf{m_i = 1, x} = \sum_{i \in \mathcal{P}([1,n])} \mathbf{m_i} \sum_{j \in i} \mathbf{v_j})\\
&\Leftrightarrow (\exists(\mathbf{x^i} : [1, n] \to [0., 1.]))(\mathbf{x = x^i v_i}) \ .
\end{aligned} \tag{2.123}$$

**Proof**

From proposition 775 (p. 457)

$$(\forall \mathbf{p})(\emptyset \in \mathcal{P}([1, \mathbf{p}])) \ .$$

Then we may set $\mathbf{m}_\emptyset = \mathbf{1} - \sum_{i \in \mathcal{P}([1,p]) - \emptyset} \mathbf{m_i}$

- proposition 502 (p. 282) is true for $\mathbf{p = 1}$:
  We may write for $\mathbf{x \in V, x = x^1 v_1}$, it gives:

$$\begin{aligned}
&(\mathbf{x^1} \in [0., 1.], \mathbf{m}_{[1,1]} = \mathbf{x^1}) \Rightarrow (\forall i \in \mathcal{P}([1, p]))(\mathbf{m_i \geq 0}, \sum_{i \in \mathcal{P}([1,p])} \mathbf{m_i = 1},\\
&\mathbf{m}_\emptyset \mathbf{0} + \mathbf{m}_{[1,1]} \sum_{j \in [1,1]} \mathbf{v_j = 0 + x^1 v_1 = x^1 v_1 = x}) \ .
\end{aligned}$$

  We have as well:

$$\begin{aligned}
&((\forall i \in \mathcal{P}([1, p]))(\mathbf{m_i \geq 0}), \sum_{i \in \mathcal{P}([1,p])} \mathbf{m_i = 1})\\
&\Rightarrow (\mathbf{x^1 = m}_{[1,1]} \in [0., 1.], \mathbf{x = x^1 v_1} \in \overline{\boldsymbol{V}}_\mathbf{b} \ .
\end{aligned}$$

- Let us suppose that it is true for $\mathbf{p} \in [1, \mathbf{n}[$.

  1. From proposition 765 (p. 454), we have:

$$\begin{aligned}
&\mathcal{P}([1, p] \cup \{p + 1\}) = \mathcal{P}([1, p + 1])\\
&= \mathbf{set_x}(\exists i \subset [1, p])(\exists j \subset \{p + 1\})(x = i \cup j)\} \ .
\end{aligned}$$

  Since

$$[1, p] \cap \{p+1\} = \emptyset, \ j \subset \{p+1\} \Leftrightarrow j = \emptyset \vee \{p+1\}, \ i \subset [1, p] \Leftrightarrow i \in \mathcal{P}([1, p])$$

  we have then for $(\mathbf{f} : [1, n] \times \mathcal{P}([1, p + 1]) \to \mathbf{E})$:

$$\begin{aligned}
&\sum_{k \in \mathcal{P}([1,p+1])} \sum_{\ell \in k} \mathbf{f}(\ell, \mathbf{k})\\
&= \sum_{k \in \mathcal{P}([1,p] \cup \{p+1\})} \sum_{\ell \in k} \mathbf{f}(\ell, \mathbf{k}) = \sum_{i \in \mathcal{P}([1,p])} \sum_{j \in \{\emptyset, \{p+1\}\}} \sum_{\ell \in i \cup j} \mathbf{f}(\ell, \mathbf{i \cup j})\\
&= \sum_{i \in \mathcal{P}([1,p])} \sum_{\ell \in i} \mathbf{f}(\ell, \mathbf{i}) + \sum_{\ell \in i \cup \{p+1\}} \mathbf{f}(\ell, \mathbf{i} \cup \{\mathbf{p + 1}\}) \ .
\end{aligned}$$

We call $\overline{\boldsymbol{V}}^{\mathbf{p}}_{\mathbf{b}}$ the inner part of $(\mathbf{v_i} : [1, p] \to \mathbf{E})$. We have:

$$\mathbf{x} \in \overline{\boldsymbol{V}}^{\mathbf{p+1}}_{\mathbf{b}} \Leftrightarrow (\exists(\mathbf{m_i} : \mathcal{P}([1, p+1]) \to \mathcal{R}^{0+}))(\mathbf{x} = \sum_{i \in \mathcal{P}([1,p+1])} \mathbf{m_i} \sum_{j \in i} \mathbf{v_j},$$
$$\sum_{i \in \mathcal{P}([1,p+1])} \mathbf{m_i} = 1) .$$

With that we have:

$$\begin{aligned}
\mathbf{x} &= \sum_{i \in \mathcal{P}([1,p])} \sum_{j \in \{\emptyset, \{p+1\}\}} \sum_{k \in i \cup j} \mathbf{m_k} \sum_{\ell \in k} \mathbf{v_\ell} \\
&= \sum_{i \in \mathcal{P}([1,p])} \mathbf{m_i} \sum_{\ell \in i} \mathbf{v_\ell} + \sum_{i \in \mathcal{P}([1,p])} \mathbf{m_{i \cup \{p+1\}}} \sum_{\ell \in i \cup \{p+1\}} \mathbf{v_\ell} \\
&= (\sum_{k \in \mathcal{P}([1,p])} \mathbf{m_k}) \sum_{i \in \mathcal{P}([1,p])} \frac{\mathbf{m_i}}{\sum_{k \in \mathcal{P}([1,p])} \mathbf{m_k}} \sum_{\ell \in i} \mathbf{v_\ell} \\
&+ (\sum_{k \in \mathcal{P}([1,p])} \mathbf{m_{k \cup \{p+1\}}}) \sum_{i \in \mathcal{P}([1,p])} \frac{\mathbf{m_{i \cup \{p+1\}}}}{\sum_{k \in \mathcal{P}([1,p])} \mathbf{m_{k \cup \{p+1\}}}} \sum_{\ell \in i} \mathbf{v_\ell} \\
&+ \sum_{i \in \mathcal{P}([1,p])} \mathbf{m_{i \cup \{p+1\}}} \mathbf{v_{p+1}} \\
&= \sum_{i \in \mathcal{P}([1,p])} \mathbf{m_i} \sum_{j \in [1,p]} \mathbf{x^j v_j} + \sum_{i \in \mathcal{P}([1,p])} \mathbf{m_{i \cup \{p+1\}}} \sum_{j \in [1,p]} \mathbf{y^j v_j} \\
&+ \sum_{i \in \mathcal{P}([1,p])} \mathbf{m_{i \cup \{p+1\}}} \mathbf{v_{p+1}} \\
&= \sum_{j \in [1,p]} ( \sum_{i \in \mathcal{P}([1,p])} \mathbf{m_i x^j} + \sum_{i \in \mathcal{P}([1,p])} \mathbf{m_{i \cup \{p+1\}} y^j}) \mathbf{v_j} \\
&+ \sum_{i \in \mathcal{P}([1,p])} \mathbf{m_{i \cup \{p+1\}}} \mathbf{v_{p+1}} .
\end{aligned}$$

From what we suppose (proposition 502 (p. 282) true up to $\mathbf{p}$), we have:

$$(\forall j \in [1, p])(\mathbf{x^j} = \in [0., 1.], \mathbf{y^j} \in [0., 1.]) .$$

We apply for that proposition 502 (p. 282) separately to the $\mathbf{m_i}$ part and to the $\mathbf{m_{i \cup \{p+1\}}}$ part. From the following relation:

$$(\forall i \in \mathcal{P}([1, p+1]))(\mathbf{m_i} > 0, \sum_{i \in \mathcal{P}([1,p+1])} \mathbf{m_i} = 1)$$

we have:

$$\sum_{i \in \mathcal{P}([1,p])} \mathbf{m_i} + \sum_{i \in \mathcal{P}([1,p])} \mathbf{m_{i \cup \{p+1\}}} = 1$$

then:

$$\begin{aligned}
&(\forall j \in [1, p])(0 \leq \sum_{i \in \mathcal{P}([1,p])} \mathbf{m_i x^j} + \sum_{i \in \mathcal{P}([1,p])} \mathbf{m_{i \cup \{p+1\}} y^j} \\
&\leq \sum_{i \in \mathcal{P}([1,p])} \mathbf{m_i} + \sum_{i \in \mathcal{P}([1,p])} \mathbf{m_{i \cup \{p+1\}}} = 1). \\
&0 \leq \sum_{i \in \mathcal{P}([1,p])} \mathbf{m_{i \cup \{p+1\}}} \leq 1 .
\end{aligned}$$

When it is read from left to right, equation (2.123) (p. 282) of proposition 502 (p. 282) is true for $\mathbf{p+1}$ if it is true for $\mathbf{p}$.

2. Supposing again that proposition 502 (p. 282) is true for $\mathbf{p}$, we have:

$$(\forall \ell \in [1,p+1])(x^\ell \in [0.,1.]), x = \sum_{\ell\in[1,p+1]} x^\ell v_\ell = x = \sum_{\ell\in[1,p]} x^\ell + x^{p+1} v_{p+1} .$$

Applying 502 (p. 282) at $\mathbf{p}$ we get:

$$x = \sum_{i\in\mathcal{P}([1,p])} m_i \sum_{\ell\in i} v_\ell + x^{p+1} v_{p+1}, \sum_{i\in\mathcal{P}([1,p])} m_i = 1$$
$$x = (\sum_{i\in\mathcal{P}([1,p])} m_i(1 - x^{p+1}) \sum_{\ell\in i} v_\ell + x^{p+1} \sum_{i\in\mathcal{P}([1,p])} m_i(v_{p+1} + \sum_{\ell\in i} v_\ell) .$$

Setting $(m'_i : \mathcal{P}([1,p+1]) \to \mathcal{R}^{+0})$ by:

$$i \in \mathcal{P}([1,p]) \Rightarrow (m'_i = m_i(1 - x^{p+1}), m'_{i\cup\{p+1\}} = x^{p+1} m_i)$$

since we have $x^{p+1} \in [0.,1.]$, we also have:

$$i \in \mathcal{P}([1,p+1]) \Rightarrow m'_i \geq 0$$

and we get:

$$\sum_{i\in\mathcal{P}([1,p])} m_i = 1 \sum_{i\in\mathcal{P}([1,p])} m_i(1 - x^{p+1}) + m_i x^{p+1} = 1$$
$$= \sum_{i\in\mathcal{P}([1,p])} m'_i + m'_{i\cup\{p+1\}} = \sum_{i\in\mathcal{P}([1,p+1])} m'_i$$

together with:

$$x = \sum_{i\in\mathcal{P}([1,p])} m'_i \sum_{\ell\in i} v_\ell + \sum_{i\in\mathcal{P}([1,p])} m'_{i\cup\{p+1\}} \sum_{\ell\in i\cup\{p+1\}} v_\ell$$
$$= \sum_{i\in\mathcal{P}([1,p+1])} m'_i \sum_{\ell\in i} v_\ell .$$

When it is read from right to left, equation (2.123) (p. 282) of proposition 502 (p. 282) is true for $\mathbf{p+1}$ if it is true for $\mathbf{p}$.

- From the recurrence principle (proposition 927 (p. 520)) it is true for any $\mathbf{p \leq n}$.

**Proposition 503** Given a vectorial space $\mathbf{E}$ built on $\mathcal{R}$ and given a set of $\mathbf{n}$ independent vectors $\mathbf{V} = (v_i : [1,n] \to E)$ with $\overline{\mathcal{V}}_b$ its box inner part, given $\mathbf{p} \in \mathcal{Z}^+$ vectors $(u_i : [1,p] \to \overline{\mathcal{V}}_b \subset E)$, given a map $(m^i : [1,p] \to [0.,1.])$ such that $\sum_{i\in[1,p]} m^i = 1.$, then:

$$\sum_{i\in[1,p]} m^i u_i \in \overline{\mathcal{V}}_b .$$

**Proof**
From definition 259 (p. 281), we have:

$$u_i = u_i^j v_j, (\forall i \in [1,p])(\forall j \in [1,n])(0 \leq u_i^j \leq 1)$$

then with

$$m^i u_i = m^i u_i^j v_j, m^i \in [0.1.]$$

we also have:

$$(1 = \sum_{i\in[1,m]} m^i \geq \sum_{i\in[1,m]} m^i u_i^j \geq 0)$$
$$\Rightarrow (\forall j \in [1,n])(1 \geq m^i u_i^j \geq 0) \Rightarrow m^i u_i = m^i u_i^j v_j \in \overline{\mathcal{V}}_b \ .$$

**Proposition 504** Given a set $\mathbf{V}$ of independent vectors of a vectorial space built on $\mathcal{R}$, their box inner part $\overline{\mathcal{V}}_b$ (definition 259 (p. 281)) is convex.

**Proof**
It is a direct consequence of propositions 503 (p. 284), and 500 (p. 281).

## 2.15 Topology on $\mathcal{R}^n$

**Proposition 505** Setting $(\|x\|_p : \mathcal{R}^n \to \mathcal{R}^{+0})$ the **p_norm** as defined in definition 243 (p. 267), given a point $x \in \mathcal{R}^n$ and $\rho \in \mathcal{R}^{0+}$, there is a point $y \in \mathcal{R}^n$ such that $\|x - y\|_p = \rho$.

**Proof**
If $n = 1$ from definition 243 (p. 267), $y = x + \rho$ verifies $\|x - y\|_p = \rho$.
If it is true for $n = n'$ it is true for $n = n' + 1$:
Having ${}^{n'+1}x \in \mathcal{R}^{n'+1}$ calling ${}^{n'}x = \prod^{e}_{i\in[1,n']} {}^{n'+1}x^i \in \mathcal{R}^{n'}$ and ${}^{n'}y = \prod^{e}_{i\in[1,n']} {}^{n'+1}y^i \in \mathcal{R}^{n'}$ a solution of $\|{}^{n'}x - {}^{n'}y\|_p = 0.5\rho$ setting ${}^{n'+1}y^{n'+1} = {}^{n'+1}x^{n'+1} + (\rho^p - (0.5\rho)^p)^{1/p}$ we have $\|{}^{n'+1}x - {}^{n'+1}y\|_p = \rho$.
We conclude using the recurrence principle 927 (p. 520).

**Definition 260 Open ball (and open cube) of $\mathcal{R}^n$.**

Using the norm $(\|x\|_p : \mathcal{R}^n \to \mathcal{R}^{+0})$ (definition 243 (p. 267)),

$$B_p(x,\rho) = \text{set}_y\{y \in \mathcal{R}^n, \|y - x\|_p < \rho\} \tag{2.124}$$

is an **open ball** centered on $x$ and of radius $\rho$. We shall use $B(x,\rho)$ and $\|\|$ when, what is written, is true for any norm and deduced Euclidean metric (see definitions 243 (p. 267) and 203 (p. 237)).

**Definition 261 Open cube.**

We call **open cube** $C(x,\rho)$, an open ball when $p = \infty$. Given an open cube $C(x,\rho)$, we call center $x = (x^1, x^2., x^n)$ and side length $2\rho$ (see definition 260 (p. 285)):

$$C(x,r) = \text{set}_y\{(y = (y^1, y^2, ..., y^n) = y^i e_i \in \mathcal{R}^n), ((\forall i)(|x^i - y^i| < r))\} = B_\infty(x,r) \tag{2.125}$$

$$C(x,r) = \prod_{i\in[1,n]} I(x^i, r) = \prod_{i\in[1,n]} ]x^i - r, x^i + r[ \ .$$

**Definition 262 Closed cube.**

One calls a closed cube centered on $\mathbf{x}$ of side $\mathbf{r}$:

$$\bar{C}(x, r) = \prod_{i \in [1,n]} \bar{I}(x^i, r) = \prod_{i \in [1,n]} [x^i - r, x^i + r] .$$

**Definition 263 Open of the $\mathcal{R}^n$ topology.**

According to proposition 263 (p. 177), those open balls define on $\mathcal{R}^n$ a topology with the set of opens $\mathcal{O}$:

$$\mathcal{O} = \text{set}_O\{(\exists e \in \mathcal{P}(\mathcal{R}^n \times \mathcal{R}^+)), O = \bigcup_{(x,\rho) \in e} B_p(x, \rho)\} .$$

**Proposition 506** The topology from cubes is the same as the topology induced by the set product. (See definition 134 (p. 153). )

**Proof**

The basic element of an open induced by the set product is the product:

$$O_x = \prod_i I(x^i, \rho^i) . \tag{2.126}$$

**Proposition 507** The topology from cubes is the same as the topology induced by any **p_norm**.

**Proof**

We have the following inequalities (see proposition 328 (p. 202)):

$$\|x\|_\infty = \max_{i \in [1,n]} |x^i| = \max_{i \in [1,n]} (|x^i|^p)^{1/p} \leq \|x\|_p \leq (n \max_{i \in [1,n]} (|x^i|^p))^{1/p} = \|x\|_\infty n^{1/p} . \tag{2.127}$$

Then we have:

$$B_\infty(x, r/n^{1/p}) \subset B_p(x, r) \subset B_\infty(x, r) \subset B_p(x, r \times n^{1/p})$$

because

$$y \in B_p(x, r) \Rightarrow \|y - x\|_\infty \leq \|y - x\|_p < r \Rightarrow y \in B_\infty(x, r)$$

and

$$y \in B_\infty(x, r) \Rightarrow \|y - x\|_p \leq \|y - x\|_\infty n^{1/p} < r \times n^{1/p} \Rightarrow y \in B_p(x, r \times n^{1/p})$$

so since $(\forall p)(\forall r > 0)(x \in B_p(x, r))$, for any open $\mathbf{O}$ of the **p_norm** and any $\mathbf{x} \in \mathbf{O}$ there is a $\mathbf{r_x}$ such that $B_p(x, r_x) \subset O$ (proposition 266 (p. 179)). Then we have:

$$O = \bigcup_{x \in O} B_p(x, r_x) = \bigcup_{x \in O} B_\infty(x, r_x/n^{1/p}) .$$

Then $\mathbf{O}$ is an open of cube topology and vice versa because if $\mathbf{O}$ is an open of cube topology and if $\mathbf{r_x}$ is such that $B_\infty(x, r_x) \subset O$

$$O = \bigcup_{x \in O} B_\infty(x, r_x) = \bigcup_{x \in O} B_p(x, r_x) .$$

**Proposition 508** The closure on an open cube (definition 261 (p. 285))

$$C(x, r) = \prod_{i \in ]1,n]} ]x^i - r, x^i + r[$$

is the closed cube $\bar{C}(x, r) = \prod_{i \in [1,n]} [x^i - r, x^i + r]$ (definition 262 (p. 286)).

**Proof**
Any point $y \in \bar{C}(x, r)$ is such that $(\forall i \in [1, n])(|y_i - x_i| \leq r)$. From 266 (p. 179), any open containing $y$ included an open ball $C(y, 2\rho)$. This ball contains a point of $C(x, r)$ because we have:

$$(\forall i \in [1, n])(|y_i - x_i| \leq r) \Rightarrow (\forall i)(\exists s_i \in \{-1, +1\})(0 \leq (y_i - x_i)s_i \leq r) .$$

Taking those $s_i$, we have used $\theta = \min(\rho, |y_i - x_i|/2)$:

$$(\forall i \in [1, n])(|y_i - s_i\theta - y_i| < 2\rho, |y_i - s_i\theta - x_i| = |(y_i - x_i)s_i - s_i^2\theta| \leq |y_i - x_i|/2 < r) .$$

Then the point $y - s\theta$ is in both $C(y, 2\rho)$ and $C(x, r)$. Then $y$ is in the closure of $C(x, r)$ (see definition 136 (p. 155)).
Since $y \in \mathcal{C}(\bar{C}(x, r))$ means there are an $i \in [1, n]$ and a $s_i \in \{-1, +1\}$ such that $r < (y_i - x_i)s_i$. It means that any point $z \in C(y, ((y_i - x_i)s_i - r)/3)$ cannot be in $C(x, r)$ because we have:

$$\begin{aligned} |z_i - x_i| &= |(z_i - y_i)s_i + (y_i - x_i)s_i| > (y_i - x_i)s_i - |(z_i - y_i)s_i| \\ &> (y_i - x_i)s_i - ((y_i - x_i)s_i - r)/3) = (2(y_i - x_i)s_i + r)/3) > r . \end{aligned}$$

Then $y$ is not in the closure of $C(x, r)$ (see definition 136 (p. 155)).

**Proposition 509** $C \subset \mathcal{R}^n$ being closed, if a point $c \in \mathcal{R}^n$ is such that any open containing $c$ contains a point of $C$, then $c$ is an element of $C$.

**Proof**
It is a direct consequence of proposition 220 (p. 158).

**Proposition 510** If $C \in \mathcal{R}$ is closed and if its **lower_limit** and its **upper_limit** exist, they belong to $C$.

**Proof**
It is a consequence of proposition 509 (p. 287) and of the properties of the **lower_limit** and **upper_limit**.

**Proposition 511** Given an open set $A \subset \mathcal{R}^n$, a map $(f : A \to \mathcal{R}^m)$, $n \in \mathcal{Z}^+$, $m \in \mathcal{Z}^+$, we have (see definition 158 (p. 183)):

$$u \in A \Rightarrow \lim_{x \to u} f(x) \in \overline{f(A)} \tag{2.128}$$

where $\overline{f(A)}$ is the closure of $f(A)$ (definition 136 (p. 155)).

**Proof**
From definition 157 (p. 182), any open $O$ containing $\lim_{x \to u} f(x)$ contains the image of an open $O_A$ containing $u$. $O_A \cap A$ is open (definition 132 (p. 151)). From proposition 266 (p. 179) $O_A \cap A$ contains an open ball $B(u, \rho)$ centered on $u$ and of radius $\rho$. From proposition 505 (p. 285), there is a point $v \neq u$ such that $v \in B(u, \rho) \subset A$. It means $f(v) \in O$. Then since $f(v) \in f(A)$, $\lim_{x \to u} f(x) \in \overline{f(A)}$.

**Proposition 512** With the same premise as proposition 511 (p. 287), if $\mathbf{C}$ is a closed subset of $\mathcal{R}^m$ such that $\mathbf{f(A)} \subset \mathbf{C}$ then $\lim_{x \to u} f(x) \in C$.

**Proof**
It is a direct consequence of proposition 511 (p. 287) and proposition 211 (p. 155).

**Proposition 513** An open ball of $\mathcal{R}^n$ is convex (definition 255 (p. 277)).

**Proof**
Since $t \in [0,1] \Rightarrow (t \geq 0, 1-t \geq 0) \Rightarrow (|t| = t, |1-t| = 1-t)$, from definitions 203 (p. 237), 243 (p. 267), propositions 472 (p. 268), 473 (p. 268), we have:

$$\begin{aligned}&(x_1 \in B_p(x,\rho), x_2 \in B_p(x,\rho), t \in [0,1]) \Rightarrow \|tx_1 + (1-t)x_2 - x\|_p\\&= \|t(x_1 - x) + (1-t)(x_2 - x)\|_p\\&\leq t\|x_1 - x\|_p + (1-t)\|x_2 - x\|_p < t\rho + (1-t)\rho = \rho\end{aligned}$$

(see definition 203 (p. 237)), proposition 475 (p. 270)). Then $tx_1 + (1-t)x_2 \in B_p(x,\rho)$.

**Proposition 514** The closure of a ball is the same ball plus the points at a distance to the center ball equal to the radius of the ball and it is a closed subset (see also proposition 268 (p. 179)).

**Proof**
Given an open ball $B(x,\rho)$ of center $x$ and radius $\rho$, for any point $y$ at a distance $\rho$ of $x$ and in any open $O_y$ including the point $y$, there is a ball $B(y,\rho') \subset O_y$ (proposition 266 (p. 179)). There are points common to both balls $B(x,\rho)$ and $B(y,\rho')$ because the distance of their centers is bigger than the difference of radius and smaller than their sum. From proposition 219 (p. 158), the closure of a ball is closed.

**Proposition 515** Given a metric space $\mathbf{E}$, considering $\mathcal{R}$ with the open interval topology (see definition 147 (p. 176), proposition 263 (p. 177)), then for all $x \in E$, the map

$$(d(x,y) = d(y,x) : E \to \mathcal{R}^{+0})$$

is continuous. $d(x,y)$ represents the distance between $x \in E$ and $y \in E$ (definition 147 (p. 176)).

**Proof**
Calling $\mathcal{O}_{\mathcal{R}}$ the set of opens of $\mathcal{R}$, by definition (see proposition 263 (p. 177)), we have:

$$O \in \mathcal{O}_{\mathcal{R}} \Leftrightarrow (\exists(\rho(x) : \mathcal{R} \to \mathcal{R}^+))(O = \bigcup_{u \in O} ]u - \rho(u), u + \rho(u)[) .$$

We have:

$$u - \rho(u) > 0 \Rightarrow ]u - \rho(u), u + \rho(u)[ = ]0, u + \rho(u)[ \cap \mathcal{C}([0, u - \rho(u)]) .$$

By definition of a ball using propositions 164 (p. 138) and 165 (p. 138)

$$\begin{aligned}&\{d^{-1}(x, ]0, u + \rho(u)[) = B(x, u + \rho(u)),\\&d^{-1}(x, \mathcal{C}([0, u - \rho(u)])) = \mathcal{C}(\bar{B}(x, u - \rho(u)))\}\\&\Rightarrow d^{-1}(x, ]u - \rho(u), u + \rho(u)[) = B(x, u + \rho(u)) \cap \mathcal{C}(\bar{B}(x, u - \rho(u))) \in \mathcal{O}_E ,\end{aligned}$$

where $\mathcal{O}_E$ is the set of opens of $\mathbf{E}$.

**Proposition 516** The map $(\|r\| : \mathcal{R}^n \to \mathcal{R}^{+0})$ is a continuous map.

**Proof**
It is a direct consequence of proposition 277 (p. 182) because $\|\mathbf{r}\| = \|\mathbf{0} - \mathbf{r}\| = \mathbf{d}(\mathbf{0}, \mathbf{r})$.

**Proposition 517** The only subsets of $\mathcal{R}^n$ which are open and closed are $\mathcal{R}^n$ itself and $\emptyset$.
**Proof**
If $\mathbf{A} \subset \mathcal{R}^n$ is open and closed with $\mathbf{A} \neq \emptyset$, $\mathbf{A} \neq \mathcal{R}^n$. Then $(\exists \mathbf{a} \in \mathcal{R}^n)(\mathbf{a} \in \mathbf{A}), (\exists \mathbf{b} \in \mathcal{R}^n)(\mathbf{b} \notin \mathbf{A})$. Then given such $\mathbf{a}$ and $\mathbf{b}$, and $\mathbf{d}(\mathbf{a}, \mathbf{b}) = \|\mathbf{a} - \mathbf{b}\|$, we call $\mathbf{f}(\lambda)$, the map $\mathbf{f}(\lambda) : [0, \mathbf{d}(\mathbf{a}, \mathbf{b})] \to \mathcal{R}^n$:

$$\mathbf{f}(\lambda) = \frac{\mathbf{a}(\mathbf{d}(\mathbf{a}, \mathbf{b}) - \lambda) + \mathbf{b}\lambda}{\mathbf{d}(\mathbf{a}, \mathbf{b})} = \mathbf{a} - \lambda \frac{\mathbf{a} - \mathbf{b}}{\mathbf{d}(\mathbf{a}, \mathbf{b})}$$

and

$$\mathbf{a} - \mathbf{f}(\lambda) = \lambda \frac{\mathbf{a} - \mathbf{b}}{\mathbf{d}(\mathbf{a}, \mathbf{b})}$$

(see equation (2.105) (p. 267)).
Using definition 203 (p. 237), 3 we have:

$$\lambda = \|\mathbf{a} - \mathbf{f}(\lambda)\|, \ \|\mathbf{f}(\lambda) - \mathbf{f}(\lambda')\| = |\lambda - \lambda'| \ . \tag{2.129}$$

We now define $\Lambda$ as the set:

$$\Lambda = \mathbf{set}_\lambda\{(\exists \mathbf{x} \in \mathbf{A}), (\lambda \in \mathcal{R}), (0 \leq \lambda \leq \mathbf{d}(\mathbf{a}, \mathbf{b})), \mathbf{x} = \mathbf{f}(\lambda)\}$$

$\Lambda \neq \emptyset$ because $0 \in \Lambda$ and $\mathbf{d}(\mathbf{a}, \mathbf{b})$ is a majoring value. So, from proposition 258 (p. 174), $\Lambda$ has a **upper_limit**:

$$\lambda_s = \mathbf{upper\_limit}(\Lambda) \ .$$

We call $\mathbf{u} = \mathbf{f}(\lambda_s)$.
Any open containing $\mathbf{u}$, contains an open ball of $\mathcal{R}^n$ centered on $\mathbf{u}$ (proposition 266 (p. 179)). Let us call its radius $\mathbf{r}$. There is from proposition 260 (p. 175) a $\lambda' \in \Lambda$ with $0 \leq \lambda_s - \lambda' < \mathbf{r}$. From equation (2.129) (p. 289) $\|\mathbf{f}(\lambda_s) - \mathbf{f}(\lambda')\| = \lambda_s - \lambda' < \mathbf{r}$, $\mathbf{f}(\lambda')$ is in that ball so also in the open and in $\mathbf{A}$. Since $\mathbf{A}$ is closed, from proposition 509 (p. 287), $\mathbf{u}$ is in $\mathbf{A}$.
Since $\mathbf{A}$ is open, there is, as before, an open ball centered on $\mathbf{u}$ in $\mathbf{A}$, $\mathbf{r}$ being its radius. From equation (2.129) (p. 289), we have $\mathbf{f}(\lambda_s + \mathbf{r}/2)$ in that ball. The ball being in $\mathbf{A}$, $\mathbf{f}(\lambda_s + \mathbf{r}/2)$ is in $\mathbf{A}$. Then $\lambda_s + \mathbf{r}/2 \in \Lambda$ and $\lambda_s$ cannot be $\mathbf{upper\_limit}(\Lambda)$.
$\mathbf{A}$ open and closed leads to an absurd situation if $\mathbf{A} \neq \emptyset, \mathbf{A} \neq \mathcal{R}^n$. From A.1.4 -8 (p. 435) we conclude that 517 (p. 289) is true.

## 2.15.1 Projection of $\mathcal{R}^n$ on $\mathcal{R}^{n-1}$

We shall deal here with some topological properties of projections as defined in definition 244 (p. 270).
**Definition 264 Projection of $\mathcal{R}^n$ on $\mathcal{R}^{n-1}$ along the axis i: $(\mathbf{Pr_i} : \mathcal{R}^n \to \mathcal{R}^{n-1})$.**

The usual notations for an element of $\mathcal{R}^n$ are

$$(\mathbf{x^k} : [1, \mathbf{n}] \to \mathcal{R}) = (\mathbf{x^1}, \mathbf{x^2}, ..., \mathbf{x^i}, ..., \mathbf{x^n}) = \mathbf{x^i e_i} = \sum_{\mathbf{i} \in [1,\mathbf{n}]} \mathbf{x^i e_i} \in \mathcal{R}^n$$

$\mathbf{e_i} = \prod^{\mathbf{e}}_{\mathbf{j} \in [1,\mathbf{n}]} \delta^{\mathbf{j}}_{\mathbf{i}} = (\delta^{\mathbf{j}}_{\mathbf{i}} : [1, \mathbf{n}] \to \mathcal{R}) \in \mathcal{R}^n$ is the reference basis of $\mathcal{R}^n$, $\delta^{\mathbf{j}}_{\mathbf{i}}$ is the Kronecker symbol (see equation (2.106) (p. 267) in definition 242 (p. 267)).
The map $(\mathbf{Pr_i} : \mathcal{R}^n \to \mathcal{R}^{n-1})$ gives for all

$$(\mathbf{x^k} : [1, \mathbf{n}] \to \mathcal{R}) = (\mathbf{x^1}, \mathbf{x^2}, ..., \mathbf{x^{i-1}}, \mathbf{x^i}, \mathbf{x^{i+1}}, ..., \mathbf{x^n}) = \mathbf{x^i e_i} \in \mathcal{R}^n$$

a term

$$(\mathbf{x^k} : [1, n] - \{i\} \to \mathcal{R}) = (\mathbf{x^1}, \mathbf{x^2}, ..., \mathbf{x^{i-1}}, \mathbf{x^{i+1}} ..., \mathbf{x^n}) \in \mathcal{R}^{n-1}$$

where the term of rank $\mathbf{i}$ has been dropped. For simplicity we shall limit the propositions and proves to the case where $\mathbf{i} = \mathbf{n}$.

**Proposition 518** The projection of an open ball of $\mathcal{R}^n$ on $\mathcal{R}^{n-1}$ along the axis $\mathbf{n}$ ($\mathbf{Pr_n}$) is a ball of the same radius.

**Proof**

$\mathbf{x_n}$ and $\mathbf{y_n}$ being in $\mathcal{R}^n$, $\mathbf{x_{n-1}}$ and $\mathbf{y_{n-1}}$ their projection on $\mathcal{R}^{n-1}$, we have

$$\mathbf{p} \geq 1 \Rightarrow \mathbf{d_p(x_n, y_n)} = \mathbf{d(x_n, y_n)} = (\sum_{i \in [1,n]} |\mathbf{x^i} - \mathbf{y^i}|^p)^{1/p} \geq (\sum_{i \in [1,n-1]} |\mathbf{x^i} - \mathbf{y^i}|^p)^{1/p}$$

so $\mathbf{d(x_n, y_n)} \leq \mathbf{r} \Rightarrow \mathbf{d(x_{n-1}, y_{n-1})} \leq \mathbf{r}$ So a ball of radius $\mathbf{r}$ centered on $\mathbf{x_n}$ projects inside a ball of the same radius centered on $\mathbf{x_{n-1}}$. The ball is fully filled since any point of the ball centered on $\mathbf{x_{n-1}}$ and of radius $\mathbf{r}$ in $\mathcal{R}^{n-1}$ is the projection of a point in the ball of $\mathcal{R}^n$ centered on $\mathbf{x_n}$ and of the same radius with the last coordinate $\mathbf{y_n} = \mathbf{x_n}$.

**Proposition 519** The projection of an open set of $\mathcal{R}^n$ on $\mathcal{R}^{n-1}$ along the axis $\mathbf{n}$ ($\mathbf{Pr_n}$) is an open set.

**Proof**

See also propositions 205 (p. 154), 506 (p. 286), 507 (p. 286). An open set $\mathbf{A} \subset \mathcal{R}^n$ can be considered as the union of all open ball centered on each point of $\mathbf{A}$ and fully in $\mathbf{A}$. From proposition 518, their projection are also open balls fully inside the projection of $\mathbf{A}$. Their union is equal to the projection of $\mathbf{A}$ so the projection of $\mathbf{A}$ is open.

**Proposition 520** The inverse (image of the) projection (definition 115 (p. 133)) of an open set of $\mathcal{R}^{n-1}$ on $\mathcal{R}^n$ along the axis $\mathbf{n}$ ($\mathbf{Pr_n}$) is an open set. The projection map is continuous.

.

**Proof**

Any point $\mathbf{x} = (\mathbf{x^1}, \mathbf{x^2}, ..., \mathbf{x^{n-1}}) = \mathbf{x^i e_i}$ of an open set of $\mathcal{R}^{n-1}$ is the center of a ball $\mathbf{B^{n-1}(x, \rho)}$, calling $\rho$ its radius. For any point $\mathbf{x'}$ of $\mathcal{R}^n$ which projects on $\mathbf{x}$, there is an $\mathbf{y} \in \mathcal{R}$ such that $\mathbf{x'} = (\mathbf{x^1}, \mathbf{x^2}, ..., \mathbf{x^{n-1}}, \mathbf{y})$. The ball $\mathbf{B^n(x', \rho)} \subset \mathcal{R}^n$ projects in $\mathbf{B^{n-1}(x, \rho)}$ so is in the inverse projection of it and of the open set of $\mathcal{R}^{n-1}$. Then the inverse projection of the open of $\mathcal{R}^{n-1}$ is the union of those open balls associated to each of its points so it is open (see definition 132). From definition 141, the projection map is continuous.

### 2.15.2 Compact set in $\mathcal{R}^n$

**Proposition 521** A closed subset of $\mathcal{R}^n$ included in a ball is compact.

**Proof**

Let us demonstrate it in the **p_norm** $= \infty$ with $(\mathbf{e_i} : [1, n] \to \mathcal{R}^n)$ as a reference basis: $\mathbf{x} \in \mathbf{B(y, \rho)} \Rightarrow \|\mathbf{x} - \mathbf{y}\|_\infty < \rho \Leftrightarrow (\forall \mathbf{i} \in [1, n])(|\mathbf{x^i} - \mathbf{y^i}| < \rho)$ and $\mathbf{x} = \mathbf{x^i e_i}, \mathbf{y} = \mathbf{y^i e_i}$. It is the norm corresponding to the set product. $\mathbf{B(y, \rho)} = \prod_{i \in [1,n]} ]\mathbf{y^i} - \rho, \mathbf{y^i} + \rho[$. The conditions of proposition 521 suppose that there is a $\mathbf{c}$ such that the closed set $\mathbf{A}$ is included in the closed ball $\bar{\mathbf{B}}(\mathbf{0}, \mathbf{c})$. From definition 262 (p. 286) $\bar{\mathbf{B}}(\mathbf{0}, \mathbf{c}) = [-\mathbf{c}, \mathbf{c}]^n$. From proposition 298 (p. 190), $[-\mathbf{c}, \mathbf{c}]$ is compact. Then from proposition 230 (p. 164), $\bar{\mathbf{B}}(\mathbf{0}, \mathbf{c}) = [-\mathbf{c}, \mathbf{c}]^n$ is also compact. From proposition 223 (p. 160), $\mathbf{A}$ is compact. Since all the topologies generated by the **p_norm** (equation (2.107) (p. 268)) are equivalent to the $\infty$**_norm** topology (proposition 507 (p. 286)), the same proposition is valid whatever is the **p_norm**.

**Proposition 522** Any compact subset of $\mathcal{R}^n$ is a closed set.
**Proof**
It is a direct application of proposition 226 (p. 160).

**Proposition 523** Any compact subset of $\mathcal{R}^n$ is a closed set enclosed in a ball.
**Proof**
It is a direct application of proposition 271 (p. 180).

**Proposition 524** Given a closed set not empty, $K \subset \mathcal{R}^n$, we consider the map: $(md_K(x) : \mathcal{R}^n \to \mathcal{R}^{0+})$ defined by (see definitions 148 (p. 176), 261 (p. 176)):

$$x \in \mathcal{R}^n \Rightarrow md_K(x) = \min_{y \in K} \|x - y\| .$$

This map is continuous and we have:

$$(\forall x \in \mathcal{R}^n)(\exists y \in K)(\|x - y\| = md_K(x)) .$$

**Proof**
Since $\|x - y\| \geq 0$, from proposition 258 (p. 174), $md(x)$ exists and is unique. It is then a map (definition 112 (p. 131)). Since $K \neq \emptyset$, $\exists z \in K$, setting $d = \|x - z\| \geq md(x)$ we have:

$$md_K(x) = \min_{y \in K} \|x - y\| = \min_{y \in K \cap \overline{B(x,d)}} \|x - y\| .$$

From proposition 277 (p. 182), $\|x - y\|$ is a continuous map. From proposition 206 (p. 154) $K \cap \overline{B(x, d)}$ is closed. From proposition 521 (p. 290) $K \cap \overline{B(x, d)}$ is compact. From proposition 297 (p. 189),

$$(\exists y \in K \cap \overline{B(x, d)})(md_K(x) = \|x - y\|) .$$

Using that, we shall prove that $md_K(x)$ is uniformly continuous:
$x_1 \in \mathcal{R}^n$, $x_2 \in \mathcal{R}^n$ calling $y_1 \in K$ and $y_2 \in K$ such as:

$$md_K(x_1) = \|x_1 - y_1\|,\ md(x_2) = \|x_2 - y_2\| ,$$

one has:

$$\|x_1 - y_1\| \leq \|x_1 - y_2\| \leq \|x_1 - x_2\| + \|x_2 - y_2\|$$

and the same exchanging 1 and 2.

$$\|x_2 - y_2\| \leq \|x_1 - x_2\| + \|x_1 - y_1\| .$$

Then if $0 < \|x_1 - x_2\| < \epsilon$ we have (proposition 317 (p. 197)):

$$\|x_1 - x_2\| < \epsilon \Rightarrow -\epsilon < \|x_1 - y_1\| - \|x_2 - y_2\| < \epsilon \Rightarrow |md_K(x_1) - md_K(x_2)| < \epsilon .$$

Then:

$$(\forall \alpha > 0)(\exists \epsilon = \alpha > 0)(\forall x \in \mathcal{R}^n)(\forall x' \in \mathcal{R}^n)(\|x - x'\| < \epsilon$$
$$\Rightarrow |md_K(x) - md_K(x')| < \alpha) .$$

**Proposition 525** Given an open $O \subset \mathcal{R}^n, O \neq \mathcal{R}^n$

$$(\forall x \in O)(\exists x_m \in \mathcal{R}^+)(((y \in \mathcal{R}^n, \|y - x\| < x_m) \Rightarrow y \in O),$$
$$(\exists y \notin O)(\|y - x\| = x_m)) .$$

**Proof**
$\mathcal{C}(O)$ is closed from definition 135 (p. 154), and not empty since $O \neq \mathcal{R}^n \Leftrightarrow \mathcal{C}(O) \neq \emptyset$. Then proposition 524 (p. 291) gives 525 (p. 291).

**Proposition 526** Given a non-empty compact set $K \subset \mathcal{R}^n$, given an open $O \subset \mathcal{R}^n$ containing $K \subset O$, we have:

$$(\exists c \in K)(\exists d \in \mathcal{C}(O))(||c - d|| = \min_{x \in K, y \in \mathcal{C}(O)} ||x - y|| > 0) . \tag{2.130}$$

**Proof**
$\mathcal{C}(O)$ is closed from definition 135 (p. 154). From proposition 524 (p. 291), the map:

$$md_{\mathcal{C}(O)}(x) = \min_{y \in \mathcal{C}(O)} ||x - y||$$

is continuous. Since $K$ is compact, from proposition 297 (p. 189),

$$(\exists c \in K)(md_{\mathcal{C}(O)}(c) = \min_{z \in K}(md_{\mathcal{C}(O)}(z))) .$$

Then from proposition 524 (p. 291),

$$(\exists d \in \mathcal{C}(O))(||c - d|| = md_{\mathcal{C}(O)}(c) = \min_{z \in K}(md_{\mathcal{C}(O)}(z)))$$

In such a case we have

$$||c - d|| = \min_{z \in K}(md_{\mathcal{C}(O)}(z)) \neq 0 .$$

Otherwise we would have $d = c \in K$ which is impossible since $K \subset O \Rightarrow \mathcal{C}(O) \cap K = \emptyset$.

**Proposition 527** Given a non-empty compact set $K \subset \mathcal{R}^n$, given an open $O \subset \mathcal{R}^n$ containing $K \subset O$, calling $c \in K$, $d \in \mathcal{C}(O)$ the two points as defined by equation (2.130) (p. 292) of proposition 526 (p. 292), for any $0 < u \leq ||c - d||$ there is a continuous map $(h_u : \mathcal{R}^n \to [0, 1])$ with a compact support such that

$$x \in K \Rightarrow h_u(x) = 1, x \notin O \Rightarrow h_u(x) = 0, (x \in \mathcal{R}^n, md_K(x) \geq u) \Rightarrow h_u(x) = 0 .$$

**Proof**
We have from proposition 524 (p. 291):

$$(\forall x \in \mathcal{R}^n)(\exists y \in \mathcal{C}(O))(\exists z \in K)(||x - y|| = md_{\mathcal{C}(O)}(x), ||x - z|| = md_K(x)) .$$

Using those $y$ and $z$ depending on $x$, using $c \in K$, and $d \in \mathcal{C}(O)$ as defined in equation (2.130) (p. 292) of proposition 526 (p. 292), setting $u \leq ||c - d|| > 0$, we have from definition 147 (p. 176) and proposition 262 (p. 176):

$$\begin{aligned}&||y - x|| \geq ||y - z|| - ||z - x|| \geq ||c - d|| - ||z - x|| \geq u - ||z - x|| \\ &= u - md_K(x)\end{aligned}$$

then

$$\begin{aligned}&md_K(x) < u \Rightarrow u - md_K(x) > 0 \\ &\Rightarrow ||y - x|| = md_{\mathcal{C}(O)}(x) > 0 \Rightarrow x \in O .\end{aligned}$$

We define a map $(h_u(x) : \mathcal{R}^n \to \mathcal{R}^{0+})$ by:

$$\begin{aligned}&1 - \frac{md_K(x)}{u} > 0 \Rightarrow h_u(x) = 1 - \frac{md_K(x)}{u}, \\ &1 - \frac{md_K(x)}{u} \leq 0 \Rightarrow h_u(x) = 0\end{aligned}$$

**$h_u(x)$ is uniformly continuous:**
If $\|x - x'\| < \epsilon$, we may label $x$ and $x'$ such that $md_K(x) \leq md_K(x')$ and, from proposition 524 (p. 291), we may set $z$ such that $z \in K, \|x - z\| = md_K(x)$. From proposition 262 (p. 176), we have:

$$|md_K(x') - md_K(x)| = md_K(x') - md_K(x) \leq \|x' - z\| - md_K(x)$$
$$= \|x' - z\| - \|z - x\| \leq \|x - x'\| < \epsilon$$

then:

$$(\frac{md_K(x)}{u} \geq 1, \frac{md_K(x')}{u} \geq 1)$$
$$\Rightarrow h_u(x) = h_u(x') = 0 \Rightarrow |h_u(x) - h_u(x')| < \frac{\epsilon}{u}$$
$$(\frac{md_K(x)}{u} \in [0, 1[, \frac{md_K(x')}{u} \in [0, 1[) \Rightarrow |h_u(x) - h_u(x')| < \frac{\epsilon}{u}$$
$$(\frac{md_K(x')}{u} \geq 1, \frac{md_K(x)}{u} \in [0, 1[)$$
$$\Rightarrow -\frac{\epsilon}{u} < 0 \leq h_u(x) - h_u(x') = 1 - \frac{md_K(x)}{u} \leq \frac{md_K(x') - md_K(x)}{u} < \frac{\epsilon}{u}$$
$$\Rightarrow |h_u(x) - h_u(x')| < \frac{\epsilon}{u} .$$

Then, in any case:

$$(\forall \alpha)(\exists \epsilon = \alpha u > 0)(\|x - x'\| < \epsilon \Rightarrow |h_u(x) - h_u(x')| < \frac{\epsilon}{u} = \alpha) .$$

Since $(\exists z \in K)(\|x - z\| = md_K(x) < u)$ and since from proposition 271 (p. 180) $(\exists r > 0)(K \subset B(0, r))$, we have (point 4 in definition 147 (p. 176)):

$$h_u(x) \neq 0 \Rightarrow \|x\| \leq \|x - z\| + \|z\| < u + r$$

$h_u(x)$ is continuous with a compact support. Proposition 527 (p. 292) is true.

**Proposition 528** Given a closed subset $S$ of $\mathcal{R}^n$, any closed subset[24] of $S$ included in a ball is compact over the induced topology (definition 133 (p. 151)). See also proposition 223 (p. 160) which is a more general version of this proposition.

**Proof**
Calling $\mathcal{O}$ the set of opens of $\mathcal{R}^n$, given a closed set $C \subset S$ over the induced topology, $C$ is supposed to be in an open ball. We call $\mathcal{O}_S$ the set of opens of $S$ in the induced topology. We have from definition 133 (p. 151) and proposition 199 (p. 151) equation (2.19):

$$O \in \mathcal{O}_S \Leftrightarrow (\exists O_{\mathcal{R}^n} \in \mathcal{O})(O = O_{\mathcal{R}^n} \cap S) .$$

Introducing, as in proposition 199 (p. 151), the injective map (definition 122 (p. 139)) $(O_{\mathcal{R}^n}(O) : \mathcal{O}_S \to \mathcal{O})$ such that:

$$O \in \mathcal{O}_S \Rightarrow (O = O_{\mathcal{R}^n}(O) \cap S, O_{\mathcal{R}^n}(O) \in \mathcal{O}) .$$

[24]From propositions 207 (p. 155) and 208 (p. 155), any closed subset under the induced topology of $S$ is closed in $\mathcal{R}^n$ and any subset of $S$ closed in $\mathcal{R}^n$ is closed under the induced topology of $S$.

From proposition 207 (p. 155). $\mathbf{C}$ is also closed in $\mathcal{R}^n$ (from proposition 208 (p. 155) the inverse is also true). From proposition 521 (p. 290), $\mathbf{C}$ is compact in $\mathcal{R}^n$ topology. Any open in the induced topology, which covers $\mathbf{C}$, is the intersection of an open of $\mathcal{R}^n$ and $\mathbf{S}$ (see proposition 798 (p. 465)):

$$\mathrm{U} \subset \mathcal{O}_\mathrm{S}, \mathrm{C} \subset \bigcup_{\mathrm{u} \in \mathrm{U}} \mathrm{u} = \bigcup_{\mathrm{u} \in \mathrm{O}_{\mathcal{R}^n}(\mathrm{U})} \mathrm{u} \cap \mathrm{S} = \mathrm{S} \cap \bigcup_{\mathrm{u} \in \mathrm{O}_{\mathcal{R}^n}(\mathrm{U})} \mathrm{u} \Rightarrow \mathrm{C} \subset \bigcup_{\mathrm{u} \in \mathrm{O}_{\mathcal{R}^n}(\mathrm{U})} \mathrm{u}\,.$$

Since $\mathbf{C}$ is compact in $\mathcal{R}^n$, there is a finite set $\mathbf{T} \subset \mathbf{O}_{\mathcal{R}^n}(\mathbf{U})$ such that $\mathbf{C} \subset \bigcup_{\mathbf{u} \in \mathbf{T}} \mathbf{u}$. From propositions 763 (p. 453), 798 (p. 465), we have $\mathbf{C} = \mathbf{C} \cap \mathbf{S} \subset \mathbf{S} \cap \bigcup_{\mathbf{u} \in \mathbf{T}} \mathbf{u} = \bigcup_{\mathbf{u} \in \mathbf{T}} \mathbf{u} \cap \mathbf{S}$. Since the map $\mathbf{O}_{\mathcal{R}^n}$ is injective and from definition 114 (p. 132), we have:

$$\mathrm{u} \in \mathrm{O}_{\mathcal{R}^n}(\mathrm{U}) \Rightarrow (\exists \mathrm{u}' \in \mathrm{U})(\mathrm{u}' = \mathrm{u} \cap \mathrm{S}) \Rightarrow \mathrm{u} \cap \mathrm{S} \in \mathrm{U}$$

$\mathbf{C}$ is compact in the induced topology.

**Proposition 529** Given a continuous injective map (definition 122 (p. 139)) $(\mathbf{f} : \mathbf{C} \subset \mathcal{R}^m \rightarrow \mathcal{R}^n)$ where $\mathbf{C}$ is a closed subset of $\mathcal{R}^m$ included in a ball $\mathbf{B}(0, \mathbf{a}) \subset \mathcal{R}^m$, the inverse map $(\mathbf{f}^{-1} : \mathbf{f}(\mathbf{C}) \rightarrow \mathbf{C} \subset \mathbf{B}(0, \mathbf{a}))$ is continuous in the induced topology.

**Proof**

From proposition 169 (p. 139), the inverse map $(\mathbf{f}^{-1} : \mathbf{f}(\mathbf{C}) \rightarrow \mathbf{C} \subset \mathbf{B}(0, \mathbf{a}))$ exists. For any open $\mathbf{O}$ of $\mathbf{C}$, there is an open $\mathbf{O}'$ of $\mathcal{R}^n$ such that $\mathbf{O} = \mathbf{O}' \cap \mathbf{C}$ then from proposition 768 (p. 455), $\mathcal{C}_\mathbf{C}(\mathbf{O}) = \mathcal{C}(\mathbf{O}') \cap \mathbf{C}$. $\mathcal{C}(\mathbf{O}') \cap \overline{\mathbf{B}(0, 2\mathbf{a})} \cap \mathbf{C}$ is closed in $\mathcal{R}^n$ topology as intersection of closed sets (proposition 206 (p. 154)), it is included in a ball then it is compact (proposition 528 (p. 293)) in the $\mathcal{R}^n$ topology. $\mathcal{C}_\mathbf{C}(\mathbf{O}) = \mathcal{C}(\mathbf{O}') \cap \overline{\mathbf{B}(0, 2\mathbf{a})} \cap \mathbf{C} = \mathcal{C}(\mathbf{O}') \cap \mathbf{C}$ is compact in $\mathbf{C}$ induced topology (proposition 227 (p. 162)). $\mathbf{f}(\mathcal{C}_\mathbf{C}(\mathbf{O}))$ is compact (proposition 240 (p. 167)) and then closed (proposition 226 (p. 160)). $\mathbf{f}^{-1}$ being a map, from propositions 164 (p. 138) and 165 (p. 138) applied to $\mathbf{f}^{-1}$ we have:

$$\mathrm{f}(\mathrm{O}) = \mathcal{C}\mathrm{f}(\mathcal{C}_\mathrm{C}(\mathrm{O})) \cap \mathrm{f}(\mathrm{C})$$

is an open of the induced topology on $\mathbf{f}(\mathbf{C})$. Then from definition 2.1.3 (p. 165) $(\mathbf{f}^{-1} : \mathbf{f}(\mathbf{C}) \rightarrow \mathbf{C})$ is continuous.

**Proposition 530** Given a closed subset of $\mathcal{R}^n$ enclosed in a ball its projection on $\mathcal{R}^{n-1}$ is also a closed subset enclosed in a ball.

**Proof**

Calling $\mathbf{C} \subset \mathcal{R}^n$ that closed set in a closed ball, from proposition 521 (p. 290) $\mathbf{C}$ is compact. From proposition 520 (p. 290) together with proposition 240 (p. 167), its projection on $\mathcal{R}^{n-1}$ is compact. From proposition 523 (p. 291), the projection of $\mathbf{C}$ is a closed set in a ball of $\mathcal{R}^{n-1}$.

**Proposition 531** Given a set $\mathbf{I}$, an open ball $\mathbf{B}(0, \mathbf{d})$ of $\mathcal{R}^n$, a map $(\mathbf{C_i} : \mathbf{I} \rightarrow \mathcal{P}(\mathbf{B}(0, \mathbf{d})))$ such that all $\mathbf{C_i}$ are not empty and such that:

$$(\forall \mathrm{i} \in \mathrm{I})(\forall \mathrm{j} \in \mathrm{I})((\mathrm{C_i} \subset \mathrm{C_j}) \vee (\mathrm{C_j} \subset \mathrm{C_i}))$$

then

$$\bigcap_{\mathrm{i} \in \mathrm{I}} \overline{\mathrm{C}}_\mathrm{j} \neq \emptyset$$

See the definition of the closure of a subset 136 (p. 155).

**Proof**

- $\mathbf{n = 1}$
We suppose that there is $\mathbf{d} \in \mathcal{R}$ with $(\forall \mathbf{i} \in \mathbf{I})(\bar{\mathbf{C}}_\mathbf{i} \subset [-\mathbf{d}, \mathbf{d}])$. Then $\mathbf{d}$ is a majoring element of all $\overline{\mathbf{C}}_\mathbf{i}$. From proposition 258 (p. 174) there is an **upper_limit** for each $\bar{\mathbf{C}}_\mathbf{i}$ and a map of $\mathbf{M_i} : \mathbf{I} \rightarrow \mathcal{R}$ where $\mathbf{M_i} = \mathbf{upper\_limit}(\bar{\mathbf{C}}_\mathbf{i})$. We have (proposition 209 (p. 155)) :
$$\mathbf{C_i} \subset \mathbf{C_j} \Rightarrow \overline{\mathbf{C}}_\mathbf{i} \subset \overline{\mathbf{C}}_\mathbf{j} \Rightarrow \mathbf{M_i} \leq \mathbf{M_j}$$
We also have $(\forall \mathbf{i} \in \mathbf{I})(\mathbf{M_i} \in \overline{\mathbf{C}}_\mathbf{i} \subset [-\mathbf{d}, \mathbf{d}])$ (proposition 276 (p. 182)). Calling:
$$\mathbf{X} = \mathbf{set_x}\{(\exists \mathbf{j} \in \mathbf{I})(\mathbf{x} = \mathbf{upper\_limit}(\bar{\mathbf{C}}_\mathbf{j}))\}$$
$\mathbf{X} \subset [-\mathbf{d}, \mathbf{d}]$ then there is an $\mathbf{m} = \mathbf{lower\_limit}(\mathbf{X})$.
For any open $\mathbf{O}$ containing $\mathbf{m}$, there is an $\mathbf{r}$ such that $]\mathbf{m} - \mathbf{r}, \mathbf{m} + \mathbf{r}[\subset \mathbf{O}$. Then $\mathcal{O}$ being the set of opens of $\mathcal{R}$ we have:
$$(\forall \mathbf{O} \in \mathcal{O})(\mathbf{m} \in \mathbf{O} \Rightarrow (\exists \mathbf{r} \in \mathcal{R})(]\mathbf{m} - \mathbf{r}, \mathbf{m} + \mathbf{r}[\subset \mathbf{O}, \\ (\exists \mathbf{i} \in \mathbf{I})(\mathbf{m} \leq \mathbf{M_i} < \mathbf{m} + \mathbf{r} \Rightarrow \mathbf{M_i} \in \mathbf{O}))) .$$
From the above formula taking the associated $\mathbf{r}$, $\mathbf{i}$ and $\mathbf{M_i} \in \overline{\mathbf{C}}_\mathbf{i}$ to each open $\mathbf{O} \in \mathcal{O}, \mathbf{m} \in \mathbf{O}$ containing $\mathbf{m}$, we have for all $\mathbf{j} \in \mathbf{I}$ either $\mathbf{C_i} \subset \mathbf{C_j}$ or $\mathbf{C_j} \subset \mathbf{C_i}$.
  - $\mathbf{C_i} \subset \mathbf{C_j}$
  $$\mathbf{C_i} \subset \mathbf{C_j} \Rightarrow \mathbf{M_i} \in \overline{\mathbf{C}}_\mathbf{j}$$
  then from proposition 727 (p. 441) $(\exists \mathbf{z})(\mathbf{z} \in \mathbf{C_j} \cap ]\mathbf{m} - \mathbf{r}, \mathbf{m} + \mathbf{r}[, \mathbf{z} \in \mathbf{O})$
  - $\mathbf{C_j} \subset \mathbf{C_i}$
  $$\mathbf{C_j} \subset \mathbf{C_i} \Rightarrow \mathbf{m} \leq \mathbf{M_j} \leq \mathbf{M_i} \Rightarrow (\mathbf{M_j} \in ]\mathbf{m} - \mathbf{r}, \mathbf{m} + \mathbf{r}[, \mathbf{M_j} \in \mathbf{C_j})$$
  and again $(\exists \mathbf{z})(\mathbf{z} \in \mathbf{C_j} \cap ]\mathbf{m} - \mathbf{r}, \mathbf{m} + \mathbf{r}[, \mathbf{z} \in \mathbf{O})$

Then since
$$(\forall \mathbf{i} \in \mathbf{I})(\forall \mathbf{j} \in \mathbf{I})(\mathbf{C_i} \subset \mathbf{C_j} \vee \mathbf{C_j} \subset \mathbf{C_i} \Rightarrow (\forall \mathbf{j})(\exists \mathbf{z})(\mathbf{z} \in \mathbf{C_j} \cap ]\mathbf{m} - \mathbf{r}, \mathbf{m} + \mathbf{r}[, \mathbf{z} \in \mathbf{O})) .$$
Then
$$(\forall \mathbf{O} \in \mathcal{O})(\mathbf{m} \in \mathbf{O} \Rightarrow (\forall \mathbf{j})(\exists \mathbf{z})(\mathbf{z} \in \bar{\mathbf{C}}_\mathbf{j} \cap \mathbf{O}))$$
$\mathbf{m}$ belongs to the closure of all $\overline{\mathbf{C}}_\mathbf{i}$ which are the $\overline{\mathbf{C}}_\mathbf{i}$ themselves and: $\mathbf{m} \in \bigcap_{\mathbf{i} \in \mathbf{I}} \bar{\mathbf{C}}_\mathbf{i}$ so $\bigcap_{\mathbf{i} \in \mathbf{I}} \overline{\mathbf{C}}_\mathbf{i} \neq \emptyset$.

- For any $\mathbf{n}$
Let us suppose that proposition 531 is true up to $\mathbf{n}$ and we prove it for $\mathbf{n} + \mathbf{1}$. We call $\mathbf{C}'^\mathbf{n}_\mathbf{i}$ the projection of $\overline{\mathbf{C}}_\mathbf{i} \in \mathcal{R}^{\mathbf{n}+\mathbf{1}}$ on $\mathcal{R}^\mathbf{n}$. From proposition 530 (p. 294) the set of $\mathbf{C}'^\mathbf{n}_\mathbf{i} = \bar{\mathbf{C}}'^\mathbf{n}_\mathbf{i}$ satisfies the condition of proposition 531. Then since we suppose that it is true for $\mathbf{n}$, there is $\mathbf{m} \in \mathcal{R}^\mathbf{n}$ belonging to all $\bar{\mathbf{C}}^\mathbf{n}_\mathbf{i}$. $\{\mathbf{m}\}$ is a closed set of $\mathcal{R}^\mathbf{n}$ because any point of $\mathbf{m} \in \mathcal{R}^\mathbf{n}$ except $\mathbf{m}$ is the center of an open ball excluding $\mathbf{m}$. Then all the points $\mathcal{R}^{\mathbf{n}+\mathbf{1}}$ projecting on $\mathbf{m}$ is a closed set of $\mathcal{R}^{\mathbf{n}+\mathbf{1}}$ because from proposition 520 (p. 290) the projection is a continuous map. The intersection with each $\overline{\mathbf{C}}_\mathbf{i}$ is closed and its projection $\mathbf{C}'^\mathbf{m}_\mathbf{i} = \bar{\mathbf{C}}'^\mathbf{m}_\mathbf{i}$ on $\mathcal{R}$ is also closed and satisfies proposition 531 and there is an $\mathbf{m}^\mathbf{1}$ common to all $\mathbf{C}'^\mathbf{m}_\mathbf{i}$. Then the point of $(\mathbf{m}, \mathbf{m}^\mathbf{1})$ of $\mathcal{R}^{\mathbf{n}+\mathbf{1}}$ is common to all $\overline{\mathbf{C}}_\mathbf{i}$.

## 2.16 Continuous map on $\mathcal{R}^n$ to $\mathcal{R}^p$

**Proposition 532** Given a map $f$ between $\mathcal{R}^n$ and $\mathcal{R}^p$, $(f : \mathcal{R}^n \to \mathcal{R}^p)$.
$f(x)$ continuous is equivalent to:

$$(\forall \alpha \in \mathcal{R}^+)(\forall x \in \mathcal{R}^n)(\exists \epsilon \in \mathcal{R}^+)(\|x - y\| < \epsilon \Rightarrow \|f(x) - f(y)\| < \alpha) . \quad (2.131)$$

**Proof**
It is a direct consequence of propositions 278 (p. 183), 281 (p. 184) and 282 (p. 184).

**Proposition 533** Given a continuous map $f$ $(f : \mathcal{R}^n \to \mathcal{R}^p)$ and a compact subset $C$ of $\mathcal{R}^n$ we have:

$$(\forall \alpha \in \mathcal{R}^+)(\exists \epsilon \in \mathcal{R}^+)(\forall (x, y) \in C \times C)(\|x - y\| < \epsilon \Rightarrow \|f(x) - f(y)\| < \alpha) . \quad (2.132)$$

**Proof**
It is a direct consequence of definition 160 (p. 184) and proposition 283 (p. 184).

**Definition 265 Functions on $\mathcal{R}^n$.**

One uses to call function a map from $\mathcal{R}^n$ to $\mathcal{R}$.

**Proposition 534** If $f(x)$ and $g(x)$ are two continuous functions from $\mathcal{R}^n$ to $\mathcal{R}$ the sum $f(x) + g(x)$ and product $f(x)g(x)$ are continuous functions from $\mathcal{R}^n$ to $\mathcal{R}$.

**Proof**
$f(x)$ and $g(x)$ are defining a map $((f, g) : \mathcal{R}^n \to \mathcal{R}^2)$ since $x \in \mathcal{R}^n$ and $(f(x), g(x)) \in \mathcal{R}^2$. $(f, g)$ is continuous according to proposition 244 (p. 168). $f(x) + g(x)$ and $f(x)g(x)$ are composite maps of $(f, g)$ and respectively of the maps sum and product $(\mathcal{R}^2 \to \mathcal{R})$ which are continuous, proposition 286 (p. 185). From proposition 241 (p. 167), $f(x) + g(x)$ and $f(x)g(x)$ are continuous.

**Proposition 535** Calling $\mathcal{F}_n$ the set of continuous $\mathcal{R}^n \to \mathcal{R}$, given $k \in \mathcal{Z}^+$ and a map $(f_i : [1, k] \to \mathcal{F}_n)$, the two maps $(\sum_{i \in [1,k]} f_i : \mathcal{R}^n \to \mathcal{R})$ and $(\prod^{\times}_{i \in [1,k]} f_i : \mathcal{R}^n \to \mathcal{R})$ (definitions 127 (p. 145), 128 (p. 145)) are continuous (they are in $\mathcal{F}_n$).

**Proof**
From proposition 750 (p. 450), $\mathcal{F}_n$ exists because any continuous map $\mathcal{R}^n \to \mathcal{R}$ is a map $\mathcal{R}^n \to \mathcal{R}$ which from proposition 787 (p. 461) forms a set. Let us suppose proposition 535 is true up to $n - 1$. From proposition 534 (p. 296), proposition 535 is true for $n$. Proposition 535 is true for $n = 1$ then from recurrence principle proposition 927 (p. 520), it is true for all $n \in \mathcal{Z}^+$.

**Proposition 536** Given $n \in \mathcal{Z}^+$, calling $\mathcal{F}$ the set of continuous map $\mathcal{R} \to \mathcal{R}$ and $\mathcal{F}_n$ the set of continuous map $\mathcal{R}^n \to \mathcal{R}$, having a map $(f_i : [1, n] \to \mathcal{F})$, the maps $(\sum_{i \in [1,n]} f_i(x^i) : \mathcal{R}^n \to \mathcal{R})$ and $(\prod^{\times}_{i \in [1,n]} f_i(x^i) : \mathcal{R}^n \to \mathcal{R})$ are continuous.

**Proof**
The maps $(\mathbf{x}^i : \mathcal{R}^n \to \mathcal{R})$ are projections (definition 121 (p. 136)). From proposition 237 (p. 166), they are continuous. From proposition 241 (p. 167), we have:

$$(\forall i \in [1,n])((\forall x \in \mathcal{R}^n)(f_i(x^i) = f_i(\mathrm{Proj}_{\{i\}}(x))) \Rightarrow (f_i(x^i) : \mathcal{R}^n \to \mathcal{R}) \in \mathcal{F}_n$$

from proposition 535 (p. 296), the maps $(\sum_{i\in[1,n]} f_i(x^i) : \mathcal{R}^n \to \mathcal{R})$ and $(\prod^{\times}_{i\in[1,n]} f_i(x^i) : \mathcal{R}^n \to \mathcal{R})$ are continuous.

**Proposition 537** Any linear transformation (definitions 224 (p. 250), 225 (p. 250)) from the vectorial space $\mathcal{R}^n$ to $\mathcal{R}^p$ is continuous.

**Proof**
Given the transformation matrix $\mathbf{T} = (\mathbf{T}^i_j : [1,n] \times [1,p] \to \mathcal{R})$, a vector $\mathbf{x} = (\mathbf{x}^i : [1,n] \to \mathcal{R}) \in \mathcal{R}^n$, we have:

$$\mathbf{T}(\mathbf{x}) = \mathbf{T}^i_j \mathbf{x}^j \in \mathcal{R}^p$$

$\mathbf{T}(\mathbf{x})$ is continuous from propositions 536 (p. 296) and 244 (p. 168).

**Proposition 538** Given a map $\mathbf{T} = (\mathbf{T}^i_j : [1,n]^2 \to \mathcal{R}) \in \mathcal{R}^{n^2}$, the map defined by the determinant of $\mathbf{T}$ (definition 227 (p. 252)): $(\det(\mathbf{T}) : \mathcal{R}^{n^2} \to \mathcal{R})$ is continuous.

**Proof**
It is a consequence of proposition 536 (p. 296) for product and sum and of proposition 244 (p. 168).

## 2.17 Sequence

**Definition 266 Sequence.**

A **sequence** in a set $\mathbf{E}$ is a map of $\mathcal{Z}^+$ into $\mathbf{E}$. $(\mathbf{U}_m : \mathcal{Z}^+ \to \mathbf{E})$

**Definition 267 Accumulation point of a sequence on a topological space.**

Given a sequence $\mathbf{S}$ on a topological space $\mathbf{E}$: $(\mathbf{U}_m : \mathcal{Z}^+ \to \mathbf{E})$. One calls an **accumulation point** of $\mathbf{S}$ a point $\mathbf{P}$ of $\mathbf{E}$ such that if $\mathbf{O}$ is any open containing $\mathbf{P}$ there is no majoring value of $\mathbf{U}^{-1}(\mathbf{O})$ (definition 380 (p. 484)).

**Proposition 539** If a sequence $\mathbf{S} = (\mathbf{U}_m : \mathcal{Z}^+ \to \mathcal{R}^n)$ is included in a closed set $\mathbf{C}$ of $\mathcal{R}^n$ $(\mathbf{U}(\mathcal{Z}^+) \subset \mathbf{C})$, any accumulation point of the sequence $\mathbf{S}$ belongs to the closed set $\mathbf{C}$.

**Proof**
Any open $\mathbf{O}$ containing the accumulation point $\mathbf{P}$ contains an element of the sequence, otherwise any integer would be a majoring element of $\mathbf{U}^{-1}(\mathbf{O})$. $\mathbf{O}$ contains an element of the closed set. From proposition 509 (p. 287), $\mathbf{P}$ is in the closed set.

**Proposition 540** Given a sequence $\mathbf{S} = (\mathbf{U}_m : \mathcal{Z}^+ \to \mathcal{R}^n)$, any point of $\mathcal{R}^n$, such that any open containing it contains the images by $\mathbf{U}$ of an infinite [25] number of integers, is an accumulation point of the sequence and vice versa.

[25] A set $\mathbf{E}$ is said to have an infinite number of elements, if Card $(\mathbf{E})$ is not an integer.

**Proof**

Let us call $\mathbf{p}$ such a point and $\mathbf{O}$ any open containing $\mathbf{p}$. If a set $\mathbf{M}$ of positive integers has a cardinal which is not an integer, it cannot have a majoring element $\mathbf{m} \in \mathcal{Z}^+$ because the cardinal of all positive integers smaller than an integer $\mathbf{m}$ is an integer (see proposition 934 (p. 523)). Then $\mathbf{Card(f^{-1}(O))}$ cannot have a majoring element. $\mathbf{P}$ is an accumulation point (definition 267 (p. 297)).

The reverse is also true: Any open containing an accumulation point contains the images of an infinite number of positive integers: If the number were finite, from proposition 941 (p. 526) in appendix A.6.2 (p. 521), there would be an integer bigger than the index of any element in that open. That is against definition 266 (p. 297).

**Proposition 541** Any sequence in a compact (definition 140 (p. 160)) of $\mathcal{R}^n$ has at least an accumulation point.

**Proof**

If it was not true, for a compact $\mathbf{C}$ in topological set $\mathbf{E}$ with a sequence $\mathbf{S = (U_m : Z^+ \to C)}$ at any point $\mathbf{p \in C}$, there would be an open $\mathbf{O_p \subset E}$ containing $\mathbf{p}$ and a finite $\mathbf{K_p}$ number of elements of $\mathbf{S}$. Since $\mathbf{C \subset \bigcup_{p \in C} O_p}$, we should have (definition 140 (p. 160)):

$$(\exists P \subset C)(\mathbf{Card}(P) \in \mathcal{Z}^+, C \subset \bigcup_{p \in P} O_p) .$$

Since $(\forall \mathbf{m} \in \mathbf{Z}^+)(\mathbf{U_m} \in \mathbf{C})$, we should have $\mathcal{Z}^+ \leq \sum_{p \in P} \mathbf{K_p}$. Since $\sum_{p \in P} \mathbf{K_p}$ is an integer (proposition 941 (p. 526)), $\mathbf{Z}^+$ would also be an integer (propositions 918 (p. 516), 919 (p. 517) and definition 397 (p. 516)). It would be in contradiction with proposition 939 (p. 525). Then proposition 541 (p. 298) is true.

**Definition 268 Limit of a sequence.**

Having a sequence $(\mathbf{U_i} : \mathcal{Z}^+ \to \mathcal{R}^n)$, it has a limit $\mathbf{L}$ if we have:

$$(\forall \alpha)(\exists m > 0)(p > m \Rightarrow \|U_p - L\| < \alpha) . \tag{2.133}$$

**Proposition 542** Given a sequence $\mathbf{S} = (\mathbf{U(i)} = \mathbf{U_i} : \mathcal{Z}^+ \to \mathcal{R}^n)$ having a limit $\mathbf{L}$, given a bijection $(\mathbf{f} : \mathcal{Z}^+ \leftrightarrow \mathcal{Z}^+)$, the sequence $\mathbf{S} = (\mathbf{U(f(i))} = \mathbf{U_{f(i)}} : \mathcal{Z}^+ \to \mathcal{R}^n)$ has a limit which is $\mathbf{L}$.

**Proof**

We have since $\mathbf{f}(\mathcal{Z}^+) = \mathcal{Z}^+$, $\mathbf{U}(\mathbf{f}(\mathcal{Z}^+)) = \mathbf{U}(\mathcal{Z}^+)$. Then given a ball $\mathbf{B(L, \alpha)}$, $\mathbf{U}(\mathbf{f}(\mathcal{Z}^+)) \cap \mathcal{C}(\mathbf{B(L, \alpha)}) = \mathbf{U}(\mathcal{Z}^+) \cap \mathcal{C}(\mathbf{B(L, \alpha)})$, from definition 268 (p. 298), we have:

$$(\exists m > 0)(p > m \Rightarrow U_p \notin \mathcal{C}(B(L, \alpha))) .$$

Taking that $\mathbf{m}$, we have:

$$U_q \in \mathcal{C}(B(L, \alpha)) \Rightarrow q \in [1, m]$$
$$\mathbf{Card}(U(\mathcal{Z}^+) \cap \mathcal{C}(B(L, \alpha))) \in [1, m]$$

(see proposition 934 (p. 523)). Then $\mathbf{Card}(\mathbf{U}(\mathcal{Z}^+) \cap \mathcal{C}(\mathbf{B(L, \alpha)})) \in \mathcal{Z}^+$ (propositions 918 (p. 516), 926 (p. 519)). From proposition 943 (p. 528),

$$(\exists m')((U \circ f)^{-1}(\mathcal{C}(B(L, \alpha))) \in [1, m'])$$

then (section A.1.4 point 7 (p. 435))

$$(\forall \alpha)(\exists m')(p \notin [-\infty, m'] \Rightarrow U(f(p)) \in B(L, \alpha)) .$$

Proposition 542 (p. 298) is true.

**Proposition 543** Any limit of a sequence is an accumulation point.

**Proof**
We have a $(\mathbf{U_i} : \mathcal{Z}^+ \to \mathcal{R}^n)$ with a limit $\mathbf{L}$. If, in an open $\mathbf{O}$ including $\mathbf{L}$, there is a majoring value $\mathbf{m'}$ to $\mathbf{U^{-1}(O)}$, we would have $\mathbf{q > m' \Rightarrow U_q \notin O}$. But, there is in $\mathbf{O}$ an open ball $\mathbf{B(L,\alpha)}$ (proposition 266 (p. 179)). That implies $\mathbf{q > m' \Rightarrow U_q \notin B(L,\alpha)}$. Definition 268 (p. 298), relation (2.133) (p. 298) means that $\mathbf{(\exists m)(p > m \Rightarrow U_p \in B(L,\alpha))}$. It would mean that, if $\mathbf{q > \max(m,m')}$, we would have both $\mathbf{U_q \in B(L,\alpha)}$ and $\mathbf{U_q \notin B(L,\alpha)}$. We conclude by A.1.4 (p. 435), 8.

**Proposition 544** Having a sequence $\mathbf{S}$ $(\mathbf{U_i} : \mathcal{Z}^+ \to \mathcal{R}^n)$ defined in a ball of $\mathcal{R}^n$, if that sequence has a unique accumulation point, it has a limit and vice versa.

**Proof**

- We suppose that $\mathbf{S}$ has a limit $\mathbf{L}$, it cannot have another accumulation point $\mathbf{L'}$, because (proposition A.1.4 (p. 435), 8) if for any $\boldsymbol{\alpha} < \|\mathbf{L - L'}\|$, we had $\mathbf{m} \in \mathcal{Z}^+$ such that $\mathbf{p > m \Rightarrow \|U_p - L\|} < \boldsymbol{\alpha}/3$, it would be in contradiction with definition 267 (p. 297) which says that there is always a $\mathbf{q > m}$ such that $\|\mathbf{U_q - L'}\| < \boldsymbol{\alpha}/3$. In that case we would have:

$$\alpha < \|\mathbf{L - L'}\| < \|\mathbf{L - U_q}\| + \|\mathbf{L' - U_q}\| < 2\alpha/3 \ .$$

- If we have a unique accumulation point $\mathbf{L}$, it is the limit:
$\mathbf{S}$ is in a ball $\mathbf{B(0,c)}$. It is also in the compact $\mathbf{C} = \overline{\mathbf{B(0,c)}}$ (proposition 219 (p. 158), 521 (p. 290)). Outside any ball $\mathbf{B(L,\alpha)}$, there is only a finite number of elements of $\mathbf{S}$. Otherwise, since (see definition 371 (p. 472)) $\mathbf{S} \cap \mathcal{C}(\mathbf{B(L,\alpha)}) \sim \mathcal{Z}^+$, from definition 371 (p. 472), we would have a sequence $\mathbf{S'} = (\mathbf{V_i} : \mathcal{Z}^+ \leftrightarrow \mathbf{S} \cap \mathcal{C}(\mathbf{B(L,\alpha)}))$. From proposition 541 (p. 298), $\mathbf{S'}$ would have an accumulation point in $\mathbf{C} \cap \mathcal{C}(\mathbf{B(L,\alpha)})$ which is closed and in a ball then compact (propositions 206 (p. 154), 514 (p. 288) and 521 (p. 290)). From proposition 540 (p. 297), that point would also be an accumulation point for $\mathbf{S}$ which is forbidden. Then $\mathbf{Card}(\mathbf{S} \cap \mathcal{C}(\mathbf{B(L,\alpha)}))$ is an integer and from proposition 943 (p. 528) $\mathbf{(\exists m)(p > m \Rightarrow U_p} \notin \mathcal{C}(\mathbf{B(L,\alpha)}))$. Then, since $\mathbf{U_p} \notin \mathcal{C}(\mathbf{B(L,\alpha)}) \Leftrightarrow \mathbf{U_p \in B(L,\alpha)}$, from definition 268 (p. 298) $\mathbf{L}$ is the limit of $\mathbf{S}$.

**Proposition 545** Given $\mathbf{a} \in [0,1[ \subset \mathcal{R}$ the sequence defined by $(\mathbf{U^n = a^n} : \mathcal{Z}^+ \to \mathcal{R})$ has a limit which is zero.

**Proof**
From proposition 320 (p. 198) we have:

$$(\forall \mathbf{i} \in \mathcal{Z}^+)(\forall \mathbf{j} \in \mathcal{Z}^+)((\mathbf{a} \in ]0,1[, \mathbf{i > j}) \Rightarrow \mathbf{0 < a^i < a^j < 1}) \ . \tag{2.134}$$

Since $\mathbf{0}$ is a minoring element of the sequence $\mathbf{a^i}$ (definition 380 (p. 484)), from proposition 258 (p. 174), it has a **lower_limit** $\mathbf{L}$ (see definition 381 (p. 485)). $\mathbf{L = 0}$ because from proposition 260 (p. 175), we have:

$$(\exists \mathbf{p} > 0)(\mathbf{a^p} \in [\mathbf{L}, \mathbf{L} + \mathbf{L} \times (1 - \mathbf{a})] \ .$$

But from proposition 318 (p. 197):

$$\mathbf{a^p} \leq \mathbf{L} + \mathbf{L} \times (1 - \mathbf{a}) \Rightarrow \mathbf{a^{p+1}} < \mathbf{aL} + \mathbf{aL} \times (1 - \mathbf{a}) < \mathbf{aL} + \mathbf{L} \times (1 - \mathbf{a}) = \mathbf{L}$$

$\mathbf{L}$ can be the **lower_limit** only if $\mathbf{L} = \mathbf{0}$. Then we have:

$$\mathbf{a} \in [0, 1[\Rightarrow (\forall \alpha > 0)(\exists \mathbf{p} > 0)(\mathbf{a}^{\mathbf{p}} < \alpha) \; .$$

From relation (2.134), we now have:

$$\mathbf{a} \in [0, 1[\Rightarrow (\forall \alpha > 0)(\exists \mathbf{p} > 0)(\mathbf{q} \geq \mathbf{p} \Rightarrow \mathbf{a}^{\mathbf{q}} < \alpha) \; .$$

**Definition 269 Cauchy sequence.**

A sequence $\mathbf{S}$ in a normed vectorial space $\mathbf{E}$: $(\mathbf{U_m} : \mathcal{Z}^+ \to \mathbf{E})$ is called **Cauchy sequence** if:

$$(\forall \alpha \in \mathcal{R}^+)(\exists \mathbf{m} \in \mathcal{Z}^+)((\mathbf{p} > \mathbf{m}, \mathbf{q} > \mathbf{m}) \Rightarrow ||\mathbf{U_p} - \mathbf{U_q}|| < \alpha) \; . \tag{2.135}$$

**Proposition 546** A Cauchy sequence in $\mathcal{R}^\mathbf{n}$ has a limit.
**Proof**
We suppose that the sequence $\mathbf{S} = (\mathbf{U_m} : \mathcal{Z}^+ \to \mathcal{R}^\mathbf{n})$ is a Cauchy sequence. Given $\alpha$, we consider the set $\mathbf{C}$ of real number $||\mathbf{U_n}|| + \alpha$ when $\mathbf{n} \leq \mathbf{m} + \mathbf{1}$. $\mathbf{m}$ is such that $||\mathbf{U_p} - \mathbf{U_q}|| < \alpha$ for $\mathbf{p}$ and $\mathbf{q}$ bigger than $\mathbf{m}$. $\mathbf{Card(C)} = \mathbf{m} + \mathbf{1}$ is an integer. Then from proposition 255 (p. 172) it has a biggest element $\mathbf{R}$ which is the **upper_limit** of that set, all the sequence is within an open ball of radius $\mathbf{R}$:

$$\mathbf{n} \leq \mathbf{m} + \mathbf{1} \Rightarrow ||\mathbf{U_n}|| \leq \mathbf{R} - \alpha < \mathbf{R}$$

and

$$\mathbf{n} > \mathbf{m} \Rightarrow ||\mathbf{U_n} - \mathbf{U_{m+1}}|| < \alpha \; .$$

Since $||\mathbf{U_n}|| - ||\mathbf{U_{m+1}}|| \leq ||\mathbf{U_n} - \mathbf{U_{m+1}}||$ (proposition 474 (p. 269)) we also have:

$$\mathbf{n} > \mathbf{m} \Rightarrow ||\mathbf{U_n}|| < \alpha + \mathbf{R}$$

The sequence has an accumulation point inside the ball $\bar{\mathbf{B}}(\mathbf{0}, \mathbf{R} + \alpha)$ (proposition 541 (p. 298) applied to $\mathbf{U}'_\mathbf{j} = \mathbf{U_{m+j}}$, see also proposition 521 (p. 290)).
It is unique because if there were two accumulation points $\mathbf{x}$ and $\mathbf{y}$, we can call $\mathbf{m}$ the index such that $(\mathbf{p} > \mathbf{m}), (\mathbf{q} > \mathbf{m}) \Rightarrow ||\mathbf{U_p} - \mathbf{U_q}|| < ||\mathbf{x} - \mathbf{y}||/4$. From the definition of an accumulation point, for any $\mathbf{m}$ there are elements of the sequence with $\mathbf{p} > \mathbf{m}$ and $\mathbf{q} > \mathbf{m}$ for which $||\mathbf{U_p} - \mathbf{x}|| < ||\mathbf{x} - \mathbf{y}||/4$ and $||\mathbf{U_q} - \mathbf{y}|| < ||\mathbf{x} - \mathbf{y}||/4$. It is impossible because, from proposition 473 (p. 268), one would have:

$$\begin{aligned} 3/4||\mathbf{x} - \mathbf{y}|| &\leq ||\mathbf{x} - \mathbf{y}|| - ||\mathbf{U_p} - \mathbf{x}|| \leq ||\mathbf{U_p} - \mathbf{y}|| \\ &\leq ||\mathbf{U_p} - \mathbf{U_q}|| + ||\mathbf{U_q} - \mathbf{y}|| \leq ||\mathbf{x} - \mathbf{y}||/2 \; . \end{aligned}$$

**Proposition 547** Every sequence on $\mathcal{R}^\mathbf{n}$ which has a limit and which is included in an open ball is a Cauchy sequence.
**Proof**
Given a sequence $\mathbf{S}$ included in an open ball with a limit $\mathbf{s}$. For any $\alpha$, there is only a finite number of elements $\mathbf{U_n}$ at a distance bigger than $\alpha/2$ otherwise from proposition 541 (p. 298) there would be a second accumulation point. Taking $\mathbf{m}$ as the maximum of the indices of those elements, we have:

$$\begin{aligned} &(\mathbf{p} > \mathbf{m}), (\mathbf{q} > \mathbf{m}) \Rightarrow ((||\mathbf{U_p} - \mathbf{s}|| < \alpha/2)(||\mathbf{U_q} - \mathbf{s}|| < \alpha/2)) \Rightarrow \\ &||\mathbf{U_p} - \mathbf{U_q}|| < \alpha \; . \end{aligned}$$

**Proposition 548** In any Cauchy sequence $\mathbf{S}$ built on a normed vectorial space $\mathbf{E}$: $(\mathbf{U_m} : \mathcal{Z}^+ \rightarrow \mathbf{E})$ and having a limit $\mathbf{L}$, for any $\alpha > 0$ there is an $\mathbf{m}$ such that $\mathbf{p} > \mathbf{m}, \mathbf{q} > \mathbf{m} \Rightarrow \|\mathbf{U_p} - \mathbf{U_q}\| < \alpha/2$ for the same couple $\mathbf{m}$ and $\alpha > 0$, we also have $\mathbf{p} > \mathbf{m} \Rightarrow \|\mathbf{U_p} - \mathbf{L}\| < \alpha$.

**Proof**

From the definition 269 to any $\alpha/3$ there is an $\mathbf{m}$ such that $\mathbf{p} > \mathbf{m}, \mathbf{q} > \mathbf{m} \Rightarrow \|\mathbf{U_p} - \mathbf{U_q}\| < \alpha/2$. Since $\mathbf{L}$ is an accumulation point from definition 267 there is an $\mathbf{m_2} > \mathbf{m}$ such that $\|\mathbf{U_{m_2}} - \mathbf{L}\| < \alpha/2$ otherwise there would be a majoring element (definition 380 (p. 484)) of the index of the points of $\mathbf{S}$ at a distance less than $\alpha/2$ of $\mathbf{L}$ which is forbidden. Now we have:

$$\mathbf{p} > \mathbf{m}, \mathbf{m_2} > \mathbf{m} \Rightarrow \|\mathbf{U_p} - \mathbf{U_{m_2}}\| < \alpha/2 \Rightarrow$$
$$\|\mathbf{U_p} - \mathbf{L}\| < \|\mathbf{U_p} - \mathbf{U_{m_2}}\| + \|\mathbf{U_{m_2}} - \mathbf{L}\| < 2\alpha/2 = \alpha\,.$$

## 2.18 Sequence in $\mathcal{R}_\infty$

**Definition 270 Accumulation point and limit for a sequence in $\mathcal{R}_\infty$.**

Given a sequence $\mathbf{S} = (\mathbf{U_m} : \mathcal{Z}^+ \rightarrow \mathcal{R}_\infty)$

- a point $\mathbf{p} \in \mathcal{R}_\infty$ (definition 144 (p. 171)) is an accumulation point of $\mathbf{S}$
  - if $\mathbf{p} \in \mathcal{R}$, $\mathbf{p}$ is an accumulation point of $\mathbf{S}$ as defined in definition 267 (p. 297).
  - if $\mathbf{p} = -\infty$ (resp. $\mathbf{p} = \infty$) (definition 144 (p. 171)), $\forall \mathbf{A} \in \mathcal{R}$ there is no majoring value (definition 380 (p. 484)) to $\mathbf{U}^{-1}([-\infty, \mathbf{A}[)$ (resp. $\mathbf{U}^{-1}(]\mathbf{A}, \infty])$).
- A point $\mathbf{p} \in \mathcal{R}_\infty$ is a limit of $\mathbf{S}$
  - if $\mathbf{p} \in \mathcal{R}$, $\mathbf{p}$ is a limit of $\mathbf{S}$ as defined in definition 268 (p. 298).
  - if $\mathbf{p} = -\infty$ (resp. $\mathbf{p} = \infty$) (definition 144 (p. 171)),

$$(\forall \mathbf{A} > 0)(\exists \mathbf{m} \in \mathcal{Z}^+)(\mathbf{p} > \mathbf{m} \Rightarrow \mathbf{U_i} < -\mathbf{A} \ \ (\text{resp.} \mathbf{A} < \mathbf{U_i}))\,.$$

**Proposition 549** Given a sequence $\mathbf{S} = (\mathbf{U(i)} = \mathbf{U_i} : \mathcal{Z}^+ \rightarrow \mathcal{R})$ having a limit $\mathbf{L} = \pm\infty$ (definition 144 (p. 171)), given a bijection $(\mathbf{f} : \mathcal{Z}^+ \leftrightarrow \mathcal{Z}^+)$, the sequence $\mathbf{S} = (\mathbf{U(f(i))} = \mathbf{U_{f(i)}} : \mathcal{Z}^+ \rightarrow \mathcal{R}^n)$ has a limit which is $\mathbf{L}$.

**Proof**

It is the same proof as proposition 542 (p. 298) with $\mathcal{R}$ for the sequence definition and replacing $\alpha$ by $\mathbf{A} > 0$ and the ball $\mathbf{B(L, \alpha)}$ by $] \leftarrow, \mathbf{A}[=] - \infty, \mathbf{A}[$ or $[\mathbf{A}, \infty[=]\mathbf{a}, \rightarrow [$ (see definition 378 (p. 484)).

**Proposition 550** Given a map $(\mathbf{u_i} : \mathcal{Z}^+ \rightarrow \mathcal{R}^+)$, considering the increasing map (definition 384 (p. 486)) or sequence $\mathbf{S} = (\mathbf{s_i} = \sum_{\mathbf{j} \in [1,\mathbf{i}]} \mathbf{u_j} : \mathcal{Z}^+ \rightarrow \mathcal{R}^+)$ this sequence has a limit in $\mathcal{R}^+_\infty$ (definition 144 (p. 171)).

**Proof**

- If $\mathbf{S}$ has a majoring element (see definition 380 (p. 484)), from proposition 258 (p. 174), it has an **upper_limit** (see definition 381 (p. 485)). From propositions 260 (p. 175) and 259 (p. 175),

$$(\forall \alpha > 0)(\exists \mathbf{s_i} \in \mathbf{S})(\mathbf{L} - \alpha < \mathbf{s_i} \leq \mathbf{L}) \Rightarrow (\exists \mathbf{i} \in \mathcal{Z}^+)(\mathbf{j} > \mathbf{i}$$
$$\Rightarrow \mathbf{L} - \alpha < \mathbf{s_i} \leq \mathbf{s_j} \leq \mathbf{L} \Rightarrow |\mathbf{s_j} - \mathbf{L}| < \alpha)$$

  $\mathbf{S}$ has a limit in $\mathcal{R}^+ \subset \mathcal{R}^+_\infty$ (definition 144 (p. 171)).

- If $\mathbf{S}$ has no majoring element,

$$(\forall A > 0)(\exists s_i \in S)(A < s_i) \Rightarrow (\exists i \in \mathcal{Z}^+)(j > i \Rightarrow A < s_i \leq s_j \Rightarrow A < s_j)$$

$+\infty$ is the limit of $\mathbf{S}$.

**Proposition 551** Given a map $(u(i) = u_i : \mathcal{Z}^+ \to \mathcal{R}^+)$, considering the increasing map (definition 384 (p. 486)) or sequence $S = (s_i = \sum_{j\in[1,i]} u_j : \mathcal{Z}^+ \to \mathcal{R}^+)$ and its limit $L$ in $\mathcal{R}^+_\infty$ (definition 144 (p. 171)). Given a bijection $(f : \mathcal{Z}^+ \leftrightarrow \mathcal{Z}^+)$, the sequence $S' = (s'_i = \sum_{j\in[1,i]} u(f(j)) : \mathcal{Z}^+ \to \mathcal{R}^+)$ has the same limit as $\mathbf{S}$.

**Proof**
We have (see definition 112 (p. 131), 4):

$$(\forall A \in ]0, L[)(\exists m(A))(s_{m(A)} \in ]A, L[) \ .$$

As in proposition 542 (p. 298), $\mathbf{Card}(\mathbf{set}_j\{f(j) \leq m\}) \in \mathcal{Z}^+$, then from proposition 943 (p. 528),

$$(\forall m > 0)(\exists h'(m))((f(j) \in [1, m] \Rightarrow j \in [1, h'(m)]), [1, m] \subset f([1, h'(m)])) \ .$$

For the same reasons:

$$(\forall m' > 0)(\exists h(m'))((i \leq m' \Rightarrow f(i) \in [1, h(m')]), f([1, m']) \subset [1, h(m')])$$

then since

$$[1, m] \subset f([1, h'(m)]) \subset f([1, j]) \Rightarrow s'_j = \sum_{k\in[1,j]} u(f(k)) \geq \sum_{k\in[1,m]} u(k)$$

$$f([1, j]) \subset [1, h(j)] \Rightarrow s'_j = \sum_{k\in[1,j]} u(f(k)) \leq \sum_{k\in[1,h(j)]} u(k)$$

$$j > h'(m) \Rightarrow A < s_m \leq s'_j \leq s_{h(j)} \leq L \ .$$

Then proposition 551 (p. 302) is true.

## 2.19 Sequence of maps

**Definition 271 Sequence of maps.**

Given a set $\mathbf{I}$, a topological set $\mathbf{E}$, we call a sequence of maps, a map $(U^i_n : \mathcal{Z}^+ \times I \to E)$.

**Definition 272 Sequence of maps converging uniformly to a map limit on a set.**

Given a set $\mathbf{I}$, a topological set $\mathbf{E}$, a sequence of maps $(U^i_n : \mathcal{Z}^+ \times I \to E)$ is said to converge uniformly to a limit $(U^i : I \to E)$ if we have:

$$(\forall \alpha)(\exists p \in \mathcal{Z}^+)(\forall i \in I)(\exists U^i \in E)(\forall m \geq p)(\|U^i_m - U^i\| < \alpha) \ . \tag{2.136}$$

**Definition 273 Cauchy sequence of maps.**

Given a set $\mathbf{I}$, a topological set $\mathbf{E}$, a sequence of maps $(\mathbf{U}^i_n : \mathcal{Z}^+ \times \mathbf{I} \to \mathbf{E})$ is said to be a Cauchy sequence of maps if:

$$(\forall i \in I)(\forall \alpha \in \mathcal{R}^+)(\exists m \in \mathcal{Z}^+)((p > m, q > m) \Rightarrow \|U^i_p - U^i_q\| < \alpha) . \qquad (2.137)$$

**Definition 274 Uniform Cauchy sequence of maps.**
With the same notation as in the above definition 273, a Cauchy sequence of maps is said to be uniform Cauchy sequence if:

$$(\forall \alpha \in \mathcal{R}^+)(\exists m \in \mathcal{Z}^+)(\forall i \in I)((p > m, q > m) \Rightarrow \|U^i_p - U^i_q\| < \alpha) . \qquad (2.138)$$

**Proposition 552** A uniform Cauchy sequence of maps (definition 274 (p. 303)) converges uniformly (definition 272 (p. 302)) to a limit map and vice versa.
**Proof**
From proposition 546 (p. 300), $(\forall i \in I)$, $U^i_k$ has a limit $U^i$. If at an $i \in I$, using the $\alpha, m(\alpha)$ of equation (2.138) (p. 303), we have $p > m(\alpha)$ and $\|U^i_p - U^i\| > \alpha$. From proposition 540 (p. 297), the set $\mathbf{Q}$ of $U^i_q$ in the ball $B(U^i, \frac{\|U^i_p - U^i\| - \alpha}{2})$ is not finite ($\approx$ not integer), $\mathbf{Card}([1,p])$ is finite ($\approx$ integer) (proposition 934 (p. 523)) as any of the subset of $[1,p]$, then $\mathbf{Q}$ is not a subset of $[1,p]$ (propositions 877 (p. 500), 918 (p. 516)).
From that, there is a $q > p$ such that $U^i_q \in B(U^i, \frac{\|U^i_p - U^i\| - \alpha}{2})$ which gives:

$$\|U^i_p - U^i_q\| > \|U^i_p - U^i\| - \|U^i - U^i_q\| \geq \frac{\|U^i_p - U^i\|}{2} + \frac{\alpha}{2} > \alpha .$$

It is in contradiction with $\|U^i_p - U^i_q\| < \alpha$. The reverse is also true because having $m(\alpha/2)$ such that $p > m(\alpha/2) \Rightarrow \|U^i_p - U^i\| < \alpha/2$, it also gives:

$$\begin{aligned}&(p > m(\alpha/2), q > m(\alpha/2))\\ &\Rightarrow \|U^i_p - U^i_q\| \leq \|U^i_p - U^i\| + \|U^i - U^i_q\| < \alpha .\end{aligned}$$

**Proposition 553** Given a sequence $\mathbf{S}$ of maps $(U_k : \mathcal{Z}^+ \times \mathcal{R}^{n_1} \to \mathcal{R}^{n_2})$ with each $U_k(x)$ continuous inside a subset $\mathbf{C}$ of $\mathcal{R}^{n_1}$. Let us suppose that for any $\mathbf{x}$ of $\mathbf{C}$ the sequence $U_k(x)$ as a limit $U(x)$ which satisfies equation (2.136) (p. 302) of the definition of uniform convergence on $\mathbf{C}$ then the limit $U(x)$ of $\mathbf{S}$ is continuous on $\mathbf{C}$.
**Proof**
We have, since for all $\mathbf{k}$, $U_k(x)$ is continuous (see again definition 112 (p. 131), 4):

$$(\forall \alpha)(\forall i \in \mathcal{Z}^+)(\exists \epsilon(\alpha, i))((\|x - x'\| < \epsilon(\alpha, i) \Rightarrow \|U_i(x) - U_i(x')\| < \alpha) .$$

Since for any $x \in C$ the sequence $U_k(x)$ converge uniformly to $U(x)$, according to equation (2.136) (p. 302), we have:

$$(\forall \alpha)(\exists q(\alpha))(\forall x \in C)(\forall i \in \mathcal{Z}^+)(i > q(\alpha) \Rightarrow \|U_i(x) - U(x)\| < \alpha) .$$

Combining the two, one gets applying proposition 473 (p. 268)

$$\begin{aligned}&(\forall \alpha)(\exists q(\alpha/3))(\exists \epsilon(\alpha/3, q(\alpha/3)))((x \in C, x' \in C, \|x - x'\| < \epsilon(\alpha/3, q(\alpha/3)))\\ &\Rightarrow (\|U_q(x) - U(x)\| < \alpha/3, \|U_q(x') - U(x')\| < \alpha/3,\\ &\|U_q(x) - U_q(x')\| < \alpha/3)\\ &\Rightarrow \|U(x) - U(x')\| \leq \|U(x) - U_q(x)\| + \|U_q(x) - U_q(x')\|\\ &+ \|U_q(x') - U(x')\| < \alpha)\end{aligned}$$

$U(x)$ is continuous on $\mathbf{C}$ according to proposition 282 (p. 184).

## 2.20 Partial derivative

**Definition 275 Partial derivative of a map, $\partial_i = \frac{\partial}{\partial x^i}$.**

Given an open $O \subset \mathcal{R}^n$ and a map $(f : O \to \mathcal{R})$, one calls partial derivative[26] $\partial_i f(x)$ or $\frac{\partial f(x)}{\partial x^i}$ of the map $f(x)$ at $x = (x^1, ..., x^n) = \prod^e_{i\in[1,n]} x^i \in O$ the derivative of the function (definition 161 (p. 192)) $(g(u) : \mathcal{R} \to \mathcal{R})$ at $u = x^i$ where $g(u)$ is defined, at $\prod^e_{j\in[1,n]-i} x^j$ which remains fixed, by

$$g(u) = f(\prod_{k<i}(x^k), u, \prod_{k>i}(x^k)) \; .$$

One uses to call $x = \prod^e_{i\in[1,n]} x^i$ the variables. If one considers a bunch of $m$ maps $(f(x) : \mathcal{R}^n \to \mathcal{R}^m)$ which can be written as $(\forall x \in \mathcal{R}^n)(f^k(x) : [1, m] \to \mathcal{R})$, each $f^k$ are called component. Most of the time, we may analyze each component separately setting $m = 1$. The conclusions could be applied at each component when $m > 1$.

**Definition 276 Derivative operator, product of derivative operators, $C^i$, $D^k$.**

We are given an open $O$ of $\mathcal{R}^n$. We call $C^0$ the set of continuous maps $(O \to \mathcal{R})$ (definition 141 (p. 165)). Subset of $C^0$, we call $C^1$, the set of maps $(O \to \mathcal{R})$ with partial derivatives over its $n$ variables (definition 275 (p. 304)) at any point of $O$. Those $n$ partial derivatives can be written as $i \in [1, n] \Rightarrow (\partial_i = \frac{\partial}{\partial x^i} : C^1 \to C^0)$. For this section $D^1$ will be the set of derivative operators $(\partial_i : [1, n] \to (C^0)^{C^1} = \prod_{i\in C^1} C^0)$ [27].

We suppose that we have defined $C^k \subset C^{k-1}$, $k \in \mathcal{Z}^+$, the set of $k$-derivable maps $(O \to \mathcal{R})$ at any point of $O$. We shall call here $D^k$, the set of $k$-derivative operator maps $C^k \to C^0$.

$$(f \in C^k, \partial \in D^k) \Rightarrow (\exists g \in C^0)(g = \partial f)$$

One calls $C^{k+1}$ the set defined by (see definitions 115 (p. 133), 333 (p. 441) and 362 (p. 462)):

$$\begin{aligned} x \in C^{k+1} &\Leftrightarrow (\exists \partial \in D^1)(\partial x \in C^k) \Leftrightarrow (\exists \partial \in D^1)(\exists y \in C^k)(\partial x = y) \\ &\Leftrightarrow (\exists \partial \in D^1)(x \in \partial^{-1}(C^k)) \Leftrightarrow x \in \bigcup_{\partial\in D^1} \partial^{-1}(C^k) \end{aligned}$$

We define the set $D^{k+1}$ by:

$$\partial^{k+1} \in D^{k+1} \Leftrightarrow (\exists i \in [1, n])(\exists \partial^k \in D^k)(\partial^{k+1} = (\partial_i \circ \partial^k : C^{k+1} \to C^0))$$

This is possible because the $\partial^{k+1}$ considered here have to be element of $\mathcal{P}(C^{k+1} \times C^0)$ which is a set (proposition 759 (p. 452), proposition 731 (p. 442), proposition 749 (p. 450)). Using the recurrence principle (proposition 927 (p. 520)), since $D^1, C^0, C^1$ are defined, $C^k$ and $D^k$ are defined for any $k \in \mathcal{Z}^{0+}$.

**Definition 277 $C^\infty$.**

Given an open $O$ of $\mathcal{R}^n$, a map $(f : O \to \mathcal{R})$ is said to be in $C^\infty$, if $(\forall k \in \mathcal{Z}^+)(f \in C^k)$.

---

[26] When $n = 1$, we suppress the down index and we use the convenient notation: $\partial f(x) = \partial_1 f(x) = \frac{df(x)}{dx}$ (see definition 164 (p. 193))

[27] $(C^0)^{C^1} = \prod_{i\in C^1} C^0 \subset \mathcal{P}(C^1 \times C^0)$ is a pedantic way to call the set of maps $C^1 \to C^0$ (see definition 372 (p. 473)).

**Proposition 554** Considering $\mathcal{R}^n$ as a vectorial space with its standard basis $(e_i = \delta_i^j : [1,n] \to \{0,1\} \subset \mathcal{R})$ (see definitions 215 (p. 245), 242 (p. 267)) and its associated dual basis $(e^j : [1,n] \to (\mathcal{R}^n)^*)$ (definition 214 (p. 245)), the following map $(e^j(x) = x^j : \mathcal{R}^n \to \mathcal{R})$ is of $C^\infty$ and we have:

$$\partial_i e^j(x) = \partial_i(x^j) = \delta_i^j \tag{2.139}$$

$\delta_j^i$ is the Kronecker symbol (definition 215 (p. 245)).

**Proof**

From propositions 308 (p. 194) and 309 (p. 194), $\partial_i e^j(x) = \delta_i^j$, from proposition 308 (p. 194), we have:

$$k > 1 \Rightarrow (\partial^k \in D^k \Rightarrow \partial^k e^j(x) = 0)\ .$$

**Proposition 555** With the notations of the above definition 276 (p. 304), there is a surjective map (definition 123 (p. 139)) between $D^k$ and $[1,n]^k$ the set of maps $([1,k] \to [1,n])$ (definition 372 (p. 473)):

$$\partial^k \in D^k \Leftrightarrow (\exists \ell \in [1,n]^k)(\partial^k = \prod_{i\in[1,k]}^{\circ} \partial_{\ell(i)} : C^k \to C^0)$$

(see definition 131 (p. 146)). One uses to write also:

$$\partial^k \in D^k \Leftrightarrow (\exists \ell \in [1,n]^k)(\partial^k = \frac{\partial^k}{\prod_{i\in[1,k]} \partial_{\ell(i)}})\ .$$

**Proof**

From definition 276 (p. 304), if proposition 555 (p. 305) is true up to $k$ it is also true for $k+1$. Then from proposition 927 (p. 520), it is true for any $k$.

**Proposition 556** $n \in \mathcal{Z}^+$, $A \subset \mathcal{R}^n$, $B \subset \mathcal{R}^n$, $O \subset A \cap B$ being an open of $\mathcal{R}^n$, two functions $(f : A \to \mathcal{R})$, $(g : B \to \mathcal{R})$, $f$ with partial derivatives to any order at any point of $O$ with $(\forall x \in O)(f(x) = g(x))$, then $g$ has also partial derivatives equal to those of $f$ to any order at any point of $O$.

**Proof**

From proposition 306 (p. 193) and definition 134 (p. 153), with $(x^i : [1,n] \to \mathcal{R}) \in \mathcal{R}^n$ it is true for $\frac{\partial g}{\partial x^j}$. If we have

$$x \in O \Rightarrow (\prod_{j\in[1,n]} \partial_j^{k_j} g = \prod_{j\in[1,n]} \frac{\partial g}{\partial^{k_j} x^j} = \prod_{j\in[1,n]} \frac{\partial f}{\partial^{k_j} x^j})$$

we also have from proposition 306 (p. 193):

$$\begin{aligned} &x \in O \Rightarrow \\ &((\exists p \in [1,n])(k' \in (\mathcal{Z})^n, (\forall i \in [1,n] - p)(k'_i = k_i), k'_p = k_p + 1) \\ &\Rightarrow \prod \frac{\partial g}{\partial^{k'_j} x^j} = \prod \frac{\partial f}{\partial^{k'_j} x^j}) \end{aligned}$$

then from proposition 928 (p. 520), proposition 556 is true.

**Proposition 557** $n \in \mathcal{Z}^+$, $A \subset \mathcal{R}^n$ and $O \subset A$ being opens of $\mathcal{R}^n$, a function $(f : A \to \mathcal{R})$, with partial derivatives to any order, the restricted map $(f_\circ : O \to \mathcal{R})$ of $f$ on $O$ (definition 118 (p. 135)) also has partial derivatives to any order.

**Proof**
Since $(\forall x \in O)(f(x) = f_o(x))$ proposition 556 (p. 305) gives proposition 557 (p. 305).
**Definition 278 Smooth map, map infinitely derivable, $C^\infty\{\mathcal{R}^m \to \mathcal{R}^n\}$, map with partial derivatives up to $i \in \mathcal{Z}^+$, $C^i$.**

If a map $(f : \mathcal{R}^m \to \mathcal{R}^n)$ has partial derivatives to any order for any component and any variable it is also said to be smooth and to belong to $C^\infty\{\mathcal{R}^m \to \mathcal{R}^n\}$ as in definition 166 (p. 200). We can also define $C^i$, $i \in \mathcal{Z}^{+0}$ as the set of maps $(f : \mathcal{R}^m \to \mathcal{R}^n)$ if any product of $i$ partial derivatives acting on any component of $f$ exists. See also definition 276 (p. 304).

**Proposition 558** Given two functions $(f : \mathcal{R} \to \mathcal{R})$, $(g : \mathcal{R} \to \mathcal{R})$ of $C^i$, $i \in \mathcal{Z}^+$ setting $\partial^0 f = f$ we have for $n \in [0, i]$

$$\partial^n fg = \sum_{p\in[0,n]} \frac{n!}{(n-p)!p!} \partial^p f \partial^{n-p} g \ . \tag{2.140}$$

**Proof**

Let us suppose that we have for $n < i$ (see proposition 467 (p. 266))

$$\partial^n fg = \sum_{p\in[0,n]} \frac{n!}{(n-p)!p!} (\partial^p f)(\partial^{n-p} g) \ . \tag{2.141}$$

From proposition 369 (p. 217), we have for $n + 1$

$$\begin{aligned}
\partial\partial^n fg &= \sum_{p\in[0,n]} \frac{n!}{(n-p)!p!} (\partial^{p+1} f \partial^{n-p} g + \partial^p f \partial^{n-p+1} g) \\
&= \sum_{p\in[1,n+1]} \frac{n!}{(n-p+1)!(p-1)!} \partial^p f \partial^{n-p+1} g + \sum_{p\in[0,n]} \frac{n!}{(n-p)!p!} \partial^p f \partial^{n-p+1} g \\
&= \sum_{p\in[1,n]} (\frac{n!}{(n-p+1)!(p-1)!} + \frac{n!}{(n-p)!p!}) \partial^p f \partial^{n-p+1} g + (\partial^{n+1} f) g + f \partial^{n+1} g \\
&= \sum_{p\in[1,n]} (\frac{n!}{(n-p+1)!(p)!} (p + n - p + 1) \partial^p f \partial^{n-p+1} g + (\partial^{n+1} f) g + f \partial^{n+1} g \\
&= \sum_{p\in[0,n+1]} \frac{n+1!}{(n+1-p)!p!} \partial^p f \partial^{n+1-p} g
\end{aligned}$$

equation (2.141) is also true for $n + 1$. Then since it is true for $n = 0$ from proposition 927 (p. 520), it is true for $n \in [0, i]$. It gives equation (2.140).

**Proposition 559** The sum and the product of an integer number of functions of $C^i$, $i \in \mathcal{Z}^+$ are in $C^i$.
**Proof**
We suppose $f \in C^i$ and $g \in C^i$

$$(\forall n \in [0, i])(\partial^n (f + g) = \partial^n f + \partial^n g) \Rightarrow f + g \in C^i$$

and from proposition 558, $fg \in C^i$. If it is true for a sum (resp. product) of $n$ maps of $C^i$ from proposition 998 (p. 556) (resp. proposition 999 (p. 556)) and from above, it is true for $n + 1$. Then we conclude using proposition 927 (p. 520).

**Proposition 560 Finite increase formula** [28].
Given an open ball $\mathbf{B^n(c,r)} \in \mathcal{R}^n$ and a map $(\mathbf{f} : \mathbf{B^n(c,r)} \to \mathcal{R})$ we suppose that all partial derivatives of $\mathbf{f}$ exist and are continuous over $\mathbf{B^n(c,r)}$. We define $\mathbf{x}$ and $\mathbf{\Delta x}$ as: $\mathbf{x} = (\mathbf{x^1, ..., x^i, ..., x^n}) = (\mathbf{x^i} : [\mathbf{1,n}] \to \mathcal{R}) \in \mathcal{R}^n$, $\mathbf{\Delta x} = (\mathbf{\Delta x^1, ..., \Delta x^i, ..., \Delta x^n}) = (\mathbf{\Delta x^i} : [\mathbf{1,n}] \to \mathcal{R}))$. (see footnote 20 (p. 239))

1. We have:

$$\begin{aligned}&(\mathbf{x} \in \mathbf{B^n(c,r)}, \mathbf{x} + \mathbf{\Delta x} \in \mathbf{B^n(c,r)}) \Rightarrow \\ &(\forall \alpha > 0)(\exists \epsilon(\alpha, \mathbf{x}) > 0)((\|\mathbf{\Delta x}\| < \epsilon(\alpha, \mathbf{x}), \\ &\mathbf{f}(\mathbf{x} + \mathbf{\Delta x}) - \mathbf{f}(\mathbf{x}) = (\partial_i \mathbf{f}(\mathbf{x}) + \eta_i)\mathbf{\Delta x^i}) \Rightarrow \|\eta\| < \alpha) .\end{aligned} \tag{2.142}$$

2. If $\mathbf{C} \subset \mathbf{B^n(c,r)}$ is closed and then compact (proposition 521 (p. 290)) we have:

$$\begin{aligned}&(\forall \alpha > 0)(\exists \epsilon(\alpha) > 0)((\mathbf{x} \in \mathbf{C}, \mathbf{x} + \mathbf{\Delta x} \in \mathbf{C}, \|\mathbf{\Delta x}\| < \epsilon(\alpha), \\ &\mathbf{f}(\mathbf{x} + \mathbf{\Delta x}) - \mathbf{f}(\mathbf{x}) = (\partial_i \mathbf{f}(\mathbf{x}) + \eta_i)\mathbf{\Delta x^i}) \Rightarrow \|\eta\| < \alpha) .\end{aligned} \tag{2.143}$$

**Proof**
Calling $\mathbf{T_i} \in \mathcal{R}^n$ such that:

$$\mathbf{u} = \mathbf{T_i} \Rightarrow (\forall \mathbf{j} \in [\mathbf{1,i}])(\mathbf{u^j} = \mathbf{\Delta x^j}), (\forall \mathbf{j} \in [\mathbf{i+1,n}])(\mathbf{u^j} = 0)$$

with:

$$\begin{aligned}\mathbf{u} = \mathbf{T_n} &\Rightarrow (\forall \mathbf{j} \in [\mathbf{1,n}])(\mathbf{u^j} = \mathbf{\Delta x^j}) \\ \mathbf{u} = \mathbf{T_0} &\Rightarrow (\forall \mathbf{j} \in [\mathbf{1,n}])(\mathbf{u^j} = 0) \\ \mathbf{f}(\mathbf{x} + \mathbf{\Delta x}) - \mathbf{f}(\mathbf{x}) &= \sum_{\mathbf{i} \in [\mathbf{1,n}]} \mathbf{f}(\mathbf{x} + \mathbf{T_i}) - \mathbf{f}(\mathbf{x} + \mathbf{T_{i-1}}) .\end{aligned}$$

Setting with $\beta^\mathbf{i} \in \mathcal{R}^n$

$$\begin{aligned}&\mathbf{j} \in [\mathbf{1, i-1}] \Rightarrow \beta^\mathbf{i}_\mathbf{j} = 1, \\ &\mathbf{j} = \mathbf{i} \Rightarrow \beta^\mathbf{i}_\mathbf{j} = \beta^\mathbf{i}_\mathbf{i} \in [0,1], \\ &\mathbf{j} \in [\mathbf{i+1,n}] \Rightarrow \beta^\mathbf{i}_\mathbf{j} = 0\end{aligned}$$

From proposition 314 (p. 196), without summation on $\mathbf{i}$, there exists a set of $\beta^\mathbf{i}_\mathbf{i} \in [0,1]$, $\mathbf{i} \in [\mathbf{1,n}]$ such as:

$$\mathbf{f}(\mathbf{x} + \mathbf{T_i}) - \mathbf{f}(\mathbf{x} + \mathbf{T_{i-1}}) = \partial_i \mathbf{f}(\mathbf{x} + \mathbf{T_{i-1}} + \beta^\mathbf{i}_\mathbf{i}\mathbf{\Delta x^i})\mathbf{\Delta x^i} .$$

It gives:

$$\mathbf{f}(\mathbf{x} + \mathbf{T_i}) - \mathbf{f}(\mathbf{x} + \mathbf{T_{i-1}}) = \partial_i \mathbf{f}(\mathbf{x} + \sum_{\mathbf{j} \in [\mathbf{1,n}]} \beta^\mathbf{i}_\mathbf{j}\mathbf{\Delta x^j})\mathbf{\Delta x^i} .$$

From that we get:

$$\begin{aligned}&\sum_{\mathbf{i} \in [\mathbf{1,n}]} \mathbf{f}(\mathbf{x} + \mathbf{T_i}) - \mathbf{f}(\mathbf{x} + \mathbf{T_{i-1}}) = \mathbf{f}(\mathbf{x} + \mathbf{T_n}) - \mathbf{f}(\mathbf{x} + \mathbf{T_0}) = \mathbf{f}(\mathbf{x} + \mathbf{\Delta x}) - \mathbf{f}(\mathbf{x}) \\ &= \sum_{\mathbf{i} \in [\mathbf{1,n}]} \partial_i \mathbf{f}(\mathbf{x} + \sum_{\mathbf{j} \in [\mathbf{1,n}]} \beta^\mathbf{i}_\mathbf{j}\mathbf{\Delta x^j})\mathbf{\Delta x^i} .\end{aligned}$$

[28] As in most part of his book, when a term $\mathbf{a}$ belongs to a product of an integer number of sets let us say $\mathbf{n}$ like $\mathcal{R}^n$ : $\mathbf{a} = (\mathbf{a^i} : [\mathbf{1,n}] \to \mathcal{R})$ (resp. $\mathbf{a_i}$), if the index is not specified, it means that we handle the whole term, if the index is specified, it means we handle the result of the map.

We may write:

$$\partial_i f(x) + \eta_i(x, \Delta x) = \partial_i f(x + \sum_{j\in[1,n]} \beta^i_j \Delta x^j) .$$

Since $\partial_i f(x)$ is continuous we have the relation:

$$(\forall\alpha)(\exists\epsilon_i)(\|\Delta x\| < \epsilon_i \Rightarrow |\eta_i(x, \Delta x)| < \alpha/n) .$$

Taking $\epsilon = \min_{i\in[1,n]}\epsilon_i$, we get equation (2.142).
If $C \subset B^n(c, r)$ is closed and then compact, from proposition 283 (p. 184) we have:

$$(\forall\alpha)(\exists\epsilon_i)((\|\Delta x\| < \epsilon_i, x \in C) \Rightarrow |\eta_i(x, \Delta x)| < \alpha/n)$$

and

$$f(x + \Delta x) - f(x) = \sum_{i\in[1,n]} (\partial_i f(x) + \eta_i(x, \Delta x))\Delta x^i .$$

Again taking $\epsilon = \min_{i\in[1,n]}\epsilon_i$, we get equation (2.143).

**Proposition 561** Given an open $O \subset \mathcal{R}^n$ and a map $(f : O \to \mathcal{R}^p)$. For any $x = (x^i : [1, n] \to \mathcal{R}) \in \mathcal{R}^n$, we quote $f(x)$ as $f(x) = (f^j(x) : [1, p] \to \mathcal{R}) \in \mathcal{R}^p$. The set of $f^j$ are called component of $f$. As in proposition 560 (p. 307) we suppose that all partial derivatives of any component of $f$: $f^j$ exists and is continuous over $O$. We define $\Delta x$ as: $\Delta x = (\Delta x^i : [1, n] \to \mathcal{R})$.

1. We have:

$$\begin{aligned}&(x \in O, x + \Delta x \in O) \Rightarrow\\ &(\forall\alpha > 0)(\exists\epsilon(\alpha, x) > 0)((\|\Delta x\| < \epsilon(\alpha, x),\\ &f(x + \Delta x) - f(x) = (\partial_i f(x) + \eta_i)\Delta x^i) \Rightarrow \|\eta\| < \alpha) .\end{aligned}$$

2. The map $(f : O \to \mathcal{R}^p)$ is continuous.

3. If we have an open ball $B^n(c, r) \subset \mathcal{R}^n$ and a closed set $C \subset O \cap B^n(c, r)$ which is then compact (proposition 521 (p. 290)), we also have:

$$\begin{aligned}&(\forall\alpha > 0)(\exists\epsilon(\alpha) > 0)(x \in C, \|\Delta x\| < \epsilon(\alpha),\\ &(f(x + \Delta x) - f(x) = (\partial_i f(x) + \eta_i)\Delta x^i \Rightarrow \|\eta\| < \alpha)) .\end{aligned} \tag{2.144}$$

**Proof**
From propositions 560 (p. 307) and 266 (p. 179), we have:

$$\begin{aligned}&(\forall j \in [1, p])(x \in O \Rightarrow\\ &(\forall\alpha > 0)(\exists\epsilon^j(\alpha, x) > 0)(\|\Delta x\| < \epsilon^j(\alpha, x), f^j(x + \Delta x) - f^j(x) = (\partial_i f^j(x) + \eta^j_i)\Delta x^i)\\ &\Rightarrow \|\eta^j\| < \alpha)\\ &(\forall j \in [1, p])(\forall\alpha > 0)(\exists\epsilon^j(\alpha) > 0)((x \in C, x + \Delta x \in C, \|\Delta x\| < \epsilon^j(\alpha),\\ &f^j(x + \Delta x) - f^j(x) = (\partial_i f^j(x) + \eta^j_i)\Delta x^i) \Rightarrow \|\eta^j\| < \alpha) .\end{aligned}$$

Then using proposition 738 (p. 444),

$$\begin{aligned}&x \in O \Rightarrow (\forall\alpha > 0)(\forall j \in [1, p])(\exists\epsilon^j(\alpha, x) > 0)\\ &((\|\Delta x\| < \epsilon^j(\alpha, x), f^j(x + \Delta x) - f^j(x) = (\partial_i f^j(x) + \eta^j_i)\Delta x^i) \Rightarrow \|\eta^j\| < \alpha)\\ &(\forall\alpha > 0)(\forall j \in [1, p])(\exists\epsilon^j(\alpha) > 0)((x \in C, x + \Delta x \in C, \|\Delta x\| < \epsilon^j(\alpha),\\ &f^j(x + \Delta x) - f^j(x) = (\partial_i f^j(x) + \eta^j_i)\Delta x^i) \Rightarrow \|\eta^j\| < \alpha) .\end{aligned}$$

From proposition 255 (p. 172), we define $\epsilon'(\alpha, x) = \min_{j\in[1,n]} \epsilon^j(\alpha, x) > 0$, $\epsilon'(\alpha) = \min_{j\in[1,n]} \epsilon^j(\alpha) > 0$. Then

$$\begin{aligned}
&x \in O \Rightarrow \\
&(\forall\alpha > 0)(\exists\epsilon'(\alpha, x) > 0)(\|\Delta x\| < \epsilon'(\alpha, x), \\
&((f(x + \Delta x) - f(x) = (\partial_i f(x) + \eta_i)\Delta x^i) \Rightarrow (\forall j \in [1, p])(\|\eta^j\| < \alpha)) \\
&(\forall\alpha > 0)(\exists\epsilon'(\alpha) > 0)(x \in C, x + \Delta x \in C, \|\Delta x\| < \epsilon(\alpha), \\
&(f(x + \Delta x) - f(x) = (\partial_i f(x) + \eta_i)\Delta x^i) \Rightarrow (\forall j \in [1, p])\|\eta^j\| < \alpha)
\end{aligned}$$

but we also have:

$$(\alpha > 0, (\forall j \in [1, p])(\|\eta^j\| < \alpha)) \Rightarrow \|\eta\| < \alpha\sqrt{p}$$

then we get proposition 561 (p. 308) points 1 and 3 setting $\epsilon = \epsilon'(\alpha/\sqrt{p})$. Having that setting $\epsilon_c = \mathrm{Min}(\frac{\alpha}{\|\partial f(x)\|+\alpha}, \epsilon)$ we have:

$$\begin{aligned}
&x \in O \Rightarrow (\forall\alpha > 0)((\exists\epsilon_c > 0)((\|\Delta x\| < \epsilon_c, x + \Delta x \in O) \Rightarrow \|f(x + \Delta x) - f(x)\| \\
&= \|(\partial_i f(x) + \eta_i)\Delta x^i\| < (\|\partial f(x)\| + \|\eta\|)\|\Delta x\| < (\|\partial f(x)\| + \alpha)\frac{\alpha}{\|\partial f(x)\| + \alpha} = \alpha) .
\end{aligned}$$

From proposition 282 (p. 184), point 2 is true.

**Proposition 562** $A \subset \mathcal{R}, B \subset \mathcal{R}, C \subset \mathcal{R}^n$ being opens, given two maps $(f : C \subset \mathcal{R}^n \to B \subset \mathcal{R})$ and $(g : A \subset \mathcal{R} \to C \subset \mathcal{R}^n)$, we suppose that:

1. All partial derivatives of $f$ exist and are continuous where $f$ is defined (the open $C \subset \mathcal{R}^n$).

2. All maps $(g^i(u) : \mathcal{R} \to \mathcal{R})$ are derivable and then continuous.

Then the derivative of the composite function $(f \circ g : A \to B)$ exists and is given by:

$$u \in A \subset \mathcal{R} \Rightarrow (f \circ g)'(u) = \sum_{i\in[1,n]} \frac{\partial f(g(u))}{\partial x^i} g'^i(u) = \partial_i f(g(u)) g'^i(u) \in \mathcal{R} \tag{2.145}$$

**Proof**
With the above notations, considering a $u \in A$, we call:

$$g(u) = \prod_{i\in[1,n]}^{e} g^i(u) = (g^1(u), g^2(u), ..., g^n(u)) \in \mathcal{R}^n$$

$$x = \prod_{i\in[1,n]}^{e} x^i = (x^1(u), x^2(u), ..., x^n(u))$$

$$f(x) = f(x^1(u), x^2(u), ..., x^n(u)) \in \mathcal{R} .$$

From proposition 560 (p. 307), having a $\rho > 0$ such that $]u - \rho, u + \rho[\subset A$ ($\rho$ exists from proposition 266 (p. 179)), considering $\Delta u \in ]0, \rho[$, we may call (see definition 241 (p. 266)):

$$\Delta g = g(u + \Delta u) - g(u)$$

$$f \circ g(u + \Delta u) - f \circ g(u) = f(g + \Delta g) - f(g) .$$

Since $\mathbf{g(u)} \in \mathbf{C}$ open of $\mathcal{R}^n$, there is a $\rho_g$ such that $\mathbf{B(g(u),\rho_g)} \subset \mathbf{C}$ (proposition 266 (p. 179)). Since $\mathbf{g(u)}$ is continuous for $\mathbf{u} \in \mathbf{A}$ $(\exists \rho_u > 0)(]u-\rho_u, u+\rho_u[ \subset g^{-1}(B(g(u),\rho_g)) \subset$ $\mathbf{A}$ (definition. 141 (p. 165)), we have from proposition 560 (p. 307):

$$(\exists \eta \in \mathcal{R}^n)(f(g+\Delta g) - f(g) = (\partial_i f(g) + \eta_i)\Delta g^i)$$
$$(\forall \alpha > 0)(\exists \epsilon^f > 0)(\|\Delta g\| < \epsilon^f < \rho_g \Rightarrow \|\eta\| < \alpha) \,. \tag{2.146}$$

From 112 (p. 131) point 4, at given $\mathbf{u}$, it defines a map $\epsilon^f(\alpha)$. Since all $\mathbf{g^i(u)}$ are continuous we have:

$$(\forall \alpha > 0)(\forall i \in [1,n])(\exists \epsilon_i^g > 0)(|\Delta u| < \epsilon_i^g \rho_u \Rightarrow |g^i(u+\Delta u) - g^i(u)| < \frac{\alpha}{n}) \,.$$

It defines, at given $\mathbf{u}$, a map $\epsilon_i^g(\alpha)$ calling

$$\epsilon^g(\alpha) = \min_{i\in[1,n]} \epsilon_i^g(\alpha)$$

$\epsilon^g(\alpha) > 0$ because from proposition 255 (p. 172) $(\forall \alpha > 0)(\exists j \in [1,n])(\epsilon^g(\alpha) = \epsilon_j^g(\alpha))$. We have:

$$|\Delta u| < \epsilon^g(\alpha) \Rightarrow \|\Delta g\| < \alpha \,. \tag{2.147}$$

Since all $\mathbf{g^i(u)}$ have a derivative at $\mathbf{u}$ (see definition 162 (p. 193)), we have:

$$(\forall \alpha > 0)(\forall i \in [1,n])(\exists \epsilon'^g_i > 0)(|\Delta u| < \epsilon'^g_i < \rho_u$$
$$\Rightarrow |\frac{g^i(u+\Delta u) - g^i(u)}{\Delta u} - g'^i(u)| < \frac{\alpha}{n}) \,. \tag{2.148}$$

It defines a map, at given $\mathbf{u}$, $\epsilon'^g_i(\alpha)$ and a $\epsilon'^g(\alpha) = \min_{i\in[1,n]}(\epsilon'^g_i(\alpha)) > 0$.
At a given $\mathbf{u}$, we have:

1. From formulae (2.147) and (2.146):

$$\Delta u < \epsilon^g(\epsilon^f(\alpha/n)) < \rho_u \Rightarrow \|\Delta g\| < \epsilon^f(\alpha/n)$$
$$\Rightarrow \|\eta\| < \alpha/n \Rightarrow (\forall i \in [1,n])(\eta_i < \alpha/n)$$

2. From formula (2.148), setting $\Delta g^i = g^i(u+\Delta u) - g^i(u)$, we have:

$$\Delta u < \epsilon'^g(\alpha) \Rightarrow (\forall i \in [1,n])(|\Delta g^i/\Delta u - g'_i| < \alpha/n) \Rightarrow \|\Delta g/\Delta u - g'\| < \alpha \,.$$

3. Combining 1 and 2, we have:

$$\Delta u < \min(\epsilon^g(\epsilon^f(\alpha_1/n)), \epsilon'^g(\alpha_2)) \Rightarrow$$
$$|\frac{f(g+\Delta g) - f(g)}{\Delta u} - \partial_i f(g) g'^i| < \alpha_1 \max_{i\in[1,n]} |g'^i(u)| + \alpha_2 \max_{i\in[1,n]} |\partial_i f(g)| + \alpha_1\alpha_2/n \,.$$

So for any $\alpha$ at any $\mathbf{u}$ choosing:

$$\alpha_1 < \min(\frac{\alpha}{3\max_{i\in[1,n]} |g'^i(u)|}, \alpha/3)$$

and

$$\alpha_2 < \min(\frac{\alpha}{3\max_{i\in[1,n]} |\partial_i f(g)|}, n)$$

we have $\epsilon = \min(\epsilon^g(\epsilon^f(\alpha_1/n)), \epsilon'^g(\alpha_2)) > 0$ such as:

$$|\Delta u| < \epsilon \Rightarrow |\frac{f(g+\Delta g) - f(g)}{\Delta u} - \partial_i f(g) g'^i| < \alpha$$

which proves equation (2.145) and proposition 562.

**Proposition 563** $(\mathbf{g}:\mathcal{R}^n \to \mathcal{R}^p)$ and $(\mathbf{f}:\mathcal{R}^p \to \mathcal{R}^q)$ being continuous maps defined at least on respectively an open of $\mathcal{R}^n$ and $\mathcal{R}^p$, both maps having all their partial derivatives defined and continuous on those opens. We have for the composite map $(\mathbf{f}\circ\mathbf{g}:\mathcal{R}^n \to \mathcal{R}^q)$ the following formula:

$$\partial_i(f\circ g(x))^j = \sum_{k\in[1,p]} \partial_k f^j(g(x))\partial_i g^k \,. \tag{2.149}$$

**Proof**

At any point of the opens where $\mathbf{g}$ and $\mathbf{f}$ are defined, there is an open cube centered on it, then, using the same argument as in proposition 561 (p. 308), we can apply proposition 562 (p. 309) on each variable of $\mathcal{R}^n$ and $\mathcal{R}^q$ using the definition of partial derivative (definition 275 (p. 304)).

## 2.20.1 Derivative of a sequence of maps

**Proposition 564** We have a sequence of maps $(U_k(x) : \mathcal{Z}^+ \times \mathcal{R}^n \to \mathcal{R})$ defined and continuous on a closed box $C(0,c)$ and converging uniformly to a limit $(U(x):\mathcal{R}^n \to \mathcal{R})$ (definition 272 (p. 302)). We suppose that the partial derivative for $x^\mu, \mu\in[1,n]$: $U'_k(x) = \frac{\partial U_k(x)}{\partial x^\mu}$ exists for all $x\in C(0,c)$ and it is such that ($\mu$ is fixed but not always quoted):

$$\begin{aligned}&(\forall\alpha)(\exists\epsilon(\alpha))(\forall x\in C(0,c))(\forall p\in\mathcal{Z}^+)\\&(|\Delta|<\epsilon(\alpha)\Rightarrow|\frac{U_p(x+e_\mu\Delta)-U_p(x)}{\Delta}-U'_p(x)|<\alpha)\end{aligned} \tag{2.150}$$

then $U'_k(x)$ is also a uniform Cauchy sequence (definition 274 (p. 303)) the limit of which is the corresponding derivative of $U(x)$: $U'(x) = \frac{\partial U(x)}{\partial x^\mu}$.
We call, in equation (2.150), $e_\mu$ the vector of $\mathcal{R}^n$ such that $e_\mu = (\delta^\nu_\mu : [1,n]\to\mathcal{R})$ where $\delta^\nu_\mu$ is the Kronecker symbol (definition 215 (p. 245)).

**Proof**

For simplicity $\mu$ is fixed but not always quoted.

1. From equation (2.150) we have:

$$\begin{aligned}&(\forall\alpha_1)(\exists\epsilon_1(\alpha_1))(\forall p)(\forall x\in C(0,c))(x+e_\mu\Delta\in C(0,c),|\Delta|<\epsilon_1(\alpha_1)\\&\Rightarrow|\frac{U_p(x+e_\mu\Delta)-U_p(x)}{\Delta}-U'_p(x)|<\alpha_1)\end{aligned}$$

2. From proposition 552 (p. 303), $U_n$ is a uniform Cauchy sequence of maps (definition 274 (p. 303)). Then from that definition we have:

$$\begin{aligned}&(\forall\alpha_2)(\forall\Delta\in\mathcal{R})(\exists m_2(\alpha_2|\Delta|)>0)(\forall p>m_2(\alpha_2|\Delta|))\\&(\forall q>m_2(\alpha_2|\Delta|))(\forall x\in C(0,c))(|U_p(x)-U_q(x)|<\alpha_2|\Delta|)\,.\end{aligned}$$

Combining both:

$$\begin{aligned}&(|\Delta|<\epsilon_1(\alpha_1),p>m_2(\alpha_2|\Delta|),q>m_2(\alpha_2|\Delta|))\Rightarrow|U'_p(x)-U'_q(x)|\\&=|U'_p(x)-\frac{U_p(x+e_\mu\Delta)-U_p(x)}{\Delta}+\frac{U_p(x+e_\mu\Delta)-U_q(x+e_\mu\Delta)}{\Delta}\\&-\frac{U_p(x)-U_q(x)}{\Delta}+\frac{U_q(x+e_\mu\Delta)-U_q(x)}{\Delta}-U'_q(x)|<2\alpha_1+2\alpha_2\,.\end{aligned}$$

With $\alpha_1 = \alpha_2 = \alpha/4, |\Delta x| < \epsilon_1(\alpha_1), m_3(\alpha) = m_2(\alpha_2|\Delta x|) = m_2(\alpha|\Delta x|/4)$, we get:

$$(\forall\alpha)(\exists m = m_3(\alpha) > 0)(\forall x \in C(0,c))(\forall p > m)(\forall q > m)(|U'_p(x) - U'_q(x)| < \alpha$$

From definition 274 (p. 303) $U'_k$ is a uniform Cauchy sequence, from proposition 552 (p. 303), it converges uniformly to a limit $U'(x)$ (definition 272 (p. 302))

$$(\forall\alpha_4)(\exists m_4(\alpha_4)(\forall x \in C(0,c))(\forall p > m_4(\alpha_4))(|U'_p(x) - U'(x)| < \alpha_4) .$$

Since $U_k$ converges also uniformly to $U(x)$ we also have

$$\begin{aligned}&(\forall\alpha_5)(\forall\Delta \in \mathcal{R})(\exists m_5(\alpha_5|\Delta|) > 0)(\forall p > m_5(\alpha_5|\Delta|))\\&(\forall x \in C(0,c))(|U_p(x) - U(x)| < \alpha_5|\Delta|)\end{aligned}$$

then $|\Delta| < \epsilon_1(\alpha_1)$ and $p > \max(m_2(\alpha_2|\Delta|), m_4(\alpha_4), m_5(\alpha_5|\Delta|))$ gives:

$$\begin{aligned}&|\frac{U_p(x+e_\mu\Delta) - U_p(x)}{\Delta} - \frac{U(x+\Delta x) - U(x)}{\Delta} - U'_p(x) + U'(x)| =\\&|\frac{U_p(x+e_\mu\Delta) - U(x+e_\mu\Delta)}{\Delta} - \frac{U_p(x) - U(x)}{\Delta} - U'_p(x) + U'(x)| < \alpha_4 + \alpha_5\end{aligned}$$

then

$$\begin{aligned}&|\frac{U(x+e_\mu\Delta) - U(x)}{\Delta} - U'(x)| =\\&|\frac{U_p(x+e_\mu\Delta) - U_p(x)}{\Delta} - \frac{U(x+\Delta x) - U(x)}{\Delta} - U'_p(x) + U'(x)\\&-(\frac{U_p(x+e_\mu\Delta) - U_p(x)}{\Delta} - U'_p(x))|\\&< \alpha_4 + \alpha_5 + |\frac{U_p(x+e_\mu\Delta) - U_p(x)}{\Delta} - U'_p(x)| < \alpha_4 + \alpha_5 + \alpha_1\end{aligned}$$

and we have set $\alpha_4 = \alpha_5 = \alpha_1 = \alpha/3$

$$\begin{aligned}&(\forall\alpha)(\exists\epsilon = \epsilon_1(\alpha/3))(|\Delta x| < \epsilon) \Rightarrow (\forall x \in C(0,c))(x + e_\mu\Delta \in C(0,c)\\&\Rightarrow |\frac{U(x+e_\mu\Delta) - U(x)}{\Delta} - U'(x)| < \alpha) .\end{aligned}$$

## 2.21 Topology on convex subsets

We shall elaborate on convex set only for what we shall use in the integral and in differentiable manifold theory.

**Proposition 565** We consider a vectorial space $E$ of dimension $n$ built on real numbers $\mathcal{R}$ with a Euclidean inner product $g$ generating a Euclidean metric. $E$ has the topology of open balls (see proposition 263 (p. 177)). $n_c \in \mathcal{Z}^+$ being an integer, any map $(p_i : [0, n_c] \to E)$ defines a closed convex (definition 255 (p. 277)) $C$ subset of $E$ by the formula (see proposition 748 (p. 449)):

$$x \in C \Leftrightarrow (\exists(m^i : [0, n_c] \to \mathcal{R}^{0+}))(\sum_{i\in[0,n]} m^i = 1, x = m^i p_i) .$$

**Proof**
C is convex from proposition 501 (p. 281). C is closed because:

- From propositions 241 (p. 167), 244 (p. 168), 286 (p. 185), 244 (p. 168), 506 (p. 286) and 507 (p. 286), the map $\mathbf{f} : \mathcal{R}^{n_c+1} \to \mathbf{E}$ defined by:

$$(\forall w \in \mathcal{R}^{n_c+1})(f(w) = w^i p_i)$$

  is continuous.

- The subset $S \subset \mathcal{R}^{n_c+1}$ defined by

$$w \in S \Leftrightarrow \sum_{i \in [0,n_c]} w_i = 1$$

  is closed:
  $\{1\} \subset \mathcal{R}$ is closed because $\mathcal{C}(\{1\}) = ]\leftarrow, 1[ \cup ]1, \to [$ is open (definitions 378 (p. 484), 135 (p. 154), 132 (p. 151) equation (2.17), proposition 275 (p. 181)). $(\sigma(w) = \sum_{i \in [0,n_c]} w_i : \mathcal{R}^{n_c+1} \to \mathcal{R})$ is a continuous map and $S = \sigma^{-1}(\{1\}) = \mathcal{C}(\sigma^{-1}(\mathcal{C}(\{1\})))$ is closed from definitions 135 (p. 154), 141 (p. 165).

- From proposition 230 (p. 164), the subset $[0,1]^{n_c} \subset \mathcal{R}^{n_c+1}$ is compact as product of an integer number of compact subsets (see proposition 298 (p. 190)) then it is closed (proposition 226 (p. 160)).

- From the above two points $S \cap [0,1]^{n_c}$ is closed (definition 135 (p. 154), equation (2.26) in proposition 206 (p. 154)). From that we may deduce: Since $S \cap [0,1]^{n_c} \subset B(0,2)$ is closed, from proposition 521 (p. 290) $S \cap [0,1]^{n_c} \subset B(0,2)$ is compact. It implies that $C = f(S \cap [0,1]^{n_c})$ is compact and closed (propositions 240 (p. 167) and 226 (p. 160)).

**Proposition 566** We have a vectorial space **E** of finite dimension **n** built on $\mathcal{R}$ with a Euclidean metric **g** and its induced open ball topology. In any convex set not empty $C \subset E$ of dimension **n** there is an open ball in it:

$$(\exists x \in C)(\exists r \in \mathcal{R}^+)(B(x,r) \subset C) .$$

**Proof**
Since $C \neq \emptyset$ is of dimension $n \in \mathcal{Z}^+$, the same as of **E**, from definition 256 (p. 280) there is a map:

$$(p_i : [0,n] \to C)$$

such that the set of vectors $(p_i - p_0 : [1,n] \to E)$ is independent. Since their number is equal to the dimension of **E**, they are a basis of **E**. Calling **S** the set such that:

$$x \in S \Leftrightarrow (x \in E, (\exists (x^i : [1,n] \to ]0,1[)) \\ (\sum_{i \in [1,n]} x^i < 1, x = x^i(p_i - p_0) + p_0 = x^i p_i + (1 - \sum_{i \in [1,n]} x^i) p_0)) .$$

From proposition 492 (p. 278) $S \subset C$. Let us prove that **S** is open.
The set $(p_i - p_0 : [1,n] \to E)$ being a basis of **E**, the map $(x^i(x) : E \to \mathcal{R}^n)$ is a continuous map:

$\mathbf{e_i} : [1, n] \to \mathbf{E}$ being an orthogonal basis of $\mathbf{E}$ defining the metric and the topology of $\mathbf{E}$, setting the matrix ${}^0p_i^j$ such that $p_i - p_0 = {}^0p_i^j e_j$ and calling $e_j^i$ its inverse which exists since the $p_i - p_0$ are a basis and we have $e_j = e_j^i(p_i - p_0)$ and

$$x = {}^e x^j e_j = {}^e x^j e_j^i (p_i - p_0) \Rightarrow x^i = {}^e x^j e_j^i \ .$$

The map $(x^i({}^e x^j) : \mathcal{R}^n \to \mathcal{R}^n)$ is continuous. For all $i \in [1, n]$, the reverse image $h^i$ of $x^i \in ]0., 1.[$ is open (definition 141 (p. 165)). Since $(\sum_{i \in [1,n]} x^i({}^e x^j) : \mathcal{R}^n \to \mathcal{R})$ is a continuous continuous map, the reverse image $H \in \mathcal{R}^n$ of $\sum_{i \in [1,n]} x^i({}^e x^j) \in ] \leftarrow, 1[$ is open. From definition 132 (p. 151) equation (2.18), $\bigcap_{i \in [1,n]} h^i \cap H$ is open. Since the map $({}^e x^j(x^i) : \mathcal{R}^n \to \mathcal{R}^n)$ is continuous, from definition 141 (p. 165), $S$ is open.
We also have:

$$\frac{\sum_{i[0,n]} p_i}{n+1} = \frac{\sum_{i[1,n]}(p_i - p_0)}{n+1} + p_0 \in S \ .$$

Then, from proposition 266 (p. 179), we have proposition 566 (p. 313).

**Proposition 567** We suppose that we have a convex set $C$ of dimension $n$ subset of $\mathcal{R}^n$ considered with the natural Euclidean metric as a Euclidean metric topological vectorial space (see definition 256 (p. 280)). In $\mathcal{R}^n$, the closure of $C$ is equal to the closure of interior of $C$ (see definitions 136 (p. 155) and 137 (p. 156)):

$$\overline{C} = \overline{\mathcal{I}(C)} \ .$$

**Proof**
Given a point $a \in \overline{C}$ and an open $O$ containing $a$, from proposition 266 (p. 179), there is an open ball $B(a, R) \subset O$ centered on $a$. Inside the open ball $B(a, R/2) \subset B(a, R)$ there is, from definition 136 (p. 155) a point, $p \in B(a, R/2) \cap C$. Calling $r_p = \|a - p\| < R/2$ we have: $B(p, R - r_p) \subset B(a, R)$. Considering $n$ points of $C$ $(p_i : [1, n] \to C)$ such that the set of the $n$ vectors $p_i - p$ are independent. Calling $d_M = \max(\max_{i \in [1,n]} \|p_i - p\|, R)$, the set of points $p + \frac{R - r_p}{2d_M}(p_i - p)$ with $i \in [1, n]$ is a subset of $B(p, R - r_p) \cap C$ because: $C$ is convex, then, since $0 < \frac{R - r_p}{2d_M} < 1$, we have:

$$(\forall i \in [0, n])(p + \frac{R - r_p}{2d_M}(p_i - p) \in C)$$

but we also have:

$$(\forall i \in [0, n])(p + \frac{R - r_p}{2d_M}(p_i - p) \in B(p, R - r_p))$$

so

$$(\forall i \in [0, n])(p + \frac{R - r_p}{2d_M}(p_i - p) \in B(p, R - r_p) \cap C) \ .$$

Since $B(p, R - r_p) \cap C$ is convex (propositions 494 (p. 278), 513 (p. 288)), calling $p_0 = p$ and $B_p$ the box generated by:

$$x \in B_p \Leftrightarrow (\exists (m^i : [0, n] \to \mathcal{R}^{+0}))$$
$$(\sum_{i \in [0,n]} m^i = 1, x \in \sum_{i \in [0,n]} m^i (p + \frac{R - r_p}{2d_M}(p_i - p)))$$

is a convex subset of $\mathbf{B}(p, \mathbf{R} - \mathbf{r_p}) \cap \mathbf{C}$ (proposition 500 (p. 281)). From proposition 566 (p. 313), there is an open ball fully in $\mathbf{B}(p, \mathbf{R} - \mathbf{r_p})) \cap \mathbf{C}$ fully in $\mathbf{B_p}$ and fully in $\mathbf{B(a, R)}$ then in $\mathbf{O}$. From that we have:

$$\overline{C} = \overline{\mathcal{I}(C)} \ .$$

**Proposition 568** In a Euclidean metric vectorial space $\mathbf{E}$ built on real numbers with the open ball topology, the interior of a convex subset of $\mathbf{E}$ is also convex.

**Proof**

With the above notation, $\mathbf{C}$ being a convex subset of $\mathbf{E}$ and $\mathcal{I}(\mathbf{C})$ its interior (definition 137 (p. 156)), given two points $\mathbf{p}, \mathbf{q}$ in $\mathcal{I}(\mathbf{C})$ and taking a $\mathbf{t} \in ]0, 1[$, we call (see definition 255 (p. 277)):

$$\mathbf{r} = (1 - \mathbf{t})\mathbf{p} + \mathbf{t}\mathbf{q} \in \mathbf{C} \ .$$

Since $\mathcal{I}(\mathbf{C})$ is open from proposition 266 (p. 179), there are two open balls centered on $\mathbf{p}$ and $\mathbf{q}$ contained in $\mathcal{I}(\mathbf{C})$:

$$(\exists \epsilon_p > 0)(\exists \epsilon_q > 0)(\mathbf{B}(\mathbf{p}, \epsilon_p) \subset \mathcal{I}(\mathbf{C}), \mathbf{B}(\mathbf{q}, \epsilon_q) \subset \mathcal{I}(\mathbf{C})) \Rightarrow (\forall \mathbf{t} \in ]0, 1[)$$
$$(\exists \epsilon > 0)(\epsilon = \min(\epsilon_p, \frac{\mathbf{t}}{1 - \mathbf{t}}\epsilon_q), \mathbf{B}(\mathbf{p}, \epsilon) \subset \mathcal{I}(\mathbf{C}), \mathbf{B}(\mathbf{q}, \frac{1 - \mathbf{t}}{\mathbf{t}}\epsilon) \subset \mathcal{I}(\mathbf{C}))$$

for any point $\mathbf{r}' \in \mathbf{B}(\mathbf{r}, \epsilon)$ one may consider a point $\mathbf{p}' \in \mathbf{B}(\mathbf{p}, \epsilon)$ and a point $\mathbf{q}' = \frac{\mathbf{t}-1}{\mathbf{t}}\mathbf{p}' + \frac{1}{\mathbf{t}}\mathbf{r}'$ we have since $\mathbf{t} \in ]0, 1[$ we have:

$$\|\mathbf{q} - \mathbf{q}'\| < (\frac{1 - \mathbf{t}}{\mathbf{t}} + \frac{1}{\mathbf{t}})\epsilon = \frac{1 - \mathbf{t}}{\mathbf{t}}\epsilon$$

then

$$\mathbf{q}' \in \mathbf{B}(\mathbf{q}, \frac{1 - \mathbf{t}}{\mathbf{t}}\epsilon) \subset \mathcal{I}(\mathbf{C}) \subset \mathbf{C}$$

since

$$\mathbf{p}' \in \mathbf{B}(\mathbf{p}, \epsilon_p) \subset \mathcal{I}(\mathbf{C}) \subset \mathbf{C} \ .$$

Since $\mathbf{C}$ is convex and $\mathbf{r}' = \mathbf{t}\mathbf{p}' + (1 - \mathbf{t})\mathbf{q}'$ with $\mathbf{t} \in [0, 1]$, we have from definition 255 (p. 277): $\mathbf{r}' \in \mathbf{C}$. From definition 137 (p. 156), we have:

$$(\forall \mathbf{r}' \in \mathbf{B}(\mathbf{r}, \epsilon))(\mathbf{r}' \in \mathbf{C}) \Rightarrow \mathbf{r} \in \mathcal{I}(\mathbf{C}) \ .$$

Then $\mathcal{I}(\mathbf{C})$ is convex (definition 255 (p. 277)).

**Proposition 569** In a Euclidean metric vectorial space $\mathbf{E}$ built on real numbers with the open ball topology, the closure of a convex subset of $\mathbf{E}$ is also convex.

**Proof**

With the above notation, $\mathbf{C}$ being a convex subset of $\mathbf{E}$, given two points $\mathbf{p}, \mathbf{q}$ in $\overline{\mathbf{C}}$ and for any $\mathbf{r}$ such that $(\exists \mathbf{t} \in [0, 1])(\mathbf{r} = \mathbf{t}\mathbf{p} + (1 - \mathbf{t})\mathbf{q})$, for any open $\mathbf{O}$ containing $\mathbf{r}$, there is an open ball $\mathbf{B}(\mathbf{r}, \epsilon) \subset \mathbf{O}, \epsilon \in \mathcal{R}^+$. By the definition 136 (p. 155) in the balls $\mathbf{B}(\mathbf{p}, \epsilon), \mathbf{B}(\mathbf{q}, \epsilon)$ there is $\mathbf{p}' \in \mathbf{B}(\mathbf{p}, \epsilon) \cap \mathbf{C}$, $\mathbf{q}' \in \mathbf{B}(\mathbf{q}, \epsilon)\mathbf{C}$ then from definition 255 (p. 277) $\mathbf{r}' = \mathbf{t}\mathbf{p}' + (1 - \mathbf{t})\mathbf{q}' \in \mathbf{C}$. We also have since $\mathbf{t} \in [0, 1]$:

$$\|\mathbf{r}' - \mathbf{r}\| = \|\mathbf{t}(\mathbf{p}' - \mathbf{p}) + (1 - \mathbf{t})(\mathbf{q}' - \mathbf{q})\| < \epsilon(\mathbf{t} + 1 - \mathbf{t}) = \epsilon \Rightarrow \mathbf{r}' \in \mathbf{B}(\mathbf{r}, \epsilon) \subset \mathbf{O} \ .$$

Then from definition 136 (p. 155), $\mathbf{r} \in \overline{\mathbf{C}}$ and from definition 255 (p. 277) $\overline{\mathbf{C}}$ is convex.

**Proposition 570** In a Euclidean metric vectorial space $\mathbf{E}$ of finite dimension $\mathbf{n}$ built on real numbers with the open ball topology, $\mathbf{C}$ being a convex of dimension $\mathbf{n}$ subset of $\mathbf{E}$ calling the interior of $\mathbf{C}$: $\mathcal{I}(\mathbf{C})$ and the closure of $\mathbf{C}$: $\overline{\mathbf{C}}$ (see definitions 137 (p. 156), 136 (p. 155)) then:

$$(\forall \mathbf{p} \in \mathcal{I}(\mathbf{C}))(\forall \mathbf{q} \in \overline{\mathbf{C}})(\forall \mathbf{t} \in [0,1[)((1-\mathbf{t})\mathbf{p} + \mathbf{t}\mathbf{q} \in \mathcal{I}(\mathbf{C})) .$$

**Proof**
With the above notation, repeating somehow the reasoning of proposition 568 (p. 315), given two points $\mathbf{p} \in \mathcal{I}(\mathbf{C})$ and $\mathbf{q} \in \overline{\mathbf{C}}$ taking a $\mathbf{t} \in [0,1[$, we call:

$$\mathbf{r} = (1-\mathbf{t})\mathbf{p} + \mathbf{t}\mathbf{q} \in \mathbf{C} .$$

Since $\mathcal{I}(\mathbf{C})$ is open from proposition 266 (p. 179) there is a ball centered on $\mathbf{p}$ contained in $\mathcal{I}(\mathbf{C})$:

$$(\exists \epsilon > 0)(\mathbf{B}(\mathbf{p},\epsilon) \subset \mathcal{I}(\mathbf{C})) .$$

Since from proposition 567 (p. 314) $\overline{\mathcal{I}(\mathbf{C})} = \overline{\mathbf{C}}$ then from definition 136 (p. 155) in any ball $\mathbf{B}(\mathbf{q},(1-\mathbf{t})\epsilon)$ there a $\mathbf{q}' \in \mathcal{I}(\mathbf{C})$, calling $\mathbf{p}' = \frac{1}{1-\mathbf{t}}\mathbf{r} - \frac{\mathbf{t}}{1-\mathbf{t}}\mathbf{q}'$, we have:

$$\|\mathbf{p}-\mathbf{p}'\| = \frac{1}{1-\mathbf{t}}\|\mathbf{q}-\mathbf{q}'\| < \epsilon \Rightarrow \mathbf{p}' \in \mathbf{B}(\mathbf{p},\epsilon) \subset \mathcal{I}(\mathbf{C}) .$$

Then since $\mathcal{I}(\mathbf{C})$ is convex (proposition 568 (p. 315)), from definition 255 (p. 277), $\mathbf{r} \in \mathcal{I}(\mathbf{C})$.

**Proposition 571** We have a Euclidean metric topological vectorial space $\mathbf{E}$ built on real numbers. Given a line $\mathcal{L} \subset \mathbf{E}$ defined by two elements $\boldsymbol{p}$, $\boldsymbol{v}$ of of $\mathbf{E}$ (definition 211 (p. 242)), such that we have:

$$\boldsymbol{x} \in \mathcal{L} \Leftrightarrow (\boldsymbol{x} \in \mathbf{E}, (\exists \lambda \in \mathcal{R})(\boldsymbol{x} = \boldsymbol{p} + \lambda \boldsymbol{v}))$$

given also a convex set $\mathbf{C} \subset \mathbf{E}$ then $\mathcal{L} \cap \mathbf{C}$ is convex and calling $\mathbf{I}$ the set of intervals of $\mathcal{R}$ (definition 378 (p. 484)), we should have:

$$(\exists \iota \in \mathbf{I})(\boldsymbol{x} \in \mathcal{L} \cap \mathbf{C} \Leftrightarrow (\exists \lambda \in \iota)(\boldsymbol{x} = \boldsymbol{p} + \lambda \boldsymbol{v})) .$$

**Proof**
It is a direct consequence of propositions 491 (p. 277), 494 (p. 278), 497 (p. 279) and 495 (p. 278).

**Proposition 572** We have a Euclidean metric topological vectorial space $\mathbf{E}$ of finite dimension $\mathbf{n}$ built on real numbers. We suppose that, there is a convex subset $\mathbf{C} \subset \mathbf{E}$ of dimension $\mathbf{n}$ inside a ball of $\mathbf{E}$, for any point $\boldsymbol{p} \in \mathcal{I}(\mathbf{C})$ ($\mathcal{I}(\mathbf{C})$ is the interior of $\mathbf{C}$) and for any nonzero vector $\boldsymbol{v} \in \mathbf{E}$, there is one and only one $\lambda \in \mathcal{R}^+$ such that $\boldsymbol{p} + \lambda\frac{\boldsymbol{v}}{\|\boldsymbol{v}\|} \in \mathcal{B}(\mathbf{C})$ where $\mathcal{B}(\mathbf{C})$ is the boundary of $\mathbf{C}$ (definition 138 (p. 157)). Furthermore we have:

$$\mu \in [0,\lambda] \Rightarrow \boldsymbol{p} + \mu\frac{\boldsymbol{v}}{\|\boldsymbol{v}\|} \in \overline{\mathbf{C}} .$$

**Proof**
Applying proposition 571 (p. 316), we have an interval $\mathbf{I}(\mathcal{I}(\mathbf{C}))$ of $\mathcal{R}$ and one $\mathbf{I}(\bar{\mathbf{C}})$ of $\mathcal{R}$ such that:

$$\lambda \in \mathbf{I}(\mathcal{I}(\mathbf{C})) \Leftrightarrow \boldsymbol{p} + \lambda\frac{\boldsymbol{v}}{\|\boldsymbol{v}\|} \in \mathcal{I}(\mathbf{C})$$

$$\lambda \in \mathbf{I}(\bar{\mathbf{C}}) \Leftrightarrow \boldsymbol{p} + \lambda\frac{\boldsymbol{v}}{\|\boldsymbol{v}\|} \in \bar{\mathbf{C}} .$$

Since we have $\mathcal{I}(C) \subset \bar{C}$, $I(\mathcal{I}(C)) \subset I(\bar{C})$. Since the map: $(p + \nu\frac{v}{\|v\|} : \mathcal{R} \rightarrow E)$ is continuous in $\nu$ (propositions 244 (p. 168)), 234 (p. 165) and 288 (p. 186)) $I(\mathcal{I}(C)) = ]\lambda_1, \lambda_2[$ is an open interval and $I(\bar{C}) = [\lambda'_1, \lambda'_2]$ a closed interval (definitions 378 (p. 484), 155 (p. 181), 156 (p. 181)). From proposition 570 (p. 316), since $p \in \mathcal{I}(C)$, we have $\lambda'_1 = \lambda_1, \lambda'_2 = \lambda_2$.

**Proposition 573** Using the same notations as in preceding proposition 572 (p. 316), for any point $p \in \mathcal{I}(C)$, for any nonzero vector $v \in E$, we have from proposition 572 (p. 316):

$$(\exists \lambda \in \mathcal{R}^+)(p + \lambda\frac{v}{\|v\|} \in \mathcal{B}(C)) .$$

Using that $\lambda$ we have:

$$\mu \in [0, 1[ \Rightarrow p + \mu\lambda\frac{v}{\|v\|} \in \mathcal{I}(C),$$
$$\mu = 1 \Rightarrow p + \mu\lambda\frac{v}{\|v\|} \in \mathcal{B}(C), \mu > 1 \Rightarrow p + \mu\lambda\frac{v}{\|v\|} \notin \overline{C} .$$

**Proof**
From proposition 572 (p. 316), $\lambda$ is the **upper_limit** of the set $\Lambda$ defined by

$$\lambda' \in \Lambda \Leftrightarrow p + \lambda'\frac{v}{\|v\|} \in \mathcal{I}(C) .$$

Then

$$\mu \in [0, 1[ \Rightarrow (\exists \lambda' \in [\mu\lambda, \lambda])(p + \lambda'\frac{v}{\|v\|} \in \mathcal{I}(C)) \Rightarrow p + \mu\lambda\frac{v}{\|v\|}$$
$$= (1 - \mu\frac{\lambda}{\lambda'})p + \mu\frac{\lambda}{\lambda'}(p + \lambda'\frac{v}{\|v\|}) .$$

Since $\mu\lambda < \lambda'$, $\mu\frac{\lambda}{\lambda'} \in [0, 1]$ and by definition 255 (p. 277) $p + \mu\lambda\frac{v}{\|v\|} \in \mathcal{I}(C)$. The same reasoning gives the last two properties of $\mu$.

**Proposition 574** In a vectorial space $E = \mathcal{R}^n$ of finite dimension $n$ built on real numbers with a Euclidean metric, we use the topology of the open balls. In $E$ we suppose that we have a convex subset $C \subset E$ of dimension $n$ inside a ball of $E$ centered on the origin. For any point $p \in \mathcal{I}(C)$ ($\mathcal{I}(C)$ is the interior of $C$). The map defined by the above proposition 572 (p. 316) between $\mathcal{I}(C) \times (\mathcal{R}^n - \{o\})$ and $\mathcal{R}^+$ such that:

$$p + \lambda(v, p)\frac{v}{\|v\|} \in \mathcal{B}(C)$$

is continuous for any nonzero vector ($v \neq o$) of $E$ and for any $p \in \mathcal{I}(C)$.

**Proof**
We have

$$p \in \mathcal{I}(C) \Rightarrow (\exists a \in ]0, \|v\|[)(B(p, 3a) \subset \mathcal{I}(C))$$

since $p + 2a\frac{v}{\|v\|} \in \mathcal{I}(C)$, $\lambda(v, p) > 2a$:

$$(\forall \alpha > 0)(\exists \Delta\lambda_i = \min(\frac{\alpha}{2}, \lambda(v, p) - a))(p + (\lambda(v, p) - \Delta\lambda_i)\frac{v}{\|v\|} \in \mathcal{I}(C),$$
$$\lambda(v, p) - \Delta\lambda_i \geq a > 0) .$$

Since $(\exists \mathrm{b} > 0)(\mathrm{C} \subset \mathrm{B}(0,\mathrm{b}))$, taking that b, we have $\|\mathrm{p}+3\mathrm{b}\frac{v}{\|v\|}\| \geq 3\mathrm{b} - \|\mathrm{p}\| \geq 2\mathrm{b}$ and we may write:

$$\mathrm{B}(\mathrm{p}+3\mathrm{b}\frac{v}{\|v\|},\mathrm{b}/2) \cap \mathrm{B}(0,\mathrm{b}) = \emptyset, \mathrm{B}(\mathrm{p}+3\mathrm{b}\frac{v}{\|v\|},\mathrm{b}/2) \cap \mathrm{C} = \emptyset\,.$$

From definition 137 (p. 156), $\mathrm{p}+3\mathrm{b}\frac{v}{\|v\|} \notin \overline{\mathrm{C}}$. From proposition 570 (p. 316) $2\mathrm{a} < \lambda(v,p) < 3\mathrm{b}$ and from proposition 572 (p. 316),

$$(\exists \Delta\lambda_{\mathrm{o}} = \min(\frac{\alpha}{2}, 3\mathrm{b} - \lambda(v,p)) > 0)(\mathrm{p} + (\lambda(v,p)+\Delta\lambda_{\mathrm{o}})\frac{v}{\|v\|} \notin \overline{\mathrm{C}})\,.$$

From proposition 219 (p. 158) and definition 135 (p. 154) $\mathcal{C}(\overline{\mathrm{C}})$ is open. from proposition 214 (p. 157) $\mathcal{I}(\mathrm{C})$ is open. Then from proposition 266 (p. 179)

$$(\exists \mathrm{r_i} > 0)(\mathrm{B}(\mathrm{p} + (\lambda(v,p) - \Delta\lambda_{\mathrm{i}})\frac{v}{\|v\|}, \mathrm{r_i}) \subset \mathcal{I}(\mathrm{C}))$$

$$(\exists \mathrm{r_o} > 0)(\mathrm{B}(\mathrm{p} + (\lambda(v,p) + \Delta\lambda_{\mathrm{o}})\frac{v}{\|v\|}, \mathrm{r_o}) \subset \mathcal{C}(\overline{\mathrm{C}}))$$

then setting

$$\epsilon_1 = 0.5\min(\frac{\mathrm{r_i}\|v\|}{\lambda(v,p)-\Delta\lambda_{\mathrm{i}}}, \frac{\mathrm{r_o}\|v\|}{\lambda(v,p)+\Delta\lambda_{\mathrm{o}}}, \mathrm{r_i}, \mathrm{r_o})\,.$$

We have for any $\Delta v \in \mathrm{E}$ and $\Delta p \in \mathrm{E}$ such that $\|\Delta v\| + \|\Delta p\| < \epsilon_1$

$$\begin{aligned}&\|\mathrm{p} + \Delta p + (\lambda(v,p) - \Delta\lambda_{\mathrm{i}})\frac{v+\Delta v}{\|v\|} - \mathrm{p} - (\lambda(v,p) - \Delta\lambda_{\mathrm{i}})\frac{v}{\|v\|}\| \\ &< \|\Delta p\| + \|(\lambda(v,p) - \Delta\lambda_{\mathrm{i}})\frac{\Delta v}{\|v\|}\| < \frac{\lambda(v,p)-\Delta\lambda_{\mathrm{i}}}{\|v\|}\epsilon_1 + \epsilon_1 \leq \mathrm{r_i}\end{aligned}$$

and

$$\begin{aligned}&\|\mathrm{p} + \Delta p + (\lambda(v,p) + \Delta\lambda_{\mathrm{o}})\frac{v+\Delta v}{\|v\|} - \mathrm{p} - (\lambda(v,p) + \Delta\lambda_{\mathrm{o}})\frac{v}{\|v\|}\| \\ &< \|\Delta p\| + \|(\lambda(v,p) + \Delta\lambda_{\mathrm{o}})\frac{\Delta v}{\|v\|}\| < \frac{\lambda(v,p)+\Delta\lambda_{\mathrm{o}}}{\|v\|}\epsilon_1 + \epsilon_1 \leq \mathrm{r_o}\,.\end{aligned}$$

Then

$$\begin{aligned}&\mathrm{p} + \Delta p + (\lambda(v,p) - \Delta\lambda_{\mathrm{i}})\frac{v+\Delta v}{\|v\|} \in \mathcal{I}(\mathrm{C}) \\ &p + \Delta p + (\lambda(v,p) + \Delta\lambda_{\mathrm{o}})\frac{v+\Delta v}{\|v\|} \in \mathcal{C}(\overline{\mathrm{C}})\,.\end{aligned}$$

Then from proposition 572 (p. 316),

$$(\lambda(v,p) - \Delta\lambda_{\mathrm{i}})\frac{\|v+\Delta v\|}{\|v\|} < \lambda(v+\Delta v, p+\Delta p) < (\lambda(v,p)+\Delta\lambda_{\mathrm{o}})\frac{\|v+\Delta v\|}{\|v\|}$$

and

$$\begin{aligned}&\|\Delta v\| + \|\Delta p\| < \epsilon_1 \Rightarrow |\lambda(v+\Delta v, p+\Delta p)\frac{\|v\|}{\|v+\Delta v\|} - \lambda(v,p)| \\ &< \max(\Delta\lambda_{\mathrm{i}}, \Delta\lambda_{\mathrm{o}}) \leq \frac{\alpha}{2}\,.\end{aligned}$$

Then since $\mathbf{0} < \lambda(v + \Delta v, p + \Delta p) < 3\mathrm{b}$, we have:

$$\|\Delta v\| + \|\Delta p\| < \epsilon_1 \Rightarrow |\lambda(v + \Delta v, p + \Delta p) - \lambda(v, p)| < \frac{\alpha}{2} + 3\mathrm{b}\frac{\|v\|}{\|v + \Delta v\|} .$$

For $v \neq 0$ $\frac{\|v\|}{\|v+\Delta v\|}$ is a continuous map in $\Delta v$ at $\Delta v = 0$ (propositions 516 (p. 288), 241 (p. 167), 286 (p. 185)). Then we have:

$$(\exists \epsilon_2 > 0)(\|\Delta v\| < \epsilon_2 \Rightarrow \frac{\|v\|}{\|v + \Delta v\|} < \frac{\alpha}{6\mathrm{b}}) .$$

Then taking $\epsilon = \min(\epsilon_1, \epsilon_2) > 0$, we have:

$$\|\Delta v\| + \|\Delta p\| < \epsilon \Rightarrow |\lambda(v + \Delta v, p + \Delta p) - \lambda(v, p)| < \alpha .$$

Then the map $(\lambda(v, p) : \mathcal{I}(\mathrm{C}) \times (\mathcal{R}^{\mathrm{n}} - \{\mathrm{o}\}) \rightarrow \mathcal{R}^+)$ is continuous for $v \neq \mathrm{o}$ and $p \in \mathcal{I}(\mathrm{C})$. From propositions 516 (p. 288), 241 (p. 167), 286 (p. 185), since $(\frac{v}{\|v\|} : \mathrm{E} \rightarrow \mathrm{E})$ is also continuous for $v \neq 0$, proposition 574 is true.

**Proposition 575** Given an open convex subset C of $\mathcal{R}^{\mathrm{n}}$ (definition 255 (p. 277)) not empty, considering the projection map $(\mathrm{proj_i} : \mathcal{R}^{\mathrm{n}} \rightarrow \mathcal{R}^{\mathrm{n-1}})$ (see definitions 121 (p. 136), 244 (p. 270)) along the coordinate $\mathrm{i} \in [1, \mathrm{n}]$, we have, calling $\mathcal{B}(\mathrm{C})$ the boundary of C (definition 138 (p. 157)):

$$\mathrm{proj_i}(\mathcal{B}(\mathrm{C})) = \overline{\mathrm{proj_i}(\mathrm{C})} = \mathrm{proj_i}(\overline{\mathrm{C}}) .$$

**Proof**

With the above notations, for any open O of $\mathcal{R}^{\mathrm{n-1}}$ containing $\mathrm{proj_i}(\mathrm{x})$ with $\mathrm{x} \in \mathcal{B}(\mathrm{C})$, there is an open ball $\mathrm{B^{n-1}}(\mathrm{proj_i}(\mathrm{x}), \mathrm{r}) \subset \mathrm{O} \subset \mathcal{R}^{\mathrm{n-1}}$ centered on $\mathrm{proj_i}(\mathrm{x})$ fully in O. From proposition 518 (p. 290), we have $\mathrm{proj_i}(\mathrm{B^n}(\mathrm{x}, \mathrm{r})) = \mathrm{B^{n-1}}(\mathrm{proj_i}(\mathrm{x}), \mathrm{r})$. From definitions 136 (p. 155) and 138 (p. 157), $\exists \mathrm{p} \in \mathrm{B^n}(\mathrm{x}, \mathrm{r}) \cap \mathrm{C}$ then we also have $\mathrm{proj_i}(\mathrm{p}) \in \mathrm{B^{n-1}}(\mathrm{proj_i}(\mathrm{x}), \mathrm{r}) \cap \mathrm{proj_i}(\mathrm{C})$ it means $\mathrm{proj_i}(\mathrm{p}) \in \overline{\mathrm{proj_i}(\mathrm{C})}$ then:

$$\mathrm{proj_i}(\mathcal{B}(\mathrm{C})) \subset \overline{\mathrm{proj_i}(\mathrm{C})} .$$

But, calling $\mathrm{e_i} = (\mathrm{v} : [1, \mathrm{n}] \rightarrow \mathcal{R}) = (\delta^{\mathrm{j}}_{\cdot}[1, \mathrm{n}] \rightarrow \mathcal{R})\mathrm{i}$ such that $\mathrm{v^j} = \delta^{\mathrm{j}}_{\mathrm{i}}$ (see definition 215 (p. 245)) and having $\mathrm{x} \in \mathrm{proj_i}(\mathrm{C})$, there is a $\mathrm{y} \in \mathrm{C}$ with $\mathrm{proj_i}(\mathrm{y}) = \mathrm{x}$. From proposition 572 (p. 316), we have:

$$(\exists \lambda \in \mathcal{R}^+)(\mathrm{y} + \lambda \mathrm{v} \in \mathcal{B}(\mathrm{C})) .$$

It gives since $\mathrm{proj_i}(\mathrm{v}) = 0$:

$$\mathrm{x} = \mathrm{proj_i}(\mathrm{y}) = \mathrm{proj_i}(\mathrm{y}) + \lambda \mathrm{proj_i}(\mathrm{v}) = \mathrm{proj_i}(\mathrm{y} + \lambda \mathrm{v}) .$$

Then we have:

$$\mathrm{proj_i}(\mathrm{C}) \subset \mathrm{proj_i}(\mathcal{B}(\mathrm{C})) .$$

Then

$$\mathrm{proj_i}(\overline{\mathrm{C}}) = \mathrm{proj_i}(\mathrm{C}) \cup \mathrm{proj_i}(\mathcal{B}(\mathrm{C})) = \mathrm{proj_i}(\mathcal{B}(\mathrm{C})) .$$

We also have (propositions 152 (p. 133) and 209 (p. 155)):

$$\mathrm{C} \subset \overline{\mathrm{C}}$$
$$\mathrm{proj_i}(\mathrm{C}) \subset \mathrm{proj_i}(\overline{\mathrm{C}})$$
$$\overline{\mathrm{proj_i}(\mathrm{C})} \subset \overline{\mathrm{proj_i}(\overline{\mathrm{C}})} .$$

Since $\overline{\mathbf{C}}$ is closed, $\mathbf{proj_i}(\overline{\mathbf{C}})$ is closed (proposition 530 (p. 294)), we have $\mathbf{proj_i}(\overline{\mathbf{C}}) = \overline{\mathbf{proj_i}(\overline{\mathbf{C}})}$ (proposition 220 (p. 158)). Combined with the above it gives:

$$\overline{\mathbf{proj_i}(\mathbf{C})} \subset \mathbf{proj_i}(\overline{\mathbf{C}}) = \mathbf{proj_i}(\mathcal{B}(\mathbf{C})) \; .$$

Since we also have $\mathbf{proj_i}(\mathcal{B}(\mathbf{C})) \subset \overline{\mathbf{proj_i}(\mathbf{C})}$, from proposition 754 (p. 451), we have:

$$\mathbf{proj_i}(\mathcal{B}(\mathbf{C})) = \overline{\mathbf{proj_i}(\mathbf{C})} \; .$$

**Definition 279** **Boxes in $\mathcal{R}^n$.**

Given two points of $\mathcal{R}^n$, $\mathbf{y_1} = (\mathbf{y_1^i} : [1,n] \to \mathcal{R}), \mathbf{y_2} = (\mathbf{y_2^i} : [1,n] \to \mathcal{R})$, one calls box $\mathbf{Bx}(\mathbf{y_1}, \mathbf{y_2})$ the following open subset of $\mathcal{R}^n$ (see definitions 378 (p. 484), 259 (p. 281) and 134 (p. 153)):

$$\mathbf{x} = (\mathbf{x^i} : [1,n] \to \mathcal{R}) \in \mathbf{Bx}(\mathbf{y_1}, \mathbf{y_2}) \Leftrightarrow (\forall \mathbf{i} \in [1,n])(\mathbf{x^i} \in ]\mathbf{y_1^i}, \mathbf{y_2^i}[) \; .$$

From definition 372 (p. 473) we have:

$$\mathbf{Bx}(\mathbf{y_1}, \mathbf{y_2}) = \prod_{\mathbf{i} \in [1,n]} ]\mathbf{y_1^i}, \mathbf{y_2^i}[ \; .$$

**Proposition 576** Boxes defined by definition 279 (p. 320) are open.

**Proof**

It is the product of open intervals. From definition 134 (p. 153), propositions 506 (p. 286) and 507 (p. 286), it is open.

**Proposition 577** The closure of an open box $\mathbf{Bx}(\mathbf{y_1}, \mathbf{y_2})$ is given by:

$$\mathbf{x} = (\mathbf{x^i} : [1,n] \to \mathcal{R}) \in \overline{\mathbf{Bx}}(\mathbf{y_1}, \mathbf{y_2}) \Leftrightarrow (\forall \mathbf{i} \in [1,n])(\mathbf{x^i} \in [\mathbf{y_1^i}, \mathbf{y_2^i}]) \; .$$

**Proof**

From definition 372 (p. 473), we have:

$$\overline{\mathbf{Bx}}(\mathbf{y_1}, \mathbf{y_2}) = \prod_{\mathbf{i} \in [1,n]} [\mathbf{y_1^i}, \mathbf{y_2^i}]$$

$\overline{\mathbf{Bx}}(\mathbf{y_1}, \mathbf{y_2})$ is the closure of the open $\mathbf{Bx}(\mathbf{y_1}, \mathbf{y_2})$ because (see definition 261 (p. 285) for $\mathbf{C}(\mathbf{x}, \mathbf{c})$):

$$\mathbf{x} \in \mathcal{R}^n, \mathbf{c} > 0, \mathbf{z} \in \mathbf{C}(\mathbf{x}, \mathbf{c}) \cap \prod_{\mathbf{i} \in [1,n]} ]\mathbf{y_1^i}, \mathbf{y_2^i}[ \Rightarrow (\forall \mathbf{i} \in [1,n])(\mathbf{z^i} \in ]\mathbf{y_1^i}, \mathbf{y_2^i}[) \; .$$

Since:

$$\mathbf{x} \in \overline{\mathbf{Bx}}(\mathbf{y_1}, \mathbf{y_2}) \Rightarrow (\forall \mathbf{i} \in [1,n])(\forall \mathbf{c} > 0)$$
$$(\exists \mathbf{z^i} \in ]\mathbf{x^i} - \mathbf{c}, \mathbf{x^i} + \mathbf{c}[ \cap ]\mathbf{y_1^i}, \mathbf{y_2^i}[) \Rightarrow \mathbf{x^i} \in [\mathbf{y_1^i}, \mathbf{y_2^i}]$$

$\overline{\mathbf{Bx}}(\mathbf{y_1}, \mathbf{y_2}) \subset \prod_{\mathbf{i} \in [1,n]} [\mathbf{y_1^i}, \mathbf{y_2^i}]$. We also have:

$$\mathbf{x} \in \prod_{\mathbf{i} \in [1,n]} [\mathbf{y_1^i}, \mathbf{y_2^i}] \Rightarrow (\forall \mathbf{c} > 0)(\exists \mathbf{z^i} \in ]\mathbf{y_1^i}, \mathbf{y_2^i}[ \cap ]\mathbf{x^i} - \mathbf{c}, \mathbf{x^i} + \mathbf{c}[) \Rightarrow$$
$$(\exists \mathbf{z} = (\mathbf{z^i} : [1,n] \to \mathcal{R})(\mathbf{z} \in \mathbf{Bx}(\mathbf{y_1}, \mathbf{y_2}) \cap \mathbf{C}(\mathbf{x}, \mathbf{c})) \; .$$

Then $\prod_{\mathbf{i} \in [1,n]} [\mathbf{y_1^i}, \mathbf{y_2^i}] \subset \overline{\mathbf{Bx}}(\mathbf{y_1}, \mathbf{y_2})$.
From proposition 754 (p. 451), $\overline{\mathbf{Bx}}(\mathbf{y_1}, \mathbf{y_2}) = \prod_{\mathbf{i} \in [1,n]} [\mathbf{y_1^i}, \mathbf{y_2^i}]$.

**Definition 280 Hybrid box.**

For integration purposes (see section 2.23 (p. 331)) we need to define a hybrid box in $\mathcal{R}^n$. Let us call $I \subset \mathcal{P}(\mathcal{R})$, the set of intervals of $\mathcal{R}$ (definition 378 (p. 484)). A hybrid box is defined by a map $(I_j : [1, n] \to I)$ with:

$$Hbx(I) = \prod_{j\in[1,n]} I_j \Leftrightarrow (x \in Hbx(I) \Leftrightarrow (\forall j \in [1, n])(x^j \in I_j)) .$$

In integration theory, it does not matter if the intervals of I are open, semi open or closed. Then in that case it is enough to specify the end points: Calling:

$$x = \prod_{j\in[1,n]}^{e} lower\_limit(I_j), y = \prod_{j\in[1,n]}^{e} upper\_limit(I_j) .$$

We may write $Hbx(x, y)$ for one of the hybrid boxes for which:

$$i \in [1, n] \Rightarrow I_i = [x_i, y_i] \vee I_i = [x_i, y_i[ \vee I_i = ]x_i, y_i] \vee I_i = ]x_i, y_i[ .$$

**Proposition 578** The union or the intersection of a finite number of hybrid boxes are the union of a finite number of disjoint hybrid boxes.

**Proof**

- $n = 1$

  It is true for one interval of $\mathcal{R}$. We suppose that it is true for p intervals (or hybrid box). Have $p + 1$: $(I_k : [1, p + 1] \to I)$, we take the p first of them: $(I_k : [1, p] \to I)$. We suppose that any combination of them with intersection or union gives q disjoint intervals $(J_k : [1, q] \to I)$, the union of them $\bigcup_{j\in[1,d]} J_j$ being equal to the combination of union or intersection of the p intervals. Considering now $I_{p+1}$, we have three possibilities:

  1. $I_{p+1} \cap \bigcup_{j\in[1,q]} J_j = I_{p+1}$.
     It means $I_{p+1} \subset \bigcup_{j\in[1,q]} J_j$. In that case proposition 578 (p. 321) is trivially true for $p + 1$ intervals since $\bigcup_{j\in[1,q]} J_j$ covers the union or intersection of the $p + 1$ intervals.
  2. $I_{p+1} \cap \bigcup_{j\in[1,q]} J_j = \emptyset$.
     In that case proposition 578 (p. 321) is also trivially true for $p + 1$ intervals since adding the last interval to the $\bigcup_{j\in[1,q]} J_j$ we have:

     $$\bigcup_{j\in[1,q]} J_j \cup I_{p+1} = \bigcup_{i\in[1,p]} I_i \cup I_{p+1}$$

     $$\emptyset = ]0, 0[ = \bigcap_{i\in[1,p]} I_i \cap I_{p+1} .$$

  3. Both cases 1, 2 are excluded.
     There is then a $k \in [1, p]$, such that $I_k \cap I_{p+1} \neq \emptyset$ in that case the union or the intersection with $I_{p+1}$ is an interval (proposition 840 (p. 484)). We set $(i \in [1, p] - k \Rightarrow H_i = I_i, H_k = I_k \cup I_{p+1})$ (versus $(i \in [1, p] - k \Rightarrow H_i = I_i, H_k = I_k \cap I_{p+1})$) it gives $\bigcup_{i\in[1,p+1]} I_i = \bigcup_{i\in[1,p]} H_i$ (versus $\bigcap_{i\in[1,p+1]} I_i = \bigcap_{i\in[1,p]} H_i$)
     Since now we have the union (versus intersection) of intersection of p intervals from what we suppose it is equal to the union of a finite number of disjoints intervals. Proposition 578 (p. 321) is true.

Then proposition 578 (p. 321) is true in any case for $\mathbf{p+1}$ we conclude using proposition 927 (p. 520).

- Any $\mathbf{n}$.
  Using proposition 832 (p. 478) and proposition 578 (p. 321) for $\mathbf{n} = \mathbf{1}$, let us show that it is also true for $\mathcal{R}^{\mathbf{n}}$: One may define a set of $\mathbf{p}$ hybrid boxes by a map $(\mathbf{Hbx^i} : [\mathbf{1}, \mathbf{p}] \to \mathcal{I}^{\mathbf{n}})$ with $\mathcal{I} \subset \mathcal{P}(\mathcal{R})$ the set of intervals of $\mathcal{R}$. Using proposition 832 (p. 478), we have from proposition 578 (p. 321) with $\mathbf{n} = \mathbf{1}$ setting $\mathbf{Hbx^i} =< \prod_{\mathbf{j}\in[1,n]} \mathbf{Hbx^i_j}$ :

$$(\exists \mathbf{q}_\ell : [1, n] \to \mathcal{Z}^+)(\forall i \in [1, n])(\exists (\mathbf{X}^i_\ell : [1, q_i] \to \mathcal{I}))$$
$$(\forall \ell \in [1, q_i])(\forall \ell' \in [1, q_i])(\mathbf{X}^i_\ell \cap \mathbf{X}^i_{\ell'} = \emptyset, \bigcup_{\ell\in[1,q_i]} \mathbf{X}^i_\ell = \bigcup_{j\in[1,p]} \mathbf{Hbx}^i_j)$$
$$(\exists \mathbf{q}_\ell : [1, n] \to \mathcal{Z}^+)(\forall i \in [1, n])(\exists (\mathbf{X}^i_\ell : [1, q_i] \to \mathcal{I}))$$
$$(\forall \ell \in [1, q_i])(\forall \ell' \in [1, q_i])((\mathbf{X}^i_\ell \cap \mathbf{X}^i_{\ell'} = \emptyset), \bigcup_{\ell\in[1,q_i]} \mathbf{X}^i_\ell = \bigcap_{j\in[1,p]} \mathbf{Hbx}^i_j) .$$

  Then using those $\mathbf{X}^i_\ell$ (separately for the union of $\mathbf{Hbx}$ and intersection of $\mathbf{Hbx}$), applying proposition 832 (p. 478), we have:

$$\bigcup_{i\in[1,p]} \prod_{j\in[1,n]} \mathbf{Hbx}^i_j = \prod_{j\in[1,n]} \bigcup_{i\in[1,p]} \mathbf{Hbx}^i_j = \prod_{j\in[1,n]} \bigcup_{\ell\in[1,q_i]} \mathbf{X}^i_\ell = \bigcup_{f\in\prod_{i\in[1,n]}[1,q_i]} \prod_{\lambda\in[1,n]} \mathbf{X}^\lambda_{f(\lambda)}$$
$$\bigcap_{i\in[1,p]} \prod_{j\in[1,n]} \mathbf{Hbx}^i_j = \prod_{j\in[1,n]} \bigcap_{i\in[1,p]} \mathbf{Hbx}^i_j = \prod_{j\in[1,n]} \bigcup_{\ell\in[1,q_i]} \mathbf{X}^i_\ell = \bigcup_{f\in\prod_{i\in[1,n]}[1,q_i]} \prod_{\lambda\in[1,n]} \mathbf{X}^\lambda_{f(\lambda)} .$$

  From proposition 829 (p. 476) and proposition 827 (p. 475), we have:

$$(f \in \prod_{i\in[1,n]} [1, q_i], f' \in \prod_{i\in[1,n]} [1, q_i], f \neq f') \Rightarrow \prod_{\lambda\in[1,n]} \mathbf{X}^\lambda_{f(\lambda)} \cap \prod_{\lambda\in[1,n]} \mathbf{X}^\lambda_{f'(\lambda)}$$
$$= \prod_{\lambda\in[1,n]} (\mathbf{X}^\lambda_{f(\lambda)} \cap \mathbf{X}^\lambda_{f'(\lambda)}) = \emptyset .$$

Then proposition 578 (p. 321) is true for $\mathbf{n} \in \mathcal{Z}^+$.

**Definition 281 Corners or summits of a box.**

One calls corner or summit of a $\mathcal{R}^{\mathbf{n}}$ closed box $\overline{\mathbf{Bx}}(\mathbf{y_1}, \mathbf{y_2})$, $\mathbf{y_1} = (\mathbf{y^i_1} : [1, n] \to \mathcal{R}), \mathbf{y_2} = (\mathbf{y^i_2} : [1, n] \to \mathcal{R})$ (definition 279 (p. 320)), a point of $\overline{\mathbf{Bx}}(\mathbf{y_1}, \mathbf{y_2})$ which is at the limit of any interval of definition: If we call $\mathbf{S}$ the set of corners or summits of the box $\overline{\mathbf{Bx}}(\mathbf{y_1}, \mathbf{y_2})$, it is defined by:

$$x = (x^i : [1, n] \to \mathcal{R}) \in S \Leftrightarrow (\forall i \in [1, n])(x^i = y^i_1 \vee x^i = y^i_2) .$$

There are $\mathbf{2^n}$ summits or corners since each of them is defined by a map:$[\mathbf{1}, \mathbf{n}] \to \{\mathbf{1}, \mathbf{2}\}$.

**Definition 282 Ridge of a box.**

One calls ridge of $\overline{\mathbf{Bx}}(\mathbf{y_1}, \mathbf{y_2})$, $\mathbf{y_1} = (\mathbf{y^i_1} : [1, n] \to \mathcal{R}), \mathbf{y_2} = (\mathbf{y^i_2} : [1, n] \to \mathcal{R})$ parallel to a direction $\mathbf{i_{rd}} \in [\mathbf{1}, \mathbf{n}]$ and associated to a map $(\mathbf{j(i)} : [1, n] - \mathbf{i_{rd}} \to \{1, 2\})$ a subset $\mathbf{rd}$ of $\overline{\mathbf{Bx}}(\mathbf{y_1}, \mathbf{y_2})$ such that:

$$x \in rd \Leftrightarrow (x = (x^i : [1, n] \to \mathcal{R}) \in \mathcal{R}^n,$$
$$(\forall i \in [1, n] - \{i_{rd}\})(x^i = y^i_{j(i)}), x^{i_{rd}} \in [y^{i_{rd}}_1, y^{i_{rd}}_2]) .$$

One says also that $\mathbf{rd}$ is parallel to the vector $\mathbf{e_{i_{rd}}}$. There are $\mathbf{2^{n-1}}$ ridges for any $\mathbf{i_{rd}}$ since any ridge is defined by a map $[\mathbf{1},\mathbf{n}] - \{\mathbf{i_{rd}}\} \to \{\mathbf{1},\mathbf{2}\}$. The total number of ridges is then $\mathbf{n} \times \mathbf{2^{n-1}}$. There are also $\mathbf{n}$ ridges per summit (one for each $\mathbf{i_{rd}} \in [\mathbf{1},\mathbf{n}]$) and two summits per ridge.

**Proposition 579** At the end of any ridge $\mathbf{rd}$ defined by $\mathbf{i_{rd}} \in [\mathbf{1},\mathbf{n}]$ and a map $(\mathbf{j(k)} : [\mathbf{1},\mathbf{n}] - \mathbf{i_{rd}} \to \{\mathbf{1},\mathbf{2}\})$ there are two summits $\mathbf{s_{rd}(1)} \in \mathcal{R}^{\mathbf{n}}, \mathbf{s_{rd}(2)} \in \mathcal{R}^{\mathbf{n}}$ at each end of the ridge:

$$\mathbf{i} \in \{\mathbf{1},\mathbf{2}\} \Rightarrow ((\forall \ell \in [\mathbf{1},\mathbf{n}] - \mathbf{i_{rd}})(\mathbf{s^{\ell}_{rd}(i)} = \mathbf{y^{\ell}_{j(\ell)}}), \mathbf{s^{i_{rd}}_{rd}(i)} = \mathbf{y^{i_{rd}}_{i}}) \ .$$

**Proof**
It comes directly from definitions.

**Proposition 580** Given a box $\mathbf{Bx(y_1,y_2)}$, calling $\mathbf{S}$ its set of summits, given a partition of $\mathbf{S}$ defined by a surjective map (definitions 123 (p. 139), 370 (p. 471), proposition 815 (p. 471)) $(\mathbf{j(k)} : \mathbf{S} \to \{\mathbf{1},\mathbf{2}\})$:

$$\mathbf{j^{-1}(1)} \cup \mathbf{j^{-1}(2)} = \mathbf{S}, \mathbf{j^{-1}(1)} \cap \mathbf{j^{-1}(2)} = \emptyset \ .$$

There is an element $\mathbf{s} \in \mathbf{j^{-1}(1)} \times \mathbf{j^{-1}(2)}$ such that $\mathbf{s_1} \in \mathbf{j^{-1}(1)}, \mathbf{s_2} \in \mathbf{j^{-1}(2)}$ are the two ends of a ridge:

$$(\exists \mathbf{s} \in \mathbf{j^{-1}(1)} \times \mathbf{j^{-1}(2)})(\exists \mathbf{i_{rd}} \in [\mathbf{1},\mathbf{n}])(\ell \in [\mathbf{1},\mathbf{n}] - \mathbf{i_{rd}} \Rightarrow \mathbf{s^{\ell}_1} = \mathbf{s^{\ell}_2},$$
$$(\mathbf{s^{i_{rd}}_1} = \mathbf{y^{i_{rd}}_1}, \mathbf{s^{i_{rd}}_2} = \mathbf{y^{i_{rd}}_2} \vee \mathbf{s^{i_{rd}}_1} = \mathbf{y^{i_{rd}}_2}, \mathbf{s^{i_{rd}}_2} = \mathbf{y^{i_{rd}}_1})) \ .$$

**Proof**
As usual, we call $(\mathbf{e_i} = \delta^{\mathbf{j}}_{\mathbf{i}} : [\mathbf{1},\mathbf{n}] \to \mathcal{R}^{\mathbf{n}})$ the reference basis of $\mathcal{R}^{\mathbf{n}}$ (definition 242 (p. 267)). We shall use the Einstein notation (see footnote 20 (p. 239)) each time this basis is mentioned and we keep the above notations. Since the map $(\mathbf{j} : \mathbf{S} \to \{\mathbf{1},\mathbf{2}\})$ is surjective, we have a summit in both $\mathbf{j^{-1}}$. From definition 281 (p. 322), we have two different maps $(\kappa_1(\mathbf{i}) : [\mathbf{1},\mathbf{n}] \to \{\mathbf{1},\mathbf{2}\})$, $(\kappa_2(\mathbf{i}) : [\mathbf{1},\mathbf{n}] \to \{\mathbf{1},\mathbf{2}\})$. Setting $\mathbf{s_1} \in \mathbf{J^{-1}(1)}$, $\mathbf{s_2} \in \mathbf{J^{-1}(2)}$, we consider the operation $\mathbf{Tr}_{\iota}, \iota \in [\mathbf{1},\mathbf{n}]$ on any summit $\mathbf{s} \in \mathbf{S}$, $(\mathbf{Tr}^{\iota}(\mathbf{s}) : \mathbf{S} \to \mathbf{S})$ defined as follows:

$$\iota \in [\mathbf{1},\mathbf{n}] \Rightarrow ((\forall \mathbf{k} \in [\mathbf{1},\mathbf{n}] - \iota)(\mathbf{Tr^k_{\iota}(s)} = \mathbf{s^k}), \mathbf{Tr^{\iota}_{\iota}(s)} = \mathbf{y^{\iota}_{\kappa_2(\iota)}}) \ .$$

We clearly have:

$$\overset{\circ}{\prod_{\mathbf{j} \in [\mathbf{1},\mathbf{n}]}} \mathbf{Tr_j(s)} = \mathbf{y^j_{\kappa_2(j)} e_j} \in \mathbf{j^{-1}(2)} \ .$$

We call $\iota_{\mathbf{m}}$ the biggest element of $[\mathbf{1},\mathbf{n}]$ for which we have [29]:

$$\iota \in [\mathbf{0},\iota_{\mathbf{m}}] \Rightarrow \overset{\circ}{\prod_{\mathbf{k} \in [\mathbf{1},\iota]}} \mathbf{Tr_k y^{\ell}_{\kappa_1(\ell)} e_{\ell}} \in \mathbf{j^{-1}(1)} \ .$$

We have $\iota_{\mathbf{m}} < \mathbf{n}$ and:

$$\overset{\circ}{\prod_{\mathbf{k} \in [\mathbf{1},\iota_{\mathbf{m}}]}} \mathbf{Tr_k y^{\ell}_{\kappa_1(\iota)} e_{\ell}} \in \mathbf{j^{-1}(1)}, \quad \prod_{\mathbf{k} \in [\mathbf{1},\iota_{\mathbf{m}}+\mathbf{1}]} \mathbf{Tr_k y^{\ell}_{\kappa_1(\iota)} e_{\ell}} \in \mathbf{j^{-1}(2)}$$

[29] When there is no k available in the above product of transformations, it means that there is no transformation: $\iota = \mathbf{0} \Rightarrow \overset{\circ}{\prod}_{\mathbf{k} \in [\mathbf{1},\iota]} \mathbf{Tr_k y^{\ell}_{\kappa_1(\ell)} e_{\ell}} = \mathbf{y^{\ell}_{\kappa_1(\ell)} e_{\ell}}$.

which defines two summits at the end of a ridge parallel to $\mathbf{e}_{\iota_m+1}$ and defined by the two maps (as above $\iota_m = 0$ gives $(\forall i \in [1,n])(\kappa(i) = \kappa_1(i))$:

$$\iota \in [1,\iota_m] \Rightarrow \kappa(\iota) = \kappa_2(\iota), \iota \in [\iota_m + 1, n] \Rightarrow \kappa(\iota) = \kappa_1(\iota)$$

$$\iota \in [1,n] - (\iota_m + 1) \Rightarrow \kappa'(\iota) = \kappa(\iota), \iota = \iota_m + 1 \Rightarrow \kappa'(\iota) = \kappa_2(\iota)\ .$$

It gives two summits at the end of a ridge:

$$y^{\iota}_{\kappa(\iota)} e_{\iota} \in j^{-1}(1), y^{\iota}_{\kappa'(\iota)} e_{\iota} \in j^{-1}(2)\ .$$

**Definition 283 Face of a box.**

One calls face of $\overline{Bx}(y_1,y_2)$, $y_1 = (y^i_1 : [1,n] \to \mathcal{R}), y_2 = (y^i_2 : [1,n] \to \mathcal{R})$ a subset $F$ of $\overline{Bx}(y_1,y_2)$ associated to an $i_f \in [1,n]$ and an $i_\ell \in [1,2]$ such as:

$$x \in F \Leftrightarrow (x = (x^i : [1,n] \to \mathcal{R}), x \in \overline{Bx}(y_1,y_2), x^{i_f} = y^{i_f}_{i_\ell})\ .$$

From that definition, there are $2n$ faces. Any summit is contained in $n$ faces. Any face contains $2^{n-1}$ summits. The two faces corresponding to the same $i_f$ are said to be parallel. They do not have any point in common.

**Proposition 581** $e_i = \prod^e_{j\in[1,n]} \delta^j_i = (\delta^j_i : [1,n] \to \mathcal{R}) \in \mathcal{R}^n$ being the reference basis of $\mathcal{R}^n$ (see definition 242 (p. 267) equation (2.106) (p. 267) in definition 242 (p. 267)), given a box $\overline{Bx}(y_1,y_2)$, $y_1 = (y^i_1 : [1,n] \to \mathcal{R}), y_2 = (y^i_2 : [1,n] \to \mathcal{R})$, the map $(z(x) = x - y_1 : \overline{Bx}(y_1,y_2) \to V_b)$ is a bijection between the inner part $V_b$ of the box generated by the basis $V = v_i = (y^i_2 - y^i_1)e_i$ and $\overline{Bx}(y_1,y_2)$.

**Proof**

It is a direct consequence of definitions 257 (p. 281), 259 (p. 281) and proposition 502 (p. 282).

**Proposition 582** Calling $S$ the set of summits of the box $\overline{Bx}(y_1,y_2)$ (definition 281 (p. 322)), considering any element of $S$ as a vector of $\mathcal{R}^n$ we have:

$$((m_s : S \to \mathcal{R}^{+0}), \sum_{s\in S} m_s = 1) \Rightarrow \sum_{s\in S} m_s s \in \overline{Bx}(y_1,y_2)$$

and

$$x \in \overline{Bx}(y_1,y_2) \Rightarrow (\exists (m_s : S \to \mathcal{R}^{+0}))(\sum_{s\in S} m_s = 1, x = \sum_{s\in S} m_s s)\ .$$

**Proof**

It is a direct consequence of propositions 581 (p. 324) and 502 (p. 282).

**Proposition 583** The boxes $Bx(y_1,y_2)$ and $\overline{Bx}(y_1,y_2)$ are convex sets.

**Proof**

It is a direct consequence of propositions 500 (p. 281) and 582 (p. 324).

**Proposition 584** The boundary of an open box $Bx(y_1,y_2)$ is given by:

$$\mathcal{B}(\overline{Bx}(y_1,y_2)) = \mathcal{B}(Bx(y_1,y_2)) = \overline{Bx}(y_1,y_2) - Bx(y_1,y_2) = \mathcal{C}_{\overline{Bx}(y_1,y_2)}(Bx(y_1,y_2))$$
$$= set_x\{x \in \mathcal{R}^n, (\exists i \in [1,n])(\exists j \in [1,2])(x^i = y^i_j), (\forall i \in [1,n])(x^i \in [y^i_1, y^i_2])\}\ .$$

**Proof**
Using proposition 577 (p. 320) and definition 279 (p. 320), we have:

$$\begin{aligned}
&\mathbf{x} \in \overline{\mathbf{Bx}}(\mathbf{y_1}, \mathbf{y_2}) - \mathbf{Bx}(\mathbf{y_1}, \mathbf{y_2}) \\
&\Leftrightarrow (\mathbf{x} \in \mathcal{R}^n, (\forall i \in [1, n])(x^i \in [y_1^i, y_2^i]), (\exists i \in [1, n])(x^i \notin ]y_1^i, y_2^i[)) \\
&\Leftrightarrow ((\forall i \in [1, n])(x^i \in [y_1^i, y_2^i]), (\exists i \in [1, n])(x^i \notin ]y_1^i, y_2^i[, x^i \in [y_1^i, y_2^i])) \\
&\Leftrightarrow (\mathbf{x} \in \mathcal{R}^n, (\exists i \in [1, n])(\exists j \in [1, 2])(x^i = y_j^i), (\forall i \in [1, n])(x^i \in [y_1^i, y_2^i])) .
\end{aligned}$$

**Proposition 585** Given a box $\mathbf{Bx}(\mathbf{y_1}, \mathbf{y_2})$ of $\mathcal{R}^n$ a convex subset $\mathbf{C}$, if we have $\overline{\mathbf{Bx}(\mathbf{y_1}, \mathbf{y_2})} \cap \mathbf{C} \neq \emptyset$ and

$$(\forall i \in [1, n])(\mathbf{proj_i}(\overline{\mathbf{Bx}(\mathbf{y_1}, \mathbf{y_2})})) \subset \mathbf{proj_i}(\mathbf{C}) .$$

There is a ridge of $\mathbf{Bx}(\mathbf{y_1}, \mathbf{y_2})$ with a point in $\mathbf{C}$.

**Proof**
As usual, we call $(\mathbf{e_i} = \delta_i^j : [1, n] \to \mathcal{R}^n)$, the reference basis of $\mathcal{R}^n$ (definition 242 (p. 267)). Proposition 585 (p. 325) is trivially true for $\mathbf{n} = \mathbf{1}$. Let us suppose that it is true for $\mathbf{n} = \mathbf{n'}$ and let us show that it is then true for $\mathbf{n} = \mathbf{n'} + \mathbf{1}$.
Having $\mathbf{n} = \mathbf{n'} + \mathbf{1}$, taking a ridge $\mathbf{rd}$ of $\mathcal{B}(\mathbf{Bx}(\mathbf{y_1}, \mathbf{y_2}))$. From what we suppose there is a point $\mathbf{z}$ of $\mathbf{C}$ on the line defined by $\mathbf{rd}$ (see definitions 211 (p. 242), 244 (p. 270)). If $\mathbf{z} \in \mathbf{rd}$ proposition 585 (p. 325) is true. Let us suppose $\mathbf{z} \notin \mathbf{rd}$. From the definition of a ridge 282 (p. 322) and proposition 584 (p. 324) $\mathbf{z}$ is not in $\mathcal{B}(\mathbf{Bx}(\mathbf{y_1}, \mathbf{y_2}))$ Since there is a point $\mathbf{x_c} \in \overline{\mathbf{Bx}(\mathbf{y_1}, \mathbf{y_2})} \cap \mathbf{C}$, because $\overline{\mathbf{Bx}(\mathbf{y_1}, \mathbf{y_2})}$ is convex (see proposition 583 (p. 324)), we have from proposition 572 (p. 316)

$$(\exists m \in [0, 1])(m\mathbf{x_c} + (1 - m)\mathbf{z} \in \mathcal{B}(\mathbf{Bx}(\mathbf{y_1}, \mathbf{y_2})) \cap \mathbf{C} = \mathcal{B}(\overline{\mathbf{Bx}(\mathbf{y_1}, \mathbf{y_2})}) \cap \mathbf{C}) .$$

Calling $\mathbf{z_m}$ that point. Since it is on $\mathcal{B}(\overline{\mathbf{Bx}(\mathbf{y_1}, \mathbf{y_2})}$ (proposition we have $\exists \mathbf{i_d} \in [1, n], \mathbf{i_s} \in \{1, 2\}$ with $z_m^{i_d} = y_{i_s}^{i_d}$. It is a point of the corresponding face. The face is in subvectorial space $\mathbf{V^{n-1}}$ which is equivalent to $\mathcal{R}^{n-1}$ because it has for basic vector:

$$(\mathbf{f_j} : [1, n] - \mathbf{i_d} \to \mathcal{R}^n), j \in [1, n] - \mathbf{i_d} \Rightarrow \mathbf{f_j} = \mathbf{e_j} .$$

The considered face is a box $\mathbf{Bx^{n-1}} = \overline{\mathbf{Bx}(\mathbf{y_1}, \mathbf{y_2})} \cap \mathbf{V^{n-1}}$ in that space all the ridges of that box are also ridge of $\overline{\mathbf{Bx}(\mathbf{y_1}, \mathbf{y_2})}$ (definition 279 (p. 320)). The space $\mathbf{C^{n-1}} = \mathbf{C} \cap \mathbf{V^{n-1}}$ is also convex (see propositions 490 (p. 277) and 494 (p. 278)). $\mathbf{Bx^{n-1}}$, $\mathbf{C^{n-1}}$ verify condition of proposition 585 (p. 325). Since proposition 585 (p. 325) is supposed to be true for $\mathcal{R}^{n-1} = \mathcal{R}^{n'}$, there is a ridge of $\mathbf{Bx^{n-1}}$ with a point of $\mathbf{C^{n-1}}$ then of $\mathbf{C}$. Then we have a ridge of $\overline{\mathbf{Bx}(\mathbf{y_1}, \mathbf{y_2})}$ which contains a point of $\mathbf{C}$. Proposition 585 (p. 325) is true for $\mathbf{n}$. Since it is trivially true for $\mathbf{n} = \mathbf{1}$, it is true for all $\mathbf{n}$ applying proposition 927 (p. 520), the recurrence principle.

**Proposition 586** Given a box $\mathbf{Bx}(\mathbf{y_1}, \mathbf{y_2})$ of $\mathcal{R}^n$ a convex subset $\mathbf{C}$, if we have $\overline{\mathbf{Bx}(\mathbf{y_1}, \mathbf{y_2})} \cap \mathbf{C} \neq \emptyset, \overline{\mathbf{Bx}(\mathbf{y_1}, \mathbf{y_2})} \cap \mathcal{C}(\mathbf{C}) \neq \emptyset$ and

$$(\forall i \in [1, n])(\mathbf{proj_i}(\overline{\mathbf{Bx}(\mathbf{y_1}, \mathbf{y_2})})) \subset \mathbf{proj_i}(\mathbf{C}) .$$

There is a ridge of $\overline{\mathbf{Bx}(\mathbf{y_1}, \mathbf{y_2})}$ with both a point in $\mathbf{C}$ and a point in $\mathcal{C}(\mathbf{C})$.

**Proof**
If proposition 586 (p. 325) was not true, all ridges which have a point in $\mathbf{C}$ would be fully in $\mathbf{C}$. It means that the summits attached to those ridges are also in $\mathbf{C}$. They cannot be all in $\mathbf{C}$ because in that case there would not be a point of $\overline{\mathbf{Bx(y_1,y_2)}}$ not in $\mathbf{C}$ (propositions 582 (p. 324), 492 (p. 278)). If there were all out of $\mathbf{C}$, from proposition 585 (p. 325), proposition 586 (p. 325) is true. If there are summits of $\overline{\mathbf{Bx(y_1,y_2)}}$ in $\mathbf{C}$ and not in $\mathbf{C}$, we have a partition of the set of summits according to the fact that there are or there are not in $\mathbf{C}$. From proposition 580 (p. 323), there is a ridge with a summit in $\mathbf{C}$ and the other one not in $\mathbf{C}$. It contradicts our hypotheses. Then from A.1.4 -8 (p. 435), proposition 586 (p. 325) is true.

## 2.22 Path connected sets

**Definition 284 Path connection relation.**

Given a topological set $\mathbf{E}$, two points $\mathbf{a}$, $\mathbf{b}$ of $\mathbf{E}$ are path connected if there is a continuous map (definition 141 (p. 165)) $(\mathbf{f}:[0,1]\subset\mathcal{R}\rightarrow\mathbf{E})$ such that $\mathbf{a}=\mathbf{f}(0)$, $\mathbf{b}=\mathbf{f}(1)$. From proposition 236 (p. 165), it is enough to say that $(\mathbf{f}:[0,1]\subset\mathcal{R}\rightarrow\mathbf{f}([0,1])\subset\mathbf{E})$ is continuous.

**Proposition 587** The path connection relation is an equivalence relation (definition 336 (p. 445)).
**Proof**

1. **Any point is connected to itself.**
For any $\mathbf{a}\in\mathbf{E}$ $(\mathbf{f(x)}=\mathbf{a}:[0,1]\rightarrow\mathbf{E})$ is continuous (proposition 234 (p. 165)).

2. **The path connection is symmetric.**
If $\mathbf{a}$ is path connected to $\mathbf{b}$, $\mathbf{b}$ is path connected to $\mathbf{a}$: If $(\mathbf{f}:[0,1]\rightarrow\mathbf{E})$ with $\mathbf{f}(0)=\mathbf{a}$, $\mathbf{f}(1)=\mathbf{b}$, $\mathbf{f}$ continuous, $(\mathbf{g(x)}=\mathbf{f}(1-\mathbf{x}):[0,1]\rightarrow\mathbf{E})$ is also continuous (propositions 241 (p. 167), 234 (p. 165), 233 (p. 165) and 286 (p. 185)) with $\mathbf{g}(0)=\mathbf{f}(1)=\mathbf{b}$, $\mathbf{g}(1)=\mathbf{f}(0)=\mathbf{a}$.

3. **The path connection is transitive.**
Let us suppose that in a topological set $\mathbf{E}$, $\mathbf{a}$, $\mathbf{b}$ are path connected as well as $\mathbf{b}$, $\mathbf{c}$. We have two continuous maps $(\mathbf{f_{ab}}:[0,1]\rightarrow\mathbf{E})$ with $\mathbf{f_{ab}}(0)=\mathbf{a}$, $\mathbf{f_{ab}}(1)=\mathbf{b}$ and $(\mathbf{f_{bc}}:[0,1]\rightarrow\mathbf{E})$ with $\mathbf{f_{bc}}(0)=\mathbf{b}$, $\mathbf{f_{bc}}(1)=\mathbf{c}$. We consider the map $(\mathbf{h}:[0,1]\rightarrow\mathbf{E})$ defined as follows:

$$\mathbf{x}\in[0,0.5]\Rightarrow\mathbf{h(x)}=\mathbf{h_{ab}(x)}=\mathbf{f_{ab}(2x)},$$
$$\mathbf{x}\in[0.5,1]\Rightarrow\mathbf{h(x)}=\mathbf{h_{bc}(x)}=\mathbf{f_{bc}(2x-1)}\ .$$

Since $\mathbf{x}=0.5\Rightarrow\mathbf{f_{ab}(2x)}=\mathbf{f_{bc}(2x-1)}=\mathbf{b}$, $\mathbf{h(x)}$ exists as defined. Given an open $\mathbf{O}$ of $\mathbf{E}$, for any point $\mathbf{x}\in\mathbf{h}^{-1}(\mathbf{O})\subset\mathbf{h}^{-1}(\mathbf{E})=[0,1]$, we have:

   - **We suppose $\mathbf{x}\in[0,0.5[$.**
Since $\mathbf{2x}\in[0,1]\in\mathbf{f_{ab}^{-1}(O)}$, we have: $\mathbf{h(x)}=\mathbf{h_{ab}(x)}=\mathbf{f_{ab}(2x)}\in\mathbf{O}$. Since $\mathbf{h_{ab}}$ is continuous (definition 2.1.3 (p. 165)) $(\exists\mathbf{O_{ab}})(\mathbf{h_{ab}^{-1}(O)}=\mathbf{O_{ab}}\cap[0,0.5])$ with $\mathbf{O_{ab}}$ an open of $\mathcal{R}$. Since we have $\mathbf{x}\in\mathbf{O_{ab}}$, from proposition 266 (p. 179), there exists $\mathbf{r}>0$ such that $]\mathbf{x}-\mathbf{r},\mathbf{x}+\mathbf{r}[\subset\mathbf{O_{ab}}$. Since $\mathbf{x}\in[0,0.5[$, we have $\mathbf{r'}=\min(\mathbf{r},0.5-\mathbf{x})>0$ and $]\mathbf{x}-\mathbf{r'},\mathbf{x}+\mathbf{r'}[\subset]\mathbf{x}-\mathbf{r},\mathbf{x}+\mathbf{r}[\subset\mathbf{O_{ab}}$. From that, we also have:

$$]\mathbf{x}-\mathbf{r'},\mathbf{x}+\mathbf{r'}[\cap[0,0.5]\subset\mathbf{O_{ab}}\cap[0,0.5]=\mathbf{h_{ab}^{-1}(O)}$$

$\mathbf{h_{ab}^{-1}(O)}$ is an open of $[0,1]$ because $]x-r', x+r'[\cap[0,0.5] = ]x-r', x+r'[\cap[0,1]$. Then $\mathbf{x}$ is in an open of $[0,1]$ fully in $\mathbf{h_{ab}^{-1}(O) \subset h^{-1}(O)}$ then fully in $\mathbf{h^{-1}(O)}$.

- **We suppose $x \in ]0.5, 1]$.**
  Transposing the same reasoning as in the above item, $\mathbf{x}$ is also in an open of $[0,1]$ fully in $\mathbf{h^{-1}(O)}$.
- **We suppose $x = 0.5$.**
  We have (definition 2.1.3 (p. 165)) two opens of $\mathcal{R}$ to which $\mathbf{x}$ belongs: $x \in h_{ab}^{-1}(O) = O_{ab} \cap [0,0.5])$ and $x \in h_{bc}^{-1}(O) = O_{bc} \cap [0.5,1])$. Then, it exists (proposition 266 (p. 179)) $r_{ab} > 0, r_{bc} > 0$ such as:

  $$]x - r_{ab}, x + r_{ab}[\subset O_{ab},\ ]x - r_{bc}, x + r_{bc}[\subset O_{bc}$$

  Taking $r = \min(r_{ab}, r_{bc}, 0.2)$. We have:

  $$]x - r, x + r[\cap[0,0.5] = ]x - r, x] \subset O_{ab} \cap [0,0.5] = h_{ab}^{-1}(O)$$

  and

  $$]x - r, x + r[\cap[0.5,1] = [x, x + r[\subset O_{bc} \cap [0.5,1] = h_{bc}^{-1}(O)\ .$$

  Then we have:

  $$]x - r, x + r[\subset h_{ab}^{-1}(O) \cup h_{bc}^{-1}(O) = h^{-1}(O)$$

  $\mathbf{x}$ is in an open of $[0,1]$ fully in $\mathbf{h^{-1}(O)}$.

Then we have:

$$(\forall x \in h^{-1}(O))(\exists r \in ]0,1])(\exists O_x = ]x - r, x + r[\cap[0,1])(O_x \subset h^{-1}(O))\ .$$

It gives:

$$h^{-1}(O) = \bigcup_{x \in h^{-1}(O))} O_x\ .$$

From definition 132 (p. 151), equation (2.17), $\mathbf{h^{-1}(O)}$ is open, then $\mathbf{h}$ is continuous (definition 141 (p. 165)) and $\mathbf{a}$ and $\mathbf{c}$ are path connected (definition 285 (p. 327)).

**Definition 285 Path connected subset.**

A subset $\mathbf{O}$ of a topological set $\mathbf{E}$ is said to be path connected [30] if any two elements of $\mathbf{O}$ are path connected (definition 284 (p. 326)). From proposition 236 (p. 165), it does not matter if we are dealing with the induced topology of $\mathbf{O}$ by $\mathbf{E}$ or with the $\mathbf{E}$ topology.

**Definition 286 Path connected part.**

A path connected subset $\mathbf{O}$ (definition 285) of a topological set $\mathbf{E}$ is said to be a path connected part if it is also an equivalence class (definition 365 (p. 469)) of the path connection equivalence relation (definition 284 (p. 326), proposition 587 (p. 326)).

**Proposition 588** If $\mathbf{O_1}$ and $\mathbf{O_1}$ are two path connected parts of a topological set $\mathbf{E}$, we have either $\mathbf{O_1 \cap O_2 = \emptyset}$ or $\mathbf{O_1 = O_2}$.

[30] In this book, since we shall never deal with connected sets which are not path connected, we shall always assume that connected set stands for path connected set.

**Proof**
It is a direct consequence of proposition 809 (p. 469).

**Proposition 589** The set of elements path connected to a point of a topological set is path connected.

**Proof**
It is a straight consequence of proposition 587 (p. 326).

**Proposition 590** Calling $(\mathbf{C_x} : \mathbf{E} \to \mathcal{P}(\mathbf{E}))$ the set of element path connected to a point $\mathbf{x}$ of a topological set, we have the following true statement:

$$(\mathbf{a} \in \mathbf{E}, \mathbf{b} \in \mathbf{E}, \mathbf{C_a} \cap \mathbf{C_b} \neq \emptyset) \Rightarrow \mathbf{C_a} = \mathbf{C_b} \ .$$

**Proof**
It is also a straight consequence of proposition 587 (p. 326).

**Proposition 591** Any convex set of $\mathcal{R}^{\mathbf{n}}$ is path connected.

**Proof**
The definition 255 (p. 277) of a convex subset $\mathbf{C}$ of $\mathcal{R}^{\mathbf{n}}$ implies for any two points $\mathbf{x}$, $\mathbf{y}$ of $\mathbf{C}$, the map $(\mathbf{xt} + (1 - \mathbf{t})\mathbf{y} : [0, 1] \to \mathcal{R}^{\mathbf{n}})$ is a map $(\mathbf{xt} + (1 - \mathbf{t})\mathbf{y} : [0, 1] \to \mathbf{C})$ since from proposition 537 (p. 297) this map is continuous, $\mathbf{C}$ is also a connected set.

**Proposition 592** Any open ball of $\mathcal{R}^{\mathbf{n}}$ is a path connected set.

**Proof**
From proposition 496 (p. 279), an open ball is convex then from proposition 591 (p. 328), it is path connected.

**Proposition 593** Given a topological set $\mathbf{E}$ path connected, for any continuous map $(\mathbf{h} : \mathbf{E} \to \mathcal{R})$ we have:

$$0 \notin \mathbf{h}(\mathbf{E}) \Rightarrow (\mathbf{h}(\mathbf{E}) \subset \mathcal{R}^- \vee \mathbf{h}(\mathbf{E}) \subset \mathcal{R}^+) \ .$$

That is to say $\mathbf{h}$ keeps its sign over whole $\mathbf{E}$.

**Proof**
If it was not true, we should have $\mathbf{x}$ and $\mathbf{y}$ such that $\mathbf{h}(\mathbf{x}) < 0 < \mathbf{h}(\mathbf{y})$ and, from definition 285 (p. 327), a continuous map $(\mathbf{f} : [0, 1] \to \mathbf{E})$ such that $\mathbf{h}(\mathbf{f}(0)) < 0 < \mathbf{h}(\mathbf{f}(1))$. From proposition 241 (p. 167), $\mathbf{h} \circ \mathbf{f}$ is continuous. From proposition 299 (p. 190), we have:

$$\mathbf{h}(\mathbf{f}([0, 1])) = [\mathbf{Min}(\mathbf{h}(\mathbf{f}([0, 1]))), \mathbf{Max}(\mathbf{h}(\mathbf{f}([0, 1])))] \ .$$

Then since $\mathbf{h}(\mathbf{f}(0)) \in \mathbf{h}(\mathbf{f}([0, 1])), \mathbf{h}(\mathbf{f}(1)) \in \mathbf{h}(\mathbf{f}([0, 1]))$, from definition 378 (p. 484), we would have:

$$[\mathbf{h}(\mathbf{f}(0)), \mathbf{h}(\mathbf{f}(1))] \subset \mathbf{h}(\mathbf{f}([0, 1])) \ .$$

Since $0 \in [\mathbf{h}(\mathbf{f}(0)), \mathbf{h}(\mathbf{f}(1))]$, it would mean:

$$0 \in \mathbf{h}(\mathbf{f}([0, 1])) \subset \mathbf{h}(\mathbf{E}) \ .$$

It is not permitted. Then we have proposition 593 (p. 328) from A.1.4 (p. 435), 8.

**Proposition 594** Given a topological set $\mathbf{E}$ path connected, $\mathbf{A}$ and $\mathbf{B}$ are two non-empty subsets of $\mathbf{E}$ and not equal to $\mathbf{E}$. We have:

$$\mathbf{A} \cup \mathbf{B} = \mathbf{E} \Rightarrow \bar{\mathbf{A}} \cap \mathbf{B} \neq \emptyset \vee \mathbf{A} \cap \bar{\mathbf{B}} \neq \emptyset \ .$$

See definition 136 (p. 155).

**Proof**
$\bar{\mathbf{A}}$ and $\bar{\mathbf{B}}$ are closed (proposition 219 (p. 158)). $\mathcal{C}(\bar{\mathbf{A}})$ and $\mathcal{C}(\bar{\mathbf{B}})$ are open (definitions 135 (p. 154), 353 (p. 454)). $\mathbf{A}$ and $\mathbf{B}$ are not empty, they cannot be equal otherwise $\mathbf{A} \cup \mathbf{B} = \mathbf{A} \neq \mathbf{E}$. Then we have:

$$(\exists \mathbf{a} \in \mathbf{A})(\exists \mathbf{b} \in \mathbf{B})(\mathbf{a} \neq \mathbf{b})$$

(see convention 26 (p. 441)). Since $\mathbf{E}$ is connected, from definition 285 (p. 327), there a continuous map $(\mathbf{f} : [0,1] \to \mathbf{E})$ with $\mathbf{f}(0) = \mathbf{a}$, $\mathbf{f}(1) = \mathbf{b}$. From definition 2.1.3 (p. 165), $\mathbf{f}^{-1}(\mathcal{C}(\bar{\mathbf{A}}))$ and $\mathbf{f}^{-1}(\mathcal{C}(\bar{\mathbf{B}}))$ are open in the $[0,1]$ induced topology (definition 133 (p. 151)). From propositions 164 (p. 138) and 767 (p. 455) we have:

$$\mathcal{C}(\mathbf{f}^{-1}(\mathcal{C}(\bar{\mathbf{A}})) = \mathbf{f}^{-1}(\bar{\mathbf{A}})),\ \mathcal{C}(\mathbf{f}^{-1}(\mathcal{C}(\bar{\mathbf{B}})) = \mathbf{f}^{-1}(\bar{\mathbf{B}}))\ .$$

Then from definition 135 (p. 154), $\mathbf{f}^{-1}(\bar{\mathbf{A}})$ and $\mathbf{f}^{-1}(\bar{\mathbf{B}})$ are closed in $[0,1]$ induced topology. Since $[0,1]$ is closed in $\mathcal{R}$, they are also closed in $\mathcal{R}$ topology (proposition 207 (p. 155)). We have (propositions 210 (p. 155) and 763 (p. 453))

$$\mathbf{E} = \mathbf{A} \cup \mathbf{B} \subset \bar{\mathbf{A}} \cup \bar{\mathbf{B}}\ .$$

Since $\bar{\mathbf{A}} \subset \mathbf{E}, \bar{\mathbf{B}} \subset \mathbf{E}$, from proposition 763 (p. 453) $\bar{\mathbf{A}} \cup \bar{\mathbf{B}} \subset \mathbf{E}$. From proposition 754 (p. 451) $\bar{\mathbf{A}} \cup \bar{\mathbf{B}} = \mathbf{E}$. From proposition 163 (p. 137),

$$[0,1] = \mathbf{f}^{-1}(\mathbf{E}) = \mathbf{f}^{-1}(\bar{\mathbf{A}}) \cup \mathbf{f}^{-1}(\bar{\mathbf{B}})\ .$$

Since $\mathbf{f}^{-1}(\bar{\mathbf{A}}) \subset [0,1]$ and $\mathbf{a} \in \mathbf{A} \subset \bar{\mathbf{A}}$, $0 = \mathbf{f}^{-1}(\mathbf{a}) \in \mathbf{f}^{-1}(\bar{\mathbf{A}})$.
Calling $\boldsymbol{\ell} = \mathbf{upper\_limit}(\mathbf{f}^{-1}(\bar{\mathbf{A}}))$ (definition 381 (p. 485) and see section 2.2.4). Since $\mathbf{f}^{-1}(\bar{\mathbf{A}})$ is closed in $\mathcal{R}$, from proposition 220 (p. 158), we have $\overline{\mathbf{f}^{-1}(\bar{\mathbf{A}})} = \mathbf{f}^{-1}(\bar{\mathbf{A}})$. From proposition 276 (p. 182)

$$\boldsymbol{\ell} \in \mathbf{f}^{-1}(\bar{\mathbf{A}}) \subset [0,1]\ .$$

If $\boldsymbol{\ell} = \mathbf{1}$ we have proposition 594.
If $\mathbf{0} \leq \boldsymbol{\ell} < \mathbf{1}$, for any open $\mathbf{O}$ of $\mathcal{R}$ containing $\boldsymbol{\ell}$, there is an $\epsilon > \mathbf{0}$ such that $]\boldsymbol{\ell}-\epsilon, \boldsymbol{\ell}+\epsilon[\subset \mathbf{O}$ (proposition 266 (p. 179)).
Because $\boldsymbol{\ell} = \mathbf{upper\_limit}(\mathbf{f}^{-1}(\bar{\mathbf{A}}))$, when taking $\epsilon' = \mathbf{min}(\epsilon, 1 - \boldsymbol{\ell}) > 0$, we have:

$$]\boldsymbol{\ell}, \boldsymbol{\ell} + \epsilon'[\cap \mathbf{f}^{-1}(\bar{\mathbf{A}}) = \emptyset\ .$$

Combined to:

$$]\boldsymbol{\ell}, \boldsymbol{\ell} + \epsilon'[\subset [0,1] = \mathbf{f}^{-1}(\bar{\mathbf{A}}) \cup \mathbf{f}^{-1}(\bar{\mathbf{B}})$$

we have, calling $\mathcal{O}$ the set of opens of $\mathcal{R}$:

$$(\forall \mathbf{O} \in \mathcal{O})(\boldsymbol{\ell} \in \mathbf{O} \Rightarrow (\exists \epsilon' > 0)(]\boldsymbol{\ell}, \boldsymbol{\ell} + \epsilon'[\subset \mathbf{f}^{-1}(\bar{\mathbf{B}})))$$

then from definition 136 (p. 155), $\boldsymbol{\ell} \in \overline{\mathbf{f}^{-1}(\bar{\mathbf{B}})}$. From proposition 220 (p. 158) $\overline{\mathbf{f}^{-1}(\bar{\mathbf{B}})} = \mathbf{f}^{-1}(\bar{\mathbf{B}})$, then $\boldsymbol{\ell} \in \mathbf{f}^{-1}(\bar{\mathbf{B}})$. Since we also have $\boldsymbol{\ell} \in [0,1] = \mathbf{f}^{-1}(\mathbf{A}) \cup \mathbf{f}^{-1}(\mathbf{B})$ we have:

$$\mathbf{f}(\boldsymbol{\ell}) \in \bar{\mathbf{A}} \cap \bar{\mathbf{B}}, \mathbf{f}(\boldsymbol{\ell}) \in \mathbf{A} \cup \mathbf{B}$$

then we have proposition 594.

**Proposition 595** Given a topological set $\mathbf{E}$ path connected, $\mathbf{A}$ and $\mathbf{B}$ two open non-empty subsets of $\mathbf{E}$ (definition 132 (p. 151)), we have:

$$\mathbf{A} \cup \mathbf{B} = \mathbf{E} \Rightarrow \mathbf{A} \cap \mathbf{B} \neq \emptyset\ .$$

**Proof**
From proposition 594 $(\exists p \in E = A \cup B)(p \in \bar{A} \cap \bar{B})$. From definition 351 (p. 453), we have $p \in A \vee p \in B$ (convention 26 (p. 441)). Let us suppose $p \in A$ (resp. $p \in B$). Since $p \in \bar{A} \cap \bar{B}$ (definition 352 (p. 453)), $p \in \bar{B}$ (resp. $p \in \bar{A}$). From definition 136 (p. 155), since $A$ (resp. $B$) is open, $A \cap B \neq \emptyset$.

**Proposition 596** Given a topological set $E$ path connected, $\Omega$ the set of open subsets of $E$, $n$ an integer, a map $(O_i : [1, n] \to \Omega)$ such that $E = \bigcup_{i \in [1,n]} O_i$, we have:

$$(\forall p \in E)((i \in [1, n], p \in O_i) \Rightarrow (\exists m \in \mathcal{Z}^+)(\exists j)(j = (j : [1, m] \to [1, n]), j(1) = i,$$
$$(\forall k \in [1, m - 1])(O_{j(k)} \cap O_{j(k+1)} \neq \emptyset, \bigcup_{k \in [1,m]} O_k = E))) .$$

**Proof**
Given $i$, we define the subset $H$ of $Z^+ \times [1, n]$ by:

$$h \in H \Leftrightarrow (\exists \ell \in \mathcal{Z}^+)(h = (h : [1, \ell] \to [1, n]), h(1) = i,$$
$$(\forall k \in [1, \ell - 1])(O_{h(k)} \cup O_{j(k+1)} \neq \emptyset)) .$$

From proposition 750 (p. 450), $H$ exists. It is not empty because $(h : [1, 1] \to \{i\} \subset [1, n]) \in H$. We call $K$, the set (see definition 390 (p. 499)) defined by:

$$k \in K \Leftrightarrow (k \in \mathcal{Z}^+, (\exists h \in H)(\exists \ell \in \mathcal{Z}^+)((h : [1, \ell] \to [1, n]), k = \mathrm{Card}(h([1, \ell))))) .$$

Since $h([1, \ell) \subset [1, n]$, we have $\mathrm{Card}(h([1, \ell)) \leq n$ (propositions 877 (p. 500), 934 (p. 523)). Then $K \subset \mathcal{Z}^+$ has a biggest element which is smaller than or equal to $n$ (proposition 924 (p. 519)). Let us call $L$ the biggest element of $K$ taking a corresponding map $h$ and $\ell$ with $\mathrm{Card}(H([1, \ell])) = L$ and let us suppose $L < n$. From proposition 796 (p. 464), we have:

$$\bigcup_{k \in h([1,\ell])} O_k \cup \bigcup_{k \in \mathcal{C}_{[1,n]}(h([1,\ell]))} O_k = \bigcup_{k \in [1,n]} O_k = E .$$

Since from definition 132 (p. 151), $\bigcup_{k \in h([1,\ell])} O_k$ and $\mathcal{C}_{[1,n]}(h([1, \ell]))O_k$ are open and not empty because $L < n$. From proposition 595 (p. 329), we have:

$$\bigcup_{k \in h([1,\ell])} O_k \cap \bigcup_{k \in \mathcal{C}_{[1,n]}(h([1,\ell]))} O_k \neq \emptyset .$$

From definitions 351 (p. 453) and 362 (p. 462), we have:

$$(\exists k_1 \in h([1, \ell]))(\exists k_2 \in \mathcal{C}_{[1,n]}(h([1, \ell])))(O_{k_1} \cap O_{k_2} = 0) .$$

We set $i_1 \in h^{-1}(k_1) \cap [1, \ell] \neq \emptyset$ (definitions 116 (p. 134), 115 (p. 133)) and we define $(h' : [1, \ell + 1] \to [1, n])$ by:

$$k \in [1, i_1] \Rightarrow h'(k) = h(k),\ h'(i_1 + 1) = k_2,\ k \in [i_1 + 1, \ell + 1] \Rightarrow h'(k) = h(k - 1)$$

we have:

$$k \in [1, i_1 - 1] \Rightarrow O_{h'(k)} \cap O_{h'(k+1)} = O_{h(k)} \cap O_{h(k+1)} \neq \emptyset, k \in [i_1 + 1, \ell]$$
$$\Rightarrow O_{h'(k)} \cap O_{h'(k+1)} = O_{h(k-1)} \cap O_{h(k)} \neq \emptyset, O_{h'(i_1)} \cap O_{h'(i_1+1)} = O_{k_1} \cap O_{k_2} \neq \emptyset .$$

Then $h' \in H$ since $h([1, \ell]) \subset (h'([1, \ell+1])$ and since $k_2 \in h'([1, \ell+1])$ and $k_2 \notin h([1, \ell])$, $L = \mathrm{Card}(h([1, \ell]) < \mathrm{Card}(h'([1, \ell + 1]))$ cannot be the biggest element of $K$ except if

$\mathbf{L} = \mathbf{n}$ which means, for the corresponding $\mathbf{h}$, $\mathbf{h}([1,\ell]) = [1,\mathbf{n}]$ (see proposition 935 (p. 523)).
In that case, $\mathbf{h}([1,\ell]) = [1,\mathbf{n}]$, applying proposition 595 (p. 329) with definitions 116 (p. 134) and 115 (p. 133)), we have:

$$\bigcup_{j\in[1,\ell]} O_{h(j)} = \bigcup_{j\in[1,\ell]} \bigcup_{k\in\{h(j)\}} O_k = \bigcup_{k\in\bigcup_{j\in[1,\ell]}\{h(j)\}} O_k = \bigcup_{k\in[1,n]} O_k = \mathbf{E}\,.$$

**Proposition 597** Given a topological set $\mathbf{E}$, $\mathbf{A}$ and $\mathbf{B}$ two path connected subsets of $\mathbf{E}$, if $\mathbf{A}\cap\mathbf{B} \neq \emptyset$ then $\mathbf{A}\cup\mathbf{B}$ is path connected.

**Proof**
We have $\mathbf{A}\cap\mathbf{B} \neq \emptyset \Rightarrow \exists \mathbf{p} \in \mathbf{A}\cap\mathbf{B}$. With convention 26 (p. 441), we have $\mathbf{a}$ path connected to $\mathbf{p}$ and $\mathbf{p}$ path connected to $\mathbf{b}$ then from proposition 587 (p. 326) $\mathbf{a}$ is path connected to $\mathbf{b}$.

**Proposition 598** Given an open set $\mathbf{E}$ of $\mathcal{R}^n$ with the open ball topology, for any point $\mathbf{a} \in \mathbf{E}$, the set of the path connected points to $\mathbf{a}$ in $\mathbf{E}$ is open.

**Proof**
We call $\mathbf{C_a}$, the set of the path connected points to $\mathbf{a}$ in $\mathbf{E}$. From proposition 808 (p. 469), $\mathbf{C_a}$ is path connected. For any point $\mathbf{x} \in \mathbf{C_a}$, there is an open ball $\mathbf{B}(\mathbf{x},\mathbf{r_x})$ centered on $\mathbf{x}$ fully in $\mathbf{E}$ (proposition 266 (p. 179)). From proposition 592 (p. 328), $\mathbf{B}(\mathbf{x},\mathbf{r_x})$ is path connected. $\mathbf{x} \in \mathbf{C_a}$ is path connected to $\mathbf{a}$. From proposition 597 (p. 331) we then have $\mathbf{B}(\mathbf{x},\mathbf{r_x}) \subset \mathbf{C_a}$. Using that we have:

$$\mathbf{C_a} = \bigcup_{x\in C_a} \mathbf{B}(\mathbf{x},\mathbf{r_x})$$

then from definition 132 (p. 151), 2.17, $\mathbf{C_a}$ is open.

## 2.23 Riemann integral of maps with compact support

**Definition 287 Map $(\mathbf{f}:\mathcal{R}^n \to \mathcal{R}^m)$ with compact support.**

We call maps $(\mathbf{f}:\mathcal{R}^n \to \mathcal{R}^m)$ with compact support, maps which are zero outside a compact subset of $\mathcal{R}^n$. From proposition 523 (p. 291), we have:

$$(\exists c \in \mathcal{R}^+)(x \notin \overline{C}(0,c) \Rightarrow f(x) = 0) \tag{2.151}$$

$\mathbf{B} = (\mathbf{e_i} : [1,\mathbf{n}] \to \mathcal{R}^n)$ being a basis of $\mathcal{R}^n$, it means that there exists a cube $\overline{\mathbf{C}}(\mathbf{0},\mathbf{c})$ centered on $\mathbf{0}$ and of half side $\mathbf{c} > \mathbf{0}$ such that:

$$x = x^i e_i \in \overline{C}(0,c) \Leftrightarrow (\forall i \in [1,n])(|x^i| \leq c)$$

and such that $\mathbf{x} \notin \overline{\mathbf{C}}(\mathbf{0},\mathbf{c}) \Rightarrow \mathbf{f}(\mathbf{x}) = \mathbf{0}$. Since $\overline{\mathbf{C}}(\mathbf{0},\mathbf{c})$ is compact (proposition 522 (p. 291)), it is equivalent to saying that $\mathbf{f}$ has a compact support or to say $(\exists \mathbf{c} > \mathbf{0})(\mathbf{x} \notin \bar{\mathbf{B}}(\mathbf{0},\mathbf{c}) \Rightarrow \mathbf{f}(\mathbf{x}) = \mathbf{0})$[31].

**Definition 288 Set of integration grids: Gr and its associated sum map $(\mathbf{gr^f} : (\mathcal{R}^m)^{\mathcal{R}^n} \times \mathbf{Gr} \to \mathcal{R}^m)$ (see definition 373 (p. 474)).**

[31] One considers here $\mathcal{R}^m$ as a vectorial space (definition 199 (p. 235)) of dimension $\mathbf{m}$ (definition 208 (p. 240)). When $\mathbf{m} > \mathbf{1}$, $\mathbf{0}$ means the null vector (neutral vector or unit element (definition 167 (p. 205), 2) for the addition group).

We take a reference basis $(\mathbf{e_i} : [1, n] \to \mathcal{R}^n)$ and a map $(\mathbf{f} : \mathcal{R}^n \to \mathcal{R}^m)$ with a compact support. We consider a set of grids $\mathbf{Gr}$.
Each element of $\mathbf{Gr}$ is a triplet $(\mathbf{p}, \mathbf{y}, \mathbf{z})$:

$$\begin{aligned}
&(p, y, z) \in \mathrm{Gr} \Leftrightarrow (p \in \mathcal{Z}^n, y \in \mathcal{Z} \times [1, n] \times \mathcal{R},\\
&(\forall k \in [1, n])(p^k \in \mathcal{Z}^+, y^k = (y^{i,k} : [0, p^k] \to \mathcal{R})),\\
&z = (z^{i,k} : \prod_{k \in [1,n]} [0, p^k[ \times [1, n] \to \mathcal{R})\\
&(\forall k \in [1, n])(2c > -y^{0,k} > c, 2c > y^{p^k,k} > c,\\
&(\forall i \in [0, p^k[)(0 < y^{i+1,k} - y^{i,k})),\\
&(\forall \ell \in \prod_{k \in [1,n]} [0, p^k[)(\forall k \in [1, n])(y^{\ell_k,k} \le z^{\ell,k} < y^{\ell_k+1,k}))
\end{aligned} \tag{2.152}$$

$\ell \in \prod_{k \in [1,n]}[0, p^k[$ is by definition a map $\ell = (\ell(k) = \ell^k : [1, n] \to \mathcal{Z}^+)$ such that $\ell^k \in [0, p^k[$ (see definitions 126 (p. 144) and 372 (p. 473)). It defines a box (see definition 279 (p. 320)), which we shall call $\mathbf{Bx}_\ell = \mathbf{Bx}(y^{\ell_i,i}e_i, y^{\ell_i+1,i}e_i) \subset \mathcal{R}^n$ [32]. The map $\mathbf{z}$, from its definition, verifies $z^{\ell,i}e_i \in \mathbf{Bx}_\ell$. They are used to evaluate the function $(\mathbf{f} : \mathcal{R}^n \to \mathcal{R}^m)$. For each map $(\mathbf{f} : \mathcal{R}^n \to \mathcal{R}^m)$, there is an associated map $(\mathbf{gr^f} : \mathrm{Gr} \to \mathcal{R}^m)$ defined by:

$$\begin{aligned}
&t = (p, y, z) \in \mathrm{Gr} \Rightarrow\\
&\mathrm{gr}^f((p, y, z)) = \mathrm{gr}^f(t) = \sum_{\ell \in \prod_{k \in [1,n]}[0,p^k[} f(z^{\ell,i}e_i) \prod_{k \in [1,n]} (y^{\ell_k+1,k} - y^{\ell_k,k}) .
\end{aligned} \tag{2.153}$$

**Definition 289 Set of grids with a mesh size smaller than a positive real number: $\mathbf{Gr_a}$, $\mathbf{a} \in \mathcal{R}^+$.**

For all $\mathbf{a} \in \mathcal{R}^+$, we also define $\mathbf{Gr_a}$ a subset of $\mathbf{Gr}$ and $\mathbf{S^f_a}$ a subset of $\mathcal{R}^m$ by:

$$\begin{aligned}
&(\forall a > 0)((t \in \mathrm{Gr_a}) \Leftrightarrow (t = (p, y, z) \in \mathrm{Gr}, (\forall k \in [0, n])(\forall i \in [0, p^k[)\\
&(y^{i+1,k} - y^{i,k}) \le a)\\
&S^f_a = \mathrm{gr}^f(\mathrm{Gr_a}) \subset \mathcal{R}^m
\end{aligned} \tag{2.154}$$

**Definition 290 Riemann integral J(f) of a map $(\mathbf{f} : \mathcal{R}^n \to \mathcal{R}^m)$ with a compact support.**

The Riemann integral $\mathbf{J(f)}$ of $\mathbf{f}$ is a vector $\mathbf{J} \in \mathcal{R}^m$ in the following statement if it is true (see definition 333 (p. 441)):

$$\begin{aligned}
&(\exists J)(J \in \mathcal{R}^m, (\forall \alpha > 0)(\exists \epsilon^f(\alpha) > 0)(0 < a < \epsilon^f(\alpha) \Rightarrow (\forall x \in S^f_a)(\|x - J\| < \alpha)))\\
&J = \tau_J(J \in \mathcal{R}^m, (\forall \alpha > 0)(\exists \epsilon^f(\alpha) > 0)(0 < a < \epsilon^f(\alpha)\\
&\Rightarrow (\forall x \in S^f_a)(\|x - J\| < \alpha))) .
\end{aligned} \tag{2.155}$$

[32] In principle when an index is free or mute and repeated up and down there is an implicit summation over it. This occurs mainly when a term like $\mathbf{e_i}$ which defines a basis of $\mathcal{R}^n$ is involved.

**Proposition 599** Given a map $(\mathbf{f} : \mathcal{R}^n \to \mathcal{R}^m)$, its integral $\mathbf{J}$, as defined in definition 290 (p. 332), has the following properties:

1. $\mathbf{J}$ is unique.

2. $\mathbf{J}$ does not depend on $\mathbf{c}$ as long as it is big enough to verify relation (2.151).

**Proof**

1. If $\mathbf{J}$ and $\mathbf{J}'$ satisfy (2.155) and $\mathbf{J} \neq \mathbf{J}'$, then taking $0 < \alpha < \|\mathbf{J} - \mathbf{J}'\|/2$ there are an $\epsilon_J$ and an $\epsilon_{J'}$ corresponding to $\alpha$ in (2.155). Taking an $\mathbf{a}$ such that $0 < \mathbf{a} < \min(\epsilon_J, \epsilon_{J'})$ we set:

$$(\forall \ell \in \prod_{k \in [1,n]} [0, p^k[)(\forall k \in [1,n])(p^k = 2c/a + 2,\ y^{l(k),k} = (l(k) - p^k/2) \times a),$$
$$z^{l,k} = (l(k) - p^k/2 + 1/2)a) .$$

It forms a term $\mathbf{t} = (\mathbf{p}, \mathbf{y}, \mathbf{z})$ of $\mathbf{Gr_a}$ because we have:

$$(\forall k \in [1,n])$$
$$(i \in [0, p^k[ \Rightarrow y^{i+1,k} - y^{i,k} = a, y^{0,k} = -c - a, y^{p,k} = c + a),$$
$$(\forall l \in \prod_{k \in [1,n]} [0, p^k[)(\forall k \in [1,n])$$
$$(l(k)a < (l(k) + 1/2)a < (l(k) + 1)a \Rightarrow y^{l_k,k} \le z^{l,k} < y^{l_k+1,k}) .$$

Using equation (2.153) (p. 332), we get an $\mathbf{x} = \mathbf{gr^f(t)} \in \mathbf{S^f_a}$. With that $\mathbf{x}$, using the relation of the integral definition (2.155) (p. 332), we should have $\|\mathbf{x} - \mathbf{J}\| < \alpha$ and $\|\mathbf{x} - \mathbf{J}'\| < \alpha$. It is incompatible with

$$\|x - J\| + \|x - J'\| = \|J - x\| + \|x - J'\| > \|J - x + x - J'\| = \|J - J'\| > 2\alpha .$$

From A.1.4 (p. 435), 8, $\mathbf{J} = \mathbf{J}'$.

2. It is a direct consequence of the definitions 287 (p. 331), 288 (p. 331), 289 (p. 332) and 290 (p. 332).

**Proposition 600** We are given a map $(\mathbf{f} : \mathcal{R}^n \to \mathcal{R}^m)$ with compact support (definition 287 (p. 331)). We call $(\mathbf{e}'_i = \delta^j_i : [1, m] \to \mathcal{R}^m)$, the standard independent basis of $\mathcal{R}^m$ ( equation (2.106) (p. 267) in definition 242 (p. 267)). If the integral $\mathbf{J(f)} = \mathbf{J^i(f)e'_i} \in \mathcal{R}^m$ exists, the integral $\mathbf{J(f^i)}$ of any component of $\mathbf{f} = \mathbf{f^i e'_i}$ exists too. The reverse is also true: if the integral $\mathbf{J(f^i)}$ of any component of $\mathbf{f} = \mathbf{f^i e'_i}$ exists, the integral $\mathbf{J(f)} = \mathbf{J^i(f)e'_i} \in \mathcal{R}^m$ exists. We have in both cases:

$$J^i(f)e'_i = J(f^i)e'_i .$$

**Proof**

We call $(\mathbf{e}_i = \delta^j_i : [1, m] \to \mathcal{R}^n)$ the standard independent basis of $\mathcal{R}^n$. From definition 287 (p. 331), if $\mathbf{f}$ is a map with compact support $\bar{\mathbf{C}}(\mathbf{0}, \mathbf{c})$ any of its component is a map $\mathcal{R}^n \to \mathcal{R}$ which can be considered as on the same compact support $\bar{\mathbf{C}}(\mathbf{0}, \mathbf{c})$. From proposition 599 (p. 333), all those integral does not depend of the exact value of $\mathbf{c}$ as soon it is big enough. we can then take a unique value of $\mathbf{c}$ for all of them. From definition 288 (p. 331) the grids in

$\mathbf{Gr_a(c)}$ of definition 288 (p. 331) $\mathbf{f}$ are also valid for any component of $\mathbf{f}$. We have (equation (2.153) (p. 332) and proposition 409 (p. 241)):

$$(\forall a \in \mathcal{R}^+)(\forall t = (p,y,z) \in Gr_a)(gr^f(t) = \sum_{\ell \in \prod_{k\in[1,n]}[0,p^k[} f(z^{\ell,i}e_i) \prod_{k\in[1,n]} (y^{\ell_k+1,k} - y^{\ell_k,k})$$

$$= \sum_{\ell \in \prod_{k\in[1,n]}[0,p^k[} f^j(z^{\ell,i}e_i)e'_j \prod_{k\in[1,n]} (y^{\ell_k+1,k} - y^{\ell_k,k}) = {}^j gr^f(t) e'_j,$$

$${}^j gr^f(t) = \sum_{\ell \in \prod_{k\in[1,n]}[0,p^k[} f^j(z^{\ell,i}e_i) \prod_{k\in[1,n]} (y^{\ell_k+1,k} - y^{\ell_k,k}) = gr^{f^j}(t)) \ .$$

Using definition 290 (p. 332) and proposition 599 (p. 333), we have if $\mathbf{J(f)}$ exists:

$$(\forall \alpha > 0)((\exists \epsilon^f(\alpha) > 0)(0 < a < \epsilon^f(\alpha) \Rightarrow (\forall x = x^j e'_j \in S^f_a \Rightarrow)(\|x - J(f)\| < \alpha))$$
$$\Rightarrow (\forall \alpha > 0)(\exists \epsilon^f(\alpha) > 0)(0 < a < \epsilon^f(\alpha) \Rightarrow (\forall j \in [1,m])(\forall x^j \in {}^j S^f_a)$$
$$(\|x^j - J^j(f)\| < \alpha))$$
$$\Rightarrow (\forall j \in [1,m])(J^j(f) = J(f^j)) \ .$$

Then $\mathbf{J(f^j)}$ integral of $\mathbf{(f^j : \mathcal{R}^n \to \mathcal{R})}$ exists and is equal the $\mathbf{j}$ component of $\mathbf{J(f)}$. On the opposite if $\mathbf{J(f^j)}$ exists for all $\mathbf{j \in [1,m]}$ we have using a $\mathbf{p}$ norm:

$$\|x = x^i e'_i\| = ( \sum_{i\in[1,m]} \prod_{\ell \in [1,p]} x^i)^{\frac{1}{p}} \ .$$

We have:

$$(\forall \alpha > 0)((\exists \epsilon(\alpha) > 0)$$
$$(0 < a < \epsilon(\frac{\alpha}{m^{1/p}}) \Rightarrow (\forall j \in [1,m])(\forall x^j \in S^{f^j}_a)(\|x^j - J(f^j)\| < \frac{\alpha}{m^{1/p}})$$
$$\Rightarrow (\forall x = x^j e'_j \in S^f_a)(\|x^j e'_j - J(f^j)e'_j\| = \|x - J(f^j)e'_j\| < \alpha)) \ .$$

Then $\mathbf{J(f)}$ exists and is equal to $\mathbf{J(f^j)e'_j}$ (definition 290 (p. 332) and proposition 599 (p. 333)).

**Definition 291 Translated grid.**

Considering in definition 287 (p. 331), the set $\mathbf{Gr_a(c)}$ relative to the cube $\mathbf{C(0,c)}$ defined by equation (2.154) (p. 332) and a translating vector $\mathbf{\Delta \in \mathcal{R}^n}$, we associate to $\mathbf{\Delta}$ a set of maps belonging to $\mathbf{(Gr_a(c + \|\Delta\|)}$. Both cubes $\mathbf{C(0,c)}$ and $\mathbf{C(0, c + \|\Delta\|)}$ are bigger than the compact support. A grid of $\mathbf{(Gr_a(c)}$ can be extended to be a grid of $\mathbf{C(0, c + \|\Delta\|)}$. Given $\mathbf{t(p,y,z) \in Gr_a(c)}$, $\mathbf{{}^\Delta t({}^\Delta p, {}^\Delta y, {}^\Delta z) \in Gr_a(c + \|\Delta\|))}$ is a term translated of $\mathbf{t}$ by $\mathbf{\Delta}$ if we have:

$$\begin{aligned}
&(\forall k \in [1,n])(\Delta^k \neq 0 \Rightarrow (({}^\Delta p^k \geq p^k, (\exists(\Delta^k_i : [0, {}^\Delta p^k - p^k] \to [0, \Delta^k]))\\
&(\Delta^k_{{}^\Delta p^k - p^k} = \Delta^k, ({}^\Delta p^k > p^k \Rightarrow \Delta^k_0 = 0, (\forall i \in [0, {}^\Delta p^k - p^k - 1[)\\
&(0 < \Delta^k_{i+1} - \Delta^k_i \leq a)),\\
&\Delta^k > 0 \Rightarrow \{(\forall i \in [0, {}^\Delta p^k - p^k])({}^\Delta y^{k,i} = y^{k,0} + \Delta^k_i),\\
&(\forall i \in [{}^\Delta p^k - p^k, {}^\Delta p^k])({}^\Delta y^{k,i} = y^{k,i-{}^\Delta p^k + p^k} + \Delta^k)\},\\
&\Delta^k < 0 \Rightarrow \{(\forall i \in [0, p^k])({}^\Delta y^{k,i} = y^{k,i} + \Delta^k),\\
&(\forall i \in [p^k, {}^\Delta p^k])({}^\Delta y^{k,i} = y^{k,p^k} + \Delta^k_{{}^\Delta p^k - i})\}),\\
&\Delta^k = 0 \Rightarrow \{{}^\Delta p^k = p^k, (\forall i \in [0, p^k])({}^\Delta y^{k,i} = y^{k,i})\}) \ .
\end{aligned} \tag{2.156}$$

Calling $(\mathcal{Y}(x) : \mathcal{R} \to \{0,1\})$ the Heaviside map (definition 216 (p. 245)), we can also defined translated grids by:

$$(\forall k \in [1,n])(\forall i \in [0,p^k])(y^{k,i} = {}^{\Delta}y^{k,\mathcal{Y}(\Delta^k)({}^{\Delta}p^k - p^k)+i} - \Delta^k)$$

or calling $(h^k = \mathcal{Y}(\Delta^k)({}^{\Delta}p^k - p^k) : [1,n] \to \mathcal{Z})$

$$(\forall k \in [1,n])(\forall i \in [h^k, p^k + h^k])({}^{\Delta}y^{k,i} = y^{k,i-h^k} + \Delta^k) ,$$

we have to add extra points at the limits as defined in relation (2.156).

**Proposition 601** The integral when they exist are independent of a translation of the grids as defined in definition 291 (p. 334).

**Proof**
It is a consequence of proposition 599 (p. 333) point 2.

**Proposition 602** We are given a map $(f : \mathcal{R}^n \to \mathcal{R}^m)$ and a partition (definition 370 (p. 471)) of $[1,m]$ defined by a map $(g : [1,m] \to [1,\ell])$, $\ell \leq m$ (see proposition 815 (p. 471)). We call $e_i = \prod^e_{j\in[1,m]} \delta^j_i = (\delta^j_i : [1,m] \to \mathcal{R}) \in \mathcal{R}^m$ the reference basis of $\mathcal{R}^m$ and $e_i, i \in g^{-1}(k)$ the reference basis of $\mathcal{R}^{\mathrm{Card}(g^{-1}(k))}$ (see equation (2.106) (p. 267) in definition 242 (p. 267)). For $k \in [1,\ell]$, we write $(f_{g^{-1}(k)} : \mathcal{R}^n \to \mathcal{R}^{\mathrm{Card}(g^{-1}(k))})$ the projection (definition 121 (p. 136)) of $f$ on $\mathcal{R}^{\mathrm{Card}(g^{-1}(k))}$ along $\mathcal{C}_{[1,m]}(g^{-1}(k))$ (see definition 117 (p. 134)):

$$f_{g^{-1}(k)} = Pr_{\mathcal{C}(g^{-1}(k))} \circ f = \sum_{i\in g^{-1}(k)} f^i e_i .$$

1. If $J(f)$ exists, $J(f_{g^{-1}(k)})$ exists and
$$J(f_{g^{-1}(k)}) = Pr_{\mathcal{C}(g^{-1}(k))}(J(f)) = \sum_{i\in g^{-1}(k)} J^i(f)e_i .$$
2. If $J(f_{g^{-1}(k)})$ exists for all $k \in [1,\ell]$, then $J(f)$ exists and
$$J(f) = \sum_{k\in[1,\ell]} \sum_{i\in g^{-1}(k)} J^i(f_{g^{-1}(k)})e_i .$$

**Proof**
Keeping the above notations (definition 290 (p. 332)), we have:

$$\sum_{i\in[1,n]} f^i e_i = \sum_{k\in[1,\ell]} \sum_{i\in g^{-1}(k)} f^i_{g^{-1}(k)} e_i$$

- point 1. If $J(f)$ exists, considering $J(f_{g^{-1}(k)})$, using equation (2.155) (p. 332), we take for $\alpha > 0$:
$$\epsilon^{f_{g^{-1}(k)}}(\alpha) = \epsilon^f(\alpha) .$$
It gives:
$$(0 < a < \epsilon^{f_{g^{-1}(k)}}(\alpha) = \epsilon^f(\alpha)) \Rightarrow (t \in Gr_a \Rightarrow \|\sum_{i\in g^{-1}(k)} (gr^f(t)^i - J^i(f))e_i\| \leq$$
$$\|\sum_{i\in[1,m]} (gr^f(t)^i - J^i(f))e_i\| < \alpha)$$

then

$$J(f_{g^{-1}(k)}) = \sum_{i \in g^{-1}(k)} J^i(f)e_i \ .$$

See proposition 478 (p. 270).

- point 2. Taking (see equation (2.155) (p. 332)):

$$\epsilon^f(\alpha) = \min_{k \in [1,\ell]} (\epsilon^{f_{g^{-1}(k)}}(\alpha/\ell))$$

we have:

$$0 < a < \epsilon^f(\alpha) \Rightarrow (t \in Gr_a \Rightarrow \| \sum_{i \in [1,m]} (gr^f(t)^i - J^i(f))e_i\|$$

$$= \| \sum_{k \in [1,\ell]} \sum_{i \in g^{-1}(k)} ((gr^f(t)^i - J^i(f))e_i)\| \leq \sum_{k \in [1,\ell]} \| \sum_{i \in g^{-1}(k)} (gr^f(t)^i - J^i(f))e_i\| < \alpha)$$

then:

$$J(f) = \sum_{k \in [1,\ell]} \sum_{i \in g^{-1}(k)} J^i(f_{g^{-1}(k)})e_i \ .$$

See again proposition 478 (p. 270).

**Proposition 603** If a map $(f : \mathcal{R}^n \to \mathcal{R})$ with a compact support (definition 287 (p. 331)) has an integral $J(f)$ (definition 290 (p. 332)) and if $(\forall x \in \mathcal{R}^n)(f(x) \geq 0)$, (resp. $f(x) \leq 0$) then $J(f) \geq 0$ (resp. $J(f) \leq 0$).

**Proof**
Keeping notations of definition 290 (p. 332), we have:

$$(\forall a \in \mathcal{R}^+)(\forall t \in Gr_a)(gr^f(t) \geq 0) \text{ (resp. } gr^f(t) \leq 0) \ .$$

If $J < 0$ (resp. $J > 0$), using equation (2.155) (p. 332), taking $\alpha = -J/2 > 0$ (resp. $\alpha = J/2 > 0$), for $0 < a < \epsilon^f(\alpha)$ and for $t \in Gr_a$, we get $|gr^f(t) - J| < -J/2$. (resp. $|gr^f(t) - J| < J/2$). From definition 154 (p. 181) and definition 424 (p. 557) we would have

$$gr^f(t) - J \leq |gr^f(t) - J| < -J/2 \text{ (resp. } J - gr^f(t) \leq |J - gr^f(t)| < J/2) \ .$$

Then

$$gr^f(t) < J - J/2 < 0 \text{ (resp. } 0 < J - J/2 \leq gr^f(t)) \ .$$

It is impossible. Therefore from A.1.4 (p. 435), 7 $J \geq 0$ (resp. $J \leq 0$).

**Proposition 604** Given a map $(f : \mathcal{R}^n \to \mathcal{R}^m)$, if $J(f)$ and $J(\|f\|)$ exist, we have:

$$\|J(f)\| \leq J(\|f\|) \ .$$

**Proof**
Keeping the above notations (definition 290 (p. 332)), using equation (2.153) (p. 332), we have with $t = (p, y, z) \in Gr$:

$$gr^{\|f\|}(t) = \sum_{\ell \in \prod_{k \in [1,n]} [0,p^k[} \|f(z^{\ell,i}e_i)\| \prod_{k \in [1,n]} (y^{\ell_k+1,k} - y^{\ell_k,k})$$

$$\geq \| \sum_{\ell \in \prod_{k \in [1,n]} [0,p^k[} f(z^{\ell,i}e_i) \prod_{k \in [1,n]} (y^{\ell_k+1,k} - y^{\ell_k,k})\| = \|gr^f(t)\| \ .$$

From equation (2.155) (p. 332), we have (see definition 203 (p. 237), propositions 404 (p. 239), 603 (p. 336)):

$$
\begin{aligned}
&(\forall \alpha > 0)(\exists \epsilon^{\|f\|}(\alpha/2) > 0)(\exists \epsilon^{f}(\alpha/2) > 0)(0 < a < \min(\epsilon^{\|f\|}(\alpha/2), \epsilon^{f}(\alpha/2)) \\
&\Rightarrow (t \in Gr_a \Rightarrow (|gr^{\|f\|}(t) - J(\|f\|)| < \alpha/2, \|gr^{f}(t) - J(f)\| < \alpha/2) \\
&\Rightarrow (gr^{\|f\|}(t) - J(\|f\|) < \alpha/2, \|J(f)\| - \|gr^{f}(t)\| < \alpha/2) \\
&\Rightarrow (J(\|f\|) + \alpha/2 > gr^{\|f\|}(t), \|gr^{f}(t)\| > \|J(f)\| - \alpha/2) \\
&\Rightarrow J(\|f\|) + \alpha/2 > gr^{\|f\|}(t) \geq \|gr^{f}(t)\| > \|J(f)\| - \alpha/2 \\
&\Rightarrow J(\|f\|) + \alpha \geq \|J(f)\|)) .
\end{aligned}
$$

Then

$$(\forall \alpha > 0)(J(\|f\|) + \alpha \geq \|J(f)\|)$$

it means

$$J(\|f\|) \geq \|J(f)\| .$$

**Proposition 605** Keeping notations of definition 290 (p. 332), if $(f : \mathcal{R}^n \to \mathcal{R})$ has a maximum value $f_M$ and if it is zero outside the box $\overline{C}(0, c)$ as shown by equation (2.151) (p. 331), we have if $J(f)$ exists:

$$J(f) \leq f_M(2c)^n .$$

**Proof**

In equation (2.153) (p. 332), for any term $t$ in $Gr_a$, we have (see propositions 290 (p. 187) and 289 (p. 186)):

$$
\begin{aligned}
gr^{f}(t) &\leq \sum_{l \in \prod_{k \in [1,n]} [0, p^k[} f_M \prod_{k \in [1,n]} (y^{l_k+1,k} - y^{l_k,k}) \\
&= f_M \sum_{l \in \prod_{k \in [1,n]} [0, p^k[} \prod_{k \in [1,n]} (y^{l_k+1,k} - y^{l_k,k}) \\
&= f_M \prod_{k \in [1,n]} \sum_{l \in \prod_{k \in [1,n]} [0, p^k[} y^{l_k+1,k} - y^{l_k,k} \leq f_M \prod_{k \in [1,n]} 2c = f_M(2c)^n
\end{aligned}
\tag{2.157}
$$

(see also propositions 307 (p. 193), 309 (p. 194)).

**Definition 292 Set of Riemann integrals with a compact support converging uniformly.**

Given a set $I$, a map $(f_i(x) : I \times \mathcal{R}^n \to \mathcal{R}^m)$ such that with notations of equation (2.155) (p. 332) considering $f_i$ as a map $\mathcal{R}^n \to \mathcal{R}^m$ we have:

$$
\begin{aligned}
&(\exists (J_i : I \to \mathcal{R}))(\forall \alpha > 0)(\exists \epsilon(\alpha) > 0)(\forall i \in I)(0 < a < \epsilon(\alpha) \Rightarrow \\
&(\forall x \in S^{f_i}_a)(\|x - J_i\| < \alpha))
\end{aligned}
$$

then the integral $J_i = J(f_i)$ is said to converge uniformly to $J_i$.

**Proposition 606 Multiple integral formula.**
We have $\mathbf{n} \in \mathcal{Z}^+$, $\mathbf{I_1} \subset [\mathbf{1}, \mathbf{n}]$. We set $\mathbf{I_2} = \mathcal{C}_{[1,n]}(\mathbf{I_1})$: $[\mathbf{1}, \mathbf{n}] = \mathbf{I_1} \cup \mathbf{I_2}$ and we use the following labeling $(\forall \ell \in [1,2])(\mathbf{n}_\ell = \mathbf{Card}(\mathbf{I}_\ell))$, $\mathbf{n} = \mathbf{n_1} + \mathbf{n_2}$. We consider a map $(\mathbf{f} : \mathcal{R}^n \to \mathcal{R}^m)$ with a compact support of $\mathcal{R}^n$:

$$(\exists c \in \mathcal{R}^+)(x \notin C(0,c) \Rightarrow f(x) = 0) .$$

In $\mathcal{R}^n$, with notations of definition 287 (p. 331), we consider the terms $(\mathbf{p}, \mathbf{y}, \mathbf{z}) \in \mathbf{Gr}$ defined in equation (2.152) (p. 332)). For those terms we can define the following variables (see definition 112 (p. 131) and proposition 153 (p. 133)):

$$(\forall \ell \in [1,2])(z_\ell \in \prod_{i \in I_\ell} \mathcal{R} \approx \mathcal{R}^{n_\ell}, y_\ell \in \prod_{i \in I_\ell} \mathcal{R}, p_\ell \in \prod_{i \in I_\ell} \mathcal{Z})$$
$$z = z_1 \times z_2,\ y = y_1 \times y_2,\ p = p_1 \times p_2 .$$

Given $\mathbf{z_2} \in \prod_{i \in I_2} \mathcal{R}$, we may define $(\mathbf{f_1^{z_2}} : \prod_{i \in I_1} \mathcal{R} \to \mathcal{R}^m)$ such that $\mathbf{f_1^{z_2}(z_1) = f(z_1 \times z_2)}$. Using the equation (2.155) (p. 332) to define, if it exists, the map $(\mathbf{F(z_2)} : \prod_{i \in I_2} \mathcal{R} \to \mathcal{R}^m)$ by $\mathbf{F(z_2) = J(f_1^{z_2})}$. We suppose that $\mathbf{J(f_1^{z_2})}$ is uniformly convergent (definition 292 (p. 337)):

$$\begin{aligned}&(\forall \alpha > 0)(\exists \epsilon^F(\alpha) > 0)(\forall z_2 \in \prod_{i \in I_2} \mathcal{R})(\exists F(z_2) \in \mathcal{R})(0 < a < \epsilon^F(\alpha) \\ &\Rightarrow (\forall s \in S_a^{f_1^{z_2}})(\|s - F(z_2)\| < \alpha))\end{aligned} \tag{2.158}$$

$\mathbf{S_a^{f_1^{z_2}}}$ is defined in equation (2.154) (p. 332) of definition 287 (p. 331).
Then if $\mathbf{J(f)}$ exists (see equation (2.155) (p. 332)), $\mathbf{J(F)}$ exists too with $\mathbf{J(F) = J(f)}$.

**Proof**

In the definition of $\mathbf{J(f)}$ by equation (2.155) (p. 332), we take $\epsilon_1 = \epsilon^f(\alpha/2)$ and in (2.158) (p. 338) defining $\mathbf{F}$, $\epsilon_2 = \epsilon^F(\frac{\alpha}{2(2c)^{n_2}})$. Using proposition 294 (p. 188), we set:

$$\begin{aligned}&(\forall j \in \{1,2\})(L_j = \prod_{k \in I_j} [0, p^k[),\ L = \prod_{j \in \{1,2\}} L_j,\\ &gr^f((p,y,z)) = \sum_{\ell \in L} f(z^{\ell,i} e_i) \prod_{k \in [1,n]} (y^{\ell_k+1,k} - y^{\ell_k,k})\\ &= \sum_{\ell_2 \in L_2} \sum_{\ell_1 \in L_1} f(z^{\ell_1 \times \ell_2, i} e_i) \prod_{k \in [1,n]} (y^{(\ell_1 \times \ell_2)_k+1,k} - y^{(\ell_1 \times \ell_2)_k,k}) .\end{aligned}$$

When choosing $\mathbf{a} \in ]\mathbf{0}, \min(\epsilon_1, \epsilon_2)[$, $\mathbf{t} = (\mathbf{p}, \mathbf{z}, \mathbf{y}) \in \mathbf{Gr_a}$ with

$$(\forall \ell \in L)(z^{\ell,k} = \frac{y^{\ell_k,k} + y^{\ell_k+1,k}}{2})$$

we have:

1. $\|\mathbf{gr^f(p,y,z) - J(f)}\| < \alpha/2$
2. but if we set:

$$\begin{aligned}&(\forall \ell_1 \times \ell_2 \in L)(z_1^{\ell_1} = \sum_{k \in I_1} \frac{y^{(\ell_1)_k,k} + y^{(\ell_1)_k+1,k}}{2} e_k,\\ &z_2^{\ell_2} = \sum_{k \in I_2} \frac{y^{(\ell_2)_k,k} + y^{(\ell_2)_k+1,k}}{2} e_k,\ z^{\ell_1 \times \ell_2} = z_1^{\ell_1} \cup z_2^{\ell_2})\\ &(\forall \ell_2 \in L_2)(gr_a^{f_1^{z_2}}(p_1, y_1, z_1) = \sum_{\ell_1 \in L_1} f(z_1^{\ell_1} \cup z_2^{\ell_2}) \prod_{k \in I_1} (y^{(\ell_1 \times \ell_2)_k+1,k} - y^{(\ell_1 \times \ell_2)_k,k})) .\end{aligned}$$

We also have:

$$(\forall \ell_2 \in L_2)(\|gr_a^{f_1^{z_2}}(p_1, y_1, z_1) - F(z^{\ell_2,i}e_i)\| < \frac{\alpha}{2(2c)^{n_2}}) .$$

Then we have from propositions 289 (p. 186) and 290 (p. 187):

$$\begin{aligned}
&\|gr^f(p, y, z) - \sum_{\ell_2 \in L_2} F(z^{\ell_2,i}e_i) \prod_{k \in I_2} (y^{(\ell_2)_k+1,k} - y^{(\ell_2)_k,k})\| \\
&= \|\sum_{\ell_2 \in L_2} \prod_{k \in I_2} (y^{(\ell_2)_k+1,k} - y^{(\ell_2)_k,k}) \sum_{\ell_1 \in L_1} \prod_{k \in I_1} (y^{(\ell_1)_k+1,k} - y^{(\ell_1)_k,k}) f(z_1^{\ell_1} \cup z_2^{\ell_2}) \\
&- \sum_{\ell_2 \in L_2} F(z^{\ell_2,i}e_i) \prod_{k \in I_2} (y^{(\ell_2)_k+1,k} - y^{(\ell_2)_k,k})\| \\
&< \sum_{\ell_2 \in L_2} \|gr_a^{f_1^{z_2}}(p_1, y_1, z_1) - F(z^{\ell_2,i}e_i)\| \prod_{k \in I_2} (y^{(\ell_2)_k+1,k} - y^{(\ell_2)_k,k}) \\
&< \frac{\alpha}{2(2c)^{n_2}} \sum_{\ell_2 \in L_2} \prod_{k \in I_2} (y^{(\ell_2)_k+1,k} - y^{(\ell_2)_k,k}) = \frac{\alpha}{2(2c)^{n_2}} \prod_{k \in I_2} \sum_{j \in [0,p^k]} (y^{j+1,k} - y^{j,k}) \\
&= \frac{\alpha}{2(2c)^{n_2}} \prod_{k \in I_2} (y^{p^k,k} - y^{0,k}) < \alpha/2
\end{aligned}$$

3. then

$$\begin{aligned}
&\|J(f) - \sum_{\ell_2 \in L_2} F(z^{\ell_2,i}e_i) \prod_{k \in I_2} (y^{(\ell_2)_k+1,k} - y^{(\ell_2)_k})\| \\
&< \|gr^f(p, y, z) - J(f)\| + \alpha/2 < \alpha .
\end{aligned}$$

**Proposition 607** If a map $(f : \mathcal{R}^n \to \mathcal{R}^m)$ with a compact support (definition 287 (p. 331)) has an integral $J(f)$ (definition 290 (p. 332)) then for any $\lambda \in \mathcal{R}$, $\lambda f$ has an integral and $J(\lambda f) = \lambda J(f)$.

**Proof**
If $\lambda = 0$, $\lambda f = 0$ everywhere in $\mathcal{R}^n$, then its integral is $0$ and equal to $\lambda J(f) = 0$. We suppose now $\lambda \neq 0$. Keeping the notations of definition 290 (p. 332), from equation (2.155) (p. 332),

$$(\exists J \in \mathcal{R})(\forall \alpha > 0)(\exists \epsilon(\alpha) > 0)(0 < a < \epsilon(\alpha) \Rightarrow (x \in S_a^f \Rightarrow \|x - J\| < \alpha))$$

then

$$\begin{aligned}
&(\exists J' = \lambda J \in \mathcal{R})(\forall \alpha > 0)(\exists \epsilon'(\alpha) = \epsilon(\alpha/|\lambda|) > 0)(0 < a < \epsilon'(\alpha) \Rightarrow \\
&(x' = \lambda x \in S_a^{\lambda f} \Rightarrow x \in S_a^f \Rightarrow \frac{\|x' - J'\|}{|\lambda|} \\
&= \|\frac{\lambda x - \lambda J}{\lambda}\| = \|x - J\| < \alpha/|\lambda| \Rightarrow \|x' - J'\| < \alpha)) .
\end{aligned}$$

Then $\lambda f$ has an integral which is $\lambda J(f)$.

**Proposition 608** If two maps $(f : \mathcal{R}^n \to \mathcal{R}^m)$ and $(g : \mathcal{R}^n \to \mathcal{R}^m)$ are both on compact support (definition 287 (p. 331)) and if the two maps have a Riemann integral (definition 290 (p. 332)), their sum also has a Riemann integral sum of the two integrals.

**Proof**
In equation (2.155) we may define $\epsilon_f(\alpha_f)$ and $\epsilon_g(\alpha_g)$.
Taking $\epsilon = \min(\epsilon_f(\alpha/2), \epsilon_g(\alpha/2))$ and considering $Gr_a$ with $a < \epsilon$ for any term $t$ in $Gr_a$ we have:

$$gr^{f+g}(t) = gr^f(t) + gr^g(t)$$

then

$$(\forall \alpha > 0)(\exists \epsilon = \min(\epsilon_f(\alpha/2), \epsilon_g(\alpha/2)))$$
$$(\forall t \in Gr_a)(\|gr^{f+g}(t) - J(f) - J(g)\| < \|gr^f(t) - J(f)\| + \|gr^g(t) - J(g)\| < \alpha)$$

then $J(f+g)$ exists and $J(f+g) = J(f) + J(g)$. $J(h)$ stands for the Riemann integral of $h$ if it exists.

**Proposition 609** If two maps $(f : \mathcal{R}^n \to \mathcal{R})$ and $(g : \mathcal{R}^n \to \mathcal{R})$ are both on compact support (definition 287 (p. 331)), if the two maps have a Riemann integral (definition 290 (p. 332)) and if $(\forall x \in \mathcal{R}^n)(f(x) \leq g(x))$ then $J(f) \leq J(g)$.

**Proof**
Keeping notation of definition 290 (p. 332), from proposition 607 (p. 339), $-g$ has an integral which is $-J(g) = J(-g)$. From proposition 608 (p. 339), $f - g$ has an integral which is $J(f - g) = J(f) - J(g)$ then from proposition 603 (p. 336), $J(f) - J(g) \leq 0$ and $J(f) \leq J(g)$.

**Proposition 610** If a map $(f : \mathcal{R}^n \to \mathcal{R}^m)$ with a compact support (definition 287 (p. 331)) has an integral $J(f)$ (definition 290 (p. 332)), the translated map by $\Delta \in \mathcal{R}^n$ defined by $(\forall x \in \mathcal{R}^n)({}^{\Delta}f(x + \Delta) = f(x))$ has an integral which is equal to $J(f)$.

**Proof**
If $f$ has a compact support in $\overline{C}(0, c)$, ${}^{\Delta}f$ has $\overline{C}(0, c + \|\Delta\|)$ as a compact support:

$$\begin{aligned} x \notin \overline{C}(0, c + \|\Delta\|) &\Rightarrow (\exists i \in [1, n])(|x^i| > c + \|\Delta\|) \\ &\Rightarrow (\exists i \in [1, n])(|x^i + \Delta^i| \geq |x^i| - |\Delta^i| \geq |x^i| - \|\Delta\| > c) \\ &\Rightarrow x + \Delta \notin \overline{C}(0, c) \Rightarrow {}^{\Delta}f(x + \Delta) = f(x) = 0 \,. \end{aligned}$$

We take for the box associated to ${}^{\Delta}f(x)$, $\overline{C}(0, c + \|\Delta\|)$ and we suppose that ${}^{\Delta}f(x)$ has an integral in that box. We may take for $f$, $\overline{C}(0, c + 2\|\Delta\|)$ as the box for determining the set ${}^{f}Gr$. It does not change the integral of $f$ and ${}^{\Delta}f(x)$, if they exist (see definition 290 (p. 332), 2). To any term $t = (p, y, z)$ of ${}^{f}Gr_a$ with $a < c$ corresponds the term translated by $\Delta$:

$${}^{\Delta}t = ({}^{\Delta}p, {}^{\Delta}y, {}^{\Delta}z)$$

Using $(\mathcal{Y}(x) : \mathcal{R} \to \{0, 1\})$, the Heaviside map (definition 216 (p. 245)), we define a map $(h^k = \mathcal{Y}(\Delta^k)({}^{\Delta}p^k - p^k) : [1, n] \to \mathcal{Z})$. Using that map, we set $\Delta z$ as in definition 291 (p. 334) by:

$$(\forall \ell \in \prod_{k \in [1,n]} [h^k, p^k + h^k[)({}^{\Delta}z^{\ell} = z^{\ell - h^k} + \Delta)$$

which is equivalent to:

$$(\forall \ell \in \prod_{k \in [1,n]} [0, p^k[)({}^{\Delta}z^{\ell + h^k} = z^{\ell} + \Delta) \,.$$

For the other value of $\ell$, we use (see definition 354 (p. 454)):

$$(\forall \ell \in \prod_{k\in[1,n]}([0,{}^{\Delta}p[-\prod_{k\in[1,n]}[h^k,p^k+h^k[)({}^{\Delta}z^{\ell}=0.5\sum_{k\in[1,n]}({}^{\Delta}y^{l_k+1,k}+{}^{\Delta}y^{l_k,k})e_k)\,.$$

Then since from definition 291 (p. 334):

$$\ell\in(\prod_{k\in[1,n]}[0,{}^{\Delta}p[-\prod_{k\in[1,n]}[h^k,p^k+h^k[)$$
$$\Rightarrow(\forall k\in[1,n])({}^{\Delta}y^{l_k+1,k}\notin C(0,c),{}^{\Delta}y^{l_k,k}\notin C(0,c))\Rightarrow{}^{\Delta}z^{\ell}-\Delta\notin C(0,c)$$

we have:

$$(\forall\ell\in(\prod_{k\in[1,n]}[0,{}^{\Delta}p[-\prod_{k\in[1,n]}[h^k,p^k+h^k[))({}^{\Delta}f({}^{\Delta}z^{\ell})=f({}^{\Delta}z^{\ell}-\Delta)=0)$$
$$(\forall\ell\in\prod_{k\in[1,n]}[h^k,p^k+h^k[)({}^{\Delta}z^{\ell}=z^{\ell-h^k}+\Delta,{}^{\Delta}f({}^{\Delta}z^{\ell})={}^{\Delta}f(z^{\ell}+\Delta)=f(z^{\ell}))$$

by construction we have:

$$\begin{aligned}&\sum_{\ell\in\prod_{k\in[1,n]}[0,p^k[}f(z^{\ell})\prod_{k\in[1,n]}(y^k(\ell^k+1)-y^k(\ell^k))\\
&=\sum_{\ell\in\prod_{k\in[1,n]}[h^k,p^k+h^k[}f({}^{\Delta}z^{\ell})\prod_{k\in[1,n]}({}^{\Delta}y^k(\ell^k+1)-{}^{\Delta}y^k(\ell^k))\\
&=\sum_{\ell\in\prod_{k\in[1,n]}[h^k,p^k+h^k[}{}^{\Delta}f({}^{\Delta}z^{\ell})\prod_{k\in[1,n]}({}^{\Delta}y^k(\ell^k+1)-{}^{\Delta}y^k(\ell^k))\\
&+\sum_{\ell\in\prod_{k\in[1,n]}[0,{}^{\Delta}p[-\prod_{k\in[1,n]}[h^k,p^k+h^k[}{}^{\Delta}f({}^{\Delta}z^{\ell})\prod_{k\in[1,n]}({}^{\Delta}y^k(\ell^k+1)-{}^{\Delta}y^k(\ell^k))\\
&=\sum_{\ell\in\prod_{k\in[1,n]}[0,{}^{\Delta}p^k[}{}^{\Delta}f({}^{\Delta}z^{\ell})\prod_{k\in[1,n]}({}^{\Delta}y^k(\ell^k+1)-{}^{\Delta}y^k(\ell^k))=gr^{{}^{\Delta}f}({}^{\Delta}p,{}^{\Delta}y,{}^{\Delta}z)\end{aligned}$$

then since $J({}^{\Delta}f)$ exists, we have:

$$(\forall\alpha>0)(\exists\epsilon(\alpha)>0)(0<a<\epsilon(\alpha),t\in Gr_a\Rightarrow\|gr^{{}^{\Delta}f}(t)-J({}^{\Delta}f)\|<\alpha)\,.$$

We also have $t\in Gr_a\Leftrightarrow{}^{\Delta}t\in Gr_a$ when ${}^{\Delta}t$ is calculated as above. Then $\|gr^f(t)-J({}^{\Delta}f)\|<\alpha$. $J(f)$ exists with $J({}^{\Delta}f)=J(f)$.
Setting $g={}^{\Delta}f$ and $\Delta'=-\Delta$, if $J({}^{\Delta'}g)=J(f)$ exists, the above proves that $J(g)=J({}^{\Delta}f)$ exists with $J(f)=J({}^{\Delta}f)$. Then proposition 610 (p. 340) is true.

## 2.24 Volume in $\mathcal{R}^n$

Otherwise mentioned, in this whole section, we refer only to integrals with $m=1$ (see definition 290 (p. 332)): $J\in\mathcal{R}$.

**Definition 293 Characteristic map of a subset of $\mathcal{R}^n$.**

A being a subset of $\mathcal{R}^n$, a map $(\phi(x):\mathcal{R}^n\to\{0,1\}\subset\mathcal{R})$ such that $x\in A\Rightarrow\phi(x)=1,x\notin A\Rightarrow\phi(x)=0$ is called the characteristic map of A: $(A=\phi^{-1}(1))$.

**Definition 294 Integration over a set.**

Given $\mathbf{A}$ a subset of $\mathcal{R}^n$ inside a cube of $\mathcal{R}^n$, calling $\phi_A(x)$ its characteristic map (definition 293 (p. 341)), given a map $(f : \mathcal{R}^n \to \mathcal{R}^m)$, one calls integral of $\mathbf{f}$ over $\mathbf{A}$ the following Riemann integral (definition 290 (p. 332)) if it exists:

$$J_A(f) = J(\phi_A \times f) .$$

**Definition 295 Volume of a subset $A \subset \mathcal{R}^n$ which is within a cube of $\mathcal{R}^n$ .**

$\mathbf{A}$ is within a cube, means also: $(\exists b \in \mathcal{R}^+)(A \subset \overline{C}(0, b))$ where $\overline{C}$ is a closed and then compact cube centered on $\mathbf{0}$ and of side $\mathbf{2b}$ (see the definition of a cube in definition 260 (p. 285)). It is also in a ball: $\overline{C}(0, b) \subset \overline{B}(0, {}^p\sqrt{n}\, b) = \overline{B}(0, n^{1/p} b)$ (see definition 243 (p. 267)). Calling $\phi$, the characteristic map of $\mathbf{A}$ (definition 293 (p. 341)), the volume of $\mathbf{A}$ is $J(\phi)$, the integral of $\phi$ if it exists (definition 287 (p. 331)).

**Proposition 611** We suppose that a subset $A \subset \mathcal{R}^n$ within a ball has a volume. We call $\phi(x)$ the characteristic map of $\mathbf{A}$ and for any term $t = (p, y, z)$ of a grid (definition 288 (p. 331)), we call $L^z(t)$ the subset of $\prod_{k\in[1,n]}[0, p^k[$ such that:

$$l \in L^z(t) \Leftrightarrow (l \in \prod_{k\in[1,n]} [0, p^k[, (\exists x \in \prod_{k\in[1,n]} [y^{l_k,k}, y^{l_k+1,k}[)(\phi(x) = 0)) .$$

We have the following relation:

$$(\forall \alpha > 0)(\exists \epsilon > 0)(0 < a < \epsilon \Rightarrow (\forall t = (p, y, z) \in Gr_a)$$
$$(0 < \sum_{l\in L^z(t)} \phi(z^l) \prod_{k\in[1,n]} (y^{l_k+1,k} - y^{l_k,k}) < \alpha))$$

with $\epsilon = \epsilon^\phi(\alpha/2)$ where $\epsilon^\phi$ is a map defined in definition 290 (p. 332) relation (2.155) (p. 332).

**Proof**

Since $\phi(x)$ has a Riemann integral $\mathbf{J}$, it verifies relation (2.155) (p. 332) which implies (see definition 290 (p. 332)):

$$(\exists \epsilon^\phi(\alpha^\phi) > 0)(0 < a < \epsilon^\phi(\alpha^\phi) \Rightarrow (\forall t \in Gr_a)(|gr^\phi(t) - J| =$$
$$|\sum_{l\in\prod_{k\in[1,n]}[0,p^k[-L^z(t)} \prod_{k\in[1,n]} (y^{l_k+1,k} - y^{l_k,k})$$
$$+ \sum_{l\in L^z(t)} \phi(z^l) \prod_{k\in[1,n]} (y^{l_k+1,k} - y^{l_k,k}) - J| < \alpha^\phi)) .$$

We can define a map $(t'(t) = (p, y, z') : Gr_a \to Gr_a)$ such that (see definitions 332 (p. 440) and 333 (p. 441)):

$$(\forall l \in \prod_{k\in[1,n]} [0, p^k[)(l \notin L^z(t) \Rightarrow z^l = z'^l,$$
$$l \in L^z(t) \Rightarrow z'^l = \tau_u(u \in \prod_{k\in[1,n]} [y^{l_k,k}, y^{l_k+1,k}[, \phi(u) = 0)) \tag{2.159}$$

then:

$$0 < a < \epsilon^{\phi}(\alpha/2) \Rightarrow (t \in Gr_a \Rightarrow (t'(t) \in Gr_a, \\ |gr^{\phi}(t'(t)) - J| < \alpha/2, |gr^{\phi}(t) - J| < \alpha/2)) .$$

Since by definition (see equation (2.153) (p. 332), (2.159) (p. 342)):

$$gr^{\phi}(t) = gr^{\phi}(t') + \sum_{l \in L^z(t)} \phi(z^l) \prod_{k \in [1,n]} (y^{l_k+1,k} - y^{l_k,k}) .$$

We have:

$$|\sum_{l \in L^z(t)} \phi(z^l) \prod_{k \in [1,n]} (y^{l_k+1,k} - y^{l_k,k})| = |gr^{\phi}(t) - J - gr^{\phi}(t') + J| \\ < |gr^{\phi}(t) - J| + |gr^{\phi}(t') - J| < \alpha .$$

Since $(\forall l \in \prod_{k \in [1,n]}[0, p^k[)(\forall k \in [1,n])(y^{l_k+1,k} - y^{l_k,k} > 0)$ we have:

$$0 < \sum_{l \in L^z(t)} \phi(z^l) \prod_{k \in [1,n]} (y^{l_k+1,k} - y^{l_k,k}) < \alpha .$$

**Proposition 612** We suppose that a subset $\mathbf{A} \subset \mathcal{R}^n$ within a ball has a volume. Then the closure $\bar{\mathbf{A}}$ (definition 136 (p. 155)) also has a volume which is the same as the volume of $\mathbf{A}$.

**Proof**
We call $(e_i : [1,n] \rightarrow \mathcal{R}^n)$ the reference basis of $\mathcal{R}^n$, $e_i = \prod_{j \in [1,n]}^{e} \delta_i^j$ (see equation (2.106) (p. 267) in definition 242 (p. 267)). From definition 290 (p. 332), $\mathbf{A}$ having a volume $\mathbf{V} > 0$, if we call $(\phi(x) : \mathcal{R}^n \rightarrow \{0,1\})$ its characteristic map, we have:

$$(\forall \alpha > 0)(\exists \epsilon_1(\alpha))(0 < a < \epsilon_1(\alpha) \Rightarrow S_a^{\phi} \subset ]V - \alpha, V + \alpha[) .$$

From proposition 611 (p. 342):

$$(\forall \alpha > 0)(\exists \epsilon_2(\alpha))(0 < a < \epsilon_2(\alpha) \Rightarrow \\ (\forall t = (p, y, z) \in Gr_a)(|\sum_{l \in L^z(t)} \phi(z^l) \prod_{k \in [1,n]} (y^{l_k+1,k} - y^{l_k,k})| < \alpha)) . \tag{2.160}$$

We call $(\bar{\phi}(x) : \mathcal{R}^n \rightarrow \{0,1\})$ the characteristic map of $\bar{\mathbf{A}}$ and we set:

$$\epsilon = \min(\epsilon_1(\alpha/2), \epsilon_2(\alpha/2)) .$$

We take $0 < a < \epsilon$. It gives:

$$(\forall t = (p, y, z) \in Gr_a)(|gr^{\phi}(t) - V| < \alpha/2) .$$

Considering the sum:

$$gr^{\bar{\phi}}(t) = \sum_{l \in \prod_{k \in [1,n]}[0,p^k[} \bar{\phi}(z^{l,i} e_i) \prod_{k \in [1,n]} (y^{l_k+1,k} - y^{l_k,k})$$

It can be split into two terms:
We call $\mathbf{B(A)} = \bar{\mathbf{A}} - \mathbf{A} = \mathcal{C}_{\bar{\mathbf{A}}}(\mathbf{A}) \subset \mathcal{B}(\mathbf{A})$ (see definitions 354 (p. 454), 138 (p. 157)) and we set:

$$\begin{aligned} \mathbf{gr_A(t)} &= \sum_{\mathbf{l} \in \prod_{\mathbf{k} \in [1,n]} [0,\mathbf{p^k}[, \mathbf{z^{l,i}e_i} \in \mathbf{A}} \bar{\phi}(\mathbf{z^{l,i}e_i}) \prod_{\mathbf{k} \in [1,n]} (\mathbf{y^{l_k+1,k}} - \mathbf{y^{l_k,k}}) \\ &= \mathbf{gr^{\phi}(t)} \in \mathbf{S^{\phi}_a} \subset [\mathbf{V} - \alpha/2, \mathbf{V} + \alpha/2] \end{aligned} \tag{2.161}$$

and

$$\mathbf{gr_{B(A)}(t)} = \sum_{\mathbf{l} \in \prod_{\mathbf{k} \in [1,n]} [0,\mathbf{p^k}[, \mathbf{z^{l,i}e_i} \in \mathbf{B(A)}} \bar{\phi}(\mathbf{z^{l,i}e_i}) \prod_{\mathbf{k} \in [1,n]} (\mathbf{y^{l_k+1,k}} - \mathbf{y^{l_k,k}})$$

with $\mathbf{gr^{\bar{\phi}}(t)} = \mathbf{gr_A} + \mathbf{gr_{B(A)}}$. Since by definition $(\forall \mathbf{k} \in [1,\mathbf{n}])(\mathbf{y^{l_k,k}} \leq \mathbf{z^{l_k,k}} < \mathbf{y^{l_k+1,k}})$, we can consider:

$$\mathbf{d} = \min_{\mathbf{l} \in \prod_{\mathbf{k} \in [1,n]} [0,\mathbf{p^k}[, \mathbf{k} \in [1,n]} \mathbf{y^{l_k+1,k}} - \mathbf{z^{l_k,k}}$$

we have $\mathbf{d} > \mathbf{0}$. Then we have:

$$(\forall \mathbf{k} \in [1,\mathbf{n}])(\mathbf{y^{l_k,k}} < \mathbf{z^{l_k,k}} + \mathbf{d}/2 < \mathbf{y^{l_k+1,k}}, \mathbf{y^{l_k,k}} - \mathbf{d}/2 < \mathbf{z^{l_k,k}} < \mathbf{y^{l_k+1,k}} - \mathbf{d}/2) \ .$$

We may consider a new term $(\mathbf{p}, \mathbf{y'}, \mathbf{z}) = (\mathbf{p}, \mathbf{y} - \mathbf{d}/2, \mathbf{z}) \in \mathbf{Gr_a}$ with:

$$(\forall \mathbf{l} \in \prod_{\mathbf{k} \in [1,n]} [0,\mathbf{p^k}])(\forall \mathbf{k} \in [1,\mathbf{n}])(\mathbf{y'^{l_k,k}} = \mathbf{y^{l_k,k}} - \mathbf{d}/2) \ .$$

We have with the corresponding sum $\mathbf{gr'_{B(A)}}$

$$\begin{aligned} &\mathbf{gr_{B(A)}(p,y',z)} = \sum_{\mathbf{l} \in \prod_{\mathbf{k} \in [1,n]} [0,\mathbf{p^k}[, \mathbf{z^{l,i}e_i} \in \mathbf{B(A)}} \bar{\phi}(\mathbf{z^{l,i}e_i}) \prod_{\mathbf{k} \in [1,n]} (\mathbf{y^{l_k+1,k}} - \mathbf{d}/2 - \mathbf{y^{l_k,k}} + \mathbf{d}/2) \\ &= \mathbf{gr_{B(A)}(p,y,z)} \ . \end{aligned}$$

Since the box $\mathbf{Bx'_l} = \prod_{\mathbf{k} \in [1,n]} ]\mathbf{y^{l_k,k}} - \mathbf{d}/2, \mathbf{y^{l_k+1,k}} - \mathbf{d}/2[$ containing $\mathbf{z^{l,i}e_i}$ is open, since $\mathbf{z^{l,i}e_i} \in \mathbf{B(A)}$ and $\mathbf{z^{l,i}e_i} \notin \mathbf{A}$ from definition 136 (p. 155), there is a $\mathbf{z'^{l,i}e_i} \in \mathbf{A} \cap \mathbf{Bx'_l}$ with $\phi(\mathbf{z'^{l,i}e_i}) = \bar{\phi}(\mathbf{z^{l,i}e_i}) = \mathbf{1}, \phi(\mathbf{z^{l,i}e_i}) = \mathbf{0}$. Then we have:

$$\mathbf{gr_{B(A)}(p,y',z')} = \sum_{\mathbf{l} \in \prod_{\mathbf{k} \in [1,n]} [0,\mathbf{p^k}[, \mathbf{z^{l,i}e_i} \in \mathbf{B(A)}} \phi(\mathbf{z'^{l,i}e_i}) \prod_{\mathbf{k} \in [1,n]} (\mathbf{y^{l_k+1,k}} - \mathbf{d}/2 - \mathbf{y^{l_k,k}} + \mathbf{d}/2)$$

from proposition 611 (p. 342) and equation (2.160),

$$\mathbf{0} < \mathbf{a} < \epsilon \Rightarrow \mathbf{0} < \mathbf{gr_{B(A)}(p,y',z)} = \mathbf{gr_{B(A)}(p,y,z)} \leq \mathbf{gr_{B(A)}(p,y',z')} < \alpha/2$$

which gives from equation (2.161) with $\mathbf{t} = (\mathbf{p}, \mathbf{y}, \mathbf{z})$:

$$\begin{aligned} &\mathbf{0} < \mathbf{a} < \epsilon \Rightarrow |\mathbf{gr^{\bar{\phi}}(t)} - \mathbf{V}| = |\mathbf{gr_A(t)} - \mathbf{V} + \mathbf{gr_{B(A)}(t)}| \\ &= |\mathbf{gr^{\phi}(t)} - \mathbf{V} + \mathbf{gr_{B(A)}(t)}| \leq |\mathbf{gr^{\phi}(t)} - \mathbf{V}| + |\mathbf{gr_{B|(A)}(t)}| < \alpha \ . \end{aligned}$$

Then proposition 612 (p. 343) is true.

**Proposition 613** Considering a set of grids $\mathbf{Gr}$ (definition 288 (p. 331), equation (2.152)) and inside it two terms $\mathbf{t} = (\mathbf{p}, \mathbf{y}, \mathbf{z}) \in \mathbf{Gr}$ and $\mathbf{t'} = (\mathbf{p'}, \mathbf{y'}, \mathbf{z'}) \in \mathbf{Gr}$. From those two terms we can deduce $(\mathbf{t''}(\mathbf{t}, \mathbf{t'}) = (\mathbf{p''}, \mathbf{y''}, \mathbf{z''}) : \mathbf{Gr} \to \mathbf{Gr})$ defined as follows:
For all $\mathbf{k} \in [\mathbf{1}, \mathbf{n}]$, there are two strictly increasing injections (definitions 386 (p. 486), 122 (p. 139)):

$$\begin{array}{c}(\mathbf{h}(\mathbf{i}) : [\mathbf{0}, \mathbf{p}] \to [\mathbf{0}, \mathbf{p''}]) \\ (\mathbf{h'}(\mathbf{i}) : [\mathbf{0}, \mathbf{p'}] \to [\mathbf{0}, \mathbf{p''}])\end{array} \tag{2.162}$$

such as[33]:

$$\begin{array}{c}\mathbf{y}(\mathbf{i}) = \mathbf{y''}(\mathbf{h}(\mathbf{i})), \mathbf{y'}(\mathbf{i}) = \mathbf{y''}(\mathbf{h'}(\mathbf{i})), (\exists \mathbf{q} \in [\mathbf{0}, \mathbf{p} + \mathbf{p'} + \mathbf{1}] = [\mathbf{0}, \mathbf{p''}]) \\ (\mathbf{h}([\mathbf{0}, \mathbf{p}]) \cup \mathbf{h}([\mathbf{0}, \mathbf{p'}]) = [\mathbf{0}, \mathbf{q}]), (\forall \mathbf{i} \in [\mathbf{0}, \mathbf{q} - \mathbf{1}])(\mathbf{y''}(\mathbf{i}) < \mathbf{y''}(\mathbf{i} + \mathbf{1}))\end{array} \tag{2.163}$$

and such that:

$$(\mathbf{t} \in \mathbf{Gr_a}, \mathbf{t'} \in \mathbf{Gr_a}) \Rightarrow \mathbf{t''} \in \mathbf{Gr_a} \;.$$

**Proof**
Considering in the set of grids $\mathbf{Gr}$ (definition 290 (p. 332), equation (2.154)) two terms $\mathbf{t} = (\mathbf{p}, \mathbf{y}, \mathbf{z}) \in \mathbf{Gr}$ and $\mathbf{t'} = (\mathbf{p'}, \mathbf{y'}, \mathbf{z'}) \in \mathbf{Gr}$. Those terms may be considered as members of $\mathbf{Gr_a}$ taking for instance $\mathbf{a}$ such that

$$\mathbf{a} > \max(\max_{\mathbf{i} \in [\mathbf{0}, \mathbf{p}-\mathbf{1}], \mathbf{k} \in [\mathbf{1}, \mathbf{n}]} (\mathbf{y^k}(\mathbf{i} + \mathbf{1}) - \mathbf{y^k}(\mathbf{i})), \max_{\mathbf{i} \in [\mathbf{0}, \mathbf{p}-\mathbf{1}], \mathbf{k} \in [\mathbf{1}, \mathbf{n}]} (\mathbf{y'^k}(\mathbf{i} + \mathbf{1}) - \mathbf{y'^k}(\mathbf{i}))) \;.$$

From those two terms we can deduce $(\mathbf{t''}(\mathbf{t}, \mathbf{t'}) = (\mathbf{p''}, \mathbf{y''}, \mathbf{z''}) : \mathbf{Gr_a^2} \to \mathbf{Gr_a})$:
We consider any $\mathbf{k} \in [\mathbf{1}, \mathbf{n}]$, and following footnote 33 (p. 345), for reducing the number of upper indices, we write in this part $\mathbf{y^{j,k}}$ as $\mathbf{y}(\mathbf{j})$, $\mathbf{p^k}$ as $\mathbf{p}$, $\mathbf{h^k}$ as $\mathbf{h}$ and the same with the quantities with $'$ and $''$. It means that all terms written are elements of $\mathcal{R}$ or $\mathcal{Z}$. We consider the set $\mathbf{Y}$ of possible values of $\mathbf{y}$ and $\mathbf{y'}$ which will be the set of values for $\mathbf{y''}$: $\mathbf{Y} = \mathbf{y}([\mathbf{0}, \mathbf{p}]) \cup \mathbf{y'}([\mathbf{0}, \mathbf{p'}])$. We have $\mathbf{p''} + \mathbf{1} = \mathbf{Card}(\mathbf{Y}) \leq \mathbf{p} + \mathbf{p'} + \mathbf{2}$. Then, from proposition 256 (p. 172), there is a strictly increasing bijection definition 386 (p. 486) between $[\mathbf{0}, \mathbf{p''}]$ and $\mathbf{Y}$: $(\mathbf{y''}(\mathbf{i}) : [\mathbf{0}, \mathbf{p''}] \to \mathbf{Y})$. Then we may set:

$$\begin{array}{c}(\mathbf{h} = \mathbf{y''^{-1}} \circ \mathbf{y} : [\mathbf{0}, \mathbf{p}] \to [\mathbf{0}, \mathbf{p''}]) \\ (\mathbf{h'} = \mathbf{y''^{-1}} \circ \mathbf{y'} : [\mathbf{0}, \mathbf{p'}] \to [\mathbf{0}, \mathbf{p''}])\end{array}$$

which are strictly increasing injections as product (definition 117 (p. 134)) of strictly increasing injections (proposition 848 (p. 486)).

**Proposition 614** Having $\mathbf{A} \subset \mathcal{R}^\mathbf{n}$ , we consider two elements $\mathbf{t} = (\mathbf{p}, \mathbf{y}, \mathbf{z}) \in \mathbf{Gr}$, $\mathbf{t'} = (\mathbf{p'}, \mathbf{y}, \mathbf{z}) \in \mathbf{Gr}$ and the deduced one $\mathbf{t''} = (\mathbf{p''}, \mathbf{y''}, \mathbf{z''}) \in \mathbf{Gr}$ as defined in proposition 613 (p. 345). We define, associated to the set $\mathbf{A}$ and to a term $\mathbf{t}(\mathbf{p}, \mathbf{y}, \mathbf{z})$, the three subsets $\mathbf{L_t^z}$, $\mathbf{L_t^{nz}}$, $\mathbf{L_t} = \mathbf{L_t^z} \cup \mathbf{L_t^{nz}}$ of $\mathbf{L} = \prod_{\mathbf{k} \in [\mathbf{1}, \mathbf{n}]} [\mathbf{0}, \mathbf{p^k}]$:

$$\begin{array}{c}\ell \in \mathbf{L_t^z} \Leftrightarrow (\ell \in \mathbf{L}, \prod_{\mathbf{k} \in [\mathbf{1}, \mathbf{n}]} [\mathbf{y^k}(\ell_\mathbf{k}), \mathbf{y^k}(\ell_\mathbf{k} + \mathbf{1})[ \cap \mathbf{A} \neq \emptyset, \\ \prod_{\mathbf{k} \in [\mathbf{1}, \mathbf{n}]} [\mathbf{y^k}(\ell_\mathbf{k}), \mathbf{y^k}(\ell_\mathbf{k} + \mathbf{1})[ \cap \mathcal{C}(\mathbf{A}) \neq \emptyset)\end{array}$$

[33] We have suppressed in this part for simplicity the upper index $\mathbf{k}$ and we have written $\mathbf{p^k}, \mathbf{y^{i,k}}, \mathbf{y'^{i,k}}, \mathbf{y''^{i,k}}$ as $\mathbf{p}, \mathbf{y}(\mathbf{i}), \mathbf{y'}(\mathbf{i}), \mathbf{y''}(\mathbf{i})$. The same remark is valid for the map $\mathbf{h}$ which should be written as $\mathbf{h^k}$.

$$\ell \in L_t^{nz} \Leftrightarrow (\ell \in L, \prod_{k\in[1,n]} [y^k(\ell_k), y^k(\ell_k+1)[\subset A)$$

and we consider the following three maps: $(s_t^z : Gr \to \mathcal{R})$ $(s_t^{nz} : Gr \to \mathcal{R})$ and $(s_t : Gr \to \mathcal{R})$

$$s_t^z = \sum_{\ell \in L_t^z} \prod_{i\in[1,n]} (y^i(l_i+1) - y^i(l_i))$$

$$s_t^{nz} = \sum_{\ell \in L_t^{nz}} \prod_{i\in[1,n]} (y^i(l_i+1) - y^i(l_i))$$

$$s_t = \sum_{\ell \in L_t} \prod_{i\in[1,n]} (y^i(l_i+1) - y^i(l_i))$$

$$s_t = s_t^z + s_t^{nz}$$

We have the following inequalities:

$$s_t^{nz} \le s_{t''}^{nz} \le s_{t''} \le s_t \tag{2.164}$$

and the same with $t'$ instead of $t$.

**Proof**

We call $\phi$ the characteristic map (definition 293 (p. 341)) of $A$. Using conclusions of proposition 613 (p. 345), we use its notations with a upper index specifying which coordinate we deal with: we use $y^k, h^k$ instead of $y, h$. We set (see definition 112 (p. 131), 4) $(z(l) : [0, n] \to \prod_{k\in[1,n]}([y^k(l_k, y^k(l_k+1)[))$ by:

$$\begin{aligned}
&\mathcal{C}(A) \cap \prod_{k\in[1,n]} ([y^k(l_k), y^k(l_k+1)[) \neq \emptyset \\
&\Rightarrow z(l) = \tau_x(x \in \mathcal{C}(A) \cap \prod_{k\in[1,n]} ([y^k(l_k), y^k(l_k+1)[)), \\
&\mathcal{C}(A) \cap \prod_{k\in[1,n]} ([y^k(l_k), y^k(l_k+1)[) = \emptyset \\
&\Rightarrow z(l) = \tau_x(x \in \prod_{k\in[1,n]} ([y^k(l_k), y^k(l_k+1)[))
\end{aligned}$$

and the same for $(z''(l) : [0, n] \to \prod_{k\in[1,n]}([y''^k(l_k), y''^k(l_k+1)[))$.
From the properties of the map $h$ and from propositions 289 (p. 186), 290 (p. 187) 1000 (p.

557), 832 (p. 478) and 797 (p. 465), we have:

$$\begin{aligned}
s_t^{nz} &= \sum_{\ell\in L} \phi(z(\ell)) \prod_{k[1,n]} (y^k(\ell_k+1) - y^k(\ell)) \\
&= \sum_{\ell\in L} \phi(z(\ell)) \prod_{k[1,n]} (y''^k(h^k(\ell_k+1)) - y''^k(h^k(\ell))) \\
&= \sum_{\ell\in L} \phi(z(\ell)) \prod_{k[1,n]} \sum_{j\in[h^k(\ell_k),h^k(\ell_k+1)[} (y''^k(j+1) - y''^k(j)) \\
&= \sum_{\ell\in L} \phi(z(\ell)) \sum_{l\in\prod_{k\in[1,n]}[h^k(\ell_k),h^k(\ell_k+1)[} \prod_{k\in[1,n]} (y''^k(l(k)+1) - y''^k(l(k))) \\
&\leq \sum_{\ell\in L} \sum_{l\in\prod_{k\in[1,n]}[h^k(\ell_k),h^k(\ell_k+1)[} \phi(z''(l)) \prod_{k\in[1,n]} (y''^k(l(k)+1) - y''^k(l(k))) \qquad (2.165) \\
&= \sum_{l\in\bigcup_{\ell\in L}\prod_{k\in[1,n]}[h^k(\ell_k),h^k(\ell_k+1)[} \phi(z''(l)) \prod_{k\in[1,n]} (y''^k(l(k)+1) - y''^k(l(k))) \\
&= \sum_{l\in\prod_{k\in[1,n]}\bigcup_{\ell\in[0,p]}[h^k(\ell),h^k(\ell+1)[} \phi(z''(l)) \prod_{k\in[1,n]} (y''^k(l(k)+1) - y''^k(l(k))) \\
&= \sum_{l\in\prod_{k\in[1,n]}[h^k(0),h^k(p)[} \phi(z''(l)) \prod_{k\in[1,n]} (y''^k(l(k)+1) - y''^k(l(k))) \\
&= \sum_{l\in\prod_{k\in[1,n]}[0,p''^k[} \phi(z''(l)) \prod_{k\in[1,n]} (y''^k(l(k)+1) - y''^k(l(k))) = s_{t''}^{nz} .
\end{aligned}$$

The sign $\leq$ is due to

$$\phi(z(\ell)) = 1 \Rightarrow (\forall u \in \prod_{k[1,n]} (y^k(\ell_k+1) - y^k(\ell)))(\phi(u) = 1) \Rightarrow$$
$$(\forall l \in \prod_{k\in[1,n]} [h^k(\ell_k), h^k(\ell_k+1)[)(\phi(z''(l)) = 1)$$

and

$$\phi(z(\ell)) = 0 \Rightarrow (\exists u \in \prod_{k[1,n]} (y^k(\ell_k+1) - y^k(\ell)))(\phi(u) = 0) \Rightarrow$$
$$(\forall l \in \prod_{k\in[1,n]} [h^k(\ell_k), h^k(\ell_k+1)[)(\phi(z''(l)) \in \{0,1\}) .$$

The last two lines come from $\bigcup_{\ell\in[0,p]}[h^k(\ell), h^k(\ell+1)[= [h^k(0), h^k(p)[$ (proposition 257 (p. 173)) and from $(l \in [0,p'']^n, l(k) \in ]p,p'']) \Rightarrow \phi(z''(l)) = 0$.
Then $s_t^{nz} \leq s_{t''}^{nz}$.
Using the same technique, we also prove:

$$s_t^{nz} \leq s_{t''}^{nz} \leq s_{t''} \leq s_t$$

and the same with $\mathbf{t}'$ instead of $\mathbf{t}$.

**Proposition 615** We have a set $\mathbf{A}$ of $\mathcal{R}^n$ included in a ball. We call $(\phi : \mathcal{R}^n \to \{0,1\})$, its characteristic map (definition 293 (p. 341)). As in proposition 611 (p. 342), for any term

$\mathbf{t} = (\mathbf{p}, \mathbf{y}, \mathbf{z})$ of a grid as defined in definition 288 (p. 331), we call $\mathbf{L^z(t)}$, the subset of $\prod_{\mathbf{k}\in[1,\mathbf{n}]}[0, \mathbf{p^k}[$ such that:

$$\mathbf{l} \in \mathbf{L^z(t)} \Leftrightarrow (\mathbf{l} \in \prod_{\mathbf{k}\in[1,\mathbf{n}]} [0, \mathbf{p^k}[, (\exists \mathbf{x} \in \prod_{\mathbf{k}\in[1,\mathbf{n}]} [\mathbf{y^{l_k,k}}, \mathbf{y^{l_k+1,k}}[)(\phi(\mathbf{x}) = 0)) \ .$$

We suppose that we have:

$$(\forall \alpha > 0)(\exists \epsilon(\alpha) > 0)(0 < \mathbf{a} < \epsilon(\alpha) \Rightarrow (\forall \mathbf{t} = (\mathbf{p}, \mathbf{y}, \mathbf{z}) \in \mathbf{Gr_a}) \\ (|\sum_{\mathbf{l}\in\mathbf{L^z(t)}} \phi(\mathbf{z^l}) \prod_{\mathbf{k}\in[1,\mathbf{n}]} (\mathbf{y^{l_k+1,k}} - \mathbf{y^{l_k,k}})| < \alpha) \ . \tag{2.166}$$

Then $\mathbf{A}$ has a volume $\mathbf{V}$ defined by the Riemann integral $\mathbf{J}(\phi)$ which exists. Equation (2.166) (p. 348) defines a map $(\epsilon(\alpha) : \mathcal{R}^+ \to \mathcal{R}^+)$ using that map, we have the following formula:

$$(0 < \mathbf{a} < \epsilon(\alpha/8), \mathbf{t} \in \mathbf{Gr_a}) \Rightarrow |\mathbf{J}(\phi) - \mathbf{gr}^\phi(\mathbf{t})| = |\mathbf{V} - \mathbf{gr}^\phi(\mathbf{t})| < \alpha \ ,$$

where $\mathbf{gr}^\phi(\mathbf{t})$ is defined by equation (2.153) (p. 332).

**Proof**

1. In a set of grids $\mathbf{Gr_a}$, we consider two terms $\mathbf{t}$ and $\mathbf{t'}$, we generate a third term $\mathbf{t''}$ as described in proposition 613 (p. 345). Associated to $\mathbf{A}$ and to any term $\mathbf{t} \in \mathbf{Gr}$ we consider the terms as described in proposition 614 (p. 345):

   $$\mathbf{s_t},\ \mathbf{L_t}\ \mathbf{s_t^z},\ \mathbf{L_t^z}\ \mathbf{s_t^{nz}},\ \mathbf{L_t^{nz}}$$

   We have from proposition 614 (p. 345) equation (2.164) (p. 346):

   $$\mathbf{s_t^{nz}} \le \mathbf{s_{t''}^{nz}} \le \mathbf{s_{t''}} \le \mathbf{s_t} \tag{2.167}$$

   and the same with $\mathbf{t'}$ instead of $\mathbf{t}$. From equation (2.166) (p. 348) we have:

   $$(\forall \alpha > 0)(\exists \epsilon(\alpha) > 0)(0 < \mathbf{a} < \epsilon(\alpha) \Rightarrow \\ (0 \le \mathbf{s_t} - \mathbf{s_t^{nz}} < \alpha, 0 \le \mathbf{s_{t'}} - \mathbf{s_{t'}^{nz}} < \alpha, 0 \le \mathbf{s_{t''}} - \mathbf{s_{t''}^{nz}} < \alpha)) \ . \tag{2.168}$$

   Setting:

   $$\mathbf{u} = \mathbf{s_{t''}^{nz}} - \mathbf{s_t^{nz}}, \mathbf{v} = \mathbf{s_t} - \mathbf{s_{t''}}, \\ \Delta\mathbf{u} = \mathbf{s_t^{nz}} - \mathbf{s_{t'}^{nz}}, \Delta\mathbf{v} = \mathbf{s_{t'}} - \mathbf{s_t}$$

   equation (2.167) gives:

   $$\mathbf{u} \ge 0, \mathbf{u} + \Delta\mathbf{u} \ge 0, \mathbf{v} \ge 0, \mathbf{v} + \Delta\mathbf{v} \ge 0$$

   equation (2.168) with the above gives for $\mathbf{a} < \epsilon(\alpha)$

   $$0 \le \mathbf{u} + \mathbf{v} < \alpha, 0 \le \mathbf{u} < \alpha, 0 \le \mathbf{v} < \alpha, \\ 0 \le \mathbf{u} + \Delta\mathbf{u} + \mathbf{v} + \Delta\mathbf{v} < \alpha, 0 \le \mathbf{u} + \Delta\mathbf{u} < \alpha, 0 \le \mathbf{v} + \Delta\mathbf{v} < \alpha \ .$$

   From the latter, we have $\mathbf{u} + \mathbf{v} \ge 0$, $\Delta\mathbf{u} \ge -\mathbf{u} > -\alpha, \Delta\mathbf{v} \ge -\mathbf{v} > -\alpha$ and $\Delta\mathbf{u} + \Delta\mathbf{v} < \alpha - (\mathbf{u} + \mathbf{v}) < \alpha$. Combining all, we get: $-\alpha < \Delta\mathbf{u} < \alpha - \Delta\mathbf{v} < 2\alpha$ and $-\alpha < \Delta\mathbf{v} < \alpha - \Delta\mathbf{u} < 2\alpha$ then $|\Delta\mathbf{u}| < 2\alpha, |\Delta\mathbf{v}| < 2\alpha$. It gives:

   $$\mathbf{a} < \epsilon(\alpha) \Rightarrow (|\mathbf{s_t^{nz}} - \mathbf{s_{t'}^{nz}}| < 2\alpha, |\mathbf{s_{t'}} - \mathbf{s_t}| < 2\alpha, |\mathbf{s_t^{nz}} - \mathbf{s_t}| < \alpha, |\mathbf{s_{t'}^{nz}} - \mathbf{s_{t'}}| < \alpha) \ .$$

2. Now calling for a term $\mathbf{t} = (\mathbf{p}, \mathbf{y}, \mathbf{z}) \in \mathbf{Gr_a}$:

$$\mathbf{J_t} = \mathbf{gr}^{\phi_A}(\mathbf{t}) = \sum_{l \in L} \phi_A(z_l) \prod_{i \in [1,n]} (y^{l_i+1,i} - y^{l_i,i})$$

where $(\phi_A : \mathcal{R}^n \to \{0,1\})$ and $\mathbf{gr}^{\phi}_A(\mathbf{t})$ is defined by equation (2.153) (p. 332) is the characteristic map of $\mathbf{A}$: $\mathbf{A} = \phi_A^{-1}(1)$. We have:

$$s_t^{nz} \leq J_t \leq s_t, \; s_{t'}^{nz} \leq J_{t'} \leq s_{t'} \; .$$

We also have:

$$J_{t'} - J_t = s_t^{nz} - J_t + s_{t'}^{nz} - s_t^{nz} + J_{t'} - s_{t'}^{nz}$$

then:

$$\begin{aligned} |J_{t'} - J_t| &\leq |s_t^{nz} - J_t| + |s_{t'}^{nz} - s_t^{nz}| + |J_{t'} - s_{t'}^{nz}| \\ &\leq |s_t^{nz} - s_t| + |s_{t'}^{nz} - s_t^{nz}| + |s_{t'} - s_{t'}^{nz}| < 4\alpha \; . \end{aligned}$$

Calling $\mathbf{t_p} = \tau_t(\mathbf{t} \in \mathbf{Gr}_{1/p})$ (see definition 112 (p. 131) point 2 and definition 332 (p. 440)), considering $\mathbf{p} \in \mathcal{Z}^+$, we have:

$$\begin{aligned} &(\forall \alpha)(\exists p' > 1/\epsilon(\alpha/4))((p1 > p', p2 > p') \Rightarrow (J_{t_{p1}} \in Gr_{1/p'}, J_{t_{p2}} \in Gr_{1/p'}) \\ &\Rightarrow |J_{t_{p1}} - J_{t_{p2}}| < \alpha) \end{aligned}$$

$\mathbf{J_{t_{p1}}} \in [0, (2c)^n]$ is a Cauchy sequence in a closed ball. It has an accumulation point $\mathbf{J} \in [0, (2c)^n]$ which is its limit (proposition 546 (p. 300)).

3. $\mathbf{J}$ is the Riemann integral of $\phi_A$:
From proposition 548 (p. 301), we have:

$$(\forall \alpha > 0)(\exists p'(\alpha))(p > p' \Rightarrow |J_{t_p} - J| < \alpha) \; .$$

Then:

$$\begin{aligned} &(\forall \alpha)(\exists \epsilon' = \epsilon(\alpha/8))(a < \epsilon' \Rightarrow \\ &(\forall t \in Gr_a)(\exists p > \max(1/\epsilon', p'(\alpha/2)))(J_{t_p} \in Gr_a, \\ &|J - J_{t_p}| < \alpha/2, |J_t - J_{t_p}| < \alpha/2) \\ &\Rightarrow (\forall t \in Gr_a)(|J - J_t| = |J - J_{t_p} - J_t + J_{t_p}| < |J - J_{t_p}| + |J_t - J_{t_p}| \leq \alpha)) \; . \end{aligned}$$

**Proposition 616** Given an interval of $\mathcal{R}$ $[\mathbf{x_1}, \mathbf{x_2}] \subset ]\mathbf{c}, -\mathbf{c}[$ the volume $\mathbf{V}$ of $[\mathbf{x_1}, \mathbf{x_2}]$ is $\mathbf{V} = |\mathbf{x_2} - \mathbf{x_1}|$. Given a term $\mathbf{t} = (\mathbf{p}, \mathbf{y}, \mathbf{z}) \in \mathbf{Gr_a}$. Calling $\phi$ the characteristic map of $[\mathbf{x_1}, \mathbf{x_2}]$, we have:

$$||x_2 - x_1| - gr^{\phi}(t)| < 2a \; .$$

**Proof**
We have $\mathbf{L} = [0, \mathbf{p}[$ calling:

$$\ell_1 = \max_{\ell \in L, y^\ell \leq x_1} \ell, \; \ell_2 = \min_{\ell \in L, y^\ell > x_2} \ell \; .$$

We have $]\mathbf{y}^{\ell_1+1}, \mathbf{y}^{\ell_2-1}] \subset [\mathbf{x_1}, \mathbf{x_2}] \subset [\mathbf{y}^{\ell_1}, \mathbf{y}^{\ell_2}[$ which means

$$y^{\ell_1} \leq x_1 < y^{\ell_1+1}, y^{\ell_2-1} \leq x_2 < y^{\ell_2}$$

which means:

$$\mathbf{y}^{\ell_2-1} - \mathbf{y}^{\ell_1+1} \leq \mathbf{x}_2 - \mathbf{x}_1 < \mathbf{y}^{\ell_2} - \mathbf{y}^{\ell_1}$$

we also have: $\mathbf{y}^{\ell_1} \leq \mathbf{z}^{\ell_1} < \mathbf{y}^{\ell_1+1}, \mathbf{y}^{\ell_2-1} \leq \mathbf{z}^{\ell_2} < \mathbf{y}^{\ell_2}$. Then we have according to the position of $\mathbf{z}^{\ell_1}$ and $\mathbf{z}^{\ell_2}$ compared to $\mathbf{x}_1, \mathbf{x}_2$

$$\mathbf{gr}^{\phi}(\mathbf{t}) \in \{\mathbf{y}^{\ell_2} - \mathbf{y}^{\ell_1}, \mathbf{y}^{\ell_2} - \mathbf{y}^{\ell_2-1}, \mathbf{y}^{\ell_2+1} - \mathbf{y}^{\ell_2}, \mathbf{y}^{\ell_2+1} - \mathbf{y}^{\ell_2-1}\}\ .$$

We have then the inequality:

$$|\mathbf{x}_2 - \mathbf{x}_1 - \mathbf{gr}^{\phi}(\mathbf{t})| < \mathbf{y}^{\ell_1+1} - \mathbf{y}^{\ell_1} + \mathbf{y}^{\ell_2} - \mathbf{y}^{\ell_2-1} < 2\mathbf{a}\ .$$

**Proposition 617** Given a convex set $\mathbf{A}$ (definition 255 (p. 277)) of $\mathcal{R}^{\mathbf{n}}$ inside a cube $\mathbf{C}^{\mathbf{n}}(\mathbf{0}, \mathbf{c}) = \mathbf{C}(\mathbf{0}, \mathbf{c}) \subset \mathcal{R}^{\mathbf{n}}$, considering in the set of grids $\mathbf{Gr_a}$ associated to $\mathbf{a} > \mathbf{0}$, a term $\mathbf{t} = (\mathbf{p}, \mathbf{y}, \mathbf{z})$ (definition by equation (2.154) (p. 332)), considering the subset ${}^{\mathbf{A}}\mathbf{L}^{\mathbf{z}}_{\mathbf{t}} \subset \prod_{\mathbf{k}\in[1,\mathbf{n}]}[\mathbf{0}, \mathbf{p}^{\mathbf{k}}[$ associated to $\mathbf{t} = (\mathbf{p}, \mathbf{y}, \mathbf{z})$ such as:

$$\mathbf{l} \in {}^{\mathbf{A}}\mathbf{L}^{\mathbf{z}}_{\mathbf{t}} \Leftrightarrow (\exists \mathbf{x}_1 \in \prod_{\mathbf{i}\in[1,\mathbf{n}]} [\mathbf{y}^{\mathbf{l}_i,\mathbf{i}}, \mathbf{y}^{\mathbf{l}_i+1,\mathbf{i}}[)(\exists \mathbf{x}_2 \in \prod_{\mathbf{i}\in[1,\mathbf{n}]} [\mathbf{y}^{\mathbf{l}_i,\mathbf{i}}, \mathbf{y}^{\mathbf{l}_i+1,\mathbf{i}}[)(\mathbf{x}_1 \in \mathbf{A}, \mathbf{x}_2 \notin \mathbf{A})$$

then calling $\mathbf{C}^{\mathbf{n}}_{\mathbf{x}}$ the set of convex sets subset of $\mathbf{C}^{\mathbf{n}}(\mathbf{0}, \mathbf{c})$

$$(\forall \alpha > 0)(\exists \epsilon > 0)(\forall \mathbf{A} \in \mathbf{C}^{\mathbf{n}}_{\mathbf{x}})(\mathbf{a} < \epsilon \Rightarrow (\forall \mathbf{t} \in \mathbf{Gr_a})(\sum_{\mathbf{l}\in{}^{\mathbf{A}}\mathbf{L}^{\mathbf{z}}_{\mathbf{t}}} \prod_{\mathbf{i}\in[1,\mathbf{n}]} (\mathbf{y}^{\mathbf{l}_i+1,\mathbf{i}} - \mathbf{y}^{\mathbf{l}_i,\mathbf{i}}) < \alpha))\ . \tag{2.169}$$

**Proof**

1. Let us call $\mathbf{e}_i = \prod^{\mathbf{e}}_{\mathbf{j}\in[1,\mathbf{n}]} \delta^{\mathbf{j}}_{\mathbf{i}} = (\delta^{\mathbf{j}}_{\mathbf{i}} : [1, \mathbf{n}] \to \mathcal{R}) \in \mathcal{R}^{\mathbf{n}}$ the reference basis of $\mathcal{R}^{\mathbf{n}}$ (see equation (2.106) (p. 267) in definition 242 (p. 267)). Let us suppose that proposition 617 (p. 350) is true for $\mathbf{n} = \mathbf{n}'$. Let us set it for $\mathbf{n} = \mathbf{n}' + 1$. We call: $\mathbf{Pr_k}{}^{\mathbf{A}}\mathbf{L}^{\mathbf{z}}_{\mathbf{t}} \subset \prod_{\mathbf{k}\in[1,\mathbf{n}-1]}[\mathbf{0}, \mathbf{p}^{\mathbf{k}}[$, the set such that:

$$\mathbf{l} \in \mathbf{Pr_k}{}^{\mathbf{A}}\mathbf{L}^{\mathbf{z}}_{\mathbf{t}} \Leftrightarrow (\exists (\mathbf{x}_1, \mathbf{x}_2) \in (\prod_{\mathbf{i}\in[1,\mathbf{n}-1]} [\mathbf{y}^{\mathbf{l}_i,\mathbf{i}}, \mathbf{y}^{\mathbf{l}_i+1,\mathbf{i}}])^2)(\mathbf{x}_1 \in \mathbf{Pr_k}(\mathbf{A}), \mathbf{x}_2 \notin \mathbf{Pr_k}(\mathbf{A}))\ .$$

From proposition 493 (p. 278), the projection $\mathbf{Pr_k}(\mathbf{A})$ (definition 121 (p. 136)) of $\mathbf{A}$ along $\mathbf{e_k}$ is convex and inside a cube $\mathbf{C}^{\mathbf{n}-1}(\mathbf{0}, \mathbf{c})$ (proposition 518 (p. 290)) of $\mathcal{R}^{\mathbf{n}-1}$. From proposition 617 (p. 350) which is supposed to be true at $\mathbf{n} = \mathbf{n}'$, we have:

$$(\forall \alpha > 0)(\exists \epsilon_{\mathbf{k}}(\alpha) > 0)(\forall \mathbf{A} \in \mathbf{C}^{\mathbf{n}}_{\mathbf{x}})(\mathbf{a} < \epsilon_{\mathbf{k}}(\alpha) \Rightarrow (\forall \mathbf{t} \in \mathbf{Gr_a})(\sum_{\mathbf{l}\in\mathbf{Pr_k}{}^{\mathbf{A}}\mathbf{L}^{\mathbf{z}}_{\mathbf{t}}} \prod_{\mathbf{i}\in[1,\mathbf{n}]-\{\mathbf{k}\}} (\mathbf{y}^{\mathbf{l}_i+1,\mathbf{i}} - \mathbf{y}^{\mathbf{l}_i,\mathbf{i}}) < \alpha))\ . \tag{2.170}$$

Since $\mathbf{y}^{\mathbf{l}_k+1,\mathbf{k}} - \mathbf{y}^{\mathbf{l}_k,\mathbf{k}} < 2\mathbf{c}$, equation (2.170) gives:

$$(\forall \alpha > 0)(\exists \epsilon_{\mathbf{k}}(\alpha) > 0)(\mathbf{a} < \epsilon_{\mathbf{k}}(\alpha) \Rightarrow (\forall \mathbf{t} \in \mathbf{Gr_a})(\sum_{\mathbf{l}\in\mathbf{Pr_k}{}^{\mathbf{A}}\mathbf{L}^{\mathbf{z}}_{\mathbf{t}}} \prod_{\mathbf{i}\in[1,\mathbf{n}]} (\mathbf{y}^{\mathbf{l}_i+1,\mathbf{i}} - \mathbf{y}^{\mathbf{l}_i,\mathbf{i}}) < 2\mathbf{c} \times \alpha))\ .$$

Calling (see definition 279 (p. 320)) $\overline{\mathbf{Bx}}(\mathbf{y}^{l_i,i}\mathbf{e}_i, \mathbf{y}^{l_i+1,i}\mathbf{e}_i) = \prod_{i\in[1,n]}[\mathbf{y}^{l_i,i}, \mathbf{y}^{l_i+1,i}]$ the box attached to $\mathbf{l}$, from our definition of ${}^{\mathbf{A}}\mathbf{L}^{\mathbf{z}}_{\mathbf{t}}$ and $\mathbf{Pr_k}{}^{\mathbf{A}}\mathbf{L}^{\mathbf{z}}_{\mathbf{t}}$, we have:

$$\begin{aligned}&(\mathbf{l} \in ({}^{\mathbf{A}}\mathbf{L}^{\mathbf{z}}_{\mathbf{t}} - \bigcup_{\mathbf{k}\in[1,n]} \mathbf{Pr_k}{}^{\mathbf{A}}\mathbf{L}^{\mathbf{z}}_{\mathbf{t}}), \mathbf{x} \in \prod_{i\in[1,n]} [\mathbf{y}^{l_i,i}, \mathbf{y}^{l_i+1,i}] = \overline{\mathbf{Bx}}(\mathbf{y}^{l_i,i}\mathbf{e}_i, \mathbf{y}^{l_i+1,i}\mathbf{e}_i)) \\ &\Rightarrow (\forall \mathbf{k} \in [1, n])( \sum_{i\in[1,n]-\{k\}} \mathbf{x}^i\mathbf{e}_i \in \mathbf{Pr_k}(\mathbf{A})) \\ &\Rightarrow (\forall \mathbf{k} \in [1, n])(\exists \mathbf{y} \in \mathcal{R})(\mathbf{y}\mathbf{e}_k + \sum_{i\in[1,n]-\{k\}} \mathbf{x}^i\mathbf{e}_i \in \mathbf{A}) .\end{aligned} \tag{2.171}$$

2. From proposition 586 (p. 325), any box associated to an $\mathbf{l} \in {}^{\mathbf{A}}\mathbf{L}^{\mathbf{z}}_{\mathbf{t}} - \bigcup_{\mathbf{k}\in[1,n]} \mathbf{Pr_k}{}^{\mathbf{A}}\mathbf{L}^{\mathbf{z}}_{\mathbf{t}}$ has at least one ridge (see definition 282 (p. 322)) with a point in $\mathbf{A}$ and a point not in $\mathbf{A}$.

3. In equation (2.169) (p. 350), we have split, in point 1, the $\sum_{\mathbf{l}\in{}^{\mathbf{A}}\mathbf{L}^{\mathbf{z}}_{\mathbf{t}}}$, into two sums: the sum over $\bigcup_{\mathbf{k}\in[1,n]} \mathbf{Pr_k}{}^{\mathbf{A}}\mathbf{L}^{\mathbf{z}}_{\mathbf{t}}$ and the sum over ${}^{\mathbf{A}}\mathbf{L}^{\mathbf{z}}_{\mathbf{t}} - \bigcup_{\mathbf{k}\in[1,n]} \mathbf{Pr_k}{}^{\mathbf{A}}\mathbf{L}^{\mathbf{z}}_{\mathbf{t}}$. Considering any direction $\mathbf{k} \in [1, n]$ ${}^{\mathbf{A}}\mathbf{L}^{\mathbf{z}}_{\mathbf{t}} - \bigcup_{\mathbf{k}\in[1,n]} \mathbf{Pr_k}{}^{\mathbf{A}}\mathbf{L}^{\mathbf{z}}_{\mathbf{t}}$ project along the direction $\mathbf{k}$ on a square of dimension $\mathbf{n} - \mathbf{1}$ and of side $\mathbf{2c}$. For any line along $\mathbf{k}$ (see definition 216 (p. 245))

$$(\forall \mathbf{l} \in [0, p]^{n-1})(\mathbf{x} \in \mathbf{h_k}(\mathbf{l}) \Leftrightarrow (\exists \lambda \in \mathcal{R})(\mathbf{x} = \sum_{j\in[1,n-1]} \mathbf{y}^{l_j+\mathcal{Y}(j-k)}\mathbf{e}_{j+\mathcal{Y}(j-k)} + \lambda\mathbf{e}_k)) .$$

This line intersects $\bar{\mathbf{A}}$ in a closed convex set which is then an interval $[\mathbf{a_k}, \mathbf{b_k}]$ of $[-\mathbf{c}, \mathbf{c}] \subset \mathcal{R}$. From what we just said in point 2, all the boxes of ${}^{\mathbf{A}}\mathbf{L}^{\mathbf{z}}_{\mathbf{t}} - \bigcup_{\mathbf{k}\in[1,n]} \mathbf{Pr_k}{}^{\mathbf{A}}\mathbf{L}^{\mathbf{z}}_{\mathbf{t}}$ have a ridge with a point in $\mathbf{A}$ and one not in $\mathbf{A}$ there is at maximum $\mathbf{2^{n-1}}$ such box for the point $\mathbf{a_k}$ or $\mathbf{b_k}$ then the sum of the box volume of the set ${}^{\mathbf{A}}\mathbf{L}^{\mathbf{z}}_{\mathbf{t}} - \bigcup_{\mathbf{k}\in[1,n]} \mathbf{Pr_k}{}^{\mathbf{A}}\mathbf{L}^{\mathbf{z}}_{\mathbf{t}}$ is below $(\mathbf{2c})^{\mathbf{n-1}} \times \mathbf{a} \times \mathbf{2} \times \mathbf{2^{n-1}}$. Taking into account the result of point 1 we have with:

$$0 < \epsilon < \min(\frac{\alpha}{2^{2n-1}c^{n-1}}, \min_{k\in[1,n]} \epsilon_k(\frac{\alpha}{4nc}))$$

$$a < \epsilon \Rightarrow \sum_{l\in{}^{A}L^{z}_{t}} \prod_{i\in[1,n]} (y^{l_i+1,i} - y^{l_i,i}) < \alpha .$$

Then proposition 617 (p. 350) is true for $\mathbf{n} - \mathbf{n}' + \mathbf{1}$ if it is true for $\mathbf{n}'$. Since proposition 617 (p. 350) is true for $\mathbf{n} = \mathbf{1}$ where $\epsilon_1(\alpha) = \alpha/2$ (proposition 616 (p. 349)) and does not depend on $\mathbf{A} \in \mathbf{Cx^1} \subset [-\mathbf{c}, \mathbf{c}]$, from proposition 927 (p. 520) it is true for any $\mathbf{n}$.

**Proposition 618** Given a set $\mathbf{I}$ and a map $(\mathbf{A_i} : \mathbf{I} \to \mathbf{Cx^n})$, where $\mathbf{Cx^n} \subset \mathcal{P}(\mathbf{C}(\mathbf{0}, \mathbf{c}))$ is the set of convex sets of $\mathcal{R}^n$ included in a ball $\mathbf{C}(\mathbf{0}, \mathbf{c})$. Each $\mathbf{A_i}$ has a volume defined by a Riemann integral which converges uniformly for $\mathbf{i} \in \mathbf{I}$ (definition 292 (p. 337)).

**Proof**
It is a direct consequence of propositions 615 (p. 347) and 617 (p. 350).

**Definition 296** ($\phi^1(\mathbf{w}) : \mathcal{R} \to \{0, 1\}$).

We set $\phi^1(\mathbf{w}), \mathbf{w} \in \mathcal{R}$ as

$$(\forall \mathbf{w} \in [0, 1])(\phi^1(\mathbf{w}) = 1), (\forall \mathbf{w} \notin [0, 1])(\phi^1(\mathbf{w}) = 0) .$$

It is the characteristic function of $[0, 1]$

**Proposition 619** Given the reference basis $\mathbf{e_i} = \prod^{\mathbf{e}}_{\mathbf{j}\in[1,\mathbf{n}]} \delta^{\mathbf{j}}_{\mathbf{i}} = (\delta^{\mathbf{j}}_{\mathbf{i}} : [1,\mathbf{n}] \to \mathcal{R}^{\mathbf{n}}) \in \mathcal{R}^{\mathbf{n}}$ and given a set $\mathbf{V}$ of $\mathbf{n}$ independent vectors $(\mathbf{v_i} : [1,\mathbf{n}] \to \mathcal{R}^{\mathbf{n}})$ with $\mathbf{v_i} = \mathbf{v^j_i e_j}$ and $\mathbf{e_i} = \mathbf{e^j_i v_j}$, the characteristic map of the inner part $\overline{\mathbf{V}}$ of $\mathbf{V}$ (definition 259 (p. 281)) is for a point $\mathbf{x^i e_i} \in \mathcal{R}^{\mathbf{n}}$:

$$\phi_{\overline{V}} = \prod_{\mathbf{j}\in[1,\mathbf{n}]} \phi^1(\mathbf{x^i e^j_i}) ,$$

where $\phi^1$ is defined in definition 296.

**Proof**
From definition 259 (p. 281), we have for $\mathbf{p} = \mathbf{x^i e_i} = \mathbf{x^i e^j_i v_j} \in \mathcal{R}^{\mathbf{n}}$:

$$\mathbf{p} \in \overline{\mathbf{V}} \Leftrightarrow (\forall \mathbf{i} \in [1,\mathbf{n}])(\mathbf{x^i e^j_i} \in [0,1]) \Leftrightarrow (\forall \mathbf{i} \in [1,\mathbf{n}])(\phi_1(\mathbf{x^i e^j_i}) = 1)$$
$$\Leftrightarrow \prod_{\mathbf{j}\in[1,\mathbf{n}]} \phi^1(\mathbf{x^i e^j_i}) = 1$$
$$\mathbf{p} \notin \overline{\mathbf{V}} \Leftrightarrow \prod_{\mathbf{j}\in[1,\mathbf{n}]} \phi^1(\mathbf{x^i e^j_i}) = 0 .$$

## 2.25 Integral of a continuous map

In order to reduce the number of indices, we shall treat here only integral of continuous maps $\mathbf{f} : \mathcal{R}^{\mathbf{n}} \to \mathcal{R}$. But since continuous maps $(\mathbf{f} : \mathcal{R}^{\mathbf{n}} \to \mathcal{R}^{\mathbf{m}})$ are also continuous in the $\mathbf{m}$ components of $\mathbf{f}$ and vice versa (see propositions 244 (p. 168), 245 (p. 169)), using proposition 600 (p. 333) one can extend immediately the conclusions of $(\mathbf{f} : \mathcal{R}^{\mathbf{n}} \to \mathcal{R})$ to $(\mathbf{f} : \mathcal{R}^{\mathbf{n}} \to \mathcal{R}^{\mathbf{m}})$.

**Proposition 620** Given a set $\mathbf{I}$, a map $(\mathbf{A_i} : \mathbf{I} \to \mathcal{P}(\mathcal{R}^{\mathbf{n}}))$ such that $\bigcup_{\mathbf{i}\in\mathbf{I}} \mathbf{A_i}$ is included in a compact set $\mathbf{V}$ of $\mathcal{R}^{\mathbf{n}}$. For any $\mathbf{i} \in \mathbf{I}$, we suppose that $\mathbf{A_i}$ has an existing volume defined by the Riemann integral of its characteristic map $(\phi_{\mathbf{i}}(\mathbf{x}) : \mathcal{R}^{\mathbf{n}} \to \{0,1\})$. We suppose that this integral is converging uniformly relative to the set $\mathbf{I}$ (see definition 292 (p. 337)). Calling $\mathcal{F} \subset \mathbf{V} \times \mathcal{R}$, the set of continuous maps of $\mathbf{V}$ to $\mathcal{R}$. From proposition 533 (p. 296), any map $\mathbf{f} \in \mathcal{F}$ verifies equation (2.132) (p. 296). We consider the maps $(\mathbf{f_i} : \mathbf{I} \to \mathcal{F})$ which satisfy:

$$(\forall \alpha \in \mathcal{R}^+)(\exists \epsilon \in \mathcal{R}^+)(\forall \mathbf{i} \in \mathbf{I})(\forall (\mathbf{x},\mathbf{y}) \in \mathbf{V} \times \mathbf{V})(\|\mathbf{x}-\mathbf{y}\| < \epsilon \Rightarrow |\mathbf{f_i}(\mathbf{x}) - \mathbf{f_i}(\mathbf{y})| < \alpha) \tag{2.172}$$

and

$$(\exists \mathbf{f_M} \in \mathcal{R})(\exists \mathbf{f_m} \in \mathcal{R})(\forall \mathbf{i} \in \mathbf{I})(\forall \mathbf{x} \in \mathbf{V})(\mathbf{f_i}(\mathbf{x}) \in [\mathbf{f_m}, \mathbf{F_M}]) . \tag{2.173}$$

In those conditions, when $\mathbf{i} \in \mathbf{I}$, the product $(\mathbf{f_i}(\mathbf{x})\phi_{\mathbf{i}}(\mathbf{x}) : \mathcal{R}^{\mathbf{n}} \to \mathcal{R})$ has an existing Riemann integral which is converging uniformly (see definition 292 (p. 337)).

**Proof**
We set $(\mathbf{e_k} = \delta^{\mathbf{i}}_{\mathbf{k}} : [1,\mathbf{n}] \to \mathcal{R}^{\mathbf{n}})$, the reference basis of $\mathcal{R}^{\mathbf{n}}$. We call $(\mathbf{Vol_i} : \mathbf{I} \to \mathcal{R})$ the volume associated to $\phi_{\mathbf{i}}(\mathbf{x})$, from proposition 522 (p. 291), we have $(\exists \mathbf{c} > 0)(\mathbf{V} \subset \overline{\mathbf{C}}(0,\mathbf{c}))$ where $\overline{\mathbf{C}}(0,\mathbf{c})$ is a cube. From definition 290 (p. 332) we have:

$$(\forall \alpha_1 > 0)(\exists \epsilon_1(\alpha_1) > 0)(\forall \mathbf{i} \in \mathbf{I})(0 < \mathbf{a} < \epsilon_1(\alpha_1) \Rightarrow$$
$$(\forall \mathbf{t} = (\mathbf{p},\mathbf{y},\mathbf{z}) \in \mathbf{Gr_a})(|\mathbf{Vol_i} - \sum_{\ell\in\prod_{\mathbf{k}\in[1,\mathbf{n}]}[0,\mathbf{p^k}[} \phi_{\mathbf{i}}(\mathbf{z}^{\ell,\mathbf{j}}\mathbf{e_j}) \prod_{\mathbf{k}\in[1,\mathbf{n}]} (\mathbf{y}^{\ell_k+1,\mathbf{k}} - \mathbf{y}^{\ell_k,\mathbf{k}})| < \alpha_1)) . \tag{2.174}$$

Using $\mathbf{f_M}$ and $\mathbf{f_m}$ of equation (2.173) (p. 352), we call $\mathbf{F_M} = \mathbf{max}(|\mathbf{f_M}|, |\mathbf{f_m}|)$. From equation (2.172) (p. 352), $\mathbf{f_i(x)}$ verifies in $\mathbf{V}$:

$$(\forall \alpha_2)(\exists \epsilon_2(\alpha_2))(\forall i \in I)(\forall x_1 \in V)(\forall x_2 \in V)$$
$$(\|x_1 - x_2\| < \epsilon_2(\alpha_2) \Rightarrow |f_i(x_1) - f_i(x_2)| < \alpha_2)\,. \tag{2.175}$$

For two terms $\mathbf{t(p,y,z)}$ and $\mathbf{t'(p',y',z')}$ of the grid $\mathbf{Gr_a}$ of equation (2.153) (p. 332), we define, as in proposition 613 (p. 345), $\mathbf{t''} = (\mathbf{p''}, \mathbf{y''}, \mathbf{z''}) \in \mathbf{Gr_a}$. As in proposition 2.164 (p. 346), associated to any set $\mathbf{A_i} \subset \mathbf{V}$ and to any term $\mathbf{t(p,y,z)}$, we define, ${}^{i}L^{z}_{t}$, ${}^{i}L^{nz}_{t}$ and ${}^{i}L_{t} = {}^{i}L^{z}_{t} \cup {}^{i}L^{nz}$ three subsets of $L_t = \prod_{k\in[1,n]}[0, p^k[$ by:

$$\ell \in {}^{i}L^{z}_{t} \Leftrightarrow (\ell \in L_t, U = \prod_{k\in[1,n]} [y^k(\ell_k), y^k(\ell_k+1)[ \Rightarrow (U \cap A_i \neq \emptyset, U \cap \mathcal{C}(A_i) \neq \emptyset))$$
$$\ell \in {}^{i}L^{nz}_{t} \Leftrightarrow (\ell \in L_t, \prod_{k\in[1,n]} [y^k(\ell_k), y^k(\ell_k+1)[ \subset A_i)$$

we call:

$$s_i = gr^{f_i\phi}(t) = \sum_{\ell\in L_t} f_i(z^{\ell_j j}e_j)\phi_i(z^{\ell_j j}e_j) \prod_{k\in[1,n]} (y^{\ell_k+1,k} - y^{\ell_k,k})$$
$$s'_i = gr^{f_i\phi}(t') = \sum_{\ell\in L_{t'}} f_i(z'^{\ell_j j}e_j)\phi_i(z'^{\ell_j j}e_j) \prod_{k\in[1,n]} (y'^{\ell_k+1,k} - y'^{\ell_k,k}) \tag{2.176}$$
$$s''_i = gr^{f_i\phi}(t'') = \sum_{\ell\in L_{t''}} f_i(z''^{\ell_j j}e_j)\phi_i(z''^{\ell_j j}e_j) \prod_{k\in[1,n]} (y''^{\ell_k+1,k} - y''^{\ell_k,k})\,.$$

From their definitions, we have ${}^{i}L^{z}_{t} \cap {}^{i}L^{nz}_{t} = \emptyset$. Each $\mathbf{s_i}$ can be split into three terms replacing in equation (2.176) the sum over $\mathbf{L_t}$ by three sums over ${}^{i}L^{z}_{t}$, ${}^{i}L^{nz}_{t}$, $\mathcal{C}_{L_t}({}^{i}L^{z}_{t} \cup {}^{i}L^{nz}_{t})$ (see proposition 293 (p. 188)). Since:

$$z \in \mathcal{C}_{L_t}({}^{i}L^{z}_{t} \cup {}^{i}L^{nz}_{t}) \Rightarrow z \notin A_i \Rightarrow \phi_i(z) = 0$$

the sum over $\mathcal{C}_{L_t}({}^{i}L^{z}_{t} \cup {}^{i}L^{nz}_{t})$ is zero. Then calling $s^{z}_{i}$, $s^{nz}_{i}$ the sums over ${}^{i}L^{z}_{t}$, ${}^{i}L^{nz}_{t}$, we have:

$$s_i = s^{z}_{i} + s^{nz}_{i}$$
$$s'_i = s'^{z}_{i} + s^{nz}_{i}$$
$$s''_i = s''^{z}_{i} + s''^{nz}_{i}\,.$$

Following the same development as in equation (2.165) (p. 347) of proposition 614 (p. 345) and using the map $\mathbf{h}$ as defined in the equations (2.162) (p. 345) and (2.163) (p. 345), we have:

$$s_i^{nz} = \sum_{\ell \in {}^iL_t^{nz}} f_i(z(\ell)) \sum_{l \in \prod_{k\in[1,n]}[h^k(\ell_k),h^k(\ell_k+1)[} \prod_{k\in[1,n]} (y''^k(l(k)+1) - y''^k(l(k)))$$

$$s_i'' = \sum_{\ell \in {}^iL_t^{nz} \cup (L_t - {}^iL_t^{nz})} \sum_{l \in \prod_{k\in[1,n]}[h^k(\ell_k),h^k(\ell_k+1)[} f_i(z''(l))\phi_i(z''(l)) \prod_{k\in[1,n]} (y''^k(l(k)+1) - y''^k(l(k)))$$

$$s_i^{nz} - s_i'' = \sum_{\ell \in {}^iL_t^{nz}} \sum_{l \in \prod_{k\in[1,n]}[h^k(\ell_k),h^k(\ell_k+1)[} (f_i(z(\ell)) - f_i(z''(l))) \prod_{k\in[1,n]} (y''^k(l(k)+1) - y''^k(l(k)))$$
$$- \sum_{\ell \in L_t - {}^iL_t^{nz}} \sum_{l \in \prod_{k\in[1,n]}[h^k(\ell_k),h^k(\ell_k+1)[} f_i(z''(l))\phi_i(z''(l)) \prod_{k\in[1,n]} (y''^k(l(k)+1) - y''^k(l(k))) .$$

We have from proposition 289 (p. 186) and equation (2.163) (p. 345):

$$\begin{aligned}
&| \sum_{\ell \in L_t - {}^iL_t^{nz}} \sum_{l \in \prod_{k\in[1,n]}[h^k(\ell_k),h^k(\ell_k+1)[} f_i(z''(l))\phi_i(z''(l)) \prod_{k\in[1,n]} (y''^k(l(k)+1) - y''^k(l(k)))| \\
&< F_M \sum_{\ell \in L_t - {}^iL_t^{nz}} \sum_{l \in \prod_{k\in[1,n]}[h^k(\ell_k),h^k(\ell_k+1)[} \prod_{k\in[1,n]} (y''^k(l(k)+1) - y''^k(l(k))) \\
&= F_M \sum_{\ell \in L_t - {}^iL_t^{nz}} \prod_{k\in[1,n]} \sum_{\lambda \in [h^k(\ell_k),h^k(\ell_k+1)[} (y''^k(\lambda+1) - y''^k(\lambda)) \\
&= F_M \sum_{\ell \in L_t - {}^iL_t^{nz}} \prod_{k\in[1,n]} (y''^k(h^k(\ell_k+1)) - y''^k(h^k(\ell_k))) \\
&= F_M \sum_{\ell \in L_t - {}^iL_t^{nz}} \prod_{k\in[1,n]} (y^k(\ell(k)+1) - y^k(\ell(k))) = F_M s_i^z .
\end{aligned}$$

From proposition 611 (p. 342),

$$0 < a < \epsilon_1(\alpha/(6F_M)) \Rightarrow 0 < F_M s_i^z < \alpha/3 .$$

We also have when $0 < a < \epsilon_2(\frac{\alpha}{3(2c)^n})$:

$$(\forall \ell \in {}^iL_t^{nz})(\forall l \in \prod_{k\in[1,n]} [h^k(\ell_k), h^k(\ell_k+1)[)(|f_i(z(\ell)) - f_i(z''(l))| < \frac{\alpha}{3(2c)^n}) .$$

Again from propositions 289 (p. 186), 290 (p. 187) and equation (2.163) (p. 345), we have:

$$
\begin{aligned}
&|\sum_{\ell\in{}^{\mathrm{i}}L_t^{nz}} \sum_{l\in\prod_{k\in[1,n]}[h^k(\ell_k),h^k(\ell_k+1)[} (f_i(z(\ell)) - f_i(z''(l))) \prod_{k\in[1,n]} (y''^k(l(k)+1) - y''^k(l(k)))| \\
&< \frac{\alpha}{3(2c)^n} \sum_{\ell\in{}^{\mathrm{i}}L_t^{nz}} \sum_{l\in\prod_{k\in[1,n]}[h^k(\ell_k),h^k(\ell_k+1)[} \prod_{k\in[1,n]} (y''^k(l(k)+1) - y''^k(l(k))) \\
&= \frac{\alpha}{3(2c)^n} \sum_{\ell\in{}^{\mathrm{i}}L_t^{nz}} \prod_{k\in[1,n]} (y^k(\ell(k)+1) - y^k(\ell(k))) \\
&\leq \frac{\alpha}{2(2c)^n} \prod_{k\in[1,n]} (y^k(p^k) - y^k(0)) < \alpha/3 \, .
\end{aligned}
$$

Then for $0 < a < \min(\epsilon_2(\frac{\alpha}{3(2c)^n}), \alpha/(6F_M))$ one gets

$$|s_i^{nz} - s_i''| < 2\alpha/3$$

but

$$|s_i^{nz} - s_i''| = |s_i - s_i'' - s_i^z| < |s_i - s_i''| + |s_i^z| < 2\alpha/3 + \alpha/3 = \alpha$$

the same is valid for $|s_i' - s_i''|$.
Then calling $\epsilon_I(\alpha) = \min(\epsilon_2(\frac{\alpha}{6(2c)^n}), \alpha/(12F_M))$, for all $\alpha$, there is $\epsilon = \epsilon_I(\alpha)$ such as $0 < a < \epsilon = \epsilon_I(\alpha)$ gives:

$$(t \in Gr_a, t' \in Gr_a) \Rightarrow (\forall i \in I)(|s_i - s_i'| < \alpha) \, .$$

We call now on $s_i(t)$ what we call above $s_i$ corresponding to $t$. For any $q \in \mathcal{Z}^+$ we have a $t_q = \tau_x(x \in Gr_{1/q})$. For any $i \in I$, $s_i(t_q)$ forms a uniform Cauchy sequence (definition 274 (p. 303)) because for $q > 1/\epsilon_I(\alpha)$ and $q' > 1/\epsilon_I(\alpha)$ $|s_i(t_q) - s_i(t'_q)| < \alpha$. From proposition 552 (p. 303), it has a uniform limit $J_i$ (see 272 (p. 302)). Then for any $\alpha$, we have a $q' > q_m = 1/\epsilon_I(\alpha/4))$ such as for any $i \in I$, $|s_i(t'_q) - J_i| < \alpha/2$. We also have $a$ with $0 < a < \epsilon_I(\alpha/2)$ such that $t \in Gr_a$, $t' \in Gr_a$ which means $|s_i(t) - s_i(t')| < \alpha/2$. Taking $q' > \max(1/\epsilon_I(\alpha/2), q_m)$, we have $t_{q'} \in Gr_a$ and $|s_i(t_{q'}) - J_i| < \alpha/2$. With that, we get:

$$
\begin{aligned}
&|J_i - s_i(t)| = |J_i - s_i(t_{q'}) + s_i(t_{q'}) - s_i(t)| < |J_i - s_i(t_{q'})| + |s_i(t_{q'}) - s_i(t)| \\
&< \alpha/2 + \alpha/2 = \alpha \, .
\end{aligned}
$$

We have then:

$$
\begin{aligned}
&(\exists (J_i : I \to \mathcal{R}))(\forall \alpha > 0)(\exists \epsilon = \epsilon_I(\alpha/2))(\forall i \in I)(0 < a < \epsilon \\
&\Rightarrow (\forall t \in Gr_a)(|s_i(t) - J_i| < \alpha))
\end{aligned}
$$

proposition 620 (p. 352) is true.

**Proposition 621** We have a continuous map $(f : V \to \mathcal{R}^n)$, where $V$ is a compact subset of $\mathcal{R}^n$. We call $(\phi : \mathcal{R}^n \to \{0,1\})$, the characteristic map of $V$. The Riemann integral of $f \times \phi$ exists.

**Proof**
It is a direct consequence of propositions 620 (p. 352), 600 (p. 333), 244 (p. 168) and 245 (p. 169).

**Proposition 622** Having a continuous map $\mathbf{f(u,x)} : \mathcal{R}^n \times \mathcal{R} \to \mathcal{R}$ with a compact support $(\mathbf{C(0,c)}\times]-\mathbf{c},\mathbf{c}[= (]-\mathbf{c},\mathbf{c}[)^{n+1})$ on $\mathcal{R}^{n+1}$:

$$(\mathbf{u} \notin \mathbf{C(0,c)}, \mathbf{x} \notin ]-\mathbf{c},\mathbf{c}[) \Rightarrow \mathbf{f(u,x)} = 0 \ ,$$

we suppose that for all $\mathbf{u}$ and for all $\mathbf{x}$, $\mathbf{f(u,x)}$ has a partial derivative $\mathbf{f'(u,x)} = \frac{\partial f}{\partial x}$ on $\mathbf{x}$ which is also continuous. Then we have:

$$\mathbf{J_u}(\frac{\partial f}{\partial x}) = \mathbf{J_u(f')} = \mathbf{J'_u(f)} = \frac{\partial \mathbf{J_u(f)}}{\partial x}$$

where $\mathbf{J_u}$ is the integral over $\mathbf{u}$ only.

**Proof**

Since $\mathbf{f(u,x)}$ and $\mathbf{f'(u,x)}$ are defined on a compact and are continuous, their integral exists (proposition 621 (p. 355)). From proposition 314 (p. 196), we have:

$$(\forall \mathbf{x} \in \mathcal{R})(\exists \lambda(\mathbf{u},\mathbf{x}) \in ]0,1[)(\frac{\mathbf{f(u,x+\Delta x)} - \mathbf{f(u,x)}}{\Delta \mathbf{x}} = \mathbf{f'(u, x + \lambda(u,x)\Delta x)}) \ .$$

Since $\mathbf{f'(u,x)}$ is continuous and zero outside a compact, it is uniformly continuous (definition 160 (p. 184) and proposition 283 (p. 184)). Then we have:

$$(\forall \alpha)(\exists \epsilon(\alpha) > 0)(\forall \mathbf{x})(\forall \mathbf{u})(|\Delta \mathbf{x}| < \epsilon(\alpha) \Rightarrow |\frac{\mathbf{f(u,x+\Delta x)} - \mathbf{f(u,x)}}{\Delta \mathbf{x}} - \mathbf{f'(u,x)}| < \alpha$$

and from propositions 604 (p. 336), 605 (p. 337), 607 (p. 339) and 608 (p. 339), we have:

$$\begin{aligned}
&(\forall \alpha')(\exists \epsilon' = \epsilon(\alpha'/((2c)^n)) > 0)(|\Delta \mathbf{x}| < \epsilon' \Rightarrow \\
&(|\mathbf{J}(\frac{\mathbf{f(u,x+\Delta x)} - \mathbf{f(u,x)}}{\Delta \mathbf{x}}) - \mathbf{J(f'(u,x))}| = |\mathbf{J}(\frac{\mathbf{f(u,x+\Delta x)} - \mathbf{f(u,x)}}{\Delta \mathbf{x}} - \mathbf{f'(u,x)})| \\
&\leq \mathbf{J}(|\frac{\mathbf{f(u,x+\Delta x)} - \mathbf{f(u,x)}}{\Delta \mathbf{x}} - \mathbf{f'(u,x)}|) < \mathbf{J}(\frac{\alpha'}{(2c)^n}\Phi_{C(0,c)}) = \alpha'))
\end{aligned}$$

From definition 161 (p. 192), we have proposition 622 (p. 356).

**Proposition 623** Given a subset included in a cube of $\mathcal{R}^n$ with a characteristic map $(\phi(\mathbf{x}) : \mathcal{R}^n \to \{0,1\})$ and with a volume which is null, the product of any continuous function $(\mathbf{f(x)} : \mathcal{R}^n \to \mathcal{R})$ by $\phi(\mathbf{x})$, $(\mathbf{f(x)}\phi(\mathbf{x}) : \mathcal{R}^n \to \mathcal{R})$ has a Riemann integral which is also null.

**Proof**

From proposition 620 (p. 352), $\mathbf{J(f\phi)}$ exists. From proposition 219 (p. 158), the closure of the cube $\overline{\mathbf{C}}(\mathbf{x},\mathbf{a})$ (definition 136 (p. 155)) is closed, it is inside a ball $\mathbf{B(x, 2a\sqrt{n})}$ (proposition 507 (p. 286) equation (2.127)), from proposition 521 (p. 290) it is compact and from proposition 210 (p. 155) it includes the cube $\mathbf{C(x,a)}$. From proposition 297 (p. 189):

$$(\exists \mathbf{M} \in \mathcal{R})(\exists \mathbf{m} \in \mathcal{R})(\forall \mathbf{x} \in \overline{\mathbf{C}}(\mathbf{x},\mathbf{a}))(\mathbf{m} \leq \mathbf{f(x)} \leq \mathbf{M}) \ .$$

Then since $\phi \geq 0$, we have:

$$(\forall \mathbf{x} \in \mathcal{R})(\mathbf{m}\phi(\mathbf{x}) \leq \mathbf{f(x)}\phi(\mathbf{x}) \leq \mathbf{M}\phi(\mathbf{x})) \ .$$

From proposition 234 (p. 165), the maps $(\mathbf{m} : \mathcal{R}^n \to \mathcal{R})$ and $(\mathbf{M} : \mathcal{R}^n \to \mathcal{R})$ are continuous maps. From proposition 620 (p. 352), their Riemann integral exists. From proposition 609 (p. 340), we have:

$$\mathbf{J(m\phi)} \leq \mathbf{J(f(x)\phi)} \leq \mathbf{J(M\phi)}$$

but from proposition 607 (p. 339), we have:

$$\mathbf{J(m\phi) = mJ(\phi) = 0, J(M\phi) = MJ(\phi) = 0}$$

then:

$$\mathbf{J(f(x)\phi) = 0}\ .$$

**Proposition 624** We have a subset $\mathbf{A} \subset \mathcal{R}^{\mathbf{n}}$ within a ball and we call $(\phi_{\mathbf{A}} : \mathcal{R}^{\mathbf{n}} \to \{0,1\})$ its characteristic map (definition 293 (p. 341)). We suppose that $\mathbf{A}$ has a volume. We have $\mathbf{f}$ which is a continuous map $(\mathbf{f} : \mathcal{R}^{\mathbf{n}} \to \mathcal{R})$. Then the integrals $\mathbf{J_A(f), J_{\bar{A}}(f)}$ exist and are equal. (see definitions 136 (p. 155), 137 (p. 156) and 138 (p. 157)).

**Proof**
From proposition 612 (p. 343), the following integrals exist and are equal: $\mathbf{J_A(1) = J_{\bar{A}}(1) = J(\phi_A) = J(\phi_{\bar{A}})}$. From proposition 620 (p. 352), $\mathbf{J(f\phi_A), J(f\phi_{\bar{A}})}$ exist. Since $\mathbf{\bar{A} \subset A}$, we have $\phi_{\bar{\mathbf{A}}-\mathbf{A}} = \phi_{\bar{\mathbf{A}}} - \phi_{\mathbf{A}}$. Then from proposition 608 (p. 339) $\mathbf{J_{\bar{A}}(1) - J_A(1) = J_{\bar{A}-A(1)} = J(\phi_{\bar{A}-A}) = 0}$ and from proposition 623 (p. 356) $\mathbf{J(f\phi_A) = J(f\phi_{\bar{A}})}$.

**Proposition 625** Using the notations of proposition 624 (p. 357), let us consider a convex set $\mathbf{A} \subset \mathcal{R}^{\mathbf{n}}$ inside a cube $\mathbf{C(0,c)}$. Let us consider the projection of $\mathbf{A}$ along a direction $\mathbf{i}$: $\mathbf{proj_iA}$ on $\mathcal{R}^{\mathbf{n-1}}$. It is convex and then has a volume. For any point $\mathbf{x} \in \mathbf{proj_iA} \subset \mathcal{R}^{\mathbf{n-1}}$, we consider the two points of $\boldsymbol{\mathcal{B}}(\mathbf{A})$ which project in $\mathbf{x}$ and their two $\mathbf{i}$ coordinate $\mathbf{x^i}$, $\mathbf{x'^i}$ with $\mathbf{x^i \leq x'^i}$, we have (see definitions 294 (p. 342), 290 (p. 332)):

$$\mathbf{J_A(1) = J_{proj_i(A)}(x'^i - x^i)}\ .$$

**Proof**
It is a direct consequence of propositions 606 (p. 338), 616 (p. 349) and 618 (p. 351).

**Proposition 626** We consider a set of $\mathbf{n}$ independent vectors $\mathbf{V} = (\mathbf{v_i} : [\mathbf{1,n}] \to \mathcal{R}^{\mathbf{n}})$ which are expressed in the reference basis $(\mathbf{e_i} : [\mathbf{1,n}] \to \mathcal{R}^{\mathbf{n}})$ of $\mathcal{R}^{\mathbf{n}}$. The volume of the inner part $\overline{\boldsymbol{V}}$ of $\mathbf{V}$ is given by:

$$\mathbf{Vol}(\overline{\boldsymbol{V}}) = \mathbf{J}(\phi_{\overline{\boldsymbol{V}}}) = |\mathbf{det(V)}|\ .$$

Where $(\phi_{\overline{\boldsymbol{V}}}(\mathbf{x}) : \mathcal{R}^{\mathbf{n}} \to \{\mathbf{0,1}\})$ is the characteristic map of $\overline{\boldsymbol{V}}$.

**Proof**
$\overline{\boldsymbol{V}}$ is a convex set. The characteristic map of $\overline{\boldsymbol{V}}$ is given by:

$$\phi_V(\mathbf{x}) = \prod_{\mathbf{i}\in[\mathbf{1,n}]} \phi^1(\mathbf{w^i_j x^j})\ ,$$

where $\mathbf{w^i_j}$ is the inverse of $\mathbf{v^i_j}$ ( $\mathbf{w^i_k v^k_j} = \delta^i_j$) and $\phi^1$ is defined in definition 296 (p. 351):

$$(\phi^1(\mathbf{u}) : \mathcal{R}^{\mathbf{n}} \to \{\mathbf{0,1}\}),\ \mathbf{u} \notin [\mathbf{0,1}] \Rightarrow \phi^1(\mathbf{u}) = \mathbf{0},\ \mathbf{u} \in [\mathbf{0,1}] \Rightarrow \phi^1(\mathbf{u}) = \mathbf{1}\ .$$

Using proposition 625 (p. 357) and the notations of proposition 624 (p. 357), we can add to $\mathbf{x^1}$, the term $-\frac{\sum_{\mathbf{k}\in[\mathbf{2,n}]} \mathbf{w^1_k x^k}}{\mathbf{w^1_1}}$ it changes $\mathbf{V}$ to $\mathbf{{}^1V} = (\mathbf{{}^1v_i} : [\mathbf{1,n}] \to \mathcal{R}^{\mathbf{n}})$ with:

$$\mathbf{j} \in [\mathbf{1,n}] \Rightarrow (\mathbf{i} \neq \mathbf{1} \Rightarrow \mathbf{{}^1w^j_i = w^j_i - w^j_1 \frac{w^1_i}{w^1_1}},\ \mathbf{{}^1w^j_1 = w^j_1})\ .$$

Since **det** is a multilinear map, we have:

$$\mathbf{det}({}^1\mathbf{w}_i^j) = \mathbf{det}(\mathbf{w}_i^j) + \sum_{\mathbf{k}\in[2,\mathbf{n}]} \mathbf{det}({}^d\mathbf{w}(\mathbf{k})_i^j)$$

setting

$$\mathbf{i} \neq \mathbf{k} \Rightarrow {}^d\mathbf{w}(\mathbf{k})_i^j = \mathbf{w}_i^j, \mathbf{i} = \mathbf{k} \Rightarrow {}^d\mathbf{w}(\mathbf{k})_i^j = -\mathbf{w}_1^j\frac{\mathbf{w}_i^1}{\mathbf{w}_1^1}$$

we have:

$$\mathbf{k} \in [2,\mathbf{n}] \Rightarrow \mathbf{det}({}^d\mathbf{w}(\mathbf{k})_i^j) = 0$$

because

$$\mathbf{k} \in [2,\mathbf{n}] \Rightarrow ({}^d\mathbf{w}(\mathbf{k})_i^j = -\mathbf{w}_1^j\frac{\mathbf{w}_k^1}{\mathbf{w}_1^1} = \lambda \times {}^d\mathbf{w}(\mathbf{k})_1^j, \lambda \in \mathcal{R})$$

$\mathbf{det}({}^d\mathbf{w}(\mathbf{k})_i^j)$ is the determinant of dependent vectors (proposition 430 (p. 250)). Doing that we have:

$$\mathrm{Vol}(V) = \mathrm{Vol}(\overline{V}) = \mathrm{J}(\phi_{\overline{V}}) = \mathrm{Vol}({}^1\overline{V}) = \mathrm{J}(\phi_{{}^1\overline{V}})$$

and

$$\mathbf{det}(\mathbf{V}) = \mathbf{det}(\mathbf{v}_j^i) = \mathbf{det}^{-1}(\mathbf{w}_j^i) = \mathbf{det}^{-1}({}^1\mathbf{w}_i^j) = \mathbf{det}({}^1\mathbf{v}_i^j) = \mathbf{det}({}^1\mathbf{V})\ .$$

We have only one term in ${}^1\mathbf{w}_i^1\mathbf{x}^i : {}^1\mathbf{w}_1^1 = \mathbf{w}_1^1$, $\mathbf{i} \in [2,\mathbf{n}] \Rightarrow {}^1\mathbf{w}_i^1 = \mathbf{0}$. One can do the same thing adding a term to $\mathbf{x}^2$ in order to cancel the other terms of the expression ${}^1\mathbf{w}_i^2\mathbf{x}^i$. It does not change the term ${}^1\mathbf{w}_i^1\mathbf{x}^i$ which has only one item ${}^1\mathbf{w}_1^1\mathbf{x}^1$ and does not contain $\mathbf{x}^2$. One can continue like that up to ${}^n\mathbf{V}$ the volume of which is still equal to $\mathbf{V}$ but the basic vectors are now ${}^n\mathbf{v}_i^i\mathbf{e}_i$ with ${}^n\mathbf{v}_i^i = \frac{1}{{}^n\mathbf{w}_i^i}$. In such a case we have ( propositions 606 (p. 338), 616 (p. 349) and 437 (p. 253)):

$$\mathrm{Vol}(\overline{V}) = \mathrm{Vol}({}^n\overline{V}) = \prod_{\mathbf{i}\in[1,\mathbf{n}]} {}^n\mathbf{v}_i^i = \frac{1}{\prod_{\mathbf{i}\in[1,\mathbf{n}]} {}^n\mathbf{w}_i^i} = \frac{1}{\mathbf{det}({}^n\mathbf{w}_j^i)} = \frac{1}{\mathbf{det}(\mathbf{w}_j^i)} = \mathbf{det}(\mathbf{v}_j^i)\ .$$

**Definition 297 Jacobian of a coordinate change.**

Taking $(\mathbf{e}_i : [1,\mathbf{n}] \to \mathcal{R}^n)$ as the reference basis of $\mathcal{R}^n$, we suppose that we have a map $(\mathbf{f}(\mathbf{x}) : \mathcal{R}^n \to \mathcal{R}^n)$ where $\mathbf{x} = \mathbf{x}^i\mathbf{e}_i$ and $\mathbf{f}(\mathbf{x}) = \mathbf{f}^i(\mathbf{x})\mathbf{e}_i$. If all partial derivatives exist at $\mathbf{x}$, one calls Jacobian at $\mathbf{x}$ the matrix $(\mathbf{Jac}(\mathbf{f}(\mathbf{x})))_j^i : [1,\mathbf{n}]^2 \to \mathcal{R})$:

$$\mathbf{Jac}(\mathbf{f}(\mathbf{x}))_j^i = \frac{\partial \mathbf{f}^i(\mathbf{x})}{\partial \mathbf{x}^j} = \partial_j\mathbf{f}^i(\mathbf{x})\ .$$

**Proposition 627** We suppose that we have a continuous injective map (definition 122 (p. 139)) $\mathbf{g}$ from an open $\mathbf{O} \subset \mathcal{R}^n$ to $\mathcal{R}^n$. $\mathbf{g}$ has a continuous Jacobian defined at each point of $\mathbf{O}$. We suppose that $\mathbf{g}^{-1}$ is continuous on $\mathbf{g}(\mathbf{O})$ and that $|\mathbf{det}(\mathbf{Jac}(\mathbf{g})(\mathbf{x}))|$ never gets null in $\mathbf{O}$:

$$(\forall \mathbf{x} \in \mathbf{O})(\mathbf{det}(\mathbf{Jac}(\mathbf{g})(\mathbf{x})) \neq 0\ .$$

Then the $\mathbf{g}^{-1}(\mathbf{x})$ has a continuous Jacobian for all $\mathbf{x} \in \mathbf{g}(\mathbf{O})$ and

$$(\mathbf{Jac}(\mathbf{g})(\mathbf{x}))^{-1} = \mathbf{Jac}(\mathbf{g}^{-1})(\mathbf{g}(\mathbf{x}))\ .$$

**Proof**
For simplifying the writing, we set $\mathbf{y(x) = g(x)}$ and $\mathbf{x(y) = g^{-1}(y)}$, with $\mathbf{x \in O}$, $\mathbf{y \in g(O) = y(O)}$. We take $(\mathbf{e_i} : [1, n] \to \mathcal{R}^n)$ as a basis of $\mathcal{R}^n$ and we write $\mathbf{x = x^i e_i}$, $\mathbf{y = y^i e_i}$.
For all $\mathbf{x \in O}$ and $\mathbf{x + \Delta x \in O}$, setting:

$$\Delta y = y(x + \Delta x) - y(x),\ \eta = \Delta y - \partial_i y^j \Delta x^i e_j \tag{2.177}$$

we have from proposition 560 (p. 307):

$$\begin{aligned}&(\forall \alpha > 0)(\exists \epsilon_x(\alpha) > 0)((\Delta x \in \mathcal{R}^n, \|\Delta x\| < \epsilon_x(\alpha))\\ &\Rightarrow \|y(x + \Delta x) - y(x) - \partial_i y^j \Delta x^i e_j\| = \|\Delta y - \partial_i y^j \Delta x^i e_j\| = \|\eta\| < \alpha)\ .\end{aligned} \tag{2.178}$$

We set $\mathbf{Jac(y)^j_i(x) = \partial_i y^j(x)}$. Since $\mathbf{det(Jac(y))}$ never gets to zero, we have (See definition 228 (p. 254) and proposition 444 (p. 255)):

$$Jac(y)^{-1}{}^j_i = Minor^j_i(Jac(y))/det(Jac(y))\ .$$

From equation (2.177), $\mathbf{Jac(y)\Delta x = \Delta y - \eta}$, then we have:

$$\Delta x = Jac(y)^{-1}(\Delta y - \eta) = Jac(y)^{-1}\Delta y - Jac(y)^{-1}\eta\ .$$

Since $\mathbf{x}$ is continuous in $\mathbf{y}$, we have:

$$(\forall \alpha > 0)(\exists \epsilon_y(\alpha) > 0)(\|\Delta y\| < \epsilon_y(\alpha) \Rightarrow \|\Delta x\| < \alpha)\ . \tag{2.179}$$

We have from proposition 472 (p. 268):

$$\| - Jac(y)^{-1}\eta\| \le \| - Jac(y)^{-1}\|\|\eta\|\ .$$

Setting $\epsilon_1(\alpha) = \epsilon_x(\frac{\alpha}{1+\|Jac(y)^{-1}(x)\|})$, from equations (2.178) and (2.179), we get:

$$\begin{aligned}&\|\Delta x\| < \epsilon_1(\alpha) \Rightarrow \| - Jac(y)^{-1}\eta\| < \alpha\\ &\|\Delta y\| < \epsilon_y(\epsilon_1(\alpha)) \Rightarrow \|\Delta x\| < \epsilon_1 \Rightarrow \| - Jac(y)^{-1}\eta\| < \alpha\end{aligned}$$

and we have:

$$\begin{aligned}&(\forall \alpha)(\forall y \in y(O))(\forall i \in [1, n])\\ &((|\Delta y^i| < \epsilon_y(\epsilon_1(\alpha)), (\forall j \in ([1, n] - \{i\}))(\Delta y^j = 0), x = x(y) = g^{-1}(y))\\ &\Rightarrow \|\Delta x - Jac(y)^{-1}(x(y))\Delta y\| = \| - Jac(y)^{-1}\eta\| \le \| - Jac(y)^{-1}\|\|\eta\|\\ &\le \|Jac(y)^{-1}\|\frac{\alpha}{\|Jac(y)^{-1}\| + 1} < \alpha\\ &\Rightarrow (\forall k \in [1, n])(|\frac{\Delta x^k}{\Delta y^i} - Jac(y)^{-1}(x(y))^k_i| < \alpha))\ .\end{aligned}$$

Then:

$$\begin{aligned}&(\forall i \in [1, n])((\forall j \in ([1, n] - \{i\}))(\Delta y^j = 0) \Rightarrow (\forall \alpha > 0)(\exists \epsilon = \epsilon_y(\epsilon_1(\alpha)) > 0)\\ &(|\Delta y^i| < \epsilon \Rightarrow (\forall k \in [1, n])(|\frac{\Delta x^k}{\Delta y^i} - Jac(y)^{-1}(x(y))^k_i| < \alpha)))\ .\end{aligned}$$

**Proposition 628** We suppose that we have a continuous injective map (definition 122 (p. 139)) $\mathbf{g}$ from an open $\mathbf{O} \subset \mathcal{R}^{\mathbf{n}}$ to $\mathcal{R}^{\mathbf{n}}$ and a continuous Jacobian defined at each point of $\mathbf{O}$. We suppose that $\mathbf{g}$ has a continuous inverse from $\mathbf{g(O)}$ to $\mathbf{O}$ with a continuous Jacobian defined at each point of $\mathbf{g(O)}$. Then we have:

$$(\mathbf{Jac(g)(x)})^{-1} = \mathbf{Jac(g^{-1})(g(x))}$$

and both Jacobians are different from zero.

**Proof**

We have $\mathbf{g^{-1}(g(x)) = x}$. From proposition 563 (p. 311) we have:

$$\begin{aligned}&\partial_{\mathrm{i}}(\mathbf{g}^{-1}(\mathbf{g}(\mathbf{x})))^{\mathrm{j}} = \partial_{\mathrm{i}}\mathbf{x}^{\mathrm{j}} = \delta_{\mathrm{i}}^{\mathrm{j}} = (\partial_{\mathrm{k}}\mathbf{g}^{-1\mathrm{j}})(\mathbf{g}(\mathbf{x}))\partial_{\mathrm{i}}\mathbf{g}(\mathbf{x})^{\mathrm{k}} \\ &= \mathbf{Jac}(\mathbf{g}^{-1})(\mathbf{g}(\mathbf{x}))^{\mathrm{j}}_{\mathrm{k}}\mathbf{Jac}(\mathbf{g})(\mathbf{x})^{\mathrm{k}}_{\mathrm{i}} \ .\end{aligned}$$

Then

$$\mathbf{Jac(g^{-1})(g(x))Jac(g)(x)} = \mathbb{1}$$

where $\mathbb{1}$ is the unitary matrix ($\mathbb{1}^{\mathrm{i}}_{\mathrm{j}} = \delta^{\mathrm{i}}_{\mathrm{j}}$). From proposition 448 (p. 257):

$$\mathbf{det(Jac(g^{-1}))(g(x))det(Jac(g)(x))} = \mathbf{det}(\mathbb{1}) = \mathbf{1} \ .$$

Then

$$\mathbf{x} \in \mathbf{O} \Rightarrow \mathbf{det(Jac(g)(x))} \neq \mathbf{0}, \mathbf{det(Jac(g^{-1}))(g(x))} \neq \mathbf{0} \ .$$

**Proposition 629** We suppose that we have a continuous injective map (definition 122 (p. 139)) from a closed box $\overline{\mathbf{C}}(\mathbf{0},\mathbf{c}) \subset \mathcal{R}^{\mathbf{n}}$ to $\mathcal{R}^{\mathbf{n}}$ with a continuous inverse and a continuous Jacobian defined at each point of $\overline{\mathbf{C}}(\mathbf{0},\mathbf{c})$: $(\mathbf{g(x)} : \overline{\mathbf{C}}(\mathbf{0},\mathbf{c}) \to \mathcal{R}^{\mathbf{n}})$ such that $|\mathbf{det(Jac(g(x)))}| > \mathbf{0}$. Taking a set $\mathbf{vol}$ having a volume, we call it characteristic map $(\phi_{\mathbf{vol}} : \mathcal{R}^{\mathbf{n}} \to \{\mathbf{0},\mathbf{1}\})$ such that $\mathbf{x} \notin \overline{\mathbf{C}}(\mathbf{0},\mathbf{c}) \Rightarrow \phi_{\mathbf{vol}}(\mathbf{x}) = \mathbf{0}$. Given a continuous map $(\mathbf{f} : \overline{\mathbf{C}}(\mathbf{0},\mathbf{c}) \to \mathcal{R})$, the Riemann integral $\phi_{\mathbf{vol}} \circ \mathbf{g}$ exists and we have the following identity:

$$\mathbf{J}(\mathbf{f}\phi_{\mathbf{vol}} \circ \mathbf{g}|\mathbf{det(Jac(g))}|) = \mathbf{J}(\mathbf{f}\phi_{\mathbf{vol}}) \ , \tag{2.180}$$

where $\mathbf{J(h)}$ is the Riemann integral of the map $(\mathbf{h} : \overline{\mathbf{C}}(\mathbf{0},\mathbf{c}) \to \mathcal{R})$ if it exists.

**Proof**

From propositions 628 and 444 (p. 255), $\mathbf{Jac(g^{-1})}$ exists, it is continuous and has a nonzero determinant. Since $\overline{\mathbf{C}}(\mathbf{0},\mathbf{c})$ is a closed subset of $\mathcal{R}^{\mathbf{n}}$ and enclosed in a ball then it is compact (proposition 521 (p. 290)). From proposition 240 (p. 167) $\mathbf{g}(\overline{\mathbf{C}}(\mathbf{0},\mathbf{c}))$ is also compact and enclosed in a closed box (proposition 523 (p. 291)): $(\exists \mathbf{c}' > \mathbf{0})(\mathbf{g}(\overline{\mathbf{C}}(\mathbf{0},\mathbf{c})) \subset \overline{\mathbf{C}}(\mathbf{0},\mathbf{c}'))$ then we have from proposition 560 (p. 307), 2:

$$\begin{aligned}&(\forall \alpha_{\mathbf{g}})(\exists \epsilon_{\mathbf{g}}(\alpha_{\mathbf{g}}) > \mathbf{0})(\forall \mathbf{z} \in \overline{\mathbf{C}}(\mathbf{0},\mathbf{c}))(\|\Delta\mathbf{z}\| < \epsilon_{\mathbf{g}}(\alpha_{\mathbf{g}}) \\ &\Rightarrow \|\mathbf{g}(\mathbf{z}+\Delta\mathbf{z}) - \mathbf{g}(\mathbf{z})\| < \alpha_{\mathbf{g}})\end{aligned} \tag{2.181}$$

$$\begin{aligned}&(\forall \alpha^{\mathcal{J}}_{\mathbf{g}})(\exists \epsilon^{\mathcal{J}}_{\mathbf{g}}(\alpha^{\mathcal{J}}_{\mathbf{g}}) > \mathbf{0})(\forall \mathbf{z} \in \overline{\mathbf{C}}(\mathbf{0},\mathbf{c}))(\|\Delta\mathbf{z}\| < \epsilon^{\mathcal{J}}_{\mathbf{g}}(\alpha^{\mathcal{J}}_{\mathbf{g}}) \\ &\Rightarrow \mathbf{g}(\mathbf{z}+\Delta\mathbf{z}) = \mathbf{g}(\mathbf{z}) + (\mathbf{Jac}(\mathbf{g}(\mathbf{z})) + \eta_{\mathbf{g}}(\mathbf{z},\Delta\mathbf{z}))\Delta\mathbf{z}, \|\eta_{\mathbf{g}}(\mathbf{z},\Delta\mathbf{z})\| < \alpha^{\mathcal{J}}_{\mathbf{g}})\end{aligned} \tag{2.182}$$

$$\begin{aligned}&(\forall \alpha_{\mathbf{grv}})(\exists \epsilon_{\mathbf{grv}}(\alpha_{\mathbf{grv}}) > \mathbf{0})(\forall \mathbf{x} \in \overline{\mathbf{C}}(\mathbf{0},\mathbf{c}'))(\|\Delta\mathbf{x}\| < \epsilon_{\mathbf{grv}}(\alpha_{\mathbf{grv}}) \\ &\Rightarrow \|\mathbf{g}^{-1}(\mathbf{x}+\Delta\mathbf{x}) - \mathbf{g}^{-1}(\mathbf{x})\| < \alpha_{\mathbf{grv}})\end{aligned} \tag{2.183}$$

$$\begin{aligned}&(\forall\alpha_{g^{\mathcal{J}}_{rv}})(\exists\epsilon_{g^{\mathcal{J}}_{rv}}(\alpha_{g^{\mathcal{J}}_{rv}})>0)(\forall x\in\overline{C}(0,c'))(\|\Delta x\|<\epsilon_{g^{\mathcal{J}}_{rv}}(\alpha_{g^{\mathcal{J}}_{rv}})\Rightarrow\\&g^{-1}(x+\Delta x)=g^{-1}(x)+(Jac(g^{-1})(x)+\eta_{g_{rv}}(x,\Delta x))\Delta x,\ \|\eta_{g_{rv}}(x,\Delta x)\|<\alpha_{g^{\mathcal{J}}_{rv}})\ .\end{aligned}\tag{2.184}$$

All components of $Jac(g)(x)$ (resp. $Jac(g^{-1})(u)$) are continuous on the compact $\overline{C}(0,c)$ (resp. $\overline{C}(0,c')$). From proposition 297 (p. 189), the absolute values of any component of $Jac(g)(x)$ (resp. $Jac(g^{-1})(u)$) and the absolute value of the determinant have a maximum which is reached in $\overline{C}(0,c)$ (resp. $\overline{C}(0,c')$). We set those maxima by:

$$\begin{aligned}g^{\mathcal{M}}&=\max_{z\in\overline{C}(0,c)}\max_{i\in[1,n]}\max_{j\in[1,n]}|Jac(g)^i_j(z)|\\g^{\mathcal{M}}_{rv}&=\max_{x\in\overline{C}(0,c')}\max_{i\in[1,n]}\max_{j\in[1,n]}|Jac(g^{-1})^i_j(x)|\\det^{\mathcal{M}}_{\mathcal{J}}&=\max_{z\in\overline{C}(0,c)}|det(Jac(g)(z)|\ .\end{aligned}\tag{2.185}$$

Given any continuous map $(f:\overline{C}(0,c')\rightarrow\mathcal{R})$, $f\circ g$ is also continuous over $\overline{C}(0,c)$ (proposition 241 (p. 167)) so both of them have a Riemann integral on any subset of $\overline{C}(0,c)$ having a volume defined by an integral (proposition 620 (p. 352)). Since $\overline{C}(0,c')$ is compact:

$$\begin{aligned}&(\forall\alpha>0)(\exists\epsilon_f(\alpha)>0)(\exists\epsilon_{fg}(\alpha)>0)(\forall x\in\overline{C}(0,c'))(x+\Delta x\in\overline{C}(0,c')\Rightarrow\\&(\|\Delta x\|<\epsilon_f(\alpha)\Rightarrow|f(x+\Delta x)-f(x)|<\alpha,\\&\|\Delta x\|<\epsilon_{fg}(\alpha)\Rightarrow|f(g(x+\Delta x))-f(g(x))|<\alpha))\\&(\exists x^{\mathcal{M}}\in\overline{C}(0,c'))(\forall x\in\overline{C}(0,c'))(|f(x)|\leq f^{\mathcal{M}}=f(x^{\mathcal{M}}))\ .\end{aligned}\tag{2.186}$$

Considering a set $V\subset\overline{C}(0,c)$ with a volume, we call $\phi_V$ its characteristic map. The characteristic map of $g(V)\subset\overline{C}(0,c')$ is $\phi_{g(V)}=\phi_V\circ g^{-1}$ since (see definitions 114 (p. 132), 174 (p. 141)) we have:

$$x\in g(V)\Leftrightarrow(\exists z\in V)(x=g(z)\Leftrightarrow z=g^{-1}(x))\Leftrightarrow\phi_V(g^{-1}(x))=1\ .$$

Let us consider an element $t=(p,y,z)$ of the grid $Gr_a$ of definition 290 (p. 332). For each $l\in\prod_{k\in[1,n]}[0,p^k[$ corresponds a characteristic map $(\phi_l:\mathcal{R}^n\rightarrow\{0,1\})$ such as:

$$\phi_l(z)=1\Leftrightarrow(\forall i\in[1,n])(0\leq z^i-y^i_{l_i}<y^i_{l_i+1}-y^i_{l_i})\Leftrightarrow z\in br_l\tag{2.187}$$

From definition 344 (p. 447) and proposition 749 (p. 450), the relation (2.187) defines $br_l$. Taking a point in $br_l$, $z'=z'^ie_i$, calling $\Delta z=(z'^i-y^i_{l_i})e_i$ and $z=z^ie_i=y^i_{l_i}e_i$, we should have $0\leq\Delta z^i/(y^i_{l_i+1}-y^i_{l_i})<1$. Calling $x=g(z)$, $\Delta x=g(z+\Delta z)-g(z)$, with:

$$\Delta x=g(z+\Delta z)-g(z)=(Jac(g)(z)+\eta_g)\Delta z$$

taking a basis made of vectors $(u_i:[1,n]\rightarrow\mathcal{R}^n)$:

$$u_i=Jac(g)(z)(e_i)=Jac(g)(z)^j_ie_j\Rightarrow e_j=Jac(g)(z)^{-1}{}^i_ju_i\ .$$

For $x+\Delta x=g(z+\Delta z)$ to be in $g(br_l)$, we should have $z+\Delta z\in br_l$. It means we can write:

$$\begin{aligned}&(\forall i\in[1,n])(\exists m^i\in[0,1[)(\Delta z=m^i(y^i_{l_i+1}-y^i_{l_i})e_i\Leftrightarrow z+\Delta z\in br_l\\&\Leftrightarrow x+\Delta x=g(z+\Delta z)\in g(br_l))\ .\end{aligned}$$

We may set:

$$\Delta x = m^i(y^i_{l_i+1} - y^i_{l_i})u_i + \eta_g m^i(y^i_{l_i+1} - y^i_{l_i})e_i = m^i(y^i_{l_i+1} - y^i_{l_i})u_i + \eta_g{}^j_i m^i(y^i_{l_i+1} - y^i_{l_i})e_j$$

$$\begin{aligned}\Delta x &= m^i(y^i_{l_i+1} - y^i_{l_i})u_i + \eta_g{}^j_i m^i(y^i_{l_i+1} - y^i_{l_i})jac^{-1}(g(z))^k_j u_k \\ &= m^i(y^i_{l_i+1} - y^i_{l_i})(u_i + \lambda^k_i u_k) = (m^i(y^i_{l_i+1} - y^i_{l_i}) + m^j(y^j_{l_j+1} - y^j_{l_j})\lambda^k_j)u_i\end{aligned}$$

with

$$\lambda^k_j = \eta_g{}^i_j Jac(g)^{-1}(z)^k_i \,.$$

We have from equation (2.181):

$$\|\Delta z\| < \epsilon^{\mathcal{J}}_g(\alpha^{\mathcal{J}}_g) \Rightarrow \max_{(i,j)\in[1,n]^2} |\eta_g{}^j_i| < \alpha^{\mathcal{J}}_g \,.$$

To simplify the writing, we shall use from now on for $\mathbf{t}$, $\mathbf{t_a} = (p_a, y_a, z_a) \in \mathbf{Gr_a}$ defined by:

$$\begin{aligned}&p_a > 2c/a,\ l \in [0, p_a]^n \Rightarrow y_a^{l^i,i} = a \times l(i) = a\,l^i = z_a^{l,i} \\ &z^l_a = a\,l^i e_i = z_a^{l,i} e_i \,.\end{aligned}$$

We have then:

$$\begin{aligned}&\Delta x = (m^i + m^j\lambda^i_j)au_i \\ &\|\Delta z\| < \epsilon^{\mathcal{J}}_g(\alpha^{\mathcal{J}}_g) \Rightarrow (\forall z \in \bar{C}(0, c))(\forall i \in [1, n]) \\ &(\forall j \in [1, n])(|\lambda^i_j(z)| < \alpha^{\mathcal{J}}_g g^{\mathcal{M}}_{rv} = \frac{\alpha_\lambda}{n},\ |m^j\lambda^i_j| < \alpha_\lambda) \,.\end{aligned}$$

And when $\mathbf{x}$ and $\mathbf{x} + \Delta\mathbf{x}$ are in $\mathbf{g(br_l)}$ with $\mathbf{t_a} = (p_a, y_a, z_a) \in \mathbf{Gr_a}$, we have:

$$\begin{aligned}&g^{-1}(x + \Delta x)^i - g^{-1}(x)^i < y_{a\,l_i+1}^{\ i} - y_{a\,l_i}^{\ i} = a \\ &\|g^{-1}(x + \Delta x) - g^{-1}(x)\| = \sqrt{\sum_{i\in[1,n]} (g^{-1}(x + \Delta x)^i - g^{-1}(x)^i)^2} \\ &< \sqrt{\sum_{i\in[1,n]} (y_{a\,l_i+1}^{\ i} - y_{a\,l_i}^{\ i})^2} = a\sqrt{n} \,.\end{aligned}$$

Then from equation (2.182) (p. 360), we have:

$$\begin{aligned}&\|g^{-1}(x + \Delta x) - g^{-1}(x)\| < a\sqrt{n} < \epsilon_g(\epsilon^{\mathcal{J}}_{g_{rv}}(\alpha^{\mathcal{J}}_{g_{rv}})) \Rightarrow \\ &\|g \circ g^{-1}(x + \Delta x) - g \circ g^{-1}(x)\| = \|\Delta x\| < \epsilon^{\mathcal{J}}_{g_{rv}}(\alpha^{\mathcal{J}}_{g_{rv}}) \Rightarrow \max_{(i,j)\in[1,n]^2} \eta_{g_{rv}}{}^j_i < \alpha^{\mathcal{J}}_{g_{rv}}\end{aligned}$$

and

$$\begin{aligned}&(a < \epsilon_g(\epsilon^{\mathcal{J}}_{g_{rv}}(\alpha^{\mathcal{J}}_{g_{rv}}))/\sqrt{n},\ \ \alpha_\lambda = \alpha^{\mathcal{J}}_g g^{\mathcal{M}}_{rv}) \Rightarrow ((x = g(a\,l^i e_i), x + \Delta x \in g(br_l)) \\ &\Rightarrow (\exists m \in [0, 1]^n)(\Delta x = a(m^i + m^j\lambda^i_j)u_i, \|m^j\lambda^i_j\| < \alpha_\lambda = n\alpha^{\mathcal{J}}_g g^{\mathcal{M}}_{rv})) \,.\end{aligned} \tag{2.188}$$

Which means that any point in $\mathbf{g(Bx_l)}$ is in the inner part of the box defined by the basis $((1 + 2\alpha_\lambda)u_i : [1, n] \to \mathcal{R}^n)$ displaced by $\mathbf{g(a\ l^i e_i)} - \mathbf{a\alpha_\lambda u}$ and that any point in the inner part of the box defined by the basis $((1 - 2\alpha_\lambda)u_i : [1, n] \to \mathcal{R}^n)$ displaced by

$\mathbf{g(a\, l^i e_i) + a\alpha_\lambda u}$ is in $\mathbf{g(Bx_l)}$. From definition 221 (p. 250), proposition 610 (p. 340) and proposition 626 (p. 357), the volume of those two inner parts is given by:

$$\begin{aligned}&\mathbf{Vol(set_{\Delta x}\{(\exists m \in [0,1[^n)(\Delta x = m^i a(1+2\alpha_\lambda)u_i - a\alpha_\lambda u)\})}\\&\mathbf{= (1+2\alpha_\lambda)^n a^n \det(u)}\\&\mathbf{Vol(set_{\Delta x}\{(\exists m \in [0,1[^n)(\Delta x = m^i a(1-2\alpha_\lambda)u_i + a\alpha_\lambda u)\})}\\&\mathbf{= (1-2\alpha_\lambda)^n a^n \det(u)\ .}\end{aligned} \quad (2.189)$$

We also have, since for $\mathbf{0.5 > \alpha_\lambda > 0}$, the map in $\boldsymbol{\alpha}$: $\mathbf{(1+2\alpha_\lambda)^n}$ (resp. $\mathbf{(1-2\alpha_\lambda)^n}$ ) is an increasing (resp. decreasing) continuous map (definition 384 (p. 486)) equal to 1 for $\boldsymbol{\alpha_\lambda = 0}$ (see proposition 328 (p. 202)). Given an $\boldsymbol{\alpha}$, we can find a upper value for $\boldsymbol{\alpha_\lambda}$ so the two above volumes differ from $\mathbf{a^n \det(u)}$ by $\boldsymbol{\alpha a^n \det(u)}$:

$$\begin{aligned}&\mathbf{0 < \alpha_\lambda < \frac{(1+\alpha)^{1/n} - 1}{2} \Rightarrow 1 < (1+2\alpha_\lambda)^n < 1+\alpha}\\&\mathbf{0 < \alpha_\lambda < \frac{1-(1-\alpha)^{1/n}}{2} \Rightarrow 1-\alpha < (1-2\alpha_\lambda)^n < 1\ .}\end{aligned} \quad (2.190)$$

Setting:

$$\boldsymbol{\epsilon_\lambda(\alpha) = \min(\frac{(1+\alpha)^{1/n}-1}{2}, \frac{1-(1-\alpha)^{1/n}}{2})}$$

we have:

$$\begin{aligned}&\mathbf{(\forall\alpha > 0)(\exists\epsilon_\lambda(\alpha) > 0)(\alpha_\lambda \le \epsilon_\lambda(\alpha) \Rightarrow (1-\alpha)a^n \det(u) \le (1-2\alpha_\lambda)^n a^n \det(u)}\\&\mathbf{< (1+2\alpha_\lambda)^n a^n \det(u) \le (1+\alpha)a^n \det(u))\ .}\end{aligned}$$

Now summing over $\mathbf{t_a}$, $\mathbf{f \times \phi_V}$, we have:

$$\begin{aligned}&\mathbf{gr_a^{f\phi_V}(t_a) = \sum_{l\in[0,p_a]^n} f(a\, l^i e_i)\phi_V(a\, l^i e_i)a^n = \sum_{l\in l^{nz}} f(a\, l^i e_i)a^n =}\\&\mathbf{\sum_{l\in l^{nz}} f(a\, l^i e_i)a^n \phi_{br_l}(a\, l^i e_i)\ .}\end{aligned}$$

Taking a similar term $\mathbf{t_b = (p_b, y_b, z_b) \in Gr_b}$ given by:

$$\mathbf{p_b > 2c/b,\ l \in [0,p_b]^n \Rightarrow y_b^{l^i,i} = b \times l^i = z_b^{l,i}}$$

and using (2.186) (p. 361), we have:

$$\begin{aligned}&\mathbf{a < \epsilon_f(\alpha)/\sqrt{n} \Rightarrow (x \in g(br_l) \Rightarrow \|x - a\, l^i e_i\| < a\sqrt{n} < \epsilon_f(\alpha) \Rightarrow}\\&\mathbf{|f(g^{-1}(x)) - f(a\, l^i e_i)| < \alpha)}\end{aligned}$$

$$\mathbf{a < \epsilon_f(\alpha)/\sqrt{n} \Rightarrow |(f(g^{-1}(x)) - f(a\, l^i e_i))\phi_{g(br_l)}(x)| \le \alpha\phi_{g(br_l)}(x)\ .} \quad (2.191)$$

Since $\mathbf{l \ne l' \Rightarrow Bx_l \cap Bx_{l'} = \emptyset}$ and since $\mathbf{g}$ is a bijection, from proposition 186 (p. 143) we have:

$$\boldsymbol{\phi_{g(V)} = \sum_{l\in[0,p_a]^n} \phi_{g(Bx_l)}}$$

$$\begin{aligned}
\mathrm{gr}_b^{f\circ g^{-1}\phi_{g(V)}}(t_b) &= \sum_{l\in[0,p_b]^n} (b^n f(g^{-1}(b\,l^i e_i)\phi_{g(V)}(b\,l^i e_i) \\
&= \sum_{l\in[0,p_b]^n} b^n f(g^{-1}(b\,l^i e_i)) \sum_{l'\in[0,p_a]^n} \phi_{g(Bx_{l'})}(b\,l^i e_i) \\
&= \sum_{l'\in[0,p_a]^n} \sum_{l\in[0,p_b]} b^n f(g^{-1}(b\,l^i e_i))\phi_{g(Bx_{l'})}(b\,l^i e_i)\ .
\end{aligned}$$

Using equation (2.191) and considering the set $l_a^{nz} \subset [0, p_a]^n$ where $\phi_V(Bx_l) = 1$ we have:

$$\begin{aligned}
&a < \epsilon_f(\alpha)/\sqrt{n} \Rightarrow \\
&|\sum_{l'\in l_a^{nz}} \sum_{l\in[0,p_b]} (b^n f(g^{-1}(b\,l^i e_i)) - f(a\,l'^i e_i)\phi_{g(Bx_{l'})}(b\,l^i e_i))| \\
&< \sum_{l'\in l_a^{nz}} \sum_{l\in[0,p_b]} b^n \alpha \phi_{g(Bx_{l'})}(b\,l^i e_i) = \sum_{l\in[0,p_b]} b^n \alpha \phi_{g(V))}(b\,l^i e_i) < (2c)^n \alpha
\end{aligned} \tag{2.192}$$

$$\begin{aligned}
&b < \epsilon_{fg_{rv}}(\alpha_{b1})/\sqrt{n} \Rightarrow \\
&|\sum_{l'\in l_a^{nz}} \sum_{l\in[0,p_b]} b^n f(g^{-1}(b\,l^i e_i))\phi_{g(Bx_{l'})}(b\,l^i e_i) - J(f(g^{-1}))| < \alpha_{b1}\ .
\end{aligned} \tag{2.193}$$

Applying equations (2.189) (p. 363) and (2.190) (p. 363), we also have:

$$\begin{aligned}
&(\forall\alpha > 0)((\alpha_\lambda < \epsilon_\lambda(\alpha), \alpha_g^{\mathcal{J}} = \frac{\alpha_\lambda}{g_{rv}^{\mathcal{M}}}, a < \epsilon_{g_{rv}}(\epsilon_g^{\mathcal{J}}(\alpha_g^{\mathcal{J}}))/\sqrt{n}, l' \in l_a^{nz}) \\
&\Rightarrow (\forall\alpha_{b2} > 0)(\exists\epsilon_{l'}(\alpha_{b2}) > 0)(b < \epsilon_{l'}(\alpha_{b2}) \Rightarrow (a^n \mathrm{Vol}(u) - \alpha_{b2})(1-\alpha) \\
&\leq \sum_{l\in[0,p_b]} b^n \phi_{g(Bx_{l'})}(b\,l^i e_i) \leq (a^n \mathrm{Vol}(u) + \alpha_{b2})(1+\alpha)) \\
&\Rightarrow |\sum_{l\in[0,p_b]} b^n \phi_{g(Bx_{l'})}(b\,l^i e_i) - a^n \mathrm{Vol}(u)| \leq \alpha_{b2}(1+\alpha) + \alpha a^n \mathrm{Vol}(u))\ .
\end{aligned}$$

Using:

$$u_i = \mathrm{Jac}(g)(a\,l'^i e_i)_i^j e_j \Rightarrow \mathrm{Vol}(u) = |\det(\mathrm{Jac}(g)(a\,l'^i e_i))|$$

and setting

$$\epsilon_{b2}(\alpha) = \min_{l'\in l^{nz}} \epsilon'_l(\alpha a^n |\det(\mathrm{Jac}(g)(a\,l'^i e_i))|/(1+\alpha))$$

we have with the above conditions:

$$\begin{aligned}
&(0 < a < \epsilon_{g_{rv}}(\epsilon_g^{\mathcal{J}}(\epsilon_\lambda(\alpha)))/\sqrt{n}, 0 < b < \epsilon_{b2}(\alpha)) \Rightarrow \\
&(|\sum_{l\in[0,p_b]} b^n \phi_{g(Bx_{l'})}(b\,l^i e_i) - a^n |\det(\mathrm{Jac}(g)(a\,l'^i e_i))|| \leq 2\alpha a^n |\det(\mathrm{Jac}(g)(a\,l'^i e_i))|, \\
&|\sum_{l'\in l_a^{nz}} \sum_{l\in[0,p_b]} f(a\,l'^i e_i)\phi_{g(Bx_{l'})}(b\,l^i e_i) b^n \\
&- \sum_{l'\in l_a^{nz}} f(a\,l'^i e_i)|\det(\mathrm{Jac}(g)(a\,l'^i e_i))|a^n| < 2\alpha(2c)^n f^{\mathcal{M}} \det_{\mathcal{J}}^{\mathcal{M}})\ .
\end{aligned} \tag{2.194}$$

Since $\mathbf{f|det(Jac(g)|}$ is continuous and $\mathbf{Vol(V)}$ exists, $\mathbf{f|det(Jac(g))|\phi_V}$ is integrable (proposition 620 (p. 352)). Then by definition, we have:

$$\forall(\alpha_{f|det(Jac(g))|})(\exists\epsilon_{f|det(Jac(g))|}(\alpha_{f|det(Jac(g))|}))(0 < a < \epsilon_{f|det(Jac(g))|}(\alpha_{f|det(Jac(g))|}) \Rightarrow$$
$$(|J(f|det(Jac(g))|\phi_V) - \sum_{l'\in I_a^{nz}} f(a\, l'^i e_i)|det(Jac(g))(a\, l'^i e_i)|a^n| < \alpha_{f|det(Jac(g))|})) \ . \tag{2.195}$$

Regrouping everything from equations (2.192), (2.193), (2.194) and (2.195), we set:

$$0 < a < \min(\epsilon_f(\alpha_f), \epsilon_{f|det(Jac(g))|}(\alpha_{f|det(Jac(g))|}), \epsilon_{g_{rv}}(\epsilon_g^{\mathcal{J}}(\epsilon_\lambda(\alpha_u))))/\sqrt{n},$$
$$0 < b < \min(\epsilon_{b2}(\alpha_u), \epsilon_{fg_{rv}}(\alpha_{b1}))$$

and we have:

$$\begin{aligned}
&|J(f|det(Jac(g))|\phi_V) - J(f(g^{-1})\phi_{g(V)})| = \\
&|J(f|det(Jac(g))|\phi_V) - \sum_{l'\in I_a^{nz}} f(a\, l'^i e_i)|det(Jac(g))(a\, l'^i e_i)|a^n + \\
&\sum_{l'\in I_a^{nz}} f(a\, l'^i e_i)|det(Jac(g))(a\, l'^i e_i)|a^n - \sum_{l'\in I_a^{nz}}\sum_{l\in[0,p_b]} f(a\, l'^i e_i)\phi_{g(Bx_{l'})}(b\, l^i e_i)b^n + \\
&\sum_{l'\in I_a^{nz}}\sum_{l\in[0,p_b]} f(a\, l'^i e_i)\phi_{g(Bx_{l'})}(b\, l^i e_i)b^n - \sum_{l'\in I_a^{nz}}\sum_{l\in[0,p_b]} b^n f(g^{-1}(b\, l^i e_i))\phi_{g(Bx_{l'})}(b\, l^i e_i) + \\
&\sum_{l'\in I_a^{nz}}\sum_{l\in[0,p_b]} b^n f(g^{-1}(b\, l^i e_i))\phi_{g(Bx_{l'})}(b\, l^i e_i) - J(f(g^{-1}))| \\
&\le |J(f|det(Jac(g))|\phi_V) - \sum_{l'\in I_a^{nz}} f(a\, l'^i e_i)|det(Jac(g))(a\, l'^i e_i)|a^n| + \\
&|\sum_{l'\in I_a^{nz}} f(a\, l'^i e_i)|det(Jac(g))(a\, l'^i e_i)|a^n - \sum_{l'\in I_a^{nz}}\sum_{l\in[0,p_b]} f(a\, l'^i e_i)\phi_{g(Bx_{l'})}(b\, l^i e_i)b^n| + \\
&|\sum_{l'\in I_a^{nz}}\sum_{l\in[0,p_b]} f(a\, l'^i e_i)\phi_{g(Bx_{l'})}(b\, l^i e_i)b^n - \sum_{l'\in I_a^{nz}}\sum_{l\in[0,p_b]} b^n f(g^{-1}(b\, l^i e_i))\phi_{g(Bx_{l'})}(b\, l^i e_i)| + \\
&|\sum_{l'\in I_a^{nz}}\sum_{l\in[0,p_b]} b^n f(g^{-1}(b\, l^i e_i))\phi_{g(Bx_{l'})}(b\, l^i e_i) - J(f(g^{-1}))| \\
&< \alpha_{f|det(Jac(g))|} + 2\alpha_u(2c)^n f^{\mathcal{M}} det_{\mathcal{J}}^{\mathcal{M}} + (2c)^n\alpha_f + \alpha_{b1} \ .
\end{aligned}$$

By setting for any $\alpha > 0$:

$$0 < \alpha_f < \frac{\alpha}{4(2c)^n}, 0 < \alpha_{b1} < \frac{\alpha}{4}, 0 < \alpha_u < \frac{\alpha}{4(2c)^n f^{\mathcal{M}} det_{\mathcal{J}}^{\mathcal{M}}}, 0 < \alpha_{f|det(Jac(g))} < \frac{\alpha}{4}$$

we can choose $\mathbf{a}$ and then $\mathbf{b}$, so the absolute difference between the two terms

$$J(f|det(Jac(g))|\phi_V)$$

and

$$J(f(g^{-1})\phi_{g(V)})$$

has to be smaller than any $\alpha > 0$; so smaller than their absolute difference itself if nonzero then from A.1.4, 8:

$$J(f|det(Jac(g))|\phi_V) = J(f(g^{-1})\phi_{g(V)}) = J(f\phi_V \circ g^{-1})$$

one gets the well-known formula setting $\mathbf{f\phi_V = F \circ g}$

$$J(F \circ g|det(Jac(g))|) = J(F) \ .$$

### 2.25.1 Some properties of integrals on an interval of $\mathcal{R}$

**Definition 298** $\int_a^b f(x)dx$.

With two real numbers $a, b$, such that $a \leq b$ we have a map $(f : [a, b] \to \mathcal{R})$. We extend $f$ to $[-c, c]$ with $c \geq \max(|a|, |b|)$ by $x \in [-c, a] \Rightarrow f(x) = f(a)$, $x \in [b, c] \Rightarrow f(x) = f(b)$. We name $(\phi(x) : \mathcal{R} \to \{0, 1\})$, the characteristic map on $[a, b]$ (definition 293 (p. 341)). With that $J(\phi \times f)$ defined by 290 (p. 332) is called, if it exists, $\int_a^b f(x)dx$. To make the formula valid in case $a > b$, one settles by definition in such a case $\int_a^b f(x)dx = -\int_b^a f(x)dx$.

**Definition 299** $\int_a^b f(x)dx = -\int_b^a f(x)dx$, $b < a$.

**Proposition 630** $a, b$ being real numbers, any continuous map $(f : [a, b] \to \mathcal{R})$ has an integral $\int_a^b f(x)dx$.

**Proof**

From proposition 496 (p. 279) $[a, b] = B((a+b)/2, (b-a)/2)$ is convex. From proposition 618 (p. 351), it has a volume. From proposition 620 (p. 352), proposition 630 (p. 366) is true.

**Proposition 631** Let us consider $\int_a^b f dx$ with $b > a$ and $g$ an injection with derivative on $[a, b]$. Then:

$$\int_a^b f dx = \int_{g^{-1}(a)}^{g^{-1}(b)} f(g(x))g'(x)dx \,. \tag{2.196}$$

**Proof**

$g$ is from proposition 301 (p. 191) a strictly either increasing or decreasing map (definition 386 (p. 486)):

$$(\forall x \in [a, b])(g'(x) > 0), (\forall x \in [a, b])(g'(x) < 0)$$

$\phi^1$ being the characteristic function of $[0, 1]$ as defined in definition 296, we have:

$$\phi_{[a,b]}(x) = \phi^1(\frac{x-a}{b-a}), \phi_{([a-b])}(g(x)) = \phi_{[g^{-1}(a),g^{-1}(b)]} = \phi^1(\frac{x-g^{-1}(a)}{g^{-1}(b)-g^{-1}(a)}) \,.$$

From proposition 629 (p. 360), with its notations, we have:

$$\int_a^b f dx = J(f\phi^1(\frac{x-a}{b-a})) = J(f\phi^1(\frac{x-a}{b-a})) \circ g \times |g'(x)|)$$
$$= J(f(g(x))|g'(x)|\phi^1(\frac{x-g^{-1}(a)}{g^{-1}(b)-g^{-1}(a)})) \,.$$

If $g'(x) > 0$, we have $g^{-1}(b) > g^{-1}(a)$ which gives:

$$J(f(g(x))|g'(x)|\phi^1(\frac{x-g^{-1}(a)}{g^{-1}(b)-g^{-1}(a)})) = \int_{g^{-1}(a)}^{g^{-1}(b)} f(g(x))g'(x)dx \,.$$

If $g'(x) < 0$, we have $g^{-1}(b) < g^{-1}(a)$ which gives:

$$J(f(g(x))|g'(x)|\phi^1(\frac{x-g^{-1}(a)}{g^{-1}(b)-g^{-1}(a)})) = -\int_{g^{-1}(b)}^{g^{-1}(a)} f(g(x))g'(x)dx$$
$$= \int_{g^{-1}(a)}^{g^{-1}(b)} f(g(x))g'(x)dx \,.$$

**Proposition 632** With $a \leq b$, if a map $(f : [a,b] \rightarrow \mathcal{R})$ has an integral and if $(\forall x \in [a,b])(f(x) \geq 0)$, then $\int_a^b f dx \geq 0$.
**Proof**
It is a direct consequence of proposition 603 (p. 336).

**Proposition 633** With $a \leq b$, $f(x)$ and $g(x)$ being two functions on $\mathcal{R}$, we have:

$$(\forall x \in [a,b])(f(x) \leq g(x)) \Rightarrow \int_a^b f(x)dx \leq \int_a^b g(x)dx .$$

**Proof**
It is a direct consequence of proposition 609 (p. 340).

**Proposition 634** If $f(x)$ has a derivative on $[d,e]$ which has an integral on $[d,e]$, then $\int_d^e f'(x)dx = f(e) - f(d)$.
**Proof**
$(\phi(x) : \mathcal{R} \rightarrow \{0,1\})$ is the characteristic map on $[d,e]$. In every subset $Gr_a^{f'}$ of definition 290 (p. 332), with $n \in \mathcal{Z}^+$, $n > (e-d)/a$, $p \in \mathcal{Z}^+$, $2cn/(e-d) > p > cn/(e-d)+1$, there is a term such that $(\forall i \in [0,p-1])(y^{i+1} - y(i) = (e-d)/n \leq a, y(0) = d, y(n) = e$ for these values

$$gr_a^{f'}(p,y,z) = \sum_{l \in [0,p]} \phi f'(z^l)(y^{l+1} - y(l)) = \sum_{l \in [0,n-1]} f'(z^l)(y^{l+1} - y^l)$$

but, from proposition 313 (p. 195), we can choose for $l \in [0,n-1]$ $z^l \in ]y^l, y(l+1)[$ such that

$$f'(z^l) = \frac{f(y^{l+1}) - f(y^l)}{y^{l+1} - y^l}$$

for that term

$$gr_a^{f'}(p,y,z) = \sum_{l \in [0,n-1]} f(y^{l+1}) - f(y^l) = f(y(n)) - f(y(0)) = f(e) - f(d)$$

Then

$$a < e - d \Rightarrow f(e) - f(d) \in gr(Gr_a^{f'}) = s_a^{f'}$$

which proves proposition 634 (see page 333, 290 -1).

**Proposition 635** $(f(x) : \mathcal{R} \rightarrow \mathcal{R})$ is a function on $\mathcal{R}$ with $a, b, c \in \mathcal{R}$. Then we have:

$$\int_a^b f(x)dx + \int_b^c f(x)dx + \int_c^a f(x)dx = 0 .$$

**Proof**
It is a consequence of proposition 608 (p. 339) together with definition 298 (p. 366) and 299 (p. 366): If $a < b < c$ and quoting $\phi_{[a,b]}$ the characteristic map of $[a,b]$ (definition 293 (p. 341)) we have:

$$\phi_{[a,b]} + \phi_{[b,c]} = \phi_{[a,c]}$$

then proposition 608 (p. 339) plus definition 299 (p. 366) give:

$$\int_a^b f(x)dx + \int_b^c f(x)dx = J(\phi_{[a,b]}f) + J(\phi_{[b,c]}f) = J(\phi_{[a,c]}f) = \int_a^c f(x)dx$$
$$= -\int_c^a f(x)dx .$$

It is proposition 635. We can always relabel $a, b, c$ to be in that case.

**Proposition 636** If $\mathbf{f(y)}$ is a continuous function on $\mathcal{R}$, the function on $\mathcal{R}$: $\mathbf{g(x)} = \int_{\mathbf{a}}^{\mathbf{x}} \mathbf{f(y)dy}$ has $\mathbf{f(x)}$ for derivative.

**Proof**
From proposition 635 (p. 367)

$$\int_{\mathbf{a}}^{\mathbf{x}+\Delta\mathbf{x}} \mathbf{f(y)dy} - \int_{\mathbf{a}}^{\mathbf{x}} \mathbf{f(y)dy} = \int_{\mathbf{x}}^{\Delta\mathbf{x}} \mathbf{f(y)dy}$$

if $\mathbf{f_m} = \mathbf{min(f([x, x + \Delta x]))}$ $\mathbf{f_M} = \mathbf{max(f([x, x + \Delta x]))}$ one has, with $\Delta\mathbf{x} > \mathbf{0}$, from propositions 309 (p. 194), 634 (p. 367), 633 (p. 367):

$$\mathbf{f_m}\Delta\mathbf{x} \leq \int_{\mathbf{x}}^{\Delta\mathbf{x}} \mathbf{f(y)dy} \leq \mathbf{f_M}\Delta\mathbf{x}$$

$$\mathbf{f_m} \leq \mathbf{f(x)} \leq \mathbf{f_M}$$

then

$$\mathbf{f_m} - \mathbf{f_M} \leq \frac{\int_{\mathbf{x}}^{\Delta\mathbf{x}} \mathbf{f(y)dy}}{\Delta\mathbf{x}} - \mathbf{f(x)} \leq \mathbf{f_M} - \mathbf{f_m}$$

which leads to:

$$|\frac{\int_{\mathbf{x}}^{\Delta\mathbf{x}} \mathbf{f(y)dy}}{\Delta\mathbf{x}} - \mathbf{f(x)}| \leq \mathbf{f_M} - \mathbf{f_m}\,.$$

Since $\mathbf{f(x)}$ is continuous, we have:

$$(\forall\alpha \in \mathcal{R}^+)(\exists\epsilon \in \mathcal{R}^+)(|\Delta\mathbf{x}| < \epsilon \Rightarrow \mathbf{f_M} - \mathbf{f_m} < \alpha)$$

so $\int_{\mathbf{x}}^{\Delta\mathbf{x}} \mathbf{f(y)dy}/\Delta\mathbf{x}$ has a limit for $\Delta\mathbf{x} = \mathbf{0}$ and this limit is $\mathbf{f(x)}$. Similar demonstration for $\Delta\mathbf{x} < \mathbf{0}$ can be done by reversing the first inequality.

**Proposition 637** $(\mathbf{f} : \mathcal{R} \to \mathcal{R})$ being a continuous map, the map $(\mathbf{F(x)} = \int_{\mathbf{a}}^{\mathbf{x}} \mathbf{f(y)dy} : \mathcal{R} \to \mathcal{R})$ is a continuous map.

**Proof**
It is a consequence of propositions 636 (p. 368) and 304 (p. 193).

**Proposition 638** We are given a continuous function defined on $[\mathbf{a}, \mathbf{b}] \subset \mathcal{R}$. There is a $\mathbf{c} \in ]\mathbf{a}, \mathbf{b}[$ such that:

$$\int_{\mathbf{a}}^{\mathbf{b}} \mathbf{f(x)dx} = \mathbf{(b - a)f(c)}$$
$$\mathbf{J}_{[\mathbf{a},\mathbf{b}]}(\mathbf{f(x)}) = |\mathbf{b} - \mathbf{a}|\mathbf{f(c)} \tag{2.197}$$

**Proof**
From proposition 621 (p. 355) $\mathbf{f}$ is having an integral for any value of $\mathbf{x} \in [\mathbf{a}, \mathbf{b}]$. Then from propositions 636 (p. 368), 313 (p. 195) and definition 298 (p. 366), we have proposition 638 (p. 368).

## 2.26 Differential equations

**Definition 300 Differential equations.**

Given $\theta \in \mathcal{R}$, $m \in \mathcal{Z}^+$, $n \in \mathcal{Z}^+$, considering the set $C^m_\theta$ of maps $([-\theta,\theta] \to \mathcal{R}^n)$ having derivative up to order $m$ (definition 164 (p. 193)) for any of its components, having also a continuous map: $(h(t, \prod^e_{i\in[1,n]} (\prod^e_{j\in[0,m]} x^{i,j})) : \mathcal{R}^{n(m+1)+1} \to \mathcal{R}^q)$, the following expression is called differential equation of order $m$:

$$x \in C^m_\theta, t \in [-\theta,\theta], h(t, x \overset{e}{\times} \prod^e_{j\in[1,m]} \frac{d^j x}{dt}) = 0$$

See definitions 119 (p. 135), 126 (p. 144) and conventions 14 (p. 135) 17 (p. 144) for $\prod^e$ and $\overset{e}{\times}$.

We shall consider here only first order differential equations: $m = 1$ with $q = n$ where $h$ can be expressed as $h = \frac{dx}{dt} - f(t,x) = 0 \in \mathcal{R}^n$.

**Proposition 639** We are given a closed ball $\bar{B}(u,r) \subset \mathcal{R}^n$, $u \in \mathcal{R}^n$, $r \in \mathcal{R}^+$ (definition 152 (p. 179)), a map between $[-\ell,\ell] \times \bar{B}(u,r) \subset \mathcal{R}^{n+1}$ and $\mathcal{R}^n$ (definition 141 (p. 165)) $(f : [-\ell,\ell] \times \bar{B}(u,r) \to \mathcal{R}^n)$ continuous (from proposition 245 (p. 169), it is also continuous in any of its components). We are also given a continuous map $(x(t) : [-\ell,\ell] \to \mathcal{R}^n)$. With all those conditions, the deduced map:

$$(x^{int}(t) = \int_0^t f(t',x(t'))dt' : [-\ell,\ell] \to \mathcal{R}^n)$$

is continuous.

**Proof**

From proposition 243 (p. 168), $f(t',x(t'))$ is continuous. From proposition 497 (p. 279), $[0,t]$ is convex. From propositions 618 (p. 351), 621 (p. 355) and 600 (p. 333), $(\int_0^t f(t',x(t'))dt' : [-\ell,\ell] \to \mathcal{R}^n)$ exists for any of its component. From proposition 636 (p. 368), its derivative for any of its component is $f(t',x(t'))$. From propositions 637 (p. 368) and 636 (p. 368), $(x^{int}(t) = \int_0^t f(t',x(t'))dt' : [-\ell,\ell] \to \mathcal{R}^n)$ is continuous for any of its component. It is then continuous from proposition 244 (p. 168).

**Definition 301** $x_p$ **sequence.**

As in proposition 639 (p. 369), we have a closed ball $\bar{B}(0,r) \subset \mathcal{R}^n$, $r \in \mathcal{R}^+$ (definition 152 (p. 179)), a map $(f : [-\ell,\ell] \times \bar{B}(u,r) \to \mathcal{R}^n)$ continuous (definition 141 (p. 165)). We call $(x_0(t) : \mathcal{R} \to \bar{B}(0,r) \subset \mathcal{R}^n)$, the map defined by:

$$t \in \mathcal{R} \Rightarrow x_0(t) = 0 \in \mathcal{R}^n$$

We set:

$$(\forall p \in \mathcal{Z}^+)(t \in [-\ell,\ell] \Rightarrow x_p(t) = \int_0^t f(t, x_{p-1}(t)) \in \mathcal{R}^n) .$$

From proposition 927 (p. 520), since it is defined for $p = 1$: $x_1(t) = \int_0^t f(t,0)$, it is defined for any $p \in \mathcal{Z}^+$.

**Proposition 640** We keep the premises of proposition 639 (p. 369). We suppose now that $\mathbf{f}$ is such that:

$$(\exists L > 0)(a \in \bar{B}(u,r), b \in \bar{B}(u,r), t \in [-\ell,\ell] \Rightarrow \|f(t,a) - f(t,b)\| \leq L\|a-b\|) . \tag{2.198}$$

We call ${}^0C^1_\theta$, the set of continuous maps with first order derivatives from $[-\theta,\theta]$ to $\mathcal{R}^n$ which give at zero $\mathbf{u} \in \mathcal{R}^n$:

$$x \in {}^0C^1_\theta \Rightarrow x(0) = u .$$

We have:

$$(\exists\theta > 0)(\exists x \in {}^0C^1_\theta)(t \in [-\theta,\theta] \Rightarrow \frac{dx(t)}{dt} = x'(t) = f(t,x(t))) . \tag{2.199}$$

**Proof**

As usual, we call $e_i = \prod^e_{j\in[1,n]} \delta^j_i = (\delta^j_i : [1,n] \to \mathcal{R}) \in \mathcal{R}^n$, the reference basis of $\mathcal{R}^n$. $\delta^j_i$ is the Kronecker symbol (see equation (2.106) (p. 267) in definition 242 (p. 267) and definition 215 (p. 245)). We have with the norms of $\mathcal{R}^n$ (definition 243 (p. 267)) $(\forall i \in [1,n])(\|e_i\| = 1)$. Using proposition 308 (p. 194) we have:

$$\frac{d(x(t)-u)}{dt} = \frac{dx(t)}{dt} = x'(t), \ \|x(t) - u - (y(t) - u)\| = \|x(t) - y(t)\| .$$

If proposition 640 (p. 370) is true for $\mathbf{u} = \mathbf{0}$, for any map satisfying equation (2.198) (p. 370), it is true for $\mathbf{v(t) = x(t) - u} \in \bar{B}(0,r)$ and $\mathbf{h(t,v) = f(t,v+u)}$ because $\mathbf{h}$ satisfies equation (2.198) (p. 370) if $\mathbf{f}$ do. Using that $\mathbf{v}$, $\mathbf{x = v + u}$ satisfies proposition 640 (p. 370) together with $\mathbf{f}$ , $\mathbf{x(0) = u}$ and $\mathbf{x(t)} \in \bar{B}(u,r)$. We can demonstrate proposition 640 (p. 370) only for $\mathbf{u = 0}$.

We call:

$$M = \max_{t\in[-\ell,\ell], x\in\bar{B}(0,r)} \|f(t,x)\| .$$

Since $[-\ell,\ell] \times \bar{B}(0,r)$ is compact (proposition 521 (p. 290)) and since $\|f(t,x)\|$ is continuous (propositions 277 (p. 182), 516 (p. 288), 241 (p. 167)), from proposition 297 (p. 189), $\mathbf{M}$ exists.

We take now an $\mathbf{x_p}$ sequence as defined in definition 301 (p. 369). From propositions 604 (p. 336) and 638 (p. 368) equation (2.197), when $\mathbf{t} \in [-\ell,\ell]$, we have (see also propositions 621 (p. 355), 600 (p. 333), 244 (p. 168)):

$$(\forall i \in [1,n])(|\int_0^t f^i(t',x_0(t'))dt'| = |J_{[0,t]}(f^i(t',x_0(t')))| = |x^i_1(t)|$$
$$\leq J_{[0,t]}(|f^i(t,x_0(t))|) \leq J_{[0,t]}(M) = |t|M) .$$

Then (see definition 243 (p. 267) and proposition 400 (p. 238)):

$$\|x_1(t)\| = \|x^i_1(t)e_i\| \leq \sum_{i\in[1,n]} |x^i_1(t)| < |t|M \times n .$$

We call $\theta = \min(\ell, \frac{r}{2Mn}, 2/L)$. We call $\mathcal{F}^r_\theta \subset {}^0C^1_\theta$, the set defined by $x \in \mathcal{F}^r_\theta \Leftrightarrow (x \in {}^0C^1_\theta, t \in [-\theta,\theta]) \Rightarrow x(t) \in \bar{B}(0,r))$ and $\mathcal{F}_\theta$, the one replacing $\bar{B}(0,r)$ by $\mathcal{R}^n$. We define a map $(\Psi : \mathcal{F}^r_\theta \to \mathcal{F}_\theta)$, the following way:

Given $\phi \in \mathcal{F}^r_\theta$, we define (see definitions 298 (p. 366), 299 (p. 366)):

$$\Psi(\phi(t)) = \int_0^t f(t,\phi(t))dt .$$

This integral is defined for $t \in [-\theta, \theta]$, because $\phi(t)$ is defined and continuous in $[-\theta, \theta]$ (propositions 624 (p. 357), 637 (p. 368) and 636 (p. 368)). We have by construction:

$$x_0 \in \mathcal{F}_\theta^r,\ x_1 = \Psi(x_0) = \int_0^t f(t, 0)dt \in \mathcal{F}_\theta^r .$$

Because:

$$(\forall t \in [-\theta, \theta])(x_0(t) = 0 \in \bar{B}(0, r), \|x_1(t)\| \leq |t|M \times n \leq \theta M \times n \leq \frac{r}{2Mn}Mn = \frac{r}{2}) .$$

Let us suppose that for $p \in \mathcal{Z}^+$ we have:

$$i \in [1, p] \Rightarrow x_i = \Psi(x_{i-1}), (\forall i \in [0, p])(x_i \in \mathcal{F}_\theta^r) .$$

Then it is also true for $p + 1$ with $x_{p+1} = \Psi(x_p)$. Because:

$$\begin{aligned}&(\forall t \in [-\theta, \theta])(x_p(t) \in \bar{B}(0, r) \Rightarrow \|x_{p+1}(t)\| = \|\Psi(x_p)\| = \|J_{[0,t]}x_p(t')\|\\ &\leq |t|M \times n \leq \theta M \times n \leq \frac{r}{2Mn}Mn = \frac{r}{2}) .\end{aligned}$$

From proposition 637 (p. 368) $\Psi(x_p)$ is continuous on $[-\theta, \theta]$. Then we have $x_{p+1} \in \mathcal{F}_\theta^r$. From the recurrence principle (proposition 927 (p. 520)), $(\forall p \in \mathcal{Z}^+)(x_p \in \mathcal{F}_\theta^r)$.

**$x_p$ is a uniform Cauchy sequence:**

From proposition 604 (p. 336), we have:

$$\begin{aligned}&i \in \mathcal{Z}^+ \Rightarrow \|x_{i+1}(t) - x_i(t)\| = \|\Psi(x_i)(t) - x_i(t)\| =\\ &\|\int_0^t (f(t, x_i(t)) - f(t, x_{i-1}(t)))dt\| \leq \int_0^t \|f(t, x_i(t)) - f(t, x_{i-1}(t))\|dt\\ &= J_{[0,t]}(\|f(t, x_i(t)) - f(t, x_{i-1}(t))\|) \leq LJ_{[0,t]}(\|x_i(t) - x_{i-1}(t)\|)\\ &\leq L|t| \sup_{\tau \in [-\theta, \theta]} \|x_i(\tau) - x_{i-1}(\tau)\| .\end{aligned} \tag{2.200}$$

Since we set $0 < L\theta < 0.5$, we get:

$$t \in [-\theta, \theta] \Rightarrow \|x_{i+1}(t) - x_i(t)\| \leq 0.5 \sup_{\tau \in [-\theta, \theta]} \|x_i(\tau) - x_{i-1}(\tau)\| .$$

Because $[-\theta, \theta]$ is closed and compact, since $x_i$ are continuous map, we have (proposition 140 (p. 160)):

$$(\exists t \in [-\theta, \theta])(\|x_{i+1}(t) - x_i(t)\| = \sup_{\tau \in [-\theta, \theta]} \|x_{i+1}(\tau) - x_i(\tau)\|)$$

then combining both:

$$\sup_{\tau \in [-\theta, \theta]} \|x_{i+1}(\tau) - x_i(\tau)\| \leq 0.5 \sup_{\tau \in [-\theta, \theta]} \|x_i(\tau) - x_{i-1}(\tau)\| .$$

It gives for $0 \leq p < q$:

$$\begin{aligned}&\sup_{\tau \in [-\theta, \theta]} \|x_p(\tau) - x_q(\tau)\| = \sup_{\tau \in [-\theta, \theta]} \|\sum_{j \in [p, q-1]} x_j(\tau) - x_{j+1}(\tau)\|\\ &\leq \sum_{j \in [p, q-1]} \sup_{\tau \in [-\theta, \theta]} \|x_{j+1}(\tau) - x_j(\tau)\| \leq \sup_{\tau \in [-\theta, \theta]} \|x_1(\tau) - x_0(\tau)\| \sum_{j \in [p, q-1]} (0.5)^j\\ &= \sup_{\tau \in [-\theta, \theta]} \|x_1(\tau) - x_0(\tau)\|((0.5)^{p-1} - (0.5)^{q-1})\\ &\leq \sup_{\tau \in [-\theta, \theta]} \|x_1(\tau) - x_0(\tau)\|(0.5)^{p-1} .\end{aligned}$$

Since

$$(\forall t \in [-\theta,\theta])(\|x_p(t)-x_q(t)\| \leq \sup_{\tau\in[-\theta,\theta]} \|x_1(\tau)-x_0(\tau)\|(0.5)^{p-1}) .$$

From proposition 545 (p. 299), definitions 268 (p. 298), 274 (p. 303), $(x_i(t) : \mathcal{Z}^{0+} \to \mathcal{F}^r_\theta)$ is a uniform Cauchy sequence of continuous maps which converges uniformly to a continuous map (propositions 552 (p. 303), 553 (p. 303)).

**The derivative of $x_p$ is also a uniform Cauchy sequence the limit of which is the derivative of the limit of $x_p$.**

We have:

$$(\forall i \in [2 \to [)(t \in [-\theta,\theta] \Rightarrow (x'_i(t) = f(t, x_{i-1}(t)), \|x_{i-1}(t)\| < r)$$
$$\Rightarrow \|x'_i(t)\| < M) .$$

From proposition 314 (p. 196), we have:

$$(\forall i \in \mathcal{Z}^+)((t \in [-\theta,\theta], t+\Delta t \in [-\theta,\theta]) \Rightarrow$$
$$(\exists\eta \in ]0,1[)(\exists\eta' \in ]0,1[)(\|\frac{x_i(t+\Delta t)-x_i(t)}{\Delta t} - x'_i(t)\| = \|x'_i(t+\eta\Delta t) - x'_i(t)\| =$$
$$\|f(x_{i-1}(t+\eta\Delta t)) - f(x_{i-1}(t))\| \leq L\|x_{i-1}(t+\eta\Delta t) - x_{i-1}(t)\|$$
$$= \eta|\Delta t|L\|x'_{i-1}(t+\eta\eta'\Delta t)\| \leq L\|f(t, x_{i-2})\||\Delta t| \leq LM|\Delta t|)) .$$

It means that we verify the conditions (2.150) (p. 311) of proposition 564 (p. 311):

$$(\forall\alpha > 0)(\exists\epsilon(\alpha) = \alpha/LM)(\forall i \in \mathcal{Z}^+)(\forall t \in [-\theta,\theta])$$
$$((|\Delta t| < \epsilon(\alpha), t+\Delta t \in [-\theta,\theta]) \Rightarrow \|\frac{x_i(t+\Delta t)-x_i(t)}{\Delta t} - x'_i(t)\| < \alpha) .$$

It means that $x'_i(t)$ is a uniform Cauchy sequence as $x_i(t)$ and the derivative of the limit $x(t)$ of $x_i(t)$ is the limit $x'(t)$ of $x'_i(t)$.

**The limit $x$ of the sequence $x_p$ is a solution of the differential equation $\frac{dx(t)}{dt} = x'(t) = f(t, x(t))$ in $[-\theta,\theta]$ (definition 300 (p. 369)).**

We have:

$$(\forall\alpha > 0)(\exists m_1 \in \mathcal{Z}^+)(\exists m_2 \in \mathcal{Z}^+)(\forall t \in [-\theta,\theta])$$
$$((i > m_1(\alpha) > 0 \Rightarrow \|x_i(t) - x(t)\| < \alpha), (i > m_2(\alpha) > 0 \Rightarrow \|x'_i(t) - x'(t)\| < \alpha)) .$$

We have $x'_i = f(t, x_i)$. Then:

$$(\forall\alpha > 0)(\forall t \in [-\theta,\theta])(\exists i > \max(m_2(\alpha/3), m_1(\alpha/(3L))))$$
$$(\|x' - f(t,x_i)\| = \|x' - x'_i\| < \alpha/3, \|f(t,x_i) - f(t,x)\| < L\|x_i(t) - x(t)\|$$
$$< L\alpha/(3L) = \alpha/3)$$
$$(\forall\alpha > 0)(\forall t \in [-\theta,\theta])(\|x'(t) - f(t,x)\| = \|x'(t) - f(t,x_i) + f(t,x_i) - f(t,x)\|$$
$$< \|x' - x'_i\| + \|f(t,x_i) - f(t,x)\| < \alpha) .$$

Then $(\forall t \in [-\theta,\theta])(\forall\alpha > 0)(\|x'(t) - f(t,x)\| < \alpha)$. It implies $x'(t) = f(t,x)$. Then:

$$(\exists\theta > 0)(\exists(x^i : [1,n] \to {}^0C^1_\theta))(t \in [-\theta,\theta] \Rightarrow \frac{dx(t)}{dt} = x'(t) = f(t, x(t))) .$$

Proposition 640 (p. 370) is true.

**Proposition 641** Keeping the notation and condition of proposition 640 (p. 370), the solution satisfying relation (2.199) (p. 370) is unique.

**Proof**

If we have another solution as:

$$(\exists\theta_y > 0)(\exists(y^i : [1,n] \to {}^0C^1_{\theta_y}))(t \in [-\theta_y, \theta_y] \Rightarrow \frac{dy(t)}{dt} = y'(t) = f(t, y(t)))$$

with $x(0) = y(0) = u = 0$, we can set $\theta_m = \min(\theta, \theta_y) \leq \min(\ell, \frac{r}{2Mn}, 2/L)$ and we shall have as for equation (2.200) (p. 371):

$$\begin{aligned}&t \in [-\theta_m, \theta_m] \Rightarrow\\ &\|\int_0^t (x'(t) - y'(t))dt\| = \|x(t) - y(t) - x(0) + y(0)\| = \|x(t) - y(t)\| =\\ &\|\int_0^t (f(t, x(t)) - f(t, y(t)))dt\| \leq J_{[0,t]}(\|f(t, x(t)) - f(t, y(t))\|)\\ &< |t|L \max_{t\in[-\theta_m,\theta_m]} \|x(t) - y(t)\| \leq 0.5 \max_{t\in[-\theta_m,\theta_m]} \|x(t)) - y(t)\| \,.\end{aligned}$$

Since $x(t)$ and $y(t)$ are continuous, $\|x(t) - y(t)\|$ is continuous (propositions 277 (p. 182), 516 (p. 288)) on the compact $[-\theta_m, \theta_m]$ (proposition 298 (p. 190)). From proposition 140 (p. 160), we have:

$$\begin{aligned}&(\exists\tau \in [-\theta_m, \theta_m])(\forall t \in [-\theta_m, \theta_m])(\|x(\tau) - y(\tau)\|\\ &= \max_{t_1\in[-\theta_m,\theta_m]} \|x(t_1) - y(t_1)\| > 2\|x(t) - y(t)\|)\end{aligned}$$

it is impossible except if $(\forall t \in [-\theta_m, \theta_m])(x(t) - y(t) = 0)$. Propositions 641 (p. 373) and 640 (p. 370) are true.

**Proposition 642** We have a continuous bijection $g$ between an open ${}^nO'$ not empty of $\mathcal{R}^n$ and an open ${}^mO'$ of $\mathcal{R}^m$: $(g : {}^nO' \to {}^mO')$. We suppose that $g$ and $g^{-1}$ are continuous and that $g$ has first and second order continuous derivatives over the open ${}^nO'$. Then we have $m = n$.

**Proof**

1. **At a point $p \in {}^nO' \subset \mathcal{R}^n$, definition of the map f.**
   Having $p \in {}^nO' \subset \mathcal{R}^n$ a point in the open defining $g$, we may use, for simplification, instead of $g$, the map $f = g(x + p) - g(p) = {}^mT_{-g(p)} \circ g \circ {}^nT_p$ where $({}^\ell T_q : \mathcal{R}^\ell \Leftrightarrow \mathcal{R}^\ell)$, $q \in \mathcal{R}^\ell$ is the translation bijection: $(\forall x \in \mathcal{R}^\ell)({}^\ell T_q(x) = x + q)$. ${}^\ell T_q$ is continuous (propositions 288 (p. 186), 242 (p. 168)) and from propositions 308 (p. 194), 309 (p. 194) and 311 (p. 194) has for derivative along the variable $x^i$: $\partial_i {}^\ell T_q = \partial_i(x + q) = \partial_i x = \delta_i^j$.
   We call ${}^nO = {}^nT_{-p}({}^nO')$ and ${}^mO = {}^mT_{g(p)}({}^mO')$, the opens defining the bijection $f$ (see definition 141 (p. 165)). From proposition 241 (p. 167), $f$ is continuous. $f^{-1}$ is also continuous for the same reason, because:

   $$f^{-1n}T_{-p} \circ g^{-1} \circ {}^mT_{g(p)} = g^{-1}(x + g(p)) - p \,.$$

   From proposition 563 (p. 311), we have:

   $$\partial_k f(x) = \partial_k g(x + p)$$

   and we have $0 = f(0)$.

2. **Choice of p.**
At any point $\mathbf{p}$ of $^n\mathbf{O}'$, we consider the set of vectors $\partial_i \mathbf{g} = \frac{\partial \mathbf{g}}{\partial \mathbf{x}^i} = \partial_i \mathbf{g}^\ell(\mathbf{x})\ {}^m\mathbf{e}_\ell \in \mathcal{R}^m$. We call $\mathbf{k_p}$ the maximum number of independent vectors of that set. We consider the set of integers $\mathbf{I}$ such as $\mathbf{i} \in \mathbf{I} \Leftrightarrow (\exists \mathbf{p} \in \mathbf{O})(\mathbf{i} = \mathbf{k_p})$. Since $\mathbf{i} \leq \mathbf{m}$ (definition 208 (p. 240)), $\mathbf{I}$ has a biggest value $\mathbf{k}$ which belongs to $\mathbf{I}$ (proposition 924 (p. 519)). From definition 208 (p. 240), we have $\mathbf{k} \leq \mathbf{m}$. From the definition of $\mathbf{I}$, there is a $\mathbf{p} \in \mathbf{O}'$ such that $\mathbf{k} = \mathbf{k_p}$ is the biggest value of $\mathbf{I}$. In order to simplify the writing, we choose $\mathbf{p}$ to define $\mathbf{f}(\mathbf{x}) = \mathbf{g}(\mathbf{x}+\mathbf{p}) - \mathbf{g}(\mathbf{p})$. We order the indices such that the set $\mathbf{S}_\partial$ of independent vectors is given by $\mathbf{s} \in \mathbf{S}_\partial \Leftrightarrow (\exists \mathbf{i} \in [1,\mathbf{k}])(\mathbf{s} = \partial_i \mathbf{f}(0))$. We call ${}^n\mathbf{e}_i = (\delta_i^j : [1,\mathbf{n}] \to \mathcal{R}^n)$, ${}^m\mathbf{e}_i = (\delta_i^j : [1,\mathbf{m}] \to \mathcal{R}^m)$ the standard basic vectors of $\mathcal{R}^n$ and $\mathcal{R}^m$ (see equation (2.106) (p. 267) in definition 242 (p. 267)). From proposition 415 (p. 244), we can complete the set $\mathbf{i} \in [1,\mathbf{k}]$, $\partial_i \mathbf{f}$ with $\mathbf{n}-\mathbf{k}$ vectors of type ${}^m\mathbf{e}_j$ making a set $\mathbf{S_e}$ such that $\mathbf{S}_\partial \cup \mathbf{S_e}$ is an independent set of $\mathbf{m}$ vectors (if $\mathbf{k} = \mathbf{m}$, $\mathbf{S_e} = \emptyset$). We reorder the indices such that we have $\mathbf{s} \in \mathbf{S_e} \Leftrightarrow (\exists \mathbf{i} \in [\mathbf{k}+1,\mathbf{n}])(\mathbf{s} = \mathbf{e}_i)$. We have $\alpha_{\mathbf{det}} = |\mathbf{det}(\mathbf{S}_\partial \cup \mathbf{S_e})| \neq 0$. When $\mathbf{x} \neq 0$, we can define $\mathbf{s}_i \in \mathbf{S}_\partial(\mathbf{x}) \Leftrightarrow \mathbf{i} \in [1,\mathbf{k}], \mathbf{s}_i = \partial_i \mathbf{f}(\mathbf{x})$, keeping invariant $\mathbf{S_e}$ $(\partial_i \mathbf{f}(\mathbf{x}) : \mathbf{O} \to \mathcal{R}^m)$ are continuous maps. $|\mathbf{det}(\mathbf{S}_\partial(\mathbf{x}) \cup \mathbf{S_e})|$ is also a continuous map $\mathbf{O} \to \mathcal{R}$. Then:

$$(\forall \alpha > 0)(\exists \epsilon(\alpha) > 0)(||\mathbf{x}|| \leq \epsilon(\alpha) \Rightarrow$$
$$||\mathbf{det}(\mathbf{S}_\partial(\mathbf{x}) \cup \mathbf{S_e})| - |\mathbf{det}(\mathbf{S}_\partial(0) \cup \mathbf{S_e})|| < \alpha\ .$$

It gives (see proposition 273 (p. 181)):

$$\alpha_{\mathbf{det}} = |\mathbf{det}(\mathbf{S}_\partial(0) \cup \mathbf{S_e})| \Rightarrow \{||\mathbf{x}|| \leq \epsilon(\alpha_{\mathbf{det}}/2) \Rightarrow |\mathbf{det}(\mathbf{S}_\partial(\mathbf{x}) \cup \mathbf{S_e})| =$$
$$|\mathbf{det}(\mathbf{S}_\partial(\mathbf{x}) \cup \mathbf{S_e}) - \mathbf{det}(\mathbf{S}_\partial(0) \cup \mathbf{S_e}) + \mathbf{det}(\mathbf{S}_\partial(0) \cup \mathbf{S_e})|$$
$$\geq |\mathbf{det}(\mathbf{S}_\partial(0) \cup \mathbf{S_e})| - |\mathbf{det}(\mathbf{S}_\partial(\mathbf{x}) \cup \mathbf{S_e}) - \mathbf{det}(\mathbf{S}_\partial(0) \cup \mathbf{S_e})| \geq \alpha_{\mathbf{det}}/2\}\ .$$

The vectors of $\mathbf{S}_\partial(\mathbf{x})$ are independent for $\mathbf{x} \in {}^n\bar{\mathbf{B}}(0,\epsilon(\alpha/2))$. $\mathbf{S}_\partial(\mathbf{x})$ contains $\mathbf{k}$ independent vectors, the maximum number of independent vectors of the set $\partial_i \mathbf{f}$, $\mathbf{i} \in [1,\mathbf{n}]$ being $\mathbf{k}$, only those are independent.

3. **k=n.**
Then if $\mathbf{k} < \mathbf{n}$, we should have:

$$(\forall \mathbf{i} \in [\mathbf{k}+1,\mathbf{n}])(\forall \mathbf{x} \in {}^n\bar{\mathbf{B}}(0,\epsilon(\alpha_{\mathbf{det}}/2)))(\forall \mathbf{j} \in [1,\mathbf{k}])(\exists \lambda_i^j(\mathbf{x}) \in \mathcal{R})$$
$$(\partial_i \mathbf{f}(\mathbf{x}) = \sum_{\mathbf{j} \in [1,\mathbf{k}]} \lambda_i^j(\mathbf{x}) \partial_j \mathbf{f}(\mathbf{x})) \tag{2.201}$$

Calling $({}_\mathbf{f}\mathbf{A} : [1,\mathbf{m}] \times [1,\mathbf{m}] \to \mathcal{R}) \in \mathcal{R}^{2m}$ the matrix built with the components of $\mathbf{S}_\partial(\mathbf{x}) \cup \mathbf{S_e}$:

$$\mathbf{j} \in [1,\mathbf{k}] \Rightarrow {}_\mathbf{f}\mathbf{A}_j^i = \partial_j \mathbf{f}^i,\ \mathbf{j} \in [\mathbf{k}+1,\mathbf{m}] \Rightarrow {}_\mathbf{f}\mathbf{A}_j^i = \delta_j^i$$

with that we can rewrite equation (2.201) (p. 374):

$$\mathbf{i} \in [\mathbf{k},\mathbf{n}] \Rightarrow \partial_i \mathbf{f}^j(\mathbf{x})\ {}^m\mathbf{e}_j = \sum_{\ell \in [1,\mathbf{k}]} \lambda_i^\ell(\mathbf{x})\, {}_\mathbf{f}\mathbf{A}_\ell^j\ {}^m\mathbf{e}_j\ .$$

Since $\mathbf{S}_\partial(\mathbf{x}) \cup \mathbf{S_e}$ are a set of independent vectors from proposition 430 (p. 250), we have:

$$\mathbf{det}({}_\mathbf{f}\mathbf{A}) \neq 0$$

then from proposition 444 (p. 255) $_{\mathbf{f}}\mathbf{A}^{-1}$ exists. From propositions 444 (p. 255) and 563 (p. 311), since we suppose that all partial derivatives of $\mathbf{g}$ (and then of $\mathbf{f}$) exist up to the second order and are continuous, all partial derivatives of $\lambda_i^j(\mathbf{x})$ exist and are continuous. From proposition 516 (p. 288), $|\lambda_i^j(\mathbf{x})|$ is continuous. Since the ball $^{\mathbf{n}}\bar{\mathbf{B}}(\mathbf{0}, \epsilon(\alpha_{\mathbf{det}}/2))$ is compact, then $|\lambda_i^j({}^{\mathbf{n}}\bar{\mathbf{B}}(\mathbf{0}, \epsilon(\alpha_{\mathbf{det}}/2)))|$ is compact (proposition 240 (p. 167)) and, from propositions 271 (p. 180) and 255 (p. 172), it has a majoring value (definition 380 (p. 484)) $\mathbf{M}' = \mathbf{M}/(2\mathbf{n}^2)$ for all $\mathbf{j} \in [1, \mathbf{k}], \mathbf{i} \in [\mathbf{k}+1, \mathbf{n}]$. From proposition 560 (p. 307), 2, we have:

$$(\exists a \in ]0, \epsilon(\alpha_{det}/4)])(\forall j \in [1,k])(\forall i \in [k+1,n])((x_1 \in {}^{n}\bar{B}(0,a), \\ x_2 \in {}^{n}\bar{B}(0,a), \lambda_i^j(x_1) - \lambda_i^j(x_2) = (\partial_\ell \lambda_i^j(x_1) + \eta_{\ell,i}^j)(x_1^\ell - x_2^\ell)) \\ \Rightarrow (\forall \ell \in [0,n])(|\eta_{\ell,i}^j| < M/(2n^2)) \Rightarrow |\lambda_i^j(x_1) - \lambda_i^j(x_2)| < \|x_1 - x_2\| M/n) \ .$$

Then we have:

$$(\exists a \in ]0, \epsilon(\alpha_{det}/4)])(\forall i \in [k+1,n])(\sum_{j\in[1,k]} |\lambda_i^j(x_1) - \lambda_i^j(x_2)| < M\|x_1 - x_2\|) \ .$$

For all $\mathbf{i} \in [\mathbf{k}+1, \mathbf{n}]$, we are in the conditions of proposition 640 (p. 370). We apply proposition 640 (p. 370) when $\mathbf{i} = \mathbf{k}+1$. From proposition 640 (p. 370), there are a $\theta > 0$ and a set of $\mathbf{k}$ continuous maps $\mathbf{j} \in [1, \mathbf{k}], (\mathbf{x}^j : [-\theta, \theta] \to \mathcal{R})$ such that:

$$(x(0) = 0, t \in [-\theta, \theta]) \Rightarrow (x^{k+1}(t) = t, (\forall j \in [1,k])(\frac{dx^j(t)}{dt} = -\lambda_{k+1}^j(x)), \\ (\forall i \in [k+2,n])(x^i(t) = 0)) \ .$$

Since $|\mathbf{x}^i(\mathbf{t})|$ are continuous (proposition 304 (p. 193)), $\theta$ can be limited (definition 141 (p. 165)) such that $\| \sum_{j\in[1,k]} \mathbf{x}^j\, {}^{\mathbf{n}}\mathbf{e}_j\| < \mathbf{a}/2$ and $\mathbf{x}^{k+1} = \mathbf{t} \in ]-\mathbf{a}/2, \mathbf{a}/2[$. It gives $\mathbf{t} \in [-\theta, \theta] \Rightarrow \| \sum_{i\in[1,k+1]} \mathbf{x}^i\, {}^{\mathbf{n}}\mathbf{e}_i\| < \mathbf{a}$. With that, we have:

$$\frac{df(x(t))}{dt} = \partial_{k+1} f(x) + \sum_{j\in[1,k]} \partial_j f(x) \frac{dx_j}{dt} = \partial_{k+1} f(x) - \sum_{j\in[1,k]} \lambda_{k+1}^j(x) \partial_j f(x) = 0 \ .$$

Then from proposition 634 (p. 367):

$$(\forall t \in [-\theta, \theta])(f(x(t)) = f(0) = 0) \ .$$

Since $\mathbf{x}^{k+1}(0) = 0$, $\mathbf{x}^{k+1}(\theta) = \theta$, it implies $\mathbf{x}(0) \neq \mathbf{x}(\theta)$ and $\mathbf{f}(\mathbf{x}(0)) = \mathbf{f}(\mathbf{x}(\theta)) = 0$. $\mathbf{f}$ has at least two inverses for $\mathbf{0}$, it cannot be an injection (definition 122 (p. 139)). Then from proposition 927 (p. 520), $\mathbf{k} = \mathbf{n}$ which means $\mathbf{n} \leq \mathbf{m}$.

4. **m=n**
If $\mathbf{n} < \mathbf{m}$, from what we demonstrate above, there is a point $\mathbf{p}$ such that there are $\mathbf{n}$ independent vectors $\partial_i \mathbf{g}(\mathbf{p})$. As above, to simplify the writing, we center our axis on $\mathbf{p}$ and, as before, we call $\mathbf{f}(\mathbf{x}) = \mathbf{g}(\mathbf{p}+\mathbf{x}) - \mathbf{g}(\mathbf{p})$ and we shall use $\mathbf{f}$ instead of $\mathbf{g}$.
We call $\mathbf{S}_\partial(\mathbf{x})$ the set of $\mathbf{n}$ vectors $\partial_i \mathbf{f}$, $\mathbf{i} \in [1, \mathbf{n}]$. From proposition 415 (p. 244), we can complete those $\mathbf{n}$ vectors of $\mathbf{S}_\partial(\mathbf{0})$ to $\mathbf{m}$ vectors with a set $\mathbf{S}_e$ of $\mathbf{m}-\mathbf{n}$ vectors of type ${}^{\mathbf{m}}\mathbf{e}_i, \mathbf{i} \in [1, \mathbf{m}]$ such that to realize an independent basis. We reorder the indices such that $\mathbf{i} \in [1, \mathbf{n}]$ are devoted to the $\mathbf{n}$ vectors of $\mathbf{S}_\partial$ and $\mathbf{i} \in [\mathbf{n}+1, \mathbf{m}]$ to the extra ${}^{\mathbf{m}}\mathbf{e}_i$ of $\mathbf{S}_e$ so $\mathbf{S}_\partial \cup \mathbf{S}_e$ makes an independent basis of $\mathcal{R}^{\mathbf{m}}$. We have:

$$\det(S_\partial(0) \cup S_e) = \det({}_fA) \neq 0 \ .$$

Then $\det({}_f A^{-1})$ exists. Since $\det({}_f A^{-1})$ is continuous, we have, as above, a closed ball ${}^n\bar{B}(0, r)$ and an $a > 0$ such as:

$$x \in {}^n\bar{B}(0, r) \Rightarrow \det({}_f A^{-1}) > a .$$

Then $S_\partial(x) \cup S_e$ is an independent basis of $\mathcal{R}^m$ when $x \in {}^n\bar{B}(0, r)$. We call those independent vectors $s_i : [1, m] \to S_\partial(x) \cup S_e$ with:

$$i \in [1, n] \Rightarrow s_i(x) = \partial_i f(x) \in S_\partial(x), i \in [n + 1, m] \Rightarrow s_i(x) = e_i \in S_e .$$

Using definition 227 (p. 252), since $(i \in [n + 1, m], j \in [n + 1, m]) \Rightarrow {}_f A^i_j = \delta^i_j$ we have:

$$\det({}_f A) = \det_{i\in[1,n],j\in[1,n]}({}_f A),$$
$$\det({}_f A^{-1}) = \det({}_f A)^{-1} = \det_{i\in[1,n],j\in[1,n]}({}_f A)^{-1} = \det_{i\in[1,n],j\in[1,n]}({}_f A^{-1}) .$$

Since $f^{-1}$ is continuous $f({}^nB(0, r))$ is open and there is a closed ball ${}^m\bar{B}(0, r')$ (proposition 266 (p. 179)) such that:

$${}^m\bar{B}(0, r') \subset {}^mB(0, 1.1r') \subset f({}^nB(0, r)) \subset f({}^n\bar{B}(0, r)) .$$

We take a point $q \neq 0, q = q^i s_i \in {}^m\bar{B}(0, r'/2)$. We consider the map $x(t) = f^{-1}(qt)$, $f(x(t)) = qt$. We have from proposition 560 (p. 307):

$$f^j(x(t + \Delta t)) - f^j(x(t)) = (\partial_i f^j(x(t)) + \eta^j_i)(x^i(t + \Delta t) - x^i(t)) = q^j \Delta t .$$

Since

$$x \in {}^n\bar{B}(0, r) \Rightarrow (\det_{[1,n]\times[1,n]}(\partial f(x(t))) = \det_{[1,n]\times[1,n]} {}_f A, \det_{[1,n]\times[1,n]} {}_f A^{-1} > a)$$

we have $(\underset{[1,n]\times[1,n]}{\partial f(x(t))} + \eta)^{-1}$ exists and is continuous at $\eta = 0$. Its determinant is also a continuous map at $0$ since ${}^n\bar{B}(0, r)$ is compact. Setting:

$$(\underset{[1,n]\times[1,n]}{\partial f(x(t))} + \eta)^{-1} = (\underset{[1,n]\times[1,n]}{\partial f(x(t))})^{-1} + \eta_q = (\underset{[1,n]\times[1,n]}{{}_f A})^{-1} + \eta_q$$

we have:

$$(\forall \alpha_q > 0)(\forall x \in {}^n\bar{B}(0, r))(\exists \epsilon_q(\alpha_q) > 0)(\|\eta\| < \epsilon_q(\alpha_q) \Rightarrow \|\eta_q\| < \alpha_q) .$$

We also have:

$$(\forall \alpha_t > 0)(\exists \epsilon_t(\alpha_t) > 0)(|\Delta t| < \epsilon_t(\alpha_t) \Rightarrow \|x(t + \Delta t) - x(t)\| < \alpha_t)$$
$$(\forall \alpha_x > 0)(\exists \epsilon_x(\alpha_x) > 0)(\|\Delta x\| < \epsilon_x(\alpha_x) \Rightarrow \|f(x + \Delta x) - f(x)\| < \alpha_x) .$$
$$(\forall \alpha_d > 0)(\exists \epsilon_d(\alpha_d) > 0)(\|\Delta x\| < \epsilon_d(\alpha_d) \Rightarrow \|\eta\| < \alpha_d)$$

Then we have:

$$(\forall i \in [1, n])(\frac{x^i(t + \Delta t) - x^i(t)}{\Delta t} = ((\underset{[1,n]\times[1,n]}{{}_f A})^{-1} q)^i + (\eta_q q)^i) .$$

Choosing for any $\alpha > 0$, $\epsilon < \epsilon_t(\epsilon_d(\epsilon_q(\alpha/\|q\|)))$, we have:

$$|\Delta t| < \epsilon \Rightarrow |\frac{x^i(t + \Delta t) - x^i(t)}{\Delta t} - ((\underset{[1,n]\times[1,n]}{{}_f A})^{-1})^i_j q^j| < \alpha .$$

It means $\frac{dx^i(t)}{dt} = (({}_{[1,n]\times[1,n]}{}^f A)^{-1})^i_j q^j$. Since $({}_{[1,n]\times[1,n]}{}^f A)^{-1}$ exists, is continuous and has derivative in ${}^m\bar{B}(0, r')$, then from propositions 2.198 (p. 370) and 560 (p. 307), there exist $\theta$ and n maps $(x^i(t) : [-\theta, \theta] \to {}^n\bar{B}(0, r))$ such that:

$$t \in [-\theta, \theta] \Rightarrow \frac{dx^i(t)}{dt} = (({}_{[1,n]\times[1,n]}{}^f A)^{-1})^i_j q^j .$$

From proposition 640 (p. 370) the solution exists and is unique for $x(0) = 0$. We should have $(\forall j \in [n+1, m])(\theta q^j = f^j(x(\theta)))$ but $x(t)$ does not depend on $q^j, j \in [n+1, m]$ then it would be the same value for another point $q'$ such as $(\forall i \in [1, n])(q^i = q'^i)$ and different for the others with $\| \sum_{i\in[n+1,m]} q'^i e_i \| < r'/2$ such that $\|q'\| < r'$. It is not possible since f is supposed to have a different inverse for $\theta q$ and $\theta q'$. Then $n = m$.

## 2.27 Lebesgue integral

We shall here give only some definitions. In that book we shall not extend Riemann integral to Lebesgue integral. For a more complete description see reference [4]. When we want to integrate on a set, we shall suppose that the set has a volume given by a Riemann integral.

**Definition 302 σ-algebra.**

Given a set A, $B \subset \mathcal{P}(A)$ (definition 350 (p. 453)) is a **σ-algebra** on A, if we have:

$$B \neq \emptyset \tag{2.202}$$

$$c \in B \Leftrightarrow \mathcal{C}(c) \in B \tag{2.203}$$

$$(\forall C \subset B)(\mathrm{Card}(C) \leq \mathrm{Card}(\mathcal{Z}^{0+}) \Rightarrow \bigcup_{c\in C} c \in B) . \tag{2.204}$$

Equations (2.203), (2.204) plus proposition 801 (p. 466) implies:

$$(\forall C \subset B)(\mathrm{Card}(C) \leq \mathrm{Card}(\mathcal{Z}^{0+}) \Rightarrow \bigcap_{c\in C} c \in B) . \tag{2.205}$$

**Proposition 643** Given a **σ-algebra** B on A,

$$A \in B \tag{2.206}$$

$$\emptyset \in B . \tag{2.207}$$

**Proof**
It is a consequence of equations (2.202) (p. 377), (2.203) (p. 377) and (2.204) (p. 377). From equation (2.202), $\exists c \in B$ then $\mathcal{C}(c) \in B$. It means from (2.204), $c \cup \mathcal{C}(c) = A \in B$ and $\mathcal{C}(A) = \emptyset \in B$.

**Proposition 644** Given a set A, a subset $C \subset \mathcal{P}(A)$ of $\mathcal{P}(A)$ (definition 350 (p. 453)), we consider the subset $B \subset \mathcal{P}(\mathcal{P}(A))$ which is the set of **σ-algebra** of A containing C. It exists from proposition 750 (p. 450). We have:

$$\mathcal{P}(A) \in B, \bigcap_{b\in B} b \in B, C \subset \bigcap_{b\in B} b .$$

**Proof**
$\mathcal{P}(\mathbf{A})$ satisfies all the condition for being a $\boldsymbol{\sigma}$**-algebra** on $\mathbf{A}$. Then $\mathbf{B} \neq \emptyset$ and, from definition 363 (p. 463), footnote 8 (p. 463), $\bigcap_{\mathbf{b}\in\mathbf{B}} \mathbf{b}$ can be defined. We have from definitions 349 (p. 451) and 363 (p. 463), $\mathbf{C} \subset \bigcap_{\mathbf{b}\in\mathbf{B}} \mathbf{b}$. Moreover,

$$(\forall \mathbf{b} \in \mathbf{B})(\emptyset \in \mathbf{b}, \mathbf{A} \in \mathbf{b}) \Rightarrow \mathbf{A} \in \bigcap_{\mathbf{b}\in\mathbf{B}} \mathbf{b}, \emptyset \in \bigcap_{\mathbf{b}\in\mathbf{B}} \mathbf{b}$$

$$\mathbf{u} \in \bigcap_{\mathbf{b}\in\mathbf{B}} \mathbf{b} \Rightarrow (\forall \mathbf{b} \in \mathbf{B})(\mathbf{u} \in \mathbf{b}, \mathcal{C}(\mathbf{u}) \in \mathbf{b}) \Rightarrow \mathcal{C}(\mathbf{u}) \in \bigcap_{\mathbf{b}\in\mathbf{B}} \mathbf{b}$$

$$(\mathbf{u} \subset \bigcap_{\mathbf{b}\in\mathbf{B}} \mathbf{b}, \mathbf{Card}(\mathbf{u}) \in \mathcal{Z}^{0+}) \Rightarrow (\forall \mathbf{b} \in \mathbf{B})(\mathbf{u} \subset \mathbf{b}, \bigcup_{\mathbf{d}\in\mathbf{u}} \mathbf{d} \in \mathbf{b}) \Rightarrow \bigcup_{\mathbf{d}\in\mathbf{u}} \mathbf{d} \subset \bigcap_{\mathbf{b}\in\mathbf{B}} \mathbf{b}$$

then equations (2.202) (p. 377), (2.207) (p. 377), (2.204) (p. 377) are satisfied. $\bigcap_{\mathbf{b}\in\mathbf{B}}$ is a $\boldsymbol{\sigma}$**-algebra**:

$$\bigcap_{\mathbf{b}\in\mathbf{B}} \mathbf{b} \in \mathbf{B}\ .$$

**Definition 303 Borel $\boldsymbol{\sigma}$-algebra.**

Given a topological set $\mathbf{A}$ with $\mathcal{O} \subset \mathcal{P}(\mathbf{A})$ its set of opens, calling $\mathbf{B} \subset \mathcal{P}(\mathbf{A})$ the set of $\boldsymbol{\sigma}$**-algebra** of $\mathbf{A}$ containing $\mathcal{O}$, one calls a Borel $\boldsymbol{\sigma}$**-algebra** on $\mathbf{A}$, the following set:

$$\bigcap_{\mathbf{b}\in\mathbf{B}} \mathbf{b}$$

which is a $\boldsymbol{\sigma}$**-algebra** from proposition 644 (p. 377).
**We shall consider here only Borel $\boldsymbol{\sigma}$-algebra.**

**Proposition 645** Given a topological set $\mathbf{A}$ with $\mathcal{O} \subset \mathcal{P}(\mathbf{A})$, its set of opens, we call $\mathcal{O}^{\mathbf{b}} \subset \mathcal{P}(\mathbf{A})$, the following set:

$$\mathbf{x} \in \mathcal{O}^{\mathbf{b}} \Leftrightarrow (\exists \mathbf{a} \in \mathcal{O})(\exists \mathbf{b} \in \mathcal{O})(\mathbf{x} = \mathbf{a} \cap \mathcal{C}(\mathbf{b}) \vee \mathbf{x} = \mathbf{a} \cup \mathcal{C}(\mathbf{b}))\ . \tag{2.208}$$

Calling $\mathbb{O} \subset \mathcal{P}(\mathbf{A})$ its Borel $\boldsymbol{\sigma}$**-algebra**, we have:

$$\mathbf{x} \in \mathbb{O} \Leftrightarrow (\exists \mathbf{K} \subset \mathcal{O}^{\mathbf{b}})(\mathbf{Card}(\mathbf{K}) \leq \mathbf{Card}(\mathcal{Z}^{0+}), \mathbf{x} = \bigcup_{\mathbf{k}\in\mathbf{K}} \mathbf{k} \vee \mathbf{x} = \bigcap_{\mathbf{k}\in\mathbf{K}} \mathbf{k})\ . \tag{2.209}$$

**Proof**
Keeping the above notation, supposing that $\mathbb{O}$ is defined by equation (2.209) (p. 378), let us prove that $\mathbb{O}$ is the Borel $\boldsymbol{\sigma}$**-algebra** of $\mathbf{A}$. For that, we demonstrate the following items:

1. $\mathcal{O}^{\mathbf{b}} \subset \mathbb{O}$.
   Using equation (2.209), we have:

$$\mathbf{x} \in \mathcal{O}^{\mathbf{b}} \Rightarrow (\exists \mathbf{K} = \{\mathbf{x}\} \subset \mathcal{O}^{\mathbf{b}})(\mathbf{Card}(\mathbf{K}) = 1 \leq \mathbf{Card}(\mathcal{Z}^{0+}),$$
$$\mathbf{x} = \bigcup_{\mathbf{u}\in\mathbf{K}} \mathbf{u}) \Rightarrow \mathbf{x} \in \mathbb{O}\ .$$

   Then from definition 349 (p. 451), $\mathcal{O}^{\mathbf{b}} \subset \mathbb{O}$.

2. $\mathbf{B} \subset \mathbf{A}$ being a **$\sigma$-algebra**, if $\mathcal{O} \subset \mathbf{B}$ then $\mathcal{O}^{\mathbf{b}} \subset \mathbf{B}$.
From equation (2.208) (p. 378), we have:

$$u \in \mathcal{O}^b \leftrightarrow (\exists a \in \mathcal{O})(\exists b \in \mathcal{O})(u = a \cap \mathcal{C}(b) \vee u = a \cup \mathcal{C}(b)) .$$

From equation (2.203) (p. 377), we have:

$$(a \in \mathcal{O} \subset B,\ b \in \mathcal{O} \subset B) \Rightarrow (\mathcal{C}(a) \in B,\ \mathcal{C}(b) \in B) .$$

From equation (2.204) (p. 377) we have: $a \cup \mathcal{C}(b) \in B$, $\mathcal{C}(a) \cup b \in B$.
From equation (2.203) (p. 377) and propositions 767 (p. 455), 768 (p. 455) we have:

$$\mathcal{C}(\mathcal{C}(a) \cup b) = a \cap \mathcal{C}(b) \in B .$$

It means that in the two cases of equation (2.208) (p. 378), we have $\mathbf{u} \in \mathbf{B}$. From 349 (p. 451), $\mathcal{O}^b \subset \mathbf{B}$.

3. $\mathcal{O} \subset \mathbb{O}$.
From definitions 349 (p. 451), 355 (p. 456) and equation (2.209) (p. 378) we have:

$$(A = \mathcal{C}(\emptyset) \in \mathcal{O}, \emptyset \in \mathcal{O}) \Rightarrow (\forall a \in \mathcal{O})(a = a \cap \mathcal{C}(\emptyset) \in \mathcal{O}^b \subset \mathbb{O}) \Rightarrow \mathcal{O} \subset \mathbb{O} .$$

4. $A \in \mathbb{O},\ \emptyset \in \mathbb{O},\ \mathbb{O} \neq \emptyset$.
From definitions 132 (p. 151) and 355 (p. 456), we have:

$$A = \mathcal{C}(\emptyset) \in \mathcal{O},\ \mathcal{C}(A) = \emptyset \in \mathcal{O} .$$

Using point 3, $\mathbf{A} \in \mathbb{O}$, $\emptyset \in \mathbb{O}$, $\mathbb{O} \neq \emptyset$. $\mathbb{O}$ satisfies equation (2.202) (p. 377).

5. $x \in \mathcal{O}^b \Rightarrow \mathcal{C}(x) \in \mathcal{O}^b$.
Using propositions 767 (p. 455) and 768 (p. 455), we have: $\mathcal{C}(a \cap \mathcal{C}(b)) = \mathcal{C}(a) \cup b = b \cup \mathcal{C}(a)$ and $\mathcal{C}(a \cup \mathcal{C}(b)) = \mathcal{C}(a) \cap b = b \cap \mathcal{C}(a)$ plus equation (2.208) (p. 378).

$$\begin{aligned} x \in \mathcal{O}^b &\Leftrightarrow (\exists a \in \mathcal{O})(\exists b \in \mathcal{O})(x = a \cap \mathcal{C}(b) \vee x = a \cup \mathcal{C}(b)) \\ &\Leftrightarrow (\exists a \in \mathcal{O})(\exists b \in \mathcal{O})(\mathcal{C}(x) = \mathcal{C}(a \cap \mathcal{C}(b)) = b \cup \mathcal{C}(a) \\ &\vee \mathcal{C}(x) = \mathcal{C}(a \cup \mathcal{C}(b)) = b \cap \mathcal{C}(a)) \\ &\Leftrightarrow \mathcal{C}(x) \in \mathcal{O}^b . \end{aligned}$$

6. $x \in \mathbb{O} \Rightarrow \mathcal{C}(x) \in \mathbb{O}$.
Given $\mathbf{K} \subset \mathcal{O}^b$, we call $\mathbf{K_c}$ the following set:

$$k \in K_c \Leftrightarrow \mathcal{C}(k) \in K$$

$(\mathcal{C} : K \to K_c)$ is a bijection (definition 124 (p. 139), proposition 790 (p. 462)). Then from definition 390 (p. 499), we have: $\mathbf{Card(K) = Card(K_c)}$.
Since, from proposition 801 (p. 466), we have:

$$\mathcal{C}(\bigcup_{k \in K} k) = \bigcap_{k \in K} \mathcal{C}(k),\ \mathcal{C}(\bigcap_{k \in K} k) = \bigcup_{k \in K} \mathcal{C}(k)$$

from equation (2.209) (p. 378) and point 5, we have:

$$\begin{aligned}
&\mathbf{x} \in \mathbb{O} \Leftrightarrow (\exists \mathbf{K} \subset \mathcal{O}^{\mathbf{b}})(\mathbf{Card}(\mathbf{K}) \leq \mathbf{Card}(\mathcal{Z}^{0+}), \mathbf{x} = \bigcup_{\mathbf{k}\in\mathbf{K}} \mathbf{k} \vee \mathbf{x} = \bigcap_{\mathbf{k}\in\mathbf{K}} \mathbf{k}) \\
&\Leftrightarrow (\exists \mathbf{K} \subset \mathcal{O}^{\mathbf{b}})(\mathbf{Card}(\mathbf{K}) \leq \mathbf{Card}(\mathcal{Z}^{0+}), (\mathcal{C}(\mathbf{x}) = \mathcal{C}(\bigcup_{\mathbf{k}\in\mathbf{K}} \mathbf{k}) = \bigcap_{\mathbf{k}\in\mathbf{K}} \mathcal{C}(\mathbf{k}) \\
&\vee \mathcal{C}(\mathbf{x}) = \mathcal{C}(\bigcap_{\mathbf{k}\in\mathbf{K}} \mathbf{k}) = \bigcup_{\mathbf{k}\in\mathbf{K}} \mathcal{C}(\mathbf{k}))) \\
&\Leftrightarrow (\exists \mathbf{K}' = \mathbf{K}_c \subset \mathcal{O}^{\mathbf{b}})(\mathbf{Card}(\mathbf{K}') = \mathbf{Card}(\mathbf{K}_c) \leq \mathbf{Card}(\mathcal{Z}^{0+}), \\
&(\mathcal{C}(\mathbf{x}) = \bigcup_{\mathbf{k}\in\mathbf{K}'} \mathbf{k} \vee \mathcal{C}(\mathbf{x}) = \bigcap_{\mathbf{k}\in\mathbf{K}'} \mathbf{k})) \\
&\Leftrightarrow \mathcal{C}(\mathbf{x}) \in \mathbb{O}\,.
\end{aligned}$$

7. Having a set $\mathbf{K} \subset \mathbb{O}$, let us prove that: $\mathbf{Card}(\mathbf{K}) \leq \mathbf{Card}(\mathcal{Z}^{0+}) \Rightarrow \bigcup_{\mathbf{k}\in\mathbf{K}} \mathbf{k} \in \mathbb{O}$. Equation (2.204) (p. 377)
We can split $\mathcal{O}^{\mathbf{b}}$ in $\mathcal{O}^{\mathbf{b}}_{\mathbf{u}}$ and $\mathcal{O}^{\mathbf{b}}_{\mathbf{i}}$ with:

$$\begin{aligned}
\mathbf{x} \in \mathcal{O}^{\mathbf{b}}_{\mathbf{u}} \Leftrightarrow (\exists \mathbf{a} \in \mathcal{O})(\exists \mathbf{b} \in \mathcal{O})(\mathbf{x} = \mathbf{a} \cup \mathcal{C}(\mathbf{b})) \\
\mathbf{x} \in \mathcal{O}^{\mathbf{b}}_{\mathbf{i}} \Leftrightarrow (\exists \mathbf{a} \in \mathcal{O})(\exists \mathbf{b} \in \mathcal{O})(\mathbf{x} = \mathbf{a} \cap \mathcal{C}(\mathbf{b}))
\end{aligned}$$

We have $\mathcal{O}^{\mathbf{b}} = \mathcal{O}^{\mathbf{b}}_{\mathbf{u}} \cup \mathcal{O}^{\mathbf{b}}_{\mathbf{i}}$. We have also:

(a)

$$(\mathbf{Card}(\mathbf{K}) \in \mathcal{Z}, \mathbf{K} \subset \mathcal{O}^{\mathbf{b}}_{\mathbf{u}}) \Rightarrow (\bigcup_{\mathbf{k}\in\mathbf{K}} \mathbf{k} \in \mathcal{O}^{\mathbf{b}}_{\mathbf{u}}) \tag{2.210}$$

Calling $\mathcal{O}^{\mathbf{c}}$ the closed sets of $\mathbf{A}$ we can split any $\mathbf{k} \in \mathbf{K}$ in two sets $\mathbf{k}^{\mathbf{o}}$ and $\mathbf{k}^{\mathbf{c}}$ with $\mathbf{k} = \mathbf{k}^{\mathbf{o}} \cup \mathbf{k}^{\mathbf{c}}$ which gives:

$$\bigcup_{\mathbf{k}\in\mathbf{K}} \mathbf{k} = \bigcup_{\mathbf{k}\in\mathbf{K}} \mathbf{k}^{\mathbf{o}} \cup \bigcup_{\mathbf{k}\in\mathbf{K}} \mathbf{k}^{\mathbf{c}} \in \mathcal{O}^{\mathbf{b}}_{\mathbf{u}}\,.$$

See definition 132 (p. 151) relation (2.17), propositions 796 (p. 464) and 206 (p. 154) relation (2.26).

(b)

$$(\mathbf{Card}(\mathbf{K}) \in \mathcal{Z}, \mathbf{K} \subset \mathcal{O}^{\mathbf{b}}) \Rightarrow (\exists \mathbf{K}' \subset \mathcal{O}^{\mathbf{b}}_{\mathbf{i}})(\mathbf{Card}(\mathbf{K}') \in \mathcal{Z}, \bigcup_{\mathbf{k}\in\mathbf{K}} \mathbf{k} = \bigcup_{\mathbf{k}\in\mathbf{K}'} \mathbf{k}) \tag{2.211}$$

Because setting $\mathbf{K}_{\mathbf{u}} = \mathbf{K} \cap \mathcal{O}^{\mathbf{b}}_{\mathbf{u}}$, $\mathbf{K}_{\mathbf{i}} = \mathbf{K} \cap \mathcal{O}^{\mathbf{b}}_{\mathbf{i}}$ we have from above with $\mathbf{a} \in \mathcal{O}$, $\mathbf{b} \in \mathcal{O}$

$$\bigcup_{\mathbf{k}\in\mathbf{K}} \mathbf{k} = \bigcup_{\mathbf{k}\in\mathbf{K}_{\mathbf{u}}} \mathbf{k} \cup \bigcup_{\mathbf{k}\in\mathbf{K}_{\mathbf{i}}} \mathbf{k} = \mathbf{a} \cup \mathcal{C}(\mathbf{b}) \cup \bigcup_{\mathbf{k}\in\mathbf{K}_{\mathbf{i}}} \mathbf{k} = (\mathbf{a} \cap \mathbf{A}) \cup (\mathbf{A} \cap \mathcal{C}(\mathbf{b})) \cup \bigcup_{\mathbf{k}\in\mathbf{K}_{\mathbf{i}}} \mathbf{k}$$

It gives $\mathbf{K}' = \{\mathbf{a} \cap \mathbf{A}\} \cup \{\mathbf{A} \cap \mathcal{C}(\mathbf{b})\} \cup \mathbf{K}_{\mathbf{i}}$ with $\mathbf{a} \cap \mathbf{A} \in \mathcal{O}^{\mathbf{b}}_{\mathbf{i}}$, $\mathbf{A} \cap \mathcal{C}(\mathbf{b}) \in \mathcal{O}^{\mathbf{b}}_{\mathbf{i}}$.

(c)

$$(\mathbf{Card}(\mathbf{K}) \in \mathcal{Z}, \mathbf{K} \subset \mathcal{O}^{\mathbf{b}}_{\mathbf{i}}) \Rightarrow (\exists \mathbf{K}' \subset \mathcal{O}^{\mathbf{b}}_{\mathbf{u}})(\mathbf{Card}(\mathbf{K}') \in \mathcal{Z}, \bigcup_{\mathbf{k}\in\mathbf{K}} \mathbf{k} = \bigcap_{\mathbf{k}\in\mathbf{K}'} \mathbf{k}) \tag{2.212}$$

Using proposition 796 (p. 464) and splitting again any $\mathbf{k} \in \mathbf{K}$ in $\mathbf{k^o}$ and $\mathbf{k^c}$ with $\mathbf{k} = \mathbf{k^o} \cap \mathbf{k^c}$, we have:

$$\bigcup_{k \in K} k = \bigcup_{k \in K} \bigcap_{i \in \{c,o\}} k^i = \bigcap_{f \in \{c,o\}^K} \bigcup_{i \in K} i^{f(i)}$$

Since $i^{f(i)} = i^{f(i)} \cup \emptyset \in \mathcal{O}^b_u$ from point 7a $\bigcup_{i \in K} i^{f(i)} \in \mathcal{O}^b_u$.

Since $\mathbf{k} \in \mathcal{O}^b_u \Rightarrow \mathcal{C}(\mathbf{k}) \in \mathcal{O}^b_i$, taking the complement (definition 353 (p. 454)) of all above equations, we have also from property 6 (p. 435) of A.1.3 and propositions 767 (p. 455), 801 (p. 466):
Using equation (2.210), we have:

$$(\mathrm{Card}(K) \in \mathcal{Z}, K \subset \mathcal{O}^b_i) \Rightarrow (\bigcap_{k \in K} k \in \mathcal{O}^b_i) \tag{2.213}$$

Using equation (2.211)

$$(\mathrm{Card}(K) \in \mathcal{Z}, K \subset \mathcal{O}^b) \Rightarrow (\exists K' \subset \mathcal{O}^b_u)(\mathrm{Card}(K') \in \mathcal{Z}, \bigcap_{k \in K} k = \bigcap_{k \in K'} k) \tag{2.214}$$

Using equation (2.212), we have:

$$(\mathrm{Card}(K) \in \mathcal{Z}, K \subset \mathcal{O}^b_u) \Rightarrow (\exists K' \subset \mathcal{O}^b_i)(\mathrm{Card}(K') \in \mathcal{Z}, \bigcup_{k \in K} \mathcal{C}(k) = \bigcap_{k \in K'} \mathcal{C}(k))$$

$$\Leftrightarrow (\exists K' \subset \mathcal{O}^b_i)(\mathrm{Card}(K') \in \mathcal{Z}, \bigcap_{k \in K} k = \bigcup_{k \in K'} k) \tag{2.215}$$

From definition 2.209 (p. 378) of $\mathbb{O}$, we can also split it in two sets: $\mathbb{O}^i$, $\mathbb{O}^u$ according to the sign $\bigcap$ or $\bigcup$ used for $\mathbf{x}$. From above relations ((2.212), (2.215)) $\mathbb{O}^i = \mathbb{O}^u$. If $x_1 = \bigcap_{k \in K_1} k \in \mathbb{O}^i$, $x_2 = \bigcap_{k \in K_2} k \in \mathbb{O}^i$, we have (proposition 803 (p. 466) )

$$x_1 \cap x_2 = \bigcap_{k \in K_1} k \cap \bigcap_{k \in K_2} k = \bigcap_{k \in K_1 \cup K_2} k \in \mathbb{O}$$

If $x_1 = \bigcup_{k \in K_1} k \in \mathbb{O}^i$, $x_2 = \bigcup_{k \in K_2} k \in \mathbb{O}^i$, we have (proposition 796 (p. 464) )

$$x_1 \cup x_2 = \bigcup_{k \in K_1} k \cap \bigcup_{k \in K_2} k = \bigcup_{k \in K_1 \cup K_2} k \in \mathbb{O}$$

Then since $\mathbb{O}^i = \mathbb{O}^u$, the condition (2.204) is filled for two elements. But if it true for $\mathbf{n}$ elements using the relation for two it is also true for $\mathbf{n+1}$. Condition (2.204) is true using proposition 927 (p. 520) (Recurrence principle).

Using all those relations, we see that $\mathbb{O}$ fulfills all the conditions for being a ***σ*-algebra** containing $\mathcal{O}$ (see definition 302 (p. 377)). Since by definition of a ***σ*-algebra** (definition 302 (p. 377)), any ***σ*-algebra** containing $\mathcal{O}$ contains $\mathbb{O}$. Since $\mathbb{O}$ is a ***σ*-algebra** and contains $\mathcal{O}$, it is the intersection of all ***σ*-algebra** containing $\mathcal{O}$. It is then the Borel ***σ*-algebra** of **A**.

**Definition 304 Measurable space.**

A set **A** is a measurable space if it has a ***σ*-algebra**. In that book, we ask for more: it means that it is a topological set on which one considers its Borel ***σ*-algebra** as its ***σ*-algebra**.

**Definition 305 Measure.**

Given a measurable space $\mathbf{A}$ and $\mathbf{B} \subset \mathcal{P}(\mathbf{A})$ its Borel **$\sigma$-algebra** (definitions 350 (p. 453), 304 (p. 381), 303 (p. 378)), one calls measure on $\mathbf{B}$ a map $(\mu : \mathbf{B} \to \mathcal{R}^{0+}_{\infty})$ (definition 144 (p. 171)), such that $\mu(\emptyset) = 0, \mu(\mathcal{R}^{n}) = \infty$ and if we have:

$$(\mathbf{C} \subset \mathcal{P}(\mathbf{B}), (\forall x \in \mathbf{C})(\forall y \in \mathbf{C})(x \neq y \Rightarrow x \cap y = \emptyset),$$
$$\mathrm{Card}(\mathbf{C}) \leq \mathrm{Card}(\mathcal{Z})) \Rightarrow \mu(\bigcup_{c \in \mathbf{C}} c) = \sum_{c \in \mathbf{C}} \mu(c)$$

with the following convention:

$$(\exists c \in \mathbf{C})(\mu(c) = \infty) \Rightarrow \sum_{c \in \mathbf{C}} \mu(c) = \infty .$$

**Definition 306 Lebesgue measure of a compact set of $\mathcal{R}^{n}$.**

Given a compact set $\mathbf{K} \subset \mathcal{R}^{n}$ included in an open set $\mathbf{O}$ of the $\mathcal{R}^{n}$ open ball topology (definition 263 (p. 286)), we call $\mathcal{F}^{\mathbf{K}}_{\mathbf{O}}$, the set of continuous maps $(f : \mathcal{R}^{n} \to [0, 1])$ with a compact support such that $f(\mathbf{K}) = \{1\}$ and $f(\mathcal{C}(\mathbf{O})) = 0$ (see definition 353 (p. 454) and convention 29 (p. 455)). From proposition 527 (p. 292), $\mathcal{F}^{\mathbf{K}}_{\mathbf{O}} \neq \emptyset$. We call $\mathcal{O}_{\mathbf{K}} \subset \mathcal{P}(\mathcal{R}^{n})$, the set of opens which contain $\mathbf{K}$. $\mathcal{O}_{\mathbf{K}} \neq \emptyset$ because from definitions 132 (p. 151), 150 (p. 177):

$$r > 0 \Rightarrow \bigcup_{x \in \mathbf{K}} B(x, r) \in \mathcal{O}_{\mathbf{K}} .$$

When it exists, we call $\mathbf{J(h)}$, the Riemann integral of $\mathbf{h}$. With that, one defines the Lebesgue measure of $\mathbf{K}$, $\mu(\mathbf{K})$ by[34]:

$$\mathbf{K} = \emptyset \Rightarrow \mu(\mathbf{K}) = 0,$$
$$\mathbf{K} \neq \emptyset \Rightarrow \mu(\mathbf{K}) = \min_{\mathbf{O} \in \mathcal{O}_{\mathbf{K}}} \max_{f \in \mathcal{F}^{\mathbf{K}}_{\mathbf{O}}} J(f) .$$

From proposition 621 (p. 355), $\mathbf{J(f)}$ always exists.

**Definition 307 Lebesgue integral of a positive continuous map over a compact set of $\mathcal{R}^{n}$.**

With the same notation as in definition 306 (p. 382), given a compact set $\mathbf{K} \subset \mathcal{R}^{n}$, one calls integral $\mu_{g}(\mathbf{K})$ of a positive continuous map $(g : \mathcal{R} \to \mathcal{R}^{0+})$ over $\mathbf{K}$ , the following quantity:

$$\mu_{g}(\mathbf{K}) = \min_{\mathbf{O} \in \mathcal{O}_{\mathbf{K}}} \max_{f \in \mathcal{F}^{\mathbf{K}}_{\mathbf{O}}} J(fg) .$$

**Proposition 646** With the same notation as above, given an open $\mathbf{O} \in \mathcal{O}_{\mathbf{K}}$, we can replace, in definitions 306 (p. 382) and 307 (p. 382), $\mathcal{F}^{\mathbf{K}}_{\mathbf{O}}$ by $\mathcal{F}_{\mathbf{O}}$: the set of continuous maps $(f : \mathcal{R}^{n} \to [0, 1])$ such that $f(\mathcal{C}(\mathbf{O})) = 0$.

[34] In all that section min and max has to be understood as in definition 146 (p. 175) and not as in definition 382 (p. 485).

**Proof**
We have $\mathcal{F}_O^K \subset \mathcal{F}_O$. Then:

$$\begin{aligned}\max_{f\in\mathcal{F}_O^K} J(f) &\leq \max_{f\in\mathcal{F}_O} J(f)\\ \max_{f\in\mathcal{F}_O^K} J(fg) &\leq \max_{f\in\mathcal{F}_O} J(fg)\ .\end{aligned} \tag{2.216}$$

With $(f_1 : \mathcal{R}^n \to \mathcal{R})$, $(f_2 : \mathcal{R}^n \to \mathcal{R})$ calling $(\max(f_1, f_2) : \mathcal{R}^n \to \mathcal{R})$ the map defined by:

$$(\forall x \in \mathcal{R}^n)(\max(f_1, f_2)(x) = \max(f_1(x), f_2(x)))\ .$$

From proposition 296 (p. 189), if $f_1$ and $f_2$ are continuous, $\max(f_1, f_2)$ is continuous. Then, from proposition 609 (p. 340), we have:

$$(\forall f \in \mathcal{F}_O)(\forall f^K \in \mathcal{F}_O^K)((f^m = \max(f, f^K) \in \mathcal{F}_O^K) \Rightarrow J(f^m) \geq J(f))\ .$$

Since $\mathcal{F}_O^K \neq \emptyset$,

$$(\forall f \in \mathcal{F}_O)(\exists f^K \in \mathcal{F}_O^K)(J(f^K) \geq J(f))\ .$$

Then from proposition 609 (p. 340) again, we have:

$$\begin{aligned}\max_{f\in\mathcal{F}_O^K} J(f) &\geq \max_{f\in\mathcal{F}_O} J(f)\\ \max_{f\in\mathcal{F}_O^K} J(fg) &\geq \max_{f\in\mathcal{F}_O} J(fg)\ .\end{aligned} \tag{2.217}$$

Combining equations (2.216) and (2.217), we have:

$$\begin{aligned}\max_{f\in\mathcal{F}_O^K} J(f) &= \max_{f\in\mathcal{F}_O} J(f)\\ \max_{f\in\mathcal{F}_O^K} J(fg) &= \max_{f\in\mathcal{F}_O} J(fg)\ .\end{aligned}$$

**Definition 308 Lebesgue measure of an open of $\mathcal{R}^n$.**

Having an open set $O \subset \mathcal{R}^n$, one considers the set $\mathcal{K}_O$ of compact subsets of O. If $O \neq \emptyset$, any point $x$ of O is the center of an open ball in O (proposition 266 (p. 179)): $(\exists r_x > 0)(B(x, r_x) \subset O)$ then $\bar{B}(x, r_x/2) \subset B(x, r_x) \subset O$. $\bar{B}(x, r_x/2)$ is compact then $\mathcal{K}_O \neq \emptyset$. One calls Lebesgue measure $(\mu(O) : \mathcal{O} \to \mathcal{R}_\infty^{0+})$ (definition 144 (p. 171)), $\mathcal{O}$ being the set of opens of $\mathcal{R}^n$, the following quantity:

$$\mu(O) = \max_{K\in\mathcal{K}_O} \mu(K)\ .$$

If there is no maximum of $J(f)$, by convention $\mu(O) = \infty$ (definition 144 (p. 171)).

**Definition 309 Lebesgue integral of a positive continuous map over an open of $\mathcal{R}^n$.**

With the same notation as in definition 308 (p. 383), given an open $O \subset \mathcal{R}^n$, one calls integral over O of a positive continuous map $(g : \mathcal{R}^n \to \mathcal{R}^{0+})$ defined at least on O, the following quantity:

$$\mu_g(O) = \max_{K\in\mathcal{K}_O} \mu_g(K)\ .$$

It exists (proposition 620 (p. 352)) if by convention $\mu_g(O) = \infty$ when there is no maximum.

**Proposition 647** We have an open $\mathbf{O} \subset \mathcal{R}^{\mathbf{n}}$. One considers the set of semi-open boxes $(\mathbf{S}^{\beta}(\mathbf{p},\mathbf{q}) : [0,1] \times \mathcal{Z}^{\mathbf{n}} \times \mathcal{Z}^{+0} \Rightarrow \mathcal{P}(\mathcal{R}^{\mathbf{n}}))$ defined by (see definition 165 (p. 197) and proposition 319 (p. 197)):

$$((\mathbf{p},\mathbf{q}) \in \mathcal{Z}^{\mathbf{n}} \times \mathcal{Z}^{+0}, \beta \in [0,1] \subset \mathcal{R}) \Rightarrow (\mathbf{x} \in \mathbf{S}^{\beta}(\mathbf{p},\mathbf{q}) \Leftrightarrow (\forall \mathbf{i} \in [1,\mathbf{n}])(\mathbf{x}^{\mathbf{i}} \in [\frac{\mathbf{p}^{\mathbf{i}}-\beta}{2^{\mathbf{q}}}, \frac{\mathbf{p}^{\mathbf{i}}+1+\beta}{2^{\mathbf{q}}}[)) . \tag{2.218}$$

Using that definition, we have:

$$(\exists((\mathbf{p_j},\mathbf{q_j}) : \mathcal{Z}^{+0} \to \mathcal{Z}^{\mathbf{n}} \times \mathcal{Z}^{+0})) ((\forall \beta \in [0,1])(\mathbf{O} = \bigcup_{\mathbf{j} \in \mathcal{Z}^{+0}} \mathbf{S}^{\beta}(\mathbf{p_j},\mathbf{q_j})), \mathbf{i} \neq \mathbf{j} \Rightarrow \mathbf{S}^0(\mathbf{p_i},\mathbf{q_i}) \cap \mathbf{S}^0(\mathbf{p_j},\mathbf{q_j}) = \emptyset) . \tag{2.219}$$

**Proof**
With the above notations, we have:

$$\mathbf{p} \neq \mathbf{p}' \Rightarrow \mathbf{S}^0(\mathbf{p},\mathbf{q}) \cap \mathbf{S}^0(\mathbf{p}',\mathbf{q}) = \emptyset, \tag{2.220}$$

$$(\mathbf{q} \in \mathcal{Z}^{+0}, \beta \in [0,1]) \Rightarrow \mathbf{O} \subset \bigcup_{\mathbf{p} \in \mathcal{Z}^{\mathbf{n}}} \mathbf{S}^{\beta}(\mathbf{p},\mathbf{q}) = \mathcal{R}^{\mathbf{n}}, \tag{2.221}$$

$$\mathbf{q} \leq \mathbf{q}' \Rightarrow \mathbf{S}^0(\mathbf{p},\mathbf{q}) = \bigcup_{\mathbf{p}' \in \prod_{\mathbf{j} \in [1,\mathbf{n}]} [(\mathbf{p}^{\mathbf{j}} 2^{\mathbf{q}'-\mathbf{q}}, (\mathbf{p}^{\mathbf{j}}+1) 2^{\mathbf{q}'-\mathbf{q}}[} \mathbf{S}^0(\mathbf{p}',\mathbf{q}'), \tag{2.222}$$

$$\mathbf{S}^0(\mathbf{p},\mathbf{q}) \cap \mathbf{S}^0(\mathbf{p}',\mathbf{q}') \neq \emptyset \Leftrightarrow \mathbf{S}^0(\mathbf{p},\mathbf{q}) \subset \mathbf{S}^0(\mathbf{p}',\mathbf{q}') . \tag{2.223}$$

From proposition 985 (p. 551), $\mathbf{Card}(\mathcal{Z}^{\mathbf{n}} \times \mathcal{Z}^{0,+}) = \mathbf{Card}(\mathcal{Z}^{+})$. We define the map $(\sigma : \mathcal{Z}^{\mathbf{n}} \times \mathcal{Z}^{0,+} \Rightarrow \{0,1\})$ by:

$$\begin{aligned} &\sigma(\mathbf{p},\mathbf{q}) = 0 \Leftrightarrow (\mathbf{S}^1(\mathbf{p},\mathbf{q}) \cap \mathcal{C}(\mathbf{O}) \neq \emptyset \\ &\vee (\exists \mathbf{q}' < \mathbf{q})(\exists \mathbf{p}' \in \mathcal{Z}^{\mathbf{n}})(\mathbf{S}^1(\mathbf{p},\mathbf{q}) \subset \mathbf{S}^1(\mathbf{p}',\mathbf{q}') \subset \mathbf{O}, \mathbf{S}^0(\mathbf{p},\mathbf{q}) \subset \mathbf{S}^0(\mathbf{p}',\mathbf{q}'))), \\ &\sigma(\mathbf{p},\mathbf{q}) = 1 \Leftrightarrow (\mathbf{S}^1(\mathbf{p},\mathbf{q}) \subset \mathbf{O}, (\forall \mathbf{q}' < \mathbf{q})(\forall \mathbf{p}' \in \mathcal{Z}^{\mathbf{n}})(\mathbf{S}^0(\mathbf{p},\mathbf{q}) \cap \mathbf{S}^0(\mathbf{p}',\mathbf{q}') = \emptyset \\ &\vee ((\mathbf{S}^1(\mathbf{p},\mathbf{q}) \subset \mathbf{S}^1(\mathbf{p}',\mathbf{q}') \Rightarrow \mathbf{S}^1(\mathbf{p}',\mathbf{q}') \cap \mathcal{C}(\mathbf{O}) \neq \emptyset))) . \end{aligned}$$

Let us prove now:

$$(\forall \mathbf{x} \in \mathbf{O})(\exists \mathbf{q} \in \mathcal{Z}^{+0})(\exists \mathbf{p} \in [-2^{\mathbf{q}}, 2^{\mathbf{q}}[^{\mathbf{n}})(\mathbf{x} \in \mathbf{S}^0(\mathbf{p},\mathbf{q}), \mathbf{S}^1(\mathbf{p},\mathbf{q}) \subset \mathbf{O}) .$$

From proposition 266 (p. 179), we have:

$$(\forall \mathbf{x} \in \mathbf{O})(\exists \mathbf{r}(\mathbf{x}) > 0)(\mathbf{C}(\mathbf{x},\mathbf{r}(\mathbf{x})) \subset \mathbf{O}) .$$

We also have:

$$(\exists \mathbf{q_r} \in \mathcal{Z}^{+})(2^{-\mathbf{q_r}} < \mathbf{r}/3),$$
$$(\forall \mathbf{q} \in \mathcal{Z}^{+0})(\mathbf{O} \subset \bigcup_{\mathbf{p} \in \mathcal{Z}^{\mathbf{n}}} \mathbf{S}^0(\mathbf{p},\mathbf{q}) = \mathcal{R}^{\mathbf{n}})$$

we have:

$$(\forall \mathbf{x} \in \mathbf{O})(\forall \mathbf{q} \in \mathcal{Z}^{+0})(\exists \mathbf{p}(\mathbf{x},\mathbf{q}))(\mathbf{x} \in \mathbf{S}^0(\mathbf{p}(\mathbf{x},\mathbf{q}),\mathbf{q})) .$$

From equation (2.218) (p. 384), it gives:

$$\begin{aligned} &(\forall \mathbf{i})(\mathbf{x}^{\mathbf{i}} - \mathbf{r}(\mathbf{x}) < \mathbf{x}^{\mathbf{i}} - \frac{2}{2^{\mathbf{q_{r(x)}}}} < \frac{\mathbf{p}^{\mathbf{i}}(\mathbf{x},\mathbf{q_{r(x)}}) - 1}{2^{\mathbf{q_{r(x)}}}} < \frac{\mathbf{p}^{\mathbf{i}}(\mathbf{x},\mathbf{q_{r(x)}})}{2^{\mathbf{q_{r(x)}}}} \leq \mathbf{x}^{\mathbf{i}} \\ &< \frac{\mathbf{p}^{\mathbf{i}}(\mathbf{x},\mathbf{q_{r(x)}}) + 1}{2^{\mathbf{q_{r(x)}}}} < \frac{\mathbf{p}^{\mathbf{i}}(\mathbf{x},\mathbf{q_{r(x)}}) + 2}{2^{\mathbf{q_{r(x)}}}} \leq \mathbf{x}^{\mathbf{i}} + \frac{2}{2^{\mathbf{q_{r(x)}}}} < \mathbf{x}^{\mathbf{i}} + \mathbf{r}(\mathbf{x})) . \end{aligned}$$

With that we have:

$$S^0(p(x, q_{r(x)}), q_{r(x)}) \subset S^1(p(x, q_{r(x)}), q_{r(x)}) \subset C(x, r(x)) \subset O \,.$$

We can consider the set $Q_x \subset [0, q_{r(x)}]$ such that

$$q \in Q_x \Leftrightarrow (q \leq q_{r(x)}, (\exists p \in \mathcal{Z}^n)(S^0(p(x, q_{r(x)}), q_{r(x)}) \subset S^0(p, q), S^1(p, q) \subset O)) \,.$$

Since $0$ is minoring value of $Q_x$, since $Q_x$ is not empty because $q_{r(x)} \in Q_x$, $Q_x$ has a smallest element $q(x) \in Q_x$ (proposition 886 (p. 504)) to which corresponds $p(x)$ such that: $x \in S^0(p(x), q(x))$ and $\sigma(p(x), q(x)) = 1$. Then we have (proposition 727 (p. 441)):

$$(\forall x \in O)(\exists (p, q) \in \mathcal{Z}^n \times \mathcal{Z}^{+0})(x \in S^0(p, q), \sigma(p, q) = 1) \,.$$

Then

$$O \subset \bigcup_{(p,q) \in \mathcal{Z}^n \times \mathcal{Z}^{+0}, \sigma(p,q)=1} S^0(p, q) \subset \bigcup_{(p,q) \in \mathcal{Z}^n \times \mathcal{Z}^{+0}, \sigma(p,q)=1} S^1(p, q) \,.$$

Since from the definition of $\sigma$, $\sigma(p, q) = 1 \Rightarrow S^1(p, q) \subset O$

$$\bigcup_{(p,q) \in \mathcal{Z}^n \times \mathcal{Z}^{+0}, \sigma(p,q)=1} S^0(p, q) \subset \bigcup_{(p,q) \in \mathcal{Z}^n \times \mathcal{Z}^{+0}, \sigma(p,q)=1} S^1(p, q) \subset O$$

then from the axiom 754 (p. 451):

$$\bigcup_{(p,q) \in \mathcal{Z}^n \times \mathcal{Z}^{+0}), \sigma(p,q)=1} S^0(p, q) = \bigcup_{(p,q) \in \mathcal{Z}^n \times \mathcal{Z}^{+0}, \sigma(p,q)=1} S^1(p, q) = O \,.$$

Calling $H = \mathrm{Set}_u\{u = (p, q), (p, q) \in \mathcal{Z}^n \times \mathcal{Z}^{+0}, \sigma(p, q) = 1\} \subset \mathcal{Z}^n \times \mathcal{Z}^{+0}$, we have (definition 392 (p. 500), proposition 938 (p. 525)):

$$\begin{aligned} \mathrm{Card}(H) &\leq \mathrm{Card}(\mathcal{Z}^n \times \mathcal{Z}^{+0}) \\ &= \mathrm{Card}(\mathcal{Z}^{+0}) \,. \end{aligned}$$

From propositions 940 (p. 525) and 255 (p. 172), we have:

$$\mathrm{Card}(H) < \mathrm{Card}(\mathcal{Z}) \Rightarrow \mathrm{Card}(H) \in \mathcal{Z}^+ \Rightarrow (\exists k \in H)(p^1_k/2^{q_k} = \min_{j \in H}(p^1_j/2^{q_j})) \,.$$

If $H$ is finite, $\mathrm{Card}(H) \in \mathcal{Z}^+$ and vice versa. Then, in that case, we may consider this $k \in H$ and the point $x = (x^i : [1, n] \to \mathcal{R})$ such that $x^i = p^i_k/2^{q_k}$. From equation (2.218) (p. 384), $x \in S^0(p_k, q_k) \subset O$, from proposition 266 (p. 179), there is a cube $C(x, r) \subset O, r > 0$. $C(x, r)$ cannot be included in the union $\bigcup_{j \in H} S^0(p_j, q_j)$ because the point $y$ such as $y^1 = x^1 - r/2, (\forall i \in [2, n])(y^i = x^i)$ which belongs to $C(x, r)$, does not belong to the union because by definition of $x$:

$$y^1 < \min_{u \in \bigcup_{j \in H} S^0(p_j, q_j)} u^1 = \min_{u \in O} u^1 \,.$$

Then it remains: $\mathrm{Card}(H) = \mathrm{Card}(\mathcal{Z}^{+0})$. We have (definition 390 (p. 499)):

$$\begin{aligned} &(\exists ((p_j, q_j) : \mathcal{Z}^{+0} \to \mathcal{Z}^n \times \mathcal{Z}^{+0})) \\ &(\forall \beta \in [0, 1])(O = \bigcup_{j \in H} S^\beta(p_j, q_j)) \,. \end{aligned} \tag{2.224}$$

We also have:

$$(p \neq p', q \neq q', \sigma(p,q) = \sigma(p',q')) \Rightarrow S^0(p,q) \cap S^0(p',q') = \emptyset \,.$$

Because:
By construction $(q = q', p \neq p') \Rightarrow S^0(p,q) \cap S^0(p',q') = \emptyset$. We have from section A.1.4 (p. 435) point 7:

$$\begin{aligned} & q > q' \Rightarrow \\ & (S^0(p',q') \cap S^0(p,q) \neq \emptyset \Rightarrow S^0(p',q') \subset S^0(p,q) \Rightarrow S^1(p',q') \subset S^1(p,q) \\ & \quad \Rightarrow \sigma(p',q') = 0) \\ & \Rightarrow (\sigma(p',q') = 1 \Rightarrow \sigma(p',q') \neq 0 \Rightarrow S^0(p',q') \cap S^0(p,q) = \emptyset) \,. \end{aligned} \tag{2.225}$$

Combining the two relations (2.224) and (2.225), proposition 647 (p. 384) is true.

**Proposition 648** If a compact set of $\mathcal{R}^n$ has a volume defined by a Riemann integral (see definition 295 (p. 342)), this volume is equal to its Lebesgue measure defined by definition 306 (p. 382).

**Proof**
Using the notation of definition 306 (p. 382), propositions 526 (p. 292) and 527 (p. 292), given a compact set $K \subset \mathcal{R}^n$, we have $r > 0$ such that $K \subset B(0,r)$ (proposition 271 (p. 180)). Taking an open $O \in \mathcal{O}_K$ ($\mathcal{O}_K$ is the set of opens which contain $K$ as in definition 306 (p. 382)), defining $c, d$ as in equation (2.130) (p. 292) ($c \in K,\ d \in \mathcal{C}(O)$) and a continuous map (definition 141 (p. 165)) ($h_u : \mathcal{R}^n \to [0,1]$) as in proposition 527 (p. 292), for any $u \in ]0, \|c-d\|[$, $h_u^{-1}(]0,2[) \subset O$ is open (definition 141 (p. 165)). We have (see definition 306 (p. 382)):

$$(O_1 \in \mathcal{O}_K, O_2 \in \mathcal{O}_K, O_1 \subset O_2) \Rightarrow \mathcal{F}^K_{O_1} \subset \mathcal{F}^K_{O_2} \Rightarrow \max_{f \in \mathcal{F}^K_{O_1}} J(f) \leq \max_{f \in \mathcal{F}^K_{O_2}} J(f)$$

$$0 < u < \|c-d\| \Rightarrow \max_{f \in \mathcal{F}^K_{h_u^{-1}(]0,2[)}} J(f) \leq \max_{f \in \mathcal{F}^K_O} J(f) \,.$$

From that we also have:

$$(\forall O \in \mathcal{O}_K)(\exists u_O > 0)(u < u_O \Rightarrow \max_{f \in \mathcal{F}^K_{h_u^{-1}(]0,2[)}} J(f) \leq \max_{f \in \mathcal{F}^K_O} J(f))$$

$$(\forall u > 0)(\exists O = B(0, r+u))(h_u^{-1}(]0,2[) \subset O, \max_{f \in \mathcal{F}^K_{h_u^{-1}(]0,2[)}} J(f) \leq \max_{f \in \mathcal{F}^K_O} J(f)) \,.$$

Then:

$$\min_{u \in \mathcal{R}^+} \max_{f \in \mathcal{F}^K_{h_u^{-1}(]0,2[)}} J(f) \leq \min_{O \in \mathcal{O}_K} \max_{f \in \mathcal{F}^K_O} J(f) \,.$$

Since:

$$(\forall u \in \mathcal{R}^+)(h_u^{-1}(]0,2[) \in \mathcal{O}_K)$$

calling $\mathcal{O}' = \bigcup_{u \in \mathcal{R}^+} \{h_u^{-1}(]0,2[)\} \subset \mathcal{O}_K$, we have:

$$\min_{u \in \mathcal{R}^+} \max_{f \in \mathcal{F}^K_{h_u^{-1}(]0,2[)}} J(f) = \min_{O \in \mathcal{O}'} \max_{f \in \mathcal{F}^K_O} J(f) \geq \min_{O \in \mathcal{O}_K} \max_{f \in \mathcal{F}^K_O} J(f) \,.$$

Then we have:

$$\min_{u \in \mathcal{R}^+} \max_{f \in \mathcal{F}^K_{h_u^{-1}(]0,2[)}} J(f) = \min_{O \in \mathcal{O}_K} \max_{f \in \mathcal{F}^K_O} J(f) \,.$$

We also have:

$$0 < u_1 < u_2 \Rightarrow h_{u_1}^{-1}(]0,2[) \subset h_{u_2}^{-1}(]0,2[) \Rightarrow \max_{f \in \mathcal{F}^K_{h_{u_1}^{-1}(]0,2[)}} J(f) \le \max_{f \in \mathcal{F}^K_{h_{u_2}^{-1}(]0,2[)}} J(f) .$$

We can then restrict $u \in ]0,r[$. Calling $(\phi_K : \mathcal{R}^n \to \{0,1\})$ the characteristic map of $K$ (definition 293 (p. 341)), using definition 290 (p. 332) and defining as in proposition 614 (p. 345), $L_a \subset \mathcal{Z}^n$, $L^Z_a \subset \mathcal{Z}^n$, $L^{NZ}_a \subset \mathcal{Z}^n$ ($L_a = L^Z_a \cup L^{NZ}_a$) and defining also an $i_a \in \mathcal{Z}^+$ associated to any $a \in \mathcal{R}^+$, we have as definition of $L^Z_a, L^{NZ}_a, L_a$: [35]

$$(\forall a \in \mathcal{R}^+)(i_a = 2r/a \Rightarrow \{$$
$$\ell \in L_a \subset [-i_a, i_a]^n \Leftrightarrow (\exists z \in \prod_{j \in [1,n]} [a\ell_j, a\ell_{j+1}[)(\phi_K(z) = 1),$$
$$\ell \in L^{NZ}_a \subset L_a \Leftrightarrow (z \in \prod_{j \in [1,n]} [a\ell_j, a\ell_{j+1}[ \Rightarrow \phi_K(z) = 1),$$
$$\ell \in L^Z_a \subset L_a \Leftrightarrow ((\exists z \in \prod_{j \in [1,n]} [a\ell_j, a\ell_{j+1}[)(\phi_K(z) = 1),$$
$$(\exists z' \in \prod_{j \in [1,n]} [a\ell_j, a\ell_{j+1}[)(\phi_K(z') = 0))\}) .$$

From definition 290 (p. 332), we have, calling $N_a = Card(L_a)$:

$$(\forall \alpha > 0)(\exists \epsilon_J(\alpha) > 0)(0 < a < \epsilon_J(\alpha), \Rightarrow |J(\phi_K) - a^n N_a| < \alpha) .$$

From proposition 615 (p. 347), we have, calling $N^Z_a = Card(L^z_a)$:

$$(\forall \alpha > 0)(\exists \epsilon_Z(\alpha) > 0)(0 < a < \epsilon_Z(\alpha), \Rightarrow a^n N^Z_a < \alpha) .$$

We also have:

$$(\forall v \in \mathcal{R}^+)(\forall x \in \mathcal{R}^n)(h_u(x) \in ]0,1] \Rightarrow (\exists y \in K)(\|x - y\| < v,$$
$$(\exists \ell \in L_a)(\forall j \in [1,n])(y_j \in [a\ell_j, a\ell_{j+1}[, x_j \in [a\ell_j - v, a\ell_{j+1} + v[))) .$$

Since by definition $f \in \mathcal{F}^K_{h_u^{-1}(]0,2[)}$ has a Riemann integral and is zero when $h_u$ is zero. Since $(\forall x \in \mathcal{R}^n)(f(x) \ge \phi_K(x))$, we have

$$J(f) \ge J(\phi_K) .$$

Let us take $u = a/p$ with $p > 2$. We define $L'_u \subset \mathcal{Z}^n$, $i'_u \in \mathcal{Z}^+$, $N'_u$ by

$$(\forall u \in \mathcal{R}^+)(i'_u = 2r/u,$$
$$\ell \in L'_u \subset [-i'_u, i'_u]^n \Leftrightarrow (\exists z \in \mathcal{R}^n)(\forall j \in [1,n])(z_j \in [u\ell_j, u\ell_{j+1}[), h_u(z) \ne 0)$$

setting:

$$N'_u = Card(L'_u), V^u_K = \bigcup_{\ell \in L'_u} \prod_{j \in [1,n]} [u\ell_j, u\ell_{j+1}[ .$$

[35]When writing $i_a = 2r/a$, it is an extension of the Euclidean division (definition 425 (p. 558)): $i_a$ is the integer such that $2r = i_a \times a + b \times a$ with $b \in [0,1[$.

From propositions 608 (p. 339) and 626 (p. 357), $V^u_K$ has a volume. Calling $\phi_{V^u_K}$ the characteristic map of $V^u_K$, we have by definition of $\mathcal{F}^K_{h_u^{-1}(]0,2[)}$: $(\forall x \in \mathcal{R}^n)(0 \leq f(x) \leq \phi_{V^u_K}(x))$. Then, from proposition 609 (p. 340), we have:

$$J(f) \leq J(\phi_{V^u_K}) = u^n N'_u .$$

Any point of $x \in V^u_K$ has at least a point of $y \in K$ at a distance less than $u$: $\|x - y\| \leq u = a/p$. Then:

$$(\forall x \in V^u_K)(\exists \ell^x \in L_a)(\exists y(x) \in \prod_{j\in[1,n]} [a\ell^x_j, a(\ell^x_j + 1)[= \prod_{j\in[1,n]} [up\ell^x_j, up(\ell^x_j + 1)[)$$
$$(\|x - y(x)\| < u = a/p) .$$

It defines two maps $(y(x) : V^u_K \to K)$ and $(\ell^x : V^u_K \to L_a)$ (see definition 112 (p. 131), point 4). With those maps, since $\|x - y(x)\| < u \Rightarrow (\forall j \in [1, n])(x_j \in ]y_j(x) - u, y_j(x) + u[)$, we have:

$$(\forall x \in V^u_K)(\exists (p^y_j : [1, n] \to [0, p[))(y(x) \in \prod_{j\in[1,n]} [u(p + p^y_j)\ell^x_j, u((p + p^y_j)\ell^x_j + 1)[,$$
$$x \in \prod_{j\in[1,n]} [u((p + p^y_j)\ell^x_j - 1), u((p + p^y_j)\ell^x_j + 1)[) .$$

From that we have:

$$N'_u \leq (p + 2)^n(N^{NZ}_a + N^Z_a) .$$

Then we have:

$$0 < u^n N'_u - a^n N^{NZ}_a \leq u^n((p + 2)^n - p^n)N^{NZ}_a + u^n(p + 2)^n N^Z_a .$$

For $p \geq 2$ we have:

$$(p + 2)^n - p^n \leq 2(2^n - 1)p^{n-1}, (p + 2)^n \leq 2^n p^n .$$

Because for $p \geq 2$, it is true for $n = 1$ and if $(p + 2)^n - p^n \leq 2(2^n - 1)p^{n-1}$ is true.

$$(p+2)^{n+1} - p^n = p(p+2)^n + 2(p+2)^n - p^n \leq 2(2^n - 1)p^n + 2(p+p)^n = 2(2^{n+1} - 1)p^n$$

is true (see proposition 927 (p. 520)). The second is true because $p \geq 2 \Rightarrow (p + 2)^n = p^n(1 + 2/p)^n \leq 2^n p^n$. Then we have:

$$-a^n N^Z_a < u^n N'_u - a^n N_a \leq \frac{2(2^n - 1)}{p} a^n N^{NZ}_a + (2^n + 1)a^n N^Z_a .$$

Using:

$$0 < \epsilon(\alpha) < \min(\epsilon_J(\frac{\alpha}{2}), \epsilon_Z(\frac{\alpha}{4(2^n + 1)}))$$

$0 < a < \epsilon$ and $p > q(\alpha) = \max(8(2^n - 1)a^n N^{NZ}_a/\alpha, 2)$, we have:

$$|u^n N'_u - a^n N_a| < \alpha/2, |a^n N_a - J(\phi_K)| < \alpha/2 .$$

Then we have:

$$|u^n N'_u - J(\phi_K)| < \alpha .$$

Then, combining everything, writing $\mathcal{F}^K_u = \mathcal{F}^K_{h_u^{-1}(]0,2[)}$, we have:

$$(\forall \alpha > 0)((a < \epsilon(\alpha), p > q(\alpha) > 2, f \in \mathcal{F}^K_{a/p}) \\ \Rightarrow J(\phi_K) \leq J(f) \leq u^n N'_u \leq J(\phi_K) + \alpha) .$$

Then using the fact that $0 < u_1 < u_2 \Rightarrow \max_{f \in \mathcal{F}^K_{u_1})} J(f) < \max_{f \in \mathcal{F}^K_{u_2})} J(f)$, we have:

$$(\forall \alpha > 0)(0 < u < \epsilon(\alpha)/q(\alpha) \Rightarrow (f \in \mathcal{F}^K_u \Rightarrow |J(\phi_K) - J(f)| < \alpha) \\ \Rightarrow |J(\phi_K) - \max_{f \in \mathcal{F}^K_u} J(f)| < \alpha) \Rightarrow J(\phi_K) - \min_{u>0} \max_{f \in \mathcal{F}^K_u} J(f) = 0 .$$

**Proposition 649** If an open set has a volume defined by a Riemann integral (see definition 295 (p. 342)), this volume is equal to its Lebesgue measure defined by definition 308 (p. 383).

**Proof**

Using the notation of definition 308 (p. 383), given an open set $O$ with a volume defined by a Riemann integral, we call $\mathcal{K}_O$ the set of compact subsets of $O$. From our definition of Riemann integral (definition 287 (p. 331)), $O$ has to be in a ball $B(0, r)$. We Call $\mu(O)$ the Lebesgue measure of $O$ (definition 308 (p. 383)) and $V(O) = J(\phi_O)$ its volume defined by the Riemann integral (definitions 290 (p. 332), 295 (p. 342)). In $J(\phi_O)$, $(\phi_O : \mathcal{R}^n \to \{0,1\})$ is the characteristic map (definition 293 (p. 341)) of $O$. From definitions 308 (p. 383), 381 (p. 485) and proposition 260 (p. 175), we have:

$$(\forall \alpha > 0)(\exists K(\alpha) \in \mathcal{K}_O)(\mu(O) - \alpha < \mu(K(\alpha)) \leq \mu(O)) .$$

We use proposition 647 (p. 384) together with its notations for $S^\beta(p, q)$ and the associated map $((p_j, q_j) : \mathcal{Z}^{+0} \to \mathcal{Z}^n \times \mathcal{Z}^{+0})$ as defined in equation (2.219) (p. 384). We call $]\![S^\beta(p, q)$ the open defined by:

$$x \in ]\![S^\beta(p, q) \Leftrightarrow (\forall i \in [1, n])(x^i \in ]\frac{p^i - \beta}{2^q}, \frac{p^i + 1 + \beta}{2^q}[) .$$

We have:

$$S^0(p, q) \subset ]\![S^{0.5}(p, q) \subset S^{0.5}(p, q) \subset \overline{S^{0.5}(p, q)} \subset S^1(p, q) .$$

Using equation (2.219) (p. 384), we get:

$$K(\alpha) \subset O = \bigcup_{j \in \mathcal{Z}^{0+}} S^0(p_j, q_j) = \bigcup_{j \in \mathcal{Z}^{0+}} ]\![S^{0.5}(p_j, q_j) = \bigcup_{j \in \mathcal{Z}^{0+}} S^{0.5}(p_j, q_j) \\ = \bigcup_{j \in \mathcal{Z}^{0+}} \overline{S^{0.5}(p_j, q_j)} = \bigcup_{j \in \mathcal{Z}^{0+}} S^1(p_j, q_j) .$$

From definition 140 (p. 160), we also have:

$$(\exists H(K(\alpha)) \subset \mathcal{Z}^{0+})(\mathrm{Card}(H(K(\alpha))) \in \mathcal{Z}^+, K(\alpha) \subset \bigcup_{j \in H(K(\alpha))} ]\![S^{0.5}(p_j, q_j) \\ \subset \bigcup_{j \in H(K(\alpha))} \overline{S^{0.5}(p_j, q_j)} \subset O \subset B(0, r)) .$$

From definition 136 (p. 155) and proposition 206 (p. 154), calling:

$$K'(\alpha) = \bigcup_{j \in H(K(\alpha))} \overline{S^{0.5}(p_j, q_j)}$$

$K'(\alpha)$ is closed. Then, from proposition 521 (p. 290), it is compact and in $\mathcal{K}_O$. Using equation (2.218) (p. 384), we have:

$$\overline{S^{0.5}(p_j, q_j)} = \prod_{i\in[1,n]} [\frac{p^i - 1/2}{2^{q_j}}, \frac{p_j^i + 1 + 1/2}{2^{q_j}}] =$$
$$\prod_{i\in[1,n]} [\frac{2p^i - 1}{2^{q_j+1}}, \frac{2p^i}{2^{q_j+1}}[\cup[\frac{2p^i}{2^{q_j+1}}, \frac{2p^i + 1}{2^{q_j+1}}[\cup[\frac{2p^i + 1}{2^{q_j+1}}, \frac{2p^i + 2}{2^{q_j+1}}[\cup[\frac{2p^i + 2}{2^{q_j+1}}, \frac{2p_j^i + 3}{2^{q_j+1}}] .$$

Then we have:

$$(\exists k'(\alpha) \in \mathcal{Z}^+)(\exists((p_j', q_j') : [0, k'(\alpha)] \to \mathcal{Z}^n \times \mathcal{Z}^{+0}))$$
$$(\bigcup_{j\in H(K(\alpha))} \overline{S^{0.5}(p_j, q_j)} = \overline{\bigcup_{j\in[0,k'(\alpha)]} S^0(p_j', q_j')}, j \neq j' \Rightarrow S^0(p_j', q_j') \cap S^0(p_{j'}', q_{j'}') = \emptyset) .$$

Applying now propositions 648 (p. 386), 612 (p. 343), 626 (p. 357), 608 (p. 339), we have:

$$\mathrm{Vol}(\bigcup_{j\in H(K(\alpha))} ]\![S^{0.5}(p_j, q_j)) = \mathrm{Vol}(\overline{\bigcup_{j\in H(K(\alpha))} ]\![S^{0.5}(p_j, q_j)}) =$$
$$\mathrm{Vol}(\bigcup_{j\in H(K(\alpha))} \overline{S^{0.5}(p_j, q_j)}) = \mathrm{Vol}(K'(\alpha)) = \mu(K'(\alpha)) = J(\phi_{K'(\alpha)}) = \sum_{j\in[1,k'(\alpha)]} 2^{-nq_j'} .$$

Applying equation (2.222) (p. 384), we also have:

$$\begin{aligned}&(q'' \leq q'(\alpha) = \min_{j\in[1,k'(\alpha)]} q_j' \Rightarrow (\exists k''(q'') \in \mathcal{Z}^+)(\exists(p_i'' : [1, k''(q'')] \to \mathcal{Z}^n))\\&(K'(\alpha) = \bigcup_{j\in[1,k''(q'')]} S^0(p_j'', q''), \mathrm{Vol}(K'(\alpha)) = k''(q'')2^{-nq''}) .\end{aligned} \quad (2.226)$$

We call:

$$O'(\alpha) = \bigcup_{j\in H(K(\alpha))} ]\![S^{0.5}(p_j, q_j) .$$

We also call $\mathcal{F}^K_{O'(\alpha)}$ as in definition 306 (p. 382). From definitions 306 (p. 382) and 308 (p. 383), from proposition 609 (p. 340), we have:

$$\mu(K(\alpha)) \leq \max_{f\in\mathcal{F}^K_{O'(\alpha)}} J(f) \leq \mathrm{Vol}(O'(\alpha)) = \mu(K'(\alpha)) \leq \mu(O) . \quad (2.227)$$

Since $O$ has a volume, we have:

$$\begin{aligned}&(\forall\alpha > 0)(\exists q^o(\alpha) > 0)\{q > q^o(\alpha) \Rightarrow (\exists P^{nz}(q) \subset \mathcal{Z}^n)\\&(p \in P^{nz}(q) \Rightarrow S^0(p, q) \subset O, p \notin P^{nz}(q) \Rightarrow S^0(p, q) \cap \mathcal{C}(O) \neq \emptyset,\\&\mathrm{Card}(P^{nz}(q)) \in \mathcal{Z}^+, 0 < \mathrm{Vol}(O) - \mathrm{Card}(P^{nz}(q))2^{-nq} < \alpha)\} .\end{aligned} \quad (2.228)$$

Choosing $q < \min(q'(\alpha), q^o(\alpha))$ common for relations (2.226) (p. 390) and (2.228) (p. 390) since in that case for any $j \in [1, k''(q)]$, $\overline{S^0(p_j'', q'')} \subset O$ and using the corresponding $P^{nz}(q))$, we have:

$$(\forall j \in [1, k''(q)])(S^0(p_j'', q'') \subset \bigcup_{p\in P^{nz}(q)} S^0(p, q) \Rightarrow p_j'' \in P^{nz}(q)) .$$

Then there is a map:

$$(k(j) : [1, k''(q)] \to P^{nz}(q))$$

and a map:

$$(k(j) : [k''(q) + 1, Card(P^{nz}(q))] \to P^{nz}(q))$$

such that:

$$K'(\alpha) = \bigcup_{j\in:[1,k''(q)]} \overline{S^0(p_{k(j)}, q)} \;.$$

When $k''(q) < Card(P^{nz}(q))$, we define $\epsilon$ such as:

$$0 < \epsilon < \min(0.5, \alpha)/(Card(P^{nz}(q)) - k''(q)) < 1 \;.$$

Using the definition of **Bx** given in definition 279 (p. 320), and setting $e_j = \delta^i_j$ (definition 215 (p. 245)), a basis of $\mathcal{R}^n$, we consider the compact set $K^R(\alpha)$ defined by:

$$k''(q) = Card(P^{nz}(q)) \Rightarrow K^R(\alpha) = K'(\alpha), k''(q) < Card(P^{nz}(q)) \Rightarrow$$
$$(K^R(\alpha) = K'(\alpha) \cup \bigcup_{j\in[k''(q)+1,Card(P^{nz}(q))]} \overline{Bx(p''^i_j 2^{-q} e_i, (p''^i_j + (1-\epsilon)^{1/n}) 2^{-q} e_i)}\;)$$

since $0 < \epsilon < 1$, we have:

$$K(\alpha) \subset K'(\alpha) \subset K^R(\alpha) \subset O$$

and we have:

$$Vol(K^R(\alpha)) = Vol(K'(\alpha)) + (Card(P^{nz}(q)) - k''(q))(1-\epsilon)2^{-nq}$$

$$= k''(q)2^{-nq} + (Card(P^{nz}(q)) - k''(q))(1-\epsilon)2^{-nq} \;.$$

Then we have:

$$0 < Vol(O) - Vol(K^R(\alpha)) =$$
$$Vol(O) - Card(P^{nz}(q))2^{-nq} + Card(P^{nz}(q))2^{-nq} - Vol(K^R(\alpha))$$
$$< \alpha + (Card(P^{nz}(q)) - k''(q))\epsilon 2^{-nq} < 2\alpha \;.$$

Since:

$$\mu(K(\alpha)) \le \mu(K^R(\alpha)) = Vol(K^R(\alpha)) \le Vol(O)$$

and

$$\mu(O) - \mu(K^R(\alpha)) < \mu(O) - \mu(K(\alpha)) < \alpha$$

we have:

$$(\forall\alpha > 0)(|\mu(O) - vol(O)| \le |\mu(O) - \mu(K^R(\alpha))| + |Vol(K^R(\alpha)) - Vol(O)|$$
$$< \alpha + 2\alpha = 3\alpha) \;.$$

Then we have:

$$\mu(O) = Vol(O) \;.$$

**Definition 310 Lebesgue measure of an element of the Borel $\sigma$-algebra of $\mathcal{R}^n$ (definition 303 (p. 378)).**

Given a set **B** member of the Borel **σ-algebra** of $\mathcal{R}^n$, calling $\mathcal{O}_B$ the set of opens containing B. The measure of **B** is:

$$\mu(B) = \min_{O \in \mathcal{O}_B} \mu(O)$$

**Definition 311 Lebesgue integral of a positive continuous map over an an element of the Borel σ-algebra of $\mathcal{R}^n$.**

With the same notation as in definition 310 (p. 391), given a set **B** member of the Borel **σ-algebra** of $\mathcal{R}^n$, one calls integral over B of a positive continuous map $(g : \mathcal{R}^n \to \mathcal{R}^{0+})$ defined at least on O, the following quantity:

$$\mu_g(O) = \min_{O \in \mathcal{O}_B} \mu_g(O) \ .$$

## 2.28 Taylor expansion of functions with derivatives

**Proposition 650** Given a closed interval C, a function $f : C \to \mathcal{R}$ and two numbers x, a in C, we suppose f has continuous derivatives up to the order n at each point of C. We have the following equalities:

$$f(x) = \sum_{i \in [0,n]} \frac{(x-a)^i}{i!} \frac{d^i f(a)}{dx^i} + \eta_n(x,a) \frac{(x-a)^n}{n!} \tag{2.229}$$

with

$$\frac{(x-a)^n}{n!} \eta_n(x,a) = [\prod_{i \in [1,n]} \int_a^x dx] (\frac{d^n f(x)}{dx^n} - \frac{d^n f(a)}{dx^n}) \tag{2.230}$$

and we also have:

$$(\exists c \in ]a,x[)(\eta_n(x,a) = n!(\frac{d^n f(c)}{dx^n} - \frac{d^n f(a)}{dx^n})) \tag{2.231}$$

and

$$(\forall \alpha)(\exists \epsilon)(\forall a \in C)(\forall x \in C)(|x-a| < \epsilon \Rightarrow |\eta_n(x,a)| < \alpha) \ . \tag{2.232}$$

**Proof**

1. Equation (2.230) (p. 392). From propositions 309 (p. 194), 326 (p. 202) and 634 (p. 367) we have:

$$[\prod_{i \in [1,n+1]} \int_a^x dx] (\frac{d^{n+1} f(x)}{dx^{n+1}} - \frac{d^{n+1} f(a)}{dx^{n+1}}) =$$
$$[\prod_{i \in [1,n]} \int_a^x dx] (\frac{d^n f(x)}{dx^n} - \frac{d^n f(a)}{dx^n} - (x-a) \frac{d^{n+1} f(a)}{dx^{n+1}}) =$$
$$[\prod_{i \in [1,n]} \int_a^x dx] (\frac{d^n f(x)}{dx^n} - \frac{d^n f(a)}{dx^n}) - \frac{(x-a)^{n+1}}{(n+1)!} \frac{d^{n+1} f(a)}{dx^{n+1}} \ .$$

Then using that relation, supposing the following is true for n:

$$f(x) - \sum_{i \in [0,n]} \frac{(x-a)^i}{i!} \frac{d^i f(a)}{dx^i} = [\prod_{i \in [1,n]} \int_a^x dx] (\frac{d^n f(x)}{dx^n} - \frac{d^n f(a)}{dx^n})$$

it is true for $\mathbf{n+1}$:

$$f(x) - \sum_{i\in[0,n]} \frac{(x-a)^i}{i!}\frac{d^i f(a)}{dx^i} =$$

$$[\prod_{i\in[1,n+1]} \int_a^x dx](\frac{d^{n+1}f(x)}{dx^{n+1}} - \frac{d^{n+1}f(a)}{dx^{n+1}}) + \frac{(x-a)^{n+1}}{(n+1)!}\frac{d^{n+1}f(a)}{dx^{n+1}} .$$

Since it is true for $\mathbf{n} = \mathbf{1}$ (proposition 634 (p. 367)), it is true for any $\mathbf{n}$ from the recurrence principle (proposition 927 (p. 520)).

2. Equation (2.231) (p. 392). We have applying proposition 638 (p. 368), there is a strictly decreasing map (definition 386 (p. 486)) $(c_i : [2,n] \rightarrow ]a,x[)$ such that:

$$\begin{aligned}\frac{(x-a)^n}{n!}\eta_n(x,a) &= \int_a^x dx \int_a^x dx[\prod_{i\in[3,n]} \int_a^x dx](\frac{d^n f(x)}{dx^n} - \frac{d^n f(a)}{dx^n})\\ &= (x-a)\int_a^{c_2} dx[\prod_{i\in[3,n]} \int_a^x dx](\frac{d^n f(x)}{dx^n} - \frac{d^n f(a)}{dx^n})\\ &= (x-a)^2\int_a^{c_3} dx[\prod_{i\in[4,n]} \int_a^x dx](\frac{d^n f(x)}{dx^n} - \frac{d^n f(a)}{dx^n})\\ &= (x-a)^n(\frac{d^n f(c_n)}{dx^n} - \frac{d^n f(a)}{dx^n})\end{aligned}$$

with $2 \leq i < j \leq n \Rightarrow a < c_i < c_j < x$. Then $c_n \in ]a,x[$. Then we have equation (2.231). Since $\frac{d^n f(x)}{dx^n}$ is continuous on $\mathbf{C}$ which is compact, it is uniformly continuous. Then:

$$(\forall\alpha)(\exists\epsilon)(\forall a \in C)(\forall x \in C)(|x-a| < \epsilon \Rightarrow n!|\frac{d^n f(x)}{dx^n} - \frac{d^n f(a)}{dx^n}| < \alpha)$$

coupled with equation (2.231), it gives equation (2.232).
Then we have proposition 650.

**Proposition 651** If in proposition 650, $\mathbf{f}$ is a function infinitely derivable that is to say of $C^\infty$ (see 277 (p. 304)), for all $\mathbf{n}$ in equation (2.230) $\eta_n(x,a)$ is also of $C^\infty$ in the interval of definition of $\mathbf{f}$.

**Proof**
Since with $\frac{d^0 f(x)}{dx^0} = f(x)$ we have:

$$(\forall q \in \mathcal{Z}^{0+})(\frac{d^q f(a)}{dx^q} \in C^\infty) .$$

In the expansion as defined in proposition 650, we may write when $x \neq a$ and $q \geq 0$:

$$\eta_n^q(x,a) = \frac{n!}{(x-a)^n}(\frac{d^q f(x)}{dx^q} - \sum_{i\in[0,n]} \frac{(x-a)^i}{i!}\frac{d^{i+q}f(a)}{dx^{i+q}}) . \tag{2.233}$$

For $x \neq a$, using proposition 325 (p. 201), 326 (p. 202) with $f(x) \in C^\infty$ there is a derivative at all order for the second member of equation (2.233). We have to verify that all the derivatives

of $\eta_n(x,a)$ are also continuous at $x = a$.
**The first order derivative of $\eta_n^q(x,a)$ is continuous at $x = a$.**
We have expanding proposition 650 (p. 392) to $n+1$:

$$\eta_n(x,a)\frac{(x-a)^n}{n!} = \frac{(x-a)^{n+1}}{(n+1)!}\frac{d^{n+1}f(a)}{dx^{n+1}} + \eta_{n+1}(x,a)\frac{(x-a)^{n+1}}{(n+1)!}\ .$$

Then we have:

$$\eta_n(x,a) = \frac{(x-a)}{n+1}\frac{d^{n+1}f(a)}{dx^{n+1}} + \eta_{n+1}(x,a)\frac{x-a}{n+1}\ . \tag{2.234}$$

Since $\eta_n(a,a) = 0$, we get:

$$\frac{\eta_n(x,a)-\eta_n(a,a)}{x-a} = \frac{1}{n+1}\frac{d^{n+1}f(a)}{dx^{n+1}} + \frac{\eta_{n+1}(x,a)}{n+1}\ .$$

It means:

$$\frac{\eta_n(x,a)-\eta_n(a,a)}{x-a} - \frac{1}{n+1}\frac{d^{n+1}f(a)}{dx^{n+1}} = \frac{\eta_{n+1}(x,a)}{n+1}\ .$$

For all $\alpha$, we have $\epsilon > 0$ such that $|x-a| < \epsilon \Rightarrow |\eta_{n+1}(x,a)| < (n+1)\alpha$. Using that we may write:

$$|x-a| < \epsilon \Rightarrow |\frac{\eta_n(x,a)-\eta_n(a,a)}{x-a} - \frac{1}{n+1}\frac{d^{n+1}f(a)}{dx^{n+1}}| < \alpha\ .$$

Then the derivative at $x = a$ of $\eta_n(x,a)$ is:

$$\frac{d\eta_n(a,a)}{dx} = \frac{1}{n+1}\frac{d^{n+1}f(a)}{dx^{n+1}}\ .$$

Since all the above reasoning is the same for $\frac{d^qf(x)}{dx^q}$, we have:

$$q \in \mathcal{Z}^{0+} \Rightarrow \frac{d\eta_n^q(a,a)}{dx} = \frac{1}{n+1}\frac{d^{n+1+q}f(a)}{dx^{n+1+q}}\ . \tag{2.235}$$

Now we have to prove that this expression is also the limit at $x = a$ of the $\frac{d\eta_n^q(x,a)}{dx}$ defined by formula (2.233) (p. 393):

$$\eta_n(x,a)\frac{(x-a)^{n-1}}{(n-1)!} + \frac{d\eta_n(x,a)}{dx}\frac{(x-a)^n}{n!} = \frac{df(x)}{dx} - \sum_{i\in[0,n-1]}\frac{(x-a)^i}{i!}\frac{d^{i+1}f(a)}{dx^{i+1}}\ .$$

But using (2.233) (p. 393) again for $\frac{df(x)}{dx}$, we have:

$$\eta_n^1(x,a)\frac{(x-a)^n}{n!} = \frac{df(x)}{dx} - \sum_{i\in[0,n-1]}\frac{(x-a)^i}{i!}\frac{d^{i+1}f(a)}{dx^{i+1}} + \frac{(x-a)^n}{n!}\frac{d^{n+1}f(a)}{dx^{n+1}}\ .$$

Then we have for $x \neq a$:

$$\eta_n(x,a) + \frac{d\eta_n(x,a)}{dx}\frac{(x-a)}{n} = (\frac{d^{n+1}f(a)}{dx^{n+1}} + \eta_n^1(x,a))\frac{(x-a)}{n}\ .$$

Using equation (2.234) (p. 394), we have:

$$\frac{(x-a)}{n+1}\frac{d^{n+1}f(a)}{dx^{n+1}} + \eta_{n+1}(x,a)\frac{x-a}{n+1} + \frac{d\eta_n(x,a)}{dx}\frac{(x-a)}{n} =$$
$$(\frac{d^{n+1}f(a)}{dx^{n+1}} + \eta_n^1(x,a))\frac{(x-a)}{n} .$$

Then, when $x \neq a$, we have:

$$\frac{d\eta_n(x,a)}{dx} = \frac{1}{n+1}\frac{d^{n+1}f(a)}{dx^{n+1}} + \eta_n^1(x,a) - \frac{n}{n+1}\eta_{n+1}(x,a) .$$

Since $\frac{d^q f(x)}{dx^q} \in C^\infty$, we may replace $f$ in the above expression by $\frac{d^q f(x)}{dx^q}$ and using the equation (2.233) (p. 393), we get:

$$\frac{d\eta_n^q(x,a)}{dx} = \frac{1}{n+1}\frac{d^{n+1+q}f(a)}{dx^{n+1+q}} + \eta_n^{q+1}(x,a) - \frac{n}{n+1}\eta_{n+1}^q(x,a) . \tag{2.236}$$

Then, setting in above $q = 0$ and applying equation (2.231) (p. 392) to $\eta_n^1(x,a)$ and to $\eta_{n+1}$, there is $c_0 \in ]a,x[, c_1 \in ]a,x[$ such that:

$$\frac{d\eta_n(x,a)}{dx} = \frac{1}{n+1}\frac{d^{n+1}f(a)}{dx^{n+1}} + n!(\frac{d^{n+1}f(c_1)}{dx^{n+1}} - \frac{d^{n+1}f(a)}{dx^{n+1}})$$
$$- n \times n!(\frac{d^{n+1}f(c_0)}{dx^{n+1}} - \frac{d^{n+1}f(a)}{dx^{n+1}})$$

or

$$\begin{aligned}\frac{d\eta_n^q(x,a)}{dx} &= \frac{1}{n+1}\frac{d^{n+1+q}f(a)}{dx^{n+1+q}} + n!(\frac{d^{n+1+q}f(c_1)}{dx^{n+1+q}} - \frac{d^{n+1+q}f(a)}{dx^{n+1+q}}) \\ &- n \times n!(\frac{d^{n+1+q}f(c_0)}{dx^{n+1+q}} - \frac{d^{n+1+q}f(a)}{dx^{n+1+q}}) .\end{aligned} \tag{2.237}$$

And since that when $x$ is going to $a$, $c_0$ and $c_1$ are also going to $a$, we have:

$$\lim_{x \to a} \frac{d\eta_n^q(x,a)}{dx} = \frac{1}{n+1}\frac{d^{n+1+q}f(a)}{dx^{n+1+q}} .$$

**All order derivatives of $\eta_n^q(x,a)$ are continuous at $x = a$.**
Deriving equation (2.236) (p. 395) $p$ times, we have for $p \geq 1$:

$$\frac{d^{p+1}\eta_n^q(x,a)}{dx^{p+1}} = \frac{d^p\eta_n^{q+1}(x,a)}{dx^p} - \frac{n}{n+1}\frac{d^p\eta_{n+1}^q(x,a)}{dx^p} . \tag{2.238}$$

Let us suppose that: for $1 \leq p' \leq p$ and $\forall n \in \mathcal{Z}^+, \forall q \in \mathcal{Z}^+$ with all $c_i^{n,p',q} \in ]a,x[$, $h_i^{n,p',q} \in \mathcal{R}$

$$\begin{aligned}\frac{d^{p'}\eta_n^q(x,a)}{dx^{p'}} &= \frac{p'!n!}{(n+p')!}\frac{d^{n+p'+q}f(a)}{dx^{n+p'+q}} \\ &+ \sum_{i \in [0,2^{p'}-1]} (\frac{d^{n+p'+q}f(c_i^{n,p',q})}{dx^{n+p'+q}} - \frac{d^{n+p'+q}f(a)}{dx^{n+p'+q}})h_i^{n,p',q} \\ \frac{d^{p'}\eta_n^q(a,a)}{dx^{p'}} &= \frac{p'!n!}{(n+p')!}\frac{d^{n+p'+q}f(a)}{dx^{n+p'+q}} = \lim_{x \to a}\frac{d^{p'}\eta_n^q(x,a)}{dx^{p'}} .\end{aligned} \tag{2.239}$$

From equation (2.237), (2.239) is true for $\mathbf{p = p' = 1}$. Using equations (2.238) and (2.239) for $\mathbf{p' = p}$, let us show that (2.239) is true for $\mathbf{p' = p + 1}$:

$$\frac{d^{p+1}\eta_n^q(x,a)}{dx^{p+1}} = \frac{p!n!}{(n+p)!}\frac{d^{n+p+q+1}f(a)}{dx^{n+p+1}} - \frac{n}{n+1}\times\frac{p!(n+1)!}{(n+p+1)!}\frac{d^{n+p+q+1}f(a)}{dx^{n+p+1}}$$

$$+\sum_{i\in[0,2^p-1]}\left(\frac{d^{n+p+q+1}f(c_i^{n,p,q+1})}{dx^{n+p+q+1}} - \frac{d^{n+p+q+1}f(a)}{dx^{n+p+q+1}}\right)h_i^{n,p,q+1}$$

$$-\frac{n}{n+1}\sum_{i\in[0,2^p-1]}\left(\frac{d^{n+p+q+1}f(c_i^{n+1,p,q})}{dx^{n+p+q+1}} - \frac{d^{n+p+q+1}f(a)}{dx^{n+p+q+1}}\right)h_i^{n+1,p,q}$$

$$\frac{d^{p+1}\eta_n^q(x,a)}{dx^{p+1}} = \frac{(p+1)!n!}{(n+p+1)!}\frac{d^{n+p+1}f(a)}{dx^{n+p+1}}$$

$$+\sum_{i\in[0,2^{p+1}-1]}\left(\frac{d^{n+p+q+1}f(c_i^{n,p+1,q})}{dx^{n+p+q+1}} - \frac{d^{n+p+q+1}f(a)}{dx^{n+p+q+1}}\right)h_i^{n,p+1,q}\,.$$

Moreover, using equations (2.238) and (2.239), we have:

$$\frac{d^{p+1}\eta_n^q(a,a)}{dx^{p+1}} = \lim_{x\to a}\frac{\frac{d^p\eta_n^q(x,a)}{dx^p} - \frac{p!n!}{(n+p)!}\frac{d^{n+p+q}f(a)}{dx^{n+p+q}}}{x-a}$$

$$= \lim_{x\to a}\frac{\frac{d^{p-1}\eta_n^{q+1}(x,a)}{dx^{p-1}} - \frac{n}{n+1}\frac{d^{p-1}\eta_{n+1}^q(x,a)}{dx^{p-1}} - \frac{p!n!}{(n+p)!}\frac{d^{n+p+q}f(a)}{dx^{n+p+q}}}{x-a}$$

$$= \lim_{x\to a}\frac{\frac{d^{p-1}\eta_n^{q+1}(x,a)}{dx^{p-1}} - \frac{(p-1)!n!}{(n+p-1)!}\frac{d^{n+p+q}f(a)}{dx^{n+p+q}} - \frac{n}{n+1}\left(\frac{d^{p-1}\eta_{n+1}^q(x,a)}{dx^{p-1}} - \frac{(p-1)!n!}{(n+p)!}\frac{d^{n+p+q}f(a)}{dx^{n+p+q}}\right)}{x-a}$$

$$= \frac{d^p\eta_n^{q+1}(a,a)}{dx^p} - \frac{n}{n+1}\frac{d^p\eta_{n+1}^q(a,a)}{dx^p}$$

$$= \left(\frac{p!n!}{(n+p)!} - \frac{n}{n+1}\frac{p!(n+1)!}{(n+p+1)!}\right)\frac{d^{n+p+q+1}f(a)}{dx^{n+p+q+1}}$$

$$= \frac{(p+1)!n!}{(n+p+1)!}\frac{d^{n+p+q+1}f(a)}{dx^{n+p+q+1}} = \lim_{x\to a}\frac{d^{p+1}\eta_n^q(x,a)}{dx^{p+1}}\,.$$

Then (2.239) is also true when $\mathbf{1 \le p' \le p + 1}$ then by the recurrence principle (proposition 927 (p. 520)) it is true for any $\mathbf{p' \ge 1}$. Proposition 651 is true.

**Proposition 652** $\mathbf{f(x,y)}$ being a continuous function with continuous derivatives up to the second order defined on a closed box of $\mathcal{R}^2$ ($\mathbf{f(x,y)} : \mathcal{R}^2 \to \mathcal{R}$), then we have $\frac{\partial^2 f(x,y)}{\partial x\partial y} = \frac{\partial^2 f(x,y)}{\partial y\partial x}$.

**Proof**
Using (2.229) with $(\mathbf{x_1, y_1})$, $(\mathbf{x_2, y_2})$ in the closed box defining $\mathbf{f(x,y)}$ and setting by convention:

$$c_{u,v}^i \in ]u,v[$$

$$f(x_2,y_2) = f(x_1,y_2) + (x_2-x_1)\frac{\partial f(x_1,y_2)}{\partial x} + \frac{(x_2-x_1)^2}{2!}\frac{\partial^2 f(c_{x_1,x_2},y_2)}{\partial x^2}$$

then applying (2.229) a second time:

$$\begin{aligned} f(x_2,y_2) &= f(x_1,y_1) + (y_2-y_1)\frac{\partial f(x_1,y_1)}{\partial x} + \frac{(y_2-y_1)^2}{2!}\frac{\partial^2 f(x_1,c_{y_1,y_2})}{\partial x^2} \\ &+ (x_2-x_1)(\frac{\partial f(x_1,y_1)}{\partial x} + (y_2-y_1)\frac{\partial^2 f(x_1,c^2_{y_1,y_2})}{\partial y\partial x}) \\ &+ \frac{(x_2-x_1)^2}{2!}\frac{\partial^2 f(c_{x_1,x_2},y_2)}{\partial x^2}\,. \end{aligned} \tag{2.240}$$

Since we can reverse the order of applying (2.229) by starting with $\mathbf{y}$, we have the equality from the two equations of type (2.240):

$$\begin{aligned} &\frac{(y_2-y_1)^2}{2!}\frac{\partial^2 f(x_1,c_{y_1,y_2})}{\partial y^2} + (x_2-x_1)(y_2-y_1)\frac{\partial^2 f(x_1,c'_{y_1,y_2})}{\partial y\partial x} + \\ &\frac{(x_2-x_1)^2}{2!}\frac{\partial^2 f(c_{x_1,x_2},y_2)}{\partial x^2} = \frac{(y_2-y_1)^2}{2!}\frac{\partial^2 f(x_1,c''_{y_1,y_2})}{\partial y^2} + \\ &(x_2-x_1)(y_2-y_1)\frac{\partial^2 f(c'_{x_1,x_2},y_1)}{\partial x\partial y} + \frac{(x_2-x_1)^2}{2!}\frac{\partial^2 f(c''_{x_1,x_2},y_2)}{\partial x^2}\,. \end{aligned}$$

Since, together with the map itself, all the derivatives of the map up to the second order are continuous on a compact subset of $\mathcal{R}^2$, they are uniformly continuous. Then for all $\alpha > 0$, it exists $\epsilon > 0$ such that:

$$\begin{aligned} &|x_1-x_2| < \epsilon, |y_1-y_2| < \epsilon \Rightarrow \\ &|\frac{\partial^2 f(x_1-y_1)}{\partial x\partial y} - \frac{\partial^2 f(x_2-y_2)}{\partial x\partial y}| < \alpha, |\frac{\partial^2 f(x_1-y_1)}{\partial y\partial x} - \frac{\partial^2 f(x_2-y_2)}{\partial y\partial x}| < \alpha, \\ &|\frac{\partial^2 f(x_1-y_1)}{\partial x^2} - \frac{\partial^2 f(x_2-y_2)}{\partial x^2}| < \alpha, |\frac{\partial^2 f(x_1-y_1)}{\partial y^2} - \frac{\partial^2 f(x_2-y_2)}{\partial y^2}| < \alpha\,. \end{aligned} \tag{2.241}$$

Setting $\mathbf{y_2 = y_1 + u, x_2 = x_1 + u}$ we have for $|\mathbf{u}| > 0$:

$$\begin{aligned} &\frac{\partial^2 f(x_1,c_{y_1,y_2})}{\partial y^2} + \frac{\partial^2 f(c_{x_1,x_2},y_2)}{\partial x^2} - \frac{\partial^2 f(x_1,c''_{y_1,y_2})}{\partial y^2} - \frac{\partial^2 f(c''_{x_1,x_2},y_2)}{\partial x^2} = \\ &2!(\frac{\partial^2 f(c'_{x_1,x_2},y_1)}{\partial y\partial x} - \frac{\partial^2 f(x_1,c'_{y_1,y_2})}{\partial y\partial x})\,. \end{aligned} \tag{2.242}$$

By definition we have:

$$|C^i_{v,w} - v| < |v-w|, |C^i_{v,w} - w| < |v-w|, |C^i_{v,w} - C^j_{v,w}| < |v-w|\,.$$

Setting $0 < |\mathbf{u}| < \epsilon$, we have from relation (2.241) and (2.242):

$$|\frac{\partial^2 f(c'_{x_1,x_2},y_1)}{\partial x\partial y} - \frac{\partial^2 f(x_1,c'_{y_1,y_2})}{\partial y\partial x}| < \alpha\,. \tag{2.243}$$

We also have, for $0 < |\mathbf{u}| < \epsilon$:

$$|\frac{\partial^2 f(c'_{x_1,x_2},y_1)}{\partial x\partial y} - \frac{\partial^2 f(x_1,y_1)}{\partial x\partial y}| < \alpha, |\frac{\partial^2 f(x_1,c'_{y_1,y_2})}{\partial y\partial x} - \frac{\partial^2 f(x_1,y_1)}{\partial y\partial x}| < \alpha\,.$$

It gives:

$$\left|\frac{\partial^2 f(c'_{x_1,x_2}, y_1)}{\partial x \partial y} - \frac{\partial^2 f(x_1, c'_{y_1,y_2})}{\partial y \partial x} + \frac{\partial^2 f(x_1, y_1)}{\partial y \partial x} - \frac{\partial^2 f(x_1, y_1)}{\partial x \partial y}\right| < 2\alpha \,.$$

It means (proposition 404 (p. 239)):

$$\left|\frac{\partial^2 f(x_1, y_1)}{\partial x \partial y} - \frac{\partial^2 f(x_1, y_1)}{\partial y \partial x}\right| - \left|\frac{\partial^2 f(c'_{x_1,x_2}, y_1)}{\partial x \partial y} - \frac{\partial^2 f(x_1, c'_{y_1,y_2})}{\partial y \partial x}\right| < 2\alpha$$

then from relation (2.243):

$$\left|\frac{\partial^2 f(x_1, y_1)}{\partial x \partial y} - \frac{\partial^2 f(x_1, y_1)}{\partial y \partial x}\right| < 3\alpha \,.$$

Since the **lower_limit** of possible $\alpha$ is zero, we have:

$$\frac{\partial f(x_1, y_1)}{\partial x \partial y} = \frac{\partial f(x_1, y_1)}{\partial y \partial x} \,.$$

**Definition 312 Product of a finite set of partial derivatives.**

A partial derivative acting on the set $\mathbf{F}$ of $\mathbf{C}^\infty$ (see 276 (p. 304)) maps $(\mathbf{f} : \mathcal{R}^n \to \mathcal{R})$, can be considered as an element of $\mathcal{F}$: the set of maps $\mathbf{F} \to \mathbf{F}$. Using definitions 131 (p. 146) and 275 (p. 304), we can define, from a map $(\mathbf{i_k} : [\mathbf{1}, \mathbf{p}] \to [\mathbf{1}, \mathbf{n}])$ with $\mathbf{p} \in \mathcal{Z}^+$, a map $(\partial_{\mathbf{i_k}} : [\mathbf{1}, \mathbf{p}] \to \mathcal{F})$ and then a product:

$$\overset{\circ}{\prod_{k \in [1,p]}} \partial_{i_k} \,.$$

**Proposition 653** Using notation of definitions 276 (p. 304) and 312 (p. 398), we have:

$$\overset{\circ}{\prod_{k \in [1,p]}} \partial_{i_k} = \overset{\circ}{\prod_{k \in [1,p]}} \partial_{i_{P(k)}}$$

where $(\mathbf{P(k)} : [\mathbf{1}, \mathbf{p}] \leftrightarrow [\mathbf{1}, \mathbf{p}])$ is any bijection (permutation) of $[\mathbf{1}, \mathbf{p}]$ on $[\mathbf{1}, \mathbf{p}]$ (see definitions 124 (p. 139), 172 (p. 207)).

**Proof**

It is a direct consequence of propositions 349 (p. 208) and 652 (p. 396).

**Convention 21** $\prod_{k \in [1,p]} \partial_{i_k}$.

Since the order does not matter, we may write:

$$\overset{\circ}{\prod_{k \in [1,p]}} \partial_{i_k} = \prod_{k \in [1,p]} \partial_{i_k} \,.$$

**Convention 22** $\partial^0_k$, $\partial^p_k$.

With $(\mathbf{h}:\mathcal{R}^{\mathbf{n}}\to\mathcal{R})\in\mathbf{C}^{\infty},\mathbf{p}\in\mathcal{Z}^{+}$ (see definition 276 (p. 304)), we set:

$$\partial_{\mathbf{k}}^{0}\mathbf{h}=\mathbf{h},\partial_{\mathbf{k}}^{\mathbf{p}}\mathbf{h}=(\prod_{\mathbf{i}\in[1,\mathbf{p}]}\partial_{\mathbf{k}})\mathbf{h}\,.$$

**Proposition 654** Using the notations of definition 276 (p. 304) and of 312 (p. 398), calling $(\mathbf{J_k}=\mathbf{i}^{-1}(\mathbf{k}):[1,\mathbf{n}]\to\mathcal{P}([1,\mathbf{p}]))$ the partition (definition 370 (p. 471)) defined by $(\mathbf{i}(\mathbf{k}):[1,\mathbf{p}]\to[1,\mathbf{n}])$, we have:

$$\prod_{\mathbf{k}\in[1,\mathbf{p}]}\partial_{\mathbf{i_k}}=\prod_{\mathbf{i}\in[1,\mathbf{n}]}\partial_{\mathbf{i}}^{\mathbf{Card}(\mathbf{J_i})}\,.$$

**Proof**
From definition 370 (p. 471), and proposition 816 (p. 471), we have:

$$\bigcup_{\mathbf{i}\in[1,\mathbf{n}]}\mathbf{J_i}=[1,\mathbf{p}],\mathbf{k}\neq\mathbf{l}\Rightarrow\mathbf{J_k}\cap\mathbf{J_l}=\emptyset\,.$$

Then there is a unique isomorphism (definition 387 (p. 487)) from $[1,\mathbf{p}]$ to $\bigcup_{\mathbf{i}\in[1,\mathbf{n}]}\mathbf{J_i}$ with the order on $\bigcup_{\mathbf{i}\in[1,\mathbf{n}]}\mathbf{J_i}$ as defined in proposition 190 (p. 147). Using proposition 653 (p. 398) together with proposition 190 (p. 147), we have proposition 654 (p. 399) using convention 22 (p. 398).

**Proposition 655** We consider a continuous function $\mathbf{f}(\mathbf{x},\mathbf{X})$: $(\mathbf{f}:\mathcal{R}\times\mathcal{R}^{\mathbf{m}}\to\mathcal{R})$ defined in a close interval $\mathbf{C_x}\subset\mathcal{R}$ for $\mathbf{x}$ and a compact set of $\mathbf{C_X}\subset\mathcal{R}^{\mathbf{m}}$ for $\mathbf{X}$. We suppose that when $\mathbf{f}(\mathbf{x},\mathbf{X})$ is defined, it is a smooth function in all variables: $\mathbf{f}(\mathbf{x},\mathbf{X})\in\mathbf{C}^{\infty}$ (see 276 (p. 304)). When the Taylor expansion is made on the first variable $\mathbf{x}$, the other $(\mathbf{X})$ remaining constant, as defined in proposition 651, the function $(\eta_{\mathbf{n}}(\mathbf{x},\mathbf{a},\mathbf{X}):\mathcal{R}^{2}\times\mathcal{R}^{\mathbf{m}}\to\mathcal{R})$ is derivable to all orders: it is in $\mathbf{C}^{\infty}$ whenever $\mathbf{f}$ is defined.

**Proof**
Given any derivative operator on $\mathbf{f}(\mathbf{x},\mathbf{X})$, from proposition 653 (p. 398), we can put the derivative over $\mathbf{x}$ first: any derivative operator on $\mathbf{x}$ and $\mathbf{X}$ can be written as (see definition 275 (p. 304)) $\partial_{\mathbf{x}}^{\mathbf{p}}\Delta_{\mathbf{X}}$, where $\Delta_{\mathbf{X}}$ is a product of derivative operators acting only on the $\mathbf{X}$ variable in $\mathbf{C_X}\subset\mathcal{R}^{\mathbf{m}}$. As for proposition 651 (p. 393), using proposition 650 (p. 392) for $\mathbf{n}$, we have:

$$\eta_{\mathbf{n}}(\mathbf{x},\mathbf{a},\mathbf{X})=\frac{\mathbf{n}!}{(\mathbf{x}-\mathbf{a})^{\mathbf{n}}}(\mathbf{f}(\mathbf{x},\mathbf{X})-\sum_{\mathbf{i}\in[0,\mathbf{n}]}\frac{(\mathbf{x}-\mathbf{a})^{\mathbf{i}}}{\mathbf{i}!}\partial_{\mathbf{x}}^{\mathbf{i}}\mathbf{f}(\mathbf{a},\mathbf{X}))\,. \tag{2.244}$$

It allows to define when $\mathbf{x}\neq\mathbf{a}$ any derivative of $\eta_{\mathbf{n}}(\mathbf{x},\mathbf{a},\mathbf{X})$ with respect to any variable at any order. They are from this fact continuous when $\mathbf{x}\neq\mathbf{a}$. From proposition 651 (p. 393) equation (2.239), we have, with all $\mathbf{c}_{\mathbf{i}}^{\mathbf{n},\mathbf{p}}(\mathbf{x},\mathbf{X})\in]\mathbf{a},\mathbf{x}[$:

$$\partial_{\mathbf{x}}^{\mathbf{p}}\Delta_{\mathbf{X}}\eta_{\mathbf{n}}(\mathbf{x},\mathbf{a},\mathbf{X})=\frac{\mathbf{p}!\mathbf{n}!}{(\mathbf{n}+\mathbf{p})!}\partial_{\mathbf{x}}^{\mathbf{n}+\mathbf{p}}\Delta_{\mathbf{X}}\mathbf{f}(\mathbf{a},\mathbf{X})$$
$$+\sum_{\mathbf{i}\in[1,2^{\mathbf{p}}]}\mathbf{h}_{\mathbf{i}}^{\mathbf{n},\mathbf{p}}(\partial_{\mathbf{x}}^{\mathbf{n}+\mathbf{p}}\Delta_{\mathbf{X}}\mathbf{f}(\mathbf{c}_{\mathbf{i}}^{\mathbf{n},\mathbf{p}}(\mathbf{x},\mathbf{X}),\mathbf{X})-\partial_{\mathbf{x}}^{\mathbf{n}+\mathbf{p}}\Delta_{\mathbf{X}}\mathbf{f}(\mathbf{a},\mathbf{X}))$$
$$\partial_{\mathbf{x}}^{\mathbf{p}}\Delta_{\mathbf{X}}\eta_{\mathbf{n}}(\mathbf{a},\mathbf{a},\mathbf{X})=\frac{\mathbf{p}!\mathbf{n}!}{(\mathbf{n}+\mathbf{p})!}\partial_{\mathbf{x}}^{\mathbf{p}}\Delta_{\mathbf{X}}\mathbf{f}(\mathbf{a},\mathbf{X})$$

$\mathbf{X}$ fixed, $\eta_n(x, a, X)$ is in $C^\infty$ (see 276 (p. 304)), $C_X \times C_x$ is compact, all $h_i^{n,p}$ do not depend on $x$ and $X$ then all $\sum_{i\in[1,2^p]} h_i^{n,p}\partial_x^{n+p}\Delta_X f(x, X)$ are uniformly continuous in $C_X \times C_x$ then

$$(\forall\alpha > 0)(\exists\epsilon' > 0)(|x - x'| < \epsilon', \|X - X'\| < \epsilon'$$
$$\Rightarrow |\sum_{i\in[1,2^p]} h_i^{n,p}(\partial_x^{n+p}\Delta_X f(x, X) - \partial_x^{n+p}\Delta_X f(x', X'))| < \alpha/2) .$$

Similarly, we have:

$$(\forall\alpha > 0)(\exists\epsilon' > 0)((a - a' < \epsilon'', \|X - X'\| < \epsilon'')$$
$$\Rightarrow |\frac{p!n!}{(n+p)!}\partial_x^{n+p}\Delta_X f(a', X') - \frac{p!n!}{(n+p)!}\partial_x^{n+p}\Delta_X f(a, X)| < \alpha/2) .$$

Then calling $\epsilon = \min(\epsilon', \epsilon'')$ for $|x - a| < \epsilon$, $\|X - X'\| < \epsilon$ since $c_i^{n,p}(x, X') \in ]a, x[$, we have:

$$|\partial_x^p\Delta_X\eta_n(x, a, X') - \partial_x^p\Delta_X\eta_n(a, a, X)|$$
$$< |\frac{p!n!}{(n+p)!}\partial_x^{n+p}\Delta_X f(a, X') - \frac{p!n!}{(n+p)!}\partial_x^{n+p}\Delta_X f(a, X)|$$
$$+ |\sum_{i\in[1,2^p]} h_i^{n,p}(\partial_x^{n+p}\Delta_X f(c_i^{n,p}(x, X'), X') - \partial_x^{n+p}\Delta_X f(a, X'))| < \alpha .$$

Proposition 655 is true.

**Proposition 656** Given a map $(f : \mathcal{R} \to \mathcal{R})$ of $C^\infty$ (see 276 (p. 304)), two positive integers $k, n$ with $k \in [1, n]$, the map $(g_k : \mathcal{R}^n \to \mathcal{R})$ defined by:

$$\forall((x_i : [1, n] \to \mathcal{R}) \in \mathcal{R}^n)(g_k(x) = f(x_k))$$

belongs to $C^\infty\{\mathcal{R}^n \to \mathcal{R}\}$.

**Proof**
Any partial derivative of $g_k$ is generated by a map $(i_j : [1, n] \to \mathcal{Z}^{0+})$: $D_i(g_k) = \prod_{j\in[1,n]} \partial_j^{i_j} g_k(x)$.

$$(\exists l \in ([1, n] - k)(i_l \neq 0) \Rightarrow D_i(g_k) = 0 .$$

From proposition 308 (p. 194) and definition 275 (p. 304), it is true for $i_k = 0$, then, from propositions 308 (p. 194) and 927 (p. 520), it is true for any $i_k$.

**Proposition 657** Given a map $(u_i : [1, n] \to \mathcal{R})$, we call $v_i = u_i\partial_i$ and for $p \in \mathcal{Z}^+$, $v_i^p = u_i^p\partial_i^p$, we consider for $q \in \mathcal{Z}^+$ the set of maps:

$$J_q = \text{set}_i\{i = (i_k : [1, n] \to \mathcal{Z}^{0+}), \sum_{k\in[1,n]} i_k = q\} \subset (\mathcal{Z}^+)^n$$

we have the identity:

$$\sum_{l\in[1,n]^q} \prod_{j\in[1,q]} v_{l_j} = q! \sum_{i\in J_q} \prod_{k\in[1,n]} \frac{v_k^{i_k}}{i_k!} .$$

**Proof**

From propositions 653 (p. 398) and 654 (p. 399), we have:

$$l \in [1,n]^q \Rightarrow (l_j : [1,q] \rightarrow [1,n]) \Rightarrow \prod_{j\in[1,q]} v_{l_j} = \prod_{k\in[1,n]} v_k^{Card(l^{-1}(k))}$$

$l^{-1}$ is the inverse (definition 115 (p. 133)) of $(l_i : [1,q] \rightarrow [1,n])$, $l^{-1}(k) = l^{-1}(\{k\})$ is the inverse set of the element $k \in \mathcal{Z}^+$ and we have (see definition 370 (p. 471), definition 372 (p. 473), proposition 815 (p. 471)).

$$\bigcup_{k\in[1,n]} l^{-1}(k) = [1,q], k \neq j \Rightarrow l^{-1}(k) \cap l^{-1}(j) = \emptyset .$$

From that we have (see proposition 887 (p. 504) and definition 393 (p. 506)):

$$\sum_{k\in[1,n]} Card(l^{-1}(k)) = Card([1,q]) = q .$$

Setting $i_k = Card(l^{-1}(k))$, we set a map $(I : [1,n]^q \rightarrow J_q)$. We have from proposition 355 (p. 210):

$$Card(I^{-1}(i)) = \frac{q!}{\prod_{k\in[1,n]} i_k!} .$$

Since $i \in J_q, I^{-1}(i)$ defines a partition of $[1,n]^q$, we have:

$$\sum_{l\in[1,n]^q} \prod_{j\in[1,q]} v_{l_j} = \sum_{i\in[1,q]} \sum_{l\in I^{-1}(i)} \prod_{j\in[1,q]} v_{l_j} = \sum_{i\in[1,q]} \sum_{l\in I^{-1}(i)} \prod_{k\in[1,n]} v_k^{Card(l^{-1}(k))} .$$

Since by definition:

$$l \in I^{-1}(i) \Rightarrow (\forall k \in [1,n])(Card(l^{-1}(k)) = i_k)$$

we have:

$$(\forall i \in [1,q])( \sum_{l\in I^{-1}(i)} \prod_{k\in[1,n]} v_k^{Card(l^{-1}(k))} = \sum_{l\in I^{-1}(i)} \prod_{k\in[1,n]} v_k^{i_k}$$
$$= Card(I^{-1}(i)) \prod_{k\in[1,n]} v_k^{i_k} = \frac{q!}{\prod_{k\in[1,n]} i_k!} \prod_{k\in[1,n]} v_k^{i_k}) .$$

That relation is proposition 657 (p. 400).

**Proposition 658** f is a continuous map $(f : \mathcal{R}^n \rightarrow \mathcal{R})$ having derivatives up to order q whatever is the variable of the partial derivative. f is defined on a closed ball B. Calling $\partial_k^i f(x) = \frac{\partial^i f(x)}{\partial^i x_k}$, we have for every $a \in B$ and $b \in B$, $a = (a_1, ..., a_n)$, $b = (b_1, ..., b_n)$:

$$f(b) = f(a) + \sum_{i\in[1,q]} \frac{1}{i!} \sum_{l\in[1,n]^i} \prod_{j\in[1,i]} (b-a)_{l_j} \partial_{l_j} f(a) + \frac{1}{q!} \sum_{l\in[1,n]^q} \eta_l \prod_{j\in[1,q]} (b-a)_{l_j} \quad (2.245)$$

the maps $\eta_l$ are continuous for a and b in the ball B. All $\eta_l$ are zero for $a = b$. If f belongs to $C^\infty$ (see 276 (p. 304)), then the $\eta_l$ belong also to $C^\infty$. Since the order of derivatives does not count: we have $\eta_l = \eta_{P(l)}$ where P is a bijective map (definition 124 (p. 139)) $(P : [1,n]^q \rightarrow [1,n]^q)$. P is a permutation acting on $[1,q]$: $P(l_i) = l_{P(i)}$ (see definition 172 (p. 207)).

**Proof**

Considering $\mathbf{J_q(n)}$, the set of maps:

$$J_q(n) = set_i\{i = (i_k : [1,n] \to \mathcal{Z}^{0+}), \sum_{k\in[1,n]} i_k = q\} \subset (\mathcal{Z}^+)^n .$$

Using the fact that the order of the derivatives does not count, let us prove by recurrence (see proposition 927 (p. 520)):

$$f(b) = \sum_{p\in[0,q]} \sum_{i\in J_p(n)} \prod_{k\in[1,n]} \frac{(b_k - a_k)^{i_k} \partial_k^{i_k}}{i_k!} f(a) + \sum_{i\in J_q(n)} \eta_i \prod_{k\in[1,n]} \frac{(b_k - a_k)^{i_k}}{i_k!} . \quad (2.246)$$

With

$$\sum_{i\in J_0(n)} \prod_{k\in[1,n]} \frac{(b_k - a_k)^{i_k} \partial_k^{i_k}}{i_k!} f(a) = \prod_{k\in[1,n]} \frac{(b_k - a_k)^0 \partial_k^0}{0!} f(a) = f(a) . \quad (2.247)$$

Let us suppose (2.246) is true for $\mathbf{n' < n}$ and considering $\mathbf{a^{(n')}}$ such that:

$$(\forall i \in [1,n'])(a_i^{(n')} = a_i), (\forall i \in [n'+1,n])(a_i^{(n')} = b_i) .$$

We suppose:

$$f(b) = \sum_{p\in[0,q]} \sum_{i'\in J_p(n')} \prod_{k\in[1,n']} \frac{(b_k - a_k)^{i'_k} \partial_k^{i'_k}}{i'_k!} f(a) + \sum_{i'\in J_q(n')} \eta_{i'} \prod_{k\in[1,n']} \frac{(b_k - a_k)^{i'_k}}{i'_k!} . \quad (2.248)$$

We can change $\mathbf{a^{(n')} = (a_1, ..., a_{n'}, b_{n'+1}, ..., b_n)}$ into $\mathbf{a^{(n')} = (a_1, ..., a_{n'}, a_{n'+1} + b_{n'+1} - a_{n'+1}, ..., b_n)}$ and expand on the variable $\mathbf{n' + 1}$ according to equation (2.230) (p. 392):

$$(\forall p \in [0,q])(\forall i' \in J_p(n'))((\prod_{k\in[1,n']} \partial_k^{i'_k}) f(a^{(n')})$$

$$= \sum_{j\in[0,l]} \frac{(b_{n'+1} - a_{n'+1})^j \partial_{n'+1}^j}{j!} (\prod_{k\in[1,n']} \partial_k^{i'_k}) f(a^{(n'+1)}) + \frac{(b_{n'+1} - a_{n'+1})^l}{l!} \eta_{i',l}(b,a)) .$$

Plugging into equation (2.248) (p. 402), we get:

$$f(b) =$$

$$\sum_{p\in[0,q]} \sum_{i'\in J_p(n')} \prod_{k\in[1,n']} \frac{(b_k - a_k)^{i'_k} \partial_k^{i'_k}}{i'_k!} \sum_{i'_{n'+1}\in[0,q-p]} \frac{(b_{n'+1} - a_{n'+1})^{i'_{n'+1}} \partial_{n'+1}^{i'_{n'+1}}}{i'_{n'+1}!} f(a^{(n'+1)})$$

$$+ \sum_{i'\in J_q(n')} \eta_{i'} \prod_{k\in[1,n']} \frac{(b_k - a_k)^{i'_k}}{i'_k!}$$

$$+ \sum_{p\in[0,q]} \sum_{i'\in J_p(n')} \prod_{k\in[1,n']} \frac{(b_k - a_k)^{i'_k}}{i'_k!} \frac{(b_{n'+1} - a_{n'+1})^{q-p}}{(q-p)!} \eta_{i',q-p} .$$

Setting $\mathbf{(\forall p \in [0,q])(A_p(n'+1) \subset (\mathcal{Z}^+)^{n'+1})}$ such that:

$$i'' \in A_p(n'+1) \Leftrightarrow$$
$$(\exists p' \in [1,p])(\exists i' \in J_{p'}(n'))(k \in [1,n'] \Rightarrow i''_k = i'_k, i''_{n'+1} = p - p')$$

we have $\mathbf{A_p(n'+1)} = \mathbf{J_p(n'+1)}$ because any term of $\mathbf{A_p(n'+1)}$ is in $\mathbf{J_p(n,+1)}$ and reciprocally. Then we have, setting $\eta_{i''} = \eta_{i',q-p}$ where $\mathbf{i'} \in \mathbf{J_p}$, $\mathbf{p} \in [0,\mathbf{q}]$ and $\mathbf{i''_{n+1}} = \mathbf{q}-\mathbf{p}$:

$$f(b) = \sum_{p\in[0,q]} \sum_{i''\in J_p(n'+1)} \prod_{k\in[1,n'+1]} \frac{(b_k-a_k)^{i''_k}\partial_k^{i''_k}}{i''_k!} f(a^{(n'+1)})$$
$$+ \sum_{i''\in J_q(n'+1)} \eta_{i''} \prod_{k\in[1,n'+1]} \frac{(b_k-a_k)^{i''_k}}{i''_k!} .$$

That way for each $\mathbf{i'} \in \mathbf{J_q(n')}$, we use an $\mathbf{l} = \mathbf{q} - \sum_{j\in[1,n']} \mathbf{i_j}$ and we generate the $\mathbf{l}$ new terms of $\mathbf{J_q(n'+1)}$ and a term of $\mathbf{I^q_m(n'+1)}$ using the $\eta_{l,n'+1}(\mathbf{b},\mathbf{a})$.
Since equation (2.246) is true for $\mathbf{n}=1$, it is true for any $\mathbf{n}$ (proposition 927 (p. 520)). Applying proposition 657 (p. 400) to equation (2.246) when $\mathbf{p} \in [1,\mathbf{q}]$ in the first sum and equation (2.247) when $\mathbf{p}=0$ one gets equation (2.245). The $\eta$ introduced here have the same property as in proposition 650 (p. 392).
If $\mathbf{f} \in \mathbf{C}^\infty$ (see 276 (p. 304)), according to proposition 655 (p. 399) the final $\eta_i$ of equation (2.246) are also in $\mathbf{C}^\infty$.

### 2.28.1 Remarks on complex functions

In general, things which are true for a mapping from a set to $\mathcal{R}$ are also true for a mapping from the same set to complex numbers $\mathcal{C}$ (see section 2.13 (p. 271)). The reverse is not generally true: A mapping from $\mathcal{R}$ to a topological set like $\mathcal{R}^n$ or $\mathcal{C}^n$ does not behave like a mapping from $\mathcal{C}$ to the same sets. For instance the derivative does not really exist for $\mathcal{C}$ except for holomorphic complex function (definition 314 (p. 403)).

**Definition 313 Complex derivative (see section 2.13 (p. 271)).**

We consider the set of complex numbers $\mathcal{C} = \mathcal{R}^2$ with the topology of open balls of $\mathcal{R}^2$ (see definitions 263 (p. 286) and 260 (p. 285)). A map $(\mathbf{f}:\mathcal{C}\to\mathcal{C})$ is said to have a derivative $\mathbf{f'(z)} \in \mathcal{C}$ at $\mathbf{z} \in \mathcal{C}$ (see definition 161 (p. 192)) if the map:

$$(h(\Delta z) = \frac{f(z+\Delta z)-f(z)}{\Delta z} : \mathcal{C}\to\mathcal{C}) \tag{2.249}$$

has a limit (definition 157 (p. 182)) $\mathbf{h(0)} \in \mathcal{C}$ for $\Delta\mathbf{z} = 0$ and then we have $\mathbf{f'(z)} = \mathbf{h(0)}$. That is to say:

$$(\forall\alpha_1>0)(\exists\epsilon_1>0)(\|\Delta z\|<\epsilon_1 \Rightarrow \|h(\Delta z)-h(0)\|<\alpha_1) . \tag{2.250}$$

**Definition 314 Holomorphic map.**

Given an open subset $\mathbf{O} \subset \mathcal{R}^2 = \mathcal{C}$ in the $\mathcal{R}^2$ topology, a map is said to be holomorphic in $\mathbf{O}$ if it has a complex derivative (definition 313 (p. 403)) at any point of $\mathbf{O}$.

**Proposition 659** $(\mathbf{f} = \mathbf{f_r} + \mathbf{i}\mathbf{f_i} : \mathbf{O}\to\mathcal{C})$ is a holomorphic map between an open $\mathbf{O}$ of $\mathcal{R}^2 = \mathcal{C}$ and $\mathcal{C}$. At any point of $\mathbf{O}$, $\mathbf{f}$ should verify, setting $\mathbf{z} = \mathbf{z_r} + \mathbf{i}\mathbf{z_i}$:

$$\begin{aligned} \frac{\partial f_r}{\partial z_r} &= \partial_r f_r(z) = \frac{\partial f_i}{\partial z_i} = \partial_i f_i(z) \\ \frac{\partial f_i}{\partial z_r} &= \partial_r f_i(z) = -\frac{\partial f_r}{\partial z_i} = -\partial_i f_r(z) . \end{aligned} \tag{2.251}$$

If equation (2.251) is true at any point of $\mathbf{O}$ then the map $\mathbf{f}$ is holomorphic. The derivative at a point $\mathbf{z} \in \mathbf{O}$ is:

$$\mathbf{f}'(\mathbf{z}) = \frac{\partial \mathbf{f}_r}{\partial \mathbf{z}_r} + \mathbf{i}\frac{\partial \mathbf{f}_i}{\partial \mathbf{z}_r} .$$

**Proof**

Writing $\mathbf{z} = \mathbf{z}_r + \mathbf{i}\mathbf{z}_i$ and $\mathbf{f}(\mathbf{z}) = \mathbf{f}_r(\mathbf{z}) + \mathbf{i}\mathbf{f}_i(\mathbf{z})$, from proposition 560 (p. 307) we have:

$$\mathbf{f}_r(\mathbf{z}+\Delta\mathbf{z}) - \mathbf{f}_r(\mathbf{z}) = (\partial_r \mathbf{f}_r(\mathbf{z}) + \eta_r^r)\Delta\mathbf{z}_r + (\partial_i \mathbf{f}_r(\mathbf{z}) + \eta_r^i)\Delta\mathbf{z}_i$$

$$\mathbf{f}_i(\mathbf{z}+\Delta\mathbf{z}) - \mathbf{f}_i(\mathbf{z}) = (\partial_r \mathbf{f}_i(\mathbf{z}) + \eta_i^r)\Delta\mathbf{z}_r + (\partial_i \mathbf{f}_i(\mathbf{z}) + \eta_i^i)\Delta\mathbf{z}_i$$

with

$$(\forall \alpha_2)(\exists \epsilon_2)(\|\Delta\mathbf{z}\| < \epsilon_2 \Rightarrow \|\eta^r\| < \alpha_2, \|\eta^i\| < \alpha_2) \tag{2.252}$$

and for any $\alpha$, we should have an $\epsilon > 0$: $\epsilon = \min(\epsilon_1, \epsilon_2)$, $\epsilon_1$ and $\epsilon_2$ being such that $\alpha_1 = \alpha/2$ in equation (2.250) and $\alpha_2 = \alpha/4$ in equation (2.252) in order to have:

$$\begin{aligned}&\|\mathbf{h}_r\Delta\mathbf{z}_r - \mathbf{h}_i\Delta\mathbf{z}_i + \mathbf{i}(\mathbf{h}_r\Delta\mathbf{z}_i + \mathbf{h}_i\Delta\mathbf{z}_r)\\ &\quad - (\partial_r\mathbf{f}_r(\mathbf{z})\Delta\mathbf{z}_r + \partial_i\mathbf{f}_r(\mathbf{z})\Delta\mathbf{z}_i + \mathbf{i}(\partial_r\mathbf{f}_i(\mathbf{z})\Delta\mathbf{z}_r + \partial_i\mathbf{f}_i(\mathbf{z})\Delta\mathbf{z}_i))\| < \alpha\|\Delta\mathbf{z}\|\end{aligned}$$

for any $\|\Delta\mathbf{z}\| < \epsilon$. $\mathbf{h}$ has the definition given in equation (2.249) of definition 313 (p. 403). But whatever $\epsilon$ even when $\Delta\mathbf{z}_r < \epsilon$ and $\Delta\mathbf{z}_i = 0$, we need to have for the equation (2.250) to be true:

$$\|\mathbf{h}_r - \partial_r\mathbf{f}_r(\mathbf{z}) + \mathbf{i}(\mathbf{h}_i - \partial_r\mathbf{f}_i(\mathbf{z}))\| < \alpha .$$

Since it should be true for any $\alpha > 0$

$$\mathbf{h}_r = \partial_r\mathbf{f}_r(\mathbf{z}), \mathbf{h}_i = \partial_r\mathbf{f}_i(\mathbf{z})$$

and similarly when $\Delta\mathbf{z}_i < \epsilon$ and $\Delta\mathbf{z}_r = 0$

$$\mathbf{h}_r = \partial_i\mathbf{f}_i(\mathbf{z}), \mathbf{h}_i = -\partial_i\mathbf{f}_r(\mathbf{z})$$

$$\begin{aligned}\partial_r\mathbf{f}_r(\mathbf{z}) &= \partial_i\mathbf{f}_i(\mathbf{z})\\ \partial_r\mathbf{f}_i(\mathbf{z}) &= -\partial_i\mathbf{f}_r(\mathbf{z}) .\end{aligned}$$

Otherwise there is no solution to the system. If $\mathbf{f}_r(\mathbf{z})$ and $\mathbf{f}_i(\mathbf{z})$ have partial derivatives at $\mathbf{z} = \mathbf{z}_r + \mathbf{i}\mathbf{z}_i$ and if the above conditions are satisfied, then equation (2.252) implies equation (2.250) with $\mathbf{h}_r = \partial_i\mathbf{f}_i(\mathbf{z}), \mathbf{h}_i = -\partial_i\mathbf{f}_r(\mathbf{z})$. $\mathbf{f}'(\mathbf{z}) = \frac{\partial \mathbf{f}_r}{\partial \mathbf{z}_r} + \mathbf{i}\frac{\partial \mathbf{f}_i}{\partial \mathbf{z}_r}$ exists.

**Proposition 660** Any complex function which has a derivative at a point is continuous at that point.

**Proof**

Having found an $\epsilon_1$ to satisfy equation (2.250) for $\alpha_1 = \alpha/2$, we take:

$$\epsilon = \min(\epsilon_1, \alpha/(2\|\mathbf{h}(0)\|))$$

and we get (proposition 404 (p. 239)):

$$\begin{aligned}&(\forall\alpha > 0)(\exists\epsilon > 0)(\|\Delta\mathbf{z}\| < \epsilon \le \epsilon_1 \Rightarrow \|\mathbf{f}(\mathbf{z}+\Delta\mathbf{z}) - \mathbf{f}(\mathbf{z})\| - \|\mathbf{h}(0)\|\|\Delta\mathbf{z}\|\\ &\le \|\mathbf{f}(\mathbf{z}+\Delta\mathbf{z}) - \mathbf{f}(\mathbf{z}) - \mathbf{h}(0)\Delta\mathbf{z}\| \le \alpha_1\Delta\mathbf{z} < \alpha_1 = \alpha/2\\ &\Rightarrow \|\mathbf{f}(\mathbf{z}+\Delta\mathbf{z}) - \mathbf{f}(\mathbf{z})\| < \alpha/2 + \|\mathbf{h}(0)\|\|\Delta\mathbf{z}\| \le \alpha)\end{aligned}$$

**Proposition 661** The map $(\prod_{i\in[1,n]} z_i : \mathcal{C}^n \to \mathcal{C})$ is continuous.

**Proof**

For $n = 2$, we have:

$$(z_1 + \Delta z_1)(z_2 + \Delta z_2) - z_1 z_2 = z_2\Delta z_1 + z_1\Delta z_2 + \Delta z_1\Delta z_2$$

then $(\forall\alpha > 0)$ choosing $0 < \epsilon < \min(\sqrt{\alpha/3}, \alpha/(3Z_1), \alpha/(3Z_2))$ we have:

$$(\forall\alpha > 0)(\exists\epsilon)(\|\Delta z_1\| < \epsilon, \|\Delta z_2\| < \epsilon$$
$$\Rightarrow \|(z_1 + \Delta z_1)(z_2 + \Delta z_2) - z_1 z_2\| < \|z_2\Delta z_1\| + \|z_1\Delta z_2\| + \|\Delta z_1\Delta z_2\| < \alpha)\,.$$

Then from proposition 282 (p. 184), proposition 661 is true for $n = 2$. If it is true up to $n$, from propositions 243 (p. 168), 836 (p. 480), definition 134 (p. 153), combined to proposition 661 applied for $n = 2$, proposition 661 is true for $n + 1$. Then from proposition 927 (p. 520) it is true for any $n$.

**Proposition 662** The map $(\sum_{i\in[1,n]} z_i : \mathcal{C}^n \to \mathcal{C})$ is continuous.

**Proof**

For $n = 2$, we have:

$$z_1 + \Delta z_1 + z_2 + \Delta z_2 - (z_1 + z_2) = \Delta z_1 + \Delta z_2$$

then taking $\forall\alpha > 0$, $0 < \epsilon < \alpha/2$, we have:

$$(\forall\alpha > 0)(\exists\epsilon)((\|\Delta z_1\| < \epsilon, \|\Delta z_2\| < \epsilon) \Rightarrow$$
$$\|z_1 + \Delta z_1 + z_2 + \Delta z_2 - (z_1 + z_2)\| < \|\Delta z_1\| + \|\Delta z_2\| < \alpha)\,.$$

From proposition 282 (p. 184), proposition 662 is true for $n = 2$. Then as in proposition 661, it is true for any $n$.

## 2.29 Exponentials

**Definition 315 Vectorial space E needed for the exponentials.**

We suppose that we have a normed vectorial space (definitions 199 (p. 235) and 203 (p. 237)) $E$ built on $\mathcal{R}$ with an internal map (an internal composition law) $(f : E \times E \to E)$ such that there is a unit element $\mathbb{1}$:

$$(\exists\mathbb{1} \in E)(\forall A \in E)(f(A, \mathbb{1}) = f(\mathbb{1}, A) = A) \tag{2.253}$$

and such that it is double linear:

$$(\forall A \in E)(\forall B \in E)(\forall C \in E)(\forall\lambda \in \mathcal{R})(\forall\mu \in \mathcal{R})$$
$$(f(\lambda A + \mu B, C) = \lambda f(A, C) + \mu f(B, C)), (f(C, \lambda A + \mu B) = \lambda f(C, A) + \mu f(C, B))\,. \tag{2.254}$$

We shall write now $f(A, B) = AB$ as commonly done.

We suppose that we also have for the norm:

$$(\forall A \in E)(\forall B \in E)(\|AB\| \leq \|A\|\|B\|)\,. \tag{2.255}$$

**Proposition 663** $\mathbf{E}$ being a vectorial space with the above properties, $\mathbb{O}$ being the unit element of $\mathbf{E}$: $\mathbf{A} \in \mathbf{E} \Rightarrow \mathbf{A} + \mathbb{O} = \mathbb{O} + \mathbf{A} = \mathbf{A}$, then $\mathbf{A} \in \mathbf{E} \Rightarrow (\|\mathbf{A}\| = 0 \Leftrightarrow \mathbf{A} = \mathbb{O})$, $\mathbf{A} \in \mathbf{E} \Rightarrow (\mathbf{A}\mathbb{O} = \mathbb{O})$.

**Proof**

$\mathbf{A} \in \mathbf{E} \Rightarrow (\|\mathbf{A}\| = 0 \Leftrightarrow \mathbf{A} = \mathbb{O})$ is a direct consequence of definition 203 (p. 237), 2. $\mathbf{A} \in \mathbf{E} \Rightarrow (\mathbf{A}\mathbb{O} = \mathbb{O})$ is a consequence of equation (2.255) (p. 405), definition 203 (p. 237), 1 and the above relation.

**Definition 316** $\mathbf{A}^{\mathbf{n}}$.

Given $\mathbf{n} \in \mathbf{\mathcal{Z}}^+$ and a normed vectorial space $\mathbf{E}$ with an internal composition law (definition 315 (p. 405)), we use the definitions 128 (p. 145) and 127 (p. 145) with a map of $[\mathbf{1}, \mathbf{n}] \to \mathbf{E}$ which gives $\mathbf{A}$ for all $\mathbf{i} \in [\mathbf{1}, \mathbf{n}]$ and we call:

$$\mathbf{A}^{\mathbf{n}} = \prod_{\mathbf{i} \in [\mathbf{1},\mathbf{n}]} \mathbf{A} \,.$$

It also defines a map $(\mathbf{A}^{\mathbf{i}} : \mathbf{\mathcal{Z}}^{0+} \to \mathbf{E})$. For $\mathbf{i} = \mathbf{0}$, one uses to set $\mathbf{A}^{\mathbf{0}} = \mathbb{1}$ where $\mathbb{1}$ is the unit element of $\mathbf{E}$ for the internal law of $\mathbf{E}$ (product).

**Proposition 664** $\mathbf{A}^{\mathbf{0}} = \mathbb{1}$

**Proof**

It comes from what we said in definition 316 (p. 406).

**Proposition 665** $\mathbf{E}$ being a normed vectorial space (definition 315 (p. 405)) with the above properties(definition 316 (p. 406)), then

$$(\mathbf{A} \in \mathbf{E}, \|\mathbf{A}\| < \mathbf{1}, \mathbf{n} \leq \mathbf{p}) \Rightarrow \|\mathbf{A}^{\mathbf{n}}\| \leq \|\mathbf{A}^{\mathbf{p}}\| \,.$$

**Proof**

Let us assume $\|\mathbf{A}^{\mathbf{p}+\mathbf{k}}\| \leq \|\mathbf{A}^{\mathbf{p}}\|$. Then we have from proposition 316 (p. 196):

$$\|\mathbf{A}\| < \mathbf{1} \Rightarrow \|\mathbf{A}^{\mathbf{p}+\mathbf{k}+\mathbf{1}}\| \leq \|\mathbf{A}^{\mathbf{p}+\mathbf{k}}\|\|\mathbf{A}\| \leq \|\mathbf{A}^{\mathbf{p}+\mathbf{k}}\| \leq \|\mathbf{A}^{\mathbf{p}}\| \,.$$

Since it is true for $\mathbf{k} = \mathbf{0}$, it is true for any $\mathbf{k} \in \mathbf{\mathcal{Z}}^{0+}$ (proposition 927 (p. 520)). Since $\mathbf{n} \geq \mathbf{p} \Rightarrow \mathbf{n} - \mathbf{p} \in \mathbf{\mathcal{Z}}^{0+}$

$$\|\mathbf{A}^{\mathbf{n}}\| = \|\mathbf{A}^{\mathbf{p}+\mathbf{n}-\mathbf{p}}\| \leq \|\mathbf{A}^{\mathbf{p}}\| \,.$$

**Proposition 666** Given a normed vectorial space $\mathbf{E}$ (definitions 315 (p. 405), 316 (p. 406)), $\forall \mathbf{A} \in \mathbf{E}$, the sequence $(\mathbf{U}_{\mathbf{n}} : \mathbf{\mathcal{Z}}^{0+} \to \mathbf{E})$ defined as

$$\mathbf{U}_{\mathbf{n}} = \sum_{\mathbf{i} \in [\mathbf{0},\mathbf{n}]} \frac{\mathbf{A}^{\mathbf{i}}}{\mathbf{i}!}$$

is a Cauchy sequence (definition 269 (p. 300)). See definition 127 (p. 145).

In fact, the value $\mathbf{m}$ entering in the formula (2.135) (p. 300) of definition 269 (p. 300) is only depending on $\|\mathbf{A}\|$. With the above conditions, the Cauchy condition which is to be satisfied can then be written as:

$$(\forall \alpha \in \mathcal{R}^+)(\forall \mathbf{h} \in \mathcal{R}^+)(\exists \mathbf{m} \in \mathbf{\mathcal{Z}}^+)(\|\mathbf{A}\| \leq \mathbf{h},\ \mathbf{p} > \mathbf{m},\ \mathbf{q} > \mathbf{m}$$
$$\Rightarrow \| \sum_{\mathbf{i} \in [\mathbf{0},\mathbf{p}]} \frac{\mathbf{A}^{\mathbf{i}}}{\mathbf{i}!} - \sum_{\mathbf{i} \in [\mathbf{0},\mathbf{q}]} \frac{\mathbf{A}^{\mathbf{i}}}{\mathbf{i}!} \| < \alpha) \,. \tag{2.256}$$

**Proof**

We have with the above notation assuming $p < q$

$$U_q - U_p = \frac{A^p}{p!}\Big(\sum_{i\in[p+1,q]} \frac{A^i}{i!}\Big) . \tag{2.257}$$

We have:

1. $(\exists p_1)(n > p_1 \Rightarrow \|\frac{A}{n}\| \leq h/n < 1/2)$ it is enough to take $p_1 > 2h \geq 2\|A\|$
2. $p_1 > 2h \geq 2\|A\| \Rightarrow (\forall p_2 > p_1)(\forall p_3 > p_2)(\frac{h^{p_3}}{p_3!} \leq \frac{h^{p_2}}{p_2!})$
   It is true for $p_3 = p_2$. From the preceding item, from proposition 316 (p. 196), if it is true for $p_3 = p_2 + k$, it is true for $p_3 = p_2 + k + 1$:
   $$\frac{h^{p_2+k+1}}{(p_2+k+1)!} = \frac{h}{(p_2+k+1)}\frac{h^{p_2+k}}{(p_2+k)!} \leq 1/2\frac{h^{p_2+k}}{(p_2+k)} < \frac{h^{p_2+k}}{(p_2+k)!} \leq \frac{h^{p_2}}{p_2!} .$$
   Then by (proposition 927 (p. 520)), it is true for all $k$.
3. $p_1 > 2h \geq 2\|A\| \Rightarrow (\forall \alpha > 0)(\exists p_2)(n > p_2 \rightarrow \|\frac{A^n}{n!}\| < \alpha)$
   Considering the subset of $\mathcal{R}^+$ made by the numbers $\frac{h^n}{n!}$ with $\|\frac{A^n}{n!}\| \leq \frac{h^n}{n!}$ when $n > p_1$, this set has $0$ as a minoring value. It has then a lower_limit $m$. We have $m = 0$ otherwise there is a $p_3 > p_1$ such that (proposition 260 (p. 175)) $0 < \frac{h^{p_3}}{p_3!} - m < m/3$ and we would have:
   $$\frac{h^{p_3+1}}{p_3+1!} = \frac{h}{p_3+1}\frac{h^{p_3}}{p_3!} < \frac{1}{2}\frac{h^{p_3}}{p_3!} < \frac{m+m/3}{2} < m .$$
   This cannot be true since $m$ is the lower_limit. We conclude $m = 0$, using appendix A.1.4, 8.
   Then, from proposition 260 (p. 175) and from point 2, we have:
   $$(\forall \alpha > 0)(\exists p_2 \in [p_1, \rightarrow [)(p_3 > p_2 \Rightarrow \|\frac{A^{p_3}}{p_3!}\| \leq \frac{h^{p_3}}{p_3!} < \frac{h^{p_2}}{p_2!} \leq \alpha/2) .$$
4. Now from equation (2.257), for $q > p > \max(p_2, p_1)$, we have:
   $$\|U_q - U_p\| \leq \sum_{i\in[p+1,q]} \|\frac{A^i}{i!}\| \leq \sum_{i\in[p+1,q]} \|\frac{h^i}{i!} \leq \alpha/2 \sum_{i\in[1,q-p]} \frac{1}{2^i}$$
   $$= \frac{\alpha}{2} \times \frac{1 - 1/2^{q-p+1}}{1 - 1/2} < \alpha$$
   so $U_n$ is a Cauchy sequence.

**Definition 317 Exponential.**

Having a normed vectorial space $E$ with the multiplicative mapping as defined at the beginning of section 2.29 (p. 405). If moreover $E$ is such that all Cauchy sequences have a limit. Then from propositions 666 and 203 (p. 237) for any $A \in E$, the sequence $U_n = \sum_{i\in[0,n]} \frac{A^i}{i!}$ has a limit. This limit is written as $e^A$ and called exponential of $A$.

**Proposition 667** $\mathbf{e^{\mathbb{O}} = \mathbb{1}}$

**Proof**
From propositions 663 (p. 406) and 664 (p. 406), we have:

$$\|\mathbf{A}\| = 0 \Rightarrow \mathbf{A} = \mathbb{O} \Rightarrow (\mathrm{i} \neq 0 \Rightarrow \frac{\mathbf{A}^{\mathrm{i}}}{\mathrm{i}!} = 0, \mathbf{A}^0 = \mathbb{1}) .$$

Then we have:

$$\|\mathbf{A}\| = 0 \Rightarrow (\forall \mathrm{n} \in \mathcal{Z}^{0+})(\mathbf{U}_{\mathrm{n}} = \mathbf{A}^0 = \mathbb{1})$$

$\mathbb{O}$ is the null vector of $\mathbf{E}$ and $\mathbb{1}$ the unit vector of the internal law. Then:

$$e^{\mathbb{O}} = \mathbb{1}$$

**Proposition 668** Having a normed vectorial space $\mathbf{E}$ with an internal composition law and with Cauchy sequences converging as defined in definitions 315 (p. 405) and 317 (p. 407) , if the law is also commutative then there is the following identity:

$$\mathrm{e}^{\mathbf{A}+\mathbf{B}} = \mathrm{e}^{\mathbf{A}}\mathrm{e}^{\mathbf{B}} .$$

**Proof**

1. We have the following equalities (see the formula (2.72) (p. 218))

$$\mathbf{U}_{\mathrm{n}} = \sum_{\mathrm{i}\in[0,\mathrm{n}]} \frac{(\mathbf{A}+\mathbf{B})^{\mathrm{i}}}{\mathrm{i}!} \tag{2.258}$$

$$= \sum_{\mathrm{i}\in[0,\mathrm{n}]}\sum_{\mathrm{j}\in[0,\mathrm{i}]} \frac{\mathbf{A}^{\mathrm{j}}}{\mathrm{j}!}\frac{\mathbf{B}^{\mathrm{i-j}}}{(\mathrm{i-j})!} = \sum_{(\mathrm{i,j})\in[0,\mathrm{n}]^2,\mathrm{j}\in[0,\mathrm{i}]} \frac{\mathbf{A}^{\mathrm{j}}}{\mathrm{j}!}\frac{\mathbf{B}^{\mathrm{i-j}}}{(\mathrm{i-j})!} \tag{2.259}$$

$$= \sum_{(\ell,\mathrm{k})\in[0,\mathrm{n}]^2,\mathrm{k}+\ell\in[0,\mathrm{n}]} \frac{\mathbf{A}^{\ell}}{\ell!}\frac{\mathbf{B}^{\mathrm{k}}}{\mathrm{k}!} \tag{2.260}$$

$$\mathbf{V}^{\mathrm{a}}_{\mathrm{p}} = \sum_{\mathrm{i}\in[0,\mathrm{p}]} \frac{\mathbf{A}^{\mathrm{i}}}{\mathrm{i}!} \tag{2.261}$$

$$\mathbf{V}^{\mathrm{b}}_{\mathrm{q}} = \sum_{\mathrm{i}\in[0,\mathrm{q}]} \frac{\mathbf{B}^{\mathrm{i}}}{\mathrm{i}!} \tag{2.262}$$

$$\mathbf{V}^{\mathrm{a}}_{\mathrm{p}}\mathbf{V}^{\mathrm{b}}_{\mathrm{q}} = (\sum_{\mathrm{i}\in[0,\mathrm{p}]} \frac{\mathbf{A}^{\mathrm{i}}}{\mathrm{i}!})(\sum_{\mathrm{i}\in[0,\mathrm{q}]} \frac{\mathbf{B}^{\mathrm{i}}}{\mathrm{i}!}) \tag{2.263}$$

$$= \sum_{(\mathrm{i,j})\in[0,\mathrm{p}]\times[0,\mathrm{q}]} \frac{\mathbf{A}^{\mathrm{i}}}{\mathrm{i}!}\frac{\mathbf{B}^{\mathrm{j}}}{\mathrm{j}!} . \tag{2.264}$$

One goes from equation (2.259) to (2.260) by setting $\ell = \mathrm{j}, \mathrm{k} = \mathrm{i}-\mathrm{j}$ which is a bijective map (definition 124 (p. 139)).
It is an injective map (definition 122 (p. 139)):

$$\ell = \mathrm{j}, \mathrm{k} = \mathrm{i} - \mathrm{j} \Leftrightarrow \mathrm{j} = \ell, \mathrm{i} = \ell + \mathrm{k} .$$

It is a surjective map (definition 123 (p. 139)):

$$((\mathbf{i},\mathbf{j}) \in [0,\mathbf{n}]^2, \mathbf{j} \in [0,\mathbf{i}], \ell = \mathbf{j}, \mathbf{k} = \mathbf{i}-\mathbf{j}) \Rightarrow (\ell = \mathbf{j} \in [0,\mathbf{n}], 0 \leq \mathbf{k} = \mathbf{i}-\mathbf{j} \leq \mathbf{i})$$
$$\Rightarrow ((\ell,\mathbf{k}) \in [0,\mathbf{n}]^2, \mathbf{k}+\mathbf{j} = \mathbf{k}+\ell = \mathbf{i} \in [0,\mathbf{n}])$$
$$((\ell,\mathbf{k}) \in [0,\mathbf{n}]^2, \mathbf{k}+\ell \in [0,\mathbf{n}], \mathbf{j} = \ell, \mathbf{i} = \ell+\mathbf{k})$$
$$\Rightarrow (\mathbf{i} \in [0,\mathbf{n}], \mathbf{j} \in [0,\mathbf{n}], \mathbf{j} = \mathbf{i}-\mathbf{k} \leq \mathbf{i})$$
$$\Rightarrow ((\mathbf{i},\mathbf{j}) \in [0,\mathbf{n}]^2, \mathbf{j} \in [0,\mathbf{n}]) .$$

Then the sums are equal (proposition 291 (p. 187)).

2. We have the following statement:

$$(\forall \alpha > 0)(\exists (\mathbf{p},\mathbf{q}) \in (\mathcal{Z}^+)^2)(\forall \mathbf{S} \subset [\mathbf{p},\rightarrow[\times[\mathbf{q},\rightarrow[)$$
$$(\mathbf{Card}(\mathbf{S}) \in \mathcal{Z}^+ \Rightarrow \sum_{(\mathbf{i},\mathbf{j})\in \mathbf{S}} \|\frac{\mathbf{A}^\mathbf{i}}{\mathbf{i}!}\frac{\mathbf{B}^\mathbf{j}}{\mathbf{j}!}\| < \alpha) . \tag{2.265}$$

We can choose $\mathbf{p}$ and $\mathbf{q}$ (see equation (2.256) (p. 406) with $\mathbf{h} = \|\mathbf{A}\|$) such that:

$$(\forall \mathbf{l} \geq \mathbf{p})(\forall \mathbf{m} \geq \mathbf{p})(\sum_{\mathbf{i}\in[\mathbf{l},\mathbf{m}]} \|\frac{\mathbf{A}^\mathbf{i}}{\mathbf{i}!}\| < \sqrt{\alpha}) \tag{2.266}$$

$$(\forall \mathbf{l} \geq \mathbf{q})(\forall \mathbf{m} \geq \mathbf{q})(\sum_{\mathbf{i}\in[\mathbf{l},\mathbf{m}]} \|\frac{\mathbf{B}^\mathbf{i}}{\mathbf{i}!}\| < \sqrt{\alpha}) \tag{2.267}$$

which is possible because the sequence $\sum_{\mathbf{i}\in[0,\mathbf{n}]} \|\frac{\mathbf{A}^\mathbf{i}}{\mathbf{i}!}\|$ and $\sum_{\mathbf{i}\in[0,\mathbf{n}]} \|\frac{\mathbf{B}^\mathbf{i}}{\mathbf{i}!}\|$ are Cauchy sequences from proposition 666. Since $\mathbf{Card}(\mathbf{S}) \in \mathcal{Z}$, from propositions 255 (p. 172) and 248 (p. 171), there is always a biggest element in both projections of $\mathbf{S}$ on $\mathcal{Z}^+$. Then we have, using proposition 328 (p. 202):

$$\|\sum_{(\mathbf{i},\mathbf{j})\in \mathbf{S}} \frac{\mathbf{A}^\mathbf{i}}{\mathbf{i}!}\frac{\mathbf{B}^\mathbf{j}}{\mathbf{j}!}\| \leq \sum_{(\mathbf{i},\mathbf{j})\in \mathbf{S}} \|\frac{\mathbf{A}^\mathbf{i}}{\mathbf{i}!}\|\|\frac{\mathbf{B}^\mathbf{j}}{\mathbf{j}!}\| \leq \sum_{(\mathbf{i},\mathbf{j})\in[\mathbf{p},\mathbf{r}]\times[\mathbf{q},\mathbf{s}]} \|\frac{\mathbf{A}^\mathbf{i}}{\mathbf{i}!}\|\|\frac{\mathbf{B}^\mathbf{j}}{\mathbf{j}!}\| < \alpha$$

where $\mathbf{r}$ and $\mathbf{s}$ are majoring of the projections of $\mathbf{S}$. Relation (2.265) is then true.

3. Calling $\mathbf{l_u}$ the limit of the $\mathbf{U_n}$ and $\mathbf{l_v^a}$, $\mathbf{l_v^b}$ the limits of $\mathbf{V_n^a}$ and $\mathbf{V_n^b}$, if $\|\mathbf{l_u} - \mathbf{l_v^a l_v^b}\| \neq 0$, we have:

$$(\exists \alpha)(\|\mathbf{l_u} - \mathbf{l_v^a l_v^b}\| > 2\alpha > 0) .$$

But for that $\alpha > 0$, supposing $\|\mathbf{l_v^b}\|$ and $\|\mathbf{l_v^a}\|$ nonzero (Otherwise if one of them is zero the second member of the corresponding $\mathbf{Min}$ could be ignored), there are $\mathbf{n_1}$, $\mathbf{p_1}$ and $\mathbf{q_1}$ such that:

$$\mathbf{i} > \mathbf{n_1} \Rightarrow \|\mathbf{U_i} - \mathbf{l_u}\| < \alpha/3$$
$$\mathbf{i} > \mathbf{p_1} \Rightarrow \|\mathbf{V_i^a} - \mathbf{l_v^a}\| < \mathbf{Min}(\sqrt{\alpha/9}, \frac{\alpha}{9\|\mathbf{l_v^b}\|})$$
$$\mathbf{i} > \mathbf{q_1} \Rightarrow \|\mathbf{V_i^b} - \mathbf{l_v^b}\| < \mathbf{Min}(\sqrt{\alpha/9}, \frac{\alpha}{9\|\mathbf{l_v^a}\|})$$

and $\mathbf{p_2} > \mathbf{p_1}$, $\mathbf{q_2} > \mathbf{q_1}$ such that
$(\forall \mathbf{S} \subset [\mathbf{p_2},\rightarrow[\times[\mathbf{q_2},\rightarrow[)(\mathbf{Card}(\mathbf{S}) \in \mathcal{Z}^+ \Rightarrow \sum_{(\mathbf{i},\mathbf{j})\in \mathbf{S}} \|\frac{\mathbf{A}^\mathbf{i}}{\mathbf{i}!}\frac{\mathbf{B}^\mathbf{j}}{\mathbf{j}!}\| < \alpha/3)$

taking $\mathbf{n_2 > p_2 + q_2}$ we have $||\mathbf{U_{n_2} - V^a_{p_2} V^b_{q_2}}|| < \alpha/3$.
Then we have:

$$\begin{aligned}||\mathbf{l_u - l^a_v l^b_v}|| =& ||\mathbf{l_u - U_{n_2} + U_{n_2} - V^a_{p_2} V^b_{q_2} + V^a_{p_2} V^b_{q_2} - l^a_v l^b_v}|| \\ &\leq ||\mathbf{l_u - U_{n_2}}|| + ||\mathbf{U_{n_2} - V^a_{p_2} V^b_{q_2}}|| + ||\mathbf{V^a_{p_2} V^b_{q_2} - l^a_v l^b_v}||\end{aligned}$$

$$\begin{aligned}&||\mathbf{V^a_{p_2} V^b_{q_2} - l^a_v l^b_v}|| = ||\mathbf{(V^a_{p_2} - l^a_v + l^a_v)(V^b_{q_2} - l^b_v + l^b_v) - l^a_v l^b_v}|| \\ &= ||\mathbf{(V^a_{p_2} - l^a_v)(V^b_{q_2} - l^b_v) + l^a_v(V^b_{q_2} - l^b_v) + (V^a_{p_2} - l^a_v) l^b_v}|| \\ &< ||\mathbf{(V^a_{p_2} - l^a_v)(V^b_{q_2} - l^b_v)}|| + ||\mathbf{l^a_v(V^b_{q_2} - l^b_v)}|| + ||\mathbf{(V^a_{p_2} - l^a_v) l^b_v}|| \\ &< \alpha/3\ .\end{aligned}$$

Then putting all together, we have:

$$||\mathbf{l_u - l^a_v l^b_v}|| < \alpha$$

which is incompatible with our assumption $||\mathbf{l_u - l^a_v l^b_v}|| > \mathbf{2}\alpha > \mathbf{0}$ then:

$$||\mathbf{l_u - l^a_v l^b_v}|| = \mathbf{0} \Rightarrow \mathbf{l_u = l^a_v l^b_v}\ .$$

Since by definition, $\mathbf{l_u = e^{A+B}},\ \mathbf{l^a_v = e^A},\ \mathbf{l^b_v = e^B}$, we get proposition 668 (p. 408).

**Proposition 669** Assuming the same condition as in the preceding proposition 668 (p. 408) we have also called $\mathbb{1}$ the unit element of the multiplication ($\mathbb{1} = \mathbf{A^0} = \mathbf{e^{\mathbb{O}}}$):

$$(\forall\alpha)(\exists\epsilon)(||\mathbf{\Delta A}|| < \epsilon \Rightarrow \frac{||\mathbf{e^{A+\Delta A} - e^A(\mathbb{1} + \Delta A)}||}{||\mathbf{\Delta A}||} < \alpha)\ .$$

**Proof**
As before considering: $\mathbf{U_n} = \sum_{\mathbf{i}\in[\mathbf{0},\mathbf{n}]} \frac{\mathbf{\Delta A^i}}{\mathbf{i!}}$ and for $\mathbf{n > 2}$

$$\mathbf{U_n} - \mathbb{1} - \mathbf{\Delta A} = \sum_{\mathbf{i}\in[\mathbf{2},\mathbf{n}]} \frac{\mathbf{\Delta A^i}}{\mathbf{i!}}$$

for $||\mathbf{\Delta A}|| < \mathbf{1/2}$ and for $\mathbf{i \geq 2}$ we have:

$$\frac{||\mathbf{\Delta A^i}||}{\mathbf{i!}} < \frac{||\mathbf{\Delta A}||^2}{\mathbf{2}} \frac{\mathbf{1}}{\mathbf{2^{i-2}}}$$

then:

$$\mathbf{n > 2} \Rightarrow ||\mathbf{U_n} - \mathbb{1} - \mathbf{\Delta A}|| < \frac{||\mathbf{\Delta A}||^2}{\mathbf{2}} \times \frac{\mathbf{1 - 1/2^{n-1}}}{\mathbf{1 - 1/2}} < ||\mathbf{\Delta A}||^2$$

then

$$||\mathbf{\Delta A}|| < \mathbf{1/2} \Rightarrow ||\mathbf{e^{\Delta A}} - \mathbb{1} - \mathbf{\Delta A}|| < ||\mathbf{\Delta A}||^2$$

then by taking $\mathbf{0} < \epsilon < \mathbf{min(1/2,} \frac{\mathbf{4}\alpha}{||\mathbf{e^A}||})$ we have:

$$||\mathbf{\Delta A}|| < \epsilon \Rightarrow \frac{||\mathbf{e^{A+\Delta A} - e^A(\mathbb{1} + \Delta A)}||}{||\mathbf{\Delta A}||} < \alpha\ .$$

## 2.29.1 Exponential of complex numbers

$\mathcal{C}$ is equivalent to the $\mathcal{R}^2$ vectorial space and using the multiplication as a map of $\mathcal{C} \times \mathcal{C} \to \mathcal{C}$ it fulfills the condition of propositions 666 (p. 406) and of 668 (p. 408). See the section 2.13 (p. 271).

**Proposition 670** For any $z \in \mathcal{C}$, the complex conjugate of $e^z$ is $e^{\bar{z}}$ where $\bar{z}$ is the complex conjugate of $z$:

$$\overline{e^z} = e^{\bar{z}} .$$

**Proof**
Using the sequence $(U_n : \mathcal{Z}^{0+} \to \mathcal{C})$ which defines $e^z$: $U_n = \sum_{k\in[0,n]} z^k/k!$, from propositions 481 (p. 273) and 485 (p. 273), we have:

$$\overline{U_n} = \sum_{k\in[0,n]} \bar{z}^k/k! .$$

Then, from proposition 483 (p. 273):

$$\begin{aligned}
&(\forall \alpha > 0)(\exists p(\alpha))(\exists p'(\alpha))\\
&((n > p(\alpha) \Rightarrow \|U_n - e^z\| < \alpha/2 \Rightarrow \|\overline{U_n - e^z}\| < \alpha/2 \Rightarrow \|\overline{U_n} - \overline{e^z}\| < \alpha/2),\\
&(n > p'(\alpha) \Rightarrow \|\overline{U_n} - e^{\bar{z}}\| < \alpha/2))\\
&n > \max(p(\alpha), p'(\alpha))\\
&\Rightarrow \|\overline{e^z} - e^{\bar{z}}\| = \|\overline{e^z} - \overline{U_n} + \overline{U_n} - e^{\bar{z}}\| \leq \|\overline{U_n} - \overline{e^z}\| + \|\overline{U_n} - e^{\bar{z}}\| < \alpha\\
&(\forall \alpha > 0)(\|\overline{e^z} - e^{\bar{z}}\| < \alpha) \Rightarrow \overline{e^z} = e^{\bar{z}} .
\end{aligned}$$

**Proposition 671** $x \in \mathcal{R} \Rightarrow e^x > 0$

**Proof**
$x \geq 0 \Rightarrow e^x > 0$ because all $U_n$ defining $e^x$ are $> 0$ when $x \geq 0$. Since $e^{-x}e^x = e^{x-x} = e^0 = 1 \Rightarrow e^x = 1/e^{-x}$ then if $x < 0$, $-x > 0 \Rightarrow e^{-x} > 0 \Rightarrow 1/e^{-x} > 0 \Rightarrow e^x > 0$.

**Proposition 672** $\|e^{x+iy}\| = e^x$

**Proof**
Since $x \in \mathcal{R} \Rightarrow e^x > 0$, using propositions 483 (p. 273) and 670 (p. 411):

$$\|e^{x+iy}\|^2 = e^{x+iy}\overline{e^{x+iy}} = e^{2x} = e^x e^x \Leftrightarrow \sqrt{\|e^{x+iy}\|^2} = e^x .$$

**Proposition 673** $\theta \in \mathcal{R} \Rightarrow \|e^{i\theta}\| = 1$

**Proof**
From proposition 672 $\|e^{i\theta}\| = \|e^{0+i\theta}\| = e^0 = 1$.

**Derivative of complex exponential**

**Proposition 674** $e^z$ is the complex derivative of $e^z$.

**Proof**

From definition 313 (p. 403) and proposition 669 (p. 410), we have:

$$(\forall\alpha > 0)(\exists\epsilon > 0)(\|\Delta z\| < \epsilon \Rightarrow \|\frac{e^{z+\Delta z} - e^z(1+\Delta z)}{\Delta z}\| < \alpha$$
$$\Rightarrow \|\frac{e^{z+\Delta z} - e^z}{\Delta z} - e^z\| < \alpha) .$$

**Proposition 675** Given a derivable map $(z(t) : \mathcal{R} \to \mathcal{C})$, setting $z(t) = x(t) + iy(t)$ and its derivative $z'(t) = x'(t) + iy'(t)$, the derivative of $Z = e^{z(t)}$ is $Z' = e^{z(t)}z'(t)$.

**Proof**

We may write:

$$\begin{aligned} Z(t+\Delta t) - Z(t) &= e^{z(t)}(z(t+\Delta t) - z(t)) + \eta_1 \times (z(t+\Delta t) - z(t)) \\ &= e^{z(t)}(z'(t) + \eta_2)\Delta t + \eta_1 \times (z'(t) + \eta_2)\Delta t \\ &= e^{z(t)}(z'(t) + \eta_2 + \eta_1 \times (z'(t) + \eta_2))\Delta t . \end{aligned}$$

From the definition of normal and complex derivatives we have:

$$(\forall\alpha > 0)(\exists\epsilon_1(\alpha) > 0)(\|z(t+\Delta t) - z(t)\| < \epsilon_1(\alpha) \Rightarrow \|\eta_1\| < \alpha)$$
$$(\forall\alpha > 0)(\exists\epsilon_2(\alpha) > 0)(|\Delta t| < \epsilon_2(\alpha) \Rightarrow \|\eta_2\| < \alpha) .$$

If $z'(t) \neq 0$, we have:

$$(\forall\alpha > 0)(\exists\epsilon(\alpha) = \min(\epsilon_1(0.5\alpha/\|z'(t)\|), \epsilon_2(0.5\alpha)) > 0)(|\Delta t| < \epsilon(\alpha) \Rightarrow$$
$$\|\eta_2 + \eta_1(z'(t) + \eta_2)\| \leq \|\eta_2\| + \|\eta_1\|\|z'(t) + \eta_2\| \leq \|\eta_2\| + \|\eta_1\|(\|z'(t)\| + \|\eta_2\|)$$
$$< 0.5\alpha + 0.5\alpha\frac{\|z'(t)\| + \|\eta_2\|}{\|z'(t)\|} \leq \alpha) .$$

If $z'(t) = 0$, we have:

$$(\forall\alpha > 0)(\exists\epsilon(\alpha) = \min(\epsilon_1(1), \epsilon_2(0.5\alpha)) > 0)(|\Delta t| < \epsilon(\alpha) \Rightarrow$$
$$\|\eta_2 + \eta_1(z'(t) + \eta_2)\| \leq \|\eta_2\| + \|\eta_1\|\|z'(t) + \eta_2\| \leq \|\eta_2\| + \|\eta_1\|\|\eta_2\|$$
$$< 0.5\alpha + 1 \times 0.5\alpha \leq \alpha) .$$

Then we have:

$$(\forall\alpha > 0)(\exists\epsilon(\alpha) > 0)(|\Delta t| < \epsilon \Rightarrow \|\frac{Z(t+\Delta t) - Z(t)}{\Delta t} - e^{z(t)}z'(t)\| < \alpha) .$$

Both the real and imaginary parts of $Z$ have to satisfy the same equation so $Z'(t) = e^z(t)z'(t)$.

**Proposition 676** When $x \in \mathcal{R}$, the derivative of $e^x$ is $e^x$ and $e^x \in C^\infty$.

**Proof**

From proposition 671 (p. 411) $x \in \mathcal{R} \Rightarrow e^x \in \mathcal{R}^+$. Setting $z = x$ in proposition 675 (p. 412), we have $\frac{de^x}{dx} = e^x$ (see definition 164 (p. 193)). It means applying proposition 927 (p. 520) that the derivative at all order of $e^x$ is $e^x$ and $e^x \in C^\infty$.

**Proposition 677** With $\theta \in \mathcal{R}$, the derivative of $e^{i\theta}$ is $ie^{i\theta}$.

**Proof**

We put $z = i\theta$ in proposition 675 (p. 412).

## 2.29.2 Trigonometry

**Definition 318 sine and cosine.**

$\theta \in \mathcal{R}$, one writes down $e^{i\theta} = \cos\theta + i\sin\theta$ with $\cos\theta \in \mathcal{R}$, $\sin\theta \in \mathcal{R}$.

**Proposition 678** $\cos\theta^2 + \sin\theta^2 = 1$ .

**Proof**
It is a direct consequence of proposition 673 (p. 411).

**Proposition 679** The derivatives of the trigonometric functions are given by:

$$\cos'\theta = -\sin\theta,\ \sin'\theta = \cos\theta\ .$$

**Proof**
It is the consequence of propositions 677 (p. 412) and 674 (p. 412). From propositions 677 (p. 412) and 674 (p. 412), we have:

$$\cos'\theta + i\sin'\theta = i(\cos\theta + i\sin\theta) = -\sin\theta + i\cos\theta\ .$$

**Proposition 680** $\theta \in \mathcal{R} \Rightarrow (i\theta)^{4n} = \theta^{4n}$.

**Proof**
It is true for $n = 0$, if it is true for $n$, it is true for $n + 1$ since: $i^4 = i^2 \times i^2 = (-1)^2$.

**Proposition 681** $(i)^{4n+1} = i, (i)^{4n+2} = -1, (i)^{4n+3} = -i$.

**Proof**
It comes from proposition 680 (p. 413) using $i^2 = -1$.

**Proposition 682** $\cos\theta$ is the limit of the sequence $U_n = \sum_{i\in[0,n]}(-1)^n\frac{\theta^{2n}}{2n!}$.
$\sin\theta$ is the limit of $V_n = \sum_{i\in[0,n]}(-1)^n\frac{\theta^{2n+1}}{2n+1!}$.

**Proof**
It comes from the definition of $e^{i\theta}$, $\cos\theta$, $\sin\theta$ and from propositions 680 (p. 413) and 681 (p. 413).

**Proposition 683** Moivre formula:

$$n \in \mathcal{Z}^+ \Rightarrow (\cos\theta + i\sin\theta)^n = e^{in\theta} = \cos n\theta + i\sin n\theta\ .$$

**Proof**
It is a direct consequence of proposition 668 (p. 408).

**Definition 319 $\pi$.**

We have from proposition 682 (p. 413) $\cos 0 = 1$ and $\sin 0 = 0$.
Let us suppose $(\forall\theta)(\cos\theta > 0)$ and let us show that it is impossible.
Since $\cos\theta$ is the derivative of $\sin\theta$, $\sin\theta$ is a strictly increasing function (definition 386 (p. 486) and proposition 307 (p. 193)). Then $\sin\theta$ would be positive for $\theta > 0$. The derivative of $\cos\theta$ is $-\sin\theta$. $\sin\theta > 0$ would imply that $\cos\theta$ is a decreasing (definition 385 (p. 486))

function in any interval $[\mathbf{a}, \mathbf{b}]$ where $\mathbf{a}$ and $\mathbf{b}$ are positive. From proposition 314 (p. 196), using again the fact that $\cos\theta$ is the derivative of $\sin\theta$ we should have:

$$(\forall\theta > 0)(\exists x \in ]0, \theta[)(\cos x = \sin\theta/\theta \leq 1/\theta)$$

because $\sin\theta \leq 1$. If we take $\theta = 2$ there is an $x$ with $\cos x \leq 1/2$ and applying proposition 683 (p. 413), we have $\cos 2x = \cos^2 x - \sin^2 x = 2\cos^2 x - 1 \leq -1/2$ which is against our first assumption $(\forall\theta)\cos\theta > 0$. So from A.1.4, 8, there is a $b$ for which $\cos b \leq 0$. $\cos\theta$ being a continuous function from proposition 299 (p. 190) we have successively true:

$$\begin{aligned}
&\cos([0, b]) = [\mathrm{Min}(\cos([0, b])), \mathrm{Max}(\cos([0, b]))] \\
&((\forall\theta)(\cos\theta \leq 1), \cos 0 = 1) \Rightarrow \mathrm{Max}(\cos([0, b]) = 1) \\
&\mathrm{Min}(\cos([0, b])) \leq \cos b \leq 0, \mathrm{Max}(\cos([0, b]) = 1 \\
&0 \in [\mathrm{Min}(\cos([0, b])), 1] = [\mathrm{Min}(\cos([0, b])), \mathrm{Max}(\cos([0, b]))] \Rightarrow 0 \in \cos([0, b]) \\
&0 \in \cos([0, b])
\end{aligned}$$

one calls $\pi$ the smallest positive value of $\theta$ for which $\cos(\theta/2) = 0$.

**Proposition 684** $i = e^{i\pi/2}$.

**Proof**
From definition 319, $\cos(\pi/2) = 0$ and $\cos[0, \pi/2] = [1, 0]$. Then $\sin$ is an increasing function in $[0, \pi/2]$. With $\sin 0 = 0$, it means $\sin\pi/2 \geq 0$ (proposition 315 (p. 196)). Then $\sin^2(\pi/2) = 1 - \cos^2(\pi/2)$, $\sin(\pi/2) = 1$ (see proposition 328 (p. 202)). Then:

$$e^{i\pi/2} = \cos(\pi/2) + i\sin(\pi/2) = i\ .$$

**Proposition 685** $e^{i\pi} = e^{-i\pi} = -1$.

**Proof**

$$e^{i\pi/2}e^{i\pi/2} = e^{i\pi} = i^2 = -1\ .$$

**Proposition 686** $e^{2i\pi} = 1$.

**Proof**

$$e^{i\pi}e^{i\pi} = e^{2i\pi} = (-1)^2 = 1\ .$$

**Proposition 687** $n \in \mathcal{Z} \Rightarrow e^{i(\theta+2n\pi)} = e^{i\theta}$.

**Proof**
It is a direct consequence of proposition 668 (p. 408) and proposition 686. We also have from that:

$$\cos(\theta + 2n\pi) = \cos\theta;\quad \sin(\theta + 2n\pi) = \sin\theta\ .$$

**Proposition 688** $e^{i(\theta+\pi)} = -e^{i\theta}$.

**Proof**
It comes from propositions 668 (p. 408) and 685. Then we have:

$$\cos(\theta+\pi) = -\cos\theta; \quad \sin(\theta+\pi) = -\sin\theta$$

$$\cos(\theta-\pi) = -\cos\theta; \quad \sin(\theta-\pi) = -\sin\theta\ .$$

**Proposition 689** $\cos(-\theta) = \cos\theta; \quad \sin(-\theta) = -\sin\theta.$
**Proof**

$$\cos(-\theta) + i\sin(-\theta) = e^{-i\theta} = \overline{e^{i\theta}} = \overline{\cos\theta + i\sin\theta} = \cos\theta - i\sin\theta\ .$$

**Proposition 690** $\cos(\pi-\theta) = -\cos\theta; \quad \sin(\pi-\theta) = \sin\theta.$
**Proof**
It is obtained by combining propositions 689 (p. 415) and 688 (p. 414).

**Proposition 691** $\cos(\pi/2-\theta) = \sin\theta; \quad \sin(\pi/2-\theta) = \cos\theta.$
**Proof**

$$e^{i(\pi/2-\theta)} = ie^{-i\theta} = \sin\theta + i\cos\theta\ .$$

**Proposition 692** $e^{i\theta}$ is a bijection between the interval $[0, 2\pi[$ of $\mathcal{R}$ and the subset of $\mathcal{C}$ which contains the complex numbers of module one.
**Proof**
From the definition of $\pi$ 319 (p. 413) the following maps:

$$(\cos\theta : [0, \pi/2] \to [0,1])$$

$$(\sin\theta : [0, \pi/2] \to [0,1])$$

are bijections. Then the map on $\mathcal{C}$:

$$e^{i\theta} : [0, \pi/2[\to \mathrm{set}_z\{z = x + iy \in \mathcal{C}, \|z\| = 1, x > 0, y \geq 0\}$$

is a bijection and from using the preceding formulae, we deduce the other 3 bijections:

$$e^{i\theta} : [\pi/2, \pi[\to \mathrm{set}_z\{z = x + iy \in \mathcal{C}, \|z\| = 1, x \leq 0, y > 0\}$$

$$e^{i\theta} : [\pi, 3\pi/2[\to \mathrm{set}_z\{z = x + iy \in \mathcal{C}, \|z\| = 1, x < 0, y \leq 0\}$$

$$e^{i\theta} : [3\pi/2, 2\pi[\to \mathrm{set}_z\{z = x + iy \in \mathcal{C}, \|z\| = 1, x \geq 0, y < 0\}\ .$$

Those 4 bijections assemble in one bijection:

$$(e^{i\theta} : [0, 2\pi[\to \mathrm{set}_z\{z \in \mathcal{C}, \|z\| = 1\})\ .$$

Using proposition 687 (p. 414), we also have the set of bijections for any value of $a \in \mathcal{R}$:

$$(e^{i\theta} : [a, 2\pi + a] \to \mathrm{set}_z\{z \in \mathcal{C}, \|z\| = 1\})\ .$$

**Proposition 693** Any $\mathbf{z}$ of $\mathcal{C}$ can be written as $\mathbf{z} = \rho e^{i\theta}$ with $\rho = ||z|| \geq 0$ and $\theta \in [0, 2\pi[$.

**Proof**
Any complex number $z = x + iy$ can be written with $\rho = \sqrt{x^2 + y^2}$ as:

$$z = \rho(x/\rho + iy/\rho)$$

we have $||x/\rho + iy/\rho|| = \sqrt{x^2 + y^2}/\rho = 1$ then from proposition 692 (p. 415)

$$(\exists \theta \in [0, 2\pi[)(x/\rho + iy/\rho = e^{i\theta})$$

then with those $\rho$, $\theta$ we have:

$$z = \rho e^{i\theta} .$$

## 2.30 Polynomials

**Definition 320 Polynomial.**

A polynomial on $\mathcal{C}$ of degree $n \in \mathcal{Z}^{0+}$ is the map $(P(z) : \mathcal{C} \to \mathcal{C})$ associated to a map $(a_k : [0, n] \to \mathcal{C})$ with $a_n \neq 0$ by the following formula:

$$P(z) = \sum_{k \in [0,n]} a_k z^k .$$

**Definition 321 Polynomial expression.**

Given a finite set $S \subset (\mathcal{Z}^{0+})^p$, one calls a polynomial expression on $\mathcal{C}^p$, a map

$$(P^S(a, Z) : \mathcal{C}^{Card(S)} \times \mathcal{C}^p \to \mathcal{C})$$

with $(a_j : S \to \mathcal{C})$ and $Z = (z_1, ..., z_p) \in \mathcal{C}^p$, defined as follows

$$P^S(a, Z) = \sum_{j=\prod_{k \in [1,p]} j_k \in S} a_j \prod_{k \in [1,p]} z_k^{j_k} .$$

**Proposition 694** The set of polynomials and the set of polynomial expressions are vector spaces (definition 199 (p. 235)) with $\mathcal{C}$ (see section 2.13 (p. 271)) as field (definition 181 (p. 217)).

**Proof**
It is clear from the definition.

**Proposition 695** The product of two polynomials (resp. polynomial expressions) is a polynomial (resp. polynomial expression).

**Proof**
It is also clear from the definition.

**Proposition 696** Given a polynomial $P(z)$ of degree $n$ defined by:
$(a_k : [0, n] \to \mathcal{C})$, $a_n \neq 0$,
we have:

$$(\forall \alpha \in \mathcal{R}^+)(\exists R \in \mathcal{R}^+)(||z|| > R \Rightarrow ||\frac{P(z) - a_n z^n}{a_n z^n}|| < \alpha) .$$

**Proof**

It is true for $\mathbf{n = 0}$. Let us suppose that it is true up to $\mathbf{n}$, let us prove it is then true for $\mathbf{n+1}$. We set $\mathbf{a} = \|\mathbf{a_{n+1}}\| > 0$. We call $\mathbf{p}$, the highest value of $\mathbf{i}$ below $\mathbf{n+1}$ for which $\mathbf{a_i \neq 0}$. If there is no such $\mathbf{p}$ proposition 696 is true since $\mathbf{P(z) - a_{n+1}z^{n+1} = 0}$. We have:

$$\frac{P(z) - a_{n+1}z^{n+1}}{a_{n+1}} = \frac{P(z) - a_{n+1}z^{n+1} + a_pz^p - a_pz^p}{a_pz^p} \times \frac{a_pz^p}{a_{n+1}z^{n+1}}$$
$$= (\frac{P(z) - a_pz^p - a_{n+1}z^{n+1}}{a_pz^p} + 1)\frac{a_pz^p}{a_{n+1}z^{n+1}} .$$

From the above, definition 247 (p. 272) and proposition 480 (p. 272), we have:

$$\|\frac{P(z) - a_{n+1}z^{n+1}}{a_{n+1}}\| < \|\frac{P(z) - a_pz^p - a_{n+1}z^{n+1}}{a_pz^p}\|\|\frac{a_pz^p}{a_{n+1}z^{n+1}}\| + \|\frac{a_pz^p}{a_{n+1}z^{n+1}}\| .$$

Having $\mathbf{R}$ verifying proposition 696 (p. 416) for $\alpha$ and $\mathbf{p}$, having $\mathbf{A > 0}$ verifying (proposition 255 (p. 172)):

$$(\forall i \in [0, n])(\|a_i\| < A)$$

we also have $\mathbf{R' = \max(R, \frac{2A}{a}, \frac{2A}{\alpha a}, 1)}$. Then for $\|\mathbf{z}\| > \mathbf{R'}$, we have successively:

$$\|z\| > R' > R \Rightarrow \|\frac{P(z) - a_pz^p - a_{n+1}z^{n+1}}{a_pz^p}\| < \alpha$$
$$\|z\| > R' > \frac{2A}{a}, \|z\| > R' > 1 \Rightarrow$$
$$\|\frac{a_pz^p}{a_{n+1}z^{n+1}}\| = \frac{\|a_p\|}{\|a_{n+1}\|\|z^{n+1-p}\|} \leq \frac{A}{\|a\|\|z\|} < 1/2$$
$$\Rightarrow \|\frac{P(z) - a_pz^p - a_{n+1}z^{n+1}}{a_pz^p}\|\|\frac{a_pz^p}{a_{n+1}z^{n+1}}\| < \alpha/2$$
$$\|z\| > R' > \frac{2A}{\alpha a}, \|z\| > R' > 1 \Rightarrow \|\frac{a_pz^p}{a_{n+1}z^{n+1}}\| = \|\frac{a_p}{a_{n+1}z^{n+1-p}}\| < \alpha/2 .$$

Then

$$\|z\| > R' \Rightarrow \|\frac{P(z) - a_{n+1}z^{n+1}}{a_{n+1}z^{n+1}}\| < \alpha .$$

**Proposition 697** Given a polynomial of degree $\mathbf{n}$ defined by:

$$(a_k : [0, n] \to \mathcal{C}), \ \|a_n\| > 0 ,$$

we have:

$$(\forall \alpha \in \mathcal{R}^+)(\exists R \in \mathcal{R}^+)(\|z\| > R \Rightarrow \|P(z)\| > \alpha,$$
$$\|P(z)\| \leq \alpha \Rightarrow \|z\| \leq R) . \tag{2.268}$$

**Proof**
Having a polynomial $\mathbf{P(z) = \sum_{k \in [0,n]} a_kz^k}$, from proposition 404 (p. 239), we have:

$$\|P(z) - a_nz^n - P(z)\| = \|a_nz^n\| < \|P(z) - a_nz^n\| + \|P(z)\| .$$

From proposition 696 (p. 416), we also have:

$$(\forall \beta)(\exists R_0)(\|z\| > R_0 \Rightarrow \|a_nz^n\| - \|P(z)\| < \|P(z) - a_nz^n\| < \beta\|a_nz^n\|) .$$

Then corresponding to $\beta = 0.5$, there is $R_0$ such that:

$$\begin{aligned}\|z\| > R_0 &\Rightarrow \|a_n z^n\| < 0.5\|a_n z^n\| + \|P(z)\| \\ &\Rightarrow 0.5\|a_n z^n\| < \|P(z)\| \\ &\Rightarrow (\forall \alpha > 0)(\exists R = \text{Max}\{(2\alpha/\|a_n\|)^{1/n}, R_0\})(\|z\| > R \Rightarrow \|P(z)\| > \alpha)\ .\end{aligned} \tag{2.269}$$

The second line of equation (2.268) (p. 417) comes from 7 (p. 435) and 23 (p. 436) in A.1.4 (p. 435).

**Proposition 698** $(\forall z)$ the derivative of $P(z) = \sum_{k\in[0,n]} a_k z^k$ is

$$P'(z) = \sum_{k\in[1,n]} k a_k z^{k-1}\ .$$

**Proof**
From proposition 369 (p. 217), we have:

$$P(z+\Delta z) - P(z) = \sum_{k\in[0,n]} a_k((z+\Delta z)^k - z^k) = \sum_{k\in[1,n]} a_k \sum_{j\in[1,k]} \frac{k!}{j!(k-j)!} z^{k-j}\Delta z^j$$

calling

$$A = \max_{k\in[1,n]} \|a_k\| \max_{j\in[2,k]} \frac{k!}{j!(k-j)!}\|z\|^{k-j}$$

we have when $\|\Delta z\| < 1$

$$\begin{aligned}&\left\|\frac{P(z+\Delta z) - P(z)}{\Delta z} - \sum_{k\in[1,n]} k a_k z^{k-1}\right\| \le A \sum_{j\in[1,n-1]} \|\Delta z\|^j \\ &\le A\|\Delta z\|\frac{1-\|\Delta z\|^{n-1}}{1-\|\Delta z\|} \le A\frac{\|\Delta z\|}{1-\|\Delta z\|}\end{aligned}$$

for $\alpha > 0$ setting $0 < \epsilon < \min(1/2, \alpha/A)$

$$(\forall \alpha > 0)(\|\Delta z\| < \epsilon \Rightarrow \left\|\frac{P(z+\Delta z) - P(z)}{\Delta z} - \sum_{k\in[1,n]} k a_k z^{k-1}\right\| < \alpha)\ . \tag{2.270}$$

**Proposition 699** $P(z)$ is a continuous map.

**Proof**
It comes directly from proposition 660 (p. 404).

**Proposition 700** Polynomial expressions (definition 321 (p. 416)) are continuous maps.

**Proof**
From proposition 699, they are sums and products of continuous maps. Then, from propositions 661 (p. 405), 662 (p. 405), 241 (p. 167) and proposition 242 (p. 168), polynomial expressions are continuous in the product of them and from proposition 246 (p. 169) in all the variables.

**Proposition 701** Given two polynomials $P(z) = \sum_{i\in[1,n]} a_i z^i$, $Q(z) = \sum_{i\in[1,n]} b_i z^i$,

$$(\forall z \in \mathcal{C})(P(z) = Q(z)) \Leftrightarrow (\forall i)(a_i = b_i)\ .$$

**Proof**
The left side of the arrow is obvious. $P(z)$ and $Q(z)$ being equal polynomials $P(z) - Q(z) = \sum_{i\in[1,n]} c_i z^i = \sum_{i\in[1,n]}(a_i - b_i)z^i$ is always a null polynomial. Calling its highest degree term $c_p z^p$, the preceding proposition gives:

$$\begin{aligned}&(\forall\alpha > 0)(\exists R > 0)\\&(\|z\| > R \Rightarrow \|c_p\|\|z^p\| = \|c_p z^p\|\\&= \|P(z) - Q(z) - c_p z^p\| < \alpha\|c_p z^p\|) \Rightarrow \|c_p\| < \alpha\|c_p\|)\\&\Rightarrow \|c_p\| = 0\end{aligned}$$

then $(\forall i \in [1,n])(c_i = 0)$.

**Proposition 702** For any polynomial $P(z)$ of degree $n \in \mathcal{Z}^+$ acting on $\mathcal{C}$ there is a $z_0 \in \mathcal{C}$ such that $P(z_0) = 0$.

**Proof**
From proposition 697 (p. 417), we have:

$$(\exists R \in \mathcal{R}^+)(\|z\| \geq R \Rightarrow \|P(z)\| > \|a_0\|)\ . \tag{2.271}$$

We consider the closed ball $\bar{B}(R, 0)$ of $\mathcal{C}$, $\|z\| \leq R$. This ball is compact. Since $P(z)$ is a continuous map, from proposition 516 (p. 288) and proposition 241 (p. 167), the following map:

$$(\|P(z)\| : \mathcal{C} \to \mathcal{R}^{+0})$$

is a continuous map.
From proposition 240 (p. 167), $\|P(\bar{B}(0,R))\|$ is a compact set of $\mathcal{R}^{+0}$. 0 is a minoring value of $\mathcal{R}^{+0} \subset \mathcal{R}$. $\|P(\bar{B}(0,R))\|$ then has a **lower_limit** $P_0$ (proposition 258 (p. 174)) which is in $\|P(\bar{B}(0,R))\|$ (propositions 226 (p. 160) and 220 (p. 158)).
There is then a $z_0 \in (\bar{B}(0,R))$ such that $\|P(z_0)\| = P_0$.

$$0 \in \bar{B}(0,R) \Rightarrow \|a_0\| = \|P(0)\| \in \|P(\bar{B}(0,R))\|\ .$$

We have then $P_0 \leq \|a_0\|$. From the preceding inequality (2.271), we also have:

$$\|z_0\| < R,\ z_0 \in B(0,R)\ .$$

From proposition 266 (p. 179), there is a $r$ such that $B(z_0, r) \subset B(0,R)$.
We can consider the map $(Z + z_0 : \mathcal{C} \to \mathcal{C})$ and the new polynomial of degree $n$:

$$P \circ (Z + z_0) = P(Z + z_0) = P_Z(Z) = \sum_{k\in[1,n]} a_k (Z + z_0)^k \tag{2.272}$$

with (see proposition 369 (p. 217)):

$$(Z + z_0)^k = \sum_{i\in[0,k]} \frac{k!}{i!(k-i)!} Z^i z_0^{k-i}\ .$$

We set $P_Z(Z) = \sum_{k\in[0,n]} b_k Z^k$ with $b_i = \sum_{k\in[0,n-i]} a_k z_0^{k-i} \frac{k!}{i!(k-i)!}$, $b_0 = P(z_0)$ and $b_n = a_n$.

Let us consider the set $\mathbf{S_b} = \mathbf{set}_{q\in[1,n]}\{\mathbf{b_q} \neq \mathbf{0}\}$, from definition 320 $\mathbf{S_b} \neq \emptyset$ and $\mathbf{S_b}$ has a smallest value $\mathbf{p} \in [\mathbf{1}, \mathbf{n}]$ (see the proposition 886 (p. 504)). We have:

$$\mathbf{P_Z(Z) = Z^n(\sum_{q\in[0,n-p]} b_{n-q}(1/Z)^q) + b_0 \ , \ b_p \neq 0 \ .}$$

From proposition 696, $\forall\alpha \in \mathcal{R}^+$ there is an $\epsilon \in ]\mathbf{0}, \mathbf{r}[\subset \mathcal{R}^+$ such that:

$$\begin{aligned} &\mathbf{\|1/Z\| > 1/\epsilon \Leftrightarrow \|Z\| < \epsilon} \\ &\Rightarrow \mathbf{\|\sum_{q\in[0,n-p]} b_{n-q}(1/Z)^q - b_p(1/Z)^{n-p}\| < \alpha(\|1/Z\|)^{n-p}} \ . \end{aligned}$$

Using that and if $\mathbf{b_0} \neq \mathbf{0}$, we have an $\epsilon \in ]\mathbf{0}, \mathbf{r}[\subset \mathcal{R}^+$ such that:

$$\begin{aligned} \mathbf{\|Z\| < \epsilon} &\Rightarrow \mathbf{\|P_Z(Z) - b_0 - b_p Z^p\| < \alpha\|Z\|^p} \\ &\Rightarrow \mathbf{\|\frac{P_Z(Z)}{b_0} - 1 - \frac{b_p}{b_0} Z^p\| < \alpha\|\frac{Z^p}{b_0}\|} \ . \end{aligned}$$

From propositions 328 (p. 202), 693 (p. 416) and also section 2.2.3 (p. 173) there is $\boldsymbol{\rho}$ and $\boldsymbol{\theta}$ such that $\mathbf{\frac{b_0}{b_p} = \rho^p e^{ip\theta}}$, $\boldsymbol{\rho} = \mathbf{\|\frac{b_0}{b_p}\|^{1/p}} \in \mathcal{R}^+$ and setting:

$$\mathbf{Z = x\rho e^{i\theta + \pi/p}, x \in \mathcal{R}^+}$$

we have:

$$\mathbf{0 < x < \min(\epsilon/\rho, 1) \Rightarrow \|\frac{P_Z(Z)}{b_0} - 1 + x\| < \alpha x/\|b_p\|}$$

choosing $\epsilon$ for $\boldsymbol{\alpha} = \mathbf{0.5\|b_p\|}$ we have

$$\mathbf{0 < x < \min(\epsilon/\rho, 1) \Rightarrow \|\frac{P_Z(Z)}{b_0} - 1 + x\| < 0.5x}$$

Then, from proposition 404 (p. 239), taking $\mathbf{x}$ such as $\mathbf{0 < x < \min(\epsilon/\rho, 1)}$ we have:

$$\begin{aligned} &\mathbf{\|\frac{P_Z(Z)}{b_0} - 1 + x - \frac{P_Z(Z)}{b_0}\| = |1 - x| = 1 - x} \\ &\mathbf{\geq \|\frac{P_Z(Z)}{b_0}\| - \|\frac{P_Z(Z)}{b_0} - 1 + x\| > \|\frac{P_Z(Z)}{b_0}\| - 0.5x} \end{aligned}$$

then:

$$\mathbf{1 > 1 - 0.5x > \|\frac{P_Z(Z)}{b_0}\|}$$

then for our choice of $\mathbf{x > 0}$, we have a $\mathbf{Z \neq 0}$ and a $\mathbf{z = Z + z_0}$ such that

$$\mathbf{P_0 = \|b_0\| > \|P_Z(Z)\| = \|P(z)\|} \ .$$

From equation (2.271) $\mathbf{z \in \bar{B}(z_0, R)}$ so $\mathbf{\|P(z_0)\|}$ is not the smallest value of

$$\mathbf{\|P(\bar{B}(z_0, R))\| \subset \mathcal{R}^{+0}}$$

except if $\mathbf{b_0 = P(z_0) = 0}$. From A.1.4 -8 (p. 435) we conclude proposition 702.

**Proposition 703** Having a polynomial on $\mathcal{C}$: $\mathbf{P^n(z) = \sum_{k\in[0,n]} a_k z^k}$ of degree $\mathbf{n \geq 1}$, if $\mathbf{z_0}$ is such that $\mathbf{P^n(z_0) = 0}$, then there is a polynomial of degree $\mathbf{n-1 \geq 0}$: $\mathbf{Q^{n-1}(z) = \sum_{k\in[0,n-1]} c_k z^k}$ such that $\mathbf{c_{n-1} = a_n}$ and $\mathbf{P^n(z) = (z - z_0)Q^{n-1}(z)}$.

**Proof**

It is enough to consider the map $(\mathbf{Z + z_0} : \mathcal{C} \to \mathcal{C})$ and to apply the expression in formula (2.272) (p. 419) and using the notations in there, we have $\mathbf{b_0 = 0}$ since $\mathbf{P^n(z_0) = 0}$ and we have:

$$\mathbf{P^n \circ (Z + z_0) = P^n(Z + z_0) = P^n_Z(Z) = \sum_{k\in[1,n]} b_k Z^k = Z \sum_{k\in[0,n-1]} b_{k+1} Z^k}$$

with $\mathbf{b_n = a_n}$. Applying first the map $(\mathbf{z - z_0} : \mathcal{C} \to \mathcal{C})$, we get:

$$\mathbf{P^n \circ (Z+z_0) \circ (z-z_0) = P^n(Z+z_0) \circ (z-z_0) = P^n(z) = (z-z_0) \sum_{k\in[0,n-1]} b_{k+1}(z-z_0)^k} .$$

Then

$$\mathbf{Q^{n-1}(z) = \sum_{k\in[0,n-1]} c_k z^k = \sum_{k\in[0,n-1]} b_{k+1}(z - z_0)^k}$$

with $\mathbf{c_n - 1 = b_n = a_n}$.

**Proposition 704** For any polynomial on $\mathcal{C}$: $\mathbf{P^n(z) = \sum_{k\in[0,n]} a_k z^k}$, there is a map $(\mathbf{z_i} : \mathbf{[1,n]} \to \mathcal{C})$ such that:

$$\mathbf{P^n(z) = a_n \prod_{i\in[1,n]} (z - z_i)} . \tag{2.273}$$

The $\mathbf{z_i}$ are called the roots of the polynomial.

**Proof**

It is true for $\mathbf{n = 1}$ with $\mathbf{z_1 = -a_0/a_1}$. If it is true for $\mathbf{n - 1}$, from propositions 702 (p. 419) and 703 (p. 421) it is true for $\mathbf{n}$. Then we conclude by the recurrence principle (proposition 927 (p. 520)).

**Definition 322 Roots of a polynomial, multiplicity of a root.**

Given a polynomial $(\mathbf{P^n(z)} : \mathcal{C} \to \mathcal{C})$ of degree $\mathbf{n}$. Any element of the set $\mathbf{(P^n)^{-1}(\{0\})}$ is called a root of $\mathbf{P^n}$ (definition 115 (p. 133)). From proposition 702 (p. 419), that set is not empty. From proposition 704 using the map $(\mathbf{z(i) = z_i} : \mathbf{[1,n]} \to \mathcal{C})$ such that $\mathbf{P^n(z) = a_n \prod_{i\in[1,n]}(z - z_i)}$ the set of the roots is also given by $\mathbf{z([1,n])} \subset \mathcal{C}$ (definition 114 (p. 132)). Given a root $\mathbf{r \in z([1,n])}$ one calls multiplicity of $\mathbf{r}$ $\mathbf{Card(z^{-1}(\{r\}))}$. Since $\mathbf{z^{-1}(z([1,n])) = z^{-1}}(\mathcal{C}) = \mathbf{[1,n]}$ (proposition 154 (p. 133)), we have:

$$\mathbf{\sum_{r\in z([1,n])} Card(z^{-1}(\{r\})) = Card([1,n]) = n} .$$

**Proposition 705** Considering a polynomial $\mathbf{P^n(a,z)}$ of degree $\mathbf{n}$, a number $\rho \in \mathcal{C}$, there is a polynomial $\mathbf{P^{n-1}(b,z)}$ such that:

$$\mathbf{P^n(a,z) = P^n(a,\rho) + (z - \rho)P^{n-1}(b,z)}$$

where the $(\mathbf{b} : \mathcal{C}^{n+1} \to \mathcal{C}^{n-1})$ are continuous maps of $\rho$ and the $\mathbf{a}$.

**Proof**
We have to identify the terms of the same degree in $\mathbf{z}$ it gives:

$$\begin{aligned}&a_n = b_{n-1}\\&(\forall k \in [1, n-1])(a_k = b_{k-1} - \rho b_k)\\&a_0 = P^n(a, \rho) - \rho b_0\end{aligned}$$

from that we get:

$$(\forall i \in [0, n-1])(b_i = \sum_{j \in [i, n-1]} a_{i+1}\rho^{j-i}) . \tag{2.274}$$

From proposition 701 (p. 419) the polynomial $P^{n-1}(b, z)$ is unique.

**Proposition 706** We consider the polynomial on $\mathcal{C}$:

$$P^n(a, z) = z^n + \sum_{i \in [0, n-1]} a_i z^i$$

of degree $\mathbf{n}$ which is a polynomial of degree $\mathbf{n}$ divided by its highest degree coefficient $a_n \neq 0$. The coefficients $(a_i : [0, n-1] \to \mathcal{C})$ are elements of $\mathcal{C}^n$. We consider also the roots $(\rho_i : [1, n] \to \mathcal{C})$ of $P^n(a, z)$. Considering a variation $\Delta a$ of $\mathbf{a}$, we have:

$$(\forall a \in \mathcal{C}^{n+1})(\forall z \in \mathcal{C})(\forall \alpha > 0)(\exists \epsilon > 0)((\Delta a \in \mathcal{C}^{n+1}, a_n = 1, \Delta a_n = 0, \|\Delta a\| < \epsilon) \Rightarrow$$
$$(\exists (\Delta\rho_i : [1, n] \to \mathcal{C}))(\max_{i \in [1,n]} \|\Delta\rho_i\| < \alpha, P_n(a + \Delta a, z) = \prod_{i \in [1,n]} (z - \rho_i - \Delta\rho_i))) .$$

**Proof**
Given a polynomial $P^n(a, z)$ of degree $\mathbf{n}$ with $a_n = 1$, we label its $\mathbf{n}$ roots by the map $(\rho_k : [1, n] \to \mathcal{C})$ (some roots can be multiple). We may call any other polynomial as $P^n(a + \Delta a, z)$ with $(\Delta a_k : [0, n] \to \mathcal{C})$ and $\Delta a_n = 0$ and call $\|\Delta a\| = \sqrt{\sum_{i \in [0,n[} \|\Delta a_i\|^2}$. Let us suppose that we have for any $\mathbf{a}$:

$$\begin{aligned}&(\forall \alpha > 0)(\exists \epsilon > 0)(\|\Delta a\| < \epsilon \Rightarrow (\exists (\Delta\rho_k : [1, n] \to \mathcal{C}))\\&((\forall k \in [1, n])(\|\Delta\rho_k\| < \alpha), P^n(a + \Delta a, z) = \prod_{k \in [1,n]} (z - \rho_k - \Delta\rho_k))) .\end{aligned} \tag{2.275}$$

Let us prove that it is also true for any polynomial of degree $n' = n + 1$. Considering a polynomial $P^{n'}(a, z)$, we label one root as $\rho_{n'}$ taking another polynomial of degree $\mathbf{n'}$ written as $P^{n'}(a + \Delta a, z)$ with $(\Delta a_i : [1, n'] \to \mathcal{C}), \Delta a_{n'} = 0$. We set $\Delta z = z - \rho_{n'} \in \mathcal{C}$ and consider the polynomial in $\Delta z$: $P^{n'}(a + \Delta a, \rho_{n'} + \Delta z)$ and its roots $(\Delta r_i : [1, n'] \to \mathcal{C})$. We have:

$$P^n(a + \Delta a, \rho_{n'} + \Delta z) = \prod_{i \in [1,n']} (\Delta z - \Delta r_i)$$

which gives:

$$(-1)^{n'} \prod_{i \in [1,n']} \Delta r_i = P^{n'}(a + \Delta a, \rho_{n'}) .$$

From proposition 255 (p. 172), we have:

$$(\exists k \in [1, n'])(\|\Delta r_k\| = \min_{i \in [1,n']} \|\Delta r_i\|) .$$

Taking that $\mathbf{k}$, we call $\mathbf{\Delta\rho_{n'}} = \mathbf{\Delta r_k}$ and have from above:

$$\|\Delta\rho_{n'}\| = \|\Delta r_k\| \leq \|P^{n'}(a+\Delta a, \rho_{n'})\|^{1/n'} .$$

Since $\|P^{n'}(a+\Delta a, \rho_{n'})\|^{1/n'}$ is continuous in $\mathbf{\Delta a}$ we have:

$$(\forall\alpha_{n'} > 0)(\exists\epsilon_{n'}(\alpha_{n'}) > 0)(\|\Delta a\| < \epsilon_{n'}(\alpha_{n'}) \Rightarrow \Delta\rho_{n'} < \alpha_{n'}) .$$

The map $(\Delta\rho_{n'}(\Delta a) : \mathcal{C}^{n'} \to \mathcal{C})$ is continuous at $\mathbf{a} = \mathbf{0}$.
Let us find now $\mathbf{b}$ and $\mathbf{\Delta b}$ such that:

$$P^{n'}(a+\Delta a, z) = (z - \rho_{n'} - \Delta\rho_{n'})P^{n}(b+\Delta b, z)$$

with $\Delta a = 0 \Rightarrow \Delta b = 0, b_n = 1, \Delta b_n = 0$. From proposition 705 and equation (2.274), we have a formula which links $\mathbf{b}$, $\mathbf{\Delta b}$ to $\mathbf{a}$, $\mathbf{\Delta a}$, $\rho_{n'}$ and $\Delta\rho_{n'}$ in a continuous way. The roots of $P^n(b+\Delta b, z)$ are the same as the roots of $P^{n'}(a+\Delta a, z)$ except may be $\rho_{n'} + \Delta\rho_{n'}$. From expression (2.275) which is supposed to be true for $\mathbf{n}$, we may set $(\Delta\rho_i : [1, n] \to \mathcal{C})$ in order to have:

$$\begin{aligned}(\forall i \in [1, n])(P^{n'}(a, \rho_i) &= P^{n'}(a+\Delta a, \rho_i + \Delta\rho_i) \\ &= P^n(b, \rho_i) = P^n(b+\Delta b, \rho_i + \Delta\rho_i) = 0)\end{aligned}$$

with:

$$(\forall\alpha_b > 0)(\exists\epsilon_b(\alpha_b) > 0)(\|\Delta b\| < \epsilon_b(\alpha_b) \Rightarrow (\forall i \in [1, n])(\|\Delta\rho_i\| < \alpha_b)) .$$

Since $\mathbf{\Delta b}$ are continuous in $\mathbf{\Delta a}$ and in $\Delta\rho_{n'}$ for $\Delta a = 0$ and $\Delta\rho_{n'} = 0$ we have:

$$(\forall\alpha_1)(\exists\epsilon_1(\alpha_1))(\exists\epsilon'_1(\alpha_1)(\|\Delta a\| < \epsilon_1(\alpha_1), \|\Delta\rho'_n\| < \epsilon'_1(\alpha_1) \Rightarrow \|\Delta b\| < \alpha_1) .$$

Setting $\alpha_1 = \epsilon_b(\alpha)$ and $\alpha_{n'} = \min(\alpha, \epsilon'_1(\epsilon_b(\alpha)))$ we have combining all:

$$(\forall\alpha > 0)(\|\Delta a\| < \min(\epsilon_b(\alpha), \epsilon_{n'}(\alpha_{n'})) \Rightarrow (\forall i \in [1, n'])(\|\Delta\rho_i\| < \alpha))$$

then if relation (2.275) is true for $\mathbf{n}$ it is true for $\mathbf{n} = \mathbf{n}+\mathbf{1}$ since it is true for $\mathbf{n} = \mathbf{1}$, it is true for all $\mathbf{n}$ (proposition 927 (p. 520)) The relation (2.275) gives proposition 706 then proposition 706 is true.

**Proposition 707** Considering the polynomial of degree $\mathbf{n}$, $P_n(a, z)$ with $a_n = 1$, if there is an open $O \subset \mathcal{R}^{n-1}$ such that all the roots of $P_n(a, z)$ are real for $a \in O$, then the roots $(\rho_i : [1, n] \to \mathcal{R})$ can be ordered such that for all $i \in [1, n]$ the roots $\rho_i$ are a continuous map of $a \in \mathcal{R}^{n-1}$.

**Proof**
Since the roots are all real, we can set a map $(\rho_n(a) : \mathcal{R}^{n-1} \to \mathcal{R})$ such that for any $a \in O$ it is the biggest root. That root exists and is unique from proposition 255 (p. 172). From proposition 706 we have:

$$\begin{aligned}&(\forall\alpha > 0)(\exists\epsilon(\alpha) > 0)((a+\Delta a \in O, \|\Delta a\| < \epsilon(\alpha)) \\ &\Rightarrow (\exists r_\Delta \in \mathcal{R})(|r_\Delta - \rho_n(a)| < \alpha, P_n(a+\Delta a, r_\Delta) = 0)) .\end{aligned}$$

If $r_\Delta \neq \rho_n(a+\Delta a)$, we have $r_\Delta < \rho_n(a+\Delta a)$. Then we have:

$$\rho_n(a) - \alpha < r_\Delta \leq \rho_n(a+\Delta a)$$

and

$$\rho_n(a) - \rho_n(a + \Delta a) < \alpha \ . \tag{2.276}$$

From proposition 706 (p. 422), keeping its notations, we have:

$$(\exists i \in [1, n])(\rho_i(a) + \Delta\rho_i(a) = \rho_n(a + \Delta a), |\Delta\rho_i(a)| < \alpha) \ .$$

Then we have:

$$(\exists r \leq \rho_n(a))(|r - \rho_n(a + \Delta a)| < \alpha, P_n(a, r) = 0) \ .$$

It gives:

$$\rho_n(a + \Delta a) - \rho_n(a) \leq \rho_n(a + \Delta a) - r < \alpha \ . \tag{2.277}$$

Combining both equations (2.276) (p. 424) and (2.277) (p. 424), we have:

$$(\forall \alpha > 0)(\exists \epsilon > 0)(\|\Delta a\| < \epsilon \Rightarrow |\rho_n(a + \Delta a) - \rho_n(a)| < \alpha) \ .$$

From proposition 282 (p. 184), $\rho_n(a)$ is continuous on $O$. Applying again proposition 705 (p. 421) and relation (2.274) (p. 422) we can factorize the root $\rho_n(a)$ and be left with a polynomial defined on another open of $\mathcal{R}^{n-2}$ which has all its roots again real applying the same reasoning recurrently (proposition 927 (p. 520)), we have proposition 707.

## 2.31 Useful smooth maps built with exponentials

In order to ease the understanding of manifold theory, we introduce a set of smooth maps which are smooth versions of the Dirac and Heaviside distributions very popular in electronics. For that we shall prove the following propositions:

**Proposition 708** The map $(x \neq 0 \Rightarrow f = e^{-1/x^2}, x = 0 \Rightarrow f = 0 : \mathcal{R} \to \mathcal{R})$ belongs to $C^\infty$ (definitions 166 (p. 200), 276 (p. 304)) and verifies:

$$x = 0 \Rightarrow (\forall i \in \mathcal{Z}^{0+})(\partial^i f = 0)$$

$\partial^i f$ stands for the $i^{th}$ derivative of $f$[36].

**Proof**

For simplicity, for all $x \in \mathcal{R}$, we identify $f$ to $e^{-1/x^2}$. From definition 317 (p. 407), since $(\forall i \in \mathcal{Z})(\forall x \in \mathcal{R})(x^{2i} > 0)$, we have when $x \neq 0$:

$$(\forall p \in \mathcal{Z}^+)(\frac{x^{-2p}}{p!} < \sum_{i\in[0,p]} \frac{x^{-2i}}{i!} < e^{x^{-2}}) \ . \tag{2.278}$$

It gives:

$$m \in \mathcal{Z} \Rightarrow (\forall p > 0)(|x^m e^{-1/x^2}| = |x^m / e^{x^{-2}}|$$
$$< |x^m| / \sum_{i\in[0,p]} \frac{x^{-2i}}{i!} < \frac{|x|^m p!}{|x|^{-2p}} = |x|^{m+2p} p!)$$

choosing $2p > -m$ and $\epsilon = (\alpha/p!)^{1/(2p+m)}$ using proposition 328 (p. 202) we have now:

$$(\forall m \in \mathcal{Z})(\forall \alpha > 0)(\exists \epsilon > 0)(0 < |x| < \epsilon \Rightarrow |x^m e^{-1/x^2}| < \alpha) \ . \tag{2.279}$$

[36] We abandon here the traditional notation $d^k f/dx^k$ for $\partial^k f = d^k f/dx^k$ (see definition 164 (p. 193)). It avoids the confusion with the differential along the $i^{th}$ coordinate as defined in definition 40 (p. 34).

We set $\mathbf{x = 0 \Rightarrow e^{x^{-2}} = 0}$. From definition 159 (p. 183) and equation (2.279) $\mathbf{\lim_{x=0} e^{x^{-2}} =}$ $\mathbf{0}$ and from proposition 282 (see definition 166 (p. 200)) $\mathbf{(\forall x \in \mathcal{R})(e^{-1/x^2} \in C^1)}$.
We have:

$$\mathbf{x \neq 0 \Rightarrow \partial e^{-1/x^2} = 2\frac{e^{-1/x^2}}{x^3}, x = 0 \Rightarrow \partial e^{-1/x^2} = 0\ .} \tag{2.280}$$

When $\mathbf{x \neq 0}$, equation (2.280) is true from propositions 562 (p. 309), 326 (p. 202) and 675 (p. 412). For $\mathbf{x = 0}$ applying, definition 161 (p. 192) and definition 157 (p. 182), $\mathbf{x = 0 \Rightarrow \partial e^{-1/x^2} = \lim_{x=0} \frac{e^{-1/\Delta x^2}}{\Delta x}}$ then from equation (2.279) and definition 157 (p. 182) $\mathbf{x = 0 \Rightarrow \partial e^{-1/x^2} = 0 = \lim_{x=0} \frac{e^{-1/x^2}}{x^3}}$. Then from definition 159 (p. 183), proposition 282 (p. 184) and equation (2.279), we have:

$$\mathbf{(\forall x \in \mathcal{R})(e^{-1/x^2} \in C^1)\ .}$$

With $\mathbf{i \in \mathcal{Z}^+}$, let us suppose that the following equations (2.281) (see definition 166 (p. 200)) are true:

$$\begin{aligned}&\mathbf{(\forall x \in \mathcal{R})(e^{-1/x^2} \in C^i), x \neq 0 \Rightarrow \partial^i(e^{-1/x^2}) = P_i(1/x)e^{-1/x^2},}\\ &\mathbf{x = 0 \Rightarrow \partial^i(e^{-1/x^2}) = 0,}\end{aligned} \tag{2.281}$$

where $\mathbf{P_i(y)}$ is a polynomial in $\mathbf{y}$ of degree $\mathbf{3i}$ (definition 320 (p. 416)).
If equation (2.281) is true for $\mathbf{i}$, it is also true for $\mathbf{i+1}$:
Applying propositions 563 (p. 311), 325 (p. 201) and 674 (p. 412), we have when $\mathbf{x \neq 0}$:

$$\mathbf{\partial(P_i(1/x)e^{-1/x^2}) = -\frac{e^{-1/x^2}}{x^2}(\partial P_i(1/x) - \frac{2}{x}P_i(1/x))\ .}$$

From proposition 698 (p. 418), $\mathbf{P_{i+1}(1/x) = \frac{1}{x^2}(\partial P_i(1/x) - \frac{2}{x}P_i(1/x))}$ is a polynomial. Relations (2.281) can be written as

$$\mathbf{\partial^i(e^{-1/x^2}) = \sum_{j \in [0,3i]} a^i_j x^{-j} e^{-1/x^2}}$$

$$\mathbf{\partial^{i+1}(e^{-1/x^2}) = \sum_{j \in [0,3i+3]} a^{i+1}_j x^{-j} e^{-1/x^2}\ .}$$

We have:

$$\mathbf{\lim_{\Delta x = 0} \frac{\partial^i e^{-1/\Delta x^2} - 0}{\Delta x} = 0}$$

and

$$\mathbf{\lim_{x=0} \partial^{i+1}(e^{-1/x^2}) = 0}$$

because, from (2.279) (p. 424): $\mathbf{(\exists \epsilon_m(\alpha))(|x| < \epsilon_m(\alpha) \Rightarrow |x^{-m}e^{-1/x^2}| < \alpha)}$, setting:

$$\mathbf{\epsilon_1 = \min_{j \in [0,3i]} \epsilon_{j+1}(\frac{\alpha}{\sum_{k \in [0,3i]} |a^i_k|})}$$

$$\mathbf{\epsilon_2 = \min_{j \in [0,3+3i]} \epsilon_j(\frac{\alpha}{\sum_{k \in [0,3i+3]} |a^{i+1}_k|})}$$

we have:

$$\mathbf{(\forall \alpha)(\exists \epsilon_1)(|\Delta x| < \epsilon_1 \Rightarrow |\frac{\partial^i e^{-1/\Delta x^2} - 0}{\Delta x}| < \alpha) \Rightarrow (x = 0 \Rightarrow \partial^{i+1} e^{-1/x^2} = 0)}$$

and

$$(\forall\alpha)(\exists\epsilon_2)(|x| < \epsilon_2 \Rightarrow |\partial^{i+1}e^{-1/x^2} - 0| < \alpha) \Rightarrow (x = 0 \Rightarrow e^{-1/x^2} \in C^{i+1})$$
$$\Rightarrow (\forall x \in \mathcal{R})(e^{-1/x^2} \in C^{i+1}) .$$

We conclude by the recurrence principle 927 (p. 520) since it is true for $\mathbf{i = 1}$.

**Proposition 709** $\mathbf{a} \in \mathcal{R}^+$, the map $((x \neq 0 \Rightarrow f = e^{-a/x^2}, x = 0 \Rightarrow f = 0) : \mathcal{R} \to \mathcal{R})$ belongs to $\mathbf{C^\infty}$ (see 276 (p. 304)) and verifies:

$$x = 0 \Rightarrow (\forall i \in \mathcal{Z}^{0+})(\partial^i f = 0) .$$

**Proof**
Setting $\mathbf{g(x)} = e^{-1/x^2}$, since $\mathbf{a} > 0$ we have: $\mathbf{f} = e^{-1/(x/\sqrt{a})^2} = g(x/\sqrt{a})$. We conclude from proposition 324 (p. 200).

**Proposition 710** Defining $(\mathcal{U} : \mathcal{R} \to \mathcal{R})$ by

$$x \in \mathcal{R}^+ \Rightarrow \mathcal{U}(x) = e^{\frac{-1}{x^2}},$$
$$x \notin \mathcal{R}^+ \Rightarrow \mathcal{U}(x) = 0$$

$\mathcal{U}(x)$ belongs to $\mathbf{C^\infty}$ (see 276 (p. 304)).

**Proof**
It belongs to $\mathbf{C^0}$: From proposition 708 (p. 424), the derivative exists for $x \in [0, \to [$ and it is $\mathbf{0}$ for $x \in ] \leftarrow, 0[$ (proposition 308 (p. 194), definition 378 (p. 484)) then the derivative of $\mathcal{U}$ exists for all $x \in \mathcal{R}$, $\mathcal{U}$ is continuous (proposition 305 (p. 193)) and we have:

$$x \in ] \leftarrow, 0[ \Rightarrow \partial\mathcal{U}(x) = 0, x \in [0, \to [ \Rightarrow \partial\mathcal{U}(x) = \partial e^{\frac{-1}{x^2}}$$

If:

$$x \in ] \leftarrow, 0[ \Rightarrow \partial^n\mathcal{U}(x) = 0, x \in [0, \to [ \Rightarrow \partial^n\mathcal{U}(x) = \partial^n e^{\frac{-1}{x^2}}$$

is true, from propositions 305 (p. 193) and 708 (p. 424), it is also true for $\mathbf{n = n + 1}$ then from proposition 927 (p. 520) it is true for any $\mathbf{n} \in \mathcal{Z}^{0+}$.

**Proposition 711** Keeping notations of proposition 710, the derivative $\mathcal{U}'$ of $\mathcal{U}$ is given by:

$$x \leq 0 \Rightarrow \mathcal{U}'(x) = 0, x > 0 \Rightarrow \mathcal{U}'(x) = \frac{2}{x^3} e^{\frac{-1}{x^2}} > 0$$

and it belongs to $\mathbf{C^\infty}$.

**Proof**
It comes from proposition 710 (p. 426).

**Proposition 712** The map $(\mathbf{h} : \mathcal{R} \times \mathcal{R}^+ \to \mathcal{R})$ defined when $x \in \mathcal{R}$, $a \in \mathcal{R}^+$ by $h(x, a) = \frac{\mathcal{U}(a^2 - x^2)}{\mathcal{U}(a^2)}$ belongs to $\mathbf{C^\infty}$ and verifies:

$$(x \notin [-a, a] \Rightarrow h(x, a) = 0), (x \in [-a, a] \Rightarrow 0 \leq h(x, a) \leq 1, h(0, a) = 1) . \quad (2.282)$$

With $\mathbf{n} \in \mathcal{Z}^+$, $\mathbf{h^n}$ and $\mathbf{h^{1/n}}$ are also in $\mathbf{C^\infty}$ and verify also the relations of (2.282).

**Proof**
From 324 (p. 200) and 710 (p. 426), $\mathbf{h}$ belongs to $\mathbf{C}^\infty$ for any $\mathbf{a} > 0$. $\mathbf{h(0,a)} = \frac{\mathcal{U}(a^2-0)}{\mathcal{U}(a^2)} = 1$. The derivative $\mathcal{U}'(\mathbf{x})$ of $\mathcal{U}(\mathbf{x})$ is given by proposition 711 (p. 426). The derivative $\mathbf{h}'$ of $\mathbf{h}$ is (propositions 562 (p. 309), 330 (p. 203)):

$$h' = -2x\frac{\mathcal{U}'(a^2 - x^2)}{\mathcal{U}(a^2)} .$$

Since $(\forall x \in \mathcal{R})(\mathcal{U}(x) \geq 0, \mathcal{U}'(x) \geq 0)$, we have:

$$x \leq 0 \Rightarrow h'(x) \geq 0, \;\; x \geq 0 \Rightarrow h'(x) \leq 0$$

then from proposition 307 (p. 193), $\mathbf{h(0) = 1}$ is the biggest value of $\mathbf{h}(\mathcal{R}) \subset \mathcal{R}$.
The last line of proposition 712 is given by proposition 709 (p. 426):

$$a^2 - x^2 > 0 \Rightarrow h^n = \frac{e^{-\frac{n}{a^2-x^2}}}{e^{-\frac{n}{a^2}}}, h^{1/n} = \frac{e^{-\frac{1}{n(a^2-x^2)}}}{e^{-\frac{1}{na^2}}}, a^2 - x^2 \leq 0 \Rightarrow h^n = h^{1/n} = 0 .$$

**Proposition 713** The map $(\mathbf{Y} : \mathcal{R} \times \mathcal{R}^+ \to \mathcal{R})$ defined by

$$(a > 0, x < -a) \Rightarrow Y(x,a) = 0,$$
$$(a > 0, -a \leq x \leq a) \Rightarrow Y(x,a) = \frac{\int_{-a}^{x} h(x,a)}{\int_{-a}^{a} h(x,a)}$$
$$(a > 0, a < x) \Rightarrow Y(x,a) = 1$$

$\mathbf{Y(x,a)}$ belongs to $\mathbf{C}^\infty$, and verifies:

$$\begin{aligned} &x \leq -a \Rightarrow Y(x,a) = 0, \\ &x \in [-a,a] \Rightarrow 0 \leq Y(x,a) \leq 1, \\ &x \geq a \Rightarrow Y(x,a) = 1 . \end{aligned} \tag{2.283}$$

With $\mathbf{n} \in \mathcal{Z}^+$, $\mathbf{Y}^n$ and $\mathbf{Y}^{1/n}$ are also in $\mathbf{C}^\infty$ and verify also the relations of (2.283).

**Proof**
It is a direct consequence of proposition 636 (p. 368) and propositions 330 (p. 203), 562 (p. 309).

**Proposition 714** Given an open of $\mathcal{R}^n$, at any point in it, there is a smooth map of $\mathbf{C}^\infty$ (see 276 (p. 304)): $(\mathbf{g} : \mathcal{R}^n \to \mathcal{R})$ which is $\mathbf{1}$ at that point and zero outside an open cube (definition 261 (p. 285)) centered also on that point.

**Proof**
Calling $(\mathbf{x}^i : [1,n] \to \mathcal{R})$ that point, there is an open cube $(\mathbf{C(x,a)}, \mathbf{a} \in \mathcal{R}^+)$ centered on $\mathbf{x}$ inside any open including $\mathbf{x}$ (see proposition 266 (p. 179)). The map is defined by:

$$g(y-x,a) = \prod_{i\in[1,n]} \frac{\mathcal{U}(a^2 - (y^i - x^i)^2)}{\mathcal{U}(a^2)}$$

$\mathcal{U}$ defined in proposition 710 (p. 426). According to proposition 710 (p. 426), $\mathbf{g(y-x,a)}$ satisfies proposition 714 (p. 427) and clearly we can also use $\mathbf{g}^{1/n}\mathbf{(y-x,a)}$.

**Proposition 715** Given an open of $\mathcal{R}^n$, at any point in it, there is a smooth map of $C^\infty$ (see 276 (p. 304)): $(g : \mathcal{R}^n \to [0,1] \subset \mathcal{R})$ which is $1$ inside an open cube centered on that point and zero outside another open cube centered on the same point and containing the previous cube.

**Proof**
Calling $(x^i : [1,n] \to \mathcal{R})$ that point, there is an open cube $C(x,b)$ centered on $x$ inside any open including $x$ (see proposition 266 (p. 179)). From its definition 261 (p. 285), any cube $C(x,a)$ with $a < b$ is contained in $C(x,b)$. The map is defined by:

$$g(y-x,a,b) = \prod_{i\in[1,n]} Y(-|y_i - x_i| + (a+b)/2, (b-a)/2) \ .$$

## 2.32 Eigenvectors of a linear transformation

We have enough propositions to treat eigenvectors of a linear transformation acting on a vectorial space $V$ of finite dimension $n$ (see section 2.11 (p. 235)).
We shall use only as field $\mathcal{C}$.
We suppose furthermore that the vectorial space has an inner product quoted as:
$< a|b > = < b|a >^*, a \in V, b \in V$ (see definition 234 (p. 261)).

**Definition 323 Eigenvectors, eigenvalues.**

Given an $n$ dimension vector space $V$ on $\mathcal{C}$, a linear transformation $T$ of $V$ to itself $(T : V \to V)$ (see definition 224 (p. 250)). A vector $v \in V$ is an eigenvector of $T$ if it is not the null vector and if there is a $r \in \mathcal{C}$ such as $Tv = rv$. $r$ is called the eigenvalue associated to the eigenvector $v$.

**Proposition 716** Any eigenvalue $r$, associated to a linear transformation $T$ acting on a finite dimension $n$ vectorial space $V$, verifies the equality: $\det(T - rI) = 0$ where $I$ is the identical transformation.

**Proof**
We have by definition $(T - rI)v = 0$. We may take as an independent basis $B_1$ of $V$ one containing $v$ among its $n$ vectors (proposition 415 (p. 244)). The basis (definition 205 (p. 239)) $B_2 = (T - rI)B_1$ of $(T - rI)V$ constructed from $B_1$ would have at maximum $n - 1$ vectors no zero. $(T - rI)$ cannot be a bijection and from proposition 451 (p. 258) $\det(T - rI) = 0$.

**Proposition 717** With the same notation as above, if $r \in \mathcal{C}$ is such that $\det(T - rI) = 0$, it is an eigenvalue and there is an eigenvector $v \in V$ such that $Tv = rv$.

**Proof**
Since $\det(T - rI) = 0$, the vectors $(T - rI)e_i$ are a dependent basis of $T(V)$ ( proposition 451 (p. 258)). Then there are $(\lambda_i : [1,n] \to \mathcal{C})$ such that:

$$\lambda_i(T - rI)e_i = 0$$

which means that $v = \lambda_i e_i$ is such that $(T - rI)v = 0$ and $v$ is an eigenvector.

**Proposition 718** $\det(T - rI)$ is a polynomial of degree $n$.

**Proof**
We call $\mathbf{PG_n}$ the permutation group of $[1, n]$, we may write in a basis $(e_i : [1, n] \to V)$ of $V$:

$$(T - rI)e_i = (T - rI)_i^j e_j = (T_i^j - r\delta_i^j)e_j \Rightarrow (T - rI)_i^j = T_i^j - r\delta_i^j$$

and

$$\det(T - rI) = \sum_{P \in PG_n} (-1)^{S(P)} \prod_{k \in [1,n]} (T_k^{P(k)} - r\delta_k^{P(k)})$$

where $S(P)$ is the signature of $P$ (definition 175 (p. 208)). It is a polynomial as the sum of polynomials. The degree of each of them is given by the number of integers in $[1, n]$ for which $P(k) = k$. The highest degree is then given by the identical permutation: $\prod_{k \in [1,n]} T_k^k - r^k$ the degree of which is $n$.

**Proposition 719** $\det(T - rI)$ is the same polynomial in any independent basis and we have the following identity:

$$\det(T - rI) = \prod_{i \in [1,n]} (r_i - r)$$

where $(r_i : [1, n] \to \mathcal{R})$ is the map given by proposition 704 (p. 421), the set of $r_i$ is the set of roots of the polynomial $\det(T - rI)$. It is also the set of eigenvalues. The root multiplicity is also the eigenvalue multiplicity (definition 322 (p. 421)).
**Proof**
From proposition 447 (p. 256), the value of $\det(T - rI)$ is the same in any basis. From proposition 701 (p. 419), we have the first part of proposition 719. From proposition 704 (p. 421) and from the last line of the proof of proposition 718, in both members of the identity, the coefficient of the highest degree is $(-1)^n$. Then we have the second part of proposition 719.

### 2.32.1 Eigenvectors for a self adjoint linear transformation

**Definition 324 A self adjoint linear transformation.**

Given a vectorial space $V$ of finite dimension $n$ on $\mathcal{C}$ with a Riemannian inner product $g$ (definition 234 (p. 261)), a linear transformation $T$: $(T : V \to V)$, is said to be self adjoint if it is equal to its adjoint (definition 239 (p. 264)). $T$ is such that:

$$T^\dagger = T \Leftrightarrow g_{ik} T_l^{*k} g^{lj} = (T^\dagger)_i^j = T_i^j$$

with $g_{ij} = g(e_i, e_j) =< e_i|e_j >$ and $g^{ik} g_{kj} = \delta_j^i$ see propositions 451 (p. 258) and 459 (p. 262). See also footnote 22 (p. 261).
Using the usual notation in which if $c = Tb$, $< c|a >$ is written as $< bT|a >$, we have:

$$< a|Tb >=< bT|a >^* =< aT^\dagger|b >=< aT|b >=< b|Ta >^* .$$

**Proposition 720** The eigenvalues of a self adjoint linear transformation are real.
**Proof**
With the notation at the beginning of the section, if $r$ is an eigenvalue of $T$ for a vector $v$, we have since $< v|v >> 0$ (see definition 234 (p. 261) equation (2.98) and definition 323 (p. 428)):

$$(< v|Tv >= r < v|v >=< vT|v >=< v|Tv >^* = r^* < v|v >) \Rightarrow r = r^* .$$

**Proposition 721** The eigenvectors of a self adjoint linear transformation corresponding to different eigenvalues are orthogonal (definition 238 (p. 263)).
**Proof**
Given $\mathbf{v_1}$ and $\mathbf{v_2}$ two eigenvectors corresponding to two different eigenvalues $\mathbf{r_1}$ and $\mathbf{r_2}$ of a self adjoint linear transformation $\mathbf{T}$, we have:

$$\begin{aligned}&(< v_1|Tv_2 >= r_2 < v_1|v_2 >=< v_2|Tv_1 >^*= r_1 < v_2|v_1 >^*= r_1 < v_1|v_2 >)\\ &\Rightarrow (r_1 \neq r_2 \Rightarrow < v_1|v_2 >= 0) .\end{aligned}$$

**Proposition 722** If a vector $\mathbf{a}$ is orthogonal to an eigenvector $\mathbf{v}$ of a self adjoint linear transformation, its transformed is also orthogonal to that eigenvector $\mathbf{v}$.
**Proof**
$\mathbf{r}$ being the eigenvalue of $\mathbf{v}$, we have:

$$< v|Ta >=< vT|a >=< a|Tv >^*= r < a|v >^*= 0 .$$

**Proposition 723** Given a self adjoint linear transformation $\mathbf{T}$ on a finite dimension $\mathbf{n}$ vector space with an inner product, there are a set of eigenvectors which is an orthogonal basis of that vector space. Any eigenvalue $\mathbf{r_i}$ is a root with a multiplicity $\mathbf{m_i} \in [1, n]$ (definition 322 (p. 421)) of the polynomial in $\mathbf{r}$ of degree $\mathbf{n}$: $\mathbf{det(T - rI)}$. Corresponding to it, there are $\mathbf{m_i}$ eigenvectors of the set of orthogonal vectors.
**Proof**
Given a self adjoint linear transformation $\mathbf{T}$ acting on a finite dimension $\mathbf{n}$ vector space $\mathbf{V}$ with a Riemannian inner product, the eigenvalue polynomial of $\mathbf{T}$ is $\mathbf{det(T - rI)}$. Let us suppose that $\mathbf{det(T - rI)}$ has $\mathbf{J}$ different roots $(r_j : [1, J] \to \mathcal{R} - \{0\})$ with a multiplicity $(m_j : [1, J] \to [1, n])$ such that $\sum_{j\in[1,J]} m_j = n$, giving the following identity (proposition 719 (p. 429)):

$$\det(T - rI) = \prod_{j\in[1,J]} (r_j - r)^{m_j} .$$

1. We suppose that we already have a set $\mathbf{O_1}$ of $\mathbf{p}$ $(0 \leq p < n)$ orthogonal eigenvectors of norm $\mathbf{1}$: $(v_i : [1, p] \to V)$. That means that there is a map $(k(i) : [1, p] \to [1, J])$ such that, for any $\mathbf{i}$, we have $Tv_i = r_{k(i)}v_i$. Considering for any integer $l \in [1, J]$, the number $n_l^p = Card(k^{-1}(\{l\}))$ of vectors having the same eigenvalue $r_l$[37], we suppose also that we have $n_l^p \leq m_l$. We should have $\sum_{l\in[1,J]} n_l^p = p$. From proposition 463 (p. 264), we can complete $\mathbf{O_1}$ with a set of orthogonal vectors $\mathbf{O_2}$:

   $$(v_i : [p + 1, n] \to V)$$

   which we can set to be normed at $\mathbf{1}$. $O = O_1 \cup O_2$ is an orthogonal basis of $\mathbf{V}$. In the $\mathbf{O}$ basis, the matrix ${}^pT$ is given by

   $${}^pTv_i = {}^pT_i^jv_j .$$

   Then we have:

   $$\begin{aligned}&i \in [1, p] \Rightarrow {}^pTv_i = r_k(i)v_i \Rightarrow\\ &(\forall j \in [1, n])({}^pT_i^j =< v_j|{}^pT_i^{j'}v_{j'} >=< v_j|Tv_i >= r_{k(i)}\delta_i^j,\\ &{}^pT_j^i =< v_i|{}^pT_j^{i'}v_{i'} >=< v_i|Tv_j >=< v_iT|v_j >= r^*_{k(i)}\delta_i^j = r_{k(i)}\delta_i^j) .\end{aligned}$$

[37] It can be 0 if there is no vector corresponding to l.

Then from proposition 719 (p. 429):

$$\mathbf{det(T - rI) = det({}^{p}T - rI) = (\prod_{l \in [1,J]} (r_l - r)^{n_l^p}) det({}^{p2}T - rI)}$$

setting:

$$\mathbf{(i \in [1, n - p], j \in [1, n - p]) \Rightarrow {}^{p2}T^i_j = {}^{p}T^{i+p}_{j+p}} \, .$$

2. The linear combinations of vectors of $\mathbf{O_1}$ or $\mathbf{O_2}$ generate two vectorial spaces $\mathbf{V_1}$ and $\mathbf{V_2}$. By construction any vector of $\mathbf{V_1}$ is orthogonal to any vector of $\mathbf{V_2}$.

3. By construction, any vector $\mathbf{x}$ orthogonal to all vectors of $\mathbf{O_1}$ is also orthogonal to all vectors of $\mathbf{V_1}$. It is also in $\mathbf{V_2}$:
   $\mathbf{O}$ is a basis of $\mathbf{V}$, $\mathbf{x \in V \Rightarrow x = \sum_{i \in [1,p]} x^i v_i + \sum_{j \in [p+1]} x^j v_j}$ then:

$$\mathbf{(\forall i \in [1, p])(< v_i | x >= 0 = x^i < v_i | v_i >) \Rightarrow (\forall i \in [1, p])(x^i = 0) \Rightarrow x \in V_2} \, .$$

Combining the last point with proposition 722, we have $\mathbf{T(V_2) = V_2}$ and clearly also $\mathbf{T(V_1) = V_1}$. Then $\mathbf{T}$ is a self adjoint linear transformation on $\mathbf{V_2}$, its eigenvalues are given by the roots of

$$\mathbf{det({}^{p2}T - rI) = \frac{det(T - rI)}{\prod_{l \in [1,J]} (r_l - r)^{n_l^p}} = \prod_{l \in [1,J]} (r_l - r)^{m_l - n_l^p}}$$

so one can choose any root of $\mathbf{det({}^{p2}T - rI)}$, it has at least one corresponding eigenvector in $\mathbf{V_2}$ which can be added to the set $\mathbf{O_1}$. It diminishes by one unit the multiplicity of the root of the next $\mathbf{V_2}$ eigenvalue polynomial and increases by one unit the multiplicity of the root of the next $\mathbf{V_1}$.
Then if $\mathbf{p < n}$ we can add one more vector to the basis $\mathbf{O_1}$ keeping the same properties changing only $\mathbf{p}$ to $\mathbf{p + 1}$. From that and applying proposition 927 (p. 520) we conclude proposition 723.

**Proposition 724** Given a linear self adjoint linear transformation $\mathbf{T}$ over a finite dimension $\mathbf{n}$ vectorial space $\mathbf{V}$ on $\mathcal{C}$ with a Riemannian inner product, we have the following identity (proposition 719 (p. 429)):

$$\mathbf{det(T - rI) = \prod_{i \in [1,n]} (r_i - r)}$$

where $\mathbf{(r_i : [1, n] \rightarrow \mathcal{R})}$ are the eigenvalues of proposition 704 (p. 421) associated to the $\mathbf{n}$ orthogonal eigenvectors making an independent basis of $\mathbf{V}$.

**Proof**
It is a consequence of proposition 723 and proposition 719 (p. 429).

# Appendix A

# Conventions, basic relations and symbols

## A.1 Logic theory

We use in this book the formalism logical theory which is well described in [1]. A logical theory consists of terms and relations **R** between terms. The basic framework of our theory is the set theory.

**Definition 325 Terms.**

Terms are usually letters or bunches of letters. The first list of terms is settled at the beginning of the theory; others can be generated following rules which can be specific of the theory (see section A.3 (p. 441)).

**Definition 326 Relations.**

Relations are terms connected by specific symbols of the theory. Those relations can connect 2 3 etc. terms ($\mathbf{R(x,y)}$, $\mathbf{R(x_1,x_2,x_3,...,x_n)}$). To write a relation, one has to follow the specific rules which are peculiar to each theory. The relations, which can be written according to the axioms of the theory, are said to be true. Part of the axioms or axiom generators are common to all logical theory and use logical symbols. It what follows, we shall not make a clear separation between axioms and the way those axioms are generated.

**Convention 23** Substitution in a relation.

**x** and **y** being terms **A** an expression. $\mathbf{(y|x)A}$ stands for the expression **A** where any time **x** occurs in **A**, it is replace by **y**.

### A.1.1 Logic symbols

Logic theories use the following specific symbols:

- non written as $\neg$ in front of the relation
- or written as $\vee$
- $\Rightarrow$, $\Leftrightarrow$
- and

Those symbols allow to generate new relations from previous ones. There are two notations for grouping relations with those symbols:

- **the polish notation**. In that notation, the symbol is put before the terms on which it acts.
- **The bracket notation**. It can be used when there are two terms in the relation. The symbol is between the terms but brackets are needed to separate various actions.

**We shall use the bracket notation which is more commonly used although the polish notation is more logical.**
In the following definitions, **A**, **B**, **C** are supposed to be relations.

**Definition 327** $\neg$ **(non).**

non **A**, which can be written as $\neg\mathbf{A}$, is a relation.

**Definition 328** $\vee$ **(or).**

$(\mathbf{A} \vee \mathbf{B}) = (\mathbf{A}\,\mathrm{or}\,\mathbf{B})$ is a relation. It is written as $\vee\mathbf{A}\,\mathbf{B}$ in polish notation.

**Definition 329 and (, ).**

(A and B) stands for non (non A or non B): $\neg(\neg\mathbf{A} \vee \neg\mathbf{B})$, $\neg \vee \neg\mathbf{A}\neg\mathbf{B}$ in polish notation.

**Convention 24** $\mathbf{A}, \mathbf{B}$.

By convention, for simplifying $\mathbf{A}, \mathbf{B}$ stands for **A** and **B**.

**Definition 330** $\Rightarrow$**.**

$\mathbf{A} \Rightarrow \mathbf{B}$ stands for $\neg\mathbf{A} \vee \mathbf{B}$.

**Definition 331** $\Leftrightarrow$.

$(\mathbf{A} \Leftrightarrow \mathbf{B})$ means:
$(\mathbf{A} \Rightarrow \mathbf{B}$ and $\mathbf{B} \Rightarrow \mathbf{A})$
It can be written as:
$\neg(\neg(\neg\mathbf{A} \vee \mathbf{B}) \vee \neg(\neg\mathbf{B} \vee \mathbf{A}))$.

### A.1.2 Demonstration

A demonstration is a list of true statements. A statement **R** can be on that list of true statement and so declared true if:

- It is an axiom of the theory or an axiom generated by the theory axiom generator rules. (see section A.1.3)
- It is a true statement of a demonstration already written.
- There is a true statement **S** of the theory and before the statement **R**, the true statement: $\mathbf{S} \Rightarrow \mathbf{R}$

**Convention 25** Implicitly true statement.

By convention, any statement which is written separately according to the above rules is implicitly a true statement.

### A.1.3 Properties of logical symbols

In this section **A**, **B**, **C** are relations. The following properties are axiom generators of the logical theory in the sense that they generate true relations from the rules of construction of the logical theories.

1. $\mathbf{A} \vee \mathbf{A} \Rightarrow \mathbf{A}$
2. $\mathbf{A} \Rightarrow (\mathbf{A} \vee \mathbf{B})$
3. $\mathbf{A} \vee \mathbf{B} \Leftrightarrow \mathbf{B} \vee \mathbf{A}$
4. $(\mathbf{A} \Rightarrow \mathbf{B}) \Rightarrow ((\mathbf{C} \vee \mathbf{A}) \Rightarrow (\mathbf{C} \vee \mathbf{B}))$
5. (See A.1.2 (p. 434)). If **A** is true and $(\mathbf{A} \Rightarrow \mathbf{B})$ true, **B** is true.
6. $=$
   **x, y, z** are terms, $\mathbf{R}(\mathbf{z})$ is any relation without **x** and **y** explicitly in it.
   $(\mathbf{y} = \mathbf{x}) \Rightarrow (\mathbf{R}(\mathbf{x}) \Leftrightarrow \mathbf{R}(\mathbf{y}))$.

### A.1.4 Deduced properties of logical symbols

In this section **A**, **B**, **C** are relations. From the rules of A.1.3 we can deduce [1]:

1. **B** true gives $(\mathbf{A} \Rightarrow \mathbf{B})$ true for any **A**
2. $\neg\mathbf{B}$ true gives $(\mathbf{B} \Rightarrow \mathbf{A})$ true for any **A**
3. $(\mathbf{B} \Rightarrow \mathbf{C}) \Rightarrow ((\mathbf{A} \Rightarrow \mathbf{B}) \Rightarrow (\mathbf{A} \Rightarrow \mathbf{C}))$
   or $\mathbf{A} \Rightarrow \mathbf{B}$ and $\mathbf{B} \Rightarrow \mathbf{C}$ being true, $\mathbf{A} \Rightarrow \mathbf{C}$ is true.
4. $(\mathbf{A} \vee \neg\mathbf{A}), (\neg\mathbf{A} \vee \mathbf{A}), (\mathbf{A} \Rightarrow \mathbf{A})$ are true.
5. $\mathbf{A} \Leftrightarrow \neg\neg\mathbf{A}$.
6. $(\mathbf{A} \Rightarrow (\mathbf{A} \Rightarrow \mathbf{B})) \Rightarrow (\mathbf{A} \Rightarrow \mathbf{B})$.
7. $(\mathbf{A} \Rightarrow \mathbf{B}) \Leftrightarrow (\neg\mathbf{B} \Rightarrow \neg\mathbf{A})$ (Reasoning by the opposite).
8. $(\neg\mathbf{A} \Rightarrow (\mathbf{B}, \neg\mathbf{B})) \Rightarrow \mathbf{A}$ [1] (Reasoning by the absurd).
9. **A** true, $\mathbf{A} \Rightarrow \mathbf{B}$ true and $(\mathbf{A}, \neg\mathbf{C}) \Rightarrow \neg\mathbf{B}$ true make **C** true.
10. $(\mathbf{A} \Leftrightarrow \mathbf{B}) \Leftrightarrow (\neg\mathbf{B} \Leftrightarrow \neg\mathbf{A})$
11. $(\mathbf{A} \Rightarrow \mathbf{B}) \Rightarrow ((\mathbf{B} \Rightarrow \mathbf{C}) \Rightarrow (\mathbf{A} \Rightarrow \mathbf{C}))$
12. $\mathbf{A} \Rightarrow ((\mathbf{B} \Rightarrow (\mathbf{A} \Rightarrow \mathbf{C})) \Rightarrow (\mathbf{B} \Rightarrow \mathbf{C}))$
13. $(\mathbf{A}, \mathbf{B}) \Leftrightarrow (\mathbf{B}, \mathbf{A}) \Rightarrow \mathbf{A}$
14. **A** and **B** being true, gives $(\mathbf{A}, \mathbf{B})$ true.
15. $\mathbf{A} \Rightarrow \mathbf{C}$ true and $\mathbf{B} \Rightarrow \mathbf{D}$ true give $(\mathbf{A} \vee \mathbf{B}) \Rightarrow (\mathbf{C} \vee \mathbf{D})$ true:
    $((\mathbf{A} \Rightarrow \mathbf{C}), (\mathbf{B} \Rightarrow \mathbf{D})) \Rightarrow ((\mathbf{A} \vee \mathbf{B}) \Rightarrow (\mathbf{C} \vee \mathbf{D}))$.

[1] We use, as in all that book, the convention 24 (p. 434)

16. $((\mathbf{A} \Rightarrow \mathbf{C}), (\mathbf{B} \Rightarrow \mathbf{C})) \Rightarrow ((\mathbf{A} \vee \mathbf{B}) \Rightarrow \mathbf{C})$ true.

17. $((\mathbf{A} \Rightarrow \mathbf{C}), (\mathbf{B} \Rightarrow \mathbf{D})) \Rightarrow ((\mathbf{A}, \mathbf{B}) \Rightarrow (\mathbf{C}, \mathbf{D}))$

18. $((\mathbf{A} \vee \mathbf{B}) \vee \mathbf{C}) \Leftrightarrow (\mathbf{A} \vee (\mathbf{B} \vee \mathbf{C}))$

19. $(\mathbf{A} \Rightarrow (\mathbf{B} \Rightarrow \mathbf{C})) \Leftrightarrow ((\mathbf{A}, \mathbf{B}) \Rightarrow \mathbf{C})$

20. $\neg(\mathbf{A}, \mathbf{B}) \Leftrightarrow (\neg\mathbf{A} \vee \neg\mathbf{B})$

21. $(\mathbf{A} \Rightarrow \neg\mathbf{A}) \Rightarrow \neg\mathbf{A}$; $(\neg\mathbf{A} \Rightarrow \mathbf{A}) \Rightarrow \mathbf{A}$

22. $\mathbf{A} \Leftrightarrow (\mathbf{A}, \mathbf{A})$

23. $(\mathbf{A}, (\neg\mathbf{A} \vee \mathbf{B})) = (\mathbf{A}, (\mathbf{A} \Rightarrow \mathbf{B})) \Leftrightarrow (\mathbf{A}, \mathbf{B})$

24. $\mathbf{B}$ true, $\mathbf{A} \Rightarrow (\mathbf{A}, \mathbf{B})$ true.

25. $((\mathbf{A} \vee \mathbf{B}), \mathbf{C}) \Leftrightarrow ((\mathbf{A}, \mathbf{C}) \vee (\mathbf{B}, \mathbf{C}))$

26. $((\mathbf{A}, \mathbf{B}), \mathbf{C}) \Leftrightarrow (\mathbf{A}, (\mathbf{B}, \mathbf{C})) \Leftrightarrow ((\mathbf{A}, \mathbf{C}), (\mathbf{B}, \mathbf{C}))$

27. $((\mathbf{A}, \mathbf{B}) \vee \mathbf{C}) \Leftrightarrow ((\mathbf{A} \vee \mathbf{C}), (\mathbf{B} \vee \mathbf{C}))$

**Proof**
The proofs may also be found in [1]. [2]

- (A.1.4, 1). From (A.1.3, 2) and (A.1.3, 3)
  $\mathbf{B} \Rightarrow (\mathbf{B} \vee \neg\mathbf{A}) \Leftrightarrow (\neg\mathbf{A} \vee \mathbf{B} = \mathbf{A} \Rightarrow \mathbf{B})$.
  Then from (A.1.3, 5):
  $(\mathbf{A} \Rightarrow \mathbf{B})$ true.
- (A.1.4, 2). From (A.1.3, 2) $\neg\mathbf{B} \Rightarrow \neg\mathbf{B} \vee \mathbf{A} = \mathbf{B} \Rightarrow \mathbf{A}$
- (A.1.4, 3). From (A.1.3, 4), $(\mathbf{B} \Rightarrow \mathbf{C}) \Rightarrow (((\neg\mathbf{A} \vee \mathbf{B}) \Rightarrow (\neg\mathbf{A} \vee \mathbf{C})) = ((\mathbf{A} \Rightarrow \mathbf{B}) \Rightarrow (\mathbf{A} \Rightarrow \mathbf{C})))$. Then when $\mathbf{B} \Rightarrow \mathbf{C}$ and $\mathbf{A} \Rightarrow \mathbf{B}$ are true, $\mathbf{A} \Rightarrow \mathbf{C}$ is true by (A.1.3, 5).
- (A.1.4, 4). From (A.1.3, 2), $(\mathbf{A} \Rightarrow \mathbf{A} \vee \mathbf{A})$ is true, from (A.1.3, 1) $(\mathbf{A} \vee \mathbf{A} \Rightarrow \mathbf{A})$ is true, then by (A.1.4, 3) $(\mathbf{A} \Rightarrow \mathbf{A}) = (\neg\mathbf{A} \vee \mathbf{A})$ is true.
- (A.1.4, 5). The $\mathbf{A} \Rightarrow \neg\neg\mathbf{A}$ part of (A.1.4, 5) comes from (A.1.4, 4) applies to $\neg\mathbf{A}$ with (A.1.3, 3):
  $\neg\mathbf{A} \Rightarrow \neg\mathbf{A} = \neg\neg\mathbf{A} \vee \neg\mathbf{A} \Leftrightarrow \neg\mathbf{A} \vee \neg\neg\mathbf{A} = \mathbf{A} \Rightarrow \neg\neg\mathbf{A}$. The reverse $(\neg\neg\mathbf{A} \Rightarrow \mathbf{A})$ will be demonstrated after (A.1.4, 12) is proved.

---

[2] The following proofs sometimes can be difficult to follow. An easy way is to apply the Boolean logic: one gives $\geq \mathbf{1}$ to any logical statement true and $\mathbf{0}$ to any wrong. The **or**$=\vee$ symbol is replaced by the additive sign: $+$. The **and** $=$**,** symbol by the multiplicative sign: $\cdot$ or $\times$. If the result is $\geq \mathbf{1}$, it is true. If it is $\mathbf{0}$, it is wrong. Then we have the logic table:

$$\mathbf{0} \vee \mathbf{0} = \mathbf{0} + \mathbf{0} = \mathbf{0}$$
$$\mathbf{0} \vee \mathbf{1} \Leftrightarrow \mathbf{1} \vee \mathbf{0} = \mathbf{1} + \mathbf{0} = \mathbf{1}$$
$$\mathbf{1} \vee \mathbf{1} = \mathbf{1} + \mathbf{1} = \mathbf{2} \approx \mathbf{1}$$
$$\mathbf{0}, \mathbf{0} = \mathbf{0} \times \mathbf{0} = \mathbf{0}$$
$$\mathbf{0}, \mathbf{1} \Leftrightarrow \mathbf{1}, \mathbf{0} = \mathbf{0} \times \mathbf{1} = \mathbf{0}$$
$$\mathbf{1}, \mathbf{1} = \mathbf{1} \times \mathbf{1} = \mathbf{1}$$

- (A.1.4, 6). We have, applying (A.1.3, 2): $\neg\mathbf{A} \Rightarrow (\neg\mathbf{A} \vee \mathbf{B})$. Then with (A.1.3, 5), (A.1.3, 4) and (A.1.3, 1):
$(\neg\mathbf{A} \vee (\mathbf{A} \Rightarrow \mathbf{B})) \Rightarrow ((\mathbf{A} \Rightarrow \mathbf{B}) \vee \neg\mathbf{A}) \Rightarrow ((\mathbf{A} \Rightarrow \mathbf{B}) \vee (\mathbf{A} \Rightarrow \mathbf{B})) \Rightarrow (\mathbf{A} \Rightarrow \mathbf{B})$
Then:
$(\mathbf{A} \Rightarrow (\mathbf{A} \Rightarrow \mathbf{B})) \Rightarrow (\mathbf{A} \Rightarrow \mathbf{B})$

- (A.1.4, 7) comes directly from (A.1.4, 3) and applying to **B** the part of (A.1.4, 5) that we have proved:
$(\mathbf{A} \Rightarrow \mathbf{B})$ and $(\mathbf{B} \Rightarrow \neg\neg\mathbf{B})$ true, give as true by (A.1.4, 3):
$(\mathbf{A} \Rightarrow \neg\neg\mathbf{B} = \neg\mathbf{A} \vee \neg\neg\mathbf{B}) \Rightarrow (\neg\neg\mathbf{B} \vee \neg\mathbf{A} = \neg\mathbf{B} \Rightarrow \neg\mathbf{A})$
applying at the end (A.1.3, 3).

- (A.1.4, 8) can be seen as:
$(\neg\mathbf{A} \Rightarrow (\mathbf{B}, \neg\mathbf{B})) \Leftrightarrow (\mathbf{A} \vee \neg(\neg\mathbf{B} \vee \mathbf{B})) = \neg\mathbf{B} \vee \mathbf{B} \Rightarrow \mathbf{A}$
Since $\neg\mathbf{B} \vee \mathbf{B}$ is true, from (A.1.4, 12) we have:
$(\neg\mathbf{A} \Rightarrow (\mathbf{B}, \neg\mathbf{B})) \Rightarrow \mathbf{A}$ is true and **A** is true.

- (A.1.4, 9). $(\mathbf{A}, \neg\mathbf{C}) \Rightarrow \neg\mathbf{B} = \neg\neg(\neg\mathbf{A} \vee \neg\neg\mathbf{C}) \vee \neg\mathbf{B} \Leftrightarrow (\neg\mathbf{A} \vee \mathbf{C}) \vee \neg\mathbf{B} = (\mathbf{B} \Rightarrow (\mathbf{A} \Rightarrow \mathbf{C}))$. Since **A** and **B** are true, **C** is true.

- (A.1.4, 10) comes from 7 applied also to $\mathbf{B} \Rightarrow \mathbf{A}$.

- (A.1.4, 11). Using (A.1.4, 7) and (A.1.3, 2)
$(\mathbf{A} \Rightarrow \mathbf{B}) \Rightarrow (\neg\mathbf{B} \Rightarrow \neg\mathbf{A}) \Rightarrow ((\mathbf{C} \vee \neg\mathbf{B}) \Rightarrow (\mathbf{C} \vee \neg\mathbf{A})$.
Since $(\mathbf{C} \vee \neg\mathbf{B}) \Leftrightarrow (\mathbf{B} \Rightarrow \mathbf{C})$ and $(\mathbf{C} \vee \neg\mathbf{A}) \Leftrightarrow (\mathbf{A} \Rightarrow \mathbf{C}))$
$(\mathbf{A} \Rightarrow \mathbf{B}) \Rightarrow ((\mathbf{B} \Rightarrow \mathbf{C}) \Rightarrow (\mathbf{A} \Rightarrow \mathbf{C}))$.

- (A.1.4, 12). From (A.1.4, 1), (A.1.4, 7): $\mathbf{A} \Rightarrow ((\mathbf{B} \Rightarrow \mathbf{A}) \Rightarrow (\neg\mathbf{A} \Rightarrow \neg\mathbf{B}))$. Then by (A.1.4, 11) $\mathbf{A} \Rightarrow (\mathbf{B} \Rightarrow \mathbf{A}) \Rightarrow ((\mathbf{A} \Rightarrow \mathbf{C}) \Rightarrow (\mathbf{B} \Rightarrow \mathbf{C}))$ and from (A.1.3, 4):
$\mathbf{A} \Rightarrow ((\neg\mathbf{B} \vee (\mathbf{A} \Rightarrow \mathbf{C})) \Rightarrow (\neg\mathbf{B} \vee (\mathbf{B} \Rightarrow \mathbf{C})))$
which is also using (A.1.4, 6):
$\mathbf{A} \Rightarrow ((\mathbf{B} \Rightarrow (\mathbf{A} \Rightarrow \mathbf{C})) \Rightarrow \mathbf{A} \Rightarrow (\mathbf{B} \Rightarrow (\mathbf{B} \Rightarrow \mathbf{C})) \Rightarrow (\mathbf{B} \Rightarrow \mathbf{C})) \Rightarrow ((\mathbf{B} \Rightarrow (\mathbf{A} \Rightarrow \mathbf{C})) \Rightarrow (\mathbf{B} \Rightarrow \mathbf{C}))$

- (A.1.4, 5). The $\neg\neg\mathbf{A} \Rightarrow \mathbf{A}$ part of (A.1.4, 5) can now be proved. From (A.1.3, 2):
$\neg\neg\mathbf{A} \Rightarrow ((\neg\neg\mathbf{A} \vee \mathbf{A}) = (\neg\mathbf{A} \Rightarrow \mathbf{A}))$.
From (A.1.3, 4) and (A.1.3, 1):
$(\neg\mathbf{A} \Rightarrow \mathbf{A}) \Rightarrow (\mathbf{A} \vee \neg\mathbf{A} \Rightarrow \mathbf{A} \vee \mathbf{A} \Rightarrow \mathbf{A})$
then: $\neg\neg\mathbf{A} \Rightarrow (\mathbf{A} \vee \neg\mathbf{A} \Rightarrow \mathbf{A})$
from (A.1.4, 12) and (A.1.4, 4) since $\mathbf{A} \vee \neg\mathbf{A} \Leftrightarrow \neg\mathbf{A} \vee \mathbf{A} = \mathbf{A} \Rightarrow \mathbf{A}$
$\neg\neg\mathbf{A} \Rightarrow \mathbf{A}$ is true.

- (A.1.4, 13). From (A.1.3, 3) $((\neg\mathbf{A} \vee \neg\mathbf{B}) \Rightarrow (\neg\mathbf{B} \vee \neg\mathbf{A})) = ((\mathbf{A}, \mathbf{B}) \vee \neg\mathbf{B} \vee \neg\mathbf{A})$. From (A.1.4, 5), (A.1.4, 11), (A.1.3, 3): $((\mathbf{A}, \mathbf{B}) \vee \neg\neg(\neg\mathbf{B} \vee \neg\mathbf{A})) \Leftrightarrow (\mathbf{B}, \mathbf{A}) \Rightarrow (\mathbf{A}, \mathbf{B})$ is true. We also have from (A.1.4, 5):
$(\neg\mathbf{A} \Rightarrow (\neg\mathbf{A} \vee \neg\mathbf{B})) = (\neg\neg\mathbf{A} \vee (\neg\mathbf{A} \vee \neg\mathbf{B})) \Leftrightarrow ((\mathbf{A}, \mathbf{B}) \Rightarrow \mathbf{A})$. Since $(\neg\mathbf{A} \Rightarrow (\neg\mathbf{A} \vee \neg\mathbf{B}))$ is true (A.1.3, 2), $((\mathbf{A}, \mathbf{B}) \Rightarrow \mathbf{A})$ is true

- (A.1.4, 14). If **A** is true, we have true:
$(\mathbf{A} \Rightarrow \neg\mathbf{B}) \Rightarrow \mathbf{A}$ from (A.1.4, 1).
$(\mathbf{A} \Rightarrow \neg\mathbf{B}) \Rightarrow \mathbf{A}) \Rightarrow ((\mathbf{A} \Rightarrow \neg\mathbf{B}) \Rightarrow ((\mathbf{A} \Rightarrow \neg\mathbf{B}) \Rightarrow \neg\mathbf{B}))$ from (A.1.4, 11).
$(\mathbf{A} \Rightarrow \neg\mathbf{B}) \Rightarrow ((\mathbf{A} \Rightarrow \neg\mathbf{B}) \Rightarrow \neg\mathbf{B})$ is true.

Then from (A.1.4, 6)
$\mathbf{(A \Rightarrow \neg B) \Rightarrow \neg B}$ is true.
Using (A.1.4, 5) and (A.1.4, 7), it is equivalent to:
$\mathbf{B \Rightarrow \neg(A \Rightarrow \neg B)}$.
Then $\mathbf{A}$ and $\mathbf{B}$ true gives:
$\mathbf{\neg(A \Rightarrow \neg B) = \neg(\neg A \vee \neg B) = (A, B)}$ true from (A.1.3, 5).

- (A.1.4, 15). From (A.1.3, 4) is true:

$$\mathbf{(B \Rightarrow D) \Rightarrow (A \vee B \Rightarrow A \vee D \Rightarrow D \vee A)}\ .$$

Then from section A.1.2 (p. 434) $\mathbf{(A \vee B \Rightarrow D \vee A)}$ is true. From (A.1.3, 4) again:

$$\mathbf{(A \Rightarrow C) \Rightarrow (D \vee A \Rightarrow D \vee C \Rightarrow C \vee D)}\ .$$

Then $\mathbf{D \vee A \Rightarrow C \vee D}$ true and from (A.1.4, 11): $\mathbf{A \vee B \Rightarrow C \vee D}$ true.

- (A.1.4, 16). That statement is the same as (A.1.4, 15) with $\mathbf{D = C}$ and using (A.1.4, 1).

- (A.1.4, 17). From (A.1.4, 7), we have:
$\mathbf{(A \Rightarrow C) \Leftrightarrow (\neg C \Rightarrow \neg A)}$; $\mathbf{(B \Rightarrow D) \Leftrightarrow (\neg D \Rightarrow \neg B)}$.
From (A.1.4, 15) and( A.1.4, 7):
$\mathbf{((\neg C \vee \neg D) \Rightarrow (\neg A \vee \neg B)) \Leftrightarrow (\neg(\neg A \vee \neg B) \Rightarrow \neg(\neg C \vee \neg D)) = ((A, B) \Rightarrow (C, D))}$.

- (A.1.4, 18). From (A.1.3, 2), (A.1.3, 3), (A.1.4, 3), we have:
$\mathbf{A \Rightarrow (A \vee B) \Rightarrow ((A \vee B) \vee C)}$; $\mathbf{(B \Rightarrow (B \vee A) \Rightarrow (A \vee B) \Rightarrow ((A \vee B) \vee C)}$;
$\mathbf{C \Rightarrow C \vee (A \vee B) \Rightarrow (A \vee B) \vee C}$.
Then from (A.1.4, 16): $\mathbf{(A \vee C) \Rightarrow ((A \vee B) \vee C)}$; $\mathbf{B \Rightarrow (A \vee B) \vee C)}$; $\mathbf{((A \vee C) \vee B) \Rightarrow ((A \vee B) \vee C)}$ and the same for $\mathbf{((A \vee B) \vee C) \Rightarrow ((A \vee C) \vee B)}$.

- (A.1.4, 19) From (A.1.4, 18), (A.1.4, 5): $\mathbf{(A \Rightarrow (B \Rightarrow C)) \Leftrightarrow (\neg A \vee (\neg B \vee C)) \Leftrightarrow ((\neg A \vee \neg B) \vee C) \Leftrightarrow (\neg\neg(\neg A \vee \neg B) \vee C) = ((A, B) \Leftrightarrow C)}$.

- (A.1.4, 20) From (A.1.4, 5): $\mathbf{\neg(A, B) = \neg\neg(\neg A, \vee\neg B) \Leftrightarrow (\neg A, \vee\neg B)}$.

- (A.1.4, 21). From (A.1.3, 1) $\mathbf{(A \Rightarrow \neg A) = (\neg A \vee \neg A) \Rightarrow \neg A}$.
From (A.1.4, 5), (A.1.4, 4) and (A.1.4, 16)
$\mathbf{(\neg A \Rightarrow A) = (\neg\neg A \vee A) \Rightarrow (A \vee A) \Rightarrow A}$.

- (A.1.4, 22). From (A.1.3, 4), the following statement:

$$\mathbf{(\neg A \Rightarrow \neg A \vee \neg A) = (\neg\neg A \vee (\neg A \vee \neg A))}$$

is true. Then from (A.1.4, 5) and (A.1.4, 15),

$$\begin{aligned}&\mathbf{(\neg\neg A \vee (\neg A \vee \neg A)) \Leftrightarrow (A \vee \neg\neg(\neg A \vee \neg A)) \Leftrightarrow (\neg\neg(\neg A \vee \neg A) \vee A)}\\ &\mathbf{= ((A, A) \Rightarrow A)}\end{aligned}$$

$\mathbf{(A, A) \Rightarrow A}$ is true.
From (A.1.3, 1, $\mathbf{((\neg A \vee \neg A) \Rightarrow \neg A)}$ is true. But $\mathbf{((\neg A \vee \neg A) \Rightarrow \neg A) \Leftrightarrow (\neg A \vee \neg(\neg A \vee \neg A)) = (A \Rightarrow (A, A))}$.
Then $\mathbf{A \Rightarrow (A, A)}$ is true.

- (A.1.4, 23) From (A.1.4, 17), $\mathbf{(A \Rightarrow A, A \Rightarrow B) \Rightarrow (A, B)}$
  $\mathbf{(A, B) \Rightarrow A, (A, B) \Rightarrow B \Rightarrow (B \vee \neg A) \Leftrightarrow (A \Rightarrow B)}$.
  From (A.1.4, 22), $\mathbf{(A, B) \Rightarrow ((A, B), (A, B))}$.
  From (A.1.4, 17), $\mathbf{((A, B), (A, B)) \Rightarrow (A, (A \Rightarrow B))}$.

- (A.1.4, 24) From (A.1.3, 1) $\mathbf{(\neg A \vee B) = (A \Rightarrow B)}$ is true, from (A.1.4, 4) $\mathbf{(\neg A \vee A) = (A \Rightarrow A)}$ is also true, then from A.1.3, 17 $\mathbf{(A, A) \Rightarrow (A, B)}$. From (A.1.3, 22) $\mathbf{A \Leftrightarrow (A, A)}$ is true, then from (A.1.3, 3), (A.1.4, 24): $\mathbf{A \Rightarrow (A, B)}$ is true.

- (A.1.4, 25). From (A.1.3, 2), (A.1.4, 4), (A.1.4, 17), (A.1.2):

$$\begin{aligned}
&\mathbf{(A \Rightarrow A \vee B), (C \Rightarrow C)} \\
&\mathbf{((A \Rightarrow A \vee B), (C \Rightarrow C)) \Rightarrow ((A, C) \Rightarrow ((A \vee B), C))} \\
&\mathbf{(A, C) \Rightarrow ((A \vee B), C)}\,.
\end{aligned}$$

The same exchanging **A** and **B** then from (A.1.4, 15),

$$\mathbf{(A, C) \vee (B, C) \Rightarrow ((A \vee B), C) \vee ((A \vee B), C) \Rightarrow ((A \vee B), C)}\,.$$

We also have from (A.1.2), (A.1.4, 13), (A.1.4, 1), (A.1.4, 15), (A.1.4, 22), (A.1.4, 18) :

$$\begin{aligned}
&\mathbf{(A \vee B), C)} \\
&\mathbf{((A \vee B), C) \Rightarrow C, ((A \vee B), C) \Rightarrow (A \vee B)} \\
&\mathbf{C} \\
&\mathbf{A \vee B} \\
&\mathbf{A \Rightarrow A;\ A \Rightarrow C;\ B \Rightarrow B;\ B \Rightarrow C} \\
&\mathbf{A \Rightarrow (A, A) \Rightarrow (A, C);\ B \Rightarrow (B, B) \Rightarrow (B, C)} \\
&\mathbf{(A \vee B) \Rightarrow ((A, C) \vee (B, C))} \\
&\mathbf{(A, C) \vee (B, C)}\,.
\end{aligned}$$

Another way to demonstrate A.1.4 (p. 435) point 25 (p. 436) is to do table (footnote 2 (p. 436)) of true and wrong on both sides of $\mathbf{((A \vee B), C) \Leftrightarrow ((A, C) \vee (B, C))}$ and to show that in the six possible situations for the relations **A, B, C** (each of them can be true (=1) or false (=0)) they are equal [3]. Some of the demonstrations in mathematics do use implicitly a table of true-wrong to prove a proposition.

- (A.1.4, 26). The only possibility to have one of these 3 relations true is when **A, B, C** are all true and when **A, B, C** are all true, all the three relations are true. All these relations are equivalent.

- (A.1.4, 27). We have true:

$$\mathbf{(A \Rightarrow A \vee C),\ (B \Rightarrow B \vee C),\ (C \Rightarrow C, C),\ (C \Rightarrow A \vee C)\ , (C \Rightarrow B \vee C)}$$

(A.1.3, 2), (A.1.4 , 22). From (A.1.4, 17), we have:

$$\mathbf{((A, B) \Rightarrow ((A \vee C), (B \vee C))),\ (C \Rightarrow (C, C) \Rightarrow ((A \vee C), (B \vee C)))}\,.$$

[3] According to footnote 2 (p. 436) 25 (p. 436) means $\mathbf{(A + B) \times C = A \times C + B \times C}$ which is true in ordinary arithmetic. It then means true whatever the state of **A, B, C**.

Then from (A.1.4 point 15 (p. 435)), :

$$((\mathbf{A},\mathbf{B}) \vee \mathbf{C}) \Rightarrow ((\mathbf{A} \vee \mathbf{C}), (\mathbf{B} \vee \mathbf{C}))$$

but from (A.1.4 point 15 (p. 435), 22 (p. 436)), we have:

$$((\neg\mathbf{A} \Rightarrow \mathbf{C}), (\neg\mathbf{B} \Rightarrow \mathbf{C})) \Rightarrow ((\neg\mathbf{A}, \neg\mathbf{B}) \Rightarrow (\mathbf{C}, \mathbf{C}) \Rightarrow \mathbf{C})$$

which is equivalent to (definition 330 (p. 434), ( A.1.4 point 5 (p. 435))):

$$(((\neg\neg\mathbf{A} \vee \mathbf{C}), (\neg\neg\mathbf{B} \vee \mathbf{C})) \Rightarrow \neg(\neg\mathbf{A}, \neg\mathbf{B}) \vee \mathbf{C})$$
$$\Leftrightarrow (((\mathbf{A} \vee \mathbf{C}), (\mathbf{B} \vee \mathbf{C})) \Rightarrow ((\mathbf{A},\mathbf{B}) \vee \mathbf{C}))\ .$$

**Proposition 725** $(\mathbf{A}, \neg\mathbf{A})$ true gives $\mathbf{B}$ true.

**Proof**

$(\neg\mathbf{A}) \Rightarrow (\neg\mathbf{A} \vee \mathbf{B}) = (\mathbf{A} \Rightarrow \mathbf{B})$ from (A.1.3, 4 )

$\neg\mathbf{A}$ true gives $(\mathbf{A} \Rightarrow \mathbf{B})$ true. Then $\mathbf{B}$ is true if $\mathbf{A}$ is also true from (A.1.3, 5).

**Remark**: If $\mathbf{A}$ is true and non $\mathbf{A}$ true, the theory is useless since any relation is true.

### A.1.5 Substitution criteria

To the logical symbols, we also have to add a substitution criterion which allows to write with symbolic letters. They are very well explained in [1] chap. 1 par. 1 nb 3. In short whenever is in an expression without any connection to a letter elsewhere it can be replaced by any other letters and the expression keep its qualities. For instance:

**Criterion 1** $\mathbf{A}$ and $\mathbf{B}$ being expression, using convention 23 (p. 433), if in the expression $\mathbf{A}$ there is the letter $\mathbf{x}$ and not $\mathbf{y}$, we have $(\mathbf{B}|\mathbf{y})(\mathbf{y}|\mathbf{x})\mathbf{A} = (\mathbf{B}|\mathbf{x})\mathbf{A}$.

## A.2 Specifics terms

### A.2.1 Terms associated with a relation

**Definition 332** $\tau_{\mathbf{x}}\mathbf{R}$ or $\tau_{\mathbf{x}}(\mathbf{R}(\mathbf{x}))$.

Given a relation $\mathbf{R}$ and a letter $\mathbf{x}$ not defined already in the theory, the letter $\mathbf{x}$ may be or not in $\mathbf{R}$, by construction of the theory $\tau_{\mathbf{x}}\mathbf{R}$ is a term of the theory. Since $\mathbf{x}$ may appear only in $\mathbf{R}$. The relation $\mathbf{R}$ is often written as $\mathbf{R}(\mathbf{x})$ and the term $\tau_{\mathbf{x}}\mathbf{R}$ as $\tau_{\mathbf{x}}(\mathbf{R}(\mathbf{x}))$. It could be a little misleading since formally $\mathbf{x}$ does not need to be in $\mathbf{R}$.

Moreover having a relation $\mathbf{R}(\mathbf{x},\mathbf{y},\mathbf{z})$, to be perfectly logic we should consider that in the term $\tau_{\mathbf{x}}\mathbf{R}(\mathbf{x},\mathbf{y},\mathbf{x})$, the letter $\mathbf{x}$ is used only as a pointer to establish a link between $\tau$ and the position of $\mathbf{x}$ in the writing of $\mathbf{R}(\mathbf{x},\mathbf{y},\mathbf{x})$ (see proposition 748 (p. 449)). $\mathbf{y}$ stand for a term which is not yet fixed.

In fact to avoid those considerations, specialists in logic write instead of $\tau_{\mathbf{x}}(\mathbf{R}(\mathbf{x},\mathbf{y},\mathbf{x}))$:

$$\tau\mathbf{R}(\square, \mathbf{y}, (\square))$$

But it is difficult to write formulae like that. We shall not do it. The reader will have to do mentally the operation and to remember that in such formulae $\mathbf{x}$ is a dummy letter.

With a term as $\mathbf{u}(\mathbf{y}) = \tau_{\mathbf{x}}(\mathbf{R}(\mathbf{x},\mathbf{y},\mathbf{x}))$ where $\mathbf{y}$ is built in (see in definition 112 (p. 131) the point 4 (p. 132)), one can build a new relation in $\mathbf{y}$: $\mathbf{R}(\mathbf{u}(\mathbf{y}),\mathbf{y},\mathbf{u}(\mathbf{y}))$ and a new term which does not contain any $\mathbf{x}$ and $\mathbf{y}$: $\tau_{\mathbf{y}}(\mathbf{R}(\mathbf{u}(\mathbf{y}),\mathbf{y},\mathbf{u}(\mathbf{y})))$. See proposition 737 (p. 444) for such a treatment.

**Proposition 726** The following relations are axioms of our theory (this book theory or a quantified theory). They are true:

$$\mathbf{R(a) \Rightarrow R(\tau_a(R(a)))} \tag{A.1}$$

$$\mathbf{\tau_a(R(a)) = \tau_x(R(x))}\ . \tag{A.2}$$

If $\mathbf{R(x),\ R'(x)}$ are two relations where $\mathbf{x}$ stands for a term of the theory, we have:

$$\mathbf{R(x) \Leftrightarrow R'(x) \Rightarrow \tau_x(R(x)) = \tau_x(R'(x))}\ . \tag{A.3}$$

In the case of $\mathbf{x}$ not explicitly in $\mathbf{R(x)}$ the relation (A.1 (p. 441)) reduced to $\mathbf{R \Rightarrow R}$, see [1].

**Proof**

They are axioms so they do not need proofs.

## A.3 Quantificators

$\mathbf{R(x)}$ is a relation where $\mathbf{x}$ stands for a (or set of adjacent) letter (s) not already used and may not be in $\mathbf{R(x)}$.

**Definition 333 ∃.**

$\mathbf{\exists x R(x)}$ stands for the relation $\mathbf{R(\tau_x(R(x)))}$.

**Proposition 727** For any term $\mathbf{a}$ and whatever relation $\mathbf{R(z)}$, $\mathbf{R(a) \Rightarrow \exists x R(x)}$ is true[4].

**Proof**

It is a direct consequence of the axiom generator defined by relation (A.1 (p. 441)).

**Proposition 728** $\mathbf{\exists x(R(x), T(x)) \Rightarrow \exists x R(x)}$

**Proof**

From definition 333 (p. 441), with $\mathbf{a = \tau_x(R(x), T(x))}$, $\mathbf{\exists x(R(x), T(x)) = R(a), T(a)}$. Then (see section A.1.4 (p. 435) point 13) from proposition 727 (p. 441), we have:

$$\mathbf{\exists x(R(x), T(x)) \Rightarrow R(a) \Rightarrow \exists x R(x)}\ .$$

**Definition 334 ∀.**

$\mathbf{\forall x R(x)}$ stand for the relation $\mathbf{\neg\neg R(\tau_x(\neg R(x))) = \neg\exists x \neg R(x)}$.

**Proposition 729** $\mathbf{\forall x R(x) \Leftrightarrow R(\tau_x(\neg R(x)))}$.

**Proof**

It is a direct consequence of definition 334 (p. 441) and point 5 (p. 435) of A.1.4.

**Convention 26** Convention with $\mathbf{x}$ when $\mathbf{\exists x R(x)}$ is written as true.

$\mathbf{R(x)}$ is a relation. $\mathbf{\exists x R(x)}$ is often written as $\mathbf{(\exists x)(R(x))}$. By convention, If the statement $\mathbf{st = (\exists x)(R(x))}$ appears in a list of statements leading to a demonstration, if $\mathbf{x}$ is not quoted before $\mathbf{st}$ and if $\mathbf{x}$ does not have any special meaning, $\mathbf{x}$ stands for $\mathbf{\tau_x(R(x))}$ for the statements following $\mathbf{st}$.

[4] In particular, we have from relation (A.1) and definition 333: $\mathbf{R(a,a) \Rightarrow (\exists y)R(a,y) \Rightarrow (\exists x)(\exists y)R(x,y)}$.

**Proposition 730** $\mathbf{R_1(x), R_2(x)}$ being two relations with $\mathbf{R_1(x) \Rightarrow R_2(x)}$, we have

$$(\exists x)\mathbf{R_1(x)} \Rightarrow (\exists x)\mathbf{R_2(x)},\ (\forall x)\mathbf{R_1(x)} \Rightarrow (\forall x)\mathbf{R_2(x)}\ .$$

**Proof**
From equation (A.1) (p. 441) in definition 332 (p. 440):

$$((\exists x)\mathbf{R_1(x)}, \mathbf{R_1(x)} \Rightarrow \mathbf{R_2(x)}) \Rightarrow (\mathbf{R_1}(\tau_x(\mathbf{R_1(x)})), \mathbf{R_2}(\tau_x(\mathbf{R_1(x)}))) \Rightarrow (\exists x)\mathbf{R_2(x)}\ .$$

point (A.1.4, 7) gives: $\mathbf{R_1(x) \Rightarrow R_2(x) \Leftrightarrow \neg R_2(x) \Rightarrow \neg R_1(x)}$ then applying again (A.1.4, 7), we have:

$$\begin{aligned}&\neg\mathbf{R_2(x)} \Rightarrow \neg\mathbf{R_1(x)} \Rightarrow (\exists x)\neg\mathbf{R_2(x)} \Rightarrow (\exists x)\neg\mathbf{R_1(x)}\\ &\Leftrightarrow \neg(\exists x)\neg\mathbf{R_1(x)} \Rightarrow \neg(\exists x)\neg\mathbf{R_2(x)} \Leftrightarrow ((\forall x)\mathbf{R_1(x)} \Rightarrow (\forall x)\mathbf{R_2(x)})\ .\end{aligned}$$

**Proposition 731** For any term $\mathbf{a}$ and whatever relation $\mathbf{R(z)}$, $\forall x\mathbf{R(x)} \Rightarrow \mathbf{R(a)}$ is true.

**Proof**
From proposition 727 (p. 441), we have the true expression:

$$\neg\mathbf{R(a)} \Rightarrow \exists x\neg\mathbf{R(x)}$$

which is equivalent to (see definition 330 (p. 434)):

$$\begin{aligned}&(\neg\neg\mathbf{R(a)} \vee \exists x\neg\mathbf{R(x)}) \Leftrightarrow (\mathbf{R(a)} \vee \exists x\neg\mathbf{R(x)}) \Leftrightarrow (\neg\neg\exists x\neg\mathbf{R(x)} \vee \mathbf{R(a)})\\ &= (\forall x\mathbf{R(x)} \Rightarrow \mathbf{R(a)})\ .\end{aligned}$$

**Proposition 732** If for any term $\mathbf{z}$ the relation $\mathbf{R(z)}$ is true, $\forall x\mathbf{R(x)}$ is true.

**Proof**
Choosing $\mathbf{z} = \tau_x\neg\mathbf{R(x)}$, we have the definition relation of $\forall$:

$$\mathbf{R}(\tau_x\neg\mathbf{R(x)}) \Leftrightarrow \neg\neg\mathbf{R}(\tau_x\neg\mathbf{R(x)}) = \neg\exists x\neg\mathbf{R(x)} = \forall x\mathbf{R(x)}\ .$$

See definition 334 (p. 441).

**Proposition 733** The following statements are true:

$$\begin{aligned}&(\forall x)(\mathbf{R(x)} \Rightarrow \mathbf{S(x)}) \Rightarrow (\exists x\mathbf{R(x)} \Rightarrow \exists x\mathbf{S(x)})\\ &(\forall x)(\mathbf{R(x)} \Rightarrow \mathbf{S(x)}) \Rightarrow (\forall x\mathbf{R(x)} \Rightarrow \forall x\mathbf{S(x)})\ .\end{aligned}$$

**Proof**
It is a consequence of propositions 730 (p. 442) and 731 (p. 442)

**Proposition 734** If $\mathbf{R(x)}$ does not contain $\mathbf{x}$ and if $\mathbf{R} = \mathbf{R(x)}$ is true, $\forall x\mathbf{R}$ and $\exists x\mathbf{R}$ are true.

**Proof**
It is a consequence of proposition 732 (p. 442) and of what was said in definition 332 (p. 440).

**Proposition 735** If $\mathbf{R(x)}$ does not contain $\mathbf{x}$, we have:

$$(\mathbf{R(x)}, \exists x\mathbf{V(x)}) \Leftrightarrow (\exists x)(\mathbf{R(x)}, \mathbf{V(x)})$$

$$(\mathbf{R(x)}, \forall x\mathbf{V(x)}) \Leftrightarrow (\forall x)(\mathbf{R(x)}, \mathbf{V(x)})\ .$$

**Proof**
From the fact that $\mathbf{R(x)}$ does not contain $\mathbf{x}$ and from proposition 727 (p. 441), we have:

$$\begin{aligned}&(\mathbf{R(x)}, \exists \mathbf{x}\mathbf{V(x)}) \Leftrightarrow (\mathbf{R(x)}, \mathbf{V}(\tau_{\mathbf{x}}(\mathbf{V(x)}))) = (\mathbf{R}(\tau_{\mathbf{x}}(\mathbf{V(x)})), \mathbf{V}(\tau_{\mathbf{x}}(\mathbf{V(x)})))\\ &\Rightarrow (\exists \mathbf{x})(\mathbf{R(x)}, \mathbf{V(x)}) \ .\end{aligned}$$

From the fact that $\mathbf{R(x)}$ does not contain $\mathbf{x}$ and from proposition 731 (p. 442) and definition 334 (p. 441), we have:

$$\begin{aligned}&(\mathbf{R(x)}, \forall \mathbf{x}\mathbf{V(x)}) \Rightarrow (\mathbf{R(x)}, \mathbf{V}(\tau_{\mathbf{x}}(\neg(\mathbf{R(x)}, \mathbf{V(x)}))))\\ &= (\mathbf{R}(\tau_{\mathbf{x}}(\neg(\mathbf{R(x)}, \mathbf{V(x)}))), \mathbf{V}(\tau_{\mathbf{x}}(\neg(\mathbf{R(x)}, \mathbf{V(x)}))))\\ &\Rightarrow (\neg\neg(\mathbf{R}(\tau_{\mathbf{x}}(\neg(\mathbf{R(x)}, \mathbf{V(x)}))), \mathbf{V}(\tau_{\mathbf{x}}(\neg(\mathbf{R(x)}, \mathbf{V(x)}))))) = \forall \mathbf{x}(\mathbf{R(x)}, \mathbf{V(x)}) \ .\end{aligned}$$

**Proposition 736** If $\mathbf{S(x)}$ and $\mathbf{R(x)}$ are relations, we have the following true statements:

$$((\forall \mathbf{y})\mathbf{S(y)}, (\exists \mathbf{x})\mathbf{R(x)}) \Rightarrow (\exists \mathbf{x})(\mathbf{R(x)}, \mathbf{S(x)}) \tag{A.4}$$

$$((\forall \mathbf{y})\mathbf{S(y)} \vee (\exists \mathbf{x})\mathbf{R(x)}) \Rightarrow (\exists \mathbf{x})(\mathbf{R(x)} \vee \mathbf{S(x)}) \tag{A.5}$$

$$((\forall \mathbf{y})\mathbf{S(y)}, (\forall \mathbf{x})\mathbf{R(x)}) \Leftrightarrow (\forall \mathbf{x})(\mathbf{R(x)}, \mathbf{S(x)}) \tag{A.6}$$

$$((\forall \mathbf{y})\mathbf{S(y)} \vee (\forall \mathbf{x})\mathbf{R(x)}) \Rightarrow (\forall \mathbf{x})(\mathbf{R(x)} \vee \mathbf{S(x)}) \tag{A.7}$$

$$(\exists \mathbf{y})\mathbf{S(y)}, (\exists \mathbf{x})\mathbf{R(x)} \Leftrightarrow (\exists \mathbf{x})(\mathbf{R(x)}, \mathbf{S(x)}) \ . \tag{A.8}$$

**Proof**
The first two are obtained by suppressing $(\exists \mathbf{x})$ and stating $\mathbf{x}$ and $\mathbf{y}$ as $\tau_{\mathbf{x}}\mathbf{R(x)}$ (see proposition 731 (p. 442)). It gives:
$\mathbf{S}(\tau_{\mathbf{x}}\mathbf{R(x)}), \mathbf{R}(\tau_{\mathbf{x}}\mathbf{R(x)})$ for A.4 and $\mathbf{S}(\tau_{\mathbf{x}}\mathbf{R(x)}) \vee \mathbf{R}(\tau_{\mathbf{x}}\mathbf{R(x)})$ for A.5
true and then $(\exists \mathbf{x})(\mathbf{R(x)}, \mathbf{S(x)})$ and $(\exists \mathbf{x})(\mathbf{R(x)} \vee \mathbf{S(x)})$ true from equation (A.1) (p. 441).
In A.6, A.7, we make $\mathbf{x}$ and $\mathbf{y}$ equal to:
$\mathbf{x} = \mathbf{y} = \tau_{\mathbf{x}}\neg(\mathbf{S(x)}, \mathbf{R(x)})$ for A.6
and
$\mathbf{x} = \mathbf{y} = \tau_{\mathbf{x}}\neg(\mathbf{S(x)} \vee \mathbf{R(x)})$ for A.7.
We obtain:

$$\neg\neg(\mathbf{S}(\tau_{\mathbf{x}}\neg(\mathbf{S(x)}, \mathbf{R(x)})), \mathbf{R}(\tau_{\mathbf{x}}\neg(\mathbf{S(x)}, \mathbf{R(x)}))) = (\forall \mathbf{x})(\mathbf{R(x)}, \mathbf{S(x)})$$

and

$$\neg\neg(\mathbf{S}(\tau_{\mathbf{x}}\neg(\mathbf{S(x)} \vee \mathbf{R(x)})) \vee \mathbf{R}(\tau_{\mathbf{x}}\neg(\mathbf{S(x)} \vee \mathbf{R(x)}))) = (\forall \mathbf{x})(\mathbf{R(x)} \vee \mathbf{S(x)})$$

(see proposition 731 (p. 442)). The reasoning works for the relation (A.6) also reversely that why there is the $\Leftrightarrow$ sign. Setting $\mathbf{x_s} = \tau_{\mathbf{x}}\neg\mathbf{S(x)}$, $\mathbf{x_r} = \tau_{\mathbf{x}}\neg\mathbf{R(x)}$, we have (A.1.4 point 17 (p. 436), 22 (p. 436)):

$$\begin{aligned}&(\forall \mathbf{x})(\mathbf{S(x)}, \mathbf{R(x)}) \Rightarrow ((\mathbf{S(x_s)}, \mathbf{R(x_s)}), (\mathbf{S(x_r)}, \mathbf{R(x_r)})) \Rightarrow (\mathbf{S(x_s)}, \mathbf{R(x_r)})\\ &\Leftrightarrow ((\forall \mathbf{x})\mathbf{S(x)}, (\forall \mathbf{y})\mathbf{R(y)}) \ .\end{aligned}$$

We get relation (A.8) by writing A.6 for $\neg\mathbf{S}$ and $\neg\mathbf{R}$ (see (A.1.4, 7), and definition 329 (p. 434)):

$$\begin{aligned}
&\{(\forall x)(\neg S(x), \neg R(x)) \Leftrightarrow ((\forall x)\neg S(x), (\forall y)\neg R(y))\}\\
&\Leftrightarrow \{\neg(\exists x)(S(x) \vee R(x)) \Leftrightarrow (\neg(\exists x)S(x), \neg(\exists x)S(x))\}\\
&\Leftrightarrow \{\neg(\exists x)(S(x) \vee R(x)) \Leftrightarrow \neg((\exists x)S(x) \vee (\exists x)S(x))\}\\
&\Leftrightarrow \{(\exists x)(S(x) \vee R(x)) \Leftrightarrow ((\exists x)S(x) \vee (\exists x)S(x))\}\ .
\end{aligned}$$

The same type of arguments also gives if $\mathbf{S}$ does not contain $\mathbf{x}$:

$$(S, (\exists x)R(x)) \Leftrightarrow (\exists x)(R(x), S) \tag{A.9}$$

$$(S \vee (\exists x)R(x)) \Leftrightarrow (\exists x)(R(x) \vee S) \tag{A.10}$$

$$(S, (\forall x)R(x)) \Leftrightarrow (\forall x)(R(x), S) \tag{A.11}$$

$$(S \vee (\forall x)R(x)) \Leftrightarrow (\forall x)(R(x) \vee S)\ . \tag{A.12}$$

**Proposition 737** $(\exists x)(\exists y)R(x, y) \Leftrightarrow (\exists y)(\exists x)R(x, y)$ .

**Proof**

Using proposition 727 (p. 441), with $u(x) = \tau_y R(x, y)$ and $v = \tau_x R(x, u(x))$, we have:

$$\begin{aligned}
(\exists x)(\exists y)R(x, y) &\Leftrightarrow (\exists x)R(x, u(x))\\
&\Leftrightarrow R(v, u(v))\\
&\Leftrightarrow (\exists x)R(x, u(v))\\
&\Leftrightarrow (\exists y)(\exists x)R(x, y)
\end{aligned}$$

applying twice proposition 727 (p. 441).

**Proposition 738** $(\forall x)(\forall y)R(x, y) \Leftrightarrow (\forall y)(\forall x)R(x, y)$ .

**Proof**

$$\begin{aligned}
(\forall x)(\forall y)R(x, y) &= \neg(\exists x)\neg\neg(\exists y)\neg R(x, y)\\
&\Leftrightarrow \neg(\exists x)(\exists y)\neg R(x, y) \Leftrightarrow \neg(\exists y)(\exists x)\neg R(x, y)\\
&\Leftrightarrow (\forall y)(\forall x)R(x, y)\ .
\end{aligned}$$

**Proposition 739** $(\exists x)(\forall y)R(x, y) \Rightarrow (\forall y)(\exists x)R(x, y)$[5].

**Proof**

Using proposition 731 (p. 442), with: $u_1(x) = \tau_y\neg R(x, y)$, $v_1 = \tau_x R(x, u_1(x))$, $v_2(y) = \tau_x R(x, y)$ and $u_2 = \tau_y(\neg R(v_2(y), y))$, we have:

$$\begin{aligned}
&((\forall y)(\exists x)R(x, y) \Leftrightarrow (\forall y)R(v_2(y), y) \Leftrightarrow R(v_2(u_2), u_2)\\
&(\exists x)(\forall y)R(x, y) \Rightarrow (\exists x)R(x, u_2) \Leftrightarrow R(v_2(u_2), u_2)\\
&(\exists x)(\forall y)R(x, y) \Rightarrow (\forall y)(\exists x)R(x, y)\ .
\end{aligned}$$

[5] Be careful the inverse relation is not true: $(\forall y)(\exists x)$ does not commute.

**Convention 27** $\mathbf{R(x)}$ stands for $\forall \mathbf{x}\mathbf{R(x)}$.

Most of the times, when one has a relation $\forall \mathbf{x}\mathbf{R(x)}$, one does not write the $\forall \mathbf{x}$ if the letter $\mathbf{x}$ does not refer to any already defined terms. For example the relation:

$$(\forall x)(((x \in \mathcal{R}) \text{ and } (0 < x < \sqrt{2})) \Rightarrow (x^2 < 2)) \tag{A.13}$$

is often written as:

$$((x \in \mathcal{R}), (0 < x < \sqrt{2})) \Rightarrow (x^2 < 2) .$$

# A.4 Specifics relations

## A.4.1 Functional relations

$\mathbf{x}$, $\mathbf{y}$ and $\mathbf{z}$ being terms, $\mathbf{R}, \mathbf{R(x)}, \mathbf{R(x,y)}$ are relations.
**Definition 335 Functional relations.**

A relation $\mathbf{R}$ is functional, if $\mathbf{y}$ and $\mathbf{z}$ are not in $\mathbf{R}$, we have true (see conventions 23 (p. 433), A.1.5 (p. 440), Criterion 1):

$$(\forall y)(\forall z)(((y|x)R, (z|x)R) \Rightarrow y = z) .$$

**Proposition 740** Given a relation $\mathbf{R(x)}$ without $\mathbf{y}$, $\mathbf{R(y)}, \mathbf{y} = \tau_{\mathbf{x}}(\mathbf{R(x)})$, is functional in $\mathbf{y}$.
**Proof**
It comes from definitions. See also proposition 726 (p. 441).
**Remark:** $\mathbf{R(x,y)}$ being relation, $\mathbf{z} = \tau_{\mathbf{y}}\mathbf{R(x,y)}$ is functional in $\mathbf{z}$ and allows to define map (see definition 112 (p. 131) point 4 (p. 132), definition 347 (p. 449)).

## A.4.2 Equivalence relation

$\mathbf{x}$, $\mathbf{y}$ and $\mathbf{z}$ being terms, $\mathbf{R(x,y)}$ is a relation.
**Definition 336 Equivalence relation.**

$\mathbf{R(x,y)}$ is an equivalence relation if $\mathbf{R}$ has the following properties:

1. $\mathbf{R(x,x)}$ is true
2. $\mathbf{R(x,y)} \Leftrightarrow \mathbf{R(y,x)}$
3. $\mathbf{R(x,y)}, \mathbf{R(y,z)} \Rightarrow \mathbf{R(x,z)}$

When in a section, $\mathbf{R(x,y)}$ is unique and when it is clear enough, $\mathbf{R(x,y)}$ is written as:

$$R(x,y) = x \sim y .$$

When it is used as in $\mathcal{Q}$ or $\mathcal{R}$, it is written as $=$. Then $=$ does not have exactly the same meaning as in (A.1.3, 6): We shall write for instance, in section A.8 (p. 551), definition 418 (p. 551), proposition 988 (p. 551):

$$3/4 = 15/20 .$$

It is more an equivalence relation than an equality. It says $(3,4)$ belongs to the same equivalence class as $(15,20)$ because $3 \times 20 = 4 \times 15$.

### A.4.3 Representative of an equivalence relation

**Definition 337 Representative of an equivalence relation.**

Given an equivalence relation $\mathbf{R}$, a term $\mathbf{s}$, one calls the representative term of the relation $\mathbf{R}$ associated to $\mathbf{s}$, the term $\tau_{\mathbf{x}}\mathbf{R}(\mathbf{x},\mathbf{s})$.

**Proposition 741** We have the following true statements:

1. $\mathbf{R}(\mathbf{s},\mathbf{s}) \Rightarrow (\exists \mathbf{x}\mathbf{R}(\mathbf{x},\mathbf{s}))$

2. $\mathbf{R}(\mathbf{t},\mathbf{s}) \Rightarrow \tau_{\mathbf{x}}\mathbf{R}(\mathbf{x},\mathbf{s}) = \tau_{\mathbf{x}}\mathbf{R}(\mathbf{x},\mathbf{t}) = \tau_{\mathbf{y}}\mathbf{R}(\mathbf{y},\mathbf{t})$

**Proof**
From proposition 727 (p. 441), $(\exists \mathbf{x}\mathbf{R}(\mathbf{x},\mathbf{s}))$ because $\mathbf{R}(\mathbf{s},\mathbf{s})$ is true.
From definition 336 (p. 445), $\mathbf{R}(\mathbf{t},\mathbf{s}) \Rightarrow (\mathbf{R}(\mathbf{x},\mathbf{s}) \Leftrightarrow \mathbf{R}(\mathbf{x},\mathbf{t}))$. From equation (A.3) (p. 441)

$$\mathbf{R}(\mathbf{t},\mathbf{s}) \Rightarrow \tau_{\mathbf{x}}\mathbf{R}(\mathbf{x},\mathbf{s}) = \tau_{\mathbf{x}}\mathbf{R}(\mathbf{x},\mathbf{t}) = \tau_{\mathbf{y}}\mathbf{R}(\mathbf{y},\mathbf{t}) = \tau_{\mathbf{y}}\mathbf{R}(\mathbf{t},\mathbf{y})$$

is true.

**Proposition 742** Given an equivalence relation $\mathbf{R}$, a term $\mathbf{y}$, the relation:

$$\mathbf{H}(\mathbf{x}) = \tau_{\mathbf{z}}\mathbf{R}(\mathbf{z},\mathbf{y}) = \tau_{\mathbf{z}}\mathbf{R}(\mathbf{y},\mathbf{z})$$

is functional (definition 335 (p. 445)).

**Proof**
It is a direct consequence of definition 335 (p. 445) and proposition 741 (p. 446).

### A.4.4 Ordering relation

**Definition 338 Ordering relation.**

A relation $\mathbf{R}(\mathbf{x},\mathbf{y})$ between two terms $\mathbf{x}$ and $\mathbf{y}$ is an order if, $\mathbf{z}$ being a third term:

1. $\mathbf{R}(\mathbf{x},\mathbf{y}) \Rightarrow (\mathbf{R}(\mathbf{x},\mathbf{x}),\mathbf{R}(\mathbf{y},\mathbf{y}))$

2. $(\mathbf{R}(\mathbf{x},\mathbf{y}),\mathbf{R}(\mathbf{y},\mathbf{x})) \Rightarrow \mathbf{x} = \mathbf{y}$

3. $(\mathbf{R}(\mathbf{x},\mathbf{y}),\mathbf{R}(\mathbf{y},\mathbf{z})) \Rightarrow \mathbf{R}(\mathbf{x},\mathbf{z})$.

**Definition 339 $\leq$.**

For convenience an ordering relation is quoted $\leq$: $\mathbf{R}(\mathbf{x},\mathbf{y})$ is quoted as $\mathbf{x} \leq \mathbf{y}$ (resp. $\mathbf{y} \geq \mathbf{x}$) and it is phrased as $\mathbf{x}$ smaller than or equal to $\mathbf{y}$.

**Definition 340 $<$.**

The relation $(\mathbf{x} \leq \mathbf{y}),(\mathbf{x} \neq \mathbf{y})$ is quoted as $(\mathbf{x} < \mathbf{y})$ and phrased as $\mathbf{x}$ smaller than $\mathbf{y}$.

**Convention 28 $\leq$, $\geq$, $<>$.**

If $\mathbf{R}(\mathbf{x},\mathbf{y})$ is an ordering relation, one uses to write:

1. $\mathbf{R}(\mathbf{x},\mathbf{y}) \Leftrightarrow \mathbf{x} \leq \mathbf{y} \Leftrightarrow \mathbf{y} \geq \mathbf{x}$

2. $\mathbf{x} < \mathbf{y} \Leftrightarrow (\mathbf{R}(\mathbf{x},\mathbf{y}),\mathbf{x} \neq \mathbf{y}) \Leftrightarrow \neg\mathbf{R}(\mathbf{y},\mathbf{x})$

3. $\mathbf{x > y \Leftrightarrow (R(y,x), x \neq y) \Leftrightarrow \neg R(x,y)}$

**Definition 341 Opposite order.**

Given an ordering relation $\mathbf{R}$, the opposite order $\mathbf{T}$ of $\mathbf{R}$ is the relation: $\mathbf{T(x,y) = R(y,x)}$.

**Proposition 743** The opposite order $\mathbf{T}$ of an ordering relation $\mathbf{R}$ is an ordering relation.

**Proof**
Points 1 and 2 of definition 338 (p. 446) are clearly satisfied. Point 3 is also satisfied (see section A.1.4 (p. 435) point 13 (p. 435)):

$$\mathbf{(T(x,y), T(y,z)) = (R(y,x), R(z,y)) \Leftrightarrow (R(z,y), R(y,x)) \Leftrightarrow R(z,x) = T(x,z)}\ .$$

# A.5 Sets

## A.5.1 Symbols of the set theory

In that section $\mathbf{A}$, $\mathbf{B}$, $\mathbf{x}$ are terms, $\mathbf{R(x)}$ a relation with $\mathbf{x}$ as a term. We also have to remember that "$\mathbf{A}$ is a set" is no different from "$\mathbf{A}$ is a term". The "$\mathbf{A}$ is a set" is only there to indicate that we shall use the sign $\in$ on the letter $\mathbf{A}$ like $\mathbf{x \in A}$ where $\mathbf{x}$ is another letter.

**Definition 342 $\in$.**

$\mathbf{x \in A}$ is a relation which means $\mathbf{x}$ belongs to $\mathbf{A}$ in plain language. In fact there is no direct definition for $\in$. It is the definition 344 (p. 447) and the following axioms which give a sense to $\in$.

**Definition 343 $\notin$.**

$\mathbf{\neg(x \in A) = x \notin A}$.

**Definition 344 $\mathbf{coll_x R(x)}$ and $\mathbf{set_x\{R(x)\}}$.**

$\mathbf{R(x)}$ is a relation, $\mathbf{A}$ a letter which is not explicitly in $\mathbf{R}$. By definition $\mathbf{coll_x R(x)}$ is the relation[6]:

$$\mathbf{coll_x R(x) = (\exists A)(\forall x)(x \in A \Leftrightarrow R(x))}\ .$$

For convenience in that book, we call $\mathbf{set_x\{R(x)\}}$ the term $\boldsymbol{\tau}_{\mathbf{A}}\mathbf{(\forall x)(x \in A \Leftrightarrow R(x))}$ which is the set of terms $\mathbf{x}$ for which $\mathbf{R(x)}$ is true, $\mathbf{set_x\{R(x)\}}$ is also called $\mathbf{(x|R(x))}$:

$$\mathbf{set_x\{R(x)\}} = \boldsymbol{\tau}_{\mathbf{A}}\mathbf{\forall x(x \in A \Leftrightarrow R(x))} = \boldsymbol{\tau}_{\mathbf{A}}\mathbf{(x \in A \Leftrightarrow R(x))}$$

$$\mathbf{\exists A(x \in A \Leftrightarrow R(x)) \Leftrightarrow (x \in set_x\{R(x)\} \Leftrightarrow R(x))}\ .$$

See convention 27 (p. 445).

**Proposition 744**

$$\mathbf{(\forall x)(\forall y)coll_z(z = x\, or\, z = y)}\ .$$

**Proof**
It is an axiom of the set theory.

**Definition 345 Set of one or two elements.**

---

[6]There is no $\mathbf{x}$ in the relation $\mathbf{coll_x R(x)}$. $\mathbf{x}$ is there to ease the reading. The same consideration is valid for the term $\mathbf{set_x\{R(x)\}}$.

Given two elements $\mathbf{x}, \mathbf{y}$, one calls $\{\mathbf{x}, \mathbf{y}\}$:

$$\{\mathbf{x}, \mathbf{y}\} = \mathbf{set_z}\{(\mathbf{z} = \mathbf{x}) \vee (\mathbf{z} = \mathbf{y})\} .$$

It exists from proposition 744 (p. 447). One calls $\{\mathbf{x}\} = \{\mathbf{x}, \mathbf{x}\}$ the set of one element.

**Proposition 745**

$$\forall \mathbf{z}(\mathbf{z} \in \{\mathbf{x}, \mathbf{y}\} \Rightarrow (\mathbf{z} = \mathbf{x}) \vee (\mathbf{z} = \mathbf{y})) .$$

**Proof**
It comes from proposition 744 (p. 447).

**Proposition 746** Given a relation $\mathbf{R}(\mathbf{x})$ two terms $\mathbf{u}, \mathbf{v}$ and the set $\mathbf{I} = \{\mathbf{u}, \mathbf{v}\}$, we have:

$$(\mathbf{R}(\mathbf{u}) \vee \mathbf{R}(\mathbf{v})) \Leftrightarrow (\exists \mathbf{i})(\mathbf{i} \in \mathbf{I}, \mathbf{R}(\mathbf{i}))$$
$$(\mathbf{R}(\mathbf{u}), \mathbf{R}(\mathbf{v})) \Leftrightarrow (\forall \mathbf{i})(\mathbf{i} \in \mathbf{I} \Rightarrow \mathbf{R}(\mathbf{i})) .$$

**Proof**
We have the following true statements:

$$(\mathbf{R}(\mathbf{u}) \vee \mathbf{R}(\mathbf{v})) \Leftrightarrow (\mathbf{R}(\mathbf{u}), \mathbf{u} \in \mathbf{I}) \vee (\mathbf{R}(\mathbf{v}), \mathbf{v} \in \mathbf{I})$$
$$\Rightarrow (\exists \mathbf{i})(\mathbf{i} \in \mathbf{I}, \mathbf{R}(\mathbf{i})) \vee (\exists \mathbf{i})(\mathbf{i} \in \mathbf{I}, \mathbf{R}(\mathbf{i})) \Leftrightarrow (\exists \mathbf{i})(\mathbf{i} \in \mathbf{I}, \mathbf{R}(\mathbf{i})) .$$

Setting $\mathbf{t} = \tau_{\mathbf{i}}(\mathbf{i} \in \mathbf{I}, \mathbf{R}(\mathbf{i}))$, we get (see point 25 (p. 436), 13 (p. 435), A.1.4, point 6 (p. 435), A.1.3):

$$(\exists \mathbf{i})(\mathbf{i} \in \mathbf{I}, \mathbf{R}(\mathbf{i})) \Leftrightarrow (\mathbf{t} \in \mathbf{I}, \mathbf{R}(\mathbf{t})) \Rightarrow ((\mathbf{t} = \mathbf{u} \vee \mathbf{t} = \mathbf{v}), \mathbf{R}(\mathbf{t}))$$
$$\Leftrightarrow ((\mathbf{t} = \mathbf{u}, \mathbf{R}(\mathbf{t})) \vee (\mathbf{t} = \mathbf{v}, \mathbf{R}(\mathbf{t}))) \Rightarrow (\mathbf{R}(\mathbf{u}) \vee \mathbf{R}(\mathbf{v})) .$$

Then we have:

$$(\mathbf{R}(\mathbf{u}) \vee \mathbf{R}(\mathbf{v})) \Leftrightarrow (\exists \mathbf{i})(\mathbf{i} \in \mathbf{I}, \mathbf{R}(\mathbf{i})) .$$

From above and definitions 329 (p. 434), 330 (p. 434), 334 (p. 441), point 5 (p. 435), A.1.3, we have:

$$(\mathbf{R}(\mathbf{u}), \mathbf{R}(\mathbf{v})) = \neg(\neg\mathbf{R}(\mathbf{u}) \vee \neg\mathbf{R}(\mathbf{v})) \Leftrightarrow \neg(\exists \mathbf{i})(\mathbf{i} \in \mathbf{I}, \neg\mathbf{R}(\mathbf{i}))$$
$$= \neg(\exists \mathbf{i})(\neg(\neg \mathbf{i} \in \mathbf{I} \vee \neg\neg\mathbf{R}(\mathbf{i}))) \Leftrightarrow \neg(\exists \mathbf{i})(\neg(\mathbf{i} \in \mathbf{I} \Rightarrow \mathbf{R}(\mathbf{i}))) = (\forall \mathbf{i})(\mathbf{i} \in \mathbf{I} \Rightarrow \mathbf{R}(\mathbf{i})) .$$

**Definition 346 Couple of two elements (x, y).**

Given two terms $\mathbf{x}, \mathbf{y}$ from the previous definitions, there are the following sets:

$$\{\mathbf{x}\},\ \{\mathbf{x}, \mathbf{y}\},\ \{\{\mathbf{x}\}, \{\mathbf{x}, \mathbf{y}\}\} . \tag{A.14}$$

One quotes:

$$(\mathbf{x}, \mathbf{y}) = \{\{\mathbf{x}\}, \{\mathbf{x}, \mathbf{y}\}\} .$$

From that definition we have:

$$(\mathbf{x}, \mathbf{y}) = (\mathbf{x}', \mathbf{y}') \Leftrightarrow ((\mathbf{x} = \mathbf{x}'), (\mathbf{y} = \mathbf{y}')) \tag{A.15}$$

because $\mathbf{x} = \mathbf{x}'$ and $\mathbf{y} = \mathbf{y}'$ is the only possibility for the set equality:

$$\{\{\mathbf{x}\}, \{\mathbf{x}, \mathbf{y}\}\} = \{\{\mathbf{x}'\}, \{\mathbf{x}', \mathbf{y}'\}\}$$

because:

$$\mathbf{x} \neq \mathbf{y} \Rightarrow (\{\mathbf{x}'\} = \{\mathbf{x}', \mathbf{x}'\} \neq \{\mathbf{x}, \mathbf{y}\}) \Rightarrow (\mathbf{x}' = \mathbf{x}, \mathbf{y}' = \mathbf{y}) \tag{A.16}$$

$\mathbf{z}$ is a couple means:

$$(\exists \mathbf{x})(\exists \mathbf{y})(\mathbf{z} = (\mathbf{x}, \mathbf{y}) = \{\{\mathbf{x}\}, \{\mathbf{x}, \mathbf{y}\}\}) .$$

**Definition 347 Projection $\mathbf{proj_1}$, $\mathbf{proj_2}$ of a couple z.**

Given a couple $\mathbf{z}$, one calls (see section A.4.1 (p. 445)):

$$\mathbf{proj_1(z) = \tau_x((\exists y)z = (x, y))}$$
$$\mathbf{proj_2(z) = \tau_y((\exists x)z = (x, y))}\ .$$

**Proposition 747** $\mathbf{z}$ being a couple (definition 346 (p. 448)), we have (see definition 347 (p. 449)):
$\mathbf{z = (proj_1(z), proj_2(z))}$ and $\mathbf{z = (x, y) \Rightarrow x = proj_1(z), y = proj_2(z)}$.
**Proof**
It comes from definition 346 (p. 448), equation (A.15) (p. 448) and definition 333 (p. 441), propositions 727 (p. 441), 726 (p. 441).
**Definition 348 Graph.**

A set $\mathbf{\Gamma}$ is call a graph if any of its elements is a couple:

$$\mathbf{\gamma \in \Gamma \Rightarrow (\exists x)(\exists y)(\gamma = (x, y))}\ .$$

**Proposition 748** $\mathbf{R}$' being a relation where $\mathbf{X}$ and $\mathbf{Y}$ do not appear, we have the axiom generator:

$$\mathbf{(\forall y)(\exists X)(\forall x)(R' \Rightarrow x \in X) \Rightarrow (\forall Y)coll_x((\exists y)((y \in Y)\ and\ R'))}\ .$$

**Remarks on proposition /axiom 748 (p. 449)**
$\mathbf{coll_x R(x)}$ cannot be always true. There is a famous example: If we accept by axiom

$$\mathbf{coll_x(x \notin x)}$$

we should have

$$\mathbf{set_x\{x \notin x\} \notin\ set_x\{x \notin x\} \Leftrightarrow set_x\{x \notin x\} \in\ set_x\{x \notin x\}}\ .$$

Then from 15 (p. 435) and (A.1.3, 1)

$$\begin{aligned}
&\mathbf{\neg(set_x\{x \notin x\} \in\ set_x\{x \notin x\}) \vee set_x\{x \notin x\} \in\ set_x\{x \notin x\}}\\
&\mathbf{\Rightarrow set_x\{x \notin x\} \in\ set_x\{x \notin x\}}\\
&\mathbf{\neg(set_x\{x \notin x\} \in\ set_x\{x \notin x\}) \vee set_x\{x \notin x\} \in\ set_x\{x \notin x\}}\\
&\mathbf{\Rightarrow set_x\{x \notin x\} \notin\ set_x\{x \notin x\}}\ .
\end{aligned}$$

Since the first member$\mathbf{\neg(set_x\{x \notin x\} \in\ set_x\{x \notin x\}) \vee set_x\{x \notin x\} \in\ set_x\{x \notin x\}}$ is always true from (A.1.4, 4), $\mathbf{coll_x(x \notin x)}$ would make the two second ones true. So we would have a statement and its opposite which are true. Applying (A.1.3, 5) and (A.1.4, 14), we would be in the case of the remark of proposition 725 (p. 440). The theory would be useless since everything is true.
Proposition 748 (p. 449) is a scheme in the set theory which limits the action $\mathbf{coll_x R(x)}$ and generates axioms which indicate when $\mathbf{coll_x R(x)}$ has to be considered as true. The following propositions are not really theorems but generators of theorems. Usual logic rules cannot be applied strictly on them because for instance the relations are not explicitly stated. Terms and relations look as treated on the same foot and they should not. As said at the beginning of this appendix a more precise description can be found in [1].

**Proposition 749** When $\mathbf{x}$ is not in the $\mathbf{U}$ definition, the expression $\mathbf{coll_x(R(x), (x \in U))}$ is a true relation.

**Proof**

In that case $\mathbf{U}$ can be inside the definition of $\mathbf{R}$. So, it is not a straightforward application of proposition 748 (p. 449). By choosing $\mathbf{R}' = (\mathbf{R}, (\mathbf{x} \in \mathbf{U}), (\mathbf{y} = \mathbf{x}))$, $\mathbf{y}$ being any term not in $\mathbf{R}$, we have from the axiom mentioned in definition 345 (p. 447) and from (A.1.3, 6):

$$\mathbf{R}' \Rightarrow \mathbf{x} \in \{\mathbf{y}\} .$$

From propositions 727 (p. 441) and 732 (p. 442), we have:

$$(\forall \mathbf{y})(\exists \mathbf{X})(\forall \mathbf{x})(\mathbf{R}' \Rightarrow \mathbf{x} \in \mathbf{X}) .$$

From proposition 748 (p. 449), we have:

$$(\forall \mathbf{Y})\mathbf{coll_x}((\exists \mathbf{y})((\mathbf{y} \in \mathbf{Y}) \,\mathbf{and}\, \mathbf{R}')) .$$

Then

$$\mathbf{coll_x}((\exists \mathbf{y})((\mathbf{y} \in \mathbf{U}) \,\mathbf{and}\, \mathbf{R}')) .$$

Since

$$(\exists \mathbf{y})(\mathbf{y} \in \mathbf{U} \,\mathbf{and}\, \mathbf{R}') \Leftrightarrow (\exists \mathbf{y})(\mathbf{y} \in \mathbf{U}, (\mathbf{x} \in \mathbf{U}, \mathbf{y} = \mathbf{x}, \mathbf{R}))$$

because $\mathbf{y}$ is not in $\mathbf{R}$ and from

$$(\mathbf{x} \in \mathbf{U}, \mathbf{R}) \Leftrightarrow (\exists \mathbf{y})(\mathbf{x} \in \mathbf{U}, \mathbf{R}, \mathbf{y} = \mathbf{x}) \Leftrightarrow (\exists \mathbf{y})(\mathbf{y} \in \mathbf{U}, \mathbf{x} \in \mathbf{U}, \mathbf{R}, \mathbf{y} = \mathbf{x})$$

we have:

$$(\exists \mathbf{y})((\mathbf{y} \in \mathbf{U}) \,\mathbf{and}\, \mathbf{R}') \Leftrightarrow \mathbf{R}, (\mathbf{x} \in \mathbf{U})$$

From (A.1.3, 6) and the axiom (A.3) (p. 441), we have:

$$\mathbf{coll_x}((\exists \mathbf{y})((\mathbf{y} \in \mathbf{U}) \,\mathbf{and}\, \mathbf{R}')) \Leftrightarrow \mathbf{coll_x}(\mathbf{R}, (\mathbf{x} \in \mathbf{U})) .$$

**Proposition 750** When $\mathbf{x}$ is not in the $\mathbf{U}$ definition, the expression $((\mathbf{R(x)} \Rightarrow (\mathbf{x} \in \mathbf{U})) \Rightarrow \mathbf{coll_x(R(x))})$ is a true relation.

**Proof**

We have $\mathbf{R(x)} \Rightarrow \mathbf{R(x)}$ true (A.1.4 (p. 435), point 4). From points 13 (p. 435), 17 (p. 436) and 22 (p. 436) of A.1.4 (p. 435), we have:

$$(\mathbf{R(x)} \Rightarrow (\mathbf{x} \in \mathbf{U})) \Rightarrow (\mathbf{R(x)} \Leftrightarrow (\mathbf{R(x)}, (\mathbf{x} \in \mathbf{U}))) .$$

We conclude using proposition 749 (p. 450).

**Proposition 751** $\mathbf{T}$ being a term, $\mathbf{x}, \mathbf{y}$ two different letters, $\mathbf{x}$ not in $\mathbf{A}$, $\mathbf{y}$ not in $\mathbf{T}$ and not in $\mathbf{A}$, the expression $\mathbf{coll_y}((\exists \mathbf{x})((\mathbf{y} = \mathbf{T}), (\mathbf{x} \in \mathbf{A})))$ is a true relation.

**Proof**

Let us call $\mathbf{R}$ the relation $\mathbf{y} = \mathbf{T}$. We have:

$$(\forall \mathbf{y})(\mathbf{R} \Rightarrow (\mathbf{y} \in \{\mathbf{T}\}))$$

then:

$$(\forall \mathbf{x})(\exists \mathbf{X})(\forall \mathbf{y})(\mathbf{R} \Rightarrow \mathbf{y} \in \mathbf{X})$$

then from proposition 748 (p. 449),

$$\mathbf{coll_y}((\exists \mathbf{x})((\mathbf{x} \in \mathbf{A}), (\mathbf{y} = \mathbf{T}))) .$$

## A.5.2 Subsets

**Definition 349** $\mathbf{A \subset B}$.

The relation $\mathbf{\forall x(x \in A \Rightarrow x \in B)}$ is written $\mathbf{A \subset B}$.

**Proposition 752** $\mathbf{x \in A \Rightarrow \{x\} \subset A}$.

**Proof**
It comes from proposition 745 (p. 448) and definitions 345 (p. 447), 349 (p. 451).

**Proposition 753** $\mathbf{(A \subset B, B \subset C) \Rightarrow A \subset C}$.

**Proof**
$\mathbf{x}$ being a term, from proposition 731 (p. 442), one has the true relation:

$$\mathbf{R_1(x) = ((A \subset B, B \subset C) \Rightarrow x \in A \Rightarrow x \in B \Rightarrow x \in C)}\ .$$

Then the following relation is true:

$$\mathbf{R(x) = ((A \subset B, B \subset C) \Rightarrow x \in A \Rightarrow x \in C)}$$

and in particular if $\mathbf{R'(x) = (x \in A \Rightarrow x \in C)}$, (see convention 23 (p. 433)), we have:

$$\mathbf{(\tau_y \neg R'(y) | x) R(x)}\ .$$

So we have:

$$\mathbf{(A \subset B, B \subset C) \Rightarrow \neg\neg R'(\tau_y \neg R'(y))}\ .$$

Since $\mathbf{\neg\neg R'(\tau_y \neg R'(y)) = \forall y R'(y)}$ (definition 334 (p. 441)), we have proposition 753 (p. 451) because from definition 349 (p. 451), we have $\mathbf{\forall y R'(y) = A \subset C}$.

**Proposition 754** $\mathbf{(A \subset B, B \subset A) \Rightarrow A = B}$ (extensionality principle).

**Proof**
It is an axiom of the set theory.

**Proposition 755** $\mathbf{A}$ being a set[7], $\mathbf{A \subset A}$.

**Proof**
From (A.1.4, 13) and proposition 731 (p. 442), $\mathbf{\forall x(x \in A \Rightarrow x \in A)}$. By definition 349 (p. 451) of $\subset$, we have $\mathbf{\forall x(x \in A \Rightarrow x \in A)}$ equivalent to $\mathbf{A \subset A}$.

**Proposition 756** $\mathbf{(A \subset B, B \subset A) \Leftrightarrow A = B}$.

**Proof**
Setting $\mathbf{R(Z) = A \subset Z}$, since $\mathbf{R(A)}$ true (proposition 755 (p. 451)) we have $\mathbf{A = B \Rightarrow R(B) = A \subset B}$ (section A.1.3 point 6 (p. 435)). For the same reason $\mathbf{A = B \Rightarrow B \subset A}$. Then from section A.1.4, 14 (p. 435), we have: $\mathbf{A = B \Rightarrow (A \subset B, B \subset A)}$. With proposition 754 (p. 451), it gives:

$$\mathbf{(A \subset B, B \subset A) \Leftrightarrow A = B}\ .$$

[7] As explained in section A.5.1, this expression does not mean anything since in a set theory any letter is a set. Even if $\mathbf{\neg(\exists a)(a \in A) = (\forall a)(a \notin A) \Leftrightarrow A = \emptyset}$, $\mathbf{A}$ is a unique set (definition 355 (p. 456)). If we admit that $\mathbf{(\exists a)(a \in A)}$ means $\mathbf{A}$ is a set, in that case since $\mathbf{A = \emptyset \vee A \neq \emptyset}$ is true (A.1.4, 4). $\mathbf{A}$ is a set whatever $\mathbf{A}$.

**Proposition 757**

$$\begin{aligned}\mathbf{A \neq B} &\Leftrightarrow \mathbf{(\exists x)((x \notin A, x \in B) \vee (x \in A, x \notin B))}\\ &\Leftrightarrow \mathbf{(\exists x)(x \notin A, x \in B) \vee (\exists x)(x \in A, x \notin B)}\ .\end{aligned}$$

**Proof**
From proposition 754 (p. 451) and definition 349 (p. 451), we have:

$$\mathbf{((\forall x)(x \in A \Rightarrow x \in B), (\forall x)(x \in B \Rightarrow x \in A)) \Leftrightarrow A = B}\ .$$

Using equation (A.6) (p. 443), we have:

$$\mathbf{(\forall x)((x \in A \Rightarrow x \in B), (x \in B \Rightarrow x \in A)) \Leftrightarrow A = B}\ .$$

From definitions 329 (p. 434), 330 (p. 434), 331 (p. 434), 333 (p. 441), 334 (p. 441) and using A.1.4 7 (p. 435), we have:

$$\begin{aligned}&\mathbf{\{(\forall x)((x \in A \Rightarrow x \in B), (x \in B \Rightarrow x \in A)) \Leftrightarrow A = B\}}\\ &\Leftrightarrow \mathbf{\{A \neq B \Leftrightarrow \neg(\forall x)((x \in A \Rightarrow x \in B), (x \in B \Rightarrow x \in A))\}}\\ &= \mathbf{\{A \neq B \Leftrightarrow \neg\neg(\exists x)\neg\neg(\neg(x \notin A \vee x \in B) \vee \neg(x \notin B \vee x \in A))\}}\\ &\Leftrightarrow \mathbf{\{A \neq B \Leftrightarrow (\exists x)((x \in A, x \notin B) \vee (x \in B, x \notin A))\}}\ .\end{aligned}$$

From equation (A.8) (p. 443), we have:

$$\begin{aligned}\mathbf{A \neq B} &\Leftrightarrow \mathbf{(\exists x)((x \in A, x \notin B) \vee (x \in B, x \notin A))}\\ &\Leftrightarrow \mathbf{((\exists x)(x \in A, x \notin B) \vee (\exists x)(x \in B, x \notin A))}\ .\end{aligned}$$

**Proposition 758**

$$\mathbf{(R(i,i), R(j,j)) \Rightarrow (R(i,j) \vee R(j,i) \Leftrightarrow (\exists k)(k \in \{i,j\}, R(i,k), R(j,k)))}\ .$$

This proposition applies in particular to $\mathbf{R(i,j) = i \subset j}$.

**Proof**
Let us suppose $\mathbf{R(i,i), R(j,j)}$ true then applying sequentially the axioms from definition 345 (p. 447), relation A.1 (p. 441) in A.2.1 (p. 440) and relations 15 (p. 435), 24 (p. 436) in A.1.4 point 1 (p. 435) in A.1.3, we have:

$$\begin{aligned}&\mathbf{(R(i,j) \vee R(j,i)) \Rightarrow (j \in \{i,j\}, R(i,j), R(j,j)) \vee (i \in \{i,j\}, R(j,i), R(i,i))}\\ &\Rightarrow \mathbf{(\exists k)(k \in \{i,j\}, R(i,k), R(j,k)) \vee (\exists k)(k \in \{i,j\}, R(j,k), R(i,k))}\\ &\Rightarrow \mathbf{(\exists k)(k \in \{i,j\}, R(i,k), R(j,k))}\ .\end{aligned}$$

We also have:

$$\begin{aligned}\mathbf{(\exists k)(k \in \{i,j\}, R(i,k), R(j,k))} &\Rightarrow \mathbf{(\exists k)((k = i \vee k = j), R(i,k), R(j,k))}\\ &\Rightarrow \mathbf{(\exists k)((k = i, R(i,k), R(j,k))}\\ &\quad\ \mathbf{\vee(k = j, R(i,k), R(j,k)))}\\ &\Rightarrow \mathbf{(\exists k)((R(i,i), R(j,i)) \vee (R(i,j), R(j,j)))}\\ &\Rightarrow \mathbf{(R(i,i), R(j,i)) \vee (R(i,j), R(j,j))}\\ &\Rightarrow \mathbf{R(i,j) \vee R(j,i)}\ .\end{aligned}$$

**Proposition 759** $\mathbf{(\forall s)(Coll_x(x \subset s))}$.

**Proof**
It is an axiom of the set theory.

**Definition 350** $\mathcal{P}(\mathrm{s})$.

s being a set, $\mathcal{P}(\mathrm{s})$ is the set of subsets of s:

$$\mathcal{P}(\mathrm{s}) = \mathrm{set}_{\mathrm{x}}\{\mathrm{x} \subset \mathrm{s}\} .$$

This set exists from proposition 759 (p. 452).

**Proposition 760** $\mathrm{t} \subset \mathrm{s} \Rightarrow \mathcal{P}(\mathrm{t}) \subset \mathcal{P}(\mathrm{s})$ .

**Proof**
From definitions 350 (p. 453), 349 (p. 451) and proposition 753 (p. 451):

$$\mathrm{t} \subset \mathrm{s} \Rightarrow (\forall \mathrm{x})(\mathrm{x} \in \mathcal{P}(\mathrm{t}) \Rightarrow \mathrm{x} \subset \mathrm{t} \subset \mathrm{s} \Rightarrow \mathrm{x} \in \mathcal{P}(\mathrm{s})) .$$

Then from definition 349 (p. 451), $\mathcal{P}(\mathrm{t}) \subset \mathcal{P}(\mathrm{s})$.

### A.5.3 Union and intersection of sets

**Definition 351** $\cup$.

$\mathrm{A} \cup \mathrm{B}$ means union of A and B:

$$(\mathrm{x} \in \mathrm{A} \cup \mathrm{B}) \Leftrightarrow (\mathrm{x} \in \mathrm{A} \vee \mathrm{x} \in \mathrm{B}) .$$

**Definition 352** $\cap$.

$\mathrm{A} \cap \mathrm{B}$ means intersection of A and B:

$$(\mathrm{x} \in \mathrm{A} \cap \mathrm{B}) \Leftrightarrow (\mathrm{x} \in \mathrm{A}, \mathrm{x} \in \mathrm{B}) .$$

**Proposition 761** **A, B** being sets, we have:

$$\mathrm{A} \subset \mathrm{A} \cup \mathrm{B}$$

$$\mathrm{A} \cap \mathrm{B} \subset \mathrm{A} .$$

**Proof**
It is a direct consequence of definitions 351 (p. 453), 352 (p. 453) and 349 (p. 451).

**Proposition 762** **A, B, C** being sets, we have:

$$\mathrm{B} \subset \mathrm{A} \Rightarrow \mathrm{B} \cup \mathrm{C} \subset \mathrm{A} \cup \mathrm{C}$$

$$\mathrm{B} \subset \mathrm{A} \Rightarrow \mathrm{B} \cap \mathrm{C} \subset \mathrm{A} \cap \mathrm{C} .$$

**Proof**
It is a direct consequence of definitions 351 (p. 453), 352 (p. 453) and 349 (p. 451).

**Proposition 763** **A, B, C, D** being sets, we have:

$$(\mathrm{A} \subset \mathrm{C}, \mathrm{B} \subset \mathrm{D}) \Rightarrow (\mathrm{A} \cup \mathrm{B} \subset \mathrm{C} \cup \mathrm{D}, \mathrm{A} \cap \mathrm{B} \subset \mathrm{C} \cup \mathrm{D}, \mathrm{A} \cap \mathrm{B} \subset \mathrm{C} \cap \mathrm{D})$$

$$\mathrm{A} \cup \mathrm{A} = \mathrm{A},\ \mathrm{A} \subset \mathrm{C} \Leftrightarrow \mathrm{A} \cap \mathrm{C} = \mathrm{A},\ \mathrm{A} \subset \mathrm{C} \Rightarrow \mathrm{A} \cup \mathrm{C} = \mathrm{C},\ \mathrm{A} \cap \mathrm{A} = \mathrm{A} .$$

It is a direct consequence of definitions 351 and 352.

**Proposition 764** Given three sets **A**, **B**, **C**, we have:

$$(\mathbf{A} \cup \mathbf{B}) \cap \mathbf{C} = (\mathbf{A} \cap \mathbf{C}) \cup (\mathbf{B} \cap \mathbf{C}) \ .$$

**Proof**
From 25 (p. 436) of A.1.4 and definitions 351 (p. 453), 352 (p. 453), we have:

$$\mathbf{x} \in (\mathbf{A} \cup \mathbf{B}) \cap \mathbf{C} \Leftrightarrow ((\mathbf{x} \in \mathbf{A} \vee \mathbf{x} \in \mathbf{B}), \mathbf{x} \in \mathbf{C}) \Leftrightarrow ((\mathbf{x} \in \mathbf{A}, \mathbf{x} \in \mathbf{C}) \vee (\mathbf{x} \in \mathbf{B}, \mathbf{x} \in \mathbf{C})) \\ \Leftrightarrow \mathbf{x} \in (\mathbf{A} \cap \mathbf{C}) \cup (\mathbf{B} \cap \mathbf{C}) \ .$$

We conclude using definition 349 (p. 451) and proposition 754 (p. 451) (extensionality principle).

**Proposition 765** Given two sets **A**, **B** (see footnote 7 (p. 451)), we have:

$$\mathcal{P}(\mathbf{A} \cup \mathbf{B}) = \mathbf{set_x}\{(\exists \mathbf{i} \subset \mathbf{A})(\exists \mathbf{j} \subset \mathbf{B})(\mathbf{x} = \mathbf{i} \cup \mathbf{j})\} \ .$$

**Proof**
We have:

$$(\mathbf{i} \subset \mathbf{A}, \mathbf{j} \subset \mathbf{B}, \mathbf{x} = \mathbf{i} \cup \mathbf{j}) \\ \Rightarrow (\forall \mathbf{y} \in \mathbf{x})(\mathbf{y} \in \mathbf{i} \subset \mathbf{A} \vee \mathbf{y} \in \mathbf{j} \subset \mathbf{B}, \mathbf{y} \in \mathbf{A} \cup \mathbf{B}) \Leftrightarrow \mathbf{x} \in \mathcal{P}(\mathbf{A} \cup \mathbf{B})) \ .$$

Then we have:

$$\mathbf{set_x}\{(\exists \mathbf{i} \subset \mathbf{A})(\exists \mathbf{j} \subset \mathbf{B})(\mathbf{x} = \mathbf{i} \cup \mathbf{j})\} \subset \mathcal{P}(\mathbf{A} \cup \mathbf{B}) \ .$$

From propositions 763 (p. 453) and 764 (p. 454), we also have:

$$\mathbf{x} \in \mathcal{P}(\mathbf{A} \cup \mathbf{B}) \Rightarrow \mathbf{x} \subset \mathbf{A} \cup \mathbf{B} \Rightarrow \mathbf{x} = \mathbf{x} \cap (\mathbf{A} \cup \mathbf{B}) = (\mathbf{x} \cap \mathbf{A}) \cup (\mathbf{x} \cap \mathbf{B}) \\ \Rightarrow (\exists \mathbf{i} = \mathbf{x} \cap \mathbf{A})(\exists \mathbf{j} = \mathbf{x} \cap \mathbf{B})(\mathbf{i} \subset \mathbf{A}, \mathbf{j} \subset \mathbf{B}, \mathbf{x} = \mathbf{i} \cup \mathbf{j}) \ .$$

Then we have:

$$\mathcal{P}(\mathbf{A} \cup \mathbf{B}) \subset \mathbf{set_x}\{(\exists \mathbf{i} \subset \mathbf{A})(\exists \mathbf{j} \subset \mathbf{B})(\mathbf{x} = \mathbf{i} \cup \mathbf{j})\} \ .$$

We conclude by proposition 754 (p. 451) (extensionality principle).

## A.5.4 Complement of a subset

**Definition 353 $\mathcal{C}$.**

**A** and **B** being sets, if $\mathbf{B} \subset \mathbf{A}$, one calls complement of **B**:

$$\mathcal{C}_\mathbf{A}(\mathbf{B}) = \mathbf{set_x}\{\mathbf{x} \in \mathbf{A}, \mathbf{x} \notin \mathbf{B}\} \ .$$

It exists from proposition 749 (p. 450).
There is another notation ($\mathcal{C}_\mathbf{A}(\mathbf{B}) = \mathbf{A} - \mathbf{B}$) which is more intuitive. But that notation may bring confusions:

**Definition 354 Subtractive notation for $\mathcal{C}$: A − B.**

When clearly $\mathbf{B} \subset \mathbf{A}$, we may also write $\mathbf{A} - \mathbf{B}$ instead of $\mathcal{C}_\mathbf{A}(\mathbf{B})$.

**Proposition 766** $\mathbf{B} \subset \mathbf{A} \Rightarrow \mathcal{C}_\mathbf{A}\mathbf{B} \cup \mathbf{B} = \mathbf{A}$.

**Proof**
From definitions 353 (p. 454) and 351 (p. 453), and since $(\mathbf{x} \in \mathbf{B} \vee \notin \mathbf{B})$ is always true, we have:

$$\begin{aligned} \mathbf{x} \in \mathcal{C}_\mathbf{A}\mathbf{B} \cup \mathbf{B} &\Leftrightarrow ((\mathbf{x} \in \mathbf{B}, \mathbf{x} \in \mathbf{A}) \vee (\mathbf{x} \notin \mathbf{B}, \mathbf{x} \in \mathbf{A})) \\ &\Leftrightarrow (\mathbf{x} \in \mathbf{A}, (\mathbf{x} \in \mathbf{B} \vee \notin \mathbf{B})) \Leftrightarrow \mathbf{x} \in \mathbf{A} \end{aligned}$$

(see A.1.3 (p. 435), 2 and A.1.4 (p. 435), 4, 13, 14, 25).

**Convention 29** $\mathcal{C}(\mathbf{B})$ for $\mathcal{C}_\mathbf{A}(\mathbf{B})$.

If in the context all the sets considered are subset of a set let say $\mathbf{A}$, given $\mathbf{B} \subset \mathbf{A}$, since there is no confusion possible, we write: $\mathcal{C}(\mathbf{B})$ for $\mathcal{C}_\mathbf{A}(\mathbf{B})$.
For instance, considering $\mathcal{R}^\mathbf{n}$ (definition 241 (p. 266)) and a subset $\mathbf{C} \subset \mathcal{R}^\mathbf{n}$, we shall always write $\mathcal{C}(\mathbf{C})$ instead of $\mathcal{C}_{\mathcal{R}^\mathbf{n}}(\mathbf{C})$. If we want to consider $\mathcal{R}^\mathbf{n}$ as a subset of $\mathcal{R}^{\mathbf{n}+1}$ and the complement of C in that bigger space, then we shall write $\mathcal{C}_{\mathcal{R}^{\mathbf{n}+1}}(\mathbf{C})$.

**Proposition 767** $\mathbf{A}$ and $\mathbf{B}$ being sets with $\mathbf{B} \subset \mathbf{A}$, we have:

$$\mathcal{C}_\mathbf{A}(\mathcal{C}_\mathbf{A}(\mathbf{B})) = \mathbf{B} \,.$$

**Proof**
From definition 353 (p. 454), we have:

$$\begin{aligned} \mathbf{x} \in \mathcal{C}_\mathbf{A}(\mathcal{C}_\mathbf{A}(\mathbf{B})) &\Leftrightarrow (\mathbf{x} \in \mathbf{A}, \mathbf{x} \notin \mathcal{C}_\mathbf{A}(\mathbf{B})) \Leftrightarrow (\mathbf{x} \in \mathbf{A}, \neg(\mathbf{x} \in \mathcal{C}_\mathbf{A}(\mathbf{B}))) \\ &\Leftrightarrow (\mathbf{x} \in \mathbf{A}, \neg\neg \mathbf{x} \in \mathbf{B}) \Leftrightarrow (\mathbf{x} \in \mathbf{B}) \,. \end{aligned}$$

**Proposition 768** $\mathbf{E}$, $\mathbf{A} \subset \mathbf{E}$ and $\mathbf{B} \subset \mathbf{E}$ being sets, we have:

$$\mathcal{C}_\mathbf{A}(\mathbf{B} \cap \mathbf{A}) = \mathcal{C}_\mathbf{E}(\mathbf{B}) \cap \mathbf{A} \,.$$

**Proof**
From definition 349 (p. 451) $\mathbf{x} \in \mathbf{A} \Rightarrow \mathbf{x} \in \mathbf{E}$, then from definition 353 (p. 454) and applying A.1.4 points 17 (p. 436), 13 (p. 435) and 26 (p. 436), we have:

$$\begin{aligned} \mathbf{x} \in \mathcal{C}_\mathbf{A}(\mathbf{B} \cap \mathbf{A}) &\Leftrightarrow (\mathbf{x} \in \mathbf{A}, \mathbf{x} \notin \mathbf{B}) \Leftrightarrow (\mathbf{x} \in \mathbf{A}, \mathbf{x} \in \mathbf{E}, \mathbf{x} \notin \mathbf{B}) \\ &\Leftrightarrow (\mathbf{x} \in \mathbf{A}, \mathbf{x} \in \mathcal{C}_\mathbf{E}(\mathbf{B}) \Leftrightarrow \mathcal{C}_\mathbf{E}(\mathbf{B}) \cap \mathbf{A} \,. \end{aligned}$$

**Proposition 769** $\mathbf{A} \subset \mathbf{B} \subset \mathbf{C} \Rightarrow \mathcal{C}_\mathbf{C}(\mathbf{B}) \subset \mathcal{C}_\mathbf{C}(\mathbf{A})$.
**Proof**
From definition 349 (p. 451), we have:

$$\begin{aligned} &\mathbf{A} \subset \mathbf{B} = (\forall \mathbf{x})((\mathbf{x} \in \mathbf{A}) \Rightarrow (\mathbf{x} \in \mathbf{B})) \\ &(\mathbf{x} \in \mathbf{A} \Rightarrow \mathbf{x} \in \mathbf{B}) \Leftrightarrow (\neg(\mathbf{x} \in \mathbf{A}) \vee \mathbf{x} \in \mathbf{B}) \\ &\Leftrightarrow (\neg\neg(\mathbf{x} \in \mathbf{B}) \vee \neg(\mathbf{x} \in \mathbf{A})) = (\mathbf{x} \notin \mathbf{B} \Rightarrow \mathbf{x} \notin \mathbf{A}) \,. \end{aligned}$$

Then we have from A.1.4 points 17 (p. 436) and definition 353 (p. 454):

$$\mathbf{A} \subset \mathbf{B} \subset \mathbf{C} \Rightarrow (\forall \mathbf{x})((\mathbf{x} \notin \mathbf{B}, \mathbf{x} \in \mathbf{C}) \Rightarrow (\mathbf{x} \notin \mathbf{A}, \mathbf{x} \in \mathbf{C})) \Leftrightarrow \mathcal{C}_\mathbf{C}(\mathbf{B}) \subset \mathcal{C}_\mathbf{C}(\mathbf{A}) \,.$$

**Proposition 770** $\neg(\exists \mathbf{x})(\mathbf{x} \in \mathcal{C}_\mathbf{A}(\mathbf{A})), (\forall \mathbf{x})(\mathbf{x} \notin \mathcal{C}_\mathbf{A}(\mathbf{A}))$.

**Proof**
The following statements are true whatever the term **x** is (see A.1.2 (p. 434), definitions 329 (p. 434), 344 (p. 447), 353 (p. 454) and (A.1.4, 4), (A.1.4, 5)):

$$\begin{aligned}
&\mathbf{x} \notin \mathbf{A} \vee \mathbf{x} \in \mathbf{A} \\
&\neg\neg(\mathbf{x} \notin \mathbf{A} \vee \mathbf{x} \in \mathbf{A}) \\
&\neg(\mathbf{x} \notin \mathbf{A}, \mathbf{x} \in \mathbf{A}) \\
&\mathbf{x} \in \mathcal{C}_\mathbf{A}(\mathbf{A}) \Leftrightarrow (\mathbf{x} \notin \mathbf{A}, \mathbf{x} \in \mathbf{A}) \Leftrightarrow \neg\neg(\mathbf{x} \notin \mathbf{A}, \mathbf{x} \in \mathbf{A}) \\
&\neg\neg(\mathbf{x} \notin \mathbf{A}, \mathbf{x} \in \mathbf{A}) \vee \neg(\mathbf{x} \in \mathcal{C}_\mathbf{A}(\mathbf{A})) = \neg(\mathbf{x} \notin \mathbf{A}, \mathbf{x} \in \mathbf{A}) \Rightarrow \neg(\mathbf{x} \in \mathcal{C}_\mathbf{A}(\mathbf{A})) \\
&\neg\mathbf{x} \in \mathcal{C}_\mathbf{A}(\mathbf{A})\ .
\end{aligned}$$

Setting $\mathbf{x} = \tau_\mathbf{x}(\mathbf{x} \in \mathcal{C}_\mathbf{A}(\mathbf{A}))$, from definitions 333 (p. 441), 334 (p. 441), 330 (p. 434), 331 (p. 434) and (A.1.4 , 5), we have:

$$\begin{aligned}
&\neg\tau_\mathbf{x}(\mathbf{x} \in \mathcal{C}_\mathbf{A}(\mathbf{A})) \in \mathcal{C}_\mathbf{A}(\mathbf{A}) \\
&\neg\tau_\mathbf{x}(\mathbf{x} \in \mathcal{C}_\mathbf{A}(\mathbf{A})) \in \mathcal{C}_\mathbf{A}(\mathbf{A}) = \neg(\exists \mathbf{x})(\mathbf{x} \in \mathcal{C}_\mathbf{A}(\mathbf{A})) \\
&\neg(\exists \mathbf{x})(\mathbf{x} \in \mathcal{C}_\mathbf{A}(\mathbf{A})) \Leftrightarrow (\forall \mathbf{x})(\mathbf{x} \notin \mathcal{C}_\mathbf{A}(\mathbf{A})) \\
&\neg(\exists \mathbf{x})(\mathbf{x} \in \mathcal{C}_\mathbf{A}(\mathbf{A})), (\forall \mathbf{x})(\mathbf{x} \notin \mathcal{C}_\mathbf{A}(\mathbf{A}))\ .
\end{aligned}$$

**Proposition 771** $(\forall \mathbf{B})(\forall \mathbf{A})(\mathcal{C}_\mathbf{A}(\mathbf{A}) = \mathcal{C}_\mathbf{B}(\mathbf{B}))$.

**Proof**
From logical rules (A.1.3) we have as true statements:

$$\begin{aligned}
(\mathbf{x} \in \mathbf{A} \vee \mathbf{x} \notin \mathbf{A}) \Leftrightarrow\ & \neg\neg(\mathbf{x} \in \mathbf{A} \vee \mathbf{x} \notin \mathbf{A}) \Leftrightarrow \neg(\mathbf{x} \notin \mathbf{A}, \mathbf{x} \in \mathbf{A}) \\
\Rightarrow\ & \neg(\mathbf{x} \in \mathbf{A}, \mathbf{x} \notin \mathbf{A}) \vee (\mathbf{x} \in \mathbf{B}, \mathbf{x} \notin \mathbf{B}) \\
=\ & (\mathbf{x} \in \mathbf{A}, \mathbf{x} \notin \mathbf{A}) \Rightarrow (\mathbf{x} \in \mathbf{B}, \mathbf{x} \notin \mathbf{B})\ .
\end{aligned}$$

Then:

$$(\mathbf{x} \in \mathbf{A}, \mathbf{x} \notin \mathbf{A}) \Leftrightarrow (\mathbf{x} \in \mathbf{B}, \mathbf{x} \notin \mathbf{B})\ .$$

Since

$$\mathbf{x} \in \mathcal{C}_\mathbf{A}(\mathbf{A}) \Leftrightarrow (\mathbf{x} \in \mathbf{A}, \mathbf{x} \notin \mathbf{A}) \Leftrightarrow (\mathbf{x} \in \mathbf{B}, \mathbf{x} \notin \mathbf{B}) \Leftrightarrow \mathbf{x} \in \mathcal{C}_\mathbf{B}(\mathbf{B})$$

we have $(\mathcal{C}_\mathbf{A}(\mathbf{A}) \subset \mathcal{C}_\mathbf{B}(\mathbf{B}), \mathcal{C}_\mathbf{B}(\mathbf{B}) \subset \mathcal{C}_\mathbf{A}(\mathbf{A}))$ so from the axiom of proposition 754 (p. 451), we have

$$\mathcal{C}_\mathbf{A}(\mathbf{A}) = \mathcal{C}_\mathbf{B}(\mathbf{B})\ . \tag{A.17}$$

As in proposition 732 (p. 442), we may set $\mathbf{A} = \tau_\mathbf{A}\neg(\mathcal{C}_\mathbf{A}(\mathbf{A}) = \mathcal{C}_\mathbf{B}(\mathbf{B}))$. It gives:

$$\begin{aligned}
&\neg\neg(\mathcal{C}_{\tau_\mathbf{A}\neg(\mathcal{C}_\mathbf{A}(\mathbf{A})=\mathcal{C}_\mathbf{B}(\mathbf{B}))}(\tau_\mathbf{A}\neg(\mathcal{C}_\mathbf{A}(\mathbf{A}) = \mathcal{C}_\mathbf{B}(\mathbf{B}))) = \mathcal{C}_\mathbf{B}(\mathbf{B})) \\
&= \neg(\exists \mathbf{A})\neg(\mathcal{C}_\mathbf{A}(\mathbf{A}) = \mathcal{C}_\mathbf{B}(\mathbf{B})) = (\forall \mathbf{A})(\mathcal{C}_\mathbf{A}(\mathbf{A}) = \mathcal{C}_\mathbf{B}(\mathbf{B}))\ .
\end{aligned}$$

Doing now in $(\forall \mathbf{A})(\mathcal{C}_\mathbf{A}(\mathbf{A}) = \mathcal{C}_\mathbf{B}(\mathbf{B}))$ the same treatment for **B**, we get proposition 771 (p. 456).

**Definition 355 Empty set $\emptyset$.**

By definition $\emptyset = \mathcal{C}_\mathbf{A}(\mathbf{A})$.

**Proposition 772**

$$\mathbf{A \subset B \Rightarrow (\neg(\exists x)(x \in B, x \in A) \Rightarrow A = \emptyset)}\ .$$

**Proof**
Using definitions 333 (p. 441), 330 (p. 434), axiom A.1 (p. 441) and keeping the above notations, we suppose $\mathbf{A \subset B}$, we have:

$$\begin{aligned}&\mathbf{\neg(\exists x)(x \in B, x \in A) = \neg(\exists x)(\neg(x \notin B \vee x \notin A)) = (\forall x)(x \in B \Rightarrow x \notin A)}\\ &\mathbf{\Rightarrow (\forall x)(x \in B \Rightarrow (x \in B, x \notin A)) = (\forall x)(x \in B \Rightarrow x \in \mathcal{C}_B(A)) \Rightarrow B \subset \mathcal{C}_B(A)}\end{aligned}$$

since $\mathbf{\mathcal{C}_B(A) \subset B}$ from proposition 754 (p. 451), $\mathbf{\mathcal{C}_B(A) = B}$ which, from proposition 767 (p. 455), gives: $\mathbf{A = \mathcal{C}_B(B) = \emptyset}$.

**Proposition 773**

$$\begin{aligned}&\mathbf{\neg(\exists x)(x \in A) = (\forall x)(x \notin A) \Leftrightarrow A = \emptyset}\\ &\mathbf{A \neq \emptyset \Leftrightarrow (\exists x)(x \in A) = \tau_x(x \in A) \in A}\ .\end{aligned}$$

**Proof**
We have $\mathbf{A \subset A}$. Using proposition 772 (p. 457), definitions 333 (p. 441) and A.1.3 (p. 435) points 4 and 22, we have:

$$\mathbf{A \subset A \Rightarrow (\neg(\exists x)(x \in A) \Rightarrow (\neg(\exists x)(x \in A, x \in A) \Rightarrow A = \emptyset)}\ .$$

Using definition 355 (p. 456) and proposition 770 (p. 456), we have:

$$\begin{aligned}&\mathbf{\neg(\exists x)(x \in \mathcal{C}_B B)}\\ &\mathbf{\neg(\exists x)(x \in \emptyset) = \neg(\exists x)(x \in \mathcal{C}_B B)}\\ &\mathbf{\neg(\exists x)(x \in \emptyset)}\ .\end{aligned}$$

Using (A.1.3 point 6 (p. 435)), from proposition 770 (p. 456) and (A.1.3, 6), we have:

$$\begin{aligned}&\mathbf{\neg(\exists x)(x \in \mathcal{C}(A))}\\ &\mathbf{A = \emptyset = \mathcal{C}(A) \Rightarrow \neg(\exists x)(x \in A) = \neg(\exists x)(x \in \emptyset)}\ .\end{aligned}$$

The second line is given by (A.1.3, 10 (p. 435)).

**Proposition 774** $\mathbf{\emptyset = \tau_X(\forall x)(x \notin X) = \tau_X(\neg\exists x)(x \in X)}$.

**Proof**
Using axioms of sections A.2.1 (p. 440) and proposition 773 (p. 457) and definition 355 (p. 456), we have:
$\mathbf{(\forall x)(x \notin C_A A)}$ true then $\mathbf{(\exists X)(\neg\exists x)(x \in X) = (\neg\exists x)(x \in \tau_X\{(\forall x)(x \notin X)\})}$ true which means $\mathbf{\emptyset = \tau_X(\forall x)(x \notin X)}$.

**Proposition 775** $\mathbf{\emptyset}$ is a subset of any set.

**Proof**
It is a consequence of relation (A.17) (p. 456) and definitions 355 (p. 456), 353 (p. 454).

**Proposition 776** $\mathbf{A \subset \emptyset \Rightarrow A = \emptyset}$.

**Proof**
From proposition 775 (p. 457), we have $(\forall A)(\emptyset \subset A)$. Then from proposition 731 (p. 442), $\emptyset \subset A$ is true. From (A.1.4 point 24 (p. 436)), $A \subset \emptyset \Rightarrow (A \subset \emptyset, \emptyset \subset A)$. From proposition 754 (p. 451), we get:

$$A \subset \emptyset \Rightarrow A = \emptyset .$$

**Proposition 777** $A \cup \emptyset = A, A \cap \emptyset = \emptyset$.

**Proof**
It is a direct consequence of propositions 763 (p. 453) and 775 (p. 457).

**Proposition 778** $(B \subset A, \mathcal{C}_A B = \emptyset) \Rightarrow A = B$.

**Proof**
We have from propositions 766 (p. 454) and 777 (p. 458):

$$B \cup \mathcal{C}_A B = A = B \cup \emptyset = B .$$

**Proposition 779** $\{\emptyset\} = \mathrm{Set}_u\{u = \emptyset\}$, $\{\emptyset, \mathcal{P}(\emptyset)\} = \mathrm{Set}_u\{u = \emptyset \vee u = \mathcal{P}(\emptyset)\}$ are sets with one and two elements. $(\emptyset, \mathcal{P}(\emptyset))$ is a couple of two elements.

**Proof**
See from definitions 345 (p. 447), 355 (p. 456), 346 (p. 448), proposition 744 (p. 447). From propositions 773 (p. 457), 774 (p. 457), 759 (p. 452), definition 350 (p. 453), the following items exist:

$$\emptyset = \tau_X((\forall x)(x \notin X), \ \mathcal{P}(\emptyset)) .$$

From proposition 754 (p. 451) (extensionality principle) $\emptyset \neq \mathcal{P}(\emptyset)$, because:

$$\emptyset \notin \emptyset, \emptyset \subset \emptyset, \emptyset \in \mathcal{P}$$

then $\{\emptyset, \mathcal{P}(\emptyset)\}$, $(\emptyset, \mathcal{P}(\emptyset))$ are set and couple of two elements.

### A.5.5 Product of two sets

**Proposition 780** Given two sets **A** and **B**, we have (see definitions 344 (p. 447), 346 (p. 448)):

$$\mathrm{Coll}_z((\exists x \in A)(\exists y \in B)(z = (x, y), (x \in A), (y \in B)) .$$

**Proof**
The relation $R(z) = (\exists x \in A)(\exists y \in B)(z = (x, y), (x \in A), (y \in B))$ verifies $\mathrm{coll}_z(R(z))$, because, applying several times the axiom generator of proposition 748 (p. 449), we have:

$$\begin{aligned}
&(\forall x)(\forall y)\forall z)(z = (x, y) \Rightarrow z \in \{(x, y)\}) \\
&\Rightarrow (\forall x)(\forall y)(\exists Z)(\forall z)(z = (x, y) \Rightarrow z \in Z) \\
&\Rightarrow (\forall x)(\forall Y)\mathrm{coll}_z((\exists y)(y \in Y, z = (x, y))) \\
&\Rightarrow (\forall x)\mathrm{coll}_z((\exists y)(y \in B, z = (x, y))) \\
&\Rightarrow (\forall x)(\exists Z)(\forall z)((\exists y)(y \in B, z = (x, y)) \Rightarrow z \in Z) \\
&\Rightarrow (\forall X)\mathrm{coll}_z((\exists x)(\exists y)((x \in X, y \in B, z = (x, y)) \\
&\Rightarrow \mathrm{coll}_z((\exists x)(\exists y)(x \in A, y \in B), z = (x, y)) .
\end{aligned}$$

**Definition 356 Product of two sets** (see also proposition 822 (p. 473)).

One calls product of two sets $\mathbf{A}$ and $\mathbf{B}$ (see definition 346 (p. 448)) the following set which exists from proposition 780 (p. 458):

$$\mathbf{A \times B = set_z\{(\exists x \in A)(\exists y \in B)(z = (x, y), x \in A, y \in B)\}}\ .$$

**Proposition 781** $\mathbf{A \times B = \emptyset \Leftrightarrow A = \emptyset \vee B = \emptyset}$.

**Proof**
We have from definition 356 (p. 459):

$$\mathbf{z \in A \times B \Rightarrow (\exists x \in A)(\exists y \in B)(z = (x, y)) \Rightarrow (\exists x \in A)(\exists y \in B)(z = (x, y))}$$
$$\mathbf{\Rightarrow (A \neq \emptyset, B \neq \emptyset)}\ .$$

Then we have:

$$\mathbf{(z \in A \times B \Rightarrow (A \neq \emptyset, B \neq \emptyset)) = (\neg(z \in A \times B) \vee \neg(A = \emptyset \vee B = \emptyset)) =}$$
$$\mathbf{((A = \emptyset \vee B = \emptyset) \Rightarrow z \notin A \times B)}\ .$$

Applying that to $\mathbf{z = \tau_u u \in A \times B}$, we have:

$$\mathbf{A = \emptyset \vee B = \emptyset \Rightarrow (\forall u)(u \notin A \times B)}\ .$$

From proposition 773 (p. 457), it means $\mathbf{A \times B = \emptyset}$

$$\mathbf{A \times B = \emptyset \Rightarrow \neg(\exists x \in A, \exists y \in B) \Rightarrow (\neg\exists x \in A \vee \neg\exists y \in B)}$$

from proposition 773 (p. 457), it means $\mathbf{A = \emptyset \vee B = \emptyset}$.

**Proposition 782** $\mathbf{A, B, C}$ being sets, $\mathbf{(A \cup B) \times C = (A \times C) \cup (B \times C)}$, $\mathbf{C \times (A \cup B) = (C \times A) \cup (C \times B)}$.

**Proof**
From (A.1.4 point 25 (p. 436)), we have:

$$\mathbf{(x, y) \in (A \cup B) \times C \Leftrightarrow (x \in A \cup B, y \in C) \Leftrightarrow ((x \in A \vee x \in B), y \in C)}$$
$$\mathbf{\Leftrightarrow (x \in A, y \in C) \vee (x \in B, y \in C) \Leftrightarrow (x, y) \in (A \times C) \cup (B \times C)}\ .$$

Reversing the order of the terms around, $\times$ we have the same proof for the second relation of proposition 782.

**Proposition 783** $\mathbf{A, B, C}$ being sets, $\mathbf{B \subset A \Rightarrow (B \times C \subset A \times C, C \times B \subset C \times A)}$.

**Proof**
Setting $\mathbf{D = \mathcal{C}_A B}$ and applying proposition 782 (p. 459), we have: $\mathbf{A \times C = (B \cup D) \times C = B \times C \cup D \times C}$. Then, from proposition 761 (p. 453), $\mathbf{B \times C \subset A \times C}$ and similarly, using the second part of proposition 761 (p. 453): $\mathbf{C \times B \subset C \times A}$.

**Proposition 784** $\mathbf{A, B, C, D}$ being sets, $\mathbf{(A \cap B) \times (C \cap D) = (A \times C) \cap (B \times D) = (A \times D) \cap (B \times C)}$.

**Proof**
From (A.1.4 point 26 (p. 436)), we have:

$$\begin{aligned}&(x, y) \in (A \cap B) \times (C \cap D) \Leftrightarrow (x \in A \cap B, y \in C \cap D)\\ &\Leftrightarrow ((x \in A, x \in B), (y \in C, y \in D)) \Leftrightarrow (x \in A, x \in B, y \in C, y \in D)\\ &\Leftrightarrow ((x \in A, y \in C), (x \in B, y \in D)) \Leftrightarrow (x, y) \in (A \times C) \cap (B \times D) .\end{aligned}$$

Clearly the same thing is true changing **C** in **D** and vice versa.

**Proposition 785**

$$B \subset C \Rightarrow \mathcal{C}_{A\times C}(A \times B) = A \times \mathcal{C}_C(B) .$$

**Proof**
From A.1.4 -25 (p. 436) -26 (p. 436) -16 (p. 436) -13 (p. 435), -4 (p. 435), section A.1.2 (p. 434) and A.1.3 -2 (p. 435), we have:

$$\begin{aligned}&x = (x_a, x_c) \in \mathcal{C}_{A\times C}(A \times B) \Leftrightarrow (x_a \in A, x_c \in C, (x_a, x_c) \notin A \times B)\\ &\Leftrightarrow (x_a \in A, x_c \in C, \neg\neg(\neg x_a \in A \vee \neg x_c \in B))\\ &\Leftrightarrow (x_a \in A, x_c \in C, (\neg x_a \in A \vee \neg x_c \in B))\\ &\Leftrightarrow ((x_a \in A, x_c \in C, \neg x_a \in A) \vee (x_a \in A, x_c \in C, \neg x_c \in B))\\ &\Leftrightarrow ((x_a \in A, \neg x_a \in A, x_c \in C) \vee (x_a \in A, x_c \in C, \neg x_c \in B))\\ &\Leftrightarrow (((x_a \in A, \neg x_a \in A) \vee (x_a \in A, x_c \in C, \neg x_c \in B)),\\ &(x_c \in C \vee (x_a \in A, x_c \in C, \neg x_c \in B)))\\ &\Leftrightarrow (((x_a \in A \vee \neg x_a \in A) \Rightarrow (x_a \in A, x_c \in C, \neg x_c \in B)),\\ &(x_c \in C \vee (x_a \in A, x_c \in C, \neg x_c \in B)))\\ &\Leftrightarrow (x_a \in A, x_c \in C, x_c \notin B) \Leftrightarrow x \in A \times \mathcal{C}_C(B) .\end{aligned}$$

**Proposition 786** **A** and **B** being sets, we have:

$$((\forall x)(\forall y)(x \neq y \Rightarrow (A \times \{x\}) \cap (B \times \{y\}) = \emptyset)$$

see definition 345 (p. 447).

**Proof**
From propositions 784 (p. 459), 781 (p. 459) and definition 345 (p. 447), we have:

$$(A \times \{x\}) \cap (B \times \{y\}) = (A \cap B) \times (\{x\} \cap \{y\}) = (A \cap B) \times \emptyset = \emptyset .$$

**Definition 357 Map, graph of a map.**

See section 2.1.1 (p. 131) for more developments. Let us consider two sets **A** and **B** and a subset **f** of the product **A** by **B** (see section A.5.1 (p. 447) and definitions 342 (p. 447), 344 (p. 447), 356 (p. 459)): $f \subset A \times B$. **f** defines a map (function, application) of **A** on **B** $(f : A \to B)$ if (see section A.3 (p. 441)):

$$(\forall x \in A)(\exists y \in B)((x, y) \in f) \tag{A.18}$$

and if:

$$(x, y) \in f, (x, y') \in f \Rightarrow y = y' .$$

See definition 112 (p. 131), 1 and also footnote 1 (p. 131). The set **f** is called graph of the map $(f : A \to B)$.

**Definition 358 f(x).**

Given a map $(\mathbf{f} : \mathbf{A} \longrightarrow \mathbf{B})$, the term $\tau_{\mathbf{y}}((\mathbf{x},\mathbf{y})) \in \mathbf{f}$ coming from relation A.18 (p. 460) is written as:

$$\mathbf{f}(\mathbf{x}) = \tau_{\mathbf{y}}((\mathbf{x},\mathbf{y}) \in \mathbf{f}) \ .$$

See point 2 of definition 112 (p. 131).

**Proposition 787** All the maps $\mathbf{A} \rightarrow \mathbf{B}$ are making a set.

**Proof**

We have from definition 357 (p. 460), $(\mathbf{f} : \mathbf{A} \longrightarrow \mathbf{B}) \Rightarrow \mathbf{f} \in \mathcal{P}(\mathbf{A} \times \mathbf{B})$. From propositions 750 (p. 450), $\mathbf{coll_f}(\mathbf{f} : \mathbf{A} \longrightarrow \mathbf{B})$.

**Definition 359 Injective map, one-to-one map.**

See definition 122 (p. 139). If a map $\mathbf{f}$ is such that:

$$\mathbf{f}(\mathbf{x}) = \mathbf{f}(\mathbf{x}') \Rightarrow \mathbf{x} = \mathbf{x}' \tag{A.19}$$

the map is said to be injective or one-to-one.

**Definition 360 Surjective map, onto map.**

See definition 123 (p. 139). If a map $(\mathbf{f} : \mathbf{A} \longrightarrow \mathbf{B})$ is such that

$$(\forall \mathbf{y} \in \mathbf{B})(\exists \mathbf{x} \in \mathbf{A})((\mathbf{x},\mathbf{y}) \in \mathbf{f}) \tag{A.20}$$

the map is said to be surjective or onto.

**Definition 361 Bijective map or bijection.**

See definition 124 (p. 139). If a map $(\mathbf{f} : \mathbf{A} \longrightarrow \mathbf{B})$ is both injective and surjective, it is said to be bijective and $\mathbf{f}$ also defines a map of $\mathbf{B}$ on $\mathbf{A}$: $(\mathbf{f}^{-1} : \mathbf{B} \longrightarrow \mathbf{A})$. In that case any couple of $\mathbf{f}$ can be written as:

$$(\mathbf{x},\mathbf{y}) = (\mathbf{x},\mathbf{f}(\mathbf{x})) = (\mathbf{f}^{-1}(\mathbf{y}),\mathbf{y}) \in \mathbf{f} \ . \tag{A.21}$$

**Proposition 788** Given two sets (see footnote 7 (p. 451)) with two different elements (definition 345 (p. 447)), there is a bijection between those two sets.

**Proof**

Let us call $\mathbf{XY} = \{\mathbf{x},\mathbf{y}\}, \mathbf{x} \neq \mathbf{y}$, $\mathbf{UV} = \{\mathbf{u},\mathbf{v}\}, \mathbf{u} \neq \mathbf{v}$ those two sets. We have (see definitions 345 (p. 447), 346 (p. 448)):

$$\mathbf{b} \in \mathbf{XY} \Rightarrow (\mathbf{b} = \mathbf{x} \vee \mathbf{b} = \mathbf{y})$$

and we set $(\mathbf{f} : \mathbf{XY} \longrightarrow \mathbf{UV})$ by:

$$\begin{aligned} &\mathbf{f} = \{(\tau_{\mathbf{x}}(\mathbf{x} \in \mathbf{XY}), \tau_{\mathbf{x}}(\mathbf{x} \in \mathbf{UV})),\\ &(\tau_{\mathbf{x}}(\mathbf{x} \in \mathbf{XY}, \mathbf{x} \neq \tau_{\mathbf{x}}(\mathbf{x} \in \mathbf{XY})), \tau_{\mathbf{x}}(\mathbf{x} \in \mathbf{UV}, \mathbf{x} \neq \tau_{\mathbf{x}}(\mathbf{x} \in \mathbf{UV})))\} \ . \end{aligned} \tag{A.22}$$

It is a bijection because it verifies definition 361 (p. 461): From the relation (A.22) (p. 461):

$$((\mathbf{x},\mathbf{y}) \in \mathbf{f}, (\mathbf{x}',\mathbf{y}') \in \mathbf{f}) \Rightarrow (\mathbf{x} = \mathbf{x}', \mathbf{y} = \mathbf{y}') \ .$$

**Proposition 789** Given two sets $\mathbf{A}$ and $\mathbf{B}$, a set of two elements $\{x, y\}, x \neq y$. There is a bijection between the set $\mathcal{F}$ of maps $(f : \{x, y\} \to A \cup B)$ such that $f(x) \in A, f(y) \in B$ and the set $A \times B$.

**Proof**
The map $(H : \mathcal{F} \to A \times B)$ defined by $f \in \mathcal{F} \Rightarrow H(f) = (f(x), f(y)) \in A \times B$. It is a bijection because from definitions 346 (p. 448), 115 (p. 133):

$$(a, b) \in A \times B \Rightarrow f \in H^{-1}\{(a, b)\}, g \in H^{-1}\{(a, b)\} \Rightarrow H(f) = H(g) = (a, b) \Rightarrow$$
$$(f(x), f(y)) = (g(x), g(y)) = (a, b) \Rightarrow f(x) = g(x) = a, f(y) = g(y) = b \Rightarrow f = g\ .$$

Then from definition 361 (p. 461), $\mathbf{H}$ is a bijection.

**Proposition 790** $\mathbf{A}$ being a set, the map (definitions 112 (p. 131), 357 (p. 460)) $\mathcal{C}_A(x) : \mathcal{P}(A) \to \mathcal{P}(A)$ is a bijection (definition 361 (p. 461)).

**Proof**
$\mathbf{A}$ being a set, $\mathcal{C}$ is a map because the set $\mathcal{C} \subset \mathcal{P}(A) \times \mathcal{P}(A)$ defined as:

$$(x, y) \in \mathcal{C} \Leftrightarrow ((x, y) \in \mathcal{P}(A) \times \mathcal{P}(A), y = \mathcal{C}(x))$$

verifies the definitions 112 (p. 131), 357 (p. 460). $\mathcal{C}$ being its own inverse from proposition 178 (p. 142) $\mathcal{C}$ is a bijection.

### A.5.6 Union and intersection of sets over a set of indices

**Definition 362** $\bigcup_{i \in I} O_i$.

$(O_i : I \to A)$ is a map (see section 2.1.1 (p. 131) and the above definition 357) from the set $\mathbf{I}$ (called set of indices) to $\mathbf{A}$ which can be qualified as a set of sets (see footnote 7 (p. 451)). One calls $\bigcup\limits_{i \in I} O_i$, the following set:

$$\bigcup_{i \in I} O_i = \text{set}_x\{(\exists i \in I)(x \in O_i)\}\ .$$

The set exists:
Setting $R(U) = (\forall x)((i \in I, x \in O_i) \Rightarrow x \in U)$, we have $R(O_i)$ true. From relation (A.1) (p. 441) and definition 333 (p. 441), the following relation is true

$$(\exists X)R(X) = (\exists X)(\forall x)((i \in I, x \in O_i) \Rightarrow x \in X)\ .$$

From propositions 748 (p. 449), 731 (p. 442), section A.1.4 (p. 435), 18 (p. 436), 22 (p. 436), we have:

$$\begin{aligned}(\forall i)(\exists X)(\forall x)((i \in I, x \in O_i) \Rightarrow x \in X) &\Rightarrow \forall Y \text{coll}_x((\exists i)(i \in Y, i \in I, x \in O_i))\\ &\Rightarrow \text{coll}_x((\exists i)(i \in I, i \in I, x \in O_i))\\ &\Rightarrow \text{coll}_x((\exists i)(i \in I, x \in O_i))\ .\end{aligned}$$

**Proposition 791** $I = \emptyset \Rightarrow \bigcup_{i \in I} O_i = \emptyset$

**Proof**
Considering a map $(\mathbf{O_i : I' \longrightarrow A})$ and its restriction to $\mathbf{I} = \mathcal{C}_{\mathbf{I'}}(\mathbf{I'}) = \emptyset$ (definition 355 (p. 456)), from A.1.2 (p. 434), A.1.3 (p. 435), A.1.4 (p. 435), propositions 736 (p. 443), 774 (p. 457), we have successively true :

$$\begin{aligned}
&\forall i)(i \notin \emptyset)\\
&(\forall i)(i \notin \emptyset) \vee (\forall i)(x \notin O_i)\\
&(\forall i)(i \notin \emptyset \vee x \notin O_i)\\
&\neg(\exists i)\neg(i \notin \emptyset \vee x \notin O_i)\\
&\neg(\exists i)(i \in \emptyset, x \in O_i)\\
&\neg(\exists i)(i \in \emptyset, x \in O_i) \vee x \in \emptyset\\
&(\exists i)(i \in \emptyset, x \in O_i) \Rightarrow x \in \emptyset\\
&x \notin \emptyset\\
&x \notin \emptyset \vee (\exists i)(i \in \emptyset, x \in O_i)\\
&x \in \emptyset \Rightarrow (\exists i)(i \in \emptyset, x \in O_i)\\
&x \in \emptyset \Leftrightarrow (\exists i)(i \in \emptyset, x \in O_i)\\
&\emptyset = \bigcup_{i \in I} O_i
\end{aligned}$$

**Convention 30** $\bigcup_{\mathbf{R(i)}}, \bigcap_{\mathbf{R(i)}}, \prod_{\mathbf{R(i)}}, \sum_{\mathbf{R(i)}}$.

When one refers explicitly in a formula to a member of a set: $\mathbf{i \in I}$, if the set $\mathbf{I}$ is the set of elements verifying a relation $\mathbf{R(i)}$ which satisfies $\mathbf{coll_i R(i)}$ or $\mathbf{I = set_i\{(R(i)\}}$ (definition 344 (p. 447)), one may replace $\mathbf{i \in I}$ by its equivalent $\mathbf{R(i)}$: $\bigcup_{i \in I} O_i = \bigcup_{R(i)} O_i$ in the above definition 362 . A common feature is when in fact the relation defines a subset: $\mathbf{I = set_i\{(R(i), i \in K\} \subset K}$. In that case because it is clearer, we shall write $\mathbf{R(i), i \in K}$ instead of $\mathbf{i \in I}$. Here we may write $\bigcup_{i \in I} O_i = \bigcup_{R(i), i \in K} O_i$

**Proposition 792** Keeping the above notations (definition 362), we have:

$$(\forall j \in I)(O_j \subset \bigcup_{i \in I} O_i) .$$

**Proof**
It is a direct consequence of definitions 362 (p. 462) and 349 (p. 451).

**Definition 363** $\bigcap_{i \in I} O_i$.

With the same notations and definitions as in definition 362 but with $\mathbf{I} \neq \emptyset$ [8], one calls:

$$\bigcap_{i \in I} O_i = \mathbf{set}_x\{(\forall i)(i \in I \Rightarrow x \in O_i)\} .$$

Since

$$I \neq \emptyset \Rightarrow (\exists i)(i \in I) = \tau_i(i \in I) \in I \Rightarrow ((\forall i)(i \in I \Rightarrow x \in O_i) \Rightarrow x \in O_{\tau_i(i \in I)}) ,$$

the set $\bigcap_{i \in I} O_i$ exists from proposition 750 (p. 450).

---

[8]That is needed because, since for any term $\mathbf{x}$ and any term $\mathbf{i}$, $\mathbf{i \notin \emptyset \vee x \in O_i}$ is true (proposition 773 (p. 457)). From proposition 732 (p. 442), $(\forall i)(i \in I \Rightarrow x \in O_i)$ is true whatever $\mathbf{x}$. Then any term would belong to $\mathbf{set_x\{(\forall i)(i \in \emptyset \Rightarrow x \in O_i)\}}$ if it exists. It cannot be because in that case from proposition 750 (p. 450), any relation would be set generating which is not true: see the remark of proposition 748 (p. 449).

**Proposition 793** We have:

$$\mathbf{A} \cup \mathbf{B} = \bigcup_{\mathbf{i} \in \{\mathbf{A},\mathbf{B}\}} \mathbf{i} \tag{A.23}$$

$$\mathbf{A} \cap \mathbf{B} = \bigcap_{\mathbf{i} \in \{\mathbf{A},\mathbf{B}\}} \mathbf{i}\,. \tag{A.24}$$

**Proof**
Using proposition 746 (p. 448),

$$\mathbf{x} \in \mathbf{A} \cup \mathbf{B} \Leftrightarrow (\mathbf{x} \in \mathbf{A} \vee \mathbf{x} \in \mathbf{B}) \Leftrightarrow (\exists \mathbf{i} \in \{\mathbf{A},\mathbf{B}\})(\mathbf{x} \in \mathbf{I}) \Leftrightarrow \bigcup_{\mathbf{i} \in \{\mathbf{A},\mathbf{B}\}} \mathbf{i}$$

$$\mathbf{x} \in \mathbf{A} \cap \mathbf{B} \Leftrightarrow (\mathbf{x} \in \mathbf{A}, \mathbf{x} \in \mathbf{B}) \Leftrightarrow (\forall \mathbf{i} \in \{\mathbf{A},\mathbf{B}\})(\mathbf{x} \in \mathbf{I}) \Leftrightarrow \bigcup_{\mathbf{i} \in \{\mathbf{A},\mathbf{B}\}} \mathbf{i}\,.$$

**Proposition 794** Keeping the above notations (definition 362),

$$\mathbf{I} \neq \emptyset \Rightarrow (\forall \mathbf{j} \in \mathbf{I})(\bigcap_{\mathbf{i} \in \mathbf{I}} \mathbf{O_i} \subset \mathbf{O_j})\,.$$

**Proof**
It is a direct consequence of definitions 363 (p. 463) and 349 (p. 451).

**Proposition 795** $\mathbf{I}$ being a set, we have:

$$\bigcup_{\mathbf{i} \in \mathbf{I}} \{\mathbf{i}\} = \mathbf{I}\,.$$

**Proof**

$$(\mathbf{j} \in \bigcup_{\mathbf{i} \in \mathbf{I}} \{\mathbf{i}\} \Rightarrow (\exists \mathbf{i} \in \mathbf{I})(\mathbf{j} \in \{\mathbf{i}\}) \Rightarrow (\exists \mathbf{i} \in \mathbf{I})(\mathbf{j} = \mathbf{i}) \Rightarrow \mathbf{j} \in \mathbf{I}) \Rightarrow \bigcup_{\mathbf{i} \in \mathbf{I}} \{\mathbf{i}\} \subset \mathbf{I}$$

(definitions 362 (p. 462), 345 (p. 447), section A.1.3 point 6 (p. 435), definition 349 (p. 451)). Setting $\mathbf{R(u)} = (\mathbf{u} = \mathbf{j}, \mathbf{u} \in \mathbf{I})$, we have:

$$\begin{aligned} &(\mathbf{j} \in \mathbf{I} \Rightarrow \mathbf{R(j)} \Rightarrow (\exists \mathbf{i})\mathbf{R(i)} \Rightarrow (\exists \mathbf{i})(\mathbf{i} \in \mathbf{I}, \mathbf{j} = \mathbf{i}) \\ &\Rightarrow (\exists \mathbf{i} \in \mathbf{I})(\mathbf{j} \in \{\mathbf{i}\}) \Rightarrow \mathbf{j} \in \bigcup_{\mathbf{i} \in \mathbf{I}} \{\mathbf{i}\}) \Rightarrow \mathbf{I} \subset \bigcup_{\mathbf{i} \in \mathbf{I}} \{\mathbf{i}\} \end{aligned}$$

(proposition 727 (p. 441), section A.1.3 point 6 (p. 435), definitions 345 (p. 447), 362 (p. 462), 349 (p. 451)). Then from proposition 754 (p. 451) (extensionality principle), we have:

$$\bigcup_{\mathbf{i} \in \mathbf{I}} \{\mathbf{i}\} = \mathbf{I}\,.$$

**Proposition 796** Given three sets (see footnote 7 (p. 451)): $\mathbf{X}$ (a set of sets), $\mathbf{I}$ and $\mathbf{J}$, two maps: $(\mathbf{X_i} : \mathbf{I} \rightarrow \mathbf{X})$ and $(\mathbf{I_j} : \mathbf{J} \rightarrow \mathcal{P}(\mathbf{I}))$, then:

$$\bigcup_{\mathbf{j} \in \mathbf{J}} \bigcup_{\mathbf{i} \in \mathbf{I_j}} \mathbf{X_i} = \bigcup_{\mathbf{i} \in \bigcup_{\mathbf{j} \in \mathbf{J}} \mathbf{I_j}} \mathbf{X_i}\,. \tag{A.25}$$

**Proof**
Using definition 362 (p. 462) and proposition 737 (p. 444), we have:

$$\begin{aligned}
& x \in \bigcup_{j\in J} \bigcup_{i\in I_j} X_i \\
& \Leftrightarrow (\exists j)(j \in J, x \in \bigcup_{i\in I_j} X_i) \\
& \Leftrightarrow (\exists j)(j \in J, (\exists i)(i \in I_j, x \in X_i)) \\
& \Leftrightarrow (\exists j)(\exists i)(j \in J, i \in I_j, x \in X_i) \\
& \Leftrightarrow (\exists i)(\exists j)(j \in J, i \in I_j, x \in X_i) \\
& \Leftrightarrow (\exists i)(i \in \bigcup_{j\in J} I_j, x \in X_i) = x \in \bigcup_{i\in \bigcup_{j\in J} I_j} X_i \, .
\end{aligned}$$

**Proposition 797** $\mathbf{I, A, B}$ being sets ($\mathbf{B}$ a set of sets), $(\mathbf{a(i) : I \to A})$ a surjective map (definition 360 (p. 461)), $(\mathbf{b(a) : A \to B})$ another map, we have:

$$\bigcup_{i\in I} b(a(i)) = \bigcup_{a\in A} b(a) \, .$$

**Proof**
We can write using definitions 117 (p. 134), 362 (p. 462) and proposition 727 (p. 441):

$$x \in \bigcup_{i\in I} b(a(i)) \Leftrightarrow (\exists i \in I)(x \in b(a(i))) \Leftrightarrow (\exists a \in A)(x \in B(a)) \Leftrightarrow x \in \bigcup_{a\in A} b(a) \, .$$

**Proposition 798**

$$A \cap \bigcup_{j\in J} B_j = \bigcup_{j\in J} A \cap B_j \, .$$

**Proof**
From proposition 736 (p. 443) equation (A.9) (p. 444)

$$\begin{aligned}
& x \in A \cap \bigcup_{j\in J} B_j \Leftrightarrow x \in A, (\exists j \in J)(x \in B_j) \Leftrightarrow (\exists j \in J)(x \in A, x \in B_j) \\
& \Leftrightarrow x \in \bigcup_{j\in J} A \cap B_j \, .
\end{aligned}$$

Then from the axiom 754 (p. 451), we conclude proposition 798 (p. 465).

**Proposition 799** $\mathbf{S}$ being a set,

$$U \subset \mathcal{P}(S) \Rightarrow \bigcup_{u\in U} u \in \mathcal{P}(S) \, .$$

**Proof**

$$x \in \bigcup_{u\in U} u \Rightarrow (\exists v \in U)(x \in v, U \subset \mathcal{P}(S), v \in \mathcal{P}(S), v \subset S) \Rightarrow x \in S \, .$$

Then

$$\bigcup_{u\in U} u \subset S, \quad \bigcup_{u\in U} u \in \mathcal{P}(S) \, .$$

**Proposition 800**

$$A \cap \bigcap_{j \in J} B_j = \bigcap_{j \in J} A \cap B_j \,.$$

**Proof**
From proposition 736 (p. 443) equation (A.11) (p. 444), we have:

$$x \in A \cap \bigcap_{j \in J} B_j \Leftrightarrow (x \in A, (\forall j \in J)(x \in B_j)) \Leftrightarrow (\forall j \in J)(x \in A, x \in B_j)$$
$$\Leftrightarrow x \in \bigcap_{j \in J} A \cap B_j \,.$$

Then from the axiom 754 (p. 451), we conclude proposition 800 (p. 466).

**Proposition 801** $A$ and $I$ being sets, $(O_i : I \to \mathcal{P}(A))$ a map, we have (see definition 353 (p. 454)):

$$\mathcal{C}_A(\bigcup_{i \in I} O_i) = \bigcap_{i \in I} \mathcal{C}_A(O_i) \,.$$

**Proof**
From definition 353 (p. 454) and proposition 735 (p. 442) and relation (A.11) (p. 444), we have:

$$\begin{array}{ll} x \in \mathcal{C}_A(\bigcup_{i\in I} O_i) & = x \in A, \neg(\exists i)(x \in O_i, i \in I) \\ \neg(\exists i)(x \in O_i, i \in I) & = \neg(\exists i)\neg(\neg(i \in I) \vee x \notin O_i) \\ & = (\forall i)(i \in I \Rightarrow x \notin O_i) = (\forall i \in I)(x \notin O_i) \\ x \in \mathcal{C}_A(\bigcup_{i\in I} O_i) & = x \in A, (\forall i \in I)(x \notin O_i) \Leftrightarrow (\forall i \in I)(x \in A, x \notin O_i) \\ & = \bigcap_{i\in I} \mathcal{C}_A(O_i) \,. \end{array}$$

**Proposition 802** With the same notation as in proposition 801 (p. 466), we have:

$$\mathcal{C}_A(\bigcap_{i \in I} O_i) = \bigcup_{i \in I} \mathcal{C}_A(O_i) \,.$$

**Proof**
Applying proposition 801 to $\mathcal{C}_A(O_i)$ and using proposition 767 (p. 455), we have:

$$\mathcal{C}_A(\bigcup_{i \in I} \mathcal{C}_A(O_i)) = \bigcap_{i \in I} \mathcal{C}_A(\mathcal{C}_A(O_i)) = \bigcap_{i \in I} O_i \,.$$

Then:

$$\bigcap_{i \in I} O_i = \mathcal{C}_A(\bigcup_{i \in I} \mathcal{C}_A(O_i)) \Leftrightarrow \mathcal{C}_A \bigcap_{i \in I} O_i = \mathcal{C}_A \mathcal{C}_A(\bigcup_{i \in I} \mathcal{C}_A(O_i)) = \bigcup_{i \in I} \mathcal{C}_A(O_i) \,.$$

**Proposition 803** Given three sets $X$, $I$ and $J$, a map $(X_i : I \to X)$, a map $(I_j : J \to \mathcal{P}(I))$ then:

$$\bigcap_{j \in J} \bigcap_{i \in I_j} X_i = \bigcap_{i \in \bigcup_{j \in J} I_j} X_i \,. \qquad \text{(A.26)}$$

**Proof**
Applying proposition A.25 (p. 464) to $\mathcal{C}\mathbf{X_i} = \mathcal{C}_{\mathbf{Y}}\mathbf{X_i}$ with $\mathbf{Y} = \bigcup_{\mathbf{i}\in\mathbf{I}} \mathbf{X_i}$, we have:

$$\bigcup_{\mathbf{j}\in\mathbf{J}}\bigcup_{\mathbf{i}\in\mathbf{I_j}} \mathcal{C}(\mathbf{X_i}) = \bigcup_{\mathbf{i}\in\bigcup_{\mathbf{j}\in\mathbf{J}}\mathbf{I_j}} \mathcal{C}(\mathbf{X_i})$$

then using proposition 802 (p. 466), we get:

$$\mathcal{C}(\bigcap_{\mathbf{j}\in\mathbf{J}}\bigcap_{\mathbf{i}\in\mathbf{I_j}} \mathbf{X_i}) = \mathcal{C}(\bigcap_{\mathbf{i}\in\bigcup_{\mathbf{j}\in\mathbf{J}}\mathbf{I_j}} \mathbf{X_i}) .$$

Using proposition 767 (p. 455), we get:

$$\bigcap_{\mathbf{j}\in\mathbf{J}}\bigcap_{\mathbf{i}\in\mathbf{I_j}} \mathbf{X_i} = \bigcap_{\mathbf{i}\in\bigcup_{\mathbf{j}\in\mathbf{J}}\mathbf{I_j}} \mathbf{X_i} .$$

**Proposition 804** $\mathbf{I}, \mathbf{A}, \mathbf{B}$ being sets ($\mathbf{B}$ a set of sets), $(\mathbf{a(i)} : \mathbf{I} \to \mathbf{A})$ a surjective map (definition 360 (p. 461)), $(\mathbf{B(a)} : \mathbf{A} \to \mathbf{B})$ another map, we have:

$$\bigcap_{\mathbf{i}\in\mathbf{I}} \mathbf{B(a(i))} = \bigcap_{\mathbf{a}\in\mathbf{A}} \mathbf{B(a)} .$$

**Proof**
From proposition 797, with $\mathcal{C} = \mathcal{C}_{\bigcup_{\mathbf{a}\in\mathbf{A}} \mathbf{B(a)}}$, we have:

$$\bigcup_{\mathbf{i}\in\mathbf{I}} \mathcal{C}(\mathbf{B(a(i))}) = \bigcup_{\mathbf{a}\in\mathbf{A}} \mathcal{C}(\mathbf{B(a)}) .$$

Applying proposition 767 (p. 455), we get proposition 804 (p. 467). Then from the axiom A.1.3, 6

$$\mathcal{C}(\bigcup_{\mathbf{i}\in\mathbf{I}} \mathcal{C}(\mathbf{B(a(i))})) = \mathcal{C}(\bigcup_{\mathbf{a}\in\mathbf{A}} \mathcal{C}(\mathbf{B(a)})) .$$

Then from propositions 801 and 767 (p. 455), we have:

$$\bigcap_{\mathbf{i}\in\mathbf{I}} \mathbf{B(a(i))} = \bigcap_{\mathbf{a}\in\mathbf{A}} \mathbf{B(a)} .$$

**Proposition 805** $\mathbf{A}$ and $\mathbf{I}$ are sets ($\mathbf{A}$ set of sets see footnote 7 (p. 451)) $(\mathbf{O_i} : \mathbf{I} \to \mathbf{A})$ is a map. We have:

$$(\forall \mathbf{i} \in \mathbf{I})(\mathbf{O_i} = \emptyset) \Leftrightarrow \bigcup_{\mathbf{i}\in\mathbf{I}} \mathbf{O_i} = \emptyset \tag{A.27}$$

$$\mathbf{I} \neq \emptyset \Rightarrow ((\exists \mathbf{i} \in \mathbf{I})(\mathbf{O_i} = \emptyset) \Rightarrow \bigcap_{\mathbf{i}\in\mathbf{I}} \mathbf{O_i} = \emptyset) . \tag{A.28}$$

**Proof**
From propositions 737 (p. 444), 735 (p. 442), 773 (p. 457) and definition 334 (p. 441), we have:

$$\begin{aligned}
&(\forall i \in I)(O_i = \emptyset) \Leftrightarrow \neg(\exists i \in I)(\neg\neg\exists x \in O_i)\\
&\Leftrightarrow \neg(\exists i \in I)(\exists x \in O_i) = \neg(\exists i)((i \in I), (\exists x)(x \in O_i)) \Leftrightarrow \neg(\exists i)(\exists x)(i \in I, x \in O_i)\\
&\Leftrightarrow \neg(\exists x)(\exists i)(i \in I, x \in O_i) \Leftrightarrow \neg(\exists x)(x \in \bigcup_{i\in I} O_i) \Leftrightarrow \bigcup_{i\in I} O_i = \emptyset\,.
\end{aligned}$$

With $j = \tau_k(k \in I, O_k = \emptyset)$ (see definition 333 (p. 441), proposition 727 (p. 441), 734 (p. 442), relation (A.11) (p. 444) of proposition 736 and convention 37 (p. 560)), we have:

$$\begin{aligned}
&(\exists i \in I)(O_i = \emptyset) = (O_j = \emptyset) \Leftrightarrow ((\forall x)(x \notin O_j), j \in I)\\
&\Leftrightarrow (\forall x)(x \notin O_j, j \in I) \Rightarrow (\forall x)(\exists i)(i \in I, x \notin O_i) \Leftrightarrow (\forall x)(\neg\neg(\exists i)(i \in I, x \notin O_i))\\
&\Leftrightarrow (\forall x)(\neg(\forall i)(\neg i \in I \vee x \in O_i)) = (\forall x)(\neg(\forall i)(i \in I \Rightarrow x \in O_i))\\
&= (\forall x)(\neg(\forall i \in I)(x \in O_i)) \Leftrightarrow (\forall x)(\neg(x \in \bigcap_{i\in I} O_i)) = (\forall x)(x \notin \bigcap_{i\in I} O_i) \Rightarrow \bigcap_{i\in I} O_i = \emptyset\,.
\end{aligned}$$

**Proposition 806** $\mathbf{A}$ and $\mathbf{I}$ are sets ($\mathbf{A}$ set of sets) $(O_i : I \to A)$ is a map. We define $I_0$ as $I_0 = \text{coll}_i\{i \in I, O_i = \emptyset\}$. We have:

$$\bigcup_{i\in I} O_i = \bigcup_{i\in \mathcal{C}_I(I_0)} O_i\,.$$

**Proof**
Calling $I_1 = \mathcal{C}_I(I_0)$ and $I_t = \{I_0, I_1\}$ we have $I = \bigcup_{j\in I_t} I_j$. Then using proposition 805 (p. 467) which gives $\bigcup_{i\in I_0} O_i = \emptyset$ we have from propositions 796 (p. 464) and 777 (p. 458):

$$\bigcup_{i\in I} O_i = \bigcup_{i\in \bigcup_{j\in I_t} j} O_i = \bigcup_{j\in I_t}\bigcup_{i\in j} O_i = (\bigcup_{i\in I_1} O_i) \cup (\bigcup_{i\in I_0} O_i) = (\bigcup_{i\in I_1} O_i) \cup \emptyset = \bigcup_{i\in I_1} O_i\,.$$

### A.5.7 Equivalence relation on sets

**Definition 364 Equivalence relation defined on a set.**

Given a set $\mathbf{A}$ and an equivalence relation $R(x, y)$ (definition 336 (p. 445)) such that:

$$(x \notin A, y \notin A \Rightarrow R(x, y)), (x \notin A, y \in A \Rightarrow \neg R(x, y)) \tag{A.29}$$

then the equivalence relation $R(x, y)$ is said to be defined on the set $\mathbf{A}$.

**Proposition 807** Given a set $\mathbf{A}$ and an equivalence relation $R(x, y)$ defined on the set A,

$$(\forall a \in A)(\exists C_a)(\forall x)(x \in C_a \Leftrightarrow R(x, a))\,.$$

**Proof**
From proposition 749 (p. 450) we have:

$$(a \in A, R(x, a)) \Rightarrow (R(x, a), x \in A) \Rightarrow (\forall a \in A)\text{Coll}_x(R(x, a))\,.$$

From definition 344 (p. 447), we have proposition 807 (p. 468).

**Definition 365 Equivalence classes on a set associated to an equivalence relation and to an element of the set.**

Given a set $\mathbf{A}$ and an equivalence relation $\mathbf{R(x,y)}$ defined on the set A, one calls the equivalence class of an element, the result of the map $(\mathbf{C_a} : \mathbf{A} \rightarrow \mathcal{P}(\mathbf{A}))$. This map is defined using proposition 807 (p. 468) with its notations together with the definition 112 point 4 (p. 132).

**Proposition 808** Given a set $\mathbf{A}$ and an equivalence relation $\mathbf{R(x,y)}$ defined on the set $\mathbf{A}$, calling $(\mathbf{C_a} : \mathbf{A} \rightarrow \mathcal{P}(\mathbf{A}))$ the equivalence class map (definition 365 (p. 469)), we have:

$$(\forall \mathbf{a} \in \mathbf{A})(\forall \mathbf{x} \in \mathbf{C_a})(\forall \mathbf{y} \in \mathbf{C_a})(\mathbf{R(x,y)}) .$$

**Proof**
From definition 365 (p. 469), we have $\mathbf{R(x,a)},\ \mathbf{R(y,a)}$. Then from definition 336 (p. 445), we have $\mathbf{R(x,y)}$.

**Definition 366 Set of equivalence classes on a set associated to an equivalence relation.**

We have a set $\mathbf{A}$ and an equivalence relation $\mathbf{R(x,y)}$ defined on the set A. We call $(\mathbf{C_a} : \mathbf{A} \rightarrow \mathcal{P}(\mathbf{A}))$, the equivalence class map (definition 365 (p. 469)). It defines for each $\mathbf{a} \in \mathbf{A}$ an element of $\mathcal{P}(\mathbf{A})$ or a subset of $\mathbf{A}$. From proposition 749 (p. 450), those subsets image of an element $\mathbf{A}$ in $\mathcal{P}(\mathbf{A})$ form a set. It is the set of equivalence classes of $\mathbf{R(x,y)}$ defined on $\mathbf{A}$. It is the image $\mathbf{C_A}$ of $\mathbf{A}$ in $\mathcal{P}(\mathbf{A})$ by the map $\mathbf{C_a}$. We have $\mathbf{C_A} \subset \mathcal{P}(\mathbf{A})$ (definition 114 (p. 132)).

**Proposition 809** Setting $\mathbf{B}$ as the set of equivalence classes (definition 366 (p. 469)) associated to an equivalence relation $\mathbf{R(x,y)}$ defined on a set $\mathbf{A}$, we have:

$$\mathbf{u} \in \mathbf{B} \Rightarrow ((\mathbf{x} \in \mathbf{u}, \mathbf{R(x,y)}) \Rightarrow \mathbf{y} \in \mathbf{u}) \tag{A.30}$$

$$\mathbf{u} \in \mathbf{B}, \mathbf{v} \in \mathbf{B} \Rightarrow (\mathbf{u} \cap \mathbf{v} = \emptyset \vee \mathbf{u} = \mathbf{v}) . \tag{A.31}$$

**Proof**
$\mathbf{B}$ being the equivalence classes associated to an equivalence relation $\mathbf{R(x,y)}$, using definitions 365 (p. 469), 330 (p. 434), 329 (p. 434), A.1.3 (p. 435), 3, A.1.4 (p. 435), 5, 18, 15, we have:

$$\begin{aligned}\mathbf{u} \in \mathbf{B} &\Rightarrow ((\mathbf{x} \in \mathbf{u}, \mathbf{y} \notin \mathbf{u}) \Rightarrow \neg\mathbf{R(x,y)})\\ &\Rightarrow (\neg\mathbf{R(x,y)} \vee (\mathbf{x} \notin \mathbf{u} \vee \mathbf{y} \in \mathbf{u}))\\ &\Rightarrow (\neg\neg(\neg\mathbf{R(x,y)} \vee \neg(\mathbf{x} \in \mathbf{u})) \vee \mathbf{y} \in \mathbf{u}) = ((\mathbf{R(x,y)}, \mathbf{x} \in \mathbf{u}) \Rightarrow \mathbf{y} \in \mathbf{u}) .\end{aligned}$$

We suppose: $\mathbf{u} \in \mathbf{B}, \mathbf{v} \in \mathbf{B}, \exists \mathbf{y} \in \mathbf{u} \cap \mathbf{v}$. From A.30, we have:

$$((\mathbf{z} \in \mathbf{v} \Rightarrow \mathbf{R(z,y)}) \Rightarrow \mathbf{z} \in \mathbf{u}) \Rightarrow \mathbf{v} \subset \mathbf{u} .$$

Exchanging $\mathbf{u}$ and $\mathbf{v}$, we also get $\mathbf{u} \subset \mathbf{v}$. Then from proposition 754 (p. 451) $\mathbf{u} = \mathbf{v}$.

**Proposition 810** Given a set $\mathbf{A}$, an equivalence relation $\mathbf{R(x,y)}$ and its associated set of equivalence classes $\mathbf{B}$, for any element of $\mathbf{x} \in \mathbf{A}$, there is one and only one $\mathbf{b} \in \mathbf{B}$ such that: $\mathbf{x} \in \mathbf{b}$.

**Proof**
It is a direct consequence of proposition 809 (p. 469).

**Definition 367 Representative of an equivalence class.**

Given a set $\mathbf{A}$ and an equivalence relation $\mathbf{R(x,y)}$ defined on the set $\mathbf{A}$, calling $\mathbf{B}$ set of equivalence classes of $\mathbf{R(x,y)}$, $\mathbf{W} \in \mathbf{B}$ an equivalence class of $\mathbf{R}$, one calls $\mathbf{w} \in \mathbf{W}$ the representative of $\mathbf{W}$, the term:

$$\mathbf{w} = \tau_{\mathbf{x}}(\mathbf{x} \in \mathbf{W}) \in \mathbf{W} \subset \mathbf{A} .$$

From propositions 809 (p. 469), A.3 (p. 441), 4 (p. 132), it defines an injective map $(\mathbf{w(W)} : \mathbf{B} \longrightarrow \mathbf{A})$ (definition 359 (p. 461)).

**Proposition 811** Keeping the setting of definition 367 (p. 470), we have the following equalities:

$$\mathbf{R(x,w)} \Leftrightarrow \mathbf{x} \in \mathbf{W}$$
$$\tau_{\mathbf{x}}\mathbf{R(x,w)} = \tau_{\mathbf{x}}(\mathbf{x} \in \mathbf{W}) = \mathbf{w}$$

$$\mathbf{y} \in \mathbf{W} \Rightarrow (\mathbf{R(y,w)}, (\mathbf{R(x,w)} \Leftrightarrow \mathbf{R(x,y)} \Leftrightarrow \mathbf{x} \in \mathbf{W}))$$
$$\mathbf{y} \in \mathbf{W} \Rightarrow \tau_{\mathbf{x}}\mathbf{R(x,w)} = \tau_{\mathbf{x}}(\mathbf{R(x,y)}) = \tau_{\mathbf{x}}\mathbf{R(y,x)} = \tau_{\mathbf{x}}(\mathbf{x} \in \mathbf{W}) = \mathbf{w} .$$

**Proof**
It is a direct consequence of the proposition (axiom) 726 (p. 441).

**Definition 368 Set of representatives of the equivalence class of an equivalence relation defined on a set.**

Given a set $\mathbf{A}$ and an equivalence relation $\mathbf{R(x,y)}$ defined on the set $\mathbf{A}$, calling $\mathbf{B}$ the set of equivalence classes of $\mathbf{R(x,y)}$, one calls $\mathbf{C}$, the set of the representatives of $\mathbf{R(x,y)}$, the following set:

$$\mathbf{C} = \mathbf{set_u}\{(\exists \mathbf{v} \in \mathbf{B})(\mathbf{u} = \tau_{\mathbf{x}}\mathbf{R}(\mathbf{x}, \tau_{\mathbf{w}}(\mathbf{w} \in \mathbf{v})))\} .$$

This is possible because $(\exists \mathbf{v} \in \mathbf{B})(\mathbf{u} = \tau_{\mathbf{x}}\mathbf{R}(\mathbf{x}, \tau_{\mathbf{w}}(\mathbf{w} \in \mathbf{v}))) \Rightarrow \mathbf{u} \in \mathbf{A}$ (proposition 750 (p. 450)). We have $\mathbf{C} \subset \mathbf{A}$.

**Proposition 812** Keeping the setting of definition 367 (p. 470), calling $\mathbf{C}$ the set of the representatives of the equivalence class. The following relation $(\mathbf{f(x)} = \tau_{\mathbf{z}}(\mathbf{z} \in \mathbf{x}) : \mathbf{B} \leftrightarrow \mathbf{C})$ defines a bijection (definition 124 (p. 139)) of $\mathbf{B}$ to $\mathbf{C}$.

**Proof**
From definition 367 (p. 470) and proposition 811 (p. 470), $\mathbf{f(x)}$ is a map (see definition 112, points 2 (p. 131), 4 (p. 132)). Since from definition 367 (p. 470), proposition 810 (p. 469) and equation (A.31) (p. 469), $\mathbf{x} \in \mathbf{B}, \mathbf{x'} \in \mathbf{B}, \mathbf{x} \neq \mathbf{x'} \Rightarrow \mathbf{x} \cap \mathbf{x'} = \emptyset$ and $\mathbf{f(x)} \in \mathbf{x}, \mathbf{f(x')} \in \mathbf{x'}$ we have $\mathbf{x} \neq \mathbf{x'} \Rightarrow \mathbf{f(x)} \neq \mathbf{f(x')}$. $\mathbf{f(x)}$ is injective (definition 122 (p. 139)). $\mathbf{C}$ being the set of representative of the equivalence relation $\mathbf{R}$ on $\mathbf{A}$ $\mathbf{C} = \mathbf{f(B)}$, $\mathbf{f}$ is surjective. Then from definition 124 (p. 139), $\mathbf{f}$ is a bijection between $\mathbf{B}$ and $\mathbf{C} = \mathbf{f(B)}$.

**Definition 369 Disjoint set of sets.**

Given a set $\mathbf{E}$, a set $\mathbf{A}$ of subset of $\mathbf{E}$ $(\mathbf{A} \subset \mathcal{P}(\mathbf{E}))$ (definition 350 (p. 453)). $\mathbf{A}$ is said to be a set of disjoint sets if:

$$(\mathbf{x} \in \mathbf{A}, \mathbf{y} \in \mathbf{A}) \Rightarrow \mathbf{x} \cap \mathbf{y} = \emptyset .$$

**Definition 370 Partition of a set.**

Given a set $\mathbf{E}$ and a set of indices $\mathbf{I}$, the map $(\mathbf{F_i} : \mathbf{I} \longrightarrow \mathcal{P}(\mathbf{E}))$ (definition 350 (p. 453)) is called partition of $\mathbf{E}$ if $\mathbf{F_i}$ has the following properties:

$$\bigcup_{\mathbf{i} \in \mathbf{I}} \mathbf{F_i} = \mathbf{E}$$

$$(\mathbf{i} \in \mathbf{I}, \mathbf{j} \in \mathbf{I}, \mathbf{i} \neq \mathbf{j}) \Rightarrow (\mathbf{F_i} \cap \mathbf{F_j} = \emptyset)$$

$$\mathbf{i} \in \mathbf{I} \Rightarrow \mathbf{F_i} \neq \emptyset .$$

**Proposition 813** $\mathbf{B}$ being the equivalence classes on a set associated to an equivalence relation $\mathbf{R(x, y)}$ defined on a set $\mathbf{A}$, $\mathbf{B}$ is a partition of $\mathbf{A}$.

**Proof**

It comes directly from proposition 809 (p. 469).

**Proposition 814** $\mathbf{B}$ being a partition of $\mathbf{A}$, the relation $\mathbf{R(x, y)}$ defined by

$$\mathbf{R(x, y)} \Leftrightarrow ((\mathbf{x} \notin \mathbf{A}, \mathbf{y} \notin \mathbf{A}) \vee (\mathbf{x} \in \mathbf{A}, \mathbf{y} \in \mathbf{A}, (\exists \mathbf{b} \in \mathbf{B})(\mathbf{x} \in \mathbf{b}, \mathbf{y} \in \mathbf{b})))$$

is an equivalence relation.

**Proof**

$\mathbf{R(x, x)}$ is always true, $\mathbf{R(x, y)} \Leftrightarrow \mathbf{R(y, x)}$ and $(\mathbf{R(x, y)},\ \mathbf{R(y, z)}) \Rightarrow \mathbf{R(x, z)}$. $\mathbf{R}$ satisfies definition 336 (p. 445).

**Proposition 815** Given two sets $\mathbf{E}$ and $\mathbf{I}$ (see footnote 7 (p. 451)), a partition :$(\mathbf{F_i} : \mathbf{I} \longrightarrow \mathcal{P}(\mathbf{E}))$, the set $(\mathbf{g} : \mathbf{E} \longrightarrow \mathbf{I}) \subset \mathbf{E} \times \mathbf{I}$ defined by

$$(\mathbf{x}, \mathbf{i}) \in \mathbf{g} \Leftrightarrow (\mathbf{x} \in \mathbf{F_i}, \mathbf{i} \in \mathbf{I})$$

is a surjective map.

**Proof**

- $(\forall \mathbf{x} \in \mathbf{E})(\exists \mathbf{i} \in \mathbf{I})(\mathbf{x} \in \mathbf{F_i} \Leftrightarrow (\mathbf{x}, \mathbf{i}) \in \mathbf{g})$
- $(\forall \mathbf{i} \in \mathbf{J})(\forall \mathbf{i} \in \mathbf{J})(\mathbf{i} \neq \mathbf{j} \Rightarrow \mathbf{F_i} \cap \mathbf{F_j} = \emptyset) \Rightarrow ((\mathbf{x}, \mathbf{i}) \in \mathbf{g}, (\mathbf{x}, \mathbf{j}) \in \mathbf{g}) \Leftrightarrow (\mathbf{x} \in \mathbf{F_i}, \mathbf{x} \in \mathbf{F_j}) \Rightarrow \mathbf{i} = \mathbf{j}$
- $(\forall \mathbf{i} \in \mathbf{I})(\mathbf{F_i} \neq \emptyset)$.

**Proposition 816** Given two sets $\mathbf{E}$ and $\mathbf{I}$ and a map $(\mathbf{g} : \mathbf{E} \longrightarrow \mathbf{I})$ (definitions 112 (p. 131), 357 (p. 460), 114 (p. 132)). The map: $(\mathbf{F_i} = \mathbf{g}^{-1} : \mathbf{g(E)} \longrightarrow \mathcal{P}(\mathbf{E}))$ defines a partition of $\mathbf{E}$ with $\mathbf{g(E)} \subset \mathbf{I}$ as the set of indices (see definitions 174 (p. 141), 350 (p. 453)).

**Proof**

From proposition 154 (p. 133), $\mathbf{g}^{-1}(\mathbf{g(E)}) = \mathbf{g}^{-1}(\mathbf{I}) = \mathbf{E}$. From definitions 112 (p. 131), 357 (p. 460), we have:

$$\mathbf{x} \in \mathbf{g}^{-1}(\mathbf{i}) \cap \mathbf{g}^{-1}(\mathbf{j}) = \mathbf{F_i} \cap \mathbf{F_j} \Rightarrow (\mathbf{x}, \mathbf{i}) \in \mathbf{g}, (\mathbf{x}, \mathbf{j}) \in \mathbf{g} \Rightarrow \mathbf{i} = \mathbf{j} .$$

Then, we have:

$$\mathbf{i} \neq \mathbf{j} \Rightarrow \mathbf{g}^{-1}(\mathbf{i}) \cap \mathbf{g}^{-1}(\mathbf{j}) = \mathbf{F_i} \cap \mathbf{F_j} = \emptyset$$

$$\mathbf{i} \in \mathbf{g(E)} \Rightarrow \mathbf{g}^{-1}(\mathbf{i}) \neq \emptyset .$$

**Definition 371 $\sim$ relation between two sets: two sets have the same number of elements.**

One defines a relation $\sim$ between two sets $\mathbf{e}$ and $\mathbf{f}$ $\mathbf{e} \sim \mathbf{f}$ phrased as "Two sets $\mathbf{e}$ and $\mathbf{f}$ have the same number of elements" or $(\mathbf{e} \sim \mathbf{f})$ if there is a map between $\mathbf{e}$ and $\mathbf{f}$ which is bijective (definition 361 (p. 461)).

**Proposition 817** The relation $\sim$ between two sets is an equivalence relation (definition 336 (p. 445)).
**Proof**
The relation $\sim$ is an equivalence relation because it verifies definition 336 (p. 445):

- Point 1 (p. 445) is satisfied by the identity map $\mathbf{I} = \mathbf{set}_{\{x,x\}}\{\mathbf{x} \in \mathbf{X}\}$ of a set $\mathbf{X}$ on itself $\mathbf{I} : \mathbf{X} \to \mathbf{X}$ which is a bijection (proposition 179 (p. 142)).

- Point 2 (p. 445) because if $\mathbf{f}$ is a bijection from $\mathbf{X}$ to $\mathbf{Y}$ there is a bijection $\mathbf{f}^{-1}$ from $\mathbf{Y}$ to $\mathbf{X}$ (see proposition 177 (p. 142)).

- Point 3 (p. 445) because if $\mathbf{f}$ is a bijection from $\mathbf{X}$ to $\mathbf{Y}$ and $\mathbf{g}$ is a bijection from $\mathbf{Y}$ to $\mathbf{Z}$ $\mathbf{f} \circ \mathbf{g}$ is a bijection from $\mathbf{X}$ to $\mathbf{Z}$ (definition 117 (p. 134), proposition 193 (p. 149)).

**Proposition 818** Having a set of maps $\mathcal{F}$ $(\mathbf{A} \to \mathbf{B})$, a bijection $(\mathbf{h} : \mathbf{B} \leftrightarrow \mathbf{C})$, we consider the set of maps $\mathcal{G}$ defined by the product of maps (definition 117 (p. 134)):

$$\mathbf{g} \in \mathcal{G} \Leftrightarrow (\exists \mathbf{f} \in \mathcal{F})(\mathbf{g} = \mathbf{h} \circ \mathbf{f} : \mathbf{A} \to \mathbf{C})$$

$\mathbf{h}$ defined a bijection between $\mathcal{F}$ and $\mathcal{G}$:

$$\mathcal{F} \sim \mathcal{G} \,.$$

**Proof**
Since $\mathbf{g} \in \mathcal{G} \Leftrightarrow (\exists \mathbf{f} \in \mathcal{F})(\mathbf{g} = \mathbf{h} \circ \mathbf{f})$, taking that $\mathbf{f}$, we have an inverse $\mathbf{h}^{-1}\mathbf{g}$ to any $\mathbf{g}$. The map is surjective (definition 123 (p. 139)). We also have:

$$\begin{aligned}&(\mathbf{f}_1 \in \mathcal{F}, \mathbf{f}_2 \in \mathcal{F}, \mathbf{f}_1 \neq \mathbf{f}_2, \mathbf{g}_1 = \mathbf{h} \circ \mathbf{f}_1, \mathbf{g}_2 = \mathbf{h} \circ \mathbf{f}_2)\\ &\Rightarrow (\exists \mathbf{x} \in \mathbf{A})(\mathbf{f}_1(\mathbf{x}) \neq \mathbf{f}_2(\mathbf{x}), \mathbf{g}_1(\mathbf{x}) = \mathbf{h} \circ \mathbf{f}_1(\mathbf{x}) \neq \mathbf{g}_2(\mathbf{x}) = \mathbf{h} \circ \mathbf{f}_2(\mathbf{x})) \Rightarrow \mathbf{g}_1 \neq \mathbf{g}_2\end{aligned}$$

then $\mathbf{g}_1 = \mathbf{g}_2 \Rightarrow \mathbf{f}_1 = \mathbf{h}^{-1} \circ \mathbf{g}_1 = \mathbf{f}_2 = \mathbf{h}^{-1} \circ \mathbf{g}_2$. The map is injective (definition 122 (p. 139)). It is bijective (definition 124 (p. 139)).

**Proposition 819** Having a set of maps $\mathcal{F}$ $(\mathbf{A} \to \mathbf{B})$, a bijection $(\mathbf{h} : \mathbf{C} \leftrightarrow \mathbf{A})$, we consider the set of maps $\mathcal{G}$ defined by the product of maps (definition 117 (p. 134)):

$$\mathbf{g} \in \mathcal{G} \Leftrightarrow (\exists \mathbf{f} \in \mathcal{F})(\mathbf{g} = \mathbf{f} \circ \mathbf{h} : \mathbf{C} \to \mathbf{B})$$

$\mathbf{h}$ defined a bijection between $\mathcal{F}$ and $\mathcal{G}$:

$$\mathcal{F} \sim \mathcal{G} \,.$$

**Proof**
The proof is the same as proposition 818 (p. 472).

**Proposition 820** Given a set $\mathbf{A}$ and a term $\mathbf{x}$, we have: $\mathbf{A} \times \{\mathbf{x}\} \sim \mathbf{A}$ .
**Proof**
Given an element $(\mathbf{a}, \mathbf{x}) \in \mathbf{A} \times \{\mathbf{x}\}$, we have $\mathbf{a} \in \mathbf{A}$. The map $(\mathbf{f}((\mathbf{a}, \mathbf{x})) = \mathbf{a} : \mathbf{A} \times \{\mathbf{x}\} \to \mathbf{A})$ is a bijection (definition 361 (p. 461)) because it is surjective (definition 360 (p. 461)): $(\forall \mathbf{a} \in \mathbf{A})(\exists (\mathbf{a}, \mathbf{x}) \in \mathbf{A} \times \{\mathbf{x}\})(\mathbf{f}((\mathbf{a}, \mathbf{x})) = \mathbf{a})$ and it is injective (definition 359 (p. 461)): $(\mathbf{b}, \mathbf{x}) = \mathbf{f}^{-1}(\mathbf{a}) = (\mathbf{a}, \mathbf{x}) \Rightarrow \mathbf{b} = \mathbf{a}$ (definition 346 (p. 448) (A.15)). Then $\mathbf{A} \times \{\mathbf{x}\} \sim \mathbf{A}$.

### A.5.8 Product of sets over a set of indices

**Definition 372 Product of sets over a set of indices.**

One has a set of indices $\mathbf{I}$ and a map of $\mathbf{I}$ to a set of sets $\mathbf{X}$, $(\mathbf{X_i} : \mathbf{I} \to \mathbf{X})$. One can always consider $\mathbf{X}$ as $\mathbf{X} \subset \mathcal{P}(\mathbf{A})$, $\mathbf{A}$ being a set which can be $\bigcup_{\mathbf{i}\in\mathbf{I}} \mathbf{X_i}$. One calls:

$$\prod_{\mathbf{i}\in\mathbf{I}} \mathbf{X_i} = \mathbf{set_u}\{(\mathbf{u} : \mathbf{I} \to \bigcup_{\mathbf{i}\in\mathbf{I}} \mathbf{X_i}), (\forall \mathbf{i} \in \mathbf{I})(\mathbf{u}(\mathbf{i}) \in \mathbf{X_i})\} .$$

It is a set from proposition 749 (p. 450). $\mathbf{u}$ being a map, by definition 112 (p. 131), 357 (p. 460), it is also a set and we have $\mathbf{u} \subset \mathbf{I} \times \bigcup_{\mathbf{i}\in\mathbf{I}} \mathbf{X_i}$.

**Proposition 821** Taking the conditions of definition 372 (p. 473), one has a set of indices $\mathbf{I}$ and a map of $\mathbf{I}$ to a set of sets $\mathbf{X}$, $(\mathbf{X_i} : \mathbf{I} \to \mathbf{X})$. One has also a set $\mathbf{J}$ and a bijection $(\mathbf{h} : \mathbf{J} \leftrightarrow \mathbf{I})$. Then we have (definition 371 (p. 472)):

$$\prod_{\mathbf{i}\in\mathbf{I}} \mathbf{X_i} \sim \prod_{\mathbf{j}\in\mathbf{J}} \mathbf{X_{h(j)}} .$$

**Proof**

Using definition 372 (p. 473), we have:

$$\prod_{\mathbf{j}\in\mathbf{J}} \mathbf{X_{h(j)}} = \mathbf{set_v}\{(\mathbf{v} : \mathbf{J} \to \bigcup_{\mathbf{j}\in\mathbf{J}} \mathbf{X_{h(j)}}), (\forall \mathbf{j} \in \mathbf{J})(\mathbf{v}(\mathbf{j}) \in \mathbf{X_{h(j)}})\} .$$

From proposition 797 (p. 465), we have $\bigcup_{\mathbf{j}\in\mathbf{J}} \mathbf{X_{h(j)}} = \bigcup_{\mathbf{i}\in\mathbf{I}} \mathbf{X_i}$. Then we have:

$$\begin{aligned}
&\mathbf{v} \in \prod_{\mathbf{j}\in\mathbf{J}} \mathbf{X_{h(j)}} \\
&\Leftrightarrow ((\mathbf{v}(\mathbf{h}^{-1}) : \mathbf{I} \to \mathbf{J} \to \bigcup_{\mathbf{j}\in\mathbf{J}} \mathbf{X_{h(j)}} = \bigcup_{\mathbf{i}\in\mathbf{I}} \mathbf{X_i}), (\forall \mathbf{i} \in \mathbf{I})(\mathbf{v}(\mathbf{h}^{-1})(\mathbf{i}) \in \mathbf{X_{h(h^{-1}(i))}} = \mathbf{X_i})) \\
&\Leftrightarrow ((\mathbf{v}(\mathbf{h}^{-1}) : \mathbf{I} \to \bigcup_{\mathbf{i}\in\mathbf{I}} \mathbf{X_i}), (\forall \mathbf{i} \in \mathbf{I})(\mathbf{v}(\mathbf{h}^{-1})(\mathbf{i}) \in \mathbf{X_i})) \Leftrightarrow \mathbf{v}(\mathbf{h}^{-1}) \in \prod_{\mathbf{i}\in\mathbf{I}} \mathbf{X_i} .
\end{aligned}$$

Then applying proposition 819 (p. 472), we have proposition 821 (p. 473).

**Proposition 822** Given two different terms $\boldsymbol{\alpha}, \boldsymbol{\beta}$, the set of them $\mathbf{I} = \{\boldsymbol{\alpha}, \boldsymbol{\beta}\}$ (see definition 345 (p. 447)) and the couple of them $\mathbf{z} = (\boldsymbol{\alpha}, \boldsymbol{\beta})$ (see definitions 346 (p. 448), 347 (p. 449)), and a map $(\mathbf{X_i} : \mathbf{I} \to \mathbf{X})$ (see footnote 7 (p. 451)). We have (definitions 356 (p. 459) and 371 (p. 472)):

$$\mathbf{X_{proj_1(z)}} \times \mathbf{X_{proj_2(z)}} \sim \prod_{\mathbf{i}\in\mathbf{I}} \mathbf{X_i} .$$

**Proof**

The couple $\mathbf{z}$ of two different terms exists (proposition 779 (p. 458)). We set (see definitions 347 (p. 449), 345 (p. 447)):

$$\mathbf{I} = \{\mathbf{Proj_1(z)}, \mathbf{Proj_2(z)}\} .$$

Defining the map

$$(\mathbf{f} : \prod_{\mathbf{i}\in\mathbf{I}} \mathbf{X_i} \to \mathbf{X_{proj_1(z)}} \times \mathbf{X_{proj_2(z)}})$$

by

$$\mathbf{u} \in \prod_{\mathbf{i}\in\mathbf{I}} \mathbf{X_i} \Rightarrow \mathbf{f}(\mathbf{u}) = (\mathbf{u_{proj_1(z)}}, \mathbf{u_{proj_2(z)}})$$

$\mathbf{f}$ is a bijection because (definition 346 (p. 448)):

$$((u'_{proj_1(z)}, u'_{proj_2(z)}) = f(u') = f(u) = (u_{proj_1(z)}, u_{proj_2(z)}))$$
$$\Rightarrow ((u'_{proj_1(z)}, u'_{proj_2(z)}) = (u_{proj_1(z)}, u_{proj_2(z)}))$$
$$\Rightarrow (u'_{proj_1(z)} = u_{proj_1(z)},\ u'_{proj_2(z)} = u_{proj_2(z)}) \Rightarrow u' = u\ .$$

**Convention 31** $\mathbf{X_aX_bX_c}$.

When dealing with elements like $\prod_{i\in I} X_i$ with $\mathbf{Card(I)} \in \mathcal{Z}^+$ (definition 390 (p. 499)), one often writes the elements side by side like $X_aX_bX_c$. It has to be understood that we consider in definition 372 (p. 473), the map $(X_i : I = [1,3] \to X)$ defined by $X_1 = X_a, X_2 = X_b, X_3 = X_c$ and we set:

$$X_aX_bX_c = \prod_{i\in[1,3]} X_i\ .$$

**Definition 373 Exponentiation of a set A by a set B:** $\mathbf{A^B}$.

Given two sets $\mathbf{A}$ and $\mathbf{B}$, in definition 372 (p. 473), we can set $I = B$ and $(\forall i \in I = B)(X_i = A)$ we set:

$$A^B = \prod_{i\in B} X_i = \prod_{i\in B} A\ .$$

From definition 372 (p. 473), it is the set of maps $B \to A$.

**Proposition 823** With notations of definition 373 (p. 474), we have:

$$B \sim C \Rightarrow A^B \sim A^C\ .$$

**Proof**
It is a direct consequence of proposition 821 (p. 473).

**Proposition 824** Given a set of indices $\mathbf{I}$ (see footnote 7 (p. 451)) and two maps from $\mathbf{I}$ to two sets of sets $\mathbf{X}$, $\mathbf{Y}$: $(X_i : I \to X)$, $(Y_i : I \to Y)$, we suppose that we have a bijection $y_i$ between $X_i$ and $Y_i$:

$$(\forall i \in I)(\exists (y_i(x_i) : X_i \leftrightarrow Y_i))\ .$$

Then there is a bijection from $\prod_{i\in I} X_i$ to $\prod_{i\in I} Y_i$:

$$(\exists (f = \prod_{i\in I} y_i(x_i) : \prod_{i\in I} X_i \leftrightarrow \prod_{i\in I} Y_i))\ .$$

**Proof**
Clearly $f = \prod_{i\in I} y_i(x_i)$ is a map $(\prod_{i\in I} X_i \to \prod_{i\in I} Y_i)$. It is a surjective map because calling $x_i(y_i) = y_i^{-1}(y_i)$, we have:

$$(\forall \prod_{i\in I} y_i \in \prod_{i\in I} Y_i)(\exists (u = \prod_{i\in I} x_i(y_i) \in \prod_{i\in I} X_i)(\prod_{i\in I} y_i = f(u) = f(\prod_{i\in I}(x_i(y_i)))$$
$$= \prod_{i\in I} y_i(x_i(y_i)))\ .$$

It is injective because to any $\prod_{i\in I} y_i$, there is only one inverse $\prod_{i\in I} x_i(y_i)$ because $y_i(x_i)$ is a bijection.

**Proposition 825** Given a set $\mathbf{XY}$ of two elements: $\mathbf{XY} = \{\mathbf{x}, \mathbf{y}\}$ with $\mathbf{x} \neq \mathbf{y}$ (definition 345 (p. 447)), and two sets $\mathbf{A^x}, \mathbf{A^y}$ which can be considered as given by a map

$$(\mathbf{A^z} : \mathbf{XY} \to \mathbf{A^x} \cup \mathbf{A^y} = \bigcup_{\mathbf{z} \in \mathbf{XY}} \mathbf{A^z}) \,.$$

We have a bijection between $\mathbf{A^x} \times \mathbf{A^y}$ defined by definition 356 (p. 459) and $\prod_{\mathbf{z} \in \mathbf{XY}} \mathbf{A^z}$ as defined by definition 372 (p. 473).

**Proof**
It is a direct consequence of proposition 789 (p. 462).

**Proposition 826** Keeping the notation of the definition 372 (p. 473), we have:

$$(\forall \mathbf{i} \in \mathbf{I})(\mathbf{A_i} \subset \mathbf{X_i}) \Rightarrow \prod_{\mathbf{i} \in \mathbf{I}} \mathbf{A_i} \subset \prod_{\mathbf{i} \in \mathbf{I}} \mathbf{X_i} \,.$$

**Proof**
It is a direct consequence of definitions 372 (p. 473) and 349 (p. 451).

**Proposition 827** Given a set of indices $\mathbf{I}$, a set of sets $\mathbf{X}$, a map: $(\mathbf{X_i} : \mathbf{I} \to \mathbf{X})$, we have:

$$(\exists \mathbf{i} \in \mathbf{I})(\mathbf{X_i} = \emptyset) \Leftrightarrow \prod_{\mathbf{i} \in \mathbf{I}} \mathbf{X_i} = \emptyset \tag{A.32}$$

$$\mathbf{A} \times \emptyset = \emptyset \,. \tag{A.33}$$

**Proof**
As from definition 333 (p. 441) with $\mathbf{i_1} = \tau_\mathbf{i}(\mathbf{i} \in \mathbf{I}, \mathbf{X_i} = \emptyset)$, we have:

$$(\exists \mathbf{i} \in \mathbf{I})(\mathbf{X_i} = \emptyset) = (\mathbf{X_{i_1}} = \emptyset) \,. \tag{A.34}$$

From propositions 731 (p. 442), 773 (p. 457), definitions 330 (p. 434), A.1.4 (p. 435) -7 we have:

$$\prod_{\mathbf{i} \in \mathbf{I}} \mathbf{X_i} \neq \emptyset \Rightarrow \exists((\mathbf{u} : \mathbf{I} \to \bigcup_{\mathbf{i} \in \mathbf{I}} \mathbf{X_i}))(\forall \mathbf{i} \in \mathbf{I})(\mathbf{u}(\mathbf{i}) \in \mathbf{X_i}) \Rightarrow \mathbf{u}(\mathbf{i_1}) \in \mathbf{X_{i_1}} \Rightarrow \mathbf{X_{i_1}} \neq \emptyset$$

we conclude from A.1.4 (p. 435), 7:

$$\mathbf{X_{i_1}} = \emptyset \Rightarrow \prod_{\mathbf{i} \in \mathbf{I}} \mathbf{X_i} = \emptyset \,.$$

Then combining with equation (A.34), we have:

$$(\exists \mathbf{i} \in \mathbf{I})(\mathbf{X_i} = \emptyset) \Rightarrow \prod_{\mathbf{i} \in \mathbf{I}} \mathbf{X_i} = \emptyset \,.$$

Moreover, we also have with $\mathbf{i_1} = \tau_\mathbf{i}(\mathbf{i} \in \mathbf{I}, \mathbf{X_i} = \emptyset)$ (see definition 333 (p. 441) and axioms A.1 (p. 441), A.2, A.3):

$$(\exists \mathbf{i} \in \mathbf{I})(\mathbf{X_i} = \emptyset) \Leftrightarrow \mathbf{X_{i_1}} = \emptyset \Leftrightarrow \neg \mathbf{X_{i_1}} \neq \emptyset \Leftrightarrow (\exists \mathbf{i} \in \mathbf{I})(\neg \mathbf{X_i} \neq \emptyset) \,.$$

Then (see definitions 334 (p. 441) and 10 (p. 435)):

$$\begin{aligned}&\neg(\exists i \in I)(X_i = \emptyset) \Leftrightarrow \neg(\exists i \in I)(\neg X_i \neq \emptyset)\\&= (\forall i \in I)(X_i \neq \emptyset) \Leftrightarrow (\forall i \in I)(\exists x \in X_i)\\&\Rightarrow (\exists(u : I \to \bigcup_{i\in I} X_i))((u(i) = \tau_x(x \in X_i) : I \to \bigcup_{i\in I} X_i), u(i) \in X_i)\\&\Rightarrow (\exists u \in \prod_{i\in I} X_i) \Rightarrow \prod_{i\in I} X_i \neq \emptyset\,.\end{aligned}$$

Then from A.1.4 (p. 435), 7, we have:

$$\prod_{i\in I} X_i = \emptyset \Rightarrow (\exists i \in I)(X_i = \emptyset)\,.$$

Equation (A.32) is true. Equation (A.33) (p. 475) is true from proposition 781 (p. 459).

**Proposition 828** Given a set of indices $\mathbf{I}$, a set of sets $\mathbf{X}$, a map: $(X_i : I \to X)$ and a deduced map $(X_i \times i : I \to X \times I)$, we have:

$$(\forall i \in I)(\forall j \in I)(i \neq j \Rightarrow X_i \times \{i\} \cap X_j \times \{j\} = \emptyset)\,.$$

**Proof**
It is a direct consequence of proposition 786 (p. 460).

**Proposition 829** Given sets $\mathbf{X}$, $\mathbf{I}$ and $\mathbf{J}$, a map $(X_{i,j} : I \times J \to \mathcal{P}(X))$, $\prod_{i\in I} \bigcap_{j\in J} X_{i,j} = \bigcap_{j\in J} \prod_{i\in I} X_{i,j}$.
**Proof**
Using proposition 738 (p. 444), one has:

$$\begin{aligned}x \in \prod_{i\in I}\bigcap_{j\in J} X_{i,j} &\Leftrightarrow (\forall i \in I)(x_i \in \bigcap_{j\in J} X_{i,j})\\&\Leftrightarrow (\forall i \in I)(\forall j \in J)(x_i \in X_{i,j})\\&\Leftrightarrow (\forall j \in J)(\forall i \in I)(x_i \in X_{i,j})\\&\Leftrightarrow x \in \bigcap_{j\in J}\prod_{i\in I} X_{i,j}\,.\end{aligned}$$

Using proposition 754 (p. 451), we conclude that the proposition 829 (p. 476) is true:

$$\prod_{i\in I}\bigcap_{j\in J} X_{i,j} = \bigcap_{j\in J}\prod_{i\in I} X_{i,j}\,.$$

**Proposition 830** $\mathbf{A}$ being a set and given a set of indices $L \neq \emptyset$ and a map $(J_\lambda : L \to \mathcal{P}(A) - \emptyset)$ and a set of maps over a set of sets $\mathbf{X}$: $(X_{\lambda,\iota} : J_\lambda \to X)$.
We call $I = \prod_{\lambda\in L} J_\lambda$. We have the following relations:

$$\begin{aligned}\bigcup_{\lambda\in L}\bigcap_{\iota\in J_\lambda} X_{\lambda,\iota} &= \bigcap_{f\in I}\bigcup_{\lambda\in L} X_{\lambda,f_\lambda}\\\bigcap_{\lambda\in L}\bigcup_{\iota\in J_\lambda} X_{\lambda,\iota} &= \bigcup_{f\in I}\bigcap_{\lambda\in L} X_{\lambda,f_\lambda}\,.\end{aligned}$$

**Proof**
Since $\lambda \in L \Rightarrow (\forall\iota \in J_\lambda X_{\lambda,\iota}) \subset X$, when we consider the operation $\mathcal{C}$, it is in fact $(\mathcal{C}_X(A) : \mathcal{P}(X) \leftrightarrow \mathcal{P}(X))$ (definitions 353 (p. 454), 350 (p. 453)).

**First relation.**

Using definition 372 (p. 473), when we have $\mathbf{f} \in \mathbf{I}$, we write $\mathbf{f} = (\mathbf{f}_\lambda : \mathbf{L} \to \bigcup_{\lambda \in \mathbf{L}} \mathbf{J}_\lambda)$ with $(\mathbf{f}_\lambda \in \mathbf{J}_\lambda)$. Using propositions 730 (p. 442) and 739 (p. 444), we have:

$$\begin{aligned}
&\mathbf{x} \in \bigcap_{\iota \in \mathbf{J}_\lambda} \mathbf{X}_{\lambda,\iota} \Leftrightarrow (\forall \iota \in \mathbf{J}_\lambda)(\mathbf{x} \in \mathbf{X}_{\lambda,\iota}) \\
&(\forall \mathbf{f})(\mathbf{f} \in \mathbf{I} \Rightarrow (\mathbf{f}_\lambda \in \mathbf{J}_\lambda) \\
&\mathbf{x} \in \bigcap_{\iota \in \mathbf{J}_\lambda} \mathbf{X}_{\lambda,\iota} \Rightarrow (\forall \mathbf{f} \in \mathbf{I})(\mathbf{x} \in \mathbf{X}_{\lambda,\mathbf{f}_\lambda}) \\
&\mathbf{x} \in \bigcup_{\lambda \in \mathbf{L}} \bigcap_{\iota \in \mathbf{J}_\lambda} \mathbf{X}_{\lambda,\iota} \Rightarrow (\exists \lambda \in \mathbf{L})(\forall \mathbf{f} \in \mathbf{I})(\mathbf{x} \in \mathbf{X}_{\lambda,\mathbf{f}_\lambda}) \\
&\Rightarrow (\forall \mathbf{f} \in \mathbf{I})(\exists \lambda \in \mathbf{L})(\mathbf{X}_{\lambda,\mathbf{f}_\lambda}) = \mathbf{x} \in \bigcap_{\mathbf{f} \in \mathbf{I}} \bigcup_{\lambda \in \mathbf{L}} \mathbf{X}_{\lambda,\mathbf{f}_\lambda} .
\end{aligned}$$

Then we have:

$$\bigcup_{\lambda \in \mathbf{L}} \bigcap_{\iota \in \mathbf{J}_\lambda} \mathbf{X}_{\lambda,\iota} \subset \bigcap_{\mathbf{f} \in \mathbf{I}} \bigcup_{\lambda \in \mathbf{L}} \mathbf{X}_{\lambda,\mathbf{f}_\lambda} . \tag{A.35}$$

**Second relation.**

Applying equation (A.35) (p. 477) to $\mathcal{C}(\mathbf{X}_{\lambda,\iota})$ (definition 353 (p. 454), using propositions 769 (p. 455), 801 (p. 466) and 802 (p. 466), we have:

$$\begin{aligned}
&\bigcup_{\lambda \in \mathbf{L}} \bigcap_{\iota \in \mathbf{J}_\lambda} \mathcal{C}(\mathbf{X}_{\lambda,\iota}) \subset \bigcap_{\mathbf{f} \in \mathbf{I}} \bigcup_{\lambda \in \mathbf{L}} \mathcal{C}(\mathbf{X}_{\lambda,\mathbf{f}_\lambda}) \Leftrightarrow \mathcal{C}(\bigcap_{\mathbf{f} \in \mathbf{I}} \bigcup_{\lambda \in \mathbf{L}} \mathbf{X}_{\lambda,\mathbf{f}_\lambda}) \subset \mathcal{C}(\bigcup_{\lambda \in \mathbf{L}} \bigcap_{\iota \in \mathbf{J}_\lambda} \mathbf{X}_{\lambda,\iota}) \\
&\Leftrightarrow \bigcup_{\lambda \in \mathbf{L}} \bigcap_{\iota \in \mathbf{J}_\lambda} \mathbf{X}_{\lambda,\iota} \subset \bigcap_{\mathbf{f} \in \mathbf{I}} \bigcup_{\lambda \in \mathbf{L}} \mathbf{X}_{\lambda,\mathbf{f}_\lambda} .
\end{aligned}$$

From proposition 754 (p. 451) (extensionality principle),

$$\bigcap_{\mathbf{f} \in \mathbf{I}} \bigcup_{\lambda \in \mathbf{L}} \mathbf{X}_{\lambda,\mathbf{f}_\lambda} = \bigcup_{\lambda \in \mathbf{L}} \bigcap_{\iota \in \mathbf{J}_\lambda} \mathbf{X}_{\lambda,\iota} . \tag{A.36}$$

Applying again relation (A.36) (p. 477) to $\mathcal{C}\mathbf{X}_{\lambda,\iota}$, using propositions 767 (p. 455), 801 (p. 466) and 802 (p. 466), we have:

$$\begin{aligned}
&\bigcup_{\lambda \in \mathbf{L}} \bigcap_{\iota \in \mathbf{J}_\lambda} \mathcal{C}(\mathbf{X}_{\lambda,\iota}) = \bigcap_{\mathbf{f} \in \mathbf{I}} \bigcup_{\lambda \in \mathbf{L}} \mathcal{C}(\mathbf{X}_{\lambda,\mathbf{f}_\lambda}) \\
&\mathcal{C}(\bigcap_{\lambda \in \mathbf{L}} \bigcup_{\iota \in \mathbf{J}_\lambda} \mathbf{X}_{\lambda,\iota}) = \mathcal{C}(\bigcup_{\mathbf{f} \in \mathbf{I}} \bigcap_{\lambda \in \mathbf{L}} \mathbf{X}_{\lambda,\mathbf{f}_\lambda}) \\
&\bigcap_{\lambda \in \mathbf{L}} \bigcup_{\iota \in \mathbf{J}_\lambda} \mathbf{X}_{\lambda,\iota} = \bigcup_{\mathbf{f} \in \mathbf{I}} \bigcap_{\lambda \in \mathbf{L}} \mathbf{X}_{\lambda,\mathbf{f}_\lambda}
\end{aligned}$$

we get the second relation.

**Proposition 831** $\mathbf{J}$ being a set, given a set of indices $\mathbf{L}$ and a map $(\mathbf{J}_\lambda : \mathbf{L} \to \mathcal{P}(\mathbf{J}))$, setting $\mathbf{I} = \prod_{\lambda \in \mathbf{L}} \mathbf{J}_\lambda$, we have:

$$(\forall \lambda \in \mathbf{L})(\exists \iota)(\iota \in \mathbf{J}_\lambda, \mathbf{R}(\iota, \lambda)) \Leftrightarrow (\exists \mathbf{f} \in \mathbf{I})(\forall \lambda \in \mathbf{L})(\mathbf{R}(\mathbf{f}_\lambda, \lambda)) \tag{A.37}$$

$$(\exists \lambda \in \mathbf{L})(\forall \iota)(\iota \in \mathbf{J}_\lambda \Rightarrow \mathbf{R}(\iota, \lambda)) \Leftrightarrow (\forall \mathbf{f} \in \mathbf{I})(\exists \lambda \in \mathbf{L})(\mathbf{R}(\mathbf{f}_\lambda, \lambda)) . \tag{A.38}$$

**Proof**
Having a relation $\mathbf{R}(\iota,\lambda)$, we define $\mathbf{R}'(\iota,\lambda) = (\iota \in \mathbf{J}_\lambda, \mathbf{R}(\iota,\lambda))$. From definition 112 (p. 131) point 4 (p. 132), since $\mathbf{R}'(\iota,\lambda) \Rightarrow \iota \in \mathbf{J}_\lambda$, we can defined a map

$$(\mathbf{f}_\lambda = \tau_\iota \mathbf{R}'(\iota,\lambda) : \mathbf{L} \to \bigcup_{\lambda\in\mathbf{L}} \mathbf{J}_\lambda) \ .$$

We have, from definition 372 (p. 473), $\mathbf{f}_\lambda \in \mathbf{I}$. Then, from equation (A.1) (p. 441) and definition 333 (p. 441), we have:

$$(\forall\lambda \in \mathbf{L})(\exists\iota)(\iota \in \mathbf{J}_\lambda, \mathbf{R}(\iota,\lambda)) = (\forall\lambda \in \mathbf{L})(\mathbf{f}_\lambda \in \mathbf{J}_\lambda, \mathbf{R}(\mathbf{f}_\lambda,\lambda)) \Rightarrow (\forall\lambda \in \mathbf{L})(\mathbf{R}(\mathbf{f}_\lambda,\lambda))$$
$$\Rightarrow (\exists\mathbf{f} \in \mathbf{I})(\forall\lambda \in \mathbf{L})(\mathbf{R}(\mathbf{f}_\lambda,\lambda)) \ .$$

From point 13 (p. 435), 18 (p. 436) of A.1.4, propositions 733 (p. 442), 739 (p. 444), equation (A.1) (p. 441), we have:

$$\begin{aligned}&(\exists\mathbf{f} \in \mathbf{I})(\forall\lambda \in \mathbf{L})(\mathbf{R}(\mathbf{f}_\lambda,\lambda)) \Rightarrow (\forall\lambda \in \mathbf{L})(\exists\mathbf{f} \in \mathbf{I})(\mathbf{R}(\mathbf{f}_\lambda,\lambda))\\ &\Rightarrow (\forall\lambda \in \mathbf{L})(\exists\mathbf{f})(\mathbf{f} \in \mathbf{I}, (\exists\iota)((\lambda,\iota) \in \mathbf{f}, \iota \in \mathbf{J}_\lambda, \mathbf{R}(\iota,\lambda))\\ &\Rightarrow (\forall\lambda \in \mathbf{L})(\exists\mathbf{f})(\exists\iota)(\iota \in \mathbf{J}_\lambda, \mathbf{R}(\iota,\lambda)) = (\forall\lambda \in \mathbf{L})(\exists\iota)(\iota \in \mathbf{J}_\lambda, \mathbf{R}(\iota,\lambda)) \ .\end{aligned}$$

The second relation (A.38) is obtained by applying (A.37) to $\neg\mathbf{R}$ and taking the opposite of both members according to point 7 of A.1.4 (p. 435):

$$\begin{aligned}&\neg(\forall\lambda \in \mathbf{L})(\exists\iota)(\iota \in \mathbf{J}_\lambda, \neg\mathbf{R}(\iota,\lambda)) \Leftrightarrow (\exists\lambda \in \mathbf{L})(\forall\iota)(\iota \notin \mathbf{J}_\lambda \vee \neg\neg\mathbf{R}(\iota,\lambda))\\ &\Leftrightarrow (\exists\lambda \in \mathbf{L})(\forall\iota)(\iota \in \mathbf{J}_\lambda \Rightarrow \mathbf{R}(\iota,\lambda))\end{aligned}$$

and the same for the other member of (A.38).

**Proposition 832** $\mathbf{J}$ being a set and given a set of indices $\mathbf{L}$ and a map $(\mathbf{J}_\lambda : \mathbf{L} \to \mathcal{P}(\mathbf{J}))$ and a set of maps over a set of sets $\mathbf{X}$: $(\mathbf{X}_{\lambda,\iota} : \mathbf{J}_\lambda \to \mathbf{X})$.
We call as before $\mathbf{I} = \prod_{\lambda\in\mathbf{L}} \mathbf{J}_\lambda$. We have the following relations:

$$\prod_{\lambda\in\mathbf{L}}(\bigcup_{\iota\in\mathbf{J}_\lambda} \mathbf{X}_{\lambda,\iota}) = \bigcup_{\mathbf{f}\in\mathbf{I}}(\prod_{\lambda\in\mathbf{L}} \mathbf{X}_{\lambda,\mathbf{f}(\lambda)})$$
$$\prod_{\lambda\in\mathbf{L}}(\bigcap_{\iota\in\mathbf{J}_\lambda} \mathbf{X}_{\lambda,\iota}) = \bigcap_{\mathbf{f}\in\mathbf{I}}(\prod_{\lambda\in\mathbf{L}} \mathbf{X}_{\lambda,\mathbf{f}(\lambda)}) \ .$$

**Proof**
Using propositions A.38 (p. 477), 831 (p. 477), definitions 372 (p. 473), 362 (p. 462), we have:

$$\begin{aligned}&\mathbf{x} \in \prod_{\lambda\in\mathbf{L}}(\bigcup_{\iota\in\mathbf{J}_\lambda} \mathbf{X}_{\lambda,\iota}) \Leftrightarrow (\mathbf{x} = (\mathbf{x}(\lambda) : \mathbf{L} \to \bigcup_{\lambda\in\mathbf{L}}\bigcup_{\iota\in\mathbf{J}_\lambda} \mathbf{X}_{\lambda,\iota}),\\ &(\forall\lambda \in \mathbf{L})(\exists\iota \in \mathbf{J}_\lambda)(\mathbf{x}(\lambda) \in \mathbf{X}_{\lambda,\iota}))\\ &\Leftrightarrow ((\mathbf{x} : \mathbf{L} \to \bigcup_{\lambda\in\mathbf{L}}\bigcup_{\iota\in\mathbf{J}_\lambda} \mathbf{X}_{\lambda,\iota}), (\exists\mathbf{f} \in \mathbf{I})(\forall\lambda \in \mathbf{L})(\mathbf{x}(\lambda) \in \mathbf{X}_{\lambda,\mathbf{f}(\lambda)}))\\ &\Leftrightarrow ((\mathbf{x} : \mathbf{L} \to \bigcup_{\lambda\in\mathbf{L}}\bigcup_{\iota\in\mathbf{J}_\lambda} \mathbf{X}_{\lambda,\iota}), (\exists\mathbf{f} \in \mathbf{I})(\mathbf{x} \in \prod_{\lambda\in\mathbf{L}} \mathbf{X}_{\lambda,\mathbf{f}(\lambda)}))\\ &\Leftrightarrow \mathbf{x} \in \bigcup_{\mathbf{f}\in\mathbf{I}}(\prod_{\lambda\in\mathbf{L}} \mathbf{X}_{\lambda,\mathbf{f}(\lambda)}) \ .\end{aligned}$$

Using propositions 730 (p. 442), 731 (p. 442), 733 (p. 442), definitions 372 (p. 473), 362 (p. 462), 363 (p. 463), we have:

$$\begin{aligned}&x \in \prod_{\lambda\in L}(\bigcap_{\iota\in J_\lambda} X_{\lambda,\iota}) \Leftrightarrow ((x : L \to \bigcup_{\lambda\in L}\bigcap_{\iota\in J_\lambda} X_{\lambda,\iota}), (\forall\lambda \in L)(x(\lambda) \in \bigcap_{\iota\in J_\lambda} X_{\lambda,\iota}))\\ &\Leftrightarrow ((x : L \to \bigcup_{\lambda\in L}\bigcap_{\iota\in J_\lambda} X_{\lambda,\iota}), (\forall\lambda \in L)(\forall\iota)(\iota \in J_\lambda \Rightarrow x(\lambda) \in X_{\lambda,\iota}))\ .\end{aligned}$$

Let us suppose that $x \in \prod_{\lambda\in L}(\bigcap_{\iota\in J_\lambda} X_{\lambda,\iota})$ and $(\exists f \in I)(\exists\lambda \in L)(x(\lambda) \notin X_{\lambda,f_\lambda})$, since $(\forall\lambda \in L)(f_\lambda \in J_\lambda)$, it would be in contradiction with $(\forall\iota)(\iota \in J_\lambda \Rightarrow x(\lambda) \in X_{\lambda,\iota})$. Then from point 9 (p. 435) of A.1.4, and definition 334 (p. 441), we have:

$$\begin{aligned}&x \in \prod_{\lambda\in L}(\bigcap_{\iota\in J_\lambda} X_{\lambda,\iota}) \Rightarrow \neg(\exists f)(f \in I, (\exists\lambda \in L)(x(\lambda) \notin X_\lambda, f_\lambda))\\ &= \neg(\exists f)(\neg(\neg f \in I \vee \neg(\exists\lambda \in L)(x(\lambda) \notin X_\lambda, f_\lambda)))\\ &= (\forall f)(f \in I \Rightarrow (\forall\lambda \in L)(x(\lambda) \in X_\lambda, f_\lambda))\\ &\Leftrightarrow \prod_{\lambda\in L}\bigcap_{\iota\in J_\lambda} X_{\lambda,\iota} \subset \bigcap_{f\in I}\prod_{\lambda\in L} X_{\lambda,f_\lambda}\ .\end{aligned}$$

Let us suppose that we have both:
$x \in \bigcap_{f\in I}\prod_{\lambda\in L} X_\lambda, f_\lambda$ and $(\exists\lambda \in L)(\exists\iota \in J_\lambda)(x(\lambda) \notin X_{\lambda,\iota})$ .
To be clearer, we call $\lambda'$ and $\iota'$ those $\lambda$ and $\iota$ (see convention 26 (p. 441)) and we set $f$ by:

$$(\forall\lambda \neq \lambda')(f_\lambda = \tau_x(x \in J_\lambda)), f_{\lambda'} = \iota'$$

that $f$ does not verify $(\forall\lambda \in L)(x(\lambda) \in X_{\lambda,f_\lambda})$. But since we have:

$$x \in \bigcap_{f\in I}\prod_{\lambda\in L} X_{\lambda,f_\lambda} \Leftrightarrow (\forall f \in I)(\forall\lambda \in L)(x(\lambda) \in X_{\lambda,f_\lambda})$$

$f$ should verify that expression (see proposition 731 (p. 442)). Then, from A.1.4 (p. 435) point 9, we have:

$$\begin{aligned}&x \in \bigcap_{f\in I}\prod_{\lambda\in L} X_{\lambda,f_\lambda} \Rightarrow \neg(\exists\lambda \in L)(\exists\iota \in J_\lambda)(x(\lambda) \notin X_{\lambda,\iota})\\ &= (\forall\lambda \in L)(\forall\iota \in J_\lambda)(x(\lambda) \in X_{\lambda,\iota})\\ &\Leftrightarrow \bigcap_{f\in I}\prod_{\lambda\in L} X_{\lambda,f_\lambda} \subset \prod_{\lambda\in L}\bigcap_{\iota\in J_\lambda} X_{\lambda,\iota}\ .\end{aligned}$$

From proposition 754 (p. 451), we have:

$$\bigcap_{f\in I}\prod_{\lambda\in L} X_{\lambda,f_\lambda} = \prod_{\lambda\in L}\bigcap_{\iota\in J_\lambda} X_{\lambda,\iota}\ .$$

**Proposition 833** Given a set of indices $K$, a set of sets $Y$, a map $(Y_k : K \to Y)$ a set $C$ (see footnote 7 (p. 451)), we have:

$$C \times \bigcup_{k\in K} Y_k = \bigcup_{k\in K} C \times Y_k\ .$$

**Proof**

It is an application of proposition 832 (p. 478), with $\mathbf{L} = \{\mathbf{a}, \mathbf{b}\}, \mathbf{a} \neq \mathbf{b}$ (see definition 345 (p. 447)), with $\mathbf{J_a} = \{\mathbf{d}\}$, $\mathbf{J_b} = \mathbf{K}$, $\mathbf{I} = \mathbf{J_a} \times \mathbf{J_b}$, $\mathbf{A} = \{\mathbf{J_a}, \mathbf{J_b}\}$, $\mathbf{X_{a,d}} = \mathbf{C}$, $(\forall \mathbf{i} \in \mathbf{J_b} = \mathbf{K})(\mathbf{X_{b,i}} = \mathbf{Y_i})$. In those conditions, proposition 832 (p. 478) gives:

$$\prod_{\lambda \in L}(\bigcup_{\iota \in J_\lambda} X_{\lambda,\iota}) = C \times \bigcup_{\iota \in K} Y_\iota = \bigcup_{f \in I}\prod_{\lambda \in L} X_{\lambda,f(\lambda)} = \bigcup_{f \in I} X_{a,f(a)} X_{b,f(b)} = \bigcup_{k \in K}(C \times Y_k) .$$

**Proposition 834** $\mathbf{A}$ and $\mathbf{X}$ being sets and given a set of indices $\mathbf{L}$ and two maps $(\mathbf{J}^+_\lambda : \mathbf{L} \to \mathcal{P}(\mathbf{A}))$, $(\mathbf{J}^-_\lambda : \mathbf{L} \to \mathcal{P}(\mathbf{A}))$ and a set of maps: $(\mathbf{X}_{\lambda,\iota} : \mathbf{J}_\lambda \to \mathcal{P}(\mathbf{X}))$ $\mathbf{J}_\lambda = \mathbf{J}^+_\lambda \cup \mathbf{J}^-_\lambda$.
We call $\mathbf{I}^+ = \prod_{\lambda \in L} \mathbf{J}^+_\lambda$ and $\mathbf{I}^- = \prod_{\lambda \in L} \mathbf{J}^-_\lambda$. We have the following relations (see definition 353 (p. 454)):

$$\prod_{\lambda \in L}(\bigcup_{\iota \in J^+_\lambda} X_{\lambda,\iota} \cap \mathcal{C}_X(\bigcup_{\iota \in J^-_\lambda} X_{\lambda,\iota})) = \prod_{\lambda \in L}(\bigcup_{\iota \in J^+_\lambda} X_{\lambda,\iota} \cap (\bigcap_{\iota \in J^-_\lambda} \mathcal{C}_X X_{\lambda,\iota}))$$
$$= \bigcup_{f \in I^+}(\prod_{\lambda \in L} X_{\lambda,f(\lambda)}) \cap \bigcap_{f \in I^-}(\prod_{\lambda \in L} \mathcal{C}_X(X_{\lambda,f(\lambda)})) .$$

**Proof**

From propositions 801 (p. 466), 802 (p. 466), 829 (p. 476) and 832 (p. 478), we have:

$$\prod_{\lambda \in L}(\bigcup_{\iota \in J^+_\lambda} X_{\lambda,\iota} \cap \mathcal{C}_X(\bigcup_{\iota \in J^-_\lambda} X_{\lambda,\iota})) = \prod_{\lambda \in L}(\bigcup_{\iota \in J^+_\lambda} X_{\lambda,\iota}) \cap \prod_{\lambda \in L}(\mathcal{C}_X(\bigcup_{\iota \in J^-_\lambda} X_{\lambda,\iota}))$$
$$= \prod_{\lambda \in L}(\bigcup_{\iota \in J^+_\lambda} X_{\lambda,\iota}) \cap \prod_{\lambda \in L}(\bigcap_{\iota \in J^-_\lambda} \mathcal{C}_X(X_{\lambda,\iota}))$$
$$= \bigcup_{f \in I^+}(\prod_{\lambda \in L} X_{\lambda,f(\lambda)}) \cap \bigcap_{f \in I^-}(\prod_{\lambda \in L} \mathcal{C}_X(X_{\lambda,f(\lambda)})) .$$

**Proposition 835** Using notation of proposition 832 (p. 478), if $\mathbf{L} \neq \emptyset$ and $(\forall \lambda \in \mathbf{L})(\mathbf{J}_\lambda \neq \emptyset)$, we have:

$$(\forall \lambda \in L)(\forall \iota \in J_\lambda)(\forall \kappa \in J_\lambda)(\iota \neq \kappa \Rightarrow X_{\lambda,\iota} \cap X_{\lambda,\kappa} = \emptyset)$$
$$\Rightarrow (\forall f \in I)(\forall g \in I)(f \neq g \Rightarrow \prod_{\lambda \in L} X_{\lambda,f(\lambda)} \cap \prod_{\lambda \in L} X_{\lambda,g(\lambda)} = \emptyset) .$$

**Proof**

It is a direct consequence of propositions 829 (p. 476) and 827 (p. 475).

**Proposition 836** We have two sets of indices $\mathbf{I}$, $\mathbf{J}$, a set of sets $\mathbf{X}$ and two maps $\mathbf{I}$ to $\mathbf{X}$, $(\mathbf{X_i} : \mathbf{I} \to \mathbf{X})$ and $\mathbf{J}$ to $\mathbf{X}$, $(\mathbf{X_j} : \mathbf{J} \to \mathbf{X})$ with $\mathbf{I} \cap \mathbf{J} = \emptyset$. We have:

$$(\prod_{i \in I} X_i) \times (\prod_{j \in J} X_j) \sim \prod_{i \in I \cup J} X_i . \tag{A.39}$$

See definition 371 (p. 472).

**Proof**
Let us have $\mathbf{u} \times \mathbf{v} \in (\prod_{i\in I} X_i) \times (\prod_{j\in J} X_j)$. From definition 357 (p. 460), we have $\mathbf{u} \subset \mathbf{I} \times \mathbf{X}, \mathbf{v} \subset \mathbf{J} \times \mathbf{X}$. Then from propositions 782 (p. 459), 763 (p. 453), we have:

$$\mathbf{u} \cup \mathbf{v} \subset (\mathbf{I} \times \mathbf{X}) \cup (\mathbf{J} \times \mathbf{X}) = (\mathbf{I} \cup \mathbf{J}) \times \mathbf{X} .$$

Since $\mathbf{I} \cap \mathbf{J} = \emptyset$, from propositions 763 (p. 453), 781 (p. 459) and 784 (p. 459), we have:

$$\mathbf{u} \cap \mathbf{v} \subset (\mathbf{I} \times \mathbf{X}) \cap (\mathbf{J} \times \mathbf{X}) = (\mathbf{I} \cap \mathbf{J}) \times \mathbf{X} \cap \mathbf{X} = (\mathbf{I} \cap \mathbf{J}) \times \mathbf{X} = \emptyset .$$

Then we have:

$$\mathbf{x} \in \mathbf{I} \Rightarrow (\exists \mathbf{y} \in \mathbf{X_x})((\mathbf{x}, \mathbf{y}) \in \mathbf{u} \subset \mathbf{u} \cup \mathbf{v}),$$
$$\mathbf{x} \in \mathbf{J} \Rightarrow (\exists \mathbf{y} \in \mathbf{X_x})((\mathbf{x}, \mathbf{y}) \in \mathbf{v} \subset \mathbf{u} \cup \mathbf{v}) .$$

Then:

$$(\forall \mathbf{x} \in \mathbf{I} \cup \mathbf{J})(\exists \mathbf{y} \in \mathbf{X})(\mathbf{y} \in \mathbf{X_x}, (\mathbf{x}, \mathbf{y}) \in \mathbf{u} \cup \mathbf{v})$$
$$((\mathbf{x}, \mathbf{y}) \in \mathbf{u} \cup \mathbf{v}, (\mathbf{x}, \mathbf{y}') \in \mathbf{u} \cup \mathbf{v}) \Rightarrow (((\mathbf{x} \in \mathbf{I}, \mathbf{x} \notin \mathbf{J}) \vee (\mathbf{x} \notin \mathbf{I}, \mathbf{x} \in \mathbf{J})),$$
$$\mathbf{y} \in \mathbf{X_x}, \mathbf{y}' \in \mathbf{X_x}) \Rightarrow (((\mathbf{x}, \mathbf{y}) \in \mathbf{u}, (\mathbf{x}, \mathbf{y}') \in \mathbf{u}) \vee ((\mathbf{x}, \mathbf{y}) \in \mathbf{v}, (\mathbf{x}, \mathbf{y}') \in \mathbf{v})) \Rightarrow \mathbf{y} = \mathbf{y}' .$$

The set $\mathbf{u} \cup \mathbf{v}$ is a map of $\prod_{i\in I\cup J} X_i$. We call $\mathbf{F} \subset (\prod_{i\in I} X_i) \times (\prod_{j\in J} X_j) \times (\prod_{i\in I\cup J} X_i)$, the map $(\mathbf{F} : \mathbf{I} \cup \mathbf{J} \to \mathbf{X}) \in \prod_{i\in I\cup J}$ defined by:

$$\mathbf{a} \in \mathbf{F} \Leftrightarrow (\mathbf{a} = ((\mathbf{u}, \mathbf{v}), \mathbf{u} \cup \mathbf{v}), ((\mathbf{u}, \mathbf{v}) \in \prod_{i\in I} X_i \times \prod_{j\in J} X_j)) .$$

From the above, $\mathbf{F}$ fills all requirement for a map $\prod_{i\in I} X_i \times \prod_{j\in J} X_j \to (\prod_{i\in I\cup J} X_i)$: see definitions 112 (p. 131) and 357 (p. 460).
**It is a bijection** (definitions 124 (p. 139), 361 (p. 461)):
Any term of $(\mathbf{t} : \mathbf{I} \cup \mathbf{J} \to \mathbf{X}) \in \prod_{i\in I\cup J} X_i$ has only one element in $\mathbf{F}^{-1}(\mathbf{t})$:

$$((\mathbf{u}, \mathbf{v}) \in \mathbf{F}^{-1}(\mathbf{t}), (\mathbf{u}', \mathbf{v}') \in \mathbf{F}^{-1}(\mathbf{t})) \Rightarrow \mathbf{u} \cup \mathbf{v} = \mathbf{u}' \cup \mathbf{v}' = \mathbf{t}$$
$$\Rightarrow (\mathbf{u} = \mathbf{t} \cap \mathbf{I} \times \mathbf{X} \in \prod_{i\in I} X_i = \mathbf{u}', \mathbf{v} = \mathbf{t} \cap \mathbf{J} \times \mathbf{X} \in \prod_{i\in J} X_i = \mathbf{v}') \Rightarrow (\mathbf{u} = \mathbf{u}, \mathbf{v} = \mathbf{v}')) .$$

Then we have equation (A.39) (p. 480).

**Proposition 837** We have two sets of indices $\mathbf{L}, \mathbf{J}$, a set of sets $\mathbf{X}$, map $\mathbf{X_j} : \mathbf{J} \to \mathbf{X}$ a map $(\boldsymbol{\lambda}_j : \mathbf{J} \to \mathbf{L})$ defining a partition (definition 370 (p. 471)) of $\mathbf{J}$ with: $(\mathbf{J}_\mu = \boldsymbol{\lambda}_\mu^{-1} : \mathbf{L} \to \mathcal{P}(\mathbf{J}))$ which means $\mathbf{j} \in \mathbf{J}_{\boldsymbol{\lambda}_j}$ and $\boldsymbol{\lambda}_{\mathbf{J}_\mu} = \boldsymbol{\mu}$ we have:

$$\prod_{\lambda\in L} \prod_{i\in J_\lambda} X_i \sim \prod_{i\in J} X_i . \tag{A.40}$$

See definition 371 (p. 472).

**Proof**
From definition 372 (p. 473), we have:

$$\mathbf{x} \in \prod_{\lambda\in L} \prod_{i\in J_\lambda} X_i \Leftrightarrow (\mathbf{x} = (\mathbf{x}_\mu : \mathbf{L} \to \bigcup_{\lambda\in L} \prod_{i\in J_\lambda} X_i), (\forall \mu \in \mathbf{L})(\mathbf{x}_\mu \in \prod_{i\in J_\mu} X_i)) .$$

We have from definition 372 (p. 473), proposition 783 (p. 459):

$$(\forall \mu \in L)(x_\mu \subset J_\mu \times \bigcup_{i\in J_\mu} X_i \subset J_\mu \times \bigcup_{i\in J} X_i \subset J \times \bigcup_{i\in J} X_i)$$

let us consider the following map (definition 112 (p. 131) and 357 (p. 460)):

$$(\forall x \in \prod_{\lambda\in L}\prod_{i\in J_\lambda} X_i)(y(x) = \bigcup_{\mu\in L} x_\mu : \prod_{\lambda\in L}\prod_{i\in J_\lambda} X_i \to J \times \bigcup_{i\in J} X_i)$$

$\bigcup_{\mu\in L} x_\mu$ belongs to $\prod_{i\in J} X_i$ because it satisfies (see definition 372 (p. 473)):

$$w \in \bigcup_{\mu\in L} x_\mu \Rightarrow (\exists \mu \in L)(w = (u, v) \in x_\mu, u \in J_\mu \subset \bigcup_{\lambda\in L} J_\lambda = J, v \in X_\mu)$$

$y(x)$ is a bijection on $\prod_{i\in J} X_i \subset J \times \bigcup_{i\in J} X_i$ because we have:

$$z \in \prod_{i\in J} X_i \Rightarrow (\forall \mu \in L)((\forall u \in J_\mu)(v = z(u) \in X_u \Leftrightarrow (u, v) \in z))$$
$$\Rightarrow (\exists (x_\mu(j) = z(j) : J_\mu \to X), (\forall j \in J_\mu)(x_\mu(j) \in X_j) .$$

Then we have:

$$(\forall z \in \prod_{i\in J} X_i)(\exists x \in \prod_{\lambda\in L}\prod_{i\in J_\lambda} X_i)(z = y(x))$$

and $z = y(x) = y(x') \Rightarrow x = x'$ because we have

$$x = \prod_{\mu\in L} x_\mu, (\forall \mu \in L)(x_\mu = \prod_{i\in J_\mu} z(i))$$

and

$$x' = \prod_{\mu\in L} x'_\mu, (\forall \mu \in L)(x'_\mu = \prod_{i\in J_\mu} z(i)) .$$

Then we have:

$$(\forall \mu \in L)(x_\mu = x'_\mu), x = \prod_{\mu\in L} x_\mu = \prod_{\mu\in L} x'_\mu = x' .$$

### A.5.9 Order on sets

**Definition 374 Ordered sets.**

A set $E$ is said to be ordered by $R(x, y)$, if $R$ is defined as:

$$R(x, y) \Leftrightarrow (R'(x, y), (x, y) \in E \times E)$$

where $R'(x, y)$ is an ordering relation (definition 338 (p. 446)). It means that $R(x, y) \Rightarrow (x \in E, y \in E)$.

**Convention 32** $\leq$, $\geq$, $<$ $>$.

We extend convention 28 (p. 446), to order on set by setting:

$$x \leq y = R(x, y) = (R'(x, y), (x, y) \in E \times E)$$

and the same way for $\geq$, $<$, $>$.

**Proposition 838** Given a set $\mathbf{E}$, the inclusion relation between two subsets of $\mathbf{E}$ defines an order on $\mathcal{P}(\mathbf{E})$.

**Proof**

If $\mathbf{x}, \mathbf{y}$ are two terms, the relation $\mathbf{R}(\mathbf{x},\mathbf{y}) = \mathbf{x} \subset \mathbf{y}, \mathbf{y} \subset \mathbf{E}$ is an ordering relation on $\mathcal{P}(\mathbf{E})$. Because first (definition 350 (p. 453) proposition 753 (p. 451)) we have:

$$\mathbf{R}(\mathbf{x},\mathbf{y}) \Rightarrow (\mathbf{x} \subset \mathbf{y},\ \mathbf{y} \subset \mathbf{E}) \Rightarrow (\mathbf{x} \subset \mathbf{E}, \mathbf{y} \subset \mathbf{E}) \Rightarrow (\mathbf{x} \in \mathcal{P}(\mathbf{E}),\ \mathbf{y} \in \mathcal{P}(\mathbf{E}))$$

and second we also have:

1. $(\mathbf{x} \subset \mathbf{y}, \mathbf{y} \subset \mathbf{E}) \Rightarrow ((\mathbf{x} \subset \mathbf{x}, \mathbf{x} \subset \mathbf{E}), (\mathbf{y} \subset \mathbf{y}, \mathbf{y} \subset \mathbf{E}))$
2. $(\mathbf{x} \subset \mathbf{y}, \mathbf{y} \subset \mathbf{E}, \mathbf{y} \subset \mathbf{x}, \mathbf{x} \subset \mathbf{E}) \Rightarrow \mathbf{x} = \mathbf{y}$
3. $(\mathbf{x} \subset \mathbf{y}, \mathbf{y} \subset \mathbf{E}, \mathbf{y} \subset \mathbf{z}, \mathbf{z} \subset \mathbf{E}) \Rightarrow (\mathbf{x} \subset \mathbf{z}, \mathbf{z} \subset \mathbf{E})$.

**Definition 375 Induced Order on a subset.**

$\mathbf{E}$ being an ordered set by $\mathbf{R}(\mathbf{x},\mathbf{y})$, the induced order of $\mathbf{R_A}(\mathbf{x},\mathbf{y})$ on a subset $\mathbf{A} \subset \mathbf{E}$ is defined as:

$$\mathbf{R_A}(\mathbf{x},\mathbf{y}) \Leftrightarrow (\mathbf{R}(\mathbf{x},\mathbf{y}), ((\mathbf{x},\mathbf{y}) \in \mathbf{A} \times \mathbf{A}))\ .$$

**Definition 376 Graph of an order on a set.**

$\mathbf{E}$ being an ordered set, the graph (definition 348 (p. 449)) $\mathbf{\Gamma} \subset \mathbf{E} \times \mathbf{E}$ of the order $\mathbf{R}(\mathbf{x},\mathbf{y})$ is:

$$\mathbf{\Gamma} = \mathbf{set}_{(\mathbf{x},\mathbf{y})}\{(\mathbf{R}(\mathbf{x},\mathbf{y}))\}$$

From proposition 749 (p. 450), this set exists because $\mathbf{R}(\mathbf{x},\mathbf{y}) \Rightarrow ((\mathbf{x},\mathbf{y}) \in \mathbf{E} \times \mathbf{E})$.

**Definition 377 Total order on a set.**

The order $\mathbf{R}(\mathbf{x},\mathbf{y})$ on a set $\mathbf{E}$ is total if:

$$(\forall \mathbf{x} \in \mathbf{E})(\forall \mathbf{y} \in \mathbf{E})(\mathbf{R}(\mathbf{x},\mathbf{y}) \vee \mathbf{R}(\mathbf{y},\mathbf{x}))\ .$$

**Proposition 839** Given a set of maps $\mathbf{M}$ (definitions 112 (p. 131), 357 (p. 460), from a subset of $\mathbf{E}$ to a subset of $\mathbf{F}$, if this set $\mathbf{M}$ is fully ordered by inclusion, then the union $\mathbf{\Gamma}$ of all these maps (graphs)(definition 357 (p. 460)) is a map from a subset of $\mathbf{E}$ to a subset of $\mathbf{F}$.

**Proof**

Calling $\mathbf{M}$ the set of maps (graphs) from a subset of $\mathbf{E}$ to a subset of $\mathbf{F}$, setting $\mathbf{\Gamma} = \bigcup_{\mathbf{f} \in \mathbf{M}} \mathbf{f}$

1. $\mathbf{a} \in \mathbf{\Gamma} \Rightarrow (\exists \mathbf{f} \in \mathbf{M})(\mathbf{a} \in \mathbf{f})$
2. Since $\mathbf{M}$ is fully ordered by inclusion, we have:

   $$(\forall \mathbf{f} \in \mathbf{M})(\forall \mathbf{g} \in \mathbf{M})(\mathbf{f} \subset \mathbf{g} \vee \mathbf{g} \subset \mathbf{f})\ .$$

   Then from proposition 758 (p. 452)

   $$(\forall \mathbf{f} \in \mathbf{M})(\forall \mathbf{g} \in \mathbf{M})(\exists \mathbf{h} \in \{\mathbf{f},\mathbf{g}\}), (\mathbf{f} \subset \mathbf{h}, \mathbf{g} \subset \mathbf{h}))\ .$$

   It gives:

   $$\begin{aligned}&(\mathbf{x},\mathbf{y}) \in \mathbf{\Gamma}, (\mathbf{x},\mathbf{y}') \in \mathbf{\Gamma}\\ &\Rightarrow (\exists \mathbf{f} \in \mathbf{M})(\exists \mathbf{g} \in \mathbf{M})\\ &((\mathbf{x},\mathbf{y}) \in \mathbf{f}, (\mathbf{x},\mathbf{y}') \in \mathbf{g} \Rightarrow (\exists \mathbf{h} \in \{\mathbf{f},\mathbf{g}\})(\mathbf{f} \subset \mathbf{h}, \mathbf{g} \subset \mathbf{h}))\\ &\Rightarrow (\exists \mathbf{h} \in \mathbf{M})((\mathbf{x},\mathbf{y}) \in \mathbf{h}, (\mathbf{x},\mathbf{y}') \in \mathbf{h} \Rightarrow \mathbf{y} = \mathbf{y}')\\ &\Rightarrow \mathbf{y} = \mathbf{y}'\ .\end{aligned}$$

Then from definitions 112 (p. 131), 357 (p. 460), $\mathbf{\Gamma}$ is a map.

**Definition 378 Intervals.**

**E** is an ordered set with the order quoted as $\leq$. With **a** and **b** two elements of **E**, one calls (see also definitions 340 (p. 446), 155 (p. 181), 156 (p. 181)):

- closed interval $[a, b]$
$$[a, b] = set_x\{a \leq x \leq b\}$$
- open interval $]a, b[$
$$]a, b[= set_x\{a < x < b\}$$
- semi-open interval $]a, b], [a, b[$
$$]a, b] = set_x\{a < x \leq b\}$$
$$[a, b[= set_x\{a \leq x < b\}$$
- $] \leftarrow, a], ] \leftarrow, a[, [a, \rightarrow [, ]a, \rightarrow [$
$$] \leftarrow, a] = set_x\{x \leq a\}$$
$$] \leftarrow, a[= set_x\{x < a\}$$
$$[a, \rightarrow [= set_x\{a \leq x\}$$
$$]a, \rightarrow [= set_x\{a < x\}$$

**Proposition 840** In a total ordered set (definition 377 (p. 483)), the intersection of two intervals is an interval. The union of two overlapping intervals is an interval.

**Proof**
It is straightforward but long by examining all the possible cases.

**Proposition 841 E** is an ordered set with the order quoted as $\leq$. We have:

$$(a \in E, b \in E) \Rightarrow ((b < a \Rightarrow [a, b] = \emptyset), (b \leq a \Rightarrow [a, b[=]a, b] =]a, b[= \emptyset)) .$$

**Proof**
It is a direct consequence of definitions 378 (p. 484), 338 (p. 446), 3.

**Definition 379 Smallest (resp. biggest) element of a set.**

An element **a** of an ordered set **E** $(a \in E)$ is said to be the smallest (resp. biggest) if

$$(\forall x \in E)(a \leq x) \; (resp(x \leq a)) .$$

**Proposition 842** The smallest (resp. biggest) element of a set is unique.

**Proof**
**E** is an ordered set, we have:

$$(a \in E, b \in E, (\forall x \in E)(a \leq x, b \leq x)) \Rightarrow (a \leq b, b \leq a) \Rightarrow a = b .$$

**Definition 380 Minoring (resp. Majoring) element.**

A subset **A** of an ordered set is said to have a minoring (resp. majoring) element $(\mathbf{a} \in \mathbf{E})$ if:

$$(\forall \mathbf{x} \in \mathbf{A})(\mathbf{a} \leq \mathbf{x})\ (\mathbf{resp}(\mathbf{x} \leq \mathbf{a}))\ .$$

**Proposition 843** If a minoring (resp. majoring) element of a subset **A** of an ordered set is an element of that subset, then it is the smallest (biggest) element of the subset (definition 379 (p. 484)).

**Proof**
Calling **a** the minoring (resp. majoring) element of a subset **A**, since $\mathbf{a} \in \mathbf{A}$, one can replace **E** by **A** in definition 379 (p. 484) which make it identical to definition 380 (p. 484).

**Definition 381 upper_limit (resp. lower_limit) of a subset.**

The **upper_limit** (resp. **lower_limit**) of a subset **A** of an ordered set is the smallest (resp. biggest) majoring (resp. minoring) element of **A**.

**Proposition 844 upper_limit** (resp. **lower_limit**) of a subset **C** of an ordered set is, if it exists, unique.

**Proof**
It comes from proposition 842 (p. 484).

**Definition 382 max or min of a subset.**

If **upper_limit** (resp. **lower_limit**) of a subset belongs to that subset. It is usually called **max** (resp. **min**). For instance **A** being a subset of an ordered set **E**:

$$\mathbf{upper\_limit}(\mathbf{A}) \in \mathbf{A} \Rightarrow \mathbf{max}(\mathbf{A}) = \max_{\mathbf{A}}(\mathbf{E}) = \mathbf{upper\_limit}(\mathbf{A})\ .$$

If we have a map $(\mathbf{x}(\mathbf{i}) = \mathbf{x_i} : \mathbf{I} \to \mathbf{E})$ from the set **I** to the set **E** defining $\mathbf{A} = \mathbf{x}(\mathbf{I})$ and if $\mathbf{upper\_limit}(\mathbf{A}) \in \mathbf{A}$, we may also write:

$$\mathbf{upper\_limit}(\mathbf{A}) = \max_{\mathbf{i} \in \mathbf{I}}(\mathbf{x_i})\ .$$

It is used very often when **I** is a set of integers.

**Proposition 845** The **max** or **min** (definition 382 (p. 485)) of a subset is also its biggest or smallest element (definition 379 (p. 484)) of that subset and vice versa.

**Proof**
The proposition direct is a consequence of proposition 843 (p. 485).
If an element **a** of an ordered subset **A** is the biggest (resp. smallest) element of **A**, it is smaller (resp. bigger) than or equal to any majoring (resp. minoring) element of **A** (definition 380 (p. 484)). Since it is also a majoring (resp. minoring) element of **A**, from definition 381 (p. 485), it is **upper_limit** (resp. **lower_limit**).

**Definition 383 maximal (resp. minimal) of an ordered set.**

An element **a** of an ordered set **E** is said to be maximal (resp. minimal), if we have the following relation true

$$(\forall \mathbf{x} \in \mathbf{E})(\mathbf{a} \leq (\mathbf{resp.} \geq)\mathbf{x} \Rightarrow \mathbf{x} = \mathbf{a})\ .$$

If the set is not fully ordered, there may exist several such elements.

**Definition 384 Increasing map.**

An increasing map from an ordered set $\mathbf{A}$ to another ordered set $\mathbf{B}$: $(\mathbf{f} : \mathbf{A} \rightarrow \mathbf{B})$ is a map which keep the order:

$$(\forall \mathbf{x} \in \mathbf{A})(\forall \mathbf{y} \in \mathbf{A})(\mathbf{x} \leq \mathbf{y} \Rightarrow \mathbf{f}(\mathbf{x}) \leq \mathbf{f}(\mathbf{y})) .$$

**Definition 385 Decreasing map.**

A decreasing map from an ordered set $\mathbf{A}$ to another ordered set $\mathbf{B}$: $(\mathbf{f} : \mathbf{A} \rightarrow \mathbf{B})$ if it is an increasing map with the opposite order is taken on $\mathbf{B}$ (definition 341 (p. 447), proposition 743 (p. 447)).

**Definition 386 Strictly increasing (resp. decreasing) map.**

With the same notations as the above definition 384 $(\mathbf{f} : \mathbf{A} \rightarrow \mathbf{B})$ is said to be a strictly increasing (resp. decreasing) map if:

$$(\forall \mathbf{x} \in \mathbf{A})(\forall \mathbf{y} \in \mathbf{A})(\mathbf{x} < \mathbf{y} \Rightarrow \mathbf{f}(\mathbf{x}) < \mathbf{f}(\mathbf{y})(\mathbf{resp.} \Rightarrow \mathbf{f}(\mathbf{x}) > \mathbf{f}(\mathbf{y}))) .$$

**Proposition 846** A strictly increasing map between two totally ordered sets is an injection (definition 359 (p. 461)).

**Proof**
$(\mathbf{f} : \mathbf{A} \rightarrow \mathbf{B})$ being a strictly increasing map between two totally ordered sets $\mathbf{A}$ and $\mathbf{B}$ we have from the above definitions:

$$\begin{aligned}
&\mathbf{x} \neq \mathbf{x}' \Leftrightarrow (\mathbf{x} > \mathbf{x}' \vee \mathbf{x}' > \mathbf{x}) \\
&\mathbf{x} > \mathbf{x}' \Rightarrow \mathbf{f}(\mathbf{x}) > \mathbf{f}(\mathbf{x}'), \mathbf{x}' > \mathbf{x} \Rightarrow \mathbf{f}(\mathbf{x}') > \mathbf{f}(\mathbf{x}) \\
&(\mathbf{f}(\mathbf{x}) > \mathbf{f}(\mathbf{x}') \vee \mathbf{f}(\mathbf{x}) > \mathbf{f}(\mathbf{x}')) \Rightarrow \mathbf{f}(\mathbf{x}) \neq \mathbf{f}(\mathbf{x}') \\
&(\mathbf{x} \neq \mathbf{x}' \Rightarrow \mathbf{f}(\mathbf{x}) \neq \mathbf{f}(\mathbf{x}')) = (\neg\neg(\mathbf{x} = \mathbf{x}') \vee \neg(\mathbf{f}(\mathbf{x}) = \mathbf{f}(\mathbf{x}'))) \\
&\Leftrightarrow (\mathbf{f}(\mathbf{x}) = \mathbf{f}(\mathbf{x}') \Rightarrow \mathbf{x} = \mathbf{x}') .
\end{aligned}$$

Then $\mathbf{f}(\mathbf{x}) = \mathbf{f}(\mathbf{x}') \Rightarrow \mathbf{x} = \mathbf{x}'$ is true and from definition 359 (p. 461) $\mathbf{f}$ is an injection.

**Proposition 847** An injection (definitions 122 (p. 139), 359 (p. 461)) which is an increasing (resp. decreasing) map between two totally ordered sets (definition 377 (p. 483)) is a strictly increasing (resp. decreasing) map.

**Proof**
From definition 122 (p. 139), $\mathbf{f}$ being an increasing (resp. decreasing) injective map, we have:

$$\mathbf{f}(\mathbf{x}) = \mathbf{f}(\mathbf{x}') \Rightarrow \mathbf{x} = \mathbf{x}' .$$

Then from section A.1.4 (p. 435) point 7, we have:

$$\mathbf{x} \neq \mathbf{x}' \Rightarrow \mathbf{f}(\mathbf{x}) \neq \mathbf{f}(\mathbf{x}') .$$

Then $\mathbf{f}$ is a strictly increasing (resp. decreasing) map (definition 386 (p. 486)).

**Proposition 848** The product of two strictly increasing maps is a strictly increasing map.

**Proof**
It comes directly from definitions 386 (p. 486) and 117 (p. 134) and point 3 (p. 446) of definition 338 (p. 446).

**Proposition 849** The inverse map of a strictly increasing map (definition 386 (p. 486)) between two totally ordered sets (definition 377 (p. 483)) is a strictly increasing map (definition 359 (p. 461)).

**Proof**
Keeping notations of proposition 846 (p. 486), from 846 (p. 486), there are two bijections (proposition 183 (p. 143)): $(\mathbf{f}:\mathbf{A}\to\mathbf{f}(\mathbf{A}))$ and $(\mathbf{f}^{-1}:\mathbf{f}(\mathbf{A})\to\mathbf{A})$. Since it is a bijection, we have $\mathbf{f}^{-1}(\mathbf{x})=\mathbf{f}^{-1}(\mathbf{x}')\Rightarrow\mathbf{x}=\mathbf{x}'$. Moreover:

$$\mathbf{x}<\mathbf{x}'\Rightarrow(\mathbf{f}^{-1}(\mathbf{x})=\mathbf{f}^{-1}(\mathbf{x}')\vee\mathbf{f}^{-1}(\mathbf{x})>\mathbf{f}^{-1}(\mathbf{x}')\vee\mathbf{f}^{-1}(\mathbf{x})<\mathbf{f}^{-1}(\mathbf{x}'))$$

$\mathbf{f}^{-1}(\mathbf{x})=\mathbf{f}^{-1}(\mathbf{x}')$ is impossible because $\mathbf{f}$ is a map and it would imply $\mathbf{x}=\mathbf{x}'$ (see definition 112 (p. 131) equation (2.2)) $\mathbf{f}^{-1}(\mathbf{x})>\mathbf{f}^{-1}(\mathbf{x}')$ is also impossible because $\mathbf{f}$ being a strictly increasing map (definition 386 (p. 486)) it would imply $\mathbf{x}>\mathbf{x}'$ which is also forbidden. It only remains $\mathbf{f}^{-1}(\mathbf{x})<\mathbf{f}^{-1}(\mathbf{x}')$ then $\mathbf{f}^{-1}$ is also a strictly increasing map (definition 386 (p. 486)).

**Definition 387 Isomorphism between two ordered sets.**

A map between two ordered sets $\mathbf{E}$ and $\mathbf{G}$ $(\mathbf{f}:\mathbf{E}\to\mathbf{G})$ is said to be an isomorphism for the orders if $\mathbf{f}$ is an increasing bijective map (definition 361 (p. 461)).

**Proposition 850** The reverse map of an isomorphism between two totally ordered sets is also an isomorphism.

**Proof**
It is a direct consequence of propositions 849 (p. 487) and 847 (p. 486).

**Proposition 851** Given two totally ordered sets $\mathbf{A}$ and $\mathbf{B}$ and an isomorphism $\mathbf{f}$ between $\mathbf{A}$ and $\mathbf{B}$ if $\mathbf{max}(\mathbf{A})$ (resp. $\mathbf{min}(\mathbf{A})$) (definition 382 (p. 485)) exists. Then $\mathbf{max}(\mathbf{B})$ (resp. $\mathbf{min}(\mathbf{B})$) exists and we have:

$$\mathbf{f}(\mathbf{max}(\mathbf{A}))=\mathbf{max}(\mathbf{B}),\mathbf{max}(\mathbf{A})=\mathbf{f}^{-1}(\mathbf{max}(\mathbf{B}))$$

resp.

$$\mathbf{f}(\mathbf{min}(\mathbf{A}))=\mathbf{min}(\mathbf{B}),\mathbf{min}(\mathbf{A})=\mathbf{f}^{-1}(\mathbf{min}(\mathbf{B}))\ .$$

**Proof**
From definitions 382 (p. 485), 379 (p. 484) and 845 (p. 485) and proposition 850 (p. 487), calling $\mathbf{a}=\mathbf{max}(\mathbf{A})$, we have:

$$(\forall\mathbf{x}\in\mathbf{A})(\mathbf{x}\leq\mathbf{a})$$
$$\mathbf{y}\in\mathbf{B}\Rightarrow\mathbf{f}^{-1}(\mathbf{y})\in\mathbf{A}\Rightarrow\mathbf{f}^{-1}(\mathbf{y})\leq\mathbf{a}\Rightarrow\mathbf{f}\circ\mathbf{f}^{-1}(\mathbf{y})\leq\mathbf{f}(\mathbf{a})\Rightarrow\mathbf{y}\leq\mathbf{f}(\mathbf{a})$$

then $\mathbf{f}(\mathbf{a})$ is the biggest element of $\mathbf{B}$ which is unique from proposition 842 (p. 484). We proceed the same way for **min** and we exchange $\mathbf{f}$ and $\mathbf{f}^{-1}$ using proposition 850 (p. 487) to demonstrate $\mathbf{max}(\mathbf{A})=\mathbf{f}^{-1}(\mathbf{max}(\mathbf{A}))$.

**Definition 388 Good order.**

A set $\mathbf{E}$ is said to have a good order, if it is an ordered set and if any subset not empty of $\mathbf{E}$ has a smallest element (definition 379 (p. 484)). $\mathbf{E}$ is also said to be well-ordered.

**Proposition 852** Any subset of a well-ordered set, is well-ordered by the induced order.
**Proof**
**A** being a subset of a well-ordered set **E**. Any subset **B** of **A** is a subset of **E**, **B** has a smallest element $\mathbf{b} \in \mathbf{B}$ for the order in **E**. Since the order in **A** is the same as in **E**, **b** is also the smallest element of **B** for the order of **E** restricted to **A**. **A** is then well-ordered.

**Proposition 853** A good order is a total order.
**Proof**
**E** being well-ordered, any subset of **E** of two elements $\mathbf{x}, \mathbf{y}$ has a smallest element. So

$$(\mathbf{x} \in \mathbf{E}), (\mathbf{y} \in \mathbf{E}) \Rightarrow (\mathbf{x} \leq \mathbf{y}) \vee (\mathbf{y} \leq \mathbf{x}) .$$

**Definition 389 Segments.**

**E** being an ordered set with the order quoted as $\mathbf{x} \leq \mathbf{y}$; the subset $\mathbf{S} \subset \mathbf{E}$ is a segment if:

$$(\forall \mathbf{x} \in \mathbf{S})(\forall \mathbf{y} \in \mathbf{E})(\mathbf{y} \leq \mathbf{x} \Rightarrow \mathbf{y} \in \mathbf{S}) .$$

**Proposition 854** **f** being an isomorphism (definition 387 (p. 487)) between two totally ordered sets (definition 377 (p. 483)) **E** and **F**, any segment (definition 389 (p. 488)) of **E** maps to a segment of **F** and vice versa for $\mathbf{f}^{-1}$.
**Proof**
**S** being a segment of **E**, from proposition 850 (p. 487) we have:

$$(\forall \mathbf{x} \in \mathbf{f}(\mathbf{S}))(\mathbf{y} \leq \mathbf{x} \Rightarrow \mathbf{f}^{-1}(\mathbf{y}) \leq \mathbf{f}^{-1}(\mathbf{x}) \in \mathbf{S} \Rightarrow \mathbf{f}^{-1}(\mathbf{y}) \in \mathbf{S} \Rightarrow \mathbf{y} \in \mathbf{f}(\mathbf{S})) .$$

The last term of the proposition comes also from proposition 850 (p. 487).

**Proposition 855** **E** being an ordered set, $\mathbf{a} \in \mathbf{E}$ an element of **E**, the intervals $]\leftarrow, \mathbf{a}[$ and $]\leftarrow, \mathbf{a}]$ (see 378 (p. 484)) are segments of E.
**Proof**
From definition 378 (p. 484),

$$\begin{aligned} &\mathbf{x} \in ]\leftarrow, \mathbf{a}[ \Leftrightarrow \mathbf{x} < \mathbf{a} \\ &\mathbf{y} \leq \mathbf{x}, \mathbf{x} \in ]\leftarrow, \mathbf{a}[ \Rightarrow \mathbf{y} \leq \mathbf{x} < \mathbf{a} \Rightarrow \mathbf{y} \in ]\leftarrow, \mathbf{a}[ . \end{aligned}$$

Same argument for $]\leftarrow, \mathbf{a}]$.

**Proposition 856** The union and the intersection of any set of segments are segments.
**Proof**
Having an ordered set **E**, a set $\mathbf{A} \subset \mathcal{P}(\mathbf{E})$ of segments of **E** (definition 389 (p. 488)), we have:

$$\begin{aligned} &(\mathbf{x} \in \bigcup_{\mathbf{S} \in \mathbf{A}} \mathbf{S}, \mathbf{y} \leq \mathbf{x}) \Rightarrow ((\exists \mathbf{S})(\mathbf{S} \in \mathbf{A}, \mathbf{x} \in \mathbf{S}), \mathbf{y} \leq \mathbf{x}) \\ &\Rightarrow (\mathbf{x} \in \tau_{\mathbf{S}}(\mathbf{S} \in \mathbf{A}, \mathbf{x} \in \mathbf{S}) \in \mathbf{A}, \mathbf{y} \leq \mathbf{x}) \Rightarrow \mathbf{y} \in \tau_{\mathbf{S}}(\mathbf{S} \in \mathbf{A}, \mathbf{x} \in \mathbf{S}) \in \mathbf{A} \Rightarrow \mathbf{y} \in \bigcup_{\mathbf{S} \in \mathbf{A}} \mathbf{S} \end{aligned}$$

$$\begin{aligned} &(\mathbf{x} \in \bigcap_{\mathbf{S} \in \mathbf{A}} \mathbf{S}, \mathbf{y} \leq \mathbf{x}) \Rightarrow ((\forall \mathbf{S})(\mathbf{S} \in \mathbf{A} \Rightarrow \mathbf{x} \in \mathbf{S}), \mathbf{y} \leq \mathbf{x}) \\ &\Rightarrow (\forall \mathbf{S})(\mathbf{S} \in \mathbf{A} \Rightarrow \mathbf{y} \in \mathbf{S}) \Rightarrow \mathbf{y} \in \bigcap_{\mathbf{S} \in \mathbf{A}} \mathbf{S} . \end{aligned}$$

Then from definition 389 (p. 488) both $\bigcap_{\mathbf{S} \in \mathbf{A}} \mathbf{S}$ and $\bigcup_{\mathbf{S} \in \mathbf{A}} \mathbf{S}$ are segments of **E**.

**Proposition 857** $\mathbf{S}$ being a segment of an ordered set $\mathbf{E}$, any segment of $\mathbf{S}$ is a segment of $\mathbf{E}$.

**Proof**
$\mathbf{S}$ is ordered with the induced order of $\mathbf{E}$ (proposition 852 (p. 488)). If $\mathbf{S'}$ is a segment of $\mathbf{S}$ segment of $\mathbf{E}$, we have:

$$\begin{aligned}(x \in S') &\Rightarrow (x \in S, x \in S') \\ &\Rightarrow (\forall y \in E)(y \leq x \Rightarrow (y \in S, y \leq x, x \in S') \Rightarrow y \in S')\end{aligned}$$

so $\mathbf{S'}$ is a segment of $\mathbf{E}$.

**Proposition 858** $\mathbf{S}$ and $\mathbf{S'}$ being segments of a totally ordered set $\mathbf{E}$ (definition 377 (p. 483)), we have:

$$S \subset S' \vee S' \subset S\ .$$

**Proof**
$S = S' \Rightarrow S \subset S'$, let us consider the other alternative $S \neq S'$. is In that case, there is an element $\mathbf{a}$ in one of the two sets and not in the other (proposition 757 (p. 452)). Let us suppose that it is $\mathbf{S}$:

$$(\exists a)(a \in S, a \notin S')\ .$$

Because $\mathbf{E}$ is totally ordered, we have:

$$b \in S' \Rightarrow b < a \tag{A.41}$$

otherwise $\mathbf{a}$ will be in $\mathbf{S'}$. Equation (A.41) and definition 389 give $\mathbf{b}$ is in $\mathbf{S}$ because $\mathbf{S}$ is a segment so:

$$(b \in S' \Rightarrow b \in S) \Rightarrow S' \subset S\ .$$

**Proposition 859** Given a well-ordered set $\mathbf{E}$ (definition 388 (p. 487)), for any segment $\mathbf{S}$ different from $\mathbf{E}$, there is an unique element $\mathbf{a}$ not in $\mathbf{S}$ which is the smallest (definition 379 (p. 484), proposition 842 (p. 484)) of all elements not in $\mathbf{S}$.

**Proof**
$\mathbf{E}$ being well-ordered, $\mathbf{S}$ being a segment (definition 389 (p. 488)) of $\mathbf{E}$ with $\mathcal{C}_E(S) \neq \emptyset$ (definition 353 (p. 454)), since $\mathcal{C}_E(S) \subset E$, there is a smallest element (definition 379 (p. 484)) $\mathbf{a}$ in $\mathcal{C}_E(S)$. Then we have:

$$((\exists a \notin S)(\forall x \in E)(x < a \Rightarrow x \notin \mathcal{C}_E(S) \Leftrightarrow x \in S)\ . \tag{A.42}$$

$\mathbf{S}$ can be written as $]\leftarrow, a[$ or $], a[$ or $S_a$ (see definition 389 (p. 488)).

**Proposition 860** In a well-ordered set $\mathbf{E}$, calling $\boldsymbol{\mathcal{S}}$ the set of segment of $\mathbf{E}$, the map $(f : \mathcal{S} \leftrightarrow E)$, where $f(S)$ is the smallest element (definition 379 (p. 484)) of $\mathcal{C}_E(S)$, is a bijection between $\boldsymbol{\mathcal{S}}$ and $\mathbf{E}$.

**Proof**
It is a map from propositions 859 (p. 489), 842 (p. 484). It is surjective (definition 360 (p. 461)), because from proposition 855 (p. 488), we have $]\leftarrow, a[\in f^{-1}(S))$ (see definition 389 (p. 488)). It is injective (definition 359 (p. 461)) because from equation (A.42), $x < f(S) \Leftrightarrow x \in S$ then $]\leftarrow, f(S)[= S$.

**Proposition 861** Ordered by inclusion ("subset of", see definition 349 (p. 451), proposition 838 (p. 483)), the order of the set of segments of a well-ordered set is a good order.

**Proof**
Given a set $\mathbf{A}$ of segments of a well-ordered set $\mathbf{E}$, if $\mathbf{A}$ contains only one segment, that segment is the smallest element of $\mathbf{A}$. If there are more than one segment, all the segments different from $\mathbf{E}$ are included in $\mathbf{E}$ so smaller than $\mathbf{E}$ which can be considered as a segment of $\mathbf{E}$. From proposition 860 (p. 489), there is a bijection $\mathbf{f}$ between all segments different from $\mathbf{E}$ and $\mathbf{E}$: it is given by the smallest element of $\mathcal{C}(\mathbf{S})$ where $\mathbf{S}$ is a segment of $\mathbf{E}$ different from $\mathbf{E}$. The set $\mathbf{f}(\mathbf{A}-\{\mathbf{E}\})$ has a smallest element $\mathbf{a}$ (definition 388 (p. 487)). The corresponding segment, which belongs to $\mathbf{A}$, is included in any segment of $\mathbf{A}$:

$$\mathrm{x} \in \mathrm{f}^{-1}(\mathrm{a}) \Rightarrow \mathrm{x} < \mathrm{a} \Rightarrow \mathrm{x} \in \mathrm{E}, (\forall \mathrm{S} \in \mathrm{A} - \{\mathrm{E}\})(\mathrm{x} < \mathrm{a} < \mathrm{f}(\mathrm{S})) \Rightarrow (\forall \mathrm{S} \in \mathrm{A})(\mathrm{x} \in \mathrm{S})$$
$$\Rightarrow (\forall \mathrm{S} \in \mathrm{A})(\mathrm{f}^{-1}(\mathrm{a}) \subset \mathrm{S}) .$$

**Proposition 862** $\mathbf{a}$ being the smallest element of a well-ordered set $\mathbf{E}$, the subset $\emptyset$ of $\mathbf{E}$ is a segment of $\mathbf{E}$ and we have $\emptyset = ], \mathrm{a}[ = \mathrm{S}_\mathrm{a}$.

**Proof**
From proposition 775 (p. 457), $\emptyset \subset \mathrm{E}$.
Since $(\exists \mathrm{a})(\forall \mathrm{y} \in \mathrm{E})(\mathrm{a} \leq \mathrm{y})$, from definitions 333 (p. 441), 353 (p. 454) and 355 (p. 456) we have:

$$\mathrm{a} = (\tau_\mathrm{x}(\forall \mathrm{y} \in \mathrm{E})(\mathrm{x} \leq \mathrm{y})) \Rightarrow \mathrm{x} < \mathrm{a} \Rightarrow (\mathrm{x} \in \mathrm{E}, \mathrm{x} \notin \mathrm{E}) \Rightarrow \mathrm{x} \in \mathcal{C}_\mathrm{E}(\mathrm{E}) = \emptyset$$
$$\Rightarrow \mathrm{x} \in \emptyset \Rightarrow \emptyset = ], \mathrm{a}[ = \mathrm{S}_\mathrm{a} .$$

**Proposition 863** $\mathbf{I}$ and $\mathbf{A}$ being two sets, if there is a map $(\mathrm{X}_\mathrm{i} : \mathrm{I} \to \mathcal{P}(\mathrm{A}))$ with the following properties:

1. $(\forall \mathrm{i})$ there is a good order (definition 388 (p. 487)) on $\mathrm{X}_\mathrm{i}$:$\mathrm{R}_\mathrm{i}(\mathrm{x}, \mathrm{y})$
2. $(\forall \mathrm{i}), (\forall \mathrm{j})(\mathrm{X}_\mathrm{i} \subset \mathrm{X}_\mathrm{j} \vee \mathrm{X}_\mathrm{j} \subset \mathrm{X}_\mathrm{i})$
3. $(\forall \mathrm{i}), (\forall \mathrm{j})(\mathrm{X}_\mathrm{i} \subset \mathrm{X}_\mathrm{j} \Rightarrow (\forall (\mathrm{x}, \mathrm{y}) \in \mathrm{X}_\mathrm{i} \times \mathrm{X}_\mathrm{i})(\mathrm{R}_\mathrm{i}(\mathrm{x}, \mathrm{y}) \Leftrightarrow \mathrm{R}_\mathrm{j}(\mathrm{x}, \mathrm{y})))$
4. $(\forall \mathrm{i}), (\forall \mathrm{j})$ $\mathrm{X}_\mathrm{i} \subset \mathrm{X}_\mathrm{j}$ implies that $\mathrm{X}_\mathrm{i}$ is a segment of $\mathrm{X}_\mathrm{j}$ for the order $\mathrm{R}_\mathrm{j}(\mathrm{x}, \mathrm{y})$.

Then there is a unique good order on $\mathrm{X} = \bigcup_{\mathrm{i} \in \mathrm{I}} \mathrm{X}_\mathrm{i}$ which induces $\mathrm{R}_\mathrm{i}(\mathrm{x}, \mathrm{y})$ for all $\mathbf{i}$.

**Proof**
Let us prove first:

$$(\mathrm{x}, \mathrm{y}) \in \mathrm{X} \times \mathrm{X} \Rightarrow (\exists \mathrm{k} \in \mathrm{I})(\mathrm{x} \in \mathrm{X}_\mathrm{k}, \mathrm{y} \in \mathrm{X}_\mathrm{k}) . \tag{A.43}$$

Taking the assumed property 2 (p. 490), we have from proposition 758 (p. 452):

$$\mathrm{X}_\mathrm{i} \subset \mathrm{X}_\mathrm{j} \vee \mathrm{X}_\mathrm{j} \subset \mathrm{X}_\mathrm{i} \Leftrightarrow (\exists (\mathrm{k} \in \{\mathrm{i}, \mathrm{j}\})(\mathrm{X}_\mathrm{i} \subset \mathrm{X}_\mathrm{k}, \mathrm{X}_\mathrm{j} \subset \mathrm{X}_\mathrm{k}) .$$

Then we have:

$$(\forall \mathrm{i} \in \mathrm{I})(\forall \mathrm{j} \in \mathrm{I})((\exists \mathrm{k} \in \{\mathrm{i}, \mathrm{j}\})((\mathrm{X}_\mathrm{i} \subset \mathrm{X}_\mathrm{k}), (\mathrm{X}_\mathrm{j} \subset \mathrm{X}_\mathrm{k}))$$
$$(\mathrm{x}, \mathrm{y}) \in \mathrm{X} \times \mathrm{X} \Rightarrow (\exists \mathrm{i})(\mathrm{x} \in \mathrm{X}_\mathrm{i}), (\exists \mathrm{j})(\mathrm{y} \in \mathrm{X}_\mathrm{j})$$
$$\Rightarrow (\exists \mathrm{i})(\exists \mathrm{j})(\exists \mathrm{k} \in \{\mathrm{i}, \mathrm{j}\})(\mathrm{x} \in \mathrm{X}_\mathrm{k}, \mathrm{y} \in \mathrm{X}_\mathrm{k})$$
$$\Rightarrow (\exists \mathrm{k})(\mathrm{x} \in \mathrm{X}_\mathrm{k}, \mathrm{y} \in \mathrm{X}_\mathrm{k}) .$$

Let us define now the order $\mathbf{R(x,y)}$ on $\mathbf{X} = \bigcup_{i\in I} \mathbf{X_i}$ as:

$$\mathbf{R(x,y) \Leftrightarrow (\exists i \in I)R_i(x,y)} \ . \tag{A.44}$$

Using the assumed property 3 (p. 490) of proposition 863 (p. 490), we also have:

$$\mathbf{(\forall i \in I)((x,y) \in X_i \times X_i \Rightarrow (R_i(x,y) \Rightarrow R(x,y)))} \ . \tag{A.45}$$

Let us verify that the formula A.44 (p. 491) defines an order by checking the definition given in definition 338 (p. 446):

1. Definition 338, 1 (p. 446) $\mathbf{R_i(x,y)}$ verifies definition 338, 1 (p. 446) and definition 374 (p. 482). Then we have:

$$\begin{aligned}\mathbf{R(x,y)} &\Rightarrow \mathbf{(\exists i \in I)R_i(x,y)} \\ &\Rightarrow \mathbf{(\exists i \in I),((x,y) \in X_i \times X_i, R_i(x,x), R_i(y,y))} \\ &\Rightarrow \mathbf{(((x,y) \in X \times X), R(x,x), R(y,y))} \ .\end{aligned}$$

2. Definition 338, 2 (p. 446). Using formula A.44 (p. 491), relation (A.43) (p. 490), the assumed property 3 (p. 490) and finally the relation (A.45) (p. 491), we have (see convention 26 (p. 441)):

$$\begin{aligned}&\mathbf{R(x,y), R(y,x)} \\ &\Rightarrow \mathbf{(\exists i \in I)(\exists j \in I)(R_i(x,y), R_j(y,x))} \\ &\Rightarrow \mathbf{((x,y) \in X_i \times X_i, (y,x) \in X_j \times X_j, (\exists k \in \{i,j\})(X_i \subset X_k, X_j \subset X_k))} \\ &\Rightarrow \mathbf{(\exists k \in \{i,j\})((x,y) \in X_k \times X_k)} \\ &\Rightarrow \mathbf{(\exists k)(R_i(x,y) \Leftrightarrow R_k(x,y), R_j(y,x) \Leftrightarrow R_k(y,x))} \\ &\Rightarrow \mathbf{(\exists k)(R_k(x,y), R_k(y,x)) \Rightarrow x = y} \\ &\Rightarrow \mathbf{x = y} \ .\end{aligned}$$

3. Definition 338, 3 (p. 446). Using again the formulae A.44 (p. 491), A.45 (p. 491) and the relations A.43 (p. 490) with the assumed property 3 (p. 490), we have:

$$\begin{aligned}\mathbf{R(x,y), R(y,z)} &\Rightarrow \mathbf{((\exists i \in I)R_i(x,y), (\exists j \in I)R_j(y,z))} \\ &\Rightarrow \mathbf{(\exists k \in I)(R_k(x,y), R_k(y,z)) \Rightarrow (\exists k \in I)R_k(x,z)} \\ &\Rightarrow \mathbf{R(x,z)}\end{aligned}$$

R is unique because if a $\mathbf{R'}$ is an order on $\mathbf{X}$ inducing the order $\mathbf{R_i}$ for all $\mathbf{i \in I}$ we have

$$\begin{aligned}\mathbf{R'(x,y)} &\Rightarrow \mathbf{(R'(x,y), (\exists i \in I)(x \in X_i), (\exists j \in I)(y \in X_j))} \\ &\Rightarrow \mathbf{(R'(x,y), (\exists k \in I)((x \in X_k), (y \in X_k)))} \\ &\Rightarrow \mathbf{(\exists k \in I)R_k(x,y)} \\ &\Rightarrow \mathbf{R(x,y)} \ .\end{aligned}$$

By exchanging $\mathbf{R'}$ and $\mathbf{R}$ in the above demonstration, we also have $\mathbf{R(x,y) \Rightarrow R'(x,y)}$. Then we have:

$$\mathbf{R'(x,y) \Leftrightarrow R'(x,y)} \ .$$

From the relation (A.43) (p. 490), we deduce that the order $\mathbf{R}$ is total since it is total for all the $\mathbf{R_i}$.

$\mathbf{R}$ is a good order for the following reasons:

We quote $\mathbf{R}$ as $\leq$. Since all $\mathbf{X_i}$ are well-ordered (property 1 (p. 490)), for any subset $\mathbf{A}$ of $\mathbf{X}$, we have:

$$\mathbf{A \subset X \Rightarrow (\exists i \in I)(A \cap X_i \neq \emptyset) \Rightarrow (\exists i \in I)(\exists a \in A \cap X_i)(\forall x \in A \cap X_i)(a \leq x)} \ .$$

Now we have:

$$\begin{aligned}\mathbf{(\forall y \in A)(\exists j \in I)(y \in X_j) \Rightarrow} \quad & \mathbf{((y \in X_i \Rightarrow (a \leq y)),} \\ & \mathbf{(y \notin X_i \Rightarrow X_i \subset X_j \Rightarrow (a \leq y)))} \ .\end{aligned}$$

The last part comes because $\mathbf{X_i}$ is by definition a segment of $\mathbf{X_j}$ when $\mathbf{X_i \subset X_j}$ (property 4 (p. 490))which means $\mathbf{y < a \Rightarrow y \in X_i}$ which is equivalent to $\mathbf{y \notin X_i \Rightarrow a \leq y}$ (A.1.4 (p. 435), 7)
So in any case:

$$\mathbf{(\forall A \subset X)(\exists a \in A)(\forall y \in A)(a \leq y)} \ .$$

So $\mathbf{R}$ is a good order on $\mathbf{X}$.

**Proposition 864** Given a set $\mathbf{E}$ and a map from $\mathbf{A} \subset \mathcal{P}(\mathbf{E})$ to E: $\mathbf{(f : A \subset \mathcal{P}(E) \to E)}$ such that $\mathbf{(\forall x \in A)(f(x) \notin x)}$ then: $\mathbf{(\exists M)(M \in \mathcal{P}(E))}$ with the following properties:

1. There is a good order on $\mathbf{M}$.

2. For any segment $\mathbf{S_a}$ of $\mathbf{M}$ not equal to $\mathbf{M}$, $\mathbf{S_a = ]\leftarrow, a[}$, we have $\mathbf{S_a \in A}$ and $\mathbf{f(S_a) = a}$.

3. $\mathbf{M \notin A}$.

**Proof**
We suppose that we have a map as described in proposition 864. On the set $\mathcal{P}(\mathbf{E}) \times \mathcal{P}(\mathbf{E})$ we consider the subsets $\boldsymbol{\mathcal{U}}$ for which $(\exists \mathbf{U} \subset \mathbf{E})$ with $(\boldsymbol{\mathcal{U}} \subset \mathbf{U} \times \mathbf{U})$ such that the relation $(\mathbf{x}, \mathbf{y}) \in \boldsymbol{\mathcal{U}}$ is a good order on $\mathbf{U}$ quoted as $\leq$:

$$(\mathbf{x}, \mathbf{y}) \in \boldsymbol{\mathcal{U}} \Leftrightarrow \mathbf{x} \leq \mathbf{y} \ .$$

According to definition 376 (p. 483), $\boldsymbol{\mathcal{U}}$ is the graph of the good order on $\mathbf{U}$. Since $\mathbf{U}$ is the projection of $\boldsymbol{\mathcal{U}}$ along one of its components and since $\boldsymbol{\mathcal{U}}$ defines the order, any $\boldsymbol{\mathcal{U}}$ define fully and in a unique way $\mathbf{U}$ and its order. Moreover we impose that any segment $\mathbf{]\leftarrow, x[}$ of $\mathbf{U}$ belongs to $\mathbf{A}$ and is such that

$$\mathbf{(\forall x \in U)(]\leftarrow, x[\in A, x = f(]\leftarrow, x[))} \ . \tag{A.46}$$

(see definition 389 (p. 488), proposition 859 (p. 489)). According to proposition 749 (p. 450), all $\boldsymbol{\mathcal{U}}$, satisfying the above conditions and in particular relation (A.46), form a set which we call $\boldsymbol{\mathcal{M}}$. This set can be empty. We also have a map defined as:

$$(\mathbf{U_i} : \boldsymbol{\mathcal{M}} \to \mathcal{P}(\mathbf{E}))$$

which sets $\mathbf{U_i} \subset \mathbf{E}$, for each graph $\mathbf{i} \in \boldsymbol{\mathcal{M}}$, the subset of $\mathbf{E}$ on which it acts: For $\mathbf{i} \in \boldsymbol{\mathcal{M}}$, we set $\boldsymbol{\mathcal{U}_i} = \mathbf{i}$ and we call $\mathbf{U_i}$ the corresponding $\mathbf{U}$ of relation (A.46) (see definition 112 (p. 131) point 4 (p. 132)). We quote $\mathbf{[a, b[_i, a \in U_i, b \in U_i}$ an interval of $\mathbf{U_i}$ for the good order $\mathbf{i}$ quoted as $\leq_\mathbf{i}$.
We have two cases:

- $\emptyset \notin A$
  If $\emptyset \notin A$, since a subset of $E$ on which there is a good order has a smallest element $a$, from proposition 862 (p. 490), the empty set $\emptyset$ is equal to $]\leftarrow, a[$. The condition A.46 cannot be satisfied for $a$. The $\mathcal{M}$ is empty ($\mathcal{M} = \emptyset$). But since the $\emptyset$ is a well-ordered subset of $E$, not in $A$, it satisfies all the condition for $M$, the trivial response $M = \emptyset$ is a solution for the subset $M$.

- $\emptyset \in A$
  Let us prove that $U_i$ and $U_j$ verify the condition 2 of proposition 863 (p. 490):

$$(i \in \mathcal{M}, j \in \mathcal{M}) \Rightarrow (U_i \subset U_j \vee U_j \subset U_i) \tag{A.47}$$

together with conditions 1 and 3 of proposition 863 (p. 490):

1. $\emptyset$ is a segment for any good order in any set (proposition 862 (p. 490)). So $\emptyset$ is a segment of any $U_i, i \in \mathcal{M}$: $]\leftarrow, u_i[_i$, $u_i$ being the smallest element of $U_i$ with the order $i$. $u_i$ exists because the order $i$ is a good order. Then from relation (A.46):

$$(\forall i \in \mathcal{M})(\exists u_i)(u_i \in U_i, \emptyset = ]\leftarrow, u_i[_i, f(]\leftarrow, u_i[_i) = f(\emptyset) = u_i \in U_i)$$

2. For any couple $i \in \mathcal{M}$, $j \in \mathcal{M}$, we may consider the set $V \subset U_i \cap U_j \subset E$ of elements $x$ in both $U$ such that the associated segment $]\leftarrow x[_i$, $]\leftarrow, x[_j$ in $U_i$ and $U_j$ are equal with the orders coming from $i$ and $j$ identical:

$$\begin{aligned} x \in V \Leftrightarrow (&]\leftarrow, x[_i = ]\leftarrow x[_j,\\ &(\forall y \in ]\leftarrow, x[_i)(\forall z \in ]\leftarrow x[_i)(y \leq_i z \Leftrightarrow y \leq_j z)) \end{aligned}$$

   From the above point 1, $f(\emptyset) \in V$ and $V \neq \emptyset$ from propositions 773 (p. 457), 727 (p. 441).

3. $V$ is a segment in both $U_i$, $U_j$ with their order:

$$(x \in V, y \in U_i, y <_i x) \Rightarrow (]\leftarrow, y[_i \subset ]\leftarrow, x[_i = ]\leftarrow, x[_j)$$

   since the order $i$ and $j$ are the same on $]\leftarrow, x[_j$ and $]\leftarrow, x[_j$ $]\leftarrow, y[_i = ]\leftarrow, y[_j$ with the order $i$ and $j$ the same on that segment. Then $y \in V$ from definition 389 (p. 488) $V$ is a segment in $U_i$. By exchanging $i$ and $j$ in the above $V$ is a segment in both $U_i$ and $U_j$.

4. We have $(V = U_i) \vee (V = U_j)$.
   If $V \neq U_i, V \neq U_j$, it means, since $V$ is a segment in both $U$, and since the order on both $U$ are good:

$$\begin{aligned} &(k \in \{i, j\} \Rightarrow (\exists x_k \in U_k)(V = ]\leftarrow x_k[, f(V) = x_k))\\ &\Rightarrow x_v = x_i = x_j \in U_i \cap U_j\,. \end{aligned}$$

   Then from the definition of $V$, $f(V) \in V$ which is forbidden. Then using A.1.4, 8 $(V = U_i) \vee (V = U_j)$ and equation (A.47) is true.

The family of sets $(U_i : \mathcal{M} \rightarrow \mathcal{P}(E))$ with $\mathcal{M}$ as the set of index verifies the conditions of proposition 863 (p. 490). Setting

$$M = \bigcup_{i \in \mathcal{M}} U_i\,.$$

From proposition 863 (p. 490), $\mathbf{M}$ has a good order. Point 1 (p. 492) is verified. Let us prove now point 2 (p. 492).
The order $\mathbf{g}$ on $\mathbf{M}$ is in $\mathcal{M}$, because since $\mathbf{M}$ has a good order, any segment of $\mathbf{M}$ not equal to $\mathbf{M}$ can be written has $]\leftarrow, \mathbf{x}[, \mathbf{x} \in \mathbf{M}$ (proposition 859 (p. 489)). Proposition 863 (p. 490) also says

$$\mathbf{x} \in \mathbf{M} \Rightarrow (\exists \mathbf{k})(\mathbf{x} \in \mathbf{U_k}) .$$

From what we just said and from proposition 863 (p. 490):

$$\mathbf{y} \leq_{\mathbf{k}} \mathbf{x} \Leftrightarrow \mathbf{y} \leq_{\mathbf{M}} \mathbf{x} .$$

It implies $]\leftarrow, \mathbf{x}[=]\leftarrow, \mathbf{x}[_{\mathbf{k}}$. Then $]\leftarrow, \mathbf{x}[\in \mathbf{A}$ and $\mathbf{f}(]\leftarrow, \mathbf{x}[) = \mathbf{x}$
Any segment $\mathbf{S}$ from $\mathbf{M}$ different of $\mathbf{M}$ is in $\mathbf{A}$ and we have $\mathbf{S} =]\leftarrow, \mathbf{f}(\mathbf{S})[$.
If $\mathbf{M} \in \mathbf{A}$, calling $\mathbf{b} = \mathbf{f}(\mathbf{M}) \notin \mathbf{M}$, we may consider $\mathbf{M}' = \mathbf{M} \cup \{\mathbf{b}\}$ and we may set an order $\mathbf{g}'$ on $\mathbf{M}'$ by:

$$(\mathbf{x} \in \mathbf{M}, \mathbf{y} \in \mathbf{M}) \Rightarrow \mathbf{g}'(\mathbf{x}, \mathbf{y}) = \mathbf{g}(\mathbf{x}, \mathbf{y}),$$
$$(\mathbf{x} = \mathbf{b}, \mathbf{y} \in \mathbf{M}) \Rightarrow \mathbf{y} < \mathbf{x} .$$

It is a good order and it verifies equation (A.46) then $\mathbf{M}' \subset \mathbf{M}$. It is impossible by construction of $\mathbf{M}$ and $\mathbf{M}'$. So using point 8 (p. 435) in A.1.4, we conclude $\mathbf{M} \notin \mathbf{A}$.

**Proposition 865** On every set, there is a good order (Zermelo theorem).

**Proof**
We use the notations of proposition 864 (p. 492). $\mathbf{E}$ being a set, let the set $\mathbf{A} \subset \mathcal{P}(\mathbf{E})$ be $\mathbf{A} = \mathbf{C}_{\mathcal{P}(\mathbf{E})}(\{\mathbf{E}\})$ so $\mathbf{E}$ is the only subset of $\mathbf{E}$ not in $\mathbf{A}$ and we take to satisfy the condition of proposition 864 (p. 492) ($\mathbf{f} : \mathbf{A} \to \mathbf{E}$) the following map (definition 353 (p. 454), proposition 726 (p. 441), equation (A.3)):

$$(\forall \mathbf{x} \in \mathbf{A})(\mathbf{f}(\mathbf{x}) = \tau_{\mathbf{a}}(\mathbf{a} \in \mathbf{E}, \mathbf{a} \notin \mathbf{x}) = \tau_{\mathbf{a}}(\mathbf{a} \in \mathcal{C}_{\mathbf{E}}(\mathbf{x}))) .$$

This map exists because $(\forall \mathbf{x} \in \mathbf{A})(\mathcal{C}(\mathbf{x}) \neq \emptyset)$ (proposition 757 (p. 452) with $\mathbf{x} \subset \mathbf{E}$, $\mathbf{x} \neq \mathbf{E}$). We satisfy the condition of proposition 864 (p. 492). The set $\mathbf{M} \subset \mathbf{E}$ which is claimed to exist, cannot be in $\mathbf{A}$, so it is $\mathbf{E}$. Since there is a good order on $\mathbf{M}$ that order exists on $\mathbf{E}$.
**Remark.** This theorem which is a crucial theorem of the set theory, validates to any set what one does to count and order ordinary set of objects:
The ordinary counting of elements of a set $\mathbf{E}$, can be described as putting $\emptyset$ in $\mathbf{A}$ and choosing at random an element $\mathbf{x_1}$ of the set $\mathbf{E}$ and naming $\mathbf{x_1}$ as the first or the smallest of $\mathbf{E}$. $\mathbf{x_1}$ can be $\mathbf{x_1} = \tau_{\mathbf{x}}(\mathbf{x} \in \mathbf{E})$. Then $\mathbf{x_1}$ is used as $\mathbf{x_1} = \mathbf{f}(\emptyset) \notin \emptyset$. The subset $\{\mathbf{x_1}\}$ is now put in $\mathbf{A}$ and $\mathbf{x_2}$ chosen. $\mathbf{x_2}$ can be $\mathbf{x_2} = \tau_{\mathbf{x}}(\mathbf{x} \in \mathcal{C}_{\mathbf{E}}(\{\mathbf{x_1}\}))$ and $\mathbf{x_2}$ is declared as $\mathbf{x_2} = \mathbf{f}(\{\mathbf{x_1}\})$. The subset $\{\mathbf{x_1}, \mathbf{x_2}\}$ is put in $\mathbf{A}$ and $\mathbf{x_3} = \tau_{\mathbf{x}}(\mathbf{x} \in \mathcal{C}_{\mathbf{E}}(\{\mathbf{x_1}, \mathbf{x_2}\}))$ and so on. The Zermelo theorem says that with the axioms of set theory, the above mechanism of counting makes the existence of a good order on any set $\mathbf{E}$.

**Proposition 866** If in an ordered set $\mathbf{E}$, any fully ordered subset (definition 377 (p. 483)) has a majoring element (definition 380 (p. 484)) in $\mathbf{E}$, then $\mathbf{E}$ has a maximal element (definition 383 (p. 485)).

**Proof**
We settle for that, the conditions of proposition 864 (p. 492). We consider the following map of a subset $\mathbf{A}$ of $\mathcal{P}(\mathbf{E})$ to $\mathbf{E}$ ($\mathbf{f} : \mathbf{A} \subset \mathcal{P}(\mathbf{E}) \to \mathbf{E}$):

- $\mathbf{A}$ is the set of the subsets $\mathbf{b} \subset \mathbf{E}$ which have a majoring element not in $\mathbf{b}$:

$$\mathbf{A} = \mathbf{set_b}\{(\mathbf{b} \in \mathcal{P}(\mathbf{E})), (\exists \mathbf{x} \in \mathcal{C}_\mathbf{E}(\mathbf{b}))(\forall \mathbf{y} \in \mathbf{b})(\mathbf{y} < \mathbf{x})\}$$

- For all $\mathbf{b}$ in $\mathbf{A}$, $\mathbf{f}(\mathbf{b}) \notin \mathbf{b}$ is one of the majoring values of $\mathbf{b}$:

$$\mathbf{f}(\mathbf{b}) = \tau_\mathbf{x}(\mathbf{x} \in \mathcal{C}_\mathbf{E}(\mathbf{b}), (\forall \mathbf{y} \in \mathbf{b})(\mathbf{y} < \mathbf{x}))$$

(see definitions 332 (p. 440), 112 (p. 131), 4)

- $\emptyset \in \mathbf{A}$ because from definition 380 (p. 484), all elements of $\mathbf{E}$ are majoring $\emptyset \subset \mathbf{E}$ and is not in $\emptyset$.

Then from proposition 864 (p. 492) there is a well-ordered set

$$\mathbf{M} \subset \mathbf{E},\ \mathbf{M} \neq \emptyset$$

not in $\mathbf{A}$, such that any segment of $\mathbf{M}$ (definition 389 (p. 488)) $\mathbf{S_x} = ]\leftarrow, \mathbf{x}[$ (see proposition 859 (p. 489)) is in $\mathbf{A}$ with $\mathbf{x} = \mathbf{f}(\mathbf{S_x})$ and $\mathbf{x}$ majoring $\mathbf{S_x}$ from the definition of $\mathbf{A}$.
Let us show that $\mathbf{M}$ has the same order as the one induced by $\mathbf{E}$. From proposition 864 (p. 492), we have:

- $(\mathbf{y} \leq \mathbf{x})_\mathbf{M} \Rightarrow (\mathbf{y} \leq \mathbf{x})_\mathbf{E}$

$$(\mathbf{y} < \mathbf{x})_\mathbf{M} \Rightarrow (\mathbf{x} \in \mathbf{M}), (\mathbf{y} \in \mathbf{M}), (\mathbf{y} \in \mathbf{S_x}), (\mathbf{f}(\mathbf{S_x}) = \mathbf{x}) \Rightarrow (\mathbf{y} < \mathbf{x})_\mathbf{E}$$

Since $\mathbf{x} = \mathbf{y}$ is the same in $\mathbf{E}$ and $\mathbf{M}$, we get:

$$(\mathbf{y} \leq \mathbf{x})_\mathbf{M} \Rightarrow (\mathbf{y} \leq \mathbf{x})_\mathbf{E}$$

- $(\mathbf{x} \in \mathbf{M}, \mathbf{y} \in \mathbf{M}, (\mathbf{y} \leq \mathbf{x})_\mathbf{E}) \Rightarrow (\mathbf{y} \leq \mathbf{x})_\mathbf{M}$.
The $\mathbf{M}$ order is good so it is total (definitions 377 (p. 483), 388 (p. 487), proposition 853 (p. 488)). If we have $\mathbf{x} \in \mathbf{M}, \mathbf{y} \in \mathbf{M}$, we have $(\mathbf{y} > \mathbf{x})_\mathbf{M} \vee (\mathbf{y} \leq \mathbf{x})_\mathbf{M}$. If $(\mathbf{y} \leq \mathbf{x})_\mathbf{E}, (\mathbf{y} > \mathbf{x})_\mathbf{M}$ was true, from the preceding paragraph we will have $(\mathbf{y} \leq \mathbf{x})_\mathbf{E}, (\mathbf{y} > \mathbf{x})_\mathbf{E}$. Since it is impossible, from appendix A.1.4 (8 (p. 435))
$(\mathbf{x} \in \mathbf{M}, \mathbf{y} \in \mathbf{M}, (\mathbf{y} \leq \mathbf{x})_\mathbf{E}) \Rightarrow (\mathbf{y} \leq \mathbf{x})_\mathbf{M}$ is true.

$\mathbf{M}$ is a subset fully ordered of $\mathbf{E}$. It then has a majoring value $\mathbf{a}$ from what it is supposed in our proposition 866 (p. 494). All majoring values of $\mathbf{M}$ are in $\mathbf{M}$, otherwise $\mathbf{M}$ will be in $\mathbf{A}$ which is against proposition 864 (p. 492). Since, if a majoring value of a subset is in the subset, it is its biggest element (proposition 843 (p. 485), definition 379 (p. 484)) which is unique (proposition 842 (p. 484)). Since all majoring elements of $\mathbf{M}$ are in $\mathbf{M}$, they are all equal to $\mathbf{a}$. Then we have:

$$\mathbf{b} \geq \mathbf{a} \Rightarrow (\forall \mathbf{x} \in \mathbf{M})(\mathbf{x} \leq \mathbf{a}, \mathbf{x} \leq \mathbf{b}) \Rightarrow (\forall \mathbf{x} \in \mathbf{M})(\mathbf{x} \leq \mathbf{b}) \Rightarrow \mathbf{a} = \mathbf{b}$$

$\mathbf{a}$ is a maximal element of $\mathbf{E}$ (see definition 383 (p. 485)). So proposition 866 (p. 494) is true.

**Proposition 867** Given two well-ordered sets (definition 388 (p. 487)) $\mathbf{E}$ and $\mathbf{F}$ and two isomorphisms (definition 387 (p. 487)) $\mathbf{f}$ and $\mathbf{g}$ from $\mathbf{E}$ each of them to a segment (definition 389 (p. 488)) of $\mathbf{F}$ we have $\mathbf{f} = \mathbf{g}$.

**Proof**
Let us suppose that there are two isomorphisms $(\mathbf{f} : \mathbf{E} \leftrightarrow \mathbf{f}(\mathbf{E}) \subset \mathbf{F})$ and $(\mathbf{g} : \mathbf{E} \leftrightarrow \mathbf{g}(\mathbf{E}) \subset \mathbf{F})$ with $\mathbf{f}(\mathbf{E})$ and $\mathbf{g}(\mathbf{E})$ segments of $\mathbf{F}$. Let us consider the set $\mathbf{A}$ (proposition 749 (p. 450)):

$$\mathbf{A} = \mathbf{set_x}\{\mathbf{x} \in \mathbf{E}, \mathbf{g}(\mathbf{x}) < \mathbf{f}(\mathbf{x})\} .$$

Let us suppose that $\mathbf{A} \neq \emptyset$. Since $\mathbf{E}$ is well-ordered, $\mathbf{A}$ has a smallest element $\mathbf{a}$. Since $\mathbf{f}(\mathbf{E})$ is a segment of $\mathbf{F}$, we have:

$$\mathbf{g}(\mathbf{a}) < \mathbf{f}(\mathbf{a}) \Rightarrow \mathbf{g}(\mathbf{a}) \in \mathbf{f}(\mathbf{E}) \Rightarrow (\exists \mathbf{z} \in \mathbf{E})(\mathbf{f}(\mathbf{z}) = \mathbf{g}(\mathbf{a})) .$$

From that we get:

$$\mathbf{a} \in \mathbf{A} \Rightarrow \mathbf{g}(\mathbf{a}) < \mathbf{f}(\mathbf{a}) \Rightarrow \mathbf{f}^{-1}(\mathbf{g}(\mathbf{a})) < \mathbf{f}^{-1}(\mathbf{f}(\mathbf{a})) \Rightarrow \mathbf{z} < \mathbf{a}$$
$$\Rightarrow \mathbf{g}(\mathbf{z}) < \mathbf{g}(\mathbf{a}) \Rightarrow \mathbf{g}(\mathbf{z}) < \mathbf{f}(\mathbf{z}) \Rightarrow \mathbf{z} \in \mathbf{A} \Rightarrow \mathbf{a} \leq \mathbf{z} .$$

It is absurd, so using the rule in appendix A.1.4, 8, we have:

$$\mathbf{A} = \emptyset .$$

For the same reason:

$$\mathbf{B} = \mathbf{set_x}\{(\mathbf{x} \in \mathbf{E}), (\mathbf{g}(\mathbf{x}) > \mathbf{f}(\mathbf{x}))\} = \emptyset .$$

Then:

$$\forall(\mathbf{x} \in \mathbf{E})(\mathbf{f}(\mathbf{x}) = \mathbf{g}(\mathbf{x})) .$$

**Proposition 868** Given two well-ordered sets $\mathbf{E}$ and $\mathbf{F}$, an isomorphism $\mathbf{f}$ (definition 387 (p. 487)) from a segment (definition 389 (p. 488)) $\mathbf{S_E} \subset \mathbf{E}$ to a segment $\mathbf{S_F} \subset \mathbf{F}$. The restriction of $\mathbf{f}$ to a segment $\mathbf{S'_E} \subset \mathbf{S_E}$ defines the unique isomorphism between $\mathbf{S'_E}$ and $\mathbf{S'_F} = \mathbf{f}(\mathbf{S'_E}) \subset \mathbf{S_F}$ which happens to be a segment of $\mathbf{F}$.

**Proof**
Keeping the above notations, following proposition 854 (p. 488), we have:

$$\mathbf{x} \in \mathbf{f}(\mathbf{S'_E}) \Rightarrow ((\mathbf{y} \in \mathbf{F}, \mathbf{y} < \mathbf{x}) \Rightarrow \mathbf{f}^{-1}(\mathbf{y}) < \mathbf{f}^{-1}(\mathbf{x}) \Rightarrow \mathbf{f}^{-1}(\mathbf{y}) \in \mathbf{S'_E} \Rightarrow \mathbf{y} \in \mathbf{f}(\mathbf{S'_E})) .$$

Then according to definition 389 (p. 488), $\mathbf{f}(\mathbf{S'_E})$ is a segment of $\mathbf{F}$. We conclude using proposition 867 (p. 495) applied to $\mathbf{S'_E}$.

**Proposition 869** $\mathbf{E}$ and $\mathbf{F}$ being two sets, considering a good order on them, which exists according to Zermelo theorem (proposition 865 (p. 494)), there is :

1. either a unique isomorphism (definition 387 (p. 487)) from $\mathbf{E}$ to a segment of $\mathbf{F}$.

2. or a unique isomorphism from $\mathbf{F}$ to a segment of $\mathbf{E}$.

**Proof**
Let us consider the set $\mathcal{F}$ of isomorphic maps (definition 387 (p. 487)) between a segment of $\mathbf{E}$ to a segment of $\mathbf{F}$. For any map $\mathbf{f} \in \mathcal{F}$, we call $\mathbf{S^f_E}$ the segment on which acts $\mathbf{f}$ and we set $\mathbf{S^f_F} = \mathbf{f}(\mathbf{S^f_E})$. From the above proposition 868 (p. 496), propositions 853 (p. 488), 152 (p. 133), and proposition 858 (p. 489), we have:

$$\mathbf{f} \in \mathcal{F}, \mathbf{g} \in \mathcal{F} \Rightarrow \mathbf{S^f_E} \subset \mathbf{S^g_E} \vee \mathbf{S^g_E} \subset \mathbf{S^f_E},$$
$$\mathbf{S^g_E} \subset \mathbf{S^f_E} \Rightarrow ((\mathbf{f}(\mathbf{S^g_E}) = \mathbf{g}(\mathbf{S^g_E}) \subset \mathbf{f}(\mathbf{S^f_E})), (\mathbf{x} \in \mathbf{S^g_E} \Rightarrow \mathbf{f}(\mathbf{x}) = \mathbf{g}(\mathbf{x}))) .$$

Then

$$x \in S_E^f \cap S_E^g \Rightarrow f(x) = g(x) . \tag{A.48}$$

From proposition 856 (p. 488), $S_E = \bigcup_{f \in \mathcal{F}} S_E^f$ and $S_F = \bigcup_{f \in \mathcal{F}} f(S_E^f) = \bigcup_{f \in \mathcal{F}} S_F^f$ are segments respectively of **E** and **F**.

**$h(x) = \tau_f(x \in S_E^f)(x)$ defines a map (definition 332 (p. 440)) ($h : S_E \to S_F$) which is the unique isomorphism h between $S_E$ and $S_F$:**

$$x \in S_E \Rightarrow (\exists f \in \mathcal{F})(x \in S_E^f) \Rightarrow h(x) = \tau_f(x \in S_E^f)(x) \in S_F^{\tau_f(x \in S_E^f)} \subset S_F .$$

We have $(\exists f)(x \in S_E^f) \Rightarrow x \in S_E^{\tau_f(x \in S_E^f)}$ (see definition 333 (p. 441)). Since $S_E$ and $S_E^{\tau_f(x \in S_E^f)}$ are segments of **E** (see definition 389 (p. 488)) with $x \in S_E^{\tau_f(x \in S_E^f)} \subset S_E$, from proposition 731 (p. 442) and equation (A.48), we have:

$$\begin{aligned} y < x, x \in S_E &\Rightarrow y \in S_E \cap S_E^{\tau_f(x \in S_E^f)} \cap S_E^{\tau_f(y \in S_E^f)} \\ &\Rightarrow \tau_f(y \in S_E^f)(y) = \tau_f(x \in S_E^f)(y) = h(y) . \end{aligned}$$

Then since $\tau_f(x \in S_E^f)$ is an isomorphism, we have:

$$y < x, x \in S_E \Rightarrow h(y) < h(x) .$$

Similarly, from equation (A.48) again, we have:

$$\begin{aligned} u \in S_F &\Rightarrow (\exists f)(u \in S_F^f) \Rightarrow ((f = \tau_f(u \in S_F^f), x = f^{-1}(u)) \\ &\Rightarrow (x \in S_E \cap S_E^{\tau_g(x \in S_E^g)} \cap S_E^f, h(x) = f \circ f^{-1}(u) = u) \\ &\Rightarrow (\forall u \in S_F)(\exists x)(x \in S_E, h(x) = u) . \end{aligned}$$

Then $h(S_E) = S_F$.
Since all **f** are isomorph and then bijective (definition 387 (p. 487)), we have:

$$(u = h(x), u = h(y), y \leq x) \Rightarrow u = \tau_f(x \in S_E^f)(x) = \tau_f(x \in S_E^f)(y) \Rightarrow x = y .$$

Then, since **E** is totally ordered, we have:

$$(u = h(x), u = h(y)) \Rightarrow (u = h(x), u = h(y), y \leq x \vee x \leq y) \Rightarrow x = y$$

**h** is a bijection (definition 361 (p. 461)) between $S_E$ and $S_F$. It keeps the order. It is an isomorphism (definition 387 (p. 487)). From the above proposition 868, it is the only one.
Let us prove that at least one of the two segments $S_E$ or $S_F$ is the full set **E** or **F**:

$$(E = S_E) \vee (F = S_F) .$$

If not since **E** and **F** are well-ordered, we have $a \in \mathcal{C}_E(S_E)$ which is bigger than any elements of $S_E$ and smaller than or equal to any elements of $\mathcal{C}_E(S_E)$ and $b \in \mathcal{C}_F(S_F)$ which is bigger than any elements of $S_F$ and smaller than or equal to any elements of $\mathcal{C}_F(S_F)$. Then $S_E \cup \{a\}$, $S_F \cup \{b\}$ are segments of **E** and **F** and the map $h'(x)$ defined by:

$$(\forall x \in S_E)(h'(x) = h(x) \in S_F), h'(a) = b$$

is an isomorphism between the two new segments $S_E \cup \{a\}$, $S_F \cup \{b\}$. It is in contradiction with the fact that $S_E$ includes all such segments. We conclude by A.1.4, 8.

**Proposition 870** Given two well-ordered sets $\mathbf{E}$, $\mathbf{F}$ with an isomorphism (definition 387 (p. 487)) $\mathbf{f}$ of $\mathbf{E}$ to a segment of $\mathbf{F}$ and an isomorphism $\mathbf{g}$ of $\mathbf{F}$ to a segment of $\mathbf{E}$ then $\mathbf{f} = \mathbf{g}^{-1}$.

**Proof**
We have $\mathbf{f(E)} \subset \mathbf{F}$, then $\mathbf{g(f(E))} \subset \mathbf{g(F)} \subset \mathbf{E}$. From proposition 868 (p. 496) since $\mathbf{f(E)}$ is a segment of $\mathbf{F}$, we have:

$$\mathbf{x \in f(E) \Rightarrow f^{-1}(x) = g(x)} \ .$$

Then $\mathbf{E} = \mathbf{f^{-1}(f(E))} = \mathbf{g(f(E))}$ which means $\mathbf{E} \subset \mathbf{g(F)} \subset \mathbf{E}$. We may write from proposition 754 (p. 451):

$$\mathbf{g(F) = E,\ f(E) = F,\ F = g^{-1}(E),\ E = f^{-1}(F)} \ .$$

Applying proposition 869 (p. 496), $\mathbf{f} = \mathbf{g}^{-1}$.

**Proposition 871** Given a well-ordered set $\mathbf{E}$ (definition 388 (p. 487)), for any subset $\mathbf{A}$ of $\mathbf{E}$, there is an isomorphism (definition 387 (p. 487)) from $\mathbf{A}$ to a segment of $\mathbf{E}$.

**Proof**
There is a good order on $\mathbf{A}$ using the induced order by the order of $\mathbf{E}$ (any subset of $\mathbf{A}$ is a subset of $\mathbf{E}$ so has a smallest element). So, from proposition 869 (p. 496), we only have to prove that there is not an isomorphism between $\mathbf{E}$ and a segment $\mathbf{S}$ of $\mathbf{A}$ (with the order on $\mathbf{A}$) different from $\mathbf{A}$[9]. If we call $\mathbf{g}$ an isomorphism between $\mathbf{E}$ and a segment $\mathbf{S}$ of $\mathbf{A}$. Let us prove that if $\mathbf{g}$ exist, we have $\mathbf{S} = \mathbf{A}$. We call:

$$\mathbf{B = set_x\{x \in E, x > g(x)\}} \ .$$

If $\mathbf{B} \neq \emptyset$, there is a $\mathbf{b}$ the smallest element of $\mathbf{B}$, since $\mathbf{b \in B}$, $\mathbf{b > g(b)}$. But we have from the isomorphism of $\mathbf{g}$:

$$\mathbf{g(b) < b \Rightarrow g(g(b)) < g(b) \Rightarrow g(b) \in B \Rightarrow b \leq g(b)} \ . \tag{A.49}$$

It is absurd. Then, from A.1.4 (8 (p. 435)), $\mathbf{B} = \emptyset$. But we also have, since $\mathbf{S}$ is a segment (definition 389 (p. 488)):

$$\mathbf{(x \in A, x \notin S) \Rightarrow (g(x) \in S, x \notin S) \Rightarrow x > g(x) \Rightarrow x \in B} \ .$$

Since $\mathbf{B}$ is empty, $\mathbf{\neg(\neg(\neg x \in A \lor \neg\neg x \in S)) \Leftrightarrow (x \in A \Rightarrow x \in S)}$ true, using $\mathbf{S \subset A}$, we have $\mathbf{S} = \mathbf{A}$ (proposition 754 (p. 451)). In that case $\mathbf{g}^{-1}$ is an isomorphism between $\mathbf{A}$ and $\mathbf{E}$ which is a segment of $\mathbf{E}$. If not from proposition 869 (p. 496), there is an isomorphism between $\mathbf{A}$ and a segment of $\mathbf{E}$. In both cases proposition 871 (p. 498) is proved.

**Proposition 872** Given two sets $\mathbf{E}$ and $\mathbf{F}$, if there is a bijection from $\mathbf{E}$ to a subset of $\mathbf{F}$ and a bijection from $\mathbf{F}$ to a subset of $\mathbf{E}$, then there is a bijection between $\mathbf{E}$ and $\mathbf{F}$.

**Proof**
Let us consider $\mathbf{H} = \mathbf{E} \cup \mathbf{F}$. From proposition 865 (p. 494), there is a good order(definition 388 (p. 487)) on $\mathbf{H}$. From proposition 871 (p. 498), there is a bijection between $\mathbf{E}$ and a segment (definition 389 (p. 488)) of $\mathbf{H}$. The same is true for $\mathbf{F}$. For both $\mathbf{E}$ and $\mathbf{F}$, one can consider the set $\mathbf{A^E} \subset \mathcal{P}(\mathbf{H})$ and the set $\mathbf{A^F} \subset \mathcal{P}(\mathbf{H})$ of segments of $\mathbf{H}$ for which there is a bijection from $\mathbf{E}$ and $\mathbf{F}$ respectively. Each element of those two sets, $\mathbf{S^E \in A^E}$ or $\mathbf{S^F \in A^F}$, can be written, if different from $\mathbf{H}$, as: $\mathbf{S^E_x = ]\leftarrow, x[}$ or $\mathbf{S^F_y = ]\leftarrow, y[}$ where $\mathbf{x}$ and $\mathbf{y}$ are the

[9] Be careful neither $\mathbf{A}$ nor $\mathbf{S}$ are segment of $\mathbf{E}$.

smallest element of respectively $\mathcal{C}(\mathbf{S^E_x})$ or $\mathcal{C}(\mathbf{S^F_y})$ and $\mathbf{S^F_y}$.
Let us consider now the sets $\mathbf{B^E}$ and $\mathbf{B^F}$:

$$\mathbf{B^E = set_x\{x \in H, S^E_x \in A^E\}}$$

$$\mathbf{B^F = set_y\{y \in H, S^F_y \in A^F\}} .$$

If both $\mathbf{B^E}$ and $\mathbf{B^F}$ are empty, it means that there is at least a bijection from $\mathbf{E}$ to $\mathbf{H}$ and from $\mathbf{F}$ to $\mathbf{H}$ so one from $\mathbf{E}$ to $\mathbf{F}$. Proposition 872 is true.
If one of them is not empty, let us say $\mathbf{B^E}$. $\exists \mathbf{x} \in \mathbf{H}$ such that there is a bijection from $\mathbf{E}$ to $\mathbf{S^E_x}$. Since from what we suppose, there is a bijection from $\mathbf{F}$ to a subset of $\mathbf{E}$ and then to a subset of $\mathbf{S^E_x}$ and from it to a segment of $\mathbf{S^E_x}$ (proposition 871 (p. 498)) which is also a segment $\mathbf{S^F_y}$ of $\mathbf{H}$ with $\mathbf{y} \leq \mathbf{x}$. Then $\mathbf{B^F} \neq \emptyset$.
Since $\mathbf{H}$ is well-ordered, $\mathbf{B^E}$ and $\mathbf{B^F}$ have each of them a smallest element $\mathbf{b^e}$ and $\mathbf{b^f}$. If they are not equal, one of them is smaller than the other. Let us suppose $\mathbf{b^e} < \mathbf{b^f}$. Then we have $\mathbf{S^E_{b^e}} \subset \mathbf{S^F_{b^f}}$. From the definitions, there is a bijection from $\mathbf{S^E_{b^e}}$ to $\mathbf{E}$ and from a subset of $\mathbf{E}$ to $\mathbf{F}$ so from $\mathbf{F}$ to a subset $\mathbf{F_E}$ of $\mathbf{S^E_{b^e}}$. Using proposition 871 (p. 498), we have a bijection from $\mathbf{F_E}$ to a segment $\mathbf{S^F_c}$ of $\mathbf{S^E_{b^e}}$ which is also a segment of $\mathbf{H}$. So from $\mathbf{F}$ to the segment $\mathbf{S^F_c}$ of $\mathbf{H}$. Since $\mathbf{S^F_c} \subset \mathbf{S^E_{b^e}} \subset \mathbf{S^F_{b^f}}$ we have $\mathbf{c} \leq \mathbf{b^e} < \mathbf{b^f}$. But, by definition, $\mathbf{c} \in \mathbf{B^F}$ which means $\mathbf{b^f} < \mathbf{c}$ which is impossible. Applying A.1.4, 8, we have $\mathbf{S^E_{b^e}} = \mathbf{S^F_{b^f}}$. Then there is a bijection from $\mathbf{E}$ to $\mathbf{S^E_{b^e}} = \mathbf{S^F_{b^f}}$ and from it to $\mathbf{F}$ so is a bijection from $\mathbf{E}$ to $\mathbf{F}$.

### A.5.10 Cardinal

**Definition 390 Card(s).**

We have by definition, $\mathbf{s}$ being a set (see definition 371 (p. 472)) :

$$\mathbf{Card(s) = \tau_x(x \sim s)}$$

$\mathbf{Card(s)}$ is a set which is unique from the properties of the operator $\tau_\mathbf{x}$ (proposition 726 (p. 441)). $\mathbf{Card(s)}$ corresponds to the number of elements of $\mathbf{s}$. $\mathbf{Card(s)}$ is the cardinal of $\mathbf{s}$.

**Definition 391 0, 1.**

We have:

$$\mathbf{0 = Card(\emptyset), 1 = Card(\mathcal{P}(0)) = Card(\mathcal{P}(\emptyset))} . \tag{A.50}$$

**Convention 33** $\mathbf{a}$ is cardinal.

The sentence $\mathbf{a}$ is cardinal stands for:

$$\mathbf{(\exists s)(a = \tau_x(x \sim s))}$$

It is considered as an added axiom and true locally which means that $\mathbf{a}$ can be replaced in all this local area whenever wanted by a term $\mathbf{T}$ which verifies $\mathbf{(\exists s)(T = \tau_x(x \sim s))}$ and the local area become a true demonstration (see section A.1.2 (p. 434)). We have to precise that because $\mathbf{coll_a(\exists s)(a = \tau_x(x \sim s))}$ is false (proposition 882 (p. 503)).

**Proposition 873** $\mathbf{a \sim b \Leftrightarrow Card(a) = Card(b)}$.

**Proof**
From proposition 817 (p. 472), $\sim$ is an equivalence relation. Applying point 3 of definition 336 (p. 445), we have:

$$a \sim b \Rightarrow (x \sim a \Leftrightarrow x \sim b) .$$

Then, from proposition 726 (p. 441), equation (A.3), we have:

$$\tau_x(x \sim a) = \tau_x(x \sim b) .$$

From proposition 179 (p. 142), definition 113 (p. 132), we have successively true (see footnote 7 (p. 451) and section A.1.2 (p. 434)):

$a \sim a$

$a \sim a \Rightarrow \tau_x(x \sim a) \sim a$

$\tau_x(x \sim a) \sim a$

$\mathrm{Card}(a) = \mathrm{Card}(b) \Leftrightarrow \tau_x(x \sim a) = \tau_x(x \sim b) \Leftrightarrow a \sim \tau_x(x \sim b) \sim \tau_x(x \sim a) \sim b$

then $\mathrm{Card}(a) = \mathrm{Card}(b) \Leftrightarrow a \sim b$

**Proposition 874** $\mathrm{Card}(a) = \mathrm{Card}(\mathrm{Card}(a))$.

**Proof**
From proposition 817 (p. 472), applying point 1 of definition 336 (p. 445), $a \sim a$.
From propositions 726 (p. 441), 873 (p. 499), we have:

$$a \sim a \Rightarrow \tau_x(x \sim a) \sim a = \mathrm{Card}(a) \sim a \Rightarrow \mathrm{Card}(\mathrm{Card}(a)) = \mathrm{Card}(a) .$$

From section A.1.2 (p. 434), we have proposition 874 (p. 500) true.

**Definition 392 Order on cardinals.**

Considering cardinals $e \leq f$ means:

$$e \leq f \Leftrightarrow (\exists s \subset f)(e \sim s) .$$

**Proposition 875** Definition 392 (p. 500) defines an ordering relation (definition 338 (p. 446)).

**Proof**
It is an ordering relation because (see 338 (p. 446)):

1. $e \leq f$ means $e$ and $f$ are cardinals. From proposition 179 (p. 142), $e \sim e$, then $e \leq e$, $f \leq f$.

2. $e \leq f, f \leq e \Rightarrow e = f$ from proposition 872 (p. 498).

3. $e \leq f, f \leq g \Rightarrow e \leq g$ from proposition 193 (p. 149).

**Proposition 876** Given a set $S$ of cardinals, the order defined by definition 392 is a total order on $S$.

**Proof**
It is a direct consequence of proposition 869 (p. 496).

**Proposition 877** Given two sets $A$ and $B$, $A \subset B \Rightarrow \mathrm{Card}(A) \leq \mathrm{Card}(B)$.

**Proof**
The identity map $(\mathbf{f(x)} : \mathbf{A} \to \mathbf{A})$: $\mathbf{f(x)} = \mathbf{x}$ is a bijection, then $\mathbf{A} \sim \mathbf{A}$ and applying definition 392 we have proposition 877.

**Proposition 878**

1. $\mathbf{A} \sim 0 \Leftrightarrow \mathbf{A} = \emptyset$
2. $\mathbf{A} \sim 1 \Leftrightarrow ((\exists \mathbf{x})(\mathbf{x} \in \mathbf{A}), (\mathbf{x} \in \mathbf{A}, \mathbf{y} \in \mathbf{A} \Rightarrow \mathbf{x} = \mathbf{y}))$
3. $\mathbf{x}$ being a term, $\mathbf{Card}(\{\mathbf{x}\}) = 1$
4. $\mathbf{A} \neq \emptyset \Rightarrow \mathbf{Card}(\mathbf{A}) \geq 1$

See definition 391 (p. 499).
**Proof**

1. From definitions 371 (p. 472), 124 (p. 139) and equation (2.1) (p. 131), we have:
$$\begin{aligned}&\mathbf{A} \sim 0 \Leftrightarrow \mathbf{A} \sim \emptyset \Rightarrow (\exists \mathbf{f})(\mathbf{f} \subset \mathbf{A} \times \emptyset, (\forall \mathbf{x})(\mathbf{x} \in \mathbf{A} \Rightarrow (\exists \mathbf{y})(\mathbf{y} \in \emptyset)),\\ &(((\mathbf{x},\mathbf{y}) \in \emptyset, (\mathbf{x},\mathbf{y}') \in \emptyset) \Rightarrow \mathbf{y} = \mathbf{y}'), (((\mathbf{x},\mathbf{y}) \in \emptyset, (\mathbf{x}',\mathbf{y}) \in \emptyset) \Rightarrow \mathbf{x} = \mathbf{x}'))\\ &\mathbf{A} \sim 0 \Rightarrow (\forall \mathbf{x})(\mathbf{x} \in \mathbf{A} \Rightarrow (\exists \mathbf{y})(\mathbf{y} \in \emptyset)) .\end{aligned}$$
Calling $\mathbf{R(x)} = (\mathbf{x} \in \mathbf{A} \Rightarrow (\exists \mathbf{y})(\mathbf{y} \in \emptyset))$, from proposition 729 (p. 441), we have:
$$\mathbf{R}(\tau_{\mathbf{x}}(\neg \mathbf{R}(\mathbf{x}))) .$$
Then, from proposition 727 (p. 441), we have:
$$\begin{aligned}&\mathbf{A} \sim \emptyset \Rightarrow \mathbf{R}(\tau_{\mathbf{x}}(\neg \mathbf{R}(\mathbf{x}))) \Rightarrow (\exists \mathbf{x})(\mathbf{x} \in \mathbf{A} \Rightarrow (\exists \mathbf{y})(\mathbf{y} \in \emptyset))\\ &\Rightarrow ((\exists \mathbf{x})(\mathbf{x} \in \mathbf{A}) \Rightarrow (\exists \mathbf{y})(\mathbf{y} \in \emptyset))\\ &\Rightarrow \neg\neg(\exists \mathbf{y})(\mathbf{y} \in \emptyset) \vee \neg(\exists \mathbf{x})(\mathbf{x} \in \mathbf{A}) \Rightarrow (\neg(\exists \mathbf{y})(\mathbf{y} \in \emptyset) \Rightarrow \neg(\exists \mathbf{x})(\mathbf{x} \in \mathbf{A})) .\end{aligned}$$
Since $\neg(\exists \mathbf{y})(\mathbf{y} \in \emptyset)$ is true $\neg(\exists \mathbf{x})(\mathbf{x} \in \mathbf{A})$ is also true and $\mathbf{A} = \emptyset$.

2. From proposition 775 (p. 457), we have $\emptyset \subset \emptyset$.
From definition 350 (p. 453) and proposition 759 (p. 452), we have: $\emptyset \in \mathcal{P}(\emptyset)$.
From propositions 773 (p. 457), 727 (p. 441) and point 1, page 501, we have:
$$\mathcal{P}(\emptyset) \neq \emptyset, \mathbf{Card}(\mathcal{P}(\emptyset)) \neq 0 .$$
From definition 371 (p. 472), we have:
$$\mathbf{A} \sim \mathcal{P}(\emptyset) \Rightarrow (\exists(\mathbf{f} : \mathbf{A} \leftrightarrow \mathcal{P}(\emptyset))) \Rightarrow (\exists \mathbf{x} \in \mathbf{A})((\mathbf{x}, \emptyset) \in \mathbf{f})$$
with $\mathbf{f} = \tau_{\mathbf{g}}((\mathbf{g} :\leftrightarrow \mathcal{P}(\emptyset)))$. Since from proposition 776 (p. 457) and definition 350 (p. 453):
$$\mathbf{x} \in \mathcal{P}(\emptyset) \Rightarrow \mathbf{x} \subset \emptyset \Rightarrow \mathbf{x} = \emptyset$$
we have using the above map $\mathbf{f}$:
$$\mathbf{x} \in \mathbf{A}, \mathbf{y} \in \mathbf{A} \Rightarrow \mathbf{f(x)} \in \mathcal{P}(\emptyset), \mathbf{f(y)} \in \mathcal{P}(\emptyset) \Rightarrow \mathbf{f(x)} = \mathbf{f(y)} = \emptyset .$$
Then from definition 361 (p. 461) $\mathbf{x} = \mathbf{y}$.

3. From definition 345 (p. 447), $\mathbf{x} \in \{\mathbf{x_1}\}, \mathbf{y} \in \{\mathbf{x_1}\} \Rightarrow \mathbf{x} = \mathbf{y} = \mathbf{x_1}$. From point 2, page 501, $\mathbf{Card}(\{\mathbf{x_1}\}) = \mathbf{1}$.

4. From propositions 773 (p. 457), 752 (p. 451), definitions 345 (p. 447), 392 (p. 500), point 3, page 501, we have:

$$\mathbf{A} \neq \emptyset \Rightarrow (\exists \mathbf{x})(\mathbf{x} \in \mathbf{A})$$
$$\mathbf{x_1} = \tau_{\mathbf{x}}(\mathbf{x} \in \mathbf{A}) \Rightarrow \mathbf{x_1} \in \mathbf{A} \Rightarrow \{\mathbf{x_1}\} \subset \mathbf{A} \Rightarrow \mathbf{1} = \mathbf{Card}(\{\mathbf{x_1}\}) \leq \mathbf{Card}(\mathbf{A}) .$$

**Proposition 879** Given a set $\mathbf{S}$ of sets, we have:

$$(\forall \mathbf{x} \in \mathbf{S})(\mathbf{Card}(\mathbf{x}) \leq \mathbf{Card}(\bigcup_{\mathbf{u} \in \mathbf{S}} \mathbf{u})) .$$

**Proof**
From definitions 362 (p. 462), 349 (p. 451) we have:

$$\mathbf{x} \in \mathbf{S} \Rightarrow \mathbf{y} \in \mathbf{x} \Rightarrow \mathbf{y} \in \bigcup_{\mathbf{u} \in \mathbf{S}} \mathbf{u} \Rightarrow \mathbf{x} \subset \bigcup_{\mathbf{u} \in \mathbf{S}} \Rightarrow \mathbf{Card}(\mathbf{x}) \leq \mathbf{Card}(\bigcup_{\mathbf{u} \in \mathbf{S}} \mathbf{u}) .$$

**Proposition 880** $\mathbf{Card}(\mathbf{s}) < \mathbf{Card}(\mathcal{P}(\mathbf{s}))$ (definitions 350 (p. 453), 390 (p. 499).
**Proof**
The application $(\mathbf{f} : \mathbf{s} \to \mathcal{P}(\mathbf{s})) \Rightarrow \mathbf{f}(\mathbf{x}) = \{\mathbf{x}\}$ is injective. Then we have:

$$\mathbf{s} \sim \mathbf{set_x}\{(\exists \mathbf{y} \in \mathbf{s}), (\mathbf{x} = \{\mathbf{y}\})\} \subset \mathcal{P}(\mathbf{s})$$

then

$$\mathbf{Card}(\mathbf{s}) \leq \mathbf{Card}(\mathcal{P}(\mathbf{s})) .$$

Let us suppose proposition 880 (p. 502) is false then from the preceding statement the only possibility left is

$$\mathbf{Card}(\mathcal{P}(\mathbf{s})) = \mathbf{Card}(\mathbf{s}) \Rightarrow \mathcal{P}(\mathbf{s}) \sim \mathbf{Card}(\mathbf{s}) .$$

There is then a map $\mathbf{g}$ bijective from $\mathbf{s}$ to $\mathcal{P}(\mathbf{s})$. Since, from proposition 749 (p. 450), we have:

$$\mathbf{coll_x}\{\mathbf{x} \in \mathbf{s}, \mathbf{x} \notin \mathbf{g}(\mathbf{x})\} .$$

By definition of $\mathcal{P}(\mathbf{s})$ (definition 350 (p. 453)) we have:

$$\mathbf{set_x}\{\mathbf{x} \in \mathbf{s}, \mathbf{x} \notin \mathbf{g}(\mathbf{x})\} \in \mathcal{P}(\mathbf{s}) .$$

Since $(\mathbf{g} : \mathbf{s} \to \mathcal{P}(\mathbf{s}))$ is a bijection (definitions 124 (p. 139), 361 (p. 461)), we have:

$$(\exists \mathbf{y})(\mathbf{g}(\mathbf{y}) = \mathbf{set_x}\{\mathbf{x} \in \mathbf{s}, \mathbf{x} \notin \mathbf{g}(\mathbf{x}),\ \mathbf{y} = \mathbf{g}^{-1}(\mathbf{set_x}\{\mathbf{x} \in \mathbf{s}, \mathbf{x} \notin \mathbf{g}(\mathbf{x})\})\}) . \tag{A.51}$$

We have from equation (A.51) second part: $\mathbf{y} \in \mathbf{g}(\mathbf{y}) \Rightarrow \mathbf{y} \notin \mathbf{g}(\mathbf{y}))$ true which means (definition 330 (p. 434)): $\mathbf{y} \notin \mathbf{g}(\mathbf{y}) \vee \mathbf{y} \notin \mathbf{g}(\mathbf{y})$ true and (point 1 (p. 435)) $\mathbf{y} \notin \mathbf{g}(\mathbf{y})$ true. But from equation (A.51) first part, $\mathbf{y} \notin \mathbf{g}(\mathbf{y}) \Rightarrow \mathbf{y} \in \mathbf{g}(\mathbf{y})$ is also true. It means (definition 330 (p. 434)) $\neg\neg\mathbf{y} \in \mathbf{g}(\mathbf{y}) \vee \mathbf{y} \in \mathbf{g}(\mathbf{y})$ and (point 5 (p. 435), 15 (p. 435)) $\mathbf{y} \in \mathbf{g}(\mathbf{y}) \vee \mathbf{y} \in \mathbf{g}(\mathbf{y})$ true then (point 1 (p. 435)) $\mathbf{y} \in \mathbf{g}(\mathbf{y})$ true. See also A.1.4 point 21 (p. 436).
This is absurd. Then from point 8 (p. 435), we have:

$$\mathbf{Card}(\mathbf{s}) < \mathbf{Card}(\mathcal{P}(\mathbf{s})) .$$

**Proposition 881** s being a term, we have $(\exists x_1)(x_1 \notin s)$ .
**Proof**
We have:

$$(\forall x_1 \in \mathcal{P}(s))(x_1 \in s) \Rightarrow \mathcal{P}(s) \subset s \Rightarrow \mathbf{Card}(\mathcal{P}(s)) \leq \mathbf{Card}(s) .$$

Since, from proposition 880 (p. 502), $\mathbf{Card}(\mathcal{P}(s)) \leq \mathbf{Card}(s)$ is false, from section A.1.4 (p. 435) point 7 (p. 435) with proposition 728 (p. 441), we have:

$$\neg(\forall x_1)(x_1 \in \mathcal{P}(s) \Rightarrow x_1 \in s) \Leftrightarrow (\exists x_1)(\neg(\neg(x_1 \in \mathcal{P}(s)) \vee \neg(x_1 \notin s)))$$
$$\Leftrightarrow (\exists x_1)(x_1 \in \mathcal{P}(s), x_1 \notin s) \Rightarrow (\exists x_1)(x_1 \notin s)$$

which gives proposition 881 (see A.3 (p. 441)).

**Proposition 882** $\mathbf{coll}_x(\exists s)(x = \mathbf{Card}(s))$ is false.
**Proof**
If $U = \mathbf{set}_x\{(\exists s)(x = \mathbf{Card}(s))\}$ exists, we have the set $S = \bigcup_{x \in U} x$ (see definition 362 (p. 462)).
Then:

$$(\forall t)(s = \mathbf{Card}(t) \Rightarrow (\exists x \subset S)((x \sim s \sim t) \Rightarrow s \leq \mathbf{Card}(S))) .$$

Then

$$(\forall t)(\mathbf{Card}(t) \leq \mathbf{Card}(S)) .$$

From proposition 731 (p. 442) we would have:

$$\mathbf{Card}(\mathcal{P}(S)) \leq \mathbf{Card}(S)$$

which is impossible from the preceding proposition 880 (p. 502). Then, from A.1.4 point 8 (p. 435), proposition 882 (p. 503) is true.

**Proposition 883** $(\forall u)(\mathbf{coll}_s((\exists x)(s = \mathbf{Card}(x), s \leq \mathbf{Card}(u))))$ .
**Proof**
Let us consider the sets $X = \mathcal{P}(u)$. We have from 751 (p. 450):

$$\mathbf{coll}_t((\exists x)(t = \mathbf{Card}(x), x \in X)) .$$

Since any cardinal smaller than $\mathbf{Card}(u)$ is equivalent to a subset of $u$, it is equivalent to a member of $X$ and then it is an element of $\mathbf{set}_t\{(\exists x)((t = \mathbf{Card}(x)), (x \in X))\}$ then:

$$(\exists x)(s = \mathbf{Card}(x), s \leq \mathbf{Card}(u)) \Rightarrow s \in \mathbf{set}_t\{(\exists x)(t = \mathbf{Card}(x), x \in X)\} .$$

Proposition 883 (p. 503) is true from proposition 750 (p. 450).

**Proposition 884** Given a set $A$ and a subset $S$ of $\mathcal{P}(A)$ we have:

$$\mathbf{coll}_x\{(\exists s \in S)(x = \mathbf{Card}(s))\} .$$

**Proof**
From definition 392 (p. 500), we have:

$$s \in S \Rightarrow \mathbf{Card}(s) \leq \mathbf{Card}(A) .$$

From proposition 883 (p. 503), $[0, \mathbf{Card}(A)]$ is a set. Then, since, from proposition 750 (p. 450), we have:

$$((\exists s \in S)(x = \mathbf{Card}(s)) \Rightarrow x \in [0, \mathbf{Card}(A)]) \Rightarrow \mathbf{coll}_x\{(\exists s \in S)(x = \mathbf{Card}(s))\}$$

proposition 884 (p. 503) is true.

**Proposition 885** Given a set $\mathbf{A}$ and a subset $\mathbf{S}$ of $\mathcal{P}(\mathbf{A})$, there is a map $(\mathbf{I(s)} = \mathbf{set_x\{(\exists s \in S)(x = Card(s))\}} : \mathbf{S} \to \mathbf{I(S)})$ such that $\mathbf{s \in S \Rightarrow I(s) = card(s)}$.

**Proof**

From proposition 884 (p. 503), the set $\mathbf{I(S) = set_x\{(\exists s \in S)(x = Card(s))\}}$ exists. From definitions 390 (p. 499) and 357 (p. 460), the set $\mathbf{I = set_i\{\exists(s \in S)(i = (s, card(s)))\}}$ is a map.

**Proposition 886** The order on cardinals is a **good order**.

**Proof**

Calling $\mathbf{E}$ a set of cardinals. Calling $\mathbf{H}$ the set:

$$\mathbf{H} = \bigcup_{\mathbf{a} \in \mathbf{E}} \mathbf{a}\,. \tag{A.52}$$

There is, from proposition 865 (p. 494), a good order on $\mathbf{H}$. We can consider together with that order, the set $\mathbf{A}$ of all segments of $\mathbf{H}$. It forms a set because they are members of the set $\mathcal{P}(\mathbf{H})$ (see definition 350 (p. 453)). On that set $\mathbf{A}$ the order coming from the inclusion: "subset of" (definition 349 (p. 451)), is a good and total order (proposition 861 (p. 490)). We consider $\mathbf{A}$ with such a good order. From equation (A.52) (p. 504), $\mathbf{a \in E \Rightarrow a \subset H}$. From proposition 871 (p. 498), there is a bijection between $\mathbf{a}$ and a segment of $\mathbf{H}$. We consider the set of segments for which there is a bijection with the cardinal $\mathbf{a}$. This set is a subset of $\mathbf{A}$. From the preceding statement that set is not empty. Since the order by inclusion is a good order, there is a unique smallest element $\mathbf{S_a}$ of that set (definition 379 (p. 484), proposition 842 (p. 484)). The set of those segment $\mathbf{S_a}$ corresponding to a cardinal $\mathbf{a \in E}$ forms also a subset of $\mathbf{a}$ which also has a smallest element $\mathbf{S_b}$ corresponding to the cardinal $\mathbf{b \in E}$. The map $(\mathbf{S_a : E \to A})$ is an injective map (definition 359 (p. 461), propositions 873 (p. 499), 874 (p. 500)):

$$\mathbf{S_u = S_v \Rightarrow u \sim S_u = S_v \sim v \Rightarrow u \sim v \Rightarrow u = v}\,.$$

It is an isomorphism because (see proposition 871 (p. 498) and A.1.4 (p. 435) point 7):

$$\begin{aligned}
&(\mathbf{S_u = S_v \Rightarrow u = v}) \Leftrightarrow (\mathbf{u \neq v \Rightarrow S_u \neq S_v})\\
&\mathbf{u < v \Rightarrow (\exists w \subset v)(u \sim w) \Rightarrow (\exists T \subset S_v)(S_u \sim T)}\\
&\;\Rightarrow (\exists \mathbf{T'} \subset \mathbf{S_v})(\mathbf{T'} \in \mathbf{A}, \mathbf{S_u} \sim \mathbf{T'})\,.
\end{aligned}$$

Since the order on $\mathbf{A}$ is total, we have $\mathbf{S_u \subset S_v \vee S_v \subset S_u}$. We cannot have $\mathbf{S_v \subset S_u}$ because it implies $(\exists \mathbf{T''} \subset \mathbf{S_u})(\mathbf{T''} \sim \mathbf{S_v})$ which, from proposition 872 (p. 498), together with $(\exists \mathbf{T'} \subset \mathbf{S_v})(\mathbf{T'} \in \mathbf{A}, \mathbf{S_u} \sim \mathbf{T'})$ means $\mathbf{S_u \sim S_v \sim u \sim v}$. It is not allowed. There remains $\mathbf{S_u \subset S_v}$, $\mathbf{S_u \neq S_v : S_u < S_v}$.

We also have $\mathbf{S_u < S_v \Rightarrow u < v}$ because:

- $\mathbf{u = v \Rightarrow S_u = S_v}$ otherwise the smallest value defining $\mathbf{S_u}$ and $\mathbf{S_v}$ is not unique.
- $\mathbf{S_u \subset S_v}$ means that there is a bijection between $\mathbf{u}$ and a subset of $\mathbf{v}$. Then $\mathbf{u \leq v}$ which gives $\mathbf{u < v}$ since $\mathbf{u = v}$ is forbidden.

Then $\mathbf{b}$ is the smallest element of $\mathbf{E}$. The order of the cardinal in $\mathbf{E}$ coming from definition 392 (p. 500) is a good order.

**Proposition 887** Given two sets $\mathbf{X}$, $\mathbf{Y}$ and a set of index $\mathbf{I}$ and two maps $(\mathbf{x_i : I \to \mathcal{P}(X)})$, $(\mathbf{y_i : I \to \mathcal{P}(Y)})$ with:

$$(\forall \mathbf{i})(\mathbf{x_i \sim y_i}, (\forall \mathbf{j})(\mathbf{i \neq j \Rightarrow x_i \cap x_j = y_i \cap y_j = \emptyset}))\,.$$

One has the following equality:

$$\mathbf{Card}(\bigcup_{i\in I} x_i) = \mathbf{Card}(\bigcup_{i\in I} y_i) .$$

**Proof**
We have (see convention 15 (p. 140)):

$$(\forall i \in I)(x_i \sim y_i \Leftrightarrow (\exists s_i)(s_i : x_i \leftrightarrow y_i)) . \tag{A.53}$$

From:

$$(\forall i)(i \neq j \Rightarrow x_i \cap x_j = y_i \cap y_j = \emptyset)$$

we defines a set $\mathbf{f}$ (resp. $\mathbf{g}$, resp. $\mathbf{y}$) by:

$$(u, i) \in f \Leftrightarrow (i \in I, u \in x_i)$$

$$(u, i) \in g \Leftrightarrow (i \in I, u \in y_i)$$

$\mathbf{f(u)}$ (resp. $\mathbf{g(u)}$) is a map (see definitions 112 (p. 131), 357 (p. 460)) $(f : \bigcup_{i\in I} x_i \to I)$, $(g : \bigcup_{i\in I} y_i \to I)$ because $i \neq j \Rightarrow x_i \cap x_j$ means:

$$(u, i) \in f, (u, j) \in f \Rightarrow i = j$$

$$(u, i) \in g, (u, j) \in g \Rightarrow i = j .$$

Using that we define $(S_x : \bigcup_{i\in I} x_i \to \bigcup_{i\in I} y_i)$ and $(S_y : \bigcup_{i\in I} y_i \to \bigcup_{i\in I} x_i)$ by:

$$u \in \bigcup_{i\in I} x_i \Rightarrow S_x(u) = s_{f(u)}(u)$$

$$v \in \bigcup_{i\in I} y_i \Rightarrow S_y(v) = s^{-1}_{g(v)}(v) .$$

One has:

$$(s_{f(u)}(u) \in y_{f(u)}, v \in y_i \Rightarrow g(v) = i) \Rightarrow g(s_{f(u)}(u)) = f(u) .$$

Then we have:

$$S_y(S_x(u)) = s^{-1}_{g(s_{f(u)}(u))}(s_{f(u)}(u)) = s^{-1}_{f(u)}(s_{f(u)}(u)) = u .$$

We have $S_y = S_x^{-1}$. From proposition 178 (p. 142) $\mathbf{S_x}$ and $\mathbf{S_y}$ are bijections (definition 361 (p. 461)). Then

$$\bigcup_{i\in I} x_i \sim \bigcup_{i\in I} y_i$$

and we have proposition 887.

**Proposition 888** Given two sets $\mathbf{X}$ and $\mathbf{Y}$, a set of index $\mathbf{I}$ and two maps: $(x_i : I \to \mathcal{P}(X)), (y_i : I \to \mathcal{P}(Y))$, we have:

$$(\forall i \in I)(x_i \sim y_i \Rightarrow \mathbf{Card}(\prod_{i\in I} x_i) = \mathbf{Card}(\prod_{i\in I} y_i)) .$$

**Proof**
Supposing the hypotheses of proposition 888, from definition 371 (p. 472), one has (see convention 15 (p. 140)):

$$(\forall i \in I)(\exists s_i)(s_i : x_i \leftrightarrow y_i) .$$

Then we define:

$$(S_x : \prod_{i\in I} x_i \to \prod_{i\in I} y_i) = (u \in \prod_{i\in I} x_i \Rightarrow (\forall i \in I)(\{S_x(u)\}(i) = s_i(u(i))))$$

$$(S_y : \prod_{i\in I} y_i \to \prod_{i\in I} x_i) = (v \in \prod_{i\in I} y_i \Rightarrow (\forall i \in I)(\{S_y(v)\}(i) = s_i^{-1}(v(i)))) .$$

We have:

$$u \in \prod_{i\in I} x_i \Rightarrow (\forall i \in I)(\{S_x(u)\}(i) = s_i(u(i)))$$
$$\Rightarrow (\forall i \in I)(\{S_y(S_x(u))\}(i) = s_i^{-1}(s_i(u(i)) = u(i))) \Rightarrow S_y(S_x(u)) = u \Rightarrow S_y = S_x^{-1}$$
$$v \in \prod_{i\in I} y_i \Rightarrow (\forall i \in I)(\{S_y(v)\}(i) = s_i^{-1}(v(i)))$$
$$\Rightarrow (\forall i \in I)(\{S_x(S_y(v))\}(i) = s_i(s_i^{-1}(v(i)) = v(i))) \Rightarrow S_x(S_y(v)) = v \Rightarrow S_x = S_y^{-1} .$$

Then $(\forall v \in \prod_{i\in I} y_i)(\exists u = S_y(v))(v = S_x(u))$, $S_x$ is surjective (definition 360 (p. 461)). From proposition 178 (p. 142), $S_x$ is a bijection (definition 361 (p. 461)) and

$$\prod_{i\in I} x_i \sim \prod_{i\in I} y_i .$$

It gives proposition 888.

**Proposition 889** Given three sets $X$, $I$ and $J$, a map $(X_i : I \to \mathcal{P}(X))$ and a partition of $I$ (definition 370 (p. 471)) $(I_j : J \to \mathcal{P}(I))$ we have:

$$\mathrm{Card}(\prod_{j\in J} \prod_{i\in I_j} X_i) = \mathrm{Card}(\prod_{i\in I} X_i) .$$

**Proof**
It is a direct consequence of proposition 837 (p. 481).

**Proposition 890** With the same notation as in proposition 889, if $g$ is a bijection (definition 361 (p. 461)) $(g : I \to J)$ we have:

$$\mathrm{Card}(\prod_{i\in I} X_i) = \mathrm{Card}(\prod_{j\in J} X_{g^{-1}(j)})$$

similarly

$$\mathrm{Card}(\prod_{j\in J} X_j) = \mathrm{Card}(\prod_{i\in I} X_{g(i)}) .$$

**Proof**
It is proposition 889 with $I_j = g^{-1}(\{j\}) = \{g^{-1}(j)\}$ and $J_i = g(\{i\}) = \{g(i)\}$.

**Definition 393 Sum of cardinals.**

Given a set of cardinals $\mathbf{A}$, a set of index $\mathbf{I}$ and a map $(\mathbf{a_i} : \mathbf{I} \longrightarrow \mathbf{A})$, one can define from that the map:

$$(\mathbf{x_i} = \mathbf{a_i} \times \{\mathbf{i}\} : \mathbf{I} \to \mathbf{A} \times \mathcal{P}(\mathbf{I})) \ .$$

From proposition 828 (p. 476), $(\forall \mathbf{i})(\forall \mathbf{j})(\mathbf{i} \neq \mathbf{j} \Rightarrow \mathbf{x_i} \cap \mathbf{x_j} = \emptyset)$. Using that one defines:

$$\sum_{\mathbf{i} \in \mathbf{I}} \mathbf{a_i} = \mathbf{Card}(\bigcup_{\mathbf{i} \in \mathbf{I}} \mathbf{x_i}) = \mathbf{Card}(\bigcup_{\mathbf{i} \in \mathbf{I}} (\mathbf{a_i} \times \{\mathbf{i}\})) \ .$$

**Proposition 891** Given a set of cardinals $\mathbf{A}$, a set of index $\mathbf{I}$ and a map $(\mathbf{a_i} : \mathbf{I} \longrightarrow \mathbf{A})$, we have:

$$\sum_{\mathbf{i} \in \mathbf{I}} \mathbf{a_i} = 0 \Leftrightarrow (\forall \mathbf{i} \in \mathbf{I})(\mathbf{a_i} = 0) \ .$$

**Proof**
From propositions 878 (p. 501) point 1, 827 (p. 475) equation (A.33), (805) (p. 467) equation (A.27), we have:

1. $(\forall \mathbf{i} \in \mathbf{I})(\mathbf{a_i} = 0)$.

$$(\forall \mathbf{i} \in \mathbf{I})(\mathbf{a_i} = 0) \Rightarrow (\forall \mathbf{i} \in \mathbf{I})(\mathbf{a_i} = \emptyset)$$
$$\Rightarrow \sum_{\mathbf{i} \in \mathbf{I}} \mathbf{a_i} = \sum_{\mathbf{i} \in \mathbf{I}} \mathbf{Card}(\emptyset) = \mathbf{Card}(\bigcup_{\mathbf{i} \in \mathbf{I}} (\emptyset \times \{\mathbf{i}\})) = \mathbf{Card}(\bigcup_{\mathbf{i} \in \mathbf{I}} (\emptyset)) = \mathbf{Card}(\emptyset) = 0 \ .$$

2. $\sum_{\mathbf{i} \in \mathbf{I}} \mathbf{a_i} = 0$.

$$\sum_{\mathbf{i} \in \mathbf{I}} \mathbf{a_i} = 0 \Rightarrow \mathbf{Card}(\bigcup_{\mathbf{i} \in \mathbf{I}} (\mathbf{a_i} \times \mathbf{i})) = 0 \Rightarrow \bigcup_{\mathbf{i} \in \mathbf{I}} (\mathbf{a_i} \times \mathbf{i}) = \emptyset$$
$$\Rightarrow (\forall \mathbf{i} \in \mathbf{I})(\mathbf{a_i} \times \mathbf{i} = \emptyset) \Rightarrow (\forall \mathbf{i} \in \mathbf{I})(\mathbf{a_i} = \emptyset) \Rightarrow (\forall \mathbf{i} \in \mathbf{I})(\mathbf{a_i} = 0) \ .$$

Then proposition 891 (p. 507) is true.

**Proposition 892** Given a set of cardinals $\mathbf{A}$, two sets of indices $\mathbf{I}, \mathbf{J}$ with $\mathbf{I} \sim \mathbf{J}$ and a map $(\mathbf{a_i} : \mathbf{I} \longrightarrow \mathbf{A})$ calling $(\mathbf{i(j)} : \mathbf{J} \leftrightarrow \mathbf{I})$ and $(\mathbf{j(i)} : \mathbf{I} \leftrightarrow \mathbf{J})$ two bijective maps with $\mathbf{i(j(k))} = \mathbf{k}$ which exists since $\mathbf{I} \sim \mathbf{J}$, we have:

$$\sum_{\mathbf{i} \in \mathbf{I}} \mathbf{a_i} = \sum_{\mathbf{j} \in \mathbf{J}} \mathbf{a_{i(j)}} \ .$$

**Proof**
It is a direct consequence of 797 (p. 465) with definition 393 (p. 506).

**Proposition 893** Given two sets $\mathbf{X}$ and $\mathbf{I}$, a map $\mathbf{x_i} : \mathbf{I} \to \mathcal{P}(\mathbf{X})$ such that $\mathbf{i} \neq \mathbf{j} \Rightarrow \mathbf{x_i} \cap \mathbf{x_j} = \emptyset$ we have:

$$\sum_{\mathbf{i} \in \mathbf{I}} \mathbf{Card}(\mathbf{x_i}) = \mathbf{Card}(\bigcup_{\mathbf{i} \in \mathbf{I}} \mathbf{x_i}) \ .$$

**Proof**
From definition 393 (p. 506), we have: $\sum_{\mathbf{i} \in \mathbf{I}} \mathbf{Card}(\mathbf{x_i}) = \mathbf{Card}(\bigcup_{\mathbf{i} \in \mathbf{I}} \mathbf{x_i} \times \{\mathbf{i}\})$. From proposition 820 (p. 472), we have: $(\forall \mathbf{i} \in \mathbf{I})(\mathbf{x_i} \times \{\mathbf{i}\} \sim \mathbf{x_i})$. From proposition 887 (p. 504), we have proposition 893. It is a direct consequence of 797 (p. 465) with definition 393 (p. 506).

**Proposition 894** Given two sets $\mathbf{X}$ and $\mathbf{I}$, a map $\mathbf{x_i : I \to \mathcal{P}(X)}$ we have:

$$\sum_{i\in I} \mathbf{Card}(x_i) \geq \mathbf{Card}(\bigcup_{i\in I} x_i) \; .$$

**Proof**
We have: $\sum_{i\in I} \mathbf{Card}(x_i) = \mathbf{Card}(\bigcup_{i\in I}(x_i \times \{i\}))$ (definition 393 (p. 506)). It leads to the following relation:

$$\begin{aligned} \alpha \in \bigcup_{i\in I}(x_i \times \{i\}) &\Rightarrow (\exists a \times i)(a \in x_i, i \in I, \alpha = a \times i) \\ &\Rightarrow (\exists a)(\exists i)(i \in I, a \in x_i, \alpha = a \times i = (a, i)) \end{aligned}$$

Using the above relation, we can define a surjective map (definition 360 (p. 461)) $(\mathbf{f} : \bigcup_{i\in I} x_i \times \{i\} \to \bigcup_{i\in I} x_i)$:

$$f(\alpha) = f(a \times i) = a \in x_i \; .$$

The map $\mathbf{f}$ is surjective, because, for any $\mathbf{a} \in \bigcup_{i\in I} x_i$, we have an $\mathbf{i}$ such that $\mathbf{a} \in x_i$ (definition 362 (p. 462)). It gives $\mathbf{a \times i} \in \bigcup_{j\in I} x_j \times \{j\}$ and $f(a \times i) = a$. Then $f(\bigcup_{i\in I} x_i \times \{i\}) = \bigcup_{i\in I} x_i$. From proposition 182 (p. 143), there is an injective map (definition 359 (p. 461)) $(\mathbf{g} : \bigcup_{i\in I} x_i \to \bigcup_{i\in I} x_i \times \{i\})$ which, from proposition 180 (p. 143), defines a bijective map (definition 361 (p. 461)) between $\bigcup_{i\in I} x_i$ and $g(\bigcup_{i\in I} x_i) \subset \bigcup_{i\in I} x_i \times \{i\}$. From definition 392 (p. 500), we have proposition 894.

**Proposition 895** Given two sets $\mathbf{X_1, X_2}$, such that $\mathbf{X_1 \cap X_2 = \emptyset}$, we have $\mathbf{Card}(X_1 \cup X_2) = \mathbf{Card}(X_1) + \mathbf{Card}(X_2)$.

**Proof**
It is a consequence of propositions 746 (p. 448), 893 (p. 507) and definition 351 (p. 453), 362 (p. 462).

**Definition 394** $\mathbf{Card(A) - Card(B)}$.

Given two sets $\mathbf{A, B}$ such that $\mathbf{B \subset A}$, one calls $\mathbf{Card(A) - Card(B)}$, $\mathbf{Card}(\mathcal{C}_A(B))$:

$$\mathbf{Card}(A) - \mathbf{Card}(B) = \mathbf{Card}(\mathcal{C}_A(B)) \; .$$

**Proposition 896** Given two sets $\mathbf{A, B}$ such that $\mathbf{B \subset A}$, we have:

$$(\mathbf{Card}(A) - \mathbf{Card}(B)) + \mathbf{Card}(B) = \mathbf{Card}(A) \; .$$

**Proof**
It is a direct consequence of propositions 895 (p. 508), 893 (p. 507).

**Proposition 897** Given three sets $\mathbf{X}$, $\mathbf{I}$ and $\mathbf{J}$, a map $(\mathbf{x_i : I \to \mathcal{P}(X)})$, a partition of $\mathbf{I}$ (definition 370 (p. 471)) $(\mathbf{I_j : J \to \mathcal{P}(I)})$, then:

$$\sum_{j\in J}\sum_{i\in I_j} \mathbf{Card}(x_i) = \sum_{i\in I} \mathbf{Card}(x_i) \; . \tag{A.54}$$

**Proof**
From proposition 796 (p. 464), we have:

$$\bigcup_{j\in J}\bigcup_{i\in I_j}(\mathbf{Card}(x_i)\times\{i\})=\bigcup_{i\in I}(\mathbf{Card}(x_i)\times\{i\})\ .$$

Since $i\neq i'\Rightarrow \mathbf{Card}(x_i)\times\{i\}\cap\mathbf{Card}(x_{i'})\times\{i'\}=\emptyset$ we have from definition 393 (p. 506):

$$\mathbf{Card}(\bigcup_{i\in I}(\mathbf{Card}(x_i)\times\{i\}))=\sum_{i\in I}\mathbf{Card}(x_i)\ .$$

We also have, from proposition 798 (p. 465):

$$j\neq j'\Rightarrow I_j\cap I_{j'}=\emptyset\Rightarrow(\bigcup_{i\in I_j}(\mathbf{Card}(x_i)\times\{i\}))\cap(\bigcup_{i\in I_{j'}}(\mathbf{Card}(x_i)\times\{i\}))=$$
$$\bigcup_{i\in I_j}\bigcup_{i'\in I_{j'}}(\mathbf{Card}(x_i)\times\{i\}\cap\mathbf{Card}(x_{i'})\times\{i'\})=\bigcup_{i\in I_j\cap I_{j'}}(\mathbf{Card}(x_i)\times\{i\})=\emptyset\ .$$

Then from proposition 892 (p. 507):

$$\mathbf{Card}(\bigcup_{j\in J}\bigcup_{i\in I_j}\mathbf{Card}(x_i)\times\{i\})=\sum_{j\in J}\mathbf{Card}(\bigcup_{i'\in I_j}\mathbf{Card}(x_{i'})\times\{i'\})=$$
$$\sum_{j\in J}\sum_{i\in I_j}\mathbf{Card}(\mathbf{Card}(x_i)\times\{i\})=\sum_{j\in J}\sum_{i\in I_j}\mathbf{Card}(x_i)\ .$$

Then since:

$$\mathbf{Card}(\bigcup_{j\in J}\bigcup_{i\in I_j}\mathbf{Card}(x_i)\times\{i\})=\mathbf{Card}(\bigcup_{i\in I}\mathbf{Card}(x_i)\times\{i\})=\sum_{i\in I}\mathbf{Card}(x_i)$$

we have proposition 897.

**Proposition 898** Given two cardinals $\mathbf{a}$, $\mathbf{b}$ we have $\mathbf{b}<\mathbf{a}\Rightarrow(\exists\mathbf{c}\neq\emptyset)(\mathbf{a}=\mathbf{b}+\mathbf{Card}(\mathbf{c}))$[10].

**Proof**
From definition 392 (p. 500), $(\exists\mathbf{a}'\subset\mathbf{b})(\mathbf{a}'\sim\mathbf{a})$. Then since $\mathbf{a}'\cup\mathcal{C}_\mathbf{b}(\mathbf{a}')=\mathbf{b},\mathbf{a}'\cap\mathcal{C}_\mathbf{b}(\mathbf{a}')=\emptyset$ from proposition 893 (p. 507),

$$\mathbf{b}=\mathbf{Card}(\mathbf{b})=\mathbf{Card}(\mathbf{a}')+\mathbf{Card}(\mathcal{C}_\mathbf{b}(\mathbf{a}'))=\mathbf{a}+\mathbf{Card}(\mathcal{C}_\mathbf{b}(\mathbf{a}'))$$

$\mathcal{C}_\mathbf{b}(\mathbf{a}')\neq\emptyset$ because otherwise $\mathbf{a}=\mathbf{b}$ (propositions 778 (p. 458), 766 (p. 454)). We conclude using proposition 727 (p. 441), with $\mathbf{c}=\mathcal{C}(\mathbf{a}')$.

**Definition 395 $\mathbf{a}-\mathbf{b}$.**

Given two cardinals $\mathbf{a}$, $\mathbf{b}$, using definition 394 (p. 508) and propositions 896 (p. 508), 898 (p. 509), if $\mathbf{a}<\mathbf{b}$ one sets:

$$\mathbf{a}-\mathbf{b}=\mathbf{Card}(\tau_x(x\neq\emptyset,\mathbf{a}=\mathbf{b}+\mathbf{Card}(x)))\ .$$

If $\mathbf{a}=\mathbf{b}$ we set $\mathbf{a}-\mathbf{b}=0$.

---

[10] Be careful we may have $\mathbf{b}<\mathbf{a}$, $\mathbf{a}=\mathbf{b}+\mathbf{c}$, $\mathbf{a}=\mathbf{b}+\mathbf{c}'$ and $\mathbf{c}\neq\mathbf{c}'$. For instance, we have: $\mathbf{Card}(\mathcal{R})>\mathbf{Card}(\mathcal{Z}),\mathbf{Card}(\mathcal{R})=\mathbf{Card}(\mathcal{Z})+\mathbf{Card}(\mathcal{R})=\mathbf{Card}(\emptyset)+\mathbf{Card}(\mathcal{R})$.

**Proposition 899** $\mathbf{a}, \mathbf{b}$ being cardinals we have (definition 395 (p. 509)): $\mathbf{a} - \mathbf{b} + \mathbf{b} = \mathbf{a}$.

**Proof**
It comes straight from definition 395 (p. 509).

**Proposition 900** Given two sets $\mathbf{X}$ and $\mathbf{I}$, a map $\mathbf{x_i} : \mathbf{I} \to \mathcal{P}(\mathbf{X})$ such that $\mathbf{i} \neq \mathbf{j} \Rightarrow \mathbf{x_i} \cap \mathbf{x_j} = \emptyset$, we define $\mathbf{I_0}$ as $\mathbf{I_0} = \mathrm{coll_i}\{\mathbf{i} \in \mathbf{I}, \mathbf{x_i} = \emptyset\}$ we have:

$$\sum_{\mathbf{i}\in\mathbf{I}} \mathbf{Card}(\mathbf{x_i}) = \sum_{\mathbf{i}\in\mathcal{C}_{\mathbf{I}}(\mathbf{I_0})} \mathbf{Card}(\mathbf{x_i}) \ .$$

**Proof**
It is a direct consequence of propositions 806 (p. 468) and 893 (p. 507).

**Definition 396 Cardinal product.**

One defines the product of cardinals $(\times^c, \underset{\mathbf{i}\in\mathbf{I}}{\times})$ by:
Given a set of cardinals $\mathbf{A}$, a set of index $\mathbf{I}$ and a map $(\mathbf{a_i} : \mathbf{I} \to \mathbf{A})$. One makes the equivalence (see definition 372 (p. 473)):

$$\underset{\mathbf{i}\in\mathbf{I}}{\times} \mathbf{a_i} = \mathbf{Card}(\prod_{\mathbf{i}\in\mathbf{I}} \mathbf{a_i}) \ .$$

**Proposition 901** Given a set $\mathbf{X}$, a set of index $\mathbf{I}$ and a map $(\mathbf{x_i} : \mathbf{I} \to \mathcal{P}(\mathbf{X}))$ we have:

$$\underset{\mathbf{i}\in\mathbf{I}}{\times} \mathbf{Card}(\mathbf{x_i}) = \mathbf{Card}(\prod_{\mathbf{i}\in\mathbf{I}} \mathbf{x_i}) \ .$$

**Proof**
From proposition 751 (p. 450), convention 36 (p. 560) and definitions 390 (p. 499), 332 (p. 440), we have:

$$(\exists \mathbf{A})(\mathbf{a} \in \mathbf{A}) \Leftrightarrow (\exists \mathbf{i} \in \mathbf{I})(\mathbf{a} = \mathbf{Card}(\mathbf{x_i})) \ .$$

From point 4 (p. 132) of definition 112 (p. 131), it defines a map $(\mathbf{a_i} : \mathbf{I} \to \mathbf{A})$ and we have: $\underset{\mathbf{i}\in\mathbf{I}}{\times} \mathbf{Card}(\mathbf{x_i}) = \underset{\mathbf{i}\in\mathbf{I}}{\times} \mathbf{a_i} = \mathbf{Card}(\prod_{\mathbf{i}\in\mathbf{I}} \mathbf{a_i})$ but $\mathbf{a_i} = \mathbf{Card}(\mathbf{x_i})$. Then $\mathbf{a_i} \sim \mathbf{x_i}$ and from proposition 888 (p. 505), we have:

$$\underset{\mathbf{i}\in\mathbf{I}}{\times} \mathbf{Card}(\mathbf{x_i}) = \underset{\mathbf{i}\in\mathbf{I}}{\times} \mathbf{a_i} = \mathbf{Card}(\prod_{\mathbf{i}\in\mathbf{I}} \mathbf{a_i}) = \mathbf{Card}(\prod_{\mathbf{i}\in\mathbf{I}} \mathbf{x_i}) \ .$$

We have proposition 901 (p. 510).

**Proposition 902** Given two sets $\mathbf{X}, \mathbf{Y}$, a set $\mathbf{I}$ (see footnote 7 (p. 451)), two maps $(\mathbf{x_i} : \mathbf{I} \to \mathcal{P}(\mathbf{X}))$, $(\mathbf{y_i} : \mathbf{I} \to \mathcal{P}(\mathbf{Y}))$ with $(\forall \mathbf{i} \in \mathbf{I})(\mathbf{Card}(\mathbf{x_i}) \leq \mathbf{Card}(\mathbf{y_i}))$, we have:

$$\underset{\mathbf{i}\in\mathbf{I}}{\times} \mathbf{Card}(\mathbf{x_i}) \leq \underset{\mathbf{i}\in\mathbf{I}}{\times} \mathbf{Card}(\mathbf{y_i}) \ .$$

**Proof**
It is a direct consequence of proposition 826 (p. 475).

**Proposition 903** We consider two terms $\boldsymbol{\alpha}, \boldsymbol{\beta}$ and the set of two elements (definition 345 (p. 447)) $\mathbf{I} = \{\boldsymbol{\alpha}, \boldsymbol{\beta}\}$. $\boldsymbol{\alpha}, \boldsymbol{\beta}$ can be considered as sets (footnote 7 (p. 451)). We have

$$\mathbf{Card}(\boldsymbol{\alpha} \times \boldsymbol{\beta}) = \mathbf{Card}(\prod_{\mathbf{i}\in\mathbf{I}} \mathbf{i}) = \underset{\mathbf{i}\in\mathbf{I}}{\times} \mathbf{Card}(\mathbf{i}) = \mathbf{Card}(\boldsymbol{\alpha}) \times^c \mathbf{Card}(\boldsymbol{\beta}) \ .$$

**Proof**
It is a direct consequence of propositions 822 (p. 473) and 901 (p. 510). The last equality can be considered as the definition of the product of two cardinals.

**Proposition 904** Given a set $\mathbf{X}$, two sets of index $\mathbf{I} \sim \mathbf{J}$ with two associated bijections, $(\mathbf{i(j)} : \mathbf{J} \leftrightarrow \mathbf{I})$, $(\mathbf{j(i)} : \mathbf{I} \leftrightarrow \mathbf{J})$, $\mathbf{i(j(k))} = \mathbf{k}$ and a map $(\mathbf{x_i} : \mathbf{I} \to \mathcal{P}(\mathbf{X}))$ we have:

$$\underset{i \in I}{\times} \mathbf{Card}(\mathbf{x_i}) = \underset{j \in J}{\times} \mathbf{Card}(\mathbf{x_{i(j)}}) .$$

**Proof**
It is a direct consequence of propositions 890 (p. 506) and 901 (p. 510).

**Proposition 905** Given three sets $\mathbf{X}$, $\mathbf{I}$ and $\mathbf{J}$, a map $(\mathbf{x_i} : \mathbf{I} \to \mathcal{P}(\mathbf{X}))$ and a partition of $\mathbf{I}$ (definition 370 (p. 471)) $(\mathbf{I_j} : \mathbf{J} \to \mathcal{P}(\mathbf{I}))$ we have:

$$\underset{j \in J}{\times} \underset{i \in I_j}{\times} \mathbf{Card}(\mathbf{x_i}) = \underset{i \in I}{\times} \mathbf{Card}(\mathbf{x_i}) .$$

**Proof**
It is a direct consequence of proposition 889 (p. 506).

**Proposition 906** Given a set $\mathbf{X}$, a set of index $\mathbf{I}$ and a map $(\mathbf{x_i} : \mathbf{I} \to \mathcal{P}(\mathbf{X}))$ we have:

$$(\exists i \in I)(\mathbf{Card}(x_i) = 0) \Leftrightarrow \mathbf{Card}(\prod_{i \in I} x_i) = 0 = \emptyset .$$

**Proof**
The $\Rightarrow$ part of proposition 906 is a consequence of proposition 827 (p. 475), 878 (p. 501)point 1.
Using propositions 878 (p. 501) and 773 (p. 457), we have:

$$\begin{aligned}
&\neg(\exists i \in I)(\mathbf{Card}(x_i) = 0) \\
&\Leftrightarrow \neg(\exists i \in I)(\neg\neg(\mathbf{Card}(x_i) = 0)) = (\forall i \in I)(\mathbf{Card}(x_i) \neq 0) \\
&\Leftrightarrow (\forall i \in I)(x_i \neq \emptyset) \Leftrightarrow (\forall i \in I)(\exists a \in x_i) .
\end{aligned}$$

From definition 2.1.1 (p. 131), point 4, $(\forall i \in I)(\exists a \in x_i)$ allows one to define a map $(\mathbf{u} : \mathbf{I} \to \mathbf{X})$ such that $(\forall i \in I)(\mathbf{u(i)} \in x_i)$. This means $(\exists u \in \prod_{i \in I} x_i)$ is true. Since from A.1.4, 7, we have:

$$(\mathbf{Card}(\prod_{i \in I} x_i) = 0 \Rightarrow \prod_{i \in I} x_i = \emptyset) \Leftrightarrow (\prod_{i \in I} x_i \neq \emptyset \Rightarrow \mathbf{Card}(\prod_{i \in I} x_i) \neq 0) .$$

We have from proposition 878 (p. 501):

$$(\exists u \in \prod_{i \in I} x_i) \Rightarrow \prod_{i \in I} x_i \neq \emptyset \Rightarrow \mathbf{Card}(\prod_{i \in I} x_i) \neq 0$$

Then we have successively true:

$$(\neg(\exists i \in I)(\mathrm{Card}(x_i) = 0)) \Rightarrow \mathrm{Card}(\prod_{i \in I} x_i) \neq 0$$
$$\neg\neg(\exists i \in I)(\mathrm{Card}(x_i) = 0) \vee \neg\mathrm{Card}(\prod_{i \in I} x_i) = 0$$
$$(\neg\neg(\exists i \in I)(\mathrm{Card}(x_i) = 0) \vee \neg\mathrm{Card}(\prod_{i \in I} x_i) = 0)$$
$$\Leftrightarrow (\mathrm{Card}(\prod_{i \in I} x_i) = 0 \Rightarrow (\exists i \in I)(\mathrm{Card}(x_i) = 0))$$
$$\mathrm{Card}(\prod_{i \in I} x_i) = 0 \Rightarrow (\exists i \in I)(\mathrm{Card}(x_i) = 0) .$$

The $\Leftarrow$ part of proposition 906 is also true.

**Proposition 907** Given two sets $\mathbf{A}$, $\mathbf{B}$

$$\mathrm{Card}(B) = 1 \Rightarrow \mathrm{Card}(A \times B) = \mathrm{Card}(A) .$$

**Proof**
Let us consider the map $(f : A \times B \to A)$ defined as $f(x, y) = x$ and the map $(g : B \leftrightarrow \mathcal{P}(\emptyset))$. Since $(\forall y \in B)(y = g^{-1}(\emptyset))$, we have

$$(\forall x \in A)((\exists((u, v) = (x, g^{-1}(\emptyset)) \in f^{-1}(x)), ((u, v) \in f^{-1}(x)), (u', v') \in f^{-1}(x)))$$
$$\Rightarrow (u = u' = x, v = v' = g^{-1}(\emptyset)) \Rightarrow (u, v) = (u', v') .$$

From definition 361 (p. 461), $\mathbf{f}$ is a bijection. Then from definitions 390 (p. 499), 371 (p. 472), we have:

$$(A \times B \sim A, \mathrm{Card}(A \times B) = \mathrm{Card}(A \times 1)) \Rightarrow \mathrm{Card}(A \times 1) = \mathrm{Card}(A) .$$

**Proposition 908** Given three sets $\mathbf{X}$, $\mathbf{Y}$ and $\mathbf{I}$, two maps $(x_i : I \to \mathcal{P}(X))$, $(y_i : I \to \mathcal{P}(Y))$ such that $(\forall i \in I)(\mathrm{Card}(x_i) \leq \mathrm{Card}(y_i))$, we have:

$$\sum_{i \in I} \mathrm{Card}(x_i) \leq \sum_{i \in I} \mathrm{Card}(y_i) .$$

**Proof**
We have from definition 392 (p. 500):

$$(\forall i \in I)(\exists s_i \subset y_i)(x_i \sim s_i) .$$

We also have, proposition 783 (p. 459), proposition 878 (p. 501) -3 and proposition 907 (p. 512):

$$(\forall i \in I)(s_i \times \{i\} \subset y_i \times \{i\}, \mathrm{Card}(s_i \times \{i\}) = \mathrm{Card}(s_i), \mathrm{Card}(y_i \times \{i\}) = \mathrm{Card}(y_i)) .$$

Then we also have:

$$(\forall i \in I)(\mathrm{Card}(x_i) \times \{i\} = \mathrm{Card}(x_i)\mathrm{Card}(s_i \times \{i\}) = \mathrm{Card}(s_i)$$
$$\leq \mathrm{Card}(y_i \times \{i\}) = \mathrm{Card}(y_i)) .$$

From proposition 893 (p. 507), we have:

$$\sum_{i\in I}\mathrm{Card}(x_i) = \sum_{i\in I}\mathrm{Card}(s_i) = \sum_{i\in I}\mathrm{Card}(s_i \times \{i\}) = \mathrm{Card}(\bigcup s_i \times \{i\})$$
$$\leq \mathrm{Card}(\bigcup(y_i \times \{i\})) = \sum_{i\in I}\mathrm{Card}(y_i)\ .$$

**Proposition 909** Given a set **X**, a set of index **I** and a map $(x_i : I \to \mathcal{P}(X))$, we have:

$$(\forall i \in I)(\mathrm{Card}(x_i) = 1) \Leftrightarrow \mathrm{Card}(\prod_{i\in I} x_i) = 1\ .$$

**Proof**

From definitions 390 (p. 499), 371 (p. 472), 361 (p. 461) and from proposition 878 (p. 501) we have:

$$(\forall i \in I)(\mathrm{Card}(x_i) = 1) \Leftrightarrow (\forall i \in I)(\exists f_i)((f_i : x_i \leftrightarrow \mathcal{P}(\emptyset)))$$

Using that $f_i$, we have $f_i^{-1}(\emptyset) \in x_i$. Then $(\forall_{i\in I})(x_i \neq \emptyset)$. The map of $(f_i^{-1} : [1,n] \to \bigcup_{i\in[1,n]} x_i)$ is an element of $\prod_{i\in I} x_i$ (definition 372 (p. 473)). Then $\prod_{i\in I} x_i \neq \emptyset$.
Given $u \in \prod_{i\in I} x_i, v \in \prod_{i\in I} x_i$ and the two associated maps $(u : I \to X)$, $(v : I \to X)$ such that:

$$i \in I \Rightarrow u(i) \in x_i, v(i) \in x_i \Rightarrow u(i) = v(i)$$

we have, using the definitions 2.1.1 (p. 131) and 357 (p. 460):

$$u \in I \times X, v \in I \times X\ .$$

It gives:

$$\begin{aligned}&(\forall(i, x_u) \in u)(\exists(i, x_v) \in v)(x_u \in x_i, x_v \in x_i)\\&\Rightarrow (\forall(i, x_u)(\exists(i, x_v) \in v)(x_u = x_v) \Rightarrow u \subset v\ .\end{aligned} \tag{A.55}$$

By switching **u** and **v**, we also have $v \subset u$. Then we have: $u = v$ (proposition 753). We can now define a map $(h : \prod_{i\in I} x_i \to \mathcal{P}(\emptyset))$ such that $(\forall u \in \prod_{i\in I} x_i)(h(u) = \emptyset)$. **h** is a bijection because:

1. It is surjective (definition 360 (p. 461))

$$(\forall z \in \mathcal{P}(\emptyset))(\exists u \in \prod_{i\in I} x_i)(h(u) = z)$$

2. It is injective (definition 359 (p. 461)) because from equation(A.55), we have:

$$(\forall z \in \mathcal{P}(\emptyset))(h(u) = h(v) = z = \emptyset \Rightarrow u \in \prod_{i\in I} x_i, v \in \prod_{i\in I} x_i \Rightarrow v = u\ .$$

Then from definition 361 (p. 461), it is bijective. $\prod_{i\in I} x_i \sim \mathcal{P}(\emptyset)$ (definition 371 (p. 472)). From definitions 390 (p. 499), 391 (p. 499) equation (A.50), we have proposition 909.

**Proposition 910** Given a set $\mathbf{X}$, a set of index $\mathbf{I}$ and a map $(\mathbf{x_i} : \mathbf{I} \to \mathcal{P}(\mathbf{X}))$ we define:

$$\mathbf{I_1} = \mathbf{set_i}\{\mathbf{i} \in \mathbf{I}, \mathbf{Card}(\mathbf{x_i}) = \mathbf{1}\},\ \mathbf{I_0} = \mathcal{C}_\mathbf{I}(\mathbf{I_1}) = \mathbf{set_i}\{\mathbf{i} \in \mathbf{I}, \mathbf{Card}(\mathbf{x_i}) \neq \mathbf{1}\} .$$

We have:

$$\mathbf{Card}(\prod_{\mathbf{i}\in\mathbf{I}} \mathbf{x_i}) = \mathbf{Card}(\prod_{\mathbf{i}\in\mathcal{C}_\mathbf{I}(\mathbf{I_1})} \mathbf{x_i}) .$$

**Proof**

It is a consequence of proposition 889 (p. 506) with $\mathbf{J} = \{\mathbf{0}, \mathbf{1}\}$, and of propositions 909 (p. 513) and 907 (p. 512).

**Convention 34** $\times_{\mathbf{i}\in\mathbf{I}}\mathbf{a_i}$ and $\prod_{\mathbf{i}\in\mathbf{I}} \mathbf{a_i}$.

Most of the time when there is no confusion, we shall forget the distinction between cardinal product $\times_{\mathbf{i}\in\mathbf{I}} \mathbf{a_i} = \mathbf{Card}(\prod_{\mathbf{i}\in\mathbf{I}} \mathbf{a_i})$ (definition 396 (p. 510)) and the product itself $\prod_{\mathbf{i}\in\mathbf{I}} \mathbf{a_i}$. See also definition 372 (p. 473).

**Proposition 911** $\mathbf{A}$ being a set, a set of indices $\mathbf{L}$, a map $(\mathbf{J}_\lambda : \mathbf{L} \to \mathcal{P}(\mathbf{A}))$ and a set of maps over a set of sets $\mathbf{X}$: $(\mathbf{x}_{\lambda,\iota} : \mathbf{J}_\lambda \to \mathbf{X})$, setting $\mathbf{I} = \prod_{\lambda\in\mathbf{L}} \mathbf{J}_\lambda$, we have:

$$\times_{\lambda\in\mathbf{L}} \sum_{\iota\in\mathbf{J}_\lambda} \mathbf{Card}(\mathbf{x}_{\lambda,\iota}) = \sum_{\mathbf{f}\in\mathbf{I}} \times_{\lambda\in\mathbf{L}} \mathbf{Card}(\mathbf{x}_{\lambda,\mathbf{f}(\lambda)})$$

which is the same as:

$$\mathbf{Card}(\prod_{\lambda\in\mathbf{L}} \sum_{\iota\in\mathbf{J}_\lambda} \mathbf{Card}(\mathbf{x}_{\lambda,\iota})) = \sum_{\mathbf{f}\in\mathbf{I}} \mathbf{Card}(\prod_{\lambda\in\mathbf{L}} \mathbf{Card}(\mathbf{x}_{\lambda,\mathbf{f}(\lambda)})) .$$

**Proof**

As in definition 393 (p. 506), we define $(\mathbf{y}_{\lambda,\iota} = \mathbf{x}_{\lambda,\iota} \times \{(\lambda, \iota)\})$. It gives:

$$\mathbf{Card}(\mathbf{y}_{\lambda,\iota}) = \mathbf{Card}(\mathbf{x}_{\lambda,\iota})$$

and

$$(\lambda', \iota') \neq (\lambda, \iota) \Leftrightarrow \mathbf{y}_{\lambda',\iota'} \cap \mathbf{y}_{\lambda,\iota} = \emptyset .$$

Then using successively propositions 893 (p. 507), 901 (p. 510), 832 (p. 478):

$$\mathbf{Card}(\prod_{\lambda\in\mathbf{L}} \sum_{\iota\in\mathbf{J}_\lambda} \mathbf{Card}(\mathbf{x}_{\lambda,\iota})) = \mathbf{Card}(\prod_{\lambda\in\mathbf{L}} \bigcup_{\iota\in\mathbf{J}_\lambda} \mathbf{y}_{\lambda,\iota}) = \mathbf{Card}(\bigcup_{\mathbf{f}\in\mathbf{I}} \prod_{\lambda\in\mathbf{L}} \mathbf{y}_{\lambda,\mathbf{f}(\lambda)}) .$$

We also have:

$$(\mathbf{f} \in \mathbf{I}, \mathbf{f}' \in \mathbf{I}) \Rightarrow (\mathbf{f} \neq \mathbf{f}' \Leftrightarrow (\exists\lambda \in \mathbf{L})(\mathbf{f}'(\lambda) \neq \mathbf{f}(\lambda)))$$

since

$$(\mathbf{f}'(\lambda) \neq \mathbf{f}(\lambda)) \Rightarrow \mathbf{y}_{\lambda,\mathbf{f}(\lambda)} \cap \mathbf{y}_{\lambda,\mathbf{f}'(\lambda)} = \emptyset .$$

From proposition 835 (p. 480), we have:

$$\mathbf{f}'(\lambda) \neq \mathbf{f}(\lambda) \Rightarrow \prod_{\lambda\in\mathbf{L}} \mathbf{y}_{\lambda,\mathbf{f}(\lambda)} \cup \prod_{\lambda\in\mathbf{L}} \mathbf{y}_{\lambda,\mathbf{f}'(\lambda)} = \emptyset .$$

Then from propositions 893 (p. 507), 901 (p. 510), we have:

$$\mathbf{Card}(\bigcup_{\mathbf{f}\in\mathbf{I}} \prod_{\lambda\in\mathbf{L}} \mathbf{y}_{\lambda,\mathbf{f}(\lambda)}) = \sum_{\mathbf{f}\in\mathbf{I}} \mathbf{Card}(\prod_{\lambda\in\mathbf{L}} \mathbf{y}_{\lambda,\mathbf{f}(\lambda)})$$

$$= \sum_{\mathbf{f}\in\mathbf{I}} \mathbf{Card}(\prod_{\lambda\in\mathbf{L}} \mathbf{Card}(\mathbf{y}_{\lambda,\mathbf{f}(\lambda)})) = \sum_{\mathbf{f}\in\mathbf{I}} \mathbf{Card}(\prod_{\lambda\in\mathbf{L}} \mathbf{Card}(\mathbf{x}_{\lambda,\mathbf{f}(\lambda)})) .$$

**Proposition 912** $\mathbf{I}$ being a set of index, $\mathbf{A}$ a set of cardinals, $(\mathbf{a_i} : \mathbf{I} \rightarrow \mathbf{A})$ a map and $\mathbf{b}$ a cardinal. We have:

$$\mathbf{b} \times \sum_{\mathbf{i} \in \mathbf{I}} \mathbf{a_i} = \sum_{\mathbf{i} \in \mathbf{I}} \mathbf{b} \times \mathbf{a_i} \ .$$

**Proof**

It is a direct application of proposition 911 (p. 514) with $\mathbf{Card(L) = 1 + 1 = 2}$ ($\mathbf{L = \{1, 2\}}$ for instance) and $\mathbf{J_1 = \{1\}}$, $\mathbf{J_2 = I}$, $\mathbf{x_{1,1} = b = Card(b)}$, $(\forall \mathbf{i} \in \mathbf{I})(\mathbf{x_{2,i} = a_i = Card(a_i)})$.

**Proposition 913** $\mathbf{s}$ being a cardinal, we have: $\mathbf{s \leq s + 1}$.

**Proof**

Using proposition 881 (p. 503), there is $\mathbf{x_1 \notin s}$. From proposition 893 (p. 507), we can write $\mathbf{s + 1 = Card(s \cup \{x_1\})}$, $\mathbf{s \subset s \cup \{x_1\}}$. From definition 392 (p. 500), the last relation gives: $\mathbf{Card(s) \leq Card(s + 1)}$, we have $\mathbf{s \leq s + 1}$.

**Proposition 914** $\mathbf{a}$, $\mathbf{b}$ being cardinals, $\mathbf{a + 1 = b + 1 \Leftrightarrow a = b}$.

**Proof**

Using proposition 881 (p. 503) $(\exists \mathbf{x_1} \in \mathcal{P}(\mathbf{a}))(\mathbf{x_1 \notin a})$, $(\exists \mathbf{x_2} \in \mathcal{P}(\mathbf{b} \cup \{\mathbf{x_1}\}))(\mathbf{x_2} \notin \mathbf{b} \cup \{\mathbf{x_1}\})$. Considering the sets $\mathbf{a' = a \cup \{(x_1\}}$ and $\mathbf{b' = b \cup \{x_2\}}$, from proposition 893 (p. 507), $\mathbf{Card(a') = a + 1}$, $\mathbf{Card(b') = b + 1}$ then $\mathbf{Card(a') = Card(b')}$. From proposition 873 (p. 499), $\mathbf{a' \sim b'}$. It implies there is a bijection (definition 361 (p. 461)) $(\mathbf{f : a' \leftrightarrow b'})$ (see definition 371 (p. 472)). Since $(\mathbf{f : a' \rightarrow b'})$ is a bijection, we have:

$$(\mathbf{f^{-1}(\{x_2\}) \in a, x_1 \notin a}) \Rightarrow \mathbf{f^{-1}(\{x_2\}) \neq x_1} \Rightarrow \mathbf{f(x_1) \neq x_2} \Rightarrow \mathbf{f(x_1) \in b}$$

Then we define a bijection $(\mathbf{g : a \leftrightarrow b})$ as:

$$\begin{aligned}
&(\mathbf{u \in a, u \neq f^{-1}(\{x_2\})}) \Rightarrow \mathbf{g(u) = f(u)} \\
&(\mathbf{u \in a, u = f^{-1}(\{x_2\})}) \Rightarrow \mathbf{g(u) = f(\{x_1\}), f(\{x_1\}) \in b} \ .
\end{aligned}$$

Since $\mathbf{g}$ exist and it is a bijection, we have $\mathbf{Card(a) = Card(b)}$. Since $\mathbf{a}$ and $\mathbf{b}$ are already cardinal, from proposition 874 (p. 500), $\mathbf{a = b}$. The reverse is also true, using a bijection $(\mathbf{g : a \rightarrow b})$ one generates a bijection $(\mathbf{f : a' \rightarrow b'})$:

$$(\forall \mathbf{u} \in \mathbf{a})(\mathbf{f(u) = g(u), f(\{x_1\}) = \{x_2\}}) \ .$$

$\mathbf{Card(a') = Card(b')}$ and $\mathbf{a + 1 = b + 1}$.

**Proposition 915** $\mathbf{I}$, $\mathbf{X}$ being sets, given a map $(\mathbf{x_i} : \mathbf{I} \rightarrow \mathcal{P}(\mathbf{X}))$ such that $(\exists \mathbf{x} \in \mathcal{P}(\mathbf{X}))(\forall \mathbf{i} \in \mathbf{I})(\mathbf{x_i = x})$, we have:

$$(\forall \mathbf{j} \in \mathbf{I})(\forall \mathbf{k} \in \mathbf{I})(\mathop{\times}_{\mathbf{i} \in \mathbf{I} - \mathbf{j}} \mathbf{x_i} = \mathop{\times}_{\mathbf{i} \in \mathbf{I} - \mathbf{k}} \mathbf{x_i}) \ .$$

**Proof**

We use the above notations. If $\mathbf{I = \emptyset}$, proposition 915 is trivially true. Calling $\mathbf{x} = \tau_{\mathbf{x}}(\forall \mathbf{i} \in \mathbf{I})(\mathbf{x_i = x})$, supposing $\mathbf{j \in I}$, $\mathbf{k \in I}$, from propositions 914 (p. 515), 893 (p. 507) we have:

$$\begin{aligned}
&\mathbf{I = (I - \{j\}) \cup \{j\} = (I - \{k\}) \cup \{k\}, (I - \{j\}) \cap \{j\} = \emptyset, (I - \{k\}) \cap \{k\} = \emptyset} \\
&\Rightarrow \mathbf{Card(I) = Card(I - \{j\}) + Card(\{j\}) = Card(I - \{k\}) + Card(\{k\})} \\
&= \mathbf{Card(I - \{j\}) + 1 = Card(I - \{k\}) + 1} \Rightarrow \mathbf{Card(I - \{j\}) = Card(I - \{k\})} \ .
\end{aligned}$$

Then there is a bijection $\mathbf{g : Card(I - \{j\}) \leftrightarrow Card(I - \{k\})}$. From definition 396 (p. 510), propositions 901 (p. 510) and 904 (p. 511), we have:

$$\mathop{\times}_{j\in I-\{k\}} x_j = \mathop{\times}_{i\in I-\{j\}} x_{g(i)} .$$

Then:

$$\mathop{\times}_{j\in I-\{k\}} x_j = \mathop{\times}_{j\in I-\{k\}} \mathrm{Card}(x_j) = \mathop{\times}_{j\in I-\{k\}} \mathrm{Card}(x)$$
$$= \mathop{\times}_{i\in I-\{j\}} x_{g(i)} = \mathop{\times}_{i\in I-\{j\}} \mathrm{Card}(x_{g(i)}) = \mathop{\times}_{i\in I-\{j\}} \mathrm{Card}(x) .$$

**Proposition 916** $\mathbf{a}$, $\mathbf{b}$ being cardinals, $\mathbf{a < b \Rightarrow a + 1 \leq b}$.

**Proof**
From definition 392 (p. 500), we have:

$$a < b \Rightarrow (\exists c)(\mathrm{Card}(c) = \mathrm{Card}(a), c \subset b, \mathcal{C}_b(c) \neq \emptyset) .$$

It means: $(\exists x \in \mathcal{C}_b(c))$.
Setting $x = \tau_x(x \in \mathcal{C}_b(c))$, we have $c \bigcup \{x\} \subset b$. From propositions 793 (p. 464) and 893 (p. 507), we have:

$$x \in \mathcal{C}_b(c) \Rightarrow x \notin c \Rightarrow \mathrm{Card}(c \bigcup \{x\}) = \mathrm{Card}(c) + 1 = a + 1 .$$

Then from definition 392 (p. 500):

$$\mathrm{Card}(c) + 1 \leq \mathrm{Card}(b) \Rightarrow a + 1 \leq b .$$

**Proposition 917** $\mathbf{a}$ being a cardinal, we have:

$$\neg(\exists s)(a < \mathrm{Card}(s) < a + 1) .$$

**Proof**
If $(\exists s)(a < \mathrm{Card}(s) < a + 1)$, using that $s = \tau_x(a < \mathrm{Card}(x) < a + 1)$ (see A.2 (p. 440), A.3 (p. 441)), we have from proposition 916 (p. 516):

$$a < \mathrm{Card}(s) \Rightarrow a + 1 \leq \mathrm{Card}(s) .$$

It is in contradiction with $\mathbf{Card(s) < a + 1}$. Then from A.1.4, 8, proposition 917 is true.

## A.6 Integers

**Definition 397 Integers.**

A cardinal $\mathbf{n}$ is an integer if $\mathbf{n + 1 > n}$.

**Proposition 918** A cardinal smaller than or equal to an integer is an integer.

**Proof**
If $\mathbf{n}$ is an integer and $\mathbf{n} = \mathbf{p}$, $\mathbf{p}$ is an integer from properties 6 (p. 435), A.1.3. If $\mathbf{n}$ is an integer and $\mathbf{p} < \mathbf{n}$, $\mathbf{p}$ being a cardinal, it means:

$$(\exists \mathbf{q})((\mathbf{q} \subset \mathbf{n}), (\mathbf{p} \sim \mathbf{q})) .$$

Then

$$\mathbf{n} = \mathbf{q} \cup \mathcal{C}(\mathbf{q}), \mathbf{q} \cap \mathcal{C}(\mathbf{q}) = \emptyset .$$

It gives (proposition 893 (p. 507)):

$$\mathbf{n} = \mathbf{Card}(\mathbf{q}) + \mathbf{Card}(\mathcal{C}(\mathbf{q})) .$$

Using proposition 881 (p. 503) with $\mathbf{x}_1 = \tau_{\mathbf{x}}(\mathbf{x} \notin \mathbf{n})$ and point 3 (p. 501) of proposition 878 (p. 501), we have:

$$\mathbf{n} + \mathbf{1} = \mathbf{Card}(\mathbf{n} \cup \{\mathbf{x}_1\}) = \mathbf{Card}(\mathbf{q} \cup \{\mathbf{x}_1\} \cup \mathcal{C}(\mathbf{q})) = (\mathbf{Card}(\mathbf{q}) + \mathbf{1}) + \mathbf{Card}(\mathcal{C}(\mathbf{q})) .$$

From proposition 913 (p. 515), $\mathbf{Card}(\mathbf{q}) \leq \mathbf{Card}(\mathbf{q}) + \mathbf{1}$.
If $\mathbf{Card}(\mathbf{q}) + \mathbf{1} = \mathbf{Card}(\mathbf{q})$, we have $\mathbf{n} + \mathbf{1} = \mathbf{n}$. It is against what we were supposing. Then from A.1.4 point 7 (p. 435): proposition 918 is true.

**Proposition 919** If $\mathbf{n}$ is an integer, $\mathbf{n} + \mathbf{1}$ is an integer.
**Proof**
From proposition 914 (p. 515), $\mathbf{n} + \mathbf{1} = \mathbf{n} + \mathbf{2} \Rightarrow \mathbf{n} = \mathbf{n} + \mathbf{1}$ is true. From A.1.4 point 7 (p. 435), proposition 919 is true.

**Proposition 920** If $\mathbf{a}$ and $\mathbf{b}$ are integers, $\mathbf{a} < \mathbf{b} \Leftrightarrow \mathbf{a} + \mathbf{1} \leq \mathbf{b}$.
**Proof**
From proposition 916 (p. 516), $\mathbf{a} < \mathbf{b} \Rightarrow \mathbf{a} + \mathbf{1} \leq \mathbf{b}$.
We suppose $\mathbf{a} + \mathbf{1} \leq \mathbf{b}$. We have from definition 397 (p. 516), $\mathbf{a} < \mathbf{a} + \mathbf{1}$. Then, from definition 338 (p. 446), $\mathbf{a} \leq \mathbf{a} + \mathbf{1}$, $\mathbf{a} + \mathbf{1} \leq \mathbf{b}$, $\mathbf{a} \leq \mathbf{b}$. If $\mathbf{a} = \mathbf{b}$ we would have $\mathbf{a} + \mathbf{1} \leq \mathbf{a}$ (6 (p. 435), A.1.3) which is forbidden by definition 397 (p. 516). Then $\mathbf{a} + \mathbf{1} \leq \mathbf{b} \Rightarrow \mathbf{a} < \mathbf{b}$ and coupling both $\mathbf{a} < \mathbf{b} \Leftrightarrow \mathbf{a} + \mathbf{1} \leq \mathbf{b}$.

**Proposition 921** If $\mathbf{a}$ and $\mathbf{b}$ are integers, $\mathbf{c}$ being an integer, we have:

$$\mathbf{a} > \mathbf{b} \Leftrightarrow (\exists \mathbf{c} \neq 0)(\mathbf{a} = \mathbf{b} + \mathbf{c})$$

$$\mathbf{a} \geq \mathbf{b} \Leftrightarrow (\exists \mathbf{c})(\mathbf{a} = \mathbf{b} + \mathbf{c})$$

with for both:

$$\mathbf{a} = \mathbf{b} + \mathbf{c}_1, \mathbf{a} = \mathbf{b} + \mathbf{c}_2 \Rightarrow \mathbf{c}_1 = \mathbf{c}_2 .$$

**Proof**
From proposition 898 (p. 509):

$$\mathbf{a} > \mathbf{b} \Rightarrow (\exists \mathbf{c} \neq 0)(\mathbf{a} = \mathbf{b} + \mathbf{c})$$

$\mathbf{c}$ is an integer because otherwise from propositions 796 (p. 464) and 893 (p. 507)

$$\mathbf{a} + \mathbf{1} = \mathbf{b} + \mathbf{c} + \mathbf{1} = \mathbf{b} + (\mathbf{c} + \mathbf{1}) = \mathbf{b} + \mathbf{c} = \mathbf{a} .$$

From proposition 878 (p. 501) point 4, $(\mathbf{c} \neq \mathbf{0} \Rightarrow \mathbf{c} \geq \mathbf{1})$. From propositions 796 (p. 464) and 893 (p. 507), we have:

$$(\mathbf{1} \leq \mathbf{c}, \mathbf{a} = \mathbf{b} + \mathbf{c}) \Rightarrow \mathbf{b} + \mathbf{1} \leq \mathbf{b} + \mathbf{c} = \mathbf{a} \Rightarrow \mathbf{b} < \mathbf{a} .$$

From proposition 876 (p. 500) $\mathbf{c_1 \neq c_2 \Rightarrow c_1 < c_2 \vee c_2 < c_1}$. Then if $\mathbf{c_1 \neq c_2}$, we may suppose $\mathbf{c_1 < c_2}$ and, from proposition 916 (p. 516), $\mathbf{c_1 + 1 \leq c_2}$. Then using proposition 913 (p. 515), we have:

$$\mathbf{b + c_2 + 1 = b + c_1 + 1 \leq b + c_2 \Rightarrow b + c_2 + 1 \leq b + c_2 \Rightarrow b + c_2 + 1 = b + c_2}$$

$\mathbf{b + c_2}$ would not be an integer which is in contradiction with $\mathbf{a = b + c_2}$ where $\mathbf{a}$ is an integer. For the second part, we have:

$$\mathbf{a \geq b \Leftrightarrow (\exists c)(a = b + c)}\ .$$

By definition 338 (p. 446), convention 28 (p. 446), we have $\mathbf{a \geq b = a > b \vee a = b}$. We then have to consider only $\mathbf{a = b}$. When $\mathbf{a = b}$ we have (propositions 900 (p. 510), 746 (p. 448)) $\mathbf{a = b + 0}$ and if $\mathbf{a = b + c}$ with $\mathbf{c \neq \emptyset}$ then (propositions 908 (p. 512), 746 (p. 448)) $\mathbf{c \geq 1}$. It would mean $\mathbf{a \geq a + 1}$ which is forbidden.

**Definition 398 $\mathbf{a - b}$ when $\mathbf{a \geq b}$.**

$\mathbf{a}$, $\mathbf{b}$ being integer with $\mathbf{a \geq b}$, by definition we call $\mathbf{a - b}$ the unique integer $\mathbf{c}$ satisfying $\mathbf{a = b + c}$ (proposition 921 (p. 517)).

**Proposition 922** If $\mathbf{a}$ and $\mathbf{b}$ are integers, $\mathbf{a < b \Leftrightarrow a \leq b - 1}$.

**Proof**

Let us suppose $\mathbf{a < b}$.

$\mathbf{a}$ being integer $\mathbf{a \geq 0}$ then $\mathbf{0 < b}$ and from proposition 920 (p. 517), $\mathbf{1 \leq b}$. From definition 398 (p. 518), there is $\mathbf{b - 1}$ as integer. If $\mathbf{a = 0}$, $\mathbf{a = \emptyset}$. Then $\mathbf{a \subset b - 1}$ which gives $\mathbf{a \leq b - 1}$. If $\mathbf{a > 0}$ $\mathbf{(\exists a' \subset b)(a \sim a')}$ and $\mathbf{(\exists x)(x \in a')}$. $\mathbf{b - 1}$ being an integer we have $\mathbf{b - 1 \neq b - 1 + 1 = b}$ (proposition 918 (p. 516)). Then $\mathbf{a' \neq b}$. From proposition 754 (p. 451), $\mathbf{\exists y \in \mathcal{C}_b(a')}$. Using that $\mathbf{y}$ (see convention 26 (p. 441)), since $\mathbf{x \in a' \subset b}$, we have $\mathbf{y \neq x}$ and $\mathbf{y \in \mathcal{C}_b(\{x\}), y \notin \mathcal{C}_{a'}(\{x\})}$ Then $\mathbf{\mathcal{C}_{a'}(\{x\}) \subset \mathcal{C}_b(\{x\}) - \{y\}}$. Since $\mathbf{\mathcal{C}_b(\{x\}) - 1}$ is an integer (proposition 918 (p. 516)), we have from definition 397 (p. 516):

$$\mathbf{Card(\mathcal{C}_{a'}(\{x\})) \leq Card(\mathcal{C}_b(\{x\})) - 1 < Card(\mathcal{C}_b(\{x\}))}$$

and we have:

$$\mathbf{a = Card(\mathcal{C}_{a'}(\{x\})) + Card(\{x\}),\ b = Card(\mathcal{C}_b(\{x\})) + Card(\{x\})}$$

and applying definition 398 (p. 518).

$$\mathbf{a - 1 = Card(\mathcal{C}_{a'}(\{x\}))\ b - 1 = Card(\mathcal{C}_b(\{x\}))}\ .$$

Then we have $\mathbf{a < b \Rightarrow a - 1 < b - 1}$. It gives from proposition 920 (p. 517):

$$\mathbf{a < b \Rightarrow a \leq b - 1}\ .$$

Let us suppose $\mathbf{a \leq b - 1}$.

$\mathbf{b - 1}$ being an integer $\mathbf{b - 1 < b - 1 + 1 = b}$, it gives from proposition 875 (p. 500):

$$\mathbf{a \leq b - 1 \Rightarrow a < b}\ .$$

Proposition 922 is true.

**Proposition 923** If two sets $\mathbf{A}$, $\mathbf{B}$ have the same cardinal which is an integer, all injections of $\mathbf{A}$ on $\mathbf{B}$ or $\mathbf{B}$ on $\mathbf{A}$ are bijections.

**Proof**

If we have an injection $\mathbf{f}$ of $\mathbf{A}$ on $\mathbf{B}$, it means (proposition 180 (p. 143)) $(\mathbf{f} : \mathbf{A} \leftrightarrow \mathbf{f(A)})$. Then from definition 390 (p. 499), propositions 893 (p. 507), 746 (p. 448) :

$$\mathbf{Card(A) = Card(f(A)) = Card(f(A)) + Card(\emptyset) = Card(B)}$$
$$\mathbf{Card(B) = Card(f(A)) + Card(C_B(f(A)))}$$
$$\mathbf{Card(f(A)) + Card(\emptyset) = Card(f(A)) + Card(C_B(f(A)))} \ .$$

From proposition 921 (p. 517) $\mathbf{Card(\mathcal{C}_B(f(A))) = Card(\emptyset)}$ and using proposition 878 (p. 501) $\mathbf{\mathcal{C}_B(f(A) = \emptyset}$. From proposition 778 (p. 458), $\mathbf{f(A) = B}$. From definitions 360 (p. 461), 361 (p. 461), $\mathbf{f}$ is a bijection.

**Definition 399 Numbers.**

One defines $\mathbf{2 = 1 + 1, 3 = 2 + 1}$ etc.

**Proposition 924** Any nonempty set of cardinals having an integer as a majoring element (definition 380 (p. 484)) has a biggest element (definition 379 (p. 484)) which is unique.

**Proof**

If it exists, it is unique from proposition 842 (p. 484).

Calling $\mathbf{A}$ a nonempty set of cardinals and $\mathbf{m}$ an integer which is strictly bigger than any element of $\mathbf{A}$: $\mathbf{(\forall x \in A)(x < m)}$. Since $\mathbf{A \neq \emptyset}$ we have $\mathbf{(\exists x \in A)(x \geq 0)}$ and $\mathbf{m > \tau_x(x \in A) \geq 0}$. From proposition 918, all the elements of $\mathbf{A}$ are integers.

Let us consider the set $\mathbf{M}$ of cardinals smaller than or equal to $\mathbf{m}$ and bigger than any element of $\mathbf{A}$:

$$\mathbf{M = set_n\{n \leq m, (\forall x \in A)(x < n)\}}$$

$\mathbf{M}$ exists from propositions 883 (p. 503) and 749 (p. 450). Since $\mathbf{m \in M}$, $\mathbf{M \neq \emptyset}$.

Since the order on cardinals is a good order (proposition 886 (p. 504)), $\mathbf{M}$ gets a smallest element $\mathbf{a}$. We have $\mathbf{a \in M}$ so $\mathbf{a \notin A}$ and $\mathbf{a > \tau_x(x \in A) \geq 0}$ which means $\mathbf{a \geq 1}$ (proposition 916 (p. 516)). Using definition 395 (p. 509), we have $\mathbf{a - 1 \geq 0}$ with $\mathbf{(a-1)+1 = a}$. From propositions 908 (p. 512), 746 (p. 448), $\mathbf{a - 1 \leq a}$. From proposition 918 (p. 516) $\mathbf{a - 1}$ is an integer. We then have $\mathbf{a - 1 < a}$ (definition 397 (p. 516)).

There is an element $\mathbf{x \in A}$ such that $\mathbf{a - 1 \leq x}$ otherwise $\mathbf{a - 1}$ would be in $\mathbf{M}$ and then bigger than or equal (see convention 28 (p. 446)) to $\mathbf{a}$ which is impossible since $\mathbf{a - 1}$ is an integer. Taking an $\mathbf{x \in A}$ with $\mathbf{a - 1 \leq x}$, we have: $\mathbf{x < a}$. But from proposition 916 (p. 516), $\mathbf{a - 1 < x}$ means $\mathbf{a \leq x}$ which is incompatible with $\mathbf{x < a}$. The only remaining case is $\mathbf{x = a - 1}$.

$\mathbf{a - 1}$ is the biggest element of $\mathbf{A}$.

**Proposition 925** There is a cardinal which is not an integer.

It is an axiom.

Such a cardinal is said to be infinite. From proposition 880 (p. 502), such cardinals as soon as they exist, they are not unique.

**Proposition 926** $\mathbf{coll_n((\exists s)(n = Card(s)), (n + 1 > n))}$.

**Proof**

Calling $\boldsymbol{\mathcal{A}}$ a cardinal such that $\boldsymbol{\mathcal{A} = \mathcal{A} + 1}$ which exists from proposition 925 (p. 519). From proposition 886 (p. 504), $\mathbf{n}$ being an integer, we have $\boldsymbol{n \geq \mathcal{A} \vee \mathcal{A} > n}$. From proposition 918 (p. 516), $\boldsymbol{n \geq \mathcal{A} \Rightarrow \mathcal{A} < \mathcal{A} + 1}$. $\boldsymbol{n \geq \mathcal{A}}$ is then forbidden by what we suppose. Then from

A.1.4 point 7 (p. 435), $\mathcal{A} > \mathbf{n}$. It means that $\mathcal{A}$ is bigger than any integer. From proposition 883 (p. 503), the integers form a set which we call $\mathcal{Z}^{0+}$ with:

$$\mathcal{Z}^{+} = \mathcal{C}_{\mathcal{Z}^{0+}}(\{0\}) = \mathcal{Z}^{0+} - 0 \ .$$

**Definition 400** $\mathcal{Z}^{+}$.

The set of positive integers is:

$$\mathcal{Z}^{+} = \mathbf{set_n}\{(\exists \mathbf{s})(\mathbf{n} = \mathbf{Card}(\mathbf{s})), (\mathbf{n}+1 > \mathbf{n}), (\mathbf{n} \neq 0)\} \ .$$

**Definition 401** $\mathcal{Z}^{0+}$.

$$\mathcal{Z}^{0+} = \mathbf{set_n}\{(\exists \mathbf{s})(\mathbf{n} = \mathbf{Card}(\mathbf{s})), (\mathbf{n}+1 > \mathbf{n})\} \ .$$

## A.6.1 Recurrence principle

**Proposition 927** $\mathbf{n} \in \mathcal{Z}^{+}$, $\mathbf{R}(\mathbf{n})$ a relation,

$$((\forall \mathbf{n} \in \mathcal{Z}^{+}), (\mathbf{R}(\mathbf{n}) \Rightarrow \mathbf{R}(\mathbf{n}+1)), \mathbf{R}(1)) \Rightarrow (\forall \mathbf{p} \in \mathcal{Z}^{+})(\mathbf{R}(\mathbf{p}))$$

is true.

**Proof**
If $\mathbf{set_q}\{(\mathbf{q} \in \mathcal{Z}^{+}), (\neg \mathbf{R}(\mathbf{q}))\} \neq \emptyset$, since, from proposition 886 (p. 504), any subset of $\mathcal{Z}^{+}$ has a smallest element, we would have:

$$(\exists \mathbf{m})(\mathbf{m} = \mathbf{lower\_limit}\ (\mathbf{set_q}\{(\mathbf{q} \in \mathcal{Z}^{+}), (\neg \mathbf{R}(\mathbf{q}))\})$$

$\mathbf{m} > 1$ since $\mathbf{R}(1)$ is true. But:

$$\mathbf{m} - 1 < \mathbf{m} \Rightarrow \mathbf{R}(\mathbf{m}-1) \Rightarrow \mathbf{R}(\mathbf{m})$$

so

$$\mathbf{m} \neq \mathbf{lower\_limit}\ (\mathbf{set_q}\{(\mathbf{q} \in \mathcal{Z}^{+}), (\neg \mathbf{R}(\mathbf{q}))\}) \ .$$

Then, from A.1.4 point 8 (p. 435), we have:

$$\mathbf{set_q}\{(\mathbf{q} \in \mathcal{Z}^{+}), (\neg \mathbf{R}(\mathbf{q}))\} = \emptyset \ .$$

**Proposition 928** $\mathbf{p} \in \mathcal{Z}^{+}, \mathbf{I} \in (\mathcal{Z}^{+})^{\mathbf{p}}$, $\mathbf{R}(\mathbf{I})$ being a relation, we have:

$$\begin{aligned}&\{(\forall \mathbf{i} \in [1,\mathbf{p}])(\mathbf{I}(\mathbf{i}) = 1) \Rightarrow \mathbf{R}(\mathbf{I}), (\forall \mathbf{I} \in (\mathcal{Z}^{+})^{\mathbf{p}})(\mathbf{R}(\mathbf{I}) \Rightarrow ((\mathbf{J} \in (\mathcal{Z}^{+})^{\mathbf{p}},\\&(\forall \mathbf{k} \in [1,\mathbf{p}])((\forall \mathbf{i} \in [1,\mathbf{p}] - \mathbf{k})(\mathbf{J}(\mathbf{i}) = \mathbf{I}(\mathbf{i})), \mathbf{J}(\mathbf{k}) = \mathbf{I}(\mathbf{k}) + 1)) \Rightarrow \mathbf{R}(\mathbf{J})))\}\\&\Rightarrow (\forall \mathbf{I} \in (\mathcal{Z}^{+})^{\mathbf{p}})\mathbf{R}(\mathbf{I}) \ .\end{aligned}$$

(Recurrence principle extended to a multi-integer variation. )

**Proof**
Calling:

$$\mathbf{T} = \{(\forall \mathbf{i} \in [\mathbf{1}, \mathbf{p}])(\mathbf{I}(\mathbf{i}) = \mathbf{1}) \Rightarrow \mathbf{R}(\mathbf{I}), (\forall \mathbf{I} \in (\mathcal{Z}^+)^{\mathbf{p}})(\mathbf{R}(\mathbf{I}) \Rightarrow ((\mathbf{J} \in (\mathcal{Z}^+)^{\mathbf{p}}, (\forall \mathbf{k} \in [\mathbf{1}, \mathbf{p}])((\forall \mathbf{i} \in [\mathbf{1}, \mathbf{p}] - \mathbf{k})(\mathbf{J}(\mathbf{i}) = \mathbf{I}(\mathbf{i})), \mathbf{J}(\mathbf{k}) = \mathbf{I}(\mathbf{k}) + \mathbf{1})) \Rightarrow \mathbf{R}(\mathbf{J})))\}$$

1. $\mathbf{T}$ true means $\mathbf{R}$ true for $\mathbf{I}(\mathbf{1}) \in \mathcal{Z}^+, (\forall \mathbf{j} \in ]\mathbf{1}, \mathbf{p}])(\mathbf{I}(\mathbf{j}) = \mathbf{1})$:
From proposition 927 (p. 520), $\mathbf{I} \in (\mathcal{Z}^+)^{\mathbf{p}}$, $\mathbf{J} \in (\mathcal{Z}^+)^{\mathbf{p}}$, we have:

$$\mathbf{T} \Rightarrow (\forall \mathbf{i} \in [\mathbf{2}, \mathbf{p}])(\mathbf{I}(\mathbf{i}) = \mathbf{J}(\mathbf{i}) = \mathbf{1}) \Rightarrow ((\mathbf{I}(\mathbf{1}) = \mathbf{1} \Rightarrow \mathbf{R}(\mathbf{I})), (\mathbf{I}(\mathbf{1}) \in \mathcal{Z}^+ \Rightarrow (\mathbf{R}(\mathbf{I}) \Rightarrow (\mathbf{J}(\mathbf{1}) = \mathbf{I}(\mathbf{1}) + \mathbf{1} \Rightarrow \mathbf{R}(\mathbf{J}))))) \Rightarrow (\forall \mathbf{I}(\mathbf{1}) \in \mathcal{Z}^+)\mathbf{R}(\mathbf{I})\,.$$

2. Now with $\mathbf{k} \in [\mathbf{1}, \mathbf{p}[$, we suppose $((\forall \mathbf{j} \in [\mathbf{1}, \mathbf{k}])(\mathbf{I}(\mathbf{j}) \in \mathcal{Z}^+), (\forall \mathbf{j} \in ]\mathbf{k}, \mathbf{p}])(\mathbf{I}(\mathbf{j}) = \mathbf{1})) \Rightarrow \mathbf{R}(\mathbf{I})$. We have, $\mathbf{T}$ being true:

$$(\forall \mathbf{j})(\forall \mathbf{i})((\mathbf{j} \in [\mathbf{1}, \mathbf{k}] \Rightarrow \mathbf{I}(\mathbf{j}) \in \mathcal{Z}^+, \mathbf{i} \in ]\mathbf{k}+\mathbf{1}, \mathbf{p}], \mathbf{I}(\mathbf{i}) = \mathbf{1}) \Rightarrow \{(\mathbf{I}(\mathbf{k}+\mathbf{1}) = \mathbf{1} \Rightarrow \mathbf{R}(\mathbf{I}), ((\forall \mathbf{I}(\mathbf{k}+\mathbf{1}) \in \mathcal{Z}^+)\mathbf{R}(\mathbf{I}) \Rightarrow ((\mathbf{J} \in (\mathcal{Z}^+)^{\mathbf{p}}, (\forall \mathbf{i} \in [\mathbf{1}, \mathbf{p}] - (\mathbf{k}+\mathbf{1}))(\mathbf{J}(\mathbf{i}) = \mathbf{I}(\mathbf{i})), \mathbf{J}(\mathbf{k}+\mathbf{1}) = \mathbf{I}(\mathbf{k}+\mathbf{1}) + \mathbf{1}) \Rightarrow \mathbf{R}(\mathbf{J}))))) \Rightarrow (\forall \mathbf{I}(\mathbf{k}+\mathbf{1}) \in \mathcal{Z}^+)\mathbf{R}(\mathbf{I})\}\,.$$

Then from proposition 927 (p. 520), $((\forall \mathbf{j} \in [\mathbf{1}, \mathbf{k}+\mathbf{1}])(\mathbf{I}(\mathbf{j}) \in \mathcal{Z}^+), (\forall \mathbf{j} \in ]\mathbf{k}+\mathbf{1}, \mathbf{p}])(\mathbf{I}(\mathbf{j}) = \mathbf{1})) \Rightarrow \mathbf{R}(\mathbf{I})$.

3. Applying again proposition 927 (p. 520), we have proposition 928.

## A.6.2 Calculus on integer

**Proposition 929** Given two integers $\mathbf{a}$ and $\mathbf{b}$, $\mathbf{a} + \mathbf{b}$ is an integer.

**Proof**
It is a direct consequence of propositions 919 (p. 517) and 927 (p. 520).

**Proposition 930** Given two integers $\mathbf{a}$ and $\mathbf{b}$, we have:

$$\mathbf{0} < \mathbf{b} \leq \mathbf{a} \Rightarrow (\exists \mathbf{r} \in [\mathbf{0}, \mathbf{b}[)(\exists \mathbf{q} \in \mathcal{Z}^{0+}) (\mathbf{a} = \mathbf{b}\mathbf{q} + \mathbf{r}, (\mathbf{a} = \mathbf{b}\mathbf{q}_1 + \mathbf{r}_1, \mathbf{r}_1 \in [\mathbf{0}, \mathbf{b}[, \mathbf{q}_1 \in \mathcal{Z}^{0+}) \Rightarrow (\mathbf{q}_1 = \mathbf{q}, \mathbf{r}_1 = \mathbf{r}))$$

(Euclidean division) (see also definition 425 (p. 558)).

**Proof**
Let us consider:

$$\mathbf{Q} = \mathbf{set}_{\mathbf{q}}\{(\exists \mathbf{q})(\exists \mathbf{r})(\mathbf{a} = \mathbf{b}\mathbf{q} + \mathbf{r}, \mathbf{q} \in \mathcal{Z}^{0+}, \mathbf{r} \in \mathcal{Z}^{0+})\}$$

$\mathbf{Q} \subset \mathcal{Z}^{0+}$ is a set from proposition 750 (p. 450). It is not empty since $\mathbf{1} \in \mathbf{Q}$ (proposition 921 (p. 517)). We may write (proposition 916 (p. 516)): $\mathbf{b} > \mathbf{0} \Rightarrow \mathbf{b} \geq \mathbf{1}$ which means (proposition 921 (p. 517)):

$$(\exists \mathbf{u})(\mathbf{b} = \mathbf{1} + \mathbf{u}, \mathbf{u} \in \mathcal{Z}^{0+})\,.$$

From propositions 912 (p. 515), 746 (p. 448), using that $\mathbf{u}$ [11], we get $\mathbf{bq} = \mathbf{q} + \mathbf{uq} \geq \mathbf{q}$ and we have:

$$\mathbf{q} > \mathbf{a} \Rightarrow \mathbf{bq} + \mathbf{r} \geq \mathbf{bq} \geq \mathbf{q} > \mathbf{a}$$

then $\mathbf{q} > \mathbf{a}$. It forbids $\mathbf{bq} + \mathbf{r} = \mathbf{a}$:

$$\mathbf{q} > \mathbf{a} \Rightarrow \mathbf{q} \notin \mathbf{Q}$$

$\mathbf{Q}$ has $\mathbf{a}$ as a majoring element (definition 380 (p. 484)). Then $\mathbf{Q}$ has an upper limit $\mathbf{q_m}$ which is in $\mathbf{Q}$ (proposition 924 (p. 519)). We have $\mathbf{a} = \mathbf{bq_m} + \mathbf{r_m}$.
$\mathbf{r_m} < \mathbf{b}$ because otherwise we would have $\mathbf{r_m} = \mathbf{b} + \mathbf{r'_m}$ (proposition 921 (p. 517)) and $\mathbf{a} = \mathbf{b}(\mathbf{q_m} + \mathbf{1}) + \mathbf{r'_m}$ (see proposition 912 (p. 515)) and $\mathbf{q_m} + \mathbf{1} \in \mathbf{Q}$ which is forbidden by definition 397 (p. 516) because $\mathbf{q_m}$ is an integer and the biggest element of $\mathbf{Q}$.
Let us suppose:

$$\mathbf{a} = \mathbf{bq} + \mathbf{r}, \mathbf{a} = \mathbf{bq_m} + \mathbf{r_m}, \mathbf{r} \in [\mathbf{0}, \mathbf{b}[\ .$$

From proposition 921 (p. 517), we have: $\mathbf{q} = \mathbf{q_m} \Rightarrow \mathbf{r} = \mathbf{r_m}$. From proposition 876 (p. 500), since $\mathbf{q_m}$ is the biggest element of $\mathbf{Q}$, we have: $\mathbf{q_m} \geq \mathbf{q}$.
Let us suppose $\mathbf{q_m} > \mathbf{q}$.
From proposition 916 (p. 516), we have: $\mathbf{q_m} \geq \mathbf{q} + \mathbf{1}$. From proposition 921 (p. 517), we have: $\mathbf{q_m} = \mathbf{q} + \mathbf{k} + \mathbf{1}, \mathbf{k} \in \mathcal{Z}^{0+}$ and from propositions 921 (p. 517), 912 (p. 515), we have:

$$\mathbf{a} = \mathbf{bq_m} + \mathbf{r_m} = \mathbf{a} = \mathbf{bq} + \mathbf{bk} + \mathbf{b} + \mathbf{r_m} \Rightarrow \mathbf{r} = \mathbf{bk} + \mathbf{b} + \mathbf{r_m} \geq \mathbf{b}\ .$$

It is against $\mathbf{r} < \mathbf{b}$. Then we have proposition 930.

**Definition 402 a/b.**

Given two integers $\mathbf{a}$, $\mathbf{b}$, using proposition 930 (p. 521), we set $\mathbf{a} = \mathbf{bq} + \mathbf{r}$, $\mathbf{q}$, $\mathbf{r} < \mathbf{b}$ being integer. One uses to write $\mathbf{q} = \mathbf{a}/\mathbf{b}$ when $\mathbf{r} = \mathbf{0}$. It is compatible with the notations described in definitions 420 (p. 552) and 419 (p. 552).

**Definition 403 Even and odd integer, parity.**

Using proposition 930 (p. 521), for any integer $\mathbf{a} \in \mathcal{Z}^{0+}$, we may write:

$$\mathbf{a} \in \mathcal{Z}^{0+} \Rightarrow (\exists \mathbf{q} \in \mathcal{Z}^{0+})(\exists \mathbf{r} \in [\mathbf{0}, \mathbf{2}[)(\mathbf{a} = \mathbf{2q} + \mathbf{r})\ .$$

We have: $\mathbf{0} < \mathbf{0} + \mathbf{1} < \mathbf{0} + \mathbf{1} + \mathbf{1} = \mathbf{2}$ (definition 399 (p. 519), proposition 900 (p. 510), 746 (p. 448)). From proposition 876 (p. 500), when $\mathbf{r} \in [\mathbf{0}, \mathbf{2}[$, we have

$$\mathbf{0} \leq \mathbf{r}, \mathbf{r} \leq \mathbf{1} < \mathbf{1} + \mathbf{1} = \mathbf{2}\ .$$

From proposition 917 (p. 516) $(\mathbf{0} \leq \mathbf{r}, \mathbf{r} \leq \mathbf{1}) \Rightarrow (\mathbf{r} = \mathbf{0} \vee \mathbf{r} = \mathbf{1})$ then

$$\mathbf{r} \in [\mathbf{0}, \mathbf{2}[ \Leftrightarrow \mathbf{r} \in \{\mathbf{0}, \mathbf{1}\}\ .$$

If $\mathbf{r} = \mathbf{0}$, the parity of $\mathbf{a}$ is said to be even, if $\mathbf{r} = \mathbf{1}$, $\mathbf{a}$ is said to be odd.

**Proposition 931** $\mathbf{a} \in \mathcal{Z}^{0+}$ being even (resp. odd), $\mathbf{a} + \mathbf{1}$ is odd (resp. even).

[11] In order to simplify, we apply frequently convention 42 (p. 560): we use the name $\mathbf{x}$ in $(\exists \mathbf{x})(\mathbf{R}(\mathbf{x}))$ in place of $\tau_{\mathbf{x}}\mathbf{R}(\mathbf{x})$ setting $\mathbf{x} = \tau_{\mathbf{x}}\mathbf{R}(\mathbf{x})$.

**Proof**
It is a direct consequence of definition 403 (p. 522):

$$\begin{aligned}&\mathbf{a = 2b \vee a = 2b+1}\\&\mathbf{a = 2b \Rightarrow a+1 = 2b+1}\\&\mathbf{a = 2b+1 \Rightarrow a+1 = 2b+1+1 = a+1 = 2b+2 = 2(b+1)}\end{aligned}$$

see A.1.4 point 16 (p. 436), definition 399 (p. 519), propositions 911 (p. 514) and 746 (p. 448).

**Proposition 932** The sum of two integers of the same parity is even and it is odd otherwise.

**Proof**
It is a direct consequence of definition 403 (p. 522) and of propositions 931 (p. 522), 912 (p. 515), 746 (p. 448):

$$\begin{aligned}&\mathbf{(a_1 = 2b_1 + r_1,\ r_1 \in \{0,1\}\ a_2 = 2b_2 + r_2,\ r_2 \in \{0,1\}) \Rightarrow}\\&\mathbf{((r_1 = 0, r_2 = 0 \Rightarrow a_1 + a_2 = 2(b_1 + b_2))}\\&\mathbf{\vee (r_1 = 0, r_2 = 1 \Rightarrow a_1 + a_2 = 2(b_1 + b_2) + 1)}\\&\mathbf{\vee (r_1 = 1, r_2 = 0 \Rightarrow a_1 + a_2 = 2(b_1 + b_2) + 1)}\\&\mathbf{\vee (r_1 = 1, r_2 = 1 \Rightarrow a_1 + a_2 = 2(b_1 + b_2) + 2 = 2(b_1 + b_2 + 1)))}\ .\end{aligned}$$

**Proposition 933** The product of an even integer by an integer is an even integer.

**Proof**
It is a direct consequence of definition 403 (p. 522), propositions 930 (p. 521), 889 (p. 506), 890 (p. 506).

**Proposition 934** $\mathbf{n}$ is an integer. We have:

$$\mathbf{n = Card(set_i\{i < n\}) = Card([0, n[) = Card([1, n])}\ .$$

**Proof**
Proposition 934 (p. 523) is true for $\mathbf{n = 0}$. Since $\mathbf{0 = Card(\emptyset) = Card(set_i\{i < Card(\emptyset)\})}$. From proposition 876 (p. 500) $\mathbf{set_i\{i < n\} \cap \{n\} = \emptyset}$. Then, since, from definitions 397 (p. 516), 338 (p. 446), 392 (p. 500), $\mathbf{set_i\{i < n\} \cup \{n\} = set_i\{i < n+1\}}$, we have from proposition 893 (p. 507):

$$\mathbf{Card(set_i\{i < n\}) + 1 = Card(set_i\{i < n\} \cup \{n\}) = Card(set_i\{i < n+1\})}$$

so

$$\mathbf{n = Card(set_i\{i < n\}) \Rightarrow n + 1 = Card(set_i\{i < n+1\})}\ .$$

Then proposition 934 (p. 523) is true from proposition 927 (p. 520).

**Proposition 935** $\mathbf{Card(a)}$ being an integer, if $\mathbf{b \subset a}$, we have:

$$\mathbf{(Card(a) = Card(b)) \Leftrightarrow (a = b)}\ .$$

**Proof**
From A.1.3 point 6 (p. 435) $(\mathbf{a} = \mathbf{b}) \Rightarrow \mathbf{Card(a)} = \mathbf{Card(b)}$.
Let us suppose $\mathbf{Card(a)} = \mathbf{Card(b)}$ and $\mathbf{a} \neq \mathbf{b}$. From proposition 754 (p. 451) (extensionality principle) we have:

$$(\mathbf{b} \subset \mathbf{a}, \mathbf{a} \neq \mathbf{b}) \Rightarrow (\exists \mathbf{c})(\mathbf{c} \in \mathbf{a}, \mathbf{c} \notin \mathbf{b})$$

because otherwise we also have $\mathbf{a} \subset \mathbf{b}$.
Then $\mathbf{b} \cup \{\mathbf{c}\} \subset \mathbf{a}$ (see definition 349 (p. 451)) and from definitions 393 (p. 506), 878 (p. 501), 397 (p. 516), propositions 918 (p. 516), 919 (p. 517), we have:

$$\mathbf{Card(b)} \leq \mathbf{Card(b \cup \{c\})} = \mathbf{Card(b)} + \mathbf{1} \leq \mathbf{Card(a)} \; .$$

Then $\mathbf{Card(b)} < \mathbf{Card(a)}$ since by definition 397 (p. 516) $\mathbf{Card(b)} \neq \mathbf{Card(b)} + \mathbf{1}$. We conclude by A.1.4 (p. 435) -7.

**Proposition 936** $\mathbf{a}$, $\mathbf{b}$ integers, $\mathbf{a} < \mathbf{b}$, we have (definition 378 (p. 484)):

1. $[\mathbf{a}, \mathbf{b}[= [\mathbf{a}, \mathbf{b} - \mathbf{1}]$, $]\mathbf{a}, \mathbf{b}] = [\mathbf{a} + \mathbf{1}, \mathbf{b}]$
2. $[\mathbf{a}, \mathbf{b}] \cup \{\mathbf{b} + \mathbf{1}\} = [\mathbf{a}, \mathbf{b} + \mathbf{1}]$, $[\mathbf{a} + \mathbf{1}, \mathbf{b}] \cup \{\mathbf{a}\} = [\mathbf{a}, \mathbf{b}]$
3. $\mathbf{Card}(]\mathbf{a}, \mathbf{b}[) = \mathbf{b} - \mathbf{a} - \mathbf{1}$, $\mathbf{Card}([\mathbf{a}, \mathbf{b}]) = \mathbf{b} - \mathbf{a} + \mathbf{1}$, $\mathbf{Card}(]\mathbf{a}, \mathbf{b}]) = \mathbf{b} - \mathbf{a} + \mathbf{1}$, $\mathbf{Card}([\mathbf{a}, \mathbf{b}[) = \mathbf{b} - \mathbf{a}$ are integers.

**Proof**

1. From definition 378 (p. 484), proposition 922 (p. 518), we have:

   $$\mathbf{x} \in [\mathbf{a}, \mathbf{b}[\Leftrightarrow (\mathbf{a} \leq \mathbf{x}, \mathbf{x} < \mathbf{b}) \Leftrightarrow (\mathbf{a} \leq \mathbf{x}, \mathbf{x} \leq \mathbf{b} - \mathbf{1}) \Leftrightarrow \mathbf{x} \in [\mathbf{a}, \mathbf{b} - \mathbf{1}] \; .$$

   From proposition 920 (p. 517), we have:

   $$\mathbf{x} \in ]\mathbf{a}, \mathbf{b}] \Leftrightarrow (\mathbf{a} < \mathbf{x}, \mathbf{x} < \mathbf{b}) \Leftrightarrow (\mathbf{a} + \mathbf{1} \leq \mathbf{x}, \mathbf{x} < \mathbf{b}) \Leftrightarrow \mathbf{x} \in [\mathbf{a} + \mathbf{1}, \mathbf{b}] \; .$$

2. From definitions 338 (p. 446), 397 (p. 516), we have:

   $$(\mathbf{a} < \mathbf{b}, \mathbf{b} < \mathbf{b} + \mathbf{1}) \Rightarrow \mathbf{a} < \mathbf{b} + \mathbf{1} \; .$$

   Then (see 2 (p. 435) of A.1.3) $\mathbf{x} = \mathbf{b} + \mathbf{1} \Rightarrow \mathbf{a} \leq \mathbf{x}$. From 16 (p. 436) of A.1.4, we have:

   $$(\mathbf{x} = \mathbf{b} + \mathbf{1} \vee \mathbf{a} \leq \mathbf{x}) \Rightarrow \mathbf{a} \leq \mathbf{x} \; .$$

   Since (see 2 (p. 435) of A.1.3) $\mathbf{a} \leq \mathbf{x} \Rightarrow \mathbf{a} \leq \mathbf{x} \vee \mathbf{x} = \mathbf{b} + \mathbf{1}$, we have (see 17 (p. 436), 22 (p. 436) of A.1.4):

   $$\mathbf{a} \leq \mathbf{x} \Leftrightarrow (\mathbf{a} \leq \mathbf{x} \vee \mathbf{x} = \mathbf{b} + \mathbf{1}) \; .$$

   From definitions 351 (p. 453), 378 (p. 484), propositions 920 (p. 517), 922 (p. 518) and A.1.4 (p. 435) point 27, we have:

   $$\begin{aligned} &[\mathbf{a}, \mathbf{b}] \cup \{\mathbf{b} + \mathbf{1}\} \Leftrightarrow ((\mathbf{a} \leq \mathbf{x}, \mathbf{x} \leq \mathbf{b}) \vee \mathbf{x} = \{\mathbf{b} + \mathbf{1}\}) \\ &\Leftrightarrow ((\mathbf{a} \leq \mathbf{x} \vee \mathbf{x} = \{\mathbf{b} + \mathbf{1}\}), (\mathbf{x} < \mathbf{b} + \mathbf{1} \vee \mathbf{x} = \{\mathbf{b} + \mathbf{1}\})) \\ &\Leftrightarrow (\mathbf{a} \leq \mathbf{x}, \mathbf{x} \leq \mathbf{b} + \mathbf{1}) \Leftrightarrow \mathbf{x} \in [\mathbf{a}, \mathbf{b} + \mathbf{1}] \; . \end{aligned}$$

   Similar proof for $[\mathbf{a} + \mathbf{1}, \mathbf{b}] \cup \{\mathbf{a}\} = [\mathbf{a}, \mathbf{b}]$.

3. From definitions 378 (p. 484), 338 (p. 446), 339 (p. 446), 349 (p. 451), 392 (p. 500), propositions 918 (p. 516), 934 (p. 523), the preceding points 1, 2, we have:

$$\mathbf{x \in [a,b] \Rightarrow a \leq x \leq b \Rightarrow 0 \leq x \leq b \Rightarrow x \in [0,b] \Rightarrow [a,b] \subset [0,b]}$$
$$\mathbf{\Rightarrow Card([a,b]) \leq Card([0,b]) = Card(b) + 1 .}$$

Then $\mathbf{Card([a,b])}$ is an integer. From definition 378 (p. 484), we have:

$$\mathbf{[0,b] = [0,a[\cup[a,b],\ [0,a[\cap[a,b] = \emptyset .}$$

From propositions 893 (p. 507), 746 (p. 448), we have:

$$\mathbf{Card([0,b]) = Card([0,a[) + Card([a,b]) .}$$

Then from propositions 934 (p. 523), 921 (p. 517), definition 398 (p. 518), we have:

$$\mathbf{b + 1 = a + Card([a,b]),\ Card([a,b]) = b - a + 1 .}$$

We have a similar proof for the other properties mentioned in point 3.

**Proposition 937** The cardinal of any segment of integers, which is not $\mathcal{Z}^{0+}$ is an integer.

**Proof**
Since the order on $\mathcal{Z}^{0+}$ is a good order, from proposition 859 (p. 489), for any segment $\mathbf{S}$ of $\mathcal{Z}^{0+}$ not $\mathcal{Z}^{0+}$, there is an integer $\mathbf{s}$ such as $\mathbf{S = [0,s[}$ so from proposition 934 (p. 523), $\mathbf{Card(S)}$ equal to $\mathbf{s}$. $\mathbf{Card(S)}$ is an integer.

**Proposition 938** The cardinal of any subset of $\mathcal{Z}^{0+}$ is either $\mathbf{Card(\mathcal{Z}^{0+})}$ or an integer.

Given $\mathbf{A}$ subset of $\mathcal{Z}^{0+}$, then $\mathbf{A}$ is equivalent to a segment of $\mathcal{Z}^{0+}$ (proposition 871 (p. 498)). Then from proposition 937 (p. 525) is equivalent to $\mathcal{Z}^{0+}$ or to an integer (proposition 937 (p. 525)).

**Proposition 939** $\mathbf{Card(\mathcal{Z}^{0+}) = Card(\mathcal{Z}^{0+}) + 1}$.

**Proof**
Let us suppose $\mathbf{Card(\mathcal{Z}^{0+})}$ is an integer. From proposition 919 (p. 517), $\mathbf{Card(\mathcal{Z}^{0+}) + 1}$ would be an integer. Setting

$$\mathbf{C = set_i\{i < Card(\mathcal{Z}^{0+}) + 1\}}$$

from proposition 918 (p. 516) any element of $\mathbf{C}$ would be an integer. Then $\mathbf{C \subset \mathcal{Z}^{0+}}$ and

$$\mathbf{Card(C) \leq Card(\mathcal{Z}^{0+}) .}$$

We should have from proposition 934 (p. 523):

$$\mathbf{Card(\mathcal{Z}^{0+}) + 1 = Card(C)}$$

and

$$\mathbf{Card(\mathcal{Z}^{0+}) + 1 \leq Card(\mathcal{Z}^{0+}) .}$$

That is impossible from proposition 913 (p. 515) if $\mathbf{Card(\mathcal{Z}^{0+}) + 1 \neq Card(\mathcal{Z}^{0+})}$. Then proposition 939 is true.

**Proposition 940** $\mathbf{Card(s) < Card(\mathcal{Z}^{0+}) \Rightarrow Card(s) \in \mathcal{Z}^{0+}}$.

**Proof**
From definition 390 (p. 499), we have:

$$(\exists \mathbf{A} \subset \mathcal{Z}^{0+})(\mathbf{A} \sim \mathbf{s}) \ .$$

From proposition 938 (p. 525), we have:

$$\mathbf{Card(s)} = \mathbf{Card}(\mathcal{Z}^{0+}) \vee \mathbf{Card(s)} \in \mathcal{Z}^{0+} \ .$$

Then since $\mathbf{Card(s)} \neq \mathbf{Card}(\mathcal{Z}^{0+})$, we have: $\mathbf{Card(s)} \in \mathcal{Z}^{0+}$.

**Proposition 941**

$$\mathbf{Card(A)} \in \mathcal{Z}^{0+} \Rightarrow (\forall \mathbf{f})((\mathbf{f} : \mathbf{A} \to \mathcal{Z}^{0+}) \Rightarrow$$
$$(\sum_{\mathbf{a}\in\mathbf{A}} \mathbf{f(a)} \in \mathcal{Z}^{0+}, (\forall \mathbf{a} \in \mathbf{A})(\mathbf{f(a)} \leq \sum_{\mathbf{i}\in\mathbf{A}} \mathbf{f(i)}), \underset{\mathbf{a}\in\mathbf{A}}{\times} \mathbf{f(a)} = \mathbf{Card}(\prod_{\mathbf{a}\in\mathbf{A}} \mathbf{f(a)}) \in \mathcal{Z}^{0+})) \ .$$

**Proof**
Let us call $\mathbf{n} = \mathbf{Card(A)} \in \mathcal{Z}^{0+}$ and $\mathbf{I} = [\mathbf{0}, \mathbf{n}[$. From definition 390 (p. 499) and proposition 934 (p. 523), there is a bijection $\mathbf{g}$ ($\mathbf{g} : \mathbf{I} \leftrightarrow \mathbf{A}$). It changes the map $\mathbf{f}$ in the map $\mathbf{f} \circ \mathbf{g} = (\mathbf{z_i} : \mathbf{I} \to \mathcal{Z}^{0+})$, $\mathbf{z_i} = \mathbf{f(g(i))}$.
Then, if proposition 941 (p. 526) is true for any finite $\mathbf{I}$ and any $\mathbf{f}$, it is also true for any finite $\mathbf{A}$ and any $\mathbf{f}$. Let us suppose $\mathbf{A} = \mathbf{I}$ for simplicity. At $\mathbf{n} = \mathbf{1}$, it gives $\mathbf{I} = \{\mathbf{0}\}$. The proposition 941 (p. 526) is trivially true.
We suppose that part 1 of proposition 941 (p. 526) is true for $\mathbf{n}$. We suppose it is also true for $\mathbf{f(n)} = \mathbf{q}$. Let us prove that it is true for $\mathbf{f(n)} = \mathbf{q} + \mathbf{1}$. From propositions 796 (p. 464), 919 (p. 517), 913 (p. 515) and definition 393 (p. 506), we have:

$$\sum_{\mathbf{i}\in[\mathbf{0},\mathbf{n}+\mathbf{1}[} \mathbf{f(i)} = \sum_{\mathbf{i}\in[\mathbf{0},\mathbf{n}[} \mathbf{f(i)} + \mathbf{f(n)} = \sum_{\mathbf{i}\in[\mathbf{0},\mathbf{n}[} \mathbf{f(i)} + \mathbf{q} + \mathbf{1}$$
$$(\sum_{\mathbf{i}\in[\mathbf{0},\mathbf{n}[} \mathbf{f(i)} + \mathbf{q} \in \mathcal{Z}^{0+}, (\forall \mathbf{i} \in [\mathbf{0},\mathbf{n}[)(\mathbf{f(i)} \leq \sum_{\mathbf{i}\in[\mathbf{0},\mathbf{n}[} \mathbf{f(i)} + \mathbf{q}, \mathbf{q} \leq \sum_{\mathbf{i}\in[\mathbf{0},\mathbf{n}[} \mathbf{f(i)} + \mathbf{q}))$$
$$\Rightarrow (\sum_{\mathbf{i}\in[\mathbf{0},\mathbf{n}[} \mathbf{f(i)} + \mathbf{q} + \mathbf{1} \in \mathcal{Z}^{0+}, (\forall \mathbf{i} \in [\mathbf{0},\mathbf{n}[)(\mathbf{f(i)} \leq \sum_{\mathbf{i}\in[\mathbf{0},\mathbf{n}[} \mathbf{f(i)} + \mathbf{q} + \mathbf{1},$$
$$\mathbf{q} + \mathbf{1} \leq \sum_{\mathbf{i}\in[\mathbf{0},\mathbf{n}[} \mathbf{f(i)} + \mathbf{q} + \mathbf{1}))$$
$$\mathbf{f(n)} = \mathbf{q} + \mathbf{1} \Rightarrow \sum_{\mathbf{i}\in[\mathbf{0},\mathbf{n}+\mathbf{1}[} \mathbf{f(i)} \in \mathcal{Z}^{0+}, (\forall \mathbf{i} \in [\mathbf{0},\mathbf{n}+\mathbf{1}[)(\mathbf{f(i)} \leq \sum_{\mathbf{i}\in[\mathbf{0},\mathbf{n}+\mathbf{1}[} \mathbf{f(i)}) \ .$$

Since proposition 941 is true for $\mathbf{q} = \mathbf{0}$ and if it is true for $\mathbf{q} \in \mathcal{Z}^{0+}$, it is also true for $\mathbf{q} + \mathbf{1}$. From proposition 927 (p. 520), it is true for any map $(\mathbf{f} : [\mathbf{0}, \mathbf{n} + \mathbf{1}[\to \mathcal{Z}^{0+})$. Applying proposition 927 (p. 520), first part of proposition 941 is true:

$$\sum_{\mathbf{a}\in\mathbf{A}} \mathbf{f(a)} \in \mathcal{Z}^{0+}, (\forall \mathbf{a} \in \mathbf{A})(\mathbf{f(a)} \leq \sum_{\mathbf{i}\in\mathbf{A}} \mathbf{f(i)}) \ .$$

We suppose now that part 2 of proposition 941 is true for $\mathbf{n}$. Let us prove it for $\mathbf{n} + \mathbf{1}$. We suppose again in that case part 2 of proposition 941 (p. 526) is true for $\mathbf{f(n)} = \mathbf{q}$. Let us prove that it is true for $\mathbf{f(n)} = \mathbf{q} + \mathbf{1}$. We have from proposition 889 (p. 506) and proposition 941 part 1:

$$\prod_{\mathbf{i}\in[\mathbf{0},\mathbf{n}+\mathbf{1}[} \mathbf{f(i)} = \prod_{\mathbf{i}\in[\mathbf{0},\mathbf{n}[} \mathbf{f(i)} \times (\mathbf{q} + \mathbf{1}) = \prod_{\mathbf{i}\in[\mathbf{0},\mathbf{n}[} \mathbf{f(i)} \times \mathbf{q} + \prod_{\mathbf{i}\in[\mathbf{0},\mathbf{n}[} \mathbf{f(i)}$$
$$\prod_{\mathbf{i}\in[\mathbf{0},\mathbf{n}[} \mathbf{f(i)} \times \mathbf{q} \in \mathcal{Z}^{0+}, \prod_{\mathbf{i}\in[\mathbf{0},\mathbf{n}[} \mathbf{f(i)} \in \mathcal{Z}^{0+} \Rightarrow \prod_{\mathbf{i}\in[\mathbf{0},\mathbf{n}[} \mathbf{f(i)} \times \mathbf{q} + \prod_{\mathbf{i}\in[\mathbf{0},\mathbf{n}[} \mathbf{f(i)} \in \mathcal{Z}^{0+} \ .$$

Then if it is true for $\mathbf{f(n) = q}$, it is true for $\mathbf{f(n) = q + 1}$. It is true for $\mathbf{q = 0}$ (proposition 827 (p. 475)) and so, from proposition 927 (p. 520), $\forall \mathbf{q}$ so for $\mathbf{f(n)}$ and again with proposition 927 (p. 520), it is true for all $\mathbf{n}$. Proposition 941 part 2 is true. Then proposition 941 is true.

**Proposition 942** $\mathbf{n \in \mathcal{Z}^+}$, given a map $(\mathbf{a_i} : [\mathbf{0, n}[\to \mathcal{Z}^{0+})$ defining a set of $\mathbf{n}$ integers of sum $\mathbf{N} = \sum_{\mathbf{i}\in[0,n[} \mathbf{a_i}$, we have:

$$(\forall \mathbf{i} \in [\mathbf{0, n}])(\mathbf{a_i} \sim [\sum_{\mathbf{k}\in[0,i[} \mathbf{a_k}, \sum_{\mathbf{k}\in[0,i+1[} \mathbf{a_k}[\sim [\mathbf{1} + \sum_{\mathbf{k}\in[0,i[} \mathbf{a_k}, \sum_{\mathbf{k}\in[0,i+1[} \mathbf{a_k}]) .$$

**Proof**

From proposition 865 (p. 494), each set $\mathbf{a_i}$ has a well-order which we quote $\leq_\mathbf{i}$, the set $[\mathbf{0, n}[$ is also well-ordered by cardinal order which we quote $\leq$. Any set $\mathbf{S} = \bigcup_{\mathbf{k}\in[0,i[} \mathbf{a_k} \times \{\mathbf{k}\}$ is well-ordered by $\mathbf{R} =\leq$ (see definition 374 (p. 482)):

$$\begin{aligned}
&\mathbf{R(x, y) = x \leq y} \Rightarrow (\exists \ell \in [\mathbf{0, n}[)(\exists \mathbf{j} \in [\mathbf{0, n}[) \\
&((\mathbf{x} = \mathbf{x}_\ell \in \mathbf{a}_\ell \times \{\ell\} \subset \mathbf{S}, \mathbf{y} = \mathbf{y_j} \in \mathbf{a_j} \times \{\mathbf{j}\} \subset \mathbf{S}), \\
&(\ell = \mathbf{j} \Rightarrow \mathbf{x}_\ell \leq_\ell \mathbf{y}_\ell), \\
&(\ell \neq \mathbf{j} \Rightarrow \ell <_\mathbf{n} \mathbf{j})) .
\end{aligned} \tag{A.56}$$

It is a good order on $\mathbf{S}$:

For any subset $\mathbf{A}$ of $\mathbf{S} = \bigcup_{\mathbf{k}\in[0,i[} \mathbf{a_k} \times \{\mathbf{k}\}$, there is a smallest term $\mathbf{j}$ in the set $\mathbf{K}$ of union index $\mathbf{k}$ of the element of $\mathbf{A}$:

$$\ell \in \mathbf{K} \Leftrightarrow (\exists \mathbf{x} \in \mathbf{A} \cap \mathbf{a}_\ell \times \{\ell\})$$

because the order on $[\mathbf{0, n}[$ is a good order.

$\mathbf{a_j}$ being also well-ordered, there is also a smallest term in $\mathbf{A} \cap \mathbf{a_j} \times \{\mathbf{j}\}$ which is the smallest of all the $\mathbf{A}$ with the above order given by equation (A.56) (p. 527).

For all $\mathbf{i} \in [\mathbf{1, n}]$, $\bigcup_{\mathbf{k}\in[0,i[} \mathbf{a_k} \times \{\mathbf{k}\}$ are segments of $\bigcup_{\mathbf{k}\in[0,n]} \mathbf{a_k} \times \{\mathbf{k}\}$ because with the order defined by A.56 (p. 527), those sets satisfy definition 389 (p. 488). From 868 (p. 496), there is a unique isomorphism (definition 387 (p. 487)) between $\bigcup_{\mathbf{k}\in[0,i[} \mathbf{a_k} \times \{\mathbf{k}\}$ and a segment of $\mathcal{Z}^{0+}$ because from proposition 941 (p. 526), we have:

$$\mathbf{Card}(\bigcup_{\mathbf{k}\in[0,i[} \mathbf{a_k} \times \{\mathbf{k}\}) < \mathbf{Card}(\mathcal{Z}^{0+}) .$$

It forbids (proposition 939 (p. 525), definition 397 (p. 516)) an isomorphism between $\mathcal{Z}^{0+}$ and a segment of that set. The only segment of $\mathcal{Z}^{0+}$ which has as cardinal $\mathbf{Card}(\bigcup_{\mathbf{k}\in[0,i[} \mathbf{a_k}\times\{\mathbf{k}\})$ is $[\mathbf{0}, \mathbf{Card}(\bigcup_{\mathbf{k}\in[0,i[} \mathbf{a_k}\times\{\mathbf{k}\})[$ (proposition 934 (p. 523)). For the same reason, the unique isomorphism of $\mathbf{Card}(\bigcup_{\mathbf{k}\in[0,i+1[} \mathbf{a_k}\times\{\mathbf{k}\})$ to $\mathcal{Z}^{0+}$ is onto the segment $[\mathbf{0}, \mathbf{Card}(\bigcup_{\mathbf{k}\in[0,i+1[} \mathbf{a_k}\times \{\mathbf{k}\})[$.

Let us call $\mathbf{f}$ that isomorphism $(\mathbf{f} : \bigcup_{\mathbf{k}\in[0,i+1[} \mathbf{a_k} \times \{\mathbf{k}\} \leftrightarrow [\mathbf{0}, \mathbf{Card}(\bigcup_{\mathbf{k}\in[0,i+1[} \mathbf{a_k} \times \{\mathbf{k}\})[)$. Since the orders on $\bigcup_{\mathbf{k}\in[0,i[} \mathbf{a_k} \times \{\mathbf{k}\}$ and on $[\mathbf{0}, \mathbf{Card}(\bigcup_{\mathbf{k}\in[0,i[} \mathbf{a_k} \times \{\mathbf{k}\})[$ are the same as the ones induced (see definition 375 (p. 483)) by the orders on $\bigcup_{\mathbf{k}\in[0,i+1[} \mathbf{a_k} \times \{\mathbf{k}\}$ and on $[\mathbf{0}, \mathbf{Card}(\bigcup_{\mathbf{k}\in[0,i+1[} \mathbf{a_k} \times \{\mathbf{k}\})[$, the restriction (definition 118 (p. 135)) of $\mathbf{f}$ to $\bigcup_{\mathbf{k}\in[0,i[} \mathbf{a_k} \times \{\mathbf{k}\}$ is also the unique isomorphism to $[\mathbf{0}, \mathbf{Card}(\bigcup_{\mathbf{k}\in[0,i[} \mathbf{a_k} \times \{\mathbf{k}\}[$. It is also an isomorphism of $\mathbf{a_i} \times \{\mathbf{i}\}$ to $\mathcal{Z}^{0+}$. Then, we have successively (see proposition 160 (p. 136),

185 (p. 143), 184 (p. 143), 186 (p. 143)):

$$a_i \times \{i\} \cup \bigcup_{k \in [0,i[} a_k \times \{k\} = \bigcup_{k \in [0,i+1[} a_k \times \{k\}$$

$$a_i \times \{i\} \cap \bigcup_{k \in [0,i[} a_k \times \{k\} = \emptyset$$

$$f(a_i \times \{i\}) \cup f(\bigcup_{k \in [0,i[} a_k \times \{k\}) = f(\bigcup_{k \in [0,i+1[} a_k \times \{k\})$$

$$f(a_i \times \{i\}) \cap f(\bigcup_{k \in [0,i[} a_k \times \{k\}) = \emptyset$$

$$f(a_i \times \{i\}) \cup [0, \mathrm{Card}(\bigcup_{k \in [0,i[} a_k \times \{k\})[= [0, \mathrm{Card}(\bigcup_{k \in [0,i+1[} a_k \times \{k\})[$$

$$f(a_i \times \{i\}) \cap [0, \mathrm{Card}(\bigcup_{k \in [0,i[} a_k \times \{k\})[= \emptyset$$

$$\mathrm{Card}(f(a_i \times \{i\})) + \mathrm{Card}([0, \mathrm{Card}(\bigcup_{k \in [0,i[} a_k \times \{k\})[)$$

$$= \mathrm{Card}([0, \mathrm{Card}(\bigcup_{k \in [0,i+1[} a_k \times \{k\})[)$$

$$\mathrm{Card}(f(a_i \times \{i\})) = \mathrm{Card}(a_i \times \{i\})$$

$$= \mathrm{Card}([0, \mathrm{Card}(\bigcup_{k \in [0,i+1[} a_k \times \{k\})[) - \mathrm{Card}([0, \mathrm{Card}(\bigcup_{k \in [0,i[} a_k \times \{k\})[)$$

$$= \mathrm{Card}([\mathrm{Card}(\bigcup_{k \in [0,i[} a_k \times \{k\}), \mathrm{Card}(\bigcup_{k \in [0,i+1[} a_k \times \{k\})[)$$

$$a_i \times \{i\} \sim [\mathrm{Card}(\bigcup_{k \in [0,i[} a_k \times \{k\}), \mathrm{Card}(\bigcup_{k \in [0,i+1[} a_k \times \{k\})[$$

$$a_i \sim [\sum_{k \in [0,i[} a_k, \sum_{k \in [0,i+1[} a_k[ \, .$$

Using proposition 936 (p. 524), we have:

$$a_i \times \{i\} \sim [1 + \sum_{k \in [0,i[} a_k, \sum_{k \in [0,i+1[} a_k] \, .$$

**Proposition 943** Any nonempty finite set of integers has a biggest element.

**Proof**
Given $\mathbf{A} \subset \mathcal{Z}^{0+}$ with $\mathbf{Card(A)} \in \mathcal{Z}^{+}$, from propositions 941 (p. 526) and 150 (p. 132), we have:

$$a \in A \Rightarrow a \leq \sum_{u \in A} u \, .$$

From proposition 924 (p. 519), we have proposition 943 (p. 528).

**Definition 404 Factorial.**

For any integer $\mathbf{n} \in \mathcal{Z}^+$, one calls factorial $\mathbf{n}$ quoted as $\mathbf{n!}$ the product of the nonzero cardinals and smaller than or equal to $\mathbf{n}$:

$$\mathbf{n!} = \prod_{\mathbf{i}\in[\mathbf{1},\mathbf{n}]} \mathbf{i}\,.$$

For convenience $\mathbf{0!} = \mathbf{1}$.

**Proposition 944** Given two integers $\mathbf{n}, \mathbf{p}$ in $\mathcal{Z}^+$ with $\mathbf{n} \geq \mathbf{p}$, the cardinal of the set of injective maps of $\mathbf{p}$ in $\mathbf{n}$ is $\mathbf{n!/(n-p)!}$ (see definition 402 (p. 522)):

$$\mathbf{Card(set_f\{(f : p \to n), i \neq j \Rightarrow f(i) \neq f(j)\}) = n!/(n-p)! = \prod_{i\in[n-p+1,n]} i}\,. \tag{A.57}$$

**Proof**

1. Proposition 944 is true for $\mathbf{p} = \mathbf{1}$.
Calling $\mathcal{F}$ the set of maps $\mathbf{(f : 1 \to n) \subset (1 \times n)}$, any element of $\mathbf{1 \times n}$ defined a subset of one element (definition 345 (p. 447)) of $\mathbf{1 \times n}$. Any such subset verifies trivially the definition 122 (p. 139). Any subset of one element of $\mathbf{1 \times n}$ is an injection. Then we have:
$$\mathbf{1 \times n \subset \bigcup_{f\in\mathcal{F}} f}\,.$$
Since by definition 112 (p. 131), point 1, any map $\mathbf{(f : 1 \to n)}$ is a subset of $\mathbf{1 \times n}$ with one element. Then:
$$\mathbf{\bigcup_{f\in\mathcal{F}} f \subset 1 \times n}\,.$$
From proposition 754 (p. 451) (extensionality principle), we have:
$$\mathbf{1 \times n = set_f\{(f : 1 \to n), i \neq j \Rightarrow f(i) \neq f(j)\}}\,.$$
The cardinal of $\mathbf{1 \times n}$ is $\mathbf{n}$ (propositions 907 (p. 512), 874 (p. 500)). Proposition 944 is true for $\mathbf{p} = \mathbf{1}$.

2. If it is true for $\mathbf{p} \in [\mathbf{1}, \mathbf{n}]$, it is true for $\mathbf{p+1}$:
There is a bijective map $\mathbf{(f_1 : [1, p{+}1] \to p{+}1)}$ and one $\mathbf{(f_2 : [1, n] \to n)}$ (proposition 934 (p. 523)). From that, let us prove that the set of injections $\mathcal{F}_{\mathbf{it}}$ (definition 359 (p. 461)) of $\mathbf{[1, p+1]}$ to $\mathbf{[1, n]}$ is equivalent to the set of injections $\mathcal{F}_{\mathbf{cd}}$ of $\mathbf{p+1}$ to $\mathbf{n}$:
If $\mathbf{h}$ is an injection $\mathbf{[1, p+1] \to [1, n]}$, it is a bijection $\mathbf{[1, p+1] \to h([1, p+1])}$ (definition 361 (p. 461), proposition 180 (p. 143)). Then $\mathbf{f_2 \circ h \circ f_1^{-1}}$ is a bijection from $\mathbf{p{+}1 \to f_2 \circ h \circ f_1^{-1}(p{+}1) \subset n}$ (proposition 193 (p. 149)). It is an injection $\mathbf{p{+}1 \to n}$ (proposition 181 (p. 143)). The couple $\mathbf{f_1}$, $\mathbf{f_2}$ defines a map $(\mathbf{F_{it,cd}} : \mathcal{F}_{\mathbf{it}} \to \mathcal{F}_{\mathbf{cd}})$. For any $\mathbf{h'} \in \mathcal{F}_{\mathbf{cd}}$ (an injection from $\mathbf{p+1 \to n}$), $\mathbf{h = f_2^{-1} \circ h' \circ f_1}$ is an injection $\mathbf{[1, p+1] \to [1, n]}$ in $\mathcal{F}_{\mathbf{it}}$ and we have:
$$\mathbf{h' = f_2 \circ h \circ f_1^{-1} = f_2 \circ f_2^{-1} \circ h' \circ f_1 \circ f_1^{-1} = F_{it,cd}(h)}$$
$\mathbf{F_{it,cd}}$ is surjective (definition 123 (p. 139)). From propositions 196 (p. 150) and 197 (p. 150), $\mathbf{F_{it,cd}}$ is injective. Then $\mathbf{F_{it,cd}}$ is a bijection: The cardinal of injections $\mathbf{[1, p{+}1] \to [1, n]}$ is equal to the cardinal of injections $\mathbf{p+1 \to n}$.
We define the map $\mathbf{(h_p : [1, p] \to h([1, p]))}$ from an injective map $\mathbf{(h : [1, p+1] \to [1, n])}$ by:
$$\mathbf{x \in [1, p] \Rightarrow h_p(x) = h(x) \in h([1, p])}\,.$$

From proposition 180 (p. 143), $\mathbf{h_p}$ is a bijection. Then from definition 390 (p. 499) and proposition 934 (p. 523), we have:

$$\mathbf{Card(h([1,p])) = Card([1,p]) = p} \ .$$

By definition of $\mathbf{h_p}$ and $\mathbf{h}$, we have:

$$\mathbf{h(p+1) \in \mathcal{C}_{[1,n]}(h_p([1,p])) = \mathcal{C}_{[1,n]}(h([1,p]))} \ .$$

From definition 353 (p. 454), we have:

$$\mathbf{\mathcal{C}_{[1,n]}(h_p([1,p])) \cup h_p([1,p]) = [1,n]} \ .$$

From definition 393 (p. 506) and proposition 793 (p. 464), we have:

$$\begin{aligned}&\mathbf{Card(\mathcal{C}_{[1,n]}(h_p([1,p]))) + Card(h_p([1,p])) = Card(\mathcal{C}_{[1,n]}(h_p([1,p]))) + p}\\&\mathbf{= Card([1,n]) = n} \ .\end{aligned}$$

From proposition 898 (p. 509) and definition 395 (p. 509), we have:

$$\mathbf{Card(\mathcal{C}_{[1,n]}(h_p([1,p]))) = n - p = Card([p+1,n])} \ .$$

From proposition 868 (p. 496) and definition 390 (p. 499), since the induced order from the order on $\mathcal{Z}$ is a good order as the one of $\mathbf{[1,n]}$, there is a unique isomorphism $\mathbf{(u : \mathcal{C}_{[1,n]}(h_p([1,p])) \rightarrow [1,n-p])}$ between $\mathbf{\mathcal{C}_{[1,n]}(h_p([1,p]))}$ and $\mathbf{[1,n-p]}$. To any injective map $\mathbf{(h : [1,p+1] \rightarrow [1,n])}$, we generate two maps:
$\mathbf{(h_p : [1,p] \rightarrow [1,n])}$ such that:

$$\mathbf{x \in [1,p] \Rightarrow h_p(x) = h(x) \in h([1,p])}$$

and $\mathbf{(h_1 : \{1\} \rightarrow [1,n-p])}$ such that:

$$\mathbf{h_1(1) = uh(p+1) \in [1,n-p]} \ .$$

We call $\mathcal{F}^{\mathrm{n}}_{\mathrm{p}}$ the set of maps $\mathbf{(h_p : [1,p] \rightarrow [1,n])}$ and $\mathcal{F}^{\mathrm{n-p}}_{1}$ the set of maps $\mathbf{(h_1 : \{1\} \rightarrow [1,n-p])}$. Seeing $\mathbf{h_p}$, $\mathbf{h_1}$ as the two members of the couple $\mathbf{(h_p, h_1)}$, that operation can be considered as a map $\mathbf{(F : \mathcal{F}^{n}_{p+1} \rightarrow \mathcal{F}^{n}_{p} \times \mathcal{F}^{n-p}_{1})}$ (definitions 346 (p. 448), 356 (p. 459)).
$\mathbf{F}$ is an injective map (definitions 122 (p. 139), 359 (p. 461)), because from definition 346 (p. 448) equation (A.15) (p. 448) and definition 347 (p. 449), propositions 757 (p. 452) and 149 (p. 132), we have:

$$\begin{aligned}&\mathbf{(h \in \mathcal{F}^{n}_{p+1}, h' \in \mathcal{F}^{n}_{p+1}, h \neq h') \Rightarrow (\exists x \in [1,p+1])(h(x) \neq h'(x))}\\&\mathbf{(x = \tau_x(h(x) \neq h'(x)), x \in [1,p]) \Rightarrow proj_1(F(h))(x) = h(x)}\\&\mathbf{\neq proj_1(F(h'))(x) = h'(x) \Rightarrow F(h) \neq F(h')}\\&\mathbf{x = \tau_x(h(x) \neq h'(x)) = p+1}\\&\mathbf{\Rightarrow proj_2(F(h))(1) = h_1(x) = u(h(p+1))}\\&\mathbf{\neq proj_2(F(h'))(1) = h'_1(x) = u(h'(p+1)) \Rightarrow F(h) \neq F(h')} \ .\end{aligned}$$

Then $\mathbf{h \neq h' \Rightarrow F(h) \neq F(h')}$. From definition 359 (p. 461), $\mathbf{F}$ is an injection.
Since for any couple of maps $\mathbf{(h_p, h_1) \in \mathcal{F}^{n}_{p} \times \mathcal{F}^{n-p}_{p}}$, we have: $\mathbf{h \in \mathcal{F}^{n-p}_{p+1}}$ such that:

$$\mathbf{x \in [1,p] \Rightarrow h(x) = h_p(x),\ x \in ]p,p+1] \Rightarrow h(x) = h(p+1) = u^{-1}(h_1(1))} \ .$$

That $\mathbf{h}$ is such that $(\mathbf{h_p}, \mathbf{h_1}) = \mathbf{F}(\mathbf{h})$. $\mathbf{F}$ is surjective (definition 123 (p. 139), 360 (p. 461)). Then $\mathbf{F}$ is a bijection (definition 124 (p. 139), 361 (p. 461)). From definition 390 (p. 499), we have:

$$\mathbf{Card}(\mathcal{F}^{n}_{p+1}) = \mathbf{Card}(\mathcal{F}^{n}_{p} \times \mathcal{F}^{n-p}_{1}) .$$

From definition 390 (p. 499) and proposition 822 (p. 473) using definition 396 (p. 510), we have:

$$\mathbf{Card}(\mathcal{F}^{n}_{p} \times \mathcal{F}^{n-p}_{1}) = \mathbf{Card}(\mathcal{F}^{n}_{p}) \times \mathbf{Card}(\mathcal{F}^{n-p}_{1}) .$$

We suppose:

$$\mathbf{Card}(\mathcal{F}^{n}_{p}) = \frac{\mathbf{n!}}{(\mathbf{n}-\mathbf{p})!} = \prod_{\mathbf{i} \in [\mathbf{n}-\mathbf{p}+1, \mathbf{n}]} \mathbf{i}$$

and from point 1, we have:

$$\mathbf{Card}(\mathcal{F}^{n-p}_{1}) = \mathbf{n} - \mathbf{p} .$$

It gives the formula (A.57) (p. 529) of proposition 944 for $\mathbf{p}+\mathbf{1}$. From proposition 927 (p. 520) (recurrence principle), equation (A.57) (p. 529) is true for any $\mathbf{p} \in [\mathbf{1}, \mathbf{n}]$.

**Proposition 945** $\mathbf{a}$ being a non-empty finite set with a total order (definition 377 (p. 483)), $\mathbf{a}$ has a smallest and a biggest elements (definition 379 (p. 484)) according that order.

**Proof**
We have $\mathbf{Card}(\mathbf{a}) \in \mathcal{Z}^{+}$.

1. Proposition 945 (p. 531) is true for $\mathbf{Card}(\mathbf{a}) = \mathbf{1}$.
   From proposition 878 (p. 501), if $\mathbf{Card}(\mathbf{a}) = \mathbf{1}$, we have $\exists \mathbf{c} \in \mathbf{a}$. Calling $\mathbf{c} = \tau_{\mathbf{c}}(\mathbf{c} \in \mathbf{a})$, we have furthermore:

   $$(\forall \mathbf{b})(\mathbf{b} \in \mathbf{a} \Rightarrow \mathbf{b} = \mathbf{c})$$
   $$(\forall \mathbf{b})(\mathbf{b} \in \mathbf{a} \Rightarrow \mathbf{b} \leq \mathbf{c}, \mathbf{c} \leq \mathbf{b})$$

   $\mathbf{c}$ fulfills the requirements of definition 379 (p. 484).

2. Let us suppose now that $\mathbf{Card}(\mathbf{a}) > \mathbf{1}$ and let us suppose that proposition 945 (p. 531) is true for $\mathbf{Card}(\mathbf{a}) = \mathbf{n} - \mathbf{1} \geq \mathbf{1}$.
   When $\mathbf{Card}(\mathbf{a}) = \mathbf{n} > \mathbf{1}$, from proposition 878 (p. 501), we have $\mathbf{a} \neq \emptyset$. From proposition 773 (p. 457), $\exists \mathbf{c} \in \mathbf{a}$. From definitions 392 (p. 500), 353 (p. 454), propositions 878 (p. 501) and 930 (p. 521), since from proposition 752 (p. 451), we have $\{\mathbf{c}\} \subset \mathbf{a}$ and $\mathbf{Card}(\{\mathbf{c}\}) = \mathbf{1} < \mathbf{Card}(\mathbf{a}) = \mathbf{n}$, we have: $\mathbf{n} = \mathbf{Card}(\mathbf{a}) = \mathbf{Card}(\mathcal{C}(\{\mathbf{c}\})) + \mathbf{Card}(\{\mathbf{c}\}) = \mathbf{Card}(\mathcal{C}_{\mathbf{a}}(\{\mathbf{c}\})) + \mathbf{1}$. Then $\mathbf{Card}(\mathcal{C}_{\mathbf{a}}(\{\mathbf{c}\})) = \mathbf{n} - \mathbf{1}$, from what we suppose $\mathcal{C}_{\mathbf{a}}(\{\mathbf{c}\})$ has a smallest element $\mathbf{m}$ and a biggest element $\mathbf{M}$. Since the order on $\mathbf{a}$ is total, we have three issues:

   - $\mathbf{c} < \mathbf{m} \leq \mathbf{M}$, $\mathbf{c}$ is the smallest element of $\mathbf{a}$ and $\mathbf{M}$ is the biggest element of $\mathbf{a}$ because we have:

     $$(\forall \mathbf{y} \in \mathbf{a})((\mathbf{y} = \mathbf{c}, \mathbf{c} \leq \mathbf{y} \leq \mathbf{M}) \vee (\mathbf{y} \in \mathcal{C}_{\mathbf{a}}(\{\mathbf{c}\}), \mathbf{c} < \mathbf{m} \leq \mathbf{y} \leq \mathbf{M})) .$$

   - $\mathbf{m} < \mathbf{c} < \mathbf{M}$, $\mathbf{m}$ (resp. $\mathbf{M}$) is the smallest (resp. the biggest) element of $\mathbf{a}$.
   - $\mathbf{M} < \mathbf{c}$, $\mathbf{c}$ is the biggest element of $\mathbf{a}$ and $\mathbf{m}$ is the smallest element of $\mathbf{a}$ because we have:

     $$(\forall \mathbf{y} \in \mathbf{a})((\mathbf{y} = \mathbf{c}, \mathbf{c} \geq \mathbf{y} \geq \mathbf{m}) \vee (\mathbf{y} \in \mathcal{C}_{\mathbf{a}}(\{\mathbf{c}\}), \mathbf{m} \leq \mathbf{y} \leq \mathbf{M} < \mathbf{c}))$$

proposition 945 (p. 531) is true for $\mathbf{Card(a) = n}$. Then proposition 945 (p. 531) is true from the recurrence principle proposition 927 (p. 520).

**Proposition 946** A total order on a finite set is a good order.

**Proof**
It is a direct consequence of propositions 945 (p. 531), 918 (p. 516) and of definitions 388 (p. 487), 392 (p. 500).

**Proposition 947** There is an isomorphism (definition 387 (p. 487)) between a finite set $\mathbf{a}$ with a total order and the set of cardinal $[\mathbf{1}, \mathbf{Card(a)}]$.

**Proof**
From proposition 869 (p. 496), there are:
Either an isomorphism between $\mathbf{a}$ and a segment of $[\mathbf{1}, \mathbf{Card(a)}]$, $[\mathbf{1}, \mathbf{k}] \subset [\mathbf{1}, \mathbf{Card(a)}]$;
or an isomorphism between $[\mathbf{1}, \mathbf{Card(a)}]$ and a segment of $\mathbf{a}$: $[\mathbf{1}, \mathbf{b}] \subset \mathbf{a}$.
Since there are bijections between $\mathbf{a}$, $[\mathbf{1}, \mathbf{Card(a)}]$ and $\mathbf{Card(a)}$, we have a bijection between $[\mathbf{1}, \mathbf{k}]$ and $\mathbf{Card(a)}$ or a bijection between $[\mathbf{1}, \mathbf{b}]$ and $\mathbf{Card(a)}$ which means from proposition 935 (p. 523) $[\mathbf{1}, \mathbf{k}] = [\mathbf{1}, \mathbf{Card(a)}]$ or $[\mathbf{1}, \mathbf{b}] = \mathbf{a}$. In both cases, it is an isomorphism between $[\mathbf{1}, \mathbf{Card(a)}]$ and $\mathbf{a}$.

## A.7 Operations on $\mathcal{Z} = \mathcal{Z}^{0+} \cup \mathcal{Z}^-$

**Definition 405** $\mathcal{Z}^-$.

One calls (see definitions 346 (p. 448) and 356 (p. 459)):

$$\mathcal{Z}^- = \{-\} \times \mathcal{Z}^+ \,.$$

**Definition 406** $\mathcal{Z}$.

$$\mathcal{Z} = \mathcal{Z}^- \cup \mathcal{Z}^{0+} \,.$$

**Definition 407 Absolute value in** $\mathcal{Z}$.

$$\begin{aligned}
&\mathbf{x} \in \mathcal{Z}^{0+} \Rightarrow |\mathbf{x}| = \mathbf{x}, \\
&\mathbf{x} \in \mathcal{Z}^- \Rightarrow \mathbf{x} \in \{-\} \times \mathcal{Z}^+ \Rightarrow (\exists \mathbf{y})(\mathbf{y} \in \mathcal{Z}^+, \mathbf{x} = -\mathbf{y}) \\
&|\mathbf{x}| = \tau_{\mathbf{y}}(\mathbf{y} \in \mathcal{Z}^+, \mathbf{x} = -\mathbf{y}) \,.
\end{aligned}$$

See definitions 332 (p. 440) and 333 (p. 441).

**Proposition 948** $(\mathbf{z_1} \in \mathcal{Z}^-, \mathbf{z_2} \in \mathcal{Z}^-) \Leftrightarrow |\mathbf{z_1}| = |\mathbf{z_2}|$

**Proof**
It is a direct consequence of definition 346 (p. 448) and equation (A.15).

**Definition 408 Minus map** $(- : \mathcal{Z} \to \mathcal{Z})$.

$$(\forall \mathbf{z} \in \mathcal{Z})(\mathbf{z} \in \mathcal{Z}^+ \Rightarrow -(\mathbf{z}) = -\mathbf{z}), -\mathbf{0} = \mathbf{0}, (\mathbf{z} \in \mathcal{Z}^- \Rightarrow -(\mathbf{z}) = |\mathbf{z}|) \,.$$

**Proposition 949** The minus map is its own inverse.

**Proof**
It comes straight from the definition: (see definition 174 (p. 141))

$$(\forall \mathbf{z} \in \mathcal{Z})(-(-\mathbf{z}) = \mathbf{z}) .$$

**Proposition 950** The minus map is a bijection: $(- : \mathcal{Z} \leftrightarrow \mathcal{Z})$.
**Proof**
It comes from the fact that $(\forall \mathbf{z} \in \mathcal{Z})(\mathbf{Card}((-)^{-1}(\{\mathbf{z}\})) = \mathbf{1})$ (definitions 174 (p. 141) and 178 (p. 142)).
**Definition 409 Order on $\mathcal{Z}$ (definition 406 (p. 532)).**

$\mathbf{z_1}$ and $\mathbf{z_2}$ are in $\mathcal{Z}$.
If $\mathbf{z_1}$ and $\mathbf{z_2}$ are both in $\mathcal{Z}^{0+}$, they are cardinals and their order is given by definition 392 (p. 500).
If they are both in $\mathcal{Z}^-$, $\mathbf{z_1} \leq \mathbf{z_2}$ if $|\mathbf{z_1}| \geq |\mathbf{z_2}|$.
If $\mathbf{z_1} \in \mathcal{Z}^-, \mathbf{z_2} \in \mathcal{Z}^{0+}$, we have: $\mathbf{z_1} < \mathbf{z_2}$.

**Proposition 951**

$$(\forall \mathbf{z_1} \in \mathcal{Z})(\forall \mathbf{z_2} \in \mathcal{Z})(\mathbf{z_1} \leq \mathbf{z_2} \Leftrightarrow -\mathbf{z_2} \leq -\mathbf{z_1}) .$$

**Proof**
It is a direct consequence of definition 409 (p. 533).

**Proposition 952** The order defined by definition 409 (p. 533) is an ordering relation (definition 338 (p. 446)).
**Proof**
Along definition 338 (p. 446), with $\mathbf{z_1}, \mathbf{z_2}, \mathbf{z_3}$ all in $\mathcal{Z}$, we have:

1. Using definition 338 (p. 446), point 2 with $\mathbf{z_1} \leq \mathbf{z_2},\ \mathbf{z_2} \leq \mathbf{z_1}$, we have:

   - If both in $\mathcal{Z}^{0+}$, the order is the same as for $\mathcal{Z}^{0+}$ so $\mathbf{z_1} \leq \mathbf{z_2}, \mathbf{z_2} \leq \mathbf{z_1} \Rightarrow \mathbf{z_1} = \mathbf{z_2}$.
   - If both in $\mathcal{Z}^-$, from definition 408 (p. 532), $-\mathbf{z_1}$ $-\mathbf{z_2}$ are both in $\mathcal{Z}^+$. From propositions 951 (p. 533) and 950 (p. 533), we have: $\mathbf{z_1} \leq \mathbf{z_2} \Rightarrow -\mathbf{z_2} \leq -\mathbf{z_1}$. From that, with equation (A.15) (p. 448), we get:

   $$\mathbf{z_1} \leq \mathbf{z_2}, \mathbf{z_2} \leq \mathbf{z_1} \Leftrightarrow -\mathbf{z_1} \leq -\mathbf{z_2}, -\mathbf{z_2} \leq -\mathbf{z_1} \Rightarrow -\mathbf{z_1} = -\mathbf{z_2} \Rightarrow \mathbf{z_1} = \mathbf{z_2} .$$

   - If $\mathbf{z_1} \in \mathcal{Z}^-,\ \mathbf{z_2} \in \mathcal{Z}^{0+}$ from definition 409 (p. 533), only $\mathbf{z_1} < \mathbf{z_2}$ is possible.

2. Using definition 338 (p. 446), point 3, supposing $\mathbf{z_1} \leq \mathbf{z_2}$, $\mathbf{z_2} \leq \mathbf{z_3}$, we have:

   - $\mathbf{z_1}, \mathbf{z_2}$ both in $\mathcal{Z}^{0+}$. $\mathbf{z_3}$ cannot be in $\mathcal{Z}^-$ for that it should be smaller than $\mathbf{z_2}$. $\mathbf{z_1}, \mathbf{z_2}, \mathbf{z_3}$ are all in $\mathcal{Z}^{0+}$, they are cardinals. From point 3 (p. 500),

   $$(\mathbf{z_1} \leq \mathbf{z_2}, \mathbf{z_2} \leq \mathbf{z_3}) \Rightarrow \mathbf{z_1} \leq \mathbf{z_3} .$$

   - $\mathbf{z_1}, \mathbf{z_2}$ both in $\mathcal{Z}^-$. If $\mathbf{z_3} \in \mathcal{Z}^-$ we have, from definition 405 (p. 532), $-\mathbf{z_1}$, $-\mathbf{z_2}$, $-\mathbf{z_3}$ all in $\mathcal{Z}^{0+}$. From proposition 951 (p. 533), we have:

   $$(\mathbf{z_1} \leq \mathbf{z_2}, \mathbf{z_2} \leq \mathbf{z_3}) \Rightarrow (-\mathbf{z_2} \leq -\mathbf{z_1}, -\mathbf{z_3} \leq -\mathbf{z_2}) \Rightarrow -\mathbf{z_3} \leq -\mathbf{z_1} \Rightarrow \mathbf{z_1} \leq \mathbf{z_3} .$$

   If $\mathbf{z_3} \in \mathcal{Z}^{0+}$, from definition 409 (p. 533), $\mathbf{z_1} < \mathbf{z_3}$.

- If $z_1 \in \mathcal{Z}^-, z_2 \in \mathcal{Z}^{0+}$, we have $z_3 \in \mathcal{Z}^{0+}$ otherwise $z_3 < z_2$, then $z_1 < z_3$ (definition 409 (p. 533)).

**Proposition 953** The order defined by definition 409 (p. 533) on $\mathcal{Z}$ (definition 406 (p. 532)) is total (definition 377 (p. 483)).

**Proof**
From proposition 876 (p. 500), the order on $\mathcal{Z}^{0+}$ is total. From definition 409 (p. 533) and proposition 951 (p. 533), the order on $\mathcal{Z}^-$ is also total since:

$$(z_1 \in \mathcal{Z}^-,\ z_2 \in \mathcal{Z}^-) \Rightarrow (-z_1 \in \mathcal{Z}^+,\ -z_2 \in \mathcal{Z}^+) \Rightarrow (-z_1 \leq -z_2 \vee -z_2 \leq -z_1)$$
$$\Rightarrow (z_2 \leq z_1 \vee z_1 \leq z_2)\ .$$

From definition 409 (p. 533), we have:

$$(z_1 \in \mathcal{Z}^-,\ z_2 \in \mathcal{Z}^{0+}) \Rightarrow z_1 < z_2 \Rightarrow z_1 \leq z_2\ .$$

**Proposition 954** Any element $z$ of $\mathcal{Z}$ has a bigger and a smaller elements in $\mathcal{Z}$.

**Proof**
If it is zero $z = 0 \in \mathcal{Z}$, one can take $1$ and $-1$ respectively because $1 > 0$ (definitions 390 (p. 499), 409 (p. 533) proposition 880 (p. 502)). If $z \in \mathcal{Z}^+,\ z > 0$ one can take $z + 1$ and $-z$ respectively (definition 397 (p. 516), proposition 877 (p. 500), definition 409 (p. 533)). If $z \in \mathcal{Z},\ z < 0$, one can take $-z$ and $z - 1$ respectively.

**Proposition 955** Any nonempty finite subset of $\mathcal{Z}$ has a smallest and a biggest elements (see definition 379 (p. 484)).

**Proof**
It is a direct consequence of propositions 953 (p. 534) and 945 (p. 531).

**Proposition 956**

$$z \in \mathcal{Z}^{0+} \Rightarrow z = \mathrm{Card}(]-z, 0]) = \mathrm{Card}([-z, 0[) = \mathrm{Card}([0, z[) = \mathrm{Card}(]0, z])\ .$$

**Proof**
Supposing $z \in \mathcal{Z}^{0+}$, $a \in \mathcal{Z}^{0+}$, from definitions 405 (p. 532), 409 (p. 533), 390 (p. 499) and proposition 934 (p. 523), we have:

$$-a \in ]-z, 0] \Leftrightarrow a \in [0, z[ \Rightarrow \mathrm{Card}(]-z, 0]) = \mathrm{Card}([0, z[) = z\ .$$

From propositions 914 (p. 515), 393 (p. 506) and definition 393 (p. 506), we have:

$$([0, z[\cup\{z\} = [0, z], [0, z[\cap\{z\} = \emptyset, ]0, z] \cup \{0\} = [0, z], ]0, z] \cap \{0\} = \emptyset)$$
$$\Rightarrow \mathrm{Card}([0, z]) = \mathrm{Card}([0, z[) + 1 = \mathrm{Card}(]0, z]) + 1$$
$$\Rightarrow \mathrm{Card}([0, z[) = \mathrm{Card}(]0, z])\ .$$

**Definition 410 Sum of two elements of $\mathcal{Z}$.**

Given a set $\mathbf{I} = \{\mathbf{i_1}, \mathbf{i_2}\}$ with $\mathbf{Card(I) = 2}$ (see definition 345 (p. 447)) and a map $(\mathbf{a_i} : \mathbf{I} \to \mathcal{Z})$ one defines the sum:

$$\sum_{i \in I} a_i \,.$$

By the following rules:

1. $(\forall i \in I)(a_i \in \mathcal{Z}^{0+})$, the sum is defined by definition 393 (p. 506).
2. $(\exists i_1 \in I)(a_{i_1} \in \mathcal{Z}^{-}), (\exists i_2 \in I)(a_{i_2} \in \mathcal{Z}^{0+})$, We use definition 392 (p. 500), propositions 876 (p. 500) and 898 (p. 509) and we define $\sum_{i \in I} a_i$ by:

$$|a_{i_1}| \leq a_{i_2} \Rightarrow (\exists x \subset a_{i_2})(x \sim |a_{i_1}|) \Rightarrow \sum_{i \in I} a_i = \mathbf{Card}(\mathcal{C}_{a_{i_2}}(x)) \tag{A.58}$$

$$|a_{i_1}| > a_{i_2} \Rightarrow (\exists x \subset |a_{i_1}|)(x \sim a_{i_2}) \Rightarrow \sum_{i \in I} a_i = -\mathbf{Card}(\mathcal{C}_{|a_{i_1}|}(x)) \,. \tag{A.59}$$

3. $(\forall i \in I)(a_i \in \mathcal{Z}^{-})$

$$\sum_{i \in I} a_i = -\sum_{i \in I} |a_i| \,.$$

**Proposition 957** Given a set $\mathbf{I} = \{\mathbf{i_1}, \mathbf{i_2}\}$ with $\mathbf{Card(I) = 2}$ (see definition 345 (p. 447)) and a map $(\mathbf{a_i} : \mathbf{I} \to \mathcal{Z})$, given a set $\mathbf{J}$ with $\mathbf{Card(J) = Card(I) = 2}$ and a bijection $(\mathbf{f} : \mathbf{J} \leftrightarrow \mathbf{I})$, we have:

$$\sum_{i \in I} a_i = \sum_{j \in J} a_{f(j)} \,.$$

This proposition is generally quoted as 'the sum of two integers is commutative'.

**Proof**

If we are in items 1 and 3 of definition 410, from proposition 731 (p. 442) and definition 112 (p. 131), 2

$$(\forall i)(i \in I \Rightarrow a_i \in B) \Rightarrow (f(j) \in I \Rightarrow a_{f(j)} \in B) \,.$$

We conclude using proposition 892 (p. 507).

In item 2, since $\mathbf{f}$ is a bijection, we have:

$$((\exists i_1 \in I)(a_{i1} \in \mathcal{Z}^{-}), (\exists i_2 \in I)(a_{i_2} \in \mathcal{Z}^{0+})) \Rightarrow$$
$$((\exists j_1 \in J)(j_1 = f^{-1}(i_1), a_{f(j_1)} \in \mathcal{Z}^{-}), (\exists j_2 \in J)(j_2 = f^{-1}(i_2), a_{f(j_2)} \in \mathcal{Z}^{0+})) \,.$$

We also have: $x \subset |a_{i_1}|, x \sim a_{i_2} \Leftrightarrow x \subset |a_{f(j_1)}|, x \sim a_{f(j_2)}$ (resp. switching $\mathbf{1}$ and $\mathbf{2}$) which gives $\mathbf{Card}(\mathcal{C}_{a_{i_2}}(x)) = \mathbf{Card}(\mathcal{C}_{a_{j(j_2)}}(x))$ (resp. $\mathbf{Card}(\mathcal{C}_{|a_{i_1}|}(x)) = -\mathbf{Card}(\mathcal{C}_{|a_{j(j_1)}|}(x))$ ). Then we have:

$$\sum_{i \in I} a_i = \sum_{j \in J} a_{f(j)} \,.$$

**Proposition 958** Given a set $\mathbf{I}$ with $\mathbf{Card(I) = 2}$ and a map $(\mathbf{a_i} : \mathbf{I} \to \mathcal{Z})$, we have:

$$\sum_{i \in I} a_i = -\sum_{i \in I} -a_i \,.$$

**Proof**

- In the case where all $a_i$ are in $\mathcal{Z}^{0+}$ (resp. $\mathcal{Z}^-$), we have, from definition 410 (p. 534), 1 (resp. 410 (p. 534), 3), $\sum_{i\in I} a_i = \sum_{i\in I} |a_i|$ (resp. $\sum_{i\in I} a_i = -\sum_{i\in I} |a_i|$).
It gives all $-a_i$ in $\mathcal{Z}^-$ (resp. $\mathcal{Z}^{0+}$) and from definition 410 (p. 534), 3 (resp. 410 (p. 534), 1, proposition 949 (p. 532)), we have:

$$\sum_{i\in I} -a_i = -\sum_{i\in I} |a_i| = -\sum_{i\in I} a_i$$

$$\text{resp. } \sum_{i\in I} -a_i = \sum_{i\in I} |a_i| = -(-\sum_{i\in I} |a_i|) = -\sum_{i\in I} a_i \;.$$

- When the $a_i$ are of different signs, we have:

$$(\exists(k,l) \in I^2)(|a_k| > |a_l|), \quad (\exists x \subset |a_k|)(x \sim |a_l| = |-a_l|) \;.$$

We have, from definition 410 (p. 534), 2, when $a_k \in \mathcal{Z}^{0+}$ (resp. $a_k \in \mathcal{Z}^-$):

$$\sum_{i\in I} a_i = \mathcal{C}_{|a_k|}x \text{ (resp. } \sum_{i\in I} a_i = -\mathcal{C}_{|a_k|}(x)) \;.$$

It gives, from definition 410 (p. 534), 2 again since $-a_k \in \mathcal{Z}^-$ (resp. $-a_k \in \mathcal{Z}^{0+}$):

$$\sum_{i\in I} -a_i = -\mathcal{C}_{|a_k|}x = -\sum_{i\in I} a_i \text{ (resp. } \sum_{i\in I} -a_i = \mathcal{C}_{|a_k|}x = -\sum_{i\in I} a_i) \;.$$

**Proposition 959** The sum of two elements of $\mathcal{Z}$ is in $\mathcal{Z}$.

**Proof**
It is a consequence of proposition 941 (p. 526).

**Definition 411 a + b a and b both in $\mathcal{Z}$.**

Given two terms **a** and **b** both in $\mathcal{Z}$, one calls:

$$a + b = b + a = \sum_{i\in\{a,b\}} i$$

and given a map $(a_i : \{1,2\} \to \{a,b\})$, we have:

$$a + b = \sum_{i\in[1,2]} a_i \;.$$

**Definition 412 a − b a and b both in $\mathcal{Z}$.**

$$a - b = a + (-b) \;.$$

**Proposition 960** a, b being in $\mathcal{Z}$ (definition 406 (p. 532)), we have (definition 407 (p. 532)):

$$|a + b| \leq |a| + |b|, \; |a| = 0 \Rightarrow a = 0 \;.$$

**Proof**
Since $\mathbf{a} \in \mathcal{Z}^- \Rightarrow |\mathbf{a}| = -\mathbf{a} \in \mathcal{Z}^+ \Rightarrow |\mathbf{a}| \neq 0$ then (point 7 (p. 435) A.1.4), we have:

$$|\mathbf{a}| = 0 \Rightarrow \mathbf{a} \notin \mathcal{Z}^- .$$

Since $\mathbf{a} \in \mathcal{Z}^- \vee \mathcal{Z}^{0+}$ and (points 4 (p. 435), 5 (p. 435), 15 (p. 435) A.1.4), we have:

$$(\mathbf{a} \in \mathcal{Z}^- \vee \mathcal{Z}^{0+}) \Leftrightarrow (\neg\neg \mathbf{a} \in \mathcal{Z}^- \vee \mathcal{Z}^{0+}) = (\mathbf{a} \notin \mathcal{Z}^- \Rightarrow \mathbf{a} \in \mathcal{Z}^{0+}) .$$

From those, we have (definition 407 (p. 532)):

$$|\mathbf{a}| = 0 \Rightarrow \mathbf{a} \in \mathcal{Z}^{0+}, |\mathbf{a}| = 0 \Rightarrow \mathbf{a} = |\mathbf{a}| = 0$$

1. $\mathbf{a} \in \mathcal{Z}^{0+}, \ \mathbf{b} \in \mathcal{Z}^{0+}$. We have successively (see A.1.2 (p. 434))

$$|\mathbf{a}| = \mathbf{a}, \ |\mathbf{b}| = \mathbf{b}; \ |\mathbf{a}| + |\mathbf{b}| = \mathbf{a} + \mathbf{b} \in \mathcal{Z}^{0+}; \ |\mathbf{a} + \mathbf{b}| = \mathbf{a} + \mathbf{b}; \ |\mathbf{a}| + |\mathbf{b}| = |\mathbf{a} + \mathbf{b}| .$$

2. $\mathbf{a} \in \mathcal{Z}^-, \ \mathbf{b} \in \mathcal{Z}^-$

$$|\mathbf{a}| = -\mathbf{a}, \ |\mathbf{b}| = -\mathbf{b}; \ \mathbf{a} + \mathbf{b} \in \mathcal{Z}^-;$$
$$|\mathbf{a} + \mathbf{b}| = -(\mathbf{a} + \mathbf{b}) = -\mathbf{a} - \mathbf{b}; \ |\mathbf{a} + \mathbf{b}| = |\mathbf{a}| + |\mathbf{b}| .$$

3. $\mathbf{a} \in \mathcal{Z}^-, \ \mathbf{b} \in \mathcal{Z}^{0+}, \ |\mathbf{a}| = -\mathbf{a}, \ |\mathbf{b}| = \mathbf{b}$

$$\begin{aligned} &\mathbf{a} + \mathbf{b} \in \mathcal{Z}^- \Rightarrow |\mathbf{a} + \mathbf{b}| = -(\mathbf{a} + \mathbf{b}) = -\mathbf{a} - \mathbf{b} = |\mathbf{a}| - |\mathbf{b}| \leq 0 \\ &\Rightarrow |\mathbf{a} + \mathbf{b}| = |\mathbf{a}| - |\mathbf{b}| \leq |\mathbf{a}| \leq |\mathbf{a}| + |\mathbf{b}|) \Rightarrow |\mathbf{a} + \mathbf{b}| \leq |\mathbf{a}| + |\mathbf{b}| \\ &\mathbf{a} + \mathbf{b} \in \mathcal{Z}^{+0} \Rightarrow |\mathbf{a} + \mathbf{b}| = \mathbf{a} + \mathbf{b} = -|\mathbf{a}| + |\mathbf{b}| \leq 0 \\ &\Rightarrow |\mathbf{a} + \mathbf{b}| = |\mathbf{b}| - |\mathbf{a}| \leq |\mathbf{b}| \leq |\mathbf{a}| + |\mathbf{b}| \ \Rightarrow |\mathbf{a} + \mathbf{b}| \leq |\mathbf{a}| + |\mathbf{b}| . \end{aligned}$$

Since we have $\mathbf{a} + \mathbf{b} \in \mathcal{Z}^- \vee \mathbf{a} + \mathbf{b} \in \mathcal{Z}^{0+}$, we have $|\mathbf{a} + \mathbf{b}| \leq |\mathbf{a}| + |\mathbf{b}|$.

4. $\mathbf{b} \in \mathcal{Z}^-, \ \mathbf{a} \in \mathcal{Z}^{0+}$. Since $\mathbf{a} + \mathbf{b} = \mathbf{b} + \mathbf{a}$, it is the same proof as point 3.

**Definition 413 Oriented intervals.**

Given two elements $\mathbf{z}_1$ and $\mathbf{z}_2$ of $\mathcal{Z}$ we define a map $(\overline{\mathbf{z}_1, \mathbf{z}_2} : \mathcal{Z}^2 \to \mathcal{Z})$ by:

$$\begin{aligned} \mathbf{z}_1 \leq \mathbf{z}_2 &\Rightarrow \overline{\mathbf{z}_1, \mathbf{z}_2} = \mathbf{Card}([\mathbf{z}_1, \mathbf{z}_2[), \\ \mathbf{z}_2 < \mathbf{z}_1 &\Rightarrow \overline{\mathbf{z}_1, \mathbf{z}_2} = -\mathbf{Card}([\mathbf{z}_2, \mathbf{z}_1[) . \end{aligned}$$

**Proposition 961** Given $\mathbf{z}$, $\mathbf{z}_1$, $\mathbf{z}_2$ and $\mathbf{z}_3$ in $\mathcal{Z}$ and using definitions 411 (p. 536) and 412 (p. 536) we have:

1. $\overline{\mathbf{z}_1, \mathbf{z}_2} = -\overline{\mathbf{z}_2, \mathbf{z}_1}$
2. $\overline{\mathbf{z}_1, \mathbf{z}_2} = -\overline{-\mathbf{z}_1, -\mathbf{z}_2}$
3. $\mathbf{z} = \overline{0, \mathbf{z}}$
4. $\overline{\mathbf{z}_1, \mathbf{z}_2} = \mathbf{z}_2 - \mathbf{z}_1$
5. $\overline{\mathbf{z}_1, \mathbf{z}_2} + \overline{\mathbf{z}_2, \mathbf{z}_3} = \overline{\mathbf{z}_1, \mathbf{z}_3}$

**Proof**

1. $\overline{z_1, z_2} = -\overline{z_2, z_1}$. It comes directly from definition 413.

2. $\overline{z_1, z_2} = -\overline{-z_1, -z_2}$. We suppose $z_1 \leq z_2$ (resp. $z_2 \leq z_1$). It means $-z_2 \leq -z_1$ (resp. $-z_1 \leq -z_2$) (proposition 951 (p. 533)). From definition 413 (p. 537), we have: $\overline{z_1, z_2} = \mathrm{Card}([z_1, z_2[)$ (resp. $\overline{z_2, z_1} = \mathrm{Card}([z_2, z_1[)$).
From proposition 951 (p. 533) and since $-$ is a bijection and its own inverse (proposition 949 (p. 532)), we have:

$$a \in [z_1, z_2[ \Leftrightarrow -a \in [-z_2, -z_1[ \text{ (resp. } a \in [z_2, z_1[ \Leftrightarrow -a \in [-z_1, -z_2[) \ .$$

Then we have:

$$[z_1, z_2[ \sim [-z_2, -z_1[ \text{(resp. } [z_2, z_1[ \sim [-z_1, -z_2[)$$

and

$$\mathrm{Card}([z_1, z_2[) = \mathrm{Card}([-z_2, -z_1[) \text{(resp. } \mathrm{Card}([z_2, z_1[) = \mathrm{Card}([-z_1, -z_2[))$$

then

$$\overline{z_1, z_2} = \overline{-z_2, -z_1} = -\overline{-z_1, -z_2}$$

resp.

$$\overline{z_1, z_2} = -\overline{z_2, z_1} = --\overline{-z_1, -z_2} = ---\overline{-z_2, -z_1} = \overline{-z_2, -z_1} \ .$$

3. $z = \overline{0, z}$. If $z \geq 0$, we have $z \in \mathcal{Z}^{0+}$. Then from proposition 934 (p. 523), we have:

$$z \geq 0 \Rightarrow z = \mathrm{Card}([0, z[) = \overline{0, z} \ .$$

If $z < 0$, we have, from the above point 1, $-z = \overline{0, -z}$. From the above point 2, $\overline{0, -z} = -\overline{0, z}$. From proposition 949 (p. 532) $z = \overline{0, z}$.

4. $\overline{z_1, z_2} = z_2 - z_1$

   (a) If $0 \leq z_1 \leq z_2$, we have $\overline{z_1, z_2} = \mathrm{Card}([z_1, z_2[) = \mathcal{C}_{[0,z_2[}([0, z_1[) = z_2 - z_1$. We also have:

   $$[0, z_2[ = [0, z_1[ \cup [z_1, z_2[, \ [0, z_1[ \cap [z_1, z_2[ = \emptyset \ .$$

   From proposition 893 (p. 507), it gives:

   $$\mathrm{Card}([0, z_2[) = \mathrm{Card}([0, z_1[) + \mathrm{Card}([z_1, z_2[) \ .$$

   Then using the fact that $|-\mathrm{Card}([0, z_1[)| = \mathrm{Card}([0, z_1[ \geq 0$, $\mathrm{Card}([0, z_2[) \geq \mathrm{Card}([0, z_1[) \geq 0$, we have from definitions 410 (p. 534), 411 (p. 536) and 412 (p. 536):

   $$\mathrm{Card}([z_1, z_2[ = \mathrm{Card}([0, z_2[) - \mathrm{Card}([0, z_1[) = z_2 - z_1 = \overline{z_1, z_2} \ .$$

   We have from point 1 (p. 537), 2 (p. 537) and proposition 958 (p. 535):

   $$\overline{z_2, z_1} = -\overline{z_1, z_2} = -(z_2 - z_1) = z_1 - z_2 \ .$$

(b) If $\mathbf{z_1} < 0 \leq \mathbf{z_2}$, Using again proposition 893 (p. 507), we have:

$$\overline{\mathbf{z_1}, \mathbf{z_2}} = \mathbf{Card}([\mathbf{z_1}, \mathbf{z_2}[) = \mathbf{Card}([\mathbf{z_1}, 0[\cup[0, \mathbf{z_2}[), [\mathbf{z_1}, 0[\cap[0, \mathbf{z_2}[= \emptyset$$
$$\mathbf{Card}([\mathbf{z_1}, \mathbf{z_2}[) = \mathbf{Card}([\mathbf{z_1}, 0[) + \mathbf{Card}([0, \mathbf{z_2}[) = \mathbf{z_2} - \mathbf{z_1}$$
$$\overline{\mathbf{z_1}, \mathbf{z_2}} = \mathbf{z_2} - \mathbf{z_1} \; .$$

We also as in above point 4a:

$$\overline{\mathbf{z_2}, \mathbf{z_1}} = -\overline{\mathbf{z_1}, \mathbf{z_2}} = -(\mathbf{z_2} - \mathbf{z_1}) = \mathbf{z_1} - \mathbf{z_2} \; .$$

(c) If $\mathbf{z_1} \leq \mathbf{z_2} < 0$, we have, applying 2, 1, 4a, definitions 411 (p. 536) and 412 (p. 536):

$$\overline{\mathbf{z_1}, \mathbf{z_2}} = -(\overline{-\mathbf{z_1}, -\mathbf{z_2}}) = \overline{-\mathbf{z_2}, -\mathbf{z_1}} = -\mathbf{z_1} - (-\mathbf{z_2}) = \mathbf{z_2} - \mathbf{z_1}$$
$$\overline{\mathbf{z_2}, \mathbf{z_1}} = \mathbf{z_1} - \mathbf{z_2} \; .$$

5. $\overline{\mathbf{z_1}, \mathbf{z_2}} + \overline{\mathbf{z_2}, \mathbf{z_3}} = \overline{\mathbf{z_1}, \mathbf{z_3}}$.
If $\mathbf{z_1} \leq \mathbf{z_2} \leq \mathbf{z_3}$, we have:

$$[\mathbf{z_1}, \mathbf{z_2}[\cup[\mathbf{z_2}, \mathbf{z_3}[= [\mathbf{z_1}, \mathbf{z_3}[ \; .$$

It gives:

$$\overline{\mathbf{z_1}, \mathbf{z_2}} + \overline{\mathbf{z_2}, \mathbf{z_3}} = \overline{\mathbf{z_1}, \mathbf{z_3}} \; .$$

Since $[\mathbf{z_2}, \mathbf{z_3}[ \subset [\mathbf{z_1}, \mathbf{z_3}[$, from definition 410 (p. 534), we have:

$$\mathbf{Card}([\mathbf{z_1}, \mathbf{z_2}[) = \mathbf{Card}([\mathbf{z_1}, \mathbf{z_3}[) - \mathbf{Card}([\mathbf{z_2}, \mathbf{z_3}[) \; .$$

Then using definition 413 (p. 537) and the already demonstrated point 2, we have:

$$\overline{\mathbf{z_1}, \mathbf{z_2}} = \overline{\mathbf{z_1}, \mathbf{z_3}} - \overline{\mathbf{z_2}, \mathbf{z_3}} = \overline{\mathbf{z_1}, \mathbf{z_3}} + \overline{\mathbf{z_3}, \mathbf{z_2}} \; .$$

Using proposition 958 (p. 535), we have:

$$\overline{\mathbf{z_2}, \mathbf{z_1}} = -\overline{\mathbf{z_1}, \mathbf{z_2}} = -\overline{\mathbf{z_1}, \mathbf{z_3}} - \overline{\mathbf{z_3}, \mathbf{z_2}} = \overline{\mathbf{z_3}, \mathbf{z_1}} + \overline{\mathbf{z_2}, \mathbf{z_3}} = \overline{\mathbf{z_2}, \mathbf{z_3}} + \overline{\mathbf{z_3}, \mathbf{z_1}} \; .$$

The same reasoning gives:

$$\overline{\mathbf{z_2}, \mathbf{z_3}} = \overline{\mathbf{z_1}, \mathbf{z_3}} + \overline{\mathbf{z_2}, \mathbf{z_1}} = \overline{\mathbf{z_2}, \mathbf{z_1}} + \overline{\mathbf{z_1}, \mathbf{z_3}}$$

$$\overline{\mathbf{z_3}, \mathbf{z_2}} = \overline{\mathbf{z_3}, \mathbf{z_1}} + \overline{\mathbf{z_1}, \mathbf{z_2}}$$
$$\overline{\mathbf{z_3}, \mathbf{z_1}} = \overline{\mathbf{z_3}, \mathbf{z_2}} + \overline{\mathbf{z_2}, \mathbf{z_1}} \; .$$

We have examined any possible combination and in any case 5 is true. We have:

$$\overline{\mathbf{z_1}, \mathbf{z_2}} + \overline{\mathbf{z_2}, \mathbf{z_3}} = \overline{\mathbf{z_1}, \mathbf{z_3}} \; .$$

**Proposition 962** $\mathbf{z_1}$, $\mathbf{z_2}$ in $\mathcal{Z}$ gives:

$$\mathbf{z_1} \leq \mathbf{z_2} \Leftrightarrow \mathbf{z_2} - \mathbf{z_1} \geq 0 \Leftrightarrow \mathbf{z_1} - \mathbf{z_2} \leq 0 \tag{A.60}$$

$$\mathbf{z_1} = \mathbf{z_2} \Leftrightarrow \mathbf{z_2} - \mathbf{z_1} = 0 \Leftrightarrow \mathbf{z_1} - \mathbf{z_2} = 0 \tag{A.61}$$

$$\mathbf{z_1} < \mathbf{z_2} \Leftrightarrow \mathbf{z_2} - \mathbf{z_1} > 0 \Leftrightarrow \mathbf{z_1} - \mathbf{z_2} < 0 \; . \tag{A.62}$$

**Proof**
From proposition 961 (p. 537) and definition 413 (p. 537), we have:

$$z_1 \geq z_2 \Leftrightarrow \overline{z_1, z_2} \geq 0 \Leftrightarrow \overline{z_2, z_1} \leq 0 \Leftrightarrow z_2 - z_1 = \overline{z_1, z_2} \geq 0 \Leftrightarrow z_1 - z_2 = \overline{z_2, z_1} \leq 0 \, .$$

Equation (A.60) is true. $z_1 = z_2 \Rightarrow z_2 - z_1 = z_1 - z_2 = \mathbf{Card}([z_1, z_2[) = \mathbf{Card}(\emptyset) = 0$

$$\begin{aligned} &z_1 \neq z_2 \Rightarrow (z_1 < z_2 \vee z_1 > z_2) \Rightarrow z_1 \in [z_1, z_2[ \vee z_1 \in [z_2, z_1[ \\ &\Rightarrow \mathbf{Card}([z_1, z_2[) \geq 1 \vee \mathbf{Card}([z_2, z_1[ \geq 1 \\ &\Rightarrow \mathbf{Card}([z_1, z_2[) \neq \emptyset = 0 \vee \mathbf{Card}([z_2, z_1[) \neq \emptyset = 0 \, . \end{aligned}$$

Then A.1.3 (p. 435), point 7 gives:

$$\begin{aligned} &\neg(\mathbf{Card}([z_1, z_2[) \neq 0 \vee \mathbf{Card}([z_2, z_1[) \neq 0) = \mathbf{Card}([z_1, z_2[) = \mathbf{Card}([z_2, z_1[ = 0 \\ &\Rightarrow \neg(z_1 \neq z_2) \Rightarrow z_1 = z_2 \, . \end{aligned}$$

It gives

$$\mathbf{Card}([z_1, z_2[) = \mathbf{Card}([z_2, z_1[) = z_1 - z_2 = z_2 - z_1 = 0 \Rightarrow z_1 = z_2 \, .$$

Equation (A.61) is true.
Since equation (A.61) is valid for $\Leftrightarrow$, applying A.1.3 (p. 435), point 7, we exclude $=$ in equation (A.60) and we get equation (A.62).

**Proposition 963** a, b, c being three terms of $\mathcal{Z}$, we have:

$$(a + b) + c = a + (b + c) \, .$$

**Proof**
From propositions 412 (p. 536), 3, 4, 5, we have:

$$(a + b) + c = \overline{-c, a + b} = \overline{-c, b} + \overline{b, a + b}$$

$$\overline{b, a + b} = (a + b) - b = \overline{-b, a} + \overline{0, -b} = \overline{0, a} = a \, .$$

Coupling both, we get:

$$(a + b) + c = \overline{-c, b} + a = a + (b + c) \, .$$

**Definition 414 Sum of a finite family of elements of $\mathcal{Z}$.**

Given a set $\mathbf{I}$ with $\mathbf{Card}(\mathbf{I}) \in \mathcal{Z}^+$ and a map: $a_i : \mathbf{I} \to \mathcal{Z}$, we call $\mathbf{I}^+ = \mathbf{set}_i\{i \in \mathbf{I}, a_i \in \mathcal{Z}^{0+}\}$ and $\mathbf{I}^- = \mathbf{set}_i\{i \in \mathbf{I}, a_i \in \mathcal{Z}^-\}$ with $\mathbf{I} = \mathbf{I}^+ \cup \mathbf{I}^-$, we set:

$$\sum_{i \in I} a_i = \sum_{i \in I^+} a_i - \sum_{i \in I^-} |a_i| \, . \tag{A.63}$$

**Proposition 964** The sum of a finite family of elements of $\mathcal{Z}$ exist and it is in $\mathcal{Z}$.

**Proof**
Using notations of definition 414 (p. 540), both sum $\sum_{i \in I^+} a_i$ and $\sum_{i \in I^-} |a_i|$ in equation (A.63) are integers (proposition 941 (p. 526)). Then $\sum_{i \in I} a_i \in \mathcal{Z}$, from proposition 959 (p. 536).

**Proposition 965** The absolute value of the sum of a finite family of elements of $\mathcal{Z}$ is smaller than or equal to the sum of the absolute value of the element of the family.

**Proof**
Using definitions 407 (p. 532), 410 (p. 534) -2, equation (A.58), proposition 960 (p. 536) and notations of definition 414 (p. 540), we have:

$$|\sum_{i \in I} a_i| = |\sum_{i \in I^+} a_i - \sum_{i \in I^-} |a_i|| \leq |\sum_{i \in I^+} |a_i|| + |\sum_{i \in I^-} |a_i|| = \sum_{i \in I} |a_i| .$$

**Proposition 966** Given two finite sets of indices $\mathbf{I}$, $\mathbf{J}$ with $\mathbf{Card(I)} = \mathbf{Card(J)}$ and a map $(a_i : I \to \mathcal{Z})$, we call $(i(j) : J \leftrightarrow I)$ and $(j(i) : I \leftrightarrow J)$ two bijective maps with $i(j(\ell)) = \ell : j^{-1}(k) = i(k)$. They exist since $I \sim J$. We have:

$$\sum_{i \in I} a_i = \sum_{j \in J} a_{i(j)} .$$

**Proof**
We call $J^+ = j(I^+)$, $J^- = j(I^-)$. Since $j^{-1}(x) = i(x)$, we have: $I^+ = i(J^+)$, $I^- = i(J^-)$. Since for all elements $z$ of $\mathcal{Z}$, $|z|$ is a cardinal, from proposition 892 (p. 507), we have $\sum_{j \in J^+} a_{i(j)} = \sum_{i \in I^+} a_i$ and $\sum_{j \in J^-} |a_{i(j)}| = \sum_{i \in I^-} |a_i|$. Then we have proposition 966.

**Proposition 967** Given two finite sets $\mathbf{I}$ and $\mathbf{J}$, a map $(a_i : I \to \mathcal{Z})$, a partition of $\mathbf{I}$ (definition 370 (p. 471)) $(I_j : J \to \mathcal{P}(I))$ then:

$$\sum_{j \in J} \sum_{i \in I_j} a_i = \sum_{i \in \bigcup_{j \in J} I_j} a_i = \sum_{i \in I} a_i . \quad \text{(A.64)}$$

**Proof**
One calls:

$$I^+ = \text{set}_i\{i \in I, a_i \in \mathcal{Z}^{0+}\},\ I^- = \text{set}_i\{i \in I, a_i \in \mathcal{Z}^-\}$$

$$J^+ = \text{set}_j\{j \in J, \sum_{i \in I_j} a_i \in \mathcal{Z}^{0+}\},\ J^- = \text{set}_j\{j \in J, \sum_{i \in I_j} a_i \in \mathcal{Z}^-\}$$

$$(\forall j \in J)(I_j^+ = \text{set}_i\{i \in I_j, a_i \in \mathcal{Z}^{0+}\},\ (\forall j \in J)(I_j^- = \text{set}_i\{i \in I_j, a_i \in \mathcal{Z}^-\})$$

Proposition 967 is trivially true for $\mathbf{Card(J)} = 1$. We suppose now proposition 967 is true when $\mathbf{Card(J)} = n$. We shall demonstrate that, in this case, it is also true for $\mathbf{Card(J)} = n + 1 > 1$. We call $(j_k : [0, n] \leftrightarrow J)$, a bijection of $[0, n]$ on $\mathbf{J}$ and we call $J_n = J - j_n = \mathcal{C}_J(j_n)$, we have $J = J_n \cup \{j_n\}$ and $\mathbf{Card}(J_n) = n$. Let us suppose:

$$\sum_{i \in j_n} a_i = (\sum_{i \in j_n^+} a_i - \sum_{i \in j_n^-} |a_i|) \in \mathcal{Z}^{0+}$$

(resp. $\sum_{i \in j_n} a_i \in \mathcal{Z}^-$). We may set $j_n^+ = j^n$, $j_n^- = \emptyset$ (resp. $j_n^+ = \emptyset$, $j_n^- = j_n$). With that we set: $J_n^+ = J^+ - \{j_n^+\}$ , $J_n^- = J^- - \{j_n^-\}$. We have: $\sum_{i \in j_n} a_i = |\sum_{i \in j_n} a_i|$ (resp. $\sum_{i \in j_n} a_i = -|\sum_{i \in j_n} a_i|$).

From propositions 897 (p. 508), 949 (p. 532), 958 (p. 535) and 963 (p. 540), we have:

$$\sum_{j\in J}\sum_{i\in I_j} a_i = \sum_{j\in J+}\sum_{i\in I_j} a_i - \sum_{j\in J-}|\sum_{i\in I_j} a_i| = (\sum_{j\in J_n^+}\sum_{i\in I_j} a_i + \sum_{i\in I_{j_n}} a_i) - \sum_{j\in J_n^-}|\sum_{i\in I_j} a_i|$$

{resp.

$$\sum_{j\in J}\sum_{i\in I_j} a_i = \sum_{j\in J+}\sum_{i\in I_j} a_i - \sum_{j\in J-}|\sum_{i\in I_j} a_i| = \sum_{j\in J_n^+}\sum_{i\in I_j} a_i - (\sum_{j\in J_n^-}|\sum_{i\in I_j} a_i| + |\sum_{i\in I_{j_n}} a_i|)\}$$

$$= \sum_{i\in I_{j_n}} a_i + (\sum_{j\in J_n^+}\sum_{i\in I_j} a_i - \sum_{j\in J_n^-}|\sum_{i\in I_j} a_i|) = (\sum_{i\in I_{j_n}^+} a_i - \sum_{i\in I_{j_n}^-} |a_i|) + \sum_{i\in(I-I_{j_n})} a_i$$

$$= \sum_{i\in I_{j_n}^+} a_i + \sum_{i\in(I-I_{j_n})} a_i - \sum_{i\in I_{j_n}^-} |a_i| =$$

$$\sum_{i\in I_{j_n}^+} a_i + \sum_{i\in(I-I_{j_n})^+} a_i - \sum_{i\in(I-I_{j_n})^-} |a_i| - \sum_{i\in I_{j_n}^-} |a_i| =$$

$$\sum_{i\in I_{j_n}^+} a_i + \sum_{i\in(I-I_{j_n})^+} a_i + (-\sum_{i\in(I-I_{j_n})^-} |a_i| - \sum_{i\in I_{j_n}^-} |a_i|) =$$

$$\sum_{i\in I_{j_n}^+} a_i + \sum_{i\in(I-I_{j_n})^+} a_i + (-(\sum_{i\in(I-I_{j_n})^-} |a_i| + \sum_{i\in I_{j_n}^-} |a_i|)) =$$

$$\sum_{i\in I_{j_n}^+} a_i + \sum_{i\in(I^+-I_{j_n}^+)} a_i - \sum_{i\in I^-} |a_i| = (\sum_{i\in I_{j_n}^+} a_i + \sum_{i\in(I^+-I_{j_n}^+)} a_i) - \sum_{i\in I^-} |a_i| =$$

$$\sum_{i\in I+} a_i - \sum_{i\in I-} |a_i| = \sum_{i\in I} a_i \,.$$

So if proposition 967 is true for $\mathbf{Card(J) = n}$ it is also true for $\mathbf{Card(J) = n+1}$. Proposition 967 being true for $\mathbf{n = 1}$, it is true for any $\mathbf{n}$ (proposition 927 (p. 520)).

**Proposition 968** Given a finite set $\mathbf{I}$ two maps $(x_i : I \to \mathcal{Z})$, $(y_i : I \to \mathcal{Z})$ such that $(\forall i \in I)(x_i \leq y_i)$, we have:

$$\sum_{i\in I} x_i \leq \sum_{i\in I} y_i \,.$$

**Proof**
Let us suppose that proposition 968 (p. 542) is true for $\mathbf{I} = [1, n]$, using proposition 967, we have for $\mathbf{I} = [1, n] \cup \{n + 1\} = [1, n + 1]$:

$$\sum_{i\in[1,n+1]} x_i = \sum_{i\in[1,n]} x_i + x_{n+1}, \quad \sum_{i\in[1,n+1]} y_i = \sum_{i\in[1,n]} y_i + y_{n+1}$$

and $\sum_{i\in[1,n]} x_i \leq \sum_{i\in[1,n]} y_i$, $x_{n+1} \leq y_{n+1}$. It means from proposition 962 (p. 539):

$$\sum_{i\in[1,n]} y_i - \sum_{i\in[1,n]} x_i \geq 0; \; y_{n+1} - x_{n+1} \geq 0 \,.$$

From propositions 908 (p. 512) and 966 (p. 541), we have:

$$\sum_{i\in[1,n]} y_i - \sum_{i\in[1,n]} x_i + y_{n+1} - x_{n+1} = \sum_{i\in[1,n+1]} y_i - \sum_{i\in[1,n+1]} x_i \geq 0$$

then applying again proposition 962 (p. 539),

$$\sum_{i\in[1,n+1]} y_i \geq \sum_{i\in[1,n+1]} x_i \,.$$

We conclude with proposition 927 (p. 520) that, since proposition 968 (p. 542) is trivially true for $I = [1,1]$, proposition 968 (p. 542) is true for $I = [1,n]$ whatever $n \in \mathcal{Z}^+$. Using proposition 966 (p. 541), proposition 968 (p. 542) is true for any finite set $I$.

**Proposition 969** $a \in \mathcal{Z}^{0+}$, $b \in \mathcal{Z}^+$ gives $a + b > 0$
**Proof**
Since $a$ and $b$ are both integers and then cardinal, we can apply proposition 891 (p. 507). It gives: $a + b = 0 \Rightarrow a = b = 0$. Since in our hypothesis $b \neq 0$ if $a + b$ was $0$, it would mean $b = 0$ which cannot be. Then we conclude, since $a + b$ is an integer (proposition 929 (p. 521)) and in $\mathcal{Z}^+$, $a + b > 0$.

**Proposition 970** Given a finite set $I$ two maps $(x_i : I \to \mathcal{Z})$, $(y_i : I \to \mathcal{Z})$ such that: $(\forall i \in I)(x_i \leq y_i)$, $(\exists k \in I)(x_k \neq y_k)$, we have:

$$\sum_{i\in I} x_i < \sum_{i\in I} y_i \,.$$

**Proof**
Let us suppose that it is true for $\mathbf{card}(I) = n$. As in proposition 968 (p. 542), we take $I = [1, n+1]$ and we have:

$$\sum_{i\in[1,n+1]} x_i = \sum_{i\in[1,n]} x_i + x_{n+1}, \quad \sum_{i\in[1,n+1]} y_i = \sum_{i\in[1,n]} y_i + y_{n+1} \,.$$

If $k \in [1,n]$, we have $\sum_{i\in[1,n]} x_i < \sum_{i\in[1,n]} y_i$, $x_{n+1} \leq y_{n+1}$. From proposition 962 (p. 539), we get: $\sum_{i\in[1,n]} y_i - \sum_{i\in[1,n]} x_i > 0$, $y_{n+1} - x_{n+1} \geq 0$. From proposition 969 (p. 543), we have:

$$\sum_{i\in[1,n]} y_i - \sum_{i\in[1,n]} x_i + y_{n+1} - x_{n+1} > 0 \,.$$

Similarly to proposition 968 (p. 542) using again proposition 962 (p. 539):

$$\sum_{i\in[1,n+1]} y_i > \sum_{i\in[1,n+1]} x_i \,.$$

If $k = n+1$ we get: $\sum_{i\in[1,n]} y_i - \sum_{i\in[1,n]} x_i \geq 0$, $y_{n+1} - x_{n+1} > 0$ and using proposition 962 (p. 539), we have:

$$\sum_{i\in[1,n+1]} y_i > \sum_{i\in[1,n+1]} x_i \,.$$

We conclude with proposition 927 (p. 520) as in proposition 968 (p. 542).
**Definition 415 Product of a finite family of elements of $\mathcal{Z}$.**

Given a set $I$ with $\mathbf{Card}(I) \in \mathcal{Z}^+$ and a map: $a_i : I \to \mathcal{Z}$, we call $I^+ = \mathbf{set}_i\{i \in I, a_i \in \mathcal{Z}^{0+}\}$ and $I^- = \mathbf{set}_i\{i \in I, a_i \in \mathcal{Z}^-\}$ with $I = I^+ \cup I^-$. We set (see definitions 399 (p. 519), 396 (p. 510)):

$$\begin{aligned} (\mathbf{Card}(I^-) = 2q) &\Rightarrow \prod_{i\in I} a_i = \prod_{i\in I} |a_i| = \mathop{\times}_{i\in I} |a_i| \\ (\mathbf{Card}(I^-) = 2q+1) &\Rightarrow \prod_{i\in I} a_i = -\prod_{i\in I} |a_i| = -\mathop{\times}_{i\in I} |a_i| \end{aligned} \tag{A.65}$$

see definition 403 (p. 522).

**Remark:**

We here start to be confused between the product of cardinals as set and the product as integers. When integers are positive, they are also cardinals and give the same results. The reader has to keep in mind that those products do not mean necessarily the same thing although they look very similar. It is impossible now to use different sign for those different operations. The reader has to realize from the context which kind of product we are dealing with.

**Definition 416 Product of two elements of $\mathcal{Z}$.**

Given two elements $\mathbf{a}, \mathbf{b}$ of $\mathcal{Z}$, one sets:

$$\mathbf{a} \times \mathbf{b} = \prod_{\mathbf{i} \in \{\mathbf{a},\mathbf{b}\}} \mathbf{i}\,.$$

**Proposition 971** Given two elements $\mathbf{a}, \mathbf{b}$ of $\mathcal{Z}$, we set $\mathbf{p}$ as the number of elements of $\{\mathbf{a}, \mathbf{b}\}$ greater than or equal to $\mathbf{0}$.
When $\mathbf{p}$ is even, $\mathbf{a} \times \mathbf{b} = |\mathbf{a}| \times |\mathbf{b}|$.
When $\mathbf{p}$ is odd, $\mathbf{a} \times \mathbf{b} = -|\mathbf{a}| \times |\mathbf{b}|$ (see definition 403 (p. 522)).

**Proof**

It is a direct consequence of definitions 416 (p. 544), 415 (p. 543) and proposition 903 (p. 510).

**Proposition 972** Given three elements $\mathbf{a}, \mathbf{z}_1, \mathbf{z}_2$ of $\mathcal{Z}$ with $\mathbf{z}_1 \leq \mathbf{z}_2$. We have:

$$\mathbf{a} \geq 0 \Rightarrow \mathbf{a} \times \mathbf{z}_1 \leq \mathbf{a} \times \mathbf{z}_2;\ \mathbf{a} \leq 0 \Rightarrow \mathbf{a} \times \mathbf{z}_1 \geq \mathbf{a} \times \mathbf{z}_2\,.$$

**Proof**

It is a consequence of proposition 902 (p. 510), definition 416 (p. 544), definition 409 (p. 533). According to those:
When $\mathbf{a} \geq \mathbf{0}$, we have:

$$\begin{aligned}
&\mathbf{z}_1 \leq 0 \leq \mathbf{z}_2 \Rightarrow \mathbf{a} \times \mathbf{z}_1 \leq 0 \leq \mathbf{a} \times \mathbf{z}_2 \Rightarrow \mathbf{a} \times \mathbf{z}_1 \leq \mathbf{a} \times \mathbf{z}_2\\
&0 \leq \mathbf{z}_1 \leq \mathbf{z}_2 \Rightarrow \mathbf{a} \times \mathbf{z}_1 \leq \mathbf{a} \times \mathbf{z}_2\\
&\mathbf{z}_1 \leq \mathbf{z}_2 \leq 0 \Rightarrow (\mathbf{z}_1 \leq \mathbf{z}_2 \leq 0, |\mathbf{z}_2| \leq |\mathbf{z}_1|) \Rightarrow (\mathbf{z}_1 \leq \mathbf{z}_2 \leq 0, \mathbf{a} \times |\mathbf{z}_2| \leq \mathbf{a} \times |\mathbf{z}_1|)\\
&\Rightarrow (\mathbf{z}_1 \leq \mathbf{z}_2 \leq 0, -\mathbf{a} \times |\mathbf{z}_1| \leq -\mathbf{a} \times |\mathbf{z}_2|) \Rightarrow \mathbf{a} \times \mathbf{z}_1 \leq \mathbf{a} \times \mathbf{z}_2\,.
\end{aligned}$$

When $\mathbf{a} \leq \mathbf{0}$, we get from above: $|\mathbf{a}| \times \mathbf{z}_1 \leq |\mathbf{a}| \times \mathbf{z}_2$. Then from definitions 416 (p. 544), 902 (p. 510) $\mathbf{a} \times \mathbf{z}_2 \leq \mathbf{a} \times \mathbf{z}_1$.

**Proposition 973** $\mathbf{I}$ being a finite set, given a map $(\mathbf{x}_\mathbf{i} : \mathbf{I} \rightarrow \mathcal{Z})$ such that $(\exists \mathbf{x} \in \mathbf{X})(\forall \mathbf{i} \in \mathbf{I})(\mathbf{x}_\mathbf{i} = \mathbf{x})$, we have:

$$(\forall \mathbf{j} \in \mathbf{I})(\forall \mathbf{k} \in \mathbf{I})(\prod_{\mathbf{i} \in \mathbf{I}-\mathbf{j}} \mathbf{x}_\mathbf{i} = \prod_{\mathbf{i} \in \mathbf{I}-\mathbf{k}} \mathbf{x}_\mathbf{i})\,.$$

**Proof**

We use the above notations. If $\mathbf{I} = \emptyset$, proposition 973 is trivially true. Calling $\mathbf{x} = \tau_\mathbf{x}(\forall \mathbf{i} \in \mathbf{I})(\mathbf{x}_\mathbf{i} = \mathbf{x})$, if $\mathbf{x} \in \mathcal{Z}^{0+}$, we have $\mathbf{I}^- = \emptyset,\ \mathbf{I}^+ = \mathbf{I}$. $\mathbf{x}_\mathbf{i}$ are cardinal. From 915 (p. 515) we have proposition 973 (p. 544). If $\mathbf{x} \in \mathcal{Z}^-$, we have $\mathbf{I}^+ = \emptyset,\ \mathbf{I}^- = \mathbf{I}$. It gives:

$$(\mathbf{j} \in \mathbf{I},\ \mathbf{k} \in \mathbf{I}) \Rightarrow \mathrm{Card}(\mathbf{I}^- - \{\mathbf{j}\}) + 1 = \mathrm{Card}(\mathbf{I}^- - \{\mathbf{k}\}) + 1 = \mathrm{Card}(\mathbf{I})\,.$$

Then $\mathbf{Card(I^- - \{j\}) = Card(I^- - \{k\})}$ (proposition 914 (p. 515)). From the above we have $\prod_{i \in I-j} |x_i| = \prod_{i \in I-k} |x_i|$. It gives using definition 415 (p. 543):

$$\prod_{i \in I-j} x_i = \prod_{i \in I-k} x_i \; .$$

**Proposition 974** Given two finite sets of index $\mathbf{I \sim J}$ with two associated bijections $\mathbf{(i(k) : J \leftrightarrow I)}$, $\mathbf{(j(k) : I \leftrightarrow J)}$, $\mathbf{i(j(k)) = k}$ and a map $\mathbf{(a_i : I \to \mathcal{Z})}$, we have:

$$\prod_{i \in I} a_i = \prod_{j \in J} a_{i(j)} \; .$$

**Proof**
From proposition 904 (p. 511), we have:

$$\prod_{i \in I} |a_i| = \prod_{j \in J} |a_{i(j)}| \; .$$

Setting $\mathbf{J^- = set_j\{j \in J, a_{i(j)} \in \mathcal{Z}^-\}}$, $\mathbf{I^- = set_i\{i \in I, a_i \in \mathcal{Z}^-\}}$, we have:

$$\begin{aligned} j \in J^- &\Rightarrow a_{i(j)} \in \mathcal{Z}^- \Rightarrow i(j) \in I^- \\ i \in I^- &\Rightarrow a_{j(i)} \in \mathcal{Z}^- \Rightarrow j(i) \in J^- \; . \end{aligned}$$

From that, $\mathbf{i(j)}$ is a map $\mathbf{J^- \to I^-}$ (definition 112 (p. 131)) which is surjective (definition 123 (p. 139)). Since it is the restriction to $\mathbf{J^-}$ of a bijective map $\mathbf{(j(i) : I \leftrightarrow J)}$, it is injective map (definition 122 (p. 139)) then it is a bijective map (definition 124 (p. 139)):

$$\mathbf{Card(I^-) = Card(J^-)} \; .$$

Then we have proposition 974 from definition 415 (p. 543).

**Definition 417** $\mathbf{(-1)^n, n \in \mathcal{Z}^{0,+}}$.

We define $\mathbf{(-1)^n : \mathcal{Z}^{0,+} \to \{-1, 1\}}$ by $\mathbf{(-1)^n = -1}$ if $\mathbf{n}$ is odd and equal $\mathbf{1}$ if $\mathbf{n}$ is even. (see definition 403 (p. 522)).

**Proposition 975**

$$(\mathbf{Card(I \in \mathcal{Z}^{0,+}), (a_i : I \to \mathcal{Z}^{0,+})) \Rightarrow (-1)^{\sum_{i \in I} a_i} = \prod_{i \in I} (-1)^{a_i}} \; . \tag{A.66}$$

**Proof**
We may assume $\mathbf{I = [0, n[}$ since there is a bijection $\mathbf{I \leftrightarrow [0, Card(I)[}$ (see propositions 966 (p. 541) and 974 (p. 545)). We suppose that proposition 975 (p. 545) is true for $\mathbf{n}$. We have $\sum_{i \in [0,n+1[} a_i = \sum_{i \in [1,n[} a_i + a_n$. From proposition 932 (p. 523), if $\mathbf{a_n}$ is even (resp. odd) $\sum_{i \in [0,n+1[} a_i$ has the same (resp. opposite) parity as $\sum_{i \in [0,n[} a_i$.
$\prod_{i \in [0,n[} (-1)^{a_i}$ is $\mathbf{1}$ (resp. $\mathbf{-1}$) if the number of $\mathbf{a_i}$ odd is even (resp. odd). This number of odd keeps (resp. changes) its parity if $\mathbf{a_n}$ is even (resp. odd).
Both members of (A.66) stay the same (resp. take the other value) if $\mathbf{a_n}$ is even (resp. odd). If (A.66) is true for $\mathbf{n}$ it is also true for $\mathbf{n+1}$. Since (A.66) is trivially true for $\mathbf{n = 0}$, it is true for any $\mathbf{n \in \mathcal{Z}^{0+}}$ (proposition 927 (p. 520)).

**Proposition 976**

$$(\mathrm{Card}(I) \in \mathcal{Z}^{0+}, (a_i : I \to \mathcal{Z}), I^- = \mathrm{set}_i\{i \in I, a_i \in \mathcal{Z}^-\})$$
$$\Rightarrow \prod_{i \in I} a_i = \prod_{i \in I^-}(-1) \prod_{i \in I} |a_i| = (-1)^{\mathrm{Card}(I^-)} \prod_{i \in I} |a_i| \ .$$

**Proof**

It is a direct consequence of definitions 415 (p. 543) and 417 (p. 545).

**Proposition 977** Given two finite sets, **I** and **J**, a map $(x_i : I \to \mathcal{Z})$ and a partition (definition 370 (p. 471)) $(I_j : J \to \mathcal{P}(I))$, we have (see proposition 889 (p. 506)):

$$\prod_{j \in J} \prod_{i \in I_j} x_i = \prod_{i \in I} x_i \ .$$

**Proof**

Calling $J^-$, the subset of **J** such that $j \in J^- \Leftrightarrow (j \in J, \prod_{i \in I_j} x_i \in \mathcal{Z}^-)$ we have (proposition 976 (p. 546)):

$$\prod_{j \in J} \prod_{i \in I_j} x_i = (-1)^{\mathrm{Card}(J^-)} \prod_{j \in J} |\prod_{i \in I_j} x_i| = (-1)^{\mathrm{Card}(J^-)} \prod_{j \in J} \prod_{i \in I_j} |x_i| \ .$$

But calling $I_j^-$, the set such that $i \in I_j^- \Leftrightarrow (i \in I_j, x_i \in \mathcal{Z}^-)$, we have (proposition 976 (p. 546)):

$$j \in J^- \Leftrightarrow (j \in J, (-1)^{\mathrm{Card}(I_j^-)} \in \mathcal{Z}^-) \ .$$

Then, from propositions 975 (p. 545) and 976 (p. 546), we get:

$$(-1)^{\mathrm{Card}(J^-)} = \prod_{j \in J}(-1)^{\mathrm{Card}(I_j^-)} = (-1)^{\sum_{j \in J} \mathrm{Card}(I_j^-)} = (-1)^{\sum_{j \in J^-} \mathrm{Card}(I_j^-)} \ .$$

Calling $I^-$, the set such that:

$$i \in I^- \Leftrightarrow (i \in I, a_i \in \mathcal{Z}^-) \Leftrightarrow (\exists j \in J)(i \in I_j, a_i \in \mathcal{Z}^-)$$
$$\Leftrightarrow (\exists j \in J)(i \in I_j^-) \Leftrightarrow i \in \bigcup_{j \in J} I_j^- \ .$$

Then $I^- = \bigcup_{j \in J} I_j^-$ (proposition 754 (p. 451)) and since $i \neq j \Rightarrow I_i^- \cap I_j^- = \emptyset$, from proposition 893 (p. 507), we have:

$$\sum_{j \in J} \mathrm{Card}(I_j^-) = \mathrm{Card}(\bigcup_{j \in J} I_j^-) = \mathrm{Card}(I^-) \ .$$

Then putting everything together:

$$\prod_{j \in J} \prod_{i \in I_j} x_i = (-1)^{\mathrm{Card}(I^-)} \prod_{j \in J} \prod_{i \in I_j} |x_i| \ .$$

But from propositions 889 (p. 506) and 905 (p. 511), we have:

$$\prod_{j \in J} \prod_{i \in I_j} |x_i| = \prod_{i \in I} |x_i| \ .$$

From definition 415 (p. 543), we have:

$$\prod_{i \in I} x_i = (-1)^{\mathrm{Card}(I^-)} \prod_{i \in I} |x_i| = (-1)^{\mathrm{Card}(I^-)} \prod_{j \in J} \prod_{i \in I_j} |x_i| = \prod_{j \in J} \prod_{i \in I_j} x_i \ .$$

**Proposition 978** $\mathbf{a, b, c}$ being in $\mathcal{Z}$ we have: $\mathbf{a(b+c) = ab + bc}$.

**Proof**
If $\mathbf{b}$, $\mathbf{c}$ are both in $\mathcal{Z}^-$ or $\mathcal{Z}^{0+}$, proposition 978 comes directly from proposition 912 (p. 515) and definitions 410 (p. 534), 415 (p. 543). If it were not the case, with for instance $\mathbf{a} > 0$, $\mathbf{b} \in \mathcal{Z}^{0+}$, $\mathbf{c} \in \mathcal{Z}^-$ with $\mathbf{b} > \mathbf{c}$, it gives (proposition 934 (p. 523)):

$$\mathbf{b} \sim [0, \mathbf{b}[, |\mathbf{c}| \sim [0, |\mathbf{c}|[$$

$$\mathbf{b} + \mathbf{c} = \mathbf{Card}(\mathcal{C}_{\mathbf{b}}([0, |\mathbf{c}|)) = \mathbf{Card}([|\mathbf{c}|, \mathbf{b}[) \ .$$

We have:

$$[0, \mathbf{b}[= [0, |\mathbf{c}|[\cup[|\mathbf{c}|, \mathbf{b}[$$
$$\mathbf{a} \times [0, \mathbf{b}[= \mathbf{a} \times ([0, |\mathbf{c}| \cup [|\mathbf{c}|, \mathbf{b}[) = \mathbf{a} \times ([0, |\mathbf{c}|[\cup \mathbf{a} \times [|\mathbf{c}|, \mathbf{b}[ \ .$$

From definition 415 $\mathbf{a} \times \mathbf{c} \in \mathcal{Z}^-$, from proposition 783 (p. 459), we have:

$$\mathbf{ab} > |\mathbf{ac}| = \mathbf{a}|\mathbf{c}| \sim \mathbf{a} \times [0, |\mathbf{c}|[ \ .$$

We have (definition 410 (p. 534) and proposition 785 (p. 460)):

$$\mathbf{ab} + \mathbf{ac} = \mathbf{Card}(\mathcal{C}_{\mathbf{a}\times[0,\mathbf{b}[}(\mathbf{a} \times [|\mathbf{c}|, \mathbf{b}[)) = \mathbf{Card}(\mathbf{a}\mathcal{C}_{[0,\mathbf{b}[}([|\mathbf{c}, \mathbf{b}[)) = \mathbf{a}(\mathbf{b}+\mathbf{c}) \ .$$

It gives:

$$\mathbf{ab} + \mathbf{ac} = \mathbf{a}(\mathbf{b}+\mathbf{c}) \ .$$

**Proposition 979** Given two integers $(\mathbf{j_i} : \{1,2\} \to \mathcal{Z}^{0+})$ two maps:

$$(\mathbf{a}_{\lambda,\iota} : [0, \mathbf{j}_\lambda[\to \mathcal{Z})$$

we have:

$$\prod_{\lambda\in\{1,2\}} \sum_{\iota\in[0,\mathbf{j}_\lambda[} \mathbf{a}_{\lambda,\iota} = \sum_{\iota\in[0,\mathbf{j}_1[} \mathbf{a}_{1,\iota} \times \sum_{\iota\in[0,\mathbf{j}_2[} \mathbf{a}_{2,\iota} = \sum_{\mathbf{f}\in\prod_{\lambda\in\{1,2\}}[0,\mathbf{j}_\lambda[} \prod_{\lambda\in\{1,2\}} \mathbf{a}_{\lambda,\mathbf{f}(\lambda)} \ .$$

**Proof**
Proposition 979 (p. 547) is true for $\mathbf{j}_1 = \mathbf{j}_2 = 1$. If it is true for $\mathbf{j}_1 = 1, \mathbf{j}_2 = \mathbf{n}_2$, it is true for $\mathbf{j}_2 = \mathbf{n}_2 + 1$ because from proposition 978 (p. 547), we have:

$$\prod_{\lambda\in\{1,2\}} \sum_{\iota\in[0,\mathbf{j}_\lambda[} \mathbf{a}_{\lambda,\iota} = \mathbf{a}_{1,0} \sum_{\iota\in[0,\mathbf{n}_2+1[} \mathbf{a}_{2,\iota} = \mathbf{a}_{1,0} \sum_{\iota\in[0,\mathbf{n}_2[} \mathbf{a}_{2,\iota} + \mathbf{a}_{1,0}\mathbf{a}_{2,\mathbf{n}_2}$$
$$= \sum_{\iota\in[0,\mathbf{n}_2[} \mathbf{a}_{1,0}\mathbf{a}_{2,\iota} + \mathbf{a}_{1,0}\mathbf{a}_{2,\mathbf{n}_2} = \sum_{\iota\in[0,\mathbf{n}_2+1[} \mathbf{a}_{1,0}\mathbf{a}_{2,\iota} = \sum_{\mathbf{f}\in\prod_{\lambda\in\{1,2\}}[0,\mathbf{j}_\lambda[} \prod_{\lambda\in\{1,2\}} \mathbf{a}_{\lambda,\mathbf{f}(\lambda)} \ .$$

If proposition 979 (p. 547) is true for $\mathbf{j}_\lambda = \mathbf{n}_\lambda, \lambda \in \{1,2\}$, let us prove that it is true also if $(\exists\mu \in \{1,2\})(\lambda = \mu \Rightarrow \mathbf{j}_\lambda = \mathbf{n}_\lambda + 1, \lambda \neq \mu \Rightarrow \mathbf{j}_\lambda = \mathbf{n}_\lambda)$ for simplicity let us suppose

$\mu = 1$ (the proof is the same for $\mu = 2$)

$$\prod_{\lambda\in\{1,2\}}\sum_{\iota\in[0,j_\lambda[} a_{\lambda,\iota} = \sum_{\iota\in[0,n_1+1[} a_{1,\iota} \times \sum_{\iota\in[0,n_2[} a_{2,\iota} = (a_{1,n_1} + \sum_{\iota\in[0,n_1[} a_{1,\iota}) \times \sum_{\iota\in[0,n_2[} a_{2,\iota}$$

$$= a_{1,n_1} \times \sum_{\iota\in[0,n_2[} a_{2,\iota} + \sum_{f\in\prod_{\lambda\in\{1,2\}}[0,n_\lambda[}\prod_{\lambda\in\{1,2\}} a_{\lambda,f(\lambda)}$$

$$= \sum_{\iota\in[0,n_2[} a_{1,n_1}a_{2,\iota} + \sum_{f\in\prod_{\lambda\in\{1,2\}}[0,n_\lambda[}\prod_{\lambda\in\{1,2\}} a_{\lambda,f(\lambda)}$$

$$= \sum_{f\in[n_1,n_1+1[\times[0,n_2[}\prod_{\lambda\in\{1,2\}} a_{\lambda,f(\lambda)} + \sum_{f\in\prod_{\lambda\in\{1,2\}}[0,n_\lambda[}\prod_{\lambda\in\{1,2\}} a_{\lambda,f(\lambda)}$$

$$= \sum_{f\in[n_1,n_1+1[\times[0,n_2[}\prod_{\lambda\in\{1,2\}} a_{\lambda,f(\lambda)} + \sum_{f\in[0,n_1[\times[0,n_2[}\prod_{\lambda\in\{1,2\}} a_{\lambda,f(\lambda)}$$

$$= \sum_{f\in[n_1,n_1+1[\times[0,n_2[\cup[0,n_1[\times[0,n_2[}\prod_{\lambda\in\{1,2\}} a_{\lambda,f(\lambda)} = \sum_{f\in[0,n_1+1[\times[0,n_2[}\prod_{\lambda\in\{1,2\}} a_{\lambda,f(\lambda)} \,.$$

We use here again propositions 978 (p. 547) and 830 (p. 476) (with $\mathbf{L} = \{1,2\}, \mathbf{J}_1 = \{1,2\}, \mathbf{J}_2 = \{1\}, \mathbf{X}_{1,1} = [n_1, n_1+1[, \mathbf{X}_{1,2} = [0, n_1[, \mathbf{X}_{2,1} = [0, n_2[$). From proposition 927 (p. 520), proposition 979 is true.

**Proposition 980** $\mathbf{A}$ and $\mathbf{L}$ are two sets with $\mathbf{Card(L)} \in \mathcal{Z}^+$. We suppose that we have a map $(\mathbf{J}_\lambda : \mathbf{L} \to \mathcal{P}(\mathbf{A}))$ with $(\forall\lambda \in \mathbf{L})(\mathbf{Card}(\mathbf{J}_\lambda) \in \mathcal{Z}^+)$ and a set of maps over $\mathcal{Z}$: $(a_{\lambda,\iota} : \mathbf{J}_\lambda \to \mathcal{Z})$. We set $\mathbf{I} = \prod_{\lambda\in L} \mathbf{J}_\lambda$. We have:

$$\prod_{\lambda\in L}\sum_{\iota\in J_\lambda} a_{\lambda,\iota} = \sum_{f\in I}\prod_{\lambda\in L} a_{\lambda,f(\lambda)} \,.$$

**Proof**

Proposition 980 (p. 548) is proposition 911 (p. 514) when we deal only with positive numbers which are cardinals. In the other cases, using definition (371 (p. 472), 390 (p. 499)), we have: $\mathbf{L} \sim \mathbf{Card(L)}$ and $\mathbf{J}_\lambda \sim \mathbf{Card}(\mathbf{J}_\lambda)$. Since they are integers, from proposition 934 (p. 523), setting $\ell = \mathbf{Card(L)}$, $j_\lambda = \mathbf{Card}(\mathbf{J}_\lambda)$, we have $\mathbf{L} \sim [0, \ell[$, $\mathbf{J}_\lambda \sim [0, j_\lambda[$ and for simplicity we can replace in proposition 980, $\mathbf{L}$ by $[0, \ell[$ and $\mathbf{J}_\lambda$ by $[0, j_\lambda[$. From the above proposition 979 (p. 547), proposition 980 (p. 548) is true for $\ell = 2$. We suppose it is true for $\ell = n$. We then have:

$$\prod_{\lambda\in[0,n+1[}\sum_{\iota\in[0,j_\lambda[} a_{\lambda,\iota} = (\sum_{\iota\in[0,j_n[} a_{n,\iota})\prod_{\lambda\in[0,n[}\sum_{\iota\in[0,j_\lambda[} a_{\lambda,\iota}$$

$$= (\sum_{\iota\in[0,j_n[} a_{n,\iota})(\sum_{f\in\prod_{\lambda\in[0,n[}[0,j_\lambda[}\prod_{\lambda\in[0,n[} a_{\lambda,f(\lambda)}) \,.$$

From proposition 979 (p. 547) replacing $[0, j_1[$ by $[0, j_n[$ and $[0, j_2[$ by $f \in \prod_{\lambda\in[0,n[}[0, j_\lambda[ \sim [0, j_f[$ setting $j_f = \mathbf{Card}(\prod_{\lambda\in[0,n[}[0, j_\lambda[)$ and $(g_\gamma : [0, j_f[ \leftrightarrow \prod_{\lambda\in[0,n[}[0, j_\lambda[)$

$$(\sum_{\iota\in[0,j_n[} a_{n,\iota})(\sum_{f\in\prod_{\lambda\in[0,n[}[0,j_\lambda[}\prod_{\lambda\in[0,n[} a_{\lambda,f(\lambda)}) = (\sum_{\iota\in[0,j_n[} a_{n,\iota})(\sum_{\gamma\in[0,j_f[}\prod_{\lambda\in[0,n[} a_{\lambda,g_\gamma(\lambda)})$$

$$= \sum_{h\in[0,j_n[\times[0,j_f[} a_{n,h(1)}\prod_{\lambda\in[0,n[} a_{\lambda,g_{h(2)}(\lambda)} = \sum_{f'\in\prod_{\lambda\in[0,n[}[0,j_\lambda[\times[0,j_n[} a_{n,f'(n)}\prod_{\lambda\in[0,n[} a_{\lambda,f'(\lambda)}$$

$$= \sum_{f'\in\prod_{\lambda\in[0,n+1[}[0,j_\lambda[}\prod_{\lambda\in[0,n+1[} a_{\lambda,f'(\lambda)} \,.$$

With $\lambda \in [0, n[ \Rightarrow f'(\lambda) = f(\lambda) = g^{-1}(\lambda), f'(n) \in [0, j_n[$.

**Proposition 981** The product of a finite family of elements of $\mathcal{Z}$ exist and it is in $\mathcal{Z}$.
**Proof**
Using notation of definition 415, from proposition 941 (p. 526), $\prod_{i \in I} |a_i|$ is an integer then we have proposition 981.

**Proposition 982** Given two sets $\mathbf{A}$ and $\mathbf{B}$ of the same cardinal: $\mathbf{Card}(\mathcal{Z}^{0+})$ (definitions 390 (p. 499), 401 (p. 520)), their product (definition 372 (p. 473)) also has as cardinal $\mathbf{Card}(\mathcal{Z}^{0+})$:

$$\mathbf{Card}(\mathcal{Z}^{0+} \times \mathcal{Z}^{0+}) = \mathbf{Card}(\mathcal{Z}^{0+}) .$$

**Proof**
Considering the map (see proposition 824 (p. 474)) $(\mathbf{f(p,q)} : \mathbf{A \times B} \sim \mathcal{Z}^{0+} \times \mathcal{Z}^{0+} \to \mathcal{Z}^{0+})$ defined by:

$$\mathbf{f(p,q)} = \frac{\mathbf{(p+q)(p+q+1)}}{\mathbf{2}} + \mathbf{q} .$$

From propositions 933 (p. 523), 931 (p. 522) $\mathbf{(p+q)(p+q+1)}$ is even and we can apply definition 402 (p. 522).
<u>$\mathbf{f(p,q)}$ **is a bijection**</u> (definition 124 (p. 139)):

1. We call $\mathbf{s = p + q}$ and $\mathbf{n = f(p,q)}$. From propositions 921 (p. 517), 902 (p. 510), 903 (p. 510), we have:

$$\frac{\mathbf{(p+q)(p+q+1)}}{\mathbf{2}} + \mathbf{q = n} \Rightarrow \frac{\mathbf{(p+q)(p+q+1)}}{\mathbf{2}} \leq \mathbf{n} .$$

Then (definition 402 (p. 522)): $\mathbf{s(s+1) \leq 2n}$.
Let us take a $\mathbf{u}$ such as : $\mathbf{s < u}$. From proposition 920 (p. 517) we have: $\mathbf{s < u \Rightarrow s + 1 \leq u}$. It means, from propositions 793 (p. 464), 941 (p. 526), 902 (p. 510):

$$\frac{\mathbf{u(u+1)}}{\mathbf{2}} \geq \frac{\mathbf{(s+1)(s+2)}}{\mathbf{2}} = \mathbf{n - q + 1 + s = n + p + 1 > n} .$$

Then calling $\mathbf{S = set_u\{u \in \mathcal{Z}^{0+}, u(u+1) \leq 2n\}}$, we have true (see A.1.4 7 (p. 435)):

$$\mathbf{u > s \Rightarrow u \notin S}$$
$$\mathbf{u \in S \Rightarrow u \leq s} .$$

From definitions 379 (p. 484), 845 (p. 485), we have:

$$\mathbf{s = p + q = \max_u u \in S} .$$

2. We look for $\mathbf{(p,q)}$ such that for an $\mathbf{n} \in \mathcal{Z}^{0+}$, we have $\mathbf{n} = \frac{\mathbf{(p+q)(p+q+1)}}{\mathbf{2}} + \mathbf{q}$. We set $\mathbf{S = set_u\{u \in \mathcal{Z}^{0+}, u(u+1) \leq 2n\}}$ and $\mathbf{s = \max_u u \in S}$.
Let us show that $\mathbf{s}$ exists.
From propositions 903 (p. 510), 906 (p. 511), 907 (p. 512), 912 (p. 515), we have:

$$\mathbf{2n \geq 0 \Rightarrow 2n \geq 0(0+1) = 0 \Rightarrow 0 \in S}$$
$$\mathbf{S \neq \emptyset} .$$

From propositions 793 (p. 464), 941 (p. 526), we have:

$$\mathbf{u > 2n \Rightarrow u(u+1) = u \times u + u > 2n \Rightarrow u \notin S}$$
$$\mathbf{u \in S \Rightarrow u \leq 2n}$$

$\mathbf{2n}$ in a majoring value of $\mathbf{S}$ (definition 380 (p. 484)). From proposition 924 (p. 519), $\mathbf{S}$ has a unique biggest value (definition 379 (p. 484)) $\mathbf{s}$ in $\mathbf{S}$.
We have:

$$q = n - \frac{s(s+1)}{2} < \frac{(s+1)(s+2)}{2} - \frac{s(s+1)}{2} = s+1 \,.$$

Then (proposition 920 (p. 517)) $0 \leq n - \frac{s(s+1)}{2} \leq s$. We have a couple $(p, q)$ given by:

$$0 \leq q = n - \frac{s(s+1)}{2} \leq s,$$
$$s \geq p = s - q \geq 0 \,.$$

This couple is unique:
Then for any $\mathbf{n} \in \mathcal{Z}^{0+}$ there is one and only one value of $\mathbf{s}$ to which it corresponds one and only one couple $(\mathbf{p}, \mathbf{q})$ such that $\mathbf{f(p,q) = n}$.

$(f(p,q) : \mathcal{Z}^{0+} \times \mathcal{Z}^{0+} \leftrightarrow \mathcal{Z}^{0+})$ is a bijection (definition 124 (p. 139)). Proposition 982 (p. 549) is true.

**Proposition 983** Considering a set of sets $\mathbf{A}$ such that $\mathbf{x} \in \mathbf{A} \Rightarrow \mathbf{Card(x)} \leq \mathbf{Card}(\mathcal{Z}^{0+})$ considering a set $\mathbf{K}$ such that $\mathbf{Card(K)} \leq \mathbf{Card}(\mathcal{Z}^{0+})$ and a map $(\mathbf{a_k} : \mathbf{K} \rightarrow \mathbf{A})$, we have:

$$\mathrm{Card}(\bigcup_{k \in K} a_k) \leq \mathrm{Card}(\mathcal{Z}^{0+}) \,.$$

**Proof**
For simplifying let us suppose that $(\forall k \in K)(\mathrm{Card}(a_k) = \mathrm{Card}(\mathcal{Z}^{0+}))$ and $\mathrm{Card}(K) = \mathrm{Card}(\mathcal{Z}^{0+})$. We also consider instead of $\mathbf{a_i}$, $\mathbf{b_i} = \mathbf{a_i} \cup \times \mathbf{i}$, we have $\mathbf{Card(b_i)} = \mathbf{Card(a_i)}$ and $\mathbf{i} \neq \mathbf{j} \Rightarrow \mathbf{b_i} \cap \mathbf{b_j} = \emptyset$ and $\mathrm{Card}(\bigcup_{k \in K} a_k) \leq \mathrm{Card}(\bigcup_{k \in K} b_k)$. For each $\mathbf{b_i}$, we have $\forall \mathbf{i} \in \mathbf{K}$ a bijection $(\mathbf{b(j)_i} : \mathcal{Z}^{0+} \leftrightarrow \mathbf{b_i})$ and another bijection (definition 124 (p. 139)):

$$(b(j)_i : b_i \leftrightarrow \mathcal{Z}^{0+} \times \{i\} \leftrightarrow \mathcal{Z}^{0+}) \,.$$

Considering the set of sets $\mathbf{B} \subset \mathbf{A}$: $\mathbf{x} \in \mathbf{B} \Leftrightarrow (\exists \mathbf{k} \in \mathbf{K})(\mathbf{x} = \mathbf{b_k})$, there is a bijection between $\mathbf{K}$ and $\mathbf{B}$. Then there is a bijection between $\mathcal{Z}^{0+}$ and $\mathbf{B}$. Altogether, it defines the bijection:

$$(b(j)_i = b(i \times j) : \mathcal{Z}^{0+} \times \mathcal{Z}^{0+} \leftrightarrow \bigcup_{b \in B} b)$$

then we have a bijection between $\mathcal{Z}^{0+} \times \mathcal{Z}^{0+} = (\mathcal{Z}^{0+})^2$ and $\bigcup_{k \in K} b_k$. Then proposition 982 (p. 549) gives proposition 983 (p. 550).

**Proposition 984** $\mathbf{Card}(\mathcal{Z}) = \mathbf{Card}(\mathcal{Z}^{+})$.

**Proof**
From proposition 949 (p. 532), there is a bijection between $\mathcal{Z}^{+}$ and $\mathcal{Z}^{-}$, since the couple $(\mathbf{0}, \mathbf{1})$ is a bijection to a subset of $\mathcal{Z}^{-}$. We have $\mathcal{Z} = \mathcal{Z}^{-} \cup \{0\} \cup \mathcal{Z}^{+}$. The $\mathcal{Z}$ is the sum of three sets each of them having a cardinal smaller than or equal to $\mathcal{Z}^{+}$. From propositions 983 (p. 550), $\mathbf{Card}(\mathcal{Z}) \leq \mathbf{Card}(\mathcal{Z}^{+})$, since we have $\mathcal{Z}^{+} \subset \mathcal{Z}$, from definition 392 (p. 500), we also have $\mathbf{Card}(\mathcal{Z}^{+}) \leq \mathbf{Card}(\mathcal{Z})$. From definition 338 (p. 446) point 2 and propositions 875 (p. 500), 872 (p. 498), we have: $\mathbf{Card}(\mathcal{Z}) = \mathbf{Card}(\mathcal{Z}^{+})$.

**Proposition 985**

$$\mathbf{n} \in \mathcal{Z}^+ \Rightarrow \mathbf{Card}((\mathcal{Z}^{0+})^{\mathbf{n}}) = \mathbf{Card}(\mathcal{Z}^{0+}) .$$

See definition 373 (p. 474).

**Proof**

Let us suppose that it is true up to $\mathbf{n} = \mathbf{n}'$. From propositions 823 (p. 474), 836 (p. 480), 888 (p. 505), 903 (p. 510) and propositions 982 (p. 549), we have:

$$(\mathcal{Z}^{0+})^{\mathbf{n}'+1} \sim (\mathcal{Z}^{0+})^{\mathbf{n}'} \times \mathcal{Z}^{0+} \sim \mathcal{Z}^{0+} \times \mathcal{Z}^{0+} \sim \mathcal{Z}^{0+} .$$

It is true for $\mathbf{n} = \mathbf{n}' + \mathbf{1}$. It is true for $\mathbf{n} = \mathbf{1}$. From propositions 927 (p. 520), it is true for any $\mathbf{n}$.

## A.8 Rational numbers

**Definition 418 Rational numbers $\mathcal{Q}$.**

We consider the equivalence relation $\mathbf{R}(\mathbf{x}, \mathbf{y})$ defined on the set $\mathcal{Z} \times (\mathcal{Z} - 0)$ (definition 364 (p. 468)) by:

$$\begin{aligned} \mathbf{R}(\mathbf{x}, \mathbf{y}) \Leftrightarrow ((&\mathbf{x} = (\mathbf{u_x}, \mathbf{v_x}) \in \mathcal{Z} \times (\mathcal{Z} - 0), \\ &\mathbf{y} = (\mathbf{u_y}, \mathbf{v_y}) \in \mathcal{Z} \times (\mathcal{Z} - 0), \mathbf{u_x}\mathbf{v_y} = \mathbf{v_x}\mathbf{u_y}) \\ &\vee (\mathbf{x} \notin \mathcal{Z} \times (\mathcal{Z} - 0), \mathbf{y} \notin \mathcal{Z} \times (\mathcal{Z} - 0))) . \end{aligned}$$

It verifies definition 336 (p. 445) and equation (A.29) (p. 468). From propositions 807 (p. 468), 808 (p. 469), an equivalence relation defined on a set determines a set of equivalence classes subset of $\mathcal{P}(\mathcal{Z} \times (\mathcal{Z} - 0))$ (definition 366 (p. 469)). In each equivalence class there is a unique representative member of $\mathcal{Z} \times (\mathcal{Z} - 0)$. Those representatives also form a subset of $\mathcal{Z} \times (\mathcal{Z} - 0)$. From proposition 812 (p. 470), the relation between an equivalence class and its representative is a bijective map (definition 124 (p. 139)). For that reason, one calls rational number either a representative of an equivalence class (definition 368 (p. 470)) either the equivalence class itself when they contain at least an element of $\mathcal{Z} \times (\mathcal{Z} - 0)$. In the first case, there are elements of $\mathcal{Z} \times (\mathcal{Z} - 0)$, in the second are elements of $\mathcal{P}(\mathcal{Z} \times \mathcal{Z})$.

**Proposition 986** When a rational $\mathbf{q}$ is considered as a representative of an equivalence class it is an element of $\mathcal{Z} \times (\mathcal{Z} - 0)$. When it is an equivalence class, it is an element of $\mathcal{P}(\mathcal{Z} \times (\mathcal{Z} - 0))$

**Proof**

It is a direct consequence of definition 418 (p. 551).

**Proposition 987** The rational numbers form a set.

**Proof**

It is a direct consequence of proposition 749 (p. 450) or 748 (p. 449) together with proposition 986 (p. 551).

**Proposition 988** Considering $\mathcal{Q}$ as the set of equivalence class of $\mathbf{R}(\mathbf{x}, \mathbf{y})$ which are rational numbers (i. e. subsets of $\mathcal{Z} \times (\mathcal{Z} - 0)$), we have:

$$(\mathbf{r_1} \in \mathcal{Q}, \mathbf{r_2} \in \mathcal{Q}, \mathbf{r_1} \cap \mathbf{r_2} \neq \emptyset) \Rightarrow \mathbf{r_1} = \mathbf{r_2} .$$

**Proof**
It is a direct consequence of propositions 809 (p. 469).

**Proposition 989** $\mathbf{Card}(\mathcal{Q}) = \mathbf{Card}(\mathcal{Z}) = \mathbf{Card}(\mathcal{Z})$.
**Proof**
There is an injective map $(\mathbf{p}/1 : \mathcal{Z} \rightarrow \mathcal{Q})$ (definition 122 (p. 139)). It is injective because $\mathbf{p} \neq \mathbf{q} \rightarrow \mathbf{p} = \mathbf{p} \times 1 \neq \mathbf{q} \times 1 = \mathbf{q}$ which means $\mathbf{p}/1 \neq /\mathbf{q}/1$. Then $\mathbf{Card}(\mathcal{Q}) \geq \mathbf{Card}(\mathcal{Z})$. There is a map $(\mathbf{p}/\mathbf{q} : \mathcal{Z} \times \mathcal{Z} \rightarrow \mathcal{Q})$ which is surjective (definitions 123 (p. 139), 418 (p. 551)). From propositions 182 (p. 143), 180 (p. 143), 903 (p. 510) and 984 (p. 550), we have:

$$\mathbf{Card}(\mathcal{Q}) \leq \mathbf{Card}(\mathcal{Z} \times \mathcal{Z}) = \mathbf{Card}(\mathcal{Z}) \ .$$

Combining with $\mathbf{Card}(\mathcal{Q}) \geq \mathbf{Card}(\mathcal{Z})$, $\mathbf{Card}(\mathcal{Q}) = \mathbf{Card}(\mathcal{Z})$ (proposition 875 (p. 500), point 3).

### A.8.1 Notation commonly used with rational numbers

**Definition 419** $\mathbf{r} = \mathbf{p}/\mathbf{q} \in \mathcal{Q}$.

One calls here $\mathcal{Q}$ the set of rationals. Any member of $\mathcal{Q}$ is defined either as a representative of an equivalence class:

$$\mathbf{r} \in \mathcal{Q} \Leftrightarrow (\exists \mathbf{p} \in \mathcal{Z})(\exists \mathbf{q} \in \mathcal{Z} - 0)(\mathbf{r} = \tau_{\mathbf{x}}(\mathbf{x} \sim (\mathbf{p}, \mathbf{q})))$$

or as the equivalence class itself:

$$\mathbf{r} \in \mathcal{Q} \Leftrightarrow (\exists \mathbf{p} \in \mathcal{Z})(\exists \mathbf{q} \in \mathcal{Z} - 0)(\mathbf{r} = \mathbf{set_u}\{\mathbf{u} \sim (\mathbf{p}, \mathbf{q})\}) \ .$$

Having $(\mathbf{p}, \mathbf{q}) \in \mathcal{Z} \times (\mathcal{Z} - 0)$, one uses to write the rational associated (the representative of the equivalence class of (p, q) or the equivalent class itself) by:

$$\mathbf{r} = \tau_{\mathbf{x}}(\mathbf{x} \sim (\mathbf{p}, \mathbf{q})) = \mathbf{p}/\mathbf{q} \text{ or } \mathbf{u} \in \mathbf{r} = \mathbf{p}/\mathbf{q} \Leftrightarrow \mathbf{u} \sim (\mathbf{p}, \mathbf{q}) \ .$$

**Definition 420 Notations associated with rational numbers.**

Given a couple of $\mathcal{Z} \times (\mathcal{Z} - 0)$, $(\mathbf{p}, \mathbf{q})$, let us call $\mathbf{r}$ the associated rational number.
We may consider it as an equivalence class of $(\mathbf{p}, \mathbf{q})$. In that case we should write:

$$((\mathbf{p}, \mathbf{q}) \in \mathbf{r}, (\mathbf{p}', \mathbf{q}') \sim (\mathbf{p}, \mathbf{q})) \Leftrightarrow (\mathbf{p}', \mathbf{q}') \in \mathbf{r} \ .$$

We may consider it as a representative of equivalence class:

$$\mathbf{r} = \tau_{\mathbf{x}}(\mathbf{x} \in \mathcal{Z} \times (\mathcal{Z} - 0), \mathbf{x} \sim (\mathbf{p}, \mathbf{q})) \ .$$

In that case, from propositions 741 (p. 446) point 2 and definition 337 (p. 446), we should have:

$$\mathbf{r} = \mathbf{p}/\mathbf{q}, (\mathbf{p}, \mathbf{q}) \sim (\mathbf{p}', \mathbf{q}') \Rightarrow \mathbf{r} \sim (\mathbf{p}, \mathbf{q}) \sim (\mathbf{p}', \mathbf{q}'), \mathbf{r} = \mathbf{p}/\mathbf{q} = \mathbf{p}'/\mathbf{q}' \ . \tag{A.67}$$

Those distinctions are rarely made and in most cases one considers $\mathbf{r} = \mathbf{p}/\mathbf{q}$ as the representative of the equivalence class to $(\mathbf{p}, \mathbf{q})$.
Since

$$(\mathbf{p} \in \mathcal{Z} - 0, \mathbf{q} \in \mathcal{Z} - 0) \Rightarrow 0 \times \mathbf{p} = 0 \times \mathbf{q} \Rightarrow (0, \mathbf{p}) \sim (0, \mathbf{q}) \ .$$

One identifies

$$0 = 0/\mathbf{p}, \mathbf{p} \in (\mathcal{Z} - 0) \ . \tag{A.68}$$

One also identifies

$$\mathbf{p}/1 = \mathbf{p} \in \mathcal{Z} \ . \tag{A.69}$$

## A.8.2 Operations on rational numbers

**Definition 421 Denominator and numerator of a rational number.**

In the expression $\mathbf{r} = \mathbf{p}/\mathbf{q}$, $\mathbf{p}$ is called the numerator and $\mathbf{q}$ the denominator.

**Proposition 990** $\mathbf{a}, \mathbf{p}, \mathbf{q}$ being in $\mathcal{Z}$, $\mathbf{p}/\mathbf{q} = (\mathbf{ap})/(\mathbf{aq})$.

**Proof**

Since $\mathbf{qap} = \mathbf{paq}$, $(\mathbf{p}, \mathbf{q}) \sim (\mathbf{ap}, \mathbf{aq})$ then from propositions 741 (p. 446) point 2 or definition 419 (p. 552), we have:

$$\mathbf{p}/\mathbf{q} = (\mathbf{ap})/(\mathbf{aq}) \,.$$

**Proposition 991** Given a set $\mathbf{I}$ with $\mathbf{Card(I)} \in \mathcal{Z}^+$, a map $(\mathbf{r_i} : \mathbf{I} \to \mathcal{Q})$ there is $\mathbf{d} \in \mathcal{Z}^+$ and a map $(\mathbf{p_i} : \mathbf{I} \to \mathcal{Z})$ such that:

$$(\forall \mathbf{i} \in \mathbf{I})(\mathbf{r_i} = \mathbf{p_i}/\mathbf{d}) \,.$$

**Proof**

From definitions 418 (p. 551) and 419 (p. 552), we have two maps $(\mathbf{a_i} : \mathbf{I} \to \mathcal{Z})$, $(\mathbf{b_i} : \mathbf{I} \to \mathcal{Z} - \mathbf{0})$ such that $(\forall \mathbf{i} \in \mathbf{I})(\mathbf{r_i} = \mathbf{a_i}/\mathbf{b_i})$. Setting $\mathbf{d} = \prod_{\mathbf{i}\in\mathbf{I}} |\mathbf{b_i}|$ and for all $\mathbf{i} \in \mathbf{I}$, $\mathbf{p_i}/\mathbf{1} = \mathbf{a_i}(\mathbf{d}/\mathbf{b_i})$, we have

$$\mathbf{r_i} = \mathbf{a_i}/\mathbf{b_i} = \mathbf{a_i d}/(\mathbf{b_i d}) = (\mathbf{a_i d}/\mathbf{b_i})/\mathbf{d} = \mathbf{p_i}/\mathbf{d} \,.$$

Then we have proposition 991 (p. 553).

**Proposition 992** Given a set $\mathbf{I}$ with $\mathbf{Card(I)} \in \mathcal{Z}^+$, a map $(\mathbf{r_i} : \mathbf{I} \to \mathcal{Q})$, with:

$$(\forall \mathbf{i} \in \mathbf{I})(\mathbf{p_i}/\mathbf{q_i} = \mathbf{p'_i}/\mathbf{q'_i} = \mathbf{r_i}) \,.$$

Calling $\mathbf{d} = \prod_{\mathbf{i}\in\mathbf{I}} \mathbf{q_i}$, $\mathbf{d'} = \prod_{\mathbf{i}\in\mathbf{I}} \mathbf{q'_i}$, we have $\mathbf{d}$ and $\mathbf{d'}$ different from zero and

$$(\sum_{\mathbf{i}\in\mathbf{I}} \mathbf{p_i}(\prod_{\mathbf{j}\in\mathbf{I}-\mathbf{i}} \mathbf{q_j}))/\mathbf{d} = (\sum_{\mathbf{i}\in\mathbf{I}} \mathbf{p'_i}(\prod_{\mathbf{j}\in\mathbf{I}-\mathbf{i}} \mathbf{q'_j}))/\mathbf{d'},$$

$$(\prod_{\mathbf{i}\in\mathbf{I}} \mathbf{p_i})/\mathbf{d} = (\prod_{\mathbf{i}\in\mathbf{I}} \mathbf{p'_i})/\mathbf{d'}.$$

**Proof**

We use the above notation. From definition 396 (p. 510), propositions 975 (p. 545), 906 (p. 511), $\mathbf{d}$ and $\mathbf{d'}$ are different from zero. We suppose that proposition 992 is true for $\mathbf{Card(I)} = \mathbf{n}$. Using propositions 892 (p. 507), we may without losing generality restrict $\mathbf{I}$ to $\mathbf{I} = [\mathbf{0}, \mathbf{n}[$. From propositions 967 (p. 541), 977 (p. 546) with $\mathbf{J} = [\mathbf{0}, \mathbf{1}]$, $\mathbf{I_0} = [\mathbf{0}, \mathbf{n}[$, $\mathbf{I_1} = [\mathbf{n}]$, we set $\mathbf{d} = \prod_{\mathbf{i}\in[\mathbf{0},\mathbf{n}]} \mathbf{q_i} = \prod_{\mathbf{i}\in[\mathbf{0},\mathbf{n}[} \mathbf{q_i q_n}$ and the same for $\mathbf{d'}$. We get:

$$\sum_{\mathbf{i}\in[\mathbf{0},\mathbf{n}]} \mathbf{p_i} \prod_{\mathbf{j}\in[\mathbf{1},\mathbf{n}]-\mathbf{i}} \mathbf{q_i} = \sum_{\mathbf{i}\in[\mathbf{0},\mathbf{n}[} \mathbf{p_i} \prod_{\mathbf{j}\in[\mathbf{0},\mathbf{n}[-\mathbf{i}} \mathbf{q_i q_n} + \mathbf{p_n} \prod_{\mathbf{j}\in[\mathbf{0},\mathbf{n}[} \mathbf{q_j} \,.$$

and the same with $\mathbf{p'}$, $\mathbf{q'}$ and, when we suppose that proposition 992 is true for $\mathbf{Card(I)} = \mathbf{n}$, we have:

$$(\sum_{\mathbf{i}\in[\mathbf{0},\mathbf{n}[} \mathbf{p_i} \prod_{\mathbf{j}\in[\mathbf{1},\mathbf{n}[-\mathbf{i}} \mathbf{q_j})\mathbf{q_n d'} = (\sum_{\mathbf{i}\in[\mathbf{0},\mathbf{n}[} \mathbf{p_i} \prod_{\mathbf{j}\in[\mathbf{1},\mathbf{n}[-\mathbf{i}} \mathbf{q_j}) \prod_{\mathbf{i}\in[\mathbf{0},\mathbf{n}[} \mathbf{q'_i q'_n q_n}$$

$$= (\sum_{\mathbf{i}\in[\mathbf{0},\mathbf{n}[} \mathbf{p'_i} \prod_{\mathbf{j}\in[\mathbf{1},\mathbf{n}[-\mathbf{i}} \mathbf{q'_j})(\prod_{\mathbf{i}\in[\mathbf{0},\mathbf{n}[} \mathbf{q_i})\mathbf{q'_n q_n} = (\sum_{\mathbf{i}\in[\mathbf{0},\mathbf{n}[} \mathbf{p'_i} \prod_{\mathbf{j}\in[\mathbf{1},\mathbf{n}[-\mathbf{i}} \mathbf{q'_j})\mathbf{d q'_n}$$

and $p_n q'_n = p'_n q_n$ we have using propositions 980 (p. 548):

$$
\begin{aligned}
&(\sum_{i\in[0,n]} p_i \prod_{j\in[1,n]-i} q_j)d' = (\sum_{i\in[0,n[} p_i \prod_{j\in[0,n[-i} q_j q_n + p_n \prod_{j\in[0,n[} q_j) \prod_{i\in[0,n[} q'_i q'_n \\
&= \sum_{i\in[0,n[} p_i \prod_{j\in[0,n[-i} q_j \prod_{i\in[0,n[} q'_i q_n q'_n + \prod_{j\in[0,n[} q_j \prod_{i\in[0,n[} q'_i p_n q'_n \\
&= \sum_{i\in[0,n[} p'_i \prod_{j\in[0,n[-i} q'_j \prod_{i\in[0,n[} q_i q_n q'_n + \prod_{j\in[0,n[} q_j \prod_{i\in[0,n[} q'_i p'_n q_n \\
&= \sum_{i\in[0,n[} p'_i \prod_{j\in[0,n[-i} q'_j q'_n \prod_{i\in[0,n[} q_i q_n + p'_n \prod_{i\in[0,n[} q'_i \prod_{j\in[0,n[} q_j q_n \\
&= \sum_{i\in[0,n[} (p'_i \prod_{j\in[0,n]-i} q'_j + p'_n \prod_{j\in[0,n]-n} q'_j) \prod_{j\in[0,n]} q_j \\
&= (\sum_{i\in[0,n]} p'_i \prod_{j\in[0,n]-i} q'_j) \prod_{j\in[0,n]} q_j = (\sum_{i\in[0,n]} p'_i \prod_{j\in[0,n]-i} q'_j)d \\
&(\sum_{i\in[0,n]} p_i \prod_{j\in[1,n]-i} q_j)/d = (\sum_{i\in[0,n]} p'_i \prod_{j\in[0,n]-i} q'_j)/d'
\end{aligned}
$$

and we also have from propositions 977 (p. 546):

$$\prod_{i\in[0,n]} p_i \prod_{i\in[0,n]} q'_i = \prod_{i\in[0,n[} p_i p_n \prod_{i\in[0,n[} q'_i q'_n = \prod_{i\in[0,n[} p'_i \prod_{i\in[0,n[} q_i p'_n q_n = \prod_{i\in[0,n]} p'_i \prod_{i\in[0,n]} q_i \,.$$

Since propositions 992 is trivially true for $I = [0,0[= \emptyset$, from the recurrence principle propositions 927 (p. 520), it is true for any $I$ with $Card(I) \in \mathcal{Z}^{0+}$.

**Definition 422 Sum and product of an integer number of rational numbers.**

Given a set $I$ with $Card(I) \in \mathcal{Z}^+$, a map $(r_i : I \to \mathcal{Q})$, we have by definition for each $r_i$ a couple $p_i, q_i$ such that $r_i = p_i/q_i$. It settles two maps:

$$(p_i : I \to \mathcal{Z})(q_i : I \to \mathcal{Z} - 0) \,.$$

Then we define using the above propositions 992 (p. 553) and calling $d = \prod_{i\in I} q_i$:

$$\sum_{i\in I} r_i = (\sum_{i\in I} p_i \prod_{j\in I-i} q_j)/d$$

$$\prod_{i\in I} r_i = \prod_{i\in I} p_i/d \,.$$

**Definition 423 Inverse of a rational number.**

Given a rational number $r = p/q \neq 0$, $p \in \mathcal{Z} - 0, q \in \mathcal{Z} - 0$, the inverse of $r$, $1/r$ is a rational such as: $r \times 1/r = 1$.

**Proposition 993** Given a rational number $r = p/q \neq 0$, $p \in \mathcal{Z} - 0, q \in \mathcal{Z} - 0$, its inverse is unique and equal to $q/p$.

**Proof**
We have (see equations (A.68) (p. 552), (A.69) (p. 552)):

$$p/q \times q/p = pq/qp = pq/pq = 1/1 = 1$$

$$p/q \times q'/p' = 1 \Rightarrow p/q(q/p - q'/p') = 0 \Rightarrow \frac{p(qp' - q'p)}{qpp'} = 0/1 = 0$$
$$\Rightarrow p(qp' - q'p) = 0 \Rightarrow qp' - q'p = 0 \Rightarrow qp' = q'p \Rightarrow q/p = q'/p' .$$

**Proposition 994** Any rational different from 0 has an inverse.
**Proof**
$(p/q)^{-1} = q/p$ .

**Proposition 995** Given a set $I$ with $Card(I) \in \mathcal{Z}^+$, a map $(r_i : I \to \mathcal{Q})$, we have:

$$(q \in \mathcal{Z} - \{0\}, (\forall i \in I)(p_i \in \mathcal{Z}, r_i = \frac{p_i}{q})) \Rightarrow \sum_{i \in I} r_i = \frac{\sum_{i \in I} p_i}{q} .$$

**Proof**
With the above notation, from proposition 992 (p. 553) we have applying definition 422 (p. 554) with $(\forall i \in I)(q_i = q)$:

$$\sum_{i \in I} r_i = \frac{\sum_{i \in I} p_i \prod_{k \in I - i} q_k}{\prod_{k \in I} q_k} .$$

Using propositions 973 (p. 544), 990 (p. 553), 977 (p. 546) and 980 (p. 548), we also have calling $j = \tau_j(j \in I)$:

$$\sum_{i \in I} r_i = \frac{\prod_{k \in I - j} q_k \sum_{i \in I} p_i}{q \prod_{k \in I - j} q_k} = \frac{\sum_{i \in I} p_i}{q} .$$

**Commutativity of sum and product**

**Proposition 996** Given two finite sets of indices $I$, $J$ with $Card(I) = Card(J)$ and a map $(a_i : I \to \mathcal{Q})$ calling $(i(j) : J \leftrightarrow I)$ and $(j(i) : I \leftrightarrow J)$ two bijective maps with $i(j(k)) = k$ which exist since $I \sim J$, we have:

$$\sum_{i \in I} a_i = \sum_{j \in J} a_{i(j)} .$$

**Proof**
Setting $a_i = p_i/q_i, p_i \in \mathcal{Z}, q_i \in \mathcal{Z}$, from proposition 974 (p. 545), we have:

$$d_I = \prod_{i \in I} q_i = d_J = \prod_{j \in J} q_{i(j)} .$$

From propositions 966 (p. 541), 974 (p. 545) and 184 (p. 143):

$$\sum_{i \in I} p_i \prod_{k \in I - i} q_k = \sum_{j \in J} p_{i(j)} \prod_{k \in J - j} q_{i(k)} .$$

Coupling both relations, we get proposition 996 (p. 555).

**Proposition 997** Given two finite sets of index $\mathbf{I} \sim \mathbf{J}$ with two associated bijections $(\mathbf{i}(\mathbf{j}) : \mathbf{J} \leftrightarrow \mathbf{I})$, $(\mathbf{j}(\mathbf{i}) : \mathbf{I} \leftrightarrow \mathbf{J})$, $\mathbf{i}(\mathbf{j}(\mathbf{k})) = \mathbf{k}$ and a map $(\mathbf{a}_\mathbf{i} : \mathbf{I} \to \mathcal{Q})$, we have:

$$\prod_{i \in I} a_i = \prod_{j \in J} a_{i(j)} \ .$$

**Proof**
Setting $a_i = p_i/q_i, p_i \in \mathcal{Z}, q_i \in \mathcal{Z}$, from proposition 974 (p. 545), definition 422 (p. 554), we have:

$$\prod_{i \in I} a_i = \frac{\prod_{i \in I} p_i}{\prod_{i \in I} q_i} = \frac{\prod_{j \in J} p_{i(j)}}{\prod_{j \in J} q_{i(j)}} = \prod_{j \in J} a_{i(j)} \ .$$

**Associativity of sum and product**

**Proposition 998** Given two finite sets $\mathbf{I}$ and $\mathbf{J}$, a map $(\mathbf{a}_\mathbf{i} : \mathbf{I} \to \mathcal{Q})$, a partition of $\mathbf{I}$ (definition 370 (p. 471)) $(\mathbf{I}_\mathbf{j} : \mathbf{J} \to \mathcal{P}(\mathbf{I}))$, we have:

$$\sum_{j \in J} \sum_{i \in I_j} a_i = \sum_{i \in \bigcup_{j \in J} I_j} a_i = \sum_{i \in I} a_i \ . \tag{A.70}$$

**Proof**
Setting $a_i = p_i/q_i, p_i \in \mathcal{Z}, q_i \in \mathcal{Z}$, $d = \prod_{i \in I} q_i$. From definition 422 (p. 554), propositions 967 (p. 541) and 977 (p. 546), we have:

$$\begin{aligned}
\sum_{j \in J} \sum_{i \in I_j} \frac{p_i}{q_i} &= \sum_{j \in J} \frac{\sum_{i \in I_j} p_i \prod_{k \in I_j - i} q_k}{\prod_{i \in I_j} q_i} = \frac{\sum_{j \in J} (\prod_{k \in J-j} \prod_{m \in I_k} q_m) \sum_{i \in I_j} p_i \prod_{k \in I_j - i} q_k}{\prod_{j \in J} \prod_{i \in I_j} q_i} \\
&= \frac{\sum_{j \in J} \sum_{i \in I_j} p_i (\prod_{m \in \bigcup_{k \in J-j} I_k} q_m) \prod_{k \in I_j - i} q_k}{\prod_{j \in J} \prod_{i \in I_j} q_i} = \frac{\sum_{j \in J} \sum_{i \in I_j} p_i \prod_{m \in (I_j - i) \cup \bigcup_{k \in J-j} I_k} q_m}{\prod_{j \in J} \prod_{i \in I_j} q_i} \\
&= \frac{\sum_{j \in J} \sum_{i \in I_j} p_i \prod_{m \in (\bigcup_{k \in J} I_k - i)} q_m}{\prod_{j \in J} \prod_{i \in I_j} q_i} = \frac{\sum_{i \in I} p_i \prod_{m \in I - i} q_m}{\prod_{i \in I} q_i} = \sum_{i \in I} a_i \ .
\end{aligned}$$

**Proposition 999** Given two finite sets $\mathbf{I}$ and $\mathbf{J}$, a map $(\mathbf{a}_\mathbf{i} : \mathbf{I} \to \mathcal{Q})$, a partition of $\mathbf{I}$ (definition 370 (p. 471)) $(\mathbf{I}_\mathbf{j} : \mathbf{J} \to \mathcal{P}(\mathbf{I}))$, we have:

$$\prod_{j \in J} \prod_{i \in I_j} a_i = \prod_{i \in \bigcup_{j \in J} I_j} a_i = \prod_{i \in I} a_i \ . \tag{A.71}$$

**Proof**
Setting $a_i = p_i/q_i, p_i \in \mathcal{Z}, q_i \in \mathcal{Z}$. From definition 422 (p. 554) and proposition 977 (p. 546), we have:

$$\begin{aligned}
\prod_{j \in J} \prod_{i \in I_j} \frac{p_i}{q_i} &= \prod_{j \in J} \frac{\prod_{i \in I_j} p_i}{\prod_{i \in I_j} q_i} \\
&= \frac{\prod_{j \in J} \prod_{i \in I_j} p_i}{\prod_{j \in J} \prod_{i \in I_j} q_i} = \frac{\prod_{i \in I} p_i}{\prod_{i \in I} q_i} = \prod_{i \in I} a_i \ .
\end{aligned}$$

**Distributivity of product with the sum.**

**Proposition 1000** J and L are two sets with $\mathbf{Card(J)} \in \mathcal{Z}^+$ and $\mathbf{Card(L)} \in \mathcal{Z}^+$. We suppose that we have a partition (definition 370 (p. 471)) of (J) $(\mathbf{J}_\lambda : \mathbf{L} \to \mathcal{P}(\mathbf{J}))$ with $(\forall \lambda \in \mathbf{L})(\mathbf{Card}(\mathbf{J}_\lambda) \in \mathcal{Z}^+)$ and a set of maps over $\mathcal{Q}$: $(\mathbf{a}_{\lambda,\iota} : \mathbf{J}_\lambda \to \mathcal{Q})$. We set $\mathbf{I} = \prod_{\lambda\in\mathbf{L}} \mathbf{J}_\lambda$. We have:

$$\prod_{\lambda\in L}\sum_{\iota\in J_\lambda} a_{\lambda,\iota} = \sum_{f\in I}\prod_{\lambda\in L} a_{\lambda,f(\lambda)} \ .$$

**Proof**

Setting $\mathbf{a}_{\lambda,\iota}\mathbf{i} = \mathbf{p}_{\lambda,\iota}/\mathbf{q}_{\lambda,\iota}, \mathbf{p}_{\lambda,\iota} \in \mathcal{Z}, \mathbf{q}_{\lambda,\iota} \in \mathcal{Z}$, from definition 422 (p. 554), propositions 980 (p. 548), 967 (p. 541), 977 (p. 546) and 995 (p. 555), we have:

$$\begin{aligned}
\prod_{\lambda\in L}\sum_{\iota\in J_\lambda} a_{\lambda,\iota} &= \prod_{\lambda\in L}\sum_{\iota\in J_\lambda}\frac{p_{\lambda,\iota}}{q_{\lambda,\iota}} = \prod_{\lambda\in L}\frac{\sum_{\iota\in J_\lambda} p_{\lambda,\iota}\prod_{\eta\in J_\lambda-\iota} q_{\lambda,\eta}}{\prod_{\iota\in J_\lambda} q_{\lambda,\iota}} \\
&= \frac{\prod_{\lambda\in L}\sum_{\iota\in J_\lambda} p_{\lambda,\iota}\prod_{\eta\in J_\lambda-\iota} q_{\lambda,\eta}}{\prod_{\lambda\in L}\prod_{\iota\in J_\lambda} q_{\lambda,\iota}} = \frac{\sum_{f\in I}\prod_{\lambda\in L} p_{\lambda,f(\lambda)}\prod_{\eta\in J_\lambda-f(\lambda)} q_{\lambda,\eta}}{\prod_{\lambda\in L}\prod_{\iota\in J_\lambda} q_{\lambda,\iota}} \\
&= \frac{\sum_{f\in I}\prod_{\lambda\in L} p_{\lambda,f(\lambda)}\prod_{\lambda\in L}\prod_{\eta\in J_\lambda-f(\lambda)} q_{\lambda,\eta}}{\prod_{\lambda\in L}\prod_{\iota\in J_\lambda} q_{\lambda,\iota}} = \sum_{f\in I}\prod_{\lambda\in L} p_{\lambda,f(\lambda)}\frac{\prod_{\lambda\in L}\prod_{\eta\in J_\lambda-f(\lambda)} q_{\lambda,\eta}}{\prod_{\lambda\in L}\prod_{\iota\in J_\lambda} q_{\lambda,\iota}} \\
&= \sum_{f\in I}\frac{\prod_{\lambda\in L} p_{\lambda,f(\lambda)}}{\prod_{\lambda\in L} q_{\lambda,f(\lambda)}} = \sum_{f\in I}\prod_{\lambda\in L}\frac{p_{\lambda,f(\lambda)}}{q_{\lambda,f(\lambda)}} = \sum_{f\in I}\prod_{\lambda\in L} a_{\lambda,f(\lambda)} \ .
\end{aligned}$$

## A.8.3 Order on $\mathcal{Q}$

**Definition 424 Order on $\mathcal{Q}$.**

By definition we have:

$$\begin{aligned}
r_1 \le r_2 \Leftrightarrow \{&r_1 \in \mathcal{Q}, r_2 \in \mathcal{Q}, (r_1 = \frac{p_1}{q_1}, r_2 = \frac{p_2}{q_2}) \\
&\Rightarrow ((q_1q_2 > 0, p_1q_2 \le q_1p_2) \vee (q_1q_2 < 0, p_1q_2 \ge q_1p_2))\} \ .
\end{aligned} \tag{A.72}$$

Or setting $\mathbf{q}_1 > 0$ and $\mathbf{q}_2 > 0$ which is always possible by propositions 990 (p. 553):

$$r_1 \le r_2 \Leftrightarrow \{(r_1 \in \mathcal{Q}, r_2 \in \mathcal{Q}, r_1 = \frac{p_1}{q_1}, q_1 > 0, r_2 = \frac{p_2}{q_2}, q_2 > 0) \Rightarrow p_1q_2 \le q_1p_2)\} \ . \tag{A.73}$$

**Proposition 1001** Definition 424 defines an order on $\mathcal{Q}$.

**Proof**

From propositions 956 (p. 534), equation (A.72) and (A.73) are equivalent. We suppose now $\mathbf{q}_1 > 0, \mathbf{q}_2 > 0$. Definition 424 defines an order because with the above notations:

- We have:

$$r_1 \le r_2, r_2 \le r_1 \Rightarrow p_1q_2 = q_1p_2 \Rightarrow r_1 = r_2 \ .$$

- Given $\mathbf{r}_1 = \mathbf{p}_1/\mathbf{q}_1$, $\mathbf{r}_2 = \mathbf{p}_2/\mathbf{q}_2$, $\mathbf{r}_3 = \mathbf{p}_3/\mathbf{q}_1$, we have:

$$r_1 = p_1q_2q_3/(q_1q_2q_3), r_2 = p_2q_1q_3/(q_1q_2q_3), r_3 = p_3q_2q_1/(q_1q_2q_3) \ .$$

  Then the order on $\mathbf{r}_1, \mathbf{r}_2, \mathbf{r}_3$ is the same as $\mathbf{p}_1\mathbf{q}_2\mathbf{q}_3, \mathbf{p}_2\mathbf{q}_1\mathbf{q}_3, \mathbf{p}_3\mathbf{q}_2\mathbf{q}_1$. From proposition 952 (p. 533) applying definition 338 (p. 446) point 3 to $\mathbf{p}_1\mathbf{q}_2\mathbf{q}_3, \mathbf{p}_2\mathbf{q}_1\mathbf{q}_3, \mathbf{p}_3\mathbf{q}_2\mathbf{q}_1$, we have:

$$r_1 \le r_2, r_2 \le r_3 \Rightarrow r_1 \le r_3 \ .$$

**Proposition 1002** The order on rational is total (see definition 377 (p. 483)).

**Proof**
With the above notation, according to proposition 876 (p. 500), it is total because $p_1q_2$ and $q_1p_2$ belong to the set of integers ($\mathcal{Z}^{0+}$).

**Proposition 1003** Given a set $I$ with $\mathbf{Card}(I) \in \mathcal{Z}^+$, a map $(r_i : I \to \mathcal{Q})$, setting according to proposition 991 (p. 553). $d \in \mathcal{Z}^+$ and the map $(p_i : I \to \mathcal{Z})$ such as:

$$(\forall i \in I)(r_i = p_i/d) .$$

The order on the $r_i$ is the same as the $p_i$.

**Proof**
It comes directly from definitions 424 (p. 557) and 409 (p. 533).

**Definition 425 Euclidean division.**

One calls Euclidean division, the map $(\mathbf{Ecd} : \mathcal{Q} \times \mathcal{Q} \to \mathcal{Z})$ defined by:

$$\mathbf{Ecd}(r_1, r_2) = r_1/r_2 - r_3 \in \mathcal{Z},$$
$$r_3 \in [0, 1[ .$$

If $r_1 = p_1/q_1$ and $r_2 = p_2/q_2$ are equivalent to integer ($q_1 = q_2 = 1, p_1 \in \mathcal{Z}, p_2 \in \mathcal{Z}$) then we are in the case of proposition 930 (p. 521) with $a = r_1$, $b = r_2$, $q = \mathbf{Ecd}(r_1, r_2) \in \mathcal{Z}$, $r = r_3$, $r_2 \in \mathcal{Z}, r < b$.

**Proposition 1004** The Euclidean division is a map.

**Proof**
We have:

$$((r_i : [1, 4] \to \mathcal{Q}), 0 \leq r_3 < 1., 0. \leq r_4 < 1, r_1/r_2 - r_3 \in \mathcal{Z}, r_1/r_2 - r_4 \in \mathcal{Z})$$
$$\Rightarrow r_3 - r_4 \in \mathcal{Z} .$$

From propositions 970 (p. 543), we have:

$$0 \leq r_3 < 1., 0 \leq r_4 \Rightarrow r_3 + 0 < r_4 + 1 \Rightarrow r_3 - r_4 < 1$$

$$0 \leq r_3, 0 \leq r_4 < 1 \Rightarrow r_4 + 0 < r_3 + 1 \Rightarrow -1 < r_3 - r_4 .$$

Then we have:

$$-1. < r_3 - r_4 < 1 \Rightarrow r_3 - r_4 = 0 \Rightarrow r_3 = r_4 .$$

**Proposition 1005** The set of rational numbers between two non-equal rational numbers has the same cardinal as $\mathcal{Z}^+$.

$r_1 = p_1/q_1 < r_2 = p_2/q_2$ being two non-equal rationals, from propositions 1003 (p. 558), 968 (p. 542), 962 (p. 539), 970 (p. 543) and 972 (p. 544), we have:

$$r_1 < r_2 \Rightarrow p_1q_2/(q_1q_2) < p_2q_1/(q_1q_2) \Rightarrow p_1q_2 < p_2q_1$$
$$n \in \mathcal{Z}^+ \Rightarrow p_1q_2n < p_1q_2(n-1) + p_2q_1 < p_2q_1(n-1) + p_2q_1 = p_2q_1n$$
$$\Rightarrow r_1 = p_1q_2n/(q_1q_2n) < (p_1q_2(n-1) + p_2q_1)/(q_1q_2n) < p_2q_1n/(q_1q_2n) = r_2$$
$$(n \in \mathcal{Z}^+, n' \in \mathcal{Z}^+, n \neq n') \Rightarrow$$
$$(p_1q_2(n-1) + p_2q_1)/(q_1q_2n) - (p_1q_2(n'-1) + p_2q_1)/(q_1q_2n')$$
$$= (p_2q_1 - p_1q_2)(n' - n)/(q_1q_2nn') \neq 0 .$$

Then the following map $\mathbf{f}$ is an injective map (definition 122 (p. 139)) between $\mathcal{Z}^+$ and $]\mathbf{r}_1, \mathbf{r}_2[$:

$$(\mathbf{f}(\mathbf{n}) = (\mathbf{p}_1\mathbf{q}_2(\mathbf{n}-1) + \mathbf{p}_2\mathbf{q}_1)/(\mathbf{q}_1\mathbf{q}_2\mathbf{n}) = \mathbf{r}_1 + (\mathbf{r}_2 - \mathbf{r}_1)/\mathbf{n} : \mathcal{Z}^+ \to ]\mathbf{r}_1, \mathbf{r}_2[) \ .$$

It means, from definition 392 (p. 500) and propositions 180 (p. 143):

$$\mathbf{Card}(\mathcal{Z}^+) \le \mathbf{Card}(]\mathbf{r}_1, \mathbf{r}_2[) \le \mathbf{Card}(\mathcal{Q}) = \mathbf{Card}(\mathcal{Z}^+) \ .$$

Then we have proposition 1005 (p. 558).

**Proposition 1006** For any segment (definition 389 (p. 488)) $\mathbf{S}$ in $\mathcal{Q}$, if there is an **upper_limit** (definition 381 (p. 485)) to $\mathbf{S}$, it is also the **lower_limit** of $\mathcal{C}(\mathbf{S})$ and vice versa.
**Proof**
If $\mathbf{s}$ is the **upper_limit** of $\mathbf{S}$ and $\mathbf{T}$, the set of majoring elements of $\mathbf{S}$, from definition 381 (p. 485), we have: $\mathbf{s} \in \mathbf{T}$ and $(\forall \mathbf{t} \in \mathbf{T})(\mathbf{s} \le \mathbf{t})$.

- We have $\mathcal{C}(\mathbf{S}) \subset \mathbf{T}$:
  If we had $\mathbf{x} \in \mathcal{C}(\mathbf{S}), \mathbf{x} \notin \mathbf{T}$, it would mean $(\exists \mathbf{y} \in \mathbf{S})(\mathbf{x} < \mathbf{y})$[12] and from definition 389 (p. 488) $\mathbf{x} \in \mathbf{S}$ which is in contradiction with $\mathbf{x} \in \mathcal{C}(\mathbf{S})$ (definition 353 (p. 454)).
- Calling $\mathbf{S}'$ the set of minoring elements of $\mathcal{C}(\mathbf{S})$, we have $\mathbf{s} \in \mathbf{S}'$ because, from definition 381 (p. 485), $\mathbf{s}$ is minoring all elements of $\mathbf{T}$ and then those of $\mathcal{C}(\mathbf{S})$.
- Using proposition 1005 (p. 558), we have:

  $$\begin{aligned}&(\mathbf{y} \in \mathbf{S}', \mathbf{y} > \mathbf{s}) \Rightarrow (\forall \mathbf{z} \in \mathbf{S})(\mathbf{y} > \mathbf{s} \ge \mathbf{z}) \Rightarrow (\mathbf{y} \notin \mathbf{S}, \mathbf{y} > \mathbf{s}) \Rightarrow (\mathbf{y} \in \mathcal{C}(\mathbf{S}), \mathbf{y} > \mathbf{s})\\ &\Rightarrow (\exists \mathbf{w} \in \mathcal{Q})(\mathbf{y} > \mathbf{w} > \mathbf{s}, \mathbf{w} \in \mathcal{C}(\mathbf{S}))\\ &\Rightarrow \mathbf{y} \notin \mathbf{S}' \ .\end{aligned}$$

  Then from point 8 (p. 435) in A.1.4, $\neg(\mathbf{y} \in \mathbf{S}', \mathbf{y} > \mathbf{s})$ is true. $\mathbf{y} \in \mathbf{S}' \Rightarrow \mathbf{y} \le \mathbf{s}$. $\mathbf{s}$ is the **lower_limit** of $\mathcal{C}(\mathbf{S})$ as the biggest element of the set of minoring element of $\mathcal{C}(\mathbf{S})$ (definition 381 (p. 485)).

Reversing the order on $\mathcal{Q}$ which is also an ordering relation (see definition 338 (p. 446) and convention 28 (p. 446)), $\mathcal{C}(\mathbf{S})$ being a segment for the reverse ordering relation $((\mathbf{x} \ge \mathbf{y}, \mathbf{y} \in \mathcal{C}(\mathbf{S})) \Rightarrow \mathbf{x} \in \mathcal{C}(\mathbf{S}))$ otherwise we would have $\mathbf{y} \in \mathcal{C}(\mathbf{S}), \mathbf{y} \in \mathbf{S}$. We can repeat exactly the same arguments and get the other conclusion.

**Proposition 1007** Any element $\mathbf{r}$ of $\mathcal{Q}$ has a bigger and smaller element in $\mathcal{Q}$.
**Proof**
Writing $\mathbf{r} = \frac{\mathbf{p}}{\mathbf{q}}$, $\mathbf{p} \in \mathcal{Z}$, $\mathbf{q} \in \mathcal{Z}^+$, from proposition 954 (p. 534), definition 424 (p. 557), we have:

$$(\exists \mathbf{p}_1 \in \mathcal{Z})(\exists \mathbf{p}_2 \in \mathcal{Z})(\mathbf{p}_1 < \mathbf{p} < \mathbf{p}_2, \frac{\mathbf{p}_1}{\mathbf{q}} < \frac{\mathbf{p}}{\mathbf{q}} < \frac{\mathbf{p}_2}{\mathbf{q}}) \ .$$

**Proposition 1008** The map $(\mathbf{n}/\mathbf{1} : \mathcal{Z} \to \mathcal{Q})$ which at any element $\mathbf{n} \in \mathcal{Z}$ associates the element $\mathbf{n}/\mathbf{1} \in \mathcal{Q}$ is a strictly increasing map (definition 386 (p. 486)) which is compatible with all the above operations:

$$\mathbf{p}, \mathbf{q} \in \mathcal{Z} \Rightarrow \{\mathbf{p}(+ - \times /)\mathbf{q}\}/1 = \mathbf{p}/1(+ - \times /)\mathbf{q}/1 \ .$$

**Proof**
It is clear from all the definitions and propositions of section A.8 (p. 551).

---

[12] $\{(\forall \mathbf{y})(\mathbf{y} \in \mathbf{S} \Rightarrow \mathbf{y} \le \mathbf{x}) \Rightarrow \mathbf{x} \in \mathbf{T}\} = \{\neg\neg(\exists \mathbf{y})\neg(\neg \mathbf{y} \in \mathbf{S} \vee \mathbf{y} \le \mathbf{x}) \vee \mathbf{x} \in \mathbf{T}\} \Leftrightarrow \{\mathbf{x} \notin \mathbf{T} \Rightarrow (\exists \mathbf{y})(\mathbf{y} \in \mathbf{S}, \mathbf{y} > \mathbf{x})\}$ .

## A.9 Conventions

$\mathbf{S}$ is a relation symbol like $\in$. $\mathbf{A}$, $\mathbf{B}$, $\mathbf{C}$ are relations. $\mathbf{x}$, $\mathbf{y}$ are terms which may enter in a relation. For simplification we adopt the conventions:

**Convention 35** A, B, C.

$\mathbf{A}, \mathbf{B}, \mathbf{C}$ means $\mathbf{A}$ and $\mathbf{B}$ and $\mathbf{C}$:
$\mathbf{A}, \mathbf{B}, \mathbf{C} = \neg(\neg\mathbf{A} \vee \neg\mathbf{B} \vee \neg\mathbf{C})$

**Convention 36** $(\exists \mathbf{x}\mathbf{S}\mathbf{y})\mathbf{B}$.

$(\exists \mathbf{x}\mathbf{S}\mathbf{y})\mathbf{B}$ stands for: $(\exists \mathbf{x})(\mathbf{x}\mathbf{S}\mathbf{y}, \mathbf{B})$
like:
$(\exists \mathbf{x} \in \mathbf{C})\mathbf{B} = (\exists \mathbf{x})(\mathbf{x} \in \mathbf{C}, \mathbf{B})$

**Convention 37** $(\forall \mathbf{x}\mathbf{S}\mathbf{y})\mathbf{B}$.

$(\forall \mathbf{x}\mathbf{S}\mathbf{y})\mathbf{B}$ stands for: $(\forall \mathbf{x})(\mathbf{x}\mathbf{S}\mathbf{y} \Rightarrow \mathbf{B})$
like:
$(\forall \mathbf{x} \in \mathbf{C})\mathbf{B} = (\forall \mathbf{x})(\mathbf{x} \in \mathbf{C} \Rightarrow \mathbf{B})$

**Convention 38** $(\exists \mathbf{x})(\exists \mathbf{y})(\mathbf{A})$.

$(\exists \mathbf{x})(\exists \mathbf{y})(\mathbf{A})$ means $\exists \mathbf{x}(\exists \mathbf{y}\ \mathbf{A})$

**Convention 39** $(\forall \mathbf{x})(\forall \mathbf{y})(\mathbf{A})$.

$(\forall \mathbf{x})(\forall \mathbf{y})(\mathbf{A})$ means $\forall \mathbf{x}(\forall \mathbf{y}\ \mathbf{A})$

**Convention 40** $(\forall \mathbf{x})(\exists \mathbf{y})(\mathbf{A})$.

$(\forall \mathbf{x})(\exists \mathbf{y})(\mathbf{A})$ means $\forall \mathbf{x}(\exists \mathbf{y}\ \mathbf{A})$

**Convention 41** $(\exists \mathbf{x})(\forall \mathbf{y})(\mathbf{A})$.

$(\exists \mathbf{x})(\forall \mathbf{y})(\mathbf{A})$ means $\exists \mathbf{x}(\forall \mathbf{y}\ \mathbf{A})$

**Convention 42** More conventions for $\exists$.

$$\begin{array}{l} (\exists \mathbf{x})(\mathbf{R}(\mathbf{x})) \\ \textbf{anything} \\ \mathbf{T}(\mathbf{x}) \end{array}$$

stands for

$$\begin{array}{l} \mathbf{R}(\tau_{\mathbf{x}}\mathbf{R}(\mathbf{x})) \\ \textbf{anything} \\ \mathbf{T}(\tau_{\mathbf{x}}\mathbf{R}(\mathbf{x})) \end{array}$$

# Bibliography

[1] Bourbaki **Eléments de mathématique, théorie des ensembles** Hermann editor Paris 1963

[2] Mikio Nakahara **Geometry, Topology and Physics** Graduate student series in Physics Editor Douglas F. Brewer Adam Hilger Bristol and New York Jersey London Hong Kong 1990

[3] John Baez, Javier P. Muniain **Gauge Fields, Knots and Gravity** World Scientific Singapore-New Jersey-London-Hong Kong 1994

[4] Walter Rubin **Real and Complex Analysis** McGraw-Hill Book Co., Singapore 1987

[5] see for instance **http://mathworld.wolfram.com/VectorSpace.html**

# List of definitions

# List of conventions

# Index

## D

## O

## P



## R



## S